The Evolution of Primate Societies

The Evolution of Primate Societies

Edited by John C. Mitani, Josep Call,
Peter M. Kappeler, Ryne A. Palombit, and Joan B. Silk

The University of Chicago Press
Chicago and London

John C. Mitani is the James N. Spuhler Collegiate Professor of Anthropology at the University of Michigan. **Josep Call** is a senior scientist and director of the Wolfgang Kohler Primate Research Centre at the Max Planck Institute for Evolutionary Anthropology. **Peter M. Kappeler** is head of the Department of Behavioral Ecology and Sociobiology/Anthropology at the University of Gottingen. **Ryne A. Palombit** is associate professor of anthropology at Rutgers, the State University of New Jersey. **Joan B. Silk** is professor in the Department of Anthropology and the Institute for Society and Genetics at the University of California, Los Angeles.

The University of Chicago Press, Chicago 60637
The University of Chicago Press, Ltd., London
© 2012 by The University of Chicago
All rights reserved. Published 2012.
Printed in the United States of America

21 20 19 18 17 16 15 14 13 12   1 2 3 4 5

ISBN-13: 978-0-226-53171-7   (cloth)
ISBN-13: 978-0-226-53172-4   (paper)
ISBN-10: 0-226-53171-6   (cloth)
ISBN-10: 0-226-53172-4   (paper)

Library of Congress Cataloging-in-Publication Data

The evolution of primate societies / edited by John C. Mitani, Josep Call, Peter M. Kappeler, Ryne A. Palombit, and Joan B. Silk.
        pages ; cm
    Includes bibliographical references and index.
    ISBN-13: 978-0-226-53171-7 (cloth : alkaline paper)
    ISBN-10: 0-226-53171-6 (cloth : alkaline paper)
    ISBN-13: 978-0-226-53172-4 (paperback : alkaline paper)
    ISBN-10: 0-226-53172-4 (paperback : alkaline paper) 1. Primates—Behavior—Evolution. 2. Primates—Evolution. 3. Primates—Ecology. 4. Animal societies. 5. Social behavior in animals. 6. Mammals—Behavior. 7. Mammals—Ecology. I. Mitani, John C. II. Call, Josep. III. Kappeler, Peter M. IV. Palombit, Ryne A. V. Silk, Joan B.
    QL737.P9E965 2012
    599.8—dc23

                                    2012007982

♾ This paper meets the requirements of ANSI/NISO Z39.48-1992 (Permanence of Paper).

Contents

Foreword

There are a few species alive today, such as the deep-sea coelacanth, without any close relatives. If there were no nonhuman primates, humans would be similarly isolated. We would have no living guides to reveal the nature of our recent evolutionary tree. Our attempts to understand the biological sources of our social behavior would therefore be relatively feeble, limited to using general principles applicable to all kinds of animals. We could only guess at how components of our exceptional mental ability had been foreshadowed, constrained, or favored in related species that had evolved within the last 50 million years.

Happily, unlike our iconic fish, we are one of a riotously diverse set of hundreds of species that offer numerous opportunities to search for the distant beginnings of the traits that make us human. The taxonomic affinities of primates are well known, including humans' place among the great apes, and primates have unusually complex social relationships and high cognitive abilities whose nature is revealed in increasing detail every year. The combination of diverse species, rich sociality, and humanlike cognition is a boon. It means that study of the primates can help show how social behavior and mental powers have evolved not merely in intelligent beings in general, but specifically in the apes, monkeys, lorisoids, and lemurs whose evolutionary biology we shared until five to six million years ago. The latter studies are important because for reasons that are only partly understood, primates show strong phylogenetic signals in many aspects of social and mental evolution. Primatology thus opens an invaluable route to understanding the origins of some of the traits that we find most interesting both in animals in general and in our own species in particular.

For most of history, the opportunities afforded by the fortunate existence of a rich Primate order lay unexploited. Although it was more than 2,000 years ago that Aristotle's dissection of a Barbary macaque began the slow movement towards a comparative anatomy of humans and other species, not until the rise of ethology, behavioral ecology, and comparative psychology in the second half of the twentieth century did primate behavior became a substantial field of study. The late bloom of behavioral primatology is illustrated by its history in the United States. While figures such as Robert Yerkes and Clarence Ray Carpenter were active researchers in the first half of the twentieth century,

no PhDs were awarded for research on primate behavior until the 1960s. Japan and Europe, the other major locales for primate study, had roughly parallel developments. By contrast, recent congresses of the International Primatological Society have attracted more than 1,000 speakers.

Efforts to survey the state of the field have therefore become rapidly more challenging. When Irven DeVore edited *Primate Behavior* in 1965 he was able to include accounts of almost all the major field studies, which covered fewer than 20 species. By the time that Barbara Smuts and colleagues edited *Primate Societies* in 1987, contributors were reviewing scores of empirical studies. Chapters in *The Evolution of Primate Societies*, by contrast, often refer to scores of reviews. They have done so in order to fulfill the editors' ambitious aim of combining a full coverage of species with careful evaluations of explanatory frameworks for all major aspects of primate societies. They have succeeded so well that this volume will surely be the definitive statement in the study of primate societies for many years.

The techniques discussed in *The Evolution of Primate Societies* range from classic field observation to multidisciplinary experimentation. Adventurous primatologists continue to penetrate remote and difficult habitats in their efforts to document the basic social attributes of little-known populations (or simply to describe new taxa: a few species or even genera of primates are still being discovered each decade). But alongside those traditional and vital forms of research, in captive settings and in a few well-habituated wild groups a combination of behavioral, genetic, physiological, neurological, and cognitive data is generating wholly new ideas about the social minds of primates, suggesting that some species think and react much more similarly to humans than they have traditionally been assumed to do. Behavioral ecology unites these investigative extremes under the same broad conceptual framework. On the one hand *The Evolution of Primate Societies* evaluates the relationships among social structures and ecological context so as to test ideas about the evolutionary origins of behavioral diversity. On the other hand, this volume discusses progress in understanding how the variety of primate mental abilities might equally be explicable from pressures arising in nature. The geneticist Theodore Dobhzhansky said that nothing in biology makes sense except in the light of evolution. *The Evolution of Primate Societies* equally affirms that nothing in primate social behavior makes sense without understanding each species' natural history.

The reasons for better understanding the natural history of the primates are not confined to understanding their social behavior. In 2010 the Primate Specialist Group of the International Union for the Conservation of Nature reported that a daunting 48% of the world's primates are threatened with extinction (Mittermeier et al. 2010). Some taxa are down to less than 100 individuals. Habitat loss is the most severe threat, a particularly challenging problem given that it emerges from human population increase and rising economic expectations. But an ever-growing awareness of the value of these extraordinary species can contribute to maintaining primate diversity. Through its revelations of the complexity and fascination of our relatives' social worlds, *The Evolution of Primate Societies* will surely accelerate the scientific study of behavioral and cognitive evolu-

tion. Let us hope that the strength of the public concern to care for and conserve the primates rises in parallel. As this book shows, the losses to our understanding of the natural world and our place in it would be irreplaceably high if we were ever to become a latter-day primate coelacanth.

Richard Wrangham

Reference

Mittermeier, R. A., Wallis, J., Rylands, A. B., Ganzhorn, J. U., Oates, J. F., Williamson, E. A., Palacios, E., Heymann, E. W., Kierulff, M. C. M., Long, Y., Supriatna, J., Roos, C., Walker, S., Cortés-Ortiz, L. & Schwitzer, C. 2009. *Primates in Peril: The World's 25 Most Endangered Primates 2008–2010.* Arlington, VA: IUCN/SSC Primate Specialist Group (PSG), International Primatological Society (IPS), and Conservation International (CI).

Preface

Robert Yerkes and Wolfgang Köhler initiated the modern study of primate behavior during the first part of the 20th century with their investigations of captive apes. Yerkes subsequently inspired Clarence Ray Carpenter, who conducted the first successful field studies of the behavior of wild primates. Carpenter's pioneering research on howler monkeys in Panama and gibbons in Thailand set the stage for the first generation of fieldwork on primate behavior following the Second World War. By the early 1960s a sizable amount of data had accumulated, leading David Hamburg and Sherwood Washburn to organize a study group at the Center for Advanced Study in the Behavioral Sciences (CASBS) in Stanford, California. The first compilation of papers on the behavior of primates emerged from that group with the publication in 1965 of *Primate Behavior: Field Studies of Monkeys and Apes*, edited by Irven DeVore. The ensuing years witnessed an explosion of interest in and research on the behavior of primates. Nearly 20 years later in 1983, Barbara Smuts, Dorothy Cheney, Robert Seyfarth, Richard Wrangham, and Tom Struhsaker formed a second study group at CASBS to take stock of current knowledge in the field. A second major collection of papers—*Primate Societies*, published in 1987—resulted from their efforts. With contributions from 46 established and aspiring scholars, *Primate Societies* became an instant classic and spawned even more research. As the field continued to grow, however, the book began to outlive its usefulness. With the 20th anniversary of its publication approaching, we decided that the time was ripe for another major synthesis of the field. After considerable planning, and with the support of the editors of *Primate Societies*, we started to solicit papers in 2008. This volume is the result.

Much has transpired in the 25 years since the publication of *Primate Societies*. The study of primate behavior has witnessed explosive growth, as ongoing research in the field and laboratory has added significantly to our understanding of these animals. Such rapid growth creates special problems for specialists and nonspecialists who are interested in keeping abreast of the latest developments in the field. The study of primate behavior has always involved researchers from a broad array of disciplines, and results are published in an equally diverse set of journals, books, and—now as we have entered the 21st century—electronic and digital media. This only adds to the problem of synthe-

sizing and integrating current knowledge. To address these issues, chapters in this volume review theories that guide research on the behavior of primates and summarize findings of recent empirical studies. Special emphasis is placed on the adaptive strategies primates adopt as they attempt to grow, survive, and reproduce in the wild. Our primary goal is to provide, in a single place, a set of synthetic and integrative reviews that will furnish readers with a comprehensive understanding of primate behavior. In doing so, we hope that this volume will serve as a standard reference for primatologists and interested researchers working in other fields, and as a text for advanced undergraduate and graduate students. Highlighting what we now know inevitably leads to questions about what we don't know, and an additional goal of this volume is to furnish a guide for future research.

The study of primate behavior continues to evolve and grow, and because of this, it is a daunting if not impossible task to attempt to synthesize current knowledge in all of its diverse forms. As a consequence, several difficult editorial decisions had to be made about what to include in and exclude from this volume. Our central concern has been to focus on theory and empirical research that bear on major questions in the study of primate behavior. We therefore made a conscious decision at the outset to limit the number of chapters devoted to describing the behavior of specific taxa. In these, we furnish readers unfamiliar with the order with background information about the ecology, life history, and social systems of different species. We recognize that technical advances in methodology and analysis have contributed in important ways to our understanding of the behavior of primates. To emphasize theory and problems over methods and analysis, however, we made a second explicit decision to not include specific chapters about the latter in this volume. Excellent treatments of some of the technical breakthroughs that have led to novel insights into the behavior of primates can be found in some other recent texts (e.g., Setchell & Curtis 2011). Third, contributors to this volume are painfully aware that habitat loss, hunting, disease outbreaks, and live capture have decimated primate populations across the world. Because proper discussion of the singularly important issue of primate conservation deserves book-length treatment by itself, we have not dealt with it here. The editors and contributors to this volume are committed to the conservation of primates in the wild, and we have all agreed to donate the royalties from this book to a fund set up for this purpose through Conservation International. Several individuals who have graciously contributed photographs to this volume have waived their fees to support this fund.

During the initial planning phases, we realized that a book that attempts to cover findings in the burgeoning field of primate behavior would require contributions from multiple authors. Not all would share the same disciplinary background, theoretical approach, or empirical research skills. To ensure consistency in style and content and to reduce overlap between chapters, we gave authors considerable but by no means complete leeway in what to write. From time to time we may have wielded an overly heavy editorial hand, resulting in real or perceived gaps in particular chapters. These omissions are likely to reflect our editorial decisions as much as the author's own choices. In compiling a volume with so many contributors, a particularly significant challenge involved the issue of overlap between chapters. Our principal aim was to produce chapters that could be read by themselves, with minimal need to refer to others. Inevitably, the theories, questions, concepts, and data in some chapters impinge upon and overlap with those in others, and in these cases we direct interested readers to relevant parts in the book through the use of cross-references.

Producing a large edited book with 5 editors, 39 contributors, and 32 chapters requires considerable work and support. We are first and foremost extremely grateful to the contributors to this volume. All have shown uncommon commitment to this project from the start. They have worked diligently to produce their chapters, all the while responding to deadlines and our editorial demands collegially and with good humor. This book represents their collective efforts. Like the primates we study, this volume has experienced an unusually long gestation, and we thank the contributors for their patience. We hope the authors agree with us that the wait has been worthwhile.

The seeds for this book were sown several years ago in discussions between Barb Smuts and John Mitani. Along with the other editors of *Primate Societies*—Dorothy Cheney, Robert Seyfarth, Richard Wrangham, and Tom Struhsaker—Barb furnished sage counsel as we planned this book. We are deeply indebted to all the editors of *Primate Societies* for helping us to conceive this volume and ushering us through to its publication. We also thank Christie Henry, Amy Krynak, and Renaldo Migaldi at the University of Chicago Press. Christie's support and encouragement as we planned and assembled this book was greatly appreciated by all of us, and we are especially grateful for Amy and Renaldo's expert assistance and patience during the final phase of production.

Several colleagues improved this book by providing reviews of various chapters. For this we thank Louise Barrett, Fred Bercovitch, Monique Bogerhoff-Mulder, Sarah Brosnan, Dick Byrne, Marie Charpentier, Tim Clutton-Brock, Lee Cronk, Robin Dunbar, Peter Fashing, Linda Fedigan, William Foley, Doree Fragazsy, Sharon Gursky, Sandy Harcourt, Kristen Hawkes, Kim Hill, Elise Huchard, Lynne Isbell, Charlie Janson, Dawn Kitchen, Andreas Koenig, Kevin Langergraber, Rich Lawler, Steve Leigh, Joe Manson, Karen McComb, Charlie Nunn, Susan Perry, Simon Reader, Martha Robbins, Laurie Santos, Gabriele Schino, Joanna Setchell, Robert Seyfarth, Eric Alden Smith, Karen Strier, Larissa Swedell, Melissa Emery Thompson, Carel van Schaik, Linda Vigilant, Andy Whiten, Mike Wilson, and Juichi Yamagiwa. We are especially grateful to Jim Moore and two anonymous reviewers for furnishing thorough and careful reviews of the entire manuscript on behalf of the University of Chicago Press.

Authors would like to thank the following individuals who provided feedback and help with various chapters: Alison Jolly, Laurie Godfrey, Stefan Merker, Melanie Dammhahn, and Claudia Fichtel (chapter 2); Lauren Chapman, Tara Harris, Kevin Potts, and Tamaini Snaith (chapter 7); Charles Janson, Katharina Peters, Carel van Schaik, and Ulrike Walbaum (chapter 8); Marlies Heesen, Charlie Nunn, Miranda Swagemakers, and Rob Walker (chapter 9); Kristen Hawkes and Janneke van Woerden (chapter 10); Andrew Burrell, Jim Dietz, Todd Disotell, John Dumbacher, Rob Fleischer, Pascal Gagneux, Cliff Jolly, Rich Lawler, and Anthony Tosi (chapter 12); Kim Hill, Jon Stieglitz, and Rob Walker (chapter 13); Karin Isler, Adrian Jaeggi, and Carel van Schaik (chapter 14); Elise Huchard, Andreas Paul, and Joanna Setchell (chapter 16); Sandy Harcourt, Dawn Kitchen, Rich Lawler, and Richard Wrangham (chapter 17); Dorothy Cheney, Lee Cronk, Sandy Harcourt, Sarah Hrdy, Robert Trivers, and Juichi Yamagiwa (chapter 19); Monique Borgerhoff Mulder, Colette Berbesque, Alexander Brewis, and Gillian Brown (chapter 20); Alicia Melis, Richard Moore, Nausicaa Pouscoulous, Thomas Scott Phillips, Amanda Seed, Michael Tomasello, and Amrisha Vaish (chapter 25); Michael Beran, Mary Beran, Stephanie Berger, Betty Chan, Christopher Elder, Lorenz Gygax, Nerida Harley, Cameron Hastie, Jan Hemmi, Megan Hoffman, Sarah Hunsberger, John Kelley, Ernst Krusi, Marion Maag, John Mustaleski, Isabel Sanchez, Ken Sayers, and Shelly Williams (chapter 27); Dick Byrne, Cristiane Cäsar, Dorothy Cheney, Juan Carlos Gomez, Cat Hobaiter, and Tobias Riede (chapter 29); Marietta Dindo, Jean-Baptiste Leca, Susan Perry, and Jason Zampol (chapter 31).

Besides the authors, the following colleagues and organizations generously provided the photos that enhanced the chapters of this book: Tyler Barry, Simon Bearder, Jacinta Beehner, Carol Berman, David Bygott, Rebecca Chancellor, Zanna Clay, Carolyn M. Crockett, Lee Cronk, Laurence Culot, Victor Dávalos, Julia Diegmann, Perry van Duijnhoven, Lynda Dunkel, Manfred Eberle, Jane Gagne, Mary Glenn, Dan Hrdy, Sarah Blaffer Hrdy, Elise Huchard, Tina Jensen, Courtney Kirkpatrick, Craig Kirkpatrick, Wiebke Lammers, Hugo van Lawick, Beth L. Leech, Rebecca J. Lewis, Petra Löttker, Lisa MacDonald, Catherine Markham, Tetsuro Matsuzawa, Stefan Merker, Ellen Meulman, Florian Möllers, Anna Nekaris, Marliyn Norconk, Susan Perry, Anne Pisor, Ulrich Reichard, Carolyn Richardson and the Georgia State University Language Research Center, Ken Sayers, Johanna van Schaik, Anna Schnoell, Dylan Schwindt, Joanna Setchell, Pascale Sicotte, Antonio Souto, Thomas Struhsaker, Larissa Swedell, Nic Thompsen, Frans de Waal, Brandon Wheeler, Richard Wrangham, Patricia Wright, Juichi Yamagiwa, Eugen Zuberbühler, the Jane Goodall Institute, the Lincoln Park Zoo, and the Max Planck Institute for Evolutionary Anthropology.

The research reported in this book has been generously supported by several public and private organizations. The authors and editors gratefully acknowledge the support of the Biotechnology and Biological Sciences Research Council, UK; the Biomedical Primate Research Centre, Rijswijk, Netherlands; the Canada Research Chairs Program; the Center for Human Evolutionary Studies, Rutgers University; Deutsche Forschungsgemeinschaft; the Economic and Social Research Council, UK; IdeaWild; the Leverhulme Trust; the L.S.B. Leakey Foundation; the National Geographic Society; the Natural Science and Engineering Research Council of Canada; the Netherlands Organisation for Scientific Research; the New York Consortium in Evolutionary Primatology; New York University; Primate Conservation, Inc.; the Royal Society; Sigma Xi; the Smithsonian Institution; the Swiss National Science Foundation; the University of Pennsylvania; the University of Wisconsin; the US National Science Foundation; the US National Institutes of Health; Utrecht University; the Wenner-Gren Foundation; the Wildlife Conservation Society; Wissenschaftskolleg zu Berlin; and the Zoological Society of San Diego.

Finally, we thank the host countries that have so graciously sponsored the field research reviewed in this volume. As the chapters in this book make clear, particularly significant challenges lie ahead for the study of primate behavior. These will only be addressed successfully with the continued support of committed scientists, conservationists, politicians, and the lay public throughout the world.

Reference

Setchell, J. & Curtis, D. 2011. *Field and Laboratory Methods in Primatology: A Practical Guide*. Second Edition. Cambridge: Cambridge University Press.

Chapter 1 Introduction

The Editors

Goals of This Book

The study of nonhuman primates and their behavior generates considerable scientific interest. Anthropologists use information about primates to make inferences about the evolution of our hominin ancestors, while psychologists and biologists study them to glean insights into the behavioral, anatomical, physiological, and genetic factors that make us uniquely human. Most importantly, there are now many primatologists who study primates because they represent a highly diverse and ecologically successful radiation of socially complex mammals with unusual life histories.

For an entire generation of students and scientists, *Primate Societies* (Smuts et al. 1987) served as the standard reference in the field of primate behavior. But in the 25 years since it was published, much has been learned. New theoretical paradigms to understand the behavioral differences between nonhuman primates and humans have been developed, debated, and tested. Field and laboratory studies of an increasing number of species continue to add to our knowledge of the diversity of adaptations within the Primate order. For example, there has been an explosion of interest in and field research on the strepsirrhine primates, and as a result, we now have a much clearer picture of their behavior. Some of the well known long-term field studies of baboons, macaques, and chimpanzees have amassed quantitative data spanning multiple generations, which are of great value in understanding patterns of life history and behavior. These have now been joined by exemplary studies of species such as mouse lemurs, capuchin monkeys, and muriquis. Experiments conducted with wild and captive animals have furnished novel insights into primate cognition. New methods, such as PCR-enabled DNA sequencing and noninvasive hormonal monitoring, have greatly amplified our understanding of the proximate and ultimate causes of behavior. While data from recent research have sometimes reinforced what we knew 25 years ago, they have also forced us to revise and alter our thinking. And as new findings accumulate, it has become increasingly difficult for researchers inside and outside the discipline to keep abreast of recent advances and to synthesize results across the entire field.

With these issues in mind, *The Evolution of Primate Societies* compiles 31 chapters that review the current state of our knowledge regarding the behavior of nonhuman primates. Like *Primate Societies*, this book has the primary goal of providing an up-to-date synthesis that will serve as a standard reference for advanced undergraduates, graduate students, and the diverse community of scientists who are interested in the behavior of our closest living relatives. As the study of primate behavior has grown, it has changed from an inductive, taxon-oriented scientific discipline into a mature problem-oriented one that relies on observations and experiments to test evolutionary hypotheses. As one measure of this change, most chapters in the first major compilation of primate field studies, *Primate Behavior: Field Studies of Monkeys and Apes* (DeVore 1965), described the behavior of specific taxa (12/18 = 67%). In contrast, the predecessor to this volume, *Primate Societies* (Smuts et al. 1987) contained considerably fewer (14/40 = 35%). We continue that trend here by minimizing the number of taxon-specific chapters and organizing this volume around a set of four broad adaptive problems primates encounter as they attempt to grow, survive, and reproduce in their natural habitats. In this fashion we move away from descriptions of the behavior of specific taxa, and instead emphasize questions, focus on concepts, and evaluate theory with empirical evidence.

A second goal of this book is to integrate research on nonhuman and human behavior through the inclusion of chapters that specifically address the behavior of the human primate. Primatology is uniquely situated at the interface between the biological and social sciences. A long series of pioneers in the field—Robert Yerkes, Wolfgang Köhler, Ray Carpenter, Kinji Imanishi, David Hamburg, and Sherwood Washburn—recognized that the findings from primate research are of central importance for understanding human behavior. As our closest living relatives, primates provide the standard for defining human uniqueness and inform us about the changes that must have taken place during the course of our own evolution. This logic provided the impetus for some of the earliest field studies of primates, and it continues to act as the intellectual glue that bonds primatology with the social sciences. By illuminating the similarities and differences between humans and our closest living relatives, primate behavioral research still has much to tell us about our place in nature.

Primatology will continue to furnish insights into human behavior only if the field remains firmly rooted within the biological sciences. Major advances in the study of primate behavior during the past 40 years have relied heavily on theoretical and methodological breakthroughs in other areas in biology. For example, the melding of evolution, behavior, and ecology into the new subfield of behavioral ecology in the 1960s and 1970s provided the theoretical rationale for many of the subsequent primate field studies that were summarized in *Primate Societies*. To maintain their vitality and rigor, modern primate field studies must continue to assimilate recent developments in the biological sciences. An effective dialogue between the two disciplines will have important consequences for the behavioral sciences.

We believe that there has never been a better time for the study of primate behavior to achieve its integrative potential and serve as a unifying force to link its sister disciplines in the social and biological sciences. By synthesizing the findings of primate behavioral research, the chapters in this volume promise to highlight the significance of primatology to several fields, including animal behavior, anthropology, cognitive science, economics, genetics, and psychology.

Contents and Organization of This Book

With the preceding goals in mind, we have divided this book into five parts. Part 1, "Primate Behavioral Diversity," will introduce readers to the animals that constitute the order. Five taxa are considered in chapters on strepsirrhines and tarsiers, platyrrhines, colobines, cercopithecines, and apes. These introductory chapters adhere to a standard format, reviewing the ecology, life history, and social systems of primates in each taxon. The aim is to summarize salient aspects of the behavior of primates across the order. Taken together, these chapters furnish taxon-specific background information for the topical chapters that follow in the rest of the book.

Parts 2 through 5 form the central part of this volume. As noted above, chapters in these sections deal with four adaptive problems primates face in the wild. Part 2, "Surviving and Growing Up in a Difficult and Dangerous World," comprises seven chapters that focus on the ways in which primates respond to ecological challenges, such as finding food and avoiding predators, and how these ecological pressures have shaped the evolution of primate sociality, life history, and development. The penultimate chapter in this section reviews how these same ecological forces, through their effects on sociality, influence the genetic composition of populations; the concluding chapter employs the comparative data furnished by primate studies to evaluate human survival and life history in evolutionary perspective.

Reproduction is the evolutionary currency of import, and part 3, "Mating and Rearing Offspring," comprises eight chapters that deal with this crucial facet of primate life. Chapters in this section examine female and male reproductive strategies, the sources and magnitude of variation in female reproduction and male mating success, female choice, parenting, and sexual conflict in the form of infanticide by males. The final chapter in part 3 reviews aspects of human reproduction, highlighting points of divergence from and convergence with nonhuman primates.

Primates are unusually social animals. As a consequence, they must interact successfully with conspecifics to grow, survive, and reproduce. Part 4, "Getting Along," considers the forces that shape the evolution of primate social behavior and social relationships. Chapters in this section highlight research on the evolution of cooperation between kin and between nonkin, the mechanisms maintaining relationships, the adaptive consequences of social bonds, the nature of social preferences, and human social behavior.

The final part of this book, "Cognitive Strategies for Coping with Life's Challenges," considers how natural selection has shaped the cognitive abilities of primates. Five chapters review how nonhuman primates have evolved cognitive solutions to problems related to ecological challenges, their knowledge of social relations and the minds of others, their communication strategies, and their social learning, traditions, and culture. Like the other parts of the book, this part concludes with a chapter devoted exclusively to humans and the issue of human cognition.

References

De Vore, I. 1965. *Primate Behavior: Field Studies of Monkeys and Apes*. New York: Holt, Rinehart, Winston.
Smuts, B., Cheney, D., Seyfarth, R., Wrangham, R. & Struhsaker, T. 1987. *Primate Societies*. Chicago: University of Chicago Press.

Part 1
Primate Behavioral Diversity

LIKE HUMANS, the nonhuman primates are unusually social animals. But the ways in which sociality manifests vary considerably between species, as well as between individuals living in groups of the same species. Similarly, nonhuman primates occupy different types of habitats where they feed on different foods and are exposed to different risks in the form of predators and other hostile forces of nature. As they strive to grow, survive, and reproduce, nonhuman primates meet the ecological and social challenges posed by their environments by adopting different life-history tactics and strategies.

Over the past 25 years we have learned a great deal about the nature and extent of behavioral, ecological, and life history variability across the Primate order. A large amount of data has accumulated for many taxa in this regard as more and more primate species have been studied in their natural habitats. The primary goal of this book is to move beyond description of this variability and to ask questions about how primates solve a set of common adaptive problems they encounter in their everyday lives. Whenever possible, we have asked authors to test hypotheses with observations collected from primates in the wild and in captivity, and to synthesize data using general theories, concepts, and principles.

We recognize that answering questions about how primates adapt to problems they face in the real world requires a basic understanding of them and their behavior. It is with this in mind that we begin this volume with five chapters regarding the major taxa that make up the order. In these, authors provide broad overviews of the ecology, life history,

social systems, and behavior of strepsirrhines and tarsiers, New World monkeys (platyrrhines), Old World monkeys (colobines and cercopithecines), and apes. It is important to recognize that these chapters are not intended to represent exhaustive summaries of what we know about the biology and behavior of individuals in these various taxa. Instead, the aim is to set the stage for the rest of the book by furnishing background information about the nonhuman primates. In this way, those who may be unfamiliar with these animals will be able to read the material in the rest of the volume in an informed manner.

Taken together, the five chapters in this section reveal the rich diversity in behavior that characterizes the order Primates. The first two chapters on strepsirrhines and tarsiers and the New World monkeys feature three taxa about which comparatively little was known 25 years ago. The situation has changed considerably since then, because of the dedicated work of a new generation of primatologists who have collected detailed behavioral observations of them in the field. As a result, we are in a much better situation to evaluate the similarities and differences that exist between all three of these taxa and some of the better known Old World monkeys and apes. Strepsirrhines were once thought to lead largely solitary existences, but as Peter M. Kappeler demonstrates in chapter 2, these primates can be quite social by living in pairs or in dispersed groups, whose individuals may not always be together in space and time. The New World monkeys radiated into a small number of ecological niches in Central and South America about 40 million years ago. In the process, they evolved an array of social systems that Eduardo Fernandez Duque, Anthony Di Fiore, and Maren Huck review in chapter 3. Recent observations of the social systems of strepsirrhines and New World monkeys furnish novel data to test hypotheses about how primates survive and grow up in a difficult and dangerous world, issues featured in part 2 of this book.

In chapter 4, Elizabeth Sterck summarizes the behavior of the Old World colobine monkeys. These monkeys were the subjects of some of the earliest primate field studies and are well known for their folivorous feeding and infanticidal behavior. These aspects of colobine behavior feature prominently in discussions of primate survival and mating strategies reviewed in parts 2 and 3 of this volume, respectively. The relatively weak dominance relationships and paucity of grooming and affiliative behavior that characterizes New World monkey and colobine monkey social relationships stands in stark contrast to that displayed by the Old World cercopithecines, the subjects of chapter 5 by Marina Cords. Until recently, much of our knowledge about the behavior of wild primates derived from studies of some of these animals that occupy relatively open habitats where they could be easily followed and observed. A few of these studies have now spanned several years and multiple generations, and have accumulated long-term data on the fitness outcomes of behavior. These data, in turn, permit us to test hypotheses about the adaptive significance of the mating and rearing strategies and social relationships that are discussed in parts 3 and 4 of this book. A fifth and final taxon, the apes, is covered by David P. Watts in chapter 6. Long-term field research on gibbons and new observations of gorillas are starting to uncover unsuspected complexity and variability in their social systems. As our closest living relatives, bonobos and chimpanzees are of special interest as they provide a means to evaluate the unique adaptations humans have evolved, which are reviewed in chapters 13, 20, 26, and 32.

As the chapters in part 1 clearly show, members of the Primate order continue to receive considerable research attention in the wild. As field researchers begin to explore relatively inaccessible geographic areas, they discover new species. In addition, molecular genetic techniques furnish novel ways to assay variation within and between species. As a consequence of these developments, the order Primates has been subject to considerable taxonomic revision in the past 10 years. In the following five chapters, authors describe the taxonomy of their subjects. We have compiled these into a single taxonomy covering the entire order, which we present in table A.1. The scientific nomenclature adopted in this composite taxonomy has been followed in the remaining chapters of this volume. By convention, primate species are denoted by their common and scientific names when they first appear in each chapter; thereafter, they are referred to by their common names only.

Table A.1. Taxonomy of living primates

Taxon	Common name	Reference
Strepsirrhini		
Lemuriformes		9
Lemuroidea		
Cheirogaleidae		
Allocebus trichotis	Hairy-eared mouse lemur	
Cheirogaleus crossleyi	Furry-eared dwarf lemur	
Cheirogaleus major	Greater dwarf lemur	
Cheirogaleus medius	Western fat-tailed dwarf lemur	
Cheirogaleus minusculus	Lesser iron-gray dwarf lemur	
Cheirogaleus sibreei	Sibree's dwarf lemur	
Microcebus berthae	Madame Berthe's mouse lemur	
Microcebus bongolavensis	Bongolava mouse lemur	
Microcebus danfossi	Danfoss' mouse lemur	
Microcebus griseorufus	Rufous-gray mouse lemur	
Microcebus jollyae	Jolly's mouse lemur	
Microcebus lehilahytsara	Goodman's mouse lemur	
Microcebus macarthurii	MacArthur's mouse lemur	
Microcebus mamiratra	Claire's mouse lemur	
Microcebus mittermeieri	Mittermeier's mouse lemur	
Microcebus murinus	Gray mouse lemur	
Microcebus myoxinus	Pygmy mouse lemur	
Microcebus ravelobensis	Lac Ravelobe mouse lemur	
Microcebus rufus	Brown mouse lemur	
Microcebus sambiranensis	Sambirano mouse lemur	

Table A.1. (continued)

Taxon	Common name	Reference
Microcebus simmonsi	Simmons' mouse lemur	
Microcebus tavaratra	Northern brown mouse lemur	
Mirza coquereli	Coquerel's dwarf lemur	
Mirza zaza	Northern dwarf lemur	
Phaner electromontis	Amber Mountain fork-marked lemur	
Phaner furcifer	Masoala fork-marked lemur	
Phaner pallescens	Pale fork-marked lemur	
Phaner parienti	Sambirano fork-marked lemur	
Daubentoniidae		
Daubentonia madagascariensis	Aye-aye	
Indriidae		
Avahi betsileo	Betsileo woolly lemur	
Avahi cleesei	Cleese's woolly lemur	
Avahi laniger	Eastern woolly lemur	
Avahi meridionalis	Southern woolly lemur	
Avahi mooreorum	The Moore's woolly lemur	
Avahi occidentalis	Western woolly lemur	
Avahi peyrierasi	Peyrieras' woolly lemur	
Avahi ramanantsoavani	Ramantsoavana's southern woolly lemur	
Avahi unicolor	Unicolor woolly lemur	
Indri indri	Indri	
Propithecus candidus	Silky sifaka	
Propithecus coronatus	Crowned sifaka	
Propithecus coquereli	Coquerel's sifaka	
Propithecus deckenii	Decken's sifaka	
Propithecus diadema	Diademed sifaka	
Propithecus edwardsi	Milne-Edward's sifaka	
Propithecus perrieri	Perrier's sifaka	
Propithecus verreauxi	Verreaux's sifaka	
Propithecus tattersalli	Tattersall's sifaka	
Lemuridae		
Eulemur albifrons	White-fronted lemur	
Eulemur cinereiceps	Gray-headed lemur	
Eulemur collaris	Red-collared lemur	
Eulemur coronatus	Crowned lemur	
Eulemur flavifrons	Sclater's lemur	
Eulemur fulvus	Brown lemur	
Eulemur macaco	Black lemur	
Eulemur mongoz	Mongoose lemur	
Eulemur rubriventer	Red-bellied lemur	
Eulemur rufifrons	Red-fronted lemur	
Eulemur rufus	Red lemur	
Eulemur sanfordi	Sanford's lemur	
Hapalemur alaotrensis	Lac Alaotra bamboo lemur	
Hapalemur aureus	Golden bamboo lemur	
Hapalemur griseus	Gray bamboo lemur	
Hapalemur meridionalis	Southern bamboo lemur	
Hapalemur occidentalis	Western bamboo lemur	
Lemur catta	Ring-tailed lemur	
Prolemur simus	Greater bamboo lemur	
Varecia variegata	Black-and-white ruffed lemur	
Varecia rubra	Red ruffed lemur	
Lepilemuridae		
Lepilemur ankaranensis	Ankarana sportive lemur	
Lepilemur aeeclis	Antafia sportive lemur	
Lepilemur betsileo	Betsileo sportive lemur	
Lepilemur dorsalis	Nosy Be sportive lemur	
Lepilemur edwardsi	Milne-Edwards' sportive lemur	

(continued)

Table A.1. (continued)

Taxon	Common name	Reference
Lepilemur fleuretae	Fleurette's sportive lemur	
Lepilemur grewcocki	The Grewcock's sportive lemur	
Lepilemur hubbardi	Hubbard's sportive lemur	
Lepilemur jamesi	James' sportive lemur	
Lepilemur leucopus	White-footed sportive lemur	
Lepilemur microdon	Small-toothed sportive lemur	
Lepilemur milanoii	Daraina sportive lemur	
Lepilemur mittermeieri	Mittermeier's sportive lemur	
Lepilemur mustelinus	Weasel lemur	
Lepilemur randrianasoli	Randrianasolo's sportive lemur	
Lepilemur otto	Otto's sportive lemur	
Lepilemur petteri	Petter's sportive lemur	
Lepilemur ruficaudatus	Red-tailed sportive lemur	
Lepilemur sahamalazensis	Sahamalaza Peninsula sportive lemur	
Lepilemur scottorum	The Scott's sportive lemur	
Lepilemur seali	Seal's sportive lemur	
Lepilemur septentrionalis	Northern sportive lemur	
Lepilemur tymerlachsoni	Hawks' sportive lemur	
Lepilemur wrighti	Wright's sportive lemur	
Lorisiformes		
Lorisoidea		
Galagidae		7
Galaginae		
Euoticus elegantulus	Southern needle-clawed galago	
Euoticus pallidus	Northern needle-clawed galago	
Galago gallarum	Somali lesser galago	
Galago matschiei	Spectacled lesser galago	
Galago moholi	Southern lesser galago	
Galago senegalensis	Northern lesser galago	
Galagoides cocos	Kenya coast galago	
Galagoides demidovii	Demidoff's dwarf galago	
Galagoides granti	Mozambique galago	
Galagoides nyasae	Malawi galago	
Galagoides orinus	Mountain dwarf galago	
Galagoides rondoensis	Rondo dwarf galago	
Galagoides thomasi	Thomas's dwarf galago	
Galagoides zanzibaricus	Zanzibar galago	
Otolemur crassicaudatus	Thick-tailed greater galago	
Otolemur garnettii	Small-eared greater galago	
Otolemur monteiri	Silver greater galago	
Sciurocheirus alleni	Allen's galago	
Sciurocheirus gabonensis	Gabon Allen's galago	
Lorisidae		3, 7
Lorisinae		
Loris lydekkerianus	Slender loris	
Loris tardigradus	Red slender loris	
Nycticebus bengalensis	Bengal slow loris	
Nycticebus coucang	Greater slow loris	
Nycticebus pygmaeus	Pygmy slow loris	
Perodicticinae		
Arctocebus calabarensis	Calabar angwantibo	
Arctocebus aureus	Golden angwantibo	
Perodicticus potto	Potto	
Haplorrhini		
Tarsiiformes		6
Tarsiodea		
Tarsiidae		
Tarsinae		
Carlito syrichta	Phillippine tarsier	
Cephalopachus bancanus	Western tarsier	
Tarsius dentatus	Dian's tarsier	

Table A.1. (continued)

Taxon	Common name	Reference
Tarsius fuscus	Brown tarsier	
Tarsius lariang	Lariang tarsier	
Tarsius pelengensis	Peleng Island tarsier	
Tarsius pumilus	Pygmy tarsier	
Tarsius sangirensis	Sangihe tarsier	
Tarsius tarsier	Spectral tarsier	
Tarsius tumpara	Siau Island tarsier	
Tarsius wallacei	Wallace's tarsier	
Platyrrhini		10, 11
Ceboidea		
Atelidae		
Alouatttinae		
Alouatta arctoidea	Ursine howler monkey	
Alouatta belzebul	Red-handed howler monkey	
Alouatta caraya	South American black howler monkey	
Alouatta discolor	Spix's red-handed howler monkey	
Alouatta guariba	Brown howler monkey	
Alouatta juara	Juruá red howler monkey	
Alouatta macconnelli	Guianan red howler monkey	
Alouatta nigerrima	Black howler monkey	
Alouatta palliata	Mantled howler monkey	
Alouatta pigra	Central American black howler monkey	
Alouatta puruensis	Purús red howler monkey	
Alouatta sara	Bolivian red howler monkey	
Alouatta seniculus	Red howler monkey	
Alouatta ululata	Maranhão red-handed howler monkey	
Atelinae		
Ateles belzebuth	White-bellied spider monkey	
Ateles chamek	Black-faced black spider monkey	
Ateles geoffroyi	Geoffroy's spider monkey	
Ateles fusciceps	Brown-headed spider monkey	
Ateles hybridus	Variegated spider monkey	
Ateles marginatus	White-whiskered spider monkey	
Ateles paniscus	Red-faced black spider monkey	
Brachyteles arachnoides	Southern muriqui	
Brachyteles hypoxanthus	Northern muriqui	
Lagothrix cana	Geoffroy's woolly monkey	
Lagothrix flavicauda	Peruvian yellow-tailed woolly monkey	
Lagothrix lagotricha	Humboldt's or common woolly monkey	
Lagothrix lugens	Colombian woolly monkey	
Lagothrix poeppigii	Poeppig's or red woolly monkey	
Cebidae		
Aotinae		
Aotus azarai	Azara's night or owl monkey	
Aotus brumbacki	Brumback's night or owl monkey	
Aotus griseimembra	Grey-legged night or owl monkey	
Aotus lemurinus	Colombian night or owl monkey	
Aotus miconax	Andean night or owl monkey	
Aotus nancymaae	Nancy Ma's night or owl monkey	
Aotus nigriceps	Black-headed night or owl monkey	
Aotus trivirgatus	Douroucouli	
Aotus vociferans	Noisy night or owl monkey	
Aotus zonalis	Panamanian night or owl monkey	
Callitrichinae		
Callibella humilis	Black-crowned dwarf marmoset	
Callimico goeldii	Goeldi's monkey or callimico	
Callithrix aurita	Buffy-tufted-ear marmoset	
Callithrix flaviceps	Buffy-headed marmoset	
Callithrix geoffroyi	Geoffroy's tufted-ear marmoset	
Callithrix jacchus	Common marmoset	

(continued)

Table A.1. (continued)

Taxon	Common name	Reference
Callithrix kuhlii	Wied's black-tufted-ear marmoset	
Callithrix penicillata	Black-tufted-ear marmoset	
Cebuella pygmaea	Pygmy marmoset	
Leontopithecus caissara	Black-faced lion tamarin	
Leontopithecus chrysomelas	Golden-headed lion tamarin	
Leontopithecus chrysopygus	Black lion tamarin	
Leontopithecus rosalia	Golden lion tamarin	
Mico acariensis	Rio Acarí marmoset	
Mico argentatus	Silvery marmoset	
Mico chrysoleucus	Golden-white tassel-ear marmoset	
Mico emiliae	Snethlage's marmoset	
Mico humeralifer	Black and white tassel-ear marmoset	
Mico intermedius	Aripuanã marmoset	
Mico leucippe	Golden-white bare-ear marmoset	
Mico manicorensis	Manicoré marmoset	
Mico marcai	Marca's marmoset	
Mico mauesi	Maués marmoset	
Mico melanurus	Black-tailed marmoset	
Mico nigriceps	Black-headed marmoset	
Mico saterei	Sateré marmoset	
Saguinus bicolor	Pied bare-face tamarin	
Saguinus fuscicollis	Saddle-back tamarin	
Saguinus geoffroyi	Geoffroy's tamarin	
Saguinus imperator	Emperor tamarin	
Saguinus inustus	Mottled-face tamarin	
Saguinus labiatus	Red-bellied tamarin	
Saguinus leucopus	Silvery-brown tamarin	
Saguinus martinsi	Martin's bare-face tamarin	
Saguinus melanoleucus	White saddle-back tamarin	
Saguinus midas	Golden-handed tamarin	
Saguinus mystax	Mustached tamarin	
Saguinus niger	Black-handed tamarin	
Saguinus nigricollis	Black-mantle tamarin	
Saguinus oedipus	Cotton-top tamarin	
Saguinus tripartitus	Golden-mantle saddle-back tamarin	
Cebinae		
Cebus albifrons	White-fronted capuchin	
Cebus apella	Tufted or brown capuchin	
Cebus capucinus	White-faced capuchin	
Cebus cay	Hooded capuchin	
Cebus flavius	Marcgraf's capuchin	
Cebus kaapori	Ka'apor capuchin	
Cebus libidinosus	Bearded capuchin	
Cebus macrocephalus	Large-headed tufted capuchin	
Cebus nigritus	Black or black-horned capuchin	
Cebus olivaceus	Wedge-capped capuchin	
Cebus robustus	Crested capuchin	
Cebus xanthosternos	Yellow-breasted capuchin	
Saimirinae		
Saimiri boliviensis	Bolivian squirrel monkey	
Saimiri oerstedii	Central American squirrel monkey	
Saimiri sciureus	Common squirrel monkey	
Saimiri ustus	Golden-backed squirrel monkey	
Saimiri vanzolinii	Vanzolini's squirrel monkey	
Pitheciidae		
Callicebinae		
Callicebus aureipalatii	Madidi titi	
Callicebus baptista	Lago do Baptista titi	
Callicebus barbarabrownae	Northern Bahian blond titi	
Callicebus bernhardi	Prince Bernhard's titi	

Taxon	Common name	Reference
Callicebus brunneus	Brown titi	
Callicebus caligatus	Chestnut-bellied titi	
Callicebus cinerascens	Ashy titi	
Callicebus coimbrai	Coimbra's titi	
Callicebus cupreus	Red titi	
Callicebus discolor	Red-crowned titi	
Callicebus donacophilus	Reed titi	
Callicebus dubius	Doubtful titi	
Callicebus hoffmannsi	Hoffmann's titi	
Callicebus lucifer	Rufous-tailed collared titi	
Callicebus lugens	Widow monkey	
Callicebus medemi	Medem's collared titi	
Callicebus melanochir	Southern Bahian masked titi	
Callicebus modestus	Beni titi	
Callicebus moloch	Orabassu or dusky titi	
Callicebus nigrifrons	Black-fronted masked titi	
Callicebus oenanthe	Andean titi	
Callicebus olallae	Ollala's titi	
Callicebus ornatus	Ornate titi	
Callicebus pallescens	Paraguayan yellow titi	
Callicebus personatus	Northern masked titi	
Callicebus purinus	Red-bellied collared titi	
Callicebus regulus	Juruá collared titi	
Callicebus stephennashi	Stephen Nash's titi	
Callicebus torquatus	White-collared titi	
Pithecinae		
Cacajao calvus	Bald-headed uacari	
Cacajao melanocephalus	Humboldt's black-headed uacari	
Cacajao ouakary	Spix's black-headed uacari	
Chiropotes albinasus	White-nosed bearded saki	
Chiropotes chiropotes	Guianan bearded saki	
Chiropotes israelita	Rio Negro bearded saki	
Chiropotes satanas	Black bearded saki	
Chiropotes utahickae	Uta Hick's bearded saki	
Pithecia aequatorialis	Equatorial saki	
Pithecia albicans	Buffy saki	
Pithecia irrorata	Bald faced saki	
Pithecia monachus	Monk saki	
Pithecia pithecia	White-faced saki	
Catarrhini		
Cercopithecoidea		
Cercopithecidae		
Cercopithecinae		4, 5, 12
Allenopithecus nigroviridis	Allen's swamp monkey	
Cercocebus agilis	Agile mangabey	
Cercocebus atys	Sooty mangabey	
Cercocebus chrysogaster	Golden-bellied mangabey	
Cercocebus galeritus	Tana river mangabey	
Cercocebus sanjei	Sanje mangabey	
Cercocebus torquatus	Collared mangabey	
Cercopithecus albogularis	Syke's monkey	
Cercopithecus ascanius	Red-tailed monkey	
Cercopithecus campbelli	Campbell's mona monkey	
Cercopithecus cephus	Moustached guenon	
Cercopithecus denti	Dent's mona monkey	
Cercopithecus diana	Diana monkey	
Cercopithecus doggetti	Silver monkey	
Cercopithecus dryas	Dryas monkey	

(continued)

Taxon	Common name	Reference
Cercopithecus erythrogaster	White-throated guenon	
Cercopithecus erythrotis	Red-eared guenon	
Cercopithecus hamlyni	Hamlyn's monkey	
Cercopithecus kandti	Golden monkey	
Cercopithecus lhoesti	L'Hoest's monkey	
Cercopithecus lowei	Lowe's mona monkey	
Cercopithecus mitis	Blue monkey	
Cercopithecus mona	Mona monkey	
Cercopithecus neglectus	DeBrazza's monkey	
Cercopithecus nictitans	Greater spot-nosed monkey	
Cercopithecus petaurista	Lesser spot-nosed monkey	
Cercopithecus pogonias	Crested mona monkey	
Cercopithecus preussi	Preuss's monkey	
Cercopithecus roloway	Roloway monkey	
Cercopithecus sclateri	Sclater's guenon	
Cercopithecus solatus	Sun-tailed monkey	
Cercopithecus wolfi	Wolf's mona monkey	
Chlorocebus aethiops	Grivet	
Chlorocebus cynosuros	Malbrouck	
Chlorocebus djamdjamensis	Bale mountains vervet	
Chlorocebus pygerythrus	Vervet monkey	
Chlorocebus sabaeus	Green monkey	
Chlorocebus tantalus	Tantalus monkey	
Erythrocebus patas	Patas monkey	
Lophocebus albigena	Gray-cheeked mangabey	
Lophocebus aterrimus	Black crested mangabey	
Lophocebus opdenboschi	Opdenbosch's mangabey	
Macaca arctoides	Stump-tailed macaque	
Macaca assamensis	Assam macaque	
Macaca cyclopis	Formosan rock macaque	
Macaca fascicularis	Crab-eating or long-tailed macaque	
Macaca fuscata	Japanese macaque	
Macaca hecki	Heck's macaque	
Macaca leonina	Northern pig-tailed macaque	
Macaca maura	Moor macaque	
Macaca mulatta	Rhesus monkey	
Macaca munzala	Arunachal macaque	
Macaca nemestrina	Southern pig-tailed macaque	
Macaca nigra	Celebes crested macaque	
Macaca nigrescens	Gorontalo macaque	
Macaca ochreata	Booted macaque	
Macaca pagensis	Pagai Island macaque	
Macaca radiata	Bonnet macaque	
Macaca siberu	Siberut macaque	
Macaca silenus	Lion-tailed macaque	
Macaca sinica	Toque macaque	
Macaca sylvanus	Barbary macaque	
Macaca thibetana	Milne-Edwards' or Tibetan macaque	
Macaca tonkeana	Tonkean macaque	
Mandrillus leucophaeus	Drill	
Mandrillus sphinx	Mandrill	
Miopithecus ogouensis	Gabon talapoin	
Miopithecus talapoin	Angolan talapoin	
Papio anubis	Olive baboon	
Papio cynocephalus	Yellow baboon	
Papio hamadryas	Hamadryas baboon	
Papio papio	Guinea baboon	
Papio ursinus	Chacma baboon	
Rungwecebus kipunji	Kipunji	
Theropithecus gelada	Gelada	

Taxon	Common name	Reference
Colobinae		3, 7
Colobus angolensis	Angola colobus	
Colobus guereza	Guereza	
Colobus polykomos	King colobus	
Colobus satanas	Black colobus	
Colobus vellerosus	White-thighed colobus	
Nasalis larvatus	Proboscis monkey	
Presbytis comata	Javan grizzled surili or leaf monkey	
Presbytis femoralis	Banded surili	
Presbytis fredericae	Javan fuscous surili	
Presbytis frontata	White-fronted surili	
Presbytis hosei	Grizzled surili	
Presbytis melalophos	Mitered surili	
Presbytis potenziani	Mentawai surili	
Presbytis rubicunda	Red surili or maroon leaf monkey	
Presbytis siamensis	Pale-thighed surili	
Presbytis thomasi	Sumatran grizzled surili or Thomas's leaf monkey	
Procolobus badius	Western red colobus	
Procolobus kirkii	Kirk's red colobus	
Procolobus gordonorum	Udzungwa red colobus	
Procolobus rufomitratus	Tana river or eastern red colobus	
Procolobus pennantii	Pennant's red colobus	
Procolobus verus	Olive colobus	
Pygathrix nemaeus	Red-shanked douc	
Pygathrix nigripes	Black-shanked douc	
Rhinopithecus avunculus	Tonkin snub-nosed monkey	
Rhinopithecus bieti	Yunnan snub-nosed monkey	
Rhinopithecus brelichi	Guizhou snub-nosed monkey	
Rhinopithecus roxellana	Golden snub-nosed monkey	
Semnopithecus entellus	Gray or Hanuman langur	
Semnopithecus johnii	Nilgiri black langur	
Semnopithecus vetulus	Purple-faced langur	
Simias concolor	Pig-tailed snub-nosed monkey	
Trachypithecus auratus	Ebony leaf monkey	
Trachypithecus barbei	Barbe's leaf monkey	
Trachypithecus delacouri	White-rumped black leaf monkey	
Trachypithecus francoisi	Black leaf monkey	
Trachypithecus geei	Golden leaf monkey	
Trachypithecus laotum	White-browed black leaf monkey	
Trachypithecus obscurus	Dusky leaf monkey	
Trachypithecus pileatus	Capped leaf monkey	
Trachypithecus poliocephalus	White-headed black leaf monkey	
Trachypithecus villosus	Silver leaf monkey	
Hominoidea		
Hylobatidae		5
Bunopithecus hoolock	Hoolock gibbon	
Hylobates agilis	Agile gibbon	
Hylobates albibarbis	Bornean white-bearded gibbon	
Hylobates klossi	Kloss gibbon	
Hylobates lar	Lar or white-handed gibbon	
Hylobates moloch	Silvery or Javan gibbon	
Hylobates muelleri	Müller's Bornean gibbon	
Hylobates pileatus	Pileated gibbon	
Nomascus concolor	Black crested gibbon	
Nomascus gabriellae	Red-cheeked gibbon	
Nomascus hainanus	Hainan gibbon	
Nomascus leucogenys	Northern white-cheeked gibbon	
Nomascus siki	Southern white-cheeked gibbon	
Symphalangus syndactylus	Siamang	

(continued)

Table A.1. (continued)

Taxon	Common name	Reference
Hominidae		
Ponginae		13
Pongo abelii	Sumatran orangutan	
Pongo pygmaeus	Bornean orangutan	
Homininae		
Gorilla gorilla	Gorilla	14
Homo sapiens	Human	8
Pan paniscus	Bonobo	1, 2
Pan troglodytes	Chimpanzee	1, 2

1. Becquet, C., Patterson, N., Stone, A., Przeworski, M. & Reich, D. 2007. Genetic structure of chimpanzee populations. *PLoS Genetics,* 3, e66.

2. Caswell, J., Mallik, S., Richter, D., Meubauer, J., Schirmer, C., Gnerre, S. & Reich, D. 2008. Analysis of chimpanzee genetic history based on genome sequence alignments. *PLoS Genetics,* 4, e100000657.

3. Brandon-Jones D., Eudey A., Geissmann T., Groves, C., Melnick, D., Morales, J., Shekelle, M. & Stewart, C-B. 2004. Asian primate classification. *International Journal of Primatology*, 25, 97–164.

4. Davenport, T., Stanley, W., Sargis, E., De Luca, D., Mpunga, N., Machaga, S., & Olson, L. 2006. A new genus of African monkey, *Rungwecebus*: Morphology, ecology, and molecular phylogenetics. *Science*, 312, 1378–1381.

5. Groves, C. 2005. Order Primates. In *Mammal Species of the World* (ed. by Wilson, D. E. & Reeder, D. M.), 111–184. Baltimore: Johns Hopkins University Press.

6. Groves, C. & Shekelle, M. 2010. The genera and species of Tarsiidae. *International Journal of Primatology*, 31, 1071–1082.

7. Grubb, P., Butynski, T., Oates, J., Bearder, S., Disotell, T., Groves, C. & Struhsaker, T. 2003. Assessment of the diversity of African primates. *International Journal of Primatology*, 24, 1301–1357.

8. Linnaeus, C. 1758. *Systema Naturae*. Stockholm: Laurentii Salvii.

9. Mittermeier, R. A., Louis Jr., E. E., Richardson, M., Schwitzer, C., Langrand, O., Rylands, A. B., Hawkins, F., Rajaobelina, S. Ratsimbazafy, J. Rasoloarison, R., Roos, C., Kappeler, P. M. & Mackinnon, J. 2010. *Lemurs of Madagascar.* Third edition. Tropical Field Guide Series, Conservation International, Arlington, VA.

10. Opazo, J. C., Wildman, D. E., Prychitko, T., Johnson, R. M. & Goodman, M. 2006. Phylogenetic relationships and divergence times among New World monkeys (Platyrrhini, Primates). *Molecular Phylogenetics and Evolution*, 40, 274–280.

11. Rylands, A. & Mittermeier, R. 2009. The diversity of the New World primates (Platyrrhini): An annotated taxonomy. In *South American Primates. Comparative Perspectives in the Study of Behavior, Ecology, and Conservation* (ed. by Garber, P. A., Estrada, A., Bicca-Marques, J.C., Heymann, E.W. & Strier, K.B.), 23–54. New York: Springer.

12. Sinha, A., Datta, A., Madhusudan, M. & Mishra, C. 2005. *Macaca munzala*: A new species from western Arunachal Pradesh, northeastern India. *International Journal of Primatology*, 26, 977 –989.

13. Steiper, E. 2006. Population history, biogeography, and taxonomy of orangutans (Genus: *Pongo*) based on a population genetic meta-analysis of multiple loci. *Journal of Human Evolution*, 50, 509–522.

14. Thalmann, O., Fischer, A., Lankaster, F., Paabo, S. & Vigilant, L. 2007. The complex evolutionary history of gorillas: Insights from genetic data. *Molecular Biology and Evolution*, 24, 146–158.

Chapter 2 The Behavioral Ecology of Strepsirrhines and Tarsiers

Peter M. Kappeler

PHYLOGENETIC ANALYSES of morphological and genetic data unequivocally identify strepsirrhines (Lemuriformes and Lorisiformes) as a monophyletic suborder of Primates (Yoder 1997). Strepsirrhines have sometimes been allied with tarsiers (Tarsiiformes) because of some plesiomorphic traits, but the corresponding taxon Prosimii is clearly paraphyletic (Schmitz et al. 2001; Roos et al. 2004) and should not be used (Fleagle 1999). Instead, tarsiers are grouped with the platyrrhines (New World monkeys) and catarrhines (Old World monkeys, apes, and humans) into the haplorrhines. However, because of some phenotypic convergences between tarsiers and nocturnal strepsirrhines, and because this numerically small primate taxon is still understudied, tarsiers will be treated here together with strepsirrhines rather than in a separate chapter.

Phenotypically, various shared primitive features of the life histories, reproductive biology, and nervous systems of strepsirrhines and tarsiers affect and constrain the way in which these basal primates have adapted to particular ecological and social challenges. An appreciation of these and other characteristic traits is therefore helpful for understanding the socioecological adaptations of these animals. Because the number of studies of the behavioral ecology of these basal primates has increased disproportionally more than that of corresponding studies of anthropoids (monkeys, apes and humans) in the past 25 years (Bearder 1987; Richard 1987), it is no longer possible to provide a complete review. The specific aim of this chapter, therefore, is to summarize selected, relevant facts about strepsirrhine and tarsier diversity, ecology, life history, and social systems to provide taxon-specific background information for the topical reviews constituting the core of this volume.

Diversity and Biogeography

Strepsirrhines and tarsiers make up about a third of all living primates. The lemurs of Madagascar (Lemuriformes) represent the largest group among strepsirrhines. Molecular methods and increased sampling in the field have resulted in a drastic increase in the number of recognized lemur species. While there are doubts about the scientific standards used to identify and describe some new species (Tattersall 2007; Markolf et al. 2011), there is an emerging consensus among taxonomic experts that lemur diversity is much higher than the 20 species recognized during most of the previous century (cf. Richard 1987). Today, there is broad consensus that the living lemurs can be divided into 5 families and 15 genera, with about 100 known species (Mittermeier et al. 2010; table 2.1). In addition, 17 species from an additional 8 genera and 3 families went extinct only during the past 500 to 2,000 years and were clearly part of the contemporary fauna (Godfrey & Jungers 2003). The remains of these animals are so recent that they have not yet fossilized; hence they are referred to as the subfossil lemurs.

Lemurs are endemic to Madagascar. About 150 million years ago, during the breakup of Gondwana, Madagascar was separated from the African mainland. It remained attached to today's Indian subcontinent for about 60 million years. Because molecular methods have estimated the time of the single successful lemur colonization of Madagascar as around 65 to 75 million years ago (Horvath et al. 2008; Matsui et al. 2009), the first lemurs presumably rafted across the Mozambique Channel (Kappeler 2000a; Roos et al. 2004; but see Stankiewicz et al. 2006 for evidence suggesting an Asian origin). Following the early divergence of the ancestral Daubentoniidae, the remaining four lineages with living representatives diverged about 24 to 40 million years ago after the emergence of the first rain forests on Madagascar (Horvath et al. 2008).

Today, lemurs inhabit a wide range of different habitats. Madagascar's north and east are dominated by wet forests, the west is covered by dry deciduous forests, and the arid south is dominated by spiny forests; only the mostly treeless central plateau is largely devoid of lemurs. All major Malagasy rivers have their source in the central highlands. They divide the island into smaller biogeographic regions, which presumably acted as natural barriers for range expansions of many species (Wilmé et al. 2006; Vences et al. 2009). Interestingly, some species have ranges covering almost the length of the island (e.g., the gray mouse lemur, *Microcebus murinus*; the western fat-tailed dwarf lemur, *Cheirogaleus medius*), whereas others have some of the smallest known ranges among living primates (e.g., Madame Berthe's mouse lemur, *Microcebus berthae*; Perrier's sifaka, *Propithecus perrieri*; the northern sportive lemur, *Lepilemur septentrionalis*). Accordingly, some lemur species with small remaining ranges are among the most threatened primates.

The Lorisiformes are found in Africa and Asia. Based on morphological and molecular data, two clades can be distinguished at the family level (Roos et al. 2004; Masters et al. 2005). The bush babies, or galagos (Galagidae), are the most diverse members of this monophyletic group, with representatives of five genera found in most suitable habitats of sub-Saharan Africa, ranging from West African rain forest to South African acacia woodlands, including secondary habitats (Nash et al. 1987; Nekaris & Bearder 2007). The true diversity of galagos remains unknown as new taxa continue to be discovered and remain undescribed (Grubb et al. 2003). It is therefore difficult to precisely characterize their geographic ranges and, hence, the ecological success of individual taxa. Some species (e.g., Rondo dwarf galago, *Galagoides rondoensis*; fig. 2.1) are known to have small, fragmented ranges, however, and are also among the most endangered primates.

Lorises (Lorisidae) are found in both Africa and Asia. One gracile and one robust genus exists on each continent: *Arctocebus* and *Perodicticus* in Africa, and *Loris* and *Nycticebus* in Asia. They remain poorly studied in the field, so that their exact ranges and local densities are not well known. According to molecular estimates, the Galagidae

Fig. 2.1. *Galago rondoensis*, a representative of an endangered and poorly studied bushbaby. Photo courtesy of Simon Bearder.

and Lorisidae separated about 35 million years ago (Matsui et al. 2009). The historical phylogeography of the Lemuriformes and Lorisiformes continues to be debated as additional molecular data and lorisid fossils emerge, along with new information about the relative positions of India and Madagascar during the breakup of Gondwana (Masters et al. 2005; Stevens & Heesy 2006; Ali & Huber 2010).

Tarsiers are an enigmatic group of Southeast Asian primates with a present-day distribution limited to the Indonesian archipelago and Phillipine islands. They are the sole survivors of a much more diverse and widespread group of Eocene tarsiiforms. As with other small nocturnal primates, new species continue to be discovered (Merker & Groves 2006), bringing the current species count to at least 10 (Shekelle et al. 2008; Groves & Shekelle 2010), some of which are separated by remarkably narrow acoustic and genetic boundaries (Merker et al. 2009). Recent analyses of morphometric and genetic data have suggested that a separation into three distinct genera is warranted (Groves & Shekelle 2010). The affiliation of tarsiers with either strepsirrhines or haplorrhines has been the topic of a controversy for more than two centuries (Gursky 2007a), but molecular analyses clearly support the monophyly of the haplorrhines—that is, the anthropoids (New and Old World monkeys, apes, and humans) and tarsiers (Schmitz et al. 2001).

Ecology and Life History

Body Size

Ecologically, lemurs, lorisiforms, and tarsiers are a highly diverse group (Charles-Dominique 1977; Richard & Dewar 1991), and much of this diversity is linked to their variation in body size. This group includes the smallest living primate (Madame Berthe's mouse lemur, 30 g) as well as some of the largest primates known to exist (*Archaeoindris fontoynonti*, more than 150 kg; table 2.1). Because the larg-

est extant lemurs (*Indri* and *Propithecus* spp.) weigh less than 10 kg, all living lemurs are considerably smaller than their recently extinct cousins. Galagids and lorisids exhibit a narrower range of body size variation than lemurs, with individual species varying between about 50 g and 1500 g, whereas tarsiers are much smaller, weighing only 50 to 130 g (table 2.1). It is noteworthy that all lemurs, including the large subfossil species, lack sexual size dimorphism. If anything, females tend to be slightly heavier than males (Kappeler 1991; Godfrey et al. 1993). Sex differences in body weight among lorisiforms and tarsiiforms are modest, with males being on average 5 to 20% heavier than females (Kappeler 1991). Despite this variability in body size and various specializations in locomotor behavior, such as vertical clinging and leaping in the Indriidae and Lepilemuri-

Table 2.1. Taxonomy and life history of strepsirrhines and tarsiers

Suborder	Family	Genus	No. of species	Common name	Activity	Body mass (g)	Social organization
Lemuriformes	Daubentoniidae	*Daubentonia*	2 (1 subfossil)	Aye-aye	Nocturnal	3,000 (14,000)	S
	Indriidae	*Avahi*	9	Woolly lemur	Nocturnal	600–1,200	P
		Propithecus	9	Sifaka	Diurnal	2,500–6,500	G
		Indri	1	Indri	Diurnal	6,500	P
	Lepilemuridae	*Lepilemur*	24	Sportive lemur	Nocturnal	500–1,000	S, P
	Cheirogaleidae	*Microcebus*	≤ 16	Mouse lemur	Nocturnal	30–70	S
		Allocebus	1	Hairy-eared dwarf lemur	Nocturnal	80	S
		Cheirogaleus	5	Dwarf lemur	Nocturnal	150–400	S, P
		Phaner	4	Fork-marked lemur	Nocturnal	300–500	P
		Mirza	2	Giant mouse lemur	Nocturnal	300	S
	Lemuridae	*Pachylemur*	2 (subfossil)	Heavy lemur	Diurnal	11–13,000	
		Hapalemur	5	Bamboo lemur	Cathemeral	700–1,600	P, G
		Prolemur	1	Greater bamboo lemur	Cathemeral	2,400	G
		Lemur	1	Ring-tailed lemur	Diurnal	2,200	G
		Eulemur	12	True lemur	Cathemeral	1,500–2,500	P, G
		Varecia	2	Ruffed lemur	Diurnal	3,000–4,500	G
	Archaeolemuridae	*Archaeolemur*	2 (subfossil)	Monkey lemur	Diurnal	18–26,000	
		Hadropithecus	1 (subfossil)	Short-jawed monkey lemur	Diurnal	35,000	
	Palaeopropithecidae	*Palaeopropithecus*	3 (subfossil)	Sloth lemur	Diurnal	25–55,000	
		Archaeoindris	1 (subfossil)	Giant sloth lemur	Diurnal	160,000	
		Babakotia	1 (subfossil)	Radofilao's sloth lemur	Diurnal	20,000	
		Mesopropithecus	3 (subfossil)	Medium-sized sloth lemur	Diurnal	9–14,000	
	Megaladapidae	*Megaladapis*	3 (subfossil)	Koala lemur	Diurnal	45–85,000	
Lorisiformes	Galagidae	*Euoticus*	2	Needle-clawed galago	Nocturnal	200–300	S
		Galago	4	Lesser galago	Nocturnal	200	S
		Galagoides	8	Dwarf galago	Nocturnal	50–150	S, P
		Sciurocheirus	2	Allen's galago	Nocturnal	250–500	S
		Otolemur	3	Greater galago	Nocturnal	600–1,000	S
	Lorisidae	*Arctocebus*	2	Angwantibo	Nocturnal	150–300	S
		Loris	2	Slender loris	Nocturnal	100–300	S
		Nycticebus	3	Slow loris	Nocturnal	300–1,400	S
		Perodicticus	1	Potto	Nocturnal	900–1,500	S
Tarsiiformes	Tarsiidae	*Tarsius*	≤ 9	Eastern tarsier	Nocturnal	60–130	S, P, G
		Carlito	1	Phillippine tarsier	Nocturnal	120	S
		Cephalopachus	1	Western tarsier	Nocturnal	120	S

S = solitary; P = pair-living; G = group-living

dae, all lemurs are or were primarily arboreal (Gebo 1987; Godfrey & Jungers 2003). Only ring-tailed lemurs (*Lemur catta*) and some *Eulemur* spend significant amounts of time on the ground (Ward & Sussman 1979). All lorisiforms and tarsiiforms are exclusively arboreal, with some galagos and tarsiers exhibiting pronounced leaping behavior (Oxnard et al. 1989; Niemitz 2010).

Brain Size

One of the functionally most important correlates of variation in body size is brain size. Even after controlling for body size, the vast majority of strepsirrhines and tarsiers have smaller brain volumes than same-sized anthropoids (Armstrong 1985; Isler et al. 2008). The brains of the largest subfossil lemurs were only about one-third the size of those of great apes with similar body size (Schwartz et al. 2005). Only the relative brain size of *Archeolemur* matches that of some anthropoids (Godfrey et al. 2006). Although lemurs tend to have slightly larger brains than lorisiforms, three lemurs (*Cheirogaleus*, *Avahi*, and *Lepilemur)* have the smallest brains, relative to body size, of any primates; the aye-aye (*Daubentonia madagascariensis*) possesses the largest brain relative to body size among strepsirrhines (Isler et al. 2008). Although some of the interspecific variation in relative brain size among all primates can be explained by differences in sensory or technical competence, the empirical evidence favors a sociocognitive explanation (Dunbar & Shultz 2007). Accordingly, the smaller relative brain size of strepsirrhines and tarsiers may reflect adaptations to the reduced demands of their less complex social lives (Kudo & Dunbar 2001), but among lemurs, interspecific variation in relative brain size appears to correlate more closely with seasonality (van Woerden et al. 2010), ecology, and diet than with group size (MacLean et al. 2009).

Diet

Dietary diversity among strepsirrhines and tarsiers is more pronounced than among all anthropoid primates taken together, and it features extreme specialists as well as many generalists. Tarsiers, for example, are exclusively faunivorous, preying upon several arthropods and small vertebrates; they are the only primates to avoid any kind of plant food (Gursky 2007a; fig. 2.2). Some lorises are almost exclusively faunivorous as well (Nekaris & Rasmussen 2003). Fork-marked lemurs (*Phaner* spp.), in contrast, specialize on tree gum and sap, spending more than 85% of their feeding time on them (Schülke 2003). Some sportive lemurs (Milne-Edwards' sportive lemur, *Lepilemur edwardsi*) and avahis (Western woolly lemur, *Avahi occidentalis*) spend up to 100% of their feeding time on leaves, buds, and flowers

(Thalmann 2001). Gentle lemurs (*Hapalemur* spp. and *Prolemur* spp.) are bamboo specialists (Tan 1999), and ruffed lemurs (*Varecia* spp.) are almost exclusively frugivorous (Britt 2000). All other species have mixed diets that combine these major components either regularly or seasonally (Ganzhorn 1988, 1989; Nekaris & Bearder 2007). The distribution and availability of different types of food have important implications for ranging patterns (chapter 7, this volume), competitive regimes (chapter 9, this volume), and life history adaptations (chapter 10, this volume).

Metabolic Rate

Because food is the primary source of energy, feeding behavior, dietary strategies and adaptations are important determinants of how much energy is available for growth, maintenance, and reproduction (Perry & Hartstone-Rose 2010; chapter 7, this volume). Basal metabolic rate describes how much energy an individual expends for the maintenance of homeostasis. In mammals it scales allometrically to body weight, and a species' deviations from expected values are explained by its diet and taxonomic affiliation (Elgar & Harvey 1987). Among primates, most strepsirrhines have lower basal metabolic rates than predicted for their body mass (Müller 1985; Richard & Nicoll 1987; Ross 1992), and a recent field study suggested that total energy expenditure of some lemurs is even lower than their predicted field metabolic rates (Simmen et al. 2010). This systematic deviation of an entire lineage is not easily explained by diet, however. Although sportive lemurs (*Lepilemur* spp.), for example, are extremely folivorous and have one of the lowest basal metabolic rates for a mammal of their body size (Schmid & Ganzhorn 1995), more frugivorous and faunivorous strepsirrhines as well as tarsiers are characterized by low metabolic rates (e.g., *Tarsius*, McNab & Wright 1987; *Eulemur*, Daniels 1984). The answer may therefore have to do with general advantages of energy savings.

Field studies of the ecophysiology of several species of *Microcebus* and *Cheirogaleus* revealed that their metabolic machinery is indeed optimized towards saving energy and water. In their pronounced seasonal habitats in southern and western Madagascar, these small lemurs are exposed to seasonal variation in food availability and to strong daily fluctuations in ambient temperature during a long cool dry season. These nocturnal cheirogaleids spend the day in tree holes, where their body temperature drops to ambient levels (as low as 7° C!) on a daily basis, so that daily torpor provides them with important energy savings (Schmid & Speakman 2000). They save additional energy by passively reheating with rising ambient temperatures in the morning. Several species of *Cheirogaleus* and *Microcebus*, as well as *Allocebus*, even hibernate for up to seven months, spending

Fig. 2.2. *Tarsius dentatus* eating a large insect. All tarsiers have an exclusively faunivorous diet. Photo courtesy of Stefan Merker.

most of the barren dry season inactive (Bourlière & Petter-Rousseau 1966; Schmid & Kappeler 1998; Atsalis 1999; Dausmann et al. 2004; Kobbe & Dausmann 2009; Schmid & Ganzhorn 2009). Interestingly, galagos use similar adaptations to comparable patterns of seasonality in southern Africa, but do so much more flexibly (Mzilikazi et al. 2006; Nowack et al. 2010). Many traits of other lemurs, such as small group size and female dominance (see below), have also been linked to energetic constraints imposed by a relatively harsh and unpredictable environment (Wright 1999; Dewar & Richard 2007). It is only during reproduction that females elevate their otherwise low metabolic rates to support the additional costs of gestation and lactation (Young et al. 1990; Kappeler 1996). Future field studies will have to identify possible specific advantages of low metabolic rates in lorisoids and tarsiers.

Life History

Life history traits, which describe age-specific schedules of reproduction and survivorship, are also scaled to body size, so that strepsirrhines generally tend to have relatively

fast life histories—with some notable exceptions, such as *Propithecus* spp. (Richard et al. 2002; Pochron et al. 2004) and the subfossil *Hadropithecus* (Catlett et al. 2010). Age at first reproduction varies between one year in the smallest species, *Microcebus*, and six years in the largest species, *Propithecus* (Richard et al. 2002; Pochron et al. 2004). Litters of two or three young are common among the Cheirogaleidae and in *Varecia* (Kappeler 1998a), and are physiologically possible in some Lemuridae in captivity or in years of exceptional resource abundance (Pereira et al. 1999). A tarsier produces a single, exceptionally large (25% of maternal weight) infant (Wright 1990). Not surprisingly, tarsier mothers park their infants for much of the night, carrying them orally between multiple caching sites as they forage (Gursky 2007a). Most galagos and lorises give birth to singletons, but twins occur regularly in *Galagoides*, *Galago*, and *Loris* (Nekaris & Bearder 2007). Infant parking is also common among lorises, cheirogaleids, lepilemurs, and aye-ayes, with species varying in the mode of infant transport (oral or active clinging) among caching sites (Kappeler 1998a; fig. 2.3). Ruffed lemurs (*Varecia* spp.) are unusual in being the only diurnal primates that park their infants

Fig. 2.3. Gray mouse lemurs, *Microcebus murinus*, are one of many nocturnal strepsirrhines that park their young infants (here a three-week-old infant in an artificial nest box in the wild). Photo courtesy of Elise Huchard.

(Morland 1990). This variation is functionally relevant because infant care patterns create conditions for paternal care and infanticide risks with far-reaching consequences for male-female relations (van Schaik & Kappeler 1997).

Compared to that of anthropoids, the infant development of strepsirrhines and tarsiers until weaning is absolutely less variable across species, ranging from several weeks in mouse lemurs and five to eight months in sifakas, aye-ayes, and indris (Feistner & Ashbourne 1994; Lee & Kappeler 2003; Godfrey et al. 2004) to possibly one to three years in the extinct *Archaeolemur* and *Hadropithecus* (Catlett et al. 2010). Care of infants during this time is chiefly provided by the mother (Doyle et al. 1969; Klopfer & Boskoff 1979; Grieser 1992; Hilgartner et al. 2008; Barthold et al. 2009), but males may groom (*Loris* spp., Nekaris 2003; *Propithecus* spp., Bastian & Brockman 2007), carry (*Tarsius* spp., Gursky 2000a), or guard infants (*Varecia* spp., Morland 1990) occasionally (summarized in Patel 2007). This relatively rapid development provides relatively little time for infant socialization (chapter 11, this volume). Data from several long-term research projects reveal that most lemur females give birth every year (Overdorff et al. 1999; Jolly et al. 2002; Gould et al. 2003). In sifakas and other large species, interbirth intervals may be two or three years, especially after droughts or cyclones (Richard et al. 2002; Pochron et al. 2004). Measures of longevity, and hence estimates of lifetime reproductive success, have been established for only a few species. Maximum lifespan was estimated at 10 years for wild gray mouse lemurs (Kappeler et al., unpublished data), 18 to 20 years for *L. catta* (Gould et al. 2003), about 20 years for diademed sifakas (*Propithecus diadema*, Pochron et al. 2004), and up to

28 years in Verreaux's sifaka (*Propithecus verreauxi*, Richard et al. 2002). Long-term field data on lorisiforms or tarsiers conducive to detailed demographic analysis are not available, but some anecdotal data indicate a lifespan of more than 10 years in some tarsiers (Shekelle & Nietsch 2008).

Mortality

Individual reproductive success is compromised by mortality. Apart from pathogens, predation looms large as the main cause of extrinsic mortality of strepsirrhines of all sizes (see chapter 8, this volume). The main predators of lemurs include several species of raptors (owls, hawks, and eagles, Goodman et al. 1993; Rasoloarison et al. 1995; Karpanty 2006), snakes (boas, Rakotoandravony et al. 1998; Burney 2002; Eberle & Kappeler 2008) and carnivores (mongooses and fossa, Wright et al. 1997; Dollar et al. 2007; Irwin et al. 2009), including introduced carnivores (dogs and cats, Brockman et al. 2008). A much larger sister species of the fossa and monkey eagles, large enough to prey upon most subfossil lemurs, went extinct along with them recently (Goodman 1994; Goodman et al. 2004). Predation rates on lemurs are among the highest across all primates (chapter 8, this volume), which is expected, given their relatively small size and the relative lack of alternative mammalian prey on Madagascar. Predation on lorisiforms and tarsiers is less well documented (see Gursky & Nekaris 2007), but known predators include snakes (Wiens und Zitzmann 1999; Gursky 2002a), viverrids (Nekaris et al. 2007), and other primates (Utami & van Hooff 1997; Poulsen & Clark 2001; Pruetz & Bertolani 2007). These taxa exhibit several antipredator strategies besides crypsis (Gursky 2007b; Nekaris et al. 2007), suggesting that they are exposed to intense predation pressure as well. Long-term data from lemur study sites without confirmed deaths due to attacks by the top predators (fossa and Harrier hawk) indicate that intrinsic mortality risks (starvation, disease) are also important, especially for infants and juveniles (Gould et al. 1999; Jolly et al. 2002; Richard et al. 2002; Pride 2005). In addition, the brief mating season also has measurable mortality costs for males in some species (Kraus et al. 2008).

Activity

The high predation risk for strepsirrhines and tarsiers may be due to increased vulnerability as a result of relatively small body size and widespread nocturnal activity: a combination of traits that is frequently associated with the use of daytime shelters (fig. 2.4). All lorises and galagos are nocturnal; only one species of galago has been observed to be

Fig. 2.4. Many nocturnal strepsirrhines (here *Lepilemur ruficaudatus*) spend the day in tree holes or other shelters. Photo courtesy of Claudia Fichtel.

cation (chapter 29, this volume), foraging (chapter 7, this volume) and predator avoidance (Fichtel 2007; chapter 8, this volume). Trichromatic color vision has recently been reported for some diurnal lemurs (Leonhardt et al. 2009) and suggested for some cathemeral ones (Veilleux & Bolnick 2009), but it remains unknown whether this originally represents an adaptation to foraging or to mate choice (Fernandez & Morris 2007). Galagos have monochromatic vision, whereas some allelic variation in opsin genes has been reported for lorises and tarsiers (Tan & Li 1999; Bradley & Mundy 2008).

Evolution of Social Systems

Together, strepsirrhines and tarsiers exhibit a stunning diversity of social systems that nearly encompasses all forms found among anthropoids. The study of lemur social systems is particularly interesting from a comparative perspective because of their long independent evolution on Madagascar: by identifying similarities and dissimilarities with anthropoids, one can discover general principles of primate social evolution (Jolly 1966a; Martin 1972; Pereira 1995). In studying social evolution, it has proven heuristically useful to distinguish among the social organization, social structure, and mating system of a species (Kappeler & van Schaik 2002; chapter 9, this volume). At the level of social organization—the size, composition, cohesion, and genetic structure of a social unit—three basic categories can be distinguished: solitary, pair-living, and group-living species. All three categories are represented among strepsirrhines and tarsiers.

Social Organization

The social organization of different primate species is highly variable, ranging from solitary individuals to richly structured aggregations of hundreds of individuals. However, most primates live in permanent bisexual groups that can be readily identified in the field. Even groups with low spatial cohesion, such as fission-fusion societies, can be identified by the fact that their members associate and interact regularly with each other, and more so than with members of neighboring groups (Struhsaker 1969). Thus, groups constitute natural social units for field primatologists, whether they have three members or hundreds. The same is true for pair-living species, in which one adult male and one adult female associate with each other and their offspring, and coordinate activities, often including defense of a joint territory. Up to a third of all primate species are solitary, however, and their social organization is more difficult to study.

active in twilight (Bearder et al. 2006). Tarsiers are secondarily nocturnal—that is, descendants of a diurnal ancestor (Martin & Ross 2005). Most extant lemurs are nocturnal as well (table 2.1). Based on comparative analyses of relative orbit size, all subfossil lemurs (except *D. robusta*) were inferred to have been diurnal (Godfrey et al. 2006). *Indri*, *Propithecus*, *Lemur*, and *Varecia* are also diurnal (Erkert & Kappeler 2004), indicating that diurnal activity has evolved independently two to four times among lemurs, depending on the phylogenetic affiliations of the subfossil lemurs (Karanth et al. 2005; Orlando et al. 2008). *Hapalemur* and *Eulemur* are unusual among primates because they are cathemeral, a pattern characterized by irregular bursts of activity around the 24-hour cycle (Tattersall 2006). The proximate and ultimate determinants of this unusual activity pattern continue to be intensively studied (Kappeler & Erkert 2003; Curtis & Rasmussen 2006; Hill 2006; Donati et al. 2009), as it has obvious implications for communi-

Solitary Species

By definition, the adults of solitary primate species are typically encountered alone during their period of activity (Charles-Dominique 1978). Because mothers may associate with their offspring, because some individuals may form sleeping groups regularly, because communication among neighbors by sounds and odors is widespread, and because individuals meet conspecifics regularly, "solitary" should not be confused or equated with "asocial." Instead, this term simply reflects the fact that, unlike those of group- and pair-living species, activities of individuals are not synchronized in a manner that results in permanent spatial association, and thus primatologists encounter solitary primates as single individuals most of the time. A solitary lifestyle is therefore a highly complex and rather challenging system to study.

A solitary social organization is ancestral for all strepsirrhines, whereas the ancestral state for tarsiers remains unresolved (Martin 1972; Kappeler 1999). Among primates, a solitary lifestyle is associated with nocturnal activity, orangutans (*Pongo* spp.) being the sole exception. Detailed field studies of individually known solitary primates are one of the last frontiers in field primatology, and methodological advances have contributed to a growing comparative database (cf. Bearder 1987; Richard 1987). Yet the behavioral ecology of most solitary primate species remains unstudied. Currently available data indicate that most lorisiforms (except Zanzibar galagos, *Galagoides zanzibaricus*), several species of *Tarsius*, *Daubentonia*, *Microcebus*, *Allocebus*, *Mirza*, and some *Cheirogaleus* and *Lepilemur* can be classified as solitary (Kappeler 1997a; Müller & Thalmann 2000; Gould & Sauther 2007; Nekaris & Bearder 2007; see table 2.1).

Two questions about the social organization of solitary species are particularly challenging. First, in a given suitable habitat, are solitary individuals evenly distributed in space, or are there areas with more or less individuals, or even none? Most field studies of nocturnal solitary primates have been conducted on local populations that were chosen either randomly or because of high density of the species of interest. Because it is difficult to observe small arboreal animals in tropical forests at night, such studies have typically been restricted either to small areas or to a handful of animals. Such hard-won data are all we currently have to characterize the social organization of some species (e.g., the hairy-eared mouse lemur, *Allocebus trichotis*, Biebouw 2009). Several recent studies with increased sampling effort have revealed substantial variation in local distribution and abundance, however. In gray mouse lemurs, for example, a continuous forest habitat that appeared uniformly suitable was found to harbor local population nuclei with high densities, areas with much lower densities, and empty tracts of forest several home range diameters (i.e., several hundred meters) wide. In more than 10 years of study, however, no mouse lemur was ever captured in the empty area despite the same sampling effort (Eberle & Kappeler 2002; Fredsted et al. 2004; Dammhahn & Kappeler 2008; Radespiel et al. 2008; see also Martin 1972). Thus, at the scale of several kilometers, solitary species can exhibit striking variation in density, which provides a caveat for generalizations about species-typical social organization based on studies of single populations, as well as a rich opportunity for studying the ecological causes and sociogenetic consequences of this variation at different spatial scales (Fredsted et al. 2005; Guschanski et al. 2007; Dammhahn & Kappeler 2008; Gligor et al. 2009).

Second, the basic social unit of a solitary species is not a priori obvious (see, e.g., Radespiel 2000). If we discount the theoretical possibility that individuals are distributed randomly in space with respect to their sex, age, or kin, we need a method to identify natural social units that is heuristically meaningful and operationally feasible. There are practical problems with describing the distribution of individuals of different age and sex classes in time and space because it is difficult to individually identify nocturnal animals, and transect censuses are useful for determining the density of species, but not for describing the distribution of individuals. Students of solitary primates have therefore resorted to field methods originally developed for the study of small mammals: live trapping and radio tracking (fig. 2.5). These methods yield useful data about the size, usage patterns, and stability of home ranges of known individuals of different age and sex classes. If all animals in an area are sampled, these data can reveal fundamental aspects of social organization, but they do not necessarily reveal the size and composition of a basic social unit. This remains more or less obscure, depending on the degree of spatial clumping at the local population level. In the case of relatively even spacing, every individual has several neighbors whose home ranges overlap to various extents, and who interact with each other with a certain probability. These neighbors themselves have neighbors, with whom a focal animal may interact with a much smaller probability. These patterns of variable mutual home range overlap among same- and opposite-sex animals have given rise to unsatisfactory classification systems (e.g., "semidispered unimalc/unifemale," Nekaris & Bearder 2007) that ignore dispersal status and the resulting kin relations, for example. Social network analyses of proximity data collected via miniaturized GPS tags (e.g., Leu et al. 2010) may reduce this problem in the future, but it has become increasingly clear that the centers

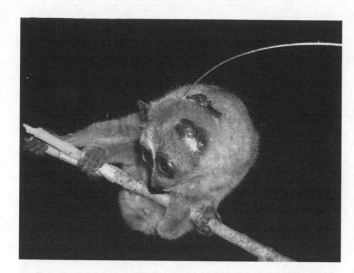

Fig. 2.5. Increased use of radio tracking of nocturnal strepsirrhines (here *Nycticebus pygmaeus*) has greatly improved our knowledge of the social systems of solitary species. Photo courtesy of Anna Nekaris.

and peripheries of social units are impossible to define with behavioral data alone.

More extensive and eventually automated sampling in space, as well as genetic analysis of population structure, is therefore required to obtain additional information on the structure of basic social units of solitary species (Radespiel et al. 2001; Kappeler et al. 2002; Wimmer et al. 2002; Weidt et al. 2004; Merker et al. 2009; chapter 12, this volume). Information about the genetic structure of populations of solitary species is currently only available for the genera *Mirza*, *Microcebus*, and *Tarsius*. These studies of lemurs revealed a genetic structure that is not obvious from behavioral data alone. Relatedness is a key factor structuring the distribution of individuals. Females exhibit positive genetic structure—that is, the genetic distance between dyads is correlated with the spatial distance between their centers of activity. As a result, closely related females cluster in space, with members of a matriline (individuals sharing a mitochondrial haplotype) occupying overlapping or adjacent home ranges. Members of different haplotypes are spatially segregated. These studies also revealed that most but not all females are philopatric, and that most but not all males disperse from their natal area. Thus, the combination of trapping, behavioral, and genetic data has provided new insights into the social organization of solitary primates. Social network analyses based on comprehensive spatial data may provide additional insights in the future.

Questions about the ultimate causes of this type of social organization also require more attention as comparative data accumulate. Much theoretical and empirical work has examined why most primates live in groups (Terborgh & Janson 1986), whereas the selective pressures favoring

solitary life have remained unclear (Kappeler 1997a). The fact that a solitary social organization is ancestral does not furnish an adaptive explanation. Similarly, constraints of activity period and small body size cannot provide sufficient explanations because there are too many exceptions among primates. The benefits of feeding on nonshareable resources and a lack of antipredator benefits favoring gregariousness seem to hold the most promising explanations for future systematic testing (Schülke & Ostner 2005; Fichtel 2007; Dammhahn & Kappeler 2009; but see Gursky 2002b).

Recognition of the rich intra- and interspecific variation in the social organization of solitary primates has been the most important finding of new field studies. This point is best illustrated by studies of mouse lemurs (*Microcebus* spp.), where both intraspecific variation within and among local populations and differences between closely related syntopic species have been documented (Radespiel 2000; Eberle & Kappeler 2002; Weidt et al. 2004; Dammhahn & Kappeler 2005; Schülke & Ostner 2005; Lahann et al. 2006). Similar variation in sociality exists between the two species of *Mirza*, which differ fundamentally in sleeping group composition and male tolerance (Kappeler 1997b; Kappeler et al. 2005; Rode 2010). These and other studies have deconstructed the myth of the uniform solitary primate on the basis of pioneering early field studies (Charles-Dominique & Martin 1970), paving the ground for more hypodeductive socioecological studies in the future.

Pair-Living Species

Pairs represent the smallest permanent cohesive social units. Pair bonding in mammals is only expected under restrictive circumstances, and is correspondingly rare (Reichard 2003; Schülke 2005). Among primates, pair-living species are twice as common (about 10% of all primates) as among other mammals (van Schaik & Kappeler 2003), and many of them are strepsirrhines and tarsiers. In fact, up to a third of all lemur species may live in pairs. This type of social organization has evolved independently in several families and independent of activity period or body size. It occurs in the genera *Phaner*, *Avahi*, and *Indri*, and also in some *Lepilemur*, *Cheirogaleus*, *Hapalemur*, and *Eulemur* species (table 2.1; Overdorff 1996; Warren & Crompton 1997; Fietz 1999; Tan 1999; Rasoloharijaona et al. 2003; Schülke & Kappeler 2003; Zinner et al. 2003). Pair living has been described for two out of four *Lepilemur* species for which data exist, thereby demonstrating that it is not strictly phylogenetically constrained. Thus, it may also be widespread among the 20 or so recently described *Lepilemur* species that remain to be studied, making pair living potentially the modal type of social organization among all lemurs.

Among lorisiforms, pair living has thus far only been reported for Zanzibar galagos (Harcourt & Nash 1986). The best evidence for pair living in tarsiers comes from a recent study of the newly described Lariang tarsier (*Tarsius lariang*, Driller et al. 2009). In some eastern tarsiers pairs are also commonly found, but, as in some lemurs (e.g., see Zinner et al. 2003; Schülke & Kappeler 2003), up to 19% of groups contain a second female (Merker 2006; Gursky 2007a; Gursky-Doyen 2010), raising the conceptually challenging problem of how to deal with intraspecific variation in social typology (Kappeler & van Schaik 2002; chapter 9, this volume). As expected for pair-living species, males and females disperse from their natal areas in western fat-tailed dwarf lemurs (Fredsted et al. 2007) as well as in the more variable spectral tarsier (*Tarsius tarsier*, Gursky 2010).

Until the 1980s, most nocturnal lemurs were considered to be solitary. Detailed studies of western fat-tailed dwarf lemurs, red-tailed sportive lemurs (*Lepilemur ruficaudatus*), and Milne-Edwards sportive lemurs (*Lepilemur edwardsi*) revealed, however, that the solitary individuals encountered were actually organized into pairs (Müller 1998; Fietz 1999; Rasoloharijaona et al. 2003; Zinner et al. 2003). In retrospect, the earlier misclassifications were based on the fact that pair partners rarely interact and sleep together. Simultaneous follows of pair partners in pale fork-marked lemurs (*Phaner pallescens*), for example, showed that males and females move largely independently of each other, with equal probabilities of being within 15 to 150 meters of each other. Direct interactions are rare, and pair partners spend on average only every third night together in the same sleeping tree (Schülke & Kappeler 2003). A very similar pattern was found in red-tailed sportive lemurs, where physical contact between partners in some pairs was observed only during the single annual night of female receptivity (Hilgartner et al. 2008). In southern woolly lemurs (*Avahi meridionalis*), adults only spend about 25% of their time within 20 meters of each other (Norsica & Borgonini Tarli 2008), and in spectral tarsiers individuals also spend most of their active period more than 20 meters apart (Gursky 2007a). In pale fork-marked lemurs (Schülke & Kappeler 2003), Milne-Edwards sportive lemurs (Méndez-Cárdenas & Zimmermann 2009), and tarsiers (Burton & Nietsch 2010), vocal duetting provides an important behavioral mechanism for synchronizing pair partners' activities also at a distance. It therefore appears that, at least among nocturnal lemurs, pair cohesion differs considerably from that displayed by other pair-living primates such as titi monkeys or gibbons (chapters 3 and 6, this volume). Nonetheless, in contrast to those of solitary species, the home ranges of pair partners in these pair-living lemurs coincide nearly perfectly, and they are defended by both sexes, so that overlap

with the ranges of other adults is minimal. Because such dispersed pairs are found in species with very different feeding ecologies, the reasons for this low cohesion remain elusive.

The ultimate reasons for pair bonding in strepsirrhines and tarsiers remain equally mysterious. Paternal care, which provides the main explanation for why other male primates and mammals associate with only one female (Wright 1990), is virtually absent in these taxa (Hilgartner et al. 2008; Driller et al. 2009). Male-female association as a means to reduce the risk of infanticide (van Schaik & Kappeler 1997) is an equally unlikely cause, due to common infant parking and high reproductive rates. The other major hypotheses proposed to explain mammalian pair living (reviewed in Schülke 2005) could neither provide obvious explanations for the other species nor be supported for *Phaner*. Perhaps the potential costs for males through missed reproductive opportunities are not very high, as indicated by the regular occurrence of extra-pair paternities (Fietz et al. 2000; Schülke et al. 2004; Driller et al. 2009). Thus, the reason why these pair-living lemur males, unlike males in solitary species, coordinate their ranges with those of a single female remains unclear for the time being.

Group-Living Species

Permanent cohesive groups with three or more adults that closely coordinate their activities have evolved at least twice independently among lemurs, once each in the Lemuridae and the Indriidae (Horvath et al. 2008), whereas lorises and galagos lack group-living species. Spectral tarsiers have been observed in small groups in some populations (Gursky 2007a), making tarsiers the most socially diverse genus of all primates apart from *Homo* (chapter 13, this volume).

Group-living lemurs (*Propithecus*, *Lemur*, *Prolemur*, *Hapalemur*, and *Varecia*) are among the largest extant species; all of them are at least partially diurnal. Group living was also presumably common in subfossil lemurs. Compared to group-living anthropoids, lemur groups exhibit three striking features. First, after controlling for body size and phylogeny, lemur groups are significantly smaller than those of anthropoids (Kappeler & Heymann 1996; fig. 2.6). Small group size presumably reflects an adaptation to feeding competition (Wright 1999; chapter 9, this volume), is not easily reconciled with their high predation risk (chapter 8, this volume), and goes hand in hand with relatively small brain size (Dunbar & Shultz 2007). Lemurs in large groups, especially those containing more males, also experience more stress, as measured through fecal cortisol (Pride 2005; Starling et al. 2010). Characteristics related to the size and composition of *Eulemur* groups were found to be rather inflexible across different habitats (Ossi & Kamilar

Fig. 2.6. Lemur groups are significantly smaller than those of anthropoids; here a complete group of redfronted lemurs (*Eulemur fulvus*) comfortably fits into a single small tree. Photo courtesy of Tina Jensen.

2006), suggesting the existence of additional phylogenetic constraints. Second, lemur groups display relatively even adult sex ratios, despite unusual intraspecific variability in group composition (Kappeler 2000b; Pochron & Wright 2003; Kappeler et al. 2009). This demographic constellation, together with pronounced breeding seasonality, generates interesting dynamics for mating strategies, especially for males (chapter 17, this volume). One hypothesis posits that even sex ratios are the result of an evolutionarily very recent transition to group living (van Schaik & Kappeler 1996). Accordingly, several pairs of previously nocturnal individuals aggregated once a diurnal life style became an option after the extinction of larger competitors and their predators. Third, certain types of social organization found among anthropoids, such as groups with just one breeding male ("harems") or one breeding female ("polyandrous groups"), have not evolved on Madagascar; neither have species with exclusive male dispersal or with multiple social layers (e.g., hamadryas baboons). Lemur groups converge with groups of many Old World monkeys, however, because female philopatry and modal male-biased natal

dispersal are widespread (Sussmann 1992; Richard et al. 1993; Wimmer & Kappeler 2002; Pochron et al. 2004; Morelli et al. 2009), thus providing conditions necessary for the establishment of kin-based social bonds despite females dispersing in some species (chapters 21 and 24, this volume). While most groups are cohesive, fission-fusion dynamics have been described for *Varecia* and some brown lemur (*Eulemur fulvus*) populations (Tattersall 1977; Vasey 2007), and large ring-tailed lemur groups fission permanently along matrilineal lines (Ichino 2006). Polyspecific associations, which are nearly obligatory in some anthropoids, are not readily observed in lemur communities. One example involves crowned (*Eulemur coronatus*) and Sanford's (*Eulemur sanfordi*) lemurs, which regularly associate for a couple of hours during the food-poor season (Freed 2006).

Mating System

The mating system of a species can be characterized by the pattern of observed mating interactions. Evolutionarily, mating success is only half the battle, however, because most primates are notoriously promiscuous (chapter 16, this volume). The reproductive system—who actually reproduces with whom—can only be determined with the help of genetic paternity analyses. It has important consequences for the genetic structure of a social unit because average genetic relatedness, and hence social tolerance, among offspring increases with increasing male reproductive skew (Schülke & Ostner 2008). Among strepsirrhines and tarsiers, various mating systems have been described, but paternity analyses have only been conducted in a few species.

Most lemur, lorisiform, and tarsier females mate every year; some mouse lemurs, lorisiforms, and tarsiforms produce two litters per year. With the exception of some lorisiforms, Bornean and Philippine tarsiers, and the aye-aye, breeding activity is highly seasonal (Wright 1999; Pullen et al. 2000; Nekaris 2003). Some lemurs in particular are noteworthy for annual periods of receptivity lasting no more than a few hours (Jolly 1967; Brockman & Whitten 1996; Eberle & Kappeler 2004a). The shortness of these periods of receptivity within a narrow seasonal mating window increases the probability of female synchrony within and between social units (Jolly 1967; Eberle & Kappeler 2004a; Mass et al. 2009; but see Pereira 1991), which in turn constrains male monopolization potential (Ostner et al. 2008a; chapter 9, this volume). Most females of species in which mating has been observed in the wild do mate with multiple males (Kappeler 1997b; Pullen et al. 2000; Radespiel et al. 2002; Nekaris 2003), but strictly polyandrous species do not exist.

Male reproductive strategies are determined by the distribution of receptive females in space and time (Emlen & Oring 1977). Because most female strepsirrhines and tarsiers are dispersed in space (solitary and pair-living species) or occur in relatively small groups, and because mating in most species is seasonal, males have a relatively small potential to monopolize access to many females. If males mate guard or associate with one female, they forgo mating opportunities elsewhere (Eberle & Kappeler 2004b; Mass et al. 2009). Passive mate guarding with copulatory plugs has been suggested as a possible solution to this dilemma (Dunham & Rudolf 2009), but plugs are expected to generate sexual conflict and female counterstrategies (Stockley 1997). In pair-living species, males may solve this dilemma by associating with one female to secure at least one reproductive opportunity. In some species, a small proportion of males associate with two females (e.g., red-tailed sportive lemur, Zinner et al. 2003; *Tarsius*, Gursky 2007a). This, in combination with detectable extrapair matings (see below), suggests that these males may mate polygynously. In solitary lemurs, as well as in the group-living Lemuridae, males apparently cannot prevent matings by rivals, and hence have relatively large testes (Kappeler 1997c), an adaptation to intense sperm competition also found among anthropoids (chapter 17, this volume). Solitary lorisoids tend to have smaller testes relative to body size than solitary lemurs, indicating a higher monopolization potential (Kappeler 1997c). In group-living *Propithecus*, which belong to the Indriidae, males have much smaller testes than expected for their body size, but they increase in size seasonally (Pochron & Wright 2002); dominant males mate guard (Mass et al. 2009) and also outperform subordinate rivals physiologically (Kraus et al. 1999; Lewis 2009). It is intriguing to imagine that Lemuridae and Indriidae, who evolved group living independently, may be relying primarily on different mechanisms to regulate male-male competition.

New genetic methods provide opportunities to measure actual male reproductive success and illuminate criteria of female choice. Genetic paternity tests permit determination of male reproductive success, mate fidelity in monogamous species, and reproductive skew in group-living species. In most promiscuous solitary species with litters, there is a potential for mixed paternities. In Coquerel's dwarf lemurs (*Mirza coquereli*) and gray mouse lemurs, mixed paternities occurred in 50% and 65% of litters respectively (Kappeler et al. 2002; Eberle & Kappeler 2004a). In gray bamboo lemurs (*Hapalemur griseus*), western fat-tailed dwarf lemurs, Lariang tarsiers, and pale fork-marked lemurs, rates of extrapair young varied between 8% and 75%, but sample sizes were small in all studies (Fietz et al. 2000; Nievergelt et al. 2002; Schülke et al. 2004; Driller et al. 2009). Ring-tailed lemurs, Verreaux's sifakas, and red-fronted lemurs (*Eulemur rufifrons*) exhibit high reproductive skew in favor of the dominant male, despite polyandrous matings by females (Pereira & Weiss 1991; Kappeler & Port 2008; Kappeler & Schäffler 2008). A paternity study in another population of Verreaux's sifaka revealed a substantial proportion of extragroup paternities (Lawler 2007), hinting at the presence of interesting intraspecific variation among group-living species. These results are difficult to reconcile with the presence of several subordinate males in such small groups containing two to five females, but subordinate males may contribute to a significant reduction in takeover risk (Port et al. 2010).

Females are not expected to mate randomly as long as there is variation in male quality. If males do not provide paternal care, females should choose mates according to some intrinsic quality (chapter 16, this volume). Immunogenetic competence is a highly adaptive trait in this context, and it can presumably be detected via olfaction in mammals; it can be assayed by determining variability in genes of the MHC complex (Penn 2002). In highly promiscuous gray mouse lemurs, females exercise postcopulatory choice among sperm of different males, preferentially fertilizing their eggs with sperm from males that differ most from them in MHC similarity (Schwensow et al. 2008a). In pair-living eastern fat-tailed dwarf lemurs, females preferred males with above-average MHC variability and heterozygosity as both social and genetic fathers for their offspring. Females also preferred males with a MHC genotype dissimilar to their own, and females that engaged in extrapair copulations shared a significantly higher number of MHC genes with their social partner than did faithful females (Schwensow et al. 2008b). Shopping for "good genes" is therefore one reason for multiple matings by females (male quality also varies in other traits; e.g., see Lawler et al. 2005). An important ultimate reason for promiscuous behavior is its diminishing effect on infanticide risk (van Schaik 2000). Even though infanticide is not expected in species with annual seasonal reproduction, it has been documented in several lemur species (ring-tailed lemurs, Hood 1994; Milne–Edwards sportive lemurs, Rasoloharijaona et al. 2000; *Propithecus* spp., Erhardt & Overdorff 1998; Lewis et al. 2003; Morelli et al. 2009) but not in wild lorisiforms (Nekaris & Bearder 2007) or tarsiers.

Social Structure

The nature and patterning of interactions define social relationships between pairs of individuals. Consistent features of these dyadic relationships can in turn be used to characterize social structure (Hinde 1976). Variation among relationships results from differences in the nature, frequency,

and intensity of affinitive, affiliative, and agonistic interactions (de Waal 1987; chapter 23, this volume). Sex has
emerged as a major organizing principle in the analysis of
social structure, with both ecological and social factors as
ultimate determinants of the observed variation in social
relationships within and among species (Kappeler & van
Schaik 2002; Watts 2010). Because the social structures of
nocturnal strepsirrhines and tarsiers remain poorly studied
and understood (but see, e.g., Gursky 2000b; Radespiel
2000; Nekaris 2006; Dammhahn & Kappeler 2009), and
because comparisons with group-living anthropoids are
most interesting from a theoretical perspective (Kappeler
1999; Erhardt & Overdorff 2008a; chapter 9, this volume),
I focus below on group-living lemurs.

Dominance

Competition for resources favors the formation of dominance relations whenever conspecifics interact regularly
because it provides a mechanism to reduce the probability
of costly fighting (chapter 9, this volume). Dominance relationships are established whenever an individual consistently exhibits submissive behavior towards a conspecific
(Hausfater 1975). The patterning of dyadic dominance
relationships can be described by a dominance hierarchy
at the group level, with linearity and transitivity being
its chief defining features. Dominance relationships exist
between group-living members of the Lemuridae and Indriidae (Pereira 1995). Ring-tailed lemurs, which form the
largest groups, exhibit linear hierarchies with few reversals;
they can be stable over several years among females (Jolly
1966b; Pereira & Kappeler 1997; Koyama et al. 2005).
Among red-fronted lemurs, most dyads do not maintain
dominance relations, and no clear hierarchy at the group
level is discernible (Pereira & Kappeler 1997; Ostner &
Kappeler 1999). In sifakas, dominance relations exist (Richard 1985), but small group size and low interaction rates
make it difficult to define group hierarchies (Pochron et al.
2003).

Male-Female Relationships

The most salient feature of lemur dominance relationships is
the widespread existence of female dominance (Jolly 1966b;
Richard 1987; Pereira et al. 1990; Jolly 1998). All adult
females are able to consistently elicit submissive behavior
from all adult males in dyadic interactions in all behavioral
contexts. In some species (e.g., ring-tailed lemurs, diademed
sifakas, gray bamboo lemurs), female dominance manifests
via spontaneous male submissive signaling in the absence
of female aggression (Pereira & Kappeler 1997; Pochron

et al. 2003; Waeber & Hemelrijk 2003), whereas in others
(e.g., gray mouse lemurs; black lemurs, *Eulemur macaco*)
female aggressive behavior is typically observed (Radespiel & Zimmermann 2001; Roeder et al. 2002; Digby &
Mclean Stevens 2007). Female dominance is facilitated by a
lack of sexual size dimorphism (Kappeler 1991) and may be
proximately linked to certain steroids (von Engelhardt et al.
2000; Drea 2007, 2011). Because it affords female feeding
priority, several ultimate explanations of female dominance
have focused on this particular competitive advantage (most
recent example: Dunham 2008), but female choice for submissive males and a by-product of low intrasexual selection
have been discussed as alternative explanations (reviewed in
Kappeler 1993a; Jolly 1998). Why virtually all members of
a suborder exhibit this behavioral trait, which is not found
in other primates and only rarely in other mammals (e.g.,
hyenas), remains unresolved.

Female-Female Relationships

Social relationships among same-sex lemur dyads also differ in several remarkable aspects from patterns described
for the better-known anthropoids. First, agonistic third-
party interventions, which are characteristic of many Old
World primates (chapter 5, this volume), are virtually absent in lemur societies (Pereira & Kappeler 1997). Despite
widespread female philopatry and the resulting coresidence
of female kin, female lemurs provide agonistic support for
close relatives only in rare cases, and there is no maternal
rank inheritance (but see Koyama et al. 2005), so that most
group-living lemurs do not easily fit into existing socioecological categories (Erhart & Overdorff 2008a). The failure
to support kin has been attributed to a lack of sufficient
visual acuity to distinguish opponents (Pereira 1995) or a
lack of social cognition due to small brain size (see above),
but it may also be an adaptive feature of their competitive
regime.

Second, female social relationships, even among close
relatives, can be exceptionally intolerant. During episodes
of targeted aggression, particular females are persistently
attacked by one or several group mates and are severely injured and/or evicted from the group (Vick & Pereira 1989),
making lack of visual acuity and limited cognitive abilities
unlikely explanations for the observed lack of kin support.
Whenever groups contain more than one matriline, targeted
aggression may lead to group fissioning (e.g., Ichino 2006).
Evicted females rarely manage to join or found new social
groups (red-fronted lemurs, Kappeler & Fichtel 2012a).
Targeted aggression peaks during the mating and birth season (Vick & Pereira 1989; Pereira & Leigh 2003), thus suggesting that it constitutes a female reproductive tactic. This

explanation is corroborated by the fact that the aggression occurs most often in groups larger than a critical size (Vick & Pereira 1989; Bayart & Simmen 2005), and because individuals in atypically large groups have the highest stress levels (Pride 2005). Moreover, females in ring-tailed lemurs and some *Eulemur* species commit infanticide under certain conditions (Jolly et al. 2000), and young female infant brown lemurs exhibit sexual mimicry, perhaps to reduce the risk of infanticide (Barthold et al. 2009). Nevertheless, reproduction in females is not physiologically suppressed (Kappeler 1989; cf. chapter 3, this volume).

Finally, the predominantly competitive nature of female social relationships is underscored by the fact that post-conflict reconciliation in lemurs is so rare that it could not be demonstrated in all species or social groups where it has been studied, or following aggression in particular social contexts (Kappeler 1993b; Palagi et al. 2005b, 2008). Thus, despite frequent affiliation and mutual grooming— preferentially among kin (Kappeler 1993c)—social relationships among lemur females are distinguished by latent competition and generally weaker networks of social bonds (chapter 24, this volume) and less pronounced cooperation (chapter 21, this volume) than in relationships among female-bonded anthropoids (but see Eberle & Kappeler 2006). Female dispersal, also in the absence of overt aggression, is therefore common in both group-living Lemuridae and Indriidae (Ehrhardt & Overdorff 2008a; Morelli et al. 2009; Kappeler & Fichtel 2012b).

Male-Male Relationships

Social relationships among lemur males are characterized by higher tolerance and affiliation than those in many anthropoids. Adult males maintain dominance relations throughout the year, and an alpha male is clearly discernible in most groups, except sometimes during brief breakdowns that occur during the mating season (Jolly 1967; Richard 1974; Pereira & Kappeler 1997). Even though lemur males compete and fight during the mating season, when their testosterone and glucocorticoid levels peak (Cavigelli & Pereira 2000; Brockman et al. 2001; Ostner et al. 2002; Fichtel et al. 2007; Ostner et al. 2008b; Starling et al. 2010), unrelated males also commonly affiliate and groom each other regularly during other times of the year (Kappeler 1993c; Gould 1997; Nakamichi & Koyama 1997; Pereira & Kappeler 1997; Port et al. 2009). Because males sometimes delay dispersal, or disperse in pairs or trios (Sussman 1992; Ostner & Kappeler 2004; Kappeler & Schäffler 2008), future genetic studies may reveal that some male dyads include related individuals (see Schoof et al. 2009).

Relationships among Social Units

Tolerance towards conspecifics among strepsirrhines and tarsiers is highly variable. Among solitary species, various patterns of within- and between-sex range overlap exist. The relationship between species differences in resource characteristics and range overlap has been studied systematically only in mouse lemurs (Dammhahn & Kappeler 2009). The degree of male range overlap ought to depend on the distribution of receptive females (Emlen & Oring 1977), but the underlying distribution of receptive females in space and time also has been studied only in mouse lemurs (Eberle & Kappeler 2002; Weidt et al. 2004). Pair-living species are much less tolerant of their neighbors, and the home ranges of individual pairs overlap very little or not at all with those of neighboring pairs (Müller 1998; Fietz et al. 2000; Schülke & Kappeler 2003; Zinner et al. 2003; Merker 2006). Among group-living lemurs, there is also considerable variation in patterns of home range overlap and sex-specific participation in agonistic intergroup encounters. For example, in Verreaux's sifaka, range overlap between neighboring groups is 30 to 70%, and only males are involved in intergroup encounters (Lawler 2007; Benadi et al. 2008), whereas in gray bamboo lemurs range overlap is much less, and both sexes participate in encounters with neighbors (Nievergelt et al. 1998). In ring-tailed lemurs, populations vary in their degree of territoriality (Sauther et al. 1999), but only females fight with their neighbors (Nunn & Deaner 2004). More detailed studies of intergroup relations, focusing on intraspecific variation across sites and seasons, are required before general conclusions about the importance of resource and mate defense in different species can be drawn.

Communication

When making broad comparisons about primate social structure, it is important to recognize that social relationships of strepsirrhines and tarsiers are mediated not only by direct interactions, but also through communication signals (chapter 29, this volume); some involve modalities unavailable to anthropoids, such as ultrasonic calls or scents (Schilling 1979; Cherry et al. 1987). Many communication signals are combined across modalities into specific displays (e.g., Mertl 1976; Pereira & Kappeler 1997; Palagi & Norsica 2009), but it is heuristically useful to separate them for functional discussions. First, in visual communication, such as the use of gestures and facial expressions, strepsirrhine primates clearly differ from anthropoids and use less variable signals (Fichtel & Kappeler 2010), but limited preliminary evidence suggests that lemurs can follow the gaze of conspecifics (Shepardt & Platt 2008; Ruiz et al. 2009). Second, all strepsirrhines and tarsiers have vocal repertoires that are on average smaller in size and complexity (Pereira et al. 1988; Macedonia 1994; Stanger 1995; Zimmermann 1995; Pereira & Kappeler 1997; Gamba & Giacoma 2007) than those of anthropoids (McComb & Semple 2005). Many species have at least one call to coordinate and maintain social cohesion and spacing (e.g., see Masters 1991; Nietsch 1999; Oda 2002; Trillmich et al. 2004; Braune et al. 2005), to warn conspecifics about predators (Pereira & Macedonia 1991; Fichtel & Kappeler 2002; Gursky 2002b, 2005; Fichtel 2007; Rahlfs & Fichtel 2010), and to facilitate species recognition in the context of mate choice (Bearder et al. 1995; Zimmermann et al. 2000; Braune et al. 2008).

Finally, chemical signals from a variety of scent glands, but also from urine and feces, serve important functions in the social lives of all lemurs, lorisiforms, and tarsiers (Schilling 1979; MacKinnon & MacKinnon 1980). These signals are deposited regularly, often accompanied by stereotypic movements (e.g., Palagi et al. 2005a), and are vigorously investigated and overmarked by conspecifics (Kappeler 1998b; Lewis 2005; Norscia et al. 2009). They encode information about species and individual identity, genetic similarity, sex, and social, reproductive, and health status, and they also function in territorial defense (Mertl 1975; Harrington 1977; Clark 1982; Schilling et al. 1984; Mertl-Millhollen 2006; Charpentier et al. 2008, 2010; Boulet et al. 2009; Crawford et al. 2010). Sex, reproductive season, and social status affect frequencies of scent marking behavior (Kappeler 1989, 1990, 1998b; Fornasieri & Roeder 1992; Gould & Overdorff 2002; Lewis 2005; Pochron et al. 2005), as well as receiver responsiveness (Kappeler 1998b). Determining how individuals use olfactory signals to compete for and attract mates represents a particularly promising area for future research on olfactory communication that goes beyond counting frequencies of scent marks (Perret 1992; Fisher et al. 2003; Palagi et al. 2004; Charpentier et al. 2008; Lewis 2009; Boulet et al. 2009; Norscia et al. 2009; Crawford et al. 2010).

Cognition

The relatively small brains of strepsirrhines and tarsiers have been suggested to reflect their less complex societies (Dunbar & Shultz 2007). Whether the cognitive performance of these primates in the domains of physical and social intelligence is actually inferior to that of anthropoids has only recently begun to be investigated (reviewed in Fichtel & Kappeler 2010; chapter 30, this volume).

The main impressions of early students of lemur cognitive abilities were eloquently summarized by Alison Jolly (1966b): "*Lemur* and *Propithecus* are both socially intelligent and socially dependent. They are, however, hopelessly stupid towards unknown inanimate objects." However, in most domains of technical cognition in which experimental tests have since been conducted, strepsirrhines seem to have the same sort of basic cognitive abilities as other primates, and their performance was in most cases not qualitatively different (Fichtel & Kappeler 2010). The number of species tested in different cognitive domains is still small, but some preliminary conclusions about the presence or strength of certain abilities in this lineage can already be drawn.

In the domain of "space and objects," lemurs have a route-based mental representation of spatial relationships, and show straight-line traveling and efficient goal-directed movements between distant sites (Joly & Zimmermann 2007; Erhart & Overdorff 2008b; Lührs et al. 2009). Some species are also able to deal with object permanence problems and solve visible, but not invisible, displacement problems successfully (Deppe et al. 2009). In discrimination learning tasks, thick-tailed greater galagos (*Otolemur crassicaudatus*) appear to be a bit slower and more error-prone (Ohta et al. 1987), but in learning-set tasks they are as skilled as other primates. Strepsirrhines also seem to have some cross-modal skills (Ward et al. 1976), and they master sorting tasks just as well as anthropoids (Merritt et al. 2007). Though their numerical discrimination skills seem to be inferior, lemurs understand the outcome of simple arithmetic operations (Lewis et al. 2005; Santos et al. 2005; Mahajan et al. 2009). Tool use and associated abilities are a striking exception to the lack of fundamental differences from anthropoids in this cognitive domain.

In the realm of social intelligence, within-group coalitions, even between mothers and daughters, are extremely rare or absent altogether; post-conflict reconciliation is also rare, but some basic exchange between grooming and other social commodities may exist (see above). Preliminary evidence suggests that some basal aspects of tactical deception exist (Genty et al. 2004, 2008), that lemurs can follow the gaze of conspecifics (Shepard & Platt 2008), and that they can co-orient their attention (Ruiz et al. 2009). Social learning abilities are more widespread among lemurs (Feldman & Klopfer 1972; Kendal et al. 2010; Stoinski et al. 2011), but true innovations of novel behaviors are apparently rare. Finally, lemurs also use functionally referential calls (Fichtel & Kappeler 2002), exhibit coordinated group movements (Erhart & Overdorff 1999; Trillmich et al. 2004), and express their emotional status in structural features of their vocalizations (Fichtel & Hammerschmidt 2002). Strepsirrhine primates use fewer and less variable visual signals, however, and in this regard they clearly differ from anthropoids (Mertl 1976; Palagi et al. 2005b; Fichtel & Kappeler 2010).

Summary and Conclusions

When studying the evolution of primate societies, it is important to acknowledge several salient aspects of the biology, natural history, and ecology of strepsirrhines and tarsiers. To the extent that generalizations across such a large and diverse group are possible or meaningful, several traits distinguish them from most anthropoids. The majority of these species are small and nocturnal; most of the larger diurnal species went extinct recently. They have smaller brains relative to body size, and faster life histories, than most anthropoids. Widespread infant parking creates different conditions for paternal care and infanticide risks with far-reaching consequences for male-female relations. These species are subject to high predation rates, and dietary specializations are common. The majority of strepsirrhines and tarsiers are solitary. Pair living is unusually common among lemurs, and many lemur pairs are characterized by a lack of strong cohesion. Group living among lemurs evolved at least twice independently from anthropoids. Lemur groups differ in several fundamental demographic aspects from groups of other primates, including size, sex ratio, and the lack of male-bonded, harem and multilevel groups. Reproduction of most species is strongly seasonal, receptivity is very brief, and promiscuity is the most common mating system. Monogamy is accompanied by high levels of extrapair paternity. Reproductive success in a lemur group is strongly skewed towards one dominant male. Social relationships among lemur females are unusually competitive, those among males unusually friendly, and those between the sexes characterized by female social dominance, which has important implications for feeding competition and mate choice. Olfactory communication is an important modality of social communication. Social cognition, social complexity, and vocal repertoires are less developed than in Old World monkeys, but not too different from those in New World monkeys.

In contrast, strepsirrhines and tarsiers also converge with many anthropoids in several fundamental aspects of their behavior and ecology. They have produced the full range of body size variation, including the smallest and some of the largest living (including subfossil) species. They evolved diurnal activity and trichromatic vision independently of anthropoids, and most of them produce single offspring, some of which are carried by the mother. They represent all major types of social organization, including the results of multiple independent evolution of pair and group living. The social units of solitary and most group-living species appear to cluster around single matrilines. Female philopatry is common and kinship facilitates cooperative behavior, such as grooming and cooperative breeding, that is relatively rare. Female social relationships reflect species-specific feeding ecologies. Sperm competition, with its attendant adaptations in promiscuous species, is similar to that in anthropoids. Strepsirrhines evolved vocal repertoires with functionally referential calls, and they solve most ecologically relevant problems requiring technical intelligence as well as do many anthropoids. Unraveling additional examples of convergence and alternative evolutionary solutions by examining levels of intra- and interspecific variation among lemur, lorisiform, and tarsier societies remains one of the most exciting challenges for primatology in this century.

References

Ali, J. R. & Huber, M. 2010. Mammalian biodiversity on Madagascar controlled by ocean currents. *Nature* 463, 653–56.

Armstrong, E. 1985. Relative brain size in monkeys and prosimians. *American Journal of Physical Anthropology* 66, 263–73.

Atsalis, S. 1999. Seasonal fluctuations in body fat and activity levels in a rain-forest species of mouse lemur, *Microcebus rufus*. *International Journal of Primatology* 20, 883–910.

Barthold, J., Fichtel, C. & Kappeler, P. M. 2009. What is it going to be? Pattern and potential function of natal coat change in sexually dichromatic redfronted lemurs (*Eulemur fulvus rufus*). *American Journal of Physical Anthropology* 138, 1–10.

Bastian, M. L. & Brockman, D. K. 2007. Paternal care in *Propithecus verreauxi coquereli*. *International Journal of Primatology* 28, 305–13.

Bayart, F. & Simmen, B. 2005. Demography, range use, and behavior in black lemurs (*Eulemur macaco macaco*) at Ampasikely, northwest Madagascar. *American Journal of Primatology* 67, 299–312.

Bearder, S. K. 1987. Lorises, bushbabies, and tarsiers: Diverse societies in solitary foragers. In *Primate Societies*, ed. by Smuts, B. B., Cheney, D. L., Seyfarth, R. M., Wrangham, R. W. & Struhsaker, T. T., 11–24. Chicago: University of Chicago Press.

Bearder, S. K., Honess, P. E. & Ambrose, L. 1995. Species diversity among galagos with special reference to mate recognition. In *Creatures of the Dark: The Nocturnal Prosimians*, ed. by Alterman, L., Doyle, G. A. & Izard, M. K., 331–52. New York: Plenum Press.

Bearder, S. K., Nekaris, K. A. I. & Curtis, D. J. 2006. A reevaluation of the role of vision in the activity and communication of nocturnal primates. *Folia Primatologia* 77, 50–71.

Benadi, G., Fichtel, C. & Kappeler, P. M. 2008. Intergroup relations and home range use in Verreaux's sifaka (*Propithecus verreauxi*). *American Journal of Primatology* 70, 956–65.

Biebouw, K. 2009. Home range size and use in *Allocebus trichotis* in Analamazaotra Special Reserve, Central Eastern Madagascar. *International Journal of Primatology* 30, 367–86.

Bourlière, F. & Petter-Rousseaux, A. 1966. Existence probable d'un rythme métabolique saisonnier chez les Cheirogaleinae (Lemuroidea). *Folia Primatologica* 4, 249–56.

Bradley, B. J. & Mundy, N. I. 2008. The primate palette: The evolution of primate coloration. *Evolutionary Anthropology* 17, 97–111.

Braune, P., Schmidt, S. & Zimmermann, E. 2005. Spacing and group coordination in a nocturnal primate, the golden brown mouse lemur (*Microcebus ravelobensis*): The role of olfactory and acoustic signals. *Behavioral Ecology and Sociobiology* 58, 587–96.

———. 2008. Acoustic divergence in the communication of cryptic species of nocturnal primates (*Microcebus* ssp.). *BMC Biology* 6, 19.

Britt, A. 2000. Diet and feeding behaviour of the black-and-white ruffed lemur (*Varecia variegata variegata*) in the Betampona Reserve, eastern Madagascar. *Folia Primatologica* 71, 133–41.

Brockman, D. K. & Whitten, P. L. 1996. Reproduction in free-ranging *Propithecus verreauxi*: Estrus and the relationship between multiple partner matings and fertilizations. *American Journal of Physical Anthropology* 100, 57–69.

Brockman, D. K., Whitten, P. L., Richard, A. F. & Benander, B. 2001. Birth season testosterone levels in male Verreaux's sifaka, *Propithecus verreauxi*: Insights into socio-demographic factors mediating seasonal testicular function. *Behavioral Ecology and Sociobiology* 49, 117–27.

Brockman, D., Godfrey, L., Dollar, L. & Ratsirarson, J. 2008. Evidence of invasive *Felis silvestris* predation on *Propithecus verreauxi* at Beza Mahafaly Special Reserve, Madagascar. *International Journal of Primatology* 29, 135–52.

Burney, D.A. 2002. Sifaka predation by a large boa. *Folia Primatologica* 73, 144–45.

Burton, J. & Nietsch, A. 2010. Geographical variation in duet songs of Sulawesi tarsiers: Evidence for new cryptic species in South and Southeast Sulawesi. *International Journal of Primatology* 31, 1123–46.

Catlett, K. K., Schwartz, G. T., Godfrey, L. R. & Jungers, W. L. 2010. "Life history space": A multivariate analysis of life history variation in extant and extinct Malagasy lemurs. *American Journal of Physical Anthropology* 142, 391–404.

Cavigelli, S. A. & Pereira, M. E. 2000. Mating season aggression and fecal testosterone levels in male ring-tailed lemurs (*Lemur catta*). *Hormones and Behavior* 37, 246–55.

Charles-Dominique, P. 1977. *Ecology and Behaviour of Nocturnal Primates*. New York: Columbia University Press.

———. 1978. Solitary and gregarious prosimians: Evolution of social structure in primates. In *Recent Advances in Primatology*, vol. 3. ed. Chivers, D. J. & Joysey, K. A., 139–49. London: Academic Press.

Charles-Dominique, P. & Martin, R. D. 1970. Evolution of lorises and lemurs. *Nature* 227, 257–60.

Charpentier, M. J. E., Boulet, M. & Drea, C. M. 2008. Smelling right: The scent of male lemurs advertises genetic quality and relatedness. *Molecular Ecology* 17, 3225–33.

Charpentier, M. J. E., Crawford, J. C., Boulet, M. & Drea, C. M. 2010. Message "scent": Lemurs detect the genetic relatedness and quality of conspecifics via olfactory cues. *Animal Behaviour* 80, 101–8.

Cherry, J. A., Izard, M. K. & Simons, E.L. 1987. Description of ultrasonic vocalizations of the mouse lemur (*Microcebus murinus*) and the fat-tailed dwarf lemur (*Cheirogaleus medius*). *American Journal of Primatology* 13, 181–85.

Clark, A. B. 1982. Scent marks as social signals in *Galago crassicaudatus*. I. Sex and reproductive status as factors in signals and responses. *Journal of Chemical Ecology* 8, 1133–51.

Crawford, J. C., Boulet, M. N. & Drea, C. M. 2010. Smelling wrong: Hormonal contraception in lemurs alters critical female odour cues. *Proceedings of the Royal Society B: Biological Sciences* 278, 122–30.

Curtis, D. J. & Rasmussen, M. A. 2006. The evolution of cathemerality in primates and other mammals: A comparative and chronoecological approach. *Folia Primatologia* 77, 178–93.

Dammhahn, M. & Kappeler, P. M. 2005. Social system of *Microcebus berthae*, the world's smallest primate. *International Journal of Primatology* 26, 407–35.

———. 2008. Small-scale coexistence of two mouse lemur species (*Microcebus berthae* and *M. murinus*) within a homogeneous competitive environment. *Oecologia* 157, 473–83.

———. 2009. Females go where the food is: Does the socioecological model explain variation in social organisation of solitary foragers? *Behavioral Ecology and Sociobiology* 63, 939–52.

Daniels, H. L. 1984. Oxygen consumption in *Lemur fulvus*: Deviation from the ideal model. *Journal of Mammalogy* 65, 584–92.

Dausmann, K. H., Glos, J., Ganzhorn, J. U. & Heldmaier, G. 2004. Hibernation in a tropical primate. *Nature* 429, 825–26.

Deppe, A. M., Wright, P. C. & Szelistowski, W. A. 2009. Object permanence in lemurs. *Animal Cognition* 12, 381–88.

De Waal, F. B. M. 1987. Dynamics of social relationships. In *Primate Societies*, ed. by Smuts, B. B., Cheney, D. L., Seyfarth, R. M., Wrangham, R. W. & Struhsaker, T. T., 421–29. Chicago: University of Chicago Press.

Dewar, R. E. & Richard, A. F. 2007. Evolution in the hypervariable environment of Madagascar. *Proceedings of the National Academy of Sciences, USA* 104, 13723–27.

Digby, L. & Mclean Stevens, A. 2007. Maintenance of female dominance in blue-eyed black lemurs (*Eulemur macaco flavifrons*) and gray bamboo lemurs (*Hapalemur griseus griseus*) under semi-free-ranging and captive conditions. *Zoo Biology* 26, 345–61.

Dollar, L., Ganzhorn, J. U. & Goodman, S. M. 2007. Primates and other prey in the seasonally variable diet of *Cryptoprocta ferox* in the dry deciduous forest of Western Madagascar. In *Primate Anti-predator Strategies*, ed. by Gursky, S. L. & Nekaris, K. A. I., 63–76. New York: Springer.

Donati, G., Baldi, N., Morelli, V., Ganzhorn, J. U. & Borgognini Tarli, S. M. 2009. Proximate and ultimate determinants of cathemeral activity in brown lemurs. *Animal Behaviour* 77, 317–25.

Doyle, G. A., Andersson, A. & Bearder S. K. 1969. Maternal behaviour in the lesser bushbaby (*Galago senegalensis moholi*) under semi-natural conditions. *Folia Primatologica* 11, 215–38.

Drea, C. M. 2007. Sex and seasonal differences in aggression and steroid secretion in *Lemur catta*: Are socially dominant females hormonally "masculinized?" *Hormones and Behavior* 51, 555–67.

———. 2011. Endocrine correlates of pregnancy in the ring-tailed lemur (*Lemur catta*): Implications for the masculinization of daughters. *Hormones and Behavior* 59, 417–27.

Driller, C., Perwitasari-Farajallah, D., Zischler, H. & Merker, S. 2009. The social system of Lariang tarsiers (*Tarsius lariang*) as revealed by genetic analyses. *International Journal of Primatology* 30, 267–81.

Dunbar, R. I. M. & Shultz, S. 2007. Evolution in the social brain. *Science* 317, 1344–47.

Dunham, A. E. 2008. Battle of the sexes: Cost asymmetry explains female dominance in lemurs. *Animal Behaviour* 76, 1435–39.

Dunham, A. E. & Rudolf, V. H. W. 2009. Evolution of sexual size monomorphism: The influence of passive mate guarding. *Journal of Evolutionary Biology* 22, 1376–86.

Eberle, M. & Kappeler, P. M. 2002. Mouse lemurs in space and time: A test of the socioecological model. *Behavioral Ecology and Sociobiology* 51, 131–39.

———. 2004a. Sex in the dark: Determinants and consequences of mixed male mating tactics in *Microcebus murinus*, a small solitary nocturnal primate. *Behavioral Ecology and Sociobiology* 57, 77–90.

———. 2004b. Selected polyandry: Female choice and intersexual conflict in a small nocturnal solitary primate (*Microcebus murinus*). *Behavioral Ecology and Sociobiology* 57, 91–100.

———. 2006. Family insurance: kin selection and cooperative breeding in a solitary primate (*Microcebus murinus*). *Behavioral Ecology and Sociobiology*, 60, 582–588.

———. 2008. Mutualism, reciprocity, or kin selection? Cooperative rescue of a conspecific from a boa in a nocturnal solitary forager, the gray mouse lemur. *American Journal of Primatology* 70, 410–14.

Elgar, M. A. & Harvey, P. H. 1987. Basal metabolic rates in mammals: Allometry, phylogeny and ecology. *Functional Ecology* 1, 25–36.

Emlen, S. T. & Oring, L. W. 1977. Ecology, sexual selection, and the evolution of mating systems. *Science* 197, 215–23.

Erhart, E. M. & Overdorff, D. J. 1998. Infanticide in *Propithecus diadema edwardsi*: An evaluation of the sexual selection hypothesis. *International Journal of Primatology* 19, 73–81.

———. 1999. Female coordination of group travel in wild *Propithecus* and *Eulemur*. *Int J Primatol* 20, 927–40.

———. 2008a. Rates of agonism by diurnal lemuroids: Implications for female social relationships. *International Journal of Primatology* 29, 1227–47.

———. 2008b. Spatial memory during foraging in prosimian primates: *Propithecus edwardsi* and *Eulemur fulvus rufus*. *Folia Primatologica* 79, 185–96.

Erkert, H. G. & Kappeler, P. M. 2004. Arrived in the light: Diel and seasonal activity patterns in wild Verreaux's sifakas (*Propithecus v. verreauxi*; Primates: Indriidae). *Behavioral Ecology and Sociobiology* 57, 174–86.

Feistner, A. T. C. & Ashbourne, C. J. 1994. Infant development in a captive-bred aye-aye (*Daubentonia madagascariensis*) over the first year of life. *Folia Primatologica* 62, 74–92.

Feldman, D. W. & Klopfer, P. H. 1972. A study of observational learning in lemurs. *Zeitschrift für Tierpsychologie* 30, 297–304.

Fernandez, A. A. & Morris, M. R. 2007. Sexual selection and trichromatic color vision in primates: Statistical support for the preexisting–bias hypothesis. *American Naturalist* 170, 10–20.

Fichtel, C. 2007. Avoiding predators at night: Antipredator strategies in red-tailed sportive lemurs (*Lepilemur ruficaudatus*). *American Journal of Primatology* 69, 611–24.

Fichtel, C. & Kappeler, P. M. 2002. Anti-predator behavior of group-living Malagasy primates: Mixed evidence for a referential alarm call system. *Behavioral Ecology and Sociobiology* 51, 262–75.

Fichtel, C. & Hammerschmidt, K. 2002. Responses of redfronted lemurs (*Eulemur fulvus rufus*) to experimentally modified alarm calls: Evidence for urgency-based changes in call structure. *Ethology* 108, 763–77.

Fichtel, C. & Kappeler, P. M. 2010. Primate plesiomorphies and human universals: Establishing the lemur baseline. In *Mind the Gap: Tracing the Origins of Human Universals*, ed. by Kappeler, P. M. & Silk, J. B., 395–426. Heidelberg: Springer.

Fichtel, C., Kraus, C., Ganswindt, A. & Heistermann, M. 2007. Influence of reproductive season and rank on fecal glucocorticoid levels in free-ranging male Verreaux's sifakas (*Propithecus verreauxi*). *Hormones and Behavior* 51, 640–48.

Fietz, J. 1999. Monogamy as a rule rather than exception in nocturnal lemurs: The case of the fat-tailed dwarf lemur, *Cheirogaleus medius*. *Ethology* 105, 259–72.

Fietz, J., Zischler, H., Schwiegk, C., Tomiuk, J., Dausmann, K. H. & Ganzhorn, J. U. 2000. High rates of extra-pair young in the pair-living fat-tailed dwarf lemur, *Cheirogaleus medius*. *Behavioral Ecology and Sociobiology* 49, 8–17.

Fisher, H. S., Swaisgood, R. R. & Fitch-Snyder, H. 2003. Countermarking by male pygmy lorises (*Nycticebus pygmaeus*): Do females use odor cues to select mates with high competitive ability? *Behavioral Ecology and Sociobiology* 53, 123–30.

Fleagle, J. G. 1999. *Primate Adaptation and Evolution*, 2nd ed. New York: Academic Press.

Fornasieri, I. & Roeder, J. J. 1992. Marking behaviour in two lemur species (*L. fulvus* and *L. macaco*): Relation to social status, reproduction, aggression and environmental change. *Folia Primatologica* 59, 137–48.

Fredsted, T., Pertoldi, C., Olesen, J. M., Eberle, M. & Kappeler, P. M. 2004. Microgeographic heterogeneity in spatial distribution and mtDNA variability of gray mouse lemurs (*Microcebus murinus*, Primates: Cheirogaleidae). *Behavioral Ecology and Sociobiology* 56, 393–403.

Fredsted, T., Pertoldi, C., Schierup, M. H. & Kappeler, P. M. 2005. Microsatellite analyses reveal fine-scale genetic structure in grey mouse lemurs (*Microcebus murinus*). *Molecular Ecology* 14, 2363–72.

Fredsted, T., Schierup, M., Groeneveld, L. & Kappeler, P. M. 2007. Genetic structure, lack of sex-biased dispersal and behavioral flexibility in the pair-living fat-tailed dwarf lemur, *Cheirogaleus medius*. *Behavioral Ecology and Sociobiology* 61, 953–54.

Freed, B. Z. 2006. Polyspecific associations of crowned lemurs and Sanford's lemurs in Madagascar. In *Lemurs: Ecology and Adaptations*, ed. by Gould L. & Sauther M. L., 111–31. New York: Springer.

Gamba, M. & Giacoma, C. 2007. Quantitative acoustic analysis of the vocal repertoire of the crowned lemur. *Ethology, Ecology and Evolution* 19, 323–43.

Ganzhorn, J. U. 1988. Food partitioning among Malagasy primates. *Oecologia* 75, 436–50.

———. 1989. Niche separation of seven lemur species in the eastern rainforest of Madagascar. *Oecologia* 79, 279–86.

Gebo, D. L. 1987. Locomotor diversity in prosimian primates. *American Journal of Primatology* 13, 271–81.

Genty, E., Foltz, J. & Roeder, J.-J. 2008. Can brown lemurs (*Eulemur fulvus*) learn to deceive a human competitor? *Animal Cognition* 11, 255–66.

Genty, E., Palmier, C. & Roeder, J.-J. 2004. Learning to suppress responses to the larger of two rewards in two species of lemurs, *Eulemur fulvus* and *E. macaco*. *Animal Behaviour* 67, 925–32.

Gligor, M., Ganzhorn, J. U., Rakotondravony, D., Ramilijaona, O., Razafimahatratra, E., Zischler, H. & Hapke, A. 2009. Hybridization between mouse lemurs in an ecological transition zone in southern Madagascar. *Molecular Ecology* 18, 520–33.

Godfrey, L. R. & Jungers, W. L. 2003. The extinct sloth lemurs of Madagascar. *Evolutionary Anthropology* 12, 252–63.

Godfrey, L. R., Lyon, S. K. & Sutherland, M. R. 1993. Sexual dimorphism in large-bodied primates: The case of the subfossil lemurs. *American Journal of Physical Anthropology* 90, 315–34.

Godfrey, L. R., Samonds, K. E., Jungers, W. L., Sutherland, M. R. & Irwin, M. T. 2004. Ontogenetic correlates of diet in Malagasy lemurs. *American Journal of Physical Anthropology* 123, 250–76.

Godfrey, L. R., Jungers, W. L. & Schwartz, G. T. 2006. Ecology and extinction of Madagascar's subfossil lemurs. In *Lemurs: Ecology and Adaptation*, ed. by Gould, L. & Sauther, M. L., 41–64. New York: Springer.

Goodman, S. M. 1994. The enigma of anti-predator behavior in lemurs: Evidence of a large extinct eagle on Madagascar. *International Journal of Primatology* 15, 129–34.

Goodman, S. M., O'Connor, S. & Langrand, O. 1993. A review of predation on lemurs: Implications for the evolution of social behavior in small, nocturnal primates. In *Lemur Social Systems and Their Ecological Basis*, ed. by Kappeler, P. M. & Ganzhorn, J. U., 51–66. New York: Plenum Press.

Goodman, S. M., Rasoloarison, R. M. & Ganzhorn, J. U. 2004. On the specific identification of subfossil *Cryptoprocta* (Mammalia, Carnivora) from Madagascar. *Zoosystema* 26, 129–43.

Gould, L. 1997. Intermale affiliative behavior in ringtailed lemurs (*Lemur catta*) at the Beza-Mahafaly Reserve, Madagascar. *Primates* 38, 15–30.

Gould, L. & Overdorff, D. J. 2002. Adult male scent-marking in *Lemur catta* and *Eulemur fulvus rufus*. *International Journal of Primatology* 23, 575–86.

Gould, L. & Sauther, M. L. 2007. Lemuriformes. In *Primates in Perspective*, ed. by Campbell, C. J., Fuentes, A., MacKinnon, K. C., Panger, M. & Bearder, S. K., 46–72. Oxford: Oxford University Press.

Gould, L., Sussman, R. W. & Sauther, M. L. 1999. Natural disasters and primate populations: The effects of a 2-year drought on a naturally occurring population of ring-tailed lemurs (*Lemur catta*) in southwestern Madagascar. *International Journal of Primatology* 20, 69–84.

———. 2003. Demographic and life-history patterns in a population of ring-tailed lemurs (*Lemur catta*) at Beza Mahafaly Reserve, Madagascar: A 15-year perpective. *American Journal of Physical Anthropology* 120, 182–94.

Grieser, B. 1992. Infant development and parental care in two species of sifakas. *Primates* 33, 305–14.

Groves, C. P. & Shekelle, M. 2010. The genera and species of Tarsiidae. *International Journal of Primatology* 31, 1071–82.

Grubb, P., Butynski, T. M., Oates, J. F., Bearder, S. K., Disotell, T. R., Groves, C. P. & Struhsaker, T. T. 2003. Assessment of the diversity of African primates. *International Journal of Primatology* 24, 1301–57.

Gursky, S. 2000a. Allocare in a nocturnal primate: Data on the spectral tarsier, *Tarsius spectrum*. *Folia Primatologica* 71, 39–54.

———. 2000b. Sociality in the spectral tarsier, *Tarsius spectrum*. *American Journal of Primatology* 51, 89–101.

———. 2002a. Predation on a wild spectral tarsier (*Tarsius spectrum*) by a snake. *Folia Primatologica* 73, 60–62.

———. 2002b. The behavioral ecology of the spectral tarsier, *Tarsier spectrum*. *Evolutionary Anthropology* 11, 226–34.

———. 2005. Predator mobbing in *Tarsius spectrum*. *International Journal of Primatology* 26, 207–21.

———. 2007a. Tarsiiformes. In: *Primates in Perspective* (Ed. by Campbell, C. J., Fuentes, A., MacKinnon, K. C., Panger, M. & Bearder, S. K.), pp. 73–85. Oxford: Oxford University Press.

———. 2007b. The response of spectral tarsiers toward avian and terrestrial predators. In: *Primate Anti-predator Strategies* (Ed. by Gursky, S. & Nekaris, K. A. I.), pp. 241–252. New York: Springer.

———. 2010. Dispersal patterns in *Tarsius spectrum*. *International Journal of Primatology*, 31, 117–131.

Gursky-Doyen, S. 2010. Intraspecific variation in the mating system of spectral tarsiers. *International Journal of Primatology* 31, 1161–73.

Gursky, S. & Nekaris, K. A. I. 2007. *Primate Anti-predator Strategies*. New York: Springer.

Guschanski, K., Olivieri, G., Funk, S. M. & Radespiel, U. 2007. MtDNA reveals strong genetic differentiation among geographically isolated populations of the golden brown mouse lemur, *Microcebus ravelobensis*. *Conservation Genetics* 8, 809–21.

Harcourt, C. S. & Nash, L. T. 1986. Social organization of galagos in Kenyan coastal forest. I. *Galago zanzibaricus*. *American Journal of Primatology* 10, 339–55.

Harrington, J. E. 1977. Discrimination between males and females by scent in *Lemur fulvus*. *Animal Behaviour* 25, 147–51.

Hausfater, G. 1975. Dominance and reproduction in baboons (*Papio cynocephalus*): A quantitative analysis. *Contributions to Primatology*, 7, 1–150.

Hilgartner, R., Zinner, D. & Kappeler, P. M. 2008. Life history traits and parental care in *Lepilemur ruficaudatus*. *American Journal of Primatology* 70, 2–11.

Hill, R. A. 2006. Why be diurnal? Or, why not be cathemeral? *Folia Primatologica* 77, 72–86.

Hinde, R. A. 1976. Interactions, relationships and social structure. *Man* 11, 1–17.

Hood, L. C. 1994. Infanticide among ringtailed lemurs (*Lemur catta*) at Berenty Reserve, Madagascar. *American Journal of Primatology* 33, 65–69.

Horvath, J. E., Weisrock, D. W., Embry, S. L., Fiorentino, I., Balhoff, J. P., Kappeler, P. M., Wray, G. A., Willard, H. F. & Yoder, A. D. 2008. Development and application of a phylogenomic toolkit: Resolving the evolutionary history of Madagascar's lemurs. *Genome Research* 18, 489–99.

Ichino, S. 2006. Troop fission in wild ring-tailed lemurs (*Lemur catta*) at Berenty, Madagascar. *American Journal of Primatology* 68, 97–102.

Irwin, M. T., Raharison, J. L. & Wright, P. C. 2009. Spatial and temporal variability in predation on rainforest primates: Do forest fragmentation and predation act synergistically? *Animal Conservation* 12, 220–30.

Isler, K., Kirk, E. C., Miller, J. M. A., Albrecht, G. A., Gelvin, B. R. & Martin, R. D. 2008. Endocranial volumes of primate species: Scaling analyses using a comprehensive and reliable data set. *Journal of Human Evolution* 55, 967–78.

Jolly, A. 1966a. Lemur social behavior and primate intelligence. *Science* 153, 501–6.

———. 1966b. *Lemur Behavior*. Chicago: University of Chicago Press.

———. 1967. Breeding synchrony in wild *Lemur catta*. In *Social Communication among Primates*, ed. by Altman, S. A., 3–14. Chicago: University of Chicago Press.

———. 1998. Pair-bonding, female aggression and the evolution of lemur societies. *Folia Primatologica* 69, 1–13.

Jolly, A., Caless, S., Cavigelli, S., Gould, L., Pereira, M. E., Pitts, A., Pride, R. E., Rabenandrasana, H. D., Walker, J. D. & Zafison, T. 2000. Infant killing, wounding and predation in *Eulemur* and *Lemur*. *International Journal of Primatology* 21, 21–40.

Jolly, A., Dobson, A., Rasamimanana, H. M., Walker, J., O'Connor, S., Solberg, M. & Perel, V. 2002. Demography of *Lemur catta* at Berenty Reserve, Madagascar: Effects of troop size, habitat and rainfall. *International Journal of Primatology* 23, 327–53.

Joly, M. & Zimmermann, E. 2007. First evidence for relocation of stationary food resources during foraging in a strepsirhine primate (*Microcebus murinus*). *American Journal of Primatology* 69, 1045–52.

Kappeler, P. M. 1989. Agonistic and grooming behavior of captive crowned lemurs (*Lemur coronatus*) during the breeding season. *Human Evolution* 4, 207–15.

———. 1990. Social status and scent marking behaviour in *Lemur catta*. *Animal Behaviour* 40, 774–76.

———. 1991. Patterns of sexual dimorphism in body weight among prosimian primates. *Folia Primatologica* 57, 132–46.

———. 1993a. Female dominance in primates and other mammals. In *Perspectives in Ethology. Vol 10: Behaviour and Evolution*, ed. by Bateson, P. P. G., Klopfer, P. H. & Thompson, N. S., 143–58. New York: Plenum Press.

———. 1993b. Reconciliation and post-conflict behaviour in ringtailed lemurs, *Lemur catta* and redfronted lemurs, *Eulemur fulvus rufus*. *Animal Behaviour* 45, 901–15.

———. 1993c. Variation in social structure: the effects of sex and kinship on social interactions in three lemur species. *Ethology* 93, 125–45.

———. 1996. Causes and consequences of life history variation among strepsirhine primates. *American Naturalist* 148, 868–91.

———. 1997a. Determinants of primate social organization: comparative evidence and new insights from Malagasy lemurs. *Biological Reviews of the Cambridge Philosophical Society* 72, 111–51.

———. 1997b. Intrasexual selection in *Mirza coquereli*: Evidence for scramble competition polygyny in a solitary primate. *Behavioral Ecology and Sociobiology* 41, 115–27.

———. 1997c. Intrasexual selection and testis size in strepsirhine primates. *Behavioral Ecology* 8, 10–19.

———. 1998a. Nests, tree holes, and the evolution of primate life histories. *American Journal of Primatology* 46, 7–33.

———. 1998b. To whom it may concern: The transmission and function of chemical signals in *Lemur catta*. *Behavioral Ecology and Sociobiology* 42, 411–21.

———. 1999. Convergence and divergence in primate social systems. In *Primate Communities*, ed. by Fleagle, J. G., Janson, C. H. & Reed, K., 158–70. Cambridge: Cambridge University Press.

———. 2000a. Lemur origins: Rafting by groups of hibernators? *Folia Primatologica* 71, 422–25.

———. 2000b. Causes and consequences of unusual sex ratios among lemurs. In *Primate Males: Causes and Consequences of Variation in Group Composition*, ed. by Kappeler, P. M., 55–63. Cambridge: Cambridge University Press.

Kappeler, P. M. & Erkert, H. G. 2003. On the move around the clock: Correlates and determinants of cathmeral activity in wild redfronted lemurs (*Eulemur fulvus rufus*). *Behavioral Ecology and Sociobiology* 54, 359–69.

Kappeler, P. M. & Fichtel, C. 2012a. Female reproductive competition in *Eulemur rufrifons*: Eviction and reproductive restraint in a plurally breeding Malagasy primate. *Molecular Ecology* 21, 685–98.

———. 2012b. A 15-year perspective on the social organization and life history of sifakas in Kirindy Forest. In *Long-term Field Studies of Primates*, ed. by Kappeler, P. M. & Watts, D. P., 101–21. Heidelberg: Springer.

Kappeler, P. M. & Heymann, E. W. 1996. Nonconvergence in the evolution of primate life history and socio-ecology. *Biological Journal of the Linnean Society* 59, 297–326.

Kappeler, P. M. & Port, M. 2008. Mutual tolerance or reproductive competition? Patterns of reproductive skew among male redfronted lemurs (*Eulemur fulvus rufus*). *Behavioral Ecology and Sociobiology* 62, 1477–88.

Kappeler, P. M. & Schäffler, L. 2008. The lemur syndrome unresolved: Extreme male reproductive skew in sifakas (*Propithecus verreauxi*), a sexually monomorphic primate with female dominance. *Behavioral Ecology and Sociobiology* 62, 1007–15.

Kappeler, P. M. & van Schaik, C. P. 2002. Evolution of primate social systems. *International Journal of Primatology* 23, 707–40.

Kappeler, P. M., Wimmer, B., Zinner, D. P. & Tautz, D. 2002. The hidden matrilineal structure of a solitary lemur: Implications for primate social evolution. *Proceedings of the Royal Society B* 269, 1755–63.

Kappeler, P. M., Rasoloarison, R. M., Razafimanantsoa, L., Walter, L. & Roos, C. 2005. Morphology, behaviour and molecular evolution of giant mouse lemurs (*Mirza* spp.) Gray, 1870, with description of a new species. *Primate Report* 71.

Kappeler, P. M., Mass, V. & Port, M. 2009. Even adult sex ratios in lemurs: Potential costs and benefits of subordinate males in Verreaux's sifaka (*Propithecus verreauxi*) in the Kirindy Forest CFPF, Madagascar. *American Journal of Physical Anthropology* 140, 487–97.

Karanth, K. P., Delefosse, T., Rakotosamimanana, B., Parsons, T. J. & Yoder, A. D. 2005. Ancient DNA from giant extinct lemurs confirms single origin of Malagasy primates. *Proceedings of the National Academy of Sciences, USA* 102, 5090–95.

Karpanty, S. M. 2006. Direct and indirect impacts of raptor predation on lemurs in southeastern Madagascar. *International Journal of Primatology* 27, 239–61.

Kendal, R. L., Custance, D. M., Kendal, J. R., Vale, G., Stoinski, T. S., Rakotomalala, N. L. & Rasamimanana, H. 2010. Evidence for social learning in wild lemurs (*Lemur catta*). *Learning & Behavior* 38, 220–34.

Klopfer, P. H. & Boskoff, K. L. 1979. Maternal behavior in prosimians. In *The Study of Prosimian Behavior*, ed. by Doyle, G. A. & Martin, R. D., 123–56. New York: Academic Press.

Kobbe, S. & Dausmann, K. 2009. Hibernation in Malagasy mouse lemurs as a strategy to counter environmental challenge. *Naturwissenschaften* 96, 1221–27.

Koyama, N., Ichino, S., Nakamichi, M. & Takahata, Y. 2005. Long-term changes in dominance ranks among ring-tailed lemurs at Berenty Reserve, Madagsacar. *Primates* 46, 225–34.

Kraus, C., Heistermann, M. & Kappeler, P. M. 1999. Physiological suppression of sexual function of subordinate males: A subtle form of intrasexual competition among male sifakas (*Propithecus verreauxi*)? *Physiology and Behavior* 66, 855–61.

Kraus, C., Eberle, M. & Kappeler, P. M. 2008. The costs of risky male behaviour: Sex differences in seasonal survival in a small sexually monomorphic primate. *Proceedings of the Royal Society B* 275, 1635–44.

Kudo, H. & Dunbar, R. I. M. 2001. Neocortex size and social network size in primates. *Animal Behaviour* 62, 711–22.

Lahann, P., Schmid, J. & Ganzhorn, J. U. 2006. Geographic variation in populations of *Microcebus murinus* in Madagascar: Resource seasonality or Bergmann's rule? *International Journal of Primatology* 27, 983–99.

Lawler, R. R. 2007. Fitness and extra-group reproduction in male Verreaux's sifaka: An analysis of reproductive success from 1989–1999. *American Journal of Physical Anthropology* 132, 267–77.

Lawler, R. R., Richard, A. F. & Riley, M. A. 2005. Intrasexual selection in Verreaux's sifaka (*Propithecus verreauxi verreauxi*). *Journal of Human Evolution* 48, 259–77.

Lee, P. C. & Kappeler, P. M. 2003. Socioecological correlates of phenotypic plasticity of primate life histories. In *Primate Life History and Socioecology*, ed. by Kappeler, P. M. & Pereira, M. E., 41–65. Chicago: University of Chicago Press.

Leonhardt, S. D., Tung, J., Camden, J. B., Leal, M. & Drea, C. M. 2009. Seeing red: Behavioral evidence of trichromatic color vision in strepsirrhine primates. *Behavioral Ecology* 20, 1–12.

Leu, S.T., Bashford, J., Kappeler, P. M. & Bull, C.M. 2010. Association networks reveal social organization in the sleepy lizard. *Animal Behaviour* 79, 217–25.

Lewis, K. P., Jaffe, S. & Brannon, E. M. 2005. Analog number representations in mongoose lemurs (*Eulemur mongoz*): Evidence from search task. *Animal Cognition* 8, 247–52.

Lewis, R. J. 2005. Sex differences in scent-marking in sifaka: Mating conflict or male services? *American Journal of Physical Anthropology* 128, 389–98.

———. 2009. Chest staining variation as a signal of testosterone levels in male Verreaux's sifaka. *Physiology and Behavior* 96, 586–92.

Lewis, R. J., Razafindrasamba, S. M. & Tolojanahary, J.-P. 2003. Observed infanticide in a seasonal breeding prosimian (*Propithecus verreauxi verreauxi*) in Kirindy Forest, Madagascar. *Folia Primatologica* 74, 101–3.

Lührs, M.-L., Dammhahn, M., Kappeler, P. M. & Fichtel, C. 2009. Spatial memory in the grey mouse lemur (*Microcebus murinus*). *Animal Cognition* 12, 599–609.

Macedonia, J. M. 1994. The vocal repertoire of the ringtailed lemur (*Lemur catta*). *Folia Primatologica* 61, 186–217.

MacKinnon, J. & MacKinnon, K. 1980. The behavior of wild spectral tarsiers. *International Journal of Primatology* 1, 361–79.

MacLean, E. L., Barrickman, N. L., Johnson, E. M. & Wall, C. E. 2009. Sociality, ecology, and relative brain size in lemurs. *Journal of Human Evolution* 56, 471–78.

Mahajan, N., Barnes, J. L., Blanco, M. & Santos, L. R. 2009. Enumeration of objects and substances in non-human primates: Experiments with brown lemurs (*Eulemur fulvus*). *Developmental Science* 12, 920–28.

Markoff, M., Brameier, M. & Kappeler, P. M. 2011. On species delimitation: Yet another lemur species or just generic variation? *BMC Evolutionary Biology*, 11, 216.

Martin, R. D. 1972. Adaptive radiation and behaviour of the Malagasy lemurs. *Philosophical Transactions of the Royal Society of London B* 264, 295–352.

Martin, R. D. & Ross, C. F. 2005. The evolutionary and ecological context of primate vision. In *The Primate Visual System: A Comparative Approach*, ed. by Kremers, J., 1–36. New York: Wiley.

Mass, V., Heistermann, M. & Kappeler, P. M. 2009. Mate-guarding as a male reproductive tactic in *Propithecus verreauxi*. *International Journal of Primatology* 30, 389–409.

Masters, J. C. 1991. Loud calls of *Galago crassicaudatus* and *G. garnettii* and their relation to habitat structure. *Primates* 32, 153–67.

Masters, J. C., Anthony, N. M., de Wit, M. J. & Mitchell, A. 2005. Reconstructing the evolutionary history of the Lorisidae using morphological, molecular, and geological data. *American Journal of Physical Anthropology* 127, 465–80.

Matsui, A., Rakotondraparany, F., Munechika, I., Hasegawa, M. & Horai, S. 2009. Molecular phylogeny and evolution of prosimians based on complete sequences of mitochondrial DNAs. *Gene* 441, 53–66.

McComb, K. & Semple, S. 2005. Coevolution of vocal communication and sociality in primates. *Biology Letters* 1, 381–85.

McNab, B. K. & Wright, P. C. 1987. Temperature regulation and oxygen consumption in the Phillipine tarsier, *Tarsius syrichta*. *Physiological Zoology* 60, 596–600.

Méndez-Cárdenas, M. G. & Zimmermann, E. 2009. Duetting: A mechanism to strengthen pair bonds in a dispersed pair-living primate (*Lepilemur edwardsi*)? *American Journal of Physical Anthropology* 139, 523–32.

Merker, S. 2006. Habitat-specific ranging patterns of Dian's tarsiers (*Tarsius dianae*) as revealed by radiotracking. *American Journal of Primatology* 68, 111–25.

Merker, S., Driller, C., Perwitasari-Farajallah, D., Pamungkas, J.

& Zischler, H. 2009. Elucidating geological and biological processes underlying the diversification of Sulawesi tarsiers. *Proceedings of the National Academy of Sciences* 106, 8459–64.

Merker, S. & Groves, C. P. 2006. *Tarsius lariang*: A new primate species from western central Sulawesi. *International Journal of Primatology* 27, 465–85.

Merritt, D., Maclean, E. L., Jaffe, S. & Brannon, E. M. 2007. A comparative analysis of serial ordering in ring-tailed lemurs (*Lemur catta*). *Journal of Comparative Psychology* 121, 363–71.

Mertl, A. S. 1975. Discrimination of individuals by scent in a primate. *Behavioral Biology* 14, 505–9.

———. 1976. Olfactory and visual cues in social interactions of *Lemur catta*. *Folia Primatologica* 26, 151–61.

Mertl-Millhollen, A. S. 2006. Scent marking as resource defense by female *Lemur catta*. *American Journal of Primatology* 68, 605–21.

Mittermeier, R. A., Louis Jr., E. E., Richardson, M., Schwitzer, C., Langrand, O., Rylands, A. B., Hawkins, F., Rajaobelina, S. Ratsimbazafy, J. Rasoloarison, R., Roos, C., Kappeler, P. M. & Mackinnon, J. 2010. *Lemurs of Madagascar*, 3rd edition. Tropical Field Guide Series. Arlington, VA: Conservation International.

Morland, H. S. 1990. Parental behavior and infant development in ruffed lemurs (*Varecia variegata*) in a northeast Madagascar rainforest. *American Journal of Primatology* 20, 253–65.

Morelli, T. L., King, S. J., Pochron, S. T. & Wright, P. C. 2009. The rules of disengagement: Takeovers, infanticide, and dispersal in a rainforest lemur, *Propithecus edwardsi*. *Behaviour* 146, 499–523.

Müller, A. E. 1998. A preliminary report on the social organisation of *Cheirogaleus medius* (Cheirogaleidae; Primates) in north-west Madagascar. *Folia Primatologica* 69, 160–66.

Müller, A. E. & Thalmann, U. 2000. Origin and evolution of primate social organisation: A reconstruction. *Biological Reviews of the Cambridge Philosophical Society* 75, 405–35.

Müller, E. F. 1985. Basal metabolic rates in primates: The possible role of phylogenetic and ecological factors. *Comparative Biochemistry and Physiology A* 81, 707–11.

Mzilikazi, N., Masters, J. C. & Lovegrove, B. G. 2006. Lack of torpor in free-ranging southern lesser galagos, *Galago moholi*: Ecological and physiological considerations. *Folia Primatologia* 77, 465–76.

Nakamichi, M. & Koyama, N. 1997. Social relationships among ring-tailed lemurs (*Lemur catta*) in two free-ranging troops at Berenty Reserve, Madagascar. *International Journal of Primatology* 18, 73–93.

Nash, L. T., Bearder, S. K. & Olson, T. R. 1989. Synopsis of *Galago* species characteristics. *International Journal of Primatology* 10, 57–80.

Nekaris, K. A. I. 2003. Observations of mating, birthing and parental behaviour in three subspecies of slender loris (*Loris tardigradus* and *Loris lydekkerianus*) in India and Sri Lanka. *Folia Primatologica* 74, 312–36.

Nekaris, K. A. I. & Bearder, S. K. 2007. The lorisiform primates of Asia and mainland Africa. In *Primates in Perspective*, ed. by Campbell, C. J., Fuentes, A., MacKinnon, K. C., Panger, M. & Bearder, S. K., 24–45. Oxford: Oxford University Press.

Nekaris, K. A. I. & Rasmussen, D. T. 2003. Diet and feeding behavior of Mysore slender lorises. *International Journal of Primatology* 24, 33–46.

Nekaris, K. A. I., Pimley, E. R. & Ablard, K. M. 2007. Predator defense by slender lorises and pottos. In *Primate Antipredator Strategies*, ed. by Gursky, S. & Nekaris, K. A. I., 220–38. New York: Springer.

Niemitz, C. T. 2010. Progreditur ordinara saltando et retrorsum . . . Normally proceeds in a leaping fashion, and backwards . . . *International Journal of Primatology* 20, 567–83.

Nietsch, A. 1999. Duet vocalizations among different populations of Sulawesi tarsiers. *International Journal of Primatology* 31, 941–57.

Nievergelt, C. M., Mutschler, T. & Feistner, A. T. C. 1998. Group encounters and territoriality in wild Alaotran gentle lemurs (*Hapalemur griseus alaotrensis*). *American Journal of Primatology* 46, 251–58.

Nievergelt, C. M., Mutschler, T., Feistner, A. T. C. & Woodruff, D. S. 2002. Social system of the Alaotran gentle lemur (*Hapalemur griseus alaotrensis*): Genetic characterization of group composition and mating system. *American Journal of Primatology* 57, 157–76.

Norscia, I. & Borgognini Tarli, S. M. 2008. Ranging behavior and possible correlates of pair-living in southeastern Avahis (Madagascar). *International Journal of Primatology* 29, 153–71.

Norscia, I., Antonacci, D. & Palagi, E. 2009. Mating first, mating more: Biological market fluctuation in a wild prosimian. *PLoS ONE* 4, e4679.

Nowack, J., Mzilikazi, N. & Dausmann, K. H. 2010. Torpor on demand: Heterothermy in the non-lemur primate *Galago moholi*. *PLoS ONE* 5, e10797.

Nunn, C. L. & Deaner, R. O. 2004. Patterns of participation and free riding in territorial conflicts among ringtailed lemurs (*Lemur catta*). *Behavioral Ecology and Sociobiology* 57, 50–61.

Oda, R. 2002. Individual distinctiveness of the contact calls of ring-tailed lemurs. *Folia Primatologica* 73, 132–36.

Ohta, H., Ishida, H. & Matano, S. 1987. Learning set formation in thick-tailed bush babies (*Galago crassicaudatus*) and comparison of learning ability among four species. *Folia Primatologica* 48, 1–8.

Orlando, L., Calvignac, S., Schnebelen, C., Douady, C., Godfrey, L. R. & Hänni, C. 2008. DNA from extinct giant lemurs links archaeolemurids to extant indriids. *BMC Evolutionary Biology* 8, 121.

Ossi, K. & Kamilar, J. M. 2006. Environmental and phylogenetic correlates of *Eulemur* behavior and ecology (Primates: Lemuridae). *Behavioral Ecology and Sociobiology* 61, 53–64.

Ostner, J. & Kappeler, P. M. 1999. Central males instead of multiple pairs in redfronted lemurs, *Eulemur fulvus rufus* (Primates, Lemuridae)? *Animal Behaviour* 58, 1069–78.

———. 2004. Male life history and the unusual adult sex ratios of redfronted lemur, *Eulemur fulvus rufus*, groups. *Animal Behaviour* 67, 249–59.

Ostner, J., Kappeler, P. M. & Heistermann, M. 2002. Seasonal variation and social correlates of androgen excretion in male redfronted lemurs (*Eulemur fulvus rufus*). *Behavioral Ecology and Sociobiology* 52, 485–95.

Ostner, J., Nunn, C. L. & Schülke, O. 2008a. Female reproductive synchrony predicts skewed paternity across primates. *Behavioral Ecology* 19, 1150–58.

Ostner, J., Kappeler, P. M. & Heistermann, M. 2008b. Androgen and glucocorticoid levels reflect seasonally occurring social challenges in male redfronted lemurs (*Eulemur fulvus rufus*). *Behavioral Ecology and Sociobiology* 62, 627–38.

Overdorff, D. J. 1996. Ecological correlates to social structure in two lemur species in Madagascar. *American Journal of Physical Anthropology* 100, 487–506.

Overdorff, D. J., Merenlender, A. M., Talata, P., Telo, A. & Forward, Z. A. 1999. Life history of *Eulemur fulvus rufus* from 1988–1998 in southeastern Madagascar. *American Journal of Physical Anthropology* 108, 295–310.

Oxnard, C. E., Crompton, R. H., & Liebermann, S. S. 1989. Animal lifestyles and anatomies: The case of the prosimian primates. Seattle: Washington University Press.

Palagi, E., Antonacci, D. & Norscia, I. 2008. Peacemaking on treetops: First evidence of reconciliation from a wild prosimian (*Propithecus verreauxi*). *Animal Behaviour* 76, 737–47.

Palagi, E., Dapporto, L. & Borgognini Tarli, S. M. 2005a. The neglected scent: On the marking function of urine in *Lemur catta*. *Behavioral Ecology and Sociobiology* 58, 437–45.

Palagi, E. & Norscia, I. 2009. Multimodal signaling in wild *Lemur catta*: Economic design and territorial function of urine marking. *American Journal of Physical Anthropology* 139, 182–92.

Palagi, E., Paoli, T. & Borgognini Tarli, S. M. 2005b. Aggression and reconciliation in two captive groups of *Lemur catta*. *International Journal of Primatology* 26, 279–94.

Palagi, E., Telara, S. & Borgognini Tarli, S. M. 2004. Reproductive strategies in *Lemur catta*: Balance among sending, receiving, and countermarking scent signals. *International Journal of Primatology* 25, 1019–31.

Patel, E. R. 2007. Non-maternal infant care in wild silky sifakas (*Propithecus candidus*). *Lemur News* 12, 39–42.

Penn, D. J. 2002. The scent of genetic compatibility: sexual selection and the major histocompatibility complex. *Ethology* 108, 1–21.

Pereira, M. E. 1991. Asynchrony within estrous synchrony among ringtailed lemurs (Primates: Lemuridae). *Physiology and Behavior* 49, 47–52.

Pereira, M. E. 1995. Development and social dominance among group-living primates. *American Journal of Primatology* 37, 143–75.

Pereira, M. E. & Kappeler, P. M. 1997. Divergent systems of agonistic relationship in lemurid primates. *Behaviour* 134, 225–74.

Pereira, M. E. & Leigh, S. R. 2003. Modes of primate development. In *Primate Life Histories and Socioecology*, ed. by Kappeler, P. M. & Pereira, M. E., 149–76. Chicago: University of Chicago Press.

Pereira, M. E. & Macedonia, J. M. 1991. Ringtailed lemur antipredator calls denote predator class, not response urgency. *Animal Behaviour* 41, 543–44.

Pereira, M. E. & Weiss, M. L. 1991. Female mate choice, male migration, and the threat of infanticide in ringtailed lemurs. *Behavioral Ecology and Sociobiology* 28, 141–52.

Pereira, M. E., Seeligson, M. L. & Macedonia, J. M. 1988. The behavioral repertoire of the black-and-white-ruffed lemur, *Varecia variegata variegata* (Primates, Lemuridae). *Folia Primatologica* 51, 1–32.

Pereira, M. E., Kaufman, R., Kappeler, P. M. & Overdorff, D. J. 1990. Female dominance does not characterize all of the Lemuridae. *Folia Primatologica* 55, 96–103.

Pereira, M. E., Strohecker, R. A., Cavigelli, S. A., Hughes, C. L. & Pearson, D. D. 1999. Metabolic strategy and social behavior in Lemuridae. In *New Directions in Lemur Studies*, ed. by Rakotosamimanana, B., Rasamimanana, H., Ganzhorn, J. U. & Goodman, S. M., 93–118. New York: Plenum Press.

Perret, M. 1992. Environmental and social determinants of sexual function in the male lesser mouse lemur (*Microcebus murinus*). *Folia Primatologica* 59, 1–25.

Perry, J. M. G. & Hartstone-Rose, A. 2010. Maximum ingested food size in captive strepsirrhine primates: Scaling and the effects of diet. *American Journal of Physical Anthropology* 142, 625–35.

Pochron, S. T. & Wright, P. C. 2002. Dynamics of testes size compensates for variation in male body size. *Evolutionary Ecology Research* 4, 577–85.

———. 2003. Variability in adult group compositions of a prosimian primate. *Behavioral Ecology and Sociobiology* 54, 285–93.

Pochron, S. T., Fitzgerald, J., Gilbert, C. C., Lawrence, D., Grgas, M., Rakotonirina, G., Ratsimbazafy, R., Rakotosoa, R. & Wright, P. C. 2003. Patterns of female dominance in *Propithecus diadema edwardsi* of Ranomafana National Park, Madagascar. *American Journal of Primatology* 61, 173–85.

Pochron, S. T., Tucker, W. T. & Wright, P. C. 2004. Demography, life history, and social structure in *Propithecus diadema edwardsi* from 1986–2000 in Ranomafana National Park, Madagascar. *American Journal of Physical Anthropology* 125, 61–72.

Pochron, S. T., Morelli, T. L., Terranova, P., Scirbona, J., Cohen, J., Kunapareddy, G., Rakotonirina, G., Ratsimbazafy, R., Rakotosoa, R. & Wright, P. C. 2005. Patterns of male scent-marking in *Propithecus edwardsi* of Ranomafana National Park, Madagascar. *American Journal of Primatology* 65, 103–15.

Port, M., Clough, D. & Kappeler, P. M. 2009. Market effects offset the reciprocation of grooming in free-ranging redfronted lemurs, *Eulemur fulvus rufus*. *Animal Behaviour* 77, 29–36.

Port, M., Johnstone, R. A. & Kappeler, P. M. 2010. Costs and benefits of multi-male associations in redfronted lemurs (*Eulemur fulvus rufus*). *Biology Letters* 6, 620–22.

Poulsen, J. R. & Clark, C. J. 2001. Predation on mammals by the grey-cheeked mangabey *Lophocebus albigena*. *Primates* 42, 391–94.

Pride, R. E. 2005. Optimal group size and seasonal stress in ringtailed lemurs (*Lemur catta*). *Behavioral Ecology* 16, 550–60.

Pruetz, J. D. & Bertolani, P. 2007. Savanna chimpanzees, *Pan troglodytes verus*, hunt with tools. *Current Biology* 17, 412–17.

Pullen, S. L., Bearder, S. K. & Dixson, A. F. 2000. Preliminary observations on sexual behavior and the mating system in free-ranging lesser Galagos (*Galago moholi*). *American Journal of Primatology* 51, 79–88.

Radespiel, U. 2000. Sociality in the gray mouse lemur (*Microcebus murinus*) in northwestern Madagascar. *American Journal of Primatology* 51, 21–40.

Radespiel, U. & Zimmermann, E. 2001. Female dominance in captive gray mouse lemurs (*Microcebus murinus*). *American Journal of Primatology* 54, 181–92.

Radespiel, U., Sarikaya, Z., Zimmermann, E. & Bruford, M. W. 2001. Sociogenetic structure in a free-living nocturnal primate population: Sex-specific differences in the grey mouse lemur (*Microcebus murinus*). *Behavioral Ecology and Sociobiology* 50, 493–502.

Radespiel, U., Dal Secco, V., Drögemüller, C., Braune, P., Labes, E. & Zimmermann, E. 2002. Sexual selection, multiple mating and paternity in grey mouse lemurs, *Microcebus murinus*. *Animal Behaviour* 63, 259–68.

Radespiel, U., Rakotondravony, R. & Chikhi, L. 2008. Natural and anthropogenic determinants of genetic structure in the largest remaining population of the endangered golden-brown mouse lemur, *Microcebus ravelobensis*. *American Journal of Primatology* 70, 860–70.

Rahlfs, M. & Fichtel, C. 2010. Anti-predator behaviour in a nocturnal primate, the grey mouse lemur (*Microcebus murinus*). *Ethology* 116, 429–39.

Rakotondravony, D., Goodman, S. M. & Soarimalala, V. 1998. Predation on *Hapalemur griseus griseus* by *Boa manditra* (Boidae) in the littoral forest of eastern Madagascar. *Folia Primatologica* 69, 405–8.

Rasoloarison, R. M., Rasolonandrasana, B. P. N., Ganzhorn, J. U. & Goodman, S. M. 1995. Predation on vertebrates in the Kirindy forest, western Madagascar. *Ecotropica* 1, 59–65.

Rasoloharijaona, S., Rakotosamimanana, B. & Zimmermann, E. 2000. Infanticide by a male Milne-Edwards' sportive lemur (*Lepilemur edwardsi*) in Ampijoroa, NW-Madagascar. *International Journal of Primatology* 21, 41–45.

Rasoloharijaona, S., Rakotosamimanana, B., Randrianambinina, B. & Zimmermann, E. 2003. Pair-specific usage of sleeping sites and their implications for social organization in a nocturnal Malagasy primate, the Milne Edwards' sportive lemur (*Lepilemur edwardsi*). *American Journal of Physical Anthropology* 122, 251–58.

Reichard, U. H. 2003. Monogamy: Past and present. In *Monogamy: Mating Strategies and Partnerships in Birds, Humans and Other Mammals*, ed. by Reichard, U. H. & Boesch, C., 3–25. Cambridge: Cambridge University Press.

Richard, A. F. 1974. Patterns of mating in *Propithecus verreauxi verreauxi*. In *Prosimian Biology*, ed. by Martin, R. D., Doyle, G. A. & Walker, A. C., 49–74. London: Duckworth.

———. 1985. Social boundaries in a Malagasy prosimian, the sifaka (*Propithecus verreauxi*). *International Journal of Primatology* 6, 553–68.

———. 1987. Malagasy prosimians: Female dominance. In *Primate Societies*, ed. by Smuts, B. B., Cheney, D. L., Seyfarth, R. M., Wrangham, R. W. & Struhsaker, T. T., 25–33. Chicago: University of Chicago Press.

Richard, A. F. & Dewar, R. E. 1991. Lemur ecology. *Annual Reviews in Ecology and Systematics* 22, 145–75.

Richard, A. F. & Nicoll, M. E. 1987. Female social dominance and basal metabolism in a Malagasy primate, *Propithecus verreauxi*. *American Journal of Primatology* 12, 309–14.

Richard, A. F., Rakotomanga, P. & Schwartz, M. 1993. Dispersal by *Propithecus verreauxi* at Beza Mahafaly, Madagascar: 1984–1991. *American Journal of Primatology* 30, 1–20.

Richard, A. F., Dewar, R. E., Schwartz, M. & Ratsirarson, J. 2002. Life in the slow lane? Demography and life histories of male and female sifaka (*Propithecus verreauxi verreauxi*). *Journal of Zoology, London* 256, 421–36.

Rode, E. J. 2010. Conservation ecology, morphology and reproduction of the nocturnal northern giant mouse lemur *Mirza zaza* in Sahamalaza National Park, Northwestern Madagascar, MSc thesis, Oxford Brooks University.

Roeder, J.-J., Fornasieri, I. & Gosset, D. 2002. Conflict and post-conflict behaviour in two lemur species with different social organizations (*Eulemur fulvus* and *Eulemur macaco*): A study on captive groups. *Aggressive Behavior* 28, 62–74.

Roos, C., Schmitz, J. & Zischler, H. 2004. Primate jumping genes elucidate strepsirrhine phylogeny. *Proceedings of the National Academy of Sciences, USA* 101, 10650–54.

Ross, C. 1992. Basal metabolic rate, body weight and diet in primates: An evaluation of the evidence. *Folia Primatologica* 58, 7–23.

Ruiz, A., Gómez, J. C., Roeder, J.-J. & Byrne, R. W. 2009. Gaze following and gaze priming in lemurs. *Animal Cognition* 12, 427–34.

Santos, L. R., Barnes, J. L. & Mahajan, N. 2005. Expectations about numerical events in four lemur species (*Eulemur fulvus*, *Eulemur mongoz*, *Lemur catta* and *Varecia rubra*). *Animal Cognition* 8, 253–62.

Sauther, M. L., Sussman, R. W. & Gould, L. 1999. The socioecology of the ringtailed lemur: Thirty-five years of research. *Evolutionary Anthropology* 8, 120–32.

Schilling, A. 1979. Olfactory communication in prosimians. In *The Study of Prosimian Behavior*, ed. by Doyle, G. A. & Martin, R. D., 461–542. New York: Academic Press.

Schilling, A., Perret, M. & Predine, J. 1984. Sexual inhibition in a prosimian primate: A pheromone-like effect. *Journal of Endocrinology* 102, 143–51.

Schmid, J. & Ganzhorn, J. U. 1995. Resting metabolic rates of *Lepilemur ruficaudatus*. *American Journal of Primatology*, 38, 169–174.

Schmid, J. & Ganzhorn, J. 2009. Optional strategies for reduced metabolism in gray mouse lemurs. *Naturwissenschaften* 96, 737–41.

Schmid, J. & Kappeler, P. M. 1998. Fluctuating sexual dimorphism and differential hibernation by sex in a primate, the gray mouse lemur (*Microcebus murinus*). *Behavioral Ecology and Sociobiology* 43, 125–32.

Schmid, J. & Speakman, J. R. 2000. Daily energy expenditure of the grey mouse lemur (*Microcebus murinus*): A small primate that uses torpor. *Journal of Comparative Physiology B* 170, 633–41.

Schmitz, J., Ohme, M. & Zischler, H. 2001. SINE insertions in cladistic analyses and the phylogenetic affiliations of *Tarsius bancanus* to other primates. *Genetics* 157, 777–84.

Schoof, V. A. M., Jack, K. M. & Isbell, L. A. 2009. What traits promote male parallel dispersal in primates? *Behaviour* 146, 701–26.

Schülke, O. 2003. To breed or not to breed: Food competition and other factors involved in female breeding decisions in the pair living nocturnal fork-marked lemur (*Phaner furcifer*). *Behavioral Ecology and Sociobiology* 55, 11–21.

———. 2005. Evolution of pair-living in *Phaner furcifer*. *International Journal of Primatology* 26, 903–19.

Schülke, O. & Kappeler, P. M. 2003. So near and yet so far: Territorial pairs but low cohesion between pair partners in a nocturnal lemur, *Phaner furcifer*. *Animal Behaviour* 65, 331–43.

Schülke, O., Kappeler, P. M. & Zischler, H. 2004. Small testes size despite high extra-pair paternity in the pair-living nocturnal primate *Phaner furcifer*. *Behavioral Ecology and Sociobiology* 55, 293–301.

Schülke, O. & Ostner, J. 2005. Big times for dwarfs: Social organization, sexual selection, and cooperation in the cheirogaleidae. *Evolutionary Anthropology* 14, 170–85.

———. 2008. Male reproductive skew, paternal relatedness, and female social relationships. *American Journal of Primatology* 70, 695–98.

Schwartz, G. T., Mahoney, P., Godfrey, L. R., Cuozzo, F. P., Jungers, W. L. & Randria, G. F. N. 2005. Dental development in *Megaladapis edwardsi* (Primates, Lemuriformes): Implications for understanding life history variation in subfossil lemurs. *Journal of Human Evolution* 49, 702–21.

Schwensow, N., Eberle, M. & Sommer, S. 2008a. Compatibility counts: MHC-associated mate choice in a wild promiscuous primate. *Proceedings of the Royal Society London, Series B* 275, 555–64.

Schwensow, N., Fietz, J., Dausmann, K. & Sommer, S. 2008b. MHC-associated mating strategies and the importance of overall genetic diversity in an obligate pair-living primate. *Evolutionary Ecology* 22, 617–36.

Shekelle, M. & Nietsch, A. 2008. Tarsier longevity: Data from a recapture in the wild and from captive animals. In *Primates of the Oriental Night*, ed. by Shekelle, M., Maryanto, I., Groves, C. P., Schulze, H. & Fitch-Snyder, H., 85–89. Cibinong: Indonesian Institute od Sciences.

Shekelle, M., Groves, C., Merker, S. & Supriatna, J. 2008. *Tarsius tumpara*: A new tarsier species from Siau Island, North Sulawesi. *Primate Conservation* 23, 55–64.

Shepherd, S. V. & Platt, M. L. 2008. Spontaneous social orienting and gaze following in ringtailed lemurs (*Lemur catta*). *Animal Cognition* 11, 13–20.

Simmen, B., Bayart, F. O., Rasamimanana, H., Zahariev, A., Blanc, S. P. & Pasquet, P. 2010. Total energy expenditure and body composition in two free-living sympatric lemurs. *PLoS ONE* 5, e9860.

Stanger, K. F. 1995. Vocalizations of some cheirogaleid prosimians evaluated in a phylogenetic context. In *Creatures of the Dark: The Nocturnal Prosimians*, ed. by Alterman, L., Doyle, G. A., Izard, M. K., 353–76. New York: Plenum Press.

Stankiewicz, J., Thiart, C., Masters, J. C. & de Wit, M. J. 2006. Did lemurs have sweepstake tickets? An exploration of Simpson's model for the colonization of Madagascar by mammals. *Journal of Biogeography* 33, 221–35.

Starling, A. P., Charpentier, M. J. E., Fitzpatrick, C., Scordato, E. S. & Drea, C. M. 2010. Seasonality, sociality, and reproduction: Long-term stressors of ring-tailed lemurs (*Lemur catta*). *Hormones and Behavior* 57, 76–85.

Stevens, N. J. & Heesy, C. P. 2006. Malagasy primate origins: Phylogenies, fossils, and biogeographic reconstructions. *Folia Primatologica* 77, 419–33.

Stockley, P. 1997. Sexual conflict resulting from adaptations to sperm competition. *Trends in Ecology and Evolution* 12, 154–59.

Stoinski, T. S., Drayton, L. A. & Price, E. E. 2011. Evidence of social learning in black-and-white ruffed lemurs (*Varecia variegata*). *Biology Letters* 7, 376–79.

Struhsaker, T. T. 1969. Correlates of ecology and social organization among African cercopithecines. *Folia Primatologica* 11, 80–118.

Sussman, R. W. 1992. Male life history and intergroup mobility among ringtailed lemurs (*Lemur catta*). *International Journal of Primatology* 13, 395–413.

Tan, C. L. 1999. Group composition, home range size, and diet of three sympatric bamboo lemur species (genus *Hapalemur*) in Ranomafana National Park, Madagascar. *International Journal of Primatology* 20, 547–66.

Tan, Y. & Li, W. H. 1999. Trichromatic vision in prosimians. *Nature* 402, 36.

Tattersall, I. 1977. Ecology and behavior of *Lemur fulvus mayottensis* (Primates, Lemuriformes). *Anthropological Papers from the American Museum of Natural History* 54, 425–82.

———. 2006. The concept of cathemerality: History and definition. *Folia Primatologica* 77, 7–14.

———. 2007. Madagascar's lemurs: Cryptic diversity or taxonomic inflation? *Evolutionary Anthropology* 16, 12–23.

Terborgh, J. & Janson, C. H. 1986. The socioecology of primate groups. *Annual Reviews in Ecology and Systematics* 17, 111–35.

Thalmann, U. 2001. Food resource characteristics in two nocturnal lemurs with different social behavior: *Avahi occidentalis* and *Lepilemur edwardsi*. *International Journal of Primatology* 22, 287–324.

Trillmich, J., Fichtel, C. & Kappeler, P. M. 2004. Coordination of group movements in wild Verreaux's sifakas (*Propithecus verreauxi*). *Behaviour* 141, 1103–20.

Utami, S. S. & van Hooff, J. A. R. A. M. 1997. Meat-eating by adult female Sumatran orangutans (*Pongo pygmaeus abelii*). *American Journal of Primatology* 43, 159–65.

Van Schaik, C. P. 2000. Social counterstrategies against male infanticide in primates and other mammals. In *Primate Males: Causes and Consequences of Variation in Group Composition*, ed. by Kappeler, P. M., 34–52. Cambridge: Cambridge University Press.

Van Schaik, C. P. & Kappeler, P. M. 1996. The social systems of gregarious lemurs: Lack of convergence with anthropoids due to evolutionary disequilibrium? *Ethology* 102, 915–41.

———. 1997. Infanticide risk and the evolution of male-female association in primates. *Proceedings of the Royal Society B* 264, 1687–94.

———. 2003. The evolution of social monogamy in primates. In *Monogamy: Mating Strategies and Partnerships in Birds, Humans and Other Mammals*, ed. by Reichard, U. H. & Boesch, C., 59–80. Cambridge: Cambridge University Press.

Van Woerden, J. T., van Schaik, C. P. & Isler, K. 2010. Effects of seasonality on brain size evolution: Evidence from strepsirrhine primates. *American Naturalist* 176, 758–67.

Vasey, N. 2007. The breeding system of wild red ruffed lemurs (*Varecia rubra*): A preliminary report. *Primates* 48, 41–54.

Veilleux, C. C. & Bolnick, D. A. 2009. Opsin gene polymorphism predicts trichromacy in a cathemeral lemur. *American Journal of Primatology* 71, 86–90.

Vences, M., Wollenberg, K. C., Vieites, D. R. & Lees, D. C. 2009. Madagascar as a model region of species diversification. *Trends in Ecology & Evolution* 24, 456–65.

Vick, L. G. & Pereira, M. E. 1989. Episodic targeting aggression and the histories of *Lemur* social groups. *Behavioral Ecology and Sociobiology* 25, 3–12.

Von Engelhardt, N., Kappeler, P. M. & Heistermann, M. 2000. Androgen levels and female social dominance in *Lemur catta*. *Proceedings of the Royal Society B* 267, 1533–39.

Waeber, P. O. & Hemelrijk, C. K. 2003. Female dominance and social structure in Alaotran gentle lemurs. *Behaviour* 140, 1235–46.

Ward, J. P., Silver, B. V. & Frank, J. 1976. Preservation of cross-modal transfer of a rate discrimination in the bushbaby (*Galago senegalensis*) with lesions of posterior neocortex. *Journal of Comparative and Physiological Psychology* 90, 520–27.

Ward, S. C. & Sussman, R. W. 1979. Correlates between loco-motor anatomy and behavior in two sympatric species of *Lemur*. *American Journal of Physical Anthropology* 50, 575–90.

Warren, R. D. & Crompton, R. H. 1997. A comparative study of the ranging behaviour, activity rythms and sociality of *Lepilemur edwardsi* (Primates, Lepilemuridae) and *Avahi occidentalis* (Primates, Indriidae) at Ampijoroa, Madagascar. *Journal of Zoology, London* 243, 397–415.

Watts, D. P. 2010. Dominance, power, and politics in nonhuman and human primates. In *Mind the Gap: Tracing the Origins of Human Universals*, ed. by Kappeler, P. M. & Silk, J. B., 109–38. Heidelberg: Springer.

Weidt, A., Hagenah, N., Randrianambinina, B., Radespiel, U. & Zimmerman, E. 2004. Social organization of the golden brown mouse lemur (*Microcebus ravelobensis*). *American Journal of Physical Anthropology* 123, 40–51.

Wiens, F. & Zitzmann, A. 1999. Predation on a wild slow loris (*Nycticebus coucang*) by a reticulated python (*Python reticulatus*). *Folia Primatologica* 70, 362–64.

Wilmé, L., Goodman, S. M. & Ganzhorn, J. U. 2006. Biogeographic evolution of Madagascar's microendemic biota. *Science* 312, 1063–65.

Wimmer, B. & Kappeler, P. M. 2002. The effects of sexual selection and life history on the genetic structure of redfronted lemur, *Eulemur fulvus rufus*, groups. *Animal Behaviour* 63, 557–68.

Wimmer, B., Tautz, D. & Kappeler, P. M. 2002. The genetic population structure of the gray mouse lemur (*Microcebus murinus*), a basal primate from Madagascar. *Behavioral Ecology and Sociobiology* 52, 166–75.

Wright, P. C. 1990. Patterns of paternal care in primates. *International Journal of Primatology* 11, 89–102.

Wright, P. C. 1999. Lemur traits and Madagascar ecology: Coping with an island environment. *Yearbook of Physical Anthropology* 42, 31–72.

Wright, P. C., Heckscher, S. K. & Dunham, A. E. 1997. Predation on Milne-Edward's sifaka (*Propithecus diadema edwardsi*) by the fossa (*Cryptoprocta ferox*) in the rain forest of southeastern Madagascar. *Folia Primatologica* 68, 34–43.

Yoder, A. D. 1997. Back to the future: A synthesis of strepsirrhine systematics. *Evolutionary Anthropology* 6, 11–22.

Young, A. L., Richard, A. F. & Aiello, L. C. 1990. Female dominance and maternal investment in strepsirhine primates. *The American Naturalist* 135, 473–88.

Zimmermann, E. 1995. Acoustic communication in nocturnal prosimians. In *Creatures of the Dark: The Nocturnal Prosimians*, ed. by Alterman, L., Doyle, G. A. & Izard, M. K., 311–30. New York: Plenum Press.

Zimmermann, E., Vorobieva, E., Wrogemann, D. & Hafen, T. 2000. Use of vocal fingerprinting for specific discrimination of gray (*Microcebus murinus*) and rufous mouse lemurs (*Microcebus rufus*). *International Journal of Primatology* 21, 837–52.

Zinner, D., Hilgartner, R. D., Kappeler, P. M., Pietsch, T. & Ganzhorn, J. U. 2003. Social organization of *Lepilemur ruficaudatus*. *International Journal of Primatology* 24, 869–88.

Chapter 3 The Behavior, Ecology, and Social Evolution of New World Monkeys

Eduardo Fernandez-Duque,
Anthony Di Fiore, and Maren Huck

RESEARCH ON the behavior and ecology of New World primates (infraorder Platyrrhini) began in the 1930s with C. R. Carpenter's pioneering work on mantled howler monkeys (*Alouatta paliatta*) and Geoffroy's spider monkeys (*Ateles geoffroyi*) in Panama (Strier 1994a, for a brief review). It was not until the late 1970s and 1980s, however, that significant work on the ecology and behavior of wild populations of platyrrhines developed (Coimbra-Filho & Mittermeier 1981; Mittermeier et al. 1981). For a long time, research on neotropical primates tended to focus more on aspects of the natural history and diversity of New World taxa than on the theoretical issues being debated by researchers focused on Old World monkeys and apes. Thus, by the mid-1980s, insufficient information was available from long-term studies of platyrrhines to contribute significantly to the canon of primate socioecological theory, or to test most hypotheses and predictions stemming from studies of Old World primates. Even by the late 1990s, most field data on New World primates had been gathered from a few genera (*Alouatta, Ateles, Cebus, Leontopithecus, Saimiri*) studied at a few research sites, or from studies of one or two social groups at a single location. In the 25 years since the publication of *Primate Societies* (Smuts et al. 1987), neotropical primatology has grown impressively. In this chapter we provide an overview of our current understanding of the behavior, ecology, and social evolution of platyrrhines.

Diversity and Biogeography

Platyrrhines occur exclusively in the tropical and subtropical Americas, from northern Mexico to northern Argentina. They represent a radiation of primates with a long evolutionary history independent from those of catarrhines and strepsirrhines. Based on several molecular studies conducted over the past decade (Schneider et al. 1993, 1996, 2001; von Dornum & Ruvolo 1999; Singer et al. 2003; Ray et al. 2005; Opazo et al. 2006; Poux et al. 2006), we now have a far better appreciation of the evolutionary relationships among the platyrrhines than we did 25 years ago. Molecular data strongly confirm that extant taxa can be divided into three major monophyletic groups: the atelids (muriquis, spider monkeys, woolly monkeys, and howlers), the pitheciids (titis, sakis, bearded sakis, and uakaris), and the cebids (marmosets and tamarins, squirrel monkeys, capuchins, and owl monkeys). The branching order among these three major groups remained unclear for many years, even after molecular data had shed light on the evolutionary relationships among genera within each of them. More recently, data from various molecular markers have provided support for the position of the pitheciids as basal within the platyrrhine radiation (Herke et al. 2007; Hodgson et al. 2009). It has also become clear that the three extant families diverged rapidly; the internode between the last common ancestor of all extant platyrrhines and the last common ancestor of the pitheciids and the atelid-cebid clade was very short, on the order of only a few million years, thus contributing to the difficulty of resolving the relationships among the major groups (Opazo et al. 2006; Hodgson et al. 2009).

Among the atelids, four genera are proposed (table 3.1). A fifth genus (*Oreonax*, Groves 2001) has been suggested, but the promotion of this taxon to a new genus has been questioned (Rosenberger & Matthews 2008). The pitheciids include four genera. The number of species is still debated, and for titi monkeys in particular the estimates vary considerably and are a topic of much debate (table 3.1 follows the classification of Rylands & Mittermeier 2009; for alternative classifications see Hershkovitz 1990; van Roosmalen et al. 2002). The third family consists of three quite distinct subfamilies: the Cebinae (including capuchins and squirrel monkeys), the Aotinae, and the Callitrichinae. This lastsubfamily traditionally included five genera (table 3.1), but some authors have proposed dividing the genus *Callithrix* (marmosets) and adding two genera, *Mico* (Amazonia marmosets, Rylands & Mittermeier 2009) and *Callibella* (black-crowned dwarf marmoset, van Roosmalen & van Roosmalen 2003). The position of *Aotus* was for a long time unclear, but it is now established within the Cebidae, even though its exact position within the family is still controversial (Opazo et al. 2006; Hodgson et al. 2009; Babb et al. 2011).

Even when the evolutionary relationships among clades are apparently resolved, the geographic and temporal origins of the platyrrhines remain topics of debate among contemporary primatologists. Phylogenetic analyses of both fossil and molecular data strongly support the position that platyrrhines are a monophyletic group that originated from migrants moving from Africa to South America (Bandoni de Oliveira et al. 2009). Additionally, coalescence analyses constrained by well-regarded fossil dates indicate that the separation of neotropical monkeys from African anthropoids occurred approximately 40 million years ago (Goodman et al. 1998; Schrago & Russo 2003). Nonetheless, there is still discussion regarding *how* stem platyrrhines moved from Africa to South America and *when* they did so. The existing evidence does not support the idea of a land bridge connecting Africa and South America, but instead indicates an oceanic dispersal sometime between 50 and 30 million years ago as the most likely explanation of the distribution of fossil and present day taxa (Bandoni de Oliveira et al. 2009).

Molecular estimates of divergence dates among the various lineages suggest a relatively rapid radiation, at least among extant taxa. The last common ancestor of living platyrrhines, for example, dates to the early Miocene, only 20 million years ago (Poux et al. 2006; Hodgson et al. 2009). The oldest fossil New World monkeys, dating to approximately 26 million years ago, are the Bolivian *Branisella boliviana* and *Szalatavus attricuspis* (Fleagle & Tejedor 2002). According to some researchers, several fossil taxa from the middle Miocene show affinities to a range of modern forms. This has led to the formulation of the "Long Lineage Hypothesis," which proposes that a preponderance of long-lived generic lineages, characterized by morphological stasis, may be a defining feature of the platyrrhine radiation during the past 15 to 20 million years (Rosenberger et al. 2009). Still others believe these fossils to belong to extinct lineages, and thus view successive radiations as crucial characteristics of the group, with a rapid radiation of the crown group of extant platyrrhines starting approximately 20 million years ago (Hodgson et al. 2009).

Ecology and Life History

A full understanding of the New World primate radiation requires knowledge of ecological conditions at the time of the colonization of South America. The first ancestors arriving on the continent would not have encountered the conditions that characterize contemporary tropical Amazonia, as the Amazon basin only began to take on its present character approximately 15 million years ago and changed profoundly during the Cenozoic (Bigarella & Ferreira 1985; Campbell et al. 2006; Hoorn et al. 2010). Due to the Andean uplift, for example, the original drainage system was reversed: the western parts of today's Amazonia harbored large areas of wetlands, shallow lakes, and swamps, changing later to fluvial systems dominated by grasses (Hoorn et al. 2010).

There is substantial evidence indicating that the radiation of New World primates occurred within a narrower range of ecological variation than the one cercopithecoids may have experienced. For example, no members of the radiation, fossil or extant, evolved to fill several comparable ecological niches occupied by fossil or extant primates in the Old World (see below). This relatively narrow ecological range available to New World monkeys is highlighted in an analysis of the ecological niche space of modern primate communities worldwide. Fleagle and Reed (1996) used a suite of variables (e.g., body size, activity pattern, locomotor pattern, diet) to characterize the members of eight well-studied primate communities, two from each of the major biogeographic regions where extant primates are found (the New World, Africa, Asia, and Madagascar). Using principal components analysis, they reduced those variables to two dimensions that maximally captured the variation in niche space across primate taxa, and examined the total "ecological space" thus covered by different primate communities.

Table 3.1. Taxonomy, number of species, body mass, diet, and brain mass of the 18 platyrrhine genera

Family[a]	Subfamily[a]/tribe	Genus	Common name	Number of species[a]	Adult female body mass (kg)[b]	Adult male body mass (kg)[b]	Diet#	Brain mass [g]/body mass [g][c]
Atelidae	Alouattini	Alouatta	Howler monkey	14	4.3–6.6	6.3–11.4	F, L, I	55.1/6550[d] 50.0/5085[e] 50.8/6400[f]
	Atelini	Ateles	Spider monkey	7	7.3–9.3	7.8–9.4	L, F, I	110.9/6000[d] 104.7/8000[d]
		Lagothrix[g]	Woolly monkey	5	4.5–7.7	7.1–9.4	L, F, I	96.4/6300[d] 92.7/7650[e] 98.9/5200[f]
		Brachyteles	Muriqui	2	8.3–8.5	9.4–10.2	L, F, I	115.5/8380[e]
Pitheciidae	Callicebinae	Callicebus	Titi monkey	29	0.81–1.4	0.85–1.3	F, S, L, I	18.6/900[f]
	Pitheciinae	Pithecia	Saki	5	1.6–2.1	1.9–3.0	S, F, L, I	34.1/1500[f]
		Chiropotes	Bearded saki	5	2.5–3.0	2.9–3.2	S, F, L, I	n.a.
		Cacajao	Uakari	3	2.7–2.9	3.2–3.5	S, F, L, I	62.6/2377[e]
Cebidae	Cebinae	Cebus	Capuchin	12	2.3–2.5	3.2–3.7	F, I, V	69.3/3100[f]
	Saimiriinae	Saimiri	Squirrel monkey	5	0.65–0.80	0.78–1.0	I, F, V	24.4/665[d]
	Aotinae	Aotus	Owl/night monkey	10	0.7–1.2	0.7–1.2	F, L, I	23.4/660[d]
	Callitrichinae	Callimico	Goeldi's monkey	1	0.36	0.37	Fu, F, I, V, Ex	18.2/960[f]
		Callithrix	Atlantic marmoset	6	0.3–0.43	0.32–0.43	Ex, F, I, V, Fu	16.8/830[f] 10.9/480[f] 7.9/300[f] 7.5/280[f]
		Callibella	Black-crowned dwarf marmoset	1	0.17	0.13	Ex, F, I, V	n.a.
		Mico	Amazonian marmoset	13	0.33–0.4	0.32–0.37	Ex, F, I, V	n.a.
		Cebuella	Pygmy marmoset	1	0.12	0.11	Ex, F, I, V	4.5/140[f]
		Saguinus	Tamarin	15	0.36–0.54	0.34–0.59	F, I, V, Ex	9.0/500[d]
		Leontopithecus	Lion tamarin	4	0.54–0.60	0.58–0.62	F, I, V, Ex	9.9/360[f] 13.0/600[e]

#: F = fruit, flowers, or nectar; Fu = fungi; S = seeds; Ex = exudates (tree gums and saps); L = leaves; I = insects; V = vertebrates.

[a] Number of species and subfamilies follows Rylands & Mittermeier (2009), except for the inclusion of Oreonax/Lagothrix flavicauda into Lagothrix; refer to the relevant chapters in Campbell et al. (2007) for other estimates of species numbers. Family designations follow Opazo et al. 2006.

[b] Body mass gives the range of the averages provided by Smith and Jungers (1997), except for Callicella (van Roosmalen and van Roosmalen 2003), Callimico (Encarnación and Heymann 1998), and Lagothrix and Brachyteles (Di Fiore et al. 2010).

[c] Values are available for only one or a few species. Body mass and brain values are reported for the same species.

[d] Schillaci (2006): Values for brain masses are average values for males and females, without stating explicitly whether wild or captive.

[e] Barrickman (2008): Values are for wild male body masses and female brain masses; for Cebus the means of three species' ratios were used.

[f] Pérez-Barbería (2007): Not explicitly stated whether samples are from males or females, or from wild or captive animals.

[g] Includes the contested genus Oreonax.

Their analysis suggested that the range of "ecological space" occupied by New World monkeys is considerably smaller than that occupied by primate communities in other major biogeographic regions. In other words, extant New World monkeys show less adaptive diversity in ecological patterns than is seen in other parts of the world and in other major primate radiations (Fleagle & Reed 1996). In contrast, the adaptive radiation of platyrrhines was accompanied by the evolution of several unique morphological and behavioral features (e.g., prehensile tails), as well as substantial variability in social systems not seen outside of the clade (see below).

Body Size and Unique Morphological Traits

The smallest New World monkeys are the pygmy marmosets (*Cebuella pygmaea*), with a body mass slightly over 100 g. The largest members, found among the atelids, can weigh more than 10 kg (table 3.1). Marmosets and tamarins apparently reduced their body size during the course of their evolution ("phyletic dwarfism," Martin 1992). The callitrichines also secondarily evolved claw-like nails (tegulae) on most digits, which enable them to use smooth vertical trunks as substrates for locomotion or feeding.

Another unique trait among some platyrrhines, prehensility in the tail (fig. 3.1), evolved not once but twice: in stem atelids and in stem capuchins (Rosenberger 1983). Again, it is not obvious what selective pressures may have led to the parallel evolution of prehensile tails in these two groups, but the tail is used in both groups to provide support and balance in a variety of suspensory postures and during locomotion, even though in *Cebus* the tail is not fully prehensile (Garber & Rehg 1999; Cant et al. 2003; Schmitt et al. 2005). In this context, it is intriguing to note that prehensile tails evolved in a variety of neotropical taxa, distributed among six mammalian, one amphibian, and two reptilian families. In contrast, their evolution has been much rarer in the paleotropics. One possible explanation is that the forest structure of the neo- and paleotropics may differ in the relative number of lianas and palm trees (Emmons & Gentry 1983).

Brain Size

There have been numerous attempts to examine the relationships that might exist between brain size and various life history traits and cognitive abilities among primates (van Schaik et al. 2006; Deaner et al. 2007; Barrickman et al. 2008; chapter 10, this volume). In both New and Old World monkeys, brain mass does not simply increase allometrically with body mass; there seems to be a clearer relationship with energy supply, suggesting an important role for basal metabolic rate (Armstrong 1985). Tradeoffs between investment in brain tissue and in growth or reproduction are examined in more detail in chapter 10. Generally, studies investigating those relationships have used different measures and methods that have hindered comparative analyses (Barrickman et al. 2008). Table 3.1 presents the ratio of brain mass to body mass for representatives of most genera (see also chapter 10, this volume). These values, however, should be considered with caution for several reasons. First, data are usually only available for one species within a genus, even when there may be considerable intrageneric variability in both of these measures. Second, the data are heterogeneous. Some are from captive individuals, who are typically larger in body mass, whereas other data are from wild animals. Some data are only from females and others are from members of both sexes. For these reasons, we include in table 3.1 the body mass of the species for which brain mass is reported. Third, although some authors maintain that total brain mass explains cognitive abilities better than brain/body ratios (van Schaik et al.

Fig. 3.1. A white-bellied spider monkey (*Ateles belzebuth*) in Amazonian Ecuador hangs by its prehensile tail. Photo courtesy of Dylan Schwindt.

2006; Deaner et al. 2007), others prefer to use specific parts of the brain (e.g., see Rilling & Insel 1999) or ratios of specific parts (de Winter & Oxnard 2001; Walker et al. 2006) to analyze potential patterns within primates. These caveats aside, there seem to be no clear patterns across taxa in the relationships among brain variables, life history traits, and cognitive abilities and, in contrast to strepsirrhines (chapter 2, this volume), no fundamental difference between catarrhines and platyrrhines (de Winter & Oxnard 2001; Oxnard 2004; Rosa & Tweedale 2005; Walker et al. 2006).

Diet

All New World monkeys have rather catholic diets (table 3.1), even if some taxa show specializations for particular kinds of food items. For example, some of the idiosyncratic structures of the callitrichines (claws and marmoset dentition) allow them to exploit food resources such as gums, saps, and embedded insect prey that are not available to many other arboreal mammals besides rodents (Garber 1992). Still, some marmosets, like the buffy-tufted-ear marmoset (*Callithrix aurita*), may devote as much as 11% of their diet to fruits and 39% to animal prey (Martins & Setz 2000). At one site in Bolivia, Goeldi's monkeys (*Callimico goeldii*) commonly consume fungi, a food source very rarely used by other primates (Porter 2001b). There are no New World primates committed to folivory, either behaviorally or morphologically, as are some other primates like Malagasy lemurs (e.g., *Propithecus, Lepilemur, Indri, Hapalemur, Prolemur*), most colobines, geladas (*Theropithecus gelada*), or mountain gorillas (*Gorilla gorilla*). Still, a significant commitment to folivory evolved twice, independently in howlers (*Alouatta* spp.) and in muriquis (*Brachyteles* spp); these taxa have evolved dental and behavioral adaptations for folivory, instead of the digestive specializations displayed by other primates (Milton 1993, 1998; Lambert 1998).

Predation

Both large and small neotropical primates are preyed upon by several animals. Predators include constricting and venomous snakes (Chapman 1986; Heymann 1987; Corrêa & Coutinho 1997; Cisneros-Heredia et al. 2005), tayras (*Eira barbara*, a mustelid species, Bezerra et al. 2009), felids (Peetz et al. 1992; Miranda et al. 2005; Bianchi & Mendes 2007; Ludwig et al. 2007) domestic animals (Oliveira et al. 2008; Raguet-Schofield 2008), raptors (Sherman 1991; Julliot 1994; Oversluijs Vásquez & Heymann 2001; Martins et al. 2005; De Luna et al. 2010), and even other monkeys (Sampaio & Ferrari 2005). Observations of unsuccessful predator attacks provide convincing evidence that New World monkeys derive benefits in terms of avoiding predation via group life and group defense (Eason 1989; Shahuano Tello et al. 2002; chapter 8, this volume).

It is plausible that the kinds of predators that platyrrhines encounter, and the antipredator strategies they might employ could differ qualitatively from those present in other primate groups. For example, some evidence suggests that platyrrhines may have radiated initially in the absence of venomous snakes, since the latter arrived in South America after the platyrrhines (Isbell 2006). Among the platyrrhines, the small-bodied tamarins, titis, and squirrel monkeys are also likely to be at risk from a somewhat different set of predators than the larger-bodied taxa. Unfortunately, our knowledge on how the risk of predation from any particular kind of predator varies with body mass, group size, or other major life-history trait is still quite limited (chapter 8, this volume). Sociality, or living in relatively larger groups, has usually been considered to decrease the risk of being preyed upon by some predators. On the other hand, it is plausible that in taxa that rely on crypsis to avoid predators, sociality may increase the risks. How animals integrate the risk posed by different predators into their decisions about whether to live with conspecifics requires additional research (chapter 8, this volume).

Locomotion and Activity Patterns

All extant neotropical primates radiated into nearly exclusively arboreal niches. While the Malagasy strepsirrhines, the cercopithecoids, and the hominoids all have various terrestrial representatives, there are no habitually terrestrial taxa among platyrrhines. Some species come to the ground occasionally to drink water or visit mineral licks (Izawa 1993; Campbell et al. 2005; Mourthé et al. 2007; Link et al. 2011), to forage for insects (Nadjafzadeh & Heymann 2008), to cross natural gaps between patches of forest (Fernandez-Duque 2009), to play (Mourthé et al. 2007), or to escape from predators (Martins et al. 2005; De Luna et al. 2010). There are in fact vast expanses of savannahs and open habitats in South America (Rosenberger et al. 2009), so a lack of open habitat cannot be the reason why none of the modern platyrrhines is habitually terrestrial.

Platyrrhines are predominantly diurnal with only one genus regularly displaying nocturnal activity: the night or owl monkeys (*Aotus*, Fernandez-Duque 2007; Erkert 2008). Owl monkeys concentrate their activities during the dark portion of the 24-hour cycle, with peaks of activity at dawn and dusk. Interestingly, our understanding of the evolution of nocturnality in the genus is further challenged by the existence of at least one owl monkey species that shows some remarkable temporal plasticity in its activity patterns. Azara's owl monkey (*Aotus azarae*) of Argentina and Paraguay is active during both day and night, like some lemurs (Wright 1989; Fernandez-Duque 2003; Fernandez-Duque & Erkert 2006; Fernandez-Duque et al. 2010). Why this species has shifted secondarily to part-daytime activity is still not completely understood. Lack of predation pressure, harsh climatic conditions, and a seasonal environment are all hypotheses that have been considered but will require further examination (Wright 1989; Engqvist & Richard 1991; Ganzhorn & Wright 1994; Overdorff & Rasmussen 1995).

Life History

Like other major primate groups, New World monkeys also have relatively slow life histories compared with other mammals. For example, age at first reproduction is considerably older than for other mammals of similar size (table 3.2). The small callitrichines (110–620 g) do not reproduce

in the wild before approximately two years of age (and usually later), and the larger woolly monkeys (*Lagothrix* spp.) and muriquis may not do so until they are nine years old (Martins & Strier 2004). These estimates should be considered with caution, since for some species there are few data available from wild individuals, and estimates of age at first reproduction tend to be younger for well-nourished captive animals. For example, golden lion tamarin (*Leontopithecus rosalia*) females in captivity mature when they are between 12 and 17 months old (review in Digby et al. 2007), whereas the average age of first reproduction for females in the wild is 3.6 years (Bales et al. 2001; table 3.2).

Gestation length ranges between four and eight months and is roughly correlated with maternal size (Hartwig 1996). Still, some of the callitrichines have quite long gestation periods given their body size, which is due to a lag phase prior to the onset of embryonic development (Oerke et al. 2002). Squirrel monkeys (*Saimiri* spp.) also have long gestation periods for their body size, resulting in relatively

Table 3.2. Social organization, mating systems, dispersals, and various life history traits. Values are typically derived from wild populations; they often come from one or a few species within the genus, and sometimes from only a single population. Mating systems refer to modal patterns.

Genus & references	Social organization[a]	Social mating system[b]	Dispersal[c]	Age at females' first reproduction in the wild (in years)	Gestation length (in days)	Interbirth interval (in months)	Allomaternal care[d]
Alouatta[1,2]	H, M(M)FF	PG	B	4–7	152–194	20	
Ateles[2]	MMFF	P	F	7	226–232	35	
Lagothrix[2,3]	MMFF	P	F, (B)	6–9	210–225	n.a.	
Brachyteles[3]	MMFF	P	F	7–9	215–219	n.a.	
Callicebus[4,5,6]	P	M	B	Average in captivity: 3.7	124–135	In captivity: 12	Yes
Pithecia[6,7]	P, MMFF (?)	M?	B	5	153	21.5	
Chiropotes[6,8,9,10]	MMFF	P?	(F?)[e]	Sex maturity (in captivity, 3)	~135–165?	24	
Cacajao[8,11]	MMFF	P?	F or B?	n.a.	n.a.	n.a.	
Cebus[12,13]	MMFF, MmFF	P, PG	M	5–7	154–162	22–26	(Yes)
Saimiri[13,14,15,16]	MMFF	P, PG	M, F, B	2.5	153–155	12–24	(Yes)
Aotus[17,18]	P	M	B	5	133–141	12	Yes
Callimico[19,20]	F(F)M(M)	PA, PG	B	Sex maturity (in captivity, 1)	147–157	6	Yes
Callithrix[20,21]	F(F),M(M)	M, PG, PA	B	Sex maturity (in captivity, 1–1.3)	143–144	6	Yes
Callibella[22]	MMFF	PG?	n.a.	n.a.	n.a.	n.a.	Yes
Mico[20]	F(F)M(M)	M?, PA, PG?	B	n.a.	n.a.	6	Yes
Cebuella[20,23,24]	F(F)M(M)	M, (PA?)	B	Sex maturity (in captivity, 1.3–1.5)	131–142	6	Yes
Saguinus[20,24,25]	F(F)MM	PA	B	Sex maturity (in captivity, 1–1.5)	140–184	(6–)12	Yes
Leontopithecus[20,26,27]	F(F)M(M)	PA,PG	B	Wild: 3.6 Sex maturity (in captivity, 1–1.5)	125	(6–)12	Yes

[a] H: harem (single male, multifemale). M(M)FF: single male or sometimes few males, multifemale. MMFF: multimale, multifemale. MmFF: multimale, multifemale, with one male clearly dominant. P: pair (with offspring of up to several generations). F(F)MM: single or few females, multimale. F(F)M(M): one to several females and one to several males.

[b] M: monogamy (extrapair copulations may occur). P: promiscuity/polygynandry. PG: polygamy (including effectively polygamous societies in which one alpha male essentially monopolizes access to group females). PA: polyandry.

[c] M: male-biased. F: female-biased. B: with dispersal by both sexes.

[d] Yes: alloparental care crucial for infant survival. (Yes): alloparental care sometimes substantial, but apparently not obligate and crucial for infant survival. No entry: no regular and intensive direct care (carrying, food provisioning) given by group members other than the mother.

[e] Based on strong male-male bonds, from which male philopatry and thus female dispersal can be suspected.

Sources: [1]Pope 1992; [2]Di Fiore et al 2010; [3]Martins and Strier 2004; [4]Anzenberger 1988; [5]Valeggia et al.1999; [6]Norconk 2007; [7]Di Fiore et al.2007; [8]Kinzey 1997; [9]Peetz 2001; [10]Silva and Ferrari 2009; [11]Bowler and Bodmer 2009; [12]Fragaszy et al.2004; [13]Jack 2007; [14]Boinski 1987; [15]Williams et al.1994; [16]Boinski et al.2005; [17]Fernandez-Duque 2002; [18]Fernandez-Duque 2007; [19]Porter 2001a; [20]Digby et al. 2007; [21]Yamamoto et al.2009; [22]van Roosmalen and van Roosmalen 2003; [23]Soini 1987; [24]Hartwig 1995; [25]Löttker et al.2004a; [26]Baker et al.1993; [27]Dietz and Baker 1993.

heavy neonates with large brains (see also above, Hartwig 1996). Unfortunately, gestation length is not known for many taxa, particularly for the larger and more recently described species.

Most callitrichines routinely give birth to twins, and occasionally litters of three or more, in a single reproductive event. This is an unusual characteristic among haplorrhine primates, and although there is no consensus regarding the evolutionary origins of twinning, it is consistently associated with small body size, male involvement in offspring care, and use of high-quality food sources (Leutenegger 1979; Goldizen 1990; Garber 1994; Ah-King & Tullberg 2000).

Interbirth intervals can be as short as half a year for some callitrichines (Soini 1987; Porter 2001a; French et al. 2002; but see Löttker et al. 2004b), and as long as three years among the atelines (table 3.2). The development of infants is usually related to maternal body mass; development occurs faster in smaller taxa than in larger ones (cf. chapter 11, this volume). There are still some notable exceptions. Capuchins (*Cebus* spp.), a medium-sized taxon, have very altricial young which are unable to completely maintain their body temperature after birth (review in Fragaszy et al. 2004).

Evolution of Social Systems

Below, we furnish a brief overview of the social organization, mating systems, and some features of the social structures in the three families of extant platyrrhines. We then focus our attention on several unique features of New World monkey social systems that have no comparable analogs among other extant primates: intensive paternal care, cooperative breeding, and cooperative mate defense.

Social Organization

A striking feature of the New World primates is the impressive range of variation in social systems, particularly in view of the comparatively narrow ecological range available to them (Fleagle & Reed 1996). Perhaps even more remarkable is the dramatic intrageneric and intraspecific variation in some taxa. In addition to the unimale-multifemale and multimale-multifemale systems that characterize many catarrhines, several platyrrhine taxa live in small monogamous and polyandrous groups. The smallest groups are found in the socially monogamous titis and owl monkeys and among the callitrichines; females who range alone only with their young, without regular contact with males, have not been described in any neotropical taxon. Even among spider monkeys, where females and their dependent offspring often travel independently of males and one another, contact between the sexes is regular and mixed-sex parties are quite common. The lack of solitary species among the platyrrhines may be linked to the paucity of nocturnal taxa, the exception being the owl monkeys. Additionally, independent and relatively persistent bachelor groups, such as those reported for many colobines and cercopithecines, are not as common, although squirrel monkeys may live for several years in all male bands before joining mixed-sex groups (Mitchell 1990, 1994), and small extragroup associations or coalitions of males have been reported for some other taxa in connection with parallel emigration or relatively brief group fissions (white-faced capuchins, *Cebus capucinus*, Jack & Fedigan 2004a; Jack & Fedigan 2004b; Lynch Alfaro 2007; Poeppig's woolly monkeys, *Lagothrix poeppigii*, Di Fiore & Fleischer 2005).

Atelids

The atelid primates (howler monkeys, woolly monkeys, spider monkeys, and muriquis) live in either unimale or multimale social groups like many Old World species. Most species of howler monkeys live in cohesive groups with fewer than 10 to 15 animals, commonly including only one adult male per group and seldom more than three. In mantled howler monkeys (*Alouatta palliata*), groups are sometimes larger (40 or more individuals) and typically contain three or more adult males and nine or more adult females (Fedigan 1986; Chapman 1988; Neville et al. 1988). Among the remaining atelids, groups are generally large, and typically contain multiple reproductive-age animals of both sexes. Woolly monkey groups, for example, may have as many as 45 individuals (Ramirez 1980, 1988; Nishimura 1990; Peres 1994; Stevenson et al. 1994; Defler 1995, 1996; Di Fiore 1997), whereas some groups of spider monkeys (*Ateles* spp.) and muriquis may contain almost twice as many (Di Fiore et al. 2010). Among woolly monkeys and northern muriquis (*Brachyteles hypoxanthus*), group members may be spread over large areas (Peres 1996). They occasionally split into separate, independently traveling subgroups (Defler 1996; Di Fiore 1997), and the spatial associations among these subgroups can be quite flexible. Nonetheless, groups tend to remain socially cohesive and to divide into discrete subgroups only infrequently (Di Fiore & Strier 2004). Spider monkeys, by contrast, typically live in "fission-fusion" societies, in which the individual members of a large community associate on a daily basis in small, flexible parties that change size and membership frequently (Klein 1972; Cant 1977; van Roosmalen 1985; McFarland 1986; Chapman 1990; Symington 1990; Di Fiore et al. 2010). In this respect they are very similar to chimpanzees (*Pan troglodytes*) and bonobos (*Pan paniscus*; Klein & Klein 1977; Symington 1990;

chapter 6, this volume). Southern muriquis (*Brachyteles arachnoides*) have also been reported to live in the same type of fission-fusion societies as spider monkeys (Torres de Assumpção 1983; Milton 1984; Coles et al. 2008).

The most significant contrast between the atelids and most Old World primate taxa living in unimale or multimale societies involves their dispersal patterns. Both natal and secondary dispersal are strongly male-biased among cercopithecoids, whereas dispersal by females and male philopatry are common in all the atelids (Di Fiore 2009; Di Fiore et al. 2009, 2010). As a result, atelid social groups are not often organized matrilineally around a core of related females like many cercopithecine groups (chapter 5, this volume). In *Ateles* and *Brachyteles*, for example, dispersal is largely or solely by females, and males become breeding adults in their natal communities when they grow up (Strier 1987, 1990, 1991; Symington 1987, 1988, 1990). In woolly monkeys, observed transfers of individuals among groups also suggest that dispersal is predominantly by females (Nishimura 1990, 2003; Stevenson et al. 1994, 2002), and genetic studies confirm that the level of female transfer is substantial (Di Fiore 2002, 2009; Di Fiore & Fleischer 2005; Di Fiore et al. 2009). Nonetheless, solitary males, including adults, have been seen in at least *Lagothrix poeppigii* (Di Fiore 2009; Di Fiore et al. 2009), suggesting some degree of male transfer as well.

The dispersal pattern of the ursine howler monkey (*Alouatta arctoidea*) population studied by Pope (1989; 1992) in Venezuela (formerly the red howler monkey, *Alouatta seniculus*) is less easily described. While only males took over established groups (male dispersal), a high proportion of females dispersed further on average than males did. Females did not enter established groups, but formed new ones. Founding females were rarely related to each other, and subsequently only the offspring of one female would stay in a group and form matrilines (female philopatry, Pope 1992). In mantled howlers, dispersal by both sexes has likewise been reported (Glander 1992).

Pitheciids

The range of variation in social systems is larger among the pitheciids (titi monkeys, sakis, bearded sakis, and uacaris) than it is among the atelids. Throughout their geographic range, titi monkeys (*Callicebus* spp.), the basal member of the clade, live in small groups, each consisting of an adult pair and two to four young (Kinzey 1981; Robinson et al. 1987; Defler 2004; Norconk 2007; Schmitt et al. 2007). The two adults in a group often coordinate their activities during feeding, resting, and travel (Mason 1966; Robinson 1979, 1981; Kinzey & Wright 1982; Wright 1985; Mendoza & Mason 1986a; Price & Piedade 2001). As might

be expected for species living in small groups, both sexes disperse (Bossuyt 2002).

The social organization of sakis is not as well understood, since there have only been a few studies of groups including identified and habituated individuals in undisturbed habitats (Setz & Gaspar 1997; Norconk 2006; Di Fiore et al. 2007). Like titis, sakis (*Pithecia* spp.) have also been reported to live in small social groups that typically include a single mating pair and a few young. Although there have also been studies reporting larger groups (Norconk 2007), many of those groups were found in island habitats that limit the dispersal possibilities of individuals (Setz & Gaspar 1997; Vié et al. 2001; Norconk 2006). Large groups have also been reported during censuses of nonhabituated individuals where the identity of groups has not always been known (Lehman et al. 2001). Preliminary data on white-faced sakis (*Pithecia pithecia*) suggest that, as in titi monkeys, both males and females disperse (M. A. Norconk pers. obs., cited in Norconk 2007).

The bearded sakis (*Chiropotes* spp.) and uacaris (*Cacajao* spp.) are the least studied and understood genera of all platyrrhines. They live in large, loosely structured multimale troops, sometimes containing more than 100 individuals, that regularly fission into smaller groups for traveling and foraging (Ayres 1986, 1989; Boubli 1994; Kinzey & Cunningham 1994; Norconk & Kinzey 1994; Barnett & Brandon Jones 1997; Defler 1999; Gregory & Norconk 2006; Boubli & Tokuda 2008; Bowler & Bodmer 2009; Silva & Ferrari 2009). These social aggregations may, in fact, represent temporary associations of smaller core social units plus peripheralized adult and subadult males (Bowler & Bodmer 2009). Genetic data regarding group structure and information on dispersal patterns are not yet available. A recent study of uacaris indicates that males affiliate more than females, and this observation has been used to suggest that the latter disperse (Bowler & Bodmer 2009). On the other hand, observations of male bachelor units at the periphery of larger groups, and of a few solitary males, suggest that males might occasionally disperse as well (Bowler & Bodmer 2009), in a pattern similar to that observed in *Lagothrix* (Martins & Strier 2004; Di Fiore et al. 2010). In black bearded sakis (*Chiropotes satanas*), observations conducted on an island that limited the possibilities for dispersal suggested that it is probably female-biased (Peetz 2001).

Cebids

The Cebids (capuchins, squirrel monkeys, owl monkeys, marmosets, and tamarins) also show significant diversity in social systems, group size, mating behavior, and dispersal patterns. Capuchins (*Cebus* spp.) usually live in multimale-multifemale social groups that range in size from 3 to 30 individuals (Janson 1984; Perry 1996, 1997, 1998; Di Bitetti 1997; Di Bitetti & Janson 2001; Jack & Fedigan 2004a, b, 2009; Jack 2007). Dispersal is predominantly by males, which would tend to reduce the opportunity for kin-based male cooperation, but parallel dispersal by pairs of males from the same social group is not uncommon (Jack & Fedigan 2004a). Female dispersal may occasionally occur, however, in the otherwise female philopatric white-faced capuchins (Jack & Fedigan 2009).

Squirrel monkeys (*Saimiri* spp.) tend to live in large groups ranging in size from 25 to 50 animals (Mitchell 1990; Boinski 1999; Jack & Fedigan 2004a; Stone 2007). Dispersal patterns vary across squirrel monkey populations and species. Females in *S. boliviensis* are philopatric, both sexes disperse in common squirrel monkeys (*S. sciureus*), and dispersal is reported to be female-biased among Central American squirrel monkeys (*S. oerstedii*, Mitchell et al. 1991; Boinski 2005; Boinski et al. 2005a, b), although a recent genetic study of *Saimiri oerstedii* found no evidence of female-biased dispersal and concluded that both males and females disperse, with males likely traveling farther than females (Blair & Melnick 2012).

Owl monkeys (*Aotus* spp.), the only nocturnal monkeys, are consistently described as socially monogamous. They live in small groups, each containing a single adult male-female pair and a few young, and defend territories. The primarily nocturnal habits of all owl monkey species have hindered the study of their social organization. However, studies of a cathemeral Azara's owl monkey population in northern Argentina have shown that both sexes disperse. Male and female dispersers may travel widely and live as solitary "floater" animals from a few weeks to several months before disappearing or successfully becoming members of adult pairs in an established group (Fernandez-Duque 2009; Huck et al. 2011).

The relatively small marmosets and tamarins (callitrichines) show highly flexible patterns of social organization and mating (Terborgh & Goldizen 1985; Heymann 2000; Baker et al. 2002; Digby et al. 2007; Porter & Garber 2009; Yamamoto et al. 2009). Most callitrichines live in small, territorial groups of 3 to 12 individuals that typically include one to three adult individuals of each sex. Animals of both sexes usually disperse, though females in some species might do so earlier or farther (Faulkes et al. 2003; Huck et al. 2007; but see Nievergelt et al. 2000). Adult-sized males commonly outnumber adult-sized females within groups of most callitrichine species (Heymann 2000).

Mating Systems

The mating systems of New World monkeys are remarkably varied. Among the atelids, spider monkeys, woolly mon-

keys, and muriquis mate promiscuously (table 3.2). Within social groups of these species, females mate multiple males and males mate multiple females with little overt aggression among males in the mating context (Di Fiore et al. 2010). Indeed, a recent genetic study revealed no significant reproductive skew among the multiple adult males in one group of northern muriquis (*Brachyteles hypoxanthus*, Strier et al. 2011). Among howler monkeys, dominance-based polygynous mating occurs in some species whereas female promiscuity, including mating with resident and nonresident males, is displayed by others (Pope 1992; Agoramoorthy & Hsu 2000; Kowalewski & Garber 2010).

Some of the pitheciids apparently fission into small, unimale-multifemale breeding groups or small groups of females defended by coalitions of affiliative males (for uacaris, see Bowler & Bodmer 2009). Sakis are assumed to be monogamous, since they have been most frequently described as living in pairs, but there are also some preliminary reports indicating the possibility of other mating systems in the genus (Norconk 2007). Little is known about the mating system of bearded sakis. Observations of large groups and of single females mating with multiple males suggest that it may be similar to that of the atelids (Peetz 2001; Norconk 2007). In contrast to the other members of the pitheciid clade, titi monkeys are socially monogamous (Kinzey 1981; Robinson et al. 1987; Defler 2004; Norconk 2007; Schmitt et al. 2007). Genetic data are not yet available to confirm whether mating is restricted to socially monogamous pairs and whether extrapair copulations occur. However, except for a few behavioral observations of extrapair copulations in Orabassu titi monkeys (*Callicebus moloch*, Mason, 1966), there are no data suggesting a high potential impact of extrapair copulations.

Mating in capuchins is promiscuous, but the degree to which dominant males monopolize matings varies across species (Fragaszy et al. 2004; Muniz et al. 2010). In white-faced and white-fronted (*Cebus albifrons*) capuchins, females sometimes copulate with lower-ranking males (Janson 1986; Fedigan 1993; Perry 1997), whereas in wedge-capped (*Cebus olivaceus*) and brown (*Cebus apella*) capuchins they apparently mate only or predominantly with alpha males (Janson 1984; Fragaszy et al. 2004). In squirrel monkeys, females usually mate promiscuously, although in Bolivean (*Saimiri boliviensis*) and Central American squirrel monkeys one male or a few may be able to monopolize the majority of matings (Boinski 1987, 2005; Boinski et al. 2005b; Jack 2007). Another curious feature of squirrel monkeys' mating system is the "fattening" of males during the mating season, a period when they may increase their body mass between 12 and 20% (Dumond & Hutchins 1967; Boinski 1987). This change seems to make them more attractive to females, who prefer to mate with the "fattest" male (Boinski 1987). Owl monkeys historically have been described as mating monogamously. The situation is actually more complex, because in Azara's owl monkey adults of either sex frequently replace same-sex residents (Fernandez-Duque 2007; Fernandez-Duque et al. 2008), resulting in serial monogamy. It is not known whether extrapair copulations occur in this species.

Callitrichines are quite unusual, even among platyrrhines, as they display an array of derived social organizational features not commonly observed in other primates and mammals. First, mating patterns within the clade are unusually variable. Monogamous, polygynous, polyandrous, and polygynandrous matings have all been reported in the different genera, sometimes within the same genus and even within the same population (Digby et al. 2007). Polyandry (fig. 3.2) is particularly noteworthy as it has only been reported outside of the callitrichines among a handful of hylobatids during relatively short study periods (Sommer & Reichard 2000; Lappan 2008; chapter 6, this volume). Second, female reproductive competition is a prominent feature of callitrichine reproductive biology. The reproductive success of females is strongly skewed within groups; breeding is typically monopolized by a single dominant female and the reproduction of subordinate females is often either physiologically or behaviorally suppressed (French et al. 1984; Abbott 1993; Snowdon et al. 1993). Continuing controversy exists over whether physiological suppression represents a by-product of captivity or exists as a general mechanism (Löttker et al. 2004b; Yamamoto et al. 2009). Except for a few callitrichine species, relatively little is known about how the social mating system translates into genetic relationships. For example, despite the clear polyandrous social mating system of moustached tamarins

Fig. 3.2: Polyandry is a common mating system among callitrichines, such as these grooming saddle-back tamarins. Photo courtesy of Petra Löttker.

(*Saguinus mystax*), paternities tend to be monopolized by one male in the group over several years, even though multiple paternities between and among litters can occur (Huck et al. 2005).

Social Structure

As with other aspects of behavior, there is also variability in the social relationships of platyrrhines. However, our understanding of how kinship influences social relationships among platyrrhines remains limited compared with our understanding of this issue in other primates. Still, some qualitative patterns distinguish the social relationships and social structure of New World monkeys from those of catarrhines and strepsirrhines.

Dominance and agonistic interactions

Clear, stable linear dominance hierarchies among either males or females have proven difficult to discern in most platyrrhines living in large multimale-multifemale social groups (e.g., *Brachyteles*, Strier 1992; *Lagothrix*, Di Fiore 1997; *Alouatta*, Wang & Milton 2003). This may be due in part to the observational challenges of distinguishing among individuals in large social groups of arboreal primates, but it is also almost certainly due to the fact that overt intrasexual competition is rare among group-living platyrrhines. Agonistic interactions, particularly severe ones with physical contact, are comparatively infrequent (Goldizen 1989; Caine 1993; Boinski 1994; Heymann 1996; Fragaszy et al. 2004). However, escalated encounters between or within groups, some leading to fatalities, have been witnessed in some taxa (Mitchell 1994; Campbell 2006; Talebi et al. 2009).

With the exception of the titi monkeys, which live in pairs and do not exhibit intrasexual dominance relationships, pitheciids have not been studied well enough to draw conclusions about the nature of dominance relationships and hierarchies within groups. In addition, low rates of aggression make it difficult to characterize dominance relationships. For instance, wild bald-headed uacaris (*Cacajao calvus*) spend about 2% of their time engaged in agonistic and display behavior, but only a small proportion of that behavior involves actual fighting with physical contact (Bowler & Bodmer 2009). Similarly, in one group of black-bearded sakis studied for more than a year, very little aggression between females or between the two males, one of which was much younger than the other, was observed (Peetz 2001).

Among cebids, patterns of within-group dominance relationships have been better documented. In some capuchins it is possible to discern a clear dominance hierarchy (e.g., brown capuchins, Janson 1985). In other populations there is a single, clear alpha male that is socially central and tends to monopolize matings; a linear hierarchy below this position, however, cannot always be determined (Izawa 1980; Robinson 1988; O'Brien 1991; Fedigan 1993; Perry 1997, 1998). Among white-faced capuchins, females can be ranked in a dominance hierarchy (Perry 1996). Male relationships in the male philopatric Central American squirrel monkeys are very peaceful, making it difficult to define their dominance relationships (Boinski 1987, 1994). By contrast, in the male-dispersing Bolivian squirrel monkeys, males may have intense aggressive interactions, with clear hierarchies forming as a consequence (Mitchell 1994). The same pattern is found among female squirrel monkeys. In species exhibiting female philopatry (Bolivian squirrel monkeys), linear dominance hierarchies have been reported. Alternatively, in species where female dispersal is common (Central American squirrel monkeys), relationships between females are more egalitarian (Mitchell et al. 1991). In callitrichines, one female usually monopolizes reproduction and is clearly dominant toward others, but even though certain males may monopolize paternity, agonistic interactions may be too infrequent to determine rank relationships (Goldizen 1989; Caine 1993; Huck et al. 2004a).

Grooming and other affiliative interactions

Allogrooming is extremely rare or nonexistent in some of the best-studied group-living platyrrhines, while in other taxa individuals may spend hours each day grooming and engaging in other sociopositive interactions. Among the atelids, female-biased dispersal and the possibility for male philopatry may limit the potential for nepotism and affiliative grooming interactions among females while setting up a unique opportunity for the kind of kin-based male bonding that among primates is elsewhere seen only in chimpanzees and bonobos (chapter 6, this volume). Among the three atelins (spider monkeys, woolly monkeys, and muriquis), males tend to be tolerant of, and in some species even affiliative with, each other and to cooperate in intergroup encounters against males from other groups (Di Fiore et al. 2010). In most species of howler monkeys, grooming is a regular activity (2 to 3% of the total activity budget), with females being much more active groomers than males (Chiarello 1995; Sánchez-Villagra et al. 1998). Mantled howler monkeys appear to be an exception in this regard, with grooming rates that are ten times lower than those for other howlers (review in Sánchez-Villagra et al. 1998). The species difference has been attributed to differences in female-female relationships; in contrast to other howler monkeys, female mantled howlers seldom form cooperative alliances or matrilines (Sánchez-Villagra et al. 1998).

Recent field studies of black-bearded sakis (Silva & Ferrari 2009) and bald-headed uacaris (*Cacajao calvus*, Bowler & Bodmer 2009) have commented on high rates of affiliative interactions among males and females (Peetz 2001). Grooming was observed regularly (3 to 5% of the activity budget) among black bearded sakis, where adult females groomed disproportionately more than males or younger individuals (Peetz 2001). Established pairs of monogamous sakis do not groom each other frequently, but newly formed pairs are much more interactive (7% of the male's activity), suggesting that grooming plays a role in establishing rather than maintaining pair bonds (Di Fiore et al. 2007). The pair mates of titi monkeys exhibit a high degree of intimacy, coordination, interdependence, and distress following separation, and the existence of a strong and specific mutual attachment or "bond" is regularly inferred (Mason 1975; Mendoza & Mason 1986b; Anzenberger 1988; Fernandez-Duque et al. 1997). Pair mates groom each other frequently (approximately 10% of daily activity), and it has been suggested that this helps to maintain social bonds (Kinzey & Wright 1982).

Among cebids, low levels of allogrooming have been reported in socially monogamous owl monkeys (Wolovich & Evans 2007); these monkeys appear to be extremely similar to titi monkeys in several aspects of their social system. Their grooming tends to be associated with sexual behavior between adults (Wolovich & Evans 2007). Among capuchins, which converge with Old World cercopithecines in many aspects of social organization (chapter 5, this volume), grooming interactions and other forms of within-group affiliation are common (e.g., 4.6% of observation time in brown capuchins, recalculated from Di Bitetti 1997). Females spend more time grooming each other than do males, and there is clear indication that grooming serves an important social function (O'Brien 1993; Perry 1996, 1998; Di Bitetti 1997). The nature of affiliative interactions in squirrel monkeys follows the reverse pattern of the aggressive interactions between same-sex partners described for them before. In the male-bonded, female-dispersing Central American squirrel monkeys, males show remarkably close associations, while females do not (Boinski 1994). In contrast, the opposite is true for female-philopatric Bolivian squirrel monkeys (Mitchell et al. 1991).

Among callitrichines, grooming is a prominent behavior observed among all combinations of individuals; it can sometimes occupy as much as 14% of the daily time budget of individual monkeys (Goldizen 1989; Heymann 1996; Lazaro-Perea et al. 2004; Löttker et al. 2007; Porter & Garber 2009). For moustached tamarins, grooming has been suggested to be a mechanism used by females to develop associations with breeding males and to induce cer-

Fig. 3.3. An owl monkey infant (*Aotus azarai*) rides dorsally on the back of his father. Photo courtesy of Victor Dávalso.

tain individuals to stay in the group and help with infant care (Löttker et al. 2007).

Paternal Care and cooperative breeding

In contrast to other primate radiations, for many platyrrhine taxa in two of the three extant families, most reproduction within groups is concentrated in a single female. This is true for titi monkeys and sakis among the pitheciids, and for owl monkeys and the callitrichines among the cebids. Associated with this pattern of female reproduction are unusual patterns of infant care. Intensive care of offspring in the form of carrying (fig. 3.3) and food sharing by the group male (i.e., the putative father) occurs in most of the taxa mentioned above. Cooperative breeding, which involves additional alloparental care, is the norm in the callitrichines.

Among titi monkeys and owl monkeys, paternal care of offspring is intensive and apparently obligate (Fernandez-Duque et al. 2009; Huck & Fernandez-Duque, in press). Both of these monkeys live in small groups that typically consist of an adult pair and two to four young (Fernandez-Duque 2007; Norconk 2007). Females give birth to a single infant each year and the male assumes the role of primary carrier for the infant soon after birth (Moynihan 1964; Wright 1981, 1994; Robinson et al. 1987; Aquino & Encarnación 1994; Kinzey 1997; Fernandez-Duque 2007). Dependent infants, carried as much as 90% of the time by their putative fathers, frequently transfer from the males' backs to their mothers for brief periods, usually for nursing (Dixson & Fleming 1981; Fragaszy et al. 1982; Wright 1984; Mendoza & Mason 1986b; Fernandez-Duque et al., in press; Huck & Fernandez-Duque, in press). In both titis and owl monkeys, males regularly play with, groom, and share food with infants (Wolovich et al. 2008; Fernandez-Duque et al. 2009). In captive titi monkeys, infants develop a preference for their fathers over their mothers, as assayed by a stronger pituitary-adrenal stress response when they are separated from their fathers rather than from their mothers (Hoffman et al. 1995). Siblings rarely help to carry titi or owl monkey infants (Fernandez-Duque et al. 2008). This contrasts to the pattern displayed by cooperatively breeding callitrichines.

Among callitrichines, parents, other relatives (e.g., older siblings), and even group members unrelated to offspring may share in the care of the offspring that are born up to twice per year. Unrelated group members may even contribute more to offspring care than the parents themselves (Tardif & Garber 1994; Bales et al. 2000; Ziegler 2000; Tardif et al. 2002; Huck et al. 2004b; Zahed et al. 2007). Dependent infants appear to be highly attractive to other group members, who often compete for the opportunity to carry them. In callitrichines, this peculiar social arrangement is associated with their habit of twinning. The combined weight of twins may require a considerable amount of care that cannot be provided by the mother alone (Tardif 1997). Outside of the callitrichines, cooperative breeding has not been reported for any other primate except humans (Gray & Anderson 2010; chapter 20, this volume) and it is relatively rare among mammals (see reviews in Solomon & French 1997).

The high level of care provided by nonmothers in titi monkeys, owl monkeys, and callitrichines is quite conspicuous and appears to be obligate. The involvement of nonmothers in the care of capuchin and squirrel monkey infants is also striking. Young capuchins and squirrel monkeys may be carried by various group members. These include males (reviews in Williams et al. 1994; Fragaszy et al. 2004), older sisters, and even unrelated females, who may nurse infants (see, e.g., O'Brien 1988; O'Brien & Robinson 1991; Williams et al. 1994; Perry 1996).

Cooperative Mate Defense

Cooperative mate defense, with quite flexible association patterns and limited overt intrasexual competition among males of the same social group, characterizes some atelids, some cebids, and perhaps even some pitheciids. Cooperative mate defense is rare among primates, occurring only in platyrrhines and chimpanzees (chapter 6, this volume). Both atelids and cebids are notable in the extent to which male group members cooperate with one another when interacting aggressively with males from other groups. Males cooperate most likely to obtain access to females and, by extension, personal reproductive opportunities (Mitchell 1994; Strier 1994b; Perry 1998). At the same time, interactions between males of the same group tend to be more tolerant or affiliative and less aggressive (see above) than is common for most Old World monkey species, like yellow baboons (*Papio cynocephalus*), where some males may form strategic coalitions with one another over consortship opportunities or in the context of intragroup conflicts with higher-ranking males, or Hanuman langurs (*Semnopithecus entellus*), which might cooperate to take over other groups but show high intrasexual competition over females

within groups (Nöe 1990; Packer 1977; Hrdy 1977; Sommer et al. 2002; Alberts et al. 2003, 2006; chapters 4 and 5, this volume). In atelids, cooperative mate defense is presumably related to the prevalence of female-biased dispersal and a greater degree of male philopatry (Strier 2008; Di Fiore et al. 2010), although males do not have to be close relatives for this system to be advantageous (Link et al. 2009). In ursine howler monkeys, coalitions of males cooperate to take over small groups of females or defend access to them from other males. Furthermore, coalitions of ursine howler monkeys composed of related males persist for longer periods of time than coalitions formed by non-relatives, suggesting a role for kin selection in the cooperative social interactions of males (Pope 1990, 1992). Among the atelins, male-male cooperation in the context of intergroup encounters has been reported for all genera (Strier 1994b, 2004; Di Fiore & Fleischer 2005; Strier 2008; Di Fiore et al. 2011), and in at least some atelin groups, adult male group members are close relatives, though this is by no means a universal pattern (Di Fiore et al. 2009). Among the cebids (e.g., white-faced capuchins), close cooperation among males may be facilitated by the high incidence of parallel dispersal, which may translate into inclusive fitness benefits as well as increased survivorship (Jack & Fedigan 2004a). In Bolivian squirrel monkeys, males emigrate together in migration alliances; alliance members support each other as they compete with males in other groups and seek entrance into new groups during immigration events (Mitchell 1994).

Summary and Conclusions

Compared with other primates, New World monkeys display relatively limited ecological variability. New World monkey anatomy and social systems, however, are extremely diverse. Several unique morphological features (e.g., claws, prehensile tails) and uncommon patterns of social organization (e.g., paternal care, cooperative breeding, female dispersal) have evolved in some platyrrhine species. Social organization and mating patterns include typical harem-like structures where mating is largely polygynous, and large multimale, multifemale groups with promiscuous mating and fission-fusion societies. In addition, some species are socially monogamous and polyandrous. Even closely related species may exhibit strikingly different social organizations, as the example of the squirrel monkeys demonstrates (Mitchell et al. 1991; Boinski et al. 2005b).

New World monkey behavior varies within species as well as between them. While the behavior of many species is known from only one study site, intriguing patterns of intraspecific variation are beginning to emerge from observations of populations that sometimes live in close proximity. For example, spider monkeys are often described as showing sex-segregated ranging behavior. Several studies show that males range farther, travel faster, and use larger areas than females, who tend to restrict their habitual ranging to smaller core areas within a group's large territory (Symington 1988; Chapman 1990; Shimooka 2005). In at least one well-studied population in Yasuní National Park, Ecuador, however, males and females both travel over the entire community home range, and different females within the community show little evidence of occupying distinct core areas (Spehar et al. 2010). Similarly, in most well-studied populations of spider monkeys, females disperse and the resident males within a group are presumed to be close relatives—a suggestion corroborated by genetic data for one local population of spider monkeys in Yasuní. Still, in a different local population, males are not closely related to one another, an unexpected pattern if significant male philopatry were common (Di Fiore 2009; Di Fiore et al. 2009). While the causes of this local variation in group genetic structure are not clear, it may be significant that the groups examined likely had different histories of contact with humans. For long-lived animals who occupy relatively small social groups, the loss of even a handful of individuals to hunting, or to any other demographic disturbance, can have a dramatic impact on a group's genetic structure. Intragroup social relationships, in turn, are likely to be influenced by patterns of intragroup relatedness and by the relative availability of social partners of different age or sex class (chapter 21, this volume). Thus, historical and demographic contingencies are likely to create conditions where considerable local, intrapopulation variation in social systems exists.

Slight changes in ecological conditions may also contribute to variation in the behavior of individuals living in a single population over time. For example, some authors have hypothesized that howler monkey populations may undergo dramatic fluctuations in size and composition in response to several ecological factors, including resource abundance, parasite and predation pressure, and climate (Milton 1982; Crockett & Eisenberg 1986; Crockett 1996; Milton 1996; Rudran & Fernandez-Duque 2003). This variability, not only among populations, but also within populations across time highlights the need for long-term studies.

In sum, our understanding of the behavior of New World monkeys has increased dramatically over the past 25 years. This understanding highlights how their behavior varies within populations over time and among populations or species across space. As our knowledge of platyrrhine behavior continues to unfold and is enriched via additional long-term studies, a central challenge will be to explain how these variations arise. It will be important to entertain adaptive explanations while acknowledging that some differences may emerge via stochastic changes in demography (Struhsaker 2008) or nongenetic, relatively short-term, nonadaptive responses to sudden ecological change.

References

Abbott, D. H. 1993. Social conflict and reproductive suppression in marmoset and tamarin monkeys. In *Primate Social Conflict* (ed. by Mason, W. A. & Mendoza, S. P.), 331–372. New York: State University of New York Press.

Agoramoorthy, G. & Hsu, M. J. 2000. Extragroup copulation among wild red howler monkeys in Venezuela. *Folia Primatologica*, 71, 147–151.

Ah-King, M. & Tullberg, B. S. 2000. Phylogenetic analysis of twinning in the Callitrichinae. *American Journal of Primatology*, 51, 135–146.

Alberts, S. C., Watts, H. E. & Altmann, J. 2003. Queuing and queue-jumping: Long-term patterns of reproductive skew in male savannah baboons, *Papio cynocephalus*. *Animal Behaviour*, 65, 821–840.

Alberts, S. C., Buchan, J. C. & Altmann, J. 2006. Sexual selection in wild baboons: From mating opportunities to paternity success. *Animal Behaviour*, 72, 1177–1196.

Anzenberger, G. 1988. The pairbond in the titi monkey (*Callicebus moloch*): Intrinsic versus extrinsic contributions of the pairmates. *Folia Primatologica*, 50, 188–203.

Aquino, R. & Encarnación, F. 1994. Owl monkey populations in Latin America: Field work and conservation. In *Aotus: The Owl Monkey* (ed. by Baer, J. F., Weller, R. E. & Kakoma, I.), 59–95. San Diego: Academic Press.

Armstrong, E. 1985. Relative brain size in monkeys and prosimians. *American Journal of Physical Anthropology*, 66, 263–273.

Ayres, J. M. 1986. Uakaris and Amazonian flooded forest. PhD dissertation, University of Cambridge.

———. 1989. Comparative feeding ecology of the uakari and bearded saki, *Cacajao* and *Chiropotes*. *Journal of Human Evolution*, 18, 697–716.

Baker, A. J., Bales, K. L. & Dietz, J. M. 2002. Mating system and group dynamics in lion tamarins. In *Lion Tamarins: Biology and Conservation* (ed. by Kleiman, D. G. & Rylands, A. B.), 188–212. Washington: Smithsonian Institution Press.

Baker, A. J., Dietz, J. M. & Kleiman, D. G. 1993. Behavioural evidence for monopolization of paternity in multi-male groups of golden lion tamarins. *Animal Behaviour*, 46, 1091–1103.

Bales, K., French, J. A. & Dietz, J. M. 2001. Reproductive and social influences on fecal cortisol levels in wild and reintroduced female golden lion tamarins. *American Journal of Primatology*, 51, 40–41.

Bales, K. L., Dietz, J. M., Baker, A. J., Miller, K. E. & Tardif, S. D. 2000. Effects of allocare-givers on fitness of infants and parents in callitrichid primates. *Folia Primatologica*, 71, 27–38.

Bandoni de Oliveira, F., Cassola Molina, E. & Marroig, G. 2009.

Paleogeography of the South Atlantic: A Route for Primates and Rodents into the New World? In *South American Primates: Comparative Perspectives in the Study of Behavior, Ecology, and Conservation* (ed. by Garber, P. A., Estrada, A., Bicca-Marques, J. C., Heymann, E. W. & Strier, K. B.), 55–68. New York: Springer.

Barnett, A. A. & Brandon Jones, D. 1997. The ecology, biogeography and conservation of the uakaris, *Cacajao* (Pitheciinae). *Folia Primatologica*, 68, 223–235.

Barrickman, N. L., Bastian, M. L., Isler, K. & van Schaik, C. P. 2008. Life history costs and benefits of encephalization: A comparative test using data from long-term studies of primates in the wild. *Journal of Human Evolution*, 54, 568–590.

Bezerra, B., Barnett, A., Souto, A. & Jones, G. 2009. Predation by the tayra on the common marmoset and the pale-throated three-toed sloth. *Journal of Ethology*, 27, 91–96.

Bianchi, R. D. C. & Mendes, S. L. 2007. Ocelot (*Leopardus pardalis*) predation on primates in Caratinga Biological Station, Southeast Brazil. *American Journal of Primatology*, 69, 1173–1178.

Bigarella, J. J. & Ferreira, A. M. M. 1985. Amazonian geology and the Pleistocene and the Cenozoic environments and paleoclimates. In *Amazonia: Key Environments Series* (ed. by Prance, G. T. & Lovejoy, T. E.), 49–71. New York: Pergamon Press.

Blair, M. E. & Melnick, D. J. 2012. Genetic evidence for dispersal by both sexes in the Central American squirrel monkey, *Saimiri oerstedii citrinellus*. *American Journal of Primatology*, 74, 37–47.

Boinski, S. 1987. Mating patterns in squirrel monkeys (*Saimiri oerstedi*): Implications for seasonal sexual dimorphism. *Behavioral Ecology and Sociobiology*, 21, 13–21.

———. 1994. Affiliation patterns among male Costa Rican squirrel monkeys. *Behaviour*, 130, 191–209.

———. 1999. The social organization of squirrel monkeys: Implications for ecological models of social evolution. *Evolutionary Anthropology*, 8, 101–112.

———. 2005. Dispersal patterns among three species of squirrel monkeys (*Saimiri oerstedii*, *S. boliviensis* and *S. sciureus*): III. Cognition. *Behaviour*, 142, 679–699.

Boinski, S., Ehmke, E., Kauffman, L., Schet, S. & Vreedzaam, A. 2005a. Dispersal patterns among three species of squirrel monkeys (*Saimiri oerstedii*, *S. boliviensis* and *S. sciureus*): II. Within-species and local variation. *Behaviour*, 142, 633–677.

Boinski, S., Kauffman, L., Ehmke, E., Schet, S. & Vreedzaam, A. 2005b. Dispersal patterns among three species of squirrel monkeys (*Saimiri oerstedii*, *S. boliviensis* and *S. sciureus*): I. Divergent costs and benefits. *Behaviour*, 142, 525–632.

Bossuyt, F. 2002. Natal dispersal of titi monkeys (*Callicebus moloch*) at Cocha Cashu, Manu National Park, Peru. *American Journal of Physical Anthropology*, Suppl. 34, 47.

Boubli, J. P. 1994. The black uakari monkey in the Pico da Neblina National Park. *Neotropical Primates*, 2, 11–12.

Boubli, J. P. & Tokuda, M. 2008. Socioecology of black uakari monkeys, *Cacajao hosomi*, in Pico da Neblina National Park, Brazil: The role of the peculiar spatial-temporal distribution of resources in the Neblina forests. *Primate Report*, 75, 3–10.

Bowler, M. & Bodmer, R. 2009. Social behavior in fission-fusion groups of red uakari monkeys (*Cacajao calvus ucayalii*). *American Journal of Primatology*, 71, 976–987.

Caine, N. G. 1993. Flexibility and co-operation as unifying themes in *Saguinus* social organization and behaviour: The role of predation pressures. In *Marmosets and Tamarins: Systematics, Behaviour, and Ecology* (ed. by Rylands, A.), 200–219. Oxford: Oxford Science Publications.

Campbell, C. J. 2006. Lethal intragroup aggression by adult male spider monkeys (*Ateles geoffroyi*). *American Journal of Primatology*, 68, 1197–1201.

Campbell, C. J., Aureli, F., Chapman, C. A., Ramos-Fernández, G., Matthews, K., Russo, S. E., Suarez, S. & Vick, L. 2005. Terrestrial behavior of *Ateles* spp. *International Journal of Primatology*, 26, 1039–1051.

Campbell, C. J., Fuentes, A., MacKinnon, K. C., Panger, M. & Beader, S. K. 2007. *Primates in Perspectives*. Oxford: Oxford University Press.

Campbell, K. E., Frailey, C. D. & Romero-Pittman, L. 2006. The Pan-Amazonian Ucayali Peneplain, late Neogene sedimentation in Amazonia, and the birth of the modern Amazon River system. *Palaeogeography, Palaeoclimatology, Palaeoecology*, 239, 166–219.

Cant, J. G. H. 1977. Ecology, locomotion, and social organization of spider monkeys (*Ateles geoffroyi*). PhD dissertation, University of California.

Cant, J. G. H., Youlatos, D. & Rose, M. D. 2003. Suspensory locomotion of *Lagothrix lagothricha* and *Ateles belzebuth* in Yasuni National Park, Ecuador. *Journal of Human Evolution*, 44, 685–699.

Chapman, C. A. 1986. Boa constrictor predation and group response in white-faced cebus monkeys. *Biotropica*, 18, 171–172.

Chapman, C. A. 1988. Patterns of foraging and range use by three species of neotropical primates. *Primates*, 29, 177–194.

Chapman, C. A. 1990. Association patterns of spider monkeys: The influence of ecology and sex on social organization. *Behavioral Ecology and Sociobiology*, 26, 409–414.

Chiarello, A. G. 1995. Grooming in brown howler monkeys, *Alouatta fusca*. *American Journal of Primatology*, 35, 73–81.

Cisneros-Heredia, D. F., Leon-Reyes, A. & Seger, S. 2005. *Boa constrictor* predation on a titi monkey, *Callicebus discolor*. *Neotropical Primates*, 13, 11–12.

Coimbra-Filho, A. & Mittermeier, R. A. 1981. *Ecology and Behavior of Neotropical Primates*. Vol. 1. Rio de Janeiro: Academia Brasileira de Ciências.

Coles, R. C., Talebi, M. G. & Lee, P. C. 2008. Fission-fusion sociality in southern muriquis (*Brachyteles arachnoides*) in the continuous Atlantic forest of Brazil. *Primate Eye*, 96 (Sp CD-ROM iss—IPS 2008), Abst #652.

Corrêa, H. K. M. & Coutinho, P. E. G. 1997. Fatal attack of a pit viper, *Bothrops jararaca*, on an infant buffy-tufted ear marmoset (*Callithrix aurita*). *Primates*, 38, 215–217.

Crockett, C. M. 1996. The relation between red howler monkey (*Alouatta seniculus*) troop size and population growth in two habitats. In *Adaptive Radiations of Neotropical Primates* (ed. by Norconk, M. A., Roseberger, A. L. & Garber, P. A.), 489–510. New York: Plenum Press.

Crockett, C. M. & Eisenberg, J. F. 1986. Howlers: Variations in group size and demography. In *Primate Societies* (ed. by Smuts, B. B., Cheney, D. L., Seyfarth, R. M., Wrangham, R. & Struhsaker, T. T.), 54–68. Chicago: University of Chicago Press.

Deaner, R. O., Isler, K., Burkart, J. & van Schaik, C. P. 2007. Overall brain size, and not encephalization quotient, best predicts cognitive ability across non-human primates. *Brain, Behavior and Evolution* ,70, 115–124.

Defler, T. R. 1995. The time budget of a group of wild woolly monkeys (*Lagothrix lagotricha*). *International Journal of Primatology*, 16, 107–120.

———. 1996. Aspects of the ranging pattern in a group of wild woolly monkeys (*Lagothrix lagothricha*). *American Journal of Primatology*, 38, 289–302.

———. 1999. Fission-fusion in the black-headed uacari (*Cacajao melanocephalus*) in eastern Colombia. *Neotropical Primates*, 7, 5–8.

———. 2004. Titi monkeys. In *Primates of Colombia*, 298–322. Bogotá: Conservation International.

De Luna, A. G., Sanmiguel, R. R., Di Fiore, A. & Fernandez-Duque, E. 2010. Predation of a red titi monkey (*Callicebus discolor*) by a harpy eagle (*Harpya harpya*) and other unsuccessful attacks by predators on pitheciines in lowland Amazonia. *Folia Primatologica*, 81, 86–95.

De Winter, W. & Oxnard, C. E. 2001. Evolutionary radiations and convergences in the structural organization of mammalian brains. *Nature*, 409, 710–714.

Di Bitetti, M. S. 1997. Evidence for an important role of allogrooming in a platyrrhine primate. *Animal Behaviour*, 54, 199–211.

Di Bitetti, M. S. & Janson, C. H. 2001. Reproductive socioecology of tufted capuchins (*Cebus apella nigritus*) in Northeastern Argentina. *International Journal of Primatology*, 22, 127–142.

Di Fiore, A. 1997. Ecology and behavior of lowland woolly monkeys (*Lagothrix lagotricha poeppigii, Atelinae*) in Eastern Ecuador. PhD dissertation, University of California.

———. 2002. Molecular perspectives on dispersal in lowland woolly monkeys (*Lagothrix lagotricha poeppigii*). *American Journal of Physical Anthropology*, supplement 34, 63.

———. 2009. Genetic approaches to the study of dispersal and kinship in New World primates. In *South American Primates: Comparative Perspectives in the Study of Behavior, Ecology, and Conservation* (ed. by Garber, P. A., Estrada, A., Bicca-Marques, J. C., Heymann, E. W. & Strier, K. B.), 211–250. New York: Springer.

Di Fiore, A., Fernandez-Duque, E. & Hurst, D. 2007. Adult male replacement in socially monogamous equatorial saki monkeys (*Pithecia aequatorialis*). *Folia Primatologica*, 78, 88–98.

Di Fiore, A. & Fleischer, R. C. 2005. Social behavior, reproductive strategies, and population genetic structure of *Lagothrix poeppigii*. *International Journal of Primatology*, 26, 1137–1173.

Di Fiore, A., Link, A. & Campbell, C. J. 2010. The atelines: Behavioral and socioecological diversity in a New World radiation. In *Primates in Perspective* (ed. by Campbell, C. J., Fuentes, A., MacKinnon, K. C., Bearder, S. K. & Stumpf, R.), 155–188. Oxford: Oxford University Press.

Di Fiore, A., Link, A., Schmitt, C. A. & Spehar, S. N. 2009. Dispersal patterns in sympatric woolly and spider monkeys: Integrating molecular and observational data. *Behaviour*, 146, 437–470.

Di Fiore, A. & Strier, K. B. 2004. Flexibility in social organisation in ateline primates. *Folia Primatologica*, 75, supplement 1, 140–141.

Dietz, J. M. & Baker, A. J. 1993. Polygyny and female reproductive success in golden lion tamarins, *Leontopithecus rosalia*. *Animal Behaviour*, 46, 1067–1078.

Digby, L. J., Ferrari, S. F. & Saltzman, W. 2007. Callitrichines: The role of competition in cooperatively breeding species. In *Primates in Perspective* (ed. by Campbell, C. J., Fuentes, A., MacKinnon, K. C., Panger, M. & Bearder, S. K.), 85–105. Oxford: Oxford University Press.

Dixson, A. F. & Fleming, D. 1981. Parental behaviour and infant development in owl monkeys (*Aotus trivirgatus griseimembra*). *Journal of Zoology, London*, 194, 25–39.

Dumond, F. V. & Hutchins, T. C. 1967. Squirrel monkey reproduction: The "fatted" male phenomenon and seasonal spermatogenesis. *Science*, 158, 1067–1070.

Eason, P. 1989. Harpy eagle attempts predation on adult howler monkey. *Condor*, 91, 469–470.

Emmons, L. H. & Gentry, A. H. 1983. Tropical forest structure and the distribution of gliding and prehensile-tailed vertebrates. *American Naturalist*, 121, 513–524.

Encarnación, F. & Heymann, E. W. 1998. Body mass of wild *Callimico goeldii*. *Folia Primatologica*, 69, 368–371.

Engqvist, A. & Richard, A. 1991. Diet as a possible determinant of cathemeral activity patterns in primates. *Folia Primatologica*, 57, 169–172.

Erkert, H. 2008. Diurnality and nocturnality in nonhuman primates: Comparative chronobiological studies in laboratory and nature. *Biological Rhythm Research*, 39, 229–267.

Faulkes, C. G., Arruda, M. F. & Monteiro da Cruz, M. A. O. 2003. Matrilineal genetic structure within and among populations of the cooperatively breeding common marmoset, *Callithrix jacchus*. *Molecular Ecology*, 12, 1101–1108.

Fedigan, L. M. 1986. Demographic trends in the *Alouatta palliata* and *Cebus capucinus* populations of Santa Rosa National Park, Costa Rica. In *Primate Ecology and Conservation* (ed. by Else, J. G. & Lee, P. C.), 285–293. New York: Cambridge University Press.

Fedigan, L. M. 1993. Sex differences and inter-sexual relations in adult white-faced capuchins (*Cebus capuchinus*). *International Journal of Primatology*, 14, 853–878.

Fernandez-Duque, E. 2002. Environmental determinants of birth seasonality in night monkeys (*Aotus azarai*) of the Argentinian Chaco. *International Journal of Primatology*, 23, 639–656.

———. 2003. Influences of moonlight, ambient temperature and food availability on the diurnal and nocturnal activity of owl monkeys (*Aotus azarai*). *Behavioral Ecology and Sociobiology*, 54, 431–440.

———. 2007. Aotinae: Social monogamy in the only nocturnal haplorhines. In *Primates in Perspective* (ed. by Campbell, C. J., Fuentes, A., MacKinnon, K. C., Panger, M. & Bearder, S. K.), 139–154. Oxford: Oxford University Press.

———. 2009. Natal dispersal in monogamous owl monkeys (*Aotus azarai*) of the Argentinean Chaco. *Behaviour*, 146, 583–606.

Fernandez-Duque, E., Di Fiore, A. & de Luna, A. G. in press. Pair-mate relationships and parenting in equatorial saki monkeys (*Pithecia aequatorialis*) and red titi monkeys (*Callicebus discolor*) of Ecuador. In *Evolutionary Biology and Conservation of Titis, Sakis and Uacaris* (ed. by Veiga, L. M. & Barnett, A. A.). Cambridge: Cambridge University Press.

Fernandez-Duque, E. & Erkert, H. G. 2006. Cathemerality and lunar periodicity of activity rhythms in owl monkeys of the Argentinian Chaco. *Folia Primatologica*, 77, 123–138.

Fernandez-Duque, E., H. de la Iglesia, et al. 2010. Moonstruck primates: owl monkeys (*Aotus*) need moonlight for nocturnal activity in their natural environment. *PLoS ONE* 5(9), e12572.

Fernandez-Duque, E., Juárez, C. & Di Fiore, A. 2008. Adult male replacement and subsequent infant care by male and siblings in socially monogamous owl monkeys (*Aotus azarai*). *Primates*, 49, 81–84.

Fernandez-Duque, E., Mason, W. A. & Mendoza, S. P. 1997. Effects of duration of separation on responses to mates and strangers in the monogamous titi monkey (*Callicebus moloch*). *American Journal of Primatology*, 43, 225–237.

Fernandez-Duque, E., Valeggia, C. R. & Mendoza, S. P. 2009. The biology of paternal care in human and nonhuman primates. *Annual Review of Anthropology*, 38, 115–130.

Fleagle, J. G. & Reed, K. E. 1996. Comparing primate communities: a multivariate approach. *Journal of Human Evolution*, 30, 489–510.

Fleagle, J. G. & Tejedor, M. F. 2002. Early platyrrhines of southern South America. In *The Primate Fossil Record* (ed. by Hartwig, W. C.), 161–173. Cambridge: Cambridge University Press.

Fragaszy, D. M., Schwarz, S. & Shimosaka, D. 1982. Longitudinal observations of care and development of infant titi monkeys (*Callicebus moloch*). *American Journal of Primatology*, 2, 191–200.

Fragaszy, D. M., Visalberghi, E. & Fedigan, L. M. 2004. *The Complete Capuchin*. Cambridge: Cambridge University Press.

French, J. A., Abbott, D. H. & Snowdon, C. T. 1984. The effect of social environment on estrogen excretion, scent marking, and sociosexual behavior in tamarins (*Saguinus oedipus*). *American Journal of Primatology*, 6, 155–167.

French, J. A., de Vleeschouwer, K., Bales, K. & Heistermann, M. 2002. Lion tamarin reproductive biology. In *Lion Tamarins: Biology and Conservation* (ed. by Kleiman, D. G. & Rylands, A. B.), 133–156. Washington: Smithsonian Institution Press.

Ganzhorn, J. U. & Wright, P. C. 1994. Temporal patterns in primate leaf eating: The possible role of leaf chemistry. *Folia Primatologica*, 63, 203–208.

Garber, P. A. 1992. Vertical clinging, small body size, and the evolution of feeding adaptations in the Callitrichinae. *American Journal of Physical Anthropology*, 88, 469–482.

Garber, P. A. 1994. Phylogenetic approach to the study of tamarin and marmoset social systems. *American Journal of Primatology*, 34, 199–219.

Garber, P. A. & Rehg, J. A. 1999. The ecological role of the prehensile tail in white-faced capuchins (*Cebus capucinus*). *American Journal of Physical Anthropology*, 110, 325–339.

Glander, K. E. 1992. Dispersal patterns in Costa Rican mantled howling monkeys. *International Journal of Primatology*, 13, 415–436.

Goldizen, A. 1989. Social relationships in a cooperatively polyandrous group of tamarins (*Saguinus fuscicollis*). *Behavioral Ecology and Sociobiology*, 24, 79–89.

Goldizen, A. W. 1990. A comparative perspective on the evolution of tamarin and marmoset social systems. *International Journal of Primatology*, 11, 63–83.

Goodman, M., Porter, C. A., Czelusniak, J., Page, S. L., Schneider, H., Shoshani, J., Gunnell, G. & Groves, C. P. 1998. Toward a phylogenetic classification of primates based on DNA evidence complemented by fossil evidence. *Molecular Phylogenetics and Evolution*, 9, 585–598.

Gray, P. B. & Anderson, K. G. 2010. *Fatherhood: Evolution and Human Paternal Behavior*. Cambridge: Harvard University Press.

Gregory, T. & Norconk, M. A. 2006. Comparative socioecology of sympatric, free-ranging bearded sakis and white-faced sakis in Brownsberg Natuurpark, Suriname. *American Journal of Primatology*, 68, 34.

Groves, C. P. 2001. *Primate Taxonomy*. Washington: Smithsonian Institution Press.

Hartwig, W. C. 1995. Effect of life history on the squirrel monkey (Platyrrhini, *Saimiri*) cranium. *American Journal of Physical Anthropology*, 97, 435–449.

———. 1996. Perinatal life history traits in New World monkeys. *American Journal of Primatology*, 40, 99–130.

Herke, S. W., Jinchuaning, Ray, D. A., Zimmerman, J. W., Cordaux, R. & Batzer, M. A. 2007. A SINE-based dichotomous key for primate identification. *Gene*, 390, 39–51.

Hershkovitz, P. 1990. Titis, New World monkeys of the genus *Callicebus* (Cebidae, Platyrrhini): A preliminary taxonomic review. *Fieldiana Zoology*, 55, 1–109.

Heymann, E. 1987. A field observation of predation on a moustached tamarin (*Saguinus mystax*) by an anaconda. *International Journal of Primatology*, 8, 193–195.

Heymann, E. W. 1996. Social behavior of wild moustached tamarins, *Saguinus mystax*, at the Estación Biológica Quebrada Blanco, Peruvian Amazonia. *American Journal of Primatology*, 38, 101–113.

———. 2000. The number of adult males in callitrichine groups and its implications for callitrichine social evolution. In *Primate Males: Causes and Consequences of Variation in Group Composition* (ed. by Kappeler, P. M.), 64–71. Cambridge: Cambridge University Press.

Hodgson, J. A., Sterner, K. N., Matthews, L. J., Burrell, A. S., Jani, R. A., Raaum, R. L., Stewart, C.-B. & Disotell, T. R. 2009. Successive radiations, not stasis, in the South American primate fauna. *Proceedings of the National Academy of Sciences USA*, 106, 5534–5539.

Hoffman, K. A., Mendoza, S. P., Hennessy, M. B. & Mason, W. A. 1995. Responses of infant titi monkeys, *Callicebus moloch*, to removal of one or both parents: evidence for paternal attachment. *Developmental Psychobiology*, 28, 399–407.

Hoorn, C., Wesselingh, F. P., ter Steege, H., Bermudez, M. A., Mora, A., Svink, J., Sanmartín, I., Sanchez-Meseguer, A., Anderson, C. L., Figueiredo, J. P., Jaramillo, C., Riff, D., Negri, F. R., Hooghiemstra, H., Lundberg, J., Stadler, T., Särkinen, T., and Antonelli, A. 2010. Amazonia through time: Andean uplift, climate change, landscape evolution, and biodiversity. *Science* 330, 927–931.

Hrdy, S. B. 1977. *The langurs of Abu: Female and male strategies of reproduction*. Cambridge, MA: Harvard University Press.

Huck, M. & Fernandez-Duque, E. in press. Building babies when dads help: Infant development of owl monkeys and other primates with allo-maternal care. In *Building Babies: Primate Development in Proximate and Ultimate Perspective* (ed. by Clancy, K., Hinde, K. & Rutherford, J.).

Huck, M., Löttker, P., Böhle, U.-R. & Heymann, E. W. 2005. Paternity and kinship patterns in polyandrous moustached tamarins (*Saguinus mystax*). *American Journal of Physical Anthropology*, 127, 449–464.

Huck, M., Löttker, P. & Heymann, E. W. 2004a. Proximate

mechanisms of reproductive monopolization in male moustached tamarins (*Saguinus mystax*). *American Journal of Primatology*, 64, 39–56.

———. 2004b. The many faces of helping: Possible costs and benefits of infant carrying and food transfer in moustached tamarins (*Saguinus mystax*). *Behaviour*, 141, 915–934.

Huck, M., Roos, C. & Heymann, E. W. 2007. Spatio-genetic population structure in moustached tamarins, *Saguinus mystax*. *American Journal of Physical Anthropology*, 132, 576–583.

Huck, M., Rotundo, M. & Fernandez-Duque E. 2011. Growth and development in wild owl monkeys (*Aotus azarai*) of Argentina. *International Journal of Primatology*, 32(5), 1133–1152.

Isbell, L. A. 2006. Snakes as agents of evolutionary change in primate brains. *Journal of Human Evolution*, 51, 1–35.

Izawa, K. 1980. Social behavior of the wild black-capped capuchin (*Cebus apella*). *Primates*, 21, 443–467.

———. 1993. Soil-eating by *Alouatta* and *Ateles*. *International Journal of Primatology*, 14, 229–242.

Jack, K. M. 2007. The cebines: Toward an explanation of variable social structure. In *Primates in Perspective* (ed. by Campbell, C. J., Fuentes, A., MacKinnon, K. C., Panger, M. & Beader, S. K.), 107–123. Oxford: Oxford University Press.

Jack, K. M. & Fedigan, L. 2004a. Male dispersal patterns in white-faced capuchins, *Cebus capucinus*, Part 1: Patterns and causes of natal emigration. *Animal Behaviour*, 67, 761–769.

———. 2004b. Male dispersal patterns in white-faced capuchins, *Cebus capucinus*, Part 2: patterns and causes of secondary dispersal. *Animal Behaviour*, 67, 771–782.

Jack, K. M. & Fedigan, L. M. 2009. Female dispersal in a female-philopatric species, *Cebus capucinus*. *Behaviour*, 146, 471–497.

Janson, C. H. 1984. Female choice and mating system of the brown capuchin monkey *Cebus apella* (Primates: Cebidae). *Zeitschrift für Tierpsychologie*, 65, 177–200.

———. 1985. Aggressive competition and individual food consumption in wild brown capuchin monkeys (*Cebus apella*). *Behavioral Ecology and Sociobiology*, 18, 125–318.

———. 1986. The mating system as a determinant of social evolution in capuchin monkeys (*Cebus*). In *Primate Ecology and Conservation* (ed. by Else, J. G. & Lee, P. C.), 169–180. Cambridge: Cambridge University Press.

Julliot, C. 1994. Predation of a young spider monkey (*Ateles paniscus*) by a crested eagle (*Morphnus guianensis*). *Folia Primatologica*, 63, 75–77.

Kinzey, W. G. 1981. The titi monkeys, genus *Callicebus*. In *Ecology and Behavior of Neotropical Primates* (ed. by Coimbra-Filho, A. F. & Mittermeier, R. A.), 241–276. Rio de Janeiro: Academia Brasileira de Ciências.

———. 1997. *Aotus*. In *New World Primates: Ecology, Evolution and Behavior* (ed. by Kinzey, W. G.), 186–191. New York: Aldine de Gruyter.

———. 1997. *New World Primates*. New York: Aldine de Gruyter.

Kinzey, W. G. & Cunningham, E. P. 1994. Variability in platyrrhine social organization. *American Journal of Primatology*, 34, 185–198.

Kinzey, W. G. & Wright, P. C. 1982. Grooming behavior in the titi monkey (*Callicebus torquatus*). *American Journal of Primatology*, 3, 267–275.

Klein, L. L. 1972. The ecology and social behavior of the spider monkey, *Ateles belzebuth*. PhD dissertation, University of California.

Klein, L. L. & Klein, D. J. 1977. Feeding behavior of the Colombian spider monkey, *Ateles belzebuth*. In *Primate Ecology: Studies of Foraging and Ranging Behaviour in Lemurs, Monkeys, and Apes* (ed. by Clutton-Brock, T. H.), 153–181. London: Academic Press.

Kowalewski, M. M. & Garber, P. A. 2010. Mating promiscuity and reproductive tactics in female black and gold howler monkeys (*Alouatta caraya*) inhabiting an island on the Parana River, Argentina. *American Journal of Primatology*, 71, 1–15.

Lambert, J. E. 1998. Primate digestion interactions among anatomy, physiology, and feeding ecology. *Evolutionary Anthropology: Issues, News, and Reviews*, 7, 8–20.

Lappan, S. 2008. Male care of infants in a siamang (*Symphalangus syndactylus*) population including socially monogamous and polyandrous groups. *Behavioral Ecology and Sociobiology*, 62, 1307–1317.

Lazaro-Perea, C., Arruda, M. D. F. & Snowdon, C. T. 2004. Grooming as a reward? Social function of grooming between females in cooperatively breeding marmosets. *Animal Behaviour*, 67, 627–636.

Lehman, S. M., Prince, W. & Mayor, M. 2001. Variations in group size in white-faced sakis (*Pithecia pithecia*): Evidence for monogamy or seasonal congregations? *Neotropical primates*, 9, 96–101.

Leutenegger, W. 1979. Evolution of litter size in primates. *The American Naturalist*, 114, 525–531.

Link, A., Di Fiore, A. & Spehar, S. N. 2009. Female-directed aggression and social control in spider monkeys. In *Sexual Coercion in Primates: An Evolutionary Perspective on Male Aggression against Females* (ed. by Muller, M. N. & Wrangham, R.), 157–183. Cambridge, MA: Harvard University Press.

Link, A., Galvis, N., Fleming, E. & Di Fiore, A. 2011. Patterns of mineral lick visitation by spider monkeys and howler monkeys in Amazonia: Are licks perceived as risky areas? *American Journal of Primatology*, 73, 386–396.

Löttker, P., Huck, M. & Heymann, E. W. 2004. Demographic parameters and events in wild moustached tamarins (*Saguinus mystax*). *American Journal of Primatology*, 64, 425–449.

Löttker, P., Huck, M., Heymann, E. W. & Heistermann, M. 2004. Endocrine correlates of reproductive status in breeding and non-breeding wild female moustached tamarins. *International Journal of Primatology* 25, 919–937.

Löttker, P., Huck, M., Zinner, D. P. & Heymann, E. W. 2007. Grooming relationships between breeding females and adult group members in cooperatively breeding moustached tamarins (*Saguinus mystax*). *American Journal of Primatology*, 69, 1–14.

Ludwig, G., Aguiar, L., Miranda, J., Teixeira, G., Svoboda, W., Malanski, L., Shiozawa, M., Hilst, C., Navarro, I. & Passos, F. 2007. Cougar predation on black-and-gold howlers on Mutum Island, Southern Brazil. *International Journal of Primatology*, 28, 39–46.

Lynch Alfaro, J. W. 2007. Subgrouping patterns in a group of wild *Cebus apella nigritus*. *International Journal of Primatology*, 28, 271–289.

Martin, R. D. 1992. Goeldi and the dwarfs: the evolutionary biology of the small New World monkeys. *Journal of Human Evolution*, 22, 367–393.

Martins, M. M. & Setz, E. Z. F. 2000. Diet of buffy tufted-eared marmosets (*Callithrix aurita*) in a forest fragment in southeastern Brazil. *International Journal of Primatology*, 21, 467–476.

Martins, S. d. S., Lima, E. M. d. & Silva Jr., J. d. S. e. 2005. Predation of a bearded saki (*Chiropotes utahicki*) by a harpy eagle (*Harpia harpyja*). *Neotropical Primates*, 13, 7–10.

Martins, W. P. & Strier, K. B. 2004. Age at first reproduction in philopatric female muriquis (*Brachyteles arachnoides hypoxanthus*). *Primates*, 45, 63–67.

Mason, W. A. 1966. Social organization of the South American monkey, *Callicebus moloch*: A preliminary report. *Tulane Studies in Zoology*, 13, 23–28.

———. 1975. Comparative studies of social behavior in *Callicebus* and *Saimiri*: Strength and specificity of attraction between male-female cagemates. *Folia Primatologica*, 23, 113–123.

McFarland, M. J. 1986. Ecological determinants of fission-fusion sociality in Ateles and Pan. In *Primate Ecology and Conservation* (ed. by Else, J. G. & Lee, P. C.), 181–190. New York: Cambridge University Press.

Mendoza, S. P. & Mason, W. A. 1986a. Constrasting responses to intruders and to involuntary separation by monogamous and polygynous New World monkeys. *Physiology and Behavior*, 38, 795–801.

———. 1986b. Parental division of labour and differentiation of attachments in a monogamous primate (*Callicebus moloch*). *Animal Behaviour*, 34, 1336–1347.

Milton, K. 1982. Dietary quality and demographic regulation in a howler monkey population. In *The Ecology of a Tropical Forest: Seasonal Rhythms and Long-Term Changes* (ed. by Leigh, E. G. J., Rand, S. A. & Windsor, D. M.), 273–289. Washington: Smithsonian Institution Press.

———. 1984. Habitat, diet, and activity patterns of free-ranging woolly spider monkeys (*Brachyteles arachnoides*, E. Geoffroy 1806). *International Journal of Primatology*, 5, 491–514.

———. 1993. Diet and primate evolution. *Scientific American*, 269, 86–93.

———. 1996. Effects of bot fly (*Alouattamyia baeri*) parasitism on a free-ranging howler monkey (*Alouatta palliata*) population in Panamá. *Journal of Zoology, London*, 239, 39–63.

———. 1998. Physiological ecology of howlers (*Alouatta*): Energetic and digestive considerations and comparison with the Colobinae. *International Journal of Primatology*, 19, 513–548.

Miranda, J. M. D., Bernardi, I. P., Abreu, K. C. & Passos, F. C. 2005. Predation on *Alouatta guariba clamitans* Cabrera (Primates, Atelidae) by *Leopardus pardalis* (Linnaeus) (Carnivora, Felidae). *Revista Brasileira de Zoologia* 22, 793–95.

Mitchell, C. L. 1990. The ecological basis for female social dominance: A behavioral study of the squirrel monkey (*Saimiri sciureus*) in the wilds. PhD dissertation, Princeton University.

———. 1994. Migration alliances and coalitions among adult male South American squirrel monkeys (*Saimiri sciureus*). *Behaviour*, 130, 169–190.

Mitchell, C. L., Boinski, S. & van Schaik, C. P. 1991. Competitive regimes and female bonding in two species of squirrel monkeys (*Saimiri oerstedi* and *S. sciureus*). *Behavioral Ecology and Sociobiology*, 28, 55–60.

Mittermeier, R. A., Coimbra-Filho, A. F. & Fonseca., G. A. B. d. 1981. *Ecology and Behavior of Neotropical Primates.*, Vol. 2. Academia Brasileira de Ciências.

Mourthé, Í. M. C., Guedes, D., Fidelis, J., Boubli, J. P., Mendes,

S. & Strier, K. B. 2007. Ground use by northern muriquis (*Brachyteles hypoxanthus*). *American Journal of Primatology*, 69, 706–712.

Moynihan, M. 1964. Some behavior patterns of playtyrrhine monkeys. I. The night monkey (*Aotus trivirgatus*). *Smithsonian Miscellaneous Collections*, 146, 1–84.

Muniz, L., S. Perry, J. H. Manson, H. Gilkenson, J. Gros-Louis and L. Vigilant 2010. Male dominance and reproductive success in wild white-faced capuchins (*Cebus capucinus*) at Lomas Barbudal, Costa Rica. *American Journal of Primatology* 72, 1118–1130.

Nadjafzadeh, M. N. & Heymann, E. W. 2008. Prey foraging of red titi monkeys, *Callicebus cupreus*, in comparison to sympatric tamarins, *Saguinus mystax* and *Saguinus fuscicollis*. *American Journal of Physical Anthropology* 135, 56–63.

Neville, M. K., Glander, K. E., Braza, F. & Rylands, A. B. 1988. The howling monkeys, genus *Alouatta*. In *Ecology and Behavior of Neotropical Primates* (ed. by Mittermeier, R. A., Rylands, A. B., Coimbra-Filho, A. F. & da Fonseca, G. A. B.), 349–453. Washington: World Wildlife Fund.

Nievergelt, C. M., Digby, L. J., Ramkrishnan, U. & Woodruff, D. S. 2000. Genetic analysis of group composition and breeding system in a wild common marmoset (*Callithrix jacchus*) population. *International Journal of Primatology*, 21, 1–20.

Nishimura, A. 1990. A sociological and behavioral study of woolly monkeys, *Lagothrix lagotricha*, in the Upper Amazon. *The Science and Engineering Review of Doshisha University*, 31, 87–121.

———. 2003. Reproductive parameters of wild female *Lagothrix lagotricha*. *International Journal of Primatology*, 24, 707–722.

Noë, R. 1990. A veto game played by baboons: A challenge to the use of the Prisoner's Dilemma as a paradigm for reciprocity and cooperation. *Animal Behaviour*, 39, 78–90.

Norconk, M. A. 2006. Long-term study of group dynamics and female reproduction in Venezuelan *Pithecia pithecia*. *International Journal of Primatology*, 27, 653–674.

———. 2007. Sakis, uakaris, and titi monkeys: Behavioral diversity in a radiation of seed predators. In *Primates in Perspective* (ed. by Campbell, C. J., Fuentes, A., MacKinnon, K. C., Panger, M. & Bearder, S. K.), 123–138. Oxford: Oxford University Press.

Norconk, M. A. & Kinzey, W. G. 1994. Challenge of neotropical frugivory: Travel patterns of spider monkeys and bearded sakis. *American Journal of Primatology*, 34, 171–183.

O'Brien, T. G. 1988. Parasitic nursing in the wedge-capped capuchin monkey (*Cebus olivaceus*). *American Journal of Primatology*, 16, 341–344.

———. 1991. Female-male social interactions in wedge-capped capuchin monkeys: Benefits and costs of group living. *Animal Behaviour*, 41, 555–567.

———. 1993. Allogrooming behaviour among adult female wedge-capped capuchin monkeys. *Animal Behaviour*, 46, 499–510.

O'Brien, T. G. & Robinson, J. G. 1991. Allomaternal care by female wedge-capped capuchin monkeys: Effects of age, rank and relatedness. *Behaviour*, 119, 30–50.

Oerke, A.-K., Heistermann, M., Küderling, I., Martin, R. D. & Hodges, J. K. 2002. Monitoring reproduction in Callitrichidae by means of ultrasonography. *Evolutionary Anthropology*, Suppl. 1, 183–185.

Oliveira, V. B. D., Linares, A. M., Correa, G. L. C. & Chiarello,

A. G. 2008. Predation on the black capuchin monkey *Cebus nigritus* (Primates: Cebidae) by domestic dogs *Canis lupus familiaris* (Carnivora: Canidae), in the Parque Estadual Serra do Brigadeiro, Minas Gerais, Brazil. *Revista Brasileira de Zoologia*, 25, 376–378.

Opazo, J. C., Wildman, D. E., Prychitko, T., Johnson, R. M. & Goodman, M. 2006. Phylogenetic relationships and divergence times among New World monkeys (Platyrrhini, Primates). *Molecular Phylogenetics and Evolution*, 40, 274–280.

Overdorff, D. J. & Rasmussen, M. A. 1995. Determinants of nightime activity in "diurnal" lemurid primates. In *Creatures of the Dark: The Nocturnal Prosimians* (ed. by Alterman, L.), 61–74. New York: Plenum Press.

Oversluijs Vásquez, M. R. & Heymann, E. W. 2001. Crested eagle (*Morphnus guianensis*) predation on infant tamarins (*Saguinus mystax* and *Saguinus fuscicollis*, Callitrichinae). *Folia Primatologica*, 72, 301–303.

Oxnard, C. E. 2004. Brain evolution: mammals, primates, chimpanzees, and humans. *International Journal of Primatology*, 25, 1127–1158.

Packer, C. 1977. Reciprocal altruism in *Papio anubis*. *Nature*, 265, 441–443.

Peetz, A. 2001. Ecology and social organization of the bearded saki, *Chiropotes satanas chiropotes* (Primates: Pitheciinae) in Venezuela. *Ecotropical Monographs*, 1, 1–170.

Peetz, A., Norconk, M. A. & Kinzey, W. G. 1992. Predation by jaguar on howler monkeys (*Alouatta seniculus*) in Venezuela. *American Journal of Primatology*, 28, 223–228.

Peres, C. A. 1994. Diet and feeding ecology of gray woolly monkeys (*Lagothrix lagotricha cana*) in Central Amazonia: Comparisons with other atelines. *International Journal of Primatology*, 15, 333–372.

———. 1996. Use of space, spatial group structure, and foraging group size of gray woolly monkeys (*Lagothrix lagotricha cana*) at Urucu, Brazil. In *Adaptive Radiations of Neotropical Primates* (ed. by Norconk, M. A., Rosenberger, A. L. & Garber, P. A.), 467–488. New York: Plenum Press.

Pérez-Barbería, F. J., Shultz, S. & Dunbar, R. I. M. 2007. Evidence for coevolution of sociality and relative brain size in three orders of mammals. *Evolution*, 61, 2811–2821.

Perry, S. 1996. Female-female social relationships in wild white-faced capuchin monkeys, *Cebus capucinus*. *American Journal of Primatology*, 40, 167–182.

———. 1997. Male-female social relationships in wild white-faced capuchins (*Cebus capucinus*). *Behaviour*, 134, 477–510.

———. 1998. Male-male social relationships in wild white-faced capuchins, *Cebus capucinus*. *Behaviour*, 135, 139–172.

Pope, T. R. 1989. The influence of mating system and dispersal patterns on the genetic structure of red howler monkey populations. PhD dissertation, University of Florida.

———. 1990. The reproductive consequences of male cooperation in the red howler monkey: Paternity exclusion in multimale and single-male troops using genetic markers. *Behavioral Ecology and Sociobiology*, 27, 439–446.

———. 1992. The influence of dispersal patterns and mating systems on genetic differentiation within and between populations of the red howler monkey (*Alouatta seniculis*). *Evolution*, 46, 1112–1128.

Porter, L. M. 2001a. Social organization, reproduction and rearing strategies of *Callimico goeldii*: New clues from the wild. *Folia Primatologica*, 72, 69–79.

———. 2001b. Dietary differences among sympatric Callitrichinae in Northern Bolivia: *Callimico goeldii, Saguinus fuscicollis* and *S. labiatus*. *International Journal of Primatology*, 22, 961–992.

Porter, L. M. & Garber, P. A. 2009. Social behavior of *Callimicos*: Mating strategies and infant care. In *The Smallest Anthropoids:The Marmoset/Callimico Radiation* (ed. by Ford, S. M., Porter, L. M. & Davis, L. C.) 87–102. New York: Springer.

Poux, C., Chevret, P., Huchon, D., De Jong, W. W. & Douzery, E. J. P. 2006. Arrival and diversification of caviomorph rodents and platyrrhine primates in South America. *Systematic Biology*, 52, 228–244.

Price, E. C. & Piedade, H. M. 2001. Ranging behavior and intraspecific relationships of masked titi monkeys (*Callicebus personatus personatus*). *American Journal of Primatology*, 53, 87–92.

Raguet-Schofield, M. 2008. The effects of human encroachment and seasonality on the risk of mantled howler monkey (*Alouatta palliate*) predation by dogs on Ometepe Island, Nicaragua. Abstract, *American Journal of Physical Anthropology*, Suppl. 46, 176.

Ramirez, M. 1980. Grouping patterns of the woolly monkey, *Lagothrix lagothricha*, at the Manu National Park , Peru. *American Journal of Physical Anthropology*, 52, 269.

———. 1988. The woolly monkeys, genus *Lagothrix*. In *Ecology and Behavior of Neotropical Primates* (ed. by Mittermeier, R. A., Rylands, A. B., Coimbra-Filho, A. F. & da Fonseca, G. A. B.), 539–575. Washington: World Wildlife Fund.

Ray, D. A., Xing, J. C., Hedges, D. J., Hall, M. A., Laborde, M. E., Anders, B. A., White, B. R., Stoilova, N., Fowlkes, J. D., Landry, K. E., Chemnick, L. G., Ryder, O. A. & Batzer, M. A. 2005. *Alu* insertion loci and platyrrhine primate phylogeny. *Molecular Phylogenetics and Evolution*, 35, 117–126.

Rilling, J. K. & Insel, T. R. 1999. The primate neocortex in comparative perspective using magnetic resonance imaging. *Journal of Human Evolution*, 37, 191–223.

Robinson, J. G. 1979. Vocal regulation of use of space by groups of titi monkeys *Callicebus moloch*. *Behavioral Ecology and Sociobiology*, 5, 1–15.

———. 1981. Vocal regulation of inter- and intragroup spacing during boundary encounters in the titi monkey, *Callicebus moloch*. *Primates*, 22, 161–172.

———. 1988. Demography and group structure in wedge-capped capuchin monkeys, *Cebus olivaceus*. *Behaviour*, 104, 202–232.

Robinson, J. G., Wright, P. C. & Kinzey, W. G. 1987. Monogamous cebids and their relatives: Intergroup calls and spacing. In *Primate Societies* (ed. by Smuts, B. B., Cheney, D. L., Seyfarth, R. M., Wrangham, R. & Struhsaker, T. T.), 44–53. Chicago: University of Chicago Press.

Rosa, M. G. P. & Tweedale, R. 2005. Brain maps, great and small: Lessons from comparative studies of primate visual cortical organization. *Philosophical Transactions of the Royal Society of London, Series B*, 360, 665–691.

Rosenberger, A. L. 1983. Tail of tails: Parallelism and prehensility. *American Journal of Physical Anthropology*, 60, 103–107.

Rosenberger, A. L. & Matthews, L. J. 2008. *Oreonax*: Not a genus. *Neotropical Primates*, 15, 8–12.

Rosenberger, A. L., Tejedor, M. F., Siobhán B. Cooke & Pekar, S. 2009. Platyrrhine ecophylogenetics in space and time. In *South American Primates: Comparative Perspectives in the Study of Behavior, Ecology, and Conservation* (ed. by Garber, P. A., Estrada, A., Bicca-Marques, J. C., Heymann, E. W. & Strier, K. B.), 69–116. New York: Springer.

Rudran, R. & Fernandez-Duque, E. 2003. Demographic changes over thirty years in a red howler population in Venezuela. *International Journal of Primatology*, 24, 925–947.

Rylands, A. B. & Mittermeier, R. A. 2009. The diversity of the New World primates (Platyrrhini): An annotated taxonomy. In *South American Primates: Comparative Perspectives in the Study of Behavior, Ecology, and Conservation* (ed. by Garber, P. A., Estrada, A., Bicca-Marques, J. C., Heymann, E. W. & Strier, K. B.), 23–54. New York: Springer.

Sampaio, D. T. & Ferrari, S. F. 2005. Predation of an infant titi monkey (*Callicebus moloch*) by a tufted capuchin (*Cebus apella*). *Folia Primatologica*, 76, 113–115.

Sánchez-Villagra, M. R., Pope, T. R. & Salas, V. 1998. Relation of intergroup variation in allogrooming to group social structure and ectoparasite loads in red howlers (*Alouatta seniculus*). *International Journal of Primatology*, 19, 473–491.

Schillaci, M. A. 2006. Sexual selection and the evolution of brain size in primates. *PLoS ONE*, 1, e62.

Schmitt, C. A., Di Fiore, A., Link, A., Matthews, L. J., Montague, M. J., Derby, A. M., Carrillo, G., Sendall, C. & Fernandez-Duque, E. 2007. Comparative ranging behavior of eight species of primates in a western Amazonian rainforest. *American Journal of Physical Anthropology*, 132, 208.

Schmitt, D., Rose, M. D., Turnquist, J. E. & Lemelin, P. 2005. Role of the prehensile tail during Ateline locomotion: Experimental and osteological evidence. *American Journal of Physical Anthropology*, 126, 435–446.

Schneider, H., Canavez, F. C., Sampaio, I., Moreira, M. Â. M., Tagliaro, C. H. & Seuánez, H. N. 2001. Can molecular data place each neotropical monkey in its own branch? *Chromosoma*, 109, 515–523.

Schneider, H., Sampaio, I., Harada, M. L., Barroso, C. M. L., Schneider, M. P. C., Czelusniak, J. & Goodman, M. 1996. Molecular phylogeny of the New World monkeys (Platyrrhini, Primates) based on two unlinked nuclear genes: IRBP intron 1 and e-globin sequences. *American Journal of Physical Anthropology*, 100, 153–179.

Schneider, H., Schneider, M. P. C., Sampaio, M. I. C., Harada, M. L., Stanhope, M., Czelusniak, J. & Goodman, M. 1993. Molecular phylogeny of the New World monkeys (Platyrrhini, Primates). *Molecular Phylogenetics and Evolution*, 2, 225–242.

Schrago, C. G. & Russo, C. A. M. 2003. Timing the origin of New World monkeys. *Molecular Biology and Evolution*, 20, 1620–1625.

Setz, E. Z. F. & Gaspar, D. A. 1997. Scent marking behaviour in free-ranging golden-faced saki monkeys *Pithecia pithecia chrysocephala*: Sex differences and context. *Journal of Zoology* 241, 603–611.

Shahuano Tello, N., Huck, M. & Heymann, E. W. 2002. *Boa constrictor* attack and successful group defence in moustached tamarins, *Saguinus mystax*. *Folia Primatologica*, 73, 146–148.

Sherman, P. T. 1991. Harpy eagle predation on a red howler monkey. *Folia Primatologica*, 56, 53–56.

Shimooka, Y. 2005. Sexual differences in ranging of *Ateles belzebuth belzebuth* at La Macarena, Colombia. *International Journal of Primatology*, 26, 385–406.

Silva, S. S. B. & Ferrari, S. F. 2009. Behavior patterns of Southern bearded sakis (*Chiropotes satanas*) in the fragmented landscape of Eastern Brazilian Amazonia. *American Journal of Primatology*, 71, 1–7.

Singer, S. S., Schmitz, J., Schwiegk, C. & Zischler, H. 2003. Molecular cladistic markers in New World monkey phylogeny (Platyrrhini, Primates). *Molecular Phylogenetics and Evolution*, 26, 490–501.

Smith, R. J. & Jungers, W. L. 1997. Body mass in comparative primatology. *Journal of Human Evolution*, 32, 523–559.

Smuts, B. B., Cheney, D. L., Seyfarth, R. M., Wrangham, R. W. & Struhsaker, T. T. 1987. *Primate Societies*. Chicago: University of Chicago Press.

Snowdon, C. T., Ziegler, T. E. & Widowski, T. M. 1993. Further hormonal suppression of eldest daughter cotton-top tamarins following birth of infants. *American Journal of Primatology*, 31, 11–21.

Soini, P. 1987. Sociosexual behavior of a free-ranging *Cebuella pygmaea* (Callitrichidae, platyrrhini) troop during postpartum estrus of its reproductive female. *American Journal of Primatology*, 13, 223–230.

Solomon, N. G. & French, J. A. 1997. *Cooperative Breeding in Mammals*. Cambridge: Cambridge University Press.

Sommer, V. & Reichard, U. 2000. Rethinking monogamy: The gibbon case. In *Primate Males. Causes and Consequences of Variation in Group Composition* (ed. by Kappeler, P. M.), 159–168. Cambridge: Cambridge University Press.

Spehar, S. N., Link, A. & Di Fiore, A. 2010. Male and female range use in a group of white-bellied spider monkeys (*Ateles belzebuth*) in Yasuní National Park, Ecuador. *American Journal of Primatology*, 72, 129–141.

Stevenson, P. R., Castellanos, M. C., Pizarro, J. C. & Garavito, M. 2002. Effects of seed dispersal by three ateline monkey species on seed germination at Tinigua National Park, Colombia. *International Journal of Primatology*, 23, 1187–1204.

Stevenson, P. R., Quiñones, M. J. & Ahumada, J. A. 1994. Ecological strategies of woolly monkeys (*Lagothrix lagotricha*) at Tinigua National Park, Colombia. *American Journal of Primatology*, 32, 123–140.

Stone, A. 2007. Responses of squirrel monkeys to seasonal changes in food availability in an eastern Amazonian forest. *American Journal of Primatology*, 69, 142–157.

Strier, K. B. 1987. Ranging behavior of woolly spider monkeys, or muriquis, *Brachyteles arachnoides*. *International Journal of Primatology*, 8, 575–591.

———. 1990. New World primates, new frontiers: Insights from the woolly spider monkey, or muriqui (*Brachyteles arachnoides*). *International Journal of Primatology*, 11, 7–19.

———. 1991. Demography and conservation of an endangered primate, *Brachyteles arachnoides*. *Conservation Biology*, 5, 214–218.

———. 1992. *Faces in the Forest: The Endangered Muriqui Monkeys of Brazil*. New York: Oxford University Press.

———. 1994a. Myth of the typical primate. *Yearbook of Physical Anthropology*, 37, 233–271.

———. 1994b. Brotherhoods among atelins: Kinship, affiliation, and competition. *Behaviour*, 130, 151–167.

————. 2004. Patrilineal kinship and primate behavior. In *Kinship and Behavior in Primates* (ed. by Chapais, B. & Berman, C. M.), 177–198. Oxford: Oxford University Press.

————. 2008. The effects of kin on primate life histories. *Annual Review of Anthopology*, 37, 21–36.

Strier, K. B., Chaves, P. B., Mendes, S. L., Fagundes, V., and Di Fiore, A. 2011. Low paternity skew and the influence of maternal kin in an egalitarian, patrilocal primate. *Proceedings of the National Academy of Sciences, USA*, 108, 18915–18919.

Struhsaker, T. 2008. Demographic variability in monkeys: Implications for theory and conservation. *International Journal of Primatology*, 29, 19–34.

Symington, M. M. 1987. Sex ratio and maternal rank in wild spider monkeys: When daughters disperse. *Behavioral Ecology and Sociobiology*, 20, 421–425.

Symington, M. M. 1988. Demography, ranging patterns, and activity budgets of black spider monkeys (*Ateles paniscus chamek*) in the Manu National Park, Peru. *American Journal of Primatology*, 15, 45–67.

Symington, M. M. 1990. Fission-fusion social organization in *Ateles* and *Pan*. *International Journal of Primatology*, 11, 47–61.

Talebi, M. G., Beltrão-Mendes, R. & Lee, P. C. 2009. Intra-community coalitionary lethal attack of an adult male southern muriqui (*Brachyteles arachnoides*). *American Journal of Primatology*, 71, 860–867.

Tardif, S. D. 1997. The bioenergetics of parental behavior and the evolution of alloparental care in marmosets and tamarins. In *Cooperative Breeding in Mammals* (ed. by Solomon, N. G. & French, J. A.), 11–33. Cambridge: Cambridge University Press.

Tardif, S. D. & Garber, P. A. 1994. Social and reproductive patterns in neotropical primates: Relation to ecology, body size, and infant care. *American Journal of Primatology*, 34, 111–114.

Tardif, S. D., Santos, C. V., Baker, A. J., Van Elsacker, L., Ruiz-Miranda, C. R., De a Moura, A. C., Passos, F. C., Price, E. C., Rapaport, L. G. & de Vleeschouwer, K. 2002. Infant care in lion tamarins. In *Lion Tamarins: Biology and Conservation* (ed. by Kleiman, D. G. & Rylands, A. B.), 213–232. Washington: Smithsonian University Press.

Terborgh, J. & Goldizen, A. W. 1985. On the mating system of the cooperatively breeding saddle-backed tamarin (*Saguinus fuscicollis*). *Behavioral Ecology and Sociobiology*, 16, 293–299.

Torres de Assumpção, C. 1983. Ecological and behavioral information on *Brachyteles arachnoides*. *Primates*, 24, 584–593.

Valeggia, C. R., Mendoza, S. P., Fernandez-Duque, E., Mason, W. A. & Lasley, B. 1999. Reproductive biology of female titi monkeys (*Callicebus moloch*) in captivity. *American Journal of Primatology*, 47, 183–195.

Van Roosmalen, M. G. M. 1985. Habitat preferences, diet, feeding strategy and social organization of the black spider monkey (*Ateles paniscus paniscus*: Linnaeus 1758) in Surinam. *Acta Amazonica*, 15, 1–238.

Van Roosmalen, M. G. M. & van Roosmalen, T. 2003. The description of a new marmoset genus, *Callibella* (Callitrichinae, Primates), including its molecular phylogenetic status. *Neotropical Primates*, 11, 1–10.

Van Roosmalen, M. G. M., van Roosmalen, T., Mittermeier, R. A. 2002. A taxonomic review of the titi monkeys, genus *Callicebus* Thomas, 1903, with the description of two new species, *Callicebus bernhardi* and *Callicebus stephennashi*, from Brazilian Amazonia. *Neotropical* Primates, 10, Supplement,1–52.

Van Schaik, C. P., Barrickman, N., Bastian, M., Krakauer, E. & van Noordwijk, M. 2006. Primate life histories and the role of brains. In *The Evolution of Human Life History* (ed. by Hawkes, K. & Paine, R.), 127–154. Santa Fe, NM: SAR Press.

Vié, J.-C., Richard-Hansen, C. & Fournier-Chambrillon, C. 2001. Abundance, use of space, and activity patterns of white-faced sakis (*Pithecia pithecia*) in French Guiana. *American Journal of Primatology*, 55, 203–221.

Von Dornum, M. & Ruvolo, M. 1999. Phylogenetic relationships of the New World monkeys (Primates, Platyrrhini) based on nuclear G6PD DNA sequences. *Molecular Phylogenetics and Evolution*, 11, 459–476.

Walker, R., Burger, O., Wagner, J. & Von Rueden, C. R. 2006. Evolution of brain size and juvenile periods in primates. *Journal of Human Evolution*, 51, 480–489.

Wang, E. & Milton, K. 2003. Intragroup social relationships of male *Alouatta palliata* on Barro Colorado Island, Republica of Panama. *International Journal of Primatology*, 24, 1227–1243.

Williams, L., Gibson, S., McDaniel, M., Bazzel, J., Barnes, S. & Abee, C. 1994. Allomaternal interactions in the Bolivian squirrel monkey (*Saimiri boliviensis boliviensis*). *American Journal of Primatology*, 34, 145–156.

Wolovich, C. K. & Evans, S. 2007. Sociosexual behavior and chemical communication of *Aotus nancymaae*. *International Journal of Primatology*, 28, 1299–1313.

Wolovich, C. K., Evans, S. & French, J. A. 2008. Dads do not pay for sex but do buy the milk: Food sharing and reproduction in owl monkeys (*Aotus* spp.). *Animal Behaviour*, 75, 1155–1163.

Wright, P. C. 1981. The night monkeys, genus *Aotus*. In *Ecology and Behavior of Neotropical Primates* (ed. by Coimbra-Filho, A. & Mittermeier, R. A.), 211–240. Rio de Janeiro: Academia Brasileira de Ciencias.

————. 1984. Biparental care in *Aotus trivirgatus* and *Callicebus moloch*. In *Female Primates: Studies by Women Primatologists* (ed. by Small, M.), 59–75. New York: Alan R. Liss, Inc.

————. 1985. The costs and benefits of nocturnality for *Aotus trivirgatus* (the night monkey). PhD dissertation, City University of New York.

————. 1989. The nocturnal primate niche in the New World. *Journal of Human Evolution*, 18, 635–658.

————. 1994. The behavior and ecology of the owl monkey. In *Aotus: The Owl Monkey* (ed. by Baer, J. F., Weller, R. E. & Kakoma, I.), 97–112. San Diego: Academic Press.

Yamamoto, M. E., Fátima Arruda, M. D., Irene Alencar, A., Sousa, M. B. C. D. & Araújo, A. 2009. Mating systems and female-female competition in the common marmoset, *Callithrix jacchus*. In *The Smallest Anthropoids: The Marmoset/Callimico Radiation* (ed. by Ford, S. M., Porter, L. M. & Davis, L. C.), 119–134. New York: Springer.

Zahed, S. R., Prudom, S. L., Snowdon, C. T. & Ziegler, T. E. 2007. Male parenting and response to infant stimuli in the common marmoset (*Callithrix jacchus*). *American Journal of Primatology*, 69, 1–15.

Ziegler, T. E. 2000. Hormones associated with non-maternal infant care: A review of mammalian and avian studies. *Folia Primatologica*, 71, 6–21.

Chapter 4 The Behavioral Ecology of Colobine Monkeys

Elisabeth H. M. Sterck

THE COLOBINES are a monophyletic group of the Cercopithecoid monkeys (Ting 2008; Ting et al. 2008). Their name derives from their relatively small or absent thumb (*kolobos* is ancient Greek for mutilated), yet their digestive system is their defining morphological feature. Colobines have a multichambered stomach that allows the fermentation of structural carbohydrates such as those found in leaves (Lambert 1998). This distinguishes the group from other leaf-eating primates, such as sportive lemurs (*Lepilemur* spp.), howler monkeys (*Alouatta* spp.), and gorillas (*Gorilla gorilla*), who ferment leaves in an enlarged caecum or colon. Apart from their digestive strategy, leaf-eating primates share several other characteristics. They often live in relatively small groups (Clutton-Brock & Harvey 1977), and many species display female dispersal (Moore 1984; Newton 1992). It has been hypothesized that within-group scramble (van Schaik 1989) or lack of competition (Isbell 1991) typifies their behavior because foliage is a noncontestable food source, but observations of colobines do not necessarily conform to these expectations (Koenig & Borries 2001; Snaith & Chapman 2005; chapters 7 and 9, this volume). Studies of leaf-eating primates have led to the "folivore paradox," the fact that group sizes are smaller than those predicted by food competition (Moore 1984; Janson & Goldsmith 1995; Steenbeek & van Schaik 2001; chapter 9, this volume). It has been proposed that social factors, such as infanticide risk, limit group sizes in folivores (Crockett & Janson 2000; Isbell 2004) and highlight how sexual strategies can affect social relationships (Sterck et al. 1997). In this chapter I outline the range and limits of colobine behavioral diversity through a review of their biogeography, ecology, and social behavior.

Diversity and Biogeography

Colobines are arboreal monkeys that inhabit forests of Africa and Asia. The exception is the Hanuman langur (*Semnopithecus entellus*), which occupies relatively dry areas and semiterrestrial terrain on the Indian subcontinent (Bennett & Davies 1994). Also the Yunnan snub-nosed monkey (*Rhinopithecus bieti*) spends considerable time on the ground (Kirkpatrick & Long 1994), but is still a mainly arboreal species (Kirkpatrick et al. 1998). Most authors divide colobines into distinct genera or subgenera. Much less agreement exists on the number of species. For purposes of this chapter, I follow two comprehensive reviews of Asian and African primate taxonomy and subdivide the subfamily Colobinae into two African genera, which comprise 11 species, and seven Asian genera, comprising 31 species (Grubb et al. 2003; Brandon-Jones et al. 2004; table 4.1).

Colobine monkeys (subfamily Colobinae) are closely related to Old World monkeys in another subfamily, the Cercopithecinae (chapter 5, this volume). These subfamilies both diverged from a common ancestor in Africa about 14.7 and 16.2 million years ago (Raaum et al. 2005; Sterner et al. 2006). The divergence between African and Asian colobines occurred around 11.5 million years ago (Raaum et al. 2005). These dates are consistent with the fossil record

Table 4.1. Colobinae subfamily classification and diversity (Grubb et al. 2003; Brandon-Jones et al. 2004).

Continent	Genus (number of species)	Species name	Continent	Genus (number of species)	Species name
Africa	*Procolobus* (6)	*P. verus*	Asia	*Semnopithecus* (3)	*S. entellus*
		P. badius			*S. johnii*
		P. gordonorum			*S. vetulus*
		P. kirkii	Asia	*Trachypithecus* (10)	*T. auratus*
		P. pennantii			*T. barbei*
		P. rufomitratus			*T. delacouri*
Africa	*Colobus* (5)	*C. angolensis*			*T. francoisi*
		C. guereza			*T. geei*
		C. polykomos			*T. laotum*
		C. satanus			*T. obscurus*
		C. vellerosus			*T. pileatus*
Asia	Presbytis (10)	*P. comata*			*T. poliocephalus*
		P. femoralis			*T. villosus*
		P. fredericae	Asia	*Nasalis* (1)	*N. larvatus*
		P. frontata	Asia	*Simias* (1)	*S. concolor*
		P. hosei	Asia	*Pygathrix* (2)	*P. nemaeus*
		P. melalophos			*P. nigripens*
		P. potenziani	Asia	*Rhinopithecus* (4)	*R. avunculus*
		P. rubicunda			*R. bieti*
		P. siamensis			*R. brelichi*
		P. thomasi			*R. roxellana*

and suggest a spread from Africa to Eurasia. Several colobine lineages went extinct, among them some large-bodied and partially terrestrial, African, and European species (Ting 2008).

The African colobines can be divided into two genera: *Colobus* and *Procolobus* (Grubb et al. 2003; table 4.1, figs. 4.1, 4.2). *Colobus* includes five species. The status of the five *Colobus* species and the olive colobus (*Procolobus verus*) are not disputed. No consensus exists, however, on the number of other *Procolobus* species, with authors distinguishing between 1 to 16 (Ting 2008). Grubb et al. (2003) tentatively propose five species. The *Colobus* group diverged from *Procolobus* around 7.5 (± 1.2) million years ago, while the remaining two subgenera diverged later about 6.4 (±1.1) million years ago (Ting 2008). All of these species are arboreal and inhabit rain forests or gallery forests in sub-Saharan Africa (Oates et al. 1994).

Two Asian colobine groups are typically distinguished: the langurs, consisting of three genera, *Semnopithecus*, *Trachypithecus*, and *Presbytis* (fig. 4.3), and the odd-nosed monkeys, consisting of four genera, *Nasalis*, *Pygathrix*, *Rhinopithecus* and *Simias* (Brandon-Jones et al. 2004; Sterner et al. 2006; table 4.1). Langurs are more successful and outnumber odd-nosed monkeys. The taxonomic status of both groups continues to generate controversy, as there is no consensus regarding their scientific and common names, the number of genera and species, and their evolutionary history.

The taxonomy of *Semnopithecus* and *Trachypithecus* is currently in flux. For example, Hanuman langurs, *Semnopithecus entellus*, may actually represent more than one species (Brandon-Jones et al. 2004; Osterholtz et al. 2008), and recent molecular analyses suggest that some species of *Trachypithecus* should be reclassified as *Semnopithecus* (Karanth et al. 2008; Osterholz et al. 2008). As a result, the taxonomy provided in table 4.1 should be viewed as provisional and is likely to be revised in the future. Most langurs are arboreal and inhabit evergreen rainforests (Oates et al. 1994). One species, the well-studied Hanuman langur, occupies several habitats ranging from dry coastal forest to montane conifer forests. They also live in and near towns and are often terrestrial.

Two odd-nosed monkeys, *Simias* and *Nasalis*, are grouped together on the basis of morphological, behavioral, and molecular data (Whittaker et al. 2006), while molecular analyses suggest that *Pygathrix* and *Rhinopithecus* represent a monophyletic clade (Sterner et al. 2006). *Simias* is confined to the Mentawai Islands near Sumatra and inhabits evergreen forests. *Nasalis* inhabits mangroves and coastal and inland forests on Borneo (Meijaard & Nijman 2000). *Pygathrix* occupies forests in Indochina, and *Rhinopithecus* inhabits forests in Vietnam and montane areas of Southern China and Nepal (fig. 4.4). Most species of both genera are arboreal, with the exception of the Yunnan snub-nosed monkey (*Rhinopithecus bieti*), which is often terrestrial (Oates et al. 1994).

Fig. 4.1. An adult male quereza (*Colobus guereza*) at Kibale, Uganda, one of the 11 species of colobines currently recognized in Africa. Photo courtesy of Colin A. Chapman.

The conventional classification of Asian colobines into langur and odd-nose groups is controversial, as some analyses place *Semnopithecus* with the latter group rather than the former (Sterner et al. 2006; Whittaker et al. 2006; Ting et al. 2008). Additional research that synthesizes molecular and anatomical data with the fossil record will be necessary to resolve this issue, along with other recurring debates about the evolutionary history and taxonomy of colobine monkeys.

Ecology and Life History

Digestive System

Easily digestible food such as ripe fruit and animal matter can be processed by enzymes produced in the vertebrate di-gestive tract. Leaves, however, contain structural carbohy-drates that require fermentation by symbiotic microorgan-isms (Lambert 2007). Moreover, efficient fermentation by symbionts requires a long processing time (Clauss et al. 2008). Fermentation takes place in specialized enlarged gut areas that can contain food for relatively long periods of time and simultaneously harbor symbiotic microorgan-isms. Two different gut specializations for fermentation can be distinguished: an enlargement and partitioning of the stomach, which results in forestomach fermentation, and an enlargement of the caecum and colon, which re-sults in caecolic fermentation (Lambert 1998). In primates, gorillas, howler monkeys, and some cercopithecines and strepsirrhines possess enlarged caecocolic regions allowing fermentation (Lambert 1998). In contrast, forestomach fer-mentation is found only in colobine monkeys (Bauchop & Martucci 1968; Lambert 1998).

Fig. 4.2.   The eastern or Tana River red colobus monkey (*Procolobus rufomitratus*). These well-studied monkeys of east Africa have contributed much to our understanding of predation on primates as well as colobine social organization, sexual strategies, infanticide, and dispersal. Photo courtesy of Tom Struhsaker.

While both gut specializations permit the fermentation of leaves, they have different characteristics. Forestomach fermentation requires an alkaline environment. When too much food with easy-to-digest sugars or acids—for example, ripe fruit—is ingested, the production of acids may increase, resulting in deadly acidosis (Kay & Davies 1994). This potential problem is likely to limit consumption of ripe fruit by colobine monkeys. Hanuman langurs nevertheless eat large amounts of ripe fruit (79% in some months: Koenig & Borries 2001). This suggests that the problem of acidosis is exaggerated or that colobines avoid the problem by selectively feeding on only a few kinds of suitable ripe fruits or their parts. Colobine monkeys benefit from forestomach fermentation in several ways. For example, it helps them to process leaves that contain many hard-to-digest structural carbohydrates, break down toxic secondary plant compounds, and absorb metabolic by-products produced by forestomach symbionts. However, forestomach fermentation is also costly, as it leads to the loss of some digestible components of food used by the symbiotic microorganisms.

In sum, forestomach fermentation in colobine monkeys allows them to subsist on a diet of leaves, unripe fruit, and seeds; it may limit their consumption of ripe fruit containing easily digestible carbohydrates. Food-processing time is correspondingly long (chapter 7, this volume). With long food retention times and high digestive efficiencies, colobines require relatively low food intake (Clauss et al. 2008). This distinctive morphological adaptation has important implications not only for colobines' diet, but for their ecology and behavior as well.

Diet

As expected by their forestomach fermentation, colobine diets typically contain a relatively large proportion of leaves (Fashing 2007; table 4.2). The proportion of leaves in the diet, as assayed by feeding time, varies from 22.5 to 87.5% in African colobines and from 20.1 to 89% in Asian colobines (table 4.2). Fruits and seeds are a second important food source, composing 4.8 to 66.7% of the diet in African colobines and 6.1 to 56.9% in Asian colobines (table 4.2).

Fig. 4.3. Thomas's leaf monkey (*Presbytis thomasi*). These monkeys are a representative example of an Asian colobine. Field research on them has added to our understanding of colobine life history, social structure, infanticide, and dispersal. Photo courtesy of Perry van Duijnhoven.

Fig. 4.4. An adult male of one of the most endangered of Asia's primates, the Yunnan snub nosed monkey of Yunnan Province, People's Republic of China. Photo courtesy of Craig Kirkpatrick.

Colobines also eat flowers (0.2 to 30%; table 4.2), animals and their products (Hanuman langurs, Srivastava 1991; Thomas's leaf monkey, *Presbytis thomasi*, Sterck 1995), and soil (Black colobus, *Colobus satanus*, Oates 1994). In addition, Angola colobus (*Colobus angolensis*) and Yunnan snub-nosed monkeys have an unusual major food source: terrestrial vegetation or lichens can form as much as 37 to 87% of their diet (Kirkpatrick et al. 1998; Vedder & Fashing 2002; Fashing 2007; Kirkpatrick 2007).

Colobines are relatively small compared to nonprimates who display forestomach fermentation. Adult females weigh between 4.2 (olive colobus, *Procolobus verus*) and 14.8 kg (Hanuman langurs), and typically range between 5 and 10 kg (Smith & Jungers 1997). While forestomach fermentation permits animals to digest plant fiber, those that weigh less than 15 kg are unable to do so efficiently enough to meet their nutritional requirements and are considered "fiber-intolerant" (Wasserman & Chapman 2003). Consequently, colobine food selection appears to be affected by food quality, as measured by the ratio of protein to fiber in mature leaves (chapter 7, this volume). This measure is considered a good indication of the digestibility of young and mature leaves (Chapman et al. 2004), and is generally a good predictor of colobine biomass (Davies 1994; Chapman et al. 2004; Fashing et al. 2007a).

Predation

Colobine monkeys are preyed upon by carnivores (Seidensticker 1985; Zuberbuehler & Jenny 2002), snakes (Sterck 1997; Koenig et al. 1998), raptors (Struhsaker & Leakey 1990, Cui 2003) and chimpanzees (*Pan troglodytes*, Uehara 1997; Stanford 1998, 2002; Boesch & Boesch-Achermann 2000; Watts & Mitani 2002). The predatory behavior of

Table 4.2. Colobine diets. Feeding is shown as a percentage of feeding activity.

Species	Site	Remark	All leaves	YL	ML	Other l	Fruit+ seed	Fruit	Ripe fruit	Unripe fruit	Seed	Flowers	Animal matter	Lichen	other	References
Procolobus verus	Tiwai Island		68	57	11		16	2			14	9			7	Davies et al. 1999
Procolobus verus	Tiwai Island		74	58.9	1.9	13.2	18.6			2.1	14.4	7.3				Oates 1988
Procolobus tholloni*	Botsima, Salonga National Park	12 mo	60.7	54.3	6.4		37.9	7.1	7.1		30.8	1.4				Maisels et al. 1994
Procolobus badius	Taï National Park		43				33					23	1			Korstjens et al. 2002
Procolobus rufomitratus	Kibale National Park		86.45	69.85	9.45	7.15	7.75	7.75	5.7	2.05		2.15			3.35	Chapman & Pavelka 2005
Procolobus rufomitratus	in and near Kibale National Park and	average (N = 4)	87.5	69.9	12.2	5.4	7.2	7.2	4.4	2.8		2			3.3	Wasserman & Chapman 2003
Procolobus badius	Tiwai Island		52	32	20		31	6			25	16			1	Davies et al. 1999
Procolobus badius	Taï National Park		30.8	23.7	7	0.1	37.1		1.9	34.3		30.1	0		2	Wachter et al. 1997
Procolobus rufomitratus	Tana River: Baomo South		47.2	43.5	1.3	2.4	25.6					26.7			0.5	Decker 1994
Procolobus rufomitratus	Tana River: Mchelelo 1986–88		63.4	56.8	2.2	4.4	21.7					13.3			1.1	Decker 1994
Procolobus rufomitratus	Tana River: Mchelelo 1973–75		63.9	36	11.5	16.4	25					6.2			4.9	Decker 1994
Procolobus tholloni*	Botsima, Salonga National Park	8 mo	27.4	21	6.4		66.7	16.8	16.8		49.9	5.9				Maisels et al. 1994
Colobus guereza	Bole Valley		53.9				28.3					1.3			16.5	Dunbar 1987
Colobus guereza	Kibale National Park		86.35	81.05	4.7	0.6	4.8	4.8	1.2	3.6		1.1			2.4	Chapman & Pavelka 2005
Colobus guereza	In and near Kibale National Park	average (N = 4)	84.8	76.3	7.2	1.3	7.9	7.9	3.4	4.5		3			1.4	Wasserman & Chapman 2003
Colobus guereza	Kakamega Forest	Top 10 species	52.8	23.7	29.1		38.6	37.4			1.2	0.5			8.2	Fashing et al. 2007a
Colobus guereza	Kibale National Park		87.7	60.4	22.7	4.6	7.5	7.3			0.2	2.2		0.01	2.54	Harris & Chapman 2007
Colobus polykomos	Taï National Park		42				56					2	0.05			Korstjens et al. 2002
Colobus polykomos	Tiwai Island		57	30	27		36	3			33	3			4	Davies et al. 1999
Colobus satanus	Makandé		37.8	34.8	3		49.9	21.8	6	15.8	28.1	11.8			0.4	Brugiere et al. 2002
Colobus satanus	Lopé Reserve		26				64.2	4.1			60.1	5.3	2.6		2.1	Tutin et al. 1997
Colous vellerosus	Boabeng-Fiema Monkey Sanctuary		87.7	60.4	36.1	4.4	10.9	5.1			4.8	2.7			2.4	Teichroeb & Sicotte 2009

Species	Location													Reference
Semnopithecus entellus	Langtang National Park	56.8	10.2	25.2	21.4	21	11.4	7.3			6.8			15.4 *	Sayers & Norconk 2008
Semnopithecus entellus	Ramnagar	64.2	8.4	39.6	16.2	15.1	21				6.3	3.1		11.4	Podzuweit 1994
Semnopithecus entellus	Kanha	51.6	3.6	34.9	13.1	24.5	15.1				9.5	2.8		10.9	Newton 1992
Semnopithecus vetulus	Panadura and Piliyandala	30.55	19.1	5.9	5.55	56.9	53.1				5.8	3.8			Dela 2007
Semnopithecus johnii	Kakachi	62.2	31.2	26.8	4.2	25.1					9.3			3.4	Oates et al 1980
Presbytis femoralis	Perawang	28.75	26.3	2.45		57.65	19.6			35.7	0.35			13.25	Megantara 1989
Presbytis melalophos	Kuala Lompat	35.7	27.7	3.2	4.3	49.5	10			25.3	11.5	1.2		2.1	Bennett 1983
Presbytis rubicunda	Sepilok	37.6	36.5	1.1		49.3	19.2			30.1	11.1			1.1	Davies 1984
Presbytis thomasi	Ketambe	44.2	30.8		13.4	36.1					3.6			9.8	Sterck 1995
Trachypithecus poliocephalus	Fusui Precious Animal Reserve	89	75.2	10.5		6.1	5.7	3.8	1.9	0.4	2.7	6.4		2.2	Li & Rogers 2006
Trachypithecus auratus	Pangandaran Nature Reserve	49.2	46.1	0.7	3.3	33.1		11.4	20.8		15.1			2.7	Kool 1989; 1993
Trachypithecus francoisi	Nonggang Nature Reserve	56.9	38.9	13.9	4.1	31.4	17.2	1.2	16	14.2	7.5			4.3	Zhou et al. 2006
Trachypithecus pileata	Madhupur	66.8	10.9	42		24.2	15.1			9.3	7	1.6			Stanford 1991c
Nasalis larvatus	Samunsan Wildlife Sanctuary	41	38	3		50	35	15			3			6	Bennett & Sebastian 1988
Nasalis larvatus	Tanjung Puting National Park	51.9				40.3					3	<1		4.7	Yeager 1989
Nasalis larvatus	Menanggul River	65.93	65.9	0.03		25.6					7.7			0.5	Matsuda et al. 2009
Rhinopithecus bieti	Tacheng, Yunnan												60		Ding & Zhao 2004
Rhinopithecus bieti	Samage; Baimaxueshan	20.05	12.4	4.1	3.55	11.55					0.15		66.85	1.65	Grueter et al. 2009

* *Procolobus* of "central assemblage," species identity unclear (Brandon-Jones et al. 2003)
All leaves, as given by author or sum of yl, ml, and ol;
YL: Young leaves
ML: Mature leaves
OL: Other leaf matter or leaves of undetermined age, including petioles
Fruit + seed: As given by author or sum of fruit and seed
Fruit: As given by author or sum of ripe fruit and unripe fruit

Fig. 4.5. An adult male chimpanzee eating a red colobus monkey it has captured at Ngogo, Uganda. Current data suggest that chimpanzee predation has contributed to the decline of the red colobus monkey population at this site (Teelen 2008). Photo courtesy of Kevin Langergraber.

chimpanzees has been studied extensively. Chimpanzee hunting appears to have had a significant impact on some colobine populations (fig. 4.5), leading to severe reductions in their population densities in some cases (Eastern red colobus, *Procolobus rufomitratus*, Teelen 2008; guereza, *Colobus guereza*, Krueger et al. 1998). African colobines may be exposed to higher predation risk than their Asian counterparts because more species prey on them; some predators strike from long distances (e.g., raptors, van Schaik & Hörstermann 1994), and some, such as chimpanzees, kill several individuals during a single hunting attempt (Stanford 1998, 2002; Boesch & Boesch-Achermann 2000; Watts & Mitani 2002).

Because primates decrease their vulnerability to predation by living in groups, individuals might benefit by doing so (chapters 8 and 9, this volume). Nevertheless, red colobus monkeys that live in large groups are hunted more often by chimpanzees and leopards than those that live with fewer conspecifics (Stanford 1998; Zuberbuehler & Jenny 2002). Two factors other than group size may influ-

ence colobine predation risk. First, African colobines may form polyspecific associations to reduce their vulnerability to predation. By associating with members of other primate species, individuals may be able to dilute their chances of being preyed upon (Olive colobus, Korstjens 2001), increase the probability of detecting predators (Western red colobus, *Procolobus badius*, Noë & Bshary 1997; Bshary & Noë 1997), and improve their ability to deter predatory attacks (Bshary & Noë 1997). Second, several observations are consistent with the hypothesis that adult males may protect group members from predation (Struhsaker & Leland 1979; Cui 2003). African colobines form multimale groups more often than do Asian langurs; one possible reason is that the former live in areas that expose them to relatively high predation risk (see above; van Schaik & Hörstermann 1994). Intraspecies variation in predation risk has also been linked to the number of males. Hanuman langurs form groups containing a relatively large number of males in areas where they are exposed to high predation risk (Treves & Chapman 1996), and large groups of male Eastern red colobus monkeys are able to deter predatory attacks by chimpanzees more successfully than are smaller groups of males (Stanford 1998, 2002).

Life History

Colobine females typically give birth to one offspring at a time. In many species the interbirth interval is about two years (table 4.3). Infant survival and food availability have been shown to influence birth intervals, which are longer when previous infants have survived than when they have died (table 4.3). Hanuman langur mothers living in provisioned populations have shorter birth intervals (16.7 months) than do females in unprovisioned groups (28.8 months, Borries et al. 2001). Similarly, mothers wean their infants relatively early in situations where food availability is high (white-headed black leaf monkey, *Trachypithecus poliocephalus*, Zhao et al. 2008b; silver leaf monkey, *Trachypithecus villosus*, Shelmidine et al. 2009), and when they receive help from allomothers (Mitani & Watts 1997). Allomothering, infant care by individuals other than the mother, is common in colobine monkeys (McKenna 1979; Newton & Dunbar 1994). In addition to improving maternal reproduction, allomothering has important fitness benefits and costs for mothers, allomothers, and infants (Stanford 1992; Mitani & Watts 1997; chapter 11, this volume). While early studies of allomothering in primates focused on colobines (e.g., Hrdy 1977), the behavior has now been investigated more widely in other taxa such as cercopithecine monkeys (chapter 5, this volume).

Female age at first reproduction also varies with the food supply and habitat, ranging from 2.9 years in captivity (Silver leaf monkey, Shelmidine et al 2009) and 3.6 years in provisioned groups (Hanuman langur, Borries et al. 2001) to about 5 to 7 years in populations occupying undisturbed habitats (table 4.3). Available data suggest that male colobines die earlier than do females (Hanuman langur, Rajpurohit et al. 1995; Thomas's leaf monkey [oldest known male, 13 years old; oldest known female, 22 years old], Wich et al. 2007).

Life history traits such as age at first reproduction, birth interval, and lifespan appear to differ between folivorous colobines and frugivorous primates of similar size. Colobines give birth for the first time at a younger age, continue to give birth more often, and live shorter lives than do frugivores. All of this suggests that colobines may possess relatively fast life histories (Leigh 1994; Wich et al. 2007). Early maturation and the associated fast life history may in turn represent an adaptation to a folivorous diet or, alternatively, may be a consequence of high protein intake (Leigh 1994; chapter 10, this volume). Additional long-term studies will be required to estimate life history traits more accurately and to test these hypotheses.

Evolution of Social Systems

Colobines differ from cercopithecines in their social organization. They live in relatively small home ranges and small groups (Clutton-Brock & Harvey 1977). While cercopithecine females form coalitions against other females in their group (chapter 21, this volume), this behavior has not been described in colobines. These differences may be due to phylogenetic inertia (Di Fiore & Rendall 1994; Chapman & Rothman 2009), to the type of food competition colobines experience (van Schaik 1989), or to the effects of grouping, such as a group-size-dependent infanticide risk (Janson & Goldsmith 1995). Despite these generalizations, colobine social relationships are variable, and this diversity is only partly understood.

Social Organization

Colobine groups vary in size and composition (table 4.4). They range from pairs with their offspring to one-male, multifemale groups to very large (> 300 individuals) multimale, multifemale groups. Dispersal patterns include male dispersal with female philopatry, female dispersal with male philopatry, and dispersal by both sexes (table 4.5). Dispersal affects the relatedness of group members (chapter 12, this volume), a factor that in turn influences which member forms social bonds with which and cooperates with which (chapter 21, this volume).

One-Male, One-Female Groups

Pair living is found in two colobines endemic to the Mentawai Islands in Southeast Asia (table 4.4). The Mentawai surili (*Presbytis potenziani*) lives in pairs (Watanabe 1981), although one-male, multifemale groups have also been recorded (Sangchantr 2003). Pair living also occurs in approximately 60% of groups in the other Mentawai colobine, the pig-tailed snub-nosed monkey (*Simias concolor*); the other 40% of groups contain one male and a few females (Tenaza & Fuentes 1995). These observations differ from those made at another, relatively undisturbed site where groups consisted of one male and multiple females (Hadi et al. 2009).

Compared with other Old World monkeys, pair living in Mentawai Island colobines is unusual and remains unexplained. Human activities, including intensive hunting and habitat degradation, have been implicated (Watanabe

Table 4.3. Colobine interbirth interval, female age of first reproduction, and infant survival

Species	Site	IBI (months)			Female age at first reproduction (year)	Infant survival to 1 year	Death due to infanticide	References
		Average	After infant death	after infant survival				
Procolobus verus	Taï Forest	19.2						Korstjens & Schippers 2003
Procolobus rufomitratus	Tana River	25.3	14–17 (N = 2)	27 (N = 2)				Marsh 1979a
Procolobus rufomitratus	Kibale	24–27	9–18			83% (N = 58)	30% (N = 10)	Struhsaker & Leland 1985
Procolobus rufomitratus	Kibale	24.4		27.5		Survival to 24 mo: 68%		Struhsaker & Pope 1991
Procolobus badius	Abuko			29.4	4.2	79% (N = 28)		Starin 1994; 2001
Colobus guereza	Bole					63% (N = 8)		Dunbar 1987
Colobus guereza	Kanyawara, Kibale	21.5	6.1 (N = 1)	21.8		46% (N = 26)		Harris & Montfort 2006
Colobus vellerosus	Boabeng-Fiema		10				71% (N = 14)	Teichroeb & Sicotte 2008
Semnopithecus entellus	Ramnagar	28	13.2 (shortest)	21.8 (shortest)	6.7	Survival to 24 mo: 50% (N = 52)	31% (N = 26)	Borries et al. 2001; Borries 1997
Semnopithecus entellus	Jodhpur		11.1	15.9				Agoramoorthy & Mohnot 1988
Semnopithecus entellus	Jodhpur	16.7	15.4	17.2	3.5	During male takeover: 59% (N = 29)		Sommer et al. 1992
Semnopithecus vetulus	Polonnaruwa					79% (N = 92)		Rudran 1973b
Semnopithecus vetulus	Horton Plains					96% (N = 49)		Rudran 1973b
Presbytis thomasi	Ketambe	22	17.7	26.8	5.4	55% (N = 111)		Wich et al. 2007
Trachypithecus villosus	Kuala Selangor					70% (N = 40)		Wolf 1984
Trachypithecus poliocephalus	Nongguan			21.3–24.5		85% (N = 53)		Zhao et al. 2008b
Trachypithecus obscurus phayre	Phu Khieo			23.2				Borries et al. 2008
Rhinopithecus bieti	Kunming Institute of Zoology; Kumming Zoo	20.5	14.1			In captivity: 70%; in wild 40%		Cui et al. 2006; Kirkpatrick et al. 1998
Rhinopithecus roxellana	Yuhuangmiao	21.9	11.6	23.3	5 to 6	78% (N = 67)		Qi et al. 2008

1981; Hadi et al. 2009). Pig-tailed snub-nosed monkeys in an area minimally affected by humans live in one-male, multifemale groups (Hadi et al. 2009), a finding consistent with this interpretation. Additional observations will be required to clarify the social organization of these elusive colobines and its causes.

One-Male, Multifemale Groups

Most colobines live in "harem" groups, consisting of one male and several females (table 4.4). Such groups may represent the ancestral group type in Asian langurs (Grueter & van Schaik 2010). It is the typical grouping pattern in African *Colobus* and Asian *Presbytis* and *Trachypithecus* (Moore 1999). Asian *Semnopithecus* are also commonly found in these groups (ibid.). One-male groups occur when a single male is able to monopolize several females (chapter 9, this volume). Such males are expected to father most if not all of the offspring. Behavioral observations are consistent with this interpretation, and the results of paternity analyses utilizing genetic data also support it (Launhardt et al. 2001). Some colobine populations consist of both one-male and multimale groups. The latter may represent age-graded groups, where multiple males are present due to the maturation of natal males (Sterck & van Hooff 2000). Alternatively, males may not be able to prevent additional males joining the group, and multi-male groups form via male immigration (Moore 1999; Borries & Koenig 2000).

In a review of colobine social organization, Newton and Dunbar (1994) considered one-male colobine groups to consist of closely bonded matrilineal relatives. The implication is that females do not disperse and remain in the same group with their relatives. The situation is more complex than this. While males always disperse (Sterck & Korstjens 2000), females can be philopatric or disperse (table 4.5). Dispersal, however, does not preclude females residing with relatives because migrating females often join groups with familiar individuals and kin (Sterck et al. 2005; Teichroeb et al. 2009). In sum, two types of social dynamics can be distinguished in one-male, multi-female colobine groups. In one type, only males disperse and females are philopatric. In another, members of both sexes disperse. This distinction does not represent a strict dichotomy as patterns of dispersal intermediate between the two types are possible (Sterck & Korstjens 2000).

Multimale, Multifemale Groups

Multimale, multifemale groups are found in *Procolobus*, *Nasalis*, *Rhinopithecus*, and in some *Colobus* and Hanuman langurs (table 4.4). Despite their superficial similarity, the composition of these groups varies. Moreover, the groups form via different patterns of dispersal.

Proboscis monkeys (*Nasalis larvatus*) and snub-nosed monkeys (*Rhinopithecus* spp.) form large aggregations that consist of several one-male units (Bennett & Sebastian 1988; Yeager 1991; Tan et al. 2007). These one-male units congregate to form larger bands. Males and females disperse in these species (Bennett & Sebastian 1988; Zhao et al. 2008a). In golden snub-nosed monkeys (*Rhinopithecus roxellana*), males leave their natal one-male unit when they mature, while females disperse between one-male units in the same or a different band (Zhao et al. 2008a; table 4.5). A similar hierarchical social organization has also been suggested to account for the extremely large groups that sometimes exceed 300 individuals in Angola colobus (Fashing et al. 2007b). The hierarchical social organization displayed by some colobines is similar to that shown by hamadryas baboons (*Papio hamadryas*) and geladas (*Theropithecus gelada*, chapter 5, this volume). These groups may form to reduce the threat of male takeover by bachelor males when food sources do not limit group size (Grueter & van Schaik 2010). The latter may apply in species that exploit slowly regenerating resources, such as lichens for the Yunnan snub-nosed monkey (Kirkpatrick et al. 1998) or terrestrial vegetation for such species as the Angola colobus (Fashing et al. 2007b).

Multimale, multifemale groups occur in several colobine species apart from the hierarchical societies described above. In these, individuals live in groups that are relatively stable over time. Group composition nevertheless changes as a function of dispersal, which itself is quite variable (table 4.5). In multimale, multifemale groups of Hanuman langurs, males disperse and females are philopatric (Koenig 2000). In contrast, multimale, multifemale red colobus groups contain philopatric males and females who disperse while immature (Struhsaker & Leland 1979; Starin 1994). Male dispersal has occasionally been described to their natal groups (Western red colobus, Starin 1994). Finally, members of both sexes disperse in multimale, multifemale olive colobus (Korstjens & Schippers 2003) and white-thighed colobus (*Colobus vellerosus*, Saj et al. 2007; Teichroeb & Sicotte 2009) groups. The causes of these observed differences in dispersal patterns remain unclear (see below).

Sexual Strategies

Male monopolization of females

For males, residency in a mixed-sex group ensures reproduction as offspring are often fathered by resident males (e.g., Hanuman langurs, Launhardt et al. 2001). Mating between females and extragroup males is observed only rarely

Table 4.4. Colobine group size, composition, sex ratio and adult/immature ratio

Species	Site	N (groups)	Group size	Adult females	Subadult females	Adult males	subadult males	Immatures[a]	Adult sex ratio (F/M)	(Sub)adult sex ratio (F/M)	Adult/subadult +immature ratio	(Sub)adult/ immature ratio	References
Procolobus verus	Taï NP	5	6.8	2.8	0	1.6	0	2.4	1.75		1.83		Korstjens & Schippers 2003
Procolobus badius	Taï NP	4	52.3	18.3	2.5	10.5	2.5	18.5	1.74	1.60	1.22	1.82	Korstjens 2001
Procolobus badius	Taï Reserve	5		8.4		3.8			2.21				Struhsaker 1975
Procolobus kirkii	Zanzibar, Jozani	5	32.0	10.7		2.7		18.6	4.03		0.72		Siex & Struhsaker 1999
Procolobus kirkii	Zanzibar, Shamba	8	23.8	9.7		1.6		12.5	5.95		0.91		Siex & Struhsaker 1999
Procolobus pennantii	Korup Reserve	5		7.6		3.4			2.24				Struhsaker 1975
Procolobus pennantii	Tana River 1987	9	10.9	5.3	0	1.0	0	4.6	5.33		1.39	1.83	Decker & Kinnaird 1992
Procolobus rufomitratus	Tana River 1973–75	13	18.1	9.7	0.5	1.5	0.1	6.4	6.63	6.60	1.61		Marsh 1979a
Procolobus rufomitratus	Tana River: Reserve	9	11.2	5.6		1.1		4.5	5.03		1.47		Decker 1994
Procolobus rufomitratus	Tana River: Reserve Area 1987–1988												Decker 1994
Procolobus rufomitratus	Tana River: Wema Area 1991–1992	11	11.4	5.6		0.9		4.8	6.20		1.36		Decker 1994
Procolobus tephrosceles[b]	Gombe	5	23.0	11.2		6.0		5.8	1.87		2.97		Stanford 1998
Procolobus rufomitratus	Kibale Forest	8		9.3		3.6			2.55				Struhsaker 1975
Procolobus rufomitratus	Kanyawara, Kibale	5	50.4	25.5c		6.8c		18.1		3.75		1.78	Treves 1998a
Procolobus gordonorum	Mwanihana	12	36.1	15.8	1.4	4.0	0.8	14.2	3.90	3.58	1.21	1.55	Struhsaker et al. 2004
Procolobus gordonorum	Kalunga	5	17.5	8.1	1.1	3.2	0	5.1	2.53	2.88	1.82	2.43	Struhsaker et al. 2004
Procolobus gordonorum	Magombera 1977	2	28.0	10.0	0.5	4.0	0	13.5	2.50	2.63	1.00	1.07	Struhsaker et al. 2004
Procolobus gordonorum	Magombera 1992	4	33.3	16.0	0.0	2.0		15.3	8.00	8.00	1.18	1.18	Struhsaker et al. 2004
Procolobus rufomitratus	Kibale National Park	9	65.2										Snaith & Chapman 2008
Colobus angolensis	Mazumbai	3	4	1.7		1.3		1	1.31		3		Groves 1973
Colobus angolensis	Nyungwe Forest	1	>300							1.03		3.94	Fashing et al. 2007b
Colobus guereza	Bole Valley	12	7.0	2.1	0.8	1.6	1.2	1.4	1.32	1.31	1.10	1.76	Dunbar 1987
Colobus guereza	Lake Shalla	6	7.8	2.0	0.8	1.0	1.2	2.8	2.00		0.62		Dunbar 1987
Colobus guereza	Arusha NP	4	5.3	2.0		1.5		1.8	1.33	2.37	2.00	4	Groves 1973
Colobus guereza	Budongo Forest	9	7.9	3.6	0.9	1.3	0.6	1.6	2.77		1.58		Marler 1969
Colobus guereza	Queen Elizabeth NP	3	5.3	2.3	0.3	1.3	0.7	8.6	1.77	1.3	0.38	0.53	Marler 1969
Colobus guereza	Kanyawara	5	11.8	3.6		2.0	0.6	6.2	1.80	1.28	0.90	1.06	Oates 1977
Colobus guereza	Kakamega Forest	5	12.8	3.7	3.7	2.3		6.2	1.61		0.88		Fashing 2001
Colobus polykomos	Taï NP	10	16.2	4.7		1.2	0.5	6.1	3.92	4.94	0.57	1.66	Korstjens 2001
Colobus satanus	Makandé	19	11.8	7.1	1.3	3.6	3.0	5.1	1.97	1.27	1.14	2.94	Brugière et al. 2002
Colobus vellerosus	Boabeng-Fiema Monkey Sanctuary	4	20.1	7.0c	1.3	3.0c	3.0			2.33			Teichroeb & Sicotte 2009
Colobus vellerosus	Boabeng-Fiema Monkey Sanctuary	2	20.0					10.0				1.00	Saj & Sicotte 2007
Presbytis comata	Patengang	5	5.8	1.4		1.2		3.2	1.17		0.81		Ruhiyat 1983
Presbytis comata	Kamojang	4	8.0	3.0		1.0		4.0	3.00		1.00		Ruhiyat 1983
Presbytis femoralis	Perawang	7	11.1	5.7	0.6	1.0	1.3	2.6	5.71	2.75	1.52	3.33	Megantara 1989
Presbytis melalophos	Kuala Lompat 1969–71	4	15.0	5.8		2.3		7.0	2.56		1.14		Curtin & Chivers 1978
Presbytis melalophos	Kuala Lompat 1981	3	15.0	7.7		1.0		6.3	7.67		1.37		Bennett 1983
Presbytis potenziani	Sarabua	6	3.2	1.0		1.0		1.2	1.00		1.71		Watanabe 1981
Presbytis potenziani	Grukna	10	3.4	1.0		1.0		1.4	1.00		1.43		Watanabe 1981
Presbytis rubicunda	Tanjung Puting	9	6.1	2.6		1.0		2.6	2.56		1.39		Supriatna et al. 1986
Presbytis thomasi	Bohorok, area 1	9	9.7	3.7	0.6	1.2	0.9	3.3	3.00	2.00	1.02	1.90	Gurmaya 1986
Presbytis thomasi	Bohorok, area 2	8	7.5	3.5	0.4	1.3	0.3	2.1	2.80	2.58	1.73	2.53	Gurmaya 1986

Taxon	Locality	N											Source
Presbytis thomasi	Bohorok, area 3	6	6.2	3.5	0.3	1.0[c]	0.2	1.2	3.50	3.29	2.70	4.29	Gurmaya 1986
Presbytis thomasi	Ketambe	7	8.9	3.6	0.9	1.0[c]	0.6	2.9	3.57	2.82	1.07	2.10	Assink & van Dijk 1990
Semnopithecus entellus	Mundanthurai Wildlife Sanctuary	3	38.3	7.7	5.3	2.7	1.0	21.7	2.88	3.55	0.37	0.77	Hohmann 1989
Semnopithecus entellus	Abu	3	16.0	7.3	1.3	1.0	1.0	5.3	7.33	4.3	1.09	2	Hrdy 1974
Semnopithecus entellus	Orcha	3	18.7	6.0	0.3	3.7	0.3	7.3	1.64	1.83	1.07	1.55	Jay 1965
Semnopithecus entellus	Kanha Meadow 1981	14	21.7	9.1	2.4	1.1	1.2	7.9	7.94	4.88	0.89	1.76	Newton 1987
Semnopithecus entellus	Kanha Meadow 1990	8	19.9	10.5	1.9	1.0	0.6	5.9	10.50	7.62	1.37	2.38	Newton 1994
Semnopithecus entellus	Deotalao	4	22.8	10.5		1.0		11.3	10.50		1.02		Newton 1987
Semnopithecus entellus	Jaipur town	8	32.4	16.3		1.0		15.1	16.25		1.14		Mathur & Manohar 1992
Semnopithecus entellus	Ambaragh RF	5	80.8	35.2		1.2		44.4	29.33		0.82		Mathur & Manohar 1992
Semnopithecus entellus	Mudanthurai WS	4	37.0	14.3	1.5	5.3	0.8	15.3	2.71	2.63	1.11	1.43	Ross 1993
Semnopithecus entellus	Dharwar 1961	37	14.9	8.1		1.3	0.3	5.3	6.21		1.67	1.84	Sugiyama 1964
Semnopithecus entellus	Dharwar 1976	18	16.6	7.6	1.2	0.9	0.2	6.7	8.00	7.85	1.06	1.46	Sugiyama & Parthasarathy 1979
Semnopithecus entellus	Jodhpur	3	21.3	10.3	0.3	1.0	[c]	9.7	10.33	10.67	1.13	1.21	Sommer 1985
Semnopithecus entellus	Simla	5	47.6	14.0	2.0	3.6	1.6	26.4	3.89	3.08	0.59	0.80	Sugiyama 1976
Semnopithecus entellus	Ramnagar	8		7.0		3.8							Borries 2000
Semnopithecus entellus	Polonnaruwa	4	24.8	8.6	3.8	2.6		7.4	3.31				Ripley 1967
Semnopithecus entellus	Southern India	94	18.6	3.4	1.2	1.2	0.4	5.2	2.83		1.51	1.19	Vasudev et al. 2008
Semnopithecus entellus	Mundanthurai Wildlife Sanctuary	5	11.4			1.2				2.88	0.68		Hohmann 1989
Semnopithecus johnii	Ootacamund	7	9.7	3.1		1.3		5.3	2.44		0.84		Poirier 1969
Trachypithecus villosus	Kuala Selangor 1965–66	4	30.6	15.6		1.3	0.6	13.1	12.50		1.23	1.33	Bernstein 1968
Trachypithecus villosus	Kuala Selangor 1977	5	30.6	14.0		1.8	0.4	14.4	7.78		1.07	1.13	Wolf 1984
Trachypithecus auratus	Pangandaran TW	6	13.5	6.8	1.3[d]	1.2	0.3[d]	3.3	5.86		0.98		Kool 1989
Trachypithecus geei	Jamduar	7	12.7	5.1		1.1	0.6	5.9	4.50		0.87	1.17	Mukherjee & Saha 1974
Trachypithecus obscurus	Kuala Lompat	5	14.2	5.0		1.6		7.6	3.13		0.82		Curtin & Chivers 1978
Trachypithecus obscurus phayrei	Phu Khieo	3	17.7	6.3		1.7		9.7	3.71				Borries et al. 2008
Trachypithecus pileatus	Madhupur NP	9	8.2	3.6		1.0		3.7	3.56		1.24		Stanford 1991a
Trachypithecus pileatus	Madhupur Forest	5	9.4	2.8		1.0	0.4	5.2	2.80		0.68	0.81	Green 1981
Nasalis larvatus	Tanjung Puting NP	7	12.4	4.7		1.0		6.7	4.71		0.85		Yeager 1990
Nasalis larvatus	Samunsam Wildlife Sanctuary	6	9.0	3.7	0.2	1.0	0.2	4.0	3.67	3.29	1.08	1.25	Bennett & Sebastian 1988
Simias concolor	Sirimuri River	4	3.8	1.0		1.0		1.8	1.00		1.14		Tilson 1977
Simias concolor	Betumonga	6	5.5	2.5		1.2		1.8	2.14		2.00		Tenaza & Fuentes 1995
Simias concolor	Sinakak	4	5.0	2.8		1.0		1.3	2.75		3.00		Tenaza & Fuentes 1995
Simias concolor	Simalegu	7	3.1	1.3		1.0		0.9	1.29		2.67		Tenaza & Fuentes 1995
Simias concolor	Loh Bajou	3	8.7	3.3	0.3	1.0	0.7	3.3	3.33		1.00	1.60	Hadi et al. 2009
Rhinopithecus bieti	Wuyapiya[e]	1	175–200	3.60		1.0			3.60				Kirkpatrick et al. 1998
Rhinopithecus bieti	Zhouzhi National Nature Reserve[e]	11	10.5	3.9	1.3	1.0		4.3	3.90	5.2	0.88	1.44	Zhao et al. 2008a
Rhinopithecus roxellana	Qinling Montains[e]	8		3.4	1.0	1.0			3.38				Zhang et al. 2008

a. Immatures, excluding subadults
b. Procolobus of "central assemblage," species identity unclear (Brandon-Jones et al. 2003)
c. Includes adults and subadults
d. N = 3 groups with all subadults sexed
e. One-male unit size provided

Table 4.5. Female dispersal and male philopatry in colobines

Species	Site	Incidence of female dispersal	Nulliparous females (N)	Total parous females (N)	Females parous without infant (N)	Females parous with infant (N)	Female eviction	Female aggression against immigrants	Male philopatry	Male dispersal	Solitary males	All-male bands	References
Procolobus verus	Tai National Park	Common	Most likely yes	8	7	1			No	MI	Yes		Korstjens & Schippers 2003
Procolobus verus	Tiwai		0	1 (+6)	1	0					Yes		Oates 1994
Procolobus badius	Tai National Park	Occurs	3	5	4	1			Yes?	Return to natal group	Yes		Korstjens 2001; Korstjens et al. 2002
Procolobus badius	Abuko N. R.	Common	23				No	Mild	Yes		Yes		Starin 1991, 1994
Procolobus rufomitratus	Kibale	Common	Several	1		0	No	No	Yes	Attempt at MI	Yes	Yes	Struhsaker & Leland 1979; 1985; Struhsaker & Pope 1991
Procolobus rufomitratus	Tana	Common	3	10	10	0			No	MTO	Yes	Yes	Marsh 1979 a,b
Colobus guereza	Kanyawara, Kibale National Park	Occurs, according to genetic data	0	0					No	MI; MTO	Yes	Yes	Oates 1977; Struhsaker & Leland 1979; Harris et al. 2009
Colobus guereza	Bole		0	1	1	0			No?	MI			Dunbar & Dunbar 1974; Dunbar 1987
Colobus polykomos	Tai National Park	Occurs	3	2	1	1		Yes					Korstjens 2001; Korstjens et al. 2002
Colobus polykomos	Tiwai	Occurs	0	1		1							Oates 1994
Colobus satanus	Lopé	Occurs	7	3	2	1	Yes and no		No	MI; FSM	Yes	Yes	Oates 1994
Colous vellerosus	Boabeng-Fiema Monkey Sanctuary		7	5	5	0					Yes	Yes	Saj & Sicotte 2005; Teichroeb et al. 2009
Presbytis femoralis	Perawang		0	4[a]	4[a]	0		Mild	No?	MTO	Yes	Yes	Megantara 1989
Presbytis melalophos	Kuala Lompat		0	1	0	1		No	No?	ME	Yes	Yes?	Bennett 1983
Presbytis rubicunda	Sepilok		2[b]	0	0	0				FSM?		Yes	Davies 1984
Presbytis thomasi	Ketambe	Common	7	23	21	2	No	Mild	No	MTO; FSM	Yes	Yes	Sterck 1997; et al. 2005; Steenbeek 1999; et al. 2000
Semnopithecus entellus	Dharwar	Rare	1 (some)	Some		3[c]			No	MTO	Yes	Yes	Sugiyama 1964; 1965; Yoshiba 1968
Semnopithecus entellus	Jaipur	Rare	3	3	0	1		No	No	MTO		Yes	Mathur & Manohar 1992
Semnopithecus entellus	Abu	Occurs	0	3		1			No	MTO; FSM	Yes	Yes	Hrdy 1974; 1977
Semnopithecus entellus	Kanha	Rare	1	6	5	1			No	MTO	Possibly	Yes	Newton 1987
Semnopithecus entellus	Jodhpur		1	0	0		No		No	MTO	Yes	Yes	Sommer 1987; Sommer & Rajurohit 1989; Sommer et al. 1992; Rajurohit et al. 1995
Semnopithecus entellus	Junbesi	Rare	2	0				(No)	No	MTO or MI	Yes	Yes	Boggess 1980
Semnopithecus entellus	Ramnagar	Rare	0	4	3	1		No	No	MI	Yes	Yes	Borries 1997; Borries & Koenig 2000
Semnopithecus johnii	Ootacamund		1	2	1	1			No	MTO?; FSM?	Yes	Yes	Poirier 1969; 1970
Semnopithecus vetulus	Polonnaruwa		1	5	6	1[d]			No	MTO	Yes	Yes	Rudran 1973a
Trachypithecus villosus	Kuala Selangor	Occurs	0	1(+4?)	6	0			No	MTO	Yes	Yes	Wolf 1984; Wolf & Fleagle 1977
Trachypithecus obscurus phayrei	Phu Khieo Wildlife Sanctuary	Occurs	0		6	0				None?			Borries et al. 2004
Trachypithecus pileata	Madhupur		0	6	6	0		No	No?	FSM	Yes	Yes	Stanford 1991a
Nasalis larvatus	Samunsam	Common	?	3	1	2	No	No	No		Yes	Yes	Bennett & Sebastian 1988
Rhinopithecus roxellana	Yuhuangmiao, Zhouzhi National Nature Reserve	Common	9	34	25	9			No	MTO; FSM	Yes	Yes	Qi et al. 2009; Zhao et al. 2008a; Zhang et al. 2008

a. One female really transferred; three females only stayed for short time with new group.

b. Dispersal of these females was likely.

c. Three females ranged with ousted male, while the other females resided with male that took over group, leading to a splitting of group. These three females later joined their former group.

d. Female with infant left group with ousted male.

e. Females dispersed between one-male units (OMU) within (N = 11) and between (N = 15) band.

Incidence of female dispersal: Common: most or all females disperse at least once; Occurs: significant proportion of the females disperses at least once; Rare: female philopatry is the norm, but occasional female dispersal has been reported.

Male dispersal: MI: male immigration into bisexual group; ME: male emigration from bisexual group; MTO: Male Take-Over, male immigrant ousts resident alpha-male; FSM: Female Split Merger: extra-group male forms group with and attracts transferring females.

(Hanuman langurs, Rajpurohit et al. 1995; Launhardt et al. 2001; olive colobus, Korstjens & Noë 2004). Males may nonetheless father offspring via surreptitious matings, and how often males mate with females may not correspond to how often they reproduce. In multimale Hanuman langur groups, for example, resident males father most offspring, but more than 20% of all infants are sired by extragroup males (Launhardt et al. 2001).

Male primates typically compete for fertilizations, and their success depends on their ability to monopolize a group of females. This, in turn, varies as a function of the temporal and spatial distribution of females (Emlen & Oring 1977). Single males are able to exclude others and form one-male groups in Hanuman langurs when groups contain few females, who mate over extended periods and have nonseasonal births (Srivastava & Dunbar 1996; Treves & Chapman 1996; Nunn 1999; chapter 9, this volume). In contrast, multimale groups form when groups contain many females, who mate during brief breeding seasons and have concentrated birth seasons (ibid.). Similarly, small female group sizes and nonseasonal breeding in *Trachypithecus* and *Presbytis* frequently lead to groups containing only one male. Some groups with multiple males appear to be age-graded, with surplus adult males still residing in their natal group (Sterck & van Hooff 2000). In multimale groups, alpha males have preferential mating access to females, but cannot always monopolize them (red colobus, Struhsaker & Leland 1979; Struhsaker & Pope 1991; Starin 1994; Hanuman langur, Borries & Koenig 2000; Launhardt et al. 2001; guereza, Harris & Monfort 2003). Population density also affects group composition and the ability of males to monopolize females (Srivastava & Dunbar 1996; Moore 1999). Intruder pressure is high in low- and high-density populations and leads to multimale groups; one-male groups form at intermediate densities.

Studies thus far reveal how males reproduce during only parts of their lives. Much less is known about their reproductive careers (chapter 18, this volume). In populations with one-male groups, males may have only one chance to achieve residency, having their best chance to do so and to reproduce as young adults (Hanuman langur, Sommer & Rajpurohit 1989; Thomas's leaf monkey, Sterck et al. 2005). In populations with multimale groups, males can disperse more than once, reside in different groups, and potentially reproduce in them (Hanuman langur, Borries & Koenig 2000; olive colobus, Korstjens & Schippers 2003). In a Hanuman langur population with both one-male and multimale groups, males who live in one-male groups nevertheless appear to obtain a distinct reproductive advantage as they produce almost three times as many offspring as alpha males in multimale groups (Launhardt et al. 2001). Whether males in multimale groups compensate by surviv-ing longer is unclear. More long-term data from multiple species will be required to address this and other questions concerning male lifetime reproductive success.

Male behavior towards extra-group males

Males who live in one-male groups are aggressive to extragroup males because they are reproductive competitors. In addition, males may temporarily leave their groups to harass neighboring conspecific groups (Thomas's leaf monkey, Steenbeek et al. 2000; king colobus, *Colobus polykomos*, Korstjens et al. 2002). Males behave aggressively in these contexts to defend their female mates who live with them (female defense polygyny) or to defend resources crucial for female reproduction and survival (resource defence polygyny, Kitchen & Beehner 2007; chapter 9, this volume). Both occur in colobines.

Known individuals are likely to represent less threatening competitors than do strange males, and resident males react more strongly to the latter than to the former (capped leaf monkey, *Trachypithecus pileatus*, Stanford 1991a; Thomas's leaf monkey, Wich et al. 2002b; Wich & Sterck 2007). Differential male responsiveness in these cases is thought to reflect female mate defense (langurs, van Schaik et al. 1992; Thomas's leaf monkey, Wich et al. 2002a, b; king colobus, Korstjens et al. 2002).

Strange individuals who move over the central part of the home range of conspecifics threaten the latter's food supply more than do animals who only skirt the periphery. Thus, when males defend food, they are likely to respond more strongly to strange conspecifics in the core of their home range than at the periphery. Such responses have been elicited from males in several colobine species (Thomas's leaf monkey, Wich et al. 2002a; guereza, Fashing 2001; Harris 2006), and have been interpreted to represent resource defense by males (e.g., guereza, Fashing 2001; Harris 2006). Alternatively, location-specific responses to intruders may be due to the threat of infanticide posed by strange males (chapter 19, this volume), which are likely to appear at the center of the range rather than the edge (Wich et al. 2002a, 2004). Resource defense by males is also expected to result in heightened male aggression near food sources. This prediction has been validated in some species (e.g., guereza, Fashing 2001; Harris 2006), but not others (e.g., Thomas's leaf monkeys, Steenbeek 1999; Wich et al. 2002a; white-thighed colobus, Sicotte & MacIntosh 2004).

Fate of surplus males

Colobine males do not always reside in mixed-sex groups (table 4.5). Solitary males are found in many species, as males typically disperse from their natal groups when they are immature. Solitary males succumb to high rates of mortality. In many species, dispersing males form all-male bands (AMB). In African colobines these males associate with other primate species (Western red colobus, Starin 1994; olive colobus, Korstjens & Schippers 2003) or begin to follow mixed-sex groups as "satellite" males (guereza, Struhsaker & Leland 1979). AMBs form when new resident males chase immature and adult males out of their groups (Hanuman langur, Treves & Chapman 1996), or when males leave voluntarily and begin to associate with each other (capped leaf monkeys, Stanford 1991b). AMBs also result after females disperse from their groups and leave their male offspring with resident males (Thomas's leaf monkey, Steenbeek et al. 2000). Living in an AMB may be costly, since these groups are often attacked by conspecifics (Thomas's leaf monkey, Steenbeek et al. 2000) and mortality rates of individuals living in them are relatively high (Hanuman langur, Rajpurohit & Sommer 1991).

Solitary males and those living in AMBs achieve residency in mixed-sex groups by taking them over aggressively (Hanuman langur, Sugiyama 1965; guereza, Struhsaker & Leland 1979) or by attracting females and forming new groups (Marsh 1979b; Stanford 1991b; Sterck 1997; table 4.5). Male takeovers can be quick, lasting days, or extended, lasting months (Rajpurohit & Sommer 1993; Teichroeb & Sicotte 2008). Forming a group by recruiting females typically takes place over an extended period of time (Thomas's leaf monkey, Sterck 1997). AMB males may form coalitions to take over groups (Rajpurohit & Sommer 1993), but the level of coordination between individuals in these is not clear, as only one of them gains residency, after which he expels other members (Rajpurohit et al. 1995). The benefits obtained by expelled AMB males are equally unclear. They may obtain matings with females for a short period. Alternatively, they may gain inclusive fitness benefits by helping relatives achieve resident status. In sum, these observations suggest that males may prefer to remain in their natal groups but cannot do so because they are evicted or deserted. In addition, dispersing males reduce the costs of solitary life by associating with members of other species or with conspecifics in AMBs.

Infanticide by males and female counterstrategies

Hanuman langurs in India were the first primate species in which male infanticide was described (Sugiyama 1965) and experimentally investigated (Sugiyama 1966). Many reports of infanticide in this and other primates followed (chapter 19, this volume). Typically, male infanticide occurs after a new male becomes the dominant male in a group because he ousts the former resident male in a harem group (Jodhpur, Sommer 1987) or, more rarely, after he enters a multimale group (Ramnagar, Borries 1997). Females mate with males following infanticide.

After the first reports of infanticide, its occurrence and significance were questioned since the infanticide was not always observed and was often deduced only from circumstantial evidence (Boggess 1979; Sussman et al. 1995). Observations are necessarily rare, as acts do not occur often and only take a few seconds (e.g. Xiang & Grueter 2007). Over time more observations of infanticide have been compiled (chapter 19, this volume), and there is now growing realization that rare infanticidal events can have a large impact on the reproductive success of female primates that reproduce relatively slowly. In several populations, infanticide by males is a large source of infant mortality (e.g., Eastern red colobus, estimated 10%–30% of infant deaths, Struhsaker & Leland 1985; Hanuman langur, 31% of infant deaths, Borries & Koenig 2000; white-thighed colobus, 71% of infant deaths, Teichroeb & Sicotte 2008).

While infanticide is costly for mothers who lose their offspring, it is advantageous for male perpetrators. Hrdy (1974) was the first to propose that infanticide represents a sexually selected male reproductive tactic in primates. Several predictions follow directly from this hypothesis: infanticidal males do not kill their own infants, infanticide reduces the birth intervals of mothers whose infants are killed, and infanticidal males subsequently father offspring born to mothers who are victims. A heated debate ensued with some authors proposing nonadaptive explanations for infanticide. For example, infants might be killed as a by-product of male-male aggression or pathological male behavior (Sussman et al. 1995; reviewed in Hrdy et al. 1995). Subsequent behavioral (van Schaik 2000) and genetic data (Borries et al. 1999a) have largely corroborated predictions of the sexual selection hypothesis, and current research focuses on variation in patterns of infanticide and countertactics females employ to decrease their vulnerability (chapter 19, this volume).

Infanticide usually occurs after a male has taken over a one-male group and becomes the sole resident. By committing infanticide, males improve their ability to reproduce (Hrdy 1974). The conditions that make infanticide beneficial for males also prevail in multimale groups, and it has also been reported in these situations (Borries 1997; Borries et al. 1999a, 1999b; chapter 19, this volume). Male colobines have also been observed to commit infanticide before taking over new groups; this strategy appears to induce females to transfer or to form new groups with infanticidal males (Sterck 1997, 1998; Sterck & Korstjens 2000).

Because infanticide can have a substantial effect on their reproductive success, females are expected to evolve countertactics to decrease their vulnerability (chapter 19, this volume). Several hypotheses have been proposed: Females solicit matings from males outside the time they ovulate, even during pregnancy (Hrdy 1979; cf. Sommer 1994); females mate promiscuously to confuse paternity (Hrdy 1979; Struhsaker & Leland 1985); females form coalitions to defend their offspring against infanticidal males (Hrdy 1974; Borries 1997; Saj et al. 2007); females wean their offspring quickly to reduce the probability of attacks (Teichroeb & Sicotte 2008); females stay with resident males, even after the group splits (Sterck 1997); and females transfer to other groups which contain other males that can serve as more efficient protectors (Sterck et al. 2005; Teichroeb et al. 2009). In sum, the threat of infanticide is real and constant and plays an important role in the lives of colobines. It affects male and female sexual strategies and the social organization of several species (Sterck et al. 1997; Treves 1998b; Isbell 2004; chapter 19, this volume).

Social Structure

Female-female social interactions

Dominance interactions between females have been a long-neglected topic in the study of colobine monkey behavior (but see Hrdy & Hrdy 1976). This is partly due to the difficulty of identifying individual animals. In addition, female dominance relationships were thought to be relatively unimportant in colobines (van Schaik 1989). Studies of individually recognized animals now indicate that females have clear dominance relationships and can be organized into dominance hierarchies (Hanuman langurs, Hrdy & Hrdy 1976; Borries et al. 1991; Borries 1993; Lu et al. 2008; Thomas's leaf monkey, Sterck & Steenbeek 1997; king colobus, Korstjens et al. 2002; Phayre's leaf monkey, *Trachypithecus obscurus phayrei*, Koenig et al. 2004). Female dominance hierarchies in colobines differ from those displayed by female cercopithecines in several ways (chapter 5, this volume). First, female hierarchies in colobines are based not on kinship, but instead on characteristics of individual animals. In some situations the youngest females often obtain the highest ranks, resulting in an age-inversed hierarchy (Borries et al. 1991). Second, rank changes occur relatively often (Borries et al. 1991). Third, females do not use coalitions to achieve or maintain their status (guereza, Dunbar & Dunbar 1976; Hanuman langur, Borries 1993; Borries et al. 1991; Koenig 2000; king colobus, Korstjens et al. 2002; white-thighed colobus, Saj et al. 2007). Whether the rarity of nepotism and coalitions is the cause or consequence of dominance is unclear. Nepotism and coalitions may be absent because dominance hierarchies are unstable over time (Koenig 2000). Alternatively, their absence may lead to unstable dominance hierarchies (Broom et al. 2009). In the latter situation, the strength and motivation of individuals determines rank relationships

and frequent changes in these factors produce instability in dominance relationships.

Colobine females compete over food (Thomas's leaf monkey, Sterck & Steenbeek 1997; Hanuman langur, Borries et al. 1991; Koenig 2000; king colobus, Korstjens et al. 2002; Phayre's leaf monkey, Koenig et al. 2004), with both fruit and leaves eliciting aggression. The quality, distribution, and availability of food are three important factors that affect the amount of contest competition for food (Sterck & Steenbeek 1997; Koenig et al. 1998; chapter 9, this volume). Colobines compete for high-quality foods, those that are clumped in space, and items that are not widely available. Through its effect on feeding competition and food intake, dominance rank can influence female reproductive success. For example, high-ranking Hanuman langur females feed in larger parties (Koenig et al. 1998) and ingest more food (Borries 1993) than do low-ranking individuals. In this same species, well-fed high-ranking females are in better body condition (Koenig 2000), and conceive earlier (Koenig et al. 1997) and more often (Borries et al. 1991) than do low-ranking animals. Despite these patterns, the impact of competition on Hanuman langur female reproduction may be unusual for colobines (Koenig & Borries 2001). Additional studies have failed to find similar relationships. For instance, observations of Thomas's leaf monkeys indicate that female dominance rank does not influence food intake, feeding effort, or reproductive success (Sterck 1995; Sterck et al. 1997).

Aggression disrupts group life (Aureli et al. 2002; chapter 23, this volume), and several mechanisms to reduce its impact have been proposed. A large body of research shows that female colobines reconcile after fights via friendly interactions (golden snub-nosed monkeys, Ren et al. 1991; guereza, Björnsdotter et al. 2000; dusky leaf monkeys, *Trachypithecus obscurus*, Arnold & Barton 2001a; Yunnan snub-nosed monkeys, Grueter 2004). These studies indicate that colobines reconcile after about half of all of their conflicts. The high frequency of reconciliation differs from that shown by other, more despotic species such as macaques, and it appears to reflect the tolerant nature of colobine relationships (Arnold & Aureli 2007). Interestingly, reconciliation occurs less often in Hanuman langurs (Sommer et al. 2002; Arnold & Aureli 2007), a species characterized by its generally intolerant nature (Lu et al. 2008). In colobines, reconciliation occurs most often between individuals with good relationships, a finding that mirrors observations in studies of other primates (Arnold & Barton 2001a; chapter 23, this volume). In addition, post-conflict affiliation with bystanders also takes place (Arnold & Barton 2001b). In sum, conflict management strategies are employed by many colobine species.

Female colobines typically groom other females more often than they groom males (Hanuman langurs, Borries et al. 1994; Western red colobus, Starin 1994; white-thighed colobus, Saj et al. 2007; golden snub-nosed monkeys, Zhang et al. 2008; Thomas's leaf monkey, Sterck unpublished data). Females can maintain close social bonds with related females via grooming when they do not disperse from their natal groups, and even after they emigrate to new groups containing some relatives (Western red colobus, Starin 1994; Thomas's leaf monkey, Sterck et al. 2005; guereza, Harris et al. 2009). Additional benefits female colobines obtain via grooming remain to be determined (Seyfarth 1977; Noë et al. 1991; Noë & Hammerstein 1994).

Male-female social interactions

In several colobine species, males police by intervening in disputes between females (Eastern red colobus, Struhsaker & Leland 1979; golden snub-nosed monkey, Ren et al. 1991; Yunnan snub-nosed monkey, Grueter 2004; Thomas's leaf monkey, Sterck & Steenbeek 1997). These interventions reduce the severity and frequency of aggression between females and the deleterious consequences of aggression such as injuries and stress. Policing may be typical of the behavior of males in one-male groups, since resident males stand to gain and improve their fitness by decreasing aggression between females. In cases where females transfer, male policing may induce females to remain in the group (Watts et al. 2000). These hypotheses, while intriguing, await empirical tests.

Male-male social interactions

In many colobine species, only one resident male is present in a group. In multimale groups, males have clear dominance hierarchies, and high-ranking males have preferential access to females (Eastern red colobus, Struhsaker & Leland 1979; Western red colobus, Starin 1994; Hanuman langur, Borries & Koenig 2000). Hanuman langur males that immigrate into new groups at Ramnagar attain all ranks, from low to high, but their social relationships have not been described (Borries 2000). In general, male-male social interactions and relationships in colobines have received surprisingly little study and require more attention.

Female Dispersal

Dispersal is a crucial feature in determining social dynamics because it potentially determines kin bonds in groups (chapter 21, this volume). However, dispersal is a conditional strategy that depends on its costs and benefits relative to philopatry (Bowler & Benton 2005; Jack & Isbell 2009). Various dispersal patterns are found in colobines (table 4.5): male dispersal and female philopatry (guereza, Harris et al. 2009; Hanuman langur, Koenig & Borries 2001), male philopatry and female dispersal (Eastern red colobus, Struhsaker & Leland 1979; Western red colobus, Starin 1994), and dispersal by both sexes (olive colobus, Korstjens & Schippers 2003; Thomas's leaf monkey, Sterck et al. 2005).

Females disperse in several colobine species and do so at different ages and reproductive states (table 4.5). In many populations, only primary dispersal by nulliparous females occurs (Eastern red colobus, Struhsaker & Leland 1979), while in others, females disperse as nulliparas and secondarily after giving birth (Eastern red colobus, Marsh 1979a, b; Thomas's leaf monkey, Sterck et al. 2005; golden snub-nosed monkey, Zhao et al. 2008a; white-thighed colobus, Teichroeb et al. 2009). Dispersal by only parous females has also been reported in capped leaf monkeys, but the relatively short duration of this study may have precluded observations of younger females dispersing as nulliparas (Stanford 1991b). Rates of dispersal vary widely. In some populations female dispersal is rare (Hanuman langurs, Koenig & Borries 2001), while in others all females disperse (Sterck & Korstjens 2000). Dispersal can be voluntary (Thomas's leaf monkey, Sterck 1997) or forced when females are evicted from their group (king colobus, Korstjens et al. 2002; reviewed in Sterck & Korstjens 2000).

Dispersal is costly, especially when individuals range alone or into unknown areas (locational dispersal, Isbell & van Vuren 1996). These costs will be lower when dispersers join adjacent social groups whose ranges overlap extensively with their own (social dispersal, ibid.). Females of most colobine species disperse socially, as they join adjacent groups or form another group within their natal home range (Isbell & van Vuren 1996). In some species, however, dispersing females move into an entirely new home range (Eastern red colobus, Marsh 1979a; Western red colobus, Starin 1994; olive colobus, Korstjens & Schippers 2003). Dispersing females do not frequently range alone. Instead, they move directly from one group into another or make the transition into a new conspecific group by following members of other species (olive colobus, Korstjens & Schippers 2003).

Resident females impose additional costs on dispersing females (table 4.5). In some species, residents direct aggression towards dispersers, preventing their immigration into the group (white-thighed colobus, Teichroeb et al. 2009). Such opposition on the part of residents, however, does not take place in several other colobine species (Sterck et al. 2005). Lastly, parous females are likely to suffer additional costs during secondary dispersal. Dispersing females with infants may be especially vulnerable to predation and infanticide by new males. As a result, most parous females disperse without dependent offspring (Sterck & Korstjens 2000). Mothers who disperse into new groups experience longer birth intervals than those who remain resident in their original groups, thus suggesting an additional cost of secondary dispersal (Thomas's leaf monkey, Sterck et al. 2005).

Several ultimate factors have been proposed to explain female dispersal in primates, including inbreeding avoidance (Clutton-Brock 1989), predation risk (Rasmussen 1981), competition for limiting resources (Dobson 1982), and infanticide (Marsh 1979b; Sterck 1997). The inbreeding avoidance hypothesis suggests that females disperse to avoid mating with their fathers. This can occur when males reside in groups for periods that last longer than the time it takes females to reach sexual maturity. Most cases of dispersal by nulliparous female colobines are consistent with this hypothesis (Sterck & Korstjens 2000; olive colobus, Korstjens & Schippers 2003; golden snub-nosed monkey, Zhao et al. 2008a; but see white-thighed colobus, Teichroeb et al. 2009). The predation risk hypothesis proposes that females emigrate to reduce their vulnerability to predation. A prediction is that females will move to large groups since individuals living in them are more effective at deterring predatory attacks than those living in smaller groups (chapter 8, this volume). Observations of colobines do not support this prediction, as females usually join groups that are similar in size or smaller than their groups of origin (Thomas's leaf monkey, Sterck 1997; olive colobus, Korstjens & Schippers 2003; golden snub-nosed monkey, Zhao et al. 2008a; white-thighed colobus, Teichroeb et al. 2009). These observations also conform to the hypothesis that females disperse to dampen the effects of food competition. Alternatively, females may move into small groups to lower their risk of infanticide, as males appear to selectively target infants living in large groups rather than those living in small groups (ursine howler monkeys, *Alouatta arctoidea*, Crockett & Janson 2000; Hanuman langur, Treves & Chapman 1996; Thomas's leaf monkey, Steenbeek & van Schaik 2001). Additional observations support the hypothesis that females disperse to avoid infanticide. Females leave their groups when extragroup males challenge resident males (Thomas's leaf monkey, Sterck 1997) or when male group membership is unstable (white-thighed colobus, Teichroeb et al. 2009). Females join males that protect them and their infants effectively; infants born after dispersals are more likely to survive than those born before their mothers disperse (Thomas's leaf monkey, Sterck et al. 2005).

Summary and Conclusion

Studies of colobines have featured prominently in the study of primate behavior. They provide a test case for socioecological models and have been central to our understanding of infanticide (chapters 9 and 19, this volume). Several colobine species have been the subjects of long-term studies, and much more information on their behavior is available now than 25 years ago. These data reveal a primate taxon that is relatively uniform in some features of its behavior: female colobines do not form coalitions, and relatively mild aggression characterizes the behavior of most species. Despite these similarities, colobines are quite variable in other aspects of their behavior; grouping and dispersal patterns vary considerably within and between species. While groups with a single male and multiple females predominate, small groups consisting of a pair of adults and immatures and large groups composed of multiple males and females and several one-male units also form. Several factors including habitat quality, food characteristics, predation pressure, and the behavior of conspecifics appear to affect this variability in grouping within and between species (Watanabe 1981; van Schaik & Hörstermann 1994; Srivastava & Dunbar 1996; Kirkpatrick et al. 1998; Moore 1999; Fashing et al. 2007b; Hadi et al. 2009; chapter 9, this volume).

Variable dispersal patterns are a conspicuous aspect of colobine monkey behavior. In some species, males disperse while females remain in their natal groups; in others, females disperse and males stay. In some species, members of both sexes disperse. Our understanding of this variation is incomplete. Specifically, the benefits of male philopatry in species such as red colobus are not understood. In addition, while the lack of between-group aggression and within-group coalitions may provide low-cost opportunities for female colobines to disperse, the benefits they accrue by doing so require further study. Inbreeding avoidance and the threat of male infanticide are frequently hypothesized to play a significant role in this regard, but more research is needed to clarify how they and other factors influence female dispersal decisions.

Field studies of colobine monkeys have shown that female grouping patterns are affected not only by ecology but also by male reproductive strategies. Females form groups when males defend food resources (Fashing 2001; Harris 2006), and male infanticide can override the effects of food competition and limit group size (Steenbeek & van Schaik 2001). When social factors affect grouping patterns, the ensuing social system can be quite variable, as social circumstances can change relatively quickly. While the effects of male reproductive strategies on female grouping patterns have been well established via studies of colobine behavior, how these interact with female reproductive strategies, especially in multimale-multifemale groups, remains to be determined.

One particularly large gap in our understanding of colobine monkey behavior concerns their life history. As is the case with most long-lived primates, scant data exist regarding development, reproduction, and longevity for most colobine species (chapters 10 and 11, this volume). Thus it is not yet possible to fully evaluate the proposal that langurs and other colobines experience a relatively fast pace of life compared to other primates (chapter 10, this volume). As more long-term studies of individually recognized animals accumulate, our knowledge of colobine life history and behavior promises to increase. Additional data on ecology, life history, and evolution will help us to unravel the factors that influence the behavioral similarities and differences that exist between colobine monkeys.

References

Agoramoorthy, G. & Mohnot, S. M. 1988. Infanticide and juvenilicide in Hanuman langurs (*Presbytis entellus*) around Jodhpur, India. *Human Evolution*, 3, 279–296.

Arnold, K. & Aureli, F. 2007. Postconflict reconciliation. In *Primates in Perspective* (ed. by C. J. Campbell, A. Fuentes, K. C. MacKinnon, M. Panger & S. K. Bearder), 592–608. New York: Oxford University Press.

Arnold, K. & Barton, R. A. 2001a. Postconflict behavior of spectacled leaf monkeys (*Trachypithecus obscurus*). I. Reconciliation. *International Journal of Primatology*, 22, 243–266.

———. 2001b. Postconflict behavior of spectacled leaf monkeys (*Trachypithecus obscurus*). II. Contact with third parties. *International Journal of Primatology*, 22, 267–286.

Assink, P. & van Dijk, I. 1990. Social organization, ranging and density of *Presbytis thomasi* at Ketambe (Sumatra), and a comparison with other *Presbytis* species at several South-east Asian locations. PhD thesis, Utrecht University, Utrecht.

Aureli, F., Cords, M. & Van Schaik, C. P. 2002. Conflict resolution following aggression in gregarious animals: A predictive framework. *Animal Behaviour*, 64, 325–343.

Bauchop, T. & Martucci, R. W. 1968. Ruminant-like digestion of the langur monkey. *Science*, 161, 698–700.

Bennett, E.L. 1983. The banded langur: Ecology of a colobine in West Malaysian rain-forest. PhD thesis, Cambridge University.

Bennett, E. L. & Davies, A. G. 1994. The ecology of Asian colobines. In *Colobine Monkeys: Their ecology, behaviour and evolution* (ed. by A. G. Davies & J. F. Oates), pp. 129–171. Cambridge: Cambridge University Press.

Bennett, E. L. & Sebastian, A. C. 1988. Social organization and ecology of proboscis monkeys (*Nasalis larvatus*) in mixed coastal forest in Sarawak. *International Journal of Primatology*, 9, 233–255.

Bernstein, I. S. 1968. The lutong of Kuala Selangor. *Behaviour*, 32, 1–16.

Bjornsdotter, M., Larsson, L. & Ljungberg, T. 2000. Postconflict affiliation in two captive groups of black-and-white guereza *Colobus guereza. Ethology*, 106, 289–300.

Boesch, C. & Boesch-Achermann, H. 2000. *The Chimpanzees of the Taï Forest: Behavioural Ecology and Evolution*. Oxford: Oxford University Press.

Boggess, J. 1979. Troop male membership changes and infant killing in langurs (*Presbytis entellus*). *Folia Primatologica*, 32, 65–107.

———. 1980. Intermale relations and troop male membership changes in langurs (*Presbytis entellus*) in Nepal. *International Journal of Primatology*, 3, 233–274.

Borries, C. 1993. Ecology of female social relationships: Hanuman langurs (*Presbytis entellus*) and the van Schaik model. *Folia Primatologica*, 61, 21–30.

———. 1997. Infanticide in seasonally breeding multimale groups of Hanuman langurs (*Presbytis entellus*) in Ramnagar (South Nepal). *Behavioral Ecology and Sociobiology*, 41, 139–150.

———. 2000. Male dispersal and mating season influxes in Hanuman langurs living in multi-male groups. In *Primate Males: Causes and Consequences of Variation in Group Composition* (ed. by P.M. Kappeler), 146–158. Cambridge: Cambridge Univiversity Press.

Borries, C. & Koenig, A. 2000. Hanuman langurs: Infanticide in multimale groups. In *Infanticide by Males and its Implications* (ed. by C. P. van Schaik & C. H. Janson), 99–122. Cambridge: Cambridge University Press.

Borries, C., Sommer, V. & Srivastava, A. 1991. Dominance, age, and reproductive success in free-ranging female Hanuman langurs (*Presbytis entellus*). *International Journal of Primatology*, 12, 231–257.

———. 1994. Weaving a tight social net: Allogrooming in free-ranging female langurs (*Presbytis entellus*). *International Journal of Primatology*, 15, 421–443.

Borries, C., Launhardt, K., Epplen, C., Epplen, J. T. & Winkler, P. 1999a. DNA analyses support the hypothesis that infanticide is adaptive in langur monkeys. *Proceedings of the Royal Society of London B*, 266, 901–904.

———. 1999b. Males as infant protectors in Hanuman langurs (*Presbytis entellus*) living in multi-male groups: Defence pattern, paternity and sexual behaviour. *Behavioral Ecology and Biology*, 46, 350–356.

Borries, C., Koenig, A. & Winkler, P. 2001. Variation of life history traits and mating patterns in female langur monkeys (*Semnopithecus entellus*). *Behavioural Ecology and Sociobiology*, 50, 391–402.

Borries, C., Larney, E., Derby, A. M. & Koenig, A. 2004. Temporary absence and dispersal in Phayre's leaf monkeys (*Trachypithecus phayrei*). *Folia Primatologica*, 75, 27–30.

Borries, C., Larney, E., Lu, A., Ossi, K. & Koenig, A. 2008. Costs of group size: Lower developmental and reproductive rates in larger groups of leaf monkeys. *Behavioral Ecology*, 19, 1186–1191.

Bowler, D. E. & Benton, T. G. 2005. Causes and consequences of animal dispersal strategies: Relating individual behaviour to spatial dynamics. *Biological Reviews*, 80, 205–225.

Brandon-Jones, D., Eudey, A. A., Geissmann, T., Groves, C.P., Melnick, D. J., Morales, J. C., Shekelle, M. & Stewart, C.-B. 2004. Asian primate classification. *International Journal of Primatology*, 25, 97–164.

Broom, M., Koenig, A. & Borries, C. 2009. Variation in dominance hierarchies among group-living animals: Modeling stability and the likelihood of coalitions. *Behavioral Ecology*, 20, 844–855.

Brugiere, D., Gautier, J.-P., Moungazi, A. & Gautier-Hion, A. 2002. Primate diet and biomass in relation to vegetation composition and fruiting phenology in a rain forest in Gabon. *International Journal of Primatology*, 23, 999–1024.

Bshary, R. & Noë, R. 1997. Anti-predation behaviour of red

colobus monkeys in the presence of chimpanzees. *Behavioural Ecology and Sociobiology*, 41, 321–333.

Chapman, C. A. & Pavelka, M. S. M. 2005. Group size in folivorous primates: Ecological constraints and the possible influence of social factors. *Primates*, 46, 1–9.

Chapman, C. A. & Rothman, J. M. 2009. Within-species differences in primate social structure: Evolution of plasticity and phylogenetic constraints. *Primates*, 50, 12–22.

Chapman, C. A., Chapman, L. J., Naughton-Treves, L., Lawes, M. J. & McDowell, L. R. 2004. Predicting folivorous primate abundance: Validation of a nutritional model. *American Journal of Primatology*, 62, 55–69.

Clauss, M., Streich, W. J., Nunn, C. L., Ortmann, S., Hohmann, G., Schwarm, A. & Hummel, J. 2008. The influence of natural diet composition, food intake level, and body size on ingesta passage in primates. *Comparative Biochemistry and Physiology, Part A*, 150, 274–281.

Clutton-Brock, T. H. 1989. Female transfer and inbreeding avoidance in social mammals. *Nature*, 337, 70–72.

Clutton-Brock, T. H. & Harvey, P. H. 1977. Primate ecology and social organization. *Journal of Zoology*, 183, 1–39.

Crockett, C. M. & Janson, C. H. 2000. Infanticide in red howlers: Female group size, male membership, and a possible link to folivory. In *Infanticide by Males and Its Implications* (ed. by C. P. van Schaik & C. H. Janson), 75–98. Cambridge: Cambridge University Press.

Cui, L. W. 2003. A note on an interaction between *Rhinopithecus bieti* and a buzzard at Baima Snow Mountain. *Folia primatologica*, 74, 51–53

Cui, L. W., Sheng, A. H., He, S. C. & Xiao, W. 2006. Birth seasonality and interbirth interval of captive *Rhinopithecus bieti*. *American Journal of Primatology*, 68, 457–463.

Curtin, S. H. & Chivers, D. J. 1978. Leaf-eating primates of peninsular Malaysia: The siamang and the dusky leaf monkey In *The Ecology of Arboreal Folivores* (ed. by G. G. Montgomery), 441–464. Washington: Smithsonian Institution Press.

Davies, A. G. 1984. An ecological study of the red leaf monkey (*Presbytis rubicunda*) in the dipterocarp forest of Northern Borneo. PhD thesis, Cambridge University.

———. 1994. Colobine populations. In *Colobine Monkeys: Their Ecology, Behaviour and Evolution* (ed. by A. G. Davies & J. F. Oates), 285–310. Cambridge: Cambridge University Press.

Davies, A. G., Oates, J. F. & Dasilva, G. L. 1999. Patterns of frugivory in three West African colobine monkeys. *International Journal of Primatology*, 20, 327–357.

Decker, B. S. 1994. Effects of habitat disturbance on the behavioural ecology and demographics of the Tana River red colobus (*Colobus badius rufomitratus*). *International Journal of Primatology*, 15, 703–737.

Decker, B. S. & Kinnaird, M. F. 1992. Tana River red colobus and crested mangabey: Results of recent censuses. *International Journal of Primatology*, 26, 47–52.

Dela, J. D. S. 2007. Seasonal food use strategies of *Semnopithecus vetulus nestor*, at Panadura and Piliyandala, Sri Lanka. *International Journal of Primatology*, 28, 607–626.

Di Fiore, A. & Rendall, D. 1994. Evolution of social organization: A reappraisal for primates by using phylogenetic methods. *Proceedings of the National Academy of Sciences*, 91, 9941–9945.

Ding, W. & Zhao, Q. K. 2004. *Rhinopithecus bieti* at Tacheng, Yunnan: Diet and daytime activities. *International Journal of Primatology*, 25, 583–598.

Dobson, F. S. 1982. Competition for mates and predominant juvenile male dispersal in mammals. *Animal Behaviour*, 30, 1183–1192.

Dunbar, R. I. M. 1987. Habitat quality, populations dynamics, and group composition in colobus monkeys (*Colobus guereza*). *International Journal of Primatology*, 8, 299–329.

Dunbar, R. I. M. & Dunbar, E. P. 1974. Ecology and population dynamics of *Colobus guereza* in Ethiopia. *Folia primatologica*, 21, 188–208.

———. 1976. Contrasts in social structure among black-and-white colobus monkey groups. *Animal Behaviour*, 24, 84–92.

Emlen, S. T. & Oring, L. W. 1977. Ecology, sexual selection, and the evolution of mating systems. *Science*, 197, 215–223.

Fashing, P. J, 2001. Male and female strategies during intergroup encounters in guerezas (*Colobus guereza*): Evidence for resource defense mediated through males and a comparison with other primates. *Behavioural Ecology and Sociobiology*, 50, 219–230.

———. 2007. African colobine monkeys: Patterns of between-group interaction. In *Primates in Perspective* (ed. by C. J. Campbell, A. Fuentes, K. C. MacKinnon, M. Panger & S. K. Bearder), 201–224. New York: Oxford University Press.

Fashing, P. J., Dierenfeld, E. S. & Mowry, C. B. 2007a. Influence of plant and soil chemistry on food selection, ranging patterns, and biomass of *Colobus guereza* in Kakamega Forest, Kenya. *International Journal of Primatology*, 28, 673–703.

Fashing, P. J., Mulindahabi, F., Gakima, J. B., Masozera, M., Mununura, I., Plumptre, A. J. & Nguyen, N. 2007b. Activity and ranging patterns of *Colobus angolensis ruwenzorii* in Nyungwe Forest, Rwanda: Possible costs of large group size. *International Journal of Primatology*, 28, 529–550.

Green, K. M. 1981. Preliminary observations on the ecology and behavior of the capped langur, *Presbytis pileatus*, in the Madhupur Forest of Bangladesh. *International Journal of Primatology*, 2, 131–151.

Groves, C. P. 1973. Notes on the ecology and behaviour of the Angola colobus (*Colobus angolensis* P. L. Sclater 1860) in N. E. Tanzania. *Folia Primatologica*, 20, 12–26.

Grubb, P., Butynski, T. M., Oates, J. F., Bearder, S. K., Disotell, T. R., Groves, C. P. & Struhsaker, T. T. 2003. Assessment of the diversity of African primates. *International Journal of Primatology*, 24, 1301–1357.

Grueter, C. C. 2004. Conflict and postconflict behaviour in captive black-and-white snub-nosed monkeys (*Rhinopithecus bieti*). *Primates*, 45, 197–200.

Grueter, C. C. & van Schaik, C. P. 2010. Evolutionary determinants of modular societies in colobines. *Behavioral Ecology*, 21, 63–71

Grueter, C. C., Li, D. Y., Ren, B. P., Wei, F., Xiang, Z. & van Schaik, C. P. 2009. Fallback foods of temperate-living primates: A case study on snub-nosed monkeys. *American Journal of Physical Anthropology*, 140, 700–715.

Gurmaya, K. J. 1986. Ecology and behavior of *Presbytis thomasi* in Northern Sumatra. *Primates*, 27, 151–172.

Hadi, S., Ziegler, T. & Hodges, J. K. 2009. Group structure and physical characteristics of simakobu monkeys (*Simias con-*

color) on the Mentawai Island of Siberut, Indonesia. *Folia Primatologica*, 80, 74–82.

Harris, T. R. 2006. Between-group contest competition for food in a highly folivorous population of black and white colobus monkeys (*Colobus guereza*). *Behavioural Ecology and Sociobiology*, 61, 317–329.

Harris, T. R. & Chapman, C. A. 2007. Variation in diet and ranging of black and white colobus monkeys in Kibale National Park, Uganda. *Primates*, 48, 208–221.

Harris, T. R. & Monfort, S. L. 2003. Behavioral and endocrine dynamics associated with infanticide in a black and white Colobus monkey (*Colobus guereza*). *American Journal of Primatology*, 61, 135–142.

———. 2006. Mating behavior and endocrine profiles of wild black and white Colobus monkeys (*Colobus guereza*): Toward an understanding of their life history and mating system. *American Journal of Primatology*, 68, 383–396.

Harris, T. R., Caillaud, D., Chapman, C. A. & Vigilant, L. 2009. Neither genetic nor observational data alone are sufficient for understanding sex-biased dispersal in a social-group-living species. *Molecular Ecology*, 18, 1777–1790.

Hohmann, G. 1989. Group fission in Nilgiri langurs (*Presbytis johnii*). *International Journal of Primatology*, 10, 441–454.

Hrdy, S. B. 1974. Male-male competition and infanticide among the langurs (*Presbytis entellus*) of Abu, Rajasthan. *Folia Primatologica*, 22, 19–58.

———. 1977. *The Langurs of Abu: Female and Male Strategies of Reproduction*. Cambridge: Harvard University Press.

———. 1979. Infanticide among animals: A review, classification, and examination of the implications for the reproductive strategies of females. *Ethology and Sociobiology*, 1, 13–40.

Hrdy, S. B. & Hrdy, D. B. 1976. Hierarchical relations among female hanuman langurs (Primates: Colobinae, *Presbytis entellus*). *Science*, 193, 913–915.

Hrdy, S. B., Janson, C. H. & van Schaik, C. P. 1995. Infanticide: Let's not throw out the baby with the bath water. *Evolutionary Anthropology*, 3, 151–154.

Isbell, L. A. 1991. Contest and scramble competition: Patterns of female aggression and ranging behavior among primates. *Behavioral Ecology*, 2, 143–155.

———. 2004. Is there no place like home? Ecological bases of dispersal in primates and their consequences for the formation of kin groups. In *Kinship and Behavior in Primates* (ed. by B. Chapais & C. Berman), 71–108. New York: Oxford University Press.

Isbell, L. & van Vuren, D. 1996.. Differential costs of locational and social dispersal and their consequences for female group-living primates. *Behaviour*, 133, 1–36.

Jack, K. M. & Isbell, L. A. 2009. Dispersal in primates: Advancing an individualized approach. Preface. *Behaviour*, 146, 429–436.

Janson, C. H. & Goldsmith, M. L. 1995. Predicting group size in primates: Foraging costs and predation risks. *Behavioral Ecology*, 6, 326–336.

Jay, P. 1965. The common langur of North India. In *Primate Behavior: Field Studies of Monkeys and Apes* (ed. by I. DeVore), 197–249. New York: Holt, Rinehart and Winston.

Karanth, K. P., Singh, L., Collura, R. V. & Stewart, C. B. 2008. Molecular phylogeny and biogeography of langurs and leaf monkeys of South Asia (Primates: *Colobinae*). *Molecular Phylogenetics and Evolution*, 46, 683–694.

Kay, R. N. B. & Davies, A. G. 1994. Digestive physiology. In *Colobine Monkeys: Their Ecology, Behaviour and Evolution* (ed. by A. G. Davies & J. F. Oates), 229–249. Cambridge: Cambridge University Press.

Kirkpatrick, R. C. 2007. The Asian colobines: Diversity among leaf-eating monkeys. In *Primates in Perspective* (ed. by C. J. Campbell, A. Fuentes, K. C. MacKinnon, M. Panger & S. K. Bearder), 186–200. New York: Oxford University Press.

Kirkpatrick, R. C. & Long, Y. C 1994. Altitudinal ranging and terrestriality in the Yunnan snub-nosed monkey (*Rhinopithecus bieti*). *Folia primatologica*, 63, 102–106.

Kirkpatrick, R. C., Long, Y. C., Zhong, T. & Xiao, L. 1998. Social organization and range use in the Yunnan snub-nosed monkey *Rhinopithecus bieti*. *International Journal of Primatology*, 19, 13–51.

Kitchen, D. M. & Beehner, J. C. 2007. Factors affecting individual participation in group-level aggression among non-human primates. *Behaviour*, 144, 1551–1581.

Koenig, A. 2000. Competitive regimes in forest-dwelling Hanuman langur females (*Semnopithecus entellus*). *Behavioural Ecology and Sociobiology*, 48, 93–109.

Koenig, A. & Borries, C. 2001. Socioecology of Hanuman langurs: The story of their success. *Evolutionary Anthropology*, 10, 122–137.

Koenig, A., Borries, C., Chalise, M. K. & Winkler, P. 1997. Ecology, nutrition, and timing of reproductive events in an Asian primate, the Hanuman langur (*Presbytis entellus*). *Journal of Zoology*, 243, 215–235.

Koenig, A., Beise, J., Chalise, M. K. and Ganzhorn, J. U. 1998. When females should contest for food: Testing hypotheses about resource density, distribution, size, and quality with Hanuman langurs (*Presbytis entellus*). *Behavioral Ecology and Sociobiology*, 42, 225–237.

Koenig, A., Larney, E., Lu, A. & Borries, C. 2004. Agonistic behavior and dominance relationships in female Phayre's leaf monkeys: Preliminary results. *American Journal of Primatology*, 64, 351–357.

Kool, K. M. 1989. Behavioural ecology of the silver leaf monkey in the Pangandaran Nature Reserve, West Java. PhD thesis, University of New South Wales, Sydney.

———. 1993. The diet and feeding behavior of the silver leaf monkey (*Trachypithecus auratus sondaicus*) in Indonesia. *International Journal of Primatology*, 14, 667–700.

Korstjens, A. H. 2001. The mob, the secret sorority and the phantoms: An analysis of the socio-ecological strategies of the three Taï colobines. PhD thesis, Utrecht University, Utrecht.

Korstjens, A. H. & Noë, R. 2004. Mating system of an exceptional primate, the olive colobus (*Procolobus verus*). *American Journal of Primatology*, 62, 261–273.

Korstjens, A. H. & Schippers, E. P. 2003. Dispersal patterns among olive colobus in Tai National Park. *International Journal of Primatology*, 24, 515–539.

Korstjens, A. H., Sterck, E. H. M. & Noë, R. 2002. How adaptive or phylogenetically inert is primate social behaviour? A test with two sympatric colobines. *Behaviour*, 139, 203–225.

Krueger, O., Affeldt, E., Brackmann, M. & Milhahn, K. 1998. Group size and composition of *Colobus guereza* in Kyambura Gorge, southwest Uganda, in relation to chimpanzee activity. *International Journal of Primatology*, 19, 287–297.

Lambert, J. E. 1998. Primate digestion: Interactions among anatomy, physiology, and feeding ecology. *Evolutionary Anthropology*, 7, 8–20.

———. 2007. Primate nutritional ecology: Feeding biology and diet at ecological and evolutionary scales. In *Primates in Perspective* (ed. by C. J. Campbell, A. Fuentes, K. C. MacKinnon, M. Panger & S. K. Bearder), 482–495. New York: Oxford University Press.

Launhardt, K., Borries, C., Hardt, C., Epplen, J. T. & Winkler, P. 2001. Paternity analysis of alternative male reproductive routes among the langurs (*Semnopithecus entellus*) of Ramnagar. *Animal Behaviour*, 61, 53–64.

Leigh, S. R. 1994. Ontogenetic correlates of diet in anthropoid primates. *American Journal of Physical Anthropology*, 94, 499–522.

Li, Z. Y. & Rogers, M. E. 2006. Food items consumed by white-headed langurs in Fusui, China. *International Journal of Primatology*, 27, 1551–1567.

Lu, A., Koenig, A. & Borries, C. 2008. Formal submission, tolerance and socioecological models: a test with female Hanuman langurs. *Animal Behaviour*, 76, 415–428.

Maisels, F., Gautier Hion, A. & Gautier J. P. 1994. Diets of 2 sympatric colobines in Zaire: More evidence on seed-eating in forests on poor soils. *International Journal of Primatology*, 15, 681–701.

Marler, P. 1969. *Colobus guereza*: Territoriality and group composition. *Science*, 163, 93–95.

Marsh, C. W. 1979a. Comparative aspects of social organization in the Tana River red colobus, *Colobus badius rufomitratus*. *Zeitschrift für Tierpsychologie*, 51, 337–362.

———. 1979b. Female transference and mate choice among Tana River red colobus. *Nature*, 281, 568–569.

Mathur, R. & Mahonar, B. R. 1992. Rate of takeovers in groups of hanuman langurs (*Presbytis entellus*) at Jaipur. *Folia primatologica*, 58, 61–71.

Matsuda, I., Tuuga, A. & Higashi, S. 2009. The feeding ecology and activity budget of proboscis monkeys. *American Journal of Primatology*, 71, 478–492

McKenna, J. J. 1979. Evolution of allomothering behavior among colobine monkeys: Function and opportunism in evolution. *Amercian Anthropologist*, 81, 818–840.

Megantara, E. N. 1989. Ecology, behavior and sociality of *Presbytis femoralis* in Eastcentral Sumatra. PhD thesis, University of Padjadjaran, Indonesia.

Meijaard, E. & Nijman, V. 2000. Distribution and conservation of the proboscis monkey (*Nasalis larvatus*) in Kalimantan, Indonesia. *Biological Conservation*, 92, 15–24.

Mitani, J. C. & Watts, D. 1997. The evolution of non-maternal caretaking in anthropoid primates: do helpers help? *Behavioural Ecology Sociobiology*, 40, 213–220.

Moore, J. 1984. Female transfer in primates. *International Journal of Primatology*, 5, 537–589.

———. 1999. Population density, social pathology, and behavioral ecology. *Primates*, 40, 1–22.

Mukherjee, R. P. & Saha, S. S. 1974. The golden langurs (*Presbytis geei* Khajuria 1956) of Assam. *Primates*, 15, 327–340.

Newton, P. N. 1987. The social organization of forest hanuman langurs (*Presbytis entellus*). *International Journal of Primatology*, 8, 199–232.

———. 1992. Feeding and ranging patterns of forest Hanuman langurs (*Presbytis entellus*). *International Journal of Primatology*, 13, 245–285.

———. 1994. Social stability and change among forest Hanuman langurs (*Presbytis entellus*). *Primates*, 35, 489–498.

Newton, P. N. & Dunbar, R. I. M. 1994. Colobine monkey society. In *Colobine Monkeys: Their Ecology, Behaviour and Evolution* (ed. by A. G. Davies & J. F. Oates), 311–346. Cambridge: Cambridge University Press.

Noë, R. & Bshary, R. 1997. The formation of red colobus-diana monkey associations under predation pressure from chimpanzees. *Proceedings of the Royal Society of London B*, 264, 253–259.

Noë, R. & Hammerstein, P. 1994. Biological markets: Supply and demand determine the effect of partner choice in cooperation, mutualism and mating. *Behavioural Ecology and Sociobiology*, 35, 1–11.

Noë, R., Schaik, C. P. van & Hooff, J. A. R. A. M. van 1991. The market effect: An explanation for pay-off asymmetries among collaborating animals. *Ethology*, 87, 97–118.

Nunn, C. L. 1999. The number of males in primate social groups: a comparative test of the socioecological model. *Behavioural Ecology and Sociobiology*, 46, 1–13.

Oates, J. F. 1977. The social life of a black-and-white colobus monkey, *Colobus guereza*. *Zeitschrift für Tierpsychologie*, 45, 1–60.

———. 1988. The diet of the olive colobus monkey (*Procolobus verus*), in Sierra Leone. *International Journal of Primatology*, 9, 457–478.

———. 1994. Africa's primates in 1992: Conservation issues and options. *American Journal of Primatology*, 34, 61–71.

Oates, J. F., Waterman, P. G. & Choo, G. M. 1980. Food selection by the South Indian leaf-monkey, *Presbytis johnii*, in relation to leaf chemistry. *Oecologia*, 45, 45–65.

Oates, J. F., Davies, A. G. & Delson, E. 1994. The natural history of African colobines. In *Colobine Monkeys: Their Ecology, Behaviour and Evolution* (ed. by A. G. Davies & J. F. Oates), 75–128. Cambridge: Cambridge University Press.

Osterholz, M., Walter, L. & Roos, C. 2008. Phylogenetic position of the langur genera *Semnopithecus* and *Trachypithecus* among Asian colobines, and genus affiliations of their species groups. *BMC Evolutionary Biology*, 8, 58 doi: 10.1186/1471-2148-8-58.

Podzuweit, D. 1994. Sozio-Ökologie weiblicher Hanuman Languren (*Presbytis entellus*) in Ramnagar, Südnepal. PhD thesis, Georg-August-Universität, Göttingen.

Poirier, F.E. 1969. The nilgiri langur (*Presbytis johnii*) troop: Its composition, structure, function and change. *Folia primatologica*, 10, 20–47.

———. 1970. The nilgiri langur (*Presbytis johnii*) of South India. In *Primate Behavior: Developments in Field and Laboratory Research* (ed. by L.A. Rosenblum), 251–383. New York, Academic Press.

Qi, X. G., Li, B. G. & Ji, W. H. 2008. Reproductive parameters of wild female *Rhihopithecus roxellana*. *American Journal of Primatology*, 70, 311–319.

Qi, X. G., Li, B. G., Garber, P. A., Ji, W. & Watanabe, K. 2009. Social dynamics of the golden snub-nosed monkey (*Rhinopithecus roxellana*): Female transfer and one-male unit succession. *American Journal of Primatology*, 71, 670–679.

Raaum, R. L., Sterner, K. N., Noviello, C. M., Stewart, C.-B. &

Disotell, T. R. 2005. Catarrhine primate divergence dates estimated from complete mitochondrial genomes: Concordance with fossil and nuclear DNA evidence. *Journal of Human Evolution*, 48, 237–257.

Rajpurohit, L. S. &Sommer, V. 1991. Sex differences in mortality among langurs (*Presbytis entellus*) of Jodhpur, Rajastan, India. *Folia Primatologica*, 56, 17–21.

———. 1993. Juvenile male emigration from natal one-male troops in Hanuman langurs. In *Juvenile Primates: Life History, Development, and Behavior* (ed. by M. E. Pereira & L. A. Fairsbanks), 86–103. Oxford: Oxford University Press.

Rajpurohit, L. S., Sommer, V. & Mohnot, S. M. 1995. Wanderers between harems and bachelor bands: Male Hanuman langurs (*Presbytis entellus*) at Jodhpur in Rajasthan. *Behaviour*, 132, 255–299.

Rasmussen, D. R. 1981. Communities of baboon troops (*Papio cynocephalus*) in Mikumi National Park, Tanzania. *Folia primatologica*, 36, 232–242.

Ren, R., Yan, K., Su, Y., Qi, H., Liang, B., Bao, W. & de Waal, F. B. M. 1991. The reconciliation bahavior of golden monkeys (*Rhinopithecus roxellanae*) in small breeding groups. *Primates*, 32, 321–327.

Ripley, S. 1967. Intertroop encounters among Ceylon gray langurs (*Presbytis entellus*). In *Social Communication among Primates* (ed. by S. A. Altmann), 237–253. Chicago: University of Chicago Press.

Ross, C. 1993. Take-over and infanticide in South Indian Hanuman langurs (*Presbytis entellus*). *American Journal of Primatology*, 30, 75–82.

Rudran, R. 1973a. Adult male replacement in one-male troops of purple-faced langurs (*Presbytis senex senex*) and its effects on population structure. *Folia Primatologica*, 19, 166–192.

———. 1973b. The reproductive cycles of two subspecies of purple-faced langurs (*Presbytis senex*) with relation to environmental factors. *Folia Primatologica*, 19, 41–60.

Ruhiyat, Y. 1983. Socio-ecological study of *Presbytis aygula* in West Java. *Primates*, 24, 334–359.

Saj, T. L. & Sicotte, P. 2005. Male takeover in *Colobus vellerosus* at boabeng-fiema monkey sanctuary, central Ghana. *Primates*, 46, 211–214

———. 2007. Scramble competition among *Colobus vellerosus* at Boabeng-Fiema, Ghana. *International Journal of Primatology*, 28, 337–355.

Saj, T. L., Marteinson, S., Chapman, C. A. & Sicotte, P. 2007. Controversy over the application of current socioecological models to folivorous primates: *Colobus vellerosus* fits the predictions. *American Journal of Physical Anthropology*, 133, 994–1003.

Sangchantr, S. 2003. Social organization and ecology of Mentawai leaf monkeys. *American Journal of Physical Anthropology Supplement*, 36, 183–183.

Sayers, K. & Norconk, M. A. 2008. Himalayan *Semnopithecus entellus* at Langtang National Park, Nepal: Diet, activity patterns, and resources. *International Journal of Primatology*, 29, 509–530.

Seidensticker, J. 1985. Primates as prey of *Panthera* cats in South Asian habitats. *American Journal of Primatology*, 8, 365–366.

Seyfarth, R. M. 1977. A model of social grooming among adult female monkeys. *Journal of theoretical Biology*, 65, 671–698.

Shelmidine, N., Borries, C. & McCann, C. 2009. Patterns of reproduction in Malayan silvered leaf monkeys at the Bronx Zoo. *American Journal of Primatology*, 71, 852–859.

Sicotte, P. & Macintosh, A. J. 2004. Inter-group encounters and male incursions in *Colobus vellerosus* in central Ghana. *Behaviour*, 141, 533–553.

Siex, K. S. & Struhsaker, T. T. 1999. Ecology of the Zanzibar red colobus monkey: Demographic variability and habitat stability. *International Journal of Primatology*, 10, 163–192.

Smith, R. J. & Jungers, W. L. 1997. Body mass in comparative primatology. *Journal of Human Evolution*, 32, 523–559

Snaith, T. V. & Chapman, C. A. 2005. Towards an ecological solution to the folivore paradox: Patch depletion as an indicator of within-group scramble competition in red colobus monkeys (*Piliocolobus tephrosceles*). *Behavioural Ecology and Sociobiology*, 59, 185–190.

———. 2008. Red colobus monkeys display alternative behavioral responses to the costs of scramble competition. *Behavioral Ecology*, 19, 1289–1296.

Sommer, V. 1985. Weibliche und Männliche Reproductionsstrategien der Hanuman Languren (*Presbytis entellus*) von Jodhpur, Rajastan/Indiën. PhD thesis, Universität Göttingen, Göttingen.

———. 1987. Infanticide among free-ranging langurs (*Presbytis entellus*) at Jodhpur (Rajasthan/India): Recent observations and a reconsideration of hypotheses. *Primates*, 28, 163–197.

———. 1994. Infanticide among the langurs of Jodhpur: Testing the sexual selection hypothesis with a long-term record. In *Infanticide and Parental Care* (ed. by S. Parmigiani & F. S. vom Saal), 155–198. Chur, Switzerland: Harwood Academic Press.

Sommer, V. & Rajpurohit, L. S. 1989. Male reproductive success in harem troops of Hanuman langurs (*Presbytis entellus*). *International Journal of Primatology*, 10, 293–317.

Sommer, V., Srivastava, A. & Borries, C. 1992. Cycles, sexuality, and conception in free-ranging female langurs (*Presbytis entellus*). *American Journal of Primatology*, 28, 1–27.

Sommer, V., Denham, A. & Little, K. 2002. Postconflict behaviour of wild Indian langur monkeys: Avoidance of opponents but rarely affinity. *Animal Behaviour*, 63, 637–648.

Srivastava, A. 1991. Insectivory and its significance to langur diets. *Primates*, 32, 237–241.

Srivastava, A. & Dunbar, R. I. M. 1996. The mating system of Hanuman langurs: A problem in optimal foraging. *Behavioural Ecology Sociobiology*, 39, 219–226.

Stanford, C. B. 1991a. Social dynamics of intergroup encounters in the capped langur (*Presbytis pileata*). *American Journal of Primatology*, 25, 35–47.

———. 1991b. *The Capped Langur in Bangladesh: Behavioral Ecology and Reproductive Tactics*. Contributions to Primatology. Vol 26. New York: Karger.

———. 1991c. The diet of the capped langur (*Presbytis pileata*) in a moist deciduous forest in Bangladesh. *International Journal of Primatology*, 12, 199–216.

———. 1992. Costs and benefits of allomothering in wild capped langurs (*Presbytis pileata*). *Behavioural Ecology and Sociobiology*, 30, 29–34.

———. 1998. *Chimpanzee and Red Colobus: The Ecology of Predator and Prey*. Cambridge, MA: Harvard University Press.

———. 2002. Avoiding predators: Expectations and evidence in primate antipredator behavior. *International Journal of Primatology*, 23, 741–757.

Starin, E. D. 1991 Socioecology of the red colobus monkey in the Gambia with particular reference to female-male differences and transfer patterns. PhD thesis, City University of New York, New York.

———. 1994. Philopatry and affiliation among red colobus monkeys. *Behaviour*, 130, 253–270.

———. 2001. Patterns of inbreeding avoidance in Temminck's red colobus. *Behaviour*, 138, 453–465.

Steenbeek, R. 1999. Tenure related changes in wild Thomas's langurs I: Between-group interactions. *Behaviour*, 136, 595–626.

Steenbeek, R. & van Schaik, C. P. 2001. Competition and group size in Thomas's langurs (*Presbytis thomasi*): The folivore paradox revisited. *Behavioural Ecology and Sociobiology*, 49, 100–110.

Steenbeek, R., Sterck, E. H. M., de Vries, H. & van Hooff, J. A. R. A. M. 2000. Costs and benefits of the one-male, age-graded and all-male phase in wild Thomas's langur groups. In *Primate Males* (ed. by P.M. Kappeler), 130–145. Cambridge: Cambridge University Press.

Sterck, E. H. M. 1995. Females, foods and fights: A socioecological comparison of the sympatric Thomas langur and long-tailed macaque. PhD thesis, Utrecht University, Utrecht.

———. 1997. Determinants of female dispersal in Thomas langurs. *American Journal of Primatology*, 42, 179–198.

———. 1998. Female dispersal, social organization, and infanticide in langurs: Are they linked to human disturbance? *American Journal of Primatology*, 44, 235–254.

Sterck, E. H. M. & Korstjens, A. H. 2000. Female dispersal and infanticide avoidance in primates. In *Infanticide by Males and Its Implications* (ed. by C. P. van Schaik & C. H. Janson), 293–321. Cambridge: Cambridge University Press.

Sterck, E. H. M. & Steenbeek, R. 1997. Female dominance relationships and food competition in the sympatric Thomas langur and long-tailed macaque. *Behaviour* 134, 749–774.

Sterck, E. H. M. & van Hooff, J. A. R. A. M. 2000. The number of males in langur groups: Monopolizability of females or demographic processes? In *Primate Males* (ed. by P. M. Kappeler), 120–129. Cambridge: Cambridge University Press.

Sterck, E. H. M., Watts, D. P. & van Schaik, C. P. 1997. The evolution of female social relationships in nonhuman primates. *Behavioural Ecology and Sociobiology*, 41, 291–309.

Sterck, E. H. M., Willems, E. P., van Hooff, J. & Wich, S. A. 2005. Female dispersal, inbreeding avoidance and mate choice in Thomas langurs (*Presbytis thomasi*). *Behaviour*, 142, 845–868.

Sterner, K. N., Raaum, R. L., Zhang, Y. P., Stewart, C.-B. & Disotell, T. R. 2006. Mitochondrial data support an odd-nosed colobine clade. *Molecular Phylogenetics and Evolution*, 40, 1–7.

Struhsaker, T. 1975. *The Red Colobus Monkey*. Chicago: University of Chicago Press.

Struhsaker, T. T. & Leland, L. 1979. Socioecology of five sympatric monkey species in the Kibale Forest, Uganda. *Advances in the Study of Behavior*, 9, 159–228.

———. 1985. Infanticide in a patrilineal society of red colobus monkeys. *Zeitschrift für Tierpsychologie*, 69, 89–132.

Struhsaker, T. T. & Leakey, M. 1990. Prey selectivity by crowned hawk-eagles on monkeys in the Kibale Forest, Uganda. *Behavioral Ecology and Sociobiology*, 26, 435–443.

Struhsaker, T. T., Marshall, A. R., Detwiler, K., Siex, K., Ehardt, C., Libjerg, D. D. & Butynski, T. M. 2004. Demographic variation among Udzungwa red colobus in relation to gross ecological and sociological parameters. *International Journal of Primatology*, 25, 615–658.

Struhsaker, T. T. & Pope, T. R. 1991. Mating system and reproductive success: A comparison of two African forest monkeys (*Colobus badius* and *Cercopithecus ascanius*). *Behaviour*, 117, 182–205.

Sugiyama, Y. 1964. Group composition, population density and some sociological observations of hanuman langurs (*Presbytis entellus*). *Primates*, 5, 7–37.

———. 1965. On the social change of Hanuman langurs (*Presbytis entellus*) in their natural habitat. *Primates*, 6, 381–418.

———. 1966. An artificial social change in a Hanuman langur troop (*Presbytis entellus*). *Primates*, 7, 41–72.

———. 1976. Characteristics of the ecology of the Himalayan langurs. *Journal of Human Evolution*, 5, 249–277.

Supriatna, J., Manullang, B. O. & Soekara, E. 1986. Group composition, home range, and diet of the maroon leaf monkey (*Presbytis rubicunda*) at Tanjung Puting Reserve, Central Kalimantan, Indonesia. *Primates*, 27, 185–190.

Sugiyama, Y. & Parthasarathy, M. D. 1979. Population change of the Hanuman langur (*Presbytis entellus*), 1961–1976, in Dharwar area, India. *Journal of the Bombay Natural History Society*, 75, 860–867.

Sussman, R. W., Cheverud, J. M. & Bartlett, T. Q. 1995. Infant killing as an evolutionary strategy: Reality or myth? *Evolutionary Anthropology*, 3, 149–151.

Tan, C. L., Guo, S. & Li, B. 2007. Population structure and ranging patterns of *Rhinopithecus roxellana* in Zhouzhi National Nature Reserve, Shaanxi, China. *International Journal of Primatology*, 28, 577–591.

Teelen, S. 2008. Influence of chimpanzee predation on the red colobus population at Ngogo, Kibale National Park, Uganda. *Primates*, 49, 41–49

Teichroeb, J. A. & Sicotte, P. 2008. Infanticide in ursine colobus monkeys (*Colobus vellerosus*) in Ghana: New cases and a test of the existing hypotheses. *Behaviour*, 145, 727–755.

———. 2009. Test of the ecological-constraints model on ursine colobus monkeys (*Colobus vellerosus*) in Ghana. *American Journal of Primatology*, 71, 49–59.

Teichroeb, J. A., Wikberg, E. C. & Sicotte, P. 2009. Female dispersal patterns in six groups of ursine colobus (*Colobus vellerosus*): Infanticide avoidance is important. *Behaviour*, 146, 551–582.

Tenaza, R. R. & Fuentes, A. 1995. Monandrous social organization of pigtailed langurs (*Simias concolor*) in the Pagai Islands, Indonesia. *International Journal of Primatology*, 16, 295–310.

Tilson, R. L. 1977. Social organization of simakobu monkeys (*Nasalis concolor*) in Siberut Island, Indonesia. *Journal of Mammalogy*, 58, 202–212.

Ting, N. 2008. Mitochondrial relationships and divergence dates of the African colobines: Evidence of Miocene origins for the living colobus monkeys. *Journal of Human Evolution*, 55, 312–325.

Ting, N., Tosi, A. J., Li, Y., Zhang, Y. P. & Disotell, T. R. 2008. Phylogenetic incongruence between nuclear and mitochondrial markers in the Asian colobines and the evolution of the langurs and leaf monkeys. *Molecular Phylogenetics and Evolution*, 46, 466–474.

Treves, A. 1998a. The influence of group size and neighbors on vigilance in two species of arboreal monkeys. *Behaviour*, 135, 453–481.

———. 1998b. Primate social systems: Conspecific threat and coercion-defense hypotheses. *Folia Primatologica*, 69, 81–88.

Treves, A. & Chapman, C. A. 1996. Conspecific threat, predation avoidance, and resource defense: Implications for grouping in langurs. *Behavioural Ecology and Sociobiology*, 39, 43–53.

Tutin, C. E. G., Ham, R. M., White, L. J. T. & Harrison, M. J. S. 1997. The primate community of the Lope Reserve, Gabon: Diets, responses to fruit scarcity, and effects on biomass. *American Journal of Primatology*, 42, 1–24.

Uehara, S. 1997. Predation on mammals by the chimpanzee (*Pan troglodytes*). *Primates*, 38, 193–214.

Van Schaik, C. P. 1989. The ecology of social relationships amongst female primates. In *Comparative Socioecology* (ed. by V. Standen & R. A. Foley), 195–218. Oxford: Blackwell.

———. 2000. Vulnerability to infanticide: Patterns among mammals. In *Infanticide by Males and Its Implications* (ed. by C. P. van Schaik & C. H. Janson), 61–71. Cambridge: Cambridge University Press.

Van Schaik, C. P. & Hörstermann, M. 1994. Predation risk and the number of adult males in a primate group: A comparative test. *Behavioral Ecology and Sociobiology*, 35, 261–272.

Van Schaik, C. P., Assink, P. R. & Salafsky, N. 1992. Territorial behavior in Southeast Asian langurs: Resource defense or mate defense? *American Journal of Primatolology*, 26, 233–242.

Vasudev, D., Kumar, A. & Sinha, A. 2008. Resource distribution and group size in the common langur *Semnopithecus entellus* in southern India. *American Journal of Primatology*, 70, 680–689.

Vedder, A. & Fashing, P. J. 2002. Diet of a 300-member Angolan colobus monkey (*Colobus angolensis*) supergroup in the Nyungwe Forest, Rwanda. *American Journal of Physical Anthropology Supplement*, 34, 159–160.

Wachter, B., Schabel, M. & Noë, R. 1997. Diet overlap and polyspecific associations of red colobus and Diana monkeys in the Tai National Park, Ivory Coast. *Ethology*, 103, 514–526.

Wasserman, M. D. & Chapman, C. A. 2003. Determinants of colobine monkey abundance: The importance of food energy, protein and fibre content. *Journal of Animal Ecology*, 72, 650–659.

Watanabe, K. 1981. Variations in group composition and population density of two sympatric Mentawaian leaf-monkeys. *Primates*, 22, 145–160.

Watts, D. P. & Mitani, J. C. 2002. Hunting behavior of chimpanzees at Ngogo, Kibale National Park, Uganda. *International Journal of Primatology*, 23, 1–28

Watts, D. P., Colmenares, F. & Arnold, K. 2000. Redirection, consolation, and male policing: How targets of aggression interact with bystanders. In *Natural Conflict Resolution* (ed. by F. Aureli & F. B. M. de Waal), 307–333. Berkeley and Los Angeles: University of California Press.

Whittaker, D. J., Ting, N. & Melnick, D. J. 2006. Molecular phylogenetic affinities of the simakobu monkey (*Simias concolor*). *Molecular Phylogenetics and Evolution*, 39, 887–892.

Wich, S. A. & Sterck E. H. M. 2007. Familiarity and threat of opponents determine variation in Thomas langur (*Presbytis thomasi*) male behaviour during between-group encounters. *Behaviour*, 144, 1583–1598.

Wich, S. A., Assink, P. R., Becher, F. & Sterck, E. H. M. 2002a. Playbacks of loud calls to wild Thomas langurs (Primates; *Presbytis thomasi*): The effect of familiarity. *Behaviour*, 139, 79–87.

———. 2002b. Playbacks of loud calls to wild Thomas langurs (Primates; *Presbytis thomasi*): The effect of location. *Behaviour*, 139, 65–78.

Wich, S. A., Assink, P. R. & Sterck, E. H. M. 2004. Thomas langurs (*Presbytis thomasi*) discriminate between calls of young solitary versus older group-living males: A factor in avoiding infanticide? *Behaviour*, 141, 41–51.

Wich, S. A., Steenbeek, R., Sterck, E. H. M., Korstjens, A. H., Willems, E. P. & van Schaik, C. P. 2007. Demography and life history of Thomas langurs (*Presbytis thomasi*). *American Journal of Primatology*, 69, 641–651.

Wolf, K. 1984. Reproductive competition among co-resident male silvered leaf monkeys (*Presbytis cristata*). PhD thesis, Yale University.

Wolf, K. E. & Fleagle, J. G. 1977. Adult male replacement in a group of silvered leaf-monkeys (*Presbytis cristata*) at Kuala Selangor, Malaysia. *Primates*, 18, 949–955.

Xiang, Z. F. & Grueter, C. C. 2007. First direct evidence of infanticide and cannibalism in wild snub-nosed monkeys (*Rhinopithecus bieti*). *American Journal of Primatology*, 69, 249–254.

Yeager, C. P. 1989. Feeding ecology of the proboscis monkey (*Nasalis larvatus*). *International Journal of Primatology*, 10, 497–530.

———. 1990. Proboscis monkey (*Nasalis larvatus*) social organization: Group structure. *American Journal of Primatology*, 20, 95–106.

———. 1991. Proboscis monkey (*Nasalis larvatus*) social organization: Intergroup patterns of association. *American Journal of Primatology*, 23, 73–86.

Yoshiba, K. 1968. Local and intertroop variability in ecology and social behavior of common Indian langurs. In *Primates: Studies in Adaptation and Variability* (ed. by P. C. Jay), 217–242. New York: Holt, Rinehart and Winston.

Zhang, P., Watanabe, K. & Li, B. G. 2008. Female social dynamics in a provisioned free-ranging band of the Sichuan snub-nosed monkey (*Rhinopithecus roxellana*) in the Qinling Mountains, China. *American Journal of Primatology*, 70, 1013–1022.

Zhao, D. P., Ji, W. H., Li, B. G. & Watanabe, K. 2008a. Mate competition and reproductive correlates of female dispersal in a polygynous primate species (*Rhinopithecus roxellana*). *Behavioural Processes*, 79, 165–170.

Zhao, Q., Tan, C. L. & Pan, W. S. 2008b. Weaning age, infant care, and behavioral development in *Trachypithecus leucocephalus*. *International Journal of Primatology*, 29, 583–591.

Zhou, Q. H., Wei, F. W., Li, M., Huang, C. M. & Luo, B. 2006. Diet and food choice of *Trachypithecus francoisi* in the Nonggang Nature Reserve, China. *International Journal of Primatology*, 27, 1441–1460.

Zhou, Q., Wei, F., Huang, C., Li,, M., Ren, B. & Luo, B. 2007. Seasonal variation in the activity patterns and time budgets of *Trachypithecus francoisi* in the Nonggang Nature Reserve, China. *International Journal of Primatology*, 28, 657–671.

Zuberbühler, K. & Jenny, D. 2002. Leopard predation and primate evolution. *Journal of Human Evolution*, 43, 873–886.

Chapter 5 The Behavior, Ecology, and Social Evolution of Cercopithecine Monkeys

Marina Cords

THE CERCOPITHECINE monkeys, subfamily Cercopithecinae, belong to the catarrhine family Cercopithecidae, along with their colobine cousins (chapter 4, this volume). The cercopithecines comprise 12 genera grouped in two tribes, the *Cercopithecini* (guenons) and *Papionini* (baboons, mangabeys, and macaques; table 5.1). Behavioral research has focused more on the papionins, most likely because several papionin taxa are common, live semiterrestrially in relatively open habitats facilitating observation, and are often kept successfully in captive groups approximating natural group sizes. Vervets (*Chlorocebus* spp.) are the exception that proves the rule in this context; they are one of two cercopithecins that live in savanna-woodland instead of forest habitats, and their behavior has been studied more thoroughly than that of other members of their tribe. Together, certain baboons, macaques, and vervets have become iconic as representatives of cercopithecine monkey behavioral biology, but they do not represent the full range of social behavior in their respective genera, much less the subfamily as a whole. In this chapter I review the behavior of cercopithecine monkeys living in natural populations.

Diversity, Biogeography, and Evolutionary History

Diversity and Biogeography

Most cercopithecine genera occur in Africa, where exceptionally diverse communities in rain forests include four or more species, along with colobines and apes. Rivers and unsuitable (often drier) habitat between forest patches constitute important biogeographic boundaries for these forest-dwelling species. A few forest dwellers, such as gray-cheeked mangabeys (*Lophocebus albigena*), blue monkeys (*Cercopithecus mitis*), and DeBrazza's monkeys (*Cercopithecus neglectus*), have, however, achieved an exceptionally broad distribution, spanning the continent within the forest zone.

Unlike the African forest-dwelling taxa, the *Papio* baboons, geladas (*Theropithecus gelada*), vervets, and patas (*Erythrocebus patas*) monkeys are adapted to life in more open and seasonal environments, mainly in drier areas surrounding the central African forest zone (Newman et al. 2004). Baboons and vervets are particularly common, likely reflecting the behavioral and ecological flexibility that predisposes them to coexist with humans (Jablonski 2002; Elton 2007). Geladas, in contrast to other open-country papionins, have a far more restricted range in high-altitude Ethiopian grasslands.

The only cercopithecines that live primarily in Asia are the macaques, a diverse and highly successful group. As a genus, *Macaca* has the broadest distribution of any nonhuman primate, from Morocco to Japan. Certain "weed" species, namely rhesus (*Macaca mulatta*), long-tailed (*Macaca fascicularis*), bonnet (*Macaca radiata*), and toque (*Macaca sinica*) macaques, have high ecological tolerance and flourish in human-altered environments, allowing them to occupy broad geographic ranges (Richard et al. 1989). Forest-dwelling species often co-occur with colobines and apes. In some forests, up to four macaque species co-occur as well (Borries et al. 2002), but many macaques do not live with other cercopithecines. The Barbary macaque (*Macaca syl-*

vanus) occurs mainly in north Africa, the only nonhuman primate north of the Sahara. This species was also introduced to Gibraltar (Modolo et al. 2005).

Cercopithecines are widely acknowledged as a taxonomically diverse branch of the primate family tree, including two of the most species-rich genera (*Cercopithecus* and *Macaca*, table 5.1). The precise degree of diversity within the clade is, however, still debated. Surprisingly, new forms previously unknown to science have been described recently. These include the kipunji (*Rungwecebus kipunji*) of Tanzania (Jones et al. 2005), distinguished as a new genus (Davenport et al. 2006), and the Arunachal macaque (*Macaca munzala*) from northeastern India (Sinha et al. 2005a; Chakraborty et al. 2007). The paraphyly of mangabeys (*Cercocebus* and *Lophocebus* genera) was also quite recently confirmed (Disotell 1994; Tosi et al. 2003; Gilbert 2007). In addition, there are ongoing debates about the taxonomic status of other forms. For example, recent recognition of a clade of terrestrial guenons, supported by both genetic and morphological evidence, suggests adjustments to the genus assignments of vervet, patas, and L'Hoest's (*Cercopithecus lhoesti*) monkeys (Tosi et al. 2003a; Xing et al. 2007; Sargis et al. 2008). Species-versus-subspecies distinctions have recently changed or remain controversial for several taxa within *Cercopithecus*, *Cercocebus*, *Macaca*, *Miopithecus*, and *Papio*. This chapter follows Groves's (2005) taxonomy while acknowledging that it is not universally accepted for these primates.

As a group, cercopithecines occupy an extremely diverse range of habitats, exceeding all other major phylogenetic groups of primates in this regard. These habitats include rain forest, gallery forest, swamps, savanna woodland, high-altitude grasslands, broadleaf deciduous forest, coniferous forest, subdesert thorn-scrub, and human-dominated landscapes including roadsides, agricultural land, towns, and cities. While most taxa live in tropical habitats, Barbary and some Japanese (*Macaca fuscata*) macaques live in temperate zones where it snows in winter. Geladas, confined to high-altitude grasslands and cliffs up to more than 4,000 meters above sea level, also experience climatic extremes with subzero temperatures and hailstorms in the wet season (Iwamoto 1983). Like Japanese macaques at northern latitudes, baboons in southern Africa face habitats with markedly seasonal variation, including significant day-length changes that may influence time budgets and feeding strategies (Hill et al. 2003).

Hunting and habitat loss have recently pushed several cercopithecines to the brink of extinction, especially where geographic range and population size were limited historically. Included among the world's 25 most endangered primates in 2009 were the Roloway monkey (*Cercopithecus diana roloway*) and kipunji (Mittermeier et al. 2009). Four species were listed as critically endangered in the 2009 IUCN Red List (kipunji; Dryas monkey, *Cercopithecus dryas*; Celebes crested macaque, *Macaca nigra*; Pagai Island macaque, *Macaca pagensis*), thus reflecting small population size or severe declines in very limited geographic ranges. Another nine species were endangered (one guenon; two *Cercocebus* mangabeys; the drill, *Mandrillus leucophaeus*; and five macaques) and still another 19 were classified as vulnerable. Altogether, 48% of 67 cercopithecine species recognized by the IUCN were threatened, and 58% are decreasing in number. Many of these species raid crops, exacerbating their conflict with humans.

Evolutionary History

The earliest known cercopithecine fossils are 7-million-year-old macaque-like monkeys from northern Africa, but molecular evidence places the cercopithecine-colobine split at about 16 million years ago, with a subsequent split between papionins and cercopithecins 9 to 11.5 million years ago (Raaum et al. 2005; Tosi et al. 2005; Steiper & Young 2006).

Among papionins, macaques diverged from Afro-papionins 6 to 11 million years ago (Harris 2000; Page & Goodman 2001; Raaum et al. 2005; Tosi et al. 2005; Steiper & Young 2006). Early macaques occurred around the Mediterranean, but by 5 million years ago they reached Asia (Delson 2007). The Barbary macaque, the most primitive living macaque, is a relict of early Mediterranean forms (Jablonski 2002), whereas all other extant macaques represent Asian lineages. Jablonski (2002) argued that the evolutionary success of macaques derives from dietary flexibility and relatively fast life histories that allowed them to thrive where contemporaneous hominins could not. Macaques comprise three to five phyletic groups, depending on placement of the Barbary and stumptail (*Macaca arctoides*) macaques (Tosi et al. 2003b). These groups derive from successive dispersals into Asia (Fooden 1976; Delson 1980), where rapid differentiation was driven by interspecific competition and climate-induced changes in forest cover (Fa 1989; Richard et al. 1989; Jablonski 2002; Thierry 2007). Life history traits, along with behavioral flexibility, temperament, and physical characteristics (terrestrial speed and agility) distinguish certain "weed" species that thrive in human-modified environments (Richard et al. 1989). The ecological tolerance of "weed" species, which do not match phylogenetic groupings, may have influenced distribution patterns and evolution via interspecific competition. Recent differentiation of modern taxa is likely related to climate change, which influenced habitat types, seawater levels, and geographic isolation of suitable habitat (Fa 1989).

In Africa, the split between *Papio/Lophocebus/Rungwe-*

Table 5.1. Cercopithecine diversity: Taxonomy, body size, and habitat

Tribe	Genus	No. of species	Common name	Adult female body mass (kg)[1]	Sexual dimorphism (M:F)[1]	Habitat	Arboreal or terrestrial?[4]
Cercopithecini	Allenopithecus	1	Swamp monkey	3.5	1.8	Forest	Semiterrestrial
	Cercopithecus	25	Forest guenon[2]	3.4	1.6	Forest	Arboreal
	Chlorocebus	6	Vervet/grivet/green monkey	3.5	1.5	Savanna woodland	Semiterrestrial
	Erythrocebus	1	Patas monkey	6.5	1.9	Savanna woodland	Semiterrestrial
	Miopithecus	2	Talapoin	1.6	1.2	Forest	arboreal
Papionini	Cercocebus	6	Ground-dwelling mangabey	5.7	1.8	Forest	Semi-terrestrial
	Lophocebus/ Rungwecebus	4[3]	Arboreal mangabey	5.9	1.4	Forest	Arboreal
	Macaca	22[3]	Macaque	6.1	1.6	Variable: forest to semidesert to human-altered	Both arboreal and semiterrestrial
	Mandrillus	2	Drill, mandrill	8.8	2.9	Forest	Semiterrestrial
	Papio	5	Baboon	11.9	1.8	Variable: forest to semi-desert to human-altered	Semi-terrestrial
	Theropithecus	1	Gelada	11.7	1.6	Mountain grassland	Terrestrial

[1] Data are averaged across species or populations as presented by Smith & Jungers 1997, excluding values designated captive or provisioned. by these authors. Values for Mandrillus come from Setchell et al. 2001 and Schaaf et al. in press.
[2] The term "guenon" is sometimes used more broadly to include all cercopithecins.
[3] New forms added from post-2005 reports of new taxa, Macaca munzala and Rungwecebus kipunji. Body mass are data are not available for these taxa.
[4] Note that many arboreal forms come down to the ground occasionally to feed.

cebus/Theropithecus and Mandrillus/Cercocebus probably followed soon after the macaques branched off (Harris 2000; Page & Goodman 2001; Raaum et al. 2005; Tosi et al. 2005; Steiper & Young 2006). In the Plio-Pleistocene, Parapapio and Theropithecus baboons were ubiquitous, inhabiting diverse habitats including forests (Elton 2007), but they were largely replaced by the end of the Pleistocene with Papio baboons. The earliest fossil form in the Cercocebus/Mandrillus lineage appeared 1.5 to 2 million years ago (Gilbert 2007), and it suggests that Cercocebus mangabeys, like baboons (Newman et al. 2004), originated in southern Africa and radiated eastward and northward as the climate dried.

Guenon fossils are rare, with poor preservation in forest habitats. Knowledge of guenon evolution therefore derives largely from molecular and morphological studies of extant forms (Tosi et al. 2003a, 2005). Molecular analyses indicate four phylogenetic units within this species-rich clade, with the Allenopithecus, Miopithecus, and terrestrial (patas, vervet, and L'Hoest's monkey) lineages diverging successively from 9.3 ± 1.0 to 4.8 ± 1.2 million years ago. Further division among the remaining arboreal guenons occurred about 3.5 million years ago. Morphological analyses do not group the three terrestrial guenons together, suggesting that after divergence from a common ancestor, terrestrial adaptations developed especially in patas and L'Hoest's monkeys (Sargis et al. 2008). Divergence dates among guenons are considerably older than previously thought, with the radiation of arboreal guenons beginning before the massive climatic and habitat changes associated with Pleistocene glaciations (Tosi et al. 2005). Fluctuations in forest cover, and presence of refugia during dry periods, were probably important in the differentiation of modern taxa (Hamilton 1988; Tosi 2008).

Ecology and Life History

Body Size

Most cercopithecines are medium-sized primates with body masses of 4 to 12 kg (table 5.1). At one extreme is the talapoin (Miopithecus talapoin), weighing less than 2 kg, about half as much as other guenons. At the other extreme, and more than 10 times larger than the talapoin, are Papio baboons, geladas (14–22 kg), and mandrills (Mandrillus sphinx, males >30 kg: Setchell et al. 2001). The heaviest cercopithecines are the most terrestrial, with baboons, mandrills, drills, and especially geladas spending much if not most of their active time on the ground. Cercocebus mangabeys and macaques also spend considerable time foraging on the ground, and are generally considered semiterrestrial. A few guenons, even forest dwellers, are also semiterrestrial, and some arboreal species occasionally feed on the ground.

Males are larger than females in all cercopithecines, with especially strong body mass dimorphism in larger species

(Plavcan 2001); males may outweigh females by a factor of two (*Papio* baboons) to three (mandrills, Setchell et al. 2001; fig. 5.1). Among most other cercopithecines, male-female body mass ratios range from 1.4 to 1.9 (table 5.1). Dimorphism arises in distinct ways in these monkeys, differentiated largely according to social organization; in species living in multimale groups, males primarily grow for longer periods (but see Schillaci and Stallmann 2005), while in species with one-male groups, males mainly grow faster than females, with a marked growth spurt (Leigh 1995).

Brain Size

Among primates as a whole, cercopithecines have large brains for their body size. Together with hominoids and noncallitrichine platyrrhines, they constitute the highest of three grades in the positive relationship between endocranial volume and body size (Isler et al. 2008). Cercopithecines have lower relative brain sizes than hominoids, but relatively larger brains than colobines.

Within cercopithecines, body size explains about ≥ 90% of the variation in endocranial volume (Isler et al. 2008). Data from Isler et al. show no systematic variation in relative endocranial volume at taxonomic levels above the species. The greatest deviation from expected values, based on female body weights in sexually dimorphic species (Isler et al. 2008), occurred in Barbary macaques and Diana monkeys (*Cercopithecus diana*), both of which are unexpectedly small-brained for their size. For Barbary macaques, Isler et al. suggested that energetic demands of life in an extraordinarily seasonal environment may have limited brain size. Additional data on brain size for this species are needed to test this hypothesis.

Life History

Cercopithecines vary in how much breeding seasonality they show. Guenons typically breed seasonally, both in forests and savanna woodland (Butynski 1988). Afro-papionins may breed throughout the year, but mangabeys and *Mandrillus* species typically show birth peaks; *Papio* and gelada baboons are the least seasonal, with no seasonality detectable in many populations (Swedell 2010). Among macaques, about half (in temperate and more seasonally varying environments) are seasonal while half (in tropical environments) are not (Bercovitch & Harvey 2004; Thierry 2007).

All cercopithecines routinely give birth to one offspring at a time (Kappeler & Pereira 2003, appendix, present summarized life history data). For females, first reproduction typically occurs in the fourth or fifth year, although for

Fig. 5.1. All cercopithecines are sexually dimorphic, with mandrills (*Mandrillus sphinx*) showing the most extreme dimorphism of all: males are three times heavier than females. Photo by Joanna Setchell.

Papio, *Cercocebus*, and blue monkeys it can occur in the sixth or seventh year (Leigh & Bernstein 2006; Cords & Chowdhury 2010), and for patas monkeys it is usually in the third year (Isbell et al. 2009). Gestation lasts approximately 160 to 180 days, and interbirth intervals are generally 12 to 24 months. Maximal lifespans for wild populations are poorly known, but females may survive into their twenties and beyond in some taxa (Alberts & Altmann 2003; Bercovitch & Harvey 2004; Cords & Chowdhury 2010).

While body size correlates with many life history variables across primates, so too do environmental variables, most likely through their effects on mortality schedules (Ross 1992a, 1998; chapter 10, this volume). Such effects are apparent among cercopithecines, whose diverse habitats drive variable exposure to predation risk and seasonal or unpredictable food supply. In eight macaque species, for example, those limited to broadleaf-evergreen forest had later ages of first female reproduction, longer interbirth intervals, and thus a lower intrinsic rate of natural increase (r_{max}) than "opportunistic" broadly sympatric species inhabiting highland areas, forest edges, and towns (Ross 1992b; Singh & Sinha 2004). Similar patterns characterize African forest vs. open-country cercopithecines (Ross 1998). Data linking these habitat-related differences to variation in mortality schedules are limited, but comparisons of forest versus open-country guenons strongly suggest that high adult mortality associated with open habitats selects for faster life histories; this one clade apparently includes both the fastest-breeding (patas) and slowest-breeding (blue monkey) cercopithecines (Isbell et al. 2009; Cords & Chowdhury 2010).

Cercopithecine mothers are the primary caretakers of offspring, and carry clinging infants for several months. Although early reviews emphasized a contrast with colobines, many of which exhibit extensive allomaternal care, many cercopithecines also exhibit considerable allocare, primarily by nonmaternal females (Chism 2000; Ross 2003). For example, infant guenons often occur in crèches, monitored by a caretaker that may not be a mother (Chism 2000). Allomaternal care by females seems to occur earlier and more extensively in guenons than in baboons and macaques, whose mothers more firmly resist the attention of other females to their infants. Differences among cercopithecine taxa are not simply related to any single variable (Paul 1999; Chism 2000), but may reflect more relaxed social relations among guenon females, making it easier for a mother to retrieve her infant and less likely that it will be roughly handled, as well as reflecting the availability of kin in smaller groups and energetic constraints on mothers, especially in species with seasonal reproduction and short interbirth intervals (Chism 2000; Ross 2003).

Some young cercopithecines also receive care from adult males. In contrast to allocare from females, however, males rarely interact directly with infants and it is primarily their vigilance and antagonism to other males that reduces predation and infanticide risk for the young (chapters 8 and 19, this volume). More direct forms of male care occur in a few species. Male *Cercocebus* and *Lophocebus* mangabeys carry infants at least occasionally (Ehardt & Butynski in press; McGraw in press; Olupot & Waser in press; Shah in press). Barbary and Thibetan (*Macaca thibetana*) macaque males extensively handle infants (fig. 5.2), but this behavior appears to benefit primarily the males and their relationships with other group members and, at least in Barbary macaques, does not involve fathers of the infants (Ogawa 1995; Kummerli & Martin 2008). Follower male hamadryas (*Papio hamadryas*) may care for young infant females in a maternal way before moving off with them years later to form their own one-male units (Kummer 1995). Male yellow baboons (*Papio cynocephalus*) support and protect their own offspring (Buchan et al. 2003; Nguyen et al. 2009; chapter 18, this volume).

There are limited comparative data on juvenile development across cercopithecines. Leigh and colleagues (Leigh 1995; Pereira & Leigh 2003; Leigh & Bernstein 2006) illustrate varied trajectories of physical development in papionins. Differences among them reflect both differential growth rates and duration of growth periods. Some variation may

Fig. 5.2. Two adult male Thibetan macaques (*Macaca thibetana*) show "bridging" behavior. Such triadic interactions between two adults and an infant occur in several papionin species, and appear to be signals of affiliation among adults. Photo by Carol Berman.

relate to differences in social behavior. For example, Pereira and Leigh contrasted the rapid acquisition of adult-sized canines in sooty mangabeys (*Cercocebus atys*) with the more gradual trajectory in Guinea baboons (*Papio papio*), which may relate to the independent acquisition of dominance status in the former versus matrilineal rank inheritance in the latter. They also contrasted *Papio* baboons, whose brain volume reaches adult size soon after birth, with mangabeys, where such growth takes an additional year (Leigh & Bernstein 2006). Such differences may reflect critical periods in socioecological development. In the case of baboons, for example, the early development of large brains and associated cognitive skills may be important for successful foraging of weanlings (Pereira & Leigh 2003; Leigh 2004). As a whole, cercopithecine immatures seem to have unusually low rates of postnatal brain growth compared to platyrhines and hominoids; this means that mothers invest most in the brain growth of their young prenatally, and likely incur relatively high metabolic costs during pregnancy (Leigh 2004).

Diets

Cercopithecines typically have diverse diets, including multiple items (fruits, seeds, flowers, leaves, shoots, bark) that come from dozens of plant species. They also consume invertebrate and occasionally vertebrate prey. Swedell (2010) emphasized the opportunistic but selective feeding of Afropapionins, which choose items based on nutritional value;

this combination of opportunism and selectivity likely characterizes all cercopithecines, which exhibit flexible diets across populations and within single groups over time (Chapman et al. 2002; Ménard 2004). One's view of diversity versus selectivity depends partly on the time frame. At one time, the diet may be dominated by a few particular foods, but over longer periods, diversity and flexibility become more conspicuous. For many species, especially those living in tropical forests, fruit is a major component of the diet when it is available, at times accounting for 70% to 90% of the diet; when fruit is scarce, other items may predominate, including seeds, leaves, stems, and invertebrates (Ménard 2004; Thierry 2007; Enstam & Isbell 2007; Swedell 2010). For *Papio* baboons and rhesus macaques outside of forests, grasses are a fallback food, while geladas feed on grasses throughout the year (Ménard 2004; Swedell 2010).

Cercocebus mangabeys and *Mandrillus* species have specific morphological adaptations for ground foraging on hard seeds and nuts. These include large thick-enameled premolars, robust jaw musculature, and forelimb features for extensive, forceful manipulation of litter on the forest floor (Fleagle & McGraw 2002). Many such features are shared with macaques, and thus are likely primitive for the papionin clade. In African communities, monkeys with these features eat dried fruits and seeds from the ground, providing a relatively nutritious alternative to the ripe fruits they also consume in the canopy. Sympatric guenons are comparatively wasteful, consuming mainly the ripest fruits from the canopy and discarding unpalatable fruit pulp and large seeds. All cercopithecines have cheek pouches which are used to store fruits and seeds, and occasionally leaves as well (fig. 5.3). Cheek pouches allow feeding monkeys to reduce their exposure to predators, but may also reduce intraspecific feeding competition (Smith et al. 2008).

Within the major groups of cercopithecines, various dietary contrasts between sets of species have been noted. Among guenons, for example, the open country vervet and patas monkeys eat far less fruit than their forest-dwelling relatives, and among the latter, semiterrestrial species tend to eat more leaves and stems than their more arboreal counterparts (Enstam & Isbell 2007). Among Afro-papionins there appears to be a gradient of dietary flexibility, from *Papio* baboons whose diets are described as so eclectic as to defy generalization (Jolly 2007), to mangabeys, mandrills, and drills, with intermediate but still considerable degrees of variation, to geladas, whose morphological and behavioral adaptations stand testament to the constant importance of grass in their diet (Swedell 2010). Among macaques, tropical forest species are the most frugivorous, but some can become primarily folivorous when fruit is scarce. Macaques living in more temperate zones—such as Barbary macaques, rhesus macaques in the Himalayan foothills, and

Fig. 5.3. All cercopithecine primates (here, a blue monkey, *Cercopithecus mitis*) have cheek pouches used for temporary storage of food. Their use minimizes the risk of predation and intraspecific competition, as the monkey can retreat to safer locations for further processing of the cheek pouch contents. Photo by Marina Cords.

Japanese macaques in northerly latitudes—have diets focused on leaves and seeds (Thierry 2007), and these may be considerably less diverse than those of tropical forest-dwelling species (Ménard 2004).

It is worth noting that while cercopithecine behavior is generally well studied, data on natural diets are quite limited. The general flexibility of feeding behavior means that characterizing species-typical patterns requires studies spanning considerable temporal and spatial extents (Chapman et al. 2002), and yet many species—especially in the two major genera *Cercopithecus* and *Macaca*—have not been studied systematically even once (Chapman et al. 2002; Ménard 2004; Enstam & Isbell 2007). Furthermore, to relate feeding behavior to social patterns, researchers must attend to the precise spatial distribution and depletion time of food resources consumed at any one time. Given the catholic and flexible diet of these monkeys, the quality and defendability of food may be highly variable, and it can even be difficult to define a food patch when members of a group are in close proximity but are feeding on several different foods at once.

Predation

Cercopithecines are vulnerable to predators including large cats, raptors, snakes, crocodiles, and, for forest-dwelling African monkeys, chimpanzees (chapter 8, this volume). For many, humans hunting bushmeat or thwarting crop raiding are the most significant predators today.

Large-bodied cercopithecines may actively confront predators. For example, baboons sometimes collectively mob, chase, and kill terrestrial predators; such behavior is often carried out by adult males (Cowlishaw 1994), but is not necessarily limited to them (Cheney et al. 2004). In small-bodied mangabeys and guenons, only adult males attack eagles (Cordeiro 2003; Arlet & Isbell 2009); in the case of one kipunji, such an attack was fatal for the bird (Jones et al. 2006). Most species have conspicuous alarm calls (chapters 8 and 29, this volume) that trigger wariness and flight in other group members and other species, but some, like mona (*Cercopithecus mona*), white-throated (*Cercopithecus erythrogaster*), and DeBrazza's guenons, flee silently instead (Gautier-Hion in press-c; Glenn et al. in press; Oates in press), a strategy that is probably most effective when groups are small. Scant data exist about the interactions of macaques and their predators, because few studies have been carried out where natural predators remain. In long-tailed macaques, large groups detect danger at a greater distance than do smaller groups (van Schaik et al. 1983), and groups are smaller in places where predators are absent (van Schaik & van Noordwijk 1985).

Predator evasion tactics vary as a function of habitat use and the degree to which animals are terrestrial or arboreal. For example, baboons and *Cercocebus* mangabeys feeding in forest trees flee on the ground, while sympatric arboreal guenons and *Lophocebus* mangabeys use the canopy. The guenons that live near rivers, such as *Allenopithecus*, *Miopithecus*, and *Cercopithecus neglectus*, may plunge into water and swim to safety (Gautier-Hion in press-a, c, d). Some macaques are also good swimmers, and, like riverine guenons, may sleep over water to increase safety at night (van Schaik et al. 1996; Ramakrishnan & Coss 2001).

Forest guenons and mangabeys also frequently form mixed-species groups. Such associations likely reduce predation risk while minimizing some costs of forming larger single-species groups that could similarly provide safety in numbers (Enstam & Isbell 2007). Among macaques, by contrast, ecological divergence among sympatric species (Fooden 1982) evidently inhibits polyspecific grouping under natural circumstances (Southwick & Southwick 1983; Wong & Ni 2000).

While smaller, more inexperienced individuals are probably especially vulnerable to predation, studies of eagle predation on African forest monkeys have failed to document selective predation on immatures (Mitani et al. 2001; McGraw et al. 2006). Similarly, Cheney et al. (2004) found that although deaths among juvenile and adult female baboons all seemed to be caused by predation, the mortality rate was higher for adults than for juveniles.

Activity and Sensory Adaptations

All cercopithecines are diurnal, and they typically intersperse periods of rest with periods of travel or feeding.

Many have routine sleeping sites, ranging from branches overhanging water for some guenons and macaques, tall *Acacia xanthophloea* trees for vervet monkeys, and other tall trees or groves for many other species. Cliffs, caves, and rocky overhangs may also be used for sleeping (Hamilton 1982; Swedell 2010).

Visual communication is important in these monkeys, which use diverse facial and postural signals (fig. 5.4) Forest guenons, as well as mandrills, are known for their bright coloration, and pelage patterns may enhance communicative displays (Kingdon 2007). In mandrills, facial coloration appears to signal male condition (Setchell et al. 2008). Vocal communication is also important, perhaps especially in forested or bushy habitats where vision is obstructed. While vocal repertoires have not been well documented for most species, it is clear that many use several calls in dif-

Fig. 5.4. A male crested black macaque (*Macaca nigra*) shows a silent bared-teeth face. In macaques from Sulawesi this expression is used in various affiliative contexts, whereas in other macaque species it indicates submission, occurring mainly in agonistic contexts. Photo by Nic Thompson.

ferent social contexts in addition to alarm calls (Gouzoules & Gouzoules 2007; chapter 29, this volume).

While cercopithecines seem to use olfactory communication less often than strepsirrhines and platyrrhines, both sniffing and scent-marking occur. Papionins and guenons sniff the mouths of feeding group mates to learn about food (Drapier et al. 2002; Laidre 2009) and males may sniff female perinea before mating or deciding not to (Dixson 1998). Cutaneous scent glands occur in some semiterrestrial species, such as *Allenopithecus, Chlorocebus, Mandrillus,* Hamlyn's monkey, *Cercopithecus hamlyni,* and DeBrazza's monkey, but the function of scent marking is not well understood, especially in the wild (Feistner 1991).

Evolution of Social Systems

Social Organization

All cercopithecines live in groups, which vary in size from less than 5 (DeBrazza's monkey, Gautier-Hion in press-c) to 845 (mandrill horde; Abernethy et al. 2002) and more than 1,200, the largest known grouping of any wild primate (gelada herd, Bergman & Beehner in press). Typically, however, groups include approximately 10 to 100 individuals. There can be considerable variation among groups within a single species (e.g., Yamagiwa & Hill 1998), with several macaques and agile mangabeys (*Cercocebus agilis*) exhibiting an order of magnitude difference between the smallest and largest groups reported (Ménard 2004; Shah in press). In contrast, membership in a single group is usually fairly stable over periods of several months to years.

In some species, group size and membership vary considerably over short time spans. Hamadryas baboons and geladas, and possibly Guinea baboons and *Mandrillus,* exhibit a hierarchical social organization in which small one-male units of approximately 10 to 20 animals are part of a larger band, comprising several such units plus unattached males (Swedell 2010). In hamadryas, an intermediate "clan" level of social organization, consisting of several one-male units whose leader males may be related, has also been reported (Abegglen 1984). These nested social units differ from simple subgrouping, which occurs in most cercopithecines in response to variable food distribution, in that the smallest one-male units have a fixed composition, and certain kinds of social interaction occur only within them. Like ordinary subgrouping, however, this hierarchical system permits considerable ecological flexibility with animals congregating in larger units at limited resources such as water holes and sleeping cliffs, or when food distribution allows it, but traveling in smaller units when food

is scattered and sparse (Kummer 1995). *Cercocebus* and *Lophocebus* mangabeys may also exhibit a hierarchical system, with groups occasionally flexibly merging into "super-groups" (Olupot & Waser in press; Shah in press) much as gelada bands can merge to form a herd. Such aggregations, however, are rarer in mangabeys.

For some species, male membership changes on a time frame of months to a year. In sooty mangabeys, some males join and leave the group for months at a time, while other males remain regular members (Range 2005). Mandrills show a similar pattern but on a seasonal basis, as adult males join the horde only when females have sexual swellings (Abernethy et al. 2002). Similarly, new males may join groups of guenons (Cords 1988) and Japanese macaques (Sprague et al. 1998) for a few months during the breeding season.

If group living provides advantages in terms of predation avoidance (chapter 8, this volume), one would expect group size to relate to predation risk; this pattern has been documented in cercopithecoids (Hill & Lee 1998) and among populations of long-tailed macaques (van Schaik & van Noordwijk 1985). In an interspecific analysis, Janson and Goldsmith (1995) confirmed that terrestrial primates, including *Papio, Theropithecus,* and some *Macaca* in their sample, lived in larger groups than their arboreal counterparts, all else being equal. Terrestriality was presumed to index predation risk because terrestrial animals have more limited escape options. Data on group sizes for African cercopithecines, however, show mixed results when related to the degree of terrestriality. Among the papionins, it is generally true that more terrestrial forms live in larger groups (geladas, 110 to 170; *Papio* baboons, typically 35 to 75; *Mandrillus,* 58 to 845; *Cercocebus* mangabeys, 20 to 30; and *Lophocebus/Rungwecebus* mangabeys, 15 to 20: Swedell 2010). Among the guenons, however, no such pattern emerges (Enstam & Isbell 2007). Terrestrial guenons lead relatively cryptic lives, however, which may preclude large groups. In addition, factors other than predation risk, such as feeding competition (chapter 7, this volume) may influence group size.

Another general characteristic of cercopithecine societies is that females remain in their natal groups for life, whereas males emigrate, typically at puberty, and may change groups several times (Pusey & Packer 1987). Thus matrilines form the stable core of the group, setting the stage for extensive cooperative bonding among female kin (chapter 21, this volume; fig. 5.5). The one notable exception to this pattern is the hamadryas baboon, in which both males and females disperse from their natal one-male units. In this species, "clans" constitute an intermediate level of social organization consisting of males that are thought to be related, along with females and juveniles in their one-male

Fig. 5.5. Cercopithecines are characterized by strong social bonds among female kin, as in this matrilineal family group of chacma baboons (*Papio ursinus*) on the Cape Peninsula, South Africa. Photo by Larissa Swedell.

units (Abegglen 1984); thus males may be more likely than females to live with adult kin (Colmenares 2004). In other species, exceptions to the general pattern of female philopatry occur as well (Moore 1984), but these are unusual.

The number of adult males in cercopithecine groups is variable, determined at least partly by the number and dispersion of females (Andelman 1986; fig. 5.6a, b). Groups with 5 females or fewer typically have one male, and those with 10 or more include multiple males, while groups with intermediate numbers of females have one or more than one male. Some species, such as most *Cercopithecus* and *Erythrocebus* guenons, typically have just one adult male per group, although breeding season influxes can temporarily increase the number of males dramatically (Cords 2000). Only three guenons (Allen's swamp monkey, *Allenopithecus nigroviridis*; vervets, and talapoins) live regularly and primarily in multi-male groups. For swamp monkeys and talapoins, large group size, seasonal reproduction, and male defense against predation may drive the multimale pattern, but too little is known of these species to be certain (Cords 2000). For vervets, whose groups are not especially large, such explanations seem insufficient (Henzi 1988; Cords 2000). Isbell et al. (2002) proposed that vervet males may

have limited dispersal options when groups are lined up along rivers, their preferred habitat, and costs of dispersal are high. Accordingly, males will often disperse into groups that contain related males, minimizing infanticide risk, and therefore increasing male-male tolerance. Intriguingly, talapoins and swamp monkeys are frequently found in riverine habitat, although DeBrazza's monkeys share that habitat preference and yet always live in one-male groups. Nothing is known of the dispersal risks or patterns for these forest-dwelling riverine taxa.

Groups of papionins are typically multimale unless they are very small; however, there are some population- or species-specific exceptions to this general pattern. For example, in contrast to most baboons, subalpine chacmas (*Papio ursinus*) live mainly in small one-male groups, apparently as a result of extreme scarcity of resources and a lack of predators (Barton 2000). In the nested social groups of hamadryas and geladas, the number of males obviously differs for the basal one-male unit and larger multimale bands; the latter, however, are typically viewed as the equivalent of other baboon groups (Swedell 2010). Finally, as noted above, mandrill hordes contain adult males only during the mating season (Abernethy et al. 2002). While the flexible

Fig. 5.6. Gelada baboons live in a multitiered society with one-male units as the most basic social unit. (A) Males compete aggressively for females. Photo by Jeff Kerby. (B) Some one-male units, however, contain one or more follower males which sometimes have amicable relations with the leader male. Here, a former leader male whose group has been taken over grooms the new leader male. Former leaders sometimes remain in a group as followers, perhaps to protect their offspring. Photo by Tyler S. Barry.

number of males during the mating season in this species is reminiscent of seasonal influxes in some guenons (Cords 2000, 2002b), the absence of fully adult males outside the breeding season has not been reported for any other cercopithecine and is not yet well understood.

In general, the adult sex ratio in cercopithecine groups is biased toward females, although the degree of bias is variable (Altmann 2000; Thierry 2007; Swedell 2010). If the sex ratio at birth is roughly 1:1 for the population as a whole, males must face higher mortality, live outside of heterosexual groups, or both. As noted above, males typically disperse in these monkeys, and the process of dispersal entails multiple risks (Alberts & Altmann 1995; Isbell & VanVuren 1996) with documented fitness costs (Alberts & Altmann 1995; Arlet et al. 2009). Aggressive competition for mates is also likely to increase male mortality selectively. These aspects of male life history, common to all cercopithecines, are likely to lead to population sex ratios that are female-biased. In some species, however, heterosexual groups are even more biased toward females than the population as a whole. In these cases, group males successfully exclude competitors, with other males living apart from females for long periods. Solitary males have been reported in macaques, guenons, mandrills, geladas, and several mangabeys (Pusey & Packer 1987; Bergman & Beehner in press; Ehardt in press; Ehardt & Butynski in press; Olupot & Waser in press; Shah in press). However, all-male bands appear to be rare among cercopithecines, and are known only in geladas (Bergman & Beehner in press), Japanese macaques (Sprague et al. 1998), and some guenons (Gartlan & Gartlan 1973; Glenn et al. 2002; Buzzard & Eckardt 2007; Rogers & Chism 2009; Lawes et al. in press). In the

latter they are loose and fluid affiliations, although some mona monkey males have remained together for two years.

Mating Systems

Cercopithecines have polygynous or polygynandrous mating systems. Although DeBrazza's monkeys living in groups with just one adult of each sex in Gabon were once classified as monogamous (Gautier-Hion & Gautier 1978), additional records suggest that one-male groups often contain two to four adult females, and sometimes many more (Gautier-Hion in press-c). Thus, this species is likely generally polygynous, as its considerable sexual size dimorphism suggests (male vs. female body mass = 1.8:2.0, Smith & Jungers 1997).

Female cercopithecines are often described as motivated to mate polyandrously (Cords 1987; Soltis 2004; Pradhan et al. 2006), even outside their own social groups (Mehlman 1986; Cords 1987; Mbora & McGrew 2002; Cooper 2004; Hayakawa 2008; Gautier-Hion in press-b). This mating pattern is usually interpreted as a strategy to create paternity confusion and thus forestall infanticide (van Noordwijk & van Schaik 2000; Paul 2002; Soltis 2002; chapter 19, this volume). Females in many species also mate at times when they cannot conceive, such as when they are pregnant or before ovulatory cycling in seasonal species. Again, this may be a strategy to reduce infanticide risk (van Noordwijk & van Schaik 2000; Setchell & Kappeler 2003; chapter 19, this volume) if males cannot distinguish fertile from infertile females (Engelhardt et al. 2007). Alternative or additional explanations may apply, however (Pazol 2003).

Many female cercopithecines have sexual swellings or

skin color changes around the time of ovulation (Dixson 1998), and some give copulation calls; these features are more common in papionins than cercopithecins, and correspond broadly with a multimale social organization (Maestripieri & Roney 2005). Among papionins, species with copulation calls also tend to show conspicuous swellings: macaques in the evolutionarily derived *sinica* group differ from most other papionins and generally exhibit neither (Maestripieri & Roney 2005). Cercopithecins do not show the same association, however; the two species with copulation calls, mona and sun-tailed (*Cercopithecus solatus*) monkeys (Glenn et al. in press), are *not* the same as those with sexual swellings (Allen's swamp monkeys and talapoins), and only monas live regularly in groups with multiple males (Glenn et al. in press).

There are numerous hypotheses for the evolution of copulation calls and swellings (Maestripieri & Roney 2005). Recently, Prahdan et al. (2006) proposed that they are signals of fertility that promote either effective mate guarding by a dominant male, thus concentrating paternity and increasing protective responses by the male, or polyandrous mating by females, thus triggering sperm competition. Which benefit accrues depends on whether mate guarding is generally successful, and ultimately on the relative benefits of infanticide protection versus postcopulatory female choice. Calls are generally more common when the partner is high-ranking, and they have been observed both to attract other males and to stimulate increased mate guarding by the male partner (Pradhan et al. 2006; chapter 16, this volume).

Even if they often tend toward polyandry, cercopithecine females do not mate indiscriminately with males. Indeed, for optimal protection of offspring at risk of infanticide, they may strike a balance between confusing paternity and concentrating it enough in relatively powerful males to promote their protective responses (van Schaik et al. 2000; chapter 19, this volume). When multiple males are present, whether in multimale groups or during breeding season influxes, females often appear to choose particular mates by persistent following and sexual solicitation, refusal of male advances, surreptitious mating, and competition with other females for mates (Cords 1988; Alberts et al. 2003; Soltis 2004; Arlet et al. 2007). Studies of macaques in which paternity was assessed have shown that such mate choice has consequences for males, because females do not necessarily base their preferences on male dominance rank. Female preferences may, in fact, have a stronger effect than male dominance rank on paternity, as Soltis et al. (1997) demonstrated in captive Japanese macaques, and as suggested also for a free-ranging group in which high-ranking males sired few offspring (Inoue & Takenaka 2008). In a smaller wild group, however, high-ranking males monopolized most re-

production despite heterogeneous female preferences, a difference attributed to lower mating synchrony (Soltis et al. 2001). In provisioned rhesus and Barbary macaques, female mate choice was also identified as influencing paternity outcomes, although it was not the most important factor (Widdig et al. 2004; Brauch et al. 2008).

Male mating strategies seem oriented toward maximizing access to fertile females. In some cercopithecine species (macaques, baboons, mandrills), males prefer to mate with females based on their likely fertility, thus targeting matings during the ovulatory phase of the female cycle and avoiding copulations with adolescent, nulliparous females (Setchell & Kappeler 2003). Male preferences for high-ranking females (Berenstain & Wade 1983) might also be related to the likelihood of successful reproduction (Setchell & Kappeler 2003).

Male mating strategies for species in which groups contain one male differ from those for species in which groups contain multiple males. In one-male groups, such as those of most guenons, males compete to monopolize the entire group. Such monopoly is difficult when intruder pressure is high, and especially when multiple females are simultaneously sexually receptive. Under these circumstances, new males may invade the group in a "multimale influx," and females are often eager to mate with them (Cords 1987, 1988; Enstam & Isbell 2007). In blue monkeys, the presence of additional males in groups is closely related, from day to day and year to year, to the number of mating females (Cords 2002b). Although the original resident may not be ousted, he is unable to repel intruders, especially whenever more than one female is mating. Fluctuations in the local density of potential intruder males may also influence the likelihood of influx breeding periods. For example, influxes appear to occur when many males are in the general area (Enstam & Isbell 2007). Only two studies to date have examined paternity in this type of mating system. They found that residents were excluded as fathers for about a third to a half of the infants born overall (Ohsawa et al. 1993; Hatcher 2006). Even in years without male influxes, residents did not fully monopolize paternity in their groups. Clearly, alternative mating strategies exist for males of these species, although such strategies may be less successful than achievement of sole resident status (Macleod et al. 2002).

In multimale cercopithecine groups, male dominance rank is often a key factor influencing mating strategies and access to females (Setchell & Kappeler 2003), but as with residence in the one-male groups, it does not guarantee a mating or paternity advantage (Cheney et al. 1988; Cowlishaw & Dunbar 1991; Alberts et al. 2003, 2006; Soltis 2004; chapter 18, this volume). High-ranking males often successfully guard fertile females, limiting access by lower-ranking males. The latter adopt alternative tactics, such as sneak matings that evade the notice of higher-ranking males and observers alike, and which lead to some paternities (Berard et al. 1994; Soltis 2004; Setchell et al. 2005; Alberts et al. 2006). Ongoing research aims to discover which factors drive variation in the relationship between male rank and reproduction in multimale groups. In various multimale cercopithecines, high-ranking males may fail to dominate actual reproduction when females are synchronously receptive (Soltis 2004; Ostner et al. 2008; but see Widdig et al. 2004), when effective mate guarding by dominants is compromised by energetic limitations or limited experience (Matsubara 2003; Alberts et al. 2006), when females prefer nondominant males (Soltis et al. 1997; Widdig et al. 2004), and when sperm competition is important (Engelhardt et al. 2006; Brauch et al. 2008). Captivity may limit opportunities for the expression of female choice and alternative male mating strategies (Bauers & Hearn 1994; Gust et al. 1998), and thus secondarily for sperm competition, and provisioning may force males to compete in unusually large groups with reduced rank-based differences in competitive ability (Kummerli & Martin 2005). Thus it will be important to address the relative effects of these factors with data from natural populations. Additional challenges arise because data relating male dominance to reproduction often come from a limited period, and dominance rank may be age-dependent or could influence the reproductive lifespan as well as the reproductive rate (Kümmerli & Martin 2005). These same challenges face studies of reproductive success of males living in one-male groups.

Social Structure

Female-female relationships

A typical cercopithecine female has her strongest social ties to close kin, and these bonds are the most enduring in her society. Close kin include offspring of both sexes, sisters, mother, and grandmother (Chapais 2001). Female relatives will be group mates for life. Close female kin typically show high rates of spatial association, grooming, co-feeding, and coalitionary support; in some species, such support is known to be critical to the development of matrilineally-based dominance hierarchies, in which mothers outrank daughters and sisters may have ranks in inverse order of age (Gouzoules & Gouzoules 1987; Chapais 1992; Silk 2002; Kapsalis 2004; chapter 21, this volume). A female's position in the dominance hierarchy often has important consequences for her fitness (Ellis 1995; Silk 2002; chapter 15, this volume). Recently, paternal kinship has also been implicated in facilitating affiliative interaction among female rhesus macaques, baboons, mandrills, and captive

sun-tailed monkeys (Widdig et al. 2001; Silk et al. 2006; Widdig et al. 2006; Charpentier et al. 2007; Charpentier et al. 2008; chapter 21, this volume).

While cercopithecine females differentiate their relationships with female group mates based on kinship and rank, there are also cases in which the entire group of females functions as a collaborative unit against external threat. In guenons, for example, group-wide female coalitions defend territories (Rowell 1988; Cords 2002a), and while not all females necessarily participate equally, they do all participate (Cords 2007). In several species, including the highly dimorphic mandrill, females may band together to counter male aggression and even exclude males from their group (Dunbar & Dunbar 1975; Cords 1984; Smuts 1987; Smuts & Smuts 1993; Setchell et al. 2006).

For decades, most data on cercopithecine female-female relationships came from a limited set of species (mainly rhesus and Japanese macaques, *Papio* baboons, and vervets), relatively easily observed in nonforested habitats. Furthermore, most studied macaque populations have been provisioned (e.g., at monkey parks, at temples, or in research colonies), which likely affects group and family size, dispersal options, and frequency of aggression, all of which may in turn influence social interaction patterns among kin and non-kin (Hill 2004). Despite these limitations, the female "cercopithecine model" of kin-based social attraction and coalitionary support that determines matrilineal dominance rank is well entrenched in the literature (Thierry 2004), and is often contrasted with female social relations in other primate (Strier 1994) and nonprimate (Holekamp et al. 2007) groups. New data from lesser-known papionin taxa studied in the wild, including mangabeys and mandrills, seem generally to support the idea that females organize affiliative and agonistic relations along kinship lines (Swedell 2010), but the importance of kinship, rank, and alliance formation may be more variable in cercopithecines generally than the classic "cercopithecine model" suggests (Cords 1997; see also below).

Female relationships in hamadryas baboons are atypical among cercopithecines. In this species, males forcibly transfer females away from their kin, and female bonds are less important than cross-sex bonds in structuring one-male units. In large one-male units, however, females develop affiliative relationships with one another, and these are not mere responses to an unavailable leader male (Swedell 2002). Where it occurs, hamadryas female bonding may vary in relation to kinship.

While female-female cooperative relationships form the core of most cercopithecine societies, females also compete with one another for mating opportunities, access to feeding sites, and safe places in their groups, all of which have consequences for relative reproductive success (Setchell & Kappeler 2003; chapter 15, this volume). In baboons and macaques, females may kill offspring of other females in their groups (Silk et al. 1981; Muroyama & Thierry 1996; Paul & Kuester 1996; Kleindorfer & Wasser 2004), although such behavior seems rare and is not well understood. More commonly, aggression and intimidation is milder but allows observers to recognize dominance relations and hierarchies, where rank position relates to access to food, social partners, the receipt of aggression, and sometimes reproduction (Silk 2002).

Male-male relationships

Because the relative parental investment of male and female mammals ensures that males are fundamentally competitors for reproductive opportunities, males often have antagonistic relations. However, they also cooperate to achieve competitive advantages. For example, low- to mid-ranking male baboons (Bercovitch 1988; Nöe 1992), geladas (Bergman & Beehner in press), sooty mangabeys (Range et al. 2007), and Tonkean (*Macaca tonkeana*), Barbary, Thibetan, rhesus, and bonnet macaques (Silk 1994; Soltis 2004; Berman et al. 2007) can form coalitions to wrest a receptive female from a higher-ranking male or to maintain dominance rank. These coalitions are typically business arrangements based on relative power (dominance rank) and partner availability; grooming and other forms of dyadic affiliation are rare. Some male macaques, however, affiliate overtly. Grooming among adult male Japanese macaques, which varies by season and population, may relate to cooperative attacks against non-troop males that compete for mates (Horiuchi 2007). Bonnet macaque males are even more exceptional, at least when living in multimale groups (see below). They are unusually tolerant, greeting, huddling, and grooming with most males in their groups much as females do (Silk 1994). Cooperation in coalitions by bonnet macaque males is based not on kinship but on a complex system of reciprocity, modulated by loyalty and revenge among the various dyads involved (Silk 1992). Because they are philopatric, one might expect a similarly developed level of affiliation and cooperation among male hamadryas baboons; however, spatial association and grooming occur only if the males have not acquired females, while males with females seem to have only rather formalized and occasionally antagonistic relations (Swedell 2006). Nonetheless, tolerance among male hamadryas in the same band is remarkable, including the "respect for possession" they exhibit toward a leader male with females, and the leader's tolerance of follower males, typically subadults, in his one-male unit (Kummer 1995). In addition, males coordinate travel patterns of their one-male units with others in their band.

In other cercopithecines, like most forest guenons and patas monkeys, adult males are largely intolerant of each other, especially when females are present. These males may associate loosely and even groom, however, when they live outside heterosexual groups (fig. 5.7); they can also be held in all-male groups in captivity (Rowell 1988). Behavioral mechanisms allowing coexistence in the multimale guenons are poorly known except in vervets, where genital signaling of relative dominance appears to play an important role (Henzi 1985). In contrast to guenons, male mandrills live alone except during the mating season (Abernethy et al. 2002). The nature of their social relations in natural circumstances is unknown, but in captivity mandrill males form marked hierarchies and have an array of visual signals to indicate relative status, as well as a tendency to avoid, appease, and ignore conflict (Setchell & Wickings 2005).

Wild bonnet macaque males have been studied in one-male and multimale groups, allowing an examination of how flexibly mechanisms of male-male tolerance can be deployed in a single population, and by a single male over time (Sinha et al. 2005b). Adult males in one-male groups were aggressive to resident subadult and juvenile males, herded their females (especially when neighboring groups were nearby), frequently joined aggressive intergroup encounters while challenging neighboring males, and effectively prevented outside males from immigrating. In contrast, bonnet

Fig. 5.7. Exceptionally for guenons, mona males form all-male bands in which males groom and huddle together. The monkey on the right is removing stored food from his cheek pouches. Photo by Mary Glenn.

macaque males in multimale groups affiliate and cooperate to a remarkable degree (see above).

Male-female relationships

With the exception of hamadryas baboons and geladas, female cercopithecines generally increase their proximity to and interactions with males when they are sexually receptive. Enduring close associations between males and receptive females have been termed "consortships," although Manson (1997), who reviewed these relationships, noted that this term has been applied so heterogeneously as to be of questionable value. Associations between males and receptive females function as mate assessment, courtship, and mate guarding, and the emphasis on different functions may differ among taxa. For example, consortships seem more clearly focused on mate guarding in baboons than in macaques (Manson 1997). Not all matings occur in the context of such enduring close relationships, and some species, such as talapoins, have been described as lacking them (Rowell & Dixson 1975). In such cases mating interactions are typically brief, and are isolated from other social exchange between the sexes.

When females are not receptive, male-female relations are more diverse. At the minimalist extreme are adult male and female mandrills, who do not even live together outside the mating season (Abernethy et al. 2002). Somewhat similarly, talapoin males and females form separate subgroups when females are not receptive, and there is very little cross-sex social interaction (Rowell & Dixson 1975). At the other extreme are the one-male units of geladas and hamadryas, in which males always stay very close to their females. In the middle are baboons and macaques, which form cross-sex "friendships" with particular partners, characterized by close proximity and grooming (Smuts & Smuts 1993; Swedell 2010; chapter 19, this volume; fig. 5.8). In baboons, a male friend protects his female and her offspring from harassment; this male is sometimes but not necessarily the infant's father (Moscovice et al. 2009; Nguyen et al. 2009). Macaque friendships seem based on mutual agonistic support, especially among high-ranking individuals (Smuts & Smuts 1993). In forest guenons and patas monkeys, males interact with females less during the non-breeding season than when females are receptive, and males are often described as socially peripheral (Lemasson et al. 2006; Buzzard & Eckardt 2007). Rowell (1988) noted an obvious but unquantified difference in the amount of male-female social interaction, especially grooming and close proximity, in these species relative to baboons. Vervets, though similar to baboons in other respects, are more like other guenons in this, with relatively weak bonds between heterosexual adults (Cheney & Seyfarth 1990).

Fig. 5.8. Social relationships between adult males and females are variable in cercopithecines, and are most affiliative among papionins. (A) Baboons (here *Papio cynocephalus*) have special "friendships" that endure outside of the mating season, providing protection for females and their young. The male friends are not necessarily fathers of the offspring. Photo by Susan Alberts. (B) In macaques, males and females show affiliation through grooming (here *Macaca thibetana*) and mutual agonistic support. Photo by Carol Berman.

High degrees of sexual dimorphism (table 5.1) in cercopithecines correspond with male dominance over females. The males act aggressively toward females, both in sexual and nonsexual contexts (Smuts 1987; Smuts & Smuts 1993), and as noted above, females may collaborate to stop such attacks by jointly chasing and attacking the male. Such female collaboration may occur even in the most dimorphic species (Smuts 1987; Setchell et al. 2006), and the threat of its occurrence may keep males from attacking females (Hall 1965). While actual killing of females is rare, the more common ultimate expression of male aggression is infanticide, which occurs in many cercopithecines (Smuts & Smuts 1993; van Schaik 2000; chapter 19, this volume). Among hamadryas baboons, male aggression toward females is also an important mechanism for maintaining one-male units.

Leader males herd their females with threats and neckbites, and females learn to follow closely (Kummer 1995).

Summary and Conclusions

Studies of cercopithecines have contributed importantly to an understanding of the role of aggressive competition and kinship in structuring primate societies, especially in the relationships of females whose philopatry makes them the core of social groups. Aggressive competition is related to kinship through the behavior of coalition formation; direct competitors can improve their power and success by joining forces with others (de Waal & Harcourt 1992). Studies of some macaques, baboons, and vervets have shown that the matrilineal rank systems of these monkeys arise from the agonistic support that young females receive from their mothers and older close kin (Chapais 1992). In these species, nepotism is common in many other sociopositive interactions as well, including grooming, infant handling, and proximity maintenance (Silk 2002; Kapsalis 2004).

While studies of cercopithecines generally implicate rank and kinship as important variables structuring female social behavior, there is growing awareness of the paucity of data for many species, and the divergence of others from what was once held to be the norm (chapter 21, this volume). For many taxa (especially *Cercocebus*, *Mandrillus*, and most forest guenons), the existence, nature, and steepness of a hierarchy have never been described in unprovisioned populations. In wild patas monkeys, hierarchies can be ambiguous and unstable, and probably not matrilineally based (Isbell & Pruetz 1998). Blue monkeys have linear hierarchies with hints of a matrilineal system, but no detectable rank-related signature on reproductive rate (Cords 2002a), and minor or inconsistent effects on behavior (Pazol & Cords 2005; Foerster et al. 2011). With the exception of vervets, it may be that the cercopithecins deal with competitive pressures differently from most papionins, perhaps by spreading out and thus avoiding direct contests (Isbell 1991; Pazol & Cords 2005). As a result, coalition formation may also be rare. The relative dearth of communicative signaling, especially of submissive signals, and the generally infrequent social interaction noted by Rowell (1988) may also relate to such differences in agonistic behavior. It will be important to gather more data, especially on guenons.

While a synthesis across all cercopithecines is not yet possible, variation in the patterns and importance of agonism, dominance, and nepotism have been documented in macaques. Thierry (2000) sorted macaques into four grades based on a covarying set of social characters including contest asymmetry and intensity, steepness of hierarchy, likelihood of post-conflict reconciliation, presence of tension-reducing behaviors and submissive signals, and kin bias in friendly behavior. The distribution of these traits among species corresponds with phylogenetic relationships (Thierry 2000; Thierry et al. 2008), but the reasons—perhaps ultimately ecological—for phylogenetic differences remain obscure (Ménard 2004). Thierry (2004) proposed that the strength of competition and its relation to the profitability of kin support may be important variables, but data to test this proposal are lacking. It will be interesting to see whether these ideas can be extended to other female-bonded cercopithecines as Thierry suggests, and important to evaluate variation within species simultaneously with differences between species (Ménard 2004).

The one aspect of the social system that shows almost no variation among cercopithecines is the fact that female kin live together for life, with hamadryas the singular exception. While the importance of kin in coalition formation has been emphasized both by those interested in the evolution of cooperation and altruism (Chapais 2001; Silk 2002; chapter 21, this volume) and by those relating social variation to ecological drivers (Sterck et al. 1997), it may be that other advantages of living with kin are more generally important. In baboons, social integration of adult females predicts both the survival of their infants (Silk et al. 2003; Silk et al. 2009) and their own longevity (Silk et al. 2010), independent of rank. Social integration was measured in terms of proximity and grooming, which are highly kin-biased in baboons and other cercopithecines (Kapsalis 2004; Silk et al. 2006). Sociality may enhance fitness through multiple mechanisms. Socially integrated animals may be better tolerated near others (Henzi & Barrett 1999), and thus improve their foraging success or diet quality (summarized for macaques by Silk 2002). Social integration may also directly enhance health, facilitate offspring survival, or allow an individual to cope better with stress (Engh et al. 2006; Shutt et al. 2007; Silk et al. 2010; chapter 24, this volume). For adult female cercopithecines, then, the primary significance of matrilineal kin may be that they are available and ready social partners that tolerate close association and are motivated to engage in social exchange.

The degree to which kinship structures social relations and influences fitness will, however, also depend on the availability of kin. Variation in demographic factors such as group size and female reproductive rates, along with different degrees of reproductive skew in males, creates the potential for considerable variability across species and populations (Chapais & Berman 2004; Hill 2004) and within a single group over time (Berman et al. 2008). Thus demography and life history variables are increasingly recognized as important contributors to social variation in cercopithecine monkeys, as well as other primates.

Twenty-five years ago, most of our information about the behavior of wild primates derived from research on cercopithecines. As summarized here, studies of cercopithecines continue to contribute in significant ways to our knowledge. Much remains to be learned, however. Long-term studies of individual species and new observations of others will ensure that cercopithecines will remain central to our theoretical and empirical understanding of primate behavior.

References

Abegglen, J. J. 1984. *On Socialization in Hamadryas Baboons.* London: Associated University Presses.

Abernethy, K. A., White, L. J. T. & Wickings, E. J. 2002. Hordes of mandrills (*Mandrillus sphinx*): Extreme group size and seasonal male presence. *Journal of Zoology*, 258, 131–137.

Alberts, S. C. & Altmann, J. 1995. Balancing costs and opportunities: Dispersal in male baboons. *American Naturalist*, 145, 279–306.

Alberts, S. C. & Altmann, J. 2003. Matrix models for primate life history analysis. In *Primate Life Histories and Socioecology* (ed. by Kappeler, P. M. & Pereira, M. E.), 66–102. Chicago: University of Chicago Press.

Alberts, S. C., Buchan, J. C. & Altmann, J. 2006. Sexual selection in wild baboons: From mating opportunities to paternity success. *Animal Behaviour*, 72, 1177–1196.

Alberts, S. C., Watts, H. E. & Altmann, J. 2003. Queuing and queue-jumping: Long-term patterns of reproductive skew in male savannah baboons, *Papio cynocephalus. Animal Behaviour*, 65, 821–840.

Altmann, J. 2000. Models of outcome and process: Predicting the number of males in primate groups. In *Primate Males* (ed. by Kappeler, P. M.), 236–247. Cambridge: Cambridge University Press.

Andelman, S. J. 1986. Ecological and social determinants of cercopithecine mating patterns. In *Ecological Aspects of Social Evolution: Birds and Mammals* (ed. by Rubenstein, D. I. & Wrangham, R. W.), 201–216. Princeton, NJ: Princeton University Press.

Arlet, M. E., Grote, M. N., Molleman, F., Isbell, L. A. & Carey, J. R. 2009. Reproductive tactics influence cortisol levels in individual male gray-cheeked mangabeys (*Lophocebus albigena*). *Hormones and Behavior*, 55, 210–216.

Arlet, M. E. & Isbell, L. A. 2009. Variation in behavioral and hormonal responses of adult male gray-cheeked mangabeys (*Lophocebus albigena*) to crowned eagles (*Stephanoaetus coronatus*) in Kibale National Park, Uganda. *Behavioral Ecology and Sociobiology*, 63, 491–499.

Arlet, M. E., Molleman, F. & Chapman, C. 2007. Indications for female mate choice in grey-cheeked mangabeys *Lophocebus albigena johnstoni* in Kibale National Park, Uganda. *Acta Ethologica*, 10, 89–95.

Barton, R. 2000. Socioecology of baboons: The interaction of male and female strategies. In *Primate Males* (ed. by Kappeler, P. M.), 97–107. Cambridge: Cambridge University Press.

Bauers, K. A. & Hearn, J. P. 1994. Patterns of paternity in relation to male social rank in the stumptailed macaque, *Macaca arctoides. Behaviour*, 129, 149–176.

Berard, J. D., Nurnberg, P., Epplen, J. T. & Schmidtke, J. 1994. Alternative reproductive tactics and reproductive success in male rhesus macaques. *Behaviour*, 129, 177–201.

Bercovitch, F. B. 1988. Coalitions, cooperation and reproductive tactics among adult male baboons. *Animal Behaviour*, 36, 1198–1209.

Bercovitch, F. B. & Harvey, N. C. 2004. Reproductive life history. In *Macaque Societies* (ed. by Thierry, B., Singh, M. & Kaumanns, W.), 61–80. Cambridge: Cambridge University Press.

Berenstain, L. & Wade, T. D. 1983. Intrasexual selection and male mating strategies in baboons and macaques. *International Journal of Primatology*, 4, 201–235.

Bergman, T. & Beehner, J. in press. Gelada *Theropithecus gelada* profile. In *The Mammals of Africa, Volume 2: Primates* (ed. by Butynski, T. M., Kingdon, J. & Kalina, J.). London: Bloomsbury.

Berman, C. M., Lonica, C. & Li, J. 2007. Supportive and tolerant relationships among male Tibetan macaques at Huangshan, China. *Behaviour*, 144, 631–661.

Berman, C. M., Ogawa, H., Ionica, C., Yin, H. B. & Li, J. H. 2008. Variation in kin bias over time in a group of Tibetan macaques at Huangshan, China: Contest competition, time constraints or risk response? *Behaviour*, 145, 863–896.

Borries, C., Larney, E., Kreetiyutanoni, K. & Koenig, A. 2002. The diurnal primate community in a dry evergreen forest in Phu Khieo Wildlife Sanctuary, northeast Thailand. *Natural History Bulletin Siam Society*, 50, 75–88.

Brauch, K., Hodges, K., Engelhardt, A., Fuhrmann, K., Shaw, E. & Heistermann, M. 2008. Sex-specific reproductive behaviours and paternity in free-ranging Barbary macaques (*Macaca sylvanus*). *Behavioral Ecology and Sociobiology*, 62, 1453–1466.

Buchan, J. C., Alberts, S. C., Silk, J. B. & Altmann, J. 2003. True paternal care in a multi-male primate society. *Nature*, 425, 179–181.

Butynski, T. M. 1988. Guenon birth seasons and correlates with rainfall and food. In *A Primate Radiation: Evolutionary Biology of the African Guenons* (ed. by Gautier-Hion, A., Bourliere, F., Gautier, J. P. & Kingdon, J.), 284–322. Cambridge: Cambridge University Press.

Buzzard, P. & Eckardt, W. 2007. The social systems of the guenons. In *Monkeys of the Tai Forest* (ed. by McGraw, W. S., Zuberbühler, K. & Noë, R.), 51–71. Cambridge: Cambridge University Press.

Chakraborty, D., Ramakrishnan, U., Panor, J., Mishra, C. & Sinha, A. 2007. Phylogenetic relationships and morphometric affinities of the Arunachal macaque *Macaca munzala*, a newly described primate from Arunachal Pradesh, northeastern India. *Molecular Phylogenetics and Evolution*, 44, 838–849.

Chapais, B. 1992. The role of alliances in social inheritance of rank among female primates. In *Coalitions and Alliances in Humans and Other Animals* (ed. by Harcourt, A. H. & de Waal, F. B. M.), 29–59. New York: Oxford University Press.

————. 2001. Primate nepotism: What is the explanatory value of kin selection? *International Journal of Primatology*, 22, 203–229.

Chapais, B. & Berman, C. M. 2004. Variation in nepotistic regimes and kin recognition: A major area for future research. In *Kinship and Behavior in Primates* (ed. by Chapais, B. & Berman, C. M.), 477–489. New York: Oxford University Press.

Chapman, C. A., Chapman, L. J., Cords, M., Gathua, J. M., Gautier-Hion, A., Lambert, J. E., Rode, K., Tutin, C. E. G. & White, L. J. T. 2002. Variation in the diets of *Cercopithecus* species: Differences within forests, among forests, and across species. In *The Guenons: Diversity and Adaptation in African Monkeys* (ed. by Glenn, M. E. & Cords, M.), 325–350. New York: Kluwer Academic/Plenum Publishers.

Charpentier, M. J. E., Deubel, D. & Peignot, P. 2008. Relatedness and social behaviors in *Cercopithecus solatus*. *International Journal of Primatology*, 29, 487–495.

Charpentier, M. J. E., Peignot, P., Hossaert-Mckey, M. & Wickings, E. J. 2007. Kin discrimination in juvenile mandrills, *Mandrillus sphinx*. *Animal Behaviour*, 73, 37–45.

Cheney, D. L. & Seyfarth, R. M. 1990. *How Monkeys See the World*. Chicago: University of Chicago Press.

Cheney, D. L., Seyfarth, R. M., Andelman, S. J. & Lee, P. C. 1988. Reproductive success in vervet monkeys. In *Reproductive Success* (ed. by Clutton-Brock, T. H.), 384–402. Chicago: University of Chicago Press.

Cheney, D. L., Seyfarth, R. M., Fischer, J., Beehner, J., Bergman, T., Johnson, S. E., Kitchen, D. M., Palombit, R. A., Rendall, D. & Silk, J. B. 2004. Factors affecting reproduction and mortality among baboons in the Okavango Delta, Botswana. *International Journal of Primatology*, 25, 401–428.

Chism, J. 2000. Allocare patterns among cercopithecines. *Folia Primatologica*, 71, 55–66.

Colmenares, F. 2004. Kinship structure and its impact on behavior in multilevel societies. In *Kinship and Behavior in Primates* (ed. by Chapais, B. & Berman, C. M.), 242–270. Oxford: Oxford University Press.

Cooper, M. A. 2004. Inter-group relationships. In *Macaque Societies* (ed. by Thierry, B., Singh, M. & Kaumanns, W.), 204–208. Cambridge: Cambridge University Press.

Cordeiro, N. J. 2003. Two unsuccessful attacks by crowned eagles (*Stephanoaetus coronatus*) on white-throated monkeys (*Cercopithecus mitis*). *African Journal of Ecology*, 41, 190–191.

Cords, M. 1984. Mating patterns and social structure in redtail monkeys (*Cercopithecus ascanius*). *Zeitschrift für Tierpsychologie*, 64, 313–329.

————. 1987. Forest guenons and patas monkeys: Male-male competition in one-male groups. In *Primate Societies* (ed. by Smuts, B. B., Cheney, D. L., Seyfarth, R. M., Wrangham, R. W. & Struhsaker, T. T.), 98–111. Chicago: University of Chicago Press.

————. 1988. Mating systems of forest guenons: A preliminary review. In *A Primate Radiation: Evolutionary Biology of the Living Primates* (ed. by Gautier-Hion, A., Bourliere, F., Gautier, J. P. & Kingdon, J.), 323–339. Cambridge: Cambridge University Press.

————. 1997. Friendships, alliances, reciprocity and repair. In *Machiavellian Intelligence II* (ed. by Whiten, A. & Byrne, R. W.), 24–49. Cambridge: Cambridge University Press.

————. 2000. The number of males in guenon groups. In *Primate Males* (ed. by Kappeler, P. M.), 84–96. Cambridge: Cambridge University Press.

————. 2002a. Friendship among adult female blue monkeys (*Cercopithecus mitis*). *Behaviour*, 139, 291–314.

————. 2002b. When are there influxes in blue monkey groups? In *The Guenons: Diversity and Adaptation in African Monkeys* (ed. by Glenn, M. E. & Cords, M.), 189–201. New York: Kluwer Academic/Plenum Publishers.

————. 2007. Variable participation in the defense of communal feeding territories by blue monkeys in the Kakamega Forest, Kenya. *Behaviour*, 144, 1537–1550.

Cords, M. & Chowdhury, S.C. 2010. Life history of blue monkeys (*Cercopithecus mitis stuhlmanni*) in the Kakamega Forest, Kenya. *International Journal of Primatology*, 31, 433–455.

Cowlishaw, G. 1994. Vulnerability to predation in baboon populations. *Behaviour*, 131, 293–304.

Cowlishaw, G. & Dunbar, R. I. M. 1991. Dominance rank and mating success in male primates. *Animal Behaviour*, 41, 1045–1056.

Davenport, T. R. B., Stanley, W. T., Sargis, E. J., De Luca, D. W., Mpunga, N. E., Machaga, S. J. & Olson, L. E. 2006. A new genus of African monkey, *Rungwecebus*: Morphology, ecology, and molecular phylogenetics. *Science*, 312, 1378–1381.

De Waal, F. B. M. & Harcourt, A. H. 1992. Coalitions and alliances: a history of ethological research. In: *Coalitions and Alliances in Humans and Other Animals* (Ed. by Harcourt, A. H. & de Waal, F. B. M.), pp. 1–19. New York: Oxford University Press.

Delson, E. 1980. Fossil macaques, phyletic relationships and a scenario of deployment. In *The Macaques: Studies in Ecology, Behavior, and Evolution* (ed. by Lindburg, D. G.), 10–30. New York. Van Nostrand Reinhold.

————. 2007. Monkey. In *McGraw-Hill Encyclopedia of Science and Technology*, 392–399. Accessed at http://www.mhest.com/index.php.

Disotell, T.R. 1994. Generic level relationships of the Papionini (Cercopithecoidea). *American Journal of Physical Anthropology*, 94, 47–57.

Dixson, A. F. 1998. *Primate Sexuality*. Oxford: Oxford University Press.

Drapier, M., Chauvin, C. & Thierry, B. 2002. Tonkean macaques (*Macaca tonkeana*) find food sources from cues conveyed by group-mates. *Animal Cognition*, 5, 159–165.

Dunbar, R. I. M. & Dunbar, P. 1975. *Social Dynamics of Gelada Baboons*. Basel: S. Karger.

Ehardt, C. L. in press. Red-capped mangabey, *Cercocebus torquatus*. In *The Mammals of Africa, Volume 2: Primates* (ed. by Butynski, T. M., Kingdon, J. & Kalina, J.). London: Bloomsbury.

Ehardt, C. L. & Butynski, T. M. in press. Sanje mangabey, *Cercocebus sanjei*. In *The Mammals of Africa, Volume 2: Primates* (ed. by Butynski, T. M., Kingdon, J. & Kalina, J.). London: Bloomsbury.

Ellis, L. 1995. Dominance and reproductive success among non-human animals: A cross-species comparison. *Ethology and Sociobiology*, 16, 257–333.

Elton, S. 2007. Environmental correlates of the cercopithecoid radiations. *Folia Primatologica*, 78, 344–364.

Engelhardt, A., Heistermann, M., Hodges, J. K., Nurnberg, P. & Niemitz, C. 2006. Determinants of male reproductive success in wild long-tailed macaques (*Macaca fascicularis*): Male monopolisation, female mate choice or post-copulatory mechanisms? *Behavioral Ecology and Sociobiology*, 59, 740–752.

Engelhardt, A., Hodges, J. K. & Heistermann, M. 2007. Post-conception mating in wild long-tailed macaques (*Macaca fascicularis*): Characterization, endocrine correlates and functional significance. *Hormones and Behavior*, 51, 3–10.

Engh, A. L., Beehner, J. C., Bergman, T. J., Whitten, P. L., Hoffmeier, R. R., Seyfarth, R. M. & Cheney, D. L. 2006. Behavioural and hormonal responses to predation in female chacma baboons (*Papio hamadryas ursinus*). *Proceedings of the Royal Society B-Biological Sciences*, 273, 707–712.

Enstam, K. L. & Isbell, L. A. 2007. The guenons (genus *Cercopithecus*) and their allies: Behavioral ecology of polyspecific associations. In *Primates in Perspective* (ed. by Campbell, C. J., Fuentes, A., MacKinnon, K. C., Panger, M. & Bearder, S. K.), 252–273. New York: Oxford University Press.

Fa, J. E. 1989. The Genus *Macaca*: A Review of Taxonomy and Evolution. *Mammal Review*, 19, 45–81.

Feistner, A. T. C. 1991. Scent marking in mandrills, *Mandrillus sphinx*. *Folia Primatologica*, 57, 42–47.

Fleagle, J. G. & McGraw, W. S. 2002. Skeletal and dental morphology of African papionins: Unmasking a cryptic clade. *Journal of Human Evolution*, 42, 267–292.

Foerster, S., Cords, M. & Monfort, S. 2011. Social behavior, foraging strategies and fecal glucocorticoids in female blue monkeys (*Cercopithecus mitis*): Potential fitness benefits of high rank in a forest guenon. *American Journal of Primatology*, 73, 870–882.

Fooden, J. 1976. Provisional classification and key to living species of macaques (Primates: *Macaca*). *Folia Primatologica*, 25, 225–236.

———. 1982. Ecogeographic segregation of macaque species. *Primates*, 23, 574–579.

Gartlan, J. S. & Gartlan, S. C. 1973. Quelques observations sur les groupes exclusivement males chez *Erythrocebus patas*. *Annales de la Faculté des Sciences du Cameroun*, 12, 121–144.

Gautier-Hion, A. in press-a. Allen's swamp monkey, *Allenopithecus nigroviridis*. In *The Mammals of Africa, Volume 2: Primates* (ed. by Butynski, T. M., Kingdon, J. & Kalina, J.). London: Bloomsbury.

———. in press-b. Black mangabey, *Cercocebus aterrimus*. In *The Mammals of Africa, Volume 2: Primates* (ed. by Butynski, T. M., Kingdon, J. & Kalina, J.). London: Bloomsbury.

———. in press-c. DeBrazza's monkey, *Cercopithecus neglectus*. In *The Mammals of Africa, Volume 2: Primates* (ed. by Butynski, T. M., Kingdon, J. & Kalina, J.). London: Bloomsbury.

———. in press-d. Northern talapoin monkey, *Miopithecus ogouensis*. In *The Mammals of Africa, Volume 2: Primates* (ed. by Butynski, T. M., Kingdon, J. & Kalina, J.). London: Bloomsbury.

Gautier-Hion, A. & Gautier, J. P. 1978. Le singe de Brazza: Une strategie originale. *Zeitschrift für Tierpsychologie*, 46, 84–104.

Gilbert, C. C. 2007. Craniomandibular morphology supporting the diphyletic origin of mangabeys and a new genus of the *Cercocebus/Mandrillus* clade, *Procercocebus*. *Journal of Human Evolution*, 53, 69–102.

Glenn, M. E., Bensen, K. J. & Goodwin, R. M. in press. Mona monkey, *Cercopithecus mona*. In *The Mammals of Africa, Volume 2: Primates* (ed. by Butynski, T. M., Kingdon, J. & Kalina, J.). London: Bloomsbury.

Glenn, M. E., Matsuda, R. & Bensen, K. J. 2002. Unique behavior of the mona mankey (*Cercopithecus mona*): All-male groups and copulation calls. In *The Guenons: Diversity and Adaptation in African Monkeys* (ed. by Glenn, M. E. & Cords, M.), 133–145. New York: Kluwer Academic/Plenum Publishers.

Gouzoules, H. & Gouzoules, S. 2007. The conundrum of communication. In: *Primates in Perspective* (ed. by Campbell, C. J., Fuentes, A., MacKinnon, K.C., Panger, M., Bearder, S.K.), 621–635. New York: Oxford University Press.

Gouzoules, S. & Gouzoules, H. 1987. Kinship. In *Primate Societies* (ed. by Smuts, B. B., Cheney, D. L., Seyfarth, R. M., Wrangham, R. W. & Struhsaker, T. T.), 299–305. Chicago: University of Chicago Press.

Groves, C. P. 2005. Order Primates. In *Mammal Species of the World* (ed. by Wilson, D. E., Reeder, DeeAnn M.), 111–184. Baltimore: Johns Hopkins University Press.

Gust, D. A., McCaster, T., Gordon, T. P., Gergits, W. F., Casna, N. J. & McClure, H. M. 1998. Paternity in sooty mangabeys. *International Journal of Primatology*, 19, 83–94.

Hall, K. R. L. 1965. Behaviour and ecology of the wild patas monkey, *Erythrocebus patas*, in Uganda. *Journal of Zoology*, 148, 15–87.

Hamilton, A. C. 1988. Guenon evolution and forest history. In *A Primate Radition: Evolutionary Biology of the African Guenons* (ed. by Gautier-Hion, A., Bourliere, F., Gautier, J. P., Kingdon, J.), 13–34. New York: Cambridge University Press.

Hamiton, W. J. 1982. Baboon sleeping site preferences and relationships to primate grouping patterns. *American Journal of Primatology*, 3, 41–53.

Harris, E. E. 2000. Molecular systematics of the Old World monkey tribe Papionini: Analysis of the total available genetic sequences. *Journal of Human Evolution*, 38, 235–256.

Hatcher, J. L. 2006. Relating paternity and population genetics in blue monkeys, *Cercopithecus mitis stuhlmanni*: Empirical results and strategies for obtaining them. Ph.D., Columbia University.

Hayakawa, S. 2008. Male-female mating tactics and paternity of wild Japanese Macaques (*Macaca fuscata yakui*). *American Journal of Primatology*, 70, 986–989.

Henzi, S. P. 1985. Genital signaling and the coexistence of male vervet monkeys (*Cercopithecus aethiops pygerythrus*). *Folia Primatologica*, 45, 129–147.

———. 1988. Many males do not a multi-male troop make. *Folia Primatologica*, 51, 165–168.

Henzi, S. P. & Barrett, L. 1999. The value of grooming to female primates. 47–59.

Hill, D. A. 2004. The effects of demographic variation on kinship structure and behavior in cercopithecines. In *Kinship and Behavior in Primates* (ed. by Chapais, B. & Berman, C. M.), 132–150. New York: Oxford University Press.

Hill, R. A., Barrett, L., Gaynor, D., Weingrill, T., Dixon, P., Payne, H. & Henzi, S. P. 2003. Day length, latitude and behavioural (in)flexibility in baboons (*Papio cynocephalus ursinus*). *Behavioral Ecology and Sociobiology*, 53, 278–286.

Hill, R. A. & Lee, P. C. 1998. Predation risk as an influence on group size in cercopithecoid primates: Implications for social structure. *Journal of Zoology*, 245, 447–456.

Holekamp, K. E., Sakai, S. T. & Lundrigan, B. L. 2007. Social intelligence in the spotted hyena (*Crocuta crocuta*). *Philosophical Transactions of the Royal Society B-Biological Sciences*, 362, 523–538.

Horiuchi, S. 2007. Social relationships of male Japanese macaques (*Macaca fuscata*) in different habitats: A comparison between Yakushima island and Shimokita peninsula populations. *Anthropological Science*, 115, 63–65.

Isbell, L. A. 1991. Contest and scramble competition: Patterns of female aggression and ranging behavior among primates. *Behavioral Ecology*, 2, 143–155.

Isbell, L. A., Cheney, D. L. & Seyfarth, R. M. 2002. Why vervet monkeys (*Cercopithecus aethiops*) live in multimale groups. In *The Guenons: Diversity and Adaptation in African Monkeys* (ed. by Glenn, M. E. & Cords, M.), 173–187. New York: Kluwer Academic/Plenum Publishers.

Isbell, L. A. & Pruetz, J. D. 1998. Differences between vervets (*Cercopithecus aethiops*) and patas monkeys (*Erythrocebus patas*) in agonistic interactions between adult females. *International Journal of Primatology*, 19, 837–855.

Isbell, L. A. & VanVuren, D. 1996. Differential costs of locational and social dispersal and their consequences for female group-living primates. *Behaviour*, 133, 1–36.

Isbell, L. A., Young, T. P., Jaffe, K. E., Carlson, A. A. & Chancellor, R. L. 2009. Demography and life histories of sympatric patas monkeys, *Erythrocebus patas*, and vervets, *Cercopithecus aethiops*, in Laikipia, Kenya. *International Journal of Primatology*, 30, 103–124.

Isler, K., Kirk, E. C., Miller, J. M. A., Albrecht, G. A., Gelvin, B. R. & Martin, R. D. 2008. Endocranial volumes of primate species: Scaling analyses using a comprehensive and reliable data set. *Journal of Human Evolution*, 55, 967–978.

Inoue, E. & Takenaka, O. 2008. The effect of male tenure and female mate choice on paternity in free-ranging Japanese macaques. *American Journal of Primatology* 70, 62–68.

Iwamoto, T., Dunbar, R.I.M. 1983. Thermoregulation, habitat quality and the behavioural ecology of gelada baboons. *Journal of Animal Ecology*, 52, 357–366.

Jablonski, N. G. 2002. Fossil Old World monkeys: The late Neogene radiation. In *The Primate Fossil Record* (ed. by Hartwig, W. C.), pp. 255–299. Cambridge: Cambridge University Press.

Janson, C. H. & Goldsmith, M. L. 1995. Predicting group-size in primates: Foraging costs and predation risks. *Behavioral Ecology*, 6, 326–336.

Jolly, C. J. 2007. Baboons, mandrills, and mangabeys: Afro-papionin socioecology in a phylogentic perspective. In *Primates in Perspective* (ed. by Campbell, C. J., Fuentes, A., MacKinnon, K. C., Panger, M. & Bearder, S. K.), 240–251. New York: Oxford University Press.

Jones, T., Ehardt, C. L., Butynski, T. M., Davenport, T. R. B., Mpunga, N. E., Machaga, S. J. & De Luca, D. W. 2005. The highland mangabey *Lophocebus kipunji*: A new species of African monkey. *Science*, 308, 1161–1164.

Jones, T., Laurent, S., Mselewa, F. & Mtui, A. 2006. Sanje mangabey *Cercocebus sanjei* kills an African crowned eagle *Stephanoaetus coronatus*. *Folia Primatologica*, 77, 359–363.

Kappeler, P. M. & Pereira, M. E. 2003. *Primate Life Histories and Socioecology*. Chicago: University of Chicago Press.

Kapsalis, E. 2004. Matrilineal kinship and primate behavior. In *Kinship and Behavior in Primates* (ed. by Chapais, B. & Berman, C. M.), 153–176. New York: Oxford University Press.

Kingdon, J. 2007. Primate visual signals in noisy environments. *Folia primatologica*, 78, 389–404.

Kleindorfer, S. & Wasser, S. K. 2004. Infant handling and mortality in yellow baboons (*Papio cynocephalus*): Evidence for female reproductive competition? *Behavioral Ecology and Sociobiology*, 56, 328–337.

Kummer, H. 1995. *In Quest of the Sacred Baboon: A Scientist's Journey*. Princeton, NJ: Princeton University Press.

Kümmerli, R. & Martin, R. D. 2005. Male and female reproductive success in *Macaca sylvanus* in Gibraltar: No evidence for rank dependence. *International Journal of Primatology*, 26, 1229–1249.

———. 2008. Patterns of infant handling and relatedness in Barbary macaques (*Macaca sylvanus*) on Gibraltar. *Primates*, 49, 271–282.

Laidre, M. E. 2009. Informative breath: Olfactory cues sought during social foraging among Old World monkeys (*Mandrillus sphinx*, *M. leucophaeus*, and *Papio anubis*). *Journal of Comparative Psychology*, 123, 34–44.

Lawes, M. J., Cords, M. & Lehn, C. in press. Blue monkey, Sykes' monkey, gentle monkey, *Cercopithecus mitis*. In *The Mammals of Africa, Volume 2: Primates* (ed. by Butynski, T. M., Kingdon, J. & Kalina, J.). London: Bloomsbury.

Leigh, S. R. 1995. Socioecology and the ontogeny of sexual size dimorphism in anthropoid primates. *American Journal of Physical Anthropology*, 97, 339–356.

———. 2004. Brain growth, life history, and cognition in primate and human evolution. *American Journal of Primatology*, 62, 139–164.

Leigh, S. R. & Bernstein, R. M. 2006. Ontogeny, life history, and maternal investment in baboons. In: *Reproduction and Fitness in Baboons* (ed. by Swedell, L. & Leigh, S. R.), 225–255. New York: Springer.

Lemasson, A., Blois-Heulin, C., Jubin, R. & Hausberger, M. 2006. Female social relationships in a captive group of Campbell's monkeys (*Cercopithecus campbelli campbelli*). *American Journal of Primatology*, 68, 1161–1170.

Macleod, M. C., Ross, C. & Lawes, M. J. 2002. Costs and benefits of alternative mating strategies of samango monkey males. In *The Guenons: Diversity and Adaptation in African Monkeys* (ed. by Glenn, M. E. & Cords, M.), 203–216. New York: Kluwer Academic/Plenum Publishers.

Maestripieri, D. & Roney, J. R. 2005. Primate copulation calls and postcopulatory female choice. *Behavioral Ecology*, 16, 106–113.

Manson, J. H. 1997. Primate consortships: A critical review. *Current Anthropology*, 38, 353–374.

Matsubara, M. 2003. Costs of mate guarding and opportunistic mating among wild male Japanese macaques. *International Journal of Primatology*, 24, 1057–1075.

Mbora, D. N. M. & McGrew, W. C. 2002. Extra-group sexual consortship in the Tana River red colobus (*Procolobus rufomitratus*). *Folia Primatologica*, 73, 210–213.

McGraw, W. S. in press. Sooty mangabey, *Cercocebus atys*. In *The Mammals of Africa, Volume 2: Primates* (ed. by Butynski, T. M., Kingdon, J. & Kalina, J.). London: Bloomsbury.

McGraw, W. S., Cooke, C. & Shultz, S. 2006. Primate remains from African crowned eagle (*Stephanoaetus coronatus*) nests in Ivory Coast's Tai Forest: Implications for primate predation and early hominid taphonomy in South Africa. *American Journal of Physical Anthropology*, 131, 151–165.

Mehlman, P. 1986. Male intergroup mobility in a wild population of the Barbary macaque (*Macaca sylvanus*), Ghomaran Rif Mountains, Morocco. *American Journal of Primatology*, 10, 67–81.

Ménard, N. 2004. Do ecological factors explain variation in social organization? In *Macaque Societies: A Model for the Study of Social Organization* (ed. by Thierry, B., Singh, M. & Kaumanns, W.), 237–262. Cambridge: Cambridge University Press.

Mitani, J. C., Sanders, W. J., Lwanga, J. S. & Windfelder, T. L. 2001. Predatory behavior of crowned hawk-eagles (*Stephanoaetus coronatus*) in Kibale National Park, Uganda. *Behavioral Ecology and Sociobiology*, 49, 187–195.

Mittermeier, R. A., Wallis, J., Rylands, A. B., Ganzhorn, J. U., Oates, J. F., Williamson, E. A., Palacios, E., Heymann, E.W., Kierulff, M. C. M., Yongcheng, L., Supriatna, J., Roos, C., Walker, S., Cortes-Ortiz, L., & Schwitzer, C. 2009. *Primates in Peril: The World's 25 Most Endangered Primates, 2008–2010*. Arlington, VA: IUCN/SSC Primate Specialist Group (PSG), International Primatological Society (IPS), and Conservation International (CI).

Modolo, L., Salzburger, W. & Martin, R. D. 2005. Phylogeography of Barbary macaques (*Macaca sylvanus*) and the origin of the Gibraltar colony. *Proceedings of the National Academy of Sciences of the United States of America*, 102, 7392–7397.

Moore, J. 1984. Female transfer in primates. *International Journal of Primatology*, 5, 537–589.

Moscovice, L. R., Heesen, M., DiFiore, A., Seyfarth, R. M., & Cheney, D. L. 2009. Paternity alone does not predict long-term investment in juveniles by male baboons. *Behavioral Ecology and Sociobiology*, 63, 1471–1482.

Muroyama, Y. & Thierry, B. 1996. Fatal attack on an infant by an adult female Tonkean macaque. *International Journal of Primatology*, 17, 219–227.

Newman, T. K., Jolly, C. J. & Rogers, J. 2004. Mitochondrial phylogeny and systematics of baboons (*Papio*). *American Journal of Physical Anthropology*, 124, 17–27.

Nguyen, N., Van Horn, R. C., Alberts, S. C. & Altmann, J. 2009. "Friendships" between new mothers and adult males: Adaptive benefits and determinants in wild baboons (*Papio cynocephalus*). *Behavioral Ecology and Sociobiology*.

Nöe, R. 1992. Alliance formation among male baboons: Shopping for profitable partners. In *Coalitions and Alliances in Humans and Other Animals* (ed. by Harcourt, A. H. & de Waal, F. B. M.), 285–321. Oxford: Oxford University Press.

Oates, J. F. in press. Red-bellied monkey, *Cercopithecus erythrogaster*. In *The Mammals of Africa, Volume 2: Primates* (ed. by Butynski, T. M., Kingdon, J. & Kalina, J.). London: Bloomsbury.

Ogawa, H. 1995. Bridging behavior and other affiliative interactions among male Tibetan macaques (*Macaca thibetana*). *International Journal of Primatology*, 16, 707–729.

Ohsawa, H., Inoue, M. & Takenaka, O. 1993. Mating strategy and reproductive success of male patas monkeys (*Erythrocebus patas*). *Primates*, 34, 533–544.

Olupot, W. & Waser, P. M. in press. Grey-cheeked mangabey, *Lophocebus albigena*. In *The Mammals of Africa, Volume 2: Primates* (ed. by Butynski, T. M., Kingdon, J. & Kalina, J.). London: Bloomsbury.

Ostner, J., Nunn, C. L. & Schulke, O. 2008. Female reproductive synchrony predicts skewed paternity across primates. *Behavioral Ecology*, 19, 1150–1158.

Page, S. L. & Goodman, M. 2001. Catarrhine phylogeny: Noncoding DNA evidence for a diphyletic origin of the mangabeys and for a human-chimpanzee clade. *Molecular Phylogenetics and Evolution*, 18, 14–25.

Paul, A. 1999. The socioecology of infant handling in primates: Is the current model convincing? *Primates*, 40, 33–46.

———. 2002. Sexual selection and mate choice. *International Journal of Primatology*, 23, 877–904.

Paul, A. & Kuester, J. 1996. Infant handling by female Barbary macaques (*Macaca sylvanus*) at Affenberg Salem: Testing functional and evolutionary hypotheses. *Behavioral Ecology and Sociobiology*, 39, 133–145.

Pazol, K. 2003. Mating in the Kakamega Forest blue monkeys (*Cercopithecus mitis*): Does female sexual behavior function to manipulate paternity assessment? *Behaviour*, 140, 473–499.

Pazol, K. & Cords, M. 2005. Seasonal variation in feeding behavior, competition and female social relationships in a forest dwelling guenon, the blue monkey (*Cercopithecus mitis stuhlmanni*), in the Kakamega Forest, Kenya. *Behavioral Ecology and Sociobiology*, 58, 566–577.

Pereira, M. E. & Leigh, S. R. 2003. Modes of primate development. In *Primate Life Histories and Socioecology* (ed. by Kappeler, P. M. & Pereira, M. E.), 149–176. Chicago: University of Chicago Press.

Plavcan, J. M. 2001. Sexual dimorphism in primate evolution. *Yearbook of Physical Anthropology*, 44, 25–53.

Pradhan, G. R., Engelhardt, A., van Schaik, C. P. & Maestripieri, D. 2006. The evolution of female copulation calls in primates: A review and a new model. *Behavioral Ecology and Sociobiology*, 59, 333–343.

Pusey, A. E. & Packer, C. 1987. Dispersal and philopatry. In *Primate Societies* (ed. by Smuts, B. B., Cheney, D. L., Seyfarth, R. M., Wrangham, R. W. & Struhsaker, T. T.), 250–266. Chicago: University of Chicago Press.

Raaum, R. L., Sterner, K. N., Noviello, C. M., Stewart, C. B. & Disotell, T. R. 2005. Catarrhine primate divergence dates estimated from complete mitochondrial genomes: Concordance with fossil and nuclear DNA evidence. *Journal of Human Evolution*, 48, 237–257.

Ramakrishnan, U. & Coss, R. G. 2001. Strategies used by bonnet macaques (*Macaca radiata*) to reduce predation risk while sleeping. *Primates*, 42, 193–206.

Range, F. 2005. Female sooty mangabeys (*Cercocebus torquatus atys*) respond differently to males depending on the male's residence status: Preliminary data. *American Journal of Primatology*, 65, 327–333.

Range, F., Förderer, T., Storrer-Meystre, Y., Benetton, C. & Fruteau, C. 2007. The structure of social relationships among sooty mangabeys in Taï. In *Monkeys of the Tai Forest* (ed. by McGraw, W. S., Zuberbühler, K. & Nöe, R.), 109–130. Cambridge: Cambridge University Press.

Richard, A. F., Goldstein, S. J. & Dewar, R. E. 1989. Weed macaques: The evolutionary implications of macaque feeding ecology. *International Journal of Primatology*, 10, 569–594.

Rogers, W. & Chism, J. 2009. Male dispersal in patas monkeys (*Erythrocebus patas*). *Behaviour*, 146, 657–676.

Ross, C. 1992a. Environmental correlates of the intrinsic rate of natural increase in primates. *Oecologia*, 90, 383–390.

———. 1992b. Life-history patterns and ecology of macaque species. *Primates*, 33, 207–215.

———. 1998. Primate life histories. *Evolutionary Anthropology*, 6, 54–63.

———. 2003. Life history, infant care strategies, and brain size in primates. In *Primate Life Histories and Socioecology* (ed. by Kappeler, P. M. & Pereira, M. E.), 266–284. Chicago: University of Chicago Press.

Rowell, T. E. 1988. The social system of guenons, compared with baboons, macaques and mangabeys. In *A Primate Radiation: Evolutionary Biology of the African Guenons* (ed. by Gautier-Hion, A., Bourliere, F., Gautier, J. P. & Kingdon, J.), 439–451. Cambridge: Cambridge University Press.

Rowell, T. E. & Dixson, A. F. 1975. Changes in social organization during the breeding season of wild talapoin monkeys. *Journal of Reproduction and Fertility*, 43, 419–434.

Sargis, E. J., Terranova, C. J. & Gebo, D. L. 2008. Evolutionary morphology of the guenon postcranium and its taxonomic implications. In *Mammalian Evolutionary Morphology: A Tribute to Frederick S. Szalay* (edited by Sargis, E. & Dagosto, M.), 361–372. Dordrecht: Springer Netherlands.

Schillaci, M. A. & Stallmann, R. R. 2005. Ontogeny and sexual dimorphism in booted macaques (*Macaca ochreata*). *Journal of Zoology*, 267, 19–29.

Setchell, J. M., Charpentier, M. & Wickings, E. J. 2005. Mate guarding and paternity in mandrills: Factors influencing alpha male monopoly. *Animal Behaviour*, 70, 1105–1120.

Setchell, J. M. & Kappeler, P. M. 2003. Selection in relation to sex in primates. *Advances in the Study of Behavior*, 33, 87–173.

Setchell, J. M., Knapp, L. A. & Wickings, E. J. 2006. Violent coalitionary attack by female mandrills against an injured alpha male. *American Journal of Primatology*, 68, 411–418.

Setchell, J. M., Lee, P. C., Wickings, E. J. & Dixson, A. F. 2001. Growth and ontogeny of sexual size dimorphism in the mandrill (*Mandrillus sphinx*). *American Journal of Physical Anthropology*, 115, 349–360.

Setchell, J. M., Smith, T., Wickings, E. J. & Knapp, L. A. 2008. Social correlates of testosterone and ornamentation in male mandrills. *Hormones and Behavior*, 54, 365–372.

Setchell, J. M. & Wickings, E. J. 2005. Dominance, status signals and coloration in male mandrills (*Mandrillus sphinx*). *Ethology*, 111, 25–50.

Shah, N. F. in press. Agile mangabey, *Cercocebus agilis*. In *The Mammals of Africa, Volume 2: Primates* (ed. by Butynski, T. M., Kingdon, J. & Kalina, J.). London: Bloomsbury.

Shutt, K., MacLarnon, A., Heistermann, M. & Semple, S. 2007. Grooming in Barbary macaques: Better to give than to receive? *Biology Letters*, 3, 231–233.

Silk, J. B. 1992. The patterning of intervention among male bonnet macaques: Reciprocity, revenge and loyalty. *Current Anthropology*, 33, 318–325.

———. 1994. Social relationships of male bonnet macaques: Male bonding in a matrilineal society. *Behaviour*, 130, 271–291.

———. 2002. Kin selection in primate groups. *International Journal of Primatology*, 23, 849–875.

Silk, J. B., Alberts, S. C. & Altmann, J. 2003. Social bonds of female baboons enhance infant survival. *Science*, 302, 1231–1234.

Silk, J. B., Altmann, J. & Alberts, S. C. 2006. Social relationships among adult female baboons (*Papio cynocephalus*) I. Variation in the strength of social bonds. *Behavioral Ecology and Sociobiology*, 61, 183–195.

Silk, J.B., Beehner, J., Bergmann, T., Crockford, C., Engh, A., Moscovice, L., Wittig, R., Seyfarth, R., & Cheney, D. 2010. Strong and consistent social bonds enhance the longevity of female baboons. *Current Biology*, 20, 1359–1361.

Silk, J. B., Clarkwheatley, C. B., Rodman, P. S. & Samuels, A. 1981. Differential reproductive success and facultative adjustment of sex ratios among captive female bonnet macaques (*Macaca radiata*). *Animal Behaviour*, 29, 1106–1120.

Singh, M. & Sinha, A. 2004. Life-history traits: Ecological adaptations or phylogenetic relics? In *Macaque Societies* (ed. by Thierry, B., Singh, M. & Kaumanns, W.), 80–83. Cambridge: Cambridge University Press.

Sinha, A., Datta, A., Madhusudan, M. D. & Mishra, C. 2005a. *Macaca munzala*: A new species from western Arunachal Pradesh, northeastern India. *International Journal of Primatology*, 26, 977–989.

Sinha, A., Mukhopadhyay, K., Datta-Roy, A. & Ram, S. 2005b. Ecology proposes, behaviour disposes: Ecological variability in social organization and male behavioural strategies among wild bonnet macaques. *Current Science*, 89, 1166–1179.

Smith, L. W., Link, A. & Cords, M. 2008. Cheek pouch use, predation risk, and feeding competition in blue monkeys (*Cercopithecus mitis stuhlmanni*). *American Journal of Physical Anthropology*, 137, 334–341.

Smith, R. J. & Jungers, W. L. 1997. Body mass in comparative primatology. *Journal of Human Evolution*, 32, 523–559.

Smuts, B. B. 1987. Gender, aggression and influence. In *Primate Societies* (ed. by Smuts, B. B., Cheney, D. L., Seyfarth, R. M., Wrangham, R. W. & Struhsaker, T. T.), 400–412. Chicago: University of Chicago Press.

Smuts, B. B. & Smuts, R. W. 1993. Male aggression and sexual coercion of females in nonhuman primates and other mammals: Evidence and theoretical implications. *Advances in the Study of Behavior*, 22, 1–63.

Soltis, J. 2002. Do primate females gain non-procreative benefits by mating with multiple males? Theoretical and empirical considerations. *Evolutionary Anthropology*, 11, 187–197.

Soltis, J. 2004. Mating systems. In *Macaque Societies* (ed. by Thierry, B., Singh, M. & Kaumanns, W.), 135–151. Cambridge: Cambridge University Press.

Soltis, J., Mitsunaga, F., Shimizu, K., Nozaki, M., Yanagihara, Y., DomingoRoura, X. & Takenaka, O. 1997. Sexual selection in Japanese macaques: 2. Female mate choice and male-male competition. *Animal Behaviour*, 54, 737–746.

Soltis, J., Thomsen, R. & Takenaka, O. 2001. The interaction of male and female reproductive strategies and paternity in wild Japanese macaques, *Macaca fuscata*. *Animal Behaviour*, 62, 485–494.

Southwick, C. H. & Southwick, K. L. 1983. Polyspecific groups of macaques on the Kowloon Peninsula, New Territories, Hong Kong. *American Journal of Primatology*, 5, 17–24.

Sprague, D. S., Suzuki, S., Takahashi, H. & Sato, S. 1998. Male life history in natural populations of Japanese macaques:

Migration, dominance rank, and troop participation of males in two habitats. *Primates*, 39, 351–363.

Steiper, M. E. & Young, N. M. 2006. Primate molecular divergence dates. *Molecular Phylogenetics and Evolution*, 41, 384–394.

Sterck, E. H. M., Watts, D. P. & van Schaik, C. P. 1997. The evolution of female social relationships in nonhuman primates. *Behavioral Ecology and Sociobiology*, 41, 291–309.

Strier, K. B. 1994. Myth of the typical primate. *Yearbook of Physical Anthropology*, 37, 233–271.

Swedell, L. 2002. Affiliation among females in wild hamadryas baboons (*Papio hamadryas hamadryas*). *International Journal of Primatology*, 23, 1205–1226.

———. 2006. *Strategies of Sex and Survival in Hamadryas Baboons: Through a Female Lens*. Upper Saddle River, NJ: Pearson Education, Inc.

———. 2010. African papionins: Diversity of social organization and ecological flexibility. In *Primates in Perspective* (ed. by Campbell, C. J., Fuentes, A., MacKinnon, K. C., Panger, M. & Bearder, S. K.), 241–277. New York: Oxford University Press.

Thierry, B. 2000. Covariation of conflict management patterns across macaque species. In *Natural Conflict Resolution* (ed. by Aureli, F. & de Waal, F. B. M.), 106–128. Berkeley: University of California Press.

———. 2004. How kinship generates dominance structures: A comparative perspective. In *Macaque Societies* (ed. by Thierry, B., Singh, M. & Kaumanns, W.), 186–204. Cambridge: Cambridge University Press.

———. 2007. The Macaques: A double-layered social organization. In *Primates in Perspective* (ed. by Campbell, C. J., Fuentes, A., MacKinnon, K. C., Panger, M., Bearder, S. K.), 224–239. New York: Oxford University Press.

Thierry, B., Aureli, F., Nunn, C. L., Petit, O., Abegg, C. & De Waal, F. B. M. 2008. A comparative study of conflict resolution in macaques: Insights into the nature of trait covariation. *Animal Behaviour*, 75, 847–860.

Tosi, A. J. 2008. Forest monkeys and Pleistocene refugia: A phylogeographic window onto the disjunct distribution of the *Chlorocebus lhoesti* species group. *Zoological Journal of the Linnean Society*, 154, 408–418.

Tosi, A. J., Detwiler, K. M. & Disotell, T. R. 2005. X-chromosomal window into the evolutionary history of the guenons (Primates : Cercopithecini). *Molecular Phylogenetics and Evolution*, 36, 58–66.

Tosi, A. J., Disotell, T. R., Morales, J. C. & Melnick, D. J. 2003a. Cercopithecine Y-chromosome data provide a test of competing morphological evolutionary hypotheses. *Molecular Phylogenetics and Evolution*, 27, 510–521.

Tosi, A. J., Morales, J. C. & Melnick, D. J. 2003b. Paternal, maternal, and biparental molecular markers provide unique windows onto the evolutionary history of macaque monkeys. *Evolution*, 57, 1419–1435.

Van Noordwijk, M. A. & van Schaik, C. P. 2000. Reproductive patterns in eutherian mammals: Adaptations against infanticide. In: *Infanticide by Males* (ed. by van Schaik, C. & Janson, C. H.), 322–360. Cambridge: Cambridge University Press.

Van Schaik, C. P. 2000. Infanticide by male primates: The sexual selection hypothesis revisted. In *Infanticide by Males and its Consequences* (ed. by van Schaik, C. P. & Janson, C. H.), 27–60. Cambridge: Cambridge University Press.

Van Schaik, C. P., Hodges, J. K. & Nunn, C. L. 2000. Paternity confusion and the ovarian cycles of female primates. In *Infanticide by Males and its Consequences* (ed. by van Schaik, C. P. & Janson, C. H.), 361–387. Cambridge: Cambridge University Press.

Van Schaik, C. P., van Amerongen, A. & van Noordwijk, M. A. 1996. Riverine refuging by wild Sumatran long-tailed macaques (*Macaca fascicularis*). In *Evolution and Ecology of Macaque Societies* (ed. by Fa, J. E. & Lindburg, D. G.), 160–181. Cambridge: Cambridge University Press.

Van Schaik, C. P. & van Noordwijk, M. A. 1985. Evolutionary effect of the absence of felids on the social organization of the macaques on the island of Simeulue (*Macaca fascicularis fusca*, Miller 1903) *Folia Primatologica*, 44, 138–147.

Van Schaik, C. P., van Noordwijk, M. A., Warsono, B. & Sutriono, E. 1983. Party size and early detection of predators in sumatran forest primates. *Primates*, 24, 211–221.

Widdig, A., Bercovitch, F. B., Streich, W. J., Sauermann, U., Nurnberg, P. & Krawczak, M. 2004. A longitudinal analysis of reproductive skew in male rhesus macaques. *Proceedings of the Royal Society of London Series B-Biological Sciences*, 271, 819–826.

Widdig, A., Nurnberg, P., Krawczak, M., Streich, W. J. & Bercovitch, F. B. 2001. Paternal relatedness and age proximity regulate social relationships among adult female rhesus macaques. *Proceedings of the National Academy of Sciences of the United States of America*, 98, 13769–13773.

Widdig, A., Streich, W. J., Nurnberg, P., Croucher, P. J. P., Bercovitch, F. B. & Krawczak, M. 2006. Paternal kin bias in the agonistic interventions of adult female rhesus macaques (*Macaca mulatta*). *Behavioral Ecology and Sociobiology*, 61, 205–214.

Wong, C. L. & Ni, I. H. 2000. Population dynamics of the feral macaques in the Kowloon Hills of Hong Kong. *American Journal of Primatology*, 50, 53–66.

Yamagiwa, J. & Hill, D.A. 1998. Intraspecific variation in the social organization of Japanese macaques: past and present scope of field studies in natural habitats. *Primates*, 39, 257–273.

Xing, J., Wang, H., Zhang, Y., Ray, D. A., Tosi, A. J., Disotell, T. R. & Batzer, M. A. 2007. A mobile element-based evolutionary history of guenons (tribe Cercopithecini). *BMC Biology*, 5, 1–10.

Chapter 6 The Apes: Taxonomy, Biogeography, Life Histories, and Behavioral Ecology

David P. Watts

Although ape taxonomic diversity is currently greatly impoverished in comparison to the Miocene (Harrison 2010) and to the diversity of extant Old and New World monkeys, research on ape behavior and ecology has had major importance in primatology. Starting with Wrangham's (1979) seminal work, research on the functional basis of the remarkable socioecological diversity within and among ape taxa has been fundamental to incorporating primatology into modern evolutionary biology (e.g., see Smuts & Smuts 1993; Langergraber et al. 2007). Also, apes have been favorite referential models for reconstructing human evolution and for comparative research on the extent and origins of human uniqueness (Moore 1996; Tomasello 2008; but see Jolly 2001; Lovejoy 2009; chapter 32, this volume).

This chapter reviews the vast literature on ape taxonomy, biogeography, life histories, feeding ecology, social organization, mating systems, and social relationships. To review the critical conservation status of apes and the enormous complexity of conservation issues in addition is impossible, but failure to address these issues in the real world will remove the material below from the discipline of animal behavior and relegate it to history.

Taxonomy and Biogeography

Recent molecular studies have revised ape taxonomy considerably. Much disagreement exists, but one consensus is that the traditionally construed family Pongidae (orangutans, gorillas, chimpanzees, and bonobos) is paraphyletic.

Extant African apes and humans shared a common ancestor since orangutans diverged, and molecular evidence supports a *Pan-Homo* clade whose members shared ancestry after gorillas diverged. Consequently, Groves (2001, 2005; see table A.1, this volume) and many others include all great apes and humans in the Hominidae, with orangutans (*Pongo* spp.) in the subfamily Ponginae and the African apes and humans in the Homininae. Chimpanzees (*Pan troglodytes*), bonobos (*Pan paniscus*), and humans are then assigned to the tribe Hominini and gorillas to the Gorillinini. The family Hylobatidae includes all gibbons.

For the generic level and below, I use the taxonomy given in table A.1, which follows Groves (2005) in recognizing 14 species in four gibbon genera. The multiple gibbon genera accord with variation in chromosome numbers: *Symphalangus* (siamangs; diploid number = 38), *Nomascus* (crested gibbons; diploid number = 44), *Bunopithecus* (hoolock gibbons; diploid number = 52), and *Hylobates* (diploid number = 50; Prouty et al. 1983a, b). Gibbons are distributed broadly from eastern India through peninsular Southeast Asia and the Sunda Shelf islands of Malaysia and Indonesia (Bartlett 2007). Most inhabit monsoon forest or lowland and mid-montane evergreen moist forest, but *Nomascus* occupies semihumid, mid-montane broadleaved evergreen forests in southern China with sharp rainfall and temperature seasonality (Fan et al. 2009).

Orangutans occur only on Sumatra and Borneo. They are restricted to evergreen rainforest with annual rainfall over 2,000 mm at altitudes below 2,000 m (McConkey 2005a, b). Groves (2001, 2005) and others divide orangutans into *Pongo pygmaeus* on Borneo and *Pongo abelii* on Sumatra. Table

Fig. 6.1. An adult male Grauer's (or eastern lowland) gorilla (*Gorilla gorilla graueri*) at Kahuzi-Biega National Park, Democratic Republic of Congo. Maturing silverback males form new groups by actively acquiring females from group fissions, a process that may account for groups that are smaller and more often unimale in structure than groups of mountain gorillas in the Vigungas. Photo courtesy of Juichi Yamagiwa.

A.1 reflects this, Steiper's (2006) meta-analysis of orangutan population history based on genetic data and results from the recently sequenced orangutan genome (Locke et al. 2011). Despite morphological differentiation between populations on the two islands, complete reproductive isolation is relatively recent, perhaps only 400,000 years ago (ibid.).

Groves (2001, 2005) used molecular and morphological data to divide gorillas into *G. gorilla*, which includes all western (or "western lowland") gorillas, and *G. beringei* ("eastern gorillas"), further divided into *G. b. graueri* (Grauer's gorillas) and *G. b. beringei* (mountain gorillas). But neither eastern nor western gorillas are genetically monophyletic (Jensen-Seaman et al. 2003). Data on multiple autosomal loci date the initial east-west split at 0.9 to 1.6 million years ago, followed by a complex history of gene flow until 80 to 20,000 years ago (Thalmann et al. 2007). Genetic divergence between western and eastern gorillas is similar to that among chimpanzee populations (Fischer et al. 2006), and Thalmann et al. (2007) concluded that gorillas are monospecific (table A.1). Gorillas have a disjunct distribution from Rwanda and Uganda in the east to Nigeria in

the west that includes diverse habitats and great altitudinal variation, from coastal evergreen forest in Gabon to open montane woodland and Afro-alpine vegetation in the Virungas (Ferris 2005; Ferris et al. 2005; Robbins 2007). Western gorillas occur in seven Central/West African countries, mountain gorillas only in the Democratic Republic of Congo (DRC), Rwanda, and Uganda, and Grauer's gorillas only in the DRC (fig. 6.1).

Disagreement exists over whether chimpanzees belong to one species, *P. troglodytes*, or to two (Groves 2001), one including "eastern" (*P. t. schweinfurthii*) and "central" chimpanzees (*P. t. troglodytes*) and the second containing "western" chimpanzees (*P. t. verus*). Recent genetic studies using microsatellites (Becquet et al. 2007) and multiple sequence alignments of millions of base pairs (Caswell et al. 2008) supported the standard division between west, central, and east, with western chimpanzees diverging first. Caswell et al. (2008) found evidence for considerable east-west migration and interbreeding since a split between east-central and west about 510,000 years ago, with divergence between east and central 273,000 years ago, and subsequent major population contraction in western chimpanzees and expansion in central chimpanzees. Both research teams considered chimpanzees monospecific (table A.1), although disagreements over species number are ultimately unresolvable (cf. Jolly 2001). Chimpanzees are the most ecologically flexible apes; they occur in lowland evergreen tropical moist forest (e.g., in the Congo basin), montane forest (e.g., in the DRC, Rwanda, and Uganda), and more open habitats with longer dry seasons (e.g., Miombo woodland/gallery forest in Tanzania and savannah/gallery forest ecotone in Senegal; Caldecott & Kapos 2005; Inskipp 2005).

Bonobos occur only in the DRC, mostly in lowland evergreen tropical rainforest, but also in drier forest and grassland to the south (Lacambra et al. 2005). Presumably they diverged from chimpanzees when a subpopulation of the last common ancestor was isolated south of the Congo River (Wrangham 1986). Genome sequence alignments place the divergence at approximately 1.29 million years ago, roughly contemporary with formation of the Congo River basin (formed 1.5 to 2 million years ago; Caswell et al. 2008).

Ecology and Life History

Body Size

Adult female body mass varies by over an order of magnitude among apes, from just over 5 kg in the smallest gibbons (e.g., Müller's gibbon, *H. muelleri*) to nearly 100 kg in gorillas (table 6.1; see also appendix 1 in Kappeler & Pereira 2003). Sex differences in body mass vary greatly (table 6.1). Hylobatids are monomorphic, with females slightly larger than males in concolor gibbons (*N. concolor*) and the reverse true in lar (*H. lar*) and hoolock (*B. hoolock*) gibbons and siamangs (*S. syndactylus*; Plavcan & van Schaik 1992). Male body mass is about 1.3 times that of females in chimpanzees, 1.2 times female mass in bonobos, 1.7 times female mass in gorillas, and 2.2 times female mass in orangutans (ibid.).

Diet and Feeding Ecology

All apes have mostly plant-based diets that vary considerably among and within taxa and populations. Except for mountain gorillas in the Virungas, fruit typically accounts for most feeding time and individuals face much spatiotemporal variation in food abundance.

Most hylobatids predominantly eat fruit, including figs; leaves are usually the second most important food category (Bartlett 2007; Elder 2009). Figs can account for more feeding time than other fruit (ibid.); they are sometimes eaten in inverse proportion to the availability of other fruit (Leighton & Leighton 1983; Fan et al. 2009) and can be important fallbacks (Marshall & Leighton 2006). Habitat variation and body size variation both contribute to interspecific dietary differences. For example, siamangs devoted 43% of

Table 6.1. Ape life-history characteristics

Taxon	Body weight ♀	Body weight ♂	Age at physical maturity (years) ♀	Age at physical maturity (years) ♂	Female age at first reproduction (years)	Interbirth interval (years)	Gestation length (days)	Max. life span (years) ♀	Max. life span (years) ♂
Hylobates lar[1,2]	5.34	5.90	8	8	11.06	3.42	190		
H. agilis[3]	5.82	5.88				3.2			
H. klossi[4]	5.67	5.92				3.3			
Symphalangus syndactylus[4]	10.7	11.9				2.6–2.8			
Nomascus hainanus[5]	7.62	7.79				2.0			
Pongo abelii[6]	35.6	77.9	≥8		15.4	9.3,[a] 8.5[b]		≥53	≥58
Pongo pygmaeus[6]	35.8	78.5			15.7	7.2,[c] 7.7[d]	244		
Gorilla g. beringei[7]	97.5	162.5	>8	15–16	10.0	4.0	260/285	42	35
Gorilla g. gorilla[8]	71.5	170.4	>10	>18					
Gorilla g. graueri	71	175.2				4.6			
Pan troglodytes[9,10]	31–35[e]	39–42[e]	12–13	15–16	13–15.4[f]	5.0–6.0	230	≥50	>40
	41.6[g]	46.3[g]							
	45.8[h]	59.7[h]							
Pan paniscus[10]	33.2	45.0			14.2	4.5	228		

[a] Ketambe; [b] Suaq Balimbing; [c] Tanjung Putting; [d] Gunung Palung; [e] *P.t. schweinfurthii;* [f] range of mean values from Bossou, Gombe, Kanyawara Mahale, and Taï; [g] *P.t. verus;* [h] *P.t. troglodytes.*

Body mass data from Smith & Jungers 1997, except values for *Pan troglodytes*, from Stumpf 2007. See also appendix in Kappeler & Pereira 2003. Other sources: [1] Reichard & Barelli 2008; [2] Brockleman et al. 1999; [3] Mitani 1990; [4] Lappan 2008; [5] Zhou et al. 2008; [6] Rodman & Mitani 1987; Wich et al. 2004; Anderson et al. 2007; [7] Watts 1991b; Watts & Pusey 1993; A. Robbins et al. 2009b; unpublished records, Karisoke Research Centre; [8] Breuer et al. 2009; [9] Boesch & Boesch-Achermann 2000; Nishida et al. 2003; Stumpf et al. 2009; [10] Stumpf 2007.

feeding time to leaves at Ketambe versus 4% for smaller sympatric lar gibbons (Palombit 1997). But black crested gibbons are also highly folivorous; their habitats have relatively low tree species diversity and pronounced seasonal fruit scarcity (Fan et al. 2009), and folivory and body size are not consistently related, although time spent eating fruit varies inversely with body size (Elder 2009).

Orangutans mostly eat fruit at varying stages of maturity and prefer ripe fruit to other foods (Knott & Kahlenberg 2007; Vogel et al. 2008). Fruit accounted for 64% of annual feeding time at Tuenan, Borneo, with a monthly range of 18% to 98% (0.6%–51.6% devoted to unripe fruit and 7.2%–67.4% to ripe fruit; Vogel et al. 2008), and for 21% to 100% of monthly feeding time at Gunung Palung, Borneo (Knott 1998). Orangutans also eat varying proportions of leaves, leaf buds, flowers, flower buds, seeds, and cambium (Delgado & van Schaik 2000; Knott & Kahlenberg 2007).

Temporal variation in fruit availability is high in orangutan habitats, especially those dominated by mast-fruiting Dipterocarpaceae (van Schaik 1986; Knott 1998; Wich & van Schaik 2000). Orangutan densities are higher in non-masting than in masting forests, and highest where soft pulp fruit or strangling figs are abundant (Delgado & van Schaik 2000). Orangutans respond behaviorally to fluctuations in fruit availability, and figs, seeds, and cambium can be important fallback foods (e.g., Rodman 1977; Sugardjito et al. 1987; Galdikas 1988). Orangutans also have apparent dental adaptations for bark stripping (e.g., broad central and small lateral incisors, Rodman 1988) and for feeding on hard objects like unripe fruit and seeds (e.g., thick molar enamel, Constantino et al. 2009). They store fat efficiently, and females in some habitats (e.g., Gunung Palung: Knott 1998) maintain positive energy balance and build fat reserves when fruit is abundant, which they then use when fruit is scarce and they are in negative energy balance and experience reduced ovarian function. In contrast, females at Ketambe (Sumatra) do not experience prolonged negative energy balance, and fruit availability affects reproductive function less (Wich et al. 2006). Orangutans may be hypometabolic (Pontzer et al. 2009), which would be advantageous when fruit is extremely scarce.

Gorilla diets span a spectrum between frugivory and folivory, largely along an altitudinal gradient. Virunga mountain gorillas have relatively narrow diets comprising almost entirely nonreproductive plant parts, especially the leaves, stems, and pith of herbaceous vegetation (Fossey & Harcourt 1977; Watts 1984). Herbaceous food abundance and quality varies with altitude and topography, and groups spend most time in relatively high-quality areas where required foraging effort is low (Watts 1988, 1991a; McNeilage 2001). Because most food sources are peren-

Fig. 6.2. Virunga mountain gorillas feeding. Adult male (right rear) eats leaves of *Droguetia iners,* while adult female (left foreground) eats shoots of bamboo (*Arundinaria alpina*). Photo courtesy of David P. Watts.

nial, seasonal dietary variation is minimal, aside from the use of bamboo shoots (Vedder 1984; Watts 1984, 1998a; NcNeilage 2001; fig. 6.2).

Western gorilla diets are broader and include considerable fruit, especially during habitat-wide fruit production peaks, when the gorillas search widely for fruit and their diets resemble those of sympatric chimpanzees more than those of Virunga mountain gorillas. Still, they use nonreproductive plant parts as year-round "staples" and employ these, and sometimes figs, as fallbacks when non-fig fruit is scarce (Rogers et al. 1990; Nishihara 1995; Kuroda et al. 1996; Watts 1996; Doran & McNeilage 1998; Doran-Sheehy et al. 2004; Rogers et al. 2004; Robbins 2007). In some habitats, western gorillas often eat aquatic plants rich in minerals, particularly sodium and calcium, in large seasonally inundated clearings or "bais" (Kuroda et al. 1996; Remis et al. 2001; Magliocca & Gautier-Hion 2002; Bermejo 2004; Doran-Sheehy et al. 2004; Rogers et al. 2004). Fruit from tree species in inundated forests can also be important, and their availability can influence habitat use (Doran-Sheehy et al. 2004).

Grauer's gorillas at Kahuzi-Biega, DRC, occupy habitats varying from lowland rainforest to montane forest/woodland/bamboo forest mosaic at altitudes overlapping the altitudinal range of Virunga gorillas. Those in lowland forest exploit seasonally abundant fruit crops while using nonreproductive plant parts as staples and fallbacks (Yamagiwa et al. 1994). Fruit is scarcer in the montane habitat, but seasonally much more abundant than in the Virungas. The gorillas eat large amounts during production peaks (ibid.). The habitat and diets of mountain gorillas in Bwindi resemble those in the Kahuzi-Biega highland sector, although bamboo shoots are seasonally important in Kahuzi-Biega (Yamagiwa & Basabose 2006), but not Bwindi, and Bwindi gorillas at lower altitudes eat more fruit from more species than those at higher altitudes. During fruit production peaks, the diets of gorillas in both populations converge with those of sympatric chimpanzees, but the gorillas use nonreproductive plant parts much more than chimpanzees when fruit is scarce (Robbins & McNeilage 2003; Stanford & Nkurunungi 2003; Rothman et al. 2004; Stanford 2006; Yamagiwa & Basabose 2006; Rothman et al. 2007).

Chimpanzee diets include hundreds of plant and animal foods. The plant foods include fruit, leaves, seeds, flowers, cambium, pith, stems, roots, wood, and tubers (reviewed in Stumpf 2007). Ripe fruit is usually the largest component, and chimpanzees prefer it to other plant foods and devote much foraging effort to it even when it is scarce, although they also eat more leaves and, in some habitats, herbaceous stems and pith at such times (Wrangham et al. 1991; Kuroda et al. 1996; Wrangham et al. 1996; Conklin-Brittain et al. 1998; Wrangham et al. 1998; Newton-Fisher 1999a; Basabose 2002; Tweheyo & Lye 2003; Preutz 2006; Yamagiwa & Basabose 2006; Potts et al. 2009). Figs, which tend to fruit nonseasonally and asynchronously, can be especially important when other fruit is scarce (Wrangham et al. 1996; Stanford 2006; Potts 2008), and they can also be year-round staples (Basabose 2002; Tweheyo & Lye 2003; Preutz 2006).

Bonobos also have highly diverse diets and mostly eat fruit (Badrian & Malenky 1984; Kano 1992; White 1996; Hohmann et al. 2006). They regularly eat herbaceous pith and stems, which may be fallbacks when fruit is scarce (fig. 6.3).

Ape nutritional ecology has been most studied in gorillas. They are not highly specialized for digesting plant structural carbohydrates, but their large size, plus hindgut fermentation facilitated by their relatively large colons, long gut retention times, and diverse gut microbe populations, allows them to use high-fiber diets efficiently (Watts 1984; Popovitch et al. 1997; Remis 2000; Remis et al. 2001; Frey et al. 2006). Fruit that gorillas eat usually is high in soluble carbohydrates, and is thus a good energy source; leaves are typically a good protein source (Waterman et al. 1983; Rogers et al. 2004; Rothman et al. 2004; Rothman et al. 2008). Gorillas feed selectively, based on foods' digestibility and abundance as well as their content of minerals, structural carbohydrates, and phenolics (Waterman et al. 1983; Rogers et al. 1990; Remis et al. 2001; Rogers et al. 2004; Rothman et al. 1996; Rothman et al. 2004; Rothman et al. 2007; Ganas et al. 2009). Similar factors presumably influence fruit and leaf choices made by other apes: for example, the fruit that chimpanzees at Gashaka and bonobos at Salonga eat has higher energy values than the fruit they ignore, and protein/fiber ratios positively influence bonobo food choice (Hohmann et al. 2009; cf. chapter 7, this volume).

All apes engage in some faunivory. Insects are minor components of hylobatid, gorilla, and orangutan diets, and orangutans occasionally prey opportunistically on vertebrates (Kuroda et al. 1996; Deblauwe et al. 2003; Bartlett 2007; Knott & Kahlenberg 2007). Only chimpanzees prey on vertebrates regularly (fig. 6.4). They opportunistically hunt small ungulates; more often they hunt monkeys, especially red colobus (Procolobus spp., Goodall 1986; Boesch & Boesch 1989; Stanford 1998; Hosaka et al. 2001; Watts & Mitani 2002; Gilby & Wrangham 2007; chapter 8, this volume). Their hunting of red colobus is highly effective (Watts & Mitani 2002) and can have major impacts on prey populations (Teelen 2007, 2008). Chimpanzees are not obligate carnivores, and hunting is uncommon in some habitats (e.g., Goualago: Morgan & Sanz 2006). But meat is probably a valuable source of macronutrients for some individuals (Mitani & Watts 2001; Watts & Mitani 2002) and a generally important source of micronutrients scarce in plant tissues (Tennie et al. 2008). Bonobos hunt duiker opportunistically (Fruth & Hohmann 2002), and their predation on guenons has been documented at Salonga (Surbeck & Hohmann 2007), but the rarity of their predation on monkeys distinguishes them from chimpanzees.

Life Histories

All apes have slow life histories. Their body mass and life history speed are not consistently related (cf. chapter 10, this volume). Females mature about as slowly in gibbons as in mountain gorillas, and faster in gorillas than in orangutans and chimpanzees (table 6.1). Hylobatids have long interbirth intervals relative to female body size; body mass is highest in siamangs, but their reproductive rates are not correspondingly low (table 6.1). Hainan gibbons (Nomascus hainanus) may have higher reproductive rates than other similar-sized gibbons (table 6.1). Female mountain gorillas have the shortest interbirth intervals relative to body size of all apes (table 6.1). Intervals reported for chimpanzees vary

Fig. 6.3. Despite relying on fruit, bonobos regularly eat the piths and stems of herbaceaous plants. Here a group of bonobos forage on aquatic shoots in a pond at Lola Ya Bonobo Sanctuary, Democratic Republic of Congo. Photo courtesy of Zanna Clay.

across populations (Emery Thompson et al. 2007), but all are longer than those from modern hunting and gathering human populations (Kaplan et al. 2000; chapter 13, this volume). At eight years or more (table 6.1), interbirth intervals in orangutans are the longest among primates.

Sex differences in growth rate and duration occur in the African apes (table 6.1; Leigh and Shea 1995). Sexual dimorphism results mostly from longer male growth in gorillas and bonobos, but from faster male growth in chimpanzees (ibid.). Male gorillas reach adult body size much later than do female gorillas (table 6.1; Watts & Pusey 1993; Breuer et al. 2009), but at about the same age as male chimpanzees, who grow much more slowly (Leigh & Shea 1995).

Data on survivorship and maximum lifespan are sparse and often come from populations subject to human-induced mortality (Hill et al. 2001; Emery Thompson et al. 2007). Adult survivorship is apparently lower, and lifespans shorter, in gorillas than in chimpanzees (table 6.1), as expected given higher investment in growth and reproduction in gorillas. Female lar gibbons apparently have short repro-

ductive lifespans relative to their late age at maturation and slow reproduction; their maximum lifetime reproductive output seems correspondingly low (Reichard 2009).

Contrasts between gorillas and other apes are consistent with the "ecological risk aversion" hypothesis, which proposes that growth rates, age at maturity, and reproductive rates vary inversely with juvenile starvation risk (Janson & van Schaik 1993; chapters 10 and 11, this volume). Gorillas face energy shortfalls infrequently because they can rely on perennially available nonreproductive plant parts; this reduces starvation risk, so that infants and juveniles can invest more energy in growth and females can mature relatively quickly and give birth more often (Janson & van Schaik 1993; Watts & Pusey 1993; Leigh & Shea 1995; Leigh & Blomquist 2007). A similar contrast may exist between Virunga mountain gorillas and more frugivorous western gorillas, which apparently mature more slowly and have lower reproductive rates (Breuer et al. 2009). The contrast between siamangs and smaller, less folivorous gibbons also seems to fit the hypothesis. The extremely low birth

Fig. 6.4. Chimpanzees regularly hunt vertebrate prey. Red colobus monkeys are their favored prey. Here an adult male chimpanzee, on the right, holds the remains of a recently captured red colobus monkey. Photo courtesy of John Mitani.

rates of orangutans probably occur because long periods of fruit scarcity present females with extreme challenges in devoting energy to reproduction (Knott 1998; Delgado & van Schaik 2000; Knott & Kahlenberg 2007).

Male orangutan life histories are unique among primates. Two morphs of sexually mature males occur (Delgado & van Schaik 2000, Utami et al 2004; Knott & Kahlenberg 2007). "Flanged" males are fully grown and have completely developed secondary sexual characteristics, including large cheek flanges, whereas "unflanged" males are not fully grown and lack these characteristics (fig. 6.5), despite being as old as some flanged males. Genetic differences could underlie these phenotypic differences, but rapid transitions from unflanged to flanged status imply that they result from a conditional strategy (Utami et al. 2004). Disappearance of locally dominant flanged males can trigger developmental switches, and the local density and identity of flanged males and of sexually active females may influ-ence strategic decisions (e.g., more males may be flanged where female density is low; Maggioncalda et al. 1999; Utami et al. 2004). In captivity, "arrested" unflanged males have lower growth hormone levels than males starting to become flanged; they also have lower sex steroid levels, but are fertile. Contrary to the hypothesis that socially induced stress suppresses development, cortisol levels are low in "arrested" males and much higher in transitional males (Maggioncalda et al. 1999; Maggioncalda et al 2000). Whether these same endocrine profiles apply in the wild is unknown.

Predation

Arboreality and moderate to large size give all apes some protection against predation (Wrangham 1979). Also, potential predators other than humans are no longer present in many ape habitats. Nevertheless, predation might have been a major selective force in ape social evolution (Har-

Fig. 6.5. Two morphs of male orangutans. (A) "Flanged" male orangutans possess secondary sexual characteristics, such as cheek flanges and throat pouches. (B) These characteristics are not possessed by "unflanged" males, who are typically much smaller than flanged males. Photo courtesy of Lynda Dunkel.

court & Stewart 2007a). Hylobatids face some risk from pythons, clouded leopards, and tigers (Brockleman 2009). Predation on orangutans has not been reported (Miller & Treves 2007), but tigers and clouded leopards are potential predators (Rijksen 1978). Adult male orangutans travel on the ground regularly on Borneo, where tigers are absent, but rarely on Sumatra, where tigers are present (Rijksen 1978; Rodman 1979). Leopards prey on chimpanzees (Boesch 1994) and bonobos (D'Amour et al. 2006) and might have preyed on gorillas in many habitats (Harcourt & Stewart 2007a, b). Chimpanzees in savanna woodland habitats like Ugalla (Tanzania) and Mount Assirik (Senegal) encounter a wider range of large mammalian carnivores, including leopards, lions, and spotted hyenas, than those in wetter forest habitats, and predation risk might influence their grouping patterns more (Moore 1996). Lions that immigrated temporarily into the Mahale Mountains killed several chimpanzees (Tsukahara 1992). Lions might also have preyed on mountain gorillas in the Virungas before their local extermination (Harcourt & Stewart 2007a, b). Joining bachelor groups may protect adolescent male gorillas against predators.

Social Organization

All apes except orangutans form social groups with stable memberships. Gibbon and gorilla groups are temporally cohesive, while chimpanzees and bonobos have fission-fusion societies (Aureli et al. 2008).

Social monogamy (sensu Kappeler & van Schaik 2002), or "pair living" (Reichard 2009), is common in gibbons (fig. 6.6); groups in most species and populations typically are male-female pairs accompanied by immature offspring that forage cohesively and are hostile to neighbors (Reichard & Barelli 2008). Most males and females disperse from natal groups when mature. It is not known whether sex differences in dispersal distance exist, but preliminary genetic data from a siamang population point toward longer female dispersal (Lappan 2007a). However, considerable variation in dispersal and social grouping exists within and between species (Bartlett 2007; Morino 2009; Reichard 2009). For example, four of five siamang groups were multimale during at least some of Lappan's (2007b) 1.5-year study, and groups with a single female and two males are common in *Nomascus* (Jiang et al. 1999; Zhou et al. 2008). Delayed dispersal is common in lar gibbons at Khao Yai, Thailand, where habitat saturation makes establishing new territories difficult. Successful dispersal often involves displacing same-sex adults in neighboring territories, and sometimes budding off of new territories from parental ones (Brockleman et al. 1998; Reichard 2009). Parents benefit by tolerating adult offspring if they help defend territories and contribute to protection and development of younger siblings (ibid.). However, some Khao Yai groups, as well as groups of other gibbon species, include unrelated same-sex adults (Reichard 2009). Occasionally one pair member deserts a group, then replaces a same-sex adult in another group either aggressively or after that adult dies or emigrates (e.g., lar gibbons and siamang at Ketambe; Palombit 1994a). Because of mateswitching and formation of multimale or multifemale groups, groups can contain immature individuals unrelated to one of the coresident adults (Brockleman et al. 1998).

Fig. 6.6. Most gibbons live in socially monogamous pairs, each consisting of a mature male and a mature female, with young who may or may not be their own. Here a male lar gibbon grooms his mate. In this species, males and females are sexually dichromatic, and can be dark brown or creamy-buff in color. Photo courtesy of Ulrich Reichard.

Large size and heavy reliance on fruit strongly constrain grouping in orangutans because of the high potential for feeding competition (Wrangham 1979). Gregariousness is lowest for adult males and for females with mid-sized or partly weaned infants, for whom the ecological costs of grouping should be highest (Rijksen 1978; Rodman 1979; Galdikas 1979, 1985a, b; van Schaik 1999). Sex differences in ecology, including longer daily feeding time for males, also constrain association between adult males and females (Rodman 1979; Sugardjito 1986). Adults mostly stay alone or, for females, with infants and "ecologically dependent" (van Andrichem et al. 2006) juveniles, but variation in fruit abundance and distribution leads to variation in gregariousness. At Ketambe, for example, feeding aggregations were rarest when fruit was scarce and few or no large patches were available, and most common when fruit was moderately abundant and mostly in scattered large patches (Sugardjito et al. 1987). Gregariousness and population density were higher at Suaq Balimbing, in the same ecosystem, but large feeding aggregations were uncommon

because large strangling figs were rare (van Schaik 1999). Female gregariousness is generally higher on Sumatra than on Borneo, presumably because of higher habitat productivity (Delgado & van Schaik 2000).

Most female orangutans have relatively small home ranges. "Resident" flanged males establish stable home ranges that encompass those of multiple females; nonresident flanged males and unflanged males have less stable home ranges and may roam widely (Galdikas 1979; Mitani 1985a; Delgado & van Schaik 2000). Female site fidelity can be high (Knott et al. 2008), although females in some populations make long seasonal movements to track fruit supplies (e.g., Ketambe; te Boekhorst et al. 1990). Male site fidelity is lower; even residents may be forced out of their home ranges by other males (Utami & Mitra Setia 1995) or leave them voluntarily, perhaps to seek mating opportunities (Galdikas 1979). Dispersal distance is apparently male-biased (Galdikas 1988; van Schaik & van Hooff 1996; Singleton & van Schaik 2002), as would be expected if ecological knowledge is particularly important for female reproductive success and mating competition drives male dispersal. Adolescent females may establish home ranges that overlap those of their mothers (Singleton & van Schaik 2002).

Adult female gorillas and their dependent offspring form stable social groups with one or more adult males. Mean group sizes vary from about 9 to 22 in western gorillas and 10 to 17 in western gorillas, although groups have exceeded 60 in Virunga mountain gorillas (Harcourt & Stewart 2007a; Stoinski et al. 2009). Adolescents of both sexes disperse from natal groups. Females disperse voluntarily and transfer directly into established groups or join solitary males; secondary transfer by nulliparous and parous female mountain gorillas is common (Watts 1996; Doran & MacNeilage 2000; Harcourt & Stewart 2007a; Robbins et al. 2009a, b). Natal transfer is not universal in the Virunga population, and many female relatives reside together because of natal philopatry or transfer to the same groups (Watts 1996; Robbins et al. 2009a). Natal transfer may be universal in western gorillas (Stokes et al. 2003; Gatti et al. 2004; Levrero et al. 2006; Douadi et al. 2007; Caillaud et al. 2008). Single-male gorilla groups usually disband after male deaths, although some or all females may stay together until males join them to form new breeding groups (Watts 1989; Yamagiwa & Kahekwa 2001; Stokes et al. 2003). Females may desert aging males (Levrero et al. 2006; Douadi et al. 2007).

Natal dispersal helps to avoid close inbreeding (Harcourt 1978), but mate choice, predation risk, and perhaps feeding competition drive secondary dispersal (Watts 1990, 1996; Harcourt & Stewart 2007a, b; Caillaud et al. 2008; Robbins et al. 2009b). Ecological costs of transfer are low for Virunga mountain gorillas because most food is distributed widely, densely, and perennially across overlapping home ranges (Watts 1998), and low potential to gain from successfully contesting access to food reduces costs of leaving female kin. Neither natal nor secondary transfer lead to reproductive delays (Watts 1996; Robbins et al. 2009b). Mountain gorilla females in Bwindi tend to disperse to areas where the vegetation resembles their natal areas, which should minimize foraging costs associated with transfer (Guschanski et al. 2008). Female western gorillas may rely on use of bais and dispersal to small, adjacent groups to limit such ecological costs (Stokes et al. 2003; Douadi et al. 2007).

Adult male gorillas cannot immigrate into breeding groups. Most dispersing male mountain gorillas become solitary (Watts 2000a; Robbins & Robbins 2005; Harcourt & Stewart 2007a; Stoinski et al. 2009b). Others reside in natal groups as adults and form breeding queues; they may eventually become dominant or single-breeding males by reversing rank with older males, replacing those who die, or

acquiring females after group fissions (Robbins 1995, 2001; Watts 2000a; Harcourt & Stewart 2006, 2007; Robbins et al. 2009b; Stoinski et al 2009b). Multimale groups and group fissions also occur in Bwindi (Nsubuga et al. 2008) and Grauer's gorillas (Yamagiwa & Kahekwa 2001). Males disperse farther than females in Bwindi, and randomly with respect to vegetation composition (Guschanski et al. 2008). Multimale groups are rare or absent in western gorillas, and male dispersal may be universal. Greater frugivory may limit western gorilla female group size more strictly, making it easier for single males to monopolize groups and increasing the probability that solitary males attract mates (Harcourt & Stewart 2007a). Also, some multimale mountain gorilla groups form when sons remain with fathers (ibid.; Stoinski et al. 2009b), but slower maturation in western gorillas may mean that males rarely survive long enough to reside with mature sons (Breuer et al. 2009). Some western gorilla males remain near their natal areas, leading to relatively high relatedness among neighboring males (Bradley et al. 2004), but dispersal is male-biased overall because of long-distance movements by other males (Douadi et al. 2007).

Successful transfer to breeding groups by immature males is rare in the Virungas, but more common in western gorillas (Douadi et al. 2007). Group dissolutions and takeovers by outside males following adult male deaths have led to all-male groups in the Virungas when immature males, unable to join breeding groups or forced to emigrate, join solitary silverbacks (Yamagiwa 1987; Watts 1989). Adolescent male western gorillas commonly transfer to all-male groups, or to groups that contain only males and immature females (Levrero et al. 2006; Douadi et al. 2007).

Chimpanzees and bonobos have classic fission-fusion social organizations. Both form multimale, multifemale communities with stable memberships, but community members form temporally unstable subgroups ("parties") that lack fixed membership, rather than foraging cohesively (Nishida 1968; Goodall 1986). Most females disperse from their natal communities at adolescence and transfer into neighboring communities, but secondary transfer is rare; males are philopatric (Pusey 1980). At some sites (e.g., Ngogo: Mitani et al. 2002), party size varies positively with fruit abundance. Presence of females with sexual swellings can also positively affect party size (ibid.). Males are more gregarious than females (Wrangham & Smuts 1980; Nishida et al. 1990; Pepper et al. 1999; Lehmann & Boesch 2005), although female gregariousness varies with food abundance and distribution and with predation risk. It is high at Taï, where predation risk, and perhaps food abundance, is high (Lehmann & Boesch 2005, 2008), and at Ngogo, a food-rich site (Wakefield 2008). Noncycling eastern chimpanzee females stay mostly in limited parts ("core areas") of their community ranges; this should decrease scramble competition and increase familiarity with food distribution (Goodall 1986; Hasegawa 1990; Pusey et al. 1997; Murray et al. 2006; Kahlenberg et al. 2008; Langergraber et al. 2009; Wakefield 2010). Core area fidelity can be high (Murray et al. 2007). Fewer sex differences in range use occur at Taï (Lehmann & Boesch 2005, 2008).

Overall gregariousness is not necessarily higher in bonobos than chimpanzees, but sex differences may be slighter and even reversed. Mean party size was higher for bonobos at Salonga than for chimpanzees at Gashaka (Hohmann et al. 2006). Bonobo female gregariousness is relatively high at Wamba (Kuroda 1980; Kano 1992), and at Lomako, females are alone less than is typical for eastern chimpanzees, and often associate with each other (Hohmann & Fruth 2002). In contrast to chimpanzee males, male bonobos at Lomako were more likely than females to be solitary when fruit was scarce (White 1988). These differences between the two *Pan* species may stem from lower scramble competition in bonobos, if fruit availability varies less in bonobo than chimpanzee habitats (Chapman et al. 1994; Hohmann et al. 2006). Also, absence of competition for fruit with gorillas may facilitate gregariousness in bonobos and in some chimpanzee populations that do not live with gorillas (Yamakoshi 2004). Whatever the explanation, relatively high female gregariousness in bonobos is associated with interspecific differences in power asymmetry between the sexes (see below).

Mating Systems and Sexual Conflict

Ape mating systems vary considerably within and between species. Sexual conflict, sometimes including infanticide by males (chapter 19, this volume), is important in all cases.

Despite prevailing social monogamy, gibbon mating can be polygynous, polyandrous, or polygynadrous. Extrapair copulations occur, and females mate with multiple males in socially polygynous groups (Reichard 1995, 2009; Reichard & Sommer 1997; Brockleman et al. 1998; Palombit 1994a, b; Barelli et al. 2008), although how this affects genetic mating systems is unknown. Van Schaik and Dunbar (1994) argued that male gibbons could defend the territories of, and monopolize mating with, several solitary females, but instead associate permanently with single females (or, occasionally, two females in socially polygynous groups) because of the risk of infanticide by male intruders. However, defense of multiple territories may not be energetically feasible (Bartlett 2009), as suggested by the association of social polyandry at Khao Yai with relatively large, poor-quality female home ranges (Savini et al. 2009).

Evidence for infanticide is scant and circumstantial (Borries et al. 2011), while males trying to control multiple female territories could face unacceptable risks of forfeiting paternity opportunities; thus, permanent male-female association may represent male mate guarding (Palombit 1999).

The orangutan mating system involves "roving male promiscuity" with resident flanged males unable to maintain exclusive access to solitary females whose home ranges overlap with theirs (van Schaik & van Hooff 1996). These males compete with unflanged males and nonresident flanged males, who routinely try to copulate forcibly with females (Rijksen 1978; Galdikas 1985a, b; Mitani 1985a; Fox 1998, 2004; Utami et al. 2002; Utami & van Hooff 2004). A small paternity sample from Ketambe (Utami et al. 2002) showed relatively low reproductive skew, raising questions about the reproductive payoffs of attaining flanged resident status.

Female orangutans are active in mating; they may benefit by mating promiscuously to confuse paternity and/or by changing their mating preferences (Delgado & van Schaik 2000). Females may prefer some males based on their genetic quality (Fox 1998; Knott & Kahlenberg 2007; chapter 16, this volume). By this "good genes" hypothesis, females should mate willingly with resident flanged males because those males have demonstrated high quality by successfully establishing their local dominance in the face of intense male-male competition. Preference switches following replacement of resident males are unexpected (Fox 1998) but could reflect repeated quality assessments by females (Knott & Kahlenberg 2007). Fox (2002) proposed that females might assess males by resisting their copulation attempts, but found no clear supporting evidence. An alternative "male protection" hypothesis suggests that females choose males on the basis of their ability to protect against forced copulations and infanticide (van Schaik 2004; Knott & Kahlenberg 2007). It predicts that (1) females mate preferentially with established resident flanged males because those males protect them and their infants from sexual coercion, including infanticide, (2) they mate willingly with other flanged males who are establishing local dominance and also with "up-and-coming" developing males, (3) forced copulation attempts by nonresidents succeed less often when females are close to resident males, and (4) females use loud calls to track the locations of flanged males and to reduce sexual coercion. Behavioral observations provide some support for all these predictions (Rijksen 1978; Schurmann 1982; Galdikas 1985a, b; Mitani 1985a; Fox 1998, 2004; Utami et al. 2002; Utami & van Hooff 2004), but a major difficulty remains: Extremely slow reproduction should make infanticide advantageous to males, but neither infanticide nor infanticide attempts have been observed. Delgado and van

Schaik (2000; cf. Harcourt & Greenberg 2001; van Schaik 2004; Knott & Kahlenberg 2007) caution that opportunities for infanticide are rare, but Beaudrot et al. (2008) argue that it should have been seen if it is a threat.

Gorilla males mate polygynously, although mating is often sequentially polyandrous from the female perspective. Genetic data from several western gorilla sites show that group males typically monopolize paternity (Bradley et al. 2004; Jeffrey et al. 2004; Gatti et al. 2004). Extragroup paternity is unknown in Virunga mountain gorillas (Bradley et al. 2006) and rare or absent in Bwindi (Nsubuga et al. 2008), but multiple within-group paternity occurs in multi-male groups, probably because dominant males who sire most, but not all, infants have "incomplete control" over subordinates (Bradley et al. 2006; Nsubuga et al. 2008; but see Stoinski et al. 2009a, suggesting that dominant males sometimes concede mating opportunities). Reproductive success is higher for Virunga males who stay in their natal groups than for those who disperse (Watts 2000a; Robbins & Robbins 2005; Stoinski et al. 2009b). However, some maturing males lose opportunities to queue because their groups dissolve following deaths of older breeding males, and others may disperse because they face long queues or are too close in age to young breeding males with long expected tenures (Watts 2000a). Females commonly copulate with multiple males, including their half or full brothers (but not putative fathers), despite efforts by dominant males to prevent copulations by subordinates, and they initiate most copulations (Harcourt et al. 1980; Watts 1991b, 2000a; Stoinski et al. 2009a). This may create paternity uncertainty—an expected response to sexual conflict, given that sexually selected infanticide is prominent in male reproductive strategies (Fossey 1984; Watts 1989, 1991b; Bradley et al. 2006).

Following infanticides, female mountain gorillas usually mate with the perpetrators, supporting the hypothesis that male ability to protect infants is an influence on mate choice and that infanticide protection helps explain permanent male-female association (Watts 1989, 1990; Harcourt & Greenberg 2001; Harcourt & Stewart 2007a). The females tend to emigrate from single-male groups and to join multi-male groups, where feeding competition is potentially high but infanticide risk is low (Watts 2000a; A. Robbins et al. 2009). Infanticide is sometimes discounted as a selective force in other gorillas because it has not been seen (western gorillas, Douadi et al. 2007), or because females sometimes transfer with unweaned infants who survive in their new groups (Grauer's gorillas, Yamagiwa 2009). However, observations of western gorillas away from bais are limited, low-cost infanticide opportunities are rare, and the sexual selection hypothesis allows for variation in infanticide frequency (van Schaik 2000). Also, other infant Grauer's gorillas have been killed after transfers (Yamagiwa 2009), and indirect evidence exists for infanticide in western gorillas (Stokes et al. 2003; Gatti et al. 2004).

Male morphological traits important for fighting ability and protection of infants seem to influence female choice in gorillas. At Lokoué, the probability that males gain mates increases with body length and muscularity, and the number of females per breeding male increases with the relative size of male sagittal crests, a proxy for bite force (Caillaud et al. 2008). Females may choose large males to obtain protection against leopards and infanticide (ibid.; Harcourt & Greenberg 2001; Harcourt & Stewart 2007a, b). Infant survival increases with group size at Mbeli Bai; antipredation benefits should also increase, but males good at protecting against infanticide may also attract many females whose infants survive well (Breuer et al. 2009).

Chimpanzees have a complex polygynandrous mating system. Cycling females develop exaggerated sexual swellings that attract male attention; parous females typically generate more interest than nulliparous females (Muller et al. 2007). The presence of swollen females can lead to large parties (e.g, Mitani et al. 2002) in which females mate promiscuously and sperm competition is important; male interest increases as ovulation approaches, and high-ranking males may progressively restrict copulation opportunities for lower-ranking and adolescent males (Tutin 1979; Tutin & McGuiness 1981; Hasegawa & Hiraiwa-Hasegawa 1990; Watts 1998b; Stumpf & Boesch 2004; Muller et al. 2007). Sometimes individual males or male coalitions monopolize swollen females despite the presence of other males (Tutin 1979; Watts 1998b). Male aggression can thus constrain female mate choice, although selective female rejection or acceptance of copulation attempts in a small community at Taï might have biased paternity probabilities towards preferred males (Stumpf & Boesch 2005, 2006). Males can also monopolize females by persuading them to go on "consorts," during which they try to avoid other males (Tutin 1979). Consort initiation attempts can be highly aggressive, but success requires female consent (Smuts & Smuts 1993); whether this represents female choice or acquiescence to coercion is unclear. Paternity data indicate that high rank confers reproductive advantages on males, but reproductive skew is relatively low and is inversely related to the number of males per community (Constable et al. 2001; Vigilant et al. 2004; Boesch et al. 2006; Inoue et al. 2008; Langergraber et al. 2010).

Between-group infanticide by males is known at most long-term chimpanzee study sites (Wilson & Wrangham 2003; Boesch et al. 2008), but is not easily explained by the sexual selection hypothesis because the mothers typically do not then join the killers' communities (chapter 19, this volume). Within-group infanticide by males also occurs; its function is unclear, although it could fit the sexual selection hypothesis if the attackers have not mated with the females during conception cycles or if it has improved their chances of siring subsequent infants (Wilson & Wrangham 2003).

Bonobo females also mate promiscuously, although high rank confers some mating (Kano 1996) and reproductive advantage to males (Gerloff et al. 1999). Proposed differences in female sexuality between bonobos and chimpanzees have been widely debated. In the wild, females of both species copulate mostly when they have full and maximally firm swellings (Furuichi 1987). These swellings last longer in bonobos, but not as a proportion of the cycle, which is also longer (ibid.). However, lactating female bonobos typically resume regular cycles much sooner than lactating chimpanzee females, although interbirth intervals are similar and most bonobo cycles are not fertile (Furuchi 1987; Kano 1992). Consequently, female bonobos have many more swelling cycles and are sexually receptive for much more of their adult lives (Wrangham 1993). Information about female fecundability and paternity is probably less reliable for bonobos (Wrangham 1993, 2002). This could lower the potential benefits of sexual coercion and of obtaining high rank for male bonobos compared to male chimpanzees, and help to explain the difference in intersexual power asymmetry and the absence of male alliances and apparent absence of infanticide in bonobos (see below).

Social Structure

Female-Female Social Relationships

Ape social relationships reflect the various ways females deal with feeding competition. The problem of competition is most serious for orangutans (Wrangham 1979). They adjust their participation in groups to minimize scramble competition (Sugardjito et al. 1987; Utami et al. 1997). Social relationships between females mostly reflect familiarity, although limited female dispersal could facilitate kinship effects. Overt affiliative interactions like grooming are rare (Rijksen 1978). Contests over access to fruit patches or feeding spots occur, and some dyads seem to have decided agonistic relationships (Ketambe, Utami et al. 1997; Gunung Palung, Knott et al. 2008). But females at Gunung Palung mostly avoid each other, and the rarity of displacements suggests that contest competition has little effect on foraging efficiency and reproduction (Knott et al. 2008).

Social relationships between adult females in multifemale gibbon and siamang groups have not been described. Jiang et al. (1999) characterized females in multifemale groups of black crested gibbons as highly tolerant of each other, but provided no quantitative data.

Most data on gorilla female-female relationships are from Virunga mountain gorillas. Most dyads have antagonistic or neutral relationships, but grooming and close proximity are relatively common, and serious aggression uncommon, between some (Harcourt 1979a; Watts 1994a, 2001, 2003). Natal philopatry and co-transfer produces kinship effects: social bonds are stronger between mothers and daughters and between maternal sisters than between other females and relatively strong between putative paternal sisters (Watts 1994a). Female-female relationships appear similar in western (Stokes 2004) and Grauer's gorillas (Yamagiwa et al. 2003), but kinship effects are probably weaker in western gorillas because of universal female dispersal.

Reliance on relatively evenly and densely distributed, perennially available food means that female gorillas gain little by contesting access to food, either individually or in alliances with kin (Harcourt & Stewart 1987a, 2006; Watts 1990, 1997). Female Virunga mountain gorillas can often be ranked in hierarchies based on displacements (Robbins et al. 2006), but aggression rates are low, feeding contests are infrequent and often undecided, and dyadic aggression is often bidirectional (Watts 1994b). Thus dominance relationships are weak, and any rank effects on reproductive success are slight (Robbins et al. 2007). Agonistic relationships between females are similar in Bwindi mountain gorillas (Robbins 2008). Coalitions are rare between unrelated females, but they occur between co-resident female relatives, especially maternal kin, at rates similar to those in some cercopithecines in which nepotism influences rank acquisition and maintenance. However, any nepotistic benefits are slight and usually outweighed by the potential advantages of transfer (Harcourt & Stewart 1988, 2006; Watts 1997). Also, males limit nepotism and level competitive asymmetries by intervening in many female-female contests to support likely losers or to make control interventions that prevent either female from winning (Watts 1991, 1994b).

In chimpanzees, immigrant females are mates for males, but competitors for females. Resident females often aggressively resist new immigrants (Nishida 1989; Boesch & Boesch-Achermann 2000; Kahlenberg et al. 2007); particularly severe aggression involves coalitionary attacks and can result in infanticide (Townsend et al. 2007; Pusey et al. 2008). Aggression may reduce feeding competition for residents by dissuading immigrants from staying or preventing them from establishing core areas in good habitat (Pusey et al. 2008; Kahlenberg et al. 2008). Males intervene in many resident-immigrant contests, usually to protect immigrants, who associate with males more often than anestrous resident females do (Nishida 1989; Boesch & Boesch-Achermann 2000; Kahlenberg et al. 2007, 2008; Townsend et al. 2007; Pusey et al. 2008). Immigrants might reduce harassment by establishing core areas that overlap minimally with those of residents, although this lowers foraging efficiency, and by avoiding other females when they give birth (Pusey et al. 2008). Once established, immigrants have more at stake and costs of aggression towards them should increase; this helps explain low rates of aggression between parous females (Murray et al. 2007). Also, most immigrants are adolescents, and costs should increase as they attain full size and their fighting ability increases.

Disagreement exists about whether female chimpanzees typically have dominance relationships, although this could reflect real variation in the costs and benefits of grouping. Wittig & Boesch (2003) found a statistically linear female dominance hierarchy at Taï based on formal signals of submission. Linear hierarchies have not been documented elsewhere (Nishida 1989; Williams et al. 2002; Wakefield 2008), but researchers sometimes assign females to rank categories and identify alpha females (Goodall 1986; Williams et al. 1997; Kahlenberg et al. 2008). Status differences apparently can influence feeding, space use, and reproduction. Most feeding contests at Taï occurred in large parties, over monopolizable food, and rank predicted contest outcomes (Wittig & Boesch 2003). Feeding contests, although rare, account for most agonistic interactions between females at Gombe (Murray et al. 2006), and consistent winners may obtain feeding advantages simply because losers avoid them (Kahlenberg et al. 2008). Kanyawara females categorized as ranking high have core areas with higher densities of important food species than those categorized as ranking low (Emery Thompsen et al. 2007; Kahlenberg et al. 2008), and those categorized as ranking high at Gombe show higher site fidelity (Murray et al. 2007), have smaller core areas, and expand their foraging areas less when fruit is scarce than do low-ranking females, thus implying that their core areas are higher quality (Murray et al. 2006). They also have shorter interbirth intervals, higher infant survival, and higher overall reproductive success than low-ranking females (Pusey et al. 1997), possibly because of better nutrition (Murray et al. 2006), although data on energetics and nutrition are not available. Reproductive success also varies positively with core area quality, and thus with relative rank, at Kanyawara (Emery Thompson et al. 2007; Kahlenberg et al. 2008).

Some female chimpanzees in the same community may rarely encounter each other, and others have antagonistic relationships, but some associate relatively often and some maintain long-term social bonds, reflected in frequent co-participation in parties, frequent proximity, and stable grooming relationships (e.g., Ngogo: Wakefield 2008;

Langergraber et al. 2009). Because many others have weak bonds, female-female relationships are more strongly differentiated than those between males (Langergraber et al. 2009). Most females in one Taï community groomed each other at least occasionally, usually bidirectionally (Wittig & Boesch 2003), and grooming between Ngogo females is common and reciprocal (Wakefield 2010). However, even Taï females groom less with each other than males do (Lehmann & Boesch 2008), and female coalitions are uncommon except during attacks on immigrants (Boesch & Boesch-Achermann 2000; Newton-Fisher 2006).

Captive female bonobos affiliate frequently and maintain closer proximity than adult males do (Parrish 1996). Lomako data initially showed more affiliation between females than between the sexes or between males (White 1988), but subsequent data (Hohmann et al. 1999) indicated that most dyads that maintained close spatial association, groomed often, and had relatively long and balanced grooming bouts were male-female. Grooming between females is moderately common at Lomako and Wamba, but less so at Wamba than male-male and male-female grooming (Furuichi 1997). Aggressive resistance to immigrant females by residents has not been seen. This apparent contrast to chimpanzees may reflect lower feeding competition in bonobos. Also unlike chimpanzees, newly immigrated bonobo females maintain close spatial association with particular resident females, who may protect them against harassment (Idani 1991).

Bonobos are renowned for high rates of sociosexual interactions between members of all age-sex classes, especially genito-genital contact between females (G-G rubbing; see chapter 29, this volume, and fig. 29.4). They use sociosexual behavior more than chimpanzees to resolve, and perhaps to avoid, conflicts (de Waal 1987, 1992). G-G rubbing facilitates reconciliation, regulates tension, and acknowledges or reaffirms relative social status (Hohmann & Fruth 2000).

Male-Male Relationships

Males in multimale lar gibbon groups at Khao Yai sometimes engage in mating contests, although aggression rates are low (Sommer & Reichard 2000). Males sometimes groom each other and jointly defend territories against neighboring groups (Reichard & Barelli 2008). Rates of aggression between males in two-male siamang groups were also low (Lappan 2007b); grooming was as common between males as between males and females, and both males participated in intergroup aggression.

Sexually mature male orangutans do not interact affiliatively. Flanged males are mutually hostile; they are less so to unflanged males, perhaps because chasing them is inefficient and energetically expensive, although unflanged males disrupt some forced copulation attempts (Rijksen 1978; Rodman & Mitani 1987; Galdikas 1988; van Schaik & van Hooff 1996; Fox 2002; Utami & van Hooff 2002).

Multimale gorilla groups can contain father-son, full-sibling, and maternal and paternal half-sibling dyads, so kinship can affect male relationships. Adult male mountain gorillas are often aggressive to maturing adolescents, but this apparently does not influence dispersal (Harcourt & Stewart 1987b; Stoinski et al. 2009b). Grooming is fairly common in some adult-adolescent dyads (Harcourt & Stewart 1987b; Watts & Pusey 1993), but rare or absent between adults (personal observation). Dominant males in multimale groups interrupt many copulation attempts by other sexually mature males, but tolerate copulations with their own putative daughters (Watts 1991b; Robbins 1996, 2007). Adults in multimale groups sometimes protect adolescent males against other adults, but coalitions between adults directed at other adults are rare in groups with three or more males (Robbins 2007; personal observation).

Relationships between male chimpanzees in the same community reflect a complex balance between competition and cooperation, and high rates of male-male affiliation and multiple forms of cooperation among males distinguish chimpanzees from other nonhuman primates (Muller & Mitani 2005; fig. 6.7). Males compete for status and form linear dominance hierarchies (Goodall 1986; Newton-Fisher 2004; Watts & de Vries 2009). Rank influences male-male association tactics (Newton-Fisher 1999b). High-ranking males can be attractive grooming partners and may receive more grooming than low-ranking males (Watts 2000b; but see Arnold & Whiten 2003). Males commonly form short-term coalitions, and some form long-term alliances. Alliances can affect rank acquisition, but they may be more important for maintenance of rank, especially of alpha status (de Waal 1982; Goodall 1986; Nishida & Hosaka 1996),

Figure 6.7. Male chimpanzees affiliate and cooperate with each other in several contexts. Here three adult male chimpanzees form a grooming "chain." Photo courtesy of John Mitani.

because many coalitions are conservative (both or all partners outrank their targets; Watts & de Vries 2009).

Grooming is more common between males than between females or between the sexes (Goodall 1986; Watts 2000c; Arnold & Whiten 2003). Reciprocity in male grooming and interchange of grooming for agonistic support have been documented (Watts 2002; Arnold & Whiten 2003). An alpha male at Kanyawara tolerated mating attempts by males who supported him (Duffy et al. 2007). Males maintain long-term social bonds, reflected in stable association patterns and grooming networks (Gilby & Wrangham 2008; Mitani 2009b). Alliance networks are more fluid because the value of others as allies varies more and changes more quickly than their value as grooming partners (de Waal 1984). Whether long-term social relationships influence male fitness and whether variation in sociality affects fitness are important unanswered questions (Mitani 2009a).

Maternal brothers are more likely than other dyads to be close associates, frequent grooming partners, and allies at Ngogo (and perhaps elsewhere), but maternal relatedness is not necessary for strong social bonds, as many allies are not brothers (Langergraber et al. 2007). Demography constrains maternal kinship effects because females will not often have consecutive surviving sons who could become valuable allies (ibid.). Concomitantly, males close in age are more likely to have strong bonds than those distant in age (Mitani et al. 2002). Paternal brothers at Ngogo do not have unusually strong bonds (Langergraber et al. 2007), but the possibility of paternal kinship effects warrants further investigation.

Arguably, cooperation among chimpanzee males extends to hunting and meat transfers (chapter 22, this volume). Boesch and Boesch-Achermann (1989, 2000) argue that Taï males use complex, collaborative hunting tactics, but others (e.g., Tomasello 2008) contend that apparent collaboration could result from purely individualistic tactics. Gilby (2006) proposed that meat transfers between males minimize harassment of meat possessors and improve their feeding efficiency, rather than representing nutritional or social exchanges, despite reciprocity in transfers and interchange of meat for grooming and agonistic support at Ngogo (Mitani & Watts 2001; Watts & Mitani 2002; Watts 2002). In fact, the likelihood that beggars obtain meat probably depends on multiple factors (Mitani 2009a), including participation in hunts (Boesch 1994) and relationship quality. The need for cooperation in intercommunity antagonism may constrain within-community competition and help to explain tolerance of high-ranking males for mating by lower-ranking males (Muller & Mitani 2005).

Male bonobos form dominance hierarchies, but not alliances. Early reports indicated that grooming among adult

bonobos at Wamba was most common in male-female dyads, followed by male-male and then female-female dyads. However, high levels of male-female grooming characterized only known or presumed mother/son dyads; otherwise, males groomed mostly with other males (Furuichi & Ihobe 1994).

Male-Female Relationships

Male-female gibbon pairs maintain long-term, although not necessarily permanent, social bonds. Male-female grooming frequency and bidirectionality vary. At Ketambe, male siamangs spent more time close to females and groomed with them more than male lar gibbons did (Palombit 1996), while grooming was the most common male-female social interaction in lar gibbons at Khao Yai (Bartlett 2003). Neither sex has unequivocal agonistic dominance to the other (Reichard 2009), but female lar gibbons at Khao Yai led most group movements and entered most food sources before males, and at trees with few fruit, males often waited while females fed (Barelli et al. 2008); thus females seem to have power (sensu Lewis 2002) over males regarding food. Male siamangs often carry infants, reducing reproductive costs for females (Chivers 1974; Lappan 2008). Males in other gibbons do not carry young, although male-infant play may aid the development of locomotor competence (Bartlett 2003; Lappan 2008).

Male-female social bonds are strong in mountain gorillas (Harcourt 1979; Watts 1992, 2003; Robbins 2001; Nsubuga et al. 2008). Males are unequivocally dominant. Male-to-female aggression is common, but rarely serious (Watts 1992), and is often followed by female-initiated reconciliation that makes further male aggression less likely (Watts 1995). Male-female grooming is usually more common than that between females, although not all dyads groom and grooming imbalances can favor either sex (Watts 1992). Most females in multimale groups spend more time close to, and groom more with, dominant than subordinate males, but subordinates may invest considerably in maintaining proximity to females and grooming them (ibid.); this can influence female residence choices if groups fission (Robbins 2001). Male-female relationships seem equally prominent in other gorillas (Stokes 2004). Resident female mountain gorillas are generally not unusually aggressive to immigrants, but those in large groups may resist transfer attempts. Dominant males protect new immigrants against harassment; this may be mating effort that helps persuade immigrants to stay (Watts 1991c).

Male chimpanzees usually groom with females much less than with other males, but grooming effort varies. Grooming is common in some mother–adult son pairs (Goodall

1986; pers. observ.). Males and females at Sonso showed grooming reciprocity (Arnold & Whiten 2003). Male-female proximity and grooming at Taï are differentiated (Stumpf & Boesch 2003). Differentiation is almost certainly typical and may be stronger where sex differences in range use and gregariousness are greater (e.g., at Gombe: Goodall 1986); whether it has fitness consequences is an important question. Male-to-female aggression is common, especially when females have sexual swellings (Muller et al. 2007). Female-male coalitionary support is common in captivity, where females cannot avoid males and have much at stake in their power maneuvering (de Waal 1982). Male-female and female-female coalitions against males are much rarer in the wild (Boesch & Boesch-Achermann 2000; Newton-Fisher 2006), where avoidance is easier, potential benefits are lower, and cost-to-benefit ratios are less favorable.

Male-female bonobo dyads are more likely than female dyads to be long-term close associates at Lomako (Hohmann et al. 1999). Close male-female associates at Wamba are mostly mother-son or half-sibling pairs (Furuichi & Ihobe 1994; Furuichi 1997). Unlike with chimpanzees, bonobo female-male alliances are common in the wild (Kano 1992). Females face low risk in supporting some males against others, because most within-community aggression between males does not involve physical contact and severe aggression is rare (Hohmann et al. 1999). Males may therefore benefit as much from alliances with females as from those with other males (Furuichi 1997; Hohmann et al. 1999). By supporting sons, females may help them to achieve higher rank and, potentially, high reproductive success (Gerloff et al. 2001), and females might support unrelated males in exchange for tolerance at food sources and withholding of aggression to other allies (Hohmann et al. 1999). However, females seem unwilling to support unrelated males against females, perhaps because potential return benefits are too low (ibid.).

Male chimpanzees are unequivocally dominant to females (Goodall 1986; Wittig & Boesch 2003). In small captive bonobo groups, some females are dominant to (i.e., win agonistic contests against) at least some males, and female coalitions against males are common (Parish 1994; Vervaecke et al. 2004). Because female coalitions at Wamba (Furuichi 1989; Kano 1992) and Lomako (Hohmann & Fruth 2002; White & Wood 2007) win some contests against males, Parish and de Waal (2000; cf. Parish 1994) argued that bonobos show "female dominance." However, wins by female coalitions, if consistent, are better seen as "derived dominance" (Lewis 2002), whereas males win most dyadic contests (Kano 1992; Hohmann & Fruth 2002, 2003; White & Wood 2007). White and Wood (2007) reported that Lomako females typically entered small food patches before

males and fed longer—behavior they labeled female "feeding priority" (or male "deference"). A more useful view is that male power has more limited "scope" in bonobos, which does not include feeding (Lewis 2002), as is shown by female ability to control partitioning of shareable, valuable food items like meat and *Treculia africana* fruit (Fruth & Hohmann 2002). Also, greater potential benefits of long-term affiliative relationships with females may constrain male aggression to females in bonobos more than in chimpanzees (Hohmann & Fruth 2003).

Such contrasts may depend on whatever ecological differences allow higher gregariousness in bonobos (Wrangham 1986), which should increase female potential to form coalitions to resist male aggression and, along with less reliable advertisement of female fertility (Furuichi 1987; Wrangham 1993, 2002), reduce the benefits of male alliances. The value of females as allies gives them leverage over males and increases the costs of male-to-female aggression (Lewis 2002). Still, this raises the question of why female chimpanzees in food-rich habitats, where female gregariousness is high, rarely form coalitions with males against other males.

Intergroup Relations

Relations between gibbon groups are hostile (Bartlett 2007, 2009). Females are most concerned with outside females and may be defending food resources; males focus mostly on outside males and may primarily be guarding mates (ibid.; Ellefson 1974; Mitani 1984, 1985b; Reichard & Sommer 1997; Bartlett 2003; Brockelman 2009; Reichard 2009). These sex differences led Reichard and Sommer (1997) to argue that hylobatids show "female resource defense" and "male mate defense," not "pairbonding" and "territoriality." However, defending access to females and food for females are mutually compatible (Reichard 2009)—many encounters occur at fruiting trees (Lappan 2007)—as are mate guarding (Palombit 1999; Sommer & Reichard 2000) and defense against infanticide (Dunbar & van Schaik 1994; Reichard 2009). Because females mate polyandrously with males in two-male groups, intergroup aggression by both males is consistent with mate guarding (Reichard & Barelli 2008; Reichard 2009). If two-male groups generally win against unimale groups, joint male participation could mean that females in two-male groups have higher-quality territories (Lappan 2007b).

Male mountain gorillas are antagonistic to outside males, who may try to induce females to transfer and are infanticide threats (Watts 1989; Sicotte 1993; Harcourt & Stewart 2007a; Robbins & Sawyer 2007). Solitary males use home ranges larger than needed for energy and nutrient needs;

they sometimes follow groups and may seek encounters (Watts 1994b) that allow females to assess prospective mates. Males sometimes mate-guard females to dissuade emigration (Sicotte 1993). Males in multimale groups participate jointly in antagonism against outside males (Robbins 1996; Sicotte 1993; Watts 2000a; Harcourt & Stewart 2007a, b), although their effort may be unequal (Robbins & Sawyer 2007). This helps explain why infanticide risk is lower in multimale than in unimale groups, although the main advantage of multimale groups for females is that male deaths do not leave them and their infants unprotected (Watts 1989, 2000a; Harcourt & Stewart 2007a, b; Robbins et al. 2007; Robbins et al. 2009). Escalated aggression is more likely in contests between group males and solitaries than those between group males (Sicotte 1993; Robbins & Sawyer 2007). Most contests do not lead to fights (ibid.), but escalated aggression can be fatal (Watts 1989). Escalated male-male contests seem less likely in western gorilla groups. Many encounters in bais are peaceful; during others, males try to attract female immigrants, to guard against emigration by mates, and perhaps to attack or protect infants (Caillaud et al. 2008).

Relations between neighboring chimpanzee communities are hostile and chimpanzees are territorial, at least in moist forest habitats (Herbinger et al. 2001), although the intensity of hostility varies with demography and ecology (Boesch et al. 2008). Coalitionary intergroup aggression can be lethal (Goodall et al. 1979; fig. 6.8). Such behavior occurs in only a few other mammals, most of which (e.g., spotted hyenas, lions, wolves, humans) have fission-fusion social systems that give large parties from one group low-risk opportunities to attack smaller parties or solitary individuals from neighboring groups (Manson & Wrangham 1991; Wrangham 1999; Wilson & Wrangham 2003). Chimpanzees increase the probability of such encounters by patrolling territory boundaries and making incursions into neighbors' territories (Goodall et al. 1979; Wrangham 1999; Boesch & Boesch-Achermann 2000; Watts & Mitani 2001). Patrolling is mostly by males, and the tendency to patrol increases with male party size (Mitani & Watts 2005). Patrollers are highly vigilant and they react to encountered neighbors largely on the basis of relative party size: they are likely to attack greatly outnumbered neighbors, but to avoid others (Manson & Wrangham 1991; Boesch & Boesch-Achermann 2000; Watts & Mitani 2001). Responses to neighbors met or heard in other contexts include attacks, calling outbursts, and flight, depending on proximity and relative male numbers (Watts & Mitani 2001; Wilson et al. 2001; Herbinger et al. 2009). Communities with few males may stop patrolling and try to avoid encounters (Boesch & Boesch-Achermann 2000),

although males in the small Mitumba community at Gombe patrolled and aggressively interacted with neighbors (Wilson et al. 2008).

All chimpanzees risk attack by neighbors, but the risks are particularly severe for adult and adolescent males and for females with unweaned infants (Manson & Wrangham 1991; Wrangham 1999; Watts et al. 2005). Cross-site variation in the amount of time individuals are alone or in small parties helps to explain variation in the frequency of lethal intergroup aggression (Boesch & Boesch-Achermann 2000; Boesch et al. 2008). The "imbalance of power" hypothesis (Manson & Wrangham 1991) accounts for why low-cost attacks on outnumbered male strangers occur, given that those males are dangerous, but does not account for the benefits of lethal coalitionary intergroup aggression (Watts & Mitani 2001). Fatal attacks by Kasakela males on all the Kahama at Gombe are consistent with the "mate attraction" hypothesis, as some Kahama females later joined the Kasakela community (Goodall et al. 1979); this hypothesis could also apply to adolescent females, but it fails to explain lethal attacks on mothers (Goodall et al. 1979; Pusey et al. 2008) and between-community infanticide, after which mothers do not transfer (Wrangham 1999; Watts & Mitani 2001; Wilson & Wrangham 2003; Williams et al. 2004). More fundamentally, territoriality and lethal intergroup aggression probably promote reproduction and survival by maintaining or increasing food availability (Manson & Wrangham 1991; Wrangham 1999; Wilson & Wrangham 2003; Mitani et al. 2010). At Gombe, infant survival has been higher when the chimpanzee community's range has been relatively large (Williams et al. 2004). The Ngogo community has expanded its territory into an area where the frequency of lethal attacks on neighbors has been especially high (Mitani et al. 2010). Thus chimpanzee territoriality seems to involve male defense of food resources and, perhaps, indirect parental care.

Interactions between bonobo communities are generally hostile, and females fear male strangers (Kano 1992; Gerloff et al. 1999). However, neither chimpanzee-like boundary patrols and incursions nor lethal coalitionary intergroup aggression are known. Differences in gregariousness cannot explain these differences, even if higher bonobo gregariousness makes achieving overwhelming power imbalances more difficult (Wrangham 1986). Gregariousness is similar for Taï chimpanzees and bonobos at Lomako (Hohmann & Fruth 2002; Lehmann & Boesch 2005), but lethal intergroup aggression occurs at Taï (Boesch et al. 2008). Male bonobos may be less inclined than male chimpanzees to form alliances in between-community competition, because they do not benefit from within-community alliances. Also, attacking outsiders may be more dangerous for bonobos be-

Figure 6.8.   Lethal intergroup coalitionary aggression is a characteristic feature of chimpanzee behavior. In this photo, a young adult male chimpanzee leaps on the corpse of an adult male from a neighboring group. The victim had been caught and killed 30 minutes before by a group of male chimpanzees on patrol. See text for further explanation. Photo courtesy of John Mitani.

cause females form alliances against males, and the slightly lower sexual dimorphism of bonobos in canine size and aspects of the skull related to bite force reduces sex differences in fighting ability (Plavcan & van Schaik 1992).

Cognition

Research on ape cognition is thriving, as part 5 of this volume indicates. Tool use, mental state attribution, episodic memory, and experimental work on cooperation and fairness deserve brief mention because of their historic or current relevance to questions about human uniqueness and ecological and social intelligence.

The discovery that Gombe chimpanzees make and use tools dispelled the prevailing academic notion that such

behavior was unique to hominins. Tool use and manufacture appear universal in chimpanzees, and tool kits vary extensively (McGrew 1993, 2004; Whiten 2010; chapter 31, this volume). Tools serve multiple purposes (e.g., hygiene), but most importantly increase nutrient gain rates via extractive foraging (McGrew 1979; Sugiyama & Koman 1979; Boesch & Boesch-Achermann 2000; Yamakoshi 2001). Some chimpanzees use stones as tools but do not deliberately modify them, a stark contrast with the percussive technology that allowed niche expansion by Oldowan hominins (Schick & Toth 2009). Tool use is rare or absent in the wild among other apes except Sumatran orangutans at Suaq Balimbing, who have tool kits for extraction of insects, honey, and the arils of *Neesia* spp. fruit (Fox et al. 1999). Efficient extraction of *Neesia* arils is particularly valuable nutritionally, and it helps explain the high population density and gregariousness of orangutans there (van Schaik 1999).

Experiments show that chimpanzees attribute some intentions to others, understand seeing and hearing as perceptual states, and know something about others' knowledge states (chapters 28 and 30, this volume; cf. Call & Tomasello 2008; Tomasello 2008). Ability to attribute perception and knowledge could have many influences on social dynamics—for example, by reducing costs of competition (Hare et al. 2001) and facilitating cooperation during hunts, territorial defense, and in other contexts (chapter 22, this volume). Experimental investigation of "fairness" and cooperation (chapters 25 and 32, this volume) addresses similar issues. Episodic memory involves conscious awareness of events from one's past, the ability to relive them mentally, and general awareness of subjective time; it allows mental "time travel" into the past and future (Tulving 2005) and is relevant to ecological and social knowledge in many ways (e.g., tracking first- and third-person social relationships in fission-fusion societies (chapter 28, this volume). Recent experiments with great apes (e.g., Osvath & Osvath 2008) challenge the claim that it is uniquely human (Tulving 2005; Suddendorf et al. 2009); this issue should generate exciting research and debate (chapter 27, this volume). Experimental results presently suggest that chimpanzees have only egocentric concerns about fairness and do not engage in true collaboration involving joint understanding of mutual goals (chapter 32, this volume; cf. Tomasello 2008). By implication, they and other apes are incapable of collective intentionality, a uniquely derived human characteristic that is the basis for human culture (ibid.; Plotkin 2003).

Summary and Conclusions

Diversity in the body size, ecology, social organization, and social systems of apes is remarkable. Research on this diversity has been fundamental to incorporating primatology into behavioral ecology and for incorporating and developing theory concerning ecological influences on female dispersion and social relationships and the interplay between female and male mating and social strategies (chapter 9, this volume). All apes have broad diets comprising mostly plant foods; fruit is important for all except some mountain gorillas. Females respond to potential feeding competition by forming small, typically single-female groups (gibbons); by adjusting gregariousness to ecological conditions, either without stable, multiadult social units (orangutans) or in fission-fusion communities (chimpanzees and bonobos); or by forming stable multifemale groups that typically have single males (gorillas). Males respond by typically forming

pairs with single females (gibbons); by establishing large ranges that encompass those of several females and/or roving widely (orangutans); by competing to attract females and control mating access to female groups (gorillas); or by forming multimale, multifemale, fission-fusion communities within which most mating is promiscuous, although male reproductive success varies as a function of dominance rank and other factors (chimpanzees and bonobos). Male orangutans are hostile to each other, as are males from different social units in the other apes. Cooperation against outside males by males in the same group or community is limited in gibbons, gorillas, and bonobos, but is pronounced in chimpanzees; at its extreme, chimpanzee coalitionary intergroup aggression is lethal. Chimpanzee males compete and cooperate in complex ways within communities, and research on chimpanzees has contributed significantly to debates about the evolution of cooperation in primates, notably regarding the effects of kinship (chapters 21 and 22, this volume). Research on apes also has contributed importantly to understanding sexual conflict and sexual coercion (chapter 19, this volume), which is especially prominent in chimpanzees, orangutans, and gorillas. Long-term research is revealing extensive socioecological variation across and within taxa and even populations. Notably, we now know that gibbons are not strictly monogamous; that gorillas are not specialized folivores; that female chimpanzees are more gregarious in some populations than in others, have highly differentiated social relationships, and compete in ways that have important fitness consequences; and that male reproductive skew is high in gorillas and moderate to low in chimpanzees, as expected, but perhaps lower than expected in orangutans. Nevertheless, prominent unanswered questions remain, especially concerning chimpanzee-bonobo differences, particularly in male-female power balances and in male-male social relationships; the role of infanticide in social evolution; the relevance of cognition experiments to social and ecological strategies in the wild; and the effects of variation in ecology and social relationships on fitness. The slow life histories of all apes make collecting necessary data difficult, but the challenges of preserving apes and their habitats to address these questions are even more difficult.

References

Arnold, K. & Whiten, A. 2003. Grooming interactions among the chimpanzees of the Budongo Forest, Uganda: A test of five explanatory models. *Behaviour*, 140, 519–552.

Aureli, F., Schaffner, C. M., Boesch, C., Bearder, S. K., Call, J., Chapman, C. A., Connor, R. A., Di Fiore, A., Dunbar, R. I. M, Henzi, S. P., Holekamp, K., Korstjens, A. H., Layton, R., Lee, P., Lehmann, J., Manson, J. H., Ramos-Fernandez, G.,

Strier, K. B. & van Schaik, C. P. 2007. Fission-fusion dynamics: New research frameworks. *Current Anthropology*, 49, 627–654.

Badrian, N. & Malenky, R. K. 1984. Feeding ecology of *Pan paniscus* in the Lomako Forest, Zaire. In *The Pygmy Chimpanzee: Evolutionary Biology and Behavior* (ed. by Susman, R. L.), 275–299. New York: Plenum Press.

Barelli, C., Boesch, C., Heistermann, M. & Reichard, U. H. 2008. Female white-handed gibbons (*Hylobates lar*) lead group movements and have feeding priority to food resources. *Behaviour*, 145, 965–981.

Bartlett, T. Q. 2003. Intragroup and intergroup social interactions in white-handed gibbons. *International Journal of Primatology*, 24, 239–259.

———. 2007. The Hylobatidae: Small apes of Asia. In *Primates in Perspective* (ed. by Campbell, C. J., Fuentes, A., Mackinnon, K. C., Panger, M. & Bearder, S. K.), 274–289. Oxford: Oxford University Press.

———. 2009. In *Gibbons: New Perspectives on Small Ape Socioecology and Population Biology* (ed. by Whittaker, D. J. & Lappan, S.), 265–275. New York: Springer-Verlag.

Basabose, K. 2002. Diet composition of chimpanzees inhabiting the montane forest of Kahuzi, Democratic Republic of Congo. *International Journal of Primatology*, 23, 1–21.

Beaudrot, L. H., Kahlenberg, S. M. & Marshall, A. J. 2009. Why male orangutans do not kill infants. *Behavioral Ecology and Sociobiology*, 63, 1549–1562.

Becquet, C., Patterson, N., Stone, A. C., Przeworski, M. & Reich, D. 2007. Genetic structure of chimpanzee populations. *PLoS Genetics*, 3, e66.

Bermejo, M. 2004. Home range use and intergroup encounters in western gorillas (*Gorilla g. gorilla*) in Lossi Forest, north Congo. *American Journal of Primatology*, 64, 223–232.

Boesch, C. 1994. Cooperative hunting in wild chimpanzees. *Animal Behaviour*, 48, 653–667.

Boesch, C. & Boesch, H. 1989. Hunting behaviour of wild chimpanzees in the Taï National Park. *American Journal of Physical Anthropology*, 78, 547–573.

Boesch, C. & Boesch-Achermann, H. 2000. *The Chimpanzees of the Tai Forest*. Oxford: Oxford University Press.

Boesch, C., Kohou, G., Néné, H. & Vigilant, L. 2006. Male competition and paternity of wild chimpanzees in the Taï Forest. *American Journal of Physical Anthropology*, 130, 103–115.

Boesch, C., Crockford, C., Herbinger, I., Wittig, R., Moebius, Y. & Normand, E. 2008. Intergroup conflicts among chimpanzees in Taï National Park: Lethal violence and the female perspective. *American Journal of Primatology*, 70, 519–532.

Borries, C., Savini, T. & Koenig, A. 2011. Social monogamy and the threat of infanticide in larger mammals. *Behavioral Ecology and Sociobiology*, 65, 685–693.

Bradley, B. B., Doran-Sheehy, D. M., Lukas, D., Boesch, C. & Vigilant, L. 2004. Dispersed male networks in western gorillas. *Current Biology*, 14, 510–514.

Bradley, B. B., Robbins, M. M., Williamson, E. A., Steklis, H. D., Gerald-Steklis, N., Eckhardt, N., Boesch, C. & Vigilant, L. 2005. Mountain gorilla tug-of-war: Silverbacks have limited control over reproduction in multi-male groups. *Proceedings of the National Academy of Sciences, USA*, 102, 9418–9423.

Breuer, J., Hokemba, M. B.-N., Olejniczak, C., Parnell, R. J., & Stokes, E. 2009. Physical maturation, life-history classes and age estimates of free-ranging western gorillas: Insights from Mbeli Bai, Republic of Congo. *American Journal of Primatology*, 71, 106–119.

Breuer, J., Robbins, A. M., Olejniczak, C., Parnell, R. J., & Stokes, E., Robbins, M. M. 2009. Variation in the male reproductive success of western gorillas: Acquiring females is just the beginning. *Behavioral Ecology and Sociobiology*, 6, 515–528.

Brockelman, W. Y. 2009. Ecology and social system of gibbons. In *Gibbons: New Perspectives on Small Ape Socioecology and Population Biology* (edited by Whittaker, D. J. & Lappan, S.), 211–239. New York: Springer-Verlag.

Brockelman, W. Y., Reichard, U., Treesucon, U. & Raemaekers, J. J. 1998. Dispersal, pair formation, and social structure in gibbons (*Hylobates lar*). *Behavioral Ecology and Sociobiology*, 42, 329–339.

Caillaud, D., Levrero, F., Gatti, S., Menard, N. & Raymond, M. 2008. Influence of male morphology on male mating status and behavior during interunit encounters in western lowland gorillas. *American Journal of Physical Anthropology*, 135, 379–388.

Caldecott, J. & Kapos. 2005. Great ape habitats: Tropical moist forests of the Old World. In *World Atlas of Great Apes and Their Conservation* (ed. by Caldecott, J. & Miles, L.), 31–42. Berkeley: University of California Press.

Call, J. & Tomasello, M. 2008. Does the chimpanzee have a theory of mind? 30 years later. *Trends in Cognitive Science*, 12,187–192.

Caswell, J. L., Mallik, S., Richter, D. J., Meubauer, J., Schirmer, C., Gnerre, S. & Reich, D. 2008. Analysis of chimpanzee genetic history based on genome sequence alignments. *PLoS Genetics*, 4, e100000657.

Chapman, C. C., Wrangham, R. W. & White, F. J. 1994. Party size in chimpanzees and bonobos. In *Chimpanzee Cultures* (ed. by Wrangham, R. W., McGrew, W. C., de Waal, F. B. M., Heltne, P. G. & Marquardt, L. A.), 41–57. Cambridge, MA: Harvard University Press.

Chivers, D. J. 1974. *The Siamang in Malaya: A Field Study of a Primate in a Tropical Rain Forest*. Basel: S. Karger.

Conklin-Brittain, N. I., Wrangham, R. W. 1998. *International Journal of Primatology*.

Constantino, P. J., Lucas, P. W., Lee, J. J.-W. & Lawn, B. R. 2009. The influence of fallback foods on great ape tooth enamel. *American Journal of Physical Anthropology*, 14, 653–660.

Constable, J. L., Ashley, M. V. Goodall, J. & Pusey, A. E. 2001. Noninvasive paternity assignment in Gombe chimpanzees. *Molecular Ecology*, 10, 1279–1300.

D'Amour, D. E., Hohmann, G. & Fruth, B. 2006. Evidence of leopard predation on bonobos (*Pan paniscus*). *Folia Primatologica*, 77, 212–217.

Deblauwe, I., Dupain, J., Nguenang, G. M., Werdenich, D. & Van Elsacker, L. 2003. Insectivory by western gorillas (*Gorilla g. gorilla*) in southeast Cameroon. *International Journal of Primatology*, 24, 493–501.

Delgado, R. & van Schaik, C. P. 2000. The behavioral ecology and conservation of the orangutan (*Pongo pygmaeus*): a tale of two islands. *Evolutionary Anthropology*, 9, 201–218.

De Waal, F. B. M. 1982. *Chimpanzee Politics*. Baltmore: Johns Hopkins University Press.

———. 1984. Sex differences in the formation of coalitions among chimpanzees. *Ethology and Sociobiology*, 5, 239–255.

Doran, D. & McNeilage, A. 1998. Gorilla ecology and behavior. *Evolutionary Anthropology*, 6, 120–131.

Doran-Sheehy, D., Greer, D., Mongo, P. & Schwindt, D. 2004. Impact of ecological and social factors on ranging in western gorillas. *American Journal of Primatology*, 64, 207–222.

Douadi, M. I., Gatti, S., Levrero, F., Duhamel, H., Bermejo, M., Vallet, D., Menard, N. & Petit, E.J. 2007. Sex-biased dispersal in western lowland gorillas. *Molecular Ecology*, 16, 2247–2255.

Duffy, K. G., Wrangham, R. W. & Silk, J. B. 2007. Male chimpanzees exchange political support for mating opportunities. *Current Biology*, 17, R586–587.

Elder, A. E. 2009. Hylobatid diets revisited: The importance of body mass, fruit availability, and interspecfic competition. In *The Gibbons, Developments in Primatology: Progress and Progress* (ed. by Whittaker, D. & Lappan, S.), 133–159. New York: Springer Verlag.

Ellefson, J. 1974. *A Natural History of White-Handed Gibbons in the Malayan Peninsula*. In *Gibbon and Siamang, Vol. 3* (ed. by Rumbaugh, D. M.), 1–136. Basel: S. Karger.

Emery Thompson, M., Kahlenberg, S. M., Gilby, I. C. & Wrangham, R. W. 2007. Core area quality is associated with variance in reproductive success among female chimpanzees at Kibale National Park. *Animal Behaviour*, 7, 501–512.

Emery Thompson, M., Jones, J. H., Pusey, A. E., Brewer-Marsden, S., Goodall, J., Marsden, D., Matsuzawa, T., Nishida, T., Reynolds, V., Sugiyama, Y. & Wrangham, R. W. 2007. Aging and fertility patterns in wild chimpanzees provide insights into the evolution of menopause. *Current Biology*, 17, 2150–2156.

Fan, P., Sun, G., Huang, B. & Jiang, X. 2009. Gibbons under seasonal stress: The diet of the black-crested gibbon (*Nomascus concolor*) on Mt. Wuliang, Central Yunnan, China. *Primates*, 50, 37–44.

Ferris, S. 2005. Western gorilla (*Gorilla gorilla*). In *World Atlas of Great Apes and Their Conservation* (ed. by Caldecott, J. & Miles, L.), 105–127. Berkeley: University of California Press.

Ferris, S., Robbins, M. M. & Williamson, E. A. 2005. Eastern gorilla (*Gorilla beringei*). In: *World Atlas of Great Apes and Their Conservation* (ed. by Caldecott, J. & Miles, L.), 129–152. Berkeley: University of California Press.

Fischer, A., Pollack. J, Thalmann O., Nickel B. & Pääbo S. 2006. Demographic history and genetic differentiation among apes. *Current Biology*, 16, 1133–1138.

Fossey, D. & Harcourt, A. H. 1977. Feeding ecology in free ranging mountain gorillas (*Gorilla gorilla beringei*). In *Primate Ecology: Studies of Feeding and Ranging Behavior in Lemurs, Monkeys, and Apes* (ed. by Clutton-Brock, T. H.), 539–556. London: Academic Press.

Fox, E. A. 1998. The function of male aggression and female resistance in wild Sumatran orangutans. PhD thesis, Duke University.

———. 2002. Female tactics to reduce sexual harassment in the Sumatran orangutan (*Pongo pygmaeus abelii*). *Behavioral Ecology and Sociobiology*, 53, 93–101.

Fox, E. A., Sitompul, A. & van Schaik, C. P. 1999. Intelligent tool use in wild orangutans. In *The Mentalities of Gorillas and Orangutans* (ed. by Parker, S. T., Mitchell, R. W.

& Miles, H. L.), 99–116. Cambridge: Cambridge University Press.

Frey, J. C., Rothman, J. M., Pell, A. N., Mizeye, J. B., Cranfield, J. K. & Angert, E. A. 2006. Fecal bacterial diversity in a wild gorilla (*Gorilla beringei*). *Applied and Environmental Microbiology*, 72, 3788–3792.

Fruth, B. & Hohmann, G. 2002. How bonobos handle hunts and harvests: Why share food? In *Behavioural Diversity in Chimpanzees and Bonobos* (ed. by Boesch, C., Hohmann, G. & Marchant, L. A.), 231–243. Cambridge: Cambridge University Press.

Furuichi, T. 1987. Sexual swelling, receptivity, and grouping of wild pygmy chimpanzee females at Wamba, Zaire. *Primates*, 28, 309–318.

———. 1989. Social interactions and the life history of female *Pan paniscus* in Wamba, Zaire. *International Journal of Primatology*, 10, 173–197.

———. 1997. Agonistic interactions and matrifocal dominance rank of wild bonobos (*Pan paniscus*) at Wamba. *International Journal of Primatology*, 18, 855–875.

Furuichi, T. & Ihobe, H. 1994. Variation in male relationships in bonobos and chimpanzees. *Behaviour*, 130, 211–228.

Galdikas, B. M. F. 1979. Orangutan adaptation at Tanjung Puting: Mating and ecology. In *The Great Apes* (ed. by Hamburg, D. A. & McCown, E. R.), 194–233. Menlo Park, CA: Benjamin-Cummings.

———. 1985a. Adult male sociality and reproductive tactics among orangutans at Tanjung Puting. *Folia Primatologica*, 45, 9–24.

———. 1985b. Subadult male orangutan sociality and reproductive behavior at Tanjung Puting. *American Journal of Primatology*, 8, 87–99.

———. 1988. Orangutan diet, range, and activity at Tanjung Puting, central Borneo. *International Journal of Primatology*, 9, 1–35.

Ganas, J., Ortmann, S. & Robbins, M. M. 2009. Food choice of the mountain gorilla in Bwindi Impenetrable National Park, Uganda: The influence of nutrients, phenolics, and availability. *Journal of Tropical Ecology*, 25, 123–134.

Gatti, S., Levrero, F., Ménard, N. & Gautier-Hion, A. 2004. Population and group structure of western lowland gorillas (*Gorilla gorilla gorilla*) at Lokoué, Republic of Congo. *American Journal of Primatology*, 63, 111–123.

Gerloff, V., Hartung, B., Fruth, B., Hohmann, G. & Tautz, D. 1999. Intracommunity relations, dispersal patterns, and paternity success in a wild-living community of bonobos (*Pan paniscus*) determined from DNA analysis of fecal samples. *Proceedings of the Royal Society, London B*, 266, 1189–1195.

Gilby, I. 2005. Meat sharing among the Gombe chimpanzees: Harassment and reciprocal exchange. *Animal. Behaviour*, 71, 953–963.

Gilby, I. & Wrangham, R. W. 2007. Risk-prone hunting by chimpanzees increases during periods of high diet quality. *Behavioral Ecology and Sociobiology*, 61, 1771–1779.

———. 2008. Association patterns among wild chimpanzees (*Pan troglodytes schweinfurthii*) reflect sex differences in cooperation. *Behavioral Ecology and Sociobiology*, 62, 1831–1842.

Goodall, J., Bandora, A., Bergmann, E., Busse, C., Matama, H. & Russ D. 1979. Inter- community interactions in the population of chimpanzees in the Gombe National Park. In *The Great Apes* (ed. by Hamburg, D. A. & McCown, E. R.), 13–54. Menlo Park, CA: Benjamin-Cummings.

———. 1986. *The Chimpanzees of Gombe*. Cambridge, MA: Harvard University Press.

Groves, C. 2001. *Primate Taxonomy*. Washington: Smithsonian Institution Press.

———. 2005. Order Primates. In *Mammal Species of the World* (ed. by Wilson, D. E. & Reeder, D. M.), 111–184. Baltimore: Johns Hopkins University Press.

Guschanski, K., Caillaud, D., Robbins, M. M. & Vigilant, L. 2008. Females shape the genetic structure of a gorilla population. *Current Biology*, 18, 1809–1814.

Harcourt, A. H. 1978. Strategies of emigration and transfer by female primates with special reference to mountain gorillas. *Zeitschrift für Tierpsychologie*, 48, 401–420.

———. 1979a. Social relationships among adult female mountain gorillas. *Animal Behaviour*, 27, 251–264.

———. 1979b. Social relationships between adult male and female mountain gorillas. *Animal Behaviour*, 27, 325–342.

Harcourt, A. H., Fossey, D., Stewart, K. J & Watts, D. P. 1980. Reproduction in wild gorillas and some comparisons with chimpanzees. *Journal of Reproduction and Fertility* Suppl. 28, 59–70.

Harcourt, A.H. & Stewart, K.J. 1987a. The influence of help in contests on dominance rank in primates: Hints from gorillas. *Animal Behaviour*, 35, 182–190.

Harcourt, A.H. & Stewart, K. J. 1987b. Gorillas: Variation in female relationships. In *Primate Societies* (ed. by Cheney, D. L., Seyfarth. R. M., Smuts, B. B., Struhsaker, T. T. & Wrangham, R. W.), 155–164. Chicago: University of Chicago Press.

Harcourt, A. H. & Greenberg, J. 2001. Do female gorillas join males to avoid infanticide? A quantitative model. *Animal Behaviour*, 62, 905–915.

Harcourt, A. H. & Stewart, K. J. 2007a. *Gorilla Society: Conflict, Compromise, and Cooperation Between the Sexes*. Chicago: University of Chicago Press.

———. 2007b. Gorilla society: What we do and don't know. *Evolutionary Anthropology*, 16, 147–158.

Hare, B., Call, J. & Tomasello, M. 2001. Do chimpanzees know what conspecifics know? *Animal Behaviour*, 61, 139–151.

Harrison, T. 2010. Apes among the tangled branches of human origins. *Science*, 327, 532–534.

Hasegawa, T. 1990. Sex differences in ranging patterns. In *The Chimpanzees of the Mahale Mountains: Sexual and Life History Strategies* (ed. by Nishida, T.), 99–114. Tokyo: University of Tokyo Press.

Hasegawa, T. & Hiraiwa-Hasegawa, M. 1990. Sperm competition and mating behavior. In *The Chimpanzees of the Mahale Mountains: Sexual and Life History Strategies* (ed. by Nishida, T.), 115–132. Tokyo: University of Tokyo Press.

Herbinger, I., Boesch, C. & H. Rothe. 2001. Territory characteristics among three neighboring chimpanzee communities. *International Journal of Primatology*, 22, 143–167.

Herbinger, I., Papworth, S., Boesch, C. & Zuberbühler, K. 2009. Vocal, gestural, and locomotor responses of wild chimpanzees to familiar and unfamiliar intruders: A playback study. *Animal Behaviour*, 78, 1389–1396.

Hill, K., Boesch, C., Goodall, J., Pusey, A., Williams, J. &

Wrangham, R. W. 2001. Mortality rates among wild chimpanzees. *Journal of Human Evolution*, 40, 437–450.

Hohmann, G. & Fruth, B. 2000. Use and function of genital contacts among female bonobos. *Animal Behaviour*, 60, 107–120.

Hohmann, G. & Fruth, B. 2002. Dynamics in social organization of bonobos (*Pan paniscus*). In *Behavioral Diversity in Chimpanzees and Bonobos* (ed. by Boesch, C., Hohmann, G. & Marchant, L. F.), 138–155. Cambridge: Cambridge University Press.

Hohmann, G. & Fruth, B. 2003. Intra-and inter-sexual aggression by bonobos in the context of mating. *Behaviour*, 140, 1389–1413.

Hohmann, G., Gerloff, U., Tautz, D. & Fruth, B. 1999. Social bonds and genetics: Kinship, association, and affiliation in a community of bonobos (*Pan paniscus*). *Behaviour*, 126, 1219–1235.

Hohmann, G., Fowler, A., Sommer, V. & Ortman, S. 2006. Frugivory and gregariousness of Salonga bonobos and Gashaka chimpanzees: The influence of abundance and nutritional quality of fruit. In *Feeding Ecology in Apes and Other Primates* (ed. by Hohmann, G., Robbins, M. & Boesch, C.), 123–159. Cambridge: Cambridge University Press.

Hosaka, K., Nishida, T., Hamai, M., Matsumata-Oda, A. & Uehara, S. 2001. Predation of mammals by the chimpanzees of the Mahale Mountains, Tanzania. In *All Apes Great and Small, Vol. I: African Apes* (ed. by Galdikas, B. M. F., Briggs, N. E., Sheeran, L.K., Shapiro, G. L. & Goodall, J.), 107–130. New York: Kluwer Academic Press.

Idani, G. 1991. Social relationships between immigrant and resident bonobo (*Pan paniscus*) females at Wamba. *Folia Primatologica*, 57, 83–95.

Inskipp, T. 2005. Chimpanzees (*Pan troglodytes*). In *World Atlas of Great Apes and Their Conservation* (ed. by Caldecott, J. & Miles, L.), 53–81. Berkeley. University of California Press.

Inouye, E., Inouye-Murayama, M., Vigilant, L., Takanaka, O. & Nishida, T. 2008. Relatedness in wild chimpanzees: Paternity, male philopatry, and demographic factors. *American Journal of Physical Anthropology*, 137, 256–262.

Itoh, N. & Nishida, T. 2007. Chimpanzee grouping patterns and food availability in Mahale Mountains National Park, Tanzania. *Primates*, 48, 87–96.

Janson, C. & van Schaik, C.P. 1993. Ecological risk aversion in juvenile primates: Slow and steady wins the race. In *Juvenile Primates: Life History, and Behavior* (ed. by Pereira, M. E. & Fairbanks, L. A.), 57–76. Chicago: University of Chicago Press.

Jensen, K., Hare, B., Call, J. & Tomasello, M. 2006. What's in it for me? Self-regard precludes altruism and spite in chimpanzees. *Proceedings of the National Academy of Sciences USA*, 273, 1013–1021.

Jensen, K., Call, J. & Tomasello, M. 2007a. Chimpanzees are vengeful, but not spiteful. *Proceedings of the National Academy of Sciences USA*, 104, 13046–13050.

Jensen, K., Call, J. & Tomasello, M. 2007b. Chimpanzees are rational maximizers in an ultimatum game. *Science*, 318, 107–109.

Jensen-Seaman, M., Deinard, A. S. & Kidd, K. K. 2003. Mitochondrial and nuclear DNA estimates of divergence between western and eastern gorillas. In *Gorilla Biology: A Multidisciplinary Perspective* (ed. by Taylor, A. & Goldsmith, M.), 247–268. Cambridge: Cambridge University Press.

Jiang, X. L., Wang, Y. X. & Wang, Q. A. 1999. Coexistance of monogamy and polygyny in black crested gibbons (*Hylobates concolor*). *Primates*, 40, 607–611.

Jolly, C. J. 2001. A proper study for mankind: analogies from the papionin monkeys and their implications for human evolution. *Yearbook of Physical Anthropology*, 44, 177–204.

Kahlenberg, S. M., Emery Thompson, M. & Wrangham, R. W. 2008. Female competition over core areas in *Pan troglodytes schweinfurthii*, Kibale National Park, Uganda. *International Journal of Primatology*, 29, 931–947.

Kahlenberg, S. M., Emery Thompson, M., Muller, M. N. & Wrangham, R. W. 2008. Immigration costs for female chimpanzees and male protection as an immigrant counterstrategy to intrasexual aggression. *Animal Behaviour*, 76, 1497–1509.

Kano, T. 1982. The social group of pygmy chimpanzees (*Pan paniscus*). *Primates*, 23, 171–188.

———. 1992. *The Last Ape: Pygmy Chimpanzee Behavior and Ecology*. Stanford, CA: Stanford University Press.

———. 1996. Male rank order and copulation rate in a unit group of bonobos at Wamba. In *Great Ape Societies* (ed. by McGrew, W. C. Marchant, L. F. & Nishida, T.), 135–145. Cambridge: Cambridge University Press.

Kaplan, H., Hill, K., Lancaster, J. & Hurtado, A. M. 2000. A theory of human life history evolution: Diet, intelligence, and longevity. *Evolutionary Anthropology*, 9, 156–185.

Kappeler, P. M. & Pereira, M. E. 2003. *Primate Life History*. Chicago: University of Chicago Press.

Kappeler, P. M. & van Schaik, C. P. 2002. Evolution of primate social systems. *International Journal of Primatology*, 23, 707–740.

Knott, C. A. 1998. Changes in orangutan diet, caloric intake, and ketones in response to fluctuating fruit availability. *International Journal of Primatology*, 19, 1061–1079.

———. 2001. Female reproductive ecology of the apes: Implications for human evolution. In *Reproductive Ecology and Human Evolution* (ed. by Ellison, P.), 429–463. New York: Aldine de Gruyter.

Knott, C., Beaudrot, L., Snaith, T., White, S., Tshcauner, H. & Pianansky, G. 2008. Female-female competition in Bornean orangutans. *International Journal of Primatology*, 29, 975–997.

Knott C. A. & Kahlenberg, S. 2007. Orangutans in perspective: Forced copulations and female mating resistance. In *Primates in Perspective* (ed. by Campbell, C. J., Fuentes, A., Mackinnon, K. S., Panger, M. & Bearder, S. K.), 321–344. Oxford: Oxford University Press.

Kuroda, S. 1980. Social behavior of the pygmy chimpanzees. *Primates*, 21, 181–197.

Kuroda, S., Nishihara, T., Suzuki, S. & Oko, R. A. 1996. Sympatric chimpanzees and gorillas in the Ndoki Forest, Congo. In *Great Ape Societies* (ed. by McGrew, W. C., Marchant, L. F. & Nishida, T.), 71–81. Cambridge: Cambridge University Press.

Lacambra, C., Thompson, J., Furuichi, T., Vervaecke, H. & Stevens, J. 2005. Bonobos (*Pan paniscus*). In *World Atlas of Great Apes and Their Conservation* (ed. by Caldecott, J. & Miles, L.), 83–96. Berkeley: University of California Press.

Langergraber, K. E., Mitani, J. C. & Vigilant, L. 2007. The limited impact of kinship on cooperation in wild chimpanzees. *Proceedings of the National Academy of Sciences*, 104, 7786–7790.

Langergraber, K. E., Vigilant, L. & Mitani, J. C. 2009. Kinship and social bonds in female chimpanzees (*Pan troglodytes*). *American Journal of Primatology*, 71, 840–851.

Langergraber, K. E., Mitani, J. C., Watts, D. P. & Vigilant, L. 2010. Male-female social relationships and reproductive success in wild chimpanzees. *American Journal of Physical Anthropology*, Suppl. 50, 151.

Lappan, S. 2007a. Patterns of dispersal in Sumatran siamangs (*Symphalangus syndactylus*): Preliminary mtDNA evidence suggests more frequent male than female dispersal to adjacent groups. *American Journal of Primatology*, 69, 692–698.

———. 2007b. Social relationships among males in multi-male siamang groups. *International Journal of Primatology*, 28, 369–387.

———. 2008. Male care of infants in a siamang (*Symphalangus syndactylus*) population including socially monogamous and polyandrous groups. *Behavioral Ecology and Sociobiology*, 62, 1307–1317.

Lehmann, J. & Boesch, C. 2005. Bisexually bonded ranging behavior in chimpanzees (*Pan troglodytes verus*). *Behavioral Ecology and Sociobiology*, 57, 525–535.

———. 2008. Sex differences in chimpanzee sociability. *International Journal of Primatology*, 29, 65–81.

Leigh, S. R. & Blomqvist, G.E . 2007. Life history. In *Primates in Perspective* (ed. by Campbell, C. J., Fuentes, A., Mackinnon, K. C., Panger, M. & Bearder, S. K.), 396–407. Oxford: Oxford University Press.

Leigh, S. & Shea, B. 1995. Ontogeny and the evolution of adult body size dimorphism in apes. *American Journal of Primatology*, 36, 37–60.

Leighton, N. & Leighton, D. 1983. Vertebrate responses to fruiting seasonality within a Bornean tropical rain forest. In *Tropical Rain Forests: Ecology and Management* (ed. by Sutton, S. L., Whitmore, T. C. & Chadwick, A. C.), 181–196. Oxford: Blackwell Scientific.

Levrero, F., Gatti, S., Menard, N., Petit, E., Caillaud, D. & Gautier-Hion, A. 2006. Living in nonbreeding groups: An alternative strategy for maturing gorillas. *American Journal of Primatology*, 68, 275–291.

Lewis, R. L. 2002. Beyond dominance: The importance of leverage. *Quarterly Review of Biology*, 77, 149–164.

Locke, D., Hillier, L., Warren, W., Worley, K., Nazareth, L., Muzny, D., Yang, S., Wang, Z., Chinwalla, A., Minx, P., Mitreva, M., Cook, L., Delehnty, K., Fronick, C., Schmidt, H., Fulton, L., Fulton, R., Nelson, J., Magrini, V., Pohl, C., Graves, T., Markovic, C., Cree, A., Dinh, H., Hume, J., Kovar, C., Fowler, G., Lunter, G., Meader, S., Heger, A., Ponting, C., MarquesBonet, T., Alkan, C., Chen, L., Cheng, Z., Kidd, J., Eichler, E., White, S., Searle, S., Vilella, A., Chen, Y., Flicek, P., Ma, J., Raney, B., Suh, B., Burhans, R., Herrero, J., Hssler, D., Faria, R., Fernando, O., Darre, F., Farre, D., Gazave, E., Oliva, M., Navarro, A., Roberto, R., Capozzi, O., Archidiacono, N., Valle, G., Purgato, S., Rocchi, M., Konkel, M., Walker, J., Ullmer, B., Batzer, M., Smit, A., Hubley, R., Casola, C., Schrider, D., Hahn, M., Quesada, V., Puente, X., Ordonez, G., LopezOtin, C., Vinar, T., Brejova, B., Ratan, A., Harris, R., Miller, W., Kosiol, C., Lawson, H., Taliwal, V., Martins, A., Siepel, A., RoyChoudhury, A., Ma, X., Degenhardt, J., Bustamante, C., Gutenkunst, R., Mailund, T., Dutheil, J., Hobolth, A., Schierup, M., Ryder, O., Yoshinaga, Y., de Jong, P., Weinstock, G., Rogers, J., Mardis, E., Gibbs,

R. & Wilson, R. 2011. Comparative and demographic analysis of orang-utan genomes. *Nature*, 469, 529–533.

Lovejoy, O. 2009. Reexamining human origins in light of *Ardipithecus ramidus*. *Science*, 326, 74e1–74e8.

Maggioncalda, A. M., Czekala, N. M. & Czekala, N. M. 2000. Growth hormone and thyroid stimulating hormone concentrations in captive male orangutans: Implications for understanding developmental arrest. *American Journal of Primatology*, 50, 67–76.

Magioncalda, A. M., Sapolsky, R. A. & Czekala, N. M. 1999. Reproductive hormone profiles in captive male orangutans: Implications for understanding developmental arrest. *American Journal of Physical Anthropology*, 109, 119–132.

Magliocca, F. & Gautier-Hion, A. 2002. Mineral content as a basis for food selection by western lowland gorillas in a forest clearing. *American Journal of Primatology*, 57, 67–77.

Manson, J. & Wrangham, R. W. 1991. Intergroup aggression in chimpanzees and humans. *Current Anthropology*, 32, 369–390.

Marshall, A. J. & Leighton, M. 2006. How does food availability limit the population density of white-bearded gibbons? In *Feeding Ecology of the Apes and Other Primates* (ed. by Hohmann, G., Robbins, M. M. & Boesch, C.), 311–333. Cambridge: Cambridge University Press.

McConkey, K. R. 2005a. The influence of gibbon primary seed shadows on post-dispersal seed fate in a lowland dipterocarp forest in Central Borneo. *Journal of Tropical Ecology*, 21, 255–262.

——. 2005b. Influence of faeces on seed removal from gibbon droppings in a dipterocarp forest in Central Borneo. *Journal of Tropical Ecology*, 21, 117–120.

McConkey, K., Ario, A., Aldy, F. & Chivers, D. J. 2003. Influence of forest seasonality on gibbon food choice in the rain forest of Barito Ulu, Central Kalimantan. *International Journal of Primatology*, 24, 19–32.

McGrew, W. C. 1979. Evolutionary implications of sex differences in chimpanzee predation and tool use. In *The Great Apes* (ed. by Hamburg, D. A. & McCown, E. R.), 440–463. Menlo Park, CA: Benjamin-Cummings.

——. 1993. *Chimpanzee Material Culture*. Cambridge: Cambridge University Press.

——. 2004. *The Cultured Chimpanzee: Reflections on Cultural Primatology*. Cambridge: Cambridge University Press.

——. 2007. Savanna chimpanzees dig for food. *Proceedings of the National Academy of Sciences*, 104, 19166–19168.

McGrew, W. C., Ham, R. M., White, L. J. T., Tutin, C. E. G. & Fernandez, M. 1997. Why don't chimpanzees in Gabon crack nuts? *International Journal of Primatology*, 18, 353–374.

McNeilage, A. 2001. Diet and habitat use of two mountain gorilla groups in contrasting habitats in the Virungas. In *Mountain Gorillas: Three Decades of Research at Karisoke* (ed. by Robbins, M. M., Sicotte, P. & Stewart, K. J.), 265–292. Cambridge: Cambridge University Press.

Miller, L. & Treves, A. 2007. Predation on primates; Past, present, and directions for the future. In: *Primates in Perspective* (ed. by Campbell, C. J., Fuentes, A., Mackinnon, K. C., Panger, M. & Bearder, S. K.), 525–543. Oxford: Oxford University Press.

Mitani, J. C. 1984. The behavioral regulation of monogamy in gibbons (*Hylobates muelleri*). *Behavioral Ecology and Sociobiology*, 15, 225–229.

——. 1985a. Mating behaviour of male orangutans in the Kutai Game Reserve, Indonesia. *Animal Behaviour*, 33, 272–283.

——. 1985b. Location-specific responses of gibbons (*Hylobates muelleri*) to male songs. *Zeitschrift für Tierpsychologie*, 70, 219–224.

——. 1990. Demography of agile gibbons (*Hylobates agilis*). *International Journal of Primatology*, 11, 411–424.

——. 2009a. Cooperation and competition in chimpanzees: current understanding and future challenges. *Evolutionary Anthropology*, 18, 215–227.

——. 2009b. Male chimpanzees form enduring and equitable social bonds. *Animal Behaviour*, 77, 633–640.

Mitani, J.C. & Watts, D.P. 1999. Demographic influences on the hunting behavior of chimpanzees. *American Journal of Physical Anthropology*, 109, 439–454.

——. 2001. Why do chimpanzees hunt and share meat? *Animal Behaviour*, 61, 915–924.

——. 2005. Correlates of territorial boundary patrol behaviour in chimpanzees. *Animal Behaviour*, 70, 1079–1086.

Mitani, J. C. & Watts, D. P. & Amsler, S. 2010. Lethal intergroup aggression leads to territorial expansion in wild chimpanzees. *Current Biology*, 20, R507–508.

Mitani, J. C., Watts, D. P. & Lwanga, J. S. 2002. Ecological and social correlates of chimpanzee party size and composition. In *Behavioral Diversity in Chimpanzees and Bonobos* (ed. by Boesch, C., Hohmann, G. & Marchant, L. F.), 102–111. Cambridge: Cambridge University Press.

Mitani, J. C., Watts, D. P., Pepper, J. W. & Merriwether, A. 2002. Demographic and social constraints on male chimpanzee behavior. *Animal Behaviour*, 64, 727–737.

Morgan, D. & Sanz, C. 2006. In *Feeding Ecology in Apes and Other Primates* (ed. by Hohmann, G., Robbins, M. & Boesch, C.), 123–159. Cambridge: Cambridge University Press.

Morino, L. 2009. Monogamy in mammals: Expanding the hylobatid perspective. In *The Gibbons, Developments in Primatology: Progress and Prospects* (ed. by Whittaker, D. & Lappan, S.), 279–311. New York: Springer Verlag.

Moore, J. 1996. Savanna chimpanzees, referential models and the last common ancestor. In *Great Ape Societies* (ed. by McGrew, W. C., Marchant, L. F. & Nishida, T.), 275–291. Cambridge: Cambridge University Press.

Mulcahy, N. J. & Call, J. (2006) Apes save tools for future use. *Science*, 312, 1038–1040.

Muller, M. N. 2002. Agonistic relations among Kanyawara chimpanzees. In *Behavioral Diversity in Chimpanzees and Bonobos* (ed. by Boesch, C., Hohmann, G. & Marchant, L. F.), 112–124. Cambridge: Cambridge University Press.

Muller, M. N., Kahlenberg, S. M., Emery Thompson, M. & Wrangham, R. W. 2007. Male coercion and costs of promiscuous mating for female chimpanzees. *Proceedings of the Royal Society of London (B)*, 274, 1009–1014.

Muller, M. N. & Mitani, J. C. 2007. Conflict and cooperation in wild chimpanzees. *Advances in the Study of Behavior*, 35, 275–331.

Murray, C. M., Eberly, L. E. & Pusey, A. E. 2006. Foraging strategies as a function of season and rank among wild female chimpanzees (*Pan troglodytes*). *Behavioral Ecology*, 17, 2010–1028.

Murray, C. M., Mane, S. V. & Pusey, A. E. 2007. Dominance

rank influences female space use in wild female chimpanzees (*Pan troglodytes*): Towards an ideal despotic distribution. *Animal Behaviour* 74, 1795–1804.

Newton-Fisher, N. E. 1999a. The diet of chimpanzees in the Budongo Forest. *African Journal of Ecology*, 34, 344–354.

———. 1999b. Association by male chimpanzees: A social tactic? *Behaviour*, 136, 705–730.

———. 2006. Female coalitions against male aggression in wild chimpanzees of the Budongo Forest. *International Journal of Primatology*, 27, 1589–1599.

Nishida, T. 1968. The social group of wild chimpanzees in the Mahale Mountains. *Primates*, 9, 167–224.

———. 1983. Alpha status and agonistic alliance in wild chimpanzees. *Primates*, 24, 318–336.

———. 1989. Social interactions between resident and immigrant female chimpanzees. In *Understanding Chimpanzees* (ed. by Heltne, P. G. & Marquardt, L. A.), 68–89. Cambridge, MA: Harvard University Press.

Nishida, T., Corp, N., Hamai, M., Hasegawa, T., Hiraiwa-Hasegawa, M., Hosaka, K., Hunt, K. D., Itoh, N., Kawanaka, K., Matsumoto-Oda, A., Mitani, J. C., Nakamura, M., Norikoshi, K., Sakamaki, T., Turner, L., Uehara, S. & Zamma, K. 2003. Demography, female life history, and reproductive profiles among the chimpanzees of Mahale. *American Journal of Primatology*, 59, 99–121.

Nishida, T., Hiraiwa-Hasegawa, M., Hasegawa, T. & Takahata, Y. 1985. Group extinction and female transfer in wild chimpanzees in the Mahale Mountains National Park, Tanzania. *Zeitschrift für Tierpsychologie*, 67, 284–301.

Nishida, T. & Hosaka, K. 1996. Coalition strategies among adult male chimpanzees of the Mahale Mountains, Tanzania. In *Great Ape Societies* (ed. by McGrew, W. C., Marchant, L. F. & Nishida, T.), 114–134. Cambridge: Cambridge University Press.

Nishihara, T. 1995. Feeding ecology of western lowland gorillas in Noubalé-Ndoki National Park, northern Congo. *Primates*, 36, 151–168.

Nsubuga, A., Robbins, M. M., Boesch, C. & Vigilant, L. 2008. Patterns of paternity and group fission in wild multi-male mountain gorilla groups. *American Journal of Physical Anthropology*, 135, 263–274.

O'Brien, T. G., Kinnaird, M. F., Nurcayo, A., Prasetyaningrum, M. & Iqbal, M. 2003. Fire, demography, and the persistence of siamang (*Symphalangus syndactylus*) in a Sumatran rain forest. *Animal Conservation*, 6, 115–121.

Osvath, M. & Osvath, H. 2008. Chimpanzee (*Pan troglodytes*) and orangutan (*Pongo abelii*) forethought: Self-control and pre-experience in the face of future tool use. *Animal Cognition*, 11, 661–674.

Palombit, R. 1994a. Dynamic pairbonds in hylobatids: Implications regarding monogamous social systems. *Behaviour*, 128, 65–101.

———. 1994b. Extra-pair copulations in a monogamous ape. *Animal Behaviour*, 47, 721–723.

———. 1996. Pairbonds in monogamous apes: A comparison of the siamang, *Hylobates syndactylus*, and the white-handed gibbon, *Hylobates lar*. *Behaviour*, 133, 321–356.

———. 1997. Inter- and intraspecific variation in the diets of sympatric siamang (*Symphalangus syndactylus*) and lar gibbons (*Hylobates lar*). *Folia Primatologica*, 68, 321–356.

———. 1999. Infanticide and the evolution of pair bonds in non-human primates. *Evolutionary Anthropology*, 7, 117–129.

Parrish, A.R. 1994. Sex and food control in the "uncommon chimpanzee": How bonobo females overcome a phylogenetic legacy of male dominance. *Ethology and Sociobiology*, 15, 157–179.

Parrish, A. R. 1996. Female relationships in bonobos (*Pan paniscus*): Evidence for bonding, dominance, and cooperation in a male philopatric species. *Human Nature*, 7, 61–96.

Parrish, A. R. & de Waal, F. B. M. 2000. The other "closest living relative": How bonobos (*Pan paniscus*) challenge traditional assumptions about females, dominance, intra- and inter-sexual interactions, and hominid evolution. *Annals of the New York Academy of Sciences*, 907, 97–113.

Plavcan, J. M. & van Schaik, C. P. 1992. Intrasexual competition and canine dimorphism in anthropoid primates. *American Journal of Physical Anthropology*, 87, 461–477.

Plotkin, H. 2003. *The Imagined World Made Real*. New Brunswick: Rutgers University Press.

Pontzer, H., Ocobock, C., Schumaker, R. W. & Raichlin, D. A. 2009. Daily energy expenditure in orangutans measured using doubly-labeled water. *American Journal of Physical Anthropology*, S48, 213.

Popovich, D. J., Jenkins, D. J. A., Kendall, C. W. C., Dierenfeld, E. S., Carroll, R. W., Tarig, N. & Videan, E. 1997. The western lowland gorilla diet has implications for the health of humans and other hominoids. *Journal of Nutrition*, 127, 2000–2005.

Potts, K. B. 2008. Habitat heterogeneity on multiple spatial scales in Kibale National Park, Uganda: Implications for chimpanzee population ecology and grouping patterns. PhD dissertation, Yale University.

Potts, K. B., Chapman, C. A. & Lwanga, J. S. 2009. Floristic heterogeneity between different sites in Kibale National Park, Uganda: Insights into the fine-scale determinants of density in a large-bodied frugivorous primate. *Journal of Animal Ecology*, 78, 1269–1277.

Povinelli, D. & Eddy, T. 1996. *What Chimpanzees Know About Seeing: Monographs of the Society for Research in Child Development*, 61 (3). Chicago: University of Chicago Press.

Preutz, J. 2006. Feeding ecology of savanna chimpanzees (*Pan troglodytes verus*) at Fongoli, Senegal. In *Feeding Ecology in Apes and Other Primates* (ed. by Hohmann, G., Robbins, M. & Boesch), 123–159. Cambridge: Cambridge University Press.

Prouty, L. A., Buchanan, P. D., Pollitzer, W. S. & Mootnick, A. 1983a. A presumptive new hylobatid subgenus with 38 chromosomes. *Cytogenetics and Cell Genetics*, 35, 141–142.

———. 1983b. *Bunopithecus*, a genus level taxon for the hoolock gibbon (*Hylobates hoolock*). *American Journal of Primatology*, 5, 83–87.

Pusey, A. E. 1980. Inbreeding avoidance in chimpanzees. *Animal Behavior*, 28, 543–552.

Pusey, A. E., Williams, J. & Goodall, J. 1997. The influence of dominance rank on the reproductive success of female chimpanzees. *Science*, 277, 828–831.

Pusey, A. E., Murray, C., Wallauer, W., Wilson, M., Wroblewski, E. & Goodall, J. 2008. Severe aggression among female *Pan troglodytes schweinfurthii* at Gombe National Park, Tanzania. *International Journal of Primatology*, 26, 3–31.

Reichard, U. H. 1995. Extra-pair copulation in a monogamous gibbon (*Hylobates lar*). *Ethology*, 100, 99–112.

———. 2009. The social organization and mating system of Khao Yai white-handed gibbons: 1992–1996. In *The Gibbons, Developments in Primatology: Progress and Progress* (ed. by Whittaker, D. & Lappan, S.), 347–384. New York: Springer Verlag.

Reichard, U. H. & Sommer, V. 1997. Group encounters in wild gibbons (*Hylobates lar*): Agonism, affiliation, and the concept of infanticide. *Behaviour*, 13, 1135–1174.

Reichard, U. H. & Barelli, C. 2008. Life history and reproductive strategies of Khao Yai *Hylobates lar*: Implications for social evolution in apes. *International Journal of Primatology*, 29, 823–844.

Remis, M. 2000. Initial studies on the contributions of body size and gastrointestinal passage times to dietary flexibility among gorillas. *American Journal of Physical Anthropology*, 112, 171–180.

Remis, M., Dierenfeld, E. S., Mowry, C. W. & Carroll, R. W. 2001. Nutritional aspects of western lowland gorilla (*Gorilla gorilla gorilla*) diet during seasons of fruit scarcity at Bai Hokou, Central African Republic. *International Journal of Primatology*, 22, 807–836.

Rijksen, H. 1978. *A Field Study on Sumatran Orangutans (Pongo pygmaeus abelli, Lessen 1827): Ecology, Behavior, and Conservation*. Wageningen: H. Veemen and Zonen BV.

Robbins, A. M., Robbins, M. M. 2005. Fitness consequences of dispersal decisions for male mountain gorillas (*Gorilla gorilla beringei*). *Behavioral Ecology and Sociobiology*, 58, 295–309.

Robbins, A. M., Stoinski, T. M., Fawcett, K. A. & Robbins, M. M. 2009a. Socioecological influences on the dispersal of wild female mountain gorillas: evidence of a second folivore paradox. *Behavioral Ecology and Sociobiology*, 63, 477–489.

Robbins, A. M., Stoinski, T. M., Fawcett, K. A. & Robbins, M. M. 2009b. Does dispersal cause reproductive delays in female mountain gorillas? *Behaviour*, 146, 525–549.

Robbins, A. M., Robbins, M. M., Gerald-Steklis, N. & Steklis, H. D. 2006. Age-related patterns of reproductive success among female mountain gorillas. *American Journal of Physical Anthropology*, 131, 511–521.

Robbins, M. M. 1995. A demographic analysis of male life history and social structure of mountain gorillas. *Behaviour*, 132, 21–47.

———. 1996. Male-male interactions in heterosexual and all-male wild mountain gorilla groups. *Ethology*, 102, 942–965.

———. 2001. Variation in the social system of mountain gorillas: The male perspective. In *The Mountain Gorillas of Karisoke* (ed. by Robbins, M. Sicotte, P. & Stewart, K. J.), 29–58. Cambridge: Cambridge University Press.

———. 2007. Gorillas: Diversity in ecology and behavior. In *Primates in Perspective* (ed. by Campbell, C. J., Fuentes, A., Mackinnon, K. C. Panger, K. C. & Bearder, S. K.), 321–344. Oxford: Oxford University Press.

———. 2008. Feeding competition and agonistic relationships among Bwindi *Gorilla beringei*. *International Journal of Primatology*, 29, 999–1018.

Robbins, M. M. & McNeilage, A. 2003. Home range and frugivory patterns of mountain gorillas in Bwindi-Impenetrable National Park, Uganda. *International Journal of Primatology*, 24, 467–491.

Robbins, M. M., Robbins, A. M., Gerald-Steklis, N. & Steklis, H. D. 2007. Socioecological influences on the reproductive success of female mountain gorillas (*Gorilla gorilla beringei*). *Behavioral Ecology and Sociobiology*, 61, 919–931.

Robbins, M. M. & Sawyer, S. C. 2007. Intergroup encounters in mountain gorillas of Bwindi Impenetrable National Park, Uganda. *Behaviour*, 144, 1497–1519.

Rodman, P. 1977. Feeding behavior of the orangutans of the Kutai Reserve, East Kalimantan. In *Primate Ecology: Studies of feeding and Ranging Behavior in Lemurs, Monkeys, and Apes* (ed. by Clutton-Brock, T. H.), 383–414. London/New York: Academic Press.

Rodman, P. S. 1979. Individual activity profiles and the solitary nature of orangutans. In *The Great Apes* (ed. by Hamburg, D. A. & McCown, E. R.), 234–255. Menlo Park, CA: Benjamin-Cummings.

———. 1984. Foraging and social systems of orangutans and chimpanzees. In *Adaptations for Foraging in Nonhuman Primates* (ed. by Rodman, P. S. & Cant, J. H.), 134–160. New York: Columbia University Press,

———. 1988. Diversity and consistency in ecology and behavior. In *Orang-utan Biology and Behavior* (ed. by Schwartz, J.), 31–51. New York: Oxford University Press.

Rodman, P. S. & Mitani, J. C. 1987. Orangutans: Sexual dimorphism in a solitary species. In *Primate Societies* (ed. by Smuts, B. B., Cheney, D. L., Seyfarth, R. M., Wrangham, R. W. & Struhsaker, T. T.), 146–152. Chicago: University of Chicago Press.

Rogers, E., Abernethy, K., Bermejo, M., Cipoletta, C., Doran, D., McFarland, K., Nishihara, T., Remis, M. & Tutin, C. E. G. 2004. Western gorilla diet: A synthesis from six sites. *American Journal of Primatology*, 64, 173–192.

Rogers, E., Maiscls, F., Williamson, E. A., Tutin, C. E. G. & Fernandez, M. 1990. Gorilla diet in the Lopé Reserve: A nutritional analysis. *Oecologia*, 84, 326–339.

Rothman, J., Dierenfeld, E., Molina, D. O., Shaw, A. V., Hintz, H. F. & Pell, A. N. 2004. Nutritional chemistry of foods eaten by gorillas in Bwindi Impenetrable National Park, Uganda. *American Journal of Primatology*, 68, 675–691.

Rothman, J., Plumptre, A. J., Dierenfeld, E. S. & Pell, A. N. 2007. Nutritional composition of the diet of gorilla (*Gorilla beringei*): A comparison between two montane habitats. *Journal of Tropical Biology*, 23, 673–682.

Savini, T., Boesch, C. & Reichard, U. 2009. Varying ecological quality influences the probability of polyandry in white-handed gibbons (*Hylobates lar*) in Thailand. *Biotropica*, 41, 503–513.

Schurmann, C. L. 1982. Mating behaviour of wild orangutans. In *The Orangutan: Its Biology and Conservation* (ed. by de Boer, L. E. M.), 269–284. The Hague: W. Junk.

Sicotte, P. 1993. Inter-group encounters and female transfer in mountain gorillas: Influence of group competition on male behavior. *American Journal of Primatology*, 30, 21–36.

Smith, R. L. & Jungers, W. L. 1997. Body mass in comparative primatology. *Journal of Human Evolution*, 32, 522–559.

Smuts, B. B. & Smuts, R. 1993. Sexual coercion in primates and other mammals. *Advances in the Study of Behavior*, 22, 1–63.

Sommer, V. & Reichard, U. F. 2000. Re-thinking monogamy: The gibbon case. In *Primate Males: Causes and Consequences of Variation in Group Composition* (ed. by Kappeler, P. M.), 159–168. Cambridge: Cambridge University Press.

Stanford, C. B. 1998. *Chimpanzees and Red Colobus: The Ecology of Predator and Prey*. Cambridge, MA: Harvard University Press.

———. 2006. The behavioral ecology of sympatric African apes: Implications for understanding fossil hominoid ecology. *Primates*, 47, 91–101.

Stanford, C. B. & Nkurunungi, J. B. 2003. Behavioral ecology of sympatric chimpanzees and gorillas in Bwindi Impenetrable National Park, Uganda: Diet. *International Journal of Primatology*, 24, 901–918.

Steiper, M. E. 2006. Population history, biogeography, and taxonomy of orangutans (Genus: *Pongo*) based on a population genetic meta-analysis of multiple loci. *Journal of Human Evolution*, 50, 509–522.

Stoinski, T. S., Rosenbaum, S. Ngaboyamahina, T., Vecellio, V., Ndagijimana, F. & Fawcett, K. 2009a. Patterns of male reproductive behaviour in multi-male groups of mountain gorillas: examining theories of reproductive skew. *Behaviour*, 146, 1193–1215.

Stoinski, T., Vecellio, V., Ngaboyamihana, T., Ndagijimana, F., Rosenbaum, S. & Fawcett, K. A. 2009b. Proximate factors influencing dispersal decisions in male mountain gorillas, *Gorilla gorilla beringei*. *Animal Behaviour*, 77, 1155–1164.

Stokes, E. J. 2004. Within-group social relationships amongst females and adult males in wild western lowland gorillas (*Gorilla gorilla gorilla*). *American Journal of Primatology*, 64, 233–246.

Stokes, E. J., Parnell, R. J. & Olejniczek, C. 2003. Female dispersal and reproductive success in wild western gorillas (*Gorilla gorilla gorilla*). *Behavioral Ecology and Sociobiology*, 54, 329–339.

Stumpf, R. 2007. Chimpanzees and bonobos: Diversity within and between species. In *Primates in Perspective* (ed. by Campbell, C. J., Fuentes, A., Mackinnon, K. C. Panger, M. & Bearder, S. K.), 321–344. Oxford: Oxford University Press.

Stumpf, R. & Boesch, C. 2005. Does promiscuous mating preclude female mate choice? Female sexual strategies in chimpanzees (*Pan troglodytes verus*) of the Taï National Park, Côte d'Ivoire. *Behavioral Ecology and Sociobiology*, 57, 511–524.

———. 2006. The efficacy of female choice in chimpanzees of the Taï Forest, Côte d'Ivoire. *Behavioral Ecology and Sociobiology*, 60, 749–765.

Stumpf, R., Emery Thompson, M., Muller, M. N. & Wrangham, R. W. 2009. The context of female dispersal in Kanyawara chimpanzees. *Behaviour*, 146, 629–656.

Suddendorf, T., Addis, D. R. & Corballis, M. C. 2009. Mental time travel and the shaping of the human mind. *Philosophical Transactions of the Royal Society B*, 364, 1317–1324.

Sugardjito, J., te Boekhoerst, I. J. A., van Hooff, J. A. R. A. M. 1987. Ecological constraints on the grouping of wild orangutans in (*Pongo pygmaeus*) in the Gunung Leuser National Park, Sumatra, Indonesia. *International Journal of Primatology*, 8, 17–41.

Sugiyama, A. & Koman, J. 1979. Tool-using and making behavior in wild chimpanzees at Bossou, Guinea. *Primates*, 20, 413–524.

Surbeck, M. & Hohmann, G. 2007. Primate hunting by bonobos at Lui Kotale, Salonga National Park. *Current Biology*, 18, R906-R907.

Takahata, Y., Hasegawa, T. & Nishida, T. 1984. Chimpanzee predation in the Mahale Mountains from August 1979 to May 1982. *International Journal of Primatology*, 5, 213–233.

Te Boekhoerst, I. J. A., Schurmann, C. & Sugardjito, J. 1990. Residential status and seasonal movements of wild orangutans in the Gunung Leuser Reserve (Sumatra, Indonesia). *Animal Behaviour*, 39, 1098–1109.

Teelen, S. 2007. Primate abundance along five different transect lines at Ngogo, Kibale National Park, Uganda. *American Journal of Primatology*, 69, 1030–1044.

———. 2008. Influence of chimpanzee predation on the red colobus population at Ngogo, Kibale National Park, Uganda. *Primates*, 49, 41–49.

Tennie, C., Gilby, I. & Mundry, R. 2008. The meat scrap hypothesis. *Behavioral Ecology and Sociobiology*, 63, 421–431.

Thalmann, O., Fischer, A., Lankaster, F., Paabo, S. & Vigilant, L. 2007. The complex evolutionary history of gorillas: Insights from genetic data. *Molecular Biology and Evolution*, 24, 146–158.

Tomasello, M. 2008. *Origins of Human Communication*. Cambridge, MA: Bradford/MIT Press.

Tomasello, M. & Call, J. 1997. *Primate Cognition*. Oxford: Oxford University Press.

Townsend, S. W., Slocombe, K. E., Emery Thompson, M. & Zuhberbuhler, K. 2007. Female-led infanticide in wild chimpanzees. *Current Biology*, 17, 355–356.

Toth, N. & Schick, K. 2009. The Oldowan: The tool making of early hominins and chimpanzees compared. *Annual Review of Anthropology*, 35, 289–305.

Tsukahara, T. Lions eat chimpanzees: The 1st evidence of predation by lions on wild chimpanzees. *American Journal of Primatology*, 29, 1–11.

Tulving, E. 2005. Episodic memory and autonoesis: Uniquely human? In *Missing Link in Cognition: Origins of Reflective Consciousness* (ed. by Terrace, H S. & Metcalfe, J.), 3–56. Oxford: Oxford University Press.

Tutin, C. E. G. 1979. Mating patterns and reproductive strategies in a community of wild chimpanzees (*Pan troglodytes schweinfurthii*). *Behavioral Ecology and Sociobiology*, 6, 29–38.

Tutin, C. &. McGinnis, P. 1981. Chimpanzee reproduction in the wild. In *Reproductive Biology of the Apes* (ed. by Graham, C. E.), 239–264. New York: Academic Press.

Tweheyo, M. & Lye, K. A. 2003. Phenology of figs in the Budongo Forest and its importance for the chimpanzee diet. *African Journal of Ecology*, 41, 306–316.

Tweheyo, M., Lye, K. A. & Weladji, R. B. 2003. Chimpanzee diet and habitat selection in the Budongo Forest Reserve. *Forest Ecology and Management*, 188, 267–278.

Utami, S. S., Wich, S. A., Sterck, E. H. M. & van Hooff, J. A. R. A. M. 1997. Food competition between wild orangutans in large fig trees. *International Journal of Primatology*, 18, 909–927.

Utami, S. S., Goosens, B., Bruford, M. W., de Ruiter, J. R. & van Hooff, J. A. R. A. M. 2002. Male bimaturism and reproductive success in Sumatran orangutans. *Behavioral Ecology and Sociobiology*, 13, 643–652.

Utami, S. S. & van Hooff, J. A. R. A. M. 2004. Alternative male reproductive strategies: Male bimaturism in orangutans. In *Sexual Selection in Primates: New and Comparative Perspec-*

tives (ed. by Kappeler, P. M. & van Schaik, P.), 196–207. Cambridge: Cambridge University Press.

Van Andrichen, G. G. J., Utami, S. S., Wich, S. A., & van Hooff, J. A. R. A. M. 2006. The development of wild immature orangutans at Ketambe. *Primates*, 47, 300–309.

Van Schaik, C. P. 1986. Phenological changes in a Sumatran rain forest. *Journal of Tropical Ecology*, 2, 327–347.

———. 1999. The socioecology of fission-fusion sociality in orangutans. *Primates*, 40, 69–87.

Van Schaik, C.P. & van Hooff, J. A. R. A. M. 1996. Toward an understanding of the orangutan social system. In *Great Ape Societies* (ed. by McGrew, W. C. Marchant, L. F. & Nishida, T.), 3–15. Cambridge: Cambridge University Press.

Vedder, A. L. 1984. Movement patterns of a free-ranging group of mountain gorillas (*Gorilla gorilla beringei*) and their relation to food availability. *American Journal of Primatology*, 7, 73–88.

Vogel, E., van Woerden, J. T., Lucas, P. W., Utami Atmoko, S. S., van Schaik, C. P. & Dominy, N. J. 2008. Functional ecology and evolution of hominoid molar enamel thickness: *Pan troglodytes schweinfurthii* and *Pongo pygmaeus wurmbii*. *Journal of Human Evolution*, 55, 60–74.

Wakefield, M. L. 2008. Grouping patterns and competition among female chimpanzees (*Pan troglodytes schweinfurthii*) at Ngogo, Kibale National Park. *International Journal of Primatology*, 29, 907–929.

———. 2010. Socioecology of female chimpanzees (*Pan troglodytes*) in the Kibale National Park, Uganda: Social relationships, association patterns, and costs and benefits of gregariousness in a fission-fusion society. PhD thesis, Yale University.

Waterman, P. G., Choo, G., Vedder, A. & Watts, D. P. 1983. Digestibility, digestion inhibitors, and nutrients from herbaceous foliage from an African montane flora and its comparison with other tropical flora. *Oecologia*, 6, 244–249.

Watts, D. P. 1984. Composition and variability of mountain gorilla diets in the central Virungas. *American Journal of Primatology*, 7, 325–356.

———. 1988. Environmental influences on mountain gorilla time budgets. *American Journal of Primatology*, 15, 295–312.

———. 1989. Infanticide in mountain gorillas: new cases and a reconsideration of the evidence. *Ethology*, 81, 1–18.

———. 1990. Ecology of gorillas and its relationship to female transfer in mountain gorillas. *International Journal of Primatology*, 11, 21–45.

———. 1991a. Strategies of habitat use by mountain gorillas. *Folia Primatologica*, 56, 1–16.

———. 1991b. Mountain gorilla reproduction and sexual behavior. *American Journal of Primatology*, 24, 211–226.

———. 1991c. Harassment of immigrant female mountain gorillas by resident females. *Ethology*, 89, 135–153.

———. 1992. Social relationships of immigrant and resident female mountain gorillas, I. Male-female relationships. *American Journal of Primatology*, 28,159–181.

———. 1994b. Social relationships of immigrant and resident female mountain gorillas, II: Relatedness, residence, and relationships between females. *American Journal of Primatology*, 32, 13–30.

———. 1994b. Agonistic relationships of female mountain gorillas. *Behavioral Ecology and Sociobiology*, 34, 347–358.

———. 1994c. The influence of male mating tactics on habitat use in mountain gorillas (*Gorilla gorilla beringei*). *Primates*, 35, 35–47.

———. 1995. Post-conflict social events in wild mountain gorillas, 1. Social interactions between opponents. *Ethology*, 100, 158–174.

———. 1996. Comparative socioecology of gorillas. In *Great Ape Societies* (ed. by McGrew, W. C., Marchant, L. F. & Nishida, T.), 16–28. Cambridge: Cambridge University Press.

———. 1997. Agonistic interventions in wild mountain gorilla groups. *Behaviour*, 134, 23–57.

———. 1998a. Long-term habitat use by mountain gorillas (*Gorilla gorilla beringei*), 1: Consistency, variation, and home range size and stability. *International Journal of Primatology*, 19, 651–680.

———. 1998b. Coalitionary mate guarding by male chimpanzees at Ngogo, Kibale National Park, Uganda. *Behavioral Ecology and Sociobiology*, 4, 43–55.

———. 2000a. Causes and consequences of variation in the number of males in mountain gorilla groups. In *The Socioecology of Primate Males* (ed. by Kappeler, P. M.), 169–180. Cambridge: Cambridge University Press.

———. 2000b. Grooming between male chimpanzees at Ngogo, Kibale National Park, Uganda. II. Male rank and priority of access to partners. *International Journal of Primatology*, 21, 211–238.

———. 2000c. Grooming between male chimpanzees at Ngogo, Kibale National Park, Uganda. I. Partner number and diversity and reciprocity. *International Journal of Primatology*, 21, 189–210.

———. 2001. Social relationships of female mountain gorillas. In *The Mountain Gorillas of Karisoke* (ed. by Robbins, M., Sicotte, P. & Stewart, K. J.), 215–240. Cambridge: Cambridge University Press.

———. 2002. Reciprocity and interchange in the social relationships of wild male chimpanzees. *Behaviour*, 139, 343–370.

———. 2003. Gorilla social relationships: An overview. In *Gorilla Biology: A Multidisciplinary Perspective* (ed. by A. Taylor & M. Goldsmith), 302–327. Cambridge: Cambridge University Press.

Watts D. P. & Mitani, J. C. 2001. Boundary patrolling and inter-community aggression in wild chimpanzees. *Behaviour*, 138, 299–327.

———. 2002. Hunting and meat sharing by chimpanzees at Ngogo, Kibale National Park, Uganda. In *Behavioral Diversity in Chimpanzees and Bonobos* (ed. by Hohmann, G. & Boesch, C.), 244–258. Cambridge: Cambridge University Press.

Watts D. P. & Pusey, A. E. 1993. Behavior of juvenile and adolescent great apes. In *Socioecology of Juvenile Primates* (ed. by Pereira, M. E. & Fairbanks, L.), 148–167. Cambridge: Cambridge University Press.

Watts, D. P. & de Vries, H. 2009. Linearity and strength of male chimpanzee dominance hierarchies at Ngogo, Kibale National Park, Uganda. *American Journal of Physical Anthropology*, Suppl. 48, 267.

Wittig, R. M. & Boesch, C. 2003. Food competition and linear dominance hierarchy among female chimpanzees of the Taí National Park. *International Journal of Primatology*, 24, 847–867.

White, F. J. 1996. Comparative socioecology of *Pan paniscus*. In *Great Ape Societies* (ed. by McGrew, W. C., Marchant, L. F., & Nishida, T.), 29–41. Cambridge: Cambridge University Press.

White, F. J. & Wood, K. D. 2007. Female feeding priority in bonobos, *Pan paniscus*, and the question of female dominance. *American Journal of Primatology*, 69, 837–850.

Whiten, A. 2010. Ape behavior and the origins of human culture. In *Mind the Gap* (ed. by Kappeler, P. & Silk, J. B.), 429–450. Heidelberg: Springer.

Wich, S. & van Schaik, C. P. E. 2000. The impact of El Nino on mast fruiting in Sumatra and elsewhere in Malesia. *Journal of Tropical Ecology*, 16, 563–577.

Wich, S., Utami Atmoko, S. S. & Setia, T. M. 2006. Dietary and energetic responses of *Pongo abelii* to fruit availability fluctuations. *International Journal of Primatology*, 53, 1535–1550.

Wich, S., Utami Atmoko, S. S., Setia, T. M., Rijksen, H. R., Schürmann, D. & van Hooff, J. A. R. A. M. 2004. Life histories of wild orangutans (*Pongo abelii*). *Journal of Human Evolution*, 47, 385–398.

Williams, J. M., Oehlert, G. W., Carlis, J. V. & Pusey, A. E. 2004. Why do male chimpanzees defend a group range? *Animal Behaviour*, 6, 523–532.

Wrangham, R. W. 1979a. On the evolution of ape social systems. *Social Science Information*, 18, 335–378.

———. (1980) An ecological model of female-bonded primate groups. *Behaviour*, 75, 262–300.

———. 1986. Ecology and social relationships in two species of chimpanzees. In *Ecological Aspects of Social Evolution: Birds and Mammals* (ed. by Rubenstein, D. I. & Wrangham, R. W.), 352–378. Princeton, NJ: Princeton University Press.

Wrangham, R. W., Chapman, C. A., Clark-Arcadi, A. P. & Isibirye-Basuta, G. 1996. Social ecology of Kanyawara chimpanzees: Implications for the costs of great ape groups. In *Great Ape Societies* (ed. by McGrew, W. C., Marchant, L. F. & Nishida, T.), 45–57. Cambridge: Cambridge University Press.

Wrangham, R. W., Conklin, N. L., Chapman, C. A., & Hunt, K. D. 1991. The significance of fibrous foods for Kibale Forest chimpanzees. *Philosophical Transactions of the Royal Society*, 334, 171–178.

Wrangham, R. W., Conklin-Brittain, N. L. & Hunt, K. D. 1998. Dietary response of chimpanzees and cercopithecines to seasonal variation in fruit abundance: I. Antifeedants. *International Journal of Primatology*, 19, 949–970.

Wrangham, R. W. & Smuts, B. B. 1980. Sex differences in the behavioral ecology of chimpanzees in the Gombe National Park, Tanzania. *Journal of Reproduction and Fertility*, Suppl. 28, 13–31.

Yamagiwa, J. 1987. Intra- and inter-group interactions of an all-male group of Virunga mountain gorillas (*Gorilla gorilla beringei*). *Primates*, 28, 1–30.

Yamagiwa, J. & Basabose, K. 2006. In *Feeding Ecology in Apes and Other Primates* (ed. by Hohmann, G., Robbins, M. & Boesch, C.), 123–159. Cambridge: Cambridge University Press.

Yamagiwa, J., Basabose, K., Kaleme, K. & Yamoto, T. 2003. Within-group feeding competition and socioecological factors influencing social organization of gorillas in the Kahuzi-Biega National Park, Democratic Republic of Congo. In *Gorilla Biology: A Multidisciplinary Perspective* (ed. by Taylor, A. B. & Goldsmith, M. L.), 328–357. Cambridge: Cambridge University Press.

Yamagiwa, J. & Kahekwa, J. 2001. Dispersal patterns, group structure, and reproductive parameters of eastern lowland gorillas at Kahuzi in the absence of infanticide. In *The Mountain Gorillas of Karisoke* (ed. by Robbins, M., Sicotte, P. & Stewart, K. J.), 89–122. Cambridge: Cambridge University Press.

Yamagiwa, J., Mwanza, N., Yunoto, T. & Maruhashi, T. 1994. Seasonal change in the composition of the diet of eastern lowland gorillas. *Primates*, 35, 1–14.

Yamakoshi, G. 2001. Ecology of tool use in wild chimpanzees: Toward reconstruction of early hominid evolution. In *Primate Origins of Human Cognition and Behavior* (ed. by Matsuzawa, T.), 537–556. Tokyo: Springer.

Yamakoshi, G. 2004. Food seasonality and socioecology in *Pan*: Are West African chimpanzees another bonobo? *African Study Monographs*, 25, 45–60.

Zhon, J., Wei, F., Li, M., Lok, C. B. P. & Wang, D. 2008. Reproductive characteristics and mating behavior of wild *Nomascus heinanus*. *International Journal of Primatology*, 29, 1037–1046.

Part 2

Surviving and Growing up in a Difficult and Dangerous World

PART 2 of this book explores several adaptive problems primates face as they struggle to survive, grow, and maintain themselves in a difficult and dangerous world. Two of these problems are quite familiar and have been the foci of considerable past and ongoing research. Like all other organisms, primates must obtain a sufficient amount of food to satisfy their nutritional requirements and avoid becoming someone else's food each and every day. In chapter 7, Colin A. Chapman, Jessica M. Rothman, and Joanna E. Lambert deal with this first issue in a discussion of primate feeding behavior. Primates must eat to grow, survive, and reproduce, and they spend a large amount of time searching for, handling, ingesting, and digesting food. Because the search for food dominates primate life and activity, feeding influences multiple aspects of the biology of primates, including their physiology, ecology, behavior, and ultimately their fitness. All of this did not go unnoticed by researchers who began to observe primates in their natural habitats in the 1960s and 1970s, and the study of feeding behavior has continued to play an integral role in the field ever since. Chapter 7 reviews our current knowledge about primate diets and feeding strategies, and furnishes synopses of some of the different ways in which feeding affects primate ecology, ranging, social behavior, and conservation strategies.

Predation is another topic that has loomed large in the study of primate behavior. Unlike feeding, which is relatively easy to observe in the wild, predation on primates occurs rarely, unpredictably, and often quickly, making it difficult to record and study. The paucity of data regarding predation led some early researchers to conclude that it had only a minimal ecological impact on primate populations. Subsequent studies that have focused on predators, rather than their primate prey, now indicate that primates in nature are subject to a constant threat of predation, and that this is likely to have been a strong selective force in their evolution. Because successful acts of predation can terminate the reproductive careers of individuals, primates have adapted in many ways to decrease their vulnerability. Over the past 25 years we have learned a considerable amount about this facet of primate life. In chapter 8, Claudia Fichtel summarizes studies that have been conducted to address how primates recognize predators and how predator recognition skills develop over time, as well as the behavioral tactics primates employ to reduce their chances of falling victim. Historically, predation on primates has been treated as a unitary phenomenon. But not all predators hunt in the same way. A growing appreciation of the fact that predators employ different hunting strategies has led to new ways of thinking about how and why primates respond behaviorally to the threat of predation and the effects it has on primate social systems.

Primates do not often face the hostile forces of nature alone as they attempt to grow up and survive in a difficult and dangerous world. A hallmark of the order is that individuals in many species are gregarious. Primate sociality, however, takes many forms and is quite variable. The twin forces of feeding competition and predation risk were invoked early on to explain the diversity of primate social systems, beginning with Crook and Gartlan's seminal paper published in 1966. Their socioecological model hypothesized that features of the habitat and ecology shaped different aspects of primate societies, including the size and composition of groups. The model has subsequently been developed and elaborated significantly through the incorporation of modern theories of foraging behavior, sexual selection, and kin selection. In chapter 9, Oliver Schülke and Julia Ostner describe the current version of the socioecological model, which places an increasing emphasis on how social as well as ecological factors affect the evolution of primate social systems. Socioecology has been a dominant theme in primate field research during the past 25 years, and Schülke and Ostner also summarize the large body of data that has accumulated to test various predictions derived from the socioecological model. Central predictions deal with several long-standing questions about why primates live in groups, why the groups vary in size and in composition, and the factors that influence who interacts with whom, in what ways, and how often.

Primates face the problems posed by feeding, predators, and conspecifics with whom they live over relatively long life spans, another unusual characteristic of the order discussed in the next two chapters on life histories and development. Life history refers to evolved features of the life cycle of organisms and their timing. Carel P. van Schaik and Karin Isler begin chapter 10 by describing two distinct patterns of the life histories of primates. First, primates, compared to other mammals, generally live their lives in the slow lane, growing up over prolonged periods and surviving for a long time. Second, they typically give birth to precocial young. Van Schaik and Isler review current theory which suggests that these patterns can be explained by the amount of unavoidable, extrinsic mortality primates suffer in their natural habitats. Predation, along with pathogens, is likely to be an important source of unavoidable mortality to primates; and the relatively low, albeit constant, threat posed by predators accounts in part for primates' long life spans. This selection pressure in turn affects primates' allocation decisions to invest in maintenance and repair mechanisms that promote long life, instead of in machinery that facilitates rapid growth and reproduction. These decisions result in characteristic rates of development and reproduction, as well as patterns of longevity in different primate species— all of which represent adaptations shaped by the process of natural selection. Chapter 10 outlines the ecological correlates of these adaptations, and examines how some of the unusual aspects of primate life history affect their behavior and brain evolution.

The events following the birth of relatively precocial young, the second general pattern of primate life histories, are taken up by Elizabeth V. Lonsdorf and Stephen R. Ross in chapter 11. Despite their precocial nature, primate young undergo prolonged periods of prereproductive dependency during which they acquire several important motor and social skills that are necessary for subsequent survival and reproduction. After summarizing various theories that have been proposed to explain this extended phase of early dependence, Lonsdorf and Ross review how different individuals influence the development of young primates, and some of the characteristic behaviors they acquire prior to starting their reproductive careers. The chapter concludes with a brief discussion of natal dispersal, the behavioral process that defines adult independence for members of the dispersing sex.

Dispersal is also featured in the final chapter about non-human primates in Part 2. The relatively stable social groups formed by many primates are not closed but open, due to dispersal. Upon reaching sexual maturity, members of one or both sexes leave their natal groups. In the past few years there has been a growing realization that dispersal and other characteristic behaviors displayed by primates both affect and are affected by the genetic structure of their populations. The idea that behavior influences genetics represents a novel way of thinking about the relationship between the two, and in chapter 12 Anthony Di Fiore outlines theoretical predictions about how behavior is expected to shape the genetic structure of populations. Di Fiore also reviews empirical studies that bear on these predictions, with special emphasis given to the effects of dispersal, reproductive skew, and new group formation.

In the last chapter in part 2, Michael Gurven examines the human primate, and how we differ from our primate relatives with regard to patterns of survival and life history. Many of these differences bear on topics treated in the preceding chapters in this section, and involve variations on a theme or an elaboration of patterns observed in the nonhuman primates: high-quality diets comprising nutrient-dense foods in large packages (chapter 7); an extremely long period of prereproductive dependence (chapter 11); low juvenile and adult mortality due to predation (chapter 8) and other sources of extrinsic mortality, leading to very long life spans (chapter 10); and social groups consisting of pairs of females and males, who cooperate between themselves via a sexual division of labor and with kin and non-kin alike (chapter 9). Chapter 13 outlines these and other differences between nonhuman and human primates, and looks at how human behavioral science attempts to explain them using different theoretical paradigms and empirical approaches.

Taken together, the chapters in this section highlight the old and the new in the study of primate behavior. The first three chapters, on feeding, predation, and socioecology, revisit and reexamine some enduring questions in the field, while the last four chapters highlight issues that deserve more theoretical and empirical attention. As other chapters elsewhere in this book emphasize, long-term data derived from studies of populations of wild primates are starting to provide the means to assess the fitness consequences of their mating and rearing strategies (part 3) and other aspects of their social relationships and behavior (part 4). Additional long-term data from a variety of species will furnish an empirical basis to describe patterns of primate development and life history and to test hypotheses formulated to explain their evolution. More genetic data will be required to evaluate the extent to which observations conform to theoretical predictions about how behavior influences the genetic structure of primate populations. And as is stressed throughout this book, more work synthesizing the observations derived from studies of nonhuman primates and the human primate are needed to understand our place in nature.

Chapter 7 Food as a Selective Force in Primates

*Colin A. Chapman, Jessica M. Rothman, and
Joanna E. Lambert*

PRIMATES MUST EAT to survive; the acquisition of food resources is thus among one of the most significant selective pressures affecting their biology. The necessity of finding food to meet nutritional requirements must be balanced with avoiding predation and securing mates, and thereby influences primate physiology, ecology, activity, movement, and social relationships. All animals, including primates, should typically avoid expending energy through unnecessary travel, and should eat the most nutritious foods available to them. In a classic study of baboons (*Papio cynocephalus*), Altmann found that the closer an individual's foraging approached the optimal amounts of protein and energy (Pyke 1984), the higher its fitness (Altmann 1991; Altmann et al. 1993; Altmann 1998). Because of the importance of food as a selective pressure for so many aspects of primate biology, it is not surprising that primate feeding ecology has received considerable attention in the literature over the past 25 years (Clutton-Brock 1977; Milton 1980; Cant & Temerin 1984; Rodman & Cant 1984; Hohmann et al. 2006).

Here we examine the feeding problems primates encounter and the strategies they employ to satisfy their nutritional needs. We also consider the consequences of adopting specific strategies. Our goal is to evaluate what is known about primate diets and to stimulate research on primate foraging strategies by identifying new ways to interpret diet and feeding data. We start by reviewing primate nutritional requirements, diets, and digestive strategies. We then address three central problems in the study of primate feeding biology: how do foraging strategies influence primates with respect to their (1) ecology and distribution, (2) movement and ranging patterns, and (3) behavior and social organization. For each problem, we present basic information and an illustrative example. We conclude with a summary of how feeding and foraging biology impacts the conservation potential of a species.

Primate Nutrition and Diet

Like all animals, primates require a full complement of carbohydrates, protein, lipids, vitamins, and minerals for growth, development, survival, and reproduction. Obtaining an adequate balance of these nutrients, while minimizing the ingestion of toxins, is challenging. For field biologists, understanding the precise requirements of different primates is difficult because it requires experimental studies that include manipulation and deprivation of key nutrients. Studies of this nature are often impractical and unethical. Nonetheless, the US National Research Council has assembled the requirements of 31 different nutrients, primarily based on experimental studies of rhesus macaques (*Macaca mulatta*), baboons (*Papio* spp.), squirrel monkeys (*Saimiri* spp.), and humans (NRC 1998).

Energy

Energy requirements are determined by three key factors: the energetic costs of baseline functions (basal metabolic rate, or BMR), the costs of activity, and the costs of specific life stage reproduction and lactation for females and growth for juveniles. These requirements are typically met by ingesting a combination of lipids and carbohydrates. The energy needs for sustaining basic metabolism and bodily function were historically estimated by Kleiber's equation, which states that the energy needed to maintain BMR is a function of body weight (BW), specifically $BW^{0.75}$ (Kleiber 1961). This general relationship means that larger animals need to consume less energy per kilogram of body mass than smaller animals. Some specific data are available for primates (Ross 1992; Genoud 2002). The Kleiber equation has been substantially revised over the past few decades to include data from more species (McNab 1988, 2002; Nagy 1994). For free-ranging mammals, the equation to estimate energy needed to maintain life during normal activities, field metabolic rate (FMR), is expressed in kilojoules as 4.63 $BW^{0.762}$ (Nagy & Milton 1979; Nagy 1994). Neither FMR nor BMR is constant for particular species. For example, in callitrichines, sleeping reduces BMR by 30% (Thompson et al. 1994; Genoud et al. 1997). Ambient temperature also significantly affects FMR and BMR estimates, particularly for cheirogaleid primates that hibernate or go into torpor (Schülke & Ostner 2007); metabolic rates during torpor were 86% of those not in torpor (Schmid et al. 2000). Numerous studies have attempted to estimate the energy requirements of primate activities (Madame Berthe's mouse lemurs, *Microcebus berthae*, Schmid et al. 2000; golden lion tamarins, *Leontopithecus rosalia*, Miller et al. 2006; mantled howlers, *Alouatta palliata*, Milton 1998; Nagy & Milton 1979; colobines, DaSilva 1994; Wasserman & Chapman 2003; brown capuchins, *Cebus apella*, Janson 1988; chimpanzees, *Pan troglodytes*, N'guessan et al. 2009). For example, Pontzer and Wrangham (2002) estimated the locomotor energy costs of chimpanzees and found that they spent much more energy on terrestrial than arboreal travel. Interestingly, orangutans (*Pongo* spp.) have an extremely low rate of energy use, which may be an evolutionary response to food shortages (Pontzer et al. 2010), which are seasonally common (Knott 1998).

Pregnancy and milk production incur additional costs for female primates, while juveniles need to meet the energy requirements of growth. Pregnant females require about 17% to 32% more energy than nonreproducing females (Robbins 1993). These increased energy requirements can influence female foraging strategies, since females that are sized similarly to males will need to eat higher-quality food than males, or increase their intake per kilogram of body mass. The energy cost incurred by primate females compared to males depends on the degree of sexual dimorphism. Where male body size exceeds female body size by 60% or more, male energy costs on an absolute basis are greater than those for females (Key & Ross 1999), and this may reduce the need for diet-based sexual segregation among primates (Kamilar & Pokempner 2008). Juvenile primates are likely the most susceptible to nutritional stress because their small body size coupled with the increased energy needed for growth means that they require more energy per kilogram of metabolic body mass than adults of both sexes. For example, captive juvenile gorillas (*Gorilla gorilla*) can accumulate as much as 19 kg of body mass per year, so that by age 10 they may reach 120 kg (Leigh & Shea 1996; Bedyaev 2002). This will require juveniles to consume more energy per kilogram of body mass than adults (Rothman et al. 2008b). Provisioned juvenile baboons grow faster and reach sexual maturity earlier than do unprovisioned baboons (Altmann & Alberts 2005), indicating that resource availability strongly affects the growth rate of juveniles.

Protein

Many studies have focused on the importance of protein in primate diets, suggesting that protein is an important criterion for food selection (Milton 1979; Oates et al. 1990; Chapman & Chapman 2002; Chapman et al. 2002a). However, primates generally have slow growth rates, produce dilute milk, and reach their full size later than other mammals, suggesting that they may not have high protein requirements and may consume more protein than needed (Oftedal 1992). Like energy requirements, protein requirements are dependent on life stages, and immature animals require the most protein per unit of metabolic body mass. The protein requirements of primates depend on protein quality and available energy. Most primates require less than three grams of protein per kilogram of metabolic body weight per day, or 6% to 8% of the dietary dry matter (NRC 2003). However, the quality of protein must be assessed. For example, tannins (Robbins et al. 1987), fiber-bound nitrogen (Conklin-Brittain et al. 1999; Rothman et al. 2008a), and amino acid imbalances can all affect the digestibility of protein. In general, the quality of protein is rarely estimated in primate diets (Rothman et al. 2008a), and very few studies have examined the amino acid profiles of foods eaten by primates (but see Milton & Dintzis 1981; Curtis 2004).

Vitamins and Minerals

Very little is known about vitamin and mineral requirements of wild primates, although recent work highlights their significance. Minerals are particularly important for juvenile growth, and deficiencies may have permanent consequences for growth and lifetime fitness. Calcium and phosphorous make up 70% of the mineral matter in animals, and are predominant in bones and teeth in a 2:1 ratio. They are necessary for growth, and a deficiency in either element markedly affects bone development and can cause rickets, osteomalacia, and osteoporosis (Robbins 1993). Since plants do not typically concentrate sodium and the soils of many tropical regions are poor in that element, tropical herbivores often have difficulty obtaining the sodium they need. Symptoms of sodium deficiency include softening of bones, reduced growth, blindness, and reproductive impairment. Sodium is the only mineral that elicits a particular hunger and drive for acquisition. Guerezas (*Colobus guereza*) visit underground caves and increase their travel distances to exploit high-sodium resources (Oates 1978; Fashing et al. 2007; Harris & Chapman 2007), and mountain gorillas eat decaying wood that contains high levels of sodium (fig. 7.1; Rothman et al. 2006). Rode and colleagues (2003) demonstrated that the sodium content of foods eaten by primates was extremely low; no single food met the guidelines set by the National Research Council (NRC 2003), and sodium intake from the typical plant diet fell well below suggested requirements throughout the year (Rode et al. 2003; Rode et al. 2006). However, infrequently eaten foods can provide important mineral sources. Gums from selected tree species, eaten infrequently by chimpanzees (Ushida et al. 2006) and by saddleback (*Saguinus fuscicollis*) and mustached (*Saguinus mystax*) tamarins (Smith 2000), are a rich source of calcium and sodium. Seemingly nutrient-deficient bark and wood can also provide needed micronutrients that are absent in frequently eaten foods (Rode et al. 2003; Rothman et al. 2006; Stephens et al. 2006; Reynolds et al. 2009), and for colobus monkeys, flowers provided a source of copper, a mineral typically deficient in the diets of monkeys (Rode et al. 2003, 2006). As more information about the mineral contents of primate diets becomes available, we will be in a better position to understand their nutritional needs and how food ultimately influences primate biomass and diversity.

Primate Foods

Over the past 65 million years, primates have adopted several strategies to solve problems associated with feeding (Fleagle & Gilbert 2006). The adoption of these different strategies means that the same food can yield different benefits to each species that feeds on it, but some general patterns do emerge. Extant primates have evolved specializations to consume fruit, seeds, leaves, insects, gums, or most often a mixture of these dietary items. Each food source exhibits significant structural and chemical differences which influ-

Fig. 7.1. Alternative means of obtaining salts through foraging: (A) a red colobus monkey eating soil and (B) mountain gorillas eating decaying wood. Photos courtesy of Jessica Rothman.

ence primate foraging strategies and feeding adaptations. In tropical forests, most tree species produce fleshy fruits thought to have coevolved to be eaten and dispersed by frugivores. Primates constitute 25% to 40% of the frugivore biomass in these forests, likely making them an essential part of this coevolutionary process (Chapman 1995). Many fruits contain sugars, encouraging primates to eat them and disperse their seeds. This relationship does not require fruits to be nutritionally balanced, and frugivorous primates typically supplement their diet with young leaves or insects, which provide protein (Janson & Chapman 1999). While there is debate regarding how such coevolutionary relationships develop, in many cases fruiting species are thought to have coevolved with specific dispersers and to be consumed by them (Howe & Smallwood 1982; Herrera 1985; Fischer & Chapman 1993). In such cases, fruits may contain secondary compounds that are toxic to some species but not others. For instance, the chemicals that make red peppers spicy to humans and other mammals do not apparently affect birds (Janson & Chapman 1999). Secondary compounds may have important consequences for primate feeding behavior. For example, the fruits of *Strychnos mitis* are laden with compounds that are toxic to some mammals; they include phenolics, terpenes, flavonol glycosides, and various alkaloids (Thepenier et al. 1990). Some cercopithecines, however, appear to have evolved means to deal with these compounds. Thus, *Strychnos mitis* fruits in Kibale are readily eaten by redtail monkeys (*Cercopithecus ascanius*, Lambert 2001) and blue monkeys (*Cercopithecus mitis*, Rudran 1978) but are ignored by chimpanzees (Lambert 1997; although different *Strychnos* species are eaten by chimpanzees at Gombe: Goodall 1986).

Unlike fruits, seeds are typically mechanically protected

with hard protective structures designed to stop primates getting to the nutritious embryo inside. As a result, only a handful of primates have evolved the dental and morphological skull features needed to break seeds (e.g., *Cacajao* spp. and *Lophocebus* spp., Kinzey 1992; Lambert et al. 2004). For these primates, seeds can be an important part of their diet. For example, the average monthly seed consumption by white-faced sakis (*Pithecia pithecia*) was 63.2% ± 32.7% (Norconk & Conklin-Brittain 2004; fig. 7.2).

Leaves provide the photosynthetic energy used for plant growth and reproduction. Thus, it is not to a plant's advantage to have its leaves eaten. Fiber, including hemicellulose and cellulose, is a polymer of sugar molecules, but without the help of microbes, primates cannot digest the structural components that may form a large portion of leaves. Only a few species (e.g., *Colobus*, *Procolobus*, *Indri*, and *Alouatta*) with a large capacity for hindgut or foregut fermentation can rely on a diet comprised mostly of leaves (Milton 1981b; Lambert 1998). Folivores typically choose specific leaf species and leaf parts (Glander 1982; Chapman

Fig. 7.2.   A relatively small number of primate species have evolved the specializations to support a diet based heavily on seeds, but white-faced sakis (*Pithecia pithecia*) are one example. Here an adult male at Brownsberg Nature Park, Suriname, feeds on the seeds of *Garcinia madruno* of the Clusiaceae family. Photo courtesy of Marilyn A. Norconk.

Fig, 7.3,   Individuals of many primate species—such as this young male white-faced capuchin monkey in Lomas Barbudal, Costa Rica—spend considerable time eating insects on a seasonal basis. Photo courtesy of Susan Perry.

& Chapman 2002). Young leaves are most often selected, as they usually have smaller concentrations of fiber than mature leaves (Milton 1979; Chapman & Chapman 2002). Eastern red colobus monkeys (*Procolobus rufomitratus*) commonly eat only leaf tips or petioles (Chapman & Chapman 2002).

Insectivory is practiced by many primates (fig. 7.3). Insects are typically high in protein and energy and are easy to digest, with the exception of their chitinous skeleton (Moir 1994; Barker et al. 1998). Some primates are well-adapted insectivores. For example, the northern lesser galago (*Galago senegalensis*) and western fat-tailed dwarf lemur (*Cheirogaleus medius*) masticate insects to a small size, which probably improves chitin digestion (Sheine & Kay 1979). The potto (*Perodicticus potto*) uses chitinolytic enzymes to digest insects (Cornelius et al. 1976), and a gene for chitinase has apparently been conserved in a variety of primates (Gianfrancesco & Musumeci 2004). Insects can be poisonous, but some nocturnal strepsirrhines may have evolved mechanisms to deal with their secondary compounds because they specialize on noxious prey (Charles-Dominique 1977).

Lastly, only a few species specialize on gums, but gummivory is prominent among strepsirrhines and callitrichines (Nash 1994). These species typically supplement their diets with insects, fruits, and young leaves. Patas monkeys (*Erythrocebus patas*) feed extensively on *Acacia* gums, which is unusual because patas monkeys are large-bodied in comparison to typical gummivores like callitrichines (Isbell 1998). Gums are nutritionally different from saps because gums require fermentation while saps do not (Nash 1994). Few primates possess diets containing large portions of sap, but slow loris diets (*Nycticebus coucang*) contain 35% phloem sap and 32% floral nectar (Weins et al. 2006).

This description of primate foods might lead one to assume that species consistently specialize on particular foods. However, there can be extreme variation in the types of foods eaten among populations of the same species. For example, Butynski (1990) studied four groups of blue monkeys at the same site and found that the amount of time feeding on fruit ranged from 22% to 35% among groups. Fruit intake of blue monkeys among populations in East and South Africa ranged from 26% to 91%, and leaf intake varied from 3% to 47% (Chapman et al. 2002b). Another population of blue monkeys relies on bamboo (*Arundinaria alpina*) for 60% of its foraging time (Twinomugisha et al. 2006). Such variability makes it extremely difficult to identify the components of a diet that lead to selection, because one population will experience one selective regime while other populations, which may interbreed with the first, will experience different selection pressures. In addition, there can be marked spatial and temporal variation in the same food item. In the Kibale National Park, Uganda, young leaves of the same tree species varied in protein content from 22% to 47% (Chapman et al. 2003), and the fat content of a single species of ripe fruit varied seasonally from 0.3% to 30% (Worman & Chapman 2005). Consequently, few generalizations can be made about the nutritional contents of foods, and classifying feeding strategies broadly (e.g., folivory, frugivory) may not be a reliable indicator of a diet's nutritional quality (Danish et al. 2006; Rothman et al. 2007).

Primate Digestive Strategies

It is often assumed that different primate species obtain similar nutritional benefits from the same food item (Cords 1986; Chapman 1988; Isbell 1991), but nutritional gains are best interpreted and understood in light of species adaptations. For example, leaves are considered low-quality "fibrous" food sources because they often contain high amounts of structural carbohydrates. Depending on their digestive anatomy and physiology, however, primates can gain substantial amounts of energy from leaves (Milton et al. 1980; Milton & McBee 1983). As a result, the nutritional value of a particular food item is often species-specific. Here we first explore the potential for significant variation in digestive strategies among species eating the same foods by considering a single digestive process: variation in food transit time. Second, we evaluate the digestive mechanisms that primates use to consume the secondary compounds in plants.

Food Transit Time

The time it takes food to clear the digestive tract is critical to evaluate primates' foraging strategies, since longer transit times are typically associated with a greater ability to digest structural carbohydrates via fermentation (Chivers & Hladik 1980; Milton 1981b; Kay & Davies 1994; Canton et al. 1999; Lambert 2002). In contrast, faster transit times are usually associated with an increase in the total amount of food that can be processed. Thus, easily digestible nutrients (e.g., simple sugars) are more accessible, and indigestible material like seeds and chitin is expelled faster (Lambert 2002).

In a classic study, Milton (1981b) compared the digestive ecology between two sympatric monkeys in Panama, the primarily frugivorous Geoffroy's spider monkey (*Ateles geoffroyi*) and the folivorous mantled howler monkey. Milton found that howler monkeys had much longer transit times than spider monkeys, and suggested that digestive constraints forced both monkeys to consume particular foods (Milton 1981b; 1993). Although both monkeys ate both fruit and leaves, the diet foundation of howler monkeys was leaves, and that of spider monkeys was fruit. Howler monkeys ate ripe fruit when it was available, but their diet was never completely fruit because their long transit time prevented them from consuming it in sufficient quantities. Conversely, spider monkeys with short transit times were committed to a frugivorous foraging strategy because they could not gain enough energy from a leaf-dominated diet since they required fermentation time. These seminal ideas on the differences in digestive ecology inspired subsequent investigations into primate digestive ecology.

Lambert (1998) documented large variations in transit time in similarly sized frugivorous primates, which imply a diversity of digestive strategies. Primarily frugivorous brown capuchins (3.5 kg) have transit times of 3.5 hr, while similarly sized frugivorous crowned mona monkeys (*Cercopithecus pogonias*, 3.75 kg) and redtail monkeys (3.6 kg) have transit times of 16.6 and 19.7 hr respectively. The slower transit times of crowned mona and redtail monkeys imply that guenons have a greater ability to gain nutrients from food items that contain structural carbohydrates, like fibrous leaves and fruits, than do capuchins. Although primates with specialized guts and longer transit times have the ability to digest fibrous foods, this does not mean they will select those foods; primates will typically prefer foods that have the most easily gained nutrients, like simple carbohydrates. In periods of preferred-food scarcity, primates that have adaptations to deal with structural carbohydrates may be better equipped to deal with food shortages than those with less flexible digestive tracts (Lambert 2007; Marshall & Wrangham 2007). The ability to digest fibrous foods has important implications for the conservation of frugivores, because animals faced with a low supply of their preferred foods often need to eat more fibrous foods like leaves instead of fruit.

Plant Secondary Compounds

Primate species can vary in their ability to obtain nutrients from plants containing toxins. Many plants contain one or more defensive compounds, including phenols, tannins, terpenes, alkaloids, cyanogenic glycosides, protease inhibitors, lectins, nonprotein amino acids, cardiac glycocides, and oxalates (Glander 1982; Seigler 1991; McNab 2002; Foley & Moore 2005). These include some well-known compounds such as strychnine, caffeine, cocaine, nicotine, and cyanide.

A classic and still very useful distinction of plant chemical defense concerns whether defenses are qualitative or quantitative (Feeny 1976). Qualitative defenses are typically found in small amounts in the plant and typically represent less than 2% of the dry weight. They usually interfere with a metabolic process and are often toxic, unless the species has evolved a detoxification mechanism that is specific to the defensive compound in question. One example of a qualitative defense is cyanide. For most primates cyanide is lethal at small doses, but the bamboo eaten by golden bamboo lemurs (*Hapalemur aureus*) contains about 10 times the amount of cyanide that would be lethal to humans (Glander et al. 1989; Ballhorn et al. 2009). It remains unclear what mechanism these lemurs or other bamboo-eating primates use to detoxify the bamboo (Glander et al. 1989; Twinomugisha et al. 2006).

A quantitative defense is typically present in substantial amounts and has properties that reduce the digestibility of cell constituents such as tannins. Tannins are common in tropical plants and can render protein inaccessible to animals (Mole & Waterman 1985; Robbins et al. 1987). Several studies have demonstrated that tannins affect primate foraging behaviors (Oates et al. 1977; McKey et al. 1981; Wrangham & Waterman 1981; Glander 1982; Marks et al. 1988), but others have ambiguous results (Barton et al. 1993; Chapman & Chapman 2002). This may be due in part to some primates having adaptations to deal with tannins. These include proline-rich salivary proteins, which have a higher than average affinity for binding with tannins and allow for uptake of plant protein in the presence of tannins (Mole et al. 1990) and microbes in the fore or hind-gut, which are able to degrade these toxic compounds (Foley & Moore 2005; Frey et al. 2006). Alternatively, methodological problems may have hindered our ability to measure the tannins accurately (Foley et al. 2005; Rautio et al. 2007; Rothman et al. 2009).

Evidence suggests that environmental conditions can influence the frequency in occurrence of some of these compounds, or the amount present in particular food items. For example, some areas of the Amazon have white, sandy, poor soils while other areas have soils that are high in organic content (Emmons 1984). Kinzey and Gentry (1979) contrasted the diet of two species of titi monkeys (*Callicebus* spp.): white-collared titi monkeys (*Callicebus torquatus*) are found living in habitats with poor white-sand soils, while dusky titis (*Callicebus moloch*) occur in forests that grow on soils rich in organic content. Kinzey and Gentry speculated that the monkeys' diets differed because plants growing on poor soils protected their leaves with quantitative defenses (see also McKey et al. 1978 for a similar example). This intriguing speculation requires further study.

There is good evidence that some species respond behaviorally to the presence of secondary compounds, varying their consumption of foods with different plant toxins. With the exception of weasel lemurs (*Lepilemur mustelinus*), most species in one lemur community selected foods with high levels of protein (Ganzhorn 1988, 1989). Indris (*Indri indri*) and eastern woolly lemurs (*Avahi laniger*) ate leaves with high levels of tannins but avoided alkaloids, while *Hapalemur* avoided both. In contrast, brown lemurs (*Eulemur fulvus*) and greater dwarf lemurs (*Cheirogaleus major*) tolerated tannins and alkaloids while weasel lemurs ate leaves high in alkaloids. Such food selection suggests that different species have evolved different physiological mechanisms to cope with these compounds.

The action of plant toxins is complex, and it is clear that we are only beginning to understand the physiological strategies that different species employ to deal with these compounds (Foley & Moore 2005; Rothman et al. 2009). Primates can either detoxify the compounds, bind substances to them to make them inoperable, or tolerate them. Each of these strategies will be a function of a species' abilities and will influence the value of a food item. For example, detoxifying a compound can have a significant cost. Thomas et al. (1988) demonstrated that meadow voles (*Microtus pennsylvanicus*) increase their metabolic rate by 14% to 24% when fed phenolic compounds. Comparable data for primates are not yet available.

Three Important Questions in Primate Foraging Ecology

Food as a Selective Force Influencing Primate Ecology and Distributions

How food resources influence the ecology and distribution of primates has been a central question of primate research since the first field studies began, and a critical problem involves temporal and spatial changes in food availability. Many researchers have documented seasonal variation in the food supply (Beeson 1989; van Schaik et al. 1993; DaSilva 1994; Remis 1997; Conklin-Brittain et al. 1998; Poulsen & Clark 2004), but food availability and distribution changes more rapidly than that. The availability of plant reproductive and vegetative parts is irregular, and it induces short-term changes in abundance and scarcity of food for consumers (Gautier-Hion 1980; van Schaik et al. 1993). Monitoring these phenological changes is often time-consuming because the fruiting and leafing patterns of food trees must be examined at least on a monthly basis. In addition, it is often not known at the beginning of a study which foods are being eaten. Primates respond to these phenological changes in a complex fashion that may involve increased travel, reliance on less nutritious foods, decreased activity, or lower reproductive and juvenile survival rates (Peres 1994; Brugiere et al. 2002).

Seasonal and monthly phenological differences are critical; however, there are also marked interannual changes in food availability (Tutin et al. 1997; Chapman et al. 2005a). Over 12 years, temporal variability in fruit availability in Kibale National Park, Uganda, was pronounced, with the proportion of trees per month with ripe fruit varying between 0.14% and 15.93% (Chapman et al. 2005b). Primates respond to this variation across years, making characterization of a primate population's diet for purposes of comparative studies difficult, unless studies encompass many years of the same population. For example, the fruits of *Bursera simaruba* were available for every dry season during a six-year study of Geoffroy's spider monkeys in Santa Rosa National Park, Costa Rica (Chapman et al. 1995), but it was only in the last year of the study that they fed heavily on this fruit. Because the fruit occurred at a very high density, a logical conclusion to be drawn from a study of just that year would have been that fruit density was higher in the dry season than in the wet season. However, since spider monkeys ignored this fruit in the previous five years of the study and instead fed on rarer foods (Chapman 1987), it seems likely that the opposite is true, and that in most dry seasons they relied on rare foods that were more profitable. In the last year, when other resources were unavailable, they fed on *B. simaruba*.

The phenology and availability of specific foods are important for particular primate species, and these foods have often been called keystone species. Keystone species are defined as those species that have a large and disproportionate impact on the community relative to their abundance such that their removal would be devastating to the entire animal community (Power et al. 1996). To determine community-wide importance, factors to consider include (1) temporal redundancy, (2) degree of consumer specificity, (3) reliability, and (4) abundance (Peres 2000). The term "keystone species" should not be used when referring to a food source that is important only to a single primate species; it must apply to a community of organisms. The importance of this concept for conservation is readily apparent: if a manager could identify keystone species in an ecosystem and conserve them, it would likely ensure the integrity of the whole community. Implementing this strategy is nonetheless difficult (Peres 2000).

Figs have been frequently hypothesized as examples of keystone plant resources in tropical forests (Terborgh 1986; Power et al. 1996), and recently textbooks have presented figs as a clear case to illustrate the keystone species concept (Bush 2000). This proposal remains to be critically examined, as few data exist demonstrating their importance to a community of species. This is worrisome because the conservation of figs has been advocated in management strategies (Primack 2006). Two studies have provided detailed analyses to determine whether figs serve as keystone species for primates. Gautier-Hion and Michaloud's (1989) study in Gabon showed that figs were infrequently eaten by most primates and other mammals, occurred at very low densities, and had unpredictable fruiting patterns. They concluded that fig fruits could not sustain most populations of frugivorous species during periods of low fruit availability, and thus were not keystone species. Similarly, a 12-year study in Kibale National Park, Uganda, determined patterns of fruit scarcity and the spatial and temporal availability of figs. Temporal variability in fruit resource availability was high; the proportion of trees (> 10 cm diameter at breast height) per month with ripe fruit varied from 0.14% to 15.93% (Chapman et al. 2005b). If figs served as keystone species for the frugivore community (or fallback foods; see below) over these 12 years, they would have had to be available during months when few other trees were fruiting. Less than 1% of the monitored fig trees fruited in the 34 months when fruit was scarce during the 149 months of the study. Figs failed to fruit in 17 of the 34 fruit-poor months, and more than 1% of the fig trees fruited during 11 of those 34 months. Accordingly, figs may provide fruit during some periods of food scarcity, but the number of trees is probably inadequate and the fruiting phenology is too inconsistent to sustain all of the frugivorous primate community in Kibale. This example demonstrates that the role of figs as keystone species (or as fallback foods) is likely scale-dependent. During periods of fruit scarcity, figs can best be exploited by highly mobile species with large home ranges who possess the ability to track fruiting figs. In Kibale, only gray-checked mangabeys (*Lophocebus albigena*) and chimpanzees have large enough home ranges to monitor

widely dispersed fruiting fig trees (Chapman et al. 2005b). For species with small feeding ranges, like redtail monkeys (Chapman et al. 2005b), figs are unlikely to be important resources during periods of fruit scarcity. Because figs could only be keystone species for 2 of the 13 primate species in Kibale, they cannot be considered keystone species for the whole community. It is our view that figs have never been demonstrated to be a keystone resource for any primate community.

Even if figs are not a keystone resource, they may be an important food source for some primates at some periods of time. The importance of figs in tropical forests is substantiated by the large number of primates and other vertebrate frugivores that eat them (Shanahan et al. 2001). If figs are important foods and are being commercially harvested by humans, this would be a concern for primate communities. We present Bolivia as an example of what may be a general trend in timber harvesting. In Bolivia, the volume of *Ficus* timber harvested has increased by at least 65% since 1999 and the export value of *Ficus* products was approximately $US 1.4 million, which represented 13% of Bolivia's total revenue from exported timber products (CFB 2006; Felton et al. unpublished manuscript). Given that the commercial harvest of *Ficus* timber is increasing dramatically as more profitable timbers have already been removed, the value of figs to different primate communities should be assessed. If figs prove to be a generally important food resource, then pressure could be placed on the timber industry to decrease the rates of their extraction.

Food as a Selective Force Influencing Movement

An important theoretical challenge in primatology and a pressing issue in primate conservation is to understand how primates are distributed with respect to the temporal and spatial variation of food resources. All primates, regardless of their diet, confront the problem of gaining sufficient food to satisfy their nutritional requirements, and the solution involves selecting an optimal diet and travel routes (Charnov 1976; Pyke 1984; Stephens & Krebs 1986; Grether et al. 1992; Altmann 1998). The temporal and spatial availability of ripe fruits and young leaves of high quality varies considerably (van Schaik et al. 1993; Worman & Chapman 2005), and faced with this variation, animals must move across their landscape and adjust their diet.

In the past decade a great deal has been learned about the travel routes of primates (see also chapter 27, this volume). While it is clear that primates follow different strategies, in general they aim to minimize search costs relative to resource gain. Accordingly, it is interesting to consider whether primates move to the nearest available resource

(Garber 1988; Janson 1998) or plan their travel routes in the most efficient way to maximize the gains of future resources (Noser & Byrne 2006; Janson 2007). Milton (1980) noted that mantled howler monkeys oriented their travel patterns towards "pivotal" trees, which were food sources visited repeatedly until depleted; these trees were used as bases, but the monkeys then moved to nearby areas where they could feed on leaves and monitor many other fruit trees. Instead of monitoring the phenological states of thousands of tropical trees, a simple routine may provide the best means to exploit available resources (Di Fiore & Suarez 2007). Sympatric white-bellied spider monkeys (*Ateles belzebuth*) and Poeppig's woolly monkeys (*Lagothrix poeppigii*) traveled the same routes for several years, using topographical features of the landscape such as ridge tops (Di Fiore & Suarez 2007). For gray-cheeked mangabeys living in forests whose phenological patterns are difficult to predict (Chapman et al. 1999; Chapman et al. 2005a), individuals rely on detailed and sophisticated knowledge of the fruiting states and locations of their fruit trees (Janmaat et al. 2006a). Furthermore, mangabeys appear to use weather as a cue to locate fruit resources when they become ripe (Janmaat et al. 2006b). Understanding how primates move within their habitats and the methods they use to evaluate their space provides important insights into primate cognition (Chapman et al. 1999; Barton 2000; Janson 2007; chapter 27, this volume) and useful data for conservation managers.

Food as a Selective Force Influencing Behavior and Social Organization

Theoretically, differing combinations of levels and types of feeding competition should lead to variation in social structure and social bonding (Wrangham 1980; van Schaik 1989; Sterck et al. 1997; chapter 9, this volume). The nature of food resources determines the level and type of feeding competition. Scramble competition involves the common depletion of food resources while contest competition, including aggression, displacement, and avoidance, involves direct contests over food and can occur within and between groups (Nicholson 1933; Janson & van Schaik 1988). Animals must compete for access to resources when they are (1) limited in availability, (2) patchy and depletable, (3) variable in quality, or (4) able to be monopolized (Janson & van Schaik 1988; van Schaik 1989; Isbell 1991; Saj et al. 2007). When both within- and between-group contest competition are absent and only scramble competition is present, or when there is no competition for food and food resources are not monopolizable (Snaith & Chapman 2007), females are not expected to engage in agonistic interactions over food. This in turn should be associated with an absence of linear domi-

nance hierarchies and infrequent coalitions (Sterck et al. 1997). These patterns should co-occur with female dispersal because coalition partners are not required in feeding competition. Female agonistic relationships are predicted to be rare, as are affiliative behaviors among females.

When food resources are limited, patchy, depletable, and monopolizable, contest competition will occur. Under these circumstances it will become advantageous for females to remain with kin on whom they rely as allies for cooperative defense of resources, either within or between groups (Sterck et al. 1997). Accordingly, female dispersal should not occur, as a female who attempted to transfer would lose access to allies. Contest competition between groups should only lead to a system in which female dominance relationships are egalitarian and individualistic, and where coalitions are rare. The presence of within-group contest competition (and its absence between groups) should lead to a nepotistic system in which females form linear and despotic dominance relationships, coalitions with kin, and mutualistic coalitions with other females to acquire and maintain their dominance rank, as the latter will be associated with priority of access to limited food resources.

Initially folivores were generally considered not to be food-limited. This idea stemmed from the assumption that leaves are superabundant in forest habitats. However, many studies have recently demonstrated that folivorous primates are very selective in what they eat, typically preferring young leaves of just a few species (Chapman & Chapman 2002; Koenig & Borries 2006). These observations suggest that leaf-eating primates may have different competitive regimes than previously thought. For example, several recent studies have found that eastern red colobus monkeys are food-limited and experience within-group scramble competition for food resources (Snaith & Chapman 2005, 2007, 2008). Large groups occupied larger home ranges than small groups, and group size was related to depletion of feeding patches. In addition, individuals in large groups suffered reduced foraging efficiency, assayed by long travel distances, more time spent feeding, less frequent feeding at preferred food sites, and more frequent feeding in small trees. Monkeys in large groups also experienced a concomitant reduction in female reproductive success. These results suggest that within-group competition occurs in red colobus monkeys. The behavioral consequences of this competition are currently under investigation.

Socioecological models rely on an understanding of the size, density, and distribution of food resources and how these variables influence a primate's ability to monopolize them (Wrangham 1980; Chapman et al. 1995; Isbell & Young 2002). However, the best way to measure these characteristics of food resources is not clear. There are two contrasting approaches. Isbell and colleagues have suggested that the behavior of the forager should be used to assess food distribution, because it is not the patchiness of the resource itself that is important, but whether the resource can be monopolized or usurped (Isbell et al. 1998; Isbell & Young 2002). This method suffers from measurement challenges for some species (e.g., how to measure such behaviors for primates feeding in dense canopy cover) and from the fact that usurpation of feeding sites may depend on a suite of non-food variables (e.g., stability of the dominance hierarchy). In cases where a resource can be monopolized, if it is abundant it may not induce competition. Other studies have attempted to quantify a direct measure of the characteristics of food resources (Chapman et al. 1995; Koenig 2000; Koenig & Borries 2006), but this approach also has limitations. What should one consider a food item? Most studies opt for an arbitrary inclusion of foods based on a set percentage of foraging effort (e.g., the top 5 or 10 of most frequently eaten foods), but typically there is no rationale presented for the percentage chosen or how it is calculated, and using different cutoff points can produce different outcomes (C.A. Chapman unpublished data). It is also not clear whether the animals respond similarly to all food items in the list of included foods. For example, if one food source is particularly preferred or provides needed nutrients, it may be more influential than another food source in determining competition and social organization. Several studies indicate that there is a strong tendency for primates to eat leaves at the end of the day, often just prior to entering the sleeping site (e.g., Chapman & Chapman 1991). Leaves can be high in fiber and difficult to digest. By choosing to feed on leaves just prior to resting at night, animals may reduce the distance travelled with a full stomach of leaves and not miss feeding opportunities on more profitable food items, as predicted by central place foraging (Orians & Pearson 1979). It is not clear whether the trees providing such leaf resources should be included in the resources that determine competitive regimes. It seems likely important to consider the "value" of particular items and how that influences the nature of the competitive regime.

One potential means of defining resource importance could be to use a geometric analysis of feeding and nutrition (Simpson & Raubenheimer 1995, 1999; Behmer & Joern 2008; Raubenheimer et al. 2009). Researchers could define functional optima (e.g., intake targets) for a specific period in which social behaviors are evaluated (e.g., a day), including a specified food, such as the leaves eaten by monkeys at the end of the day. Subsequently, if a food item is of questionable importance, the optima can be redefined excluding that specific item and the outcome can be reevaluated. Both the inclusion and exclusion of this item can be presented. The value of food items that are rarely selected but provide particularly important resources also requires consideration, but how they influence competitive regimes is largely unknown. For example, when species select particular food items such as decaying wood or soils to obtain salts (fig. 7.1; Rode et al. 2003; Rothman et al. 2006), the nature of the competitive regimes has not been quantified.

Food and the Conservation Potential of a Species

Current threats to primate populations have increased our need to understand these complex animals, their nutrient requirements, and how food resources act as a selective force. According to the Food and Agricultural Organization's Global Forest Resources Assessment (FAO 2005), forest lost between 2000 and 2005 was about 7.3 million hectares per year or approximately 200 km^2 of forest per day. Between 1990 and 2005, forest cover in Africa decreased by 21 million hectares (FAO 2005; Chapman et al. 2006). An understanding of the importance of specific food resources (i.e., how foods meet nutritional needs), dietary flexibility in primates, the importance of fallback foods, and nutritional requirements for different species will provide information vital for conservation planning. For example, foods that provide critical nutrients or are important fallback foods should not be cut in selective logging operations or, more realistically, harvested trees can be directionally felled away from species that bear these critical foods. This could potentially reduce the decline of some primate species caused by selective logging.

To address the issue of whether food influences the conservation potential of a species, it is important to consider whether primates are food-limited. The question of whether primate populations are limited by food resource availability has been evaluated previously (Janson and Chapman 1999). In considering the importance of minerals (see also Rode et al. 2003, 2006), food species diversity, food productivity, and food quality, a general conclusion is that many primates face food shortages—either seasonally, in the case of frugivores, or in terms of the rarity of high-quality leaves for folivores. There are, of course, exceptions to these generalizations. For example, extensive chimpanzee predation affects the eastern red colobus population size at Ngogo in Kibale National Park (Mitani et al. 2000; Lwanga 2006; Teelen 2007). Disease can also be an important limiting factor and can clearly cause short-term reductions in population size (Collias & Southwick 1952; Work et al. 1957; Milton 1996; Nunn & Altizer 2006), likely below carrying capacity. For instance, yellow fever accounted for a 50% decline in the mantled howler monkey population on Barro Colorado Island, Panama, between 1933 and 1951 (Collias & Southwick 1952).

To use nutritional ecology data in conservation efforts, one could study the nutritional needs of each endangered species, but that would be time consuming and often the required information would only become available years after action was needed. As discussed previously, in many primates there is considerable variation between the diets of different populations in the types of foods eaten, making it difficult to define their nutritional needs. Thus, there have been few direct tests of general hypotheses proposed to account for this variation. Notable exceptions are studies of folivorous primates. Milton (1979) proposed that the protein-to-fiber ratio was a good predictor of leaf choice. By measuring overall mature-leaf acceptability as the ratio of protein to fiber, several subsequent studies have found positive correlations between colobine biomass and this index of leaf quality at local (Chapman & Chapman 2002; Chapman et al. 2002a, 2004; Ganzhorn 2002) and regional scales (Waterman et al. 1988; Oates et al. 1990; fig. 7.4). Milton (1979; 1998) and Milton and colleagues (1980) proposed a physiological explanation for the importance of protein-to-fiber ratios in regulating population densities.

The protein/fiber ratio appears to be a useful predictor of the biomass of folivorous primates and a useful conservation tool. With this knowledge one can define critical food resources. Unfortunately, however, it is unclear what is driving this relationship. Most primates require about 4% to 7% protein for growth and maintenance, and 8% to 10% protein for reproduction if the protein is of high quality (Oftedal 1992; NRC 2003). At some sites where this relationship has been examined, the mean mature-leaf protein of the abundant trees is about 17% dry matter (Chapman et al. 2002a), and so it would seem that protein is not limiting (Oftedal 1992). Therefore it is unclear whether high amounts of protein are unavailable for digestion because they are bound to fiber (Rothman et al. 2008a) or tannins (Robbins et al. 1987), or if the amino acids within the leaf protein are imbalanced. Alternatively, other factors may be driving this relationship, such as energy. In a recent study, Harris et al. (2010) found that guerezas regularly excrete ketones in their urine, an indication that they could be using excess-protein foods as energy (fig. 7.5). Unraveling the mechanisms behind the protein-to-fiber model is an important priority for future studies of folivorous primates.

Geometric analysis of feeding and nutrition promises to identify targeted nutrients in foraging strategies and to yield generalizations about primate feeding behavior (Felton et al. 2009; Raubenheimer et al. 2009). This modeling technique allows identification of the intake targets of a species, and it could help in evaluating how animals direct their foraging efforts. With this information, one can make reasonable hypotheses that trees providing high levels of particular nutrients will be the most critical for specific species. Efforts can then be made to protect those types of plants or to restore them in a degraded habitat (Chapman and Chapman 2002; Felton et al. 2009).

In addition to studies concerning the protein-fiber ratio for folivores, there are three key concepts that need further consideration when evaluating how food resources act as selection pressures to influence a species conservation potential: ideal free distribution, ecological sinks, and fallback foods.

Ideal Free Distribution

Evaluating the variation in primate density as a function of habitat disturbance (an index of habitat quality) should be considered in terms of population distribution theory. If individuals or groups are free to select among habitats, then their distribution should represent an ideal free distribution (Fretwell & Lucas 1969; Fretwell 1972), with the density of animals being proportional to the resources available in the area. There is evidence that this is often not the case. These situations have not been adequately investigated, particularly in primates. A Web of Science search conducted on January 13, 2011, using the term "ideal free distribution primates" yielded 86 papers, but only three had to do with

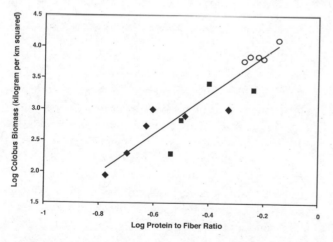

Fig. 7.4. The relationship between mature leaf chemistry and colobine biomass at rainforest sites in Africa and Asia. Chemical values are weighted by mean percentages of dry mass, standardized to the species basal area to account for different proportions of the flora being sampled at each site. The weighted values were calculated from $\Sigma(P_i + X_i)/\Sigma P_i$, where P_i is the proportion of the basal area contributed by species i and X_i is the chemical measure for species i. This figure is standardized to 100%. Diamonds denote sites around the world (Oates et al. 1990; Fashing et al. 2007), squares denote forest sites within Kibale National Park, Uganda (Chapman et al. 2002a), and open circles denote the forest fragments (Chapman et al. 2007).

Fig. 7.5. A young guereza feeds on the young leaves of *Celtis durandii* (note the small, unripe fruit also evident). *C. durandii* is one of the major dietary items of this colobine species, and its young leaves have one of the highest protein-to-fiber ratios that have been documented at Kibale National Park, Uganda. Photo courtesy of Colin A. Chapman.

nonhuman primate behavior. If, in contrast, the term "primates" was dropped from the search, 1,471 papers were listed. Work by Olupot (2000) illustrates the value of considering ideal free distributions for primate conservation. This study found that gray-cheeked mangabey males in unlogged forest in Kibale National Park, Uganda, were significantly heavier than males in logged forests. The mangabeys in the logged forest had almost 30 years to recover from the logging and reach their population equilibrium. Thus, this result is not easily understood, particularly since males frequently move among groups (Olupot & Waser 2005; Janmaat et al. 2009). According to the expectations of an ideal free distribution, the animals should distribute their density in proportion to food availability and should have equal body mass. It may be that this situation represents instead an ideal despotic distribution (Fretwell & Lucas 1969; Fretwell 1972), where mangabeys in the bet-

ter habitat exclude other mangabeys from entering. Movement patterns of male mangabeys, however, do not support this idea (Olupot & Waser 2005, Chapman unpublished data). The fact that there is a discrepancy between what theory would predict and what was observed in this study of gray-cheeked mangabeys clearly illustrates the need to evaluate the natural history of primates in situations like this to determine the missing pieces of information needed to understand such situations.

Ecological Sinks

Source-sink patterns occur when populations occupying low-quality habitats ("sinks"), where mortality rates exceed reproductive output, are sustained by immigration from populations inhabiting nearby high-quality habitats ("sources"), where reproductive output exceeds mortality and a surplus of individuals are produced. Identifying sources and sinks is difficult because it requires long-term demographic data from diverse habitats with different food resources, as well as data on dispersal patterns. Thus, the source-sink framework has rarely received explicit consideration in studies of primates, despite its usefulness for understanding population dynamics and identifying populations of high conservation priority (i.e., sources). Pulliam (1988) and Holt (1985) first formalized the theory of source-sink dynamics and highlighted the importance of explicitly considering habitat-specific demographic trends and dispersal patterns when assessing habitat quality and interpreting species' foraging adaptations (Holt 1992; Dias & Blondel 1996; Kawecki & Holt 2002). The theory of source-sink dynamics has since become a robust paradigm in the field of ecology for understanding population dynamics, selective pressures, and adaptations. Siex and Struhsaker (1999) suggested that the population dynamics of Kirk's red colobus (*Procolobus kirkii*) populations in Zanzibar represented a source-sink dynamic. If a study is conducted in a sink habitat, it is difficult to understand how food resources influence anatomical or behavioral adaptations because the observed patterns are not in response to the selective pressures that have favored them. This highlights the need to study a species in a variety of settings to obtain the information needed to construct informed management plans.

Fallback Foods

Building on a wealth of ecological information on diets and optimal foraging theory (Altmann 1998; MacArthur & Pianka 1966), there has recently been interest in the ecological and evolutionary significance of fallback foods (Conklin-Brittain et al. 1998; Wrangham et al. 1998; Lambert et al. 2004; Hemingway & Bynum 2005; Lambert 2007; Marshall & Wrangham 2007). Fallback foods are those used when preferred food items are not available, and as such are negatively correlated with the availability of the latter (Lambert 2007, 2009; Marshall & Wrangham 2007; Marshall et al. 2009). Because they help sustain populations during periods of food scarcity, fallback foods are ecologically significant components of primate diets. Information about them is also necessary to evaluate the ability of a species to respond to anthropogenic habitat disturbance.

Variation in fallback foods will have important consequences for primates, who rely on them. It seems likely that for species with flexible diets the nature of fallback foods will be highly variable. For example, when fruit was scarce, Geoffroy's spider monkeys at Santa Rosa National Park, Costa Rica, fed on young leaves during 86.3% of their feeding time in one month. In another food-poor month, how-

ever, they fed on the dry fruits of *Bursera simaruba* during 68.2% of the time (Chapman 1987; C. Chapman unpublished data). These different fallback foods will place very different selection pressures on spider monkey anatomy and social behavior. Evaluating changes in anatomy and behavior due to variation in fallback foods will require additional information about how other populations of spider monkeys respond to times of food scarcity, but such data are currently lacking.

Most primates live across a range of interconnected forest types or habitats. As a consequence they are not likely to rely on specific food items as fallback foods, but instead will exploit classes of foods with certain traits— for example, large fruits with thick exocarps, or plants with high levels of specific secondary compounds. Thus it is not surprising that cercopithecines have adapted to deal with secondary compounds or structural properties, such as hardness, rather than to specific types of foods such as fruits or leaves (Wrangham et al. 1998; Lambert et al. 1999; Balcomb & Chapman 2003; Lambert 2007). From a conservation perspective, the ability of a species to be flexible regarding the types and diversity of fallback foods will greatly increase its ability to tolerate habitat destruction and the potential removal of preferred foods.

Summary and Conclusions

Primates face the difficult challenge of obtaining a nutritious diet in an energetically efficient manner while avoiding plant toxins. To meet this challenge, different primate species have evolved a diversity of strategies and adaptations to satisfy their nutritional requirements. Foraging strategies affect virtually every aspect of primate biology, including their (1) ecology and distribution, (2) movement and ranging patterns, and (3) behavior and social organization. Most primates are endangered due to extensive deforestation and high hunting pressure, and an understanding of primate foraging strategies and their diversity is necessary to construct informed conservation management plans.

It is clear that food resources act as strong selective forces along several dimensions, including *physiology, ecology, behavior, and social organization*. We suggest that a way forward in primate foraging ecology must involve a shift in perspectives on food and primate biology. New methods in nutritional ecology, in addition to still untapped classic methods, are available to primatologists though they remain largely underused (e.g., Rothman et al. 2009). Geometric analyses, new methods in nutritional assays, and the application of ideal-free and sink-source models will be useful in interpreting not only what primates consume but why they consume it. In particular, investigations into the nutritional composition of the specific food items primates eat (e.g., a specific food item from a particular individual tree of a particular species), instead of types of foods (e.g., fruits) or single samples of food items, will shed light on how food acts as a selective pressure.

References

Altmann, J. & Alberts, S. C. 2005. Growth rates in a wild primate population: Ecological influences and maternal effects. *Behavioral Ecology and Sociobiology*, 57, 490–501.

Altmann, J., Schoeller, D., Altmann, S. A., Muruthi, P. & Sapolsky, R. M. 1993. Body size and fatness of free-living baboons reflect food availability and activity levels. *American Journal of Primatology*, 30, 149–161.

Altmann, S. A. 1991. Diets of yearling female primates (*Papio cynocephalus*) predict lifetime fitness. *Proceedings of the National Academy of Sciences of the United States of America*, 88, 420–423.

———. 1998. *Foraging for Survival: Yearling Baboons in Africa*. Chicago: University of Chicago Press.

Balcomb, S. R. & Chapman, C. A. 2003. Bridging the gap: Influence of seed deposition on seedling recruitment in a primate-tree interaction. *Ecological Monographs*, 73, 625–642.

Ballhorn, D. J., Kautz, S. & Rakotoarivelo, F. P. 2009. Quantitative variability of cyanogenesis in *Cathariostachys madagascarensis*: The main food plant of lemurs in Southeastern Madagascar. *American Journal of Primatology*, 71, 305–315.

Barker, D., Fitzpatrick, M. P. & Dierenfeld, E. S. 1998. Nutrient composition of selected whole invertebrates. *Zoo Biology*, 17, 123–134.

Barton, R. A. 2000. Primate brain evolution: Cognitive demands of foraging or of social life? In *On the Move: How and Why Animals Travel in Groups* (ed. by Boinski, S. & Garber, P. A.), 204–237. Chicago: University of Chicago Press.

Barton, R. A., Whiten, A., Byrne, R. W. & English, M. 1993. Chemical composition of baboon plant foods: Implications for the interpretation of intraspecific and interspecific differences in diet. *Folia Primatologica*, 61, 1–20.

Bedyaev, A. V. 2002. Growing apart: An ontogenetic perspective on the evolution of sexual size dimorphism. *Trends in Ecology and Evolution*, 17, 369–378.

Beeson, M. 1989. Seasonal dietary stress in a forest monkey (*Cercopithecus mitis*). *Oecologia*, 78, 565–570.

Behmer, S. T. & Joern, A. 2008. Coexisting generalists herbivores occupy unique nutritional feeding niches. *Proceeding of the National Academy of Science*, 105, 1977–1982.

Brugiere, D., Gautier, J.-P., Moungazi, A. & Gautier-Hion, A. 2002. Primate diet and biomass in relation to vegetation composition and fruiting phenology in a rain forest in Gabon. *International Journal of Primatology*, 23, 999–1023.

Bush, M. B. 2000. *Ecology of a Changing Planet*. Upper Saddle River: Prentice Hall.

Butynski, T. M. 1990. Comparative ecology of blue monkeys (*Cercopithecus mitis*) in high- and low-density subpopulations. *Ecological Monographs*, 60, 1–26.

Cámara Forestal de Bolivia. 2006a. Exportación de productos forestales de Bolivia, Gestiones: 1998—2006. *Anuario estadístico del sector forestal de Bolivia 2006*. Cámara Forestal de Bolivia.

Cant, J. G. H. & Temerin, L. A. 1984. A conceptual approach to foraging adaptations in primates. In *Adaptations for Foraging in Nonhuman Primates* (ed. by Rodman, P. S. & Cant, J. G. H.), 304–342. New York: Columbia University Press.

Canton, J. M., Hume, I. D., Hill, D. M. & Harper, P. 1999. Digesta retention in the gastro-intestinal tract of the orang utan (*Pongo pygmaeus*). *Primates*, 40, 551–558.

Chapman, C. A. 1987. Flexibility in diets of three species of Costa Rican primates. *Folia Primatologica*, 49, 90–105.

———. 1988. Patterns of foraging and range use by three species of neotropical primates. *Primates*, 29, 177–194.

———. 1995. Primate seed dispersal: Coevolution and conservation implications. *Evolutionary Anthropology*, 4, 74–82.

Chapman, C. A. & Chapman, L. J. 1991. The foraging itinerary of spider monkeys: When to eat leaves. *Folia Primatologica*, 56, 162–166.

———. 2002. Foraging challenges of red colobus monkeys: Influence of nutrients and secondary compounds. *Comparative Biochemistry and Physiology: Part A, Physiology*, 133, 861–875.

Chapman, C. A., Chapman, L. J., Bjorndal, K. A. & Onderdonk, D. A. 2002a. Application of protein-to-fiber ratios to predict colobine abundance on different spatial scales. *International Journal of Primatology*, 23, 283–310.

Chapman, C. A., Chapman, L. J., Cords, M., Gauthua, M., Gautier-Hion, A., Lambert, J. E., Rode, K. D., Tutin, C. E. G. & White, L. J. T. 2002b. Variation in the diets of Cercopithecus species: Differences within forests, among forests, and across species. In *The Guenons: Diversity and Adaptation in African Monkeys* (ed. by Glenn, M. & Cords, M.), 319–344. New York: Plenum.

Chapman, C. A., Chapman, L. J., Naughton-Treves, L., Lawes, M. J. & McDowell, L. R. 2004. Predicting folivorous primate abundance: Validation of a nutritional model. *American Journal of Primatology*, 62, 55–69.

Chapman, C. A., Chapman, L. J., Rode, K. D., Hauck, E. M. & McDowell, L. R. 2003. Variation in the nutritional value of primate foods: Among trees, time periods, and areas. *International Journal of Primatology*, 24, 317–333.

Chapman, C. A., Chapman, L. J., Struhsaker, T. T., Zanne, A. E., Clark, C. J. & Poulsen, J. R. 2005a. A long-term evaluation of fruiting phenology: Importance of climate change. *Journal of Tropical Ecology*, 21, 31–45.

Chapman, C. A., Chapman, L. J., Zanne, A. E., Poulsen, J. R. & Clark, C. J. 2005b. A 12-year phenological record of fruiting: Implications for frugivore populations and indicators of climate change. In *Tropical Fruits and Frugivores* (ed. by Dew, J. L. & Boubli, J. P.), 75–92. Dordrecht, Netherlands: Springer.

Chapman, C. A., Lawes, M. J. & Eeley, H. A. C. 2006. What hope for African primate diversity? *African Journal of Ecology*, 44, 1–18.

Chapman, C. A., Naughton-Treves, L., Lawes, M. J., Wasserman, M. D. & Gillespie, T. R. 2007. The conservation value of forest fragments: Explanations for population declines of the colobus of western Uganda. *International Journal of Primatology*, 23, 513–578.

Chapman, C. A., Wrangham, R. W. & Chapman, L. J. 1995. Ecological constraints on group size: an analysis of spider monkey and chimpanzee subgroups. *Behavioural Ecology and Sociobiology*, 36, 59–70.

Chapman, C. A., Wrangham, R. W., Chapman, L. J., Kennard, D. K. & Zanne, A. E. 1999. Fruit and flower phenology at two sites in Kibale National Park, Uganda. *Journal of Tropical Ecology*, 15, 189–211.

Charles-Dominique, P. 1977. *Ecology and Behaviour of Nocturnal Primates: Prosimians of Equatorial West Africa*. London: Columbia University Press.

Charnov, E. L. 1976. Optimal foraging: The marginal value theorem. *Theoretical Population Biology*, 9, 129–136.

Chivers, D. J. & Hladik, C. M. 1980. Morphology of the gastrointestinal tract in primates: Comparisons with other mammals in relation to diet. *Journal of Morphology*, 166, 337–386.

Clutton-Brock, T. H. 1977. *Primate ecology*. New York: Academic Press.

Collias, N. & Southwick, C. 1952. A field study of population density and social organization in howling monkeys. *Proceedings of the National Academy of Science*, 96, 143–156.

Conklin-Brittain, N. L., Dierenfeld, E. S., Wrangham, R. W., Norconk, M. & Silver, S. C. 1999. Chemical protein analysis: A comparison of Kjeldahl crude protein and total ninhydrin protein from wild, tropical vegetation. *Journal of Chemical Ecology*, 25, 2601–2622.

Conklin-Brittain, N. L., Wrangham, R. W. & Hunt, K. 1998. Dietary response of chimpanzees and cercopithecines to seasonal variation in fruit abundance: II. Macronutrients. *International Journal of Primatology*, 19, 971–997.

Cords, M. 1986. Interspecific and intraspecific variations in the diet of two forest guenons, *Cercopithecus ascanius* and *C. mitis*. *Journal of Animal Ecology*, 55, 811–827.

Cornelius, C., Dandrifosse, G. & Jeuniaux, C. 1976. Chitinolytic enzymes of gastric mucosa of *Perodicticus potto* (Primate Prosimian): Enzyme purification and enzyme specificity. *International Journal of Biochemistry and Cell Biology*, 7, 445–448.

Curtis, D. J. 2004. Diet and nutrition in wild mongoose lemurs (*Eulemur mongoz*) and their implications for the evolution of female dominance and small group size in lemurs. *American Journal of Physical Anthropology*, 124, 234–247.

Danish, L., Chapman, C. A., Hall, M. B., Rode, K. D. & Worman, C. 2006. The role of sugar in diet selection in redtail and red colobus monkeys. In *Feeding Ecology in Apes and Other Primates: Ecological, Physical, and Behavioural Aspects* (ed. by Hohmann, G., Robbins, M. M. & Boesch, C.), 471–487. Cambridge: Cambridge University Press.

DaSilva, G. L. 1994. Diet of *Colobus polykomos* on Tiwai Island: Selection of food in relation to its seasonal abundance and nutritional quality. *International Journal of Primatology*, 15, 1–26.

Di Fiore, A. & Suarez, S. A. 2007. Route-based travel and shared routes in sympatric spider and woolly monkeys: Cognitive and evolutionary implications. *Animal Cognition*, 10, 317–329.

Dias, P. C. & Blondel, J. 1996. Local specialization and maladaptation in the Mediterranean blue tit (*Parus caeruleus*). *Oecologia*, 107, 79–86.

Emmons, L. H. 1984. Geographic variation in densities and diversity of non-flying mammals in Amazonia. *Biotropica*, 16, 210–222.

Fashing, P. J., Dierenfeld, E. & Mowry, C. B. 2007. Influence of plant and soil chemistry on food selection, ranging patterns, and biomass of *Colobus guereza* in Kakamega Forest, Kenya. *International Journal of Primatology*, 28, 673–703.

Feeny, P. P. 1976. Plant apparency and chemical defense. *Recent Advances in Phytochemistry*, 10, 1–40.

Felton, A. M., Felton, A., Lindenmayer, D. B. & Foley, W. J. 2009. Nutritional goals of wild primates. *Functional Ecology*, 23, 70–78.

Felton, A.M., Felton, A., Campbell, C. J., Chapman, C. A., Pena-Claros, M., Rumiz, D. I., Shimooka, Y., Stevenson, P. R., Wallace, R. B. & Lindermayer, D. B. Commercial harvesting of *Ficus* timber: An emerging threat to frugivorous wildlife and sustainable forestry. *Conservation Letters* (submitted).

Fischer, K. E. & Chapman, C. A. 1993. Frugivores and fruit syndromes: Differences in patterns at the genus and species level. *Oikos*, 66, 472–482.

Foley, W. J. & Moore, B. D. 2005. Plant secondary metabolites and vertebrate herbivores: From physiological regulation to ecosystem function. *Current Opinion in Plant Biology*, 8, 430–435.

Food and Agricultural Organization. 2005. *Global Forest Resources Assessment 2005: Progress towards sustainable forest management*. Rome. FAO Forestry Paper 147.

Fretwell, S. D. 1972. *Populations in a Seasonal Environment*. Princeton, NJ: Princeton University Press.

Fretwell, S. D. & Lucas, H. L. 1969. On territorial behavior and other factors influencing habitat distribution in birds. 1. Theoretical development. *Acta Biotheoretica*, 19, 16–36.

Frey, J. C., Rothman, J. M., Pell, A. N., Nizeyi, J. B., Cranfield, M. R. & Angert, E. A. 2006. Fecal bacterial diversity in a wild gorilla (*Gorilla beringei*). *Applied and Environmental Microbiology*, 72, 3788–3792.

Ganzhorn, J. U. 1988. Food partitioning among Malagasy primates. *Oecologia*, 75, 436–450.

———. 1989. Primate species separation in relation to secondary plant chemicals. *Human Evolution*, 4, 125–132.

———. 2002. Distribution of a folivorous lemur in relation to seasonally varying food resources: Integrating quantitative and qualitative aspects of food characteristics. *Oecologia*, 131, 427–435.

Garber, P. A. 1988. Foraging decisions during nectar feeding by tamarin monkeys (*Saguinus mystax* and *Saguinus fuscicollis*, Callitrichidae, Primates) in Amazonian Peru. *Biotropica*, 20, 100–106.

Gautier-Hion, A. 1980. Seasonal variations of diets related to species and sex in a community of *Cercopithecus* monkeys. *Journal of Animal Ecology*, 49, 237–269.

Gautier-Hion, A. & Michaloud, G. 1989. Are figs always keystone resources for tropical frugivorous vertebrates? A test in Gabon. *Ecology*, 70, 1826–1833.

Genoud, M. 2002. Comparative studies of basal rate of metabolism in primates. *Evolutionary Anthropology*, 11, 108–111.

Genoud, M., Martin, R. D. & Glaser, D. 1997. Rate of metabolism in the smallest simian primate, the pygmy marmoset (*Cebuella pygmaea*). *American Journal of Primatology*, 41, 229–245.

Gianfrancesco, F. & Musumeci, S. 2004. The evolutionary conservation of the human chitotriosidase gene in rodents and primates. *Animal Cytogenetics and Comparative Mapping*, 105, 54–56.

Glander, K. E. 1982. The impact of plant secondary compounds on primate feeding behavior. *Yearbook of Physical Anthropology*, 25, 1–18.

Glander, K. E., Wright, P. C., Seigler, D. S., Randrianasolo, V. & Randrianasolo, B. 1989. Consumption of cyanogenic bamboo by a newly discovered species of bamboo lemur. *American Journal of Primatology*, 18, 119–124.

Goodall, J. 1986. *The Chimpanzees of Gombe: Patterns of Behavior*. Cambridge, MA: Harvard University Press.

Grether, G. F., Palombit, R. A. & Rodman, P. S. 1992. Gibbon foraging decisions and the marginal value model. *International Journal of Primatology*, 13, 1–17.

Harris, T. R. & Chapman, C. A. 2007. Variation in the diet and ranging behavior of black-and-white colobus monkeys: Implications for theory and conservation. *Primates*, 28, 208–221.

Harris, T. R., Chapman, C. A. & Monfort, S. L. 2010. Small folivorous primate groups exhibit behavioral and physiological effects of food scarcity. *Behavioral Ecology* 21, 46–56.

Hemingway, C. A. & Bynum, N. 2005. The influence of seasonality on primate diet and ranging. In *Seasonality in Primates* (ed. by Brockman, D. K. & van Schaik, C. P.), 57–104. Cambridge: Cambridge University Press.

Herrera, C. 1985. Determinants of plant-animal coevolution: The case of mutualistic dispersal of seeds by vertebrates. *Oikos*, 44, 132–141.

Hohmann, G., Robbins, M. M. & Boesch, C. 2006. *Feeding Ecology in Apes and Other Primates: Ecological, Physiological and Behavioural Aspects*. Cambridge: Cambridge University Press.

Holt, R. D. 1985. Population dynamics in 2-patch environments: Some anomalous consequences of an optimal habitat distribution. *Theoretical Population Biology*, 28, 181–208.

———. 1992. Analysis of adaptation in heterogeneous landscapes: Implications for the evolution of fundamental niches. *Evolutionary Ecology*, 6, 433–447.

Howe, H. F. & Smallwood, J. 1982. Ecology of seed dispersal. *Annual Review of Ecology and Systematics*, 13, 201–228.

Isbell, L. A. 1991. Contest and scramble competition: Patterns of female aggression and ranging behaviour among primates. *Behavioral Ecology*, 2, 143–155.

———. 1998. Diet for a small primate: Insectivory and gummivory in the (large) patas monkey (*Erythrocebus patas pyrrhonotus*). *American Journal of Primatology*, 45, 381–398.

Isbell, L. A., Pruetz, J. D. & Young, T. P. 1998. Movements of vervets (*Cercopithecus aethiops*) and patas monkeys (*Erythrocebus patas*) as estimators of food resource size, density, and distribution. *Behavioural Ecology and Sociobiology*, 42, 123–133.

Isbell, L. A. & Young, T. P. 2002. Ecological models of female social relationships in primates: Similarities, disparities and some directions for future clarity. *Behaviour*, 139, 177–202.

Janmaat, K. R. L., Byrne, R. W. & Zuberbuhler, K. 2006a. Evidence for a spatial memory of fruiting states of rainforest trees in wild mangabeys. *Animal Behaviour*, 72, 797–807.

———. 2006b. Primates take weather into account when searching for fruits. *Current Biology*, 16, 1232–1237.

Janmaat, K. R. L., Olupot, W., Chancellor, R. L., Arlet, M. E. & Waser, P. M. 2009. Long-term site fidelity and individual home range shifts in *Lophocebus albigena*. *International Journal of Primatology*, 30, 443–466.

Janson, C. H. 1988. Food competition in brown capuchin monkeys (*Cebus apella*): Quantitative effects of group size and tree productivity. *Behaviour*, 105, 53–79.

———. 1998. Experimental evidence for spatial memory in foraging wild capuchin monkeys, *Cebus apella*. *Animal Behaviour*, 55, 1229–1243.

———. 2007. What wild primates know about resources: Opening up the black box. *Animal Cognition*, 10, 357–367.

———. 2007. Experimental evidence for route integration and strategic planning in wild capuchin monkeys. *Animal Cognition*, 10, 341–356.

Janson, C. H. & Chapman, C. A. 1999. Resources and the determination of primate community structure. In *Primate Communities* (ed. by Fleagle, J. G., Janson, C. H. & Reed, K. E.), 237–267. Cambridge: Cambridge University Press.

Janson, C. H. & van Schaik, C. P. 1988. Recognizing the many faces of primate food competition: Methods. *Behaviour*, 105, 165–186.

Kamilar, J. M. & Pokempner, A. A. 2008. Does body mass dimorphism increase male-female dietary niche separation? A comparative study of primates. *Behaviour*, 145, 1211–1234.

Kawecki, T. J. & Holt, R. D. 2002. Evolutionary consequences of asymmetric dispersal rates. *American Naturalist*, 130, 333–347.

Kay, R. F. & Davies, A. G. 1994. Digestive physiology. In *Colobine Monkeys* (ed. by Davies, A. G. & Oates, J. F.). Cambridge: Cambridge University Press.

Key, C. & Ross, C. 1999. Sex differences in energy expenditure in non-human primates. *Proceedings of the Royal Society of London Series B-Biological Sciences*, 266, 2479–2485.

Kinzey, W. G. 1992. Dietary and dental adaptations in the Pitheciinae. *American Journal of Physical Anthropology*, 88, 499–514.

Kinzey, W. G. & Gentry, A. H. 1979. Habitat utilization in two species of *Callicebus*. In *Primate Ecology: Problem-Oriented Field Studies* (ed. by Sussman, R. W.), 89–100. New York: Wiley.

Kleiber, M. 1961. *The Fire of Life: An Introduction to Animal Energetics*. New York: Wiley Publishers.

Knott, C.A. 1998. Changes in orangutan caloric intake, energy balance, and ketones in response to fluctuating fruit availability. *International Journal of Primatology* 19, 1061–1079.

Koenig, A. 2000. Competitive regimes in forest-dwelling Hanuman langur females (*Semnopithecus entellus*). *Behav Ecol Sociobiol*, 48, 93–109.

Koenig, A. & Borries, C. 2006. The predictive power of socioecological models: A reconsideration of resource characteristics, agonism and dominance hierarchies. In *Feeding Ecology in Apes and Other Primates* (ed. by Hohmann, G., Robbins, M. M. & Boesch, C.), 263–284. Cambridge: Cambridge University Press.

Lambert, J. E. 1997. *Fruit Processing and Seed Dispersal by Chimpanzees (Pan troglodytes schweinfurthii) and Redtail Monkeys (Cercopithecus ascanius schmidti) in the Kibale National Park, Uganda*. Urbana: University of Illinois Press.

———. 1998. Primate digestion: Interactions among anatomy, physiology, and feeding ecology. *Evolutionary Anthropology*, 7, 7–20.

————. 2001. Red-tailed guenons (*Cercopithecus ascanius*) and *Strychnos mitis*: Evidence for plant benefits beyond seed dispersal. *International Journal of Primatology*, 22, 189–201.

————. 2002. Digestive retention times in forest guenons (*Cercopithecus* spp.) with reference to chimpanzees (*Pan troglodytes*). *International Journal of Primatology*, 23, 1169–1185.

————. 2007. Seasonality, fallback strategies, and natural selection: A chimpanzee and cercopithecoid model for interpreting the evolution of hominin diet. In *Evolution of Human Diet: The Known, the Unknown, and the Unknowable* (ed. by Ungar, P. S.), 324–343. Oxford: Oxford University Press.

————. 2009. Summary to the Symposium Issue: Primate fallback strategies as adaptive phenotypic plasticity: Scale, process, and pattern. *American Journal of Physical Anthropology*, 140, 759–766

Lambert, J. E., Chapman, C. A., Wrangham, R. W. & Conklin-Brittain, N. L. 2004. Hardness of cercopithecine foods: Implications for the critical function of enamel thickness in exploiting fallback foods. *American Journal of Physical Anthropology*, 125, 363–368.

Leigh, S. R. & Shea, B. T. 1996. Ontogeny of body size variation in African apes. *American Journal of Physical Anthropology*, 99, 43–65.

Lwanga, J. S. 2006. Spatial distribution of primates in a mosaic of colonizing and old growth forest at Ngogo, Kibale National Park, Uganda. *Primates*, 47, 230–238.

MacArthur, R. & Pianka, E. R. 1966. One optimal use of a patchy environment. *American Naturalist*, 100, 603–609.

Marks, D. L., Swain, T., Goldstein, S., Richards, A. F. & Leighton, M. 1988. Chemical correlates of rhesus monkey food choice: The influence of hydrolyzable tannins. *Journal of Chemical Ecology*, 14, 213–235.

Marshall, A. J., Boyko, C. M., Feilen, K. L., Boyko, R. H. & Leighton, M. 2009. Defining fallback foods and assessing their importance in primate ecology and evolution. *American Journal of Physical Anthropology*, 140, 603–614.

Marshall, A. J. & Wrangham, R. W. 2007. Evolutionary consequences of fallback foods. *International Journal of Primatology*, 28, 1219–1235.

McKey, D. B., Gartlan, J. S., Waterman, P. G. & Choo, G. M. 1981. Food selection by black colobus monkeys (*Colobus satanas*) in relation to plant chemistry. *Biological Journal of the Linnean Society*, 16, 115–146.

McKey, D. B., Waterman, P. G., Mbi, C. N., Gartlan, J. S. & Struhsaker, T. T. 1978. Phenolic content of vegetation in two African rain forests: Ecological implications. *Science*, 202, 61–64.

McNab, B. K. 1988. Food habits and the basal rate of metabolism in birds. *Oecologia*, 77, 343–349.

————. 2002. *The Physiological Ecology of Vertebrates: A View from Energetics*. Ithaca, NY: Cornell University Press.

Miller, K. E., Bales, K. L., Ramos, J. H. & Dietz, J. M. 2006. Energy intake, energy expenditure, and reproductive costs of female wild golden lion tamarins (*Leontopithecus rosalia*). *American Journal of Primatology*, 68, 1037–1053.

Milton, K. 1979. Factors influencing leaf choice by howler monkeys: A test of some hypotheses of food selection by generalist herbivores. *American Naturalist*, 114, 363–378.

————. 1980. *The Foraging Strategies of Howler Monkeys: A Study in Primate Economics*. New York: Columbia University Press.

————. 1981a. Distribution patterns of tropical plant foods as an evolutionary stimulus to primate mental development. *American Anthropologist*, 78, 534–548.

————. 1981b. Food choice and digestive strategies of two sympatric primate species. *American Naturalist*, 117, 496–505.

————. 1996. Effects of bot fly (*Alouattamyia baeri*) parasitism on a free-ranging howler (*Alouatta palliata*) population in Panama. *Journal of Zoology* (London) 239, 39–63.

————. 1998. Physiological ecology of howlers (*Alouatta*): Energetic and digestive considerations and comparison with the Colobinae. *International Journal of Primatology*, 19, 513–548.

Milton, K. & Dintzis, F. 1981. Nitrogen-to-protein conversion factors for tropical plant samples. *Biotropica*, 12, 177–181.

Milton, K. & McBee, R. H. 1983. Rates of fermentative digestion in the howler monkey, *Alouatta palliata* (Primates, Ceboidea). *Comparative Biochemistry and Physiology A-Molecular & Integrative Physiology*, 74, 29–31.

Milton, K., Van Soest, P. & Robertson, J. B. 1980. Digestive efficiencies of wild howler monkeys. *Physiological Zoology*, 53, 402–409.

Mitani, J. C., Struhsaker, T. T. & Lwanga, J. S. 2000. Primate community dynamics in old growth forest over 23.5 years at Ngogo, Kibale National Park, Uganda: Implications for conservation and census methods. *International Journal of Primatology*, 21, 269–286.

Moir, R. J. 1994. The "carnivorous herbivore." In *Digestive Systems in Mammals* (ed. by Chivers, D. J. & Langer, P.), 87–102. Cambridge: Cambridge University Press.

Mole, S., Butler, L. G. & Iason, G. 1990. Defense against dietary tannin in herbivores: A survey for proline rich salivary proteins in mammals. *Biochemistry, Systematics and Ecology*, 18, 287–293.

Mole, S. & Waterman, P. G. 1985. Stimulatory effects of tannins and cholic acid on tryptic hydrolysis of proteins: Ecological implications. *Journal of Chemical Ecology*, 11, 1323–1332.

N'guessan, A. K., Ortmann, S. & Boesch, C. 2009. Daily energy balance and protein gain among *Pan troglodytes verus* in the Tai National Park, Cote d'Ivoire. *International Journal of Primatology*, 30, 481–496.

Nagy, K. A. 1994. Field bioenergetics of mammals: What determines field metabolic rate? *Australian Journal of Zoology*, 42, 43–53.

Nagy, K. A. & Milton, K. 1979. Energy metabolism and food consumption by wild howler monkeys (*Alouatta palliata*). *Ecology*, 60, 475–480.

National Research Council. 2003. *Nutrient Requirements of Nonhuman Primates*, Second Revised Edition edn. Washington: National Academic Press.

Nicholson, A. J. 1933. The balance of animal populations. *Journal of Animal Ecology*, 2, 132–178.

Norconk, M. A. & Conklin-Brittain, N. L. 2004. Variation on frugivory: The diet of Venezuelan white-faced sakis (*Pithecia pithecia*). *International Journal of Primatology*, 25, 1–25.

Noser, R. & Byrne, R. W. 2006. Travel routes and planning of visits to out-of-sight resources in wild chacma baboons, *Papio ursinus*. *Animal Behavior*, 73, 257–266.

Nunn, C. L. & Altizer, S. 2006. *Infectious Diseases in Primates:*

Behavior, Ecology and Evolution. Oxford: Oxford University Press.

Oates, J. F. 1978. Water, plant and soil consumption by guereza monkeys (*Colobus guereza*): Relationship with minerals and toxins in the diet. *Biotropica*, 10, 241–253.

Oates, J. F., Swain, T. & Zantovska, J. 1977. Secondary compounds and food selection by colobus monkeys. *Biochemical Systematics and Ecology*, 5, 317–321.

Oates, J. F., Whitesides, G. H., Davies, A. G., Waterman, P. G., S.M., G., DaSilva, G. L. & Mole, S. 1990. Determinants of variation in tropical forest primate biomass: New evidence from West Africa. *Ecology*, 71, 328–343.

Oftedal, O. T. 1992. The nutritional consequences of foraging in primates: The relationship of nutrient intakes to nutrient requirements. *Philosophical Transactions of the Royal Society of London Series B-Biological Sciences*, 334, 161–170.

Olupot, W. 2000. Mass differences among male mangabey monkeys inhabiting logged and unlogged forest compartments. *Conservation Biology*, 14, 833–843.

Olupot, W. & Waser, P. M. 2005. Patterns of male residency and intergroup transfer in gray-cheeked mangabeys (*Lophocebus albigena*). *American Journal of Primatology*, 66, 331–349.

Orians, G. H. & Pearson, N. E. 1979. On the theory of central place foraging. In *Analysis of Ecological Systems* (ed. by Horn, D. J., Mitchell, R. D. & Stairs, G. R.), 155–177. Columbus: Ohio University Press.

Peres, C. A. 1994. Primate responses to phenological change in an Amazonian terra firme forest. *Biotropica*, 26, 98–112.

———. 2000. Identifying keystone plant resources in tropical forests: The case of gums from *Parkia* pods. *Journal of Tropical Ecology*, 16, 287–317.

Pontzer, H. & Wrangham, R. W. 2002. Climbing and the daily energy cost of locomotion in wild chimpanzees: Implications for hominoid locomotor evolution. *Journal of Human Evolution*, 46, 315–333.

Pontzer, H, Raichlen, D. A., Shumaker, R. W., Ocobock, C., & Wich, S. A. 2010. Metabolic adaptation for low energy throughput in orangutans. *Proceedings of the National Academy of the Sciences*. 107, 14048–14052.

Poulsen, J. R. & Clark, C. J. 2004. Densities, distributions, and seasonal movements of gorillas and chimpanzees in swamp forest in northern Congo. *International Journal of Primatology*, 25, 285–306.

Power, M. E., Tilman, D., Estes, J. A., Menge, B. A., Bond, W. J., Mills, L. S., Daily, G., Castilla, J. C., Lubchenco, J. & Paine, R. T. 1996. Challenges in the quest for keystones. *Bioscience*, 46, 609–620.

Primack, R. B. 2006. *Essentials of Conservation Biology*, 4th edn. New York: Sinauer.

Pulliam, H. R. 1988. Sources, sinks, and population regulation. *American Naturalist*, 132, 652–661.

Pyke, G. H. 1984. Optimal foraging theory: A critical review. *Annual Review of Ecology and Systematics*, 15, 525–575.

Raubenheimer, D., Simpson, S. J. & Mayntz, D. 2009. Nutrition, ecology and nutritional ecology: Towards an integrated framework. *Functional Ecology*, 23, 4–16.

Rautio, P., Bergvall, U.A., Karonen, M., & Salminen, J. P. 2007. Bitter problems in ecological feeding experiments: Commercial tannin preparations and common methods for tan-

nin quantifications. *Biochemical Systematics and Ecology* 35,257–262.

Remis, M. J. 1997. Western lowland gorillas (*Gorilla gorilla gorilla*) as seasonal frugivores: Use of variable resources. *American Journal of Primatology*, 43, 87–109.

Reynolds, V., Lloyd, A. W., Babweteera, F. & English, C. J. 2009. Decaying *Raphia farinifera* palm trees provide sodium for wild chimpanzees in the Budongo forest, Uganda. *PLoS Biology*, 4, e6194.

Robbins, C. T. 1993. *Wildlife Feeding and Nutrition*. New York: Academic Press.

Robbins, C. T., Hanley, T. A., Hagerman, A. E., Hjeljord, O., Baker, D. L., Schwartz, C. C. & Mautz, W. W. 1987. Role of tannins in defending plants against ruminants: Reduction in protein availability. *Ecology*, 68, 98–107.

Rode, K. D., Chapman, C. A., Chapman, L. J. & McDowell, L. R. 2003. Mineral resource availability and consumption by colobus in Kibale National Park, Uganda. *International Journal of Primatology*, 24, 541–573.

Rode, K. D., Chapman, C. A., McDowell, L. R. & Stickler, C. 2006. Nutritional correlates of population density across habitats and logging intensities in redtail monkeys (*Cercopithecus ascanius*). *Biotropica*, 38, 625–634.

Rodman, P. S. & Cant, J. G. H. 1984. *Adaptations for Foraging in Nonhuman Primates*. New York: Columbia University Press.

Ross, C. 1992. Basal metabolic rate, body weight and diet in primates: An evaluation of the evidence. *Folia Primatologica*, 58, 7–23.

Rothman, J. M., Chapman, C. A., Hansen, J. L., Cherney, D. J. & Pell, A. N. 2009. Rapid assessment of the nutritional value of mountain gorilla foods: Applying near-infrared reflectance spectroscopy to primatology. *International Journal of Primatology*, 30, 729–742.

Rothman, J. M., Chapman, C. A. & Pell, A. N. 2008a. Fiber-bound protein in gorilla diets: Implications for estimating the intake of dietary protein by primates. *American Journal of Primatology*, 70, 690–694.

Rothman, J. M., Dierenfeld, E. S., Hintz, H. F. & Pell, A. N. 2008b. Nutritional quality of gorilla diets: Consequences of age, sex and season. *Oecologia*, 155, 111–122.

Rothman, J. M., Dusinberre, K. & Pell, A. N. 2009. Condensed tannins in the diets of primates: A matter of methods? *American Journal of Primatology*, 71, 70–76.

Rothman, J. M., Plumptre, A. J., Dierenfeld, E. S. & Pell, A. N. 2007. Nutritional composition of the diet of the gorilla (*Gorilla beringei*): A comparison between two montane habitats. *Journal of Tropical Ecology*, 23, 673–682.

Rothman, J. M., Van Soest, P. J. & Pell, A. N. 2006. Decaying wood is a sodium source for mountain gorillas. *Biology Letters*, 2, 321–324.

Rudran, R. 1978. Intergroup dietary comparisons and folivorous tendencies of two groups of blue monkeys (*Cercopithecus mitis stuhlmanni*). In *The Ecology of Arboreal Folivores* (ed. by Montgomery, G. G.), 483–504. Washington: Smithsonian Institution Press.

Saj, T. L., Marteinson, S., Chapman, C. A. & Sicotte, P. 2007. Controversy over the application of current socioecological model to folivorous primates: *Colobus vellerosus* fits the predictions. *American Journal of Physical Anthropology*, 133, 994–1003.

Schmid, J., Ruf, T. & Heldmaier, G. 2000. Metabolism and temperature regulation during daily torpor in the smallest primate, the pygmy mouse lemur (*Microcebus myoxinus*) in Madagascar. *Journal of Comparative Physiology B: Biochemical, Systematic and Environmental Physiology*, 170, 59–68.

Schülke, O. & Ostner, J. 2007. Physiological ecology of cheirogaleid primates: Variation in hibernation and torpor. *Acta Ethologica*, 10, 13–21.

Seigler, D. A. 1991. Cyanide and cyanogenic glycosides. In *Herbivores: Their Interactions with Secondary Plant Metabolites, Volume 1: The Chemical Participants*. (ed. by Rosenthal, G. A. & Berebaum, M. R.), 35–77. New York: Academic Press.

Shanahan, M., So, S., Compton, S.G. & Corlett, R. 2001. Fig-eating by vertebrate frugivores: A global review. *Biological Reviews* 76, 529–572.

Sheine, W. S. & Kay, R. F. 1979. An analysis of chewed food particle size and it relationship to molar structure in the primates *Cheirogaleus medius* and *Galago senegalensis* and the Insectivoran *Tupaia glis*. *American Journal of Physical Anthropology*, 47, 15–20.

Siex, K. S. & Struhsaker, T. T. 1999. Ecology of the Zanzibar red colobus monkey: Demographic variability and habitat stability. *International Journal of Primatology*, 20, 163–192.

Simpson, S. J. & Raubenheimer, D. 1995. The geometric analysis of feeding and nutrition: A user's guide. *Journal of Insect Physiology*, 41, 545–553.

———. 1999. Assuaging nutritional complexity: A geometrical approach. *Proceedings of the Nutrition Society*, 58, 779–789.

Smith, A. C. 2000. Composition and proposed nutritional importance of exudates eaten by saddleback (*Saguinus fuscicollis*) and mustached (*Saguinus mystax*) tamarins. *International Journal of Primatology*, 21, 69–83.

Snaith, T. V. & Chapman, C. A. 2005. Towards an ecological solution to the folivore paradox: Patch depletion as an indicator of within-group scramble competition in red colobus. *Behavioural Ecology and Sociobiology*, 59, 185–190.

———. 2007. Primate group size and socioecological models: Do folivores really play by different rules? *Evolutionary Anthropology*, 16, 94–106.

———. 2008. Red colobus monkeys display alternative behavioural responses to the costs of scramble competition. *Behavioural Ecology*, 19, 1289–1296.

Stephens, D. W. & Krebs, J. R. 1986. *Foraging Theory*. Princeton, NJ: Princeton University Press.

Stephens, S. A., Salas, L. A. & Dierenfeld, E. S. 2006. Bark consumption by the painted ringtail (*Pseudochirulus forbesti larvatus*) in Papua New Guinea. *Biotropica*, 38, 617–624.

Sterck, E. H. M., Watts, D. P. & van Schaik, C. P. 1997. The evolution of female social relationships in nonhuman primates. *Behavioral Ecology and Sociobiology*, 41, 291–309.

Teelen, S. 2007. Primate abundance along five transect lines at Ngogo, Kibale National Park, Uganda. *American Journal of Primatology*, 69, 1–15.

Terborgh, J. 1986. Keystone plant resources in the tropical forest. In *Conservation Biology: The Science of Scarcity and Diversity* (ed. by Soule, M. E.), 330–344. Sunderland: Sinauer.

Thepenier, P., Jacquier, M. J., Massiot, G., Men-Olivier, L. L. & Delaude, C. 1990. Alkaloids from seeds of *Strychnos variabilis* and *S. longicaudata*. *Phytochemistry*, 29, 686–687.

Thomas, D. W., Samson, C. & Bergeron J, M. 1988. Metabolic costs associated with the ingestion of plant phenolics by *Microtus pennsylvanicus*. *Journal of Mammalogy*, 69, 512–515.

Thompson, S. D., Power, M. L., Rutledge, C. E. & Kleinman, D. G. 1994. Energy metabolism and thermoregulation in the golden lion tamarin (*Leontopithecus rosalia*). *Folia Primatologica*, 63, 121–143.

Tutin, C. E. G., Ham, R. M., White, L. J. T. & Harrison, M. J. S. 1997. The primate community of the Lope Reserve, Gabon: Diets, responses to fruit scarcity, and effects on biomass. *American Journal of Primatology*, 42, 1–24.

Twinomugisha, D., Chapman, C. A., Lawes, M. J., Worman, C. & Danish, L. 2006. How does the golden monkey of the Virungas cope in a fruit scarce environment? In *Primates of Western Uganda* (ed. by Newton-Fisher, N., Notman, H., Reynolds, V. & Patterson, J. D.), 45–60. New York: Springer.

Ushida, K., Fujita, S. & Ohashi, G. 2006. Nutritional significance of the selective ingestion of *Albizia zygia* gum exudate by wild chimpanzees in Bossou, Guinea. *American Journal of Primatology*, 68, 143–151.

Van Schaik, C. P., Terborgh, J. W. & Wright, S. J. 1993. The phenology of tropical forests: Adaptive significance and consequences for primary consumers. *Annual Review of Ecology and Systematics*, 24, 353–377.

Van Schaik, C. P. 1989. The ecology of social relationships amongst female primates. In *Comparative Socioecology: The Behavioural Ecology of Humans and Other Mammals* (ed. by Standen, V. & Foley, R. A.), 195–218. Boston: Blackwell Scientific Publications.

Wasserman, M. D. & Chapman, C. A. 2003. Determinants of colobine monkey abundance: The importance of food energy, protein and fibre content. *Journal of Animal Ecology*, 72, 650–659.

Waterman, P. G., Ross, J. A. M., Bennet, E. L. & Davies, A. G. 1988. A comparison of the floristics and leaf chemistry of the tree flora in two Malaysian rain forests and the influence of leaf chemistry on populations of colobine monkeys in the old world. *Biological Journal of the Linnean Society*, 34, 1–32.

Weins, F., Zitzmann, A. & Nor-Azman, H. 2006. Fast food for slow lorises: Is low metabolism related to secondary compounds in high energy diets? *Journal of Mammalogy*, 87, 790–798.

Work, T. H., Trapido, H., Murthy, D. P. N., Rao, R. L., Bhatt, R. N. & Kulkarni, K. G. 1957. Kyasanur forest disease. III. A preliminary report on the nature of the infection and clinical manifestations in human being. *Indian Journal of Medical Sciences*, 11, 619–645.

Worman, C. O. & Chapman, C. A. 2005. Seasonal variation in the quality of a tropical ripe fruit and the response of three frugivores. *Journal of Tropical Ecology*, 21, 689–697.

Wrangham, R. W. 1980. An ecological model of female-bonded primate groups. *Behaviour*, 75, 262–300.

Wrangham, R. W., Conklin-Brittain, N. L. & Hunt, K. 1998. Dietary response of chimpanzees and cercopithecines to seasonal variation in fruit abundance I. antifeedents. *International Journal of Primatology*, 19, 949–970.

Wrangham, R. W., Crofoot, M., Lundy, R. & Gilby, I. 2007. Use of overlap zones among group-living primates: A test of the risk hypothesis. *Behaviour*, 144, 1599–1619.

Wrangham, R. W. & Waterman, P. G. 1981. Feeding behaviour of vervet monkeys on *Acacia tortilis* and *Acacia xanthophloea* with special reference to reproductive strategies and tannin production. *Journal of Animal Ecology*, 50, 715–731.

Chapter 8 Predation
Claudia Fichtel

THE EXTREME fitness costs incurred by successful predation events can drive the evolution and maintenance of elaborate and costly antipredator behavior in prey species (Caro 2005). Because predation occurs quickly and is performed by animals that hunt by stealth, it is rarely observed and difficult to study. As a result, two decades ago we knew little about predation on primates. The absence of data led some to conclude that predation played only a negligible (Isbell 1994) or minimal role (Cheney & Wrangham 1987) in the evolution of primate social systems. Subsequent research has revealed, however, that predation acts as a strong selective force, with primates adapting behaviorally to this risk in many different ways (van Schaik 1983; Terborgh & Janson 1986; Dunbar 1988).

In general, predation imposes two costs on prey. These include the direct costs of mortality and the chronic indirect costs of employing behaviors that reduce mortality risk. To quantify these indirect costs, it is necessary to identify behaviors that affect fitness (Lima 1998), such as vigilance, predator-sensitive foraging, and sleeping site selection. However, predation acts not only on behavioral adaptations, but also on physiology and anatomy, as indicated by recent investigations of how predation affects hormones (Engh et al. 2006), the immune system (Semple et al. 2002), and brain size (Shultz & Dunbar 2006).

In this chapter I begin by describing the kinds of animals that prey on primates and how they do so. Second, I address the perceptual mechanisms primates use to recognize predators, and how they adapt to predation behaviorally. Because primates exhibit considerable interspecific variation in antipredator behavior that seems to be adapted to the different hunting styles of predators, I emphasize this aspect throughout the chapter. Behavioral adaptations can be broadly classified into tactics used to avoid encountering predators and those employed after predator detection (Stanford 2002). Steps taken before encountering predators include predator-sensitive foraging, appropriate sleeping site selection, increased vigilance, grouping, and the formation of mixed-species associations. Tactics that take effect during or after encounters with predators include the selfish herd, confusion and dilution effects, alarm calling, and various escape and defense tactics. Because alarm calling appears to be costly, I also discuss its benefits and evolution. I consider all of these aspects for nocturnal and diurnal species separately, and conclude with a discussion of the importance of predation as a selective force shaping the social system of primates.

Kinds of Primate Predators

Quantifying the magnitude of predation is necessary to assess the influence of predation on primate evolution (Terborgh & Janson 1986). Predation, however, occurs rarely, unpredictably, and often at night, and is therefore difficult to observe. In addition, predators are typically wary of human observers, whose presence deters their attacks. As a result, observations are frequently anecdotal, thus precluding a systematic evaluation of predation on primates. Almost all primates, ranging from the smallest (*Microcebus*) to the largest (*Gorilla*) species, have been observed to fall victim to predators, and their remains have been identified in predator scats, pellets, and nests (Hart 2000; table 8.1). Felids, canids, hyenids, additional carnivores, raptors, reptiles, and primates prey on primates. In Africa, felids and raptors account for most predation on primates. In Madagascar, raptors are the most successful predators, together with carnivores (fig. 8.1; Hart 2007). In Asia, primates frequently succumb to predation by felids, including leopards and tigers. In contrast, raptors account for most predation events in the neotropics, although felids and reptiles also figure prominently (Hart 2007). These patterns appear to be due to variation in the geographic distributions of predators and the size distributions of primate prey. For instance, Asia lacks large monkey-eating eagles outside the Philippines, thus accounting for the paucity of predation by raptors there; the relatively small size of primates in Madagascar and the neotropics may explain the major role of raptors in these regions.

Some nonhuman primates also prey on other primates (fig. 8.2). These include chimpanzees (*Pan troglodytes*), orangutans (*Pongo* spp.), baboons (*Papio* spp.), blue monkeys (*Cercopithecus mitis*), capuchin moneys (*Cebus* spp.), red lemurs (*Eulemur rufus*) and Coquerel's dwarf lemurs (*Mirza coquereli)* (Hart 2000; table 8.1). Chimpanzees at Fongoli in Senegal have recently been observed to use tools to kill northern lesser galagos (*Galago senegalensis*) as they sleep or hide in cavities of branches and tree trunks (Pruetz & Bertolani 2007). Because primates do not rely on vertebrate prey as a major source of food (chapter 7, this volume), they are not likely to have an important impact on primate populations. One exception involves chimpanzees who prey on red colobus monkeys (*Procolobus* spp.) (Busse 1977; Nishida 1983; Boesch & Boesch 1989; Wrangham & Bergmann-Riss 1990; Stanford et al. 1994; Uehara 1997). Watts & Mitani (2002) reported that chimpanzees at Ngogo in Kibale National Park, Uganda, killed 6% to 12% of the red colobus there annually (see figs. 4.5 and 6.4, this volume). More recent observations suggest that this high rate of chimpanzee predation is not sustainable (Teelen 2008).

Humans hunt primates for subsistence and medicine, and to trade as pets and bushmeat. Human hunting has a considerable impact not only on primates but also on many other animals all over the world (Fa et al. 2002; Milner-Gulland et al. 2003). Hunting wildlife for bushmeat is currently a major contributor to the loss of biodiversity in some regions such as Africa (Wilkie & Carpenter 1999; Fa et al. 2003), particularly during periods of political instability (Barrett & Ratsimbazafy 2009). Because hunting and butchering of wild animals can lead to the cross-transmission of diseases between primates and humans, pathogens may represent an additional driver of population declines (Peeters et al. 2002; LeBreton et al. 2006; Leendertz et al. 2006; Nunn & Alitzer

2006). Successful solutions to the bushmeat crisis will be difficult; they must include multidisciplinary approaches, integrating the conservation of natural resources into development agendas at the local, national, and international levels (Milner-Gulland et al. 2003).

Hunting Tactics

To understand the antipredator behavior of primates, it is important to take the various hunting strategies of predators into account. Primates are exposed to predators that hunt in different ways. Birds of prey use two distinct hunting techniques. While some actively search and hunt on the wing, others are sit-and-wait predators (Jaksić & Carothers 1985). The latter are ambush hunters, who rely on an element of surprise to capture their prey. After detecting prey, avian predators swoop in, sometimes pursuing their prey (Jaksić & Carothers 1985). Because most raptors rely on visual or acoustic cues to detect their prey, moving animals are perceived more readily than stationary ones (Rice 1983). Nocturnal birds of prey such as the barn owl (*Tyto alba*), however, use acoustic cues alone to locate and capture their prey in total darkness (Payne 1971). Some raptors, including the Madagascar harrier hawk (*Polyboroides radiatus*), have evolved specialized morphological adaptations that allow them to bend their legs to probe into narrow tree cavities and extract nocturnal lemurs (fig. 8.3; Goodman et al. 1993b).

Among terrestrial predators, many felids, small carnivores, and snakes stalk their prey or are sit-and-wait predators. Because such predators employ stealth to hunt successfully, prey that detect them before they attack typically evade capture (Schaller 1968, 1972; Slip & Shine 1988; Ayers & Shine 1997; Clark 2005). Large canids and some hyenids are pursuit predators, who hunt cooperatively. These predators use hunting tactics that include trailing prey, running them down, and attacking from the rear (Estes & Goddard 1967). Finally, some small carnivores, such as mongooses, hunt opportunistically and rapidly (Waser 1980). In summary, primates are exposed to many different types of predators who do not hunt them uniformly but can impact their survival. As a consequence, primates must adapt to the risk posed by predators appropriately and flexibly.

Predator Recognition

Which Cues Do Primates Use to Recognize Predators?

Many studies have investigated how primates respond to predators (see below), but few have focused on the cues they use to recognize predators. In this context it is important

Table 8.1. Predators of primates. Data were compiled by a literature search in PrimateLit (http://primatelit.library.wisc.edu/) with the keyword "predation." Summarized were observed events and cases in which remains of primates were identified in predator scats and pellets.

Suborder	Genus	Birds	Carnivores	Reptiles	Primates
Strepsirrhini	Cheirogaleus	Buteo brachypterus [1, 7, 63] Polyboroides radiatus [3] Asio madagascariensis [5]	Galidea elegans [1, 2, 63] Cryptoprocta ferox [4, 5, 6] Mungotictis decemlineata [6]	Sanzinia madagascariensis [2, 63]	
	Microcebus	Asio madagascariensis [5, 1, 8, 15, 16] Tyto alba [1, 5, 11] Polyboroides radiatus [9, 1, 12, 13] Accipiter henstii [10] Tyto soumagnei [14] Vanga curvirostris [1]	Cryptoprocta ferox [5, 17] Mungotictis decemlineata [1, 6, 18] Galidia elegans [19, 63, 97] Canis lupus familiaris [1]	Sanzinia madagascariensis [6] Ithycyphis miniatus [20]	Mirza coquereli [21]
	Mirza	Buteo brachypterus [1, 7] Asio madagascariensis [5]	Cryptoprocta ferox [5, 22] Mungotictis decemlineata [6, 22]		
	Phaner	Buteo brachypterus [1, 7] Aviceda madagascariensis [23]	Cryptoprocta ferox [5, 6]		
	Daubentonia		Cryptoprocta ferox [19]		
	Avahi	Accipiter henstii [1, 12, 36] Eutriorchis astur [12] Asio madagascariensis [1] Buteo brachypterus [63]			
	Propithecus	Polyboroides radiatus [1, 9, 26] Buteo brachypterus [1]	Cryptoprocta ferox [1, 5, 12, 19, 20, 24, 25, 93, 96, 98]	Acrantophis madagascariensis [64]	
	Eulemur	Accipiter henstii [1, 27]	Cryptoprocta ferox [1, 5, 12, 24]	Crocodylus niloticus [28] Acrantophis madagascariensis [1]	
	Hapalemur	Asio madagascariensis [1]	Cryptoprocta ferox [12, 16, 24, 29]	Sanzinia madagascariensis [1, 30]	
	Lemur	Polyboroides radiatus [1, 6] Accipiter francesii [31]	Viverricula indicia [1] Felis silvestris [1, 32]		Eulemur fulvus [33, 103]
	Varecia		Cryptoprocta ferox [34]		
	Lepilemur	Buteo brachypterus [35] Tyto alba [1] Accipiter henstii [36] Asio madagascariensis [5] Polyboroides radiatus [1, 13, 37]	Cryptoprocta ferox [4, 5] Mungotictis decemlineata [6]	Acrantophis sp. [105]	
	Galago		Genetta tigrina [61]	Bitis gabonica [67]	Papio cynocephalus [61] Pan paniscus [91] Pan troglodytes [99]
	Galagoides				
	Perodicticus	Stephanoaetus coronatus [86, 90]	Nandinia binotata [60]	Dendroaspis jamesonii [47]	Pan troglodytes [90]
Haplorrhini Platyrrhini	Tarsius		Felis catus [55]	Python reticulatus [81]	
	Alouatta	Harpia harpyja [69] Caracara plancus [71]	Leopardus pardalis [76, 102] Panthera onca [78] Puma concolor [94]		
	Ateles		Panthera onca [83] Felis concolor [83]		
	Brachyteles		Leopardus pardalis [76]		
	Callithrix	Athene cunicularia [59]	Eira barbara [92]	Bothrops jararaca [52]	
	Saguinus	Micrastur ruficollis [45] Spizaetus ornatus [56] Accipiter bicolor [56] Morphnus guianensis [65]	Eira barbara [59]	Eunectes murinus [51] Boa constrictor [82]	
	Cebus	Harpia harpyja [56]	Leopardus pardalis [76] Canis lupus familiaris [85]	Boa constrictor [50]	
	Saimiri	Leucopternis albicollis [43] Daptrius americanus [43] Micrastur semitorquatus [43] Spizaetus ornatus [43, 56] Ramphastos swainsonii [43] Buteo nitidus [43] Buteo magnirostris [43]			
	Callicebus	Harpia harpyja [67]	Eira barbara [67]	Boa constrictor [72]	Cebus apella [66]

(continued)

Table 8.1. (continued)

Suborder	Genus	Birds	Carnivores	Reptiles	Primates
Catarrhini	Chiropotes	Harpia harpyja[73]		Boa constrictor[74]	
	Pithecia	Morphnus guianensis[70]			
	Cercocebus	Stephanoaetus coronatus[86, 90]	Panthera pardus[90]		Pan troglodytes[90]
	Cercopithecus	Stephanoaetus coronatus[40, 86, 88, 89, 90]	Panthera pardus[61, 90]		Pan paniscus[91, L]
	Chlorocebus	Stephanoaetus coronatus[46, 57]	Panthera pardus[41, 57]		Papio cynocephalus[42,61]
		Polemaetus bellicosus[48]			Pan troglodytes[99]
		Aquila verreauxii[44]			
					Pan troglodytes[101]
	Erythrocebus				Pan paniscus[75, 104]
	Lophocebus	Stephanoaetus coronatus[88, 89]			Pan paniscus[91]
	Macaca	Pithecophaga jefferyi[53]	Canis lupus familiaris[58]	Tomistoma schlegeli[54]	
	Papio	Stephanoaetus coronatus[88]	Panthera pardus[38, 61]		Pan troglodytes[100]
			Panthera leo[38, 61, 80]		
	Colobus	Stephanoaetus coronatus[86, 88, 89, 90]	Panthera pardus[90]	Tomistoma schlegeli[48, 49]	Pan troglodytes[90]
	Nasalis		Neofelis diardi[84]		
	Presbytis		Canis aureus[77]		
	Procolobus	Stephanoaetus coronatus[86, 88, 89, 90]	Panthera padus[90]		Pan troglodytes[87, 90]
					Pan paniscus[104]
	Semnopithecus		Canis lupus familiaris[58]		
	Trachypithecus	Bubo nipalensis[58]			
	Pan		Panthera pardus[68, 90]		
			Panthera leo[79]		

Sources: [1]Goodman et al. 1993a; [2]Wright & Martin 1995; [3]Gilbert & Tingay 2001; [4]Goodman et al. 1997; [5]Rasoloarison et al. 1995; [6]Goodman 2003; [7]Goodman & Langrand 1996; [8]Goodman et al. 1993b; [9]Karpanty & Goodman 1999; [10]Rand 1935; [11]Goodman et al. 1993c; [12]Wright 1998; [13]Thorstrom & La Marca 2000; [14]Goodman & Thorstrom 1998; [15]Goodman et al. 1993b; [16]Sterling & McFadden 2000; [17]Russel 1977; [18]Albignac 1976; [19]Petter et al. 1977; [20]Richard 1978; [21]Schliehe-Diecks et al. 2010; [22]Kappeler 1997; [23]Charles-Dominique & Petter 1980; [24]Wright et al. 1997; [25]Richard et al. 1991; [26]Rasoanindrainy 1985; [27]Schwab 1999; [28]Wilson et al. 1989; [29]Goodman & Pidgeon 1999; [30]Rakotondravony et al. 1998b; [31]Koyama 1992; [32]Ratsirarson et al. 2001; [33]Pitts 1995; [34]Britt et al. 2000; [35]Ratsirarson & Emady 1999; [36]Goodman et al. 1998; [37]Schülke & Ostner 2001; [38]Busse 1980; [40]Brown 1971; [42]Hausfater 1976; [43]Boinski 1987; [44]Gargett 1971; [45]Izawa 1978; [46]Struhsaker 1967b; [47]Struhsaker 1970; [48]Galdikas 1985; [49]Yeager 1991; [50]Chapman 1986; [51]Heymann 1987; [52]Corrêa & Coutinho 1997; [53]Kennedy 1977; [54]Galdikas & Yeager 1984; [55]MacKinnon & MacKinnon 1980; [56]Terborgh 1983; [57]Cheney et al. 1988; [58]Dittus 1975; [59]Stafford & Ferreira 1995; [60]Charles-Dominique 1977; [61]Kingdon 1974; [62]Jones 1969; [63]Wright et al. 1998; [64]Burney 2002; [65]Oversluijs & Heymann 2001; [66]Sampaio & Ferrari 2005; [67]de Luna et al. 2010; [68]D'Amour et al. 2006; [69]Peres 1990; [70]Gilbert 2000; [71]McKinney 2009; [72]Cisneros-Heredia et al. 2005; [73]Martins et al. 2005; [74]Ferrari et al. 2004; [75]Surbeck et al. 2009; [76]Bianchi & Mendes 2007; [77]Stanford 1989; [78]Peetz et al. 1992; [79]Tsukahara 1993; [80]Condit & Smith 1994; [81]Gursky 2002; [82]Tello et al. 2002; [83]Matsuda & Izawa 2008; [84]Matsuda et al. 2008; [85]Oliveira et al. 2008; [86]McGraw et al. 2006; [87]Teelen 2008; [88]Mitani et al. 2001; [89]Struhsaker & Leakey 1990; [90]Shultz et al. 2004; [91]Hofreiter et al. 2010; [92]Bezerra et al. 2009; [93]Irwin et al. 2009; [94]Ludwig et al. 2007; [95]Génin 2010; [96]Lührs & Dammhahn 2010; [97]Deppe et al. 2008; [98]Patel 2005; [99]Gaspersic & Pruetz 2004; [100]Nishie 2004; [101]Pruetz & Marshack 2009; [102]Miranda et al. 2005; [103]Jolly et al. 2000; [104]Surbeck & Hohmann 2008; [105]Hilgartner et al. 2008.

Fig. 8.1. The carcass of an adult female sifaka (*Propithecus verreauxi*) (A) that has been preyed upon by a fossa (*Cryptoprocta ferox*) (B) in Kirindy, western Madagascar. The fossa, Madagascar's largest carnivore, which weighs up to 12 kilograms, is a major predator on lemurs. Photos courtesy of Claudia Fichtel.

Fig. 8.2. An adult male chacma baboon of the Okavango Delta, Botswana, feeds on a vervet monkey it has killed minutes earlier. Photo courtesy of Ryne A. Palombit.

Fig. 8.3. Even when resting in a tree hole during the day (A), red-tailed sportive lemurs (*Lepilemur ruficaudatus*) at Kirindy, western Madagascar, are vulnerable to predation by raptors such as the harrier hawk (*Polyboroides radiatus*) (B). Photos courtesy of Claudia Fichtel and Anna Viktoria Schnöll.

to consider perceptual modalities separately. Primates use pelage and skin patterns to recognize carnivores and snakes. For example, bonnet macaques (*Macaca radiata*) detected spotted leopard models earlier than dark brown control models, and responded more strongly with antipredator behavior to the former than to the latter (Coss & Ramakrishnan 2000; Coss et al. 2005). The characteristic pelage of tigers also seems to be a salient feature that Thomas's leaf monkeys (*Presbytis thomasi*) and Sumatran orangutans (*Pongo abelii*, Wich & Sterck 2003; Wich & van Schaik pers. comm.) use to recognize these predators. Similarly, bonnet macaques identify predatory snakes by their skin patterns (Ramakrishnan et al. 2005). Ring-tailed lemurs (*Lemur catta*) perceived hawk-shaped silhouettes as more threatening than goose-shaped silhouettes, suggesting that a diamond shape with its relatively short and pointed leading edge was a salient predator recognition feature (Macedonia & Polak 1989).

Several primates identify predators by acoustic cues alone. In areas where primates are regularly preyed upon by several predators, calls of raptors, carnivores, and other predators elicit antipredator responses from both diurnal and nocturnal primates (Hauser & Wrangham 1990; Zuberbühler et al. 1999a; Zuberbühler 2000a,b,d; Karpanty

& Grella 2001; Fichtel & Kappeler 2002; Arnold & Zuberbühler 2006; Fichtel 2007). However, primates that live in habitats where they are not often exposed to predators respond less strongly or not at all to such calls, suggesting that learning plays a role in the recognition of predator vocalizations (Macedonia & Yount 1991; Fichtel & van Schaik 2006). The role that acoustic cues play in predator avoidance is also demonstrated by the fact that many primates respond in adaptive ways to alarm calls emitted by other species (table 8.2). Because exposure to calling heterospecifics seems to influence responses (Hauser 1988b) and because primates respond only to alarm calls of sympatric species (Fichtel 2004), the recognition of heterospecific alarm calls also seems to be based on experience and learning.

What Do Young and Predator-naïve Primates Know about Predators?

Visual predator recognition

Visual predator recognition has been studied by presenting predator-naive individuals with stuffed models, silhouettes of predators, or live predators. Most primates respond with antipredator behavior to raptor, carnivore, and snake models (table 8.3). Captive-born rhesus monkeys (*Macaca mu-*

Table 8.2.   Recognition of heterospecific alarm calls by primates

Species that displays recognition	Species recognized	Type of heterospecific alarm call(s)	Reference
Ringtailed lemur (*Lemur catta*)	Verreaux's sifaka (*Propithecus verreauxi*)	Aerial and terrestrial predator alarm calls	1
Verreaux's sifaka (*Propithecus verreauxi*)	Redfronted lemur (*Eulemur rufifrons*)	Aerial predator and general alarm calls	2
Redfronted lemur (*Eulemur rufifrons*)	Verreaux's sifaka (*Propithecus verreauxi*)	Aerial predator and general alarm calls	2
Saddleback tamarin (*Saguinus fuscicollis*)	Moustached tamarin (*Saguinus mystax*)	Aerial and terrestrial predator alarm calls	3
Moustached tamarin (*Saguinus mystax*)	Saddleback tamarin (*Saguinus fuscicollis*)	Aerial predator and general alarm calls	3
Bonnet macaque (*Macaca radiata*)	Nilgril langur (*Semnopithecus johnii*)	Terrestrial predator alarm call	4
	Hanuman langur (*Semnopithecus entellus*)	Terrestrial predator alarm call	
	Samba deer (*Cervus unicolor*)	Terrestrial predator alarm call	
Diana monkey (*Cercopithecus diana*)	Chimpanzee (*Pan troglodytes*)	Alarm screams given in response to leopards	5
	Campbell's mona monkey (*Cercopithecus campbelli*)	Aerial predator and general alarm calls	6
	Crested guinea fowl (*Guttera pucherani*)	Terrestrial predator alarm call	7
Vervet monkey (*Chlorocebus pygerythrus*)	Superb starling (*Sprea superbus*)	Terrestrial predator alarm call	8

Sources: [1] Oda 1998; [2] Fichtel 2004; [3] Kirchhof & Hammerschmidt 2006; [4] Ramakrishnan & Coss 2000; [5, 6] Zuberbühler 2000b, c; [7] Rainey & Zuberbühler 2004; [8] Cheney & Seyfarth 1985a.

Table 8.3. Responses of captive or predator-naïve individuals to raptor, carnivore or snake models, or live snakes

Species	Raptor model	Carnivore model	Snake model/live snake	Reference
Black tufted-eared marmoset (*Callithrix penicillata*)	+	+	+	Barros et al. 2002
Cotton-top tamarin (*Saguinus oedipus*)	n.a.	n.a.	—	Hayes & Snowdon 1990
Brown capuchin (*Cebus apella*)	n.a.	n.a.	(Discrimination between live snake and live rat) +	Vitale et al. 1991
Vervet monkey (*Chlorocebus pygerythrus*)	—	+	+	Brown et al. 1992
Long-tailed macaque (*Macaca fascicularis*)	n.a.	n.a.	+	Vitale et al. 1991
Rhesus macaque (*Macaca mulatta*)	n.a.	n.a.	—	Mineka et al. 1984
Sumatran orangutan (*Pongo abelii*)	n.a.	n.a.	+	Wich & van Schaik pers. comm.

+ Individuals showed antipredator behavior.
— Individuals did not show antipredator behavior.
n.a. Type of predator model was not presented.

latta), however, do not respond with antipredator behavior to snakes. Instead, monkeys learn to fear snakes after observing wild-caught macaques reacting fearfully to them (Mineka et al. 1984; but see Campbell & Snowdon 2009). Because they could not be conditioned to display fear responses to other novel objects (Cook & Mineka 1989), macaques appear predisposed to learn only those stimuli that pose an actual threat to them in their natural environment. Moreover, Japanese macaques (*Macaca fuscata*) that are naive to snakes responded faster to pictures of them than to pictures of flowers. This suggests that they may possess an enhanced ability to visually detect evolutionarily relevant cues (Shibasaki & Kawai 2009), which in turn might facilitate predator recognition via observational learning.

Studies of wild primates indicate that young animals rely on a combination of innate mechanisms and experience for visual predator recognition. For example, young vervet monkeys (*Chlorocebus pygerythrus*) produce aerial alarm calls in response to falling leaves during the first months of life, but also to any kind of bird silhouette passing by. Terrestrial predator alarm calls are given in response to a variety of terrestrial mammals, whereas snake alarm calls are given to any snakelike object. Moreover, young vervets do not produce aerial alarm calls in response to terrestrial predators, or vice versa. These findings suggest that vervets learn to differentiate between threatening and nonthreatening species, but that they also possess an innate ability to recognize members of different types of predators (Seyfarth & Cheney 1980, 1986). In sum, these studies suggest that primates may possess a general predator template that must be reinforced by observing the responses of experienced individuals.

Acoustic predator recognition

In contrast to the findings derived from studies of visual predator recognition, primates appear to learn the vocalizations of their predators. For example, pig-tailed snub-nosed monkeys (*Simias concolor*) on the Mentawai Islands have been isolated from felids for more than half a million years, and are currently hunted heavily by humans. In this situation they discriminate between the calls of former predators (felids), current predators (humans), unfamiliar species (elephants), and familiar species (birds and pigs); they responded most strongly to humans and less so to felids and elephants (Yorzinski & Ziegler 2007). Captive-born cotton-top tamarins (*Saguinus oedipus*) do not differentiate between calls produced by predators (hawks and carnivores) and nonpredators (songbirds and other primates, Friant et al. 2008). In contrast, captive-born Geoffroy's tufted-ear marmosets (*Callithrix geoffroyi*) that had access to outdoor enclosures and the opportunity to form associations between red-tailed hawks (*Buteo jamaicensis*) and their calls, responded with antiraptor behavior to playbacks of hawk calls (Searcy & Caine 2003). Similarly, mantled howler monkeys (*Alouatta palliata*) on an island where harpy eagles (*Harpia harpyja*) had gone extinct more than 50 years earlier rapidly acquired alarm responses to calls of newly introduced harpy eagles (Gil-da-Costa et al. 2003), suggesting a link between experience and predator recognition.

Primates also appear to learn how to respond to different alarm calls produced by conspecifics. For example, three- to four-month-old vervet monkeys run to their mothers upon hearing playbacks of conspecific alarm calls. By six months of age, monkeys discriminate calls given to different predators, displaying adult-like response behavior (Seyfarth

& Cheney 1980, 1986). Chacma baboons (*Papio ursinus*), common squirrel monkeys (*Saimiri sciureus*), and Verreaux's sifakas (*Propithecus verreauxi*) responded appropriately to their own alarm calls and those given by members of other species at the same age (Fischer et al. 2000; McCowan & Newman 2000; Fichtel 2008), despite differences among species in physical rates of development. Interestingly, infants in all species showed adult-like escape responses by the time they had become physically independent from their mothers. Hence, the time span in which infants start to move around independently seems to represent an important developmental milestone in the development of alarm call recognition (Fichtel 2008). In summary, these studies suggest that learning plays a role in the development of the recognition of alarm calls.

Olfactory predator recognition

Primates also use their sense of smell to recognize predators. Studies of laboratory-reared cotton-top and red-bellied tamarins (*Saguinus labiatus*) as well as gray mouse lemurs (*Microcebus murinus*) revealed that these animals possess an innate ability to recognize the scents of predators from their scat (Caine & Weldon 1989; Buchanan-Smith et al. 1993; Sündermann et al. 2008). As suggested for mammals in general (reviewed in Apfelbach et al. 2005), primates may recognize predators through their urine and feces via metabolites produced by the digestion of meat. The production of a consistent olfactory cue may lead to selection favoring the ability of prey species to recognize predators through such cues. Because seeing or hearing a predator also provides additional cues that allow primates to classify an animal as a predator, smell alone rarely does so. Hence, recognition of predators by smell should be independent of any learning.

In conclusion, primates use different sensory cues to recognize predators, and this recognition appears to be mediated by different mechanisms that vary as a function of sensory modality. Most studies indicate that primates are genetically predisposed to recognize visual cues of predators; this predisposition must be reinforced by observing experienced individuals responding to relevant cues appropriately. In contrast, primates appear to learn the calls produced by predators. Because the calls of predators are typically more variable in their acoustic structure than in their visual gestalt, a learned recognition mechanism might be more flexible, and thus adaptive, because it allows primates to develop antipredator responses only for those predators that do in fact pose a threat. Finally, primates may possess a genetic predisposition to recognize the scents of predators in their scat.

While the ability to recognize predators has obvious and important fitness consequences for the animals themselves, knowledge regarding how primates do so takes on practical significance as more captive primates are reintroduced into the wild for the purpose of conservation. The success of such programs depends on animals possessing the ability to recognize naturally occurring predators, and efficient "predator-recognition-training" paradigms can be developed only with knowledge about how primate prey acquire their predator recognition skills.

Behavioral Tactics to Avoid Predation

The shift from an originally solitary lifestyle to a gregarious one has been considered to be one of the major evolutionary transitions (Maynard Smith & Szathmáry 1995). The benefits of grouping have been clearly shown in gray-cheeked mangabeys (*Lophocebus albigena*) and yellow baboons (*Papio cynocephalus*), where solitary males are more likely to be preyed upon than males in groups (Alberts & Altmann 1995; Olupot & Waser 2001). Whether alone or in groups, the antipredator behavior of individuals can be divided into tactics employed before and after encountering predators. Tactics employed before encountering predators include predator-sensitive foraging, the choosing of safe sleeping sites, vigilance, and the forming of mixed-species associations. Tactics employed during or after encountering predators include selfish herding, confusion and dilution effects, the emitting of alarm calls, fleeing, and the confronting and mobbing of predators. Primates may employ different tactics against different predators, and their success may vary as a function of predator types and hunting styles.

Tactics Employed before Encountering Predators

Predator-sensitive foraging

Because predators move unpredictably in space and time (Sih 1997), animals face a trade-off between finding food and avoiding becoming food for others (Lima & Dill 1990). Predator- or threat-sensitive foraging has been documented in many primates (reviewed in Miller 2002). Several variables, such as age, the presence of infants, and habitat use may affect an individual's vulnerability to predation. These variables give rise to a perceived level of risk that should in turn influence an individual's foraging decisions. Because juvenile primates are assumed to be at greater risk of predation than adults, this difference may result in age-related antipredator behavior (Janson & van Schaik 1993; Struhsaker & Leakey 1990; Stanford et al. 1994; Watts & Mitani 2002).

Because of their relatively long life span (chapter 10, this volume), juvenile primates are expected to take fewer risks than adults, to maximize their chances of surviving to adulthood. Juvenile behavior thus reflects choices that enhance safety from predators. For example, juvenile brown capuchin monkeys (*Cebus apella*) forage more often in the center of the group than do adults (Janson 1990). Wedge-capped capuchins (*Cebus olivaceus*) juveniles avoid open habitats more often than do adults (Fragaszy 1990), and juvenile long-tailed macaques (*Macaca fascicularis*) and squirrel monkeys remain in cohesive subgroups more than do adults (van Noordwijk et al. 1993; Stone 2007). Juvenile squirrel monkeys take predation risk and food availability into account by feeding less often in risky places when resources are abundant (Stone 2007). Moreover, females with infants avoid potentially dangerous situations that might expose them to predation, such as foraging on the ground (Hauser 1988a; Sauther 2002; Treves et al. 2003).

Animals use their habitats strategically as another means to avoid predation. Assuming that predation risk declines with distance from the ground, capuchin monkeys, red colobus monkeys and long-tailed macaques adjust their behavior to changing predation risk by feeding higher in trees when in smaller parties (van Schaik & van Noordwijk 1983, 1989; Stanford 1995; Cowlishaw 1997a). However, recent findings suggest an alternative explanation for this observation. Because highest densities of fruit pulps are in the upper canopy, smaller parties are more likely to feed higher in the tree, since large parties may not fit in the optimal feeding zone (Houle et al. 2010). Females in large groups of wedge-capped capuchins spend more time feeding on the ground and ingest more food than females in small groups (Miller 2002). Similarly, females in small groups of ring-tailed lemurs avoid feeding on the ground when predation pressure is high (Sauther 2002). Patterns of habitat use by chacma baboons appear to reflect the trade-off between foraging demands and the constant risk of predation. Animals spend considerable time feeding and traveling in relatively safe habitats, and spend the most time resting and grooming in the safest habitats (Cowlishaw 1997b, c). Vervet monkeys also adjust their ranging behavior according to perceived predation risk in different habitats (Enstam Jaffe & Isbell 2009).

In nocturnal primates, the lunar phase influences activity patterns, perhaps because it affects their perceived predation risk. Cathemeral brown (*Eulemur fulvus*), black (*Eulemur macaco*) and mongoose (*Eulemur mongoz*) lemurs, as well as nocturnal white-footed sportive lemurs (*Lepilemur leucopus*) and Colombian owl monkeys (*Aotus lemurinus*), are relatively active during periods of waxing moon (Colquhoun 1998; Curtis et al. 1999; Kappeler & Erkert 2003; Fernandez-Duque & Erkert 2006; Nash 2007). Slender lorises (*Loris lydekkerianus*) increased their foraging activities during moonlit nights, and southern lesser galagos (*Galago moholi*) avoided traveling on the ground during nights without moonlight, presumably because small carnivores represent a significant risk to them (Bearder et al. 2002).

Selecting safe sleeping sites

Individual primates spend more than half of their lives sleeping, and deciding where to sleep therefore represents a significant problem. Sleeping sites must be carefully chosen to avoid predators and to deal with other travails of life, including thermoregulation, parasites, feeding, and territory defense (Anderson 1998). Primates exhibit considerable interspecific variation with regard to sleeping site preferences. Most primates use branches, cliffs, and other natural structures, and some strepsirrhines and apes build nests. Some strepsirrhines and New World monkeys also use tree holes or other cavities (reviewed in Kappeler 1998). Physical characteristics of sleeping sites, their patterns of use, the timing of retirement, and preretirement behavior could all influence predation risk. By choosing sleeping sites with certain characteristics, primates may reduce the likelihood of their being attacked. For example, sleeping sites preferred by baboons (cliff faces and emergent trees) provide better protection against predators than nonpreferred sites (closed canopy without emergent trees and open woodland; Hamilton 1982). Diurnal, group-living primates usually sleep on branches or in tree hollows. They prefer to sleep in tall, emergent trees that lie along water and possess large trunks and crowns with many horizontal branches and forks. These features make it difficult for terrestrial predators to climb up into the trees (Hamilton 1982; Anderson & McGrew 1984; Di Bitetti et al. 2000; Ramakrishnan & Coss 2001a, b; Hankerson et al. 2007; Matsuda et al. 2008b). Infants may also influence the choice of sleeping sites, as suggested by observations of lar gibbon (*Hylobates lar*) mothers with infants, who select relatively high places to sleep (Reichard 1998).

Several primates also construct nests to sleep, with nest-building behavior having evolved independently six times in the order (Kappeler 1998). In diurnal primates, only apes construct nests. Nest site choice is associated with the availability of comfortable substrates in chimpanzees (Stewart et al. 2007; Stanford & O'Malley 2008), but also predator avoidance in chimpanzees and orangutans (Baldwin et al. 1981; Sugardjito 1983; Prasetyo et al. 2009). Gorillas (*Gorilla gorilla*) typically construct their nests on the ground. Nevertheless, juveniles and even adults occasionally sleep above the ground, making them less vulnerable to predation (Yamagiwa 2001).

Behavioral adaptations such as rapid movement toward sleeping sites, cryptic behavior near them, and increased vigilance when departing the shelters may reduce the vulnerability of individuals to predation (Caine 1987; Heymann 1995; Reichard 1998; Hankerson & Caine 2004; Franklin et al. 2007; Smith et al. 2007; Qihai et al. 2009). In general, primates sleep in a restricted array of places (Altmann 1974) and avoid fixed sleeping sites that can be easily located by predators. They switch areas to sleep, and avoid sleeping in the same spot on several consecutive nights, possibly to avoid parasites (Reichard 1998; Di Bitetti et al. 2000; Nunn & Heymann 2005; Hankerson et al. 2007; Smith et al. 2007).

Nocturnal primates usually sleep and rest in tree holes or nests in dense vegetation. These may serve to protect them from diurnal predators (Goodman et al. 1993b; Schülke & Ostner 2001), but they may also perform other functions, including thermoregulation, especially for young infants (Martin 1972; Schmid & Kappeler 1998; Kappeler 1998; Dausmann et al. 2004). Some nocturnal primates simply sleep in dense vegetation, an observation consistent with the hypothesis that the use of tree holes and nests performs functions other than to mitigate predation (Kappeler 1998; Bearder et al. 2003; Dammhahn & Kappeler 2005). Nevertheless, it has been suggested that sleeping holes located higher up in a tree with smaller entrances and thicker walls may provide better protection against predation (Fietz 1998; Radespiel et al. 1998). Nocturnal primates use their sleeping sites in ways similar to those of diurnal species, and they switch between preferred areas frequently (Wright 1994; Schmid & Kappeler 1998; Radespiel et al. 1998, 2003; Schülke & Kappeler 2003; Zinner et al. 2003; Rasoloharijaona et al. 2008; Dammhahn & Kappeler 2005; Eberle & Kappeler 2006; Génin 2010).

Reduced activity has also been suggested as another behavioral strategy for avoiding predators (Caro 2005). For example, gray mouse lemur females are inactive and fall into torpor during the dry season, while males remain active. These sex differences in activity have been hypothesized as a strategy to protect females against predation during periods of low food availability and high predation risk (Rasoloarison et al. 1995; Schmid & Kappeler 1998; Schmid 1999). A 10-year capture-mark-recapture study revealed no direct survival benefits of hibernation, however, with females and males surviving equally well during the dry season (Kraus et al. 2008). In summary, tree holes and nests likely serve multiple functions, one of them being protection against predators. Few data exist about nocturnal primate cryptic behavior, vigilance, and movements around sleeping sites prior to retiring, and whether variation in these behaviors reduces predation risk.

Vigilance

Early detection of approaching predators substantially enhances the survival chances of prey because it gives them time to observe the predator at a distance, to hide, to alert other conspecifics, or, in the case of sit-and-wait predators, to deter attacks. Antipredator vigilance has been referred to as scanning for predators, sometimes precluding other activities such as foraging. In many birds and mammals a decrease in individual vigilance with increasing group size has been noted (de Ruiter 1986; Elgar 1989; Lima 1995; Caro 2005). As a result, individuals enjoy increased safety in larger groups as well as more time for other fitness-related activities. However, in primates and some other taxa (reviewed in Caro 2005) this relationship has not been found (Cords 1990; Rose & Fedigan 1995; Treves 1998; Cowlishaw 1998; Treves et al. 2001). To account for this disparity, it has been suggested that the number and proximity of nearest neighbors may be better predictors of individual vigilance than total group size (reviewed in Treves 2000; Caro 2005), because with restricted visibility, the effectiveness of vigilance may decline with distance (Janson 1998). Accordingly, individual levels of vigilance decrease when party size or the number of nearby individuals increase (van Schaik & van Noordwijk 1983, 1989; Rose & Fedigan 1995; Treves 1998, 1999; Steenbeek et al. 1999; Treves et al. 2001; but see Hirsch 2002; Kutsukake 2006). Because primates also monitor conspecifics regularly (social vigilance), it is often difficult to discriminate between the social and antipredatory function of vigilance. Moreover, individuals may also gain information about the presence of a predator through social vigilance—that is, by monitoring the gaze direction or body posture of group members (Treves 2000; Partan & Marler 2002; Fichtel 2004)—thus making it even more difficult to clearly separate both functions.

Many studies clearly show that individuals increase their levels of vigilance after detecting predators or predator models, or after hearing predator calls or alarm calls (van Schaik & van Noordwijk 1989; Baldellou & Henzi 1992; König 1998; Hirsch 2002; Fichtel 2004, 2007). Several ecological and social factors, such as characteristics of the habitat and the degree of canopy cover, as well as an individual's activity, sex, age, dominance rank, spatial position, or presence of infants, can influence individual levels of vigilance. Many studies report that habitat openness, canopy cover, height in trees, terrestriality, and travel position (the center or periphery of a group) influence vigilance behavior (Rose & Fedigan 1995; van Schaik & van Noordwijk 1989; Janson 1990; Cowlishaw 1998; Steenbeck et al. 1999; Hirsch 2002; Kutsukake 2006; Stojan-Dolar & Heymann 2010a, b). Nonetheless, activity has the strongest influence on vigilance, with some individuals becoming less vigilant while feeding and allogrooming (van Schaik & van Noordwijk 1989; Baldellou & Henzi 1992; Cords 1995; Rose & Fedigan 1995; Hirsch 2002; Kutsukake 2006; Stojan-Dolar & Heymann 2010a, b). In some primates sex and rank affect vigilance, with males and high-ranking individuals being more vigilant than females and low-ranking animals, respectively (van Schaik & van Noordwijk 1989; Baldellou & Henzi 1992; Rose & Fedigan 1995; Hirsch 2002; Kutsukake 2006). In addition, vigilance by females increases after they give birth, when the probability of predatory attacks on infants is high (Hauser 1988a; Maestripieri 1993; Treves 1999; Treves et al. 2003; Kutsutake 2006) or infanticide is likely (Steenbeek et al. 1999; chapter 19, this volume).

Vigilance by nocturnal primates has not been studied in detail, most likely because it is difficult for human observers to record the behavior at night. Playback experiments with pair-living red-tailed sportive lemurs (*Lepilemur ruficaudatus)* and solitary foraging gray mouse lemurs revealed that both increase their scanning rates after playbacks of species-typical alarm calls or predator calls compared with control calls (Fichtel 2007; Rahlfs & Fichtel 2010). Sportive lemurs also increase their scanning rates after hearing the calls of their pair mates (Fichtel & Hilgartner, in press), indicating that vigilance in this nocturnal species also involves monitoring of conspecifics similar to social vigilance of diurnal primates.

In summary, all primates rely on their sense of sight to scan for predators. The extent to which they visually scan for predators versus conspecifics is unclear. Although vigilance using visual cues has been studied in detail, vigilance using acoustic cues that may indicate the presence of predators, such as wing flaps or footfalls on the forest floor, has not yet been investigated.

Mixed-species associations

Some primates associate with other species, resulting in increased effective group size. Such mixed-species groups usually consist of members of related taxa in the same genus or family (but see Newton 1989). Because primates in mixed-species groups usually eavesdrop on each other's alarm calls (see above), members of both species benefit mutually through association by detecting predators. Accordingly, mixed-species associations are mainly found in areas where primates are subject to predation by monkey-eating raptors (harpy eagles in the neotropics and crowned-hawk eagles in Africa) (Gautier-Hion et al. 1983; Terborgh 1990). Moreover, monkeys living in areas devoid of predators do not form mixed-species associations more often than would be expected on the basis of chance (Holenweg et al. 1996; Whitesides 1989).

Individuals may benefit by reducing their levels of vigilance when residing in large mixed-species groups compared with smaller single-species groups (Cords 1990; Hardie & Buchanan-Smith 1997). Moreover, the antipredator behavior of members of different species may complement each other. For example, primates who typically move in relatively high parts of the canopy might be more likely to detect raptors than those residing in lower parts of trees, thus suggesting that the former may be more vigilant than the latter (Gautier-Hion et al. 1983; Heymann & Buchanan-Smith 2000). At one site, mixed-species groups of saddleback (*Saguinus fuscicollis*) and moustached (*Saguinus mystax*) tamarins show the expected division of labor with respect to vigilance (Peres 1993); observations of these same species in another area, however, did not reveal the same pattern (Stojan-Dollar & Heymann 2010b). In West Africa, sooty mangabeys (*Cercocebus atys*), western red colobus (*Procolobus badius*), Diana monkeys (*Cercopithecus diana*), and Campbell's mona monkeys (*Cercopithecus campbelli*) forage at different heights and broaden their feeding niches when associating with each other. They consequently are vigilant for different predators and presumably reduce their risk of predation by forming mixed-species groups (Bshary & Noë 1997a, b; McGraw & Bshary 2002; Wolters & Zuberbühler 2003).

Tactics Employed after Predator Detection

Once a predator has been detected, prey can respond in several ways. They can give acoustic warning signals that might be directed to predators or conspecifics. Alternatively, they might flee or even confront predators. These tactics appear to vary as a function of the different hunting strategies employed by predators.

Alarm calls in diurnal primates

Acoustic warning signals are a widespread form of antipredator behavior. Almost all primates have been reported to emit at least one type of alarm call, and many species appear to have several alarm calls in their vocal repertoires. What kind of information is conveyed by these calls? Pioneering studies by Struhsaker (1967a) and Seyfarth and colleagues (1980a, b) revealed that vervet monkeys give three acoustically distinct alarm calls in response to leopards, eagles, and snakes. Each call elicits different escape strategies by conspecifics, which in turn seem to be adapted to the hunting strategies of the different predators. Vervet monkeys run out of trees, head for cover, and scan the sky in response to eagle alarm calls. In contrast, leopard alarm calls prompt them to climb trees and scan the ground, while they stand bipedally and scan the ground after hearing snake alarms. Playback experiments demonstrated that alarm calls alone elicit predator-specific escape responses, suggesting that vervets associate these calls with the predators themselves. Such alarm calls have been termed functionally referential when conspecific receivers respond to them as if they denote a specific external object or event in their environment. Such functional reference requires that calls are given in the same context (production specificity), elicit a specific response (perceptual specificity), and are acoustically distinct (Marler et al. 1992; Macedonia & Evans 1993; Evans 1997).

Following this influential study of vervet monkey alarm calls, functionally referential alarm calls have been found in many birds and mammals (Slobodchikoff et al. 1991; Evans et al. 1994; Manser et al. 2001; Griesser 2008). Functionally referential calls, less specific alarm calls, a combination of both, as well as multipurpose alarm calls have been described in several primate species (table 8.3). Many diurnal, group-living species have been shown to emit functionally referential alarm calls to different types of predators (Seyfarth et al. 1980b; Pereira & Macedonia 1991; Zuberbühler et al. 1999a; Zuberbühler 2001; Kirchhof & Hammerschmidt 2006) but not all species do so. First, several primates combine alarm calls or call notes into predator-specific sequences (Clarke et al. 2006; Arnold et al. 2008; Schel et al. 2009; Ouattara et al. 2009a, b; chapter 29, this volume). Second, other primates display a mixed alarm call system characterized by functionally referential alarm calls for aerial predators and general alarm calls for both aerial and terrestrial predators (Fichtel & Kappeler 2002; Digweed et al. 2005; Fichtel et al. 2005; Fichtel & van Schaik 2006; Kirchhof & Hammerschmidt 2006; Papworth et al. 2008; Wheeler 2008, 2009). General alarm calls are less specific, and are also given during encounters with other conspecifics.

Studies of redfronted lemurs and squirrel monkeys revealed that different levels of perceived threat are encoded in the acoustic structure of their general alarm calls and that conspecifics respond to these differences (Fichtel & Hammerschmidt 2002, 2003). Similarly, differences in the acoustic structure of bonnet macaques' leopard and snake alarm calls appear to reflect different levels of perceived threat (Coss et al. 2007). Alarm calls of barbary macaques (*Macaca sylvanus*), however, appear to vary according to predator type (humans vs. dogs, Fischer et al. 1995; Fischer 1998). Moreover, some species, such as black-and-white ruffed lemurs (*Varecia variegata*) and mandrills (*Mandrillus sphinx*), express different levels of threat perceived by aerial or terrestrial predators by giving different numbers or variants of alarm calls (Macedonia 1990; Yorzinski & Vehrencamp 2008). Other primates that are predominantly

exposed to terrestrial predators, such as chacma baboons or Thomas's langurs, do not possess discrete alarm calls. Instead, they use a variant of the species-typical long-distance call after detecting predators (Fischer et al. 2001, 2002; Wich et al. 2003). Taken together, these studies suggest that many primates give alarm calls in response to terrestrial predators that encode information about the level of perceived threat; such calls may lack referential specificity (see also Cheney & Seyfarth 1990).

Escape responses by diurnal primates

Most diurnal primates live in groups (van Schaik 1983; chapter 9, this volume). By doing so, they reduce their likelihood of being preyed upon after detecting predators (Hamilton 1971). For example, individuals find safety in numbers and dilute the chances that they are eaten by living with others (Hamilton 1971). This advantage is automatically accrued by all gregarious primates. Two other benefits represent escape responses to predators. First, prey may flee using synchronized or tightly coordinated movements. Such action may confuse predators, preventing them from isolating and capturing any single victim (Miller 1922). Second, individuals may benefit through the selfish herd effect; by moving as close as possible to the center of their group after detecting predators, prey reduce the danger to themselves (Hamilton 1971).

Alarm calls typically elicit startle or escape behavior in recipients. Animals can either flee in any direction or exhibit predator-specific escape tactics adapted to the hunting strategies of predators. For example, arboreal prey may climb down when attacked by aerial predators or climb up in response to encountering terrestrial predators (Cheney & Seyfarth 1990; Macedonia & Evans 1993; Fichtel & Kappeler 2002). After these initial flight responses, some primates become inactive and rely on crypsis (Karpanty & Wright 2007).

Primates respond differently to predators that adopt different hunting strategies. For example, leopards are ambush hunters while chimpanzees are pursuit predators. Leopards are more likely to give up their ambush hunting position after Diana monkeys alarm call to them (Zuberbühler et al. 1997). In contrast, Diana monkeys remain silent and move away after detecting chimpanzees (ibid.). Diana monkeys adapt to human hunting within a few generations, responding to poachers with fewer alarm calls than conspecifics living in areas with no hunting (Bshary 2001). Similarly, Diana monkeys are sensitive to cues indicating the presence of humans; they remained silent following playbacks of alarm calls of crested guinea fowls (*Guttera pucherani*), which are hunted by humans (Rainey et al. 2004). Finally, tantalus monkeys (*Chlorocebus tantalus*) hide after seeing lone dogs in areas where human farmers accompanied by dogs threaten them (Kavanagh 1980). Alternatively, they give alarm calls in response to dogs in other places where dogs are not associated with humans (ibid.).

Primates also adapt their escape responses to their risk of predation due to different habitats and predator assemblages. Patas (*Erythrocebus patas*) and vervet monkeys living in relatively, open, exposed, non-riverine habitats display flexible flight responses towards terrestrial predators (Enstam & Isbell 2002); these responses differ from those shown by conspecifics living in less risky riverine habitats (ibid.). Alternatively, responses to alarm calls by semi-free-ranging Coquerel's sifakas differ from those shown by wild Coquerel's sifakas that live with a full complement of predators (Fichtel & van Schaik 2006; Fichtel & Kappeler 2011). In addition, Verreaux's sifakas, which live in different populations with different predators, respond differently to conspecific alarm calls; in this case the signal content of alarm calls may have changed as a function of varying predator assemblages, leading to differences in the comprehension and usage of calls (Fichtel & Kappeler 2011; chapter 29, this volume). Similarly, Diana monkeys' leopard alarm calls in leopard-free areas differ from those given in areas where leopards occur (Stephan & Zuberbühler 2008). In summary, primates adapt their escape strategies and associated behaviors flexibly to varying habitats and sets of predators, and thus, to their perceived risk of predation.

Alarm calls and escape responses in nocturnal primates

Although about one-third of all primates are nocturnal and small-bodied, and thus presumably face a high predation risk (van Noordwijk et al. 1993; Isbell 1994; Hart 2000, 2007; Janson 2003), few data exist regarding the alarm-calling behavior and antipredator strategies of nocturnal primates. Small body size and nocturnality have been suggested to represent adaptations to reduce predation risk (Clutton-Brock & Harvey 1977), but Terborgh and Janson (1986) pointed out that diurnal and nocturnal species may differ in their antipredator strategies. While the former rely on early detection to ward off predators, nocturnal species may depend on reduced detectability and crypsis.

Recent studies reveal that not all nocturnal primates necessarily lead solitary lives (chapter 2, this volume) and that they mob predators (Gursky 2002; Schülke 2001; Bearder et al. 2002; Eberle & Kappeler 2008). These observations have triggered several studies of antipredator behavior in nocturnal primates. As a consequence, we now know that nocturnal primates display various antipredator behaviors that appear to be adapted to the hunting tactics of predators. For example, gray mouse lemurs behave cryptically and freeze in response to raptors that rely on acoustic cues

to locate prey. Alternatively, they mob snakes (Eberle & Kappeler 2008) and remain vigilant in confrontations with ambush-hunting terrestrial predators (Rahlfs & Fichtel 2010). Additional studies indicate that red-tailed sportive lemurs also show escape responses that correspond to the hunting strategies of predators (Fichtel 2007); they do not alarm call to alert conspecifics to the presence of predators, but instead direct these calls to the predators themselves, perhaps as a means to deter them when they enter their sleeping sites (Fichtel 2007). Thus, alarm calls of these nocturnal primates may primarily serve to communicate with the predator.

Spectral tarsiers (*Tarsius tarsier*), which are secondarily nocturnal (Martin & Ross 2005), produce two different alarm calls in response to aerial predator models and several terrestrial predators (Gursky 2006, 2007). Thick-tailed greater galago (*Otolemur crassicaudatus*) and southern lesser galagos (*Galago moholi*) produce a series of graded and discrete calls after encountering predators; these calls have been hypothesized to reflect different levels of perceived risk and serve to deter predators (Bearder 2007). The characteristic slow, non-saltatory locomotion and cryptic coloration of pottos and lorises have been suggested to camouflage them (Charles-Dominique 1977). Slender lorises mostly ignore predators but occasionally whistle at them, whereas pottos (*Perodicticus potto*) remain silent and either flee from predators (Nekaris et al. 2007) or attack them using their hands and bony neck (Charles-Dominique 1977). Slow lorises also have a brachial gland that produces oily exudates with a repellent odor, which they spread over their body while autogrooming. Although these exudates do not contain toxic compounds, humans suffer severe effects from loris bites. These observations suggest that exudates may play a role in deterring predators (Hagey et al. 2007).

In conclusion, solitary foraging and crypsis remain effective predator avoidance strategies for many nocturnal primates who are small, inconspicuously colored, and range by themselves. However, recent studies clearly indicate that several species have evolved specialized antipredator behaviors that are well adapted to the different hunting styles of their predators and that alarm calls may serve to communicate with predators. Because nocturnal strepsirrhines are the most evolutionarily basal primates living today, their alarm calling behavior might represent the ancestral form for the entire order (Fichtel 2007).

Confronting and mobbing predators

Primates may also actively confront predators. For example, baboons drive predators away by attacking them, which sometimes results in the death of the predator (Cowlishaw 1994). Mobbing is a widespread form of antipredator behavior in birds and mammals (Curio 1978; Bartecki & Heymann 1987; Tamura 1989; Fichtel et al. 2005; Templeton et al. 2005). It may serve to deter, confuse, or discourage predators from undertaking attacks (see below; Curio 1978). Primates have been reported to mob snakes and carnivores as well as raptors (Fichtel et al. 2005; Wheeler 2008; Ouattara et al. 2009c). Some species even attack snakes to rescue groupmates (van Schaik et al. 1983; Tello et al. 2002; Perry et al. 2003; Eberle & Kappeler 2008). Observations such as these indicate that confronting and mobbing predators are two viable primate antipredator strategies.

Benefits of alarm calls

Alarm calls contribute to some of the proposed antipredatory benefits of group living, including vigilance and early warning of predator presence (Lima 1995) as well as cooperative defense (Curio 1978). Alternatively, alarm calling can be costly because vocalizing in the presence of a predator might attract the predator's attention to the caller (Caro 1995) and additional predators (Mougeot & Bretagnolle 2000; Fichtel 2009). Although alarm calls may represent reflexive expression of emotions, they might also be produced strategically, as is suggested by observations of animals that adjust their alarm calling in the presence and absence of audiences (Cheney & Seyfarth 1990; Wich & Sterck 2003; reviewed in Fichtel & Manser 2010). Strategically used alarm calls may be directed at predators to discourage pursuit (Woodland et al. 1980) or to conspecifics (Maynard Smith 1965; Sherman 1977). Alarm calls directed to conspecific receivers may reduce the caller's own vulnerability (Trivers 1971) via the selfish herd or confusion effect (Hamilton 1971; Charnov & Krebs 1975), or they may warn kin (Hamilton 1964; Maynard Smith 1965).

There is evidence that primate alarm calls might be directed towards predators themselves. Observations and experiments suggest that brown capuchins, Diana monkeys, and Campbell's monkeys use alarm calls to drive predators away (Zuberbühler et al. 1999b; Wheeler 2008; Ouattara et al. 2009c). Similar findings have been reported for alarm calls emitted by several nocturnal primates (Schülke 2001; Gursky 2006; Fichtel 2007; Eberle & Kappeler 2008). Because leopards and snakes move away from ambush-hunting sites after hearing alarm calls and/or being subject to mobbing displays (Zuberbühler et al. 1997; Clark 2005), it is likely that alarm calls may also serve to deter ambush-hunting predators (chapter 29, this volume).

Individuals that give alarm calls might receive direct or indirect fitness benefits by doing so. Alarm callers may benefit via the selfish herd effect if their calling induces group members to coalesce nearby (Hamilton 1971). Although there is no direct evidence that this actually occurs, long-tailed macaques and red colobus monkeys decrease group spread and nearest-neighbor distances after detecting predators or hearing alarm calls (van Schaik & Mitrasetia 1990; Shultz et al. 2003), thus suggesting that aggregation might be beneficial (James et al. 2004). In some primates, males give alarm calls more often than females after detecting predators. Males also emit alarm calls more frequently when they are with females, and are more likely than females to attack predators. Taken together, these observations suggest that males use alarm calls to protect female mates (Anderson 1986; Cheney & Seyfarth 1985b; Cheney & Wrangham 1987; Cowlishaw 1994; van Schaik & Hörstermann 1994; van Schaik & van Noordwijk 1989). Juveniles and infants, in addition to females, are likely to benefit from male protection against predators, and alarm calls and predator defense might also be driven by kin selection. Alternatively, reciprocal altruism may explain alarm calling, because the majority of primates form long-lasting social relationships (Trivers 1971). In some species not only resident males but also recently immigrated males, which are unlikely to have sired offspring, have been observed to produce alarm calls. In such cases the alarm calling might be selfish and may serve as a commodity to pay for group membership (van Schaik & van Noordwijk 1989).

In some species such as mouse lemurs, tufted capuchin monkeys, Geoffroy's spider monkeys (*Ateles geoffroyi*), and Kloss's gibbons (*Hylobates klossi*), it has been suggested that alarm calling is driven by kin selection (Tenaza & Tilson 1977; Chapman et al. 1990; Eberle & Kappeler 2008; Wheeler 2008). Similarly, vervet monkeys alarm call more often in the presence of offspring, but most often to those species to which they themselves are most vulnerable, rather than to predators to whom their kin are most vulnerable, thus indicating that both individual and kin selection play a role in the evolution of alarm calls (Cheney & Seyfarth 1981). The adjustment of calling rates to the presence and identity of others, the so-called audience effect, is a widespread phenomenon among animals, and it has also been shown for vocalizations other than alarm calls (Sherman 1977; Marler et al. 1986; Evans & Marler 2002; Ridley et al. 2007; Townsend et al. 2008; see Fichtel & Manser 2010 for a review).

In conclusion, nocturnal primates do not appear to direct alarm calls toward conspecifics, but instead use them to communicate with predators. In contrast, diurnal species direct their alarm calls to predators as well as conspecifics. Alarm calls in the Primate order may have evolved initially as a means to communicate with predators; they may have taken on the added function of warning conspecifics after the evolution of group living. Because the potential functions of alarm calls and other defensive behaviors have been addressed in only a few studies, additional research is needed to gain a better understanding of this facet of alarm call behavior.

Evolutionary Effects of Predation on Primate Social Systems

The effect of predation on primate social systems has been the subject of a long-standing debate. The predation avoidance hypothesis proposes that primates live in groups to reduce the risk of predation, despite increased costs of within-group feeding competition (Alexander 1974; van Schaik & van Hooff 1983; van Schaik 1983; Terborgh & Janson 1986). Accordingly, primates that are exposed to high predation risk are expected to live in relatively large groups with many males (van Schaik 1983; van Schaik & van Noordwijk 1985, 1989; Nunn & van Schaik 2000). Some evidence supports these predictions. For example, long-tailed macaques on a predator-free island live in smaller groups than do conspecifics who are exposed to predators (van Schaik & van Noordwijk 1985). Alternatively, red colobus groups consist of more males in habitats where chimpanzee predators are present than in habitats where chimpanzees are absent (Stanford 1998).

Attempts to relate observed predation rates to group size and composition across species have yielded conflicting results, with positive (Anderson 1986), negative (Isbell 1994; Hill & Dunbar 1998), and no relationships (Cheney & Wrangham 1987; Majolo et al. 2008) reported. The inconsistent patterns may be partly due to a failure to distinguish between predation rate and risk across different populations (Hill & Dunbar 1998; Janson 1998). Predation rate is the annual mortality within a population that is directly attributable to predation. It represents the level of successful predator attacks that the animals are unable to control after they have implemented their antipredator strategies. Predation risk represents the expected predation rate that the animal would suffer under standardized levels of antipredator behavior (Janson 1998). The distinction is important because for a given individual or population, rare events tend to occur once or not at all, so that observed rates often have very high variances and are unreliable estimates of the expected rate. To measure the expected rate accurately requires very large samples or very long time periods. Predation risk is intended to reflect potential selection for antipredatory traits, whereas current predation rate is viewed as a result of past selection for such traits (Janson 1998).

By operationalizing predation risk according to predator presence, density, and attack rates, Hill & Lee (1998) found that populations exposed to high predation risk live in relatively large groups and tend to consist of more males. An additional analysis that assayed predation risk using habitat type and substrate use yielded similar findings (Nunn & van Schaik 2000). Reliable predation rates might be obtained by studying predators directly and also by examining the compositions of their diets (Cheney & Wrangham 1987; Struhsaker & Leakey 1990; Mitani et al. 2001). These data have been used to investigate the relationship between predation rates of African crowned eagles, leopards, and chimpanzees and the behavior and ecology of primates in the Taï forest, West Africa (Shultz et al. 2004). Overall, there was no clear relationship between group size and predation rate. For eagle predation, however, there was a strong negative relationship between predation rate and group size for arboreal but not terrestrial prey. In contrast, there was no relationship between rates of predation by leopards and group size for either terrestrial or arboreal prey. When predation rates were summed over all three predators, terrestrial species suffered higher predation rates than arboreal species and, within each prey category, predation rates declined with increasing prey group size and decreasing density of groups in the habitat (Shultz et al. 2004). Interpreting these results is difficult, as prey are likely to respond differently to predators who exhibit different hunting tactics. For example, eagles and leopards are ambush hunters, while chimpanzees are pursuit predators. Thus, prey might form large groups to obtain benefits associated with increased vigilance and alarm calling to the former, but not the latter (Zuberbühler et al. 1997; Shultz & Noë 2002).

Feeding competition is another confounding variable that complicates our understanding of the relationship between predation and group size. Individuals in large groups suffer relatively high levels of feeding competition, and observed group sizes represent a trade-off between predation risk and feeding competition (Janson & Goldsmith 1995; Dunbar 1988; chapter 9, this volume).

Large body size is an additional factor hypothesized to represent an adaptation to predation. Large-bodied primates usually live in areas with more or larger predators (Anderson 1986), and observed predation rates decrease with increasing body size (Cheney & Wrangham 1987; Isbell 1994). Because current antipredator traits may alter present-day predation risk (Abrams 1993), correlations across species between existing social traits and levels of individual predation risk assessed by a population's predation rate are difficult to interpret (Hill & Dunbar 1998). To circumvent this problem, Janson (1998, 2003) introduced the concept of "intrinsic predation risk," which considers predator densities, attack rates of individual predators, attack success, and individual prey vulnerability. Because life history traits may influence the relationship between intrinsic predation risk and the strength of selection on antipredator traits (Boinski & Chapman 1995), they also represent relevant variables. For example, a given predation risk of X% may have quite different consequences for the fitness of small species than for that of larger species (Janson 2003). By including life history traits such as body size, life span, and fecundity, an alternative hypothesis suggests that larger and longer-lived species will suffer the cost of high predation rates over more years than smaller species, and thus will gain a greater fitness benefit by reducing predation rates to low values. This in turn, will favor increased group size in larger species even if intrinsic predation risk declines slightly with body mass (Janson 2003).

Summary and Conclusions

The past 25 years of research have revealed that predation is a strong selective force, shaping the evolution of several primate behavioral adaptations. Primates ranging from the smallest to the largest species are exposed to many different types of predators that do not hunt uniformly and can have a large impact on their survival. Primates have therefore evolved many flexible behavioral adaptations to reduce this risk.

Primates rely on different sensory modalities to recognize predators. The preceding review suggests that primates use visual, auditory, and olfactory cues for this purpose, and that their use develops differently over time, implementing different mechanisms in the process. However, more work is necessary to understand the mechanisms underlying predator recognition and its development.

In contrast, recent studies have dramatically increased our knowledge of the tactics primates employ to evade predators. Several different types of predators attack primates, and primates accordingly employ various behaviors to avoid encountering them. Primates use a different set of tactics to decrease the risk of predation after they encounter predators. The observations presented here indicate that they alter their antipredator behavior as a function of their perceived predation risk and differences in the hunting strategies of predators. The activity pattern strongly influences how primates respond to the threat of predation, and here our understanding of the antipredator behavior of nocturnal primates continues to grow. Additional studies of these animals represent a particularly promising area for future comparative research.

Our current understanding of the evolution of primate alarm calls suggests that they may primarily have evolved as a means to communicate with predators, and in combination with the evolution of group living, to warn conspecifics. The potential selfish versus nepotistic function of alarm calls and other defensive behaviors is still poorly understood, and it represents another promising area for future research.

Continuing debate exists regarding the influence of predation on the evolution of primate social systems. Currently available data suggest that primates exposed to high predation risk live in larger groups than do those exposed to lower risk. Interpreting this result is not straightforward, however, as several variables, including variation in habitats, predators, and predator hunting strategies, potentially confound the relationship. As is the case with other aspects of predation reviewed here, more research will be necessary to untangle the effect of predation on the evolution of primate social systems.

References

Abrams, P. 1993. Why predation rate should not be proprational to predator density. *Ecology*, 74, 726–733.

Alberts, S. C. & Altmann, J. 1995. Balancing costs and opportunities: Dispersal in male baboons. *American Naturalist*, 145, 279–306.

Albignac, R. 1976. L'écologie du *Mungotictis decemlineata* dans les forêts decidues de l'ouest de Madagascar. *Revue d'Ecologie (Terre et la Vie)*, 30, 347–376.

Alexander, R. D. 1974. The evolution of social behavior. *Annual Review of Ecology and Systematics*, 5, 325–383.

Altmann, S. A. 1974. Baboons, space, time, and energy. *American Zoologist*, 14, 221–248.

Anderson, C. M. 1986. Predation and primate evolution. *Primates*, 27, 15–39.

Anderson, J. R. 1998. Sleep, sleeping sites, and sleep-related activities: awakening to their significance. *American Journal of Primatology*, 46, 63–75.

Anderson, J. R. & McGrew, W. C. 1984. Guinea baboons (*Papio papio*) at a sleeping site. *American Journal of Primatology*, 6, 1–14.

Apfelbach, R., Blanchard, C. D., Blanchard, R. J., Hayes, R. & McGregor, I. S. 2005. The effects of predator odors in mammalian prey species: A review of field and laboratory studies. *Neuroscience and Biobehavioral Reviews*, 29, 1123–1144.

Arnold, K. & Zuberbühler, K. 2006. The alarm-calling system of adult male putty-nosed monkeys, *Cercopithecus nictitans martini*. *Animal Behaviour*, 72, 643–653.

Arnold, K., Pohlner, Y. & Zuberbühler, K. 2008. A forest monkey's alarm call series to predator models. *Behavioral Ecology and Sociobiology*, 62, 549–559.

Ayers, D. Y. & Shine, R. 1997. Thermal influences on foraging ability: Body size, posture and cooling rate of an ambush predator, the python *Morelia spilota*. *Functional Ecology*, 11, 342–347.

Baldellou, M. & Henzi, S. P. 1992. Vigilance, predator detection and the presence of supernumery males in vervet monkey troops. *Animal Behaviour*, 43, 451–461.

Baldwin, P. J., Sabater Pi, J., McGrew, W. C. & Tutin, C. E. G. 1981. Comparisons of nests made by different populations of chimpanzees (*Pan troglodytes*). *Primates*, 22, 474–486.

Barret, M. A. & Ratsimbazafy, J. 2009. Luxury bushmeat trade threatens lemur conservation. *Nature*, 461, 470.

Barros, M., Boere, V., Mello, E. L. & Tomasz, C. 2002. Reactions to potential predators in captive-born marmosets (*Callithrix penicillata*). *International Journal of Primatology*, 23, 443–454.

Bartecki, U. & Heymann, E. W. 1987. Field observation of snake-mobbing in a group of saddle-back tamarins, *Saguinus fuscicollis nigrifrons*. *Folia Primatologica*, 48, 199–202.

Bearder, S. K. 2007. A comparison of calling patterns in two nocturnal primates, *Otolemur crassicaudatus* and *Galago moholi* as a guide to predation risk. In *Primate Anti-Predator Strategies* (ed. by S. L. Gursky & K. A. I. Nekaris), 206–221. New York: Springer.

Bearder, S. K., Nekaris, K. A. I. & Buzzell, C. A. 2002. Dangers in the night: Are some nocturnal primates afraid of the dark? In *Eat or Be Eaten: Predator Sensitive Foraging among Primates* (ed. by L. E. Miller), 21–40. Cambridge: Cambridge University Press.

Bearder, S. K., Ambrose, L., Harcourt, C., Honess, P., Perkin, A., Pimley, E., Pullen, S. & Svoboda, N. 2003. Species-typical patterns of infant contact, sleeping site use and social cohesion among nocturnal primates in Africa. *Folia Primatologica*, 74, 337–354.

Bezerra, B. M., Barnett, A. A., Souto, A. & Jones, G. 2009. Predation by the tayra on the common marmoset and the pale-throated three-toed sloth. *Journal of Ethology*, 27, 91–96.

Bianchi, R. C. & Mendes, S. I. 2007. Ocelot (*Leopardus pardalis*) predation on primates in Caratinga Biological Station, Southeast Brazil. *American Journal of Primatology*, 69, 1173–1178.

Boesch, C. & Boesch, H. 1989. Hunting behavior of wild chimpanzees in the Taï National Park. *American Journal of Physical Anthropology*, 78, 547–573.

Boinski, S. 1987. Birth synchrony in squirrel monkeys (*Saimiri oerstedi*): A strategy to reduce neonatal predation. *Behavioural Ecology and Sociobiology*, 21, 393–400.

Boinski, S. & Chapman, C. A. 1995. Predation on primates: Where are we and what's next? *Evolutionary Anthropology*, 4, 1–3.

Britt, A., Welch, C. & Katz, A. 2000. Ruffed lemur re-stocking and conservation program update. *Lemur News*, 5, 36–38.

Brown, L. H. 1971. The relations of the crowned eagle, *Stephanoaetus coronatus*, and some of its prey animals. *Ibis*, 113, 240–243.

Brown, M. M., Kreiter, N. A., Maple, J. T. & Sinnott, J. M. 1992. Silhouettes elicit alarm calls from captive vervet monkeys (*Cercopithecus aethiops*). *Journal of Comparative Psychology*, 106, 350–359.

Bshary, R. 2001. Diana monkeys, *Cercopithecus diana*, adjust their anti-predator response behaviour to human hunting strategies. *Behavioral Ecology and Sociobiology*, 50, 251–256.

Bshary, R. & Noë, R. 1997a. Anti-predation behaviour of red colobus monkeys in the presence of chimpanzees. *Behavioral Ecology and Sociobiology*, 41, 321–333.

———. 1997b. Red colobus and Diana monkeys provide mutual protection against predators. *Animal Behaviour*, 54, 1461–1474.

Buchanan-Smith, H. M., Anderson, D. A. & Ryan, C. W. 1993. Responses of cotton-top tamarins (*Saguinus oedipus*) to faecal scents of predators and non-predators. *Animal Welfare*, 2, 17–32.

Burney, D. A. 2002. Sifaka predation by a large boa. *Folia Primatologica*, 73, 144–145.

Busse, C. D. 1977. Chimpanzee predation as a possible factor in the evolution of red colobus monkey social organization. *Evolution*, 31, 907–911.

———. 1980. Leopard and lion predation upon chacma baboons living in the Moremi Wildlife Reserve. *Botswana Notes and Records*, 12, 15–21.

Caine, N. G. 1987. Vigilance, vocalizations, and cryptic behavior at retirement in captive groups of red-bellied tamarins (*Saguinus labiatus*). *American Journal of Primatology*, 12, 241–250.

Caine, N. G. & Weldon, P. J. 1989. Responses by red-bellied tamarins (*Saguinus labiatus*) to fecal scents of predatory and non-predatory neotropical mammals. *Biotropica*, 2, 186–189.

Campbell, M. W. & Snowdon, C. T. 2009. Can auditory playback condition predator mobbing in captive-reared *Saguinus oedipus*? *International Journal of Primatology*, 30, 93–102.

Caro, T. M. 1995. Pursuit-deterrence revisited. *Trends in Ecology & Evolution*, 10, 500–503.

———. 2005. *Antipredator Defenses in Birds and Mammals*. Chicago: University of Chicago Press.

Chapman, C. A. 1986. *Boa constrictor* predation and group response in white-faced cebus monkeys. *Biotropica*, 18, 171–172.

Chapman, C. A., Chapman, L. J. & Lefebvre, L. 1990. Spider monkey alarm calls: Honest advertisement or warning kin? *Animal Behaviour*, 39, 197–198.

Charles-Dominique, P. 1977. *Ecology and Behaviour of Nocturnal Primates: Prosimians of Equatorial West Africa*. London: Duckworth.

Charles-Dominique, P. & Petter, J. J. 1980. Ecology and social life of *Phaner furcifer*. In *Nocturnal Malagasy Primates: Ecology, Physiology and Behavior* (ed. by P. Charles-Dominique, H. M. Cooper, A. Hladik, C. M. Hladik, E. Pages, G. F. Pariente, A. Petter-Rousseaux, J. J. Petter & A. Schilling), 75–95. New York: Academic Press.

Charnov, E. L. & Krebs, J. R. 1975. The evolution of alarm calls: Altruism or manipulation? *American Naturalist*, 109, 107–112.

Cheney, D. L. & Seyfarth, R. M. 1981. Selective forces affecting the predator alarm calls of vervet monkeys. *Behaviour*, 76, 25–61.

———. 1985a. Social and non-social knowledge in vervet monkeys. *Philosophical Transactions of the Royal Society of London B*, 308, 187–201.

———. 1985b. Vervet monkey alarm calls: Manipulation through shared information? *Behaviour*, 94, 150–166.

———. 1990. *How Monkeys See the World*. Chicago: University of Chicago Press.

Cheney, D. L. & Wrangham, R. W. 1987. Predation. In *Primate Societies* (ed. by B. B. Smuts, D. L. Cheney, R. W. Wrangham & T. T. Struhsaker), 227–239. Chicago: University of Chicago Press.

Cheney, D. L., Seyfarth, R. M., Andelman, S. J. & Lee, P. C. 1988. Reproductive success in vervet monkeys. In *Reproductive Success: Studies of Individual Variation in Contrasting*

Breeding Systems (ed. by T. H. Clutton-Brock), 384–402. Chicago: University of Chicago Press.

Cisneros-Heredia, D. F., León-Reyes, A. & Seger, S. 2005. *Boa constrictor* predation on a titi monkey, *Callicebus discolor*. *Neotropical Primates*, 13, 11–12.

Clark, R. W. 2005. Pursuit-deterrent communication between prey animals and timber rattlesnakes (*Crotalus horridus*): The response of snakes to harrassment displays. *Behavioral Ecology and Sociobiology*, 59, 258–261.

Clarke, E., Reichard, U. H. & Zuberbühler, K. 2006. The syntax and meaning of wild gibbon songs. *PLoS ONE*, 1, e73, doi:10.1371/journal.pone.0000073.

Clutton-Brock, T. H. & Harvey, P. H. 1977. Primate ecology and social organization. *Journal of Zoology, London*, 183, 1–39.

Colquhoun, I. C. 1998. Cathemeral behavior of *Eulemur macaco macaco* at Ambato Massif, Madagascar. *Folia Primatologica*, 69, 22–34.

Condit, V. K. & Smith, E. O. 1994. Predation on a yellow baboon (*Papio cynocephalus cynocephalus*) by a lioness in the Tana River National Primate Reserve, Kenya. *American Journal of Primatology*, 33, 57–64.

Cook, M. & Mineka, S. 1989. Observational conditioning of fear to fear-relevant versus fear-irrelevant stimuli in rhesus monkeys. *Journal for Abnormal Psychology*, 98, 448–459.

Cords, M. 1990. Vigilance and mixed-species association of some East African forest monkeys. *Behavioral Ecology and Sociobiology*, 26, 297–300.

———. 1995. Predator vigilance costs of allogrooming in wild blue monkeys. *Behaviour*, 132, 559–569.

Corrêa, H. K. M. & Coutinho, P. E. G. 1997. Fatal attack of a pit viper, *Bothrops jararaca*, on an infant buffy-tufted ear marmoset (*Callithrix aurita*). *Primates*, 38, 215–217.

Coss, R. G. & Ramakrishnan, U. 2000. Perceptual aspects of leopard recognition by wild bonnet macaques (*Macaca radiata*). *Behaviour*, 137, 315–335.

Coss, R. G., Ramakrishnan, U. & Schank, J. 2005. Recognition of partially concealed leopards by wild bonnet macaques (*Macaca radiata*): the role of the spotted coat. *Behavioural Processes*, 68, 145–163.

Coss, R. G., McCowan, B. & Ramakrishnan, U. 2007. Threat-related acoustical differences in alarm calls by wild bonnet macaques (*Macaca radiata*) elicited by python and leopard models. *Ethology*, 113, 352–367.

Cowlishaw, G. 1994. Vulnerability to predation in baboon populations. *Behaviour*, 131, 293–304.

———. 1997a. Alarm calling and implications for risk perception in a desert baboon population. *Ethology*, 103, 384–394.

———. 1997b. Refuge use and predation risk in a desert baboon population. *Animal Behaviour*, 54, 241–253.

———. 1997c. Trade-offs between foraging and predation risk determine habitat use in a desert baboon population. *Animal Behaviour*, 53, 667–686.

———. 1998. The role of vigilance in the survival and reproductive strategies of desert baboons. *Behaviour*, 135, 431–452.

Curio, E. 1978. The adaptive significance of avian mobbing. I. Teleonomic hypothesis and predictions. *Zeitschrift für Tierpsychologie*, 48, 175–183.

Curtis, D. J., Zaramody, A. & Martin, R. D. 1999. Cathemerality in the mongoose lemur, *Eulemur mongoz*. *American Journal of Primatology*, 47, 279–298.

Dammhahn, M. & Kappeler, P. M. 2005. Social system of *Microcebus berthae*, the world's smallest primate. *International Journal of Primatology*, 26, 407–435.

D'Amour D. E., Hohmann, G. & Fruth, B. 2006. Evidence of leopard predation on bonobos (*Pan paniscus*). *Folia Primatologica*, 77, 212–217.

Dausmann, K. H., Glos, J., Ganzhorn, J. U. & Heldmaier, G. 2004. Hibernation in a tropical primate. *Nature*, 429, 825–826.

De Luna, A. G., Sanmiguel, R., Di Fiore, A. & Fernandez-Duque, E. 2010. Predation and predation attempts on red titi monkeys (*Callicebus discolor*) and equatorial sakis (*Pithecia aequatorialis*) in Amazonian Ecuador. *Folia Primatologica*, 81, 86–95.

Deppe, A. M., Randriamiarisoa, M., Kasprak, A. H. & Wright, P. C. 2008. Predation on the brown mouse lemur (*Microcebus rufus*) by a diurnal carnivore, the ring-tailed mongoose (*Galidia elegans*). *Lemur News*, 13, 17–18.

De Ruiter, J. R. 1986. The influence of group size on predator scanning and foraging behaviour of wedgecapped capuchin monkeys (*Cebus olivaceus*). *Behaviour*, 98, 240–258.

Di Bitetti, M. S., Vidal, E. M. L., Baldovino, M. C. & Benesovsky, V. 2000. Sleeping site preferences in tufted capuchin monkeys (*Cebus apella nigritus*). *American Journal of Primatology*, 50, 257–274.

Digweed, S. M., Fedigan, L. M. & Rendall, D. 2005. Variable specificity in the anti-predator vocalizations and behaviour of the white-faced capuchin, *Cebus capucinus*. *Behaviour*, 142, 997–1021.

Dittus, W. P. J. 1975. Population dynamics of the Ceylon toque monkey, *Macaca sinica*. In *Socioecology and Psychology of Primates* (ed. by R. H. Tuttle), 125–152. The Hague: Mouton Publishers.

Dunbar, R. I. M. 1988. *Primate Social Systems*. Ithaca, NY: Cornell University Press.

Eberle, M. & Kappeler, P. M. 2006. Family insurance: Kin selection and cooperative breeding in a solitary primate (*Microcebus murinus*). *Behavioral Ecology and Sociobiology*, 60, 582–588.

———. 2008. Mutualism, reciprocity, or kin selection? Cooperative rescue of a conspecific from a boa in a nocturnal solitary forager the gray mouse lemur. *American Journal of Primatology*, 70, 410–414.

Elgar, M. A. 1989. Predator vigilance and group size in mammals and birds: A critical review of the empirical evidence. *Biological Reviews*, 64, 13–33.

Engh, A. L., Beehner, J. C., Bergman, T. J., Whitten, P. L., Hoffmeier, R. R., Seyfarth, R. M. & Cheney, D. L. 2006. Behavioural and hormonal responses to predation in female chacma baboons (*Papio hamadryas ursinus*). *Proceedings of the Royal Society B*, 273, 707–712.

Enstam, K. L. & Isbell, L. A. 2002. Comparison of responses to alarm calls by patas (*Erythrocebus patas*) and vervet (*Cercopithecus aethiops*) monkeys in relation to habitat structure. *American Journal of Physical Anthropology*, 119, 3–14.

Enstam Jaffe, K. & Isbell, L. A. 2009. After the fire: Benefits of reduced ground cover for vervet monkeys (*Cercopithecus aethiops*). *American Journal of Primatology*, 71, 252–260.

Estes, R. & Goddard, J. 1967. Prey selection and hunting behavior of the African wild dog. *Journal of Wildlife Management*, 31, 52–70.

Evans, C. S. 1997. Referential signals. In *Perspectives in Ethology: Communication* (ed. by D. H. Owings, M. D. Beecher & N. S. Thompson), 99–143. New York: Plenum Press.

Evans, C. S. & Marler, P. 2002. Food calling and audience effects in male chickens, *Gallus gallus*: their relationships to food availability, courtship and social facilitation. *Animal Behaviour*, 47, 1159–1170.

Evans, C. S., Evans, L. & Marler, P. 1994. On the meaning of alarm calls: Functional reference in an avian vocal system. *Animal Behaviour*, 46, 23–38.

Fa, J. E., Peres, C. A. & Meeuwig, J. 2002. Bushmeat exploitation in tropical forests: An intercontinental comparison. *Conservation Biology*, 16, 232–237.

Fa, J. E., Currie, D. & Meeuwig, J. 2003. Bushmeat and food security in the Congo Basin: Linkages between wildlife and people's future. *Environmental Conservation*, 30, 71–78.

Fernandez-Duque, E. & Erkert, H. G. 2006. Cathemerality and lunar periodicity of activity rhythms in owl monkeys of the Argentinian Chaco. *Folia Primatologica*, 77, 123–138.

Ferrari, S. F., Pereira, W. L. A., Santos, R. R. & Veiga, L. M. 2004. Fatal attack of a *Boa constrictor* on a bearded saki (*Chiropotes satanas utahicki*). *Folia Primatologica*, 75, 111–113.

Fichtel, C. 2004. Reciprocal recognition in sifaka (*Propithecus verreauxi verreauxi*) and redfronted lemur (*Eulemur fulvus rufus*) alarm calls. *Animal Cognition*, 7, 45–52.

———. 2007. Avoiding predators at night: Antipredator strategies in red-tailed sportive lemurs (*Lepilemur ruficaudatus*). *American Journal of Primatology*, 69, 611–624.

———. 2008. Ontogeny of conspecific and heterospecific alarm call recognition in wild Verreaux's sifakas (*Propithecus verreauxi verreauxi*). *American Journal of Primatology*, 70, 127–135.

———. 2009. Costs of alarm calling: Lemur alarm calls attract fossas. *Lemur News*, 14, 53–55.

Fichtel, C. & Hammerschmidt, K. 2002. Responses of redfronted lemurs (*Eulemur fulvus rufus*) to experimentally modified alarm calls: Evidence for urgency-based changes in call structure. *Ethology*, 108, 763–777.

———. 2003. Responses of squirrel monkeys to their experimentally modified mobbing calls. *Journal of the Acoustical Society of America*, 113, 2927–2932.

Fichtel, C. & Hilgartner, R. in press. Noises in the dark: Vocal communication in nocturnal pair-living primates. In *Leaping Ahead: Advances in Prosimian Biology. Developments in Primatology* (ed. by J. Master, M. Gamba & F. Génin). New York: Springer.

Fichtel, C. & Kappeler, P. M. 2002. Anti-predator behavior of group-living Malagasy primates: Mixed evidence for a referential alarm call system. *Behavioral Ecology and Sociobiology*, 51, 262–275.

———. 2011. Variation in the meaning of alarm calls in Verreaux's and Coquerel's sifakas (*Propithecus verreauxi, P. coquereli*). *International Journal of Primatology* 32, 346–361.

Fichtel, C. & Manser, M. B. 2010. Vocal communication in social groups. In *Behaviour: Evolution and Mechanisms* (ed. by P. M. Kappeler), 29–54. Berlin: Springer.

Fichtel, C. & van Schaik, C. P. 2006. Semantic differences in sifaka (*Propithecus verreauxi*) alarm calls: a reflection of genetic or cultural variants? *Ethology*, 112, 839–849.

Fichtel, C., Perry, S. & Gros-Louis, J. 2005. Alarm calls of white-faced capuchin monkeys: An acoustic analysis. *Animal Behaviour*, 70, 165–176.

Fietz, J. 1998. Body mass in wild *Microcebus murinus* over the dry season. *Folia Primatologica*, 69, 183–190.

Fischer, J. 1998. Barbary macaques categorize shrill barks into two call types. *Animal Behaviour*, 55, 799–807.

Fischer, J., Hammerschmidt, K. & Todt, D. 1995. Factors affecting acoustic variation in Barbary macaque (*Macaca sylvanus*) disturbance calls. *Ethology*, 101, 51–66.

Fischer, J., Cheney, D. L. & Seyfarth, R. M. 2000. Development of infant baboons' responses to graded bark variants. *Proceedings of the Royal Society B*, 267, 2317–2321.

Fischer, J., Hammerschmidt. K,, Cheney, D. L. & Seyfarth, R. M. 2001. Acoustic features of female chacma baboon barks. *Ethology*, 107, 33–54.

———. 2002. Acoustic features of male baboon loud calls: Influences of context, age, and individuality. *Journal of the Acoustical Society of America*, 111, 1465–1474.

Fragaszy, D. M. 1990. Sex and age differences in the organization of behavior in wedge-capped capuchins, *Cebus olivaceus*. *Behavioural Ecology*, 1, 81–94.

Franklin, S. P., Hankerson, S. J., Baker, A. J. & Dietz, J. M. 2007. Golden lion tamarin sleeping-site use and pre-retirement behavior during intense predation. *American Journal of Primatology*, 69, 325–335.

Friant, S. C., Campbell, M. W. & Snowdon, C. T. 2008. Captive-born cotton-top tamarins (*Saguinus oedipus*) respond similarly to vocalizations of predators and sympatric non-predators. *American Journal of Primatology*, 70, 707–710.

Galdikas, B. M. F. 1985. Crocodile predation on a proboscis monkey in Borneo. *Primates*, 26, 495–496.

Galdikas, B. M. F. & Yeager, C. P. 1984. Crocodile predation on a crab-eating macaque in Borneo. *American Journal of Primatology*, 6, 49–51.

Gargett, V. 1971. Some observations on black eagles in the Matopos, Rhodesia. *Ostrich Supplement*, 9, 91–124.

Gašperšič, M. & Pruetz, J. D. 2004. Predation on a monkey by savanna chimpanzees at Fongoli, Senegal. *Pan Africa News*, 11, 8–10.

Gautier-Hion, A., Quris, R. & Gautier, J.-P. 1983. Monospecific vs polyspecific life: A comparative study of foraging and antipredatory tactics in a community of *Cercopithecus* monkeys. *Behavioral Ecology and Sociobiology*, 12, 325–335.

Génin, F. 2010. Who sleeps with whom? Sleeping association and socio-territoriality in *Microcebus griseorufus*. *Journal of Mammalogy*, 91, 942–951.

Gilbert, K. A. 2000. Attempted predation on a white-faced saki in Central Amazon. *Neotropical Primates*, 8, 103–104.

Gilbert, M. & Tingay, R. E. 2001. Predation of a fat-tailed dwarf lemur *Cheirogaleus medius* by a Madagascar harrier-hawk *Polyboroides radiatus*: An incidental observation. *Lemur News*, 6, 6.

Gil-da-Costa, R., Palleroni, A., Hauser, M. D., Touchton, J. & Kelley, J. P. 2003. Rapid acquisition of an alarm response by a neotropical primate to a newly introduced avian predator. *Proceedings of the Royal Society B*, 270, 605–610.

Goodman, S. M. 2003. Predation in lemurs. In *The National History of Madagascar* (ed. by S. M. Goodman & J. P. Benstead), 1221–1228. Chicago: University of Chicago Press.

Goodman, S. M. & Langrand, O. 1996. Food remains found in a nest of the Madagascar buzzard (*Buteo brachypterus*) in the Vohibasia Forest. *Working Group on Birds in the Madagascar Region Newsletter*, 6, 13–14.

Goodman, S. M. & Pidgeon, M. 1999. Carnivora of the Réserve Naturelle Intégrale d'Andohahela, Madagascar. *Fieldiana Zoology*, 94, 259–268.

Goodman, S. M. & Thorstrom, R. 1998. The diet of the Madagascar red owl (*Tyto soumagnei*) on the Masoala Peninsula, Madagascar. *Wilson Bulletin*, 110, 417–421.

Goodman, S. M., O'Connor, S. & Langrand, O. 1993a. A review of predation on lemurs: Implications for the evolution of social behavior in small, nocturnal primates. In *Lemur Social Systems and Their Ecological Basis* (ed. by P. M. Kappeler & J. U.Ganzhorn), 51–66. New York: Plenum Press.

Goodman, S. M., Langrand, O. & Raxworthy, C. J. 1993b. Food habits of the Madagascar long-eared owl *Asio madagascariensis* in two habitats in southern Madagascar. *Ostrich*, 64, 79–85.

——. 1993c. The food habits of the barn owl *Tyto alba* at three sites on Madagascar. *Ostrich*, 64, 160–171.

Goodman S. M., Langrand, O. & Rasolonandrasana, B. P. N. 1997. The food habits of *Cryptoprocta ferox* in the high mountain zone of the Andringitra Massif, Madagascar (Carnivora, Viverridae). *Mammalia*, 61, 185–192.

Goodman, S. M., de Roland, L. A. R. & Thorstrom, R. 1998. Predation on the eastern wooly lemur (*Avahi laniger*) and other vertebrates by Henst's goshawk (*Accipiter henstii*). *Lemur News*, 3, 14–15.

Griesser, M. 2008. Referential calls signal predator behavior in a group-living bird species. *Current Biology*, 18, 69–73.

Gursky, S. L. 2002. Predation on a wild spectral tarsier (*Tarsius spectrum*) by a snake. *Folia Primatologica*, 73, 60–62.

——. 2006. Function of snake mobbing in spectral tarsiers. *American Journal of Physical Anthropology*, 129, 601–608.

——. 2007. The response of spectral tarsiers toward avian and terrestrial predators. In *Primate Anti-Predator Strategies* (ed. by S. L. Gursky & K. A. I. Nekaris), 241–252. New York: Springer.

Hagey, L. R., Fry, B. G. & Fitch-Snyder, H. 2007. Talking defensively, a dual use for the brachial gland exudate of slow and pygmy lorises. In *Primate Anti-Predator Strategies* (ed. by S. L. Gursky & K. A. I. Nekaris), 253–272. New York: Springer.

Hamilton, W. D. 1964. The genetical evolution of social behavior. I and II. *Journal of Theoretical Biology*, 7, 1–52.

——. 1971. Geometry of the selfish herd. *Journal of Theoretical Biology*, 31, 295–311.

Hamilton, W. J. III. 1982. Baboon sleeping site preferences and relationships to primate grouping patterns. *American Journal of Primatology*, 3, 41–53.

Hankerson, S. J. & Caine, N. G. 2004. Pre-retirement predator encounters alter the morning behavior of captive marmosets (*Callithrix geoffroyi*). *American Journal of Primatology*, 63, 75–85.

Hankerson, S. J., Franklin, S. P. & Dietz, J. M. 2007. Tree and forest characteristics influence sleeping site choice by golden lion tamarins. *American Journal of Primatology*, 69, 976–988.

Hardie, S. M. & Buchanan-Smith, H. M. 1997. Vigilance in single- and mixed-species groups of tamarins (*Saguinus labiatus* and *Saguinus fuscicollis*). *International Journal of Primatology*, 18, 217–234.

Hart, D. L. 2000. Primates as prey: Ecological, morphological and behavioral relationships between primate species and their predators. PhD thesis, Washington University.

——. 2007. Predation on primates: A biogeographical analysis. In *Primate Anti-Predator Strategies* (ed. by S. L. Gursky & K. A. I. Nekaris), 27–59. New York: Springer.

Hauser, M. D. 1988a. Variation in maternal responsiveness in free-ranging vervet monkeys: A response to infant mortality risk? *American Naturalist*, 131, 573–587.

——. 1988b. How infant vervet monkeys learn to recognize starling alarm calls. *Behaviour*, 105, 187–201.

Hauser, M. D. & Wrangham, R. W. 1990. Recognition of predator and competitor calls in nonhuman primates and birds: A preliminary report. *Ethology*, 86, 116–130.

Hausfater, G. 1976. Predatory behavior of yellow baboons. *Behaviour*, 56, 44–67.

Hayes, S. L. & Snowdon, C. T. 1990. Predator recognition in cotton-top tamarins (*Saguinus oedipus*). *American Journal of Primatology*, 20, 283–291.

Heymann, E. W. 1987. A field observation of predation on a moustached tamarin (*Saguinus mystax*) by an anaconda. *International Journal of Primatology*, 8, 193–195.

Heymann, E. W. 1995. Sleeping habits of tamarins, *Saguinus mystax* and *Saguinus fuscicollis* (Mammalia; Primates; Callitrichidae), in north-eastern Peru. *Journal of Zoology, London*, 237, 211–226.

Heymann, E. W. & Buchanan-Smith, H. M. 2000. The behavioural ecology of mixed-species troops of callitrichine primates. *Biological Reviews*, 75, 169–190.

Hilgartner, R., Zinner, D. & Kappeler, P. M. 2008. Life history traits and parental care in *Lepilemur ruficaudatus*. *American Journal of Primatology*, 69, 1–15.

Hill, R. A. & Dunbar, R. I. M. 1998. An evaluation of the roles of predation rate and predation risk as selective pressures on primate grouping behaviour. *Behaviour*, 135, 411–430.

Hill, R. A. & Lee, P. C. 1998. Predation risk as an infuence on group size in cercopithecoid primates: implications for social structure. *Journal of Zoology, London*, 245, 447–456.

Hirsch, B. T. 2002. Social monitoring and vigilance behavior in brown capuchin monkeys (*Cebus apella*). *Behavioral Ecology and Sociobiology*, 52, 458–464.

Hofreiter, M., Kreuz, E., Eriksson, J., Schubert, G. & Hohmann, G. 2010. Vertebrate DNA in fecal samples from bonobos and gorillas: Evidence for meat consumption or artefact? *PLoS ONE*, 5, e9419, doi:10.1371/journal.pone.0009419.

Holenweg, A.-K., Noë, R. & Schabel, M. 1996. Waser's gas model applied to associations between red colobus and Diana monkeys in the Taï National Park, Ivory Coast. *Folia Primatologica*, 67, 125–136.

Houle, A., Chapman, C. A. & Vickery, W. L. 2010. Intra-tree vertical variation of fruit density and the nature of contest competition in frugivores. *Behavioral Ecology and Sociobiology*, 64, 429–441.

Irwin, M. T., Raharison, J.-L. & Wright P. C. 2009. Spatial and temporal variability in predation on rainforest primates: Do forest fragmentation and predation act synergistically? *Animal Conservation*, 12, 220–230.

Isbell, L. A. 1994. Predation on primates: Ecological patterns and evolutionary consequences. *Evolutionary Anthropology*, 3, 61–71.

Izawa, K. 1978. A field study of the ecology and behavior of the black-mantle tamarin (*Saguinus nigricollis*). *Primates*, 19, 241–274.

Jaksić, F. M. & Carothers, J. H. 1985. Ecological, morphological, and bioenergetic correlates of hunting modes in hawks and owls. *Ornis Scandinavica*, 16, 165–172.

James, R., Bennett, P. G. & Krause, J. 2004. Geometry for mutualistic and selfish herds: The limited domain of danger. *Journal of Theoretical Biology*, 228, 107–113.

Janson, C. H. 1990. Social correlates of individual spatial choice in foraging groups of brown capuchin monkeys, *Cebus apella*. *Animal Behaviour*, 40, 910–921.

———. 1998. Testing the predation hypothesis for vertebrate sociality: Prospects and pitfalls. *Behaviour*, 135, 389–410.

———. 2003. Puzzles, predation, and primates: Using life history to understand selection pressures. In *Primate Life Histories and Socioecology* (ed. by P. M. Kappeler & M. E. Pereira), 103–131. Chicago: University of Chicago Press.

Janson, C. H. & Goldsmith, M. L. 1995. Predicting group size in primates: Foraging costs and predation risks. *Behavioral Ecology*, 6, 326–336.

Janson, C. H. & van Schaik, C. P. 1993. Ecological risk aversion in juvenile primates: Slow and steady wins the race. In *Juvenile Primates: Life History, Development and Behavior* (ed. by M. E. Pereira & L. A. Fairbanks), 57–74. Chicago: University of Chicago Press.

Jolly, A., Caless, S., Cavigelli, S., Gould, L., Pereira, M. E., Pitts, A., Pride, R. E., Rabenandrasana, H. D., Walker, J. D. & Zafison, T. 2000. Infant killing, wounding and predation in *Eulemur* and *Lemur*. *International Journal of Primatology*, 21, 21–40.

Jones, C. 1969. Notes on ecological relationship of four species of lorisids in Rio Muni, West Africa. *Folia Primatologica*, 11, 255–267.

Kappeler, P. M. 1997. Intrasexual selection in *Mirza coquereli*: Evidence for scramble competition polygyny in a solitary primate. *Behavioral Ecology and Sociobiology*, 45, 115–127.

———. 1998. Nests, tree holes, and the evolution of primate life histories. *American Journal of Primatology*, 46, 7–33.

Kappeler, P. M. & Erkert, H. G. 2003. On the move around the clock: Correlates and determinants of cathemeral activity in wild redfronted lemurs (*Eulemur fulvus rufus*). *Behavioral Ecology and Sociobiology*, 54, 359–369.

Karpanty, S. M. & Goodman, S. M. 1999. Diet of the Madagascar harrier-hawk, *Polyboroides radiatus*, in southeastern Madagascar. *Journal of Raptor Research*, 33, 313–316.

Karpanty, S. M. & Grella, R. 2001. Lemur responses to diurnal raptor calls in Ranomafana National Park, Madagascar. *Folia Primatologica*, 172, 100–103.

Karpanty, S. M. & Wright, P. C. 2007. Predation on lemurs in the rainforest of Madagascar by multiple predator species: observations and experiments. In *Primate Anti-Predator Strategies* (ed. by S. L. Gursky & K. A. I. Nekaris), 77–99. New York: Springer.

Kavanagh, M. 1980. Invasion of the forest by an African savannah monkey: Behavioural adaptations. *Behaviour*, 73, 238–260.

Kennedy, R. 1977. Notes on the biology and population status of the monkey-eating eagle of the Philippines. *Wilson Bulletin*, 89, 1–20.

Kingdon, J. 1974. *East African Mammals: An Atlas of Evolution in Africa* (Vol. 3). Chicago: University of Chicago Press.

Kirchhof, J. & Hammerschmidt, K. 2006. Functionally referential alarm calls in tamarins (*Saguinus fuscicollis* and *Saguinus mystax*): Evidence from playback experiments. *Ethology*, 112, 346–354.

Koenig, A. 1998. Visual scanning by common marmosets (*Callithrix jacchus*): Functional aspects and the special role of adult males. *Primates*, 39, 85–90.

Koyama, N. 1992. Some demographic data of ring-tailed lemurs (*Lemur catta*) at Berenty, Madagascar. In *Social Structure of Madagascar Higher Vertebrates in Relation to Their Adaptive Radiation* (ed. by S. Yamagishi), 10–16. Osaka: Osaka University Press.

Kraus, C., Eberle, M. & Kappeler, P. M. 2008. The costs of risky male behaviour: Sex differences in seasonal survival in a small sexually monomorphic primate. *Proceedings of the Royal Society B*, 275, 1635–1644.

Kutsukake, N. 2006. The context and quality of social relationships affect vigilance behaviour in wild chimpanzees. *Ethology*, 112, 581–591.

Langrand, O. 1990. *Guide to the Birds of Madagascar*. New Haven: Yale University Press.

LeBreton, M., Prosser, A. T., Tamoufe, U., Sateren, W., Mpoudi-Ngole, E., Diffo, J. L. D., Burke, D. S. & Wolfe, N. D. 2006. Patterns of bushmeat hunting and perceptions of disease risk among central African communities. *Animal Conservation*, 9, 357–363.

Leendertz, F. H., Pauli, G., Maetz-Rensing, K., Boardman, W., Nunn, C. L., Ellerbrok, H., Jensen, S. A., Junglen, S. & Boesch, C. 2006. Pathogens as drivers of population declines: The importance of systematic monitoring in great apes and other threatened mammals. *Biological Conservation*, 131, 325–337.

Lima, S. L. 1995. Back to the basics of anti-predatory vigilance: The group-size effect. *Animal Behaviour*, 49, 11–20.

———. 1998. Stress and decision making under the risk of predation: Recent developments from behavioral, reproductive, and ecological perspectives. *Advances in the Study of Behavior*, 27, 215–290.

Lima, S. L. & Dill, L. M. 1990. Behavioral decisions made under the risk of predation: A review and prospectus. *Canadian Journal of Zoology*, 68, 616–640.

Ludwig, G., Aguiar, L. M., Miranda, J. M. D., Teixeira, G. M., Svoboda, W. K., Malanski, L. S., Shiozawa, M. M., Hilst, C. L. S., Navarro, I. T. & Passos, F. C. 2007. Cougar predation on black-and-gold howlers on Mutum Island, southern Brazil. *International Journal of Primatology*, 28, 39–46.

Lührs, M.-L. & Dammhahn, M. 2010. An unusual case of cooperative hunting in a solitary carnivore. *Journal of Ethology*, 28, 379–383.

Macedonia, J. M. 1990. What is communicated in the antipredator calls of lemurs: Evidence from playback experiments with ringtailed and ruffed lemurs. *Ethology*, 86, 177–190.

Macedonia, J. M. & Evans, C. S. 1993. Variation among mammalian alarm call systems and the problem of meaning in animal signals. *Ethology*, 93, 177–197.

Macedonia, J. M. & Polak, J. F. 1989. Visual assessment of

avian threat in semi-captive ringtailed lemurs (*Lemur catta*). *Behaviour*, 111, 291–304.

Macedonia, J. M. & Yount, P. L. 1991. Auditory assessment of avian predator threat in semi-captive ringtailed lemurs (*Lemur catta*). *Primates*, 32, 169–182.

MacKinnon, J. & MacKinnon, K. 1980. The behavior of wild spectral tarsiers. *International Journal of Primatology*, 1, 361–379.

Maestripieri, D. 1993. Vigilance costs of allogrooming in macaque mothers. *American Naturalist*, 141, 744–753.

Majolo, B., de Bortoli Vizioli, A. & Schino, G. 2008. Costs and benefits of group living in primates: Group size effects on behaviour and demography. *Animal Behaviour*, 76, 1235–1247.

Manser, M. B., Bell, M. B. & Fletcher, L. B. 2001. The information that receivers extract from alarm calls in suricates. *Proceedings of the Royal Society B*, 268, 2485–2491.

Marler, P., Dufty, A. & Pickert, R. 1986. Vocal communication in the domestic chicken: II. Is a sender sensitive to the presence and nature of a receiver? *Animal Behaviour*, 34, 194–198.

Marler, P., Evans, C. S. & Hauser, M. D. 1992. Animal signals: Motivational, referential, or both? In *Nonverbal Vocal Communication: Comparative and Developmental Approaches* (ed. by H. Papousek, U. Jürgens, M. Papousek), 66–86. Cambridge: Cambridge University Press.

Martin, R. D. 1972. Adaptive radiation and behaviour of the Malagasy lemurs. *Philosophical Transactions of the Royal Society of London B*, 264, 295–352.

Martin, R. D. & Ross, C. F. 2005. The evolutionary and ecological context of primate vision. In *The Primate Visual System: A Comparative Approach* (ed. by J. Kremers), 1–36. Chichester: John Wiley.

Martins, S. S., Lima, E. M. & Silva J. S. Jr. 2005. Predation of a bearded saki (*Chiropotes utahicki*) by a harpy eagle (*Harpia harpyja*). *Neotropical Primates*, 13, 7–10.

Matsuda, I. & Izawa, K. 2008. Predation of wild spider monkeys at La Macarena, Colombia. *Primates*, 49, 65–68.

Matsuda, I., Tuuga, A. & Higashi, S. 2008a. Clouded leopard (*Neofelis diardi*) predation on proboscis monkeys (*Nasalis larvatus*) in Sabah, Malaysia. *Primates*, 49, 227–231.

Matsuda, I., Tuuga, A., Akiyama, Y. & Higashi, S. 2008b. Selection of river crossing location and sleeping site by proboscis monkeys (*Nasalis larvatus*) in Sabah, Malaysia. *American Journal of Primatology*, 70, 1097–1101.

Maynard Smith, J. 1965. The evolution of alarm calls. *American Naturalist*, 49, 59–63.

Maynard Smith, J. & Szathmáry, E. 1995. *The Major Transitions in Evolution*. Oxford: Oxford University Press.

McCowan, B. & Newman, J. D. 2000. The role of learning in chuck call recognition by squirrel monkeys (*Saimiri sciureus*). *Behaviour*, 137, 279–300.

McGraw, W. S. & Bshary, R. 2002. Association of terrestrial mangabeys (*Cercocebus atys*) with arboreal monkeys: Experimental evidence for the effects of reduced ground predator pressure on habitat use. *International Journal of Primatology*, 23, 311–325.

McGraw, W. S., Cooke, C. & Shultz, S. 2006. Primate remains from African crowned eagle (*Stephanoaetus coronatus*) nests in Ivory Coast's Tai Forest: Implications for primate predation and early hominid taphonomy in South Africa. *American Journal of Physical Anthropology*, 131, 151–165.

McKinney, T. 2009. Anthropogenic change and primate predation risk: Crested caracaras (*Caracara plancus*) attempt preda-

tion on mantled howler monkeys (*Alouatta palliata*). *Neotropical Primates*, 16, 24–27.

Miller, R. C. 1922. The significance of the gregarious habit. *Ecology*, 3, 122–126.

Miller, L. E. 2002. *Eat or Be Eaten: Predator Sensitive Foraging among Primates*. Cambridge: Cambridge University Press.

Milner-Gulland, E. J., Bennett, E. L. & the SCB 2002 Annual Meeting Wild Meat Group 2003. Wild meat: The bigger picture. *Trends in Ecology & Evolution*, 18, 351–357.

Mineka, S., Davidson, M., Cook, M. & Keir, R. 1984. Observational conditioning of snake fear in rhesus monkeys. *Journal of Abnormal Psychology*, 93, 355–372.

Miranda, J. M. D., Bernardi, I. P., Abreu, K. C. & Passos, F. C. 2005. Predation on *Alouatta guariba clamitans* Cabrera (Primates, Atelidae) by *Leopardus paradalis* (Linnaeus) (Carnivora, Felidae). *Revista Brasileira de Zoologia*, 22, 793–795.

Mitani, J. C., Sanders, W. J., Lwanga, J. S. & Windfelder, T. L. 2001. Predatory behavior of crowned hawk-eagles (*Stephanoaetus coronatus*) in Kibale National Park, Uganda. *Behavioral Ecology and Sociobiology*, 49, 187–195.

Mougeot, F. & Bretagnolle, V. 2000. Predation as a cost of sexual communication in nocturnal seabirds: An experimental approach using acoustic signals. *Animal Behaviour*, 60, 647–656.

Nash, L. 2007. Moonlight and behavior in nocturnal and cathemeral primates, especially *Lepilemur leucopus*: Illuminating possible anti-predator efforts. In *Primate Anti-Predator Strategies* (ed. by S. L. Gursky & K. A. I. Nekaris), 173–205. New York: Springer.

Nekaris, K. A. I., Pimley, E. R. & Ablard, K. M. 2007. Predator defense by slender lorises and pottos. In *Primate Anti-Predator Strategies* (ed. by S. L. Gursky & K. A. I. Nekaris), 222–240. New York: Springer.

Newton, P. N. 1989. Associations between langur monkeys (*Presbytis entellus*) and chital deer (*Axis axis*): Chance encounters or a mutualism? *Ethology*, 83, 89–120.

Nishida, T. 1983. Alpha status and agonistic alliance in wild chimpanzees (*Pan troglodytes schweinfurthii*). *Primates*, 24, 318–336.

Nishie, H. 2004. Increased hunting of yellow baboons (*Papio cynocephalus*) by M group chimpanzees at Mahale. *Pan Africa News*, 11, 10–12.

Nunn, C. L. & Alitzer, S. M. 2006. *Infectious Diseases in Primates: Behavior, Ecology and Evolution*. Oxford: Oxford University Press.

Nunn, C. L. & Heymann, E. W. 2005. Malaria infection and host behavior: A comparative study of Neotropical primates. *Behavioral Ecology and Sociobiology*, 59, 30–37.

Nunn, C. L. & van Schaik, C. P. 2000. Social evolution in primates: The relative roles of ecology and intersexual conflict. In *Infanticide by Males and Its Implications* (ed. by C. P. van Schaik & C. H. Janson), 388–420. Cambridge: Cambridge University Press.

Oda, R. 1998. The responses of Verreaux's sifakas to anti-predator alarm calls given by sympatric ring-tailed lemurs. *Folia Primatologica*, 69, 357–360.

Oliveira, V. B., Linares, A. M., Correa, G. L. C. & Chiarello, A. G. 2008. Predation on the black capuchin monkey *Cebus nigritus* (Primates: Cebidae) by domestic dogs *Canis lupus familiaris* (Carnovora: Canidae), in the Parque Estadual Serra do Brigadeiro, Minas Gerais, Brazil. *Revista Brasileira de Zoologia*, 25, 376–378.

Olupot, W. & Waser, P. M. 2001. Activity patterns, habitat use and mortality risks of mangabey males living outside social groups. *Animal Behaviour*, 61, 1227–1235.

Ouattara, K., Zuberbühler, K., N'goran, E. K., Gombert, J.-E. & Lemasson, A. 2009a. The alarm call system of female Campbell's monkeys. *Animal Behaviour*, 78, 35–44.

Ouattara, K., Lemasson, A. & Zuberbühler, K. 2009b. Campbell's monkeys concatenate vocalizations into context-specific call sequences. *Proceedings of the National Academy of Sciences of the United States of America*, 51, 22026–22031.

Ouattara, K., Lemasson, A. & Zuberbühler, K. 2009c. Anti-predator strategies of free-ranging Campbell's monkeys. *Behaviour*, 146, 1687–1708.

Oversluijs Vasquez, M. R. & Heymann, E. W. 2001. Crested eagle (*Morphnus guianensis*) predation on infant tamarins (*Saguinus mystax* and *Saguinus fuscicollis*, Callithrichinae). *Folia Primatologica*, 72, 301–303.

Papworth, S., Böse, A.-S., Barker, J., Schel, A. M. & Zuberbühler, K. 2008. Male blue monkeys alarm call in response to danger experienced by others. *Biology Letters*, 4, 472–475.

Partan, S. & Marler, P. 2002. The Umwelt and its relevance to animal communication: Introduction to special issue. *Journal of Comparative Psychology*, 116, 116–119.

Patel, E. R. 2005. Silky sifaka predation (*Propithecus candidus*) by a fossa (*Cryptoprocta ferox*). *Lemur News*, 10, 25–27.

Payne, R. S. 1971. Acoustic location of prey by barn owls (*Tyto alba*). *Journal of Experimental Biology*, 54, 535–573.

Peeters, M., Courgnaud, V., Abela, B., Auzel, P., Pourrut, X., Bibollet-Ruche, F., Loul, S., Liegeois, F., Butel, C., Koulagna, D., Mpoudi-Ngole, E., Shaw, G. M., Hahn, B. H. & Delaporte, E. 2002. Risk to human health from a plethora of simian immunodeficiency viruses in primate bushmeat. *Emerging Infectious Diseases*, 8, 451–457.

Peetz, A., Norconk, M. A. & Kinzey, W. G. 1992. Predation by jaguar on howler monkeys (*Alouatta seniculus*) in Venezuela. *American Journal of Primatology*, 28, 223–228.

Pereira, M. E. & Macedonia, J. M. 1991. Ringtailed lemur anti-predator calls denote predator class, not response urgency. *Animal Behaviour*, 41, 543–544.

Peres, C. A. 1990. A harpy eagle successfully captures an adult male red howler monkey. *Wilson Bulletin*, 102, 560–561.

——. 1993. Anti-predation benefits in a mixed-species group of Amazonian tamarins. *Folia Primatologica*, 61, 61–76.

Perry, S., Manson, J. H., Dower, G. & Wikberg, E. 2003. White-faced capuchins cooperate to rescue a groupmate from a *Boa constrictor*. *Folia Primatologica*, 74, 109–111.

Petter, J.-J., Albignac, R. & Rumpler, Y. 1977. *Mammifères Lémuriens (Primates Prosimiens)*. Vol. 44 of Faune de Madagascar. Paris: ORSTOM/CNRS.

Pitts, A. 1995. Predation by *Eulemur fulvus rufus* on an infant *Lemur catta* at Berenty, Madagascar. *Folia Primatologica*, 65, 169–171.

Prasetyo, D., Ancrenaz, M., Morrogh-Bernard, H. C., Utami Atmoko, S. S., Wich, S. A. & van Schaik, C. P. 2009. Nest building in orangutans. In *Orangutans: Geographic Variation in Behavioral Ecology and Conservation* (ed. by S. A. Wich, S. S. Utami Atmoko, T. Mitra Setia & C. P. van Schaik), 269–277. New York: Oxford University Press.

Pruetz, J. D. & Bertolani, P. 2007. Savanna chimpanzees, *Pan troglodytes verus*, hunt with tools. *Current Biology*, 17, 412–417.

Pruetz, J. D. & Marshack, J. L. 2009. Savanna chimpanzees (*Pan troglodytes verus*) prey on patas monkeys (*Erythrocebus patas*) at Fongoli, Senegal. *Pan Africa News*, 16, 15–17.

Qihai, Z., Chengming, H., Ming, L. & Fuwen, W. 2009. Sleeping site use by *Trachypithecus francoisi* at Nonggang Nature Reserve, China. *International Journal of Primatology*, 30, 353–365.

Radespiel, U., Cepok, S., Zietemann, V. & Zimmermann, E. 1998. Sex-specific usage patterns of sleeping sites in grey mouse lemurs (*Microcebus murinus*) in northwestern Madagascar. *American Journal of Primatology*, 46, 77–84.

Radespiel, U., Ehresmann, P. & Zimmermann, E. 2003. Species-specific usage of sleeping sites in two sympatric mouse lemur species (*Microcebus murinus* and *M. ravelobensis*) in northwestern Madagascar. *American Journal of Primatology*, 59, 139–151.

Rahlfs, M. & Fichtel, C. 2010. Anti-predator behaviour in a nocturnal primate, the grey mouse lemur (*Microcebus murinus*). *Ethology*, 116, 429–439.

Rainey, H. J., Zuberbühler, K. & Slater, P. J. B. 2004. Hornbills can distinguish between primate alarm calls. *Proceedings of the Royal Society B*, 271, 755–759.

Rakotondravony, D., Goodman, S. M. & Soarimalala, V. 1998b. Predation on *Hapalemur griseus griseus* by *Boa manditra* (Boidae) in the littoral forest of eastern Madagascar. *Folia Primatologica*, 69, 405–408.

Ramakrishnan, U. & Coss, R. G. 2000. Recognition of heterospecific alarm vocalizations by bonnet macaques (*Macaca radiata*). *Journal of Comparative Psychology*, 114, 3–12.

——. 2001a. Strategies used by bonnet macaques (*Macaca radiata*) to reduce predation risk while sleeping. *Primates*, 42, 193–206.

——. 2001b. A comparison of the sleeping behavior of three sympatric primates: A preliminary report. *Folia Primatologica*, 72, 51–53.

Ramakrishnan, U., Coss, R. G., Schank, J., Dharawat, A. & Kim, S. 2005. Snake species discrimination by wild bonnet macaques (*Macaca radiata*). *Ethology*, 111, 337–356.

Rand, A. L. 1935. On the habits of some Madagascar mammals. *Journal of Mammalogy*, 16, 89–104.

Rasoanindrainy, J. M. 1985. Population dynamics of *Propithecus verreauxi* [in Russian]. *Nauchnye Doklady Vyssh. Shk. Biologicheskie. Nauki*, 4, 43.

Rasoloarison, R. M., Rasolonandrasana, B. P. N., Ganzhorn, J. U. & Goodman, S. M. 1995. Predation on vertebrates in the Kirindy forest, western Madagascar. *Ecotropica*, 1, 59–65.

Rasoloarison, R. M., Goodman, S. M. & Ganzhorn, J. U. 2000. Taxonomic revision of mouse lemurs (*Microcebus*) in western portions of Madagascar. *International Journal of Primatology*, 21, 963–1019.

Rasoloharijaona, S., Randrianambinina, B. & Zimmermann, E. 2008. Sleeping site ecology in a rain-forest dwelling nocturnal lemur (*Lepilemur mustelinus*): implications for sociality and conservation. *American Journal of Primatology*, 70, 247–253.

Ratsirarson, J. & Emady, R. 1999. Predation de *Lepilemur leucopus* par *Buteo brachypterus* dans la forêt galerie de Beza Mahafaly. *Working Group on Birds in the Madagascar Region Newsletter*, 9, 14–15.

Ratsirarson, J., Randrianarisoa, J., Ellis, E., Emady, R. J., Efitroarany, Ranaivonasy, J., Razanajaonarivalona, E. H. & Richard, A. F. 2001. Bezà Mahafaly: Écologie et réalités socio-économique. *Recherches pour le Développement B*, 18, 1–104.

Reichard, U. 1998. Sleeping sites, sleeping places, and presleep behavior of gibbons (*Hylobates lar*). *American Journal of Primatology*, 46, 35–62.

Rice, W. R. 1983. Sensory modality: An example of its effect on optimal foraging behavior. *Ecology*, 64, 403–406.

Richard, A. F. 1978. *Behavioral Variation: Case Study of a Malagasy Lemur*. Lewisburg, PA: Bucknell University Press.

Richard, A. F., Rakotomanga, P. & Schwartz, M. 1991. Demography of *Propithecus verreauxi* at Beza Mahafaly, Madagascar: Sex ratio, survival, and fertility, 1984–1988. *American Journal of Physical Anthropology*, 84, 307–322.

Ridley, A. R., Child, M. F. & Bell, M. B. V. 2007. Interspecific audience effects on the alarm-calling behaviour of a kleptoparasitic bird. *Biology Letters*, 3, 589–591.

Rose, L. M. & Fedigan, L. M. 1995. Vigilance in white-faced capuchins, *Cebus capucinus*, in Costa Rica. *Animal Behaviour*, 49, 63–70.

Russel, R. J. 1977. The behavior, ecology, and environmental physiology of a nocturnal primate, *Lepilemur mustelinus* (Strepsirhini, Lemuriformes, Lepilemuridae). PhD thesis, Duke University.

Sampaio, D. T. & Ferrari, S. F. 2005. Predation of an infant titi monkey (*Callicebus moloch*) by a tufted capuchin (*Cebus apella*). *Folia Primatologica*, 76, 113–115.

Sauther, M. L. 2002. Group size effects on predation sensitive foraging in wild ring-tailed lemurs (*Lemur catta*). In *Eat or Be Eaten: Predator Sensitive Foraging Among Primates* (ed. by L. E. Miller), 107–125. New York: Cambridge University Press.

Schaller, G. B. 1968. Hunting behavior of the cheetah in the Serengeti National Park, Tanzania. *East African Wildlife Journal*, 6, 95–100.

———. 1972. *The Serengeti Lion: A Study of Predator-Prey Relations*. Chicago: University of Chicago Press.

Schel, A. M. & Zuberbühler, K. 2009. Responses to leopards are independent of experience in Guereza colobus monkeys. *Behaviour*, 146, 1709–1737.

Schliehe-Diecks, S., Markolf, M. & Huchard, E. 2010. When big lemurs swallow up small ones: The Coquerel's dwarf lemur as a predator of grey mouse lemurs and endemic rodents. *Lemur News*, 15, 13–14.

Schmid, J. 1999. Sex-specific differences in activity patterns and fattening in the gray mouse lemur (*Microcebus murinus*) in Madagascar. *Journal of Mammalogy*, 80, 749–757.

Schmid, J. & Kappeler, P. M. 1998. Fluctuating sexual dimorphism and differential hibernation by sex in a primate, the gray mouse lemur (*Microcebus murinus*). *Behavioral Ecology and Sociobiology*, 43, 125–132.

Schülke, O. 2001. Social anti-predator behaviour in a nocturnal lemur. *Folia Primatologica*, 172, 332–334.

Schülke, O. & Kappeler, P. M. 2003. So near and yet so far: Territorial pairs but low cohesion between pair partners in a nocturnal lemur, *Phaner furcifer*. *Animal Behaviour*, 65, 331–343.

Schülke, O. & Ostner, J. 2001. Predation on *Lepilemur* by a harrier hawk and implications for sleeping site quality. *Lemur News*, 6, 5.

Schwab, D. 1999. Predation on *Eulemur fulvus* by *Accipiter henstii* (Henst's Goshawk). *Lemur News*, 4, 34.

Searcy, Y. M. & Caine, N. G. 2003. Hawk calls elicit alarm and defensive reactions in captive Geoffroy's marmosets (*Callithrix geoffroyi*). *Folia Primatologica*, 74, 115–125.

Semple, S., Cowlishaw, G. & Bennett, P. M. 2002. Immune system evolution among anthropoid primates: Parasites, injuries and predators. *Proceedings of the Royal Society B*, 269, 1031–1037.

Seyfarth, R. M. & Cheney, D. L. 1980. The ontogeny of vervet monkey alarm calling behavior: A preliminary report. *Zeitschrift für Tierpsychologie*, 54, 37–56.

———. 1986. Vocal development in vervet monkeys. *Animal Behaviour*, 34, 1640–1658.

Seyfarth, R. M., Cheney, D. L. & Marler, P. 1980a. Vervet monkey alarm calls: Semantic communication in a free-ranging primate. *Animal Behaviour*, 28, 1070–1094.

———. 1980b. Monkey responses to three different alarm calls: Evidence of predator classification and semantic communication. *Science*, 210, 801–803.

Sherman, P. W. 1977. Nepotism and the evolution of alarm calls. *Science*, 197, 1246–1253.

Shibasaki, M. & Kawai, N. 2009. Rapid detection of snakes by Japanese monkeys (*Macaca fuscata*): An evolutionarily predisposed visual system. *Journal of Comparative Psychology*, 123, 131–135.

Shultz, S. & Dunbar, R. I. M. 2006. Chimpanzee and felid diet composition is influenced by prey brain size. *Biology Letters*, 2, 505–508.

Shultz, S. & Noë, R. 2002. The consequences of crowned eagle central-place foraging on predation risk in monkeys. *Proceedings of the Royal Society B*, 269, 1797–1802.

Shultz, S., Faurie, C. & Noë, R. 2003. Behavioural responses of Diana monkeys to male long-distance calls: changes in ranging, association patterns and activity. *Behavioral Ecology and Sociobiology*, 53, 238–245.

Shultz, S., Noë, R., McGraw, W. S. & Dunbar, R. I. M. 2004. A community-level evaluation of the impact of prey behavioural and ecological characteristics on predator diet composition. *Proceedings of the Royal Society B*, 271, 725–732.

Sih, A. 1997. To hide or not to hide? Refuge use in a fluctuating environment. *Trends in Ecology & Evolution*, 12, 375–376.

Slip, D. J. & Shine, R. 1988. Feeding habits of the diamond python, *Morelia s. spilota*: Ambush predation by a boid snake. *Journal of Herpetology*, 22, 323–330.

Slobodchikoff, C. N., Kiriazis, J., Fischer, C. & Creef, E. 1991. Semantic information distinguishing individual predators in the alarm calls of Gunnison's prairie dogs. *Animal Behaviour*, 42, 713–719.

Smith, A. C., Knogge, C., Huck, M., Löttker, P., Buchanan-Smith, H. M. & Heymann, E. W. 2007. Long-term patterns of sleeping site use in wild saddleback (*Saguinus fuscicollis*) and mustached tamarins (*S. mystax*): Effects of foraging, thermoregulation, predation, and resource defense constraints. *American Journal of Physical Anthropology*, 134, 340–353.

Stafford, B. J. & Ferreira, F. M. 1995. Predation attempts on callitrichids in the Atlantic coastal rain forest of Brazil. *Folia Primatologica*, 65, 229–233.

Stanford, C. B. 1989. Predation on capped langurs (*Presbytis pileata*) by cooperatively hunting jackals (*Canis aureus*). *American Journal of Primatology*, 19, 53–56.

———. 1995. The influence of chimpanzee predation on group size and anti-predator behaviour in red colobus monkeys. *Animal Behaviour*, 49, 577–587.

————. 1998. Predation and male bonds in primate societies. *Behaviour*, 135, 513–533.

————. 2002. Avoiding predators: Expectations and evidence in primate antipredator behavior. *International Journal of Primatology*, 23, 741–757.

Stanford, C. B. & O'Malley, R. C. 2008. Sleeping tree choice by Bwindi chimpanzees. *American Journal of Primatology*, 70, 642–649.

Stanford, C. B., Wallis, J., Matama, H. & Goodall, J. 1994. Patterns of predation by chimpanzees on red colobus monkeys in Gombe National Park, 1982–1991. *American Journal of Physical Anthropology*, 94, 213–228.

Steenbeek, R., Piek, R. C., van Buul, M. & van Hooff, J. A. R. A. M. 1999. Vigilance in wild Thomas's langurs (*Presbytis thomasi*): The importance of infanticide risk. *Behavioral Ecology and Sociobiology*, 45, 137–150.

Stephan, C. & Zuberbühler, K. 2008. Predation increases acoustic complexity in primate alarm calls. *Biology Letters*, 4, 641–644.

Sterling, E. J. & McFadden, K. 2000. Rapid census of lemur populations in the Parc National de Marojejy, Madagascar. *Fieldiana Zoology*, 97, 265–274.

Stewart, F. A., Pruetz, J. D. & Hansell, M. H. 2007. Do chimpanzees build comfortable nests? *American Journal of Primatology*, 69, 930–939.

Stojan-Dolar, M. & Heymann, E. W. 2010a. Vigilance in mustached tamarins of single-species and mixed-species groups: The influence of group composition. *Behavorial Ecology and Sociobiology*, 64, 325–335.

————. 2010b. Vigilance in a cooperatively breeding primate. *International Journal of Primatology*, 31, 95–116.

Stone, A. I. 2007. Ecological risk aversion and foraging behaviors of juvenile squirrel monkeys (*Saimiri sciureus*). *Ethology*, 113, 782–792.

Struhsaker, T. T. 1967a. Auditory communication among vervet monkeys (*Cercopithecus aethiops*). In *Social Communication among Primates* (ed. by S. A. Altmann), 281–324. Chicago: University of Chicago Press.

————. 1967b. Ecology of vervet monkeys (*Cercopithecus aethiops*) in the Masai-Amboseli Game Reserve, Kenya. *Ecology*, 48, 891–904.

————. 1970. Notes on *Galagoides demidovii* in Cameroon. *Mammalia*, 34, 207–211.

Struhsaker, T. T. & Leakey, M. 1990. Prey selectivity by crowned hawk-eagles on monkeys in the Kibale Forest, Uganda. *Behavioral Ecology and Sociobiology*, 26, 435–443.

Sündermann, D., Scheumann, M. & Zimmermann, E. 2008. Olfactory predator recognition in predator-naïve gray mouse lemurs (*Microcebus murinus*). *Journal of Comparative Psychology*, 122, 146–155.

Sugardjito, J. 1983. Selecting nest-sites of Sumatran orang-utans, *Pongo pygmaeus abelii*, in the Gunung Leuser National Park, Indonesia. *Primates*, 24, 467–474.

Surbeck, M. & Hohmann, G. 2008. Primate hunting by bonobos at LuiKotale, Salonga National Park. *Current Biology*, 18, R906–R907.

Surbeck, M., Fowler, A., Deimel, C. & Hohmann, G. 2009. Evidence for the consumption of arboreal, diurnal primates by bonobos (*Pan paniscus*). *American Journal of Primatology*, 71, 171–174.

Tamura, N. 1989. Snake-directed mobbing by the Formosan squirrel *Callosciurus erythraeus thaiwanensis*. *Behavioral Ecology and Sociobiology*, 24, 175–180.

Teelen, S. 2008. Influence of chimpanzee predation on the red colobus population at Ngogo, Kibale National Park, Uganda. *Primates*, 49, 41–49.

Tello, N. S., Huck, M. & Heymann, E. W. 2002. *Boa constrictor* attack and successful group defence in moustached tamarins, *Saguinus mystax*. *Folia Primatologica*, 73, 146–148.

Templeton, C. N., Greene, E. & Davis, K. 2005. Allometry of alarm calls: Black-capped chickadees encode information about predator size. *Science*, 308, 1934–1937.

Tenaza, R. R. & Tilson, R. L. 1977. Evolution of long-distance alarm calls in Kloss's gibbon. *Nature*, 268, 233–235.

Terborgh, J. 1983. *Five New World Primates: A Study in Comparative Ecology*. Princeton, NJ: Princeton University Press.

————. 1990. Mixed flocks and polyspecific associations: Costs and benefits of mixed groups to birds and monkeys. *American Journal of Primatology*, 21, 87–100.

Terborgh, J. & Janson, C. H. 1986. The socioecology of primate groups. *Annual Review of Ecology and Systematics*, 17, 111–135.

Thorstrom, R. & La Marca, G. 2000. Nesting biology and behavior of the Madagascar harrier-hawk (*Polyboroides radiatus*) in northeastern Madagascar. *Journal of Raptor Research*, 34, 120–125.

Townsend, S. W., Deschner, T. & Zuberbühler, K. 2008. Female chimpanzees use copulation calls flexibly to prevent social competition. *PLoS ONE*, 3, e2431, doi:10.1371/journal.pone.0002431.

Treves, A. 1998. The influence of group size and neighbors on vigilance in two species of arboreal monkeys. *Behaviour*, 135, 453–481.

————. 1999. Has predation shaped the social systems of arboreal primates? *International Journal of Primatology*, 20, 35–67.

————. 2000. Theory and method in studies of vigilance and aggregation. *Animal Behaviour*, 60, 711–722.

Treves, A., Drescher, A. & Ingrisano, N. 2001. Vigilance and aggregation in black howler monkeys (*Alouatta pigra*). *Behavioral Ecology and Sociobiology*, 50, 90–95.

Treves, A., Drescher, A. & Snowdon, C. T. 2003. Maternal watchfulness in black howler monkeys (*Alouatta pigra*). *Ethology*, 109, 135–146.

Trivers, R. 1971. The evolution of reciprocal altruism. *Quarterly Review of Biology*, 46, 35–57.

Tsukahara, T. 1993. Lions eat chimpanzees: The first evidence of predation by lions on wild chimpanzees. *American Journal of Primatology*, 29, 1–11.

Uehara, S. 1997. Predation on mammals by the chimpanzee (*Pan troglodytes*). *Primates*, 38, 193–214.

Van Noordwijk, M. A., Hemelrijk, C. K., Herremans, L. A. M. & Sterck, E. H. M. 1993. Spatial position and behavioral sex differences in juvenile long-tailed macaques. In *Juvenile Primates* (ed. by M. E. Pereira & L. A. Fairbanks), 77–85. New York: Oxford University Press.

Van Schaik, C. P. 1983. Why are diurnal primates living in groups? *Behaviour*, 87, 120–144.

Van Schaik, C. P. & Hörstermann, M. 1994. Predation risk and the number of adult males in a primate group: A comparative test. *Behavioral Ecology and Sociobiology*, 35, 261–272.

Van Schaik, C. P. & Mitrasetia, T. 1990. Changes in the behaviour of wild long-tailed macaques (*Macaca fascicularis*) after encounters with a model python. *Folia Primatologica*, 55, 104–108.

Van Schaik, C. P. & van Hooff, J. A. R. A. M. 1983. On the ultimate causes of primate social systems. *Behaviour*, 85, 91–117.

Van Schaik, C. & van Noordwijk, M. A. 1985. Evolutionary effect of the absence of felids on the social organization of the macaques on the island of Simeulue (*Macaca fascicularis fusca*, Miller 1903). *Folia Primatologica*, 44, 138–147.

———. 1989. The special role of male *Cebus* monkeys in predation avoidance and its effect on group composition. *Behavioral Ecology and Sociobiology*, 24, 265–276.

Van Schaik, C. P., van Noordwijk, M. A., Warsono, B. & Sutriono, E. 1983. Party size and early detection of predators in Sumatran forest primates. *Primates*, 24, 211–221.

Vitale, A. F., Visalberghi, E. & De Lillo, C. 1991. Responses to a snake model in captive crab-eating macaques (*Macaca fascicularis*) and captive tufted capuchins (*Cebus apella*). *International Journal of Primatology*, 12, 277–286.

Waser, P. M. 1980. Small nocturnal carnivores: Ecological studies in the Serengeti. *African Journal of Ecology*, 18, 167–185.

Watts, D. P. & Mitani, J. C. 2002. Hunting behavior of chimpanzees at Ngogo, Kibale National Park, Uganda. *International Journal of Primatology*, 23, 1–28.

Wheeler, B. C. 2008. Selfish or altruistic? An analysis of alarm call function in wild capuchin monkeys, *Cebus apella nigritus*. *Animal Behaviour*, 76, 1465–1475.

———. 2009. Monkeys crying wolf? Tufted capuchin monkeys use anti-predator calls to usurp resources from conspecifics. *Proceedings of the Royal Society B*, 276, 3013–3018.

Whitesides, G. H. 1989. Interspecific associations of Diana monkeys, *Cercopithecus diana*, in Sierra Leone, West Africa: Biological significance or chance? *Animal Behaviour*, 37, 760–776.

Wich, S. A. & Sterck, E. H. M. 2003. Possible audience effect in Thomas langurs (Primates; *Presbytis thomasi*): An experimental study on male loud calls in response to a tiger model. *American Journal of Primatology*, 60, 155–159.

Wich, S. A., Koski, S., de Vries, H. & van Schaik, C. P. 2003. Individual and contextual variation in Thomas langur male loud calls. *Ethology*, 109, 1–13.

Wilkie, D. S. & Carpenter, J. F. 1999. Bushmeat hunting in the Congo Basin: An assessment of impacts and options for mitigation. *Biodiversity and Conservation*, 8, 927–955.

Wilson, J. M., Stewart, P. D., Ramangason, G.-S., Denning, A. M. & Hutchings, M. S. 1989. Ecology and conservation of the crowned lemur, *Lemur coronatus*, at Ankarana, N. Madagascar: With notes on Sanford's lemur, other sympatrics and subfossil lemurs. *Folia Primatologica*, 52, 1–26.

Wolters, S. & Zuberbühler, K. 2003. Mixed-species associations of Diana and Campbell's monkeys: The costs and benefits of a forest phenomenon. *Behaviour*, 140, 371–385.

Woodland, D. J., Jaafar, Z. & Knight, M.-L. 1980. The "pursuit-deterrent" function of alarm signals. *American Naturalist*, 115, 748–753.

Wrangham, R. W. & van Zinnicq Bergmann Riss, E. 1990. Rates of predation on mammals by Gombe chimpanzees, 1972–1975. *Primates*, 31, 157–170.

Wright, P. C. 1994. The nocturnal primate niche in the New World. *Journal of Human Evolution*, 18, 635–658.

———. 1998. Impact of predation risk on the behaviour of *Propithecus diadema edwardsi* in the rain forest of Madagascar. *Behaviour*, 135, 483–512.

Wright, P. C. & Martin, L. B. 1995. Predation, pollination and torpor in two nocturnal prosimians (*Cheirogaleus major* and *Microcebus rufus*) in the rain forest of Madagascar. In *Creatures of the Dark: The Nocturnal Prosimians* (ed. by L. Alterman, G. A. Doyle & M. K. Izard), 45–60. New York: Plenum.

Wright, P. C., Heckscher, S. K. & Dunham, A. E. 1997. Predation on Milne-Edward's sifaka (*Propithecus diadema edwardsi*) by the fossa (*Cryptoprocta ferox*) in the rain forest of southeastern Madagascar. *Folia Primatologica*, 68, 34–43.

Wright, P. C., Heckscher, S. & Dunham, A. 1998. Predation on rain forest prosimians in Ranomafana National Park, Madagascar. *Folia Primatologica*, 69, Suppl. 1, 401.

Yamagiwa, J. 2001. Factors influencing the formation of ground nests by eastern lowland gorillas in Kahuzi-Biega National Park: Some evolutionary implications of nesting behavior. *Journal of Human Evolution*, 40, 99–109.

Yeager, C. P. 1991. Possible antipredator behavior associated with river crossings by proboscis monkeys (*Nasalis larvatus*). *American Journal of Primatology*, 24, 61–66.

Yorzinski, J. L. & Vehrencamp, S. L. 2008. Preliminary report: Antipredator behaviors of mandrills. *Primate Report*, 75, 11–18.

Yorzinski, J. L. & Ziegler, T. 2007. Do naïve primates recognize the vocalizations of felid predators? *Ethology*, 113, 1219–1227.

Zinner, D., Hilgartner, R. D., Kappeler, P. M., Pietsch, T. & Ganzhorn, J. U. 2003. Social organization of *Lepilemur ruficaudatus*. *International Journal of Primatology*, 24, 869–888.

Zuberbühler, K. 2000a. Causal cognition in a non-human primate: Field playback experiments with Diana monkeys. *Cognition*, 76, 195–207.

———. 2000b. Causal knowledge of predators' behaviour in wild Diana monkeys. *Animal Behaviour*, 59, 209–220.

———. 2000c. Interspecies semantic communication in two forest primates. *Proceedings of the Royal Society B*, 267, 713–718.

———. 2000d. Referential labelling in Diana monkeys. *Animal Behaviour*, 59, 917–927.

———. 2001. Predator-specific alarm calls in Campbell's monkeys, *Cercopithecus campbelli*. *Behavioral Ecology and Sociobiology*, 50, 414–422.

Zuberbühler, K., Noë, R. & Seyfarth, R. M. 1997. Diana monkey long-distance calls: messages for conspecifics and predators. *Animal Behaviour*, 53, 589–604.

Zuberbühler, K., Cheney, D. L. & Seyfarth, R. M. 1999a. Conceptual semantics in a nonhuman primate. *Journal of Comparative Psychology*, 113, 33–42.

Zuberbühler, K., Jenny, D. & Bshary, R. 1999b. The predator deterrence function of primate alarm calls. *Ethology*, 105, 477–490.

Chapter 9 Ecological and Social Influences on Sociality

Oliver Schülke and Julia Ostner

RIMATE SOCIALITY has many faces. Some callitrichines form small groups with just one reproducing female and others who help care for her young. Hamadryas baboons (*Papio hamadryas*) live in small harems that aggregate into larger units containing several hundred individuals. Ring-tailed lemur (*Lemur catta*) males fight fiercely to dominate each other, but unconditionally submit to all females in their group. An entire group of Thomas's langurs (*Presbytis thomasi*) disintegrates when the females emigrate to seek a stronger protector for their infants. It proves useful to structure this diversity into three aspects of sociality that constitute the social system: social organization (who lives with whom), social structure emerging from dyadic social relationships (who interacts with whom, how often, and how), and mating systems (who mates and reproduces with whom). In the first part of this chapter we provide an overview of primate social systems, defining the terminology that is relevant for their study.

Efforts to explain the diversity in primate sociality based on differences in ecology have a long history (reviewed in Janson 2000), dating back to Crook and Gartlan's (1966) comparative analysis of data derived from early primate field studies. The socioecological model in figure 9.1 summarizes several hypotheses about the relationships between ecological and social factors and primate societies. Since mammalian females, whose potential reproductive rates are lower than those of their male conspecifics, compete mainly over food resources (Trivers 1972), they form groups if group living improves their access to food. Fe-

males also form groups if group life reduces their risk of predation or disease. Alternatively, females may associate with males if the males protect them against sexual coercion. Characteristics of food resources will then determine how females compete for food, which in turn determines their social structure. Two fundamentally different kinds of competition have been identified, namely scramble (indirect) and contest (direct) competition (Bradbury & Vehrencamp 1976; Vehrencamp 1983). In scramble competition, any individual influences the net energy gain of all other group members by using some portion of the food available (Nicolson 1954). If the resource is contested, a subset of (dominant) individuals will restrict other (subordinate) individuals' access to the resource.

Males primarily compete over access to fertile females because of those females' lower potential reproductive rate. Consequently, male distribution is constrained by and maps onto the distribution of females, which in turn strongly influences the mating system. Individual males usually strive to monopolize access to as many females as possible, and different factors determine whether they succeed. In the second, third, and fourth parts of this chapter we review theory and evidence for the different aspects of the socioecological model dealing separately with ecological influences on social organization, social structure, and mating systems. In a fifth section we review alternative explanations for the diversity of primate societies. We conclude by summarizing the strengths and weaknesses of socioecological models and suggesting a few avenues for future research.

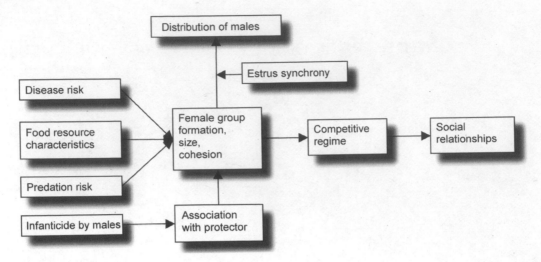

Fig. 9.1. The socioecological synthesis. Females form groups of varying size and cohesion as a response to the distribution of risks and resources in the environment. Alternatively, they gather around a strong protector against male harassment and infanticide. Predation risk sets the minimum group size, while within-group scramble competition for food or infanticide risk sets the upper limit. Varying numbers of males associate with females, depending on female group size and estrous synchrony as assayed by seasonality in reproduction. The competitive regime is characterized by modes and strengths of different types of feeding competition, as determined by food resource characteristics, and it influences female social relationships (Wrangham 1980; van Schaik 1989; Sterck et al. 2002). For the sake of simplicity, feed-back loops (e.g., between male distribution and estrus synchrony, a female strategy) are not shown.

Structuring the Diversity of Primate Societies

Social organization, social structure, and mating system all need to be clearly defined if we are to seek evolutionary explanations for their variation (Struhsaker 1969; Rowell 1979; Kappeler & van Schaik 2002). Monogamy, for example, remains an ambiguous concept (Wickler & Seibt 1983), and its evolution is still debated (Reichard 2003). Various authors use the term to describe a form of social organization (Fietz 1999; Fuentes 1999), a mating system (Ribble 1991), or a mixture of both (Brotherton et al. 1997). This has led to considerable confusion (Sommer & Reichard 2000), as all three of these aspects can be interrelated.

Social Organization

Social organization can be defined according to the size, sex composition, and fission-fusion dynamics of social units, as well as by their variability in a population or species (Kappeler & van Schaik 2002; Aureli et al. 2008). To constitute a *social unit*, animals that share a common range must also interact more frequently with each other than with other conspecifics. In other words, spatial overlap has to coincide with social boundaries (Schülke & Kappeler 2003). Each male and female either lives by themselves or with one or several others, thus yielding five types of social organization (solitary, pair, one-male–multifemale group, multimale-multifemale group, multimale–one-female group: Kappeler & van Schaik 2002). These five types of social organization

vary along two additional axes: the extent of variation in spatial and temporal cohesion (fission-fusion dynamics: Aureli et al. 2008) and the degree to which modal social organization types are realized in a given population (uniform or variable: van Schaik & Kappeler 2003).

In some species, groups split (fission) into two or more parties and merge (fusion) over time. *Fission-fusion dynamics* are described by the temporal variation in (1) spatial cohesion between party members, (2) size of parties, and (3) composition of parties. We use the term "party" instead of "subgroup" because a party can consist of single individuals who may or may not be members of a social unit. While the first axis relates to a distinction between *cohesive* (cf. Müller & Thalmann 2000, *associated* sensu van Schaik & Kappeler 2003) and *dispersed* social units (Eisenberg et al. 1972; Müller & Thalmann 2000), the second and third dimensions add significant detail to the classification of the social organization type (Aureli et al. 2008).

To characterize the specific social organization of a species or population, one usually refers to the modal social organization type—that is, the type exhibited by the majority of social units. Considerable variation exists, however, within species, populations, and even units (Strier 1994; Hill & Dunbar 2002; Thierry 2007; fig. 9.2). To understand species differences with respect to the degree of this variation (Lee & Kappeler 2003), it has been useful to distinguish between *uniformly* occurring social organization types, where more than 90% of social units resemble one type, and *variably* occurring types, where the majority but

Fig. 9.2. Folivorous Hanuman langurs, *Semnopithecus entellus,* from Ramnagar, Nepal. Across the Indian subcontinent monkeys of this species exhibit considerable flexibility in social organization, living either in one-male–multifemale groups or in multimale-multifemale groups. Photo courtesy of Oliver Schülke.

less than 90% of social units are of one type (see van Schaik & Kappeler 2003 and Fuentes 1999 for variable and uniform pair living in primates).

Mating Systems

The *mating system* distinguishes mating behavior ("social mating system") from its reproductive outcome ("genetic mating system"). Primates display most of the mammalian mating systems defined by Clutton-Brock (1989). These include *monogamy*, in which males and females only mate with one member of the opposite sex; *polyandry*, in which a female mates with several males but each male only with one female; *polygyny*, in which a male mates with several females and each female only with one male; and *polygynandry* (also called *promiscuity*) in which both males and females mate with several mating partners. While mating patterns are readily observable, reproductive success has been measured in many ways (chapters 14, 15, 17, and 18, this volume). Beyond this classification, there is tremendous variation in primate mating behavior that lies beyond the scope of this chapter (Dixson 1998).

The distributions of mating and reproductive activity across members of a social unit are primary features of mating systems. In callitrichines one breeding female systematically excludes other females from reproduction (Dietz 2004; Hager & Jones 2009). Skew in mating and repro-

ductive success among primate males living in multimale-multifemale groups varies greatly between populations and species, from absolute concentration in one male to more even distributions (Kutsukake & Nunn 2006; Ostner et al. 2008a). While males generally pursue sexual strategies to obtain exclusive access to fertile females (i.e., prefer polygyny), females frequently adopt strategies designed to enable them to mate with several males (chapter 16, this volume), thereby confusing paternity for various benefits (Birkhead 2000). Mating may not be confined to the social unit and may lead to extrapair paternity (Fietz et al. 2000; Schülke et al. 2004a) or extragroup paternity (Ostner et al. 2008a). Consequently, polygynous and polygynandrous mating systems are overrepresented in primates.

Social Structure

Social structure emerges from a bottom-up process via dyadic social interactions that determine social relationships whose patterns yield the social structure of a society (Hinde 1976; Kappeler & van Schaik 2002; chapter 23, this volume). A social interaction between two individuals of the same social unit may be *affinitive* (indicating a spatial association), *affiliative* (friendly), or *agonistic* (including aggressive or submissive behaviors). When two individuals interact repeatedly and the outcome of past interactions influences future ones, these individuals are said to have formed a *social relationship* (Hinde 1983). Social relationships vary as a function of the frequency, intensity, and quality of social interactions (de Waal 1986, 1989; chapter 23, this volume).

Dominance

Watts (2010) provides a detailed review of dominance in primates. Here we summarize some issues relevant to our discussion of primate socioecology and other chapters in this volume. *Dominance* is an emergent property that results from repeated agonistic conflicts. Dominance hierarchies describe asymmetries in agonistic relationships between individuals (de Waal & Luttrell 1989) and can be constructed in several ways. Ranking is most commonly based on (1) the I&SI method (de Vries 1995), (2) the David's score (David 1987, 1988; de Vries et al. 2006) or (3) self-reinforcing winner-loser effects (Dugatkin 1997; Hemelrijk 1999). All of these methods force individuals into a linear rank order and assume dyadic dominance relationships to be transitive (if *A* dominates *B* and *B* dominates *C*, then *A* must dominate *C*). The resulting hierarchy can be characterized in terms of its linearity, transitivity, and steepness (Appelby 1983; de Vries & Appelby 2000; Leinfelder et al. 2001).

The relationship between dominance status and stress hormones can also be viewed as another aspect of social structure that may vary between groups or species (Abbott et al. 2003; Goymann & Wingfield 2004). While short-term elevations of glucocorticoids are an adaptive response to temporary stressors and help to maintain homeostasis, a chronic increase due to prolonged exposure to stressors may have detrimental effects on an individual's fitness (Sapolsky 2002). The costs associated with rank acquisition and maintenance in dominants, the amount of aggression received, and the support available to subordinates are some factors that affect differences in stress hormone levels between individuals of different ranks (Abbott et al. 2003; Goymann & Wingfield 2004). As a result, dominance may be positively, negatively, or not at all related to glucocorticoid levels.

Recent research has introduced the nonsynonymous concepts of *power* (Lewis 2002) and *social power* (Flack & de Waal 2004) into research on dominance in primates. Lewis (2002) argued that power is a more comprehensive description of asymmetry in dyadic relationships because it includes two components. These include differences in resource holding potential or fighting ability as well as asymmetries based on other factors, such as the fighting ability of coalition partners, the possession of valuable commodities, and asymmetry in knowledge, which is called leverage (Hand 1986). Social power is defined as the "degree of implicit agreement or consensus among group members about whether an individual is capable of successfully introducing force into polyadic agonistic situations" (Flack & de Waal 2004, p. 169). The concept of social power may play a crucial role for understanding agonistic asymmetries in species with ritualized status signaling outside of agonistic encounters, such as chimpanzees (*Pan troglodytes*), macaques (*Macaca* spp.) and baboons (*Papio* spp.), as such signaling allows others to assess an individual's hidden power.

Different aspects of dominance and social power are summarized in the term *dominance style*: meaning the degree of expressed asymmetry in social power (de Waal 1989), which varies widely among species along three axes (van Schaik 1989; Sterck et al. 1997). Dominance relationships vary (1) from *individualistic* to *nepotistic*, (2) from *egalitarian* to *despotic* (Vehrencamp 1983), and (3) from *intolerant* to *tolerant*. Nepotistic relationships occur when genetic relatives support each other in agonistic conflict and hierarchies are substructured into matrilines (see below for a treatment of paternal relatedness). In egalitarian societies, dominance relationships between individuals are either absent or poorly defined, and dominance relationships and hierarchies are unclear and nonlinear. In despotic societies, individuals have clearly defined agonistic relationships and can be ordered along a linear dominance hierarchy.

Tolerant relationships are associated with increasing levels of counteraggression and reconciliation as well as with decreasing severity of aggression. Nepotistic and tolerant dominance styles can only be realized in despotic systems (Sterck et al. 1997).

If a single agonistic interaction involves more than two individuals it becomes polyadic, and if two or more individuals direct aggression against a common target it can be called an agonistic *coalition*. Coalitions can be categorized based on the ranks of the allies relative to the target and on their outcome. Coalitions can be *all-up* (*revolutionary*) if the target ranks higher than all coalition members, *all-down* (*conservative*) if the target ranks lower than the coalition members, or *bridging* if the target ranks between them (Chapais 1995; van Schaik et al. 2006). Coalitions may increase access to a limiting resource without changing the dominance rank of the individuals involved (*leveling*). Alternatively, a coalition may improve the ranks of the allies and consequently enhance their resource access (*rank changing*; van Schaik et al 2004a).

Affiliative Relationships

Like agonistic or dominance relationships, *affiliative* relationships can be classified according to the degree of asymmetry expressed within and between dyads. Affiliation may not be equally or randomly distributed across all group members; instead, it is commonly biased towards kin or higher-ranking individuals (Seyfarth 1977; chapter 24, this volume). Asymmetry may also exist within dyads—for example, when one partner is primarily responsible for maintaining the bond (Hinde & Atkinson 1970) or when reciprocation of grooming is low (Schino & Aureli 2008b). Biological market theory suggests that grooming can be exchanged for itself or interchanged for other commodities like agonistic support (Noë & Hammerstein 1995). Sometimes certain dyads form special relationships or friendships that stand out from those of other dyads (Silk 2002a). Friends frequently affiliate and support each other during conflicts. Most of these bonds form between same-sex individuals, but male-female friendships are common in some species of lemurs (Pereira & McGlynn 1997; Ostner & Kappeler 1999) and baboons (Smuts 1985; Palombit 1999). Apart from the frequency of friendly interactions, the quality of behavior may also vary. Some dyads reconcile after conflicts by affiliating, often with highly ritualized behavior (Bertrand 1969). Relationships may also be regulated via appeasement, consolation, or agonistic buffering (Aureli & de Waal 2000). Recently, social network analyses have been applied as an alternative means to evaluate social structure (Wey et al. 2007). While some local network measures have exact equivalents in classical sociometry, network analyses have the potential to assess properties of social structure beyond the individual up to the group level. These quantified features may be particularly useful for comparing social structure across groups or species.

Dispersal

Dispersal patterns have a strong influence on social structure because they determine the availability of social partners. For example, if all maturing females migrate from their natal group, adult maternal relatives will not reside together and matrilines cannot structure female society. Thus, dispersal regimes have been integrated into the study of the evolution of social relationships (Wrangham 1980). Dispersal regimes include male philopatry and female philopatry; in these, individuals of one sex remain in the natal group and members of the other sex disperse. Where both sexes disperse, dispersal is typically sex-biased, with one sex dispersing more frequently than the other. When individuals leave their natal home ranges and social groups, they experience different ecological and social costs and benefits (Isbell & van Vuren 1996; Sterck 1998; Sterck & Korstjens 2000).

The preceding discussion reveals the multifaceted nature of each of the three main aspects of primate sociality. In the next section we review theory and evidence for ecological influences on primate societies.

Ecological Influences on Social Organization

Evolution of Group Living

Socioecological theory posits that particular combinations of risks and resources in the environment promote the evolution of cohesive group living over other forms of social organization. Food, sexual coercion, predators, and pathogens have all been suggested as important evolutionary agents in this context. First, Wrangham (1980) hypothesized that group living evolved because gregarious females are more successful in finding and defending access to food resources if they cooperate against conspecifics in other groups. This between-group feeding competition hypothesis suggests that during evolution of group life, per capita net energy intake increased with group size. However, group size rarely exerts the predicted positive effect on female feeding efficiency (Janson 2000; Koenig 2002), thus suggesting that groups form for reasons other than individuals gaining benefits via between-group competition for food.

Second, Wrangham (1979) suggested that gregariousness arises when females face a risk of male sexual coer-cion and consequently associate with males who protect them from such aggression (see also Wrangham 1987; van Schaik 1996; Sterck et al. 1997). Some evidence accords with this hypothesis, as female gorillas (*Gorilla gorilla*) and some Asian colobines form groups to obtain male protection against infanticide (Watts 1989; Steenbeck et al. 2000; Steenbeck & van Schaik 2001; chapter 19, this volume).

A third hypothesis to explain group-living in primates proposes that females obtain increased protection from predators by living in cohesive groups (van Schaik 1983; Terborgh 1983; Dunbar 1988; van Schaik 1989). Predation is a rare event and its effect on primate behavior is therefore difficult to study (chapter 8, this volume). To complicate things more, it is neither the observable predation rate nor the risk of predation (i.e., the expected rate of predation or predation rate in unbiased samples that drives social evolution in this context. Instead, intrinsic predation risk, which may be high despite low observed rates when antipredation strategies are efficient, determines the adaptive value of countermeasures against predation (Janson 1998). In keeping with this idea, comparative analyses have revealed that primates exposed to higher intrinsic predation risk live in larger groups than those exposed to lower risk (Nunn & van Schaik 2000).

Female-female association in diurnal primates is thought to be an evolutionary response to high intrinsic predation risk (Hill & Lee 1998; Janson 1998). At equal encounter rates, group-living females may benefit by reducing the per capita likelihood of falling prey to a predator through the dilution, selfish herd, and confusion effects (Hamilton 1971; Charnov & Krebs 1975; Cheney & Wrangham 1987; chapter 8, this volume). Group-living animals may increase the rate at which they detect predators through increased vigilance and alarm calls, and may also launch cooperative counterattacks to drive predators away (chapter 8, this volume). An important trade-off exists, however, between investing in predator avoidance strategies and other fitness relevant behaviors like foraging (Cowlishaw 1997; Miller 2002; chapter 8, this volume). The variety of behavioral adaptations to avoid predation is testament to their evolutionary importance.

Because almost all diurnal primates live in groups or pairs, it is puzzling that nocturnal primates usually forage by themselves (Kappeler & van Schaik 2002). Recent studies have shown, however, that socioecological theory can be used to explain their social organization (Schülke 2003; Gursky 2005; Schülke & Ostner 2005; Dammhahn & Kappeler 2009). Consonant with the previous definition of a social unit—one whose social and spatial boundaries coincide—cheirogaleids alone display all types of group composition (Schülke & Ostner 2005), including truly solitary,

Fig. 9.3. Although pale forkmarked lemurs, *Phaner pallescens*, at Kirindy, Madagascar, live in heterosexual pairs, the individuals spend two of every three days sleeping on their own, away from their pair partners. Photo courtesy of Julia Ostner.

dispersed pairs and dispersed groups of all kinds (fig. 9.3). Solitary foraging individual females may form groups whose members interact frequently to derive benefits via kin-selected cooperative breeding (Eberle & Kappeler 2006) and between-group contest competition for sleeping sites (Dammhahn & Kappeler 2009). Alternatively, they may forage by themselves at night but aggregate with specific other individuals during the day, because sleeping in groups reduces predation risk and furnishes thermoregulatory benefits (Kappeler 1998; Radespiel et al. 1998; Schülke & Ostner 2005). The benefits accrued from group living are balanced against its costs, and females will not form dispersed groups if indirect competition for food is too strong (Schülke 2003, 2005; Dammhahn & Kappeler 2009).

In sum, the currently available evidence suggests that primates primarily live in groups because it decreases their vulnerability to predation and, for some primates, to risks posed by conspecifics. The question of whether feeding competition acts as a selective agent favoring the evolution of group life in primates continues to be debated. The role of pathogens in primate social evolution remains obscure (Nunn & Altizer 2006).

Constraints on Group Size

Indirect feeding competition

While the benefits of group living have long been debated, increased feeding competition was identified early on as an inevitable cost of gregariousness (Alexander 1974; Hladik 1975; Waser 1977; Leighton & Leighton 1982; Terborgh 1983). Whenever individuals live and feed closely together to reduce predation risk (Boinski & Garber 2000), they will

affect each other's foraging efficiency. When food is distributed in patches, it will be depleted faster if the groups are larger. Consequently, larger groups will leave patches earlier than smaller groups (patch depletion model, Chapman 1990; Chapman et al. 1995). If group members feed on more evenly distributed resources and slowly move forward while foraging, individuals traveling in the middle and the back of the group will arrive at feeding sites or search areas that have been visited and depleted by others, and will therefore push the group forward. From both scenarios it follows that larger groups must travel farther than smaller ones to satisfy the energy requirements of their members. Several studies, reviewed below, have tested this prediction across primates.

The first comprehensive comparative analysis of ranging costs (Wrangham et al. 1993) used mean daily travel distance as a proxy for the ecological costs of large group size. Janson and Goldsmith (1995) subsequently argued that fitness changes due to increasing costs should be viewed relative to the total costs. A small increase in travel may not be as costly in groups whose individuals travel far as it is for groups that move only little. Thus, Janson and Goldsmith evaluated feeding competition by relative ranging cost, assayed by regressing daily path length on group size and dividing the slope by the intercept (Janson & Goldsmith 1995). Both of these studies supported the predicted positive relationship between group size and ranging costs. Similarly, a recent meta-analysis confirmed that relationship and suggested that contradictory or inconclusive results from some studies may be due to their small sample size (Majolo et al. 2008; fig. 9.4). Increasing travel costs in

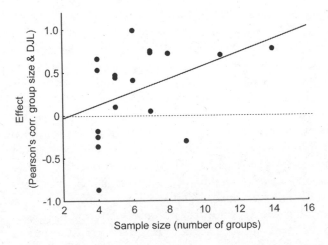

Fig. 9.4. Ecological costs of group size. As a consequence of increased feeding competition, large groups of primates have longer day journey lengths (DJL) than smaller ones (data from Majolo et al. 2008). This graph summarizes results from a meta-analysis by Majolo et al.; the larger the sample size, the more likely a positive relationship between group size and day journey length.

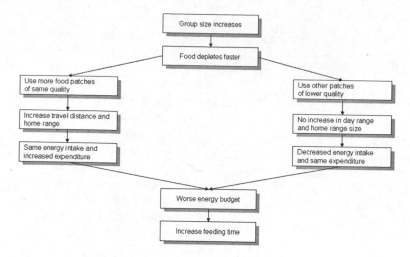

Fig. 9.5. Mechanisms of the patch depletion model of ecological constraints on group size. This flowchart illustrates the initial options for large groups when faced with faster food patch depletion. Individuals in large groups may try to compensate for the effects by increasing their feeding time.

larger groups may affect female energy budgets and translate into reduced female fecundity, but this relationship is less well supported (Majolo et al. 2008).

Increasing group size may not always lead to reduced fecundity because individuals can respond differently to the rapid depletion of food patches (fig. 9.5). When leaving a patch that has been depleted quickly, individuals in larger groups have two choices. On the one hand, they may travel to the next patch of similar food quality, which will again deplete relatively quickly, thus prompting another decision on where to feed. Using more patches of the same quality leads to increased travel distances but similar energy intake, and, in turn, to an unfavorable energy budget for individuals in larger groups. Alternatively, the individuals in these larger groups may use interspersed nearby food patches of lower food quality (in terms of energy intake rates: Schülke et al. 2006). This option will not increase their travel distance, but when feeding for the same amount of time per day, the individuals will have a lower energy intake due to inferior food quality, and consequently they will have an overall worse energy budget and lower fecundity than individuals living in smaller groups. This shows, therefore, that considering travel costs alone yields an incomplete picture.

If individuals in larger groups increase their feeding time, they may be able to compensate for, or at least reduce, the effect of lower average food quality and/or increased ranging costs (Stacey 1986; Janson 1988a; Koenig 2002; fig. 9.5). Brown capuchins (*Cebus apella*), for example, travel longer when in larger groups, and offset these costs by staying active and feeding longer in the afternoon, thus yielding similar energy budgets across group sizes (Janson 1988a). Increasing patch residence and giving-up times (Charnov 1976; Snaith & Chapman 2007) will reduce energy gain

rates, which will have the same effect as using also lower-quality patches without increasing travel effort (the path shown on the right in fig. 9.5). The model by Snaith and Chapman (2007) provides an alternative: a more mechanistic framework that seems to devalue strategic decisions involved with food selection and food choice, and treats the quality and distribution of food resources as invariant.

Predictions about the effects of increased group size hold only when groups of variable size live in the same habitat. When habitat quality varies across space, some groups may live in areas with increased food patch density or high average food quality. Day journey length may then be the same for both large and small groups because the larger group lives in a better habitat and therefore does not suffer the costs of increased size. Our review of studies that have tested the ecological constraints model suggests that unassayed variation in habitat quality alone may explain much of the contradictory evidence. Studies that ignored the effect of differences in habitat quality between groups often did not support the model or generated inconclusive results (13 of 39). Of the five studies that controlled for variation in habitat quality, four supported predictions of the patch depletion model.

Folivorous primates were long thought to be unaffected by indirect feeding competition because leaves were assumed to be a more evenly distributed, and generally more abundant, food resource than fruits (Hladik 1975; Isbell 1991). The fact that folivores do not generally live in larger groups than frugivores of similar body size is known as the folivore paradox (fig. 9.6). The folivore paradox has been resolved on ecological grounds by showing that folivores are selective feeders, their food resources may be patchily distributed, they deplete their food patches, and the abun-

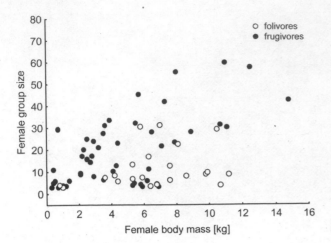

Fig. 9.6. Resolving the folivore paradox of group size. Assuming that feeding competition is more relaxed in folivores because of higher abundance, more even distribution, and lower quality of their food resources, and assuming that predation avoidance is more effective in large groups, folivores should live in larger groups than frugivores of comparable body size. This plot shows, however, that folivore groups are not generally larger in species of a given female body size. This has been termed "the folivore paradox" (data from Nunn & Barton 2001, supplemented with additional information on diet). It has now been demonstrated that the assumption of relaxed food competition in folivores is not justified, and that alternative mechanisms also constrain group size in some folivorous species.

dance of their resources may be restricted (reviewed in Snaith & Chapman 2007; chapter 7, this volume; but see chapter 19, this volume). In sum, many folivores seem to compete over food just as frugivores do.

Other constraints on group size

Ultimately, group size is constrained by the ability of individuals to compensate for increased travel costs. Digestive efficiency may limit daily energy gain and constrain individual energy budgets (Milton 1984; Schülke et al. 2004b) or time in general may be short (Dunbar 1992). Such time constraints have been demonstrated in comparative studies of baboons (*Papio* spp.). Baboons living in more temperate zones cannot compensate for the effects of indirect feeding competition during short winter days, and consequently are forced to live in small groups (Hill & Dunbar 2002). Time-constraint models have also shown that the distribution of rainfall and temperature, two predictors of habitat productivity and food resource abundance, affect the geographic distribution of species and group sizes (Korstjens et al. 2006; Lehmann et al. 2007).

Whether animals in larger groups can compensate for the costs of increased group size may also depend on their cognitive abilities. The ability to anticipate and remember what to find, where, and when will determine the effectiveness of strategic route planning (Cunningham & Janson 2007a). Comparative analyses revealed that group size

in lemurs is systematically smaller than in Asian, African, and neotropical primates (Kappeler & Heymann 1996). It has been argued that sophisticated cognitive abilities enable haplorrhines to integrate different characteristics of food resources, which allow them to offset the costs of living in large cohesive groups (Cunningham & Janson 2007b). A recent review did not reveal systematic differences in ecological intelligence between lemurs and haplorrhines (Fichtel & Kappeler 2010). Thus, small group size in lemurs remains unexplained (chapter 2, this volume).

Alternatively, lemur and other females may try to keep their groups small not because of increased ranging costs but because males immigrate preferentially into groups with many females to maximize their breeding opportunities. In this case, females in large groups might experience a higher risk of losing their offspring to infanticidal males than females in smaller groups (chapter 19, this volume). In response, the females may attempt to limit group size by expelling others when their groups grow too large (Vick & Pereira 1989). Alternatively, they may disperse preferentially into smaller groups, which will effectively constrain group size (Janson & Goldsmith 1995; Crockett & Janson 2000; Steenbeck & van Schaik 2001). Many folivores are female-dispersal species, which may explain why they often live in small groups (Moore 1984; fig. 9.6). But because dispersal is often male-biased (chapters 11 and 12, this volume) group-size-related infanticide risk is unlikely to constrain group size in most primate taxa, although it may prevent rapid increase in group size.

Group size may also be limited by contagious infections, parasites, and vector-borne diseases (Altizer et al. 2003). Under most circumstances, disease risk can be expected to increase with increasing sociality. Large groups facilitate more frequent social contacts, which increase the speed of parasite spread and thus individual parasite loads and general parasite diversity (Nunn & Altizer 2006). In support of this prediction, an increase in malaria prevalence with increasing group size has been found across New World primates (Nunn & Heymann 2005). As a consequence of increased disease risk, individuals may leave larger groups to disperse into smaller ones. The potential for disease risk to limit group size has been evaluated in few field studies, and comparative studies so far have yielded mixed results. Future investigations need to take into account more parameters of the host-parasite-environment interaction, to isolate the effects of disease ecology on social organization (Nunn & Altizer 2006).

In sum, group size is an important aspect of social systems. In large groups the number of potential social partners and competitors increases, individuals spend more time grooming (Lehmann & Dunbar 2009), and the probability that multiple males live together is increased (see below). Variation in group size is driven principally by variation in ecology. Comparative analyses indicate that maximum group size is limited by indirect feeding competition. Predation is likely to limit minimum group size, while infanticide and disease risk have the potential to constrain the upper and lower limits of group size.

Ecological Influences on Group Composition

Permanent associations between males and females make primates unusual among mammals. These associations have been linked to the high risk of infanticide in many primate species (van Schaik & Kappeler 1997, chapter 19, this volume). Males remain with females after the mating season to protect their offspring against attacks from other infanticidal males. As a consequence, much of the variation in the composition of primate groups is due to variation in the number and proportion of males that coreside with groups of females (fig. 9.7; Kappeler 2000). The socioecological model proposes that adult males will distribute themselves according to the spatiotemporal distribution of fertile females (Emlen & Oring 1977; Clutton-Brock 1989; Altmann 1990, fig. 9.1). The number of males living in a group has been causally linked to female variation in space (Andelman 1986; Altmann 1990; Mitani et al. 1996), in time (Ridley 1986), and in an interaction of both factors (Nunn 1999). Female spatial variation has been typically assayed by variation in female group size, while female temporal variation is commonly assessed by female estrous synchrony or its proxy, breeding seasonality.

Comparative analyses indicate that males indeed go where the females are. Across the phylogeny of primates, changes in male numbers have been preceded by changes in female numbers (Lindenfors et al. 2004). Additional comparative studies indicate that both male number and the probability of males monopolizing groups of females are positively and independently related to the effects of female group size and female estrous synchrony (Nunn 1999). By synchronizing their estrous periods, females may be able to influence the number of males per group. The shift from one-male to multimale units may be especially sensitive to the influence of female strategies (Kappeler 1999). This hypothesis is supported by the observation that in Nunn's (1999) sample all species with positive residual synchrony ($n = 11$) live in multimale groups, while some species with negative residuals (asynchronous estrus) form one-male units (three out of nine).

Both female group size and estrous synchrony are strongly influenced by the ecology of a species. We discussed ecological factors that limit group size above. Estrous synchrony

Fig. 9.7. Redfronted lemurs, *Eulemur rufifrons*, at Kirindy, Madagascar, live in multimale-multifemale groups with an even adult sex ratio—an arrangement that benefits both sexes in avoiding infanticide. Photo courtesy of Julia Ostner.

is largely driven by the number of females and reproductive seasonality, which in turn is strongly influenced by environmental factors, such as latitude and continent (Janson & Verdolin 2005), that constrain the options for female strategic synchronization. Thus, the environment ultimately affects variation in group composition and the number of males in primate groups.

While the positive relationship between the number of males and female group size is robust across species, residual variation may be due to differences in disease risk among species or populations. If disease risk from directly transmitted parasites is high, it may be beneficial to live in small groups (Freeland 1976). This may select for low female synchrony, which will increase monopolizability and keep male numbers low. Consequently, many males will float between groups (Nunn & Altizer 2006). Group composition may also affect disease risk. A recent model suggests that individuals in one-male groups may be more vulnerable than those living in multimale groups. If resident males succumb to disease, infected females will disperse to new groups and spread the disease (Nunn et al. 2008).

A conflict of interest between males and females is likely to exist over group composition (Kappeler 1999; van Schaik et al. 2004b; Pradhan & van Schaik 2008). Although living in groups with more males comes with costs, primate females will prefer to do so because the reduction in infanticide risk outweighs the reproductive costs from increased feeding competition. In contrast, males will generally attempt to monopolize females and exclude rival males from their groups. Under some circumstances, however, it may pay males to accept extra males in their group, and this benefit might override the cost of male-male competition (e.g., see Kappeler 1999). Comparative data across primates support the hypothesis that males may group together to reduce

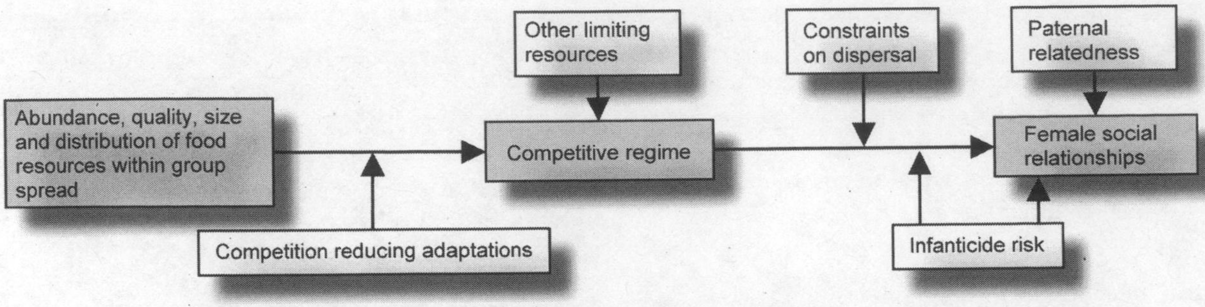

Fig. 9.8. Current socioecological theory suggests that female social relationships are affected by several factors in addition to characteristics of food resources and competitive regimes. These factors include adaptations that reduce competition, competition for limiting resources other than food, constraints on dispersal, paternal relatedness, and the influence of infanticide risk on dispersal and social tolerance. Meaningful tests of the model's internal consistency require measures of the competitive regime—that is, the relationship between net energy gain or fertility and rank and group size. Neither jumping from resource characteristics to social relationships nor using proxies of feeding competition types as predictors of social relationship characteristics can falsify the theory. The grey boxes represent the ecological model of van Schaik (1989).

the effects of predation (van Schaik & Hörstermann 1994). Living with a large number of resident males may additionally benefit an alpha male if it effectively (1) enhances safety for females and young (van Schaik & van Noordwijk 1989), (2) increases the number of coalition partners in territorial defense (Wrangham 1979; Muller & Mitani 2005), (3) increases the reproductive success of kin (Pope 1990), or (4) decreases the risk of aggressive takeovers by males outside the group (Crockett & Janson 2000; Fedigan & Jack 2004; Ostner & Kappeler 2004).

In conclusion, variation in group composition is driven to a large extent by male efforts to directly or indirectly monopolize large numbers of females. Factors determining the costs and benefits of excluding other males—such as female group size (Majolo et al. 2008), female reproductive seasonality (Janson & Verdolin 2005), ability to defend food resources (Fashing 2001), male power differentials (Schülke 2001), and infanticide risk (Janson & van Schaik 2000)—appear to be significantly affected by ecological factors and less so by sexual selection.

Ecological Influences on Social Structure

Food Resources and Modes of Competition

The distribution of risks and resources in the environment has been causally linked not only to the formation of groups but also to the patterning of female social relationships (figs. 9.1 and 9.8). The main driver is thought to be feeding competition that can occur either *within groups* (WG) or *between groups* (BG), and which will be either by *contest* (C) or by *scramble* (S). The resulting four types of feeding competition (WGC, WGS, BGC, BGS) can vary independently in their strength, yielding the specific *competitive*

regime a group, population, or species experiences (van Schaik 1989). BGS is thought to have no effect on social relationships and will not be discussed further. Different types of feeding competition have been related to their ecological causes, which are characteristics of the limited resources (fig. 9.9), and to their consequences, namely, different types of social relationships (van Schaik 1989; Sterck et al. 1997).

Contest competition is thought to arise when resources are scarce, defendable, and worth defending. Within-group contest competition (WGC) should occur when food is monopolizable and occurs in patches that can be defended by an individual or a subset of group members (van Schaik 1989), or when food sites can be usurped easily and cannot be depleted quickly (Isbell et al. 1998). Contest competition will be high if the costs of being excluded from a patch are also high (van Schaik 1989). These opportunity costs vary as a function of the relative quality (in terms of nutrient gain rates; Schülke et al. 2006) of the patch in relation to alternative patches that are close enough for an expelled individual to reach without leaving the group and experiencing increased predation risk. Patch size, quality, and distribution all need to be assessed from the primates' perspective—that is, from the perspective of the group (Janson 1988b; Isbell & Young 2002) within the area over which it can spread during feeding. If all food patches are similar with regard to food quality, exclusion of rivals from a patch is not worth the effort. Likewise, WGC is also reduced when patch density (or food abundance) is high and distances between alternative patches do not exceed the typical spread of the group. It is important to recognize that WGC can also occur over nonfood resources. If predation risk is high, individuals may directly compete to remain closer to the group's center, as this position increases one's safety (van Schaik 1989; Janson 1990; chapter 8, this volume). If food quality is high, and dispersed patches are large enough

Fig. 9.9. A female Assamese macaque (*Macaca assamensis*) at Phu Khieo Wildlife Sanctuary Thailand picks seeds slowly from a pod. Because handling time influences food intake, it is an important feature of foraging for understanding how primates compete over food resources. Photo courtesy of Oliver Schülke.

dividual net energy gain on group size and dominance rank (Janson & van Schaik 1988; Koenig 2002). If WGC is strong, net energy gain will be positively related to power or dominance rank. If WGS is strong, net energy gain will be negatively related to group size. If BGC is strong the relationship will be positive, because larger groups should have a competitive advantage over smaller ones. Net energy gain can be assessed from observations of feeding behavior combined with nutritional analyses of the foods ingested (Shopland 1987; Barton & Whiten 1994; Koenig et al. 1997; Schülke 2001; chapter 7, this volume), data on differential physical condition (Koenig et al. 1997; Schülke 2003), information on the energy balance of individuals assayed via c-peptides in urine (Emery-Thompson et al. 2009), or reproductive rates (Cheney & Seyfarth 1987; fig. 9.8).

Measures of process that tap into the proximate mechanisms of different forms of feeding competition (proxies of feeding competition in Koenig 2002) include the relationships between patch size and number of individuals co-feeding in a patch (e.g., Koenig et al. 1998; Schülke 2003), food site residence times (Isbell et al. 1998; Korstjens et al. 2007), dominance rank as an organizing feature in the composition of feeding parties (Koenig et al. 1998), food-related aggression rates (Sterck & Steenbeck 1997), and female participation in between-group agonistic encounters (Cooper 2004; Harris 2006). These measures are not direct indicators of feeding competition, though. While high rates of food-related aggression, for example, may suggest strong WGC or BGC (Vogel & Janson 2007), an absence or low rate of such aggression may result from effective avoidance of confrontation (Sterck & Steenbeck 1997; Harris 2007). In addition, individual reaction to increased WGC may be condition-dependent. For example, aggression rate may be triggered by hunger (Janson & Vogel 2006). Whenever processes of BGC are studied, other explanations (male defense or infant defense) for the occurrence of individual participation in aggression should be considered in a multivariate analysis (Harris 2007). Snaith and Chapman (2007) have compiled a list of behavioral indicators of food competition. This list needs to be treated with caution because it does not differentiate between (1) the ultimate consequences of different types of feeding competition in terms of differential energy budgets and reproductive performance, (2) the proxies of feeding competition types, and (3) aspects of the social structure that really result from different types of feeding competition. A more structured overview of hypotheses on the relationship between food resource characteristics and female social relationships, predictor and response variables, and data needed for tests is provided by Koenig and Borries (2009).

Future studies testing the ecology of female social rela-

to accommodate the entire group and can be economically defended against other groups, BGC occurs. Groups will exclude others from access to those large patches, thus leading to differences in net energy gain of individuals that live in different groups which vary in their ability to dominate others (van Schaik 1989; Harris 2006).

Within-group feeding competition may also occur indirectly without interference between members via scramble competition, in which the foraging activity of each member of a group affects the foraging efficiency of every other member, irrespective of the power differences between individuals. WGS is predicted to occur when resources are of low quality, are highly dispersed, or are very large (van Schaik 1989). Its mechanisms have been discussed above as a major constraint on group size.

Feeding competition can be measured by regressing in-

tionships should aim at using measures of differential net energy gain within and between groups, as well as the above-mentioned proxies, to evaluate the degree of different types of competition experienced in certain groups. Behavioral adaptations that compensate for the costs of feeding competition, such as cheek pouch use in cercopithecoids (Lambert 2005; Buzzard 2006), extension of activity period (Janson 1988a), fission-fusion dynamics (Symington 1988; Chapman et al. 1995; Dias & Strier 2003; Asensio et al. 2008), or changes in group spread (Di Fiore 2004; Snaith & Chapman 2008) need to be considered in these tests. Finally, constraints on behavioral change that results from the distribution of risks in the habitat require attention (Cowlishaw 1997; Boinski & Garber 2000; fig. 9.8).

Evidence for the proposed relationships between resource characteristics and competitive regimens comes from detailed studies of individual species including nocturnal lemurs, neotropical primates, colobines, cercopithecines, and apes (review in Koenig 2002; Schülke 2003; Robbins et al. 2007). Where two or a few species or populations of the same species have been compared, predictions about the relationship between food resource characteristics and competitive regimes have also been supported (Mitchell et al. 1991; Koenig 1998; Boinski et al. 2002; Korstjens et al. 2002; Dammhahn & Kappeler 2009). Only a few studies have tested the link between resource characteristics and BGC effects (review in Koenig 2002, but see Harris 2006).

Comparative analyses are hampered by the lack of a broadly applicable measure of the ecological WGC and BGC potential that results from food resource characteristics. We suggest first investigating the relationship between resource characteristics and competitive regimes within species, to determine whether inequalities in energy budgets and reproductive performance (according to rank and group size) result from feeding competition (e.g., see Robbins et al. 2007). If this is the case, in a second step the degree of inequality—the slope of net energy gain regressed against dominance rank and group size—can be used as a continuous measure of the strength of competition, and can then be related to aspects of social structure in comparative analyses.

Feeding Competition and Social Relationships

Feeding competition is thought to influence social relationships in three ways. First, as WGC increases, aggression rates also increase and the formation of dominance hierarchies and coalitions is promoted (Janson 1985; Chapais et al. 1991). Second, as WGS increases, females benefit by dispersing into smaller groups (Wrangham 1980; Dunbar 1988). Finally, as BGC increases, dominance relationships become more relaxed to improve cooperation against other

groups (van Schaik 1989; fig. 9.8). All other behavioral correlates follow from this rationale.

WGC, aggression, dominance relationships, and hierarchies If WGC is strong, selection favors traits that increase a female's access to high-quality resources. The tendency to engage in aggressive interactions will be selected for because there is a lot to gain and a lot to lose (Janson 1985). There is sound empirical support for the prediction that increased WGC leads to increased rates of aggression in many species (e.g., Isbell & Pruetz 1998; Koenig 2000; Vogel 2005; Robbins 2008). Since every fight carries energetic costs and even powerful individuals risk injury, dominance hierarchies commonly form as a means to regulate access to resources in lieu of constant aggression (de Waal & Lutrell 1989). Hence, the dominance hierarchy itself, as well as its relationship to resource access, should be expressed more clearly in species that experience strong WGC. Moreover, the evolution of formalized signals of subordination or dominance should be promoted to avoid the escalation of aggression (de Waal 1986). If WGC is strong and food patches support more than one individual at a time, selection should favor the formation of coalitions in agonistic conflicts, promoted by direct and/or indirect fitness benefits (van Schaik 1989, but see Broom et al. 2009).

Under strong WGC, high-ranking individuals have a lot to lose in a conflict. Hence, irrespective of aggression rates, most conflicts should be decided, counteraggression should be rare, and high-ranking individuals should rarely ignore aggression directed toward themselves because it represents a challenge to their status. Although these predictions follow directly from the logic of WGC, counteraggression rates are often interpreted as measures of tolerance (de Waal & Lutrell 1989; Thierry et al. 2000). On the relationship level, the directionality of these decided conflicts should be high: within a given dyad, the winner should always be the same individual. The directionality index is a robust measure that should not be affected by sample and group size or by the number of unknown relationships in a group (Koenig & Borries 2006).

Other instruments used to measure the social consequences of WGC are the linearity index h' (de Vries 1995), the time needed to find a dominance hierarchy (Isbell & Young 2002), and the steepness of the hierarchy (van Schaik 1989; de Vries et al. 2006). All of these instruments have methodological difficulties that hamper their usefulness in comparative analyses. Measurements are affected by group size and sample size (number of observations) and/or the number of empty cells in the interaction matrix (Koenig & Borries 2006; Richter et al. 2009). Sample size may be influenced by the choice of behaviors used to assess dominance, such as displacements (Barton 1993), submissive signals (Ostner et al. 2008a), and outcome of decided conflicts (Ostner & Kappeler 1999). Comparative analyses have used multiple regression methods that controlled for the effect of group size when steepness of the hierarchy was an independent factor (Schino & Aureli 2008a; Richter et al. 2009).

WGC, kin support, and dispersal

Female philopatry and the formation of long-term alliances between females have been proposed to be interrelated consequences of strong within-group contest competition, because inclusive fitness benefits facilitate alliance formation among relatives (Wrangham 1980; Moore 1984; van Schaik 1989; Sterck et al. 1997; Silk 2002b, but see Broom et al. 2009). Once kin alliances become an effective component of feeding competition between females, the females cannot easily afford to disperse from their matrilines (Sterck et al.

1997). Female philopatry is indeed a prerequisite for frequent nepotistic alliances because it is highly unlikely that many closely related females will end up living in the same group once they have dispersed from their natal group. This prompts a question: When did female dispersal evolve in primates? As a general rule, there is always at least one sex leaving its natal group to avoid inbreeding, so philopatry of both sexes is unlikely. In contrast to other attempts that investigate the relationship between dispersal patterns and feeding competition (Sterck et al. 1997; Di Fiore & Rendall 1994; Isbell 2004), here we classify dispersal according to its social aspect only and distinguish three regimes: dispersal by males only, dispersal by females only, and dispersal by both sexes away from their natal group (fig. 9.10).

Our phylogenetic reconstruction shows that in general, dispersal is a highly conserved trait that changes only rarely. This is especially true for two clades: the cercopithecines

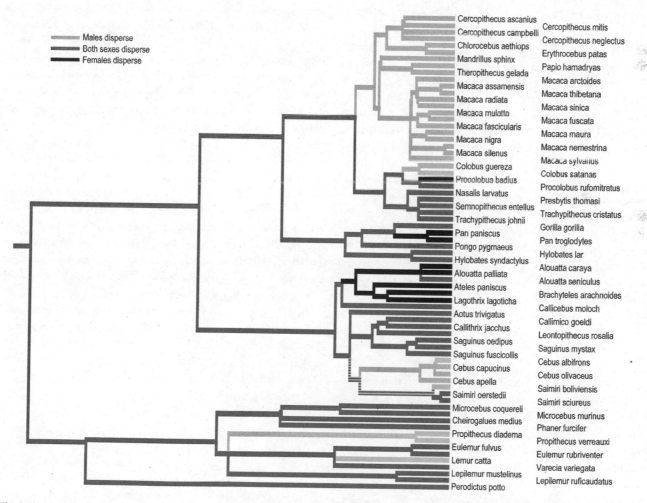

Fig. 9.10.   Patterns of social dispersal are highly conserved in primates. All cercopithecines show male-only dispersal (light gray); all but one ateline show female-only dispersal (black). In the ancestral condition, both sexes leave their natal group (medium gray). Regardless of benefits derived from WGC, once males are philopatric, females cannot switch to philopatry in one evolutionary step; this is perhaps to avoid inbreeding. Character reconstruction in Mesquite (version 1.12, Maddison & Maddison 2006). Phylogeny and branch lengths from Bininda-Emonds et al. (2008).

and the atelines. For the cercopithecines the original argument may hold that once nepotistic alliances became an important aspect of their social structure, females could not afford to leave their natal groups, even if the competitive regime would facilitate dispersal. Kin support and male dispersal in cercopithecines may therefore be regarded as highly derived (Di Fiore & Rendall 1994). The case of the atelines highlights another problem of the socioecological model. If, in a situation of male philopatry and female dispersal, WGC increases, females will benefit from being with their relatives. Staying in their natal group is not an option for females, however, because males fail to disperse. The system must go through a transitional phase of both sexes dispersing before females can switch from dispersal to philopatry. Hence, the reaction of females to long-term changes in feeding competition is strongly constrained. Our phylogenetic reconstruction of social dispersal regimes in 71 primate species supports this claim. The tree shows no character change directly from female-only dispersal to male-only dispersal. Thus, dispersal regimes cannot easily be predicted by competition for food resources and consequently should be omitted in the socioecological model (sensu Sterck et al. 1997).

An alternative yet still underexplored proposal about the relationship between strategies of food acquisition, dispersal, and kin support is provided by Isbell (2004). Her stepwise model predicts that females will disperse (1) if dispersal costs are low and male behavior permits dispersal or (2) if dispersal costs are high and mothers evict their daughters because they cannot make room in their own home ranges to accommodate additional foragers. If dispersal costs are high and mothers can extend their home range, the degree of goal-directed travel in food acquisition will determine whether females form kin groups or not. In this view, several routes lead to female philopatry, but feeding competition plays a role in only some of them. Testing this model will require a measure of dispersal costs that combines costs due to increased predation risk and decreased foraging efficiency.

BGC and social tolerance

Strong between-group contest competition may have two different consequences, depending on the strength of WGC in the competitive regime (Sterck et al. 1997). If coupled with weak WGC, strong BGC has been suggested to select for female philopatry because only related females can overcome the collective action problem of cooperatively defending resources, and the resulting dominance relationships should thus be egalitarian (Sterck et al. 1997).

If both BGC and WGC are strong, however, females will develop clear within-group dominance relationships, and

subordinates will be less inclined to partake in resource defense because their resulting benefit will be smaller. In these situations, dominant females are thought to tolerate unrelated subordinates' access to food resources and social partners (Sterck et al. 1997). Past studies related proxies of BGC, such as female participation in between-group encounters or degree of home range overlap, to social tolerance, which was measured by the relative frequency of counteraggression received by dominants, and the overall frequency and the intensity of aggression between females. No clear relationship between tolerance and BGC was found (Matsumura 1998; Koenig 2002; Lu et al. 2008). Unfortunately, no study has yet measured BGC as differential fertility or net energy gain in relation to group rank. We have argued above that high counteraggression (if coupled with a high percentage of undecided conflicts) may be an indicator of weak WGC instead of strong BGC.

The prediction that strong BGC will lead to greater tolerance of subordinates at contested resources might be questioned theoretically. Following the model's internal logic, subordinates should have ready access to resources because those generating BGC will be distributed in large, high-quality patches that cannot be easily monopolized by a few individuals, but instead can accommodate an entire group (Sterck et al. 1997). Hence, subordinates may not require an extra incentive to participate in BGC, and it remains unclear why they should be tolerated by dominants and gain access to resources that generate WGC. BGC may nonetheless overcompensate for the effects of strong WGS in very large groups and may keep non-nepotistic females from dispersing into smaller groups (Koenig 2000).

In contrast to the effects of WGS and WGC, the social consequences of BGC and the sources of social tolerance are poorly understood. Moreover, social tolerance may be affected by factors other than feeding competition. For example, females might tolerate each other if they must cooperate to defend their offspring against infanticide by males (Treves 1998; Lu et al. 2008; chapter 19, this volume). Accordingly, social tolerance in species with strong WGC should vary with infanticide risk (Treves 1998; Nunn & van Schaik 2000). Alternatively, social tolerance among females may be a function of their paternal relatedness (Schülke & Ostner 2008).

Paternal Relatedness, Social Tolerance, Dispersal, and Nepotism

It has recently been argued that social tolerance in despotic species may result from close paternal relatedness among females (Schülke & Ostner 2008). Measures of social tolerance used by prior researchers (e.g., Sterck et al. 1997;

Thierry et al. 2000) are sensitive to the degree of relatedness among females and the number of female relatives in a group. Yet only maternal kin have been considered thus far. Observed cases of frequent coalitions and reconciliation between non-kin that have previously been interpreted as examples of social tolerance may actually represent interactions between paternal relatives (Schülke & Ostner 2008).

At constant dispersal rates, variation in paternal relatedness is generated by variation in male reproductive skew. High levels of skew increase the number of paternal siblings within a group, but leave the number of maternal kin unaffected (Widdig et al. 2004). Simulations suggest that adult philopatric females will then live with more paternal female kin than maternal kin in most demographic scenarios (Schülke et al. unpublished data). In species with high male reproductive skew, age-mates are likely to be paternally related but never maternally related (Altmann 1979). Tolerance between age-mates has been reported in many primate species (reviewed in Widdig 2007), and observed alliances involving individuals of different matrilines (e.g., see de Waal & Ren 1988; Chapais et al. 1991; Aureli et al. 1997; Demaria & Thierry 2001) may include participants who are paternally related.

If kin relationships among female group mates run vertically (i.e., matrilineally) as well as horizontally (patrilineally), the social structure of such a group will be classified as highly tolerant. The effect's magnitude will be influenced by whether or not individuals are actually able to distinguish paternal kin from non-kin—an ability that has been shown in some baboon and macaque species (Widdig 2007), but not in white-faced capuchin monkeys (*Cebus capucinus*, Perry et al. 2008). But even if an individual's kin recognition is imperfect, it will not benefit by interfering aggressively against other individuals that have some chance of being its paternal kin (Widdig et al. 2006). This will lead to tolerant social systems as defined by the socioecological model (Schülke & Ostner 2008). Because animals are likely to discriminate maternal relatives more easily than paternal relatives, females may form strong bonds with the latter only when the former are unavailable (Silk et al. 2006).

In support of this "tolerance through paternal relatedness hypothesis," a review of reconciliation across primates stated that "Kinship was *not* found to influence reconciliation in . . . spectacled leaf monkeys . . . [or] in immature long-tailed macaques"(Arnold & Aureli 2007, p. 599). Both of these species, however, are characterized by a high degree of male reproductive skew, thus leading to situations in which immatures will often be paternal kin. Comparative data on eight macaque species show the predicted positive relationship between paternity concentration and female tolerance (fig. 9.11). In further support of this hypothesis,

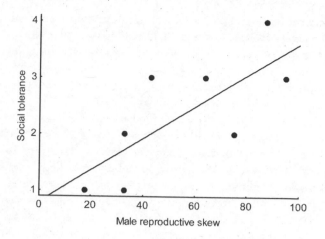

Fig. 9.11. A test of the "tolerance through paternal relatedness hypothesis" (Schülke & Ostner 2008). Social tolerance of female macaques increases with increasing male reproductive skew (Spearman's rho = 0.77, $N = 8$ species, $p < 0.05$). High paternity concentration produces female paternal relatives bound together across matrilines. Tolerance is based on Thierry's (2007) assessment of social style grades in macaques, and male reproductive skew is the proportion of paternity concentrated in the alpha male (Ostner et al. 2008, plus data on Assamese macaques [Schülke et al. 2010] and Tonkean macaques [Thierry et al. 1994]).

mathematical modeling revealed a strong influence of the variance in relatedness among group members upon cooperative behavior. If variance is low because all individuals in the group are closely related—for example, due to high paternal relatedness—the benefits and thus the occurrence of coalitions will be reduced (Broom et al. 2009), and individuals will tolerate each other.

In sum, determining whether paternal relatedness affects social tolerance in primates will require additional studies that investigate the relationships between the distribution of nepotism between paternal and maternal relatives, the regimes of dispersal, and the structure of dominance hierarchies. Since paternal relatives are most likely age-mates, any age structure in the dominance hierarchy may result from tolerance or support among paternal siblings.

Predation, Disease, and Social Relationships

Both predation and disease risk may affect the dispersal regimes of primates. Predation is thought to be a major cost of dispersal (Isbell & van Vuren 1996). Dispersal is dangerous because individuals have to move through unfamiliar areas, unaware of locations that predators might frequent and without knowledge of hiding places and potential escape routes. These individuals are usually alone and without the safety of a group (Isbell & van Vuren 1996). Therefore, dispersal decisions will result from trade-offs between costs and benefits of predation risk and feeding competition. High (intrinsic) predation risk may override the effects of feeding competition when females start competing for safety (van Schaik 1989). Female gray mouse lemurs (*Microcebus murinus*) compete for access to safe tree holes where they spend the day sleeping (Dammhahn & Kappeler 2009). Thus, dispersal and social relationships among females may be influenced by predation risk.

Likewise, disease risk may influence social relationships because disease is spread through social contact and social behavior may evolve to counter disease risk (Nunn et al. 2008). The way in which group members spread out, their interaction frequency, and their sleeping habits (in large huddles versus spread out, at ever-changing places versus fixed sleeping sites, in shelters or out in the open) may be influenced by the risk of contagious disease (Nunn & Altizer 2006).

To conclude, social structure is still perhaps the aspect of the social system in which the action of different social and ecological factors is least understood. Feeding competition features prominently in theoretical models of the evolution of female social relationships, and its influence on competitive regimes is well supported. It is less clear, however, how different competitive regimes influence female social structure. Dispersal patterns seem constrained to a degree that cannot be predicted from the type of feeding competition experienced. But characteristics of dominance hierarchies, coalitions, and social tolerance may be understood if future revisions of socioecological models include the effects of paternal relatedness among females and competition for other limiting resources.

Ecological Influences on Mating Systems

Mating systems are shaped by an array of ecological influences including feeding competition, predation, and disease. As an alternative to monopolizing exclusive access to females, a male can also engage in indirect mate defense by defending food resources for a group of females (resource defense polygyny: Fashing 2001; Harris 2005). Whether males defend resources depends on the characteristics of food resources: their abundance (since lack of food should limit female reproduction), defensibility, and quality. Apart from these ecological factors, males should defend resources only when it provides them with reproductive access to a particular group of females. Resource defense should also occur only if males are able to monopolize fertile females reproductively, and if the females cannot desert them for other males with better territories. Hence, multimale species may engage in resource defense only rarely (Fashing 2001; Harris 2006, 2007).

Male time budgets may be constrained by their diets and

food acquisition strategies, which can have far-reaching consequences for sexual strategies. Monopolization of fertile females may be easier for folivores, whose digestive physiology constrains the time they can use to ingest food and hence frees up time for other activities (Janson & van Schaik 2000). Primates with physiological adaptations to a high-fiber diet exhibit higher male reproductive skew than do other species (figure 9.12). Because this pattern may be explained by the priority-of-access model alone (chapter 18, this volume) based on differences in female estrous overlap (Ostner et al. 2008a), a crucial test will require the demonstration that frugivorous males experience higher ecological costs of mate guarding than do folivores (Alberts et al. 1996).

While parental care, mate guarding, and infanticide may have contributed to the evolution of cohesive pairs (van Schaik & Kappeler 2002; chapters 3 and 6, this volume), feeding competition may explain the evolution of dispersed pair living (Schülke 2005). Derived from a truly solitary ancestral condition, the first step towards dispersed pair living is the development of female range exclusivity, which has been shown to result from intense competition for food (Schülke 2003; Isbell 2004). Despite small home ranges and long travel distances, males fail to move economically over several ranges to exclude other males, because by doing so they suffer reduced foraging efficiency. This effect is largest when foraging movements are highly goal-directed and food patches deplete rapidly (Schülke 2005) or foraging paths cannot be backtracked (Isbell 2004). While males are

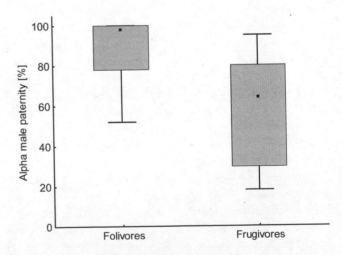

Fig. 9.12. Folivores tend to have higher male reproductive skew than frugivores (Mann-Whitney U-test: $N_{folivores} = 6$, $N_{frugivores} = 23$, $U = 18$, exact $p = 0.0042$), even when only multi-male groups are considered (Mann Whitney U-test, $N_{folivores} = 4$, $N_{frugivores} = 23$, $U = 18.0$, exact $p = 0.058$, data from Ostner et al. 2008, plus single-male groups of *Alouatta arctoidea* and *Semnopithecus entellus*). Folivores have been hypothesized to have more flexible time budgets due to their longer digestion times, and to suffer fewer constraints when monopolizing access to females (Janson & van Schaik 2000).

unable to monopolize multiple females spatially for ecological reasons, species living in dispersed pairs generally show high levels of extrapair paternity (Schülke & Ostner 2005), even among nonprimates (Munshi-South 2007), thus generating considerable variation in male reproductive success.

Mating systems and disease risk are also tightly interrelated. Their interaction generates variability in the filial generation of the host, with parasites subsequently counteradapting in an evolutionary arms race (Andersson 1994). Sexual selection theory predicts that females choose mates according to their immunocompetence or their immunocompatibility. Cues they use in the process may include ornaments that are costly to develop and maintain (Setchell & Kappeler 2003). Evidence for female choice in primates is nevertheless rare (Paul 2002; chapter 16, this volume).

Variation in mating patterns (number of mating partners, number of matings per partner, duration of mating), mating outside the social unit, and mating skew will influence the spread of sexually transmitted disease (STDs: Nunn 2002; Nunn & Altizer 2006). Mating with just one partner may be advantageous for a female in terms of avoiding STDs, but if many females choose the same infected male—for example, for his superior genetic quality—STDs may spread quickly. Because females obtain several benefits by mating polyandrously (see above), the limiting of their mating activity to a single male carries costs that may counteract the benefits they obtain via reduced STD transmission.

Self-Organization and Correlated Character Evolution: Alternatives or Integral Parts of Socioecological Theory?

Self-organization and correlated character evolution represent two potential alternative explanations for variation in primate societies (Thierry 2007). Computer simulations of individual primates exchanging agonistic behavior in a self-reinforcing manner in an artificial world revealed that several behavioral characteristics of egalitarian and despotic social relationship patterns can be generated simply by changing the intensity or frequency of aggression (Hemelrijk 1999). Given the decision rules implemented in the simulation, variance in dominance success, attack symmetry, attack rates, cohesion, and frequency of nonaggressive interactions originate from self-organization processes in the absence of selection.

Similarly, comparative studies have shown that entire suites of social behaviors that can be categorized into four different social style grades covary between macaques in different lineages (Thierry et al. 2000), which has been interpreted as evidence against the flexible reaction of social traits to changes in ecological factors. Subsequent analyses have reassigned many species to different social style grades (Thierry 2007), but the covariation hypothesis has not been reevaluated statistically using the entire data set. Recently it has been shown that even after controlling for similarity that results from common origin, four social traits remained highly correlated (Thierry et al. 2008), thus suggesting that there are tight constraints on social evolution (Thierry 2008). Yet the reasons underlying this pattern of correlated character evolution remain obscure. They may ultimately be ecological, resulting from all four traits being covariants of a hidden fifth ecological variable.

Correlated character evolution by self-organization or by the social environment has been posed as a challenge or competing alternative to the adaptationist view of socioecology (Thierry 2007). Adaptive models do not assume, however, that all characters are free to change independently, and the idea of covariation is deeply ingrained in socioecological thinking. In fact, socioecological models frequently propose that certain environmental conditions favor the evolution of whole sets of characters, as in models for female social relationships (e.g., Wrangham 1980; van Schaik 1989; Sterck et al. 1997).

While it has to be stressed that not all behavior may be adaptive (Janson 2000; Thierry 2008), the covariation hypothesis and the theory of self-organization offer no original evolutionary explanation for observed patterns. Assuming a nonrandom set of behavioral rules, frequency and intensity of aggression generate all aspects of despotic and egalitarian societies (Hemelrijk 1999). However, it remains to be explained how variation in aggressiveness comes about in the first place, especially since aggression is an energetically costly behavior that always carries the risk of injury or fatality. Socioecological theory provides a rationale by explaining how aggressive contests for limited high-quality food resources may influence differential female reproductive success, and in turn affect other aspects of social life. Finally, the most serious problem with the theory of phylogenetic constraints is that it is inevitably post hoc (Koenig & Borries 2009). Constraints on character evolution cannot be predicted, but can only be assessed from observed variation. Thus, the hypothesis cannot be tested prospectively.

Summary and Conclusions

Socioecology, if defined as the study of the relationships between social traits and the ecological and social environment, is an enormous field that cannot be fully reviewed in a single chapter. Its massive scope is its most important strength and its greatest weakness. Socioecology strives to unite research areas, such as feeding ecology, kin selection, and sexual selection, that would happily coexist in the absence of integration. Their integration via socioecology, however, has led to some of the most significant advances in our current understanding of primate social behavior (Janson 2000). It remains highly unlikely that one factor or a few can explain the entire spectrum of social systems.. But most primates are small to medium in body size, live relatively long lives, and reproduce slowly; they occupy tropical forests, subtropical forests, and savannas and feed on fruits, leaves, and invertebrates. The search for socioecological principles stems from the insight that species have adapted to survive and reproduce in similar environmental conditions in nonrandom ways.

The socioecological model for the evolution of female social relationships (sensu Sterck et al. 1997) has been criticized on several grounds and has recently been buried (Thierry 2008) and resurrected (Koenig & Borries 2009). Despite claims to the contrary (Matsumura 1999; Menard 2004; Thierry 2008), the model has never been evaluated in a large-scale comparative test investigating all relationships from resource characteristics and competitive regimes to resulting social relationships. Socioecological models are problematic because they are often verbal, conceptual models. They describe relationships between factors and responses often without incorporating actual measurements into formal models. This makes falsification difficult. Verbal models are nonetheless useful as heuristic devices that can lead to improvements in methods that can then be implemented in empirical tests. In this chapter we have offered suggestions on how to generate comparative data on contestability of resources and resulting feeding competition to test aspects of the socioecological model.

It was only in the decade following the publication of *Primate Societies* in 1987 that many of the most prominent conceptual models in primate socioecology were formulated (van Schaik 1989; Clutton-Brock 1989; Chapman 1990; Isbell 1991; van Schaik 1996; Sterck et al. 1997; Hemelrijk et al. 1999; Thierry et al. 2000). While research in the 1980s emphasized the role of ecology, it is now well recognized that social factors have profound and subtle effects on social organization and structure as well. Our understanding of the life, size, and composition of groups has improved considerably via the integration of infanticide and predation risk into socioecological theory.

Future research will have to focus on resolving puzzles about social relationships. It remains unclear to what extent the strength and mode of feeding competition influences female social relationships. It is equally unclear whether a comprehensive model of dispersal regimes can be developed that incorporates its multiple predictors, including feeding competition, predation, infanticide, disease, relatedness, and inbreeding avoidance. Dispersal is basic to a model

of social relationships; a dispersal regime, along with the genetic mating system and group size, will determine how group members are related (e.g., maternally or paternally), how many relatives they possess, and the degree to which they are related. Kin biases in behavior structure social relationships and have been reported across the Primate order (Chapais & Berman 2004; Silk 2009; chapter 21, this volume). The adaptive value and temporal stability of these biases remain relatively underexplored, which is problematic because kin biases are not equivalent to kin-selected cooperation. Biases in behavior may result from reciprocity or mutualism, and ultimate benefits may not accrue continuously in small increments but sporadically in stressful circumstances (Silk 2009; chapter 22, this volume). Thus, philopatric females need not necessarily form matrilineal dominance hierarchies even if WGC for food is strong, but behavioral kin bias (e.g., in spatial proximity) may nevertheless prevail. Understanding the way males and females associate, affiliate, and cooperate will require additional data on the ultimate causes of the observed behaviors. Genetic data are needed to uncover overall levels and patterns of relatedness among the dispersing sex as well as among members of different matrilines. In sum, multiple challenges and avenues for research remain for the next 25 years of socioecological research.

References

Abbott, D. H., Keverne, E. B., Bercovitch, F. B., Shively, C. A., Mendoza, S. P., Saltzman, W., Snowdon, C. T., Ziegler, T. E., Banjevic, M., Garland, T. J. & Sapolsky, R. M. 2003. Are subordinates always stressed? A comparative analysis of rank differences in cortisol levels among primates. *Hormones and Behavior*, 43, 67–82.

Alberts, S., Altmann, J. & Wilson, M. L. 1996. Mate guarding constrains foraging activity of male baboons. *Animal Behaviour*, 51, 1269–1277.

Alexander, R. D. 1974. The evolution of social behavior. *Annual Review of Ecology and Systematics*, 5, 325–383.

Altizer, S., Nunn, C. L., Thrall, P. H. Gittleman, J. L., Antonovics, J., Cunningham, A. A., Dobson, A. P., Ezenwa, V., Jones, K. E., Pedersen, A. B., Poss, M. & Pulliam, J. R. C. 2003. Social organization and parasite risk in mammals: Integrating theory and empirical studies. *Annual Review of Ecology and Systematics*, 34, 517–547.

Altmann, J. 1979. Age cohorts as paternal sibships. *Behavioral Ecology and Sociobiology*, 6, 161–164.

———. 1990. Primate males go where the females are. *Animal Behaviour*, 39, 193–195.

Andelman, S. J. 1986. Ecological and social determinants of cercopithecine mating patterns. In *Ecological Aspects of Social Evolution* (ed. by Rubenstein, D. I. & Wrangham, R. W.), 201–216. Princeton, NJ: Princeton University Press.

Anderson, M. B. 1994. Sexual selection. Princeton, NJ: Princeton University Press.

Appleby, M. C. 1983. The probability of linearity in hierarchies. *Animal Behaviour*, 31, 600–608.

Arnold, K. & Aureli, F. 2007. Postconflict reconciliation. In *Primates in Perspective* (ed. by Campbell, C., Fuentes, A., MacKinnon, K., Panger, M. & Bearder, S.), 592–608. New York: Oxford University Press.

Asensio, N., Korstjens, A. H., Schaffner, C. M. & Aureli, F. 2008. Intragroup aggression, fission-fusion dynamics and feeding competition in spider monkeys. *Behaviour*, 145, 983–1001.

Aureli, F., Das, M. & Veenema, H. C. 1997. Differential kinship effect on reconciliation in three species of macaques (*Macaca fascicularis*, *M. fuscata* and *M. sylvanus*). *Journal of Comparative Psychology*, 111, 91–99.

Aureli, F. & De Waal, F. 2000. *Natural Conflict Resolution*. Berkley: University of California Press.

Aureli, F, Schaffner, C. M., Boesch, C. Bearder, S. K., Call, J. Chapman, C. A. Connor, R., Di Fiore, A., Dunbar, R. I. M., Henzi, S. P., Holekamp, K., Korstjens, A. H., Layton, R. , Lee, P. C., Lehmann, J., Manson, J. H., Ramos-Fernández, G, Strier. K. B. &. van Schaik, C. P. 2008. Fission-fusion dynamics: New research frameworks. *Current Anthropology*, 49, 627–654.

Barton, R., Byrne, R. & Whiten, A. 1996. Ecology, feeding competition and social structure in baboons. *Behavioral Ecology and Sociobiology*, 38, 321- 329.

Barton, R. A. 1993. Sociospatial mechanisms of feeding competition in female olive baboons, *Papio anubis*. *Animal Behaviour*, 46, 791–802.

Barton, R. A. & Whiten, A. 1994. Reducing complex diets to simple rules: Food selection by olive baboons. *Behavioral Ecology and Sociobiology*, 35, 283–293.

Bertrand, M. 1969. *The Behavioral Repertoire of the Stumptail Macaque*. Basel: S. Karger.

Birkhead, T. 2000. *Promiscuity*. London: Faber.

Boinski, S. & Garber, P.A. 2000. *On the Move: How and Why Animals Travel in Groups*. Chicago: University of Chicago Press.

Boinski, S., Sughrue, K., Selvaggi, L., Quatrone, R., Henry, M. & Cropp, S. 2002. An expanded test of the ecological model of primate social evolution: Competitive regimes and female bonding in three species of squirrel monkeys (*Saimiri oerstedii, S. bolivensis, and S. sciureus*). *Behaviour*, 139, 227–261.

Borries, C. 1993. Ecology of female social relationships: Hanuman langurs (*Presbytis entellus*) and the van Schaik model. *Folia Primatologica*, 61, 21–30.

Bradbury, J. & Vehrencamp, S. 1976. Social organization and foraging in emballonurid bats. II. A model for the determination of group size. *Behavioral Ecology and Sociobiology*, 1, 383–404.

Broom, M. Koenig, A. & Borries, C. 2009. Variation in dominance hierarchies among group-living animals: Modelling stability and the likelihood of coalitions. *Behavioral Ecology and Sociobiology*, 20, 844–855.

Brotherton, P. N. M., Pemberton, J. M., Komers, P. E. & Malarky, G. 1997. Genetic and behavioural evidence for monogamy in a mammal, Kirk's dik-dik (*Madoqua kirkii*). *Proceedings of the Royal Society London* B, 264, 675–681.

Buzzard, P. J. 2006. Cheek pouch use in relation to interspecific competition and predator risk for three guenon monkeys (*Cercopithecus* spp.). *Primates*, 47, 336—341.

Chapais, B. 1995. Alliances as means of competition in primates: Evolutionary, developmental and cognitive aspects. *Yearbook of Physical Anthropology*, 38, 115–136.

Chapais, B. & Berman C. M. 2004. *Kinship and Behavior in Primates*. Oxford: Oxfrod University Press.

Chapais, B., Girard, M. & Primi, G. 1991. Non-kin alliances, and the stability of matrilineal dominance relations in Japanese macaques. *Animal Behaviour*, 41, 481–491.

Chapman, C. A. 1990. Ecological constraints on group size in three species of neotropical primates. *Folia Primatologica* 55, 1–9.

Chapman, C. A., Wrangham, R. W. & Chapman, L. J. 1995. Ecological constraints on group size: An analysis of spider monkey and chimpanzee subgroups. *Behavioral Ecology and Sociobiology*, 36, 59–70.

Charnov, E. L. 1976. Optimal foraging: the marginal value theorem. *Theoretical Population Biology*, 9, 129–136.

Charnov E. & Krebs J. 1975. The evolution of alarm calls: Altruism or manipulation? *American Naturalist*, 109, 107–112.

Cheney, D. L. & Seyfarth, R. M. 1987. The influence of intergroup competition on the survival and reproduction of female vervet monkeys. *Behavioral Ecology and Sociobiology*, 21, 375–386.

Cheney, D. L. & Wrangham, R. W. 1987. Predation. In *Primate Societies* (ed. by Smuts, B. B., Cheney, D. L., Seyfarth, R. M., Wrangham, R. W. & Struhsaker, T. T.), 227–239. Chicago: University of Chicago Press.

Clutton-Brock, T. H. 1989. Mammalian mating systems. *Proceedings of the Royal Society London* B, 236, 339–372.

Cooper, M. A., Aureli, F. & Singh, M. 2004. Between-group encounters among bonnet macaques (*Macaca radiata*). *Behavioral Ecology and Sociobiology*, 56, 217–227.

Cowlishaw, G. 1997. Trade-offs between foraging and predation risk determine habitat use in a desert baboon population. *Animal Behaviour*, 53, 667–686.

Crockett, C. M. & Janson, C. H. 2000. Infanticide in red howlers: Female group size, male membership, and a possible link to folivory. In *Infanticide by Males and Its Implications* (ed. by van Schaik, C. P. & Janson, C. H.), 75–98. Cambridge: Cambridge University Press.

Crook, J. H. & Gartlan, J. C. 1966. Evolution of primate societies. *Nature*, 210, 1200–1203.

Cunningham, E. & Janson, C. H. 2007a. A socioecological perspective on primate cognition. *Animal Cognition* 10, special issue.

———. 2007b. A socioecological perspective on primate cognition, past and present. *Animal Cognition* 10, 273–281.

Dammhahn, M. & Kappeler, P. M. 2009. Females go where the food is: Does the socio-ecological model explain variation in social organisation of solitary foragers? *Behavioral Ecology and Sociobiology*, 63, 939–952.

David, H.A. 1987. Ranking from unbalanced paired-comparison data. *Biometrika*, 74, 432–436.

———. 1988. *The Method of Paired Comparisons*. London: C. Griffin.

De Vries, H. 1995. An improved test of linearity in dominance hierarchies containing unknown or tied relationships. *Animal Behaviour*, 50, 1375–1389.

De Vries, H. & Appleby, M.C. 2000. Finding an appropriate order for a hierarchy: A comparison of the I&SI method and the BBS method. *Animal Behaviour*, 59, 239–245.

De Vries, H., Stevens, J. & Vervaecke, H. 2006. Measuring and testing the steepness of dominance hierarchies. *Animal Behaviour*, 71, 585–592.

De Waal, F. 1986. The integration of dominance and social bonding in primates. *Quarterly Review of Biology*, 61, 459–479.

De Waal, F. B. M. 1989. Dominance "style" and primate social organization. In *Comparative Socioecology: The Behavioural Ecology of Humans and Other Mammals* (ed. by Standen, V. & Foley, R. A.), 243–263. Oxford: Blackwell Scientific Publications.

De Waal, F. & Ren, R. 1988. Comparison of the reconciliation behavior of stumptail and rhesus macaques. *Ethology*, 78, 129–142.

De Waal, F. B. M. & Luttrell, L. M. 1989. Toward a comparative socioecology of the genus *Macaca*: Different dominance styles in rhesus and stumptailed macaques. *American Journal of Primatology*, 19, 83–109.

Demaria, C. & Thierry, B. 2001. A comparative study of reconciliation in rhesus and Tonkean macaques. *Behaviour*, 138, 397–410.

Di Fiore, A. 2004. Diet and feeding ecology of woolly monkeys in a western Amazonian rain forest. *International Journal of Primatology*, 25, 767–801.

Di Fiore, A. & Rendall, D. 1994. Evolution of social organization: A reappraisal for primates by using phylogenetic methods. *Proceedings of the National Academy of Science*, 91, 9941–9945.

Dias, L. D. & Strier, K. B. 2003. Effects of group size on ranging patterns in *Brachyteles arachnoides hypoxanthus*. *International Journal of Primatology*, 24, 209–221.

Dietz, J. 2004. Kinship structure and reproductive skew in cooperatively breeding primates. In *Kinship and Behavior in Primates* (ed. by Chapais, B. & Berman, C.), 223–241. Oxford: Oxford University Press.

Dixson, A. F. 1998. *Primate Sexuality*. Oxford: Oxford University Press.

Drews, C. 1993. The concept and definition of dominance in animal behaviour. *Behaviour*, 125, 283–313

Dugatkin, L.A. 1997. Winner and loser effects and the structure of dominance hierarchies. *Behavioral Ecology*, 8, 583–587.

Dunbar, R. I. M. 1988. *Primate Social Systems*. Ithaca, NY: Cornell University Press.

———. 1992. Time: A hidden constraint on the behavioural ecology of baboons. *Behavioral Ecology and Sociobiology*, 31, 35–49.

Eberle, M. & Kappeler, P. M. 2006. Family insurance: Kin selection and cooperative breeding in a solitary primate (*Microcebus murinus*). *Behavioral Ecology and Sociobiology*, 60, 582–588.

Eisenberg, J. F., Muckenhirn, N. A. & Rudran, R. 1972. The relation between ecology and social structure in primates. *Science*, 176, 863–874.

Emery Thompson, M., Muller, M. N., Wrangham, R. W. Lwanga, J. S. & Potts, K. B. 2009. Urinary C-peptide tracks seasonal and individual variation in energy balance in wild chimpanzees. *Hormones & Behavior*, 55, 299—305.

Emlen, S. T. & Oring, L. W. 1977. Ecology, sexual selection, and the evolution of mating systems. *Science*, 197, 215–223.

Fashing, P. J. 2001. Male and female strategies during intergroup encounters in guerezas (*Colobus guereza*): Evidence for resource defense mediated through males and a comparison with other primates. *Behavioral Ecology and Sociobiology*, 50, 219–230.

Fedigan, L. & Jack, K. 2004. The demographic and reproductive context of male replacements in *Cebus capucinus*. *Behaviour*, 141, 755–775.

Fichtel, C. & Kappeler, P. M. 2010. Human universals and primate symplesiomorphies: Establishing the lemur baseline. In *Mind the Gap* (ed. by Kappeler, P. M. & Silk, J. B.), 395–426. Berlin: Springer.

Fietz, J. 1999. Monogamy as a rule rather than exception in nocturnal lemurs: The case of the fat-tailed dwarf lemur (*Cheirogaleus medius*). *Ethology*, 105, 259–272.

Fietz, J., Zischler, H., Schiegk, C., Tomuik, Dausmann, K. D., Gardner, S. 2000. High rates of extra-pair young in the pair-living fat-tailed dwarf lemur, *Cheirogaleus medius*. *Behavioral Ecology and Sociobiology*, 49, 8–17.

Flack, J. & de Waal, F. 2004. Dominance style, social power, and conflict management: A conceptual framework. In *Macaque Societies: A Model for the Study of Social Organization* (ed. by Thierry, B., Singh, M. & Kaumanns, W.), 157–181. Cambridge: Cambridge University Press.

Freeland, W. J. 1976. Pathogens and the evolution of primate sociality. *Biotropica*, 8, 12–24.

Fuentes, A. 1999. Re-evaluating primate monogamy. *American Anthropologist*, 100, 890–907.

Goymann, W. & Wingfield, J. 2004. Allostatic load, social status and stress hormones: The costs of social status matter. *Animal Behaviour*, 67, 591–602.

Gursky, S. 2005. Associations between adult spectral tarsiers. *American Journal of Physical Anthropology*, 128, 74–83.

Hager, R. & Jones, C. 2009. *Reproductive Skew in Vertebrates: Proximate and Ultimate Causes*. Cambridge: Cambridge University Press.

Hamilton, W. D. 1971. Geometry of the selfish herd. *Journal of Theoretical Biology*, 31, 295–311

Hand, J. L. 1986. Resolution of social conflicts: Dominance, egalitarism, spheres of dominance, and game theory. *Quarterly Review of Biology*, 61, 201–220.

Harris, T. 2005. Roaring, intergroup aggression, and feeding competition in wild black and white colobus monkeys (*Colobus guereza*) at Kanyawara, Kibale National Park, Uganda. PhD thesis, Yale University.

———. 2006. Between group contest competition for food in a highly folivorous population of black and white colobus monkeys (*Colobus guereza*). *Behavioral Ecology and Sociobiology*, 61, 317–329.

———. 2007. Testing mate, resource and infant defence functions of intergroup aggression in non-human primates: Issues and methodology. *Behaviour*, 144, 1521–1535.

Hemelrijk, C. 1999. An individual-oriented model of the emergence of despotic and egalitarian societies. *Proceedings of the Royal Society London B*, 266, 361–369.

Heymann, E. W. 1999. Primate behavioral ecology and disease: Some perspective for a future primatology. *Primate Report* 55, 53–65.

Hill, R. A. & Dunbar, R. I. M. 2002. Climatic determinants of diet and foraging behaviour in baboons. *Evolutionary Ecology*, 16, 579–593.

Hill, R. A. & Lee, P. C. 1998. Predation risk as an influence on group size in cercopithecoid primates: Implications for social structure. *Journal of Zoology*, London, 245, 447–456.

Hinde, R. A. 1976. Interactions, relationships and social structure. *Man*, 11, 1–17.

———. 1983. A conceptual framework. In *Primate Social Relationships: An Integrated Approach* (ed. by Hinde, R.A.), 1–7. Oxford: Blackwell Scientific Publications.

Hinde, R. A. & Atkinson, S. 1970. Assessing the roles of social partners in maintaining mutual proximity as exemplified by mother-infant relations in rhesus monkeys. *Animal Behaviour*, 18, 169–176.

Hladik, C. M. 1975. Ecology, diet and social patterns in Old and New World primates. In *Socioecology and Psychology of Primates* (ed. by Tuttle, R. H), 3–35. The Hague, Mouton.

Isbell, L. 1991. Contest and scramble competition: Patterns of female aggression and ranging behavior in primates. *Behavioral Ecology*, 2, 143–155.

———. 2004. Is there no place like home? Ecological bases of female dispersal and philopatry and their consequences for the formation of kin groups. In *Kinship and Behavior in Primates* (ed. by Chapais, B. & Berman, C.), 71–108. Oxford: Oxford University Press.

Isbell, L. A. & van Vuren, D. 1996. Differential costs of locational and social dispersal and their consequences for female group-living primates. *Behaviour*, 133, 1–36.

Isbell, L. & Pruetz, J. 1998. Differences between vervets (*Cercopithecus aethiops*) and patas monkeys (*Erythrocebus patas*) in agonistic interactions between adult females. *International Journal of Primatology*, 19, 837–855.

Isbell, L. & Young, T. P. 1998. Ecological models of female social relationships in primates: Similarities, disparities, and some directions for future clarity. *Behaviour*, 139, 177–202.

Isbell, L., Pruetz, J. & Young, T. 1998. Movements of vervets (*Cercopithecus aethiops*) and patas monkeys (*Erythrocebus patas*) as estimators of food resource size, density, and distribution. *Behavioral Ecology and Sociobiology*, 42, 123–133.

Janson, C. H. 1985. Aggressive competition and individual food consumption in wild brown capuchin monkeys (*Cebus apella*). *Behavioral Ecology and Sociobiology*, 18, 125–138.

———. 1988a. Food competition in brown capuchin monkeys (*Cebus apella*): Quantitative effects of group size and tree productivity. *Behaviour*, 105, 53–76.

———. 1988b. Intraspecific food competition and primate social structure: A synthesis. *Behaviour*, 105, 1–17.

———. 1990. Ecological consequences of individual spatial choice in foraging groups of brown capuchin monkeys, *Cebus apella*. *Animal Behaviour*, 40, 922–934.

———. 1998. Testing the predation hypothesis for vertebrate sociality: Prospects and pitfalls. *Behaviour*, 135, 389–410.

———. 2000. Primate socio-ecology: The end of a golden age. *Evolutionary Anthropology*, 9, 73–86.

Janson, C. H. & Goldsmith, M. L. 1995. Predicting group size in primates: Foraging costs and predation risks. *Behavioral Ecology*, 6, 326–336.

Janson, C. H. & van Schaik, C. P. 1988. Recognizing the many faces of primate food competition: Methods. *Behaviour*, 105, 165–186.

———. The behavioral ecology of infanticide by males. In *Infanticide by Males and Its Implications* (ed. by van Schaik, C. P.

& Janson, C. H.), 469–494. Cambridge: Cambridge University Press.

Janson, C. H. & Verdolin, J. 2005. Seasonality of primate births in relation to climate. In *Seasonality in Primates* (ed. by Brockman, D. & van Schaik, C.), 307–350. Cambridge: Cambridge University Press.

Janson, C. H. & Vogel, E. 2006. Hunger and aggression in capuchin monkeys. In *Feeding Ecology in Apes and Other Primates* (ed. by Hohmann, G., Robbins, M. M. & Boesch, C.), 285–312. Cambridge: Cambridge University Press.

Kappeler, P. M. 1998. Nests, tree holes, and the evolution of primate life histories. *American Journal of Primatology* 46, 7–33.

———. 1999. Primate socioecology: New insights from males. *Naturwissenschaften*, 86, 18–29.

———. 2000 *Primate Males: Causes and Consequences of Variation in Group Composition*. Cambridge: Cambridge University Press.

Kappeler, P. M. & Heymann, E. W. 1996. Non-convergence in the evolution of primate life-history and socioecology. *Biological Journal of the Linnean Society*, 59, 297–326.

Kappeler, P. M. & Port, M. 2008. Mutual tolerance or reproductive competition? Patterns of reproductive skew among male redfronted lemurs (*Eulemur fulvus rufus*). *Behavioral Ecology and Sociobiology*, 62, 1477–1488.

Kappeler, P. M. & van Schaik, C. P. 2002. The evolution of primate social systems. *International Journal of Primatology*, 23, 707–740.

Koenig, A. 2000. Competitive regimes in forest-dwelling Hanuman langur females (*Semnopithecus entellus*). *Behavioral Ecology and Sociobiology*, 48, 93–109.

———. 2002. Competition for resources and its behavioral consequences among female primates. *International Journal of Primatology*, 23, 759–783.

Koenig, A. & Borries, C. 2006. The predictive power of socioecological models: A reconsideration of resource characteristics, agonism, and dominance hierarchies. In *Feeding Ecology in Apes and Other Primates: Ecological, Physical, and Behavioral Aspects* (ed. by Hohmann, G., Robbins, M. & Boesch, C.), 263–284. Cambridge: Cambridge University Press.

———. 2009. The lost dream of ecological determinism: Time to say goodbye? . . . or a white queen's proposal? *Evolutionary Anthropology*, 18, 166–174.

Koenig, A., Borries, C., Chalise, M. K. & Winkler, P. 1997. Ecology, nutrition, and timing of reproductive events in an Asian primate, the Hanuman langur (*Presbytis entellus*). *Journal of Zoology*, 243, 215–235.

Koenig, A., Beise, J., Chalise, M. & Ganzhorn, J. 1998. When females should contest for food: Testing hypotheses about resource density, distribution, size, and quality with hanuman langurs (*Presbytis entellus*). *Behavioral Ecology and Sociobiology*, 42, 225–237.

Korstjens, A., Sterck, E. H. M. & Noë, R. 2002. How adaptive or phylogenetically inert is primate social behavior? A test with two sympatric colobines. *Behaviour*, 139, 203–225.

Korstjens, A., Lugo Verhoeckx, I. & Dunbar, R. 2006. Time as a constraint on group size in spider monkeys. *Behavioral Ecology and Sociobiology*, 60, 683–694.

Korstjens, A., Bergmann, K., Deffernez, C., Krebs, M., Nijssen, E., van Oirschot, B., Paukert, C. & Schippers, E. 2007. How

small-scale differences in food competition lead to different social systems in three closely related sympatric colobines. In *The Monkeys of the Tai Forest: An African Primate Community* (ed. by McGraw, S., Zuberbühler, K. & Noe, R.), 72–108. Cambridge: Cambridge University Press.

Kutsukake, N. & Nunn, C. 2006. Comparative tests of reproductive skew in male primates: The roles of demographic factors and incomplete control. *Behavioral Ecology and Sociobiology*, 60, 695–706.

Lambert, J. 2005. Competition, predation, and the evolutionary significance of the cercopithecine cheek pouch: The case of *Cercopithecus* and *Lophocebus*. *American Journal of Physical Anthropology*, 126, 183–192.

Lee, P. C. & Kappeler, P. M. 2003. Socioecological correlates of phenotypic plasticity of primate life histories. In *Primate Life Histories and Socioecology* (ed. by Kappeler, P. M. & Pereira, M. E.), 41–65. Chicago: University of Chicago Press.

Lehmann, J. & Dunbar, R. I. M. 2009. Network cohesion, group size and neocortex size in female bonded old world monkeys. *Proceeding of the Royal Society London B*, 276, 4417–4422.

Lehmann, J., Korstjens, A. H. & Dunbar, R. I. M. 2007. Fission-fusion social systems as a strategy for coping with ecological constraints: A primate case. *Evolutionary Ecology*, 21, 613–634.

Leighton M. & Leighton, D. R. 1982. The relationship of size of feeding aggregate to size of food patch: Howler monkeys (*Alouatta palliata*) feeding in *Trichilia cipo* fruit trees on Barro Colorado Island. *Biotropica* 14, 81–90.

Leinfelder, I., de Vries, H., Deleu, R. & Nelissen, M. 2001. Rank and grooming reciprocity among females in a mixed-sex group of captive Hamadryas baboons. *American Journal of Primatology*, 55, 25–42.

Lewis, R. J. 2002. Beyond dominance: The importance of leverage. *Quarterly Review of Biology*, 77, 149–164.

Lindenfors, P., Fröberg, L. & Nunn, C. L. 2004. Females drive primate social evolution. *Proceedings of the Royal Society London B*, Suppl., 271, S101–S103.

Lu, A., Koenig, A. & Borries, C. 2008. Formal submission, tolerance and socioecological models: A test with female Hanuman langurs. *Animal Behaviour*, 76, 415–428.

Lührs, M.-L., Dammhahn, M., Kappeler, P. M. & Fichtel, C. 2009. Spatial memory in the grey mouse lemur (*Microcebus murinus*). *Animal Cognition*, 12, 599–609.

Maddison, W. & Maddison, D. 2006. Mesquite: A modular system for evolutionary analysis. Available at http://mesquite project. org/mesquite_folder/ docs/mesquite/whyMesquite .html.

Majolo, B., de Bortoli Vizioli, A. & Schino, G. 2008. Costs and benefits of group living in primates: Group size effects on behaviour and demography. *Animal Behaviour*, 76, 1235–1247.

Matsumura, S. 1998. Relaxed dominance relations among female moor macaques (*Macaca maurus*) in their natural habitat, South Sulawesi, Indonesia. *Folia Primatologica*, 69, 346-356.

Menard, N. 2004. Do ecological factors explain variation in social organization? In *Macaque Societies: A Model for the Study of Social Organization* (ed. by Thierry, B., Singh, M. & Kaumanns, W.), 235–261. Cambridge: Cambridge University Press.

Miller, L. E. 2002. *Eat or Be Eaten: Risk Sensitive Foraging among Primates*. Cambridge: Cambridge University Press.

Milton K. 1984. Habitat, diet and activity patterns of free-ranging woolly spider monkeys (*Brachyteles arachnoides* E. Geoffroy 1806). *International Journal of Primatology*, 5, 491–514.

Mitani, J. C., Gros-Louis, J. & Manson, J. H. 1996. Number of males in primate groups: Comparative tests of competing hypotheses. *American Journal of Primatology*, 38, 315–332.

Mitchell, C. L., Boinski, S. & van Schaik, C. P. 1991. Competitive regimes and female bonding in two species of squirrel monkeys (*Saimiri oerstedi* and *S. sciureus*). *Behavioral Ecology and Sociobiology*, 28, 55–60.

Moore, J. 1984. Female transfer in primates. *International Journal of Primatology*, 5, 537–589.

Mooring, M. S. & Hart, B. L. 1992. Animal grouping for protection from parasites: Selfish herd and encounter-dilution effects. *Behaviour*, 123, 173–193.

Muller, M. N. & Mitani, J. C. 2005. Conflict and cooperation in wild chimpanzees. *Advances in the Study of Behavior*, 35, 275–331.

Müller, A. E. & Thalmann, U. 2000. Origin and evolution of primate social organisation: A reconstruction. *Biological Reviews*, 75, 405–435.

Munshi-South, J. 2007. Extra-pair paternity and the evolution of testis size in a behaviorally monogamous tropical mammal, the large treeshrew (*Tupaia tana*). *Behavioral Ecology and Sociobiology* 62, 201–212.

Nicholson, A. J. 1954. An outline of the dynamics of animal populations. *Australian Journal of Zoology*, 2, 9–65.

Noë, R. & Hammerstein, P. 1995. Biological markets. *Trends in Ecology and Evolution*, 10, 336–339.

Nunn, C. L. 1999. The number of males in primate groups: A comparative test of the socioecological model. *Behavioral Ecology and Sociobiology*, 46, 1–13.

———. 2002. A comparative study of leucocyte counts and disease risk in primates. *Evolution*, 56, 177–190.

Nunn, C. L. & Altizer, S. 2006. *Infectious Disease in Primates: Behavior, Ecology and Evolution*. Oxford: Oxford University Press.

Nunn, C. L. & Heymann, E. W. 2005. Malaria infection and host behavior: A comparative study of Neotropical primates. *Behavioral Ecology and Sociobiology* 59, 30–37.

Nunn, C. L., Thrall, P. H., Stewart, K. & Harcourt, A. H. 2008. Emerging infectious disease and animal social systems. *Evolutionary Ecology* 22, 519–543.

Nunn, C. L. & van Schaik, C. P. 2000. Social evolution in primates: The relative roles of ecology and intersexual conflict. In *Infanticide by Males and Its Implications* (ed. by van Schaik, C. P. & Janson, C. H.), 388–422. Cambridge: Cambridge University Press.

Ostner, J. & Kappeler, P. M. 1999. Central males instead of multiple pairs in redfronted lemurs, *Eulemur fulvus rufus* (Primates, Lemuridae)? *Animal Behaviour*, 58, 1069–1078.

———. 2004. Male life history and the unusual adult sex ratio in redfronted lemurs (*Eulemur fulvus rufus*) groups. *Animal Behaviour*, 67, 249–259.

Ostner, J., Nunn, C. L. & Schülke, O. 2008a. Female reproductive synchrony predicts skewed paternity across primates. *Behavioral Ecology*, 19, 1150–1158.

Ostner, J., Heistermann, M. & Schülke, O. 2008b. Dominance, aggression, and physiological stress in wild male Assamese macaques (*Macaca assamensis*). *Hormones and Behavior*, 54, 613–619.

Palombit, R. A. 1999. Infanticide and the evolution of pair bonds in nonhuman primates. *Evolutionary Anthropology*, 7, 117–129.

Paul, A. 2002. Sexual selection and mate choice. *International Journal of Primatology*, 23, 887–904

Pereira, M. E. & McGlynn, C. A. 1997. Special relationships instead of female dominance for redfronted lemurs, *Eulemur fulvus rufus*. *American Journal of Primatology*, 43, 239–258.

Perry, S., Manson, J. H., Muniz, L., Gros-Louis, J. & Vigilant, L. 2008. Kin-biased social behaviour in wild adult female white-faced capuchins, *Cebus capucinus*. *Animal Behaviour*, 76, 187–199.

Pradhan, G. R. & van Schaik, C. 2008. Infanticide-driven intersexual conflict over matings in primates and its effects on social organization. *Behaviour*, 145, 251–275.

Pope, T. 1990. The reproductive consequences of male cooperation in the red howler monkey: Paternity exclusion in multi-male and single-male troops using genetic markers. *Behavioral Ecology and Sociobiology*, 27, 439–446.

Preuschoft, S. & Paul, A. 2000. Dominance, egalitarianism, and stalemate: An experimental approach to male-male competition in Barbary macaques. In *Primate Males: Causes and Consequences of Variation in Group Composition* (ed. by Kappeler, P.), 205–216. Cambridge: Cambridge University Press.

Pusey, A. E. & Packer, C. 1987. Dispersal and philopatry. In *Primate Societies* (ed. by Smuts, B. B., Cheney, D. L., Seyfarth, R. M., Wragham, R. W., Struhsaker, T. T.), 250–266. Chicago: University of Chicago Press.

Radespiel, U., Cepok, S., Zietemann, V. & Zimmermann, E. 1998. Sex-specific usage patterns of sleeping sites in grey mouse lemurs (*Microcebus murinus*) in Northwestern Madagascar. *American Journal of Primatology*, 46, 77–84.

Reichard, U. 2003. Monogamy: Past and present. In *Monogamy: Mating Strategies and Partnerships in Birds, Humans and Other Mammals* (ed. by Reichard, U. & Boesch, C.), 3–26. Cambridge: Cambridge University Press.

Ribble, D. O. 1991. The monogamous mating system of *Peromyscus californicus* as revealed by DNA fingerprinting. *Behavioral Ecology and Sociobiology*, 29, 161–166.

Richter, C., Mevis, L., Malaivijitnond, S., Schülke, O. & Ostner, J. 2009. Social relationships in free-ranging male *Macaca arctoides*. *International Journal of Primatology*, 30, 625–642.

Ridley, M. 1986. The number of males in a primate troop. *Animal Behaviour*, 34, 1848–1858.

Robbins, M. M. 2008. Feeding competition and agonistic relationships among Bwindi *Gorilla beringei*. *International Journal of Primatology*, 29, 999–1019.

Robbins, M. M., Robbins, A. M., Gerald-Stekli, N. Steklis, D. 2007. Socioecological influences on the reproductive success of female mountain gorillas (*Gorilla beringei beringei*). *Behavioral Ecology and Sociobiology*, 61, 919–931.

Rowell, T. E. 1979. How would we know if social organization were not adaptive? In *Primate Ecology and Social Organization* (ed. by Bernstein, I. S. & Smith, E. O.), 1–22. New York: Garland.

Sapolsky, R. 2002. Endocrinology of the stress-response. In *Behavioral Endocrinology* (ed. by Becker, J., Breedlove, S., Crews, D. & McCarthy, M.), 409–450. Cambridge: MIT Press.

Schino, G. & Aureli, F. 2008a. Trade-offs in primate grooming reciprocation: Testing behavioural flexibility and correlated evolution. *Biological Journal of the Linnean Society*, 95, 439–446.

Schino, G. & Aureli, F. 2008b. Grooming reciprocation among female primates: A meta-analysis. *Biology Letters*, 4, 9–11.

Schülke, O. 2001. Differential energy budget and monopolization potential of harem holders and bachelors in Hanuman langurs (*Semnopithecus entellus*): Preliminary results. *American Journal of Primatology*, 55, 57–63.

———. 2003. To breed or not to breed: Food competition and other factors influencing female reproductive decisions in the pair-living nocturnal fork-marked lemur (*Phaner furcifer*). *Behavioral Ecology and Sociobiology*, 55, 11–21.

———. 2005. Evolution of pair-living in *Phaner furcifer*. *International Journal of Primatology*, 26, 903–919.

Schülke, O. & Kappler, P.M. 2003. So near and yet so far: Territorial pairs but low cohesion between pair partners in a nocturnal lemur, *Phaner furcifer*. *Animal Behaviour*, 65, 331–343.

Schülke, O. & Ostner, J. 2005. Big times for dwarfs: Social organization, sexual selection, and cooperation in the Cheirogaleidae. *Evolutionary Anthropology*, 14, 170–185.

———, 2008. Male reproductive skew, paternal relatedness and female social relationships. *American Journal of Primatology*, 70, 1–4.

Schülke, O. Kappeler, P. M. & Zischler, H. 2004a. Small testes size despite high extra-pair paternity in the pair-living nocturnal primate *Phaner furcifer*. *Behavioral Ecology and Sociobiology*, 55, 293–301.

Schülke, O., Chalise, M. K., Nikolei, J., Podzuweit, D., Ganzhorn, J. U., Borries, C., Koenig, A. 2004b. Does digestion time limit group size in folivorous primates? *American Journal of Physical Anthropology*, 123, 175–176.

Schülke, O., Chalise, M. & Koenig, A. 2006. The importance of ingestion rates for estimating food quality and energy intake. *American Journal of Primatology*, 68, 951–965.

Schülke, O., Bhagavatula, J., Vigilant, L. & Ostner, J. 2010. Social bonds enhance reproductive success in male macaques. *Current Biology*, 20, 2207–2210.

Setchell, J. M. & Kappeler, P. M. 2003. Selection in relation to sex in primates. *Advances in the Study of Behavior*, 33, 87–173.

Seyfarth, R. 1977. A model of social grooming among female monkeys. *Journal of Theoretical Biology*, 65, 671–698.

Shopland, J. M. 1987. Food quality, spatial development, and the intensity of feeding interference in yellow baboons (*Papio cynocephalus*). *Behavioral Ecology and Sociobiology*, 21, 149–156.

Silk, J. B. 2002a. Using the 'F-word' in primatology. *Behaviour*, 139, 421–446.

———. 2002b. Kin selection in primate groups. *International Journal of Primatology*, 23, 849–875.

Silk, J. B., Altmann, J. & Alberts, S. 2006. Social relationships among adult female baboons (*Papio cynocephalus*) I. Variation in the strength of social bonds. *Behavioral Ecology and Sociobiology*, 6, 183–195.

Smuts, B. B. 1985. *Sex and Friendship in Baboons*. New York: Aldine.

Smuts, B. B., Cheney, D. L., Seyfarth, R. M., Wrangham, R. W. & Struhsaker, T. T. 1987. *Primate Societies*. Chicago: University of Chicago Press.

Snaith, T. & Chapman, C. 2007. Primate group size and interpreting socioecological models: Do folivores really play by different rules? *Evolutionary Anthropology*, 16, 94–106.

———. Red colobus monkeys display alternative behavioral responses to the costs of scramble competition. *Behavioral Ecology*, 19, 1289–1296.

Sommer, V. & Reichard, U. 2000. Rethinking monogamy: The gibbon case. In *Primate Males: Causes and Consequences of Variation in Group Composition* (ed. by Kappeler, P. M.), 159–168. Cambridge: Cambridge University Press.

Stacey, P. B. 1986. Group size and foraging efficiency in yellow baboons. *Behavioral Ecology and Sociobiology*, 18, 175–187.

Steenbeek, R. & van Schaik, C. P. 2001. Competition and group size in Thomas's langurs (*Presbytis thomasi*): The folivore paradox revisited. *Behavioral Ecology and Sociobiology*, 49, 100–110.

Steenbeek, R., Sterck, E. H. M., de Vries, H. & van Hooff, J. A. R. A. M. 2000. Costs and benefits of the one-male, age-graded, and all-male phases in wild Thomas's langur groups. In *Primate Males: Causes and Consequences of Variation in Group Composition* (ed. by Kappeler, P. M.), 130–145. Cambridge: Cambridge University Press.

Sterck, E. H. M. 1998. Female dispersal, social organization, and infanticide in langurs: Are they linked to human disturbance? *American Journal of Primatology*, 44, 235–254.

Sterck, E. & Korstjens, A. 2000. Female dispersal and infanticide avoidance in primates. In *Infanticide by Males and Its implications* (ed. by van Schaik, C. & Janson, C.), 293–321. Cambridge: Cambridge University Press.

Sterck, E. H. M. & Steenbeck, R. 1997. Female dominance relationships and food competition in the sympatric Thomas langur and long-tailed macaque. *Behaviour*, 134, 749–774.

Sterck, E. H. M., Watts, D. P. & van Schaik, C. P. 1997. The evolution of female social relationships in nonhuman primates. *Behavioral Ecology and Sociobiology*, 41, 291–309.

Strier, K. B. 1994. Myth of the typical primate. *Yearbook of Physical Anthropology*, 37, 233–271.

Struhsaker, T. T. 1969. Correlates of ecology and social organization among African cercopithecines. *Folia Primatologica*, 11, 80–118.

Symington, M. M. 1988. Food competition and foraging party size in the black spider monkey (*Ateles paniscus chamek*). *Behaviour*, 105, 117–184.

Terborgh, J. 1983. *Five New World Primates: A Study in Comparative Ecology*. Princeton, NJ: Princeton University Press.

Thierry, B. 2007. Unity in diversity. Lessons from macaque societies. *Evolutionary Anthropology*, 16, 225–238.

———. 2008. Primate socioecology, the lost dream of ecological determinism. *Evolutionary Anthropology* 17, 93–96.

Thierry, B., Iwaniuk, A. N. & Pellis, S. M. 2000. The influence of phylogeny on the social behaviour of macaques (Primates: Cercopithecidae, genus *Macaca*). *Ethology*, 106, 713–728.

Thierry, B., Aureli, F., Nunn, C. L., Petit, O., Abegg, C. & De Waal, F. 2008. A comparative study of conflict resolution in macaques: Insights into the nature of trait co-variation. *Animal Behaviour*, 75, 847–860.

Trivers, R. L. 1972. Parental investment and sexual selection. In *Sexual Selection and the Descent of Man, 1871–1971* (ed. by Campbell, B.), 136–179. London: Heinemann.

Treves A. 1998. The influence of group size and neighbors on vigilance in two species of arboreal monkeys. *Behaviour*, 135, 453–481.

Van Hooff, J. A. R. A. M. & van Schaik, C. P. 1994. Male bonds: Affiliative relationships among non-human primate males. *Behaviour*, 130, 309–337.

Van Noordwijk, M. A. & van Schaik, C. P. 1987. Competition among female long-tailed macaques, *Macaca fascicularis*. *Animal Behaviour*, 35, 577–589.

Van Noordwijk, M. A. & van Schaik, C. P. 2000. Reproductive patterns in eutherian mammals: Adaptations against infanticide? In *Infanticide by Males and Its Implications* (ed. by van Schaik, C. P. & Janson, C. H.), 322–360. Cambridge: Cambridge University Press.

Van Schaik, C. P. 1983. Why are diurnal primates living in groups? *Behaviour*, 87, 120–144.

———. 1989. The ecology of social relationships amongst female primates. In *Comparative Socioecology. The Behavioural Ecology of Humans and Other Mammals* (ed. by Standen, V. & Foley, R. A.), 195–218. Oxford: Blackwell Scientific Publications.

———. 1996. Social evolution in primates: The role of ecological factors and male behaviour. *Proceedings of the British Academy*, 88, 9–31.

———. 2000. Vulnerability to infanticide by males: Patterns among mammals. In *Infanticide by Males and Its Implications* (ed. by van Schaik, C. P. & Janson, C. H.), 61–71. Cambridge: Cambridge University Press.

Van Schaik, C. P. & Hörstermann, M. 1994. Predation risk and the number of males in a primate group: A comparative test. *Behavioral Ecology and Sociobiology*, 35, 261–272.

Van Schaik, C. & Kappeler, P. M. 2003. The evolution of social monogamy in primates. In *Monogamy: Mating Strategies and Partnerships in Birds, Humans and Other Mammals* (ed. by Reichard, U. & Boesch, C.), 59–80. Cambridge: Cambridge University Press.

Van Schaik, C. P. & van Noordwijk, M. A. 1988. Scramble and contest in feeding competition among female long-tailed macaques (*Macaca fascicularis*). *Behaviour*, 105, 77–98.

———. 1989. The special role of male *Cebus* monkeys in predation avoidance and its effect on group composition. *Behavioral Ecology and Sociobiology*, 24, 265–276.

Van Schaik, C., Pandit, S. & Vogel, E. 2004a. A model for within-group coalitionary aggression among males. *Behavioral Ecology and Sociobiology*, 57, 101–109.

Van Schaik, C. P., Pradhan, G. R. & van Noordwijk, M. A. 2004b. Mating conflict in primates: Infanticide, sexual harrassment and female sexuality. In *Sexual Selection in Primates: New and Comparative Perspectives* (ed. by Kappeler, P. M. & van Schaik, C. P.), 131–150. Cambridge: Cambridge University Press.

Van Schaik, C., Pandit, S. & Vogel, E. 2006. Toward a general model for male-male coalitions in primate groups. In *Cooperation in Primates and Humans* (ed. by Kappeler, P. & van Schaik, C.), 151–172. Heidelberg: Springer.

Vehrencamp, S. L. 1983. A model for the evolution of despotic versus egalitarian societies. *Animal Behaviour*, 31, 667–682.

Vogel, E. R. 2005. Rank differences in energy intake rates in white-faced capuchin monkeys, *Cebus capucinus*: The effects of contest competition. *Behavioral Ecology and Sociobiology*, 58, 333–344.

Vogel, E. R. & Janson, C. H. 2007. Predicting the frequency of food-related agonism in white-faced capuchin monkeys (*Cebus capucinus*), using a novel focal-tree method. *American Journal of Primatology*, 69, 1–18.

Vick, L. G. & Pereira, M. E. 1989. Episodic targeting aggression and the histories of Lemur social groups. *Behavioral Ecology and Sociobiology*, 25, 3–12.

Waser, P. 1977. Feeding, ranging, and group size in the mangabey *Cercocebus albigena*. In *Primate Ecology* (ed. by Clutton-Brock, T. H.), 182–222. London: Academic Press.

Watts, D. P. 1989. Infanticide in mountain gorillas: New cases and a reconsideration of the evidence. *Ethology* 81, 1–18.

———. 2010. Dominance, power, and politics in non-human and human primates. In *Mind the Gap: Tracing the Origins of Human Universals* (ed. by Kappeler, P. M. & Silk, J. B.), 109–138. Berlin: Springer.

Wey, T, Blumstein, D.T., Shen, W. & Jordán, F. 2008. Social network analysis of animal behaviour: A promising tool for the study of sociality. *Animal Behaviour*, 75, 333–344.

Wickler, W. & Seibt, U. 1983. Monogamy: An ambiguous concept. In *Mate Choice: Sexual Behaviour in Animals* (ed. by Bateson, P.), 33–50. Cambridge: University of Cambridge Press.

Widdig, A. 2007. Paternal kin discrimination: The evidence and likely mechanisms. *Biological Reviews*, 82, 319–334.

Widdig, A., Bercovitch, F., Streich, W. J., Sauermann, U., Nürnberg, P. & Krawczak, M. 2004. A longitudinal analysis of reproductive skew in male rhesus macaques. *Proceedings of the Royal Society London B*, 271, 819–826.

Widdig, A., Streich, W. J., Nürnberg, P., Croucher, P. J. P., Bercovitch, F. & Krawczak, M. 2006. Paternal kin bias in the agonistic interventions of adult female rhesus macaques (*Macaca mulatta*). *Behavioral Ecology and Sociobiology*, 61, 205–214.

Wilson, K., Knell, R., Boots, M. & Koch-Osborne, J. 2003. Group-living and investment in immune defence: An interspecific analysis. *Journal of Animal Ecology*, 72, 133–143.

Wrangham, R. W. 1979. On the evolution of ape social systems. *Social Sciences Information*, 18, 335–368.

———. 1980. An ecological model of female-bonded primate groups. *Behaviour*, 75, 262–300.

———. 1987. Evolution of social structure. In *Primate Societies* (ed. by Smuts, B. B., Cheney, D. L., Seyfarth, R. M., Wragham, R. W. & Struhsaker, T. T.), 282–298. Chicago: University of Chicago Press.

Wrangham, R. W., Gittleman, J. L. & Chapman, C. A. 1993. Constraints on group size in primates and carnivores: Population density and day-range as assays of exploitation competition. *Behavioral Ecology and Sociobiology*, 32, 199–209.

Wright, P. 1999. Lemur traits and Madagascar ecology: Coping with an island environment. *Yearbook of Physical Anthropology*, 42, 31–72.

Chapter 10 Life-History Evolution in Primates

Carel P. van Schaik and Karin Isler

B<small>Y</small> "<small>LIFE HISTORY</small>," we mean "the probabilities of survival and the rates of reproduction at each age in the life-span" (Partridge & Harvey 1988) or "features of the life cycle and their timing" (Stearns 1992). Fortunately, compared with the complex life cycles of many other organisms, such as parasitic protozoans or metamorphosing insects or amphibians, homeothermic vertebrates (birds and mammals) have relatively simple life cycles. Moreover, in contrast to reptiles and fish, most mammals do not exhibit indeterminate growth. They therefore show a fairly clean separation between the immature period of growth followed by sexual maturation and the period of reproduction when growth usually has stopped (chapter 11, this volume).

To characterize and quantify mammalian life history, we measure the durations of life phases or count offspring per unit time. Thus, mammalian life-history variables include age at weaning, age at first reproduction, life expectancy at birth (defined as predicted mean age at death as reckoned from birth), life expectancy at maturity, birth rates (or birth intervals after surviving infants), infant sex ratios, litter size, and maximum life span or longevity (fig. 10.1). Size and state of development at birth, as well as size at weaning and at first reproduction, are also generally considered life-history variables, and consequently so are rates of growth and development. While many life-history parameters are means of rates or durations of phases, some are extremes, such as maximum rate of growth and reproduction and maximum lifespan or longevity. Table 10.1 provides a compilation of the best currently available life-history data for primates.

Life-history parameters show characteristic mean values or ranges within any given species (table 10.1). Thus, a female gray mouse lemur (*Microcebus murinus*) will tend to become sexually mature well before her first birthday when she weighs about 65 grams, will have up to two litters of two infants a year, and will often die in her second or third year, although under the best possible conditions some surprisingly can live until they are 17 years old (Kappeler & Rasoloarison 2003). At the other extreme, a female Sumatran orangutan (*Pongo abelii*) will have her first infant at around age 15 when she weighs around 35 kg, give birth to another infant every eight or nine years, and die when approaching age 60 (Wich et al. 2004). Even species similar in body size may differ considerably in their life histories. Female patas monkeys (*Erythrocebus patas*) grow rapidly to around 6.5 kilograms and start reproduction at age three, giving birth almost every year; most females die before age 15 (Nakagawa et al. 2003), with maximum observed lifespan in captivity at 28 years. In contrast, the slightly smaller female lar gibbon (*Hylobates lar*) starts reproduction at age 11 (Reichard & Barelli 2008) and keeps giving birth about every 3.5 years for decades, with a maximum observed lifespan of over 50 years.

No matter how much they are pampered, it is impossible to keep a patas monkey alive for 50 years or to make a gibbon start to reproduce at three years of age and produce larger litters. This constancy indicates that the life-history "package" of a species—its characteristic rate of development and reproduction and its maximum life span—is an adaptation honed by natural selection. Life history is there-

Fig. 10.1. An adult female chacma baboon of the Okavango Delta, Botswana sits with her one-month-old infant and her 18-month-old juvenile offspring. The "interbirth interval" between consecutive surviving offspring is a commonly used measure of female life history and reproductive performance. Photo courtesy of Ryne A. Palombit.

and use means or extremes, such as mean age at first reproduction or maximum life span, to estimate the life-history trait of that particular genotype. This procedure would then be replicated in different conditions so that one could estimate the extent of plasticity.

In practice, of course, we have no alternative but to estimate life-history parameters by distilling them from demographic data. We therefore study wild and captive populations, and derive mean values or extremes from these samples. This procedure is not without its problems. First, individuals and populations may vary genetically in their life-history traits, but unless the differences are dramatic these are difficult to detect. Second, and importantly, these estimates will vary because ecological conditions vary in space and time, and the extent of this variation will depend on the degree of plasticity. These two problems combine to make it difficult to collect accurate estimates of the life-history traits of a population or species. Mixing wild and captive data on development and reproduction introduces problems when one examines interspecific patterns in life history or tests predictions of life-history theory with comparative data. Issues of data quality are therefore always present. In table 10.1 we mainly use data from wild populations for most of the variables, except gestation and maximum life span.

The relationship between life history and demography goes beyond methods of estimation. In the here and now, life history inevitably strongly affects demography: the age structure and rates of mortality and reproduction of a mouse lemur population will necessarily look very different from that of an orangutan. In the long run, however, the causal arrow is reversed; demographic processes are the major selective agent for evolutionary changes in life-history parameters. These topics will be developed in detail below in the section on the evolution of life histories.

In this chapter, we place the life history of primates into a broad mammalian perspective. We first describe the typical patterns in mammalian and primate life history, and then discuss how life-history theory tries to explain their evolution. The main theoretical framework contends that life histories are primarily an adaptation to rates of unavoidable extrinsic mortality. We then discuss the ecological correlates of mammalian life histories and how life history affects behavior. We end with a discussion of the relationship between life history and brain size, which turn out to have undergone extensive correlated evolution. Life-history theory is complex and much debated, so we can only give a partial overview of the current state of the art. Harvey & Purvis (1999) and Leigh & Blomquist (2007) provide useful overviews of the history of life-history theory. Stearns (2000) discusses additional theory beyond the simple optimization approach followed here.

fore essentially an individual trait amounting to a design for rates of growth, development, and reproduction, as well as maximum life span (Stearns 1992, 2000). This design is genetically based, although it is subject to a certain amount of plasticity: each primate can grow or age faster or more slowly depending on conditions.

Even though life-history parameters are individual characters, to estimate them we necessarily use demographic data derived from the study of the age structure and dynamics of populations. The reason is obvious. We cannot possibly record all relevant life-history information by documenting the fate of single individuals. After all, some individuals may die while still very young. Ideally, one would collect information on the life events for a large sample of genetically identical clones held under standard conditions,

Table 10.1. Life-history variables of primates

Species	Body mass (g)	Endocranial volume (cm³)	Neonatal mass (g)	Offspring per litter	Interbirth interval (d)	Gestation (d)	Lactation (d)	Age of female at first reproduction (y)	Maximum lifespan (y)	r_{max}
Lemuriformes										
1. Cheirogaleus major*	400	5.6	18.1	2.25	365	70	70	1	13.4	0.754
2. Cheirogaleus medius*	140	2.5	11.8	2.14	365	62	45	2[w]	23.3	0.500
3. Daubentonia madagascariensis	2,555	44.8	115.3	1	913	164[c]	205[c]	3.5	23.3	0.127
4. Eulemur fulvus*	2,292	24.8	78.6	1.08	547	121[w]	183	2.66	35.5	0.222
5. Eulemur macaco	2,390	22.7	61	1.06	365	126	135	2.18	36.2	0.312
6. Eulemur mongoz	1,212	17.9	58.2	1.21	365[w]	128	152	2.51	36.2	0.318
7. Hapalemur griseus*	709	13.8	45.2	1.4	365[w]	138	121[w]	2.73	23.3	0.332
8. Lemur catta	2,210	22.1	81.2	1.3	432[w]	136	179	3[w]	37.3	0.275
9. Lepilemur mustelinus*	804	8.3	27	1	365[w]	135	75	1.63	12	0.334
10. Microcebus murinus*	65	1.6	7.1	1.9	365[w]	60[w]	40	1.75[w]	18.2	0.502
11. Mirza coquereli*	312	5.8	17.5	1.7	365[w]	86	86	1.92	17.4	0.447
12. Propithecus diadema*	6,130	38.3	145	1	657[w]	179[w]	183	5.33[w]	21	0.131
13. Propithecus verreauxi	2,955	26.2	99.7	1.17	624[w]	159	183	6[w]	30.5	0.148
14. Varecia variegata	3,575	30.3	86.1	2.23	365	102[w]	146	2.72	36	0.428
Lorisiformes										
15. Arctocebus calabarensis	309	7.0	30.3	1	183[w]	130	115	1.12	13	0.655
16. Galago moholi	148	3.7	11.8	2	183[w]	124	84	1.21[s]	16.6	0.715
17. Galago senegalensis	194	3.9	19	1.81	200	141	84	1.64[s]	17.1	0.565
18. Galagoides demidovii	75	2.6	10.2	1.57	365[w]	113	45	1.06[s]	13.4	0.968
19. Loris tardigradus*	193	5.5	11	1.25	182.5	171	135	1.39[s]	19.3	0.674
20. Nycticebus coucang*	653	10.0	51.3	1.07	365[w]	193	180	1.54	25.8	0.364
21. Otolemur crassicaudatus	1,150	11.8	43.2	1.24	365	136	134	1.75	19.6	0.383
22. Otolemur garnettii	1,188	10.9	52	1.76	365[w]	132	140	1.63[s]	18.3	0.497
23. Perodicticus potto	835	12.5	39.1	1.03	354	197	212	0.91[s]	26.3	0.440
Tarsiiformes										
24. Cephalopachus bancanus*	111	3.2	24.4	1	365[w]	178	76	1.57[s]	16.3	0.343
25. Tarsius tarsier*	117	3.3	24	1	397[w]	191[w]	78[w]	1.42	16	0.335
Platyrrhini										
26. Alouatta caraya	4,240	49.2	262	1	483[w]	187	325	3.71	32.4	0.199
27. Alouatta palliata*	5,350	51.2	318	1.1	684[w]	186[w]	365[w]	4[w]	24	0.161
28. Alouatta seniculus*	5,210	55.4	263	1.2	517[w]	191	380[w]	5.17[w]	25	0.179
29. Aotus trivirgatus*	933	15.7	90.2	1.01	341	133[c]	76	4[w]	30.1	0.235
30. Ateles geoffroyi	7,290	103.4	490.3	1	1,055[w]	228	790[w]	7[w]	47.1	0.093
31. Ateles paniscus	8,070	110.9	452.5	1	1,459[w]	228	1,094[w]	5	46	0.083
32. Callicebus cupreus*	1,120	16.6	74	1.17	336[w]	129[c]	140	5[w]	26.4	0.227
33. Callimico goeldii	484	11.4	54.1	1	274	149[c]	67	1.49[c,s]	22.2	0.431
34. Callithrix jacchus	320	8.2	30.2	2.22	217[c]	144[c]	77	1.67	22.8	0.755
35. Cebuella pygmaea	116	4.2	16.3	2.06	213	142	91	1.9[w]	18.6	0.675
36. Cebus albifrons	2,290	66.1	232.6	1	548	155	264	4	40.4	0.178
37. Cebus apella	2,501	64.2	208.9	1	587[w]	152[w]	265	6.63[w]	46	0.135
38. Cebus capucinus	2,437	69.2	240	1	836[w]	161	517	6[w]	54	0.115
39. Lagothrix lagotricha*	7,020	89.3	463.2	1	1,116[w]	222	548	9[w]	32	0.073

40. Leontopithecus rosalia	609	12.8	57.1	2.06	311[w]	125	91	3.6[w]	31.6	0.376
41. Mico argentatus	345	8.0	32	1.91	213[w]	144	120	1.67	16.5	0.704
42. Saguinus fuscicollis*	401	8.3	39.0	1.97	258[w]	147	91[w]	1.92[w]	24.9	0.592
43. Saguinus geoffroyi	517	10.2	48.0	1.99	243	150	75[w]	2[w]	20.5	0.599
44. Saguinus midas	425	9.8	40	1.8	240	154[c]	70	2	21	0.572
45. Saguinus oedipus	431	9.7	43.3	1.90	333[w]	160	76	1.89	26.2	0.508
46. Saimiri sciureus*	763	23.5	109.5		365	171	183	3.86	30.3	0.230
Cercopithecinae										
47. Cercocebus atys	6,200	86.8	492.3	1	505	167	126	4.71	32.5	0.173
48. Cercopithecus ascanius	2,901	56.7	371	1	499[w]	172	183	5	31.2	0.080
49. Cercopithecus cephus	2,880	60.7	340	1	833[c]	170	365	5[c]	36	0.169
50. Cercopithecus mitis	4,628	65.9	398.5	1	1,429[w]	176	316	5.42[c]	37.8	0.123
51. Cercopithecus neglectus	4,130	60.9	452.5	1	700[c]	168	365	5[c]	30.8	0.137
52. Chlorocebus aethiops*	3,576	59.9	335.4	1	520.8[w]	163[c]	259	5.06[w]	30.8	0.164
53. Erythrocebus patas	6,500	88.9	504.5	1	438[w]	167	228	3.04[w]	28.3	0.231
54. Lophocebus albigena	6,010	90.7	500	1	1,004	175	213	6	29	0.096
55. Macaca arctoides	8,400	99.8	499.5	1.01	699[c]	178[c]	309[c]	4.9[c]	29.2	0.150
56. Macaca fascicularis	3,518	60.9	339.6	1.01	511[w]	163	231[j]	5.23[w]	38.5	0.148
57. Macaca fuscata	8,030	96.8	535.4	1	817[w]	175	195	6.1[w]	38.5	0.115
58. Macaca maura	6,050	87.4	389.5	1	733[w]	176	508	6.5[w]	33	0.118
59. Macaca mulatta	5,670	84.3	478.3	1.01	547[w]	167	192	5[w]	34	0.161
60. Macaca nemestrina	6,539	98.1	463.7	1	486[c]	171	237	3.92	37.6	0.193
61. Macaca nigra*	5,470	80.3	461	1	540	176	214	5.44	34	0.155
62. Macaca radiata	3,856	70.5	398	1.01	468[c]	168	365	4[c]	30	0.197
63. Macaca silenus	6,100	78.0	476	1	504	172	365	4.92	40	0.170
64. Macaca sylvanus	9,625	92.9	450	1	669[w]	165	213	4.75	29.1	0.144
65. Mandrillus leucophaeus	12,500	126.3	722	1	547[c]	181	471[c]	5	39	0.161
66. Mandrillus sphinx	12,800	136.0	898	1.07	525[c]	175	307[c]	5.04[w]	40	0.171
67. Miopithecus talapoin*	1,560	34.8	177.5	1	365	167	183	4.46[s]	27	0.213
68. Papio anubis	13,300	155.3	947.5	1	757[w]	180	420	6.92	31.6	0.111
69. Papio cynocephalus	12,000	149.5	770	1	707[w]	175	456	5.99[w]	34.1	0.126
70. Papio hamadryas	10,300	138.3	727.5	1	730[c]	170	365	6.1[w]	37.5	0.123
71. Papio ursinus	14,800	165.0	850	1	882[w]	187	502[w]	3.67[s]	32.6	0.133
72. Theropithecus gelada	11,700	124.5	464	1	781[w]	170	450	4[w]	35.5	0.141
Colobinae										
73. Colobus guereza	7,506	72.6	572.5	1	659[w]	158[w]	334	4.75	35.3	0.145
74. Colobus polykomos	6,709	71.0	597	1	584[w]	170	365[w]	5.5[w]	33.9	0.147
75. Nasalis larvatus	9,730	84.8	600	1	549	166	281	4.5	25.1	0.166
76. Pygathrix nemaeus	8,440	94.9	435.2	1	495	210	330	4	26	0.188
77. Semnopithecus entellus	11,340	95.2	500	1.14	511[w]	199[w]	389[w]	3.5[w]	29	0.214
78. Semnopithecus vetulus	5,103	59.9	447	1	714[w]	200	228	4	26.1	0.147
79. Trachypithecus obscurus	6,765	59.3	383	1	731	152[c]	365	4	33.9	0.147
Hominoidea										
80. Gorilla gorilla	71,500	433.5	2,123.5	1.0	1,826[c]	257	1,278	10.2[w]	55	0.054
81. Hylobates lar	5,595	101.5	406.0	1	1,251[c]	210	745	10[w]	52	0.071
82. Hylobates syndactylus	11,295	123.5	553.4	1	986[c]	232	639	9	43	0.085
83. Pan paniscus	33,200	326.3	1447	1	1,751[w]	231	1,094	14.2[w]	54.5	0.047
84. Pan troglodytes	40367	356.8	1,845.5	1.1	1,985[c]	235[w]	1,460	13.25[w]	59.4	0.049

(continued)

Table 10.1. (continued)

Species	Body mass (g)	Endocranial volume (cm³)	Neonatal mass (g)	Offspring per litter	Interbirth interval (d)	Gestation (d)	Lactation (d)	Age of female at first reproduction (y)	Maximum lifespan (y)	r_{max}
85. *Pongo abelii*	41,148	346.1	1,969	1	3,397^w	243^c	2,009^w	15.4^w	59	0.025
86. *Pongo pygmaeus*	36,948	337.7	1,968.1	1	2,685^w	250	1,936^w	15.7^w	56.3	0.031
87. *Homo sapiens*	56,700	1,212.7	3,319	1.01	1,167	270	720	19.5	105	0.052**

^{w}Wild; ^{c}captive; ^{i} interbirth interval minus gestation length; ^{s}sexual maturity plus gestation length.

*These species underwent recent taxonomic revision, and the given values may be derived from several of the species listed in table A.1.

**Due to midlife menopause in humans, maximum lifespan was set to 60 years to calculate this r_{max} value. Using a more realistic age for last reproduction of 47 years would yield 0.047 for *H. sapiens* and 0.045 for *P. troglodytes*.

This table lists life-history variables for 87 primate species for which complete sets of variables are available. Body mass and endocranial volume (brain size) data were taken from Isler et al. (2008) and complemented with data from Janneke van Woerden (personal communication), using female values for dimorphic species, and species means otherwise. Neonate mass (averaged for both sexes) was taken with very few exceptions from Smith and Leigh (1998), who list original sources and sample sizes. Maximum life span data were taken from the compilation by Weigl (2005), which comprises mostly records from captivity. Unconfirmed records are not included here.

For the remaining life-history variables, preference was given to long-term studies of wild populations. Values from wild (^{w}) or captive (^{c}) conditions are labeled accordingly, if known. Values from several long-term studies of the same species were averaged. If original sources could not be retrieved, we took data from previous compilations. We preferentially included data from compilations that do not present averaged values but instead are probably based on original sources, although no individual references are provided. If different values were listed in several compilations of equal credibility, the median value was taken.

The earliest compilation containing each listed value is cited. The compilations of Rudder (1979), Harvey and Clutton-Brock (1985), Harvey et al. (1987), Martin and Maclarnon (1990), and Ross and Jones (1999) unfortunately do not list sources, but are based on a thorough literature review of B. Rudder, A. Maclarnon, C. Ross, and R. D. Martin in the late 1970s and 1980s. Those values are used if no other sources were found.

In primates, weaning is a gradual process influenced by a variety of environmental and social factors. Thus, weaning age is notoriously difficult to pinpoint. In the present compilation, weaning age represents the average age end of lactation or lactational amenorrhea in wild primates. Some authors (e.g., Barrickman et al. 2008; Lee 1999; Lee et al. 1991) estimate weaning age as interbirth interval minus gestation length (^{i}), which can easily be calculated from our table if needed.

Age at first reproduction (AFR) of females is averaged for a population. Thus, seasonal breeders can still exhibit a value between full years if, for example, some females start to reproduce at two years but others do so only at tree years. Usually the actual AFR is underestimated if it is calculated from age at sexual maturity plus gestation length (^{s}).

Time units were converted as follows: 1 year = 12 × 30.4 days.

Sources (see reference list on page XX): 1. Chapman et al. 1990; Nowak 1999; Petter-Rousseaux 1964; Ross 2003. 2. Ernest 2003; Petter 1978; Petter-Rousseaux 1964. 3. Feistner and Ashbourne 1994; Garbutt 1999; Gould & Sauther 2007; Ross 2003. 4. Ernest 2003; Glander 1994; Godfrey et al. 2001; Ostner & Heistermann 2003; Ross 2003. 6. Colquhoun 1993; Ernest 2003; Lindenfors 2002; Ross 2003. 7. Campbell 2007; Chapman et al. 1990; Kappeler & Pereira 2003; Mutschler & Tan 2003; Wright 1990. 8. Chapman et al. 1990; Godfrey et al. 2001; Koyama et al. 2001; Van Horn & Eaton 1979. 9. Godfrey et al. 2001; Martin & MacLarnon 1990; Ross 2003. 10. Kaplan et al. 2000; Kappeler & Pereira 2003; Kappeler & Rasoloarison 2003; 11. Ernest 2003; Kappeler & Rasoloarison 2003; Richard 1987; Smith et al. 2007. 12. Godfrey et al. 2001; Radespiel 2000; Wright 1995. 13. Ernest 2003; Godfrey et al. 2001; Petter-Rousseaux 1964; Pochron et al. 2004. 14. Ernest 2003; Harvey & Clutton-Brock 1985; Morland 1990; Richard et al. 2002. 15. Bearder et al. 2003; Godfrey et al. 2001; Martin & Maclarnon 1990; Ross 2003. 16. Godfrey et al. 2001; Nekaris & Bearder 2007; Ross 2003. 17. Bearder 1987; Chapman et al. 1990; Harvey & Clutton-Brock 1985. 18. Ernest 2003; Nekaris & Bearder 2007. 19. Bearder 1987; Nekaris & Bearder 2007. 20. Ernest 2003; Nekaris 2003; Nekaris & Bearder 1985; Lee 1999; Whitten & Brockman 2001. 25. Formerly Ross 2003. 22. Nekaris & Bearder 2007. 23. Bearder 1987; Ernest 2003; Nekaris & Bearder 2007. 24. Bearder 1987; Gursky 2007; Izard & Rasmussen 1985; Lee 1999; Whitten & Brockman 2001. 25. Formerly known as *Tarsius spectrum*, Fogden 1974. 26. Di Fiore & Campbell 2007; Godfrey et al. 2001; Ross 2003. 27. Clarke & Glander 1984; Di Fiore & Campbell 2007; Glander 1980; Godfrey et al. 2001. 28. Crockett & Pope 1993; Di Fiore & Campbell 2007; Ernest 2003; Lee 1999. 29. Digby et al. 2007; Hunter et al. 1979; Lee 1999; Lee et al. 1991. 30. Crockett & Sekulic 1982; Di Fiore & Campbell 2007; Lee 1999. 31. Di Fiore & Campbell 2007; Fedigan & Rose 1995; Ross 2003. 32. Ernest 2003; Fernandez-Duque 2007; Norconk 2007; Valeggia et al. 1999. 33. Dettling & Pryce 1999; Gursky 2007; Lee 1999. 34. Godfrey et al. 2001; Hearn & Lunn 1975; Stevenson & Rylands 1988. 35. Chapman et al. 1990; Smucny et al. 2004; Soini 1988. 36. Godfrey et al. 2001; Kaplan et al. 2000; Martin & MacLarnon 1990. 37. Lee 1999; Ross 2003. 38. Di Bitetti & Janson 2001; Fedigan 2003; Fedigan & Rose 1995; Lee 1999. 39. Harvey & Clutton-Brock 1985; Ramirez 1988. 40. Bales et al. 2001; Chapman et al. 1990; Digby et al. 2007; Kleiman et al. 1988; Martin & MacLarnon 1990. 41. Formerly known as *Callithrix argentata*, Carroll 1993; Chapman et al. 1990; Hampton & Hampton 1977. 42. Chapman et al. 1990; Dietz et al. 1994; Snowdon & Soini 1988; Soini 1987. 43. Chapman et al. 1990; Goldizen 1987; Harvey & Clutton-Brock 1985; Rasmussen 1989. 44. Christen 1974; Ernest 2003; Harvey & Clutton-Brock 1985; Martin & MacLarnon 1990. 45. Chapman et al. 1990; Cleveland & Snowdon 1984; Ross 2003; Snowdon & Soini 1988. 46. Campbell 2007; Fragaszy et al. 2004; Harvey & Clutton-Brock 1985; Ross 2003. 47. Gust et al. 1990; Nishimura 2003; Schlee & Labejof 1994. 48. DeSilva & Lesnik 2008; Harvey & Clutton-Brock 1985; Harvey & Pereira 2003. 53. Cheney et al. Ross 2003. 49. Cords 1988; Ross 2003. 50. Cords 1988; Ernest 2003. 51. Cords 1988; Glenn & Cords 2002; Ross 2003. 52. Bramblett et al. 1975; Fedigan & Fedigan 1988; Kappeler & Pereira 2003. 53. Cheney et al. 1988; Sly et al. 1983. 54. Harvey & Cluttonbrock 1985; Lee 1999; Nakagawa et al. 2003; Ross 2003. 55. Chapman et al. 1990; Lee et al. 1991; Trollope & Blurton Jones 1975. 56. Bercovitch & Harvey 2004; Chapman et al. 1990; Lee et al. 1991; Nieuwenhuijsen et al. 1985. 57. Lee 1999; Melnick & Pearl 1987; van Noordwijk & van Schaik 1999. 58. Bercovitch & Harvey 2004; Ernest 2003; Takahata et al. 1998. 59. Bercovitch & Harvey 2004; Chapman et al. 2004; Chapman et al. 1990; Lee 1999; Okamoto et al. 2000. 60. Kaplan et al. 2000; Lee 1999; Melnick 1981. 61. Bercovitch & Harvey 2004; Silk 1988. 64. Lee 1999; Lindenfors 2002. 65. Ernest 2003; Mehlman 1989. 66. Boer 1987; Ernest 2003; Kaplan et al. 2000; Lee 1990; Hayssen et al. 1993; Lindenfors 2002; Ross 2003. 63. Bercovitch & Harvey 2004; Silk 1988. 64. Lee 1999; Lindenfors 2002. 65. Ernest 2003; Mehlman 1989. 66. Boer 1987; Ernest 2003; Kaplan et al. 2000; Lee 1999; Ross 2003. 67. Harvey & Clutton-Brock 1985; Gautier-Hion & Gautier 1976; Rowell & Dixon 1975. 68. Feistner et al. 1992; Harvey & Clutton-Brock 1985; Higham et al. 2009; Kappeler & Pereira 2003; Lindenfors 2002; Melnick & Pearl 1987. 69. Altmann et al. 1985; Bentley-Condit & Smith 1997; Lee et al. 1991; Melnick & Pearl 1987; Ross 2003. 73. Dunbar 1984; Godfrey et al. 2001; Lee 1999; Lindenfors 2002; Lippold 1977. 77. Ernest 2003; Hrdy 1977; Kirkpatrick 2007; Lycett 1994; Lycett et al. 1998; Sigg et al. 1982. 72. Harvey & Clutton-Brock 1985; Melnick & Pearl 1987; Ross 2003. 73. Dunbar 1984; Godfrey et al. 2001; Lee 1999; Lindenfors 2002; Lippold 1977. 77. Ernest 2003; Hrdy 1977; Kirkpatrick 2007; Lee 2005; Korstjens 2001. 75. Godfrey et al. 2001; Gorzitze 1996; Harris & Monfort 2006. 76. Harvey & Cluttonbrock 1985; Lee 1999; Lindenfors 2002; Lippold 1977. 77. Ernest 2003; Badham 1967; Godfrey et al. 2001; Lee 1999; Lippold 1981; Rajpurohit & Mohnot 1991; Sommer et al. 1992. 78. Formerly known as *Trachypithecus vetulus*, Godfrey et al. 2001; Lee 1999; Rudran 1973. 82. Godfrey et al. 2001; Reichard 2003; Ross Struhsaker 1987. 80. Ernest 2003; Godfrey et al. 2001; Harvey & Cluttonbrock 1985; O'Brien et al. 2003; Robbins 2007. 81. Jablonski et al. 2000; Lee 1999; Rudran 1973. 82. Godfrey et al. 2001; Reichard 2003; Ross 2003. 83. Lee et al. 1991; Stumpf 2007; Tutin 1994. 84. Ernest 2003; Wich et al. 2004. 85. Markham 1990; Stumpf 2007; Wich et al. 2004. 86. Knott et al. 2009; Ross 2003; Wich et al. 2004. 87. Hill & Hurtado 1996; 2003. 83. Lee et al. 1991; Lee 1999. Maximum life span: Weigl 2005. Neonate body mass: Smith & Leigh 1998. Female body mass and endocranial volume: Isler et al. 2008; van Woerden et al. 2010 and personal communication. Various variables: Ross & Jones 1999.

Patterns in Mammalian Life Histories

Empirical research has shown that there are two major patterns in mammalian life histories (Stearns 1983; Bielby et al. 2007). The first is the gradient from a fast to a slow pace of life history, which is found among many vertebrates (e.g., see Read & Harvey 1989). The second pattern is the somewhat more qualitative variation between altricial and precocial state at birth, explained below.

Life-History Pace

Some birds or mammals tend to have slower-paced life histories than others. They take longer to pass through all the phases of life, from gestation through immaturity to adulthood, and at the same time have lower rates of reproduction as adults. Figure 10.2.a illustrates this for the primate species whose data are compiled in table 10.1. This pattern is clearly linked to body size, as is illustrated for primates in figure 10.2.b. At first blush, this size effect may largely seem to be an inevitable outcome of fundamental metabolic laws. Obviously, larger animals must grow larger bodies, but growth rates show negative allometry, which means that the increase in body mass is less than directly proportional to the actual mass. Overall, among mammals, growth rates scale roughly as $G - aW^{0.75}$, where a is a constant, G is growth in grams per day, and W is weight in grams (Case 1978). This is not unexpected if growth largely reflects metabolic turnover, because the latter also shows negative allometry with body mass with the same exponent of about 0.75 (Kleiber's law: Lindstedt & Calder 1981). Thus, a larger animal must take longer to grow to adult size than a smaller one, although there is much variation among lineages in the size-corrected growth rates (e.g., see Mumby & Vinicius 2008). Moreover, the slower pace is maintained among adults, where life span is generally longer in large animals, although the allometric constant is weaker (usually scaling to $W^{0.25}$; see Lindstedt & Calder 1981). Finally, an even stronger negative allometry prevails in reproduction, where fertility tends to scale to $W^{-0.25}$ (Lindstedt & Calder 1981). Thus, a larger animal produces fewer offspring per unit of time than a smaller animal. Primates adhere to the general mammalian trend in that all their life-history variables, whether they estimate durations (stages) or probabilities (rates), show negative allometries (fig. 10.3).

Are these relationships between body size and life-history variables an inevitable by-product of basic physiological processes, or do they also, or even exclusively, reflect allocation decisions that are merely correlated with body size? The latter possibility is plausible, given that the fast-slow continuum is maintained when the effect of body size is

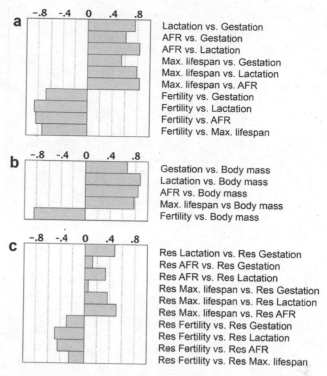

Fig. 10.2. Life-history correlations in nonhuman primates ($N = 86$ species; data from table 10.1): (a) pairwise correlations between life-history variables; (b) correlations between life-history variables and body mass; (c) pairwise correlations between residuals of life-history variables. In (a) and (b) all correlations are highly significant ($p < 0.0001$), in (c) most correlations are significant except residual maximum life span versus residual gestation length ($p = 0.635$) and residual AFR (age at first reproduction of females) versus residual gestation ($p = 0.250$). Fertility is defined as annual number of young produced per female per year.

removed statistically by calculating the residuals of a regression of the life-history trait of interest—say, gestation length—on body size. The various residuals still show the characteristic pattern of positive correlations among stage parameters and negative ones between the residuals and fertility (and also fecundity, or potential fertility). Figure 10.2.c illustrates this for primates, but it is also found for other mammals (Read & Harvey 1989; Fisher et al. 2001) and birds (Isler & van Schaik, unpublished data); the pattern does not necessarily hold in fish or reptiles (Jeschke & Kokko 2009), perhaps because many of those animals experience indeterminate growth and extremely high immature mortality. Thus, some species live fast lives and others slow lives for their respective body size. This pattern suggests that to explain the fast-slow continuum, we must look beyond the effect of body size. Indeed, the observed body-size allometries may be caused partially or entirely by the correlation between body size and the factor affecting life-history variables (Harvey & Purvis 1999), unavoidable extrinsic mortality. We will examine this issue below, in the section on evolution of life histories.

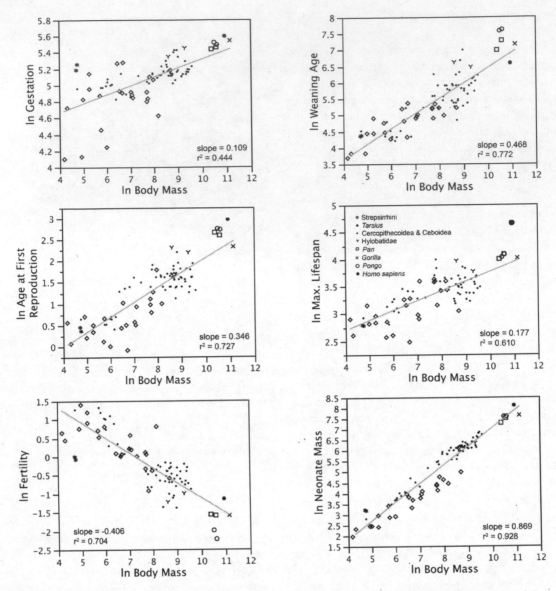

Fig. 10.3. Bivariate plots of life-history variables versus body mass in primates (*N* = 87 species). *Homo sapiens* was excluded for calculation of the least-squares regression lines. Note that the slopes for the stage parameters of life history are systematically less than 1, indicating negative allometry.

Developmental State at Birth

The second main feature of the life histories of birds and eutherian mammals is the spectrum between an altricial and a precocial state of development at birth (Martin 1990; Derrickson 1992). Altricial young are less developed at birth, being relatively small and often naked, usually with eyes still closed, and generally are unable to hold onto their mothers or move much beyond reaching the nipple. Young of precocial species, in contrast, are relatively large, furred or feathered, with eyes open; they can move around, and to some extent they can also thermoregulate. Not surprisingly, altricial young are usually born after significantly

shorter periods of gestation, and they also tend to have smaller relative brain size at birth than precocial young (Martin 1990). With few exceptions, altricial mammals are born in larger litters and have faster postnatal growth rates than precocial ones, which are generally born as singletons (Derrickson 1992). There are clearly more fast-paced taxa that have altricial young and more slow-paced ones that have precocial young among eutherian mammals, but in pattern-recognition studies using principal component analysis, developmental state at birth emerged as an independent source of variation in life history, distinct from the slow-fast continuum (Stearns 1983; Bielby et al. 2007).

Primates in general are at the precocial end of the mammalian spectrum in that young are generally born fully furred and able to cling to their mothers, but there is variation among strepsirrhines. As expected from the pattern among mammals in general, strepsirrhines that leave their young in nests or tree holes give birth to relatively altricial young. A phylogenetic analysis suggested that the ancestral state for primates is singleton births, with infants initially left in a shelter but carried around and parked soon after birth and therefore unlikely to be very altricial (Kappeler 1998). Among haplorrhines, infants are rather precocial and can cling to their mothers or other caretakers very soon after birth. Virtually all haplorrhines thus carry their infants and don't park them or leave them in nests, with two curious exceptions: the pig-tailed snub-nosed monkey (*Simias concolor*) and the Mentawai surili (*Presbytis potenziani*), both of which live on the Mentawai islands off Sumatra (see Kappeler 1998). Humans do not follow the haplorrhine trend, as our newborns have secondarily become somewhat more altricial, almost certainly due to the problems associated with giving birth through a narrow pelvic canal to infants with large heads, which in turn is a result of the evolution of bipedal locomotion (Schultz 1969).

Figure 10.4 illustrates these trends for primates. In a principal-components analysis of the life-history variables compiled in table 10.1, the first principal component (PC1) reflects the fast-slow continuum, which explains more than 80% of the variation. PC1 is also correlated with body size, but the fast-slow continuum is retained after body mass is statistically controlled (cf. fig. 10.2c). The second principal component reflects the precocial-altricial continuum, which explains only about 8% of the variation because the range of that variation is compressed in primates. The lowest scores are for various lemurs that give birth to multiple small young after relatively short periods of gestation, and some platyrrhines that receive extensive allomaternal help. The high-scoring taxa are *Tarsius* and some lorisids, which each give birth to one large young after a very long gestation, and a taxonomically diverse set of folivores. In all these cases, the unusually long gestations may be due to low metabolic rates (Martin & MacLarnon 1990; cf. Isler et al. 2008).

The Evolution of Mammalian Life Histories

Two traits show a tradeoff when there is a negative functional interaction between them (Zera & Harshman 2001). The basic tenet of life-history theory is that a species' life history reflects the optimization of a number of trade-offs between allocation targets such as growth, reproduction, maintenance, and repair—and thus survival. For a female organism——say, a mammal—all allocation targets of energy, as well as all other physiological functions, compete for the same limited pool of energy, which means that increased allocation to any one target reduces the organism's ability to allocate energy to one or more of the others (fig. 10.5). This principle also applies to trade-offs between investment in current and future needs.

There is ample empirical evidence for each of these trade-offs. Decisive evidence requires experimentation because variation in the quality of individuals causes heterogeneity in observed samples within a population, which may produce positive correlations rather than the negative correlations expected under trade-offs (see van Noordwijk &

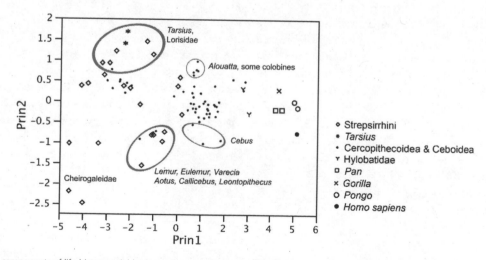

Fig. 10.4. Principal components of life-history variables in nonhuman primates (*N* = 86 species). Prin1 explains 81.5% of variation, Prin2 only 8.4% (eigenvalues 4.891 and 0.505, respectively). Prin1 loads positively on all stage parameters (gestation 0.357, lactation 0.429, age at first reproduction 0.420, and maximum life span 0.394) and negatively on fertility (−0.417). Prin2 loads positively on gestation (0.822) and negatively on maximum life span (−0.473).

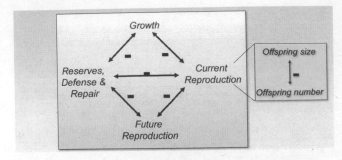

Fig. 10.5. Illustration of the trade-offs between the various functions of an organism. All these trade-offs have been shown experimentally (Stearns 1992; Mangel & Stamps 2001).

de Jong 1986). Experimental manipulation of the reproductive effort of animals shows that current reproductive effort compromises survival or future reproductive effort (Roff 1992; Stearns 1992; for a primate example, see Blomquist 2009). The trade-off between current and future reproductive effort is also the cornerstone of the theory of parent-offspring conflict (Trivers 1974). Likewise, it can be shown that high juvenile growth rates compromise future survival and thus life span (Mangel 2003). These tradeoffs arise because growth and reproduction compete with the capacity to repair oxidative damage to DNA, proteins and lipids, and thus potential life span (Mangel 2003). It is less widely appreciated that these trade-offs may also involve the immune system (Sheldon & Verhulst 1996) because increased reproductive effort tends to affect immunocompetence (antibody responsiveness). The same may hold for endocrinological processes. Whenever the body produces a sustained stress response, this takes energy away from other vital functions, so that chronic stress suppresses reproductive performance and immune function (Sapolsky 1994). Likewise, an increase in testosterone leads to more aggressive responses to social challenges and greater success in competition, but it does so at the expense of increased metabolic turnover and diminished immune function, and thus reduces survival (Ketterson & Nolan 1999). Variation in individual quality ensures that some animals can afford such increases better than others. Later in this chapter we will examine a similar trade-off with brain size.

Classic life-history theory assumes that the shifting energy allocations to each of these targets are optimized during the course of life in the face of these trade-offs. Assuming a stable population size, selection is expected to maximize lifetime reproductive success (R_0):

$$R_0 = \sum_x l_x \times m_x$$

where l_x is survival up to age class x, and m_x is birth rate (by convention, counting only daughters) during age class

x. If the data are available to construct this measure, it is also possible to calculate the well-known intrinsic rate of natural increase, r (e.g., see Begon et al. 1986). Another measure that reflects maximum possible lifetime reproductive success under optimum conditions is r_{max}, which is calculated as

$$1 = e^{-r} + be^{-ra} + be^{-r(w+1)}$$

where a = age at first reproduction, w = age at last reproduction or maximum life span, and b = birth rate (of female offspring) per year. The advantage of using r_{max} is that its calculation does not require detailed life tables, although it does require numerical methods employing iterative estimation. Estimates of r_{max} for primates are listed in table 10.1.

Figure 10.6 shows the l_x and m_x curves for the Ketambe population of long-tailed macaques (*Macaca fascicularis*: van Noordwijk & van Schaik 1999). In females, mortality (the rate of decline in the survival curve) is high during the first year and then stabilizes, rising again after about age 20. In contrast, male mortality remains high after juvenility. In this species, females typically start to reproduce at around age 5, and reach a peak birth rate between the ages of 10 and 20, after which senescence sets in and rates decline. This age profile of survival and reproduction is typical for haplorrhine primates, although specific details vary from species to species.

Any change in an energy-using process that changes the values of l_x and m_x is evaluated by natural selection with respect to its overall impact through various trade-off effects on lifetime reproductive success. Selection acts to maximize R_0 by shifting patterns of allocations over the course of a lifetime. When populations grow or shrink in size, this varying population growth rate must be incorporated into the estimates of lifetime reproductive success (see, e.g., Stearns 1992), but natural populations cannot sustain consistently positive or negative growth rates for long.

To understand the wide variation in mammalian life histories, we need to discover the optimality criteria used by natural selection. In other words, we seek to identify the major trade-offs in nature and the causes underlying their variation. Unfortunately, there is as yet no universally accepted theory that can reproduce all known life-history patterns (review in Harvey & Purvis 1999). Perhaps the best-known effort is that of Charnov (1993, 2001, 2004), who used basic trade-offs and some empirically observed allometries to derive allometric production equations that predict optimal body size—the size at which the organism should switch from growth to reproduction—and, as a by-product, the optimum age at first reproduction. From this set of equations Charnov subsequently derived a range

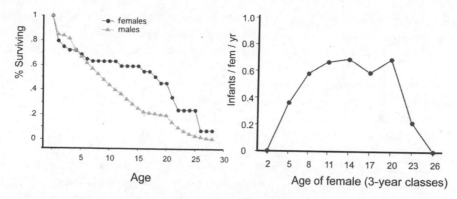

Fig. 10.6. Survival (l_x) and birth rate (m_x) curves for a wild primate, *Macaca fascicularis*, at Ketambe, Sumatra (largely after van Noordwijk & van Schaik 1999). The m_x curve here refers to births of all infants, not just females.

of invariants or dimensionless numbers, such as the ratio of average adult life span to average age at first reproduction. This approach reproduced the basic fast-slow continuum, and it also correctly predicted some invariants. However, it has failed to account for variation in rates of growth and reproduction (see below). Hawkes (2006) and Leigh & Blomquist (2007) provide detailed discussions of Charnov's theory and variations on it.

Despite the lack of a unified theory, all approaches agree that life-history variation across species is largely driven by one external variable: the level of unavoidable, extrinsic mortality. Unavoidable mortality is mortality due to predation, disease, or accidents that cannot be reduced any further by lifestyle modification. The effect of unavoidable mortality on life history has been shown by various selection experiments that manipulated longevity or mortality (Kirkwood & Austad 2000). For instance, Stearns et al. (2000) created two lines of *Drosophila melanogaster*: a low-mortality control line and a high-mortality experimental line in which the majority of adult flies were killed every week. As predicted by theory, after several generations of selection, flies in the high-mortality line had evolved faster development (earlier eclosion) and smaller adult size, higher reproductive rates during early adulthood, shorter maximum life span, and higher intrinsic mortality rates than the flies in the control group.

This theory makes intuitive sense when we think of life history as a coadapted set of allocation decisions in the face of trade-offs. Whenever an animal's life expectancy remains short regardless of what it does, the animal should forego investing limited resources into structures and processes meant to make a long life possible, and instead invest these resources into production: growth and reproduction. On the other hand, when life expectancy is long, perhaps in part due to lifestyle choices such as group living, investment in defense and repair is needed to fulfill the environmental

potential of long life, which is achieved at the expense of growth and reproduction. Thus, we expect that intrinsic mortality, and therefore the various life-history parameters, will evolve to reflect the level of unavoidable external mortality to which the species is naturally exposed. This is why a population's life table (Cole 1954), and thus its age structure (i.e., its demographic structure), reflects its life history.

The trade-off perspective neatly accounts for senescence, the "post-maturation decline in survivorship and fecundity that accompanies advancing age" (Rose & Charlesworth 1980). The main explanation for senescence is antagonistic pleiotropy: genes that have a positive effect on fitness early in life may have a negative effect later. Selection against these genes in old age is not strong enough to cancel out their positive early effects, because the animal in question is much less likely to still be alive (this is inevitably true for even the most long-lived species). We can therefore view senescence as the product of the trade-off between investment in the structures and processes that bring about growth, reproduction, and longevity along with the inevitable reduction in the strength of selection as cohort size declines with age. A second explanation for senescence is that deleterious mutations accumulate over life and thus inevitably cause aging, but empirical evidence for this is more equivocal (Kirkwood & Austad 2000; Hughes & Reynolds 2004).

Finally, the trade-off perspective makes sense of many of the intraspecific sex differences we see in life history. In many species, males can achieve mating success only if they have been successful in direct contests with others, and in species with prominent male contest competition, the males tend to be larger than females (chapter 17, this volume). They can grow larger because they grow faster or for a longer time than females. The first phenomenon, faster growth, characterizes polygynous species living in single-male groups, whereas males of species with multimale groups tend to show bimaturism: they grow for a longer period of time (Leigh 1995). Once they reach adulthood, males face tradeoffs different from those faced by females. In single-male groups with high male reproductive skew, individuals tend to have relatively short reproductive careers (van Noordwijk et al. 2004), which in turn may explain why they can afford to grow faster as immatures. As a result, interspecific analyses of life-history patterns tend to focus on females, which are often characterized as the "ecological" sex (see table 10.1).

Other historically influential ideas regarding the determinants of life history enjoy much less empirical support than the extrinsic mortality theory. The theory of *r* and *K*-selection (MacArthur & Wilson 1967; *r* stands for the population's intrinsic rate of growth [as above] and *K* for its carrying capacity) argues that the life-history continuum is linked to a shift from an emphasis on rapid reproduction in unstable environments with frequent bouts of density-independent mortality toward a strategy of "staying power" and hence longevity in stable, high-density populations. This idea is intuitively appealing and still remarkably popular, perhaps because the description of the *r/K* continuum, which is tantamount to the empirically determined fast-slow continuum, is conflated with explanation. However, it is now generally rejected, due to problems with both its logic and its predictions (Stearns 1992; Hawkes 2006). For example, *K*-strategists are not usually restricted to stable environments. Moreover, artificial selection experiments do not yield the predicted shifts in life history when density is systematically increased or reduced.

A second historically popular idea, the "rate of living" hypothesis, proposes that increased metabolic turnover limits life span because of high oxidative damage. Thus, individuals can live only until they have burned their maximum total number of calories (reviews in Lindstedt & Calder 1981; Hofman 1993; Deaner et al. 2003). This hypothesis ignores the possibility that individuals might invest in biochemical repair mechanisms (Kirkwood 2002), and thus it fails to explain why species with similar metabolic rates and body size show conspicuous differences in longevity (Harvey et al. 1991).

Beyond Extrinsic Mortality

Although this mortality-based perspective reproduces the fast-slow continuum, it still leaves some variation unexplained. In some species, growth trajectories deviate markedly from those expected by Charnov's model (Leigh 2004), although the overall pace of their life history fits the fast-slow scheme. Many other species are faster or slower than would be expected on the basis of the mortality they face—that is, they have very different productivity (growth and re-

production) than expected for their body size by Charnov's model (Mumby & Vinicius 2008), thus suggesting that additional factors influence the evolution of life-history parameters. Perhaps the main factor is brain size, which plays a major role in primates and will be discussed extensively below, but other ideas have also been suggested.

One hypothesis invokes bet hedging—the idea that when immature mortality fluctuates greatly and reproduction is costly, selection favors longer adult life, bought at the expense of lower reproductive rates (Stearns 1992). Bet hedging is thus a variation on the general theme of trade-offs and mortality-based life histories. Some evidence supports this hypothesis among primates. Verreaux's sifakas (*Propithecus verreauxi*), living in highly seasonal southwestern Madagascar with high interannual variability in food abundance, show slower development and longer life spans than expected for their body size (Richard et al. 2002). Similarly, orangutans in Southeast Asia, where mast fruiting produces pronounced interannual variation in food abundance (van Schaik & Pfannes 2005), have slower development and live longer than African great apes (fig. 10.7; Wich et al. 2004). In both cases, however, hypometabolism is equally plausible as an explanation. It reduces the risk of starvation, but at the cost of reduced growth and development (Richard & Nicoll 1987; Pontzer et al. 2010).

A second hypothesis, the juvenile risk aversion hypothesis (Janson & van Schaik 1993) argues that in species where group living produces serious feeding competition but also reduces mortality risk, individuals benefit by slowing down their growth so as to simultaneously reduce the risk of starvation and predation. While some aspects of ape life history

Fig. 10.7. Relatively long life spans characterize many primate species, particularly the great apes. This adult female orangutan at Tuanan, Borneo, is probably more than 40 years old and is still reproducing. Photo courtesy of Lynda Dunkel.

appear to be consistent with this hypothesis (chapter 6, this volume), it fails to explain why species that grow slowly in youth are also generally slow in other aspects of their life history as adults. Moreover, empirical tests do not always support this hypothesis (Ross & Jones 1999; but see below).

A third hypothesis invokes lifestyle effects, specifically ease of resource acquisition (Sibly & Brown 2007). It predicts that taxa relying on abundant and reliable foods should be more productive than expected for their body size, whereas those with reduced death rates should be less so. To date, these predictions have not been tested in primates.

These ecological hypotheses complement the demographic perspective of classic life-history theory by specifying how particular ecological conditions affect the trade-offs that form the basis of life history. We expect much theoretical and empirical work in coming years that should integrate these frameworks.

Developmental State at Birth

Newborns of altricial mammals must spend most of their time in nests, tree holes, or dens because they are unable to move or hold onto fur. Obviously they would be quite vulnerable to predators were it not for their being hidden in protective shelters. Precocial young can move independently, as in ungulates, or cling to a parent, as in most primates, and this makes them much less vulnerable to predators. However, a trade-off exists between developmental state and reproductive capacity. For any given level of investment, a female can give birth to a greater number of altricial than precocial young. In altricial species, a mother can forage unencumbered by dependent offspring, although in some species the nests or dens are guarded by the male (e.g., in *Varecia*) and in others the parked young are moved frequently (e.g., in *Tarsius*). In precocial primates, however, mothers must carry their young at all times, and the slower postnatal growth rates of primates may partly reflect this handicap. Thus, if there is a safe shelter for the young, females might often be better off producing altricial young. Lacking safe shelters, females must complete more of the development in utero and give birth to more precocial young, which inevitably limits litter size. This constraint may explain the correlation between permanent gregariousness and precocial young.

Another, possibly secondary, advantage of altriciality is that if a larger proportion of growth occurs after birth, mothers can offload some of the parenting effort onto others (i.e. allomothers). Indeed, the presence of cooperative breeding is clearly associated with the presence of altricial young in birds and mammals. The few cooperative breeders that give birth to precocial young, as in the callitrichine primates, tend to have larger litters.

Life History and Ecology

If we ignore the ecology of the developmental state at birth, the life histories of birds and mammals are largely shaped by the degree of unavoidable or extrinsic mortality. Mortality can explain why the fast-slow gradient is retained when body size effects are controlled: mortality rates tend to be negatively correlated with body size, but also have additional, independent sources. Among mammals the most obvious independent source of unavoidable mortality will be predation, which is affected by choice of habitat or intrinsic features that reduce risk, such as ability to fly, possession of armor, or unpalatability (chapter 8, this volume). Cognitive adaptations, discussed below, represent a second strategy for reducing predation rates (Zuberbühler & Byrne 2006).

Applying this argument to primates, we expect that arboreality should affect life-history pace, because arboreal animals face fewer predators than terrestrial animals of the same size and also have more refuges and escape routes. This idea finds much support among mammals and birds generally (fig. 10.8, based on our own data set; see also Prothero & Jürgens 1987, Austad & Fischer 1992). Thus, flying birds and mammals (bats) live longer lives for a given body mass than their nonflying counterparts, and primates are also longer-lived than other mammals. Because not all primates are arboreal, one can also examine variation within the order. Indeed, within the anthropoids, the primate radiation with the most variation in substrate use, the

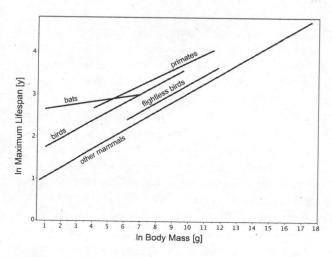

Fig. 10.8.   Off-the-ground living and maximum life span in homeothermic vertebrates (*N* = 1,322 species of eutherian mammals and birds, least-squares regression lines; Isler & van Schaik, unpublished data). Note that for each lineage, the lines cover the known size range.

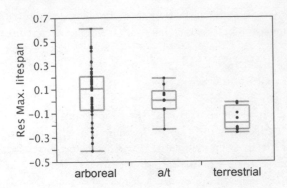

Fig. 10.9. Arboreality and maximum life span in simian primates (N = 58 species, difference between groups significant: Kruskal-Wallis test, DF = 2; chi square = 8.782; p = 0.012). See also van Schaik & Deaner (2003) for another test.

relationship between slower life-history pace and arboreality also holds (fig. 10.9).

Several studies have searched for other ecological correlates of life history in various taxa, including primates. Of the numerous correlations, only a few hold up when examined more closely or in different lineages, and they turn out to be linked to rates of unavoidable mortality. For instance, primates living in open areas tend to have faster life histories than those in forests (Ross 1988)—not surprisingly, given their higher exposure to predators. This helps to explain the contrast between patas monkeys and gibbons mentioned in the introduction to this chapter. With some exceptions (e.g., sifakas; see above), folivores experience a faster pace of life history than other primates (Wich et al. 2007; chapter 4, this volume), but there is no obvious explanation for this.

Plasticity

Many of an organism's features can vary with ecological conditions (i.e., when genetic background is held constant), and life-history traits often show remarkable plasticity, even exceeding that of behavior or physiology (Mousseau & Roff 1987). This is not surprising, since rates of reproduction and survival show enormous variation in time and space, and growth also generally responds to local conditions. More surprising, perhaps, is that animals in favorable conditions not only grow and mature faster, but also reproduce more and often live longer. Theory even predicts the form this plasticity should take. A classic model by Stearns and Koella (1986) predicted that when food is scarce, animals should try to begin reproduction neither at the same age nor at the same size as when food is abundant, but should compromise with respect to both age and size. However, provisioned Japanese macaques (*Macaca fuscata*) and yellow baboons (*Papio cynocephalus*) grew faster

than unprovisioned animals, yet started reproduction at the same size (Mori 1979; Sugiyama & Ohsawa 1982; Altmann et al. 1993). Also largely unexplored is the extent to which species differ in their degree of plasticity. For instance, do species vary in the extent to which immature growth rate increases when food is available ad libitum? And do low-ranking, poorly fed, or small mothers wean their infants at the same size as others in all species (Johnson 2003; Garcia et al. 2009)? These and other questions concerning plasticity in primate life history remain to be answered.

Life History and Behavior

Although Stearns (1992, p. 210) noted that the constraints imposed by life history on behavioral evolution "are not surprising enough to catalyse interest," it is nonetheless instructive to examine them briefly, especially with respect to primates (for reviews see Kappeler et al. 2003; van Schaik et al. 2006). First, consider the altricial-precocial contrast. As infants became more precocial during primate evolution, the young could be carried around permanently. Consequently, mothers were freed to become nomadic foragers and track resources such as ripe fruit over larger areas, albeit at the expense of slower infant growth. Infant carrying also made it possible for females to live in permanent groups, as is also suggested by phylogenetic reconstructions (cf. Kappeler 1998).

Next, consider the even more profound effects of a slower life-history pace. This could systematically impact behavior by increasing developmental periods and life expectancy, reducing demographic turnover, and, indirectly, by changing the relative duration of developmental phases. Thus, lengthening the duration of life stages or of total lifespan could (1) induce conservatism and lead to stable social groups and social relationships, (2) increase the probability of encountering drastic environmental change, and (3) lead to a longer period of immaturity. The changes in the relative proportion of life stages (4) alter the relative duration of gestation and lactation, whereas the lower rates of reproduction (5) produce more male-biased operational sex ratios.

Slow life history should affect behavioral style. Long life expectancy will almost inevitably lead to risk aversion and conservatism, with a number of possible consequences that have hardly been explored, such as willingness to queue for breeding slots rather than engage in risky escalated contests. Greater risk aversion may favor a greater ability to inhibit prepotent responses, an ability often seen as an important precondition for cognitive abilities (Geary 2005). Because slower-paced organisms are keen to reduce predation risk to a lower value than are faster-paced organisms, they are more likely to live in groups, or tend to live in larger groups. This tendency may explain the otherwise counterintuitive positive correlation between body size and group size among primates (Janson 2003). Living in cohesive groups causes food competition, thus reducing growth and reproduction, but it also enables individuals to develop long-term cooperative relationships. Longer life span also increases the likelihood that the individual will face extreme conditions to which a plastic response would be adaptive, as it ensures that the animal can survive these extreme conditions and fulfill its potential lifespan. A longer period of immaturity due to slow life history is accompanied by more time for the offspring to learn vital skills before maturity, and because all primate taxa with slow life histories experience prolonged periods of parent-offspring association, this learning can be social and therefore more efficient (Whiten & van Schaik 2007; chapter 31, this volume).

As primate life history slows down, gestation becomes an increasingly small fraction of the total prereproductive period (van Schaik et al. 2006). As a result, the lactation-gestation ratio increases (van Schaik & Deaner 2003). This ratio is a good predictor of risk of infanticide, both across orders of mammals and among primates (van Schaik 2000), because a value greater than 1 indicates that an infant's death advances the time when the mother will be fertile again, thus creating a benefit for the infanticidal male. The risk of infanticide is likely to have selected for year-round male-female association in primates (van Schaik and Kappeler 1997), as well as for female sexual counterstrategies (van Schaik et al. 1999, 2004), both of which may have had implications for cognitive evolution.

Finally, as life history slows down, female reproductive rates decline. Because the number of males is not necessarily affected by life-history pace, sexually attractive females become relatively rarer, leading to more male-biased operational sex ratios. Harassment and sexual coercion of females (Smuts & Smuts 1993) and female counterstrategies should thus be more pronounced in species with very slow life history, although we are not aware of any formal tests of this prediction.

It is important that virtually all these consequences of a slower life-history pace have a cognitive dimension; they create conditions in which greater cognitive abilities can be favored by natural selection—namely the condition of living in permanent mixed-sex groups. In these groups, long-term cooperative exchange relationships can form, behavior becomes more conservative, the young can begin to acquire skills by social learning, and social strategies are useful for coping with harassment and infanticide. All these aspects of slow life history may thus indirectly favor cognitive evo-

lution, and hence increased brain size. It is, therefore, not entirely coincidental that many long-lived organisms are large-brained. We now turn to the possible link between brain size and life history.

Life History and Brains

Although for more than a century many scholars have suggested that slow-paced life history and large brain size are correlated, the modern era began with Sacher (1959), who used a broad array of mammals to show that brain size was more tightly correlated with maximum life span than was body size. However, because body mass is quite variable, brain mass may actually be a better estimate of body size than is body mass, thus casting doubt on Sacher's conclusion. The artifact interpretation was lent credibility by the observation that the weight of some other organs, such as the liver, was also more tightly correlated with life span in this sample than was body mass (Economos 1980). Hence, the newly ascendant demographic approaches to life-history evolution (e.g., Harvey et al. 1989) could downplay the possibility of correlated evolution of life history and brain size. This resulted in a paradigm shift away from considering life-history evolution a product of allometric effects or brain-driven mechanisms, and toward emphasizing the role of unavoidable mortality. Recent studies, however, have vindicated the relationship between brain size and life span, as it has now been documented among mammals at all taxonomic levels (Barton 1999; Ross & Jones 1999; van Schaik & Deaner 2003; Ross 2003, 2004; Isler & van Schaik 2009). Many of these studies have controlled for the "Economos problem" (sometimes by examining the relationship between brain size and the size of other organs) as well as for phylogenetic nonindependence (Nunn & Barton 2001). Some of them have used carefully assembled high-quality data sets (review: Barrickman et al. 2008).

Because there is no longer any doubt that unavoidable mortality drives life-history evolution, this leaves the role of brains unclear. What processes can account for the correlated evolution of brain size and life-history traits? Theoretically, there are three non–mutually exclusive possibilities: (1) life history affects brain size, directly or indirectly; (2) brain size affects life history, directly or indirectly; and (3) another factor, such as arboreality, affects both brain size and life history. Because the third of these possibilities is unlikely to produce the same systematic effect in a variety of lineages, we consider only the first two possibilities here. In the previous section we reviewed many reasons why a slow-paced life history might favor the evolution of larger brains. However, some organisms have evolved slow life history without concomitant increases in brain size. Bats, in particular, are very long-lived but also small-brained for their body size (van Schaik & Deaner 2003). Thus, organisms with a slow life-history pace can also evolve large brains, but they do not necessarily and universally do so (contra van Schaik & Deaner 2003).

The second possibility is that a change in brain size directly or indirectly causes a change in life history. Sacher (1975) proposed two hypotheses. First, brains actively stabilize the "life processes of the organism," such that larger brains allow bodies to stave off senescence for a longer time. Second, increases in brain size have immediate metabolic and developmental costs, which can only be compensated for by an increase in reproductive span through improved survival—and hence, over evolutionary time, lifespan. While there is no support for the first possibility (Sacher 1975; review: Deaner et al. 2003), the second, which invokes allocation trade-offs, has recently been revived under the "expensive brain" framework.

The point of departure for the "expensive brain" framework is that brains are unusually expensive per unit of weight in terms of energy use. As a result, in many mammals the brain usurps 5% to 10% of total energy use during rest (Mink et al. 1981). In our own, highly encephalized species, this figure even approaches 25% (Holliday 1971). There are two reasons to think that this burden is even higher for juveniles. First, juveniles have adult-sized brains but relatively small bodies. This implies that they allocate a larger proportion of energy to their brains than adults do, despite their poorer experience and often lower social status. Second, differentiating brains require approximately 50% more energy per unit of weight, even after total size has been achieved (Kennedy & Sokoloff 1957).

As a result, any evolutionary increase in brain size has greater metabolic consequences than a similar increase in most other anatomical features, and we expect that natural selection had to choose from among a small number of options to make the extra energy available. Two non–mutually exclusive possibilities exist (fig. 10.10). Organisms could either make more energy available overall, allocate the available energy differently so that a larger proportion goes to the brain, or possibly do a bit of both. Indeed, empirical evidence indicates that larger-brained organisms tend to have higher overall metabolic turnover, as indexed by the basal metabolic rate and predicted by the *direct metabolic constraints* hypothesis (Armstrong 1983; Isler & van Schaik 2006b; Isler et al. 2008). In addition, cooperatively breeding mammals and birds, in which reproducing females receive energy subsidies, tend to have larger brains than their independently breeding sister taxa (Isler et al. 2012).

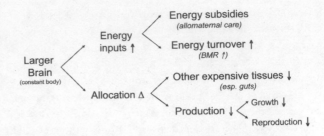

Fig. 10.10.　The expensive brain framework: how organisms could pay for brain size increases, assuming no change in body size. Increased energy turnover is predicted by the direct metabolic costs hypothesis, decreased size of other expensive tissues by the expensive tissue hypothesis (see text). Note that the logic also works in reverse: we can trace the consequences of energy gain from brain size reduction.

The second of these hypotheses, which suggests changes in energy allocation, is of particular interest because the changes reflect trade-offs, critical elements of the broader theory of life-history evolution. Trade-offs with other energy-hungry organs, such as the gastrointestinal tract, have received much attention from the *expensive tissue* hypothesis (Aiello & Wheeler 1995; Aiello et al. 2001; but see Hladik et al. 1999). However, this trade-off has not been found in birds, bats, or mammals in general (Jones & MacLarnon 2004; Isler & van Schaik 2006a; Navarrete et al. 2011).

Trade-offs with production (i.e., growth and reproduction) would have a direct effect on life history. If the higher energy needs of larger brains are paid for by reducing the allocation to production, we should expect various consequences for precocial mammals with single young (table 10.2). First, large-brained organisms should have relatively slow rates of growth and development, and thus take a long time to reach maturity (defined as age at first reproduction). Second, to lighten the heavy burden on juveniles (see above), large-brained organisms produce relatively large neonates. This may reduce annual juvenile mortality to a point below what it would be if neonate size had not increased, but whether this reduction will outweigh the overall effect of a longer immature period among larger-brained taxa is not clear. A further consequence of having large neonates is that this will, ceteris paribus, reduce reproductive rates.

These direct costs must be compensated for by increases in other fitness components if increases in brain size are to be favored by selection. Thus, a third possible cause for the correlation between brain size and life-history traits is that adult fitness components improve due to increased brain size. Large-brained organisms may increase their odds of survival by staving off starvation, turning to extracted foods hidden inside an inedible matrix (Byrne 1997), or by becoming more effective at avoiding predation (Shultz & Dun-

Table 10.2. Predicted life-history consequences of increasing brain size (assuming no change in body size) in organisms with relatively large brains, flowing from the allocation-changes version of the expensive brain framework. Major predictions are in bold.

Primary consequence	Secondary consequence
Slower development	**Delay in age at first reproduction**
Larger neonates	Reduced reproductive rates?
	Increased juvenile survival?
Improved adult performance	**Increased adult survival, hence longevity**
	Increased reproductive rates?

bar 2006; Zuberbühler & Byrne 2006). When large brain size improves survival, it leads to selection for a slower-paced life history, probably through selection on a physiology that makes longer life possible, thus creating a direct link between large brains and slow life history. In addition, large-brained organisms may increase their reproductive success by enhancing their social competence, from being more Machiavellian to being better cooperators or social learners (Dunbar 2003; Whiten & van Schaik 2007). Alternatively, augmented reproduction might result from improved foraging efficiency that counteracts some of the suppressing effects on reproduction created by larger neonates. The net effect on reproductive rates will depend on the balance of these two processes.

The primary predictions listed in table 10.2 are supported by empirical data from precocial mammals (Isler & van Schaik 2009). They are also supported among primates when we control statistically for the effects of body mass (fig. 10.11). Development is slower, adult life span is longer, and neonates are larger in larger-brained species, while reproductive rates tend to be slightly lower. For altricial mammals with large litters, the most important prediction is that litter size should be reduced; this prediction is also met (Isler & van Schaik 2009). Overall, then, the predictions of the allocation version of the expensive brain framework are well supported. The allocation-tradeoffs version is also compatible with the *maternal energy* hypothesis (Martin 1998; but see Jones & MacLarnon 2004), if we regard maternal energy flux as the mechanism through which optimal brain size is achieved during ontogeny, rather than as a constraint on the evolution of brain size.

The expensive brain framework encompasses various existing hypotheses (such as the *direct metabolic costs* and the *expensive tissue* hypotheses; fig. 10.10). Through the allocation-tradeoffs version, it provides a straightforward explanation for the link between brain size and life history. Furthermore, it explains why cooperatively breeding mammals and birds, in which females receive energy subsidies,

tend to have larger brains (Isler et al. 2012), and why unavoidable periods of starvation tend to be associated with smaller brains in orangutans (Taylor & van Schaik 2007) and lemurs (van Woerden et al. 2010). It is also consistent with several other hypotheses proposed earlier to explain the correlation between brain size and selected aspects of life history (see discussion in Deaner et al. 2003). Thus, the relatively slow development of large-brained mammals is consistent with (1) the *maturational constraints* hypothesis, also known as the *needing-to-learn* hypothesis (Ross & Jones 1999; Ross 2003), and (2) the *brain malnutrition* hypothesis (Deaner et al. 2003), which is a modification of Janson & van Schaik's (1993) starvation avoidance hypothesis. The latter idea, building on the assumption that growing brains that do not have their energy needs met will show irreversible cognitive deficits, focused on setting rates of brain growth and development to survive times of food scarcity. Long adult life spans in large-brained mammals are consistent with (3) the *cognitive buffer* hypothesis, the idea that larger-brained organisms can deal better with unusual circumstances (Allman 1999), and (4) the *delayed benefits* hypothesis, which proposes that investing in costly brain tissue during immaturity, when no reproductive benefits are possible, is worthwhile only if the period during which the reproductive benefits are generated is long enough (Dukas 1998).

Above we have noted the existence of taxonomic variation in production (i.e., growth and reproduction), which is left unexplained in Charnov's (1993) model. Although Charnov and Berrrigan (1993) felt that primate biologists seemed "to be obsessed with the benefits of large brains," the allocation-tradeoffs version of the expensive brain hypothesis explains much of this variation. Further work will be necessary to determine how much variation in production is due to brain size and how much to other factors, such as bet hedging or ecology in the form of consistently low productivity.

In sum, there is much evidence for correlated evolution between brains and life history in primates and other precocial mammals. Taxa with slow life histories are exposed to various pressures that favor the evolution of large brains, whereas selection favors the evolution of large brains only if they improve the combination of juvenile and adult survival and reproduction by a large enough margin to pay for the inevitable developmental costs. This phenomenon can explain many of the unusual cognitive abilities of primates in general and of humans in particular. Especially promising as an explanatory framework is the notion that brains are energetically expensive, and that changes in brain size affect trade-offs with other major functions, particularly production.

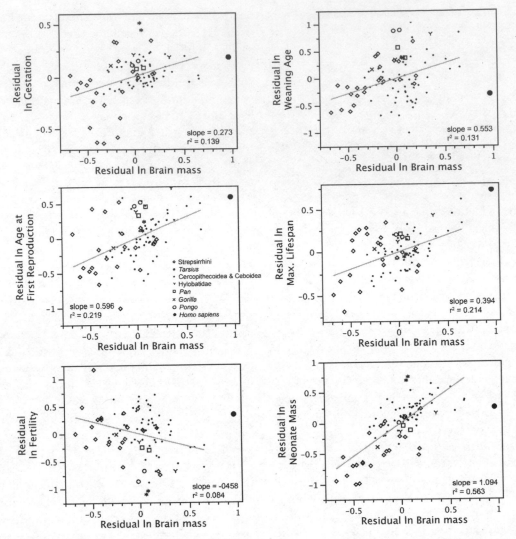

Fig. 10.11. Bivariate plots of residuals of life-history variables versus residuals of brain mass in primates ($N = 87$ species). All correlations are highly significant, and remain significant after applying independent contrasts. *Homo sapiens* was excluded for calculations of residuals and least-squares regression lines, but is always indicated to identify our place relative to the other primates.

Summary and Conclusions

Life history is essentially a design for rates of growth, development, and reproduction, as well as maximum life span. There are two major patterns in mammalian life histories: (1) a gradient from a fast to a slow pace of life history, and (2) more qualitative variation between altricial and precocial state at birth. Primates generally have a slow-paced life history for their body size and give birth to precocial young, usually singletons. Critical to understanding the evolution of life histories is the concept of trade-off. A species' life history reflects the optimization of a number of trade-offs between allocation targets, such as growth, reproduction, maintenance, and repair (and thus survival). Various theoretical approaches to the evolution of primate life histo-

ries agree that life-history variation across species is largely driven by one external variable: the level of unavoidable, extrinsic mortality. Accordingly, arboreal primates have a slower pace of life history than terrestrial ones. The evolution of precociality may be linked to the absence of safe shelters for infants. Precocial primates carry their infants at all times. Precociality limits litter size, but its evolution may have allowed the nomadic lifestyle and permanent gregariousness of most primates.

There are various interesting links between life history and behavior. Slow-paced life history generally leads to longer periods of lactation relative to gestation, which in turn has given rise to the risk of infanticide by males. Another very important link involves cognition: life history and brain size have undergone correlated evolution in primates.

This can be explained by the expensive brain hypothesis, which proposes that an increase in brain size is often paid for by reduced allocation to production—i.e., growth and reproduction.

Although the basic outlines of primate life history are well understood, there is some remaining variation not easily explained by the existing focus on extrinsic mortality. Least understood is the varying degree to which life history within a species is plastic. Although primates may not be the easiest organisms in which to examine this issue, their obvious relevance for human variation and public health makes such examination an effort well worth the investment.

References

Aiello, L. C., Bates, N. & Joffe, T. 2001. In defense of the Expensive Tissue Hypothesis. In *Evolutionary Anatomy of the Primate Cerebral Cortex* (ed. by Falk, D. & Gibson, K. R.), 57–78. Cambridge: Cambridge University Press.

Aiello, L. C. & Wheeler, P. 1995. The expensive-tissue hypothesis: The brain and the digestive system in human and primate evolution. *Current Anthropology*, 36, 199–221.

Allman, J. M. 1999. *Evolving Brains*. New York: Scientific American Library.

Altmann, J., Schoeller, D., Altmann, S. A., Muruthi, P. & Sapolsky, R. M. 1993. Body size and fatness of free-living baboons reflect food availability and activity levels. *American Journal of Primatology*, 10, 149–161.

Armstrong, E. 1983. Relative brain size and metabolism in mammals. *Science*, 220, 1302–1304.

Austad, S. N. & Fischer, K. E. 1992. Primate longevity: Its place in the mammalian scheme. *American Journal of Primatology*, 28, 251–261.

Barrickman, N. L., Bastian, M. L., Isler, K. & van Schaik, C. P. 2008. Life history costs and benefits of encephalization: A comparative test using data from long-term studies of primates in the wild. *Journal of Human Evolution*, 54, 568–590.

Barton, R. A. 1999. The evolutionary ecology of the primate brain. In *Comparative Primate Socioecology* (ed. by Lee, P. C.), 167–203. Cambridge: Cambridge University Press.

Begon, M., Harper, J. L. & Townsend, C. R. 1986. *Ecology: Individuals, Populations and Communities*. Oxford: Blackwell Scientific Publications.

Bielby, J., Mace, G. M., Bininda-Emonds, O. R. P., Cardillo, M., Gittleman, J. L., Jones, K. E., Orme, C. D. L. & Purvis, A. 2007. The fast-slow continuum in mammalian life history: An empirical reevaluation. *American Naturalist*, 169, 748–757.

Blomquist, G. E. 2009. Trade-off between age of first reproduction and survival in a female primate. *Biology Letters*, 5, 339–342.

Byrne, R. W. 1997. The technical intelligence hypothesis: An additional evolutionary stimulus to intelligence. In *Machiavellian Intelligence II* (ed. by Whiten, A. & Byrne, R.), 289–311. Cambridge: Cambridge University Press.

Case, T. J. 1978. On the evolution and adaptive significance of post-natal growth rates in terrestrial vertebrates. *Quarterly Review of Biology*, 53, 243–282.

Charnov, E. L. 1993. *Life History Invariants: Some explorations of symmetry in evolutionary ecology*. Oxford: Oxford University Press.

———. 2001. Evolution of mammal life histories. *Evolutionary Ecology Research*, 3, 521–535.

———. 2004. The optimal balance between growth rate and survival in mammals. *Evolutionary Ecology Research*, 6, 307–313.

Charnov, E. L. & Berrigan, D. 1993. Why do female primates have such long lifespans and so few babies? or, Life in the slow lane. *Evolutionary Anthropology*, 1, 191–194.

Cole, L. C. 1954. The population consequences of life history phenomena. *Quarterly Review of Biology*, 29, 103–137.

Deaner, R. O., Barton, R. A. & van Schaik, C. P. 2003. Primate brains and life histories: Renewing the connection. In *Primate Life Histories and Socioecology* (ed. by Kappeler, P. M. & Pereira, M. E.), 233–265. Chicago: University of Chicago Press.

Derrickson, E. M. 1992. Comparative reproductive strategies of altricial and precocial eutherian mammals. *Functional Ecology*, 6, 57–65.

Dukas, R. 1998. *Cognitive Ecology: The Evolutionary Ecology of Information Processing and Decision Making*. Chicago: University of Chicago Press.

Dunbar, R. I. M. 2003. The social brain: Mind, language, and society in evolutionary perspective. *Annual Review of Anthropology*, 32, 163–181.

Economos, A. C. 1980. Brain-life span conjecture: A reevaluation of the evidence. *Gerontology*, 26, 82–89.

Fisher, D. O., Owens, I. P. F. & Johnson, C. H. 2001. The ecological basis of life history variation in Marsupials. *Ecology*, 82, 3531–3540.

Garcia, C., Lee, P. C. & Rosetta, L. 2009. Growth in colony living anubis baboon infants and its relationship with maternal activity budgets and reproductive status. *American Journal of Physical Anthropology*, 138, 123–135.

Geary, D. C. 2005. *The Origin of Mind: Evolution of Brain, Cognition, and General Intelligence*. Washington: American Psychological Association.

Harvey, P., Pagel, M. & Rees, J. 1991. Mammalian metabolism and life histories. *American Naturalist*, 137, 556–566.

Harvey, P. H., Promislow, D. E. L. & Read, A. F. 1989. Causes and correlates of life history differences among mammals. In *Comparative Socioecology* (ed. by Standen, V. & Foley, R. A.), 305–318. Oxford: Blackwell.

Harvey, P. H. & Purvis, A. 1999. Understanding the ecological and evolutionary reasons for life history variation: Mammals as a case study. In *Advanced Ecological Theory: Principles and Applications* (ed. by McGlade, J.), 232–248. Oxford: Blackwell Science.

Hawkes, K. 2006. Life history theory and human evolution. In *The Evolution of Human Life History* (ed. by Hawkes, K. & Paine, R. R.), 45–93. Santa Fe: School of American Research Press.

Hladik, C. M., Chivers, D. J. & Pasquet, P. 1999. On diet and gut size in non-human primates: Is there a relationship to brain size? *Current Anthropology*, 40, 695–697.

Hofman, M. A. 1993. Encephlization and the evolution of longevity in mammals. *Journal of Evolutionary Biology*, 6, 209–227.

Holliday, M. A. 1971. Metabolic rate and organ size during

growth from infancy to maturity and during late gestation and early infancy. *Pediatrics*, 47, 169–179.

Hughes, K. A. & Reynolds, R. M. 2004. Evolutionary and mechanistic theories of aging. *Annual Review of Entomology*, 50, 421–445.

Isler, K., Kirk, E. C., Miller, J. M. A., Albrecht, G. A., Gelvin, B. R. & Martin, R. D. 2008. Endocranial volumes of primate species: scaling analyses using a comprehensive and reliable data set. *Journal of Human Evolution*, 55, 967–978.

Isler, K. & van Schaik, C. P. 2006a. Costs of encephalization: The energy trade-off hypothesis tested on birds. *Journal of Human Evolution*, 51, 228–243.

———. 2006b. Metabolic costs of brain size evolution. *Biology Letters*, 2, 557–560.

———. 2009. The expensive brain: A framework for explaining evolutionary changes in brain size. *Journal of Human Evolution*, 57, 392–400.

———. 2012. Allomaternal care, life history and brain size evolution in mammals. *Journal of Human Evolution*, doi: 10.1016/j.jhevol.2012.03.009.

Janson, C. H. 2003. Puzzles, predation, and primates: Using life history to understand selection pressures. In *Primate Life Histories and Socioecology* (ed. by Kappeler, P. M. & Pereira, M. E.), 103–131. Chicago: University of Chicago Press.

Janson, C. H. & van Schaik, C. P. 1993. Ecological risk aversion in juvenile primates: Slow and steady wins the race. In *Juvenile Primates. Life History, Development, and Behavior* (ed. by Pereira, M. E. & Fairbanks, L. A.), 57–74. New York: Oxford University Press.

Jeschke, J. M. & Kokko, H. 2009. The roles of body size and phylogeny in fast and slow life histories. *Evolutionary Ecology*, 23, 867–878.

Johnson, S. E. 2003. Life history and the competitive environment: Trajectories of growth, maturation, and reproductive output among chacma baboons. *American Journal of Physical Anthropology*, 23, 83–98.

Kappeler, P. M. 1998. Nests, tree holes, and the evolution of primate life histories. *American Journal of Primatology*, 46, 7–33.

Kappeler, P. M., Pereira, M. E. & van Schaik, C. P. 2003. Primate life histories and socioecology. In *Primate Life Histories and Socioecology* (ed. by Kappeler, P. M. & Pereira, M. E.), 1–20. Chicago: University of Chicago Press.

Kappeler, P. M. & Rasoloarison, R. M. 2003. *Microcebus*, mouse lemurs, tsidy. In *The Natural History of Madagascar* (ed. by Goodman, S. M. & Benstead, J. P.), 1310–1315. Chicago: University of Chicago Press.

Kennedy, C. & Sokoloff, L. 1957. An adaptation of the nitrous oxide method to the study of cerebral circulation in children: Normal values for cerebral blood flow and cerebral metabolic rate in childhood. *Journal of Clinical Investigation*, 36, 1130–1137.

Ketterson, E. D. & Nolan, V., Jr. 1999. Adaptation, exaptation, and constraint: A hormonal perspective. *American Naturalist*, 154, S4-S25.

Kirkwood, T. B. L. & Austad, S. N. 2000. Why do we age? *Nature*, 408, 233–238.

Leigh, S. R. 1995. Socioecology and the ontogeny of sexual dimorphism in anthropoid primates. *American Journal of Physical Anthropology*, 97, 339–356.

———. 2004. Brain growth, life history, and cognition in primates and human evolution. *American Journal of Primatology*, 62, 139–164.

Leigh, S. R. & Blomquist, G. E. 2007. Life history. In *Primates in Perspective* (ed. by Campbell, C. J., Fuentes, A., MacKinnon, K. C., Panger, M. & Bearder, S. K.), 396–407. New York: Oxford University Press.

Lindstedt, S. L. & Calder, W. A., III. 1981. Body size, physiological time, and longevity of homeothermic mammals. *Quarterly Review of Biology*, 56, 1–16.

MacArthur, R. H. & Wilson, E. O. 1967. *The Theory of Island Biogeography*. Princeton, NJ: Princeton University Press.

Mangel, M. 2003. Environment and longevity: The demography of the growth rate. In *Life Span: Evolutionary, Ecological and Demographic Perspectives* (ed. by Carey, J. R. & Tuljapurkar, S.), 57–70. New York: Population Council.

Martin, R. D. 1990. *Primate Origins and Evolution. A Phylogenetic Reconstruction*. London: Chapman & Hall.

———. 1996. Scaling of the mammalian brain: The maternal energy hypothesis. *News in Physiological Sciences*, 11, 149–156.

Martin, R. D. & MacLarnon, A. M. 1990. Reproductive patterns in primates and other mammals: The dichotomy between altricial and precocial offspring. In *Primate Life History and Evolution* (ed. by DeRousseau, C. J.), 47–79. New York: Wiley-Liss.

Mink, J. W., Blumenschine, R. J. & Adams, D. B. 1981. Ratio of central nervous system to body metabolism in vertebrates: Its constancy and functional basis. *American Journal of Physiology*, 241, R203–212.

Mori, A. 1979. Analysis of population changes by measurement of body weight in the Koshima troop of Japanese monkeys. *Primates*, 20, 371–397.

Mousseau, T. A. & Roff, D. A. 1987. Natural selection and the heritability of fitness components. *Heredity*, 59, 181–197.

Mumby, H. & Vinicius, L. 2008. Primate growth in the slow lane: A study of interspecies variation in the growth constant A. *Evolutionary Biology*, 35, 287–295.

Nakagawa, N., Ohsawa, H. & Muroyama, Y. 2003. Life-history paramaters of a wild group of West African patas monkeys (*Erythrocebus patas patas*). *Primates*, 44, 281–290.

Navarrete, A. R., van Schaik, C. P. & Isler, K. 2011. Energetics and the evolution of human brain size. *Nature*, 480, 91–93.

Nunn, C. L. & Barton, R. A. 2001. Comparative methods for studying primate adaptation and allometry. *Evolutionary Anthropology*, 10, 81–98.

Partridge, L. & Harvey, P. H. 1988. The ecological context of life history evolution. *Science*, 241, 1449–1455.

Pontzer, H., Raichlen, D. A., Shumaker, R. W., Ocobock, C. & Wich, S. A. 2010. Metabolic adaptation for low energy throughput in orangutans. *Proceedings of the National Academy of Sciences USA*, 107, 14048–14052.

Prothero, J. & Jürgens, K. D. 1987. Scaling of maximal lifespan in mammals: A review. In *Evolution of Longevity in Animals: A Comparative Approach* (ed. by Woodhead, A. D. & Thompson, K. H.). New York: Plenum Press.

Read, A. F. & Harvey, P. H. 1989. Life history differences among the eutherian radiations. *Journal of Zoology, London*, 219, 329–353.

Reichard, U. H. & Barelli, C. 2008. Life history and reproductive strategies of Khao Yai *Hylobates lar*: Implications for social evolution in apes. *International Journal of Primatology*, 29, 823–844.

Richard, A. F., Dewar, R. E., Schwartz, M. & Ratsirarson, J. 2002. Life in the slow lane? Demography and life histories of male and female sifaka (*Propithecus verreauxi verreauxi*). *Journal of Zoology, London*, 256, 421–436.

Richard, A. F. & Nicoll, M. E. 1987. Female social dominance and basal metabolism in a Malagasy primate, *Propithecus verreauxi*. *American Journal of Primatology*, 12, 309–314.

Roff, D. A. 1992. *The Evolution of Life Histories: Theory and Analysis*. New York: Chapman & Hall.

Rose, M. & Charlesworth, B. 1980. A test of evolutionary theories of senescence. *Nature*, 287, 141–142.

Ross, C. 1988. The intrinsic rate of natural increase and reproductive effort in primates. *Journal of Zoology, London*, 214, 199–219.

———. 2003. Life history, infant care strategies, and brain size in primates. In *Primate Life Histories and Socioecology* (ed. by Kappeler, P. M. & Pereira, M. E.), 266–284. Chicago: University of Chicago Press.

——— 2004. Life histories and the evolution of large brain size in great apes. In *The Evolution of Thought: Evolutionary Origins of Great Ape Intelligence* (ed. by Russon, A. E. & Begun, D. R.), 122–139. Cambridge: Cambridge University Press.

Ross, C. & Jones, K. E. 1999. Socioecology and the evolution of primate reproductive rates. In *Comparative Primate Socioecology* (ed. by Lee, P. C.), 73–110. Cambridge: Cambridge University Press.

Sacher, G. A. 1959. Relation of lifespan to brain weight and body weight in mammals. In *CIBA Foundation Colloquia on Ageing. Volume 5: The Lifespan of Animals* (ed. by Wolstenholme, G. E. W. & O'Connor, M.), 115–141. Boston: Little, Brown and Company.

———. 1975. Maturation and longevity in relation to cranial capacity in hominid evolution. In *Primate Functional Morphology and Evolution* (ed. by Tuttle, R. H.), 417–441. The Hague: Mouton Publishers.

Sapolsky, R. 1994. *Why Zebras Don't Get Ulcers: A Guide to Stress, Stress-Related Diseases and Coping*. New York: W. H. Freeman.

Schultz, A. H. 1969. *The Life of Primates*. London: Weidenfeld & Nicolson.

Sheldon, B. C. & Verhulst, S. 1996. Ecological immunology: Costly parasite defences and trade-offs in evolutionary ecology. *Trends in Ecology and Evolution*, 11, 317–321.

Shultz, S. & Dunbar, R. I. M. 2006. Chimpanzee and felid diet composition is influenced by prey brain size. *Biology Letters*, 2, 505–508.

Sibly, R. M. & Brown, J. H. 2007. Effects of body size and lifestyle on evolution of mammal life histories. *Proceedings of the National Academy of Sciences (USA)*, 104, 17707–17712.

Smuts, B. B. & Smuts, R. W. 1993. Male aggression and sexual coercion of females in nonhuman primates and other mammals: Evidence and theoretical implications. *Advances in the Study of Behavior*, 22, 1–63.

Stearns, S. C. 1983. The influence of size and phylogeny on patterns of covariation among life-history traits in the mammals. *Oikos*, 41, 173–187.

———. 1992. *The Evolution of Life Histories*. Oxford: Oxford University Press.

———. 2000. Life history evolution: Successes, limitations, and prospects. *Naturwissenschaften*, 87, 476–486.

Stearns, S. C., Ackermann, M., Doebeli, M. & Kaiser, M. 2000. Experimental evolution of aging, growth, and reproduction in fruitflies. *Proceedings of the National Academy of Sciences (USA)*, 97, 3309–3313.

Stearns, S. C. & Koella, J. C. 1986. The evolution of phenotypic plasticity in life-history traits: Predictions of reaction norms for age and size at maturity. *Evolution*, 40, 893–915.

Sugiyama, Y. & Ohsawa, H. 1982. Population dynamics of Japanese monkeys with special reference to the effect of artificial feeding. *Folia Primatologica*, 39, 238–263.

Taylor, A. B. & van Schaik, C. P. 2007. Variation in brain size and ecology in *Pongo*. *Journal of Human Evolution*, 52, 59–71.

Trivers, R. L. 1974. Parent-offspring conflict. *American Zoologist*, 14, 249–264.

Van Noordwijk, A. J. & de Jong, G. 1986. Acquisition and allocation of resources: their influence on variation in life history tactics. *American Naturalist*, 128, 137–142.

Van Noordwijk, M. A. & van Schaik, C. P. 1999. The effects of dominance rank and group size on female lifetime reproductive success in wild long-tailed macaques, *Macaca fascicularis*. *Primates*, 40, 105–130.

———. 2004. Sexual selection and the careers of primate males: Paternity concentration, dominance acquisition tactics and transfer decisions. In *Sexual Selection in Primates* (ed. by Kappeler, P. M. & van Schaik, C. P.), 208–229. Cambridge: Cambridge University Press.

Van Schaik, C. P. 2000. Vulnerability to infanticide: patterns among mammals. In *Infanticide by Males and its Implications* (ed. by van Schaik, C. P. & Janson, C. H.), 61–71. Cambridge: Cambridge University Press.

Van Schaik, C. P. & Deaner, R. O. 2003. Life history and cognitive evolution in primates. In *Animal Social Complexity* (ed. by de Waal, F. B. M. & Tyack, P. L.), 5–25. Cambridge, MA.: Harvard University Press.

Van Schaik, C. P. & Kappeler, P. M. 1997. Infanticide risk and the evolution of male-female association in primates. *Proceedings of the Royal Society London, Series B*, 264, 1687–1694.

Van Schaik, C. P. & Pfannes, K. R. 2005. Tropical climates and phenology: a primate perspective. In *Seasonality in Primates: Studies of Living and Extinct Human and Non-Human Primates* (ed. by Brockman, D. K. & van Schaik, C. P.), 23–54. Cambridge: Cambridge University Press.

Van Schaik, C. P., Pradhan, G. R. & van Noordwijk, M. A. 2004. Mating conflict in primates: Infanticide, sexual harassment and female sexuality. In *Sexual Selection in Primates* (ed. by Kappeler, P. M. & van Schaik, C. P.), 131–150. Cambridge: Cambridge University Press.

Van Schaik, C. P., Barrickman, N., Bastian, M. L., Krakauer, E. B. & van Noordwijk, M. A. 2006. Primate life histories and the role of brains. In *The Evolution of Human Life History* (ed. by Hawkes, K. & Paine, R. R.), 127–154. Santa Fe: School of American Research Press.

Van Schaik, C. P., van Noordwijk, M. A. & Nunn, C. L. 1999. Sex and social evolution in primates. In *Comparative Primate Socioecology* (ed. by Lee, P. C.), 204–240. Cambridge: Cambridge University Press.

Van Woerden, J. T., van Schaik, C. P. & Isler, K. 2010. Effects of seasonality on brain size evolution: Evidence from strepsirrhine primates. *American Naturalist*, 176, 758–767.

Whiten, A. & van Schaik, C. P. 2007. The evolution of animal "cultures" and social intelligence. *Philosophical Transactions of the Royal Society B: Biological Sciences*, 362, 603–620.

Wich, S. A., Steenbeek, R., Sterck, E. H. M., Korstjens, A. H., Willems, E. P. & van Schaik, C. P. 2007. Demography and life history of Thomas langurs (*Presbytis thomasi*). *American Journal of Primatology*, 69, 641–651.

Wich, S. A., Utami-Atmoko, S. S., Mitra Setia, T., Rijksen, H. D., Schurmann, C., van Hooff, J. A. R. A. M. & van Schaik, C. P. 2004. Life history of wild Sumatran orangutans (*Pongo abelii*). *Journal of Human Evolution*, 47, 385–398.

Zera, A. J. & Harshman, L. J. 2001. The physiology of life history trade-offs in animals. *Annual Review of Ecology and Systematics*, 32, 95–126.

Zuberbühler, K. & Byrne, R. W. 2006. Social cognition. *Current Biology*, 16, R786–790.

References for Table 10.1

Altmann, J., Hausfater, G. & Altmann, S. A. 1985. Demography of Amboseli baboons, 1963–1983. *American Journal of Primatology*, 8, 113–125.

Badham, M. 1967. A note on breeding the spectacled leaf monkey *Presbytis obscura* at Twycross Zoo. *International Zoo Yearbook*, 7, 89.

Bales, K., O'Herron, M., Baker, A. J. & Dietz, J. M. 2001. Sources of variability in numbers of live births in wild golden lion tamarins (*Leontopithecus rosalia*). *American Journal of Primatology*, 54, 211–221.

Barrickman, N. L., Bastian, M. L., Isler, K. & van Schaik, C. P. 2008. Life history costs and benefits of encephalization: A comparative test using data from long-term studies of primates in the wild. *Journal of Human Evolution*, 54, 568–590.

Bearder, S. K. 1987. Lorises, bushbabies, and tarsiers: Diverse societies in solitary foragers. In *Primate Societies* (ed. by Smuts, B. B., Cheney, D. L., Seyfarth, R. M., Wrangham, R. W. & Struhsaker, T. T.), 11–24. Chicago: University of Chicago Press.

Bearder, S. K., Ambrose, L., Harcourt, C., Honess, P., Perkin, A., Pimley, E., Pullen, S. & Svoboda, N. 2003. Species-typical patterns of infant contact, sleeping site use and social cohesion among nocturnal primates in Africa. *Folia Primatologica*, 74, 337–354.

Bentley-Condit, V. K. & Smith, E. O. 1997. Female reproductive parameters of Tana river yellow baboons. *International Journal of Primatology*, 18, 581–596.

Bercovitch, F. B. & Harvey, N. 2004. Reproductive life history. In *Macaque Societies: A Model for the Study of Social Organization* (ed. by Thierry, B., Singh, M. & Kaumanns, W.), 61–79. Cambridge: Cambridge University Press.

Boer, M. 1987. Observations on reproduction and behavior of captive drills (*Mandrillus leucophaeus* Ritgen, 1824) in the Hannover Zoo. *Zeitschrift für Säugetierkunde*, 52, 265–281.

Bramblett, C. A., Pejaver, L. D. & Drickman, D. J. 1975. Reproduction in captive vervet and Sykes monkeys. *Journal of Mammalogy*, 56, 940–946.

Campbell, C. J. 2007. Primate sexuality and reproduction. In *Primates in Perspective* (ed. by Campbell, C. J., Fuentes, A., MacKinnon, K. C., Panger, M. A. & Bearder, S. K.), 423–437. Oxford: Oxford University Press.

Carroll, J. 1993. The captive behaviour and reproduction of Goeldi's monkey *Callimico goeldii*. *Dodo Journal of the Wildlife Preservation Trust*, 29, 171–172.

Chapman, C. A., Walker, S. & Lefebvre, L. 1990. Reproductive strategies of primates: The influence of body size and diet on litter size. *Primates*, 31, 1–13.

Cheney, D. L., Seyfarth, R. M., Andelman, S. J. & Lee, P. C. 1988. Reproductive success in vervet monkeys. In *Reproductive Success: Studies of Individual Variation in Contrasting Breeding Systems* (ed. by Clutton-Brock, T. H.), 384–402. Chicago: University of Chicago Press.

Christen, A. 1974. Reproductive biology and behavior of *Cebuella pygmaea* and *Tamarin tamarin* (Primates, Platyrrhina, Callitrichidae). *Fortschritte der Verhaltensforschung*, 14, 1–80.

Clarke, M. & Glander, K. E. 1984. Female reproductive success in a group of free-ranging howling monkeys (*Alouatta palliata*) in Costa Rica. In *Female Primates: Studies by Women Primatologists* (ed. by Small, M.), 111–126. New York: Alan R. Liss, Inc.

Cleveland, J. & Snowdon, C. T. 1984. Social development during the first 20 weeks in the cotton-top tamarin (*Saguinus o. oedipus*). *Animal Behaviour*, 32, 432–444.

Colquhoun, I. C. 1993. The socioeceology of *Eulemur macaco*: A preliminary report. In *Lemur Social Systems and Their Ecological Basis* (ed. by Kappeler, P. M. & Ganzhorn, J. U.), 11–23. New York: Plenum Press.

Cords, M. 1988. Mating systems of forest guenons: A preliminary review. In *A Primate Radiation: Evolutionary Biology of the African Guenons* (ed. by Gauthier-Hion, A., Bourliere, F., Gauthier, J. P. & Kingdon, J.), 323–339. Cambridge: Cambridge University Press.

Crockett, C. M. & Pope, T. R. 1993. Consequences of sex differences in dispersal for juvenile red howler monkeys. In *Juvenile Primates. Life History, Development, and Behavior* (ed. by Pereira, M. E. & Fairbanks, L. A.), 104–118. New York: Oxford University Press.

Crockett, C. M. & Sekulic, R. 1982. Gestation length in red howler monkeys. *American Journal of Primatology*, 3, 291–294.

DeSilva, J. M. & Lesnik, J. J. 2008. Brain size at birth throughout human evolution: A new method for estimating neonatal brain size in hominins. *Journal of Human Evolution*, 55, 1064–1074.

Dettling, A. & Pryce, C. R. 1999. Hormonal monitoring of age at sexual maturation in female Goeldi's monkeys (*Callimico goeldii*) in their family groups. *American Journal of Primatology*, 48, 77–83.

Di Bitetti, M. S. & Janson, C. H. 2001. Reproductive socioecology of tufted capuchins (*Cebus apella nigritus*) in northeastern Argentina. *International Journal of Primatology*, 22, 127–142.

Dietz, J. M., Baker, A. J. & Miglioretti, D. 1994. Seasonal variation in reproduction, juvenile growth, and adult body mass in golden lion tamarins (*Leontopithecus rosalia*). *American Journal of Primatology*, 34, 115–132.

Di Fiore, A. & Campbell, C. J. 2007. The atelines: Variations in ecology, behavior, and social organization. In *Primates in Perspective* (ed. by Campbell, C. J., Fuentes, A., MacKinnon, K. C., Panger, M. A. & Bearder, S. K.), 155–185. Oxford: Oxford University Press.

Digby, L. J., Ferrari, S. & Saltzmann, W. 2007. Callitrichines: The role of competition in cooperatively breeding species. In *Primates in Perspective* (ed. by Campbell, C. J., Fuentes, A.,

MacKinnon, K. C., Panger, M. A. & Bearder, S. K.). 155–185. Oxford: Oxford University Press.

Dunbar, R. I. M. 1984. *Reproductive Decisions: An Economic Analysis of Gelada Baboon Social Strategies*. Princeton, NJ: Princeton University Press.

Ernest, S. K. M. 2003. Life history characteristics of placental nonvolant mammals. *Ecology*, 84, 3402–3402.

Fedigan, L. & Fedigan, L. M. 1988. *Cercopithecus aethiops*: A review of field studies. In *A Primate Radiation: Evolutionary Biology of the African Guenons* (ed. by Gautier-Hion, A., Bourliere, F., Gautier, J.-P. & Kingdon, J.), 389–411. Cambridge: Cambridge University Press.

Fedigan, L. M. 2003. Impact of takeovers on infant deaths, births and conceptions in *Cebus capucinus* at Santa Rosa, Costa Rica. *International Journal of Primatology*, 24, 723–741.

Fedigan, L. M. & Rose, L. M. 1995. Interbirth interval variation in three sympatric species of neotropical monkey. *American Journal of Primatology*, 37, 9–24.

Feistner, A. T. C. & Ashbourne, C. J. 1994. Infant development in a captive-bred aye-aye (*Daubentonia madagascariensis*) over the first year of life. *Folia Primatologica*, 62, 74–92.

Feistner, A. T. C., Cooper, R. W. & Evans, S. 1992. The establishment and reproduction of a group of semifree-ranging mandrills. *Zoo Biology*, 11, 385–395.

Fernandez-Duque, E. 2007. Aotinae: Social monogamy in the only nocturnal haplorhines. In *Primates in Perspective* (ed. by Campbell, C. J., Fuentes, A., MacKinnon, K. C., Panger, M. A. & Bearder, S. K.), 139–154. Oxford: Oxford University Press.

Fogden, M. 1974. A preliminary field study of the western tarsier, *Tarsius bancanus* Horsefield. In *Prosimian Biology* (ed. by Martin, R. D., Doyle, G. A. & Walker, A. C.), 151–165. London: Duckworth.

Fragaszy, D. M., Fedigan, L. M. & Visalberghi, E. 2004. *The Complete Capuchin: The Biology of the Genus Cebus*. Cambridge: Cambridge University Press.

Garbutt, N. 1999. *Mammals of Madagascar*. New Haven: Yale University Press.

Glander, K. E. 1980. Reproduction and population growth in free-ranging mantled howling monkeys. *American Journal of Physical Anthropology*, 53, 25–36.

———. 1994. Morphometrics and growth in captive aye-ayes (*Daubentonia madagascariensis*). *Folia Primatologica*, 62, 108–114.

Glenn, M. & Cords, M. 2002. *The Guenons: Diversity and Adaptation in African Monkeys*. New York: Kluwer Academic/Plenum.

Godfrey, L. R., Samonds, K. E., Jungers, W. L. & Sutherland, M. R. 2001. Teeth, brains, and primate life histories. *American Journal of Physical Anthropology*, 114, 192–214.

Goldizen, A. W. 1987. Tamarins and marmosets: Communal care of offspring. In *Primate Societies* (ed. by Smuts, B. B., Cheney, D. L., Seyfarth, R. M., Wrangham, R. W. & Strusaker, T. T.), 34–43. Chicago: University of Chicago Press.

Gorzitze, A. B. 1996. Birth-related behaviors in wild proboscis monkeys (*Nasalis larvatus*). *Primates*, 37, 75–78.

Gould, L. & Sauther, M. 2007. Lemuriformes. In *Primates in Perspective* (ed. by Campbell, C. J., Fuentes, A., MacKinnon, K. C., Panger, M. A. & Bearder, S. K.), 46–72. Oxford: Oxford University Press.

Gursky, S. L. 2007. Tarsiiformes. In *Primates in Perspective* (ed. by Campbell, C. J., Fuentes, A., MacKinnon, K. C., Panger, M. A. & Bearder, S. K.), 73–85. Oxford: Oxford University Press.

Gust, D. A., Busse, C. D. & Gordon, T. P. 1990. Reproductive parameters in the sooty mangabey (*Cercocebus torquatus atys*). *American Journal of Primatology*, 22, 241–250.

Ha, J. C., Robinette, R. L. & Sackett, G. P. 2000. Demographic analysis of the Washington Regional Primate Research Center pigtailed macaque colony, 1967–1996. *American Journal of Primatology*, 52, 187–198.

Hampton, S. & Hampton, J. J. 1977. Detection of reproductive cycles and pregnancy in tamarins (*Saguinus spp.*). In *The Biology and Conservation of the Callitrichidae* (ed. by Kleiman, D.), 173–179. Washington: Smithsonian Institution Press.

Harris, T. R. & Monfort, S. L. 2006. Mating behavior and endocrine profiles of wild black and white colobus monkeys (*Colobus guereza*): Toward an understanding of their life history and mating system. *American Journal of Primatology*, 68, 383–396.

Harvey, P. H. & Clutton Brock, T. H. 1985. Life-history variation in primates. *Evolution*, 39, 559–581.

Harvey, P. H., Martin, R. D. & Clutton Brock, T. H. 1987. Life histories in a comparative perspective. In *Primate Societies* (ed. by Smuts, B. B., Cheney, D. L., Seyfarth, R. M., Wrangham, R. W. & Struhsaker, T. T.), 181–196. Chicago: University of Chicago Press.

Hayssen, V., van Tienhoven, A. & van Tienhoven, A. 1993. *Asdell's Patterns of Mammalian Reproduction: A Compendium of Species-Specific Data*. Ithaca, NY: Comstock/Cornell University Press.

Hearn, J. P. & Lunn, S. F. 1975. The reproductive biology of the marmoset monkey, *Callithrix jacchus*. In *The Breeding of Simians and Their Uses in Development Biology* (ed. by Perkins, F. T. & O'Donoghue, P. N.), 191–202. London: Laboratory Animals Ltd.

Higham, J. P., Warren, Y., Adanu, J., Umaru, B. N., MacLarnon, A. M., Sommer, V. & Ross, C. 2009. Living on the edge: Life-history of olive baboons at Gashaka-Gumti National Park, Nigeria. *American Journal of Primatology*, 71, 293–304.

Hill, K. & Hurtado, A. M. 1996. *Ache Life History: The Ecology and Demography of a Foraging People*. Hawthorne, NY: Aldine de Gruyter.

Hrdy, S. B. 1977. *The Langurs of Abu. Female and Male Strategies of Reproduction*. Cambridge, MA: Harvard University Press.

Hunter, J., Martin, R., Dixson, A. & Rudder, B. 1979. Gestation and interbirth intervals in the owl monkey (*Aotus trivirgatus griseimembra*). *Folia Primatologica*, 31, 165–175.

Isler, K., Kirk, E. C., Miller, J. M. A., Albrecht, G. A., Gelvin, B. R. & Martin, R. D. 2008. Endocranial volumes of primate species: Scaling analyses using a comprehensive and reliable data set. *Journal of Human Evolution*, 55, 967–978.

Izard, M. K. & Rasmussen, D. T. 1985. Reproduction in the slender loris, *Loris tardigradus malabaricus*. *American Journal of Primatology*, 8, 153–166.

Jablonski, N. G., Whitfort, M. J., Roberts-Smith, N. & Xu, Q. Q. 2000. The influence of life history and diet on the distribution of catarrhine primates during the Pleistocene in eastern Asia. *Journal of Human Evolution*, 39, 131–157.

Kaplan, H., Hill, K., Lancaster, J. & Hurtado, A. M. 2000. A theory of human life history evolution: Diet, intelligence, and longevity. *Evolutionary Anthropology*, 9, 156–185.

Kappeler, P. & Pereira, M. 2003. *Primate Life Histories and Socioecology*. Chicago: University of Chicago Press.

Kappeler, P. & Rasoloarison, R. 2003. *Microcebus*, mouse lemurs, tsidy. In *The Natural History of Madagascar* (ed. by Goodman, S. M. & Benstad, J.), 1310–1315. Chicago: University of Chicago Press.

Kirkpatrick, R. 2007. The Asian colobines: Diversity among leaf-eating monkeys. In *Primates in Perspective* (ed. by Campbell, C. J., Fuentes, A., MacKinnon, K. C., Panger, M. A. & Bearder, S. K.), 186–200. Oxford: Oxford University Press.

Kleiman, D., Hoage, R. & Green, K. 1988. The lion tamarins, genus *Leontopithecus*. In *Ecology and Behavior of Neotropical Primates* (ed. by Mittermeier, R., Rylands, A., Coimbra-Filho, A. & da Fonseca, G.), 299–347. Washington: World Wildlife Fund.

Knott, C. D., Emery Thompson, M. & Wich, S. 2009. The ecology of female reproduction in wild orangutans. In *Orangutans: Geographic Variation in Behavioral Ecology and Conservation* (ed. by Wich, S., Utami-Atmoko, S. S., Mitra Setia, T. & van Schaik, C. P.). New York: Oxford University Press.

Koyama, N., Nakamichi, M., Oda, R., Miyamoto, N., Ichino, S. & Takahata, Y. 2001. A ten-year summary of reproductive parameters for ring-tailed lemurs at Berenty, Madagascar. *Primates*, 42, 1–14.

Lee, P. C. 1999. Comparative ecology of postnatal growth and weaning among haplorhine primates. In *Comparative Primate Socioecology* (ed. by Lee, P. C.), 111–139. Cambridge: Cambridge University Press.

Lee, P. C., Majluf, P. & Gordon, I. J. 1991. Growth, weaning and maternal investment from a comparative perspective. *Journal of Zoology*, 225, 99–114.

Lekagul, B. & McNeely, J. 1977. *Mammals of Thailand*. Bangkok: Association for the Conservation of Wildlife.

Lindenfors, P. 2002. Sexually antagonistic selection on primate size. *Journal of Evolutionary Biology*, 15, 595–607.

Lippold, L. 1977. The Douc langur: A time for conservation. In *Primate Conservation* (ed. by Prince Rainier III & Bourne, G.), 513–538. New York: Academic Press.

———. 1981. Monitoring female reproductive status in the Douc langur *Pygathrix nemaeus* at San Diego Zoo. *International Zoo Yearbook*, 21, 184–187.

Lycett, J. E. 1994. The developmental behavioural ecology of infant baboons (*Papio cynocephalus ursinus*), PhD thesis, Natal University.

Lycett, J. E., Henzi, S. P. & Barrett, L. 1998. Maternal investment in mountain baboons and the hypothesis of reduced care. *Behavioral Ecology and Sociobiology*, 42, 49–56.

Markham, R. J. 1990. Breeding orangutans at Perth Zoo: Twenty years of appropriate husbandry. *Zoo Biology*, 9, 171–182.

Martin, R. D. & MacLarnon, A. M. 1990. Reproductive patterns in primates and other mammals: The dichotomy between altricial and precocial offspring. In *Primate Life History and Evolution* (ed. by DeRousseau, C. J.), 47–79. New York: Wiley-Liss.

Mehlman, P. 1989. Comparative density, demography, and ranging behavior of Barbary macaques (*Macaca sylvanus*) in marginal and prime conifer habitats. *International Journal of Primatology*, 10, 269–292.

Melnick, D. J. 1981. Microevolution in a population of Himalayan rhesus monkeys (*Macaca mulatta*), PhD thesis, Yale University.

Melnick, D. J. & Pearl, M. 1987. Cercopithecines in multimale groups: Genetic diversity and population structure. In *Primate Societies* (ed. by Smuts, B. B., Cheney, D. L., Seyfarth, R. M., Wrangham, R. W. & Struhsaker, T. T.), 121–134. Chicago: University of Chicago Press.

Morland, H. S. 1990. Parental behavior and infant development in ruffed lemurs (*Varecia variegata*) in a Northeast Madagascar rainforest. *American Journal of Primatology*, 20, 253–265.

Mutschler, T. & Tan, C. 2003. *Hapalemur*, bamboo or gentle lemurs. In *The Natural History of Madagascar* (ed. by Goodman, S. M. & Benstead, J.), 1324–1329. Chicago: University of Chicago Press.

Nakagawa, N., Ohsawa, H. & Muroyama, Y. 2003. Life-history parameters of a wild group of West African patas monkeys (*Erythrocebus patas patas*). *Primates*, 44, 281–290.

Nekaris, K. 2003. Observations of mating, birthing and parental behaviour in three subspecies of slender loris (*Loris tardigradus* and *Loris lydekkerianus*) in India and Sri Lanka. *Folia Primatologica*, 74, 312–336.

Nekaris, K. & Bearder, S. K. 2007. The lorisiform primates of Asia and mainland Africa: Diversity shrouded in darkness. In *Primates in Perspective* (ed. by Campbell, C. J., Fuentes, A., MacKinnon, K. C., Panger, M. A. & Bearder, S. K.), 24–45. Oxford: Oxford University Press.

Nieuwenhuijsen, K., Lammers, A. J. J. C., De Neef, K. J. & Slob, A. K. 1985. Reproduction and social rank in female stumptail macaques (*Macaca arctoides*). *International Journal of Primatology*, 6, 77–99.

Nishimura, A. 2003. Reproductive parameters of wild female *Lagothrix lagotricha*. *International Journal of Primatology*, 24, 707–722.

Norconk, M. 2007. Sakis, uakaris, titi monkeys: Behavioral diversity in a radiation of primate seed predators. In *Primates in Perspective* (ed. by Campbell, C. J., Fuentes, A., MacKinnon, K. C., Panger, M. A. & Bearder, S. K.), 123–138. Oxford: Oxford University Press.

Nowak, R. 1999. *Walker's Mammals of the World*, Sixth edition. Baltimore and London: Johns Hopkins University Press.

O'Brien, T., Kinnaird, M., Nurcahyo, A., Prasetyaningrum, M. & Iqbal, M. 2003. Fire, demography and the persistence of siamang (*Symphalangus syndactylus*: Hylobatidae) in a Sumatran rainforest. *Animal Conservation*, 6, 115–121.

Okamoto, K., Matsumura, S. & Watanabe, K. 2000. Life history and demography of wild moor macaques (*Macaca maurus*): Summary of ten years of observations. *American Journal of Primatology*, 52, 1–11.

Ostner, J. & Heistermann, M. 2003. Endocrine characterization of female reproductive status in wild redfronted lemurs (*Eulemur fulvus rufus*). *General and Comparative Endocrinology*, 131, 274–283.

Petter, J. 1978. Ecological and physiological adaptations of five sympatric nocturnal lemurs to seasonal variations in food production. In *Recent Advances in Primatology, Vol. 1: Behaviour* (ed. by Chivers, D. & J, H.), 211–223. New York: Academic Press.

Petter-Rousseaux, A. 1964. Reproductive physiology and behavior of the Lemuroidea. In *Evolutionary and Genetic Biology of Primates* (ed. by Buettner-Janusch, J.), 91–132. New York: Academic Press.

Pochron, S. T., Tucker, W. T. & Wright, P. C. 2004. Demography, life history, and social structure in *Propithecus diadema edwardsi* from 1986–2000 in Ranomafana National Park, Madagascar. *American Journal of Physical Anthropology*, 125, 61–72.

Radespiel, U. 2000. Sociality in the gray mouse lemur (*Microcebus murinus*) in Northwestern Madagascar. *American Journal of Primatology*, 51, 21–40.

Rajpurohit, L. S. & Mohnot, S. M. 1991. The progress of weaning in Hanuman langurs *Presbytis entellus entellus*. *Primates*, 32, 213–218.

Ramirez, M. 1988. The woolly monkey, genus *Lagothrix*. In *Ecology and Behavior of Neotropical Primates* (ed. by Mittermeier, R., Rylands, A., Coimbra-Filho, A. & da Fonseca, G.), 539–575. Washington: World Wildlife Fund.

Rasmussen, D. T. 1989. Social ecology and conservation of the Panamanian tamarin. *Anthroquest*, 40, 12–15.

Reichard, U. 2003. Social monogamy in gibbons: The male perspective. In *Monogamy: Mating Strategies and Partnerships in Birds, Humans, and Other Mammals*, 190–213. Cambridge: Cambridge University Press.

Rhine, R. J., Norton, G. W. & Wasser, S. K. 2000. Lifetime reproductive success, longevity, and reproductive life history of female yellow baboons (*Papio cynocephalus*) of Mikumi National Park, Tanzania. *American Journal of Primatology*, 51, 229–241.

Richard, A. F. 1987. Malagasy prosimians: Female dominance. In *Primate Societies* (ed. by Smuts, B.B., Cheney, D. L., Seyfarth, R. M., Wrangham, R. W. & Struhsaker, T. T.), 25–33. Chicago: University of Chicago Press.

Richard, A. F., Dewar, R. E., Schwartz, M. & Ratsirarson, J. 2002. Life in the slow lane? Demography and life histories of male and female sifaka (*Propithecus verreauxi verreauxi*). *Journal of Zoology, London*, 256, 421–436.

Robbins, M. M. 2007. Gorillas: Diversity in ecology and behavior. In *Primates in Perspective* (ed. by Campbell, C. J., Fuentes, A., MacKinnon, K. C., Panger, M. A. & Bearder, S. K.), 305–321. Oxford: Oxford University Press.

Ross, C. 2003. Life history, infant care strategies, and brain size in primates. In *Primate Life Histories and Socioecology* (ed. by Kappeler, P. M. & Pereira, M. E.), 266–284. Chicago: University of Chicago Press.

Ross, C. & Jones, K. E. 1999. Socioecology and the evolution of primate reproductive rates. In *Comparative Primate Socioecology* (ed. by Lee, P. C.), 73–110. Cambridge: Cambridge University Press.

Rudder, B. 1979. The allometry of primate reproductive parameters, PhD thesis, University College London.

Rudran, R. 1973. The reproductive cycles of two subspecies of purple-faced langurs (*Presbytis senex*) with relation to environmental factors. *Folia Primatologica*, 19, 41–60.

Schlee, M. & Labejof, L. 1994. Management and early development of infant behaviour in the white-crowned mangabey *Cercocebus torquatus lunulatus* at the Paris Menagerie. *International Zoo Yearbook*, 33, 228–234.

Sigg, H., Stolba, A., Abegglen, J. J. & Dasser, V. 1982. Life history of hamadryas baboons: Physical development, infant mortality, reproductive parameters and family relationships. *Primates*, 23, 473–487.

Silk, J. B. 1988. Social mechanisms of population regulation in a captive group of bonnet macaques (*Macaca radiata*). *American Journal of Primatology*, 14, 111–124.

Sly, D. L., Harbaugh, S. W., London, W. T. & Rice, J. M. 1983. Reproductive performance of a laboratory breeding colony of Patas monkeys (*Erythrocebus patas*). *American Journal of Primatology*, 4, 23–32.

Smith, R. J. & Leigh, S. R. 1998. Sexual dimorphism in primate neonatal body mass. *Journal of Human Evolution*, 34, 173–201.

Smith, T., Alport, L., Burrows, A., Bhatnagar, K., Dennis, J., Tuladhar, P. & Morrison, E. 2007. Perinatal size and maturation of the olfactory and vomeronasal neuroepithelia in lorisoids and lemuroids. *American Journal of Primatology*, 69, 74–85:

Smucny, D. A., Abbott, D. H., Mansfield, K. G., Schultz-Darken, N., Yamamoto, M. E. & Tardif, S. D. 2004. Sources of variation in reproductive output of captive common marmosets (*Callithrix jacchus*). *American Journal of Primatology*, 62, 149.

Smuts, B. B. & Nicolson, N. 1989. Reproduction in wild female olive baboons. *American Journal of Primatology*, 19, 229–246.

Snowdon, C. T. & Soini, P. 1988. The tamarins, genus *Saguinus*. In *Ecology and Behavior of Neotropical Primates* (ed. by Mittermeier, R., Rylands, A., Coimbra-Filho, A. & da Fonseca, G.), 223–298. Washington: World Wildlife Fund.

Soini, P. 1987. Ecology of the saddle-back tamarin *Saguinus fuscicollis illigeri* on the Rio Pacaya, Northeastern Peru. *Folia Primatologica*, 49, 11–32.

——— 1988. The pygmy marmoset, genus *Cebuella*. In *Ecology and Behavior of Neotropical Primates* (ed. by Mittermeier, R., Rylands, A., Coimbra-Filho, A. & da Fonseca, G.), 79–129. Washington: World Wildlife Fund.

Sommer, V., Srivastava, A. & Borries, C. 1992. Cycles, sexuality, and conception in free-ranging langurs. *American Journal of Primatology*, 28, 1–28.

Stevenson, M. & Rylands, A. 1988. The marmosets, genus *Callithrix*. In *Ecology and Behavior of Neotropical Primates* (ed. by Mittermeier, R., Rylands, A., Coimbra-Filho, A. & da Fonseca, G.), 131–222. Washington: World Wildlife Fund.

Struhsaker, T. T., and Lysa Leland. 1987. Colobines: Infanticide by adult males. In *Primate Societies* (ed. by Smuts, B. B., Cheney, D. L., Seyfarth, R. M., Wrangham, R. W. & Strusaker, T. T.), 83–97. Chicago and London: University of Chicago Press.

Stumpf, R. M. 2007. Chimpanzees and bonobos: Diversity within and between species. In *Primates in Perspective* (ed. by Campbell, C. J., Fuentes, A., MacKinnon, K. C., Panger, M. A. & Bearder, S. K.), 321–344. Oxford: Oxford University Press.

Takahata, Y., Suzuki, S., Okayasu, N., Sugiura, H., Takahashi, H., Yamagiwa, J., Izawa, K., Agetsuma, N., Hill, D., Saito, C., Sato, S., Tanaka, T. & Sprague, D. 1998. Does troop size of wild Japanese macaques influence birth rate and infant mortality in the absence of predators? *Primates*, 39, 245–251.

Trollope, J. & Blurton Jones, N. G. 1975. Aspects of reproduction and reproductive behaviour in *Macaca arctoides*. *Primates*, 16, 191–205.

Tutin, C. 1994. Reproductive success story: Variability among chimpanzees and comparisons with gorillas. In *Chimpanzee Cultures* (ed. by Wrangham, R. W., McGrew, W. C. & de Waal, F.), 181–194. Cambridge, MA: Harvard University Press.

Valeggia, C. R., Mendoza, S. P., Fernandez-Duque, E., Mason, W. A. & Lasley, B. 1999. Reproductive biology of female titi monkeys (*Callicebus moloch*) in captivity. *American Journal of Primatology*, 47, 183–195.

Van Horn, R. N. & Eaton, G. G. 1979. Reproductive physiology and behavior in prosimians. In *The Study of Prosimian Behavior* (ed. by Doyle, G. A. & Martin, R. D.), 79–122. New York: Academic Press.

Van Noordwijk, M. A. & van Schaik, C. P. 1999. The effects of dominance rank and group size on female lifetime reproductive success in wild long-tailed macaques, *Macaca fascicularis*. *Primates*, 40, 105–130.

Van Woerden, J. T., van Schaik, C. P. & Isler, K. 2010. Effects of seasonality on brain size evolution: Evidence from strepsirrhine primates. *American Naturalist*, 176, 758–767.

Weigl, R. 2005. *Longevity of Mammals in Captivity: From the Living Collections of the World*. Stuttgart: Schweizerbart.

Whitten, P. & Brockman, D. 2001. Strepsirrhine reproductive ecology. In *Reproductive Ecology and Human Evolution* (ed. by Ellison, P.), 321–350. New York: Aldine de Gruyter.

Wich, S. A., Utami-Atmoko, S. S., Setia, T. M., Rijksen, H. D., Schurmann, C. & van Schaik, C. 2004. Life history of wild Sumatran orangutans (*Pongo abelii*). *Journal of Human Evolution*, 47, 385–398.

Wright, P. C. 1990. Patterns of paternal care in primates. *International Journal of Primatology*, 11, 89–102.

———. 1995. Demography and life history of free-ranging *Propithecus diadema edwardsi* in Ranomafana National Park, Madagascar. *International Journal of Primatology*, 16, 835–854.

Chapter 11 Socialization and Development of Behavior

Elizabeth V. Lonsdorf and Stephen R. Ross

IN PRIMATES, early experiences have a profound impact on the behavior of offspring later in life. Compared to other mammals, primates experience long periods of prereproductive dependency. During this time, they acquire an astonishing array of skills, including those for finding food, avoiding predators, and, in group-living species, navigating complex social situations. Variability abounds in primate behavioral development. The social systems of primates range from relatively simple systems to the complex and ever-changing dynamics of fission-fusion groups (chapter 9, this volume). Their diets range from those of food specialists such as lorises, who have morphological adaptations to feed on gum, to those of chimpanzees (*Pan troglodytes*), which encompass hundreds of different species of plants, insects, and other animals.

The variety of socioecological niches occupied by primates results in widely differing infant care strategies. Strepsirrhines illustrate this variability with more than 100 species in seven families, including species that are nocturnal and diurnal, social and solitary, and monogamous and promiscuous (chapter 2, this volume). In general, the diurnal lemurs tend to be social and carry their infants, while nocturnal species are solitary and "park" or "nest" their young in protected locations while the adults forage (Kappeler 1998; Ross 2001). Lorises typically give birth to single offspring or twins, and may have more than one litter per year (Nekaris & Bearder 2007). Nocturnal lemurs show variation in litter size, while diurnal lemurs typically produce singletons (Gould & Sauther 2007). Most haplorrhines are diurnal and give birth to single offspring, which they carry

throughout the early stages of development rather than park or nest. An exception is the callitrichines, for which twinning is the norm (chapter 3, this volume). Ross (2003) and van Schaik and Isler (chapter 10, this volume) discuss how variation in life history influences morphological development and infant care strategies. Moreover, van Noordwijk (chapter 14, this volume) provides an overview of maternal investment, including challenges to primate mothers and strategies they employ to balance the competing demands of reproduction and offspring care. In this chapter, we take the perspective of offspring and review the factors that shape primate behavioral and social development.

Some stages of primate development are more easily defined than others. Infancy is relatively simple to classify as the stage that begins at birth and continues while offspring are nutritionally dependent on mothers. It ends when offspring can provide for themselves nutritionally, typically upon weaning, and can therefore survive the death of their mothers (Pereira & Altmann 1985; Setchell & Lee 2004). During weaning, major physical, social, and behavioral changes occur as infants make the transition to nutritional independence. Here we follow Pereira and Altmann (1985) and define juveniles as weaned individuals, which have not yet reached puberty or started to reproduce. Adolescence is harder to define precisely, but it involves the social and physical maturation necessary to become reproductively mature (Pereira & Altmann 1985). That is, by the end of adolescence, individuals have attained most if not all of their adult body size and are physically if not socially capable of reproduction (Setchell & Lee 2004). During or towards the end of adolescence, males and/or females may disperse from their natal groups.

Organizationally, we divide our discussion into four parts. We first outline theoretical explanations for the prolonged period of dependence in primates. Next, we review what is known about the influence of particular individuals or classes of individuals on the behavioral development of young primates. Third, we provide a summary of characteristic behaviors that develop during this period and theories regarding their acquisition. Fourth, we end with a brief review of dispersal.

Theoretical Explanations for a Prolonged Developmental Period

An extended period of juvenility and adolescence is the life history feature that best distinguishes the order Primates among mammals (Schultz 1956; Tanner 1962; Harvey & Clutton-Brock 1985; Pereira & Fairbanks; 1993; Kappeler et al. 2003). The length of the postweaning period is likely tied to several physical, social, and ecological factors (chapters 10 and 13, this volume). Kappeler and Pereira (2002) provide an extensive discussion of primate life histories and Leigh and Blomquist (2007) conduct a comprehensive comparative analysis with these data. All told, there appear to be some associations between variables historically reported as functionally critical, such as duration of gestation, neonatal period, weaning age, and life span (Pagel & Harvey 1993; Leigh & Blomquist 2007). Nonetheless, the prolonged period between weaning and sexual maturity across primate species has received considerable attention and is of particular relevance to understanding the nature and significance of behavioral development. Here we review various theoretical explanations for delayed maturation in primates, which include trade-offs between growth and reproduction, ecological risk aversion, and acquisition of foraging and social skills.

Growth Rate and Reproductive Trade-offs

Early explanations of mammalian development relied heavily on the ideas that growth laws and trends in physical development are the primary factors influencing the duration of the juvenile period. These ideas are more fully explored in chapter 10 of this volume, but to summarize, during the early period of development the biological apparatus necessary for adult reproductive activities is under construction, and the rate of growth and development is likely physiologically constrained by the upper limits of biological growth. This framework minimizes the evolutionary significance of the extended juvenile period in favor of physiological constraints. We might predict however, an evolutionary friction between the slow rate of growth and the fitness benefits accrued through an earlier start to reproduction. Pagel and Harvey (1993) provide a comprehensive review of the potential trade-offs between maximizing lifetime fecundity and growth rate constraints; they conclude that size-related variation in life histories are likely due to the effects of body size on adult mortality, and that delayed maturation arises automatically from natural selection for larger body sizes. Consideration of the relationships between body size, growth rates, and key life-history variables such as fecundity provides worthwhile insight into the evolution of extended juvenility, but below we consider additional hypotheses that are based more on the organism's behavioral development and learning of skills.

Ecological Risk Hypothesis

An ecological risk hypothesis has been proposed to explain the relatively extended juvenile period (Janson & van Schaik 1993; chapters 10 and 13, this volume). According to this hypothesis, slow growth in juveniles is an adaptive response to ecological risks such as predation and starvation. Because of their lower foraging efficiency, juveniles spend more time acquiring food than do adults. Juveniles are also more vulnerable to predators than are adults (Struthsaker 1973; Dittus 1977; Altmann 1980; Seyfarth & Cheney 1980, 1986; Robinson 1986; van Schaik & van Noordwijk 1989; see also chapter 8, this volume). Ongoing debate revolves around whether size, experience, or both contribute to foraging competence (see below). Body size effects may be especially significant in species that feed on food items that are particularly difficult to access or extract (Boesch & Boesch 1983; Terborgh 1983; Visalberghi & Neel 2003; Moura & Lee 2004; Visalberghi et al. 2005, 2009; Gunst et al. 2008), but less evident in species that use small or easily manipulated items (Boinski & Fragaszy 1989).

Though the ecological risk aversion hypothesis provides a way to understand the extended period of juvenility, it does not necessarily fit data across all primate taxa (Garber & Leigh 1997; Godfrey et al. 2004). This hypothesis assumes that juvenile primates do not forage as efficiently as adults, but this is not always the case (Fragaszy 1986; Hanya 2003). For instance, Stone (2007) found no evidence that juvenile common squirrel monkeys (*Saimiri sciureus*) sacrificed access to food for predator protection, and adults did not obtain preferential access to fruit patches. Thus, predation or starvation risk does not appear to influence slow growth and prolonged juvenility in common squirrel monkeys in the way predicted by the risk aversion hypothesis.

Needing-to-Learn Hypothesis

An alternate though not mutually exclusive hypothesis for explaining prolonged juvenility is the "needing-to-learn" hypothesis. This hypothesis proposes that, rather than being a way of avoiding starvation or predation, the extended developmental period shown by primates results from a need to acquire large amounts of information and skills. Experience can improve juvenile foraging skills through direct observation (Nicholson & Demment 1982; Terborgh 1983; Mineka et al. 1984; Boinski & Fragaszy 1989; Wiens & Zitzmann 2003; Lonsdorf et al. 2004), trial and error (Post 1984; Watts 1985; Whitehead 1986) and practice (Goodall 1986; Boinski & Fragaszy 1989), but the relationship between maturity and food acquisition proficiency remains equivocal. Gunst et al. (2008) demonstrated that wild brown capuchins (*Cebus apella*) achieved adult levels of extractive foraging proficiency several years after weaning, and at about the same time as they reached sexual maturity. However, many young primates master adult foraging techniques well before maturity (Hauser 1987; Boinski & Fragaszy 1989).

Social Skills and Competition

In addition to acquiring foraging expertise, young primates must learn several social skills during juvenility and adolescence so that they develop into socially and reproductively competent adults. This is not a simple task, given the concomitant reduction in maternal support after weaning. Young primates also experience changes in their relationships with other group members including possible fathers, other family members, and non-kin. Pereira and Fairbanks (1993) suggest that "cooperation and conflict in juvenile-adult social relations contribute fundamentally to the organization of primate groups and populations." In particular, increasing adult support for weanlings has been critical in permitting the evolutionary extension of juvenility among anthropoid primates (Pereira & Fairbanks 1993). In many primate societies, males and females often undergo rank changes around the time of puberty. Adolescents either move out of their natal group or integrate themselves into their present adult social world, but in both cases this is a transitory developmental period. For example, macaque (*Macaca* spp.) females eventually attain a rank just under their mothers, but during puberty, they may rank below other adults in lower-ranking matrilines (Berman 1982). Young long-tailed macaque (*Macaca fascicularis*) males

emigrate to new social groups in a variety of ways, including doing so "unobtrusively" and assuming a low rank, or "bluffing" their way in and taking over the highest rank (van Noordwijk & van Schaik 1985). Therefore, it is hypothesized that primates require this complex and transitory period of delayed maturation so that they can acquire the necessary social skills.

The extended period of development that is characteristic of primates has been a fertile area of study for several decades. While several theories have been proposed to explain it, the competing hypotheses are not necessarily mutually exclusive. It is likely that the effects of body size, social learning, competition, and ecological risk act in concert to affect primate development. As young primates continue to grow, the influence of external factors—namely those individuals with which they are most likely to interact—becomes increasingly important in shaping adult-level characteristics that will define their behavioral and reproductive competencies. We turn now to a discussion of the role of these social factors on primate development.

Filial and Other Social Relationships: Their Role in Offspring Development

The Mother-Infant Relationship

The care and environment provided by a primate mother exerts a powerful influence on offspring development. These influences begin in utero, as the primate fetus develops (see chapter 14, this volume), and can continue through adulthood via maternal presence and/or support. Periods of postnatal infant dependence can be as short as a few weeks, as in strepsirrhines, or as long as seven to eight years, as in great apes (see table 10.1 in chapter 10, this volume; fig. 11.1). During infancy the mother and maternal environment (both social and physical) provide the experiential background in which newborn primates first interact with the world and begin to acquire the information they need to grow up and survive. Therefore, suboptimal mothering can show lasting and negative impact throughout development and into adulthood. The precise nature of these effects is often difficult to measure, due to the necessity of intensive and long-term studies, but we summarize the data that exist below.

Mother-offspring relationships have been studied in the field and in captivity. In the late 1960s, Hinde and colleagues used rhesus monkeys (*Macaca mulatta*) as a model to investigate the relative roles of the mother and the infant in maintaining contact or close proximity to one another, and how those roles change through development (Hinde

Fig. 11.1. A western lowland gorilla mother cradles her newborn. She will be the primary source of nutritional support until weaning at three to five years of age, and will continue to provide social support for several years thereafter. Photo courtesy of the Lincoln Park Zoo.

Disruptions to this typical trajectory can have profound negative effects on offspring development. Considerable research demonstrates that the role the mother plays in providing "comfort" to her offspring is as important to normal behavioral and social development as the nutrition and protection she provides. Harlow (1974) conducted a series of pioneering experiments in which rhesus monkeys were separated from their mothers shortly after birth and subjected to surrogates that provided either comfort (a cloth figure) or food (a baby bottle attached to a wire figure). When given a choice between the two options, the infants overwhelmingly spent more time with the cloth surrogates. When cloth surrogates were unavailable, the infants showed afflictions such as diarrhea, which were interpreted to indicate psychological stress. Infants raised in total social isolation showed profound negative psychological consequences, suggesting that proper socialization at an early age was critical for normal primate behavioral development.

Since the 1950s, numerous studies have shown how maternal attachment and separation affect behavioral development in captive primates. Infants that experience early maternal separations display many physiological, developmental, and behavioral abnormalities, including neurobiological dysfunction (Ichise et al. 2006), increased fearfulness and anxiety, social and sexual dysfunction, and aberrant behaviors such as self-injurious behavior (Ljungberg & Westlund 2000; reviewed in Sanchez et al. 2001). Access to peers and other conspecifics may counteract some of the negative outcomes, but peer-reared individuals remain reluctant to explore novel situations, are highly reactive and impulsive, and often attain low dominance rank compared to mother-reared monkeys (Suomi 1997, 1999). In addition, peer-reared females typically show neglectful and abusive behavior towards their own infants when they become mothers, thus suggesting multigenerational effects of deficient mothering. In extreme cases of maternal separation (i.e., loss of the mother), orphaned infants in captivity and in the wild typically do not survive without supplementary care. Infants who lose their mothers after weaning also display lasting behavioral effects (chapter 14, this volume).

The juvenile period is characterized by one of the most significant shifts in social relations for a young primate, namely weaning. Primate mothers provide their infants with protection, nutrition, warmth, and transportation. During weaning and beyond, the mother-infant relationship radically changes to one involving considerably less maternal support and investment. These changes result in extensive alteration of the weanling's filial relationships, including increases in agonism reminiscent of classic parent-offspring conflict (Trivers 1974; reviewed in Fairbanks 2003). There is growing evidence that variation in the quality of care

& Spencer-Booth 1967; Hinde & Atkinson 1970). Several field and captive studies have since shown that most diurnal primate mother-infant interactions follow a similar developmental trajectory. In vervet monkeys (*Chlorocebus pygerythrus*), for example, mothers and infants remain in constant contact shortly after birth, with the former responsible for maintaining proximity to the latter (Fairbanks & McGuire 1987). As the infants mature, they break contact to explore the area immediately around their mothers. Contact continues to decline predictably as the infants age and begin to use their mothers as a "base" for exploration, making and breaking contacts regularly. The mothers soon shift toward actively promoting their offspring's independence by breaking contact and leaving them. Weaning conflict then begins as mothers reject their offspring's attempts to ride or nurse, and then the offspring display varying amounts of distress (see below).

given to infants influences maternal ability to invest in subsequent offspring (Fairbanks 1988; Fairbanks & McGuire 1995) and that weaning conflict affects maternal mating opportunities and the timing of conception (Berman et al. 1993). Infant behavior may also actively influence conflict. For instance, Gore (1986) showed that young long-tailed macaques interfered frequently with their mothers' mating attempts, which had an effect on the mothers' subsequent reproductive success. Alternatively, others argue that it would be more advantageous to mothers and offspring to agree over the allocation and timing of parental investment (Barrett & Henzi 2000), and to cooperate rather than compete. In practice, parent-offspring interactions are likely to involve both conflict and cooperation but there is little doubt that the period ripest for conflict is the time when the mother begins refusing to nurse her infant.

Variability among Mothers and Its Consequences

As described above, primate mothers critically influence infant development. Despite their importance, there is considerable variability in the quality of care they provide. Females have to balance their investment in providing care with several other demands on their energy, including feeding, giving birth to other offspring, cultivating social relationships, and competing for status within the social group. How mothers accomplish this balance is a fascinating area of research. We briefly highlight key findings on this topic below.

Stress, Environmental Variability, and Social Status

The endocrine system plays a significant role in maternal behavior, and stress hormones have been linked to maternal behavior and offspring development. Sources of stress for the mother can come from both the physical and the social environment, and they can exert influence prenatally or after parturition. For example, Schneider and Moore (2000) found that rhesus monkey infants born to mothers who had been repeatedly exposed to loud noises during pregnancy showed delayed motor development and reduced attention, along with negative social behaviors that continued through adolescence. Post-partum stress is also correlated with negative maternal behaviors in both nonhuman primates and humans. Southern pig-tailed macaque (*Macaca nemestrina*) females were more abusive towards their infants during stressful situations (Maestripieri 1994), and infant rejection correlated positively with postpartum maternal cortisol levels in Japanese macaques (*Macaca fuscata*) and rhesus monkeys (Bardi & Huffman 2005). Similarly, western gorilla (*Gorilla gorilla gorilla*) mothers who experienced considerable postpartum stress carried their infants less frequently in a ventral position and spent less time in ventro-ventral contact with them (Bahr et al. 1998). Common marmosets (*Callithrix jacchus*) treated with cortisol also show reduced infant carrying when compared to controls (Saltzman & Abbott 2009). Studies have also demonstrated that maternal care affects the subsequent stress levels of offspring (e.g., nonhuman primates, Dettling et al. 1998; humans, Hane & Fox 2006). For example, juvenile baboons (*Papio* spp.) reared by mothers that displayed high levels of stress-related behaviors such as scratching showed relatively high cortisol levels (Bardi et al. 2005). More recent work, however, suggests that inoculation to early stress is more important for the development of stress resistance in infants than the amount or quality of maternal care (Parker et al. 2006).

Environmental stress may be particularly important to primate mothers, and one well-studied example involves variation in resource availability. In a series of experiments, Rosenblum and colleagues manipulated the foraging regimes of bonnet macaque (*Macaca radiata*) mothers and found that mothers with unpredictable access to food became anxious, erratic, and less responsive (Rosenblum & Paully 1984). Infants raised in this condition show decreases in exploratory and social behavior, hyperactive stress responses, and neurological and immunological alterations (reviewed in Champagne & Curley 2009). Studies of humans have also found that children in "risky" families characterized by aggression, coldness, or parental extremes, often have an overreactive stress response (reviewed in Repetti et al. 2002).

Dominance rank relationships permeate the lives of many female primates (chapter 15, this volume) and variations in maternal social status have a significant impact on the development of offspring during infancy and beyond. In yellow baboons (*Papio cynocephalus*), high-ranking females produce relatively large juveniles (Altmann & Alberts 2005) and subadult males with low stress hormone concentrations (Onyango et al. 2008). In chimpanzees, daughters of high-ranking females give birth for the first time at a younger age and experience shorter birth intervals than do females born to low-ranking mothers (Pusey et al. 1997). The mechanisms underlying these differences are yet to be determined. One possibility is that high-ranking mothers feed their infants better because they have priority of access to food (Murray et al. 2006, 2007). Alternatively, high-ranking females may be able to provide improved maternal care because they experience relatively low levels of social stress.

Maternal Styles and the Quality of the Mother-Infant Relationship

Many studies of maternal behavior in primates have combined specific behaviors into broader maternal styles. In cercopithecine monkeys, most variation in maternal behavior falls along the two independent axes of rejection and protection—for example, in Japanese macaques (Schino et al. 1995; Bardi & Huffman 2002), vervet monkeys (Fairbanks & McGuire 1995), and southern pig-tailed macaques (Maestripieri 1998). The combination of these dimensions results in four different maternal styles: rejecting, controlling, laissez-faire, and protective (table 11.1). Most studies have focused on the factors that influence maternal styles (e.g., Fairbanks & McGuire 1993, 1995), rather than on the specific outcomes for offspring behavioral development, as prolonged periods of infancy and long life spans generally preclude gathering of the relevant data. Data that do exist indicate that infants that experience high rates of rejection by mothers become independent at an earlier age than protected individuals—for example, in Japanese macaques (Schino et al. 2001; Bardi & Huffman 2002). In vervet monkeys, maternal protectiveness correlates with increased offspring timidity when the offspring enter novel situations (Fairbanks & McGuire 1988), and infants raised by laissez-faire mothers were more exploratory and independent than those raised by mothers with different styles (Fairbanks 1996). In terms of social development, amicable mother-infant relationships in southern pig-tailed and bonnet macaques resulted in offspring that coped more effectively with later social challenges (Weaver et al. 2004).

Research on nonhuman primate maternal styles shows parallels with work on humans. A similar model has been proposed whereby maternal style varies along two independent axes of demandingness (the willingness to supervise children) and responsiveness (attunement to the child's needs) (Darling & Steinberg 1993). These dimensions combine into the following parenting styles: authoritarian, authoritative, indulgent, and neglecting (table 11.1). In humans, several studies have proposed that the authoritative style results in positive outcomes, producing children that are mature, independent, friendly, and have high self-esteem (e.g., Baumrind 1971). In contrast, emotional neglect and inconsistent discipline result in anxiety disorders (Holmes

& Robbins 1988), and children whose parents are rejecting and anxious tend to worry a lot (Muris et al. 2000).

How external factors, such as the environmental and social stressors described above, interact with maternal style and quality is an exciting area for current and future research. In vervet monkeys, low-ranking females face more threats and are more protective than high-ranking individuals, resulting in the former's infants spending considerable time in close physical contact with their mothers (Fairbanks 2000a). High-ranking mothers tend to be more rejecting and to have infants that spend relatively more time away from them and other conspecifics. Japanese macaque and rhesus monkey females that experience high peripartum cortisol levels and display low levels of maternal responsiveness give birth to infants that are relatively anxious but nonetheless independent with respect to exploration and locomotion (Bardi & Huffman 2005). These interactions may also have consequences for adult behavior, as mothers tend to be consistent in their style across offspring, and styles tend to run in families. For example, daughters adopt the mothering style of their mothers (Fairbanks 1989; Maestripieri et al. 1997, 2006). The question of whether these intergenerational similarities are due to heritable genetic predispositions, temperament, environmental influences, or social learning remains open and in need of additional research.

Paternal Care and Relationships with Adult Males

True or obligate paternal care, behavior that directly affects an infant's chances of survival, occurs in only a few primates. In some socially monogamous species, the combination of paternity certainty and relatively large infant size at birth results in fathers that provide direct paternal care (Dunbar 1988; Wright 1990). In several New World monkeys such as marmosets, tamarins, owl monkeys, and titis (fig. 11.2), males contribute substantial amounts of care in the form of carrying, grooming, and other supportive behaviors (chapter 3, this volume). They provide comfort to infants and serve as the primary attachment figures. For

Table 11.1. Comparison of nonhuman primate and human maternal styles

Nonhuman primate maternal styles			Human maternal styles		
	Protectiveness			Responsiveness	
Rejection	Low	High	Demandingness	Low	High
Low	Laissez-faire	Protective	Low	Neglecting	Indulgent
High	Rejection	Controlling	High	Authoritarian	Authoritative

Fig. 11.2. A young male titi monkey (*Callicebus donacophilus*) nestles on the back of an adult male. Unlike many primate species, titi monkey males contribute substantial amounts of parental care including carrying, grooming, and agonistic support. Photo courtesy of the Lincoln Park Zoo.

example, in Azara's owl monkeys (*Aotus azarae*), infants prefer to stay in social groups with male caretakers after their mothers have been evicted by new females (Fernandez-Duque 2004). In cotton-top tamarins (*Saguinus oedipus*), infants may seek their fathers, rather than their mothers, when confronted by aversive stimuli (Kostan & Snowdon 2002). Similarly, infant dusky titi monkeys (*Callicebus moloch*) react to separation from their fathers (but not their mothers) with a strong cortisol response (Hoffman 1998).

Paternal care is rare across most of the rest of the primate order. Among strepsirrhines, male mongoose (*Eulemur mongoz*) and red-bellied lemurs (*Eulemur rubriventer*) regularly carry infants (Patel 2007). Paternal behavior in the form of infant holding and grooming has also been reported for Coquerel's sifaka (*Propithecus coquereli*, Bastian & Brockman 2007). However, it is still unknown whether these behaviors contribute substantially to offspring fitness. In Old World monkeys, male yellow baboons preferentially support their own offspring in aggressive conflicts (Buchan et al. 2003). In these same baboons, the presence of fathers

appears to have a direct effect on offspring fitness; females whose fathers remain in their groups for long periods reach reproductive maturity earlier than do individuals whose fathers experience relatively short group tenures (Charpentier et al. 2008). Moreover, mother-infant pairs receive a direct benefit by associating with male "friends," who are the fathers of the infants in nearly half the cases and who protect them against harassment by other females (Nguyen et al. 2009). Siamangs (*Symphalangus syndactylus*) are the only apes that show direct paternal care, with fathers carrying infants most of the time after the first year of life (Chivers 1974). In summary, more research is needed to evaluate the extent and importance of paternal care in primates (see chapters 17 and 18, this volume). Questions ripe for future research relate to the previous discussion of variability in maternal care. Does suboptimal paternal care have long-term effects on offspring development? Are there deleterious consequences for offspring development if deficiencies in paternal comforting and attachment occur while infants are still properly nourished by mothers?

Male-Offspring Interactions

Outside of direct paternal care, adult males and infants interact in different ways which can be functionally positive, neutral, or negative from the standpoint of the offspring. Interactions include males tolerating the close proximity of infants, playing with and grooming them, and adopting infants after the loss of their mothers. In multimale primate groups, associations between adult males and infants usually do not match father-offspring relationships, probably because paternity is often uncertain (Snowdon & Suomi 1982; Smuts 1985), but recent data show some surprising exceptions to this rule (see Widdig 2007 for a review). In mandrills (*Mandrillus sphinx*), juveniles and their fathers affiliate more than do juveniles and unrelated males (Charpentier et al. 2007). In vervet monkeys, fathers interact differently with their juvenile offspring as a function of their rank, age, and sex; these interactions also depend on whether males are currently under challenge for troop leadership (Horrocks & Hunte 1993). Low-ranking juveniles receive less aggression from resident adult males than do higher-ranking peers, and the relationships that juveniles form with males affect their foraging and social success. For instance, low-ranking juveniles may wait for their fathers to arrive at a food source before feeding, to obtain protection against aggression by females in high-ranking matrilines (ibid.). Likewise, immature mountain and western lowland gorillas (*Gorilla gorilla*) are strongly attracted to dominant silverback males, and spend considerable time with them (Yamagiwa 1983; Stewart & Harcourt 1986). Orphaned immature gorillas will sometimes treat silverbacks as their primary caretakers following the loss of their mothers (Fossey 1979; Stewart 1981).

In macaques and baboons, adult males interact extensively with infants (Deag & Crook 1971; Packer 1980). Such interactions were first described in Barbary macaques (*Macaca sylvanus*) by Deag and Crook (1971), who hypothesized that males use infants as "agonistic buffers" to protect themselves against attack by their opponents (reviewed in Paul et al. 2000). Alternative explanations are that male-infant interactions are a form of direct paternal care (i.e., males interact with their own offspring) or a form of mating effort in which males attempt to gain favor with the infants' mother for future matings. Paul et al. (1996) used genetic data and behavioral observations to reject both the paternal and mating effort hypotheses in a captive colony of Barbary macaques. Observations by Menard et al. (2001) of wild Barbary macaques, however, support the mating effort hypothesis. More recent data suggest that males use infants as "social tools" in their relationships with other males (Henkel et al. 2010); those individuals that carried infants had stronger social relationships with other males than those

that did not. Intriguingly, Henkel et al. (2010) also found evidence that this behavior is potentially costly, since infant carriers displayed relatively high stress hormone levels. Menard and colleagues emphasize that the social, paternal, and reproductive explanations are not mutually exclusive and may interact with each other and additional ecological factors in complex ways. Apart from the motivations of males that carry infants, the immediate and long-term consequences of these interactions for infant behavioral and social development are likely to vary across species and situations and to require further study (Smuts 1985).

In sum, new observations give us a different and more complete picture of the extent and types of male-offspring interactions than was available 25 years ago (Nicholson 1987). Mackinnon (2007, see table 35.4) provides a recent and concise summary of the types of male-infant interactions observed across the Primate order. But despite the wide range of affiliative interactions now documented between young primates and adult males, their long-term effects on social and behavioral development of the immatures remain unclear.

Relationships with Others: Alloparents, Siblings, and Peers

Most primate infants grow up and socialize within complex groups. Mothers and, less often, fathers are significant influences during this period. However, other conspecifics also play important roles in the behavioral and social development of primates. These include individuals who perform actual caretaking behaviors (i.e. alloparents), as well as siblings and peers.

Alloparenting

Alloparenting is defined as nonmaternal care of infants, and it can be performed by relatives and nonrelatives alike. Allocare includes infant carrying and transport, guarding, food sharing, grooming, play, and nursing. Alloparenting occurs in many primate taxa, and it has been hypothesized to provide several potential benefits for parents, infants, and caretakers. These include (1) reduction of energetic costs for mothers, (2) increased fitness of the infants who are targets of care, and (3) developing the parenting skills of caretakers (chapters 14 and 15, this volume). In contrast, allocare may sometimes be costly, resulting in decreased survival of infants (Hrdy 1976). Infants targeted for allocare can vary on the basis of their mother's social status or the infant's own preferences (reviewed in Mackinnon 2007). An exhaustive review of the species, circumstances, and hypotheses for the evolution of alloparental care is outside the scope of this chapter. For more information we refer readers to chapter 14 in this volume, Hrdy (2009), and reviews by Mitani and

Watts (1997) and Ross and MacLarnon (2000) for anthropoid primates; to Chism (2000) for cercopithecines; and to Vasey (2007) and Patel (2007) for strepsirrhines. For the purposes of this chapter, we note that the effects of allocare on offspring social and behavioral development, apart from increased survival (Mitani & Watts 1997), have not been well documented (see also chapter 14, this volume).

Siblings

Siblings may play particularly important roles in infant behavioral development and socialization, as they are likely to be frequent social partners (fig. 11.3). Indeed, infants often play with and groom their siblings in many primate species (reviewed in Nicholson 1987). Few data exist, however, regarding how these interactions affect infant behavior. Suomi (1982) found that in the laboratory, infant rhesus monkeys reared with their siblings became independent from their mothers and developed a more complete behavioral repertoire at a younger age than did infants raised without siblings. This acceleration to independence has also been found in free-ranging rhesus monkeys (Berman 1982) and Japanese macaques (Hiraiwa 1981). In a field study on wild chimpanzees, infants with siblings displayed accelerated independence by interacting with their mothers less and with other group members more than did infants without siblings (Brent et al. 1997). In general, there is a paucity of data on how siblings interact with each other, and on the consequences of these interactions for infant development.

Peers

The social network concept developed by Hinde (1976) proposes that every individual in a social group plays a specific role in the socialization of infants. For an infant, the social network typically starts with close kin and develops

Fig. 11.3. A mother rhesus macaque (*Macaca mulatta*) holds her infant and his older sibling. Both the mother and the sibling will play an important role in the offspring's behavioral development. Photo courtesy of Stephen R. Ross.

along kin lines. Berman et al. (1997) found that infant social networks preferentially included related individuals even when demographic changes in the group occurred resulting in more unrelated individuals with whom to interact. How the frequency, form, and quality of peer relationships influence infant development has not been investigated extensively, although some studies suggest that these relationships are important for appropriate socialization. Laboratory studies found that contact with peers, even for as little as 20 minutes per day, reduced some of the negative effects for infants of being reared without their mothers in bonnet (Boccia et al. 1997) and southern pig-tailed macaques (Sackett 1982; Worlein & Sackett 1997). Recent research by Kemps et al. (2008) expands on these findings by comparing rhesus monkeys reared with mothers and no other social partners and those reared in natural social groups. In this study, the subadult females reared only with their mothers displayed several social incompetencies, including higher rates of submissive and stereotypic behaviors and inappropriate responses to nonthreatening situations. These results suggest that peer relationships play an important role in the development of appropriate social behavior.

As described above, primate youngsters experience a relatively long period of dependency and may have multiple social influences during this time. The presence and quality of these relationships contribute to their behavioral and social maturation. We now turn our attention to some of the characteristic behaviors acquired during ontogeny and what is known about their developmental trajectory.

Development of Characteristic Behaviors

To maximize their reproductive potential, young primates must develop specific skills, including the abilities to forage and to detect predators. Young primates must also develop skills necessary to maintain social relationships and status. The mechanisms by which these skills are acquired and the extent to which they are learned remain an important area of research. Several studies have been conducted to unravel specific types of learning, with the underlying assumption that this provides a window into the cognitive abilities of the subjects. In the last decade, research on social learning has exploded (chapter 31, this volume). The most basic and inclusive definition of social learning is that "individual B learns some or part of the behavior from individual A." This is presumed to draw on more complicated cognitive processes than individual trial-and-error learning (Whiten & Ham 1992). In addition, social learning is argued to be key to the development and maintenance of animal "traditions" or "culture," although the existence of culture in animals remains a matter of great debate (chapter 31, this volume). Most research has used experiments in captivity, since carrying out longitudinal developmental studies in the wild is typically precluded by time and cost. However, as more long-term field studies progress, new findings in this area continually come to light. As we detail the development of specific behaviors below, the influence of others on offspring learning will be described where appropriate.

Not all young primates will develop the same skills in the same way, due to differences in social systems, ecological influences, and maternal factors (e.g., dominance rank). In addition, sexual selection theory predicts that sex differences in behavior will exist, and that behavioral development will proceed along sex-specific trajectories, with females showing a higher frequency of behaviors associated with caretaking (e.g., affiliation) and males showing a greater propensity for behaviors associated with male-male competition (e.g., aggression). The relative contributions of physiological factors, such as differences in the prenatal hormonal environment (Wallen 1996) versus post-natal influences, such as social factors, remain an area of considerable interest and debate. Thus far, it is not clear whether mothers or others socialize male and female infants differently. Most studies have shown few if any differences in the behavior of mothers towards sons versus daughters (e.g., macaques, Brown & Dixon 2000; vervets, Fairbanks 1996; chimpanzees, Lonsdorf 2006; northern lesser galagos, *Galago senegalensis*, Nash 2003) despite the theoretical assertion that mothers should invest more in the sex that will deliver higher lifetime reproductive success through grandchildren (chapter 14, this volume). Throughout our discussion below, we will specify sex differences in development where they have been reported.

Foraging

Primate infants begin their lives with the simplest of diets: milk from their mothers. The period during which youngsters nurse exclusively varies depending on a combination of life history (chapter 10, this volume) and maternal (chapter 14, this volume) factors, but the proportion of the diet made up by milk gradually reduces until weaning. During this time and depending on the species, a youngster may acquire foraging skills for hundreds of different types of plants and animals (chapter 7, this volume). The variety of primate diets and the potentially difficult processing necessitated by some foods have greatly influenced hypotheses regarding the prolonged development of primates, including those invoking ecological risks and the need to learn (see above).

A young primate's first interaction with solid food often occurs when infants ingest parts of food items during cofeeding events with the primary caregiver (e.g., Japanese macaques, Ueno 2005; mountain gorillas, *Gorilla gorilla beringei*, Watts 1985). Caregivers may also share food with the infants and tolerate their scrounging for food items. Such sharing may benefit the sharer by providing the infant with nutrients that may speed the time to weaning (the nutritional hypothesis) or may allow the infant to learn about the food's palatability or processing (the informational hypothesis) (reviewed in Brown et al. 2004). These hypotheses are not mutually exclusive and are likely both at work to promote the evolution of food sharing and the development of foraging behaviors. Typically, adult tolerance for offspring scrounging decreases with age as the youngster gradually acquires the adult diet.

Synchronicity between adults and offspring feeding suggests an important role for social learning in terms of which foods are edible or preferred, and how to access food items that need processing. In a recent review, Rapaport and Brown (2008) summarize and describe the potential role of social learning in the development of foraging across the Primate order. For strepsirrhines, very little is known about the role of social versus individual-based learning in the ontogeny of foraging. Among New World monkeys, examples abound (Rapaport & Brown 2008), but perhaps the most intriguing is that of the complex stone tool use exhibited by wild capuchins (*Cebus* spp.) while foraging on nuts. There is good evidence that social learning plays a key role in the acquisition of this behavior, leading to different foraging traditions displayed by different capuchin populations (reviewed in Ottoni & Izar 2008). Among the apes, several studies of chimpanzees have now focused on how young develop complex, tool-assisted foraging techniques. These include studies of nut cracking (reviewed in Biro et al. 2003), termite fishing (Lonsdorf 2005), and ant dipping (Humle et al. 2009), all of which present strong evidence that social learning is critical for the development of these behaviors (see fig. 11.4). Similar evidence is beginning to come to light in Bornean orangutans (*Pongo pygmaeus*) as well (Jaeggi et al. 2010). Finally, there are some data for black-horned capuchins (*Cebus nigritus*, Agostini & Visalberghi 2005) and chimpanzees (Lonsdorf 2005) that suggest that youngsters exhibit sex differences in the development of foraging skills consistent with the differences in diet eventually shown by adults of the species.

Antipredator Behavior

Young primates begin life being carried full-time by their primary caregivers or parked in relatively safe locations. As they attain locomotor independence, they must be able

Fig. 11.4. An adult chimpanzee uses stones to crack nuts at Bossou, Guinea, while a youngster observes closely. Complex foraging behaviors such as stone tool use suggest an important role for social learning in the development of foraging. Photo courtesy of Tetsuro Matsuzawa.

to detect and avoid predators (chapter 8, this volume). Predation risk is arguably one of the most important selective pressures driving the evolution of group living (van Schaik & van Hooff 1983), and most of what we know about the ontogeny of antipredator behavior comes from group-living diurnal primates. In contrast, relatively little is known about the acquisition of predator avoidance behaviors in solitary and nocturnal taxa. In the past, crypsis has been presumed to be the main antipredator strategy in these species (Terborgh & Janson 1986), but more recent research has described complex strategies such as mobbing and alarm calling (Fichtel 2007; chapter 8, this volume). It remains to be determined whether the development of these behaviors differs from diurnal species.

Research on the ontogeny of antipredator behavior has focused largely on the production, comprehension, and response to alarm calls given by conspecifics. In a pioneering study, Seyfarth and Cheney (1980, 1986) used observations and playback experiments to study the development of alarm calls in vervet monkeys, which produce predator-specific calls that elicit appropriate escape responses. For example, a "leopard" call causes group members to run up into trees, while a "snake" call elicits bipedal scanning. Youngsters develop these skills gradually, with adult-like production and responsiveness occurring after four years. Prior to this age, infants call to a variety of nonpredators, showing error rates approaching 80%, and display maladaptive responses, such as bipedal scanning in response to a "leopard" call. Additional studies have since been conducted on chacma baboons (*Papio ursinus*, Fischer et al.

2000), common squirrel monkeys (McCowan et al. 2001), and Verreaux's sifakas (*Propithecus verreauxi*, Fitchel 2008); all describe young animals that gradually develop the ability to produce and respond to alarm calls adaptively. In vervet monkeys there appears to be a strong social learning component; adults respond more strongly to a correct call produced by infants, and infants tend to respond appropriately to a call if they look at an adult first.

Play Behavior

Social play is a distinctive activity displayed by all primate taxa (Fagen 1993), with high rates exhibited during late infancy and early juvenility (fig. 11.5). Play steadily declines during the late juvenile period and through adolescence to adulthood (e.g., see Hinde & Spencer-Booth 1967; Goodall 1968; Chivers 1974; Pereira 1984; Hayaki 1985; Caine 1986; Fagen 1993). Fagen (1993) suggested that age-related differences in the costs and benefits of play may explain these shifts. For instance, declining maternal investment in weanlings could amplify the relatively high energetic costs of play. Play is not only energetically costly but also incurs risks such as injury (e.g., falling during arboreal play) and increased vulnerability to predation (Baldwin 1986; Sacks et al. 1989; Fagen 1993; Pellis & Iwaniuk 1999). Biben et al. (1989) experimentally examined these risks in Bolivian squirrel monkeys (*Saimiri boliviensis*) and demonstrated that during play, adults become more vigilant and young alert them by emitting calls.

The benefits that individuals derive from play remain unresolved (Burghardt 2005). The idea that play helps develop social and motor skills needed in adult life has been the most commonly cited explanation (Bekoff 1974, 1988, 1989; Fagen 1976, 1981; Symons 1978; Smith 1982), and recent studies have extended this idea. The "training for the unexpected" theory proposes that play allows individuals to perform behavioral sequences under safe conditions in which animals purposefully lose control and learn improvisational tactics (Spinka et al. 2001). Petru et al. (2008) reported that Hanuman langurs (*Semnopithecus entellus*) engage in head-rotational movements that temporarily disorient the individual and may serve to create unexpected situations during play. Likewise, juvenile chimpanzees seek relatively safe environments in which to "practice" losing control by initiating play with younger individuals and using handicapping movements to signal that they will not use their full strength (Mendoza-Granados & Sommer 1995).

Byers and Walker (1995) point out that for some nonprimate species, high rates of play occur during periods when cerebellar synaptogenesis and muscle-fiber differentiation are most malleable, thus suggesting a functional associa-

Fig. 11.5. A juvenile female and juvenile male mangabey (*Lophocebus albigena*) of Kibale, Uganda, play boisterously with one another. The benefits of play for youngsters are still not entirely clear, but they may be significant in light of the empirical evidence for the costs incurred by play behavior. Photo courtesy of Rebecca Chancellor.

tion. Fairbanks (2000b) provides data consistent with this idea: in vervet monkeys, various types of play peak when neural plasticity is maximal in brain regions associated with play. Together, these results support the hypothesis that play functions to promote motor coordination, fighting ability, and food-handling skills through its effects on neural and physical development. Thus, play may occur frequently at young ages when underlying brain structures and skeletal and muscular systems are still developing and can be permanently affected.

In general, juvenile males tend to play more than juvenile females; this matches what has been found in infancy (Fagen 1993), with some notable exceptions (baboons, Owens 1975, Pereira 1984; chimpanzees, Hayaki 1985; bonobos, *Pan paniscus*, Becker 1984; common marmosets, Stevenson & Poole 1982; cotton-top tamarins, Cleveland & Snowdon 1984). In addition to amount of play, the type of play differs between males and females. For example, Brown & Dixson (2000) studied infant rhesus monkeys in the first six months of life and found that males initiated play and engaged in rough-and-tumble, stationary, and chasing play more than females. Males also exhibit more rough-and-tumble play in patas monkeys (*Erythrocebus patas*), Japanese macaques, and olive baboons (*Papio anubis*) (reviewed in Brown & Dixson 2000). Conversely, female macaques engage in more solitary play than males across all ages (Deag & Cook 1971; Ehardt & Bernstein 1987), and show early interest in infants that peaks in juvenility and adolescence. Sex differences in play parenting (e.g., attempting to carry other infants) are also widespread, as female infants show intense interest in young infants and begin trying to interact with and carry them even before they are weaned themselves (e.g., rhesus monkeys, Lovejoy & Wallen 1988). In chimpanzees, an intriguing new study from one wild community provides evidence that young female chimpanzees play with sticks in a manner similar to human doll play, while young males' stick use resembles the toy weapon play seen in human boys (Kahlenberg & Wrangham 2010). As mentioned above, these sex differences in play are likely to reflect patterns of mating competition and parental investment.

Grooming

Grooming is a characteristic primate behavior thought to function as a social tool to develop, reinforce, or restore social bonds (see chapters 23 and 24, this volume). In most species, the amount of time mothers spend grooming their infants decreases as the latter ages, mirroring a reduction in parental investment. As a result, offspring change from passively receiving maternal grooming in infancy to actively soliciting it after weaning. However, mothers groom their adult offspring at high rates in many species, which suggests that grooming represents a form of maternal investment when offspring are young, but then shifts to be used as a social commodity or mechanism to maintain relationships when offspring are mature (Muroyama 1995).

Sex differences in the development of grooming behavior have been extensively studied in rhesus monkeys (reviewed in Roney & Maestripieri 2003) and appear to reflect the differing social roles of males and females. Rhesus monkey society is based around stable groups of related females organized in matrilines whose individuals display strong bonds. In contrast, males disperse from their natal groups (Melnick & Pearl 1987). Female infants groom others more than males in the first year of life, and grooming continues to develop along sex-specific social lines through puberty and early adulthood. That is, young females groom their female kin the most, and male youngsters groom other adult males the most. Similar sex biases and social trajectories in the development of grooming have been found in other species. For example, in golden snub-nosed monkeys (*Rhinopithecus roxellana*, Wang et al. 2007) and baboons (Young et al. 1982), female youngsters consistently groomed more than did males. In Japanese macaques (Eaton et al. 1986) and chimpanzees (Nishida 1988; Pusey 1990), the increase in grooming shown by juvenile females is specifically directed at the mother; male offspring groom more frequently outside the maternal unit. As pointed out by Roney & Maestripieri (2003), sex differences in grooming by infants may reflect and be adapted to the particular social structure of the species, whereas sex differences in rough play and play parenting are likely to be a more widespread primate or mammalian adaptation, reflecting the ubiquity of male competition and parental care by females.

Aggression, Rank, and Reconciliation

As young primates spend more time away from their mothers and increase their interactions with peers and adults, conflict arises as weanlings deal with aggression and acquire their status in the group. During aggressive play, young primates use and inhibit aggression, learning to whom they can direct aggression and dominance behavior and from whom they might expect it. Whether early rates of aggression correlate with adult rates has seldom been investigated. Studies of captive rhesus monkeys showed that impulsive aggression early in life predicted aggressive tendencies as adults (Westergaard 2003; Howell et al. 2007).

While infants are rarely attacked by conspecifics (Horrocks & Hunte 1983; Bernstein & Ehardt 1985; but see chapter 19, this volume), juveniles are often involved in aggressive interactions (Silk et al. 1981; Bernstein & Ehardt 1985; Pereira 1988a, b; Pusey 1990). Juvenile rhesus monkeys show heightened rates of aggression, with female aggression leveling off during juvenility and male aggression continuing to increase through adolescence. Adolescent male rhesus monkeys are the instigators and recipients of the highest rates of aggression of all age classes (Bernstein & Ehardt 1985). In addition, they are most likely to respond to threats and to receive mobbing and prolonged agonism than are others (ibid.). Adolescent males also actively form alliances with kin and high-ranking non-kin. Although males will become dominant to females when they reach adult size, maternal rank strongly influences agonistic encounters involving immatures, especially between females. In mountain gorillas, the severity of aggressive encounters increases as immatures grow older. Silverbacks rarely attack juveniles, but direct aggression to adolescent males occurs at rates higher than among females (Watts & Pusey 1993). Demography also influences the expression of adolescent agonism. Ross et al. (2009) found that groups of captive chimpanzees that contained more than one adult male, as is typical in the wild, quelled social disturbances involving adolescent males more effectively than groups with only one adult male.

The relationships between reproduction, social rank, and the onset of puberty continue to be debated in most primates, especially for females. Some studies of macaques have shown earlier ages of first reproduction in dominant females (Drickamer 1974; Sade 1976; Paul & Thommen 1984; Bercovitch & Berard 1993; Pusey et al. 1997), but in Japanese and stump-tailed macaques (*Macaca arctoides*) there appears to be no effect of rank on age of first ovulation or parturition (Gouzoules et al. 1982; Nieuwenhuijsen et al. 1985). Males follow a more consistent pattern across taxa in which rank is correlated with testicular descent and volume as well as with testosterone levels (Glick 1979; Bercovitch 1993; Alberts & Altmann 1995a; Dixson & Nevison 1997; Mann et al. 1998).

Because juveniles are so frequently involved in agonistic encounters, coping strategies such as appeasement, redirection, and reconciliation may be particularly important for young primates to learn quickly. The development of conflict resolution in nonhuman primates remains understudied, but in some species these skills are acquired at a relatively young age (long-tailed macaques, Cords 1988, Cords & Aureli 1993; rhesus monkeys, Judge et al. 1997; Japanese macaques, Schino et al. 1998; brown capuchins, Weaver & de Waal 2000). Although conflict between adults and juveniles is less often reconciled than conflict between members of the same age class (Japanese macaques, Schino et al. 1998; Javan grizzled leaf monkeys, *Presbytis comata*, Arnold & Barton 2001; chimpanzees, Arnold & Whiten 2001), young primates alter their conflict-resolution behavior on the basis of previous interactions with others (Cords & Thurnheer 1993) and by observing third parties (de Waal & Johanowicz 1993). In an analysis of reconciliation among juvenile long-tailed macaques, Cords & Aureli (1993) found that juveniles develop the ability to reconcile at rates similar to those of adult victims of aggression, and that this allows juvenile victims to restore tolerance and individual tension levels quickly to baseline levels. No sex differences in reconciliation were found overall, but when conflicts with unrelated adult females were examined, juvenile females reconciled more than juvenile males, which may be due to individuals in some dyads having a mutual interest in restoring relations. A second sex difference involved males reconciling with same-sex juvenile peers more than females. In a similar analysis of juvenile rhesus monkey reconciliation, de Waal and Luttrell (1986) concluded that sex differences resulted from females reconciling selectively with particular opponents.

Sexual Behavior

At puberty, hormonal changes facilitate reproductive capabilities, with associated growth spurts in most species (Leigh 1992, 1995; cf. Bogin 1999). While physiological changes make reproduction possible, sexual maturity may still be years away behaviorally. The timing of puberty is likely strongly associated with several factors, including resource availability. Young females raised in ecologically harsh conditions may delay puberty, which in turn affects juvenile mortality and ultimately reproductive fitness (Cheney et al. 1988; Rubenstein 1993). Some behaviors do not solely affect reproductive function but are also involved in dominance interactions. For instance, presenting and mounting by juveniles are a component of dominance behavior, rather than being strictly sexual in function. Furthermore, many sex differences in juvenile behavior are related to attaining social competence in adulthood. In a study of wild bonobos, sexual behavior occurred in almost all age-sex combinations, but most genital contacts between juveniles and mature males occurred during agonistic interactions to reduce tension (Hashimoto 1997). Males steadily increased the frequency of their sexual contacts through juvenility, including with mature females. Upon reaching adolescence, however, the frequency of this behavior dropped drastically, as it often resulted in aggression from adult males.

Male and female primates show differences in their interest in sexual behavior even before weaning. Male infants display more mounting behavior than females in several species of baboons and macaques (reviewed in Brown & Dixson 2000). Males, and to a lesser extent females, harass mating pairs: this has been hypothesized to be a form of parent-offspring conflict (Niemeyer & Anderson 1983). Alternatively, harassment may enable subordinate weanlings to approach and threaten dominant males when they are unlikely to retaliate effectively. Sexual interest grows through adolescence in many species, and both males and females engage in sexual behavior long before they are reproductively viable. In some macaques, juvenile males direct most of their sexual activity to their mothers, but this behavior becomes less frequent as they age (Missakian 1973; Hanby & Brown 1974). Adolescent male yellow baboons increasingly target estrous females as sexual partners and are able to copulate with them successfully (Hausfater 1975). Juvenile female baboons show low levels of sexual behavior prior to puberty, but their interest increases rapidly in adolescence as they solicit adult males, who typically show little interest in them until they have completed several cycles (Altmann et al. 1977; Scott 1984).

In some callitrichines and Old World monkeys, adolescent sexual behavior is suppressed by the presence of adult conspecifics (Manson 1993). While behavioral suppression occurs in some species (e.g., Koford 1963; Chaoui & Hasler-Gallusser 1999), physiological suppression takes place in some callitrichines as well (e.g., Snowdon et al. 1993). In male orangutans (*Pongo* spp.), the timing of development of secondary sexual characteristics (e.g., cheek flanges) may occur immediately following adolescence or only after many years (Wich et al. 2004). This arrested development was originally thought to be a product of chronic stress imposed by resident males, but additional data suggest that it might be due to males adopting an alternative reproductive strategy (Maggioncalda et al. 1999, 2002; chapter 6, this volume).

Adolescents mate more often than juveniles, but less often than adults (Hanby & Brown 1974). In most primate species, dominant males do not tolerate sexual behavior by adolescents, and as such, adolescents may use surreptitious or "sneaky" strategies to mate (Dunbar 1984; Ohsawa et al. 1993; Berard et al. 1994; Manson 1996; Soltis et al. 2001; Setchell 2003). When females achieve sexual maturity, their behavior becomes more overt. In some species that display external signs of sexual receptivity, adolescent females possess exaggerated sexual swellings in comparison to adult females (Anderson & Bielert 1994). This may relate to the fact that adolescent females undergo a period of adolescent sterility during which they experience several anovulatory menstrual cycles (Hartman 1931; Smith & Rubenstein 1940; Resko et al. 1982) or have insufficient progesterone (Foster 1977). During this prereproductive period, adult males may be less attracted to them, though there are differing opinions about whether the size of the swelling affects female fitness (Domb & Pagel 2001; Zinner et al. 2002).

Early studies found that adolescent male primates learned intromission behavior, and that with experience, they were more successful at achieving intromission with fewer mount attempts (Michael & Wilson 1973; Wolfe 1978). Before adolescent males were able to achieve intromissions, they mounted male and female partners in approximately equal proportions; but after experiencing intromission, they almost always restricted their mounting to females (Wallen 2000, 2001). This suggests that males learn the rewarding stimulus properties of intromission, which in turn promotes the selection of proper mating partners.

Dispersal

One of the final behavioral patterns on the path to independence is dispersal: the process whereby individuals leave their natal groups. This has proven to be a difficult phenomenon to study, because field primatologists do not normally record observations of individuals after they leave their study groups. Most data on dispersal are derived from observations of individuals that immigrate into known groups, or of those that don't disperse at all. In many primate species, members of one sex remain in their natal group throughout their lives while members of the other sex emigrate at sexual maturity. Male-biased dispersal occurs in many social strepsirrhines and in several Old World monkeys, with the exception of hamadryas baboons (*Papio hamadryas*) and some colobines. Female-biased dispersal occurs in chimpanzees, bonobos, and some atelines. In New World monkeys, both male-biased and female-biased dispersal occur, with members of both sexes dispersing occasionally (see Strier 2008 for a review). When both males and females disperse (e.g., most callitrichines, ursine howler monkeys [*Alouatta arctoidea*], mantled howler monkeys [*Alouatta palliata*], gorillas, hamadryas baboons, and some populations of red colobus [*Procolobus* spp.]), individuals of one sex may move farther away from their natal groups than do members of the other sex (Pusey & Packer 1987). Following dispersal, some primate males may spend time as solitary floaters or join all-male groups before taking over

or creating new social groups that include females (e.g., mountain gorillas, Robbins 1995; gray-cheeked mangabeys, *Lophocebus albigena*, Oluput & Waser 2001).

While there is considerable variety in the forms of primate dispersal, the timing of dispersal also varies. In most species, male emigration occurs at puberty, although hamadryas baboons leave their natal group before puberty and savannah baboons, sakis, and titis leave several years after puberty (Bossuyt 2002). Female emigration usually occurs following the female's first sexual cycle, although there are exceptions to this as well (e.g., howler monkeys, hamadryas baboons). In species where females disperse, the females have few opportunities to interact with mature kin, and matrilineal bonds are less likely to develop. Socially monogamous owl monkeys disperse at various ages before reaching full sexual maturity, but never before they are 26 months old (Juarez et al. 2003; Fernandez-Duque 2004). Some stay in their natal groups until over five years of age (Fernandez-Duque & Huntington 2002). Finally, dispersal is not limited to transfers from the natal group in adolescence or early adulthood. Individuals may also exhibit secondary dispersal by joining another group after dispersing for the first time. In these cases they rarely return to their natal group (Drickamer & Vessey 1973; Sugiyama 1976; Packer & Pusey 1979; Cheney 1983; Jack & Fedigan 2004).

Natal dispersal involves several costs. Individuals spend considerable time traveling and incur ecological and social costs that exacerbate the stress of leaving familiar group mates (Alberts & Altmann 1995b). Ursine howler females suffer high mortality rates while traveling between social groups (Crockett & Pope 1993). Individuals sometimes gain a competitive advantage by dispersing with others. For example, multiple females are more easily integrated into groups of common squirrel monkeys and white-faced capuchin (*Cebus capucinus*) monkeys (Mitchell 1994; Jack & Fedigan 2004). Likewise, males may be more successful integrating into groups that contain familiar individuals who have dispersed earlier (Cheney & Seyfarth 1983; van Noordwijk & van Schaik 1985; Fuentes 2000; Bradley et al, 2004).

Why do primates leave their natal group and forego the potential benefits of long-term access to kin who can provide social support? One hypothesis suggests that individuals disperse to avoid the deleterious consequences of inbreeding (Pusey & Packer 1987). An alternate hypothesis argues that primates disperse to increase their mating opportunities (Moore 1992). A third hypothesis proposes that females disperse after group takeovers to avoid potentially infanticidal males (Jack & Fedigan 2009; chapter 19, this volume). All of these hypotheses are ultimate, functional explanations of dispersal. Others have proposed proximate explanations for dispersal that invoke ecological and social factors. Here, individuals might disperse because local habitats are saturated (Sterck 1997, 1998) or because of unfavorable sex ratios (Alberts & Altmann 1995; Sugiyama 1999; Hohmann 2001; Jack & Fedigan 2004; Strier et al. 2006). In the latter case, unbalanced sex ratios may serve as the proximate cue that causes the dispersal that results in increased mating opportunities. Alternatively, other social cues, such as the prolonged presence of males in groups, may cause young females to disperse, and to thus avoid inbreeding with their fathers (Clutton-Brock 1989). Clearly the forms and functions of dispersal are variable and require further study. For the primates themselves, the process serves as the final bridge between behavioral development and adult independence.

Summary and Future Directions

Primates exhibit an extended period of juvenility and adolescence in comparison to other mammals. During this time they acquire the skills and relationships they need to survive and grow into a competent adult of their species. Several theories exist as to why the developmental period is so long in primates, including those based on biological constraints and advantages in terms of learning and behavioral development. In most maturing primates, the mother is the primary caregiver and the single most important influence; deficits in maternal care or ability can have long-lasting negative effects on infants. The roles of fathers, peers, and siblings are less well understood, especially with regard to how they influence the social and behavioral development of youngsters. The final step on the path to independence is often dispersal, when a primate leaves its natal group as it reaches adulthood.

Nearly every behavior displayed by an adult primate is shaped by early experiences that may include the interaction of several genetic, parental, ecological, and social factors. However, despite the importance of understanding the ontogenetic trajectory of behaviors such as predator avoidance and foraging, the study of primate development in the wild has lagged behind studies of adult primate behavior. This is especially true in understanding adult outcomes as they relate to an individual's developmental history. The reasons for this are obvious: the long period of prereproductive dependence necessitates a researcher's substantial long-term commitment to follow individuals' developmental trajectory from infancy to adulthood. Much has been and will continue to be learned in the laboratory, but long-term field studies are also needed to fully understand the complex influences on primate development.

Some specific issues present opportunities to expand our knowledge of primate development and evolution. First, studies of the development of nocturnal primates, while exceedingly difficult to carry out, will fill a major gap in our understanding of the evolution and complexity of the primate order. Second, the mother-infant relationship continues to be studied in terms of the consequences for development of offspring, but the roles of others (e.g., males, siblings, peers) remain understudied. Future research might focus on whether infants who experience secure relationships with males attain high rank in adulthood, or whether interactions with adult males are particularly important for learning male-specific behaviors, such as those involved in dominance or mating competition. Finally, the use of new technologies to sequence DNA will allow us to make strides in understanding kin discrimination, male-offspring interactions, and how genes interact with the environment to affect the expression of maternal behavior. In sum, the combination of long-term field data, cross-sectional and multigenerational studies in captivity, and new technologies portends a rich and fascinating field of study for decades to come.

References

Alberts, S. C. & Altmann, J. 1995a. Preparation and activation: Determinants of age at reproductive maturity in male baboons. *Behavioral Ecology and Sociobiology*, 36, 397–406.

———. 1995b. Balancing costs and opportunities: Dispersal in male baboons. *American Naturalist*, 145, 279–306.

Agostini, I. & Visalberghi, E. 2005. Social influences on the acquisition of sex-typical foraging patterns by juveniles in a group of wild tufted capuchin monkeys (*Cebus nigritus*). *American Journal of Primatology*, 65, 335–351.

Altmann, J. 1980. *Baboon Mothers and Infants*. Cambridge, MA: Harvard University Press.

Altmann, J. & Alberts, S. C. 2005. Growth rates in a wild primate population: Ecological influences and maternal effects. *Behavioral Ecology and Sociobiology*, 57, 490–501.

Altmann, J., Altmann, S., Hausfater, G. & McCuskey, S. A. 1977. Life history of yellow baboons: Physical development, reproductive parameters, and infant mortality. *Primates*, 18, 315–330.

Anderson, C. M. & Bielert, C. F. 1994. Adolescent exaggeration in female catarrhine primates. *Primates*, 35, 283–300.

Arnold, K. & Barton, R. A. 2001. Postconflict behavior of spectacled leaf monkeys (*Trachypithecus obscurus*). I. Reconciliation. *International Journal of Primatology*, 22, 243–266.

Arnold, K. & Whiten, A. 2001. Post-conflict behaviour of wild chimpanzees (*Pan troglodytes schweinfurthii*) in the Budongo Forest, Uganda. *Behaviour*, 138, 649–690.

Bahr N. I., Pryce, C. R. P., Döbeli, M. & Martin, R. D. M. 1998. Evidence from urinary cortisol that maternal behavior is related to stress in gorillas. *Physiology and Behavior*, 64, 429–437.

Baldwin, J. D. 1986. Behavior in infancy: Exploration and play.

In *Comparative Primate Biology*, Volume 2; Part A: *Behavior, Conservation and Ecology* (ed. by G. Mitchell & J. Ervin), 295–326. New York: Liss.

Bardi, M. & Huffman, M. A. 2002. Effects of maternal style on infant behavior in Japanese macaques (*Macaca fuscata*). *Developmental Psychobiology*, 41, 364–372.

———. 2005. Maternal behavior and maternal stress are associated with infant behavioral development in macaques. *Developmental Psychobiology*, 48, 1–9.

Bardi, M., Bode, A. E., Ramirez, S. M. & Brent, L. Y. 2005. Maternal care and development of stress responses in baboons. *American Journal of Primatology*, 66, 263–278.

Barrett, L. & Henzi, S. P. 2000. Are baboon infants Sir Philip Sydney's offspring? *Ethology*, 106, 645–658.

Bastian, M. L. & Brockman, D. K. 2007. Paternal care in *Propithecus verreauxi coquereli*. *International Journal of Primatology*, 28, 305–313.

Baumrind, D. 1971. Current patterns of parental authority. *Developmental Psychology Monographs*, 4, 1–103.

Becker, C. 1984. *Orang-Utans und Bonobos im Spiel*. Munich: Profil Verlag.

Bekoff, M. 1974. Social play and play-soliciting by infant canids. *American Zoologist*, 14, 323–340.

———. 1988. Motor training and physical fitness: Possible short- and long-term influences on the development of individual differences. *Developmental Psycholobiology*, 21, 601–612.

———. 1989. Social play and physical training: When "not enough" may be plenty. *Ethology*, 80, 330–333.

Berard, J. D., Nunberg, P., Epplen, J. T. & Schmidtke, J. 1994. Alternative reproductive tactics and reproductive success in male rhesus macaques. *Behaviour*, 129, 177–201.

Bercovitch, F. B. 1993. Dominance rank and reproductive maturation in male rhesus macaques (*Macaca mulatta*). *Journal of Reproduction and Fertility*, 99, 113–120.

Bercovitch, F. B. & Berard, J. D. 1993. Life history costs and consequences of rapid reproductive maturation in female rhesus macaques. *Behavioral Ecology and Sociobiology*, 32, 103–109.

Berman, C. M. 1982. The ontogeny of social relationships with group companions among free-ranging infant rhesus monkeys, 1: Social networks and differentiation. *Animal Behaviour*, 28, 860–873.

Berman, C. M., Rasmussen, K. L. R. & Suomi, S. J. 1993. Reproductive consequences of maternal care patterns during estrus among free-ranging rhesus monkeys. *Behavioral Ecology and Sociobiology*, 32, 391–399.

———. 1997. Group size, infant development and social networks in free-ranging rhesus monkeys. *Animal Behaviour*, 53, 405–421.

Bernstein, I. S. & Ehardt, C. L. 1985. Agonistic aiding: Kinship, rank, age and sex influences. *American Journal of Primatology*, 8, 37–52.

Biben, M., Symmes, D. & Bernhards, D. 1989. Vigilance during play in squirrel monkeys. *American Journal of Primatology*, 17, 41–49.

Biro, D., Inoue-Nakamura, N., Tonooka, R., Yamakoshi, G., Sousa, C. & Matsuzawa, T. 2003. Cultural innovation and transmission of tool use in wild chimpanzees: Evidence from field experiments. *Animal Cognition*, 6, 213–223.

Boccia, M. L., Scanlan, J. M., Laudenslager, M. L., Berger, C. L., Hijazi, A. S. & Reite, M. L. 1997. Juvenile friends, behavior, and immature responses to separation in bonnet macaque infants. *Physiology and Behavior*, 61, 191–198.

Boesch, C. & Boesch, H. 1983. Optimisation of nut-cracking with natural hammers by wild chimpanzees. *Behaviour*, 83, 265–286.

Bogin, B. 1999. Evolutionary perspective on human growth. *Annual Review of Anthropology*, 28, 109–153.

Boinski, S. & Fragaszy, D. M. 1989. The ontogeny of foraging in squirrel monkeys, *Saimiri oestedi*. *Animal Behaviour*, 37, 415–428.

Bossuyt, F. 2002. Natal dispersal of titi monkeys (*Callicebus moloch*) at Cocha Cashu, Manu National Park, Peru. *American Journal of Physical Anthropology*, 34, 47.

Bradley B. J., Doran-Sheehy, D. M., Lukas, D., Boesch, C. & Vigilant, L. 2004. Dispersed male networks in western gorillas. *Current Biology*, 14, 510–513.

Brent, L., Bramblett, C. A., Bard, K. A., Bloomsmith, M. A. & Blangero, J. 1997. The influence of siblings on wild chimpanzee social interaction. *Behaviour*, 134, 1189–1210.

Brown, G. R., Almond, R. E. A. & van Bergen, Y. 2004. Begging, stealing and offering: Food transfer in non-human primates. *Advances in the Study of Behavior*, 34, 265–295.

Brown, G. R. & Dixson, A. F. 2000. The development of behavioural sex differences in infant rhesus macaques (*Macaca mulatta*). *Primates*, 41, 63–77.

Buchan, J. C., Alberts, S. C., Silk, J. B. & Altmann, J. 2003. True paternal care in a multi-male primate society. *Nature*, 425, 179–181.

Burghardt, G. M. 2005. *The Genesis of Animal Play*. Cambridge: MIT Press.

Byers, J. A. & Walker, C. 1995. Refining the motor training hypothesis for the evolution of play. *American Naturalist*, 146, 25–40.

Caine, N. G. 1986. Behavior during puberty and adolescence. In *Comparative Primate Biology*, Volume 2; Part A: *Behavior, Conservation, and Ecology* (ed. by G. Mitchell & J. Ervin), 327–362. New York: Liss.

Champagne, F. A., & Curley, J. P. 2009. Epigenetic mechanisms mediating the long-term effects of maternal care on development. *Neuroscience and Biobehavioral Reviews*, 33, 593–600.

Chaoui, N. J. & Hasler-Gallusser, S. 1999. Incomplete sexual suppression in *Leontopithecus chrysomelas*: A behavioural and hormonal study in a semi-natural environment. *Folia Primatologica*, 70, 47–54.

Charpentier, M. J. E., Peignot, P., Hossaert-McKey, M. & Wickings, J. 2007. Kin discrimination in juvenile mandrills, Mandrillus sphinx. *Animal Behaviour*, 73, 37–45.

Charpentier, M. J. E., van Horn, R. C., Altmann, J. & Alberts, S. C. 2008. Paternal effects on offspring survival in a multi-male primate society. *Proceedings of the National Academy of Sciences*, 105, 1988–1992.

Cheney, D. L. 1983. Extra-familial alliances among vervet monkeys. In *Primate Social Relationships. An Integrated Approach* (ed. by R. A. Hinde), 278–285. Oxford: Blackwell.

Cheney D. L. & Seyfarth, R. M. 1983. Non-random dispersal in free-ranging vervet monkeys: Social and genetic consequences. *American Naturalist*, 122, 392–412.

Cheney, D. L., Seyfarth, R. M., Andelman, S. J. & Lee, P. C. 1988. Reproductive success in vervet monkeys. In *Reproductive success* (ed. by T. H. Clutton-Brock), 384–402. Chicago: University of Chicago Press.

Chism, J. 2000. Allocare patterns among cercopithecines. *Folia primatologia*, 71, 55–66.

Chivers, D. J. 1974. The siamang in Malaya: A field study of a primate in tropical rain forest. *Contributions to Primatology*, Volume 4. Basel: Karger.

Cleveland, J. & Snowdon, C. T. 1984. Social development during the first twenty weeks in the cotton-top tamarin (*Saguinus oedipus*). *Animal Behaviour*, 32, 432–444.

Clutton-Brock T. H. 1989. Female transfer and inbreeding avoidance in social mammals. *Nature*, 337, 70–72.

Cords, M. 1988. Resolution of aggressive conflicts by immature long-tailed macaques, *Macaca fascicularis*. *Animal Behaviour*, 36, 1124–1135.

Cords, M. & Aureli, F. 1993. Patterns of reconciliation among juvenile long-tailed macaques. In *Juvenile Primates: Life History, Development, and Behavior* (ed. by M. E. Periera & L. A. Fairbanks), 271–284. New York: Oxford University Press.

Cords, M. & Thurnheer, S. 1993. Reconciliation with valuable partners by long-tailed macaques. *Ethology*, 93, 315–325.

Crockett, C. M. & Pope T. R. 1993. Consequences of sex differences in dispersal for juvenile red howler monkeys. In *Juvenile Primates: Life History, Development and Behavior* (ed. by M.E. Pereira & Fairbanks), 104–118. New York: Oxford Press.

Darling, N. & Steinberg, L. 1993. Parenting styles as context: An integrative model. *Psychological Bulletin*, 113, 487–496.

De Waal, F. B. M. 1986. The integration of dominance and social bonding in primates. *Quarterly Review of Biology*, 61, 459–479.

De Waal, F. B. M. & Johanowicz, D. L. 1993. Modification of reconciliation behavior through social experience: An experiment with two macaque species. *Child Development*, 64, 897–908.

De Waal, F. & Luttrell, L. 1986. The similarity principle underlying social bonding among female rhesus monkeys. *Folia Primatologica*, 46, 215–234.

Deag, J. M. & Crook, J. H. 1971. Social behaviour and "agonistic buffering" in the wild barbary macaque *Macaca sylvana*. *Folia Primatologica*, 15, 183–200.

Dettling, A., Pryce, C. R., Martin, R. D. & Dobeli, M. 1998. Physiological responses to parental separation and a strange situation are related to parental care received in Goeldi's monkeys (*Callimico goeldii*). *Developmental Psychobiology*, 33, 21–31.

Dittus, W. P. J. 1977. The social regulation of population density and age-sex distribution in the toque monkey. *Behaviour*, 63, 281–322.

Dixson, A. F. & Nevison, C. M. 1997. The socioendocrinology of adolescent development in male rhesus monkeys (*Macaca mulatta*). *Hormones and Behavior*, 31, 126–135.

Domb, L.G. & Pagel, M. 2001. Sexual swellings advertise female quality in wild baboons. *Nature*, 410, 204–206.

Drickamer, L. C. 1974. Social rank, observability and sexual behavior of male rhesus monkeys (*Macaca mulatta*). *Journal of Reproduction and Fertility*, 37, 117–120.

Drickhamer, L. C. & Vessey, S. H. 1973. Group changing in free-ranging male rhesus monkeys. *Primates*, 14, 359–368.

Dunbar, R. I. M. 1984. Infant-use by male gelada in agonistic contexts: Agonistic buffering, progeny protection or soliciting support? *Primates*, 25, 485–506.

———. 1988. *Primate Social Systems*. Ithaca, NY: Comstock Publishing Associates.

Eaton, G. C., Johnson, D. F., Glick, B. B. & Worlein, J. M. 1986. Japanese macaques (*Macaca fuscata*) social development: Sex difference in juvenile behavior. *Primates*, 24, 141–150.

Ehardt, C. L. & Bernstein, I. S. 1987. Patterns of affiliation among immature rhesus monkeys (*Macaca mulatta*). *American Journal of Primatology*, 13, 255–269.

Fagen, R. M. 1976. Exercise, play, and physical training in animals. In *Perspectives in Ethology*, Vol. 2 (ed. by P. P. G. Bateson & P. H. Klopfer), 188–219. New York: Plenum.

Fagen, R. 1981. *Animal Play Behavior*. New York: Oxford University Press.

———. 1993. Primate juveniles and primate play. In *Juvenile Primates: Life History, Development, and Behavior* (ed. by M. E. Pereira & L. A. Fairbanks), 183–196. New York: Oxford University Press.

Fairbanks, L. A. 1988. Mother-infant behavior in vervet monkeys: Response to failure of last pregnancy. *Behavioral Ecology and Sociobiology*, 23, 157–165.

———. 1989. Early experience and cross-generational continuity of mother infant contact in vervet monkeys. *Developmental Psychobiology*, 22, 669–681.

———. 1993. Juvenile vervet monkeys: Establishing relationships and practicing skills for the future. In *Juvenile Primates: Life History, Development and Behavior* (ed. by M. E. Pereira & L. A. Fairbanks), 211–227. New York: Oxford University Press.

———. 1996. Individual differences in maternal style. *Advances in the Study of Behavior*, 25, 579–611.

———. 2000a. Dominance in vervets: Social, reproductive and developmental correlates of rank. *American Journal of Primatology Supplement*, 51, 34.

———. 2000b. The developmental timing of primate play: A neural selection model. In *Biology, Brains, and Behavior* (ed. by S. T. Parker, J. L. Langer & M. McKinney), 31–158. Santa Fe: SAR Press.

Fairbanks, L. A. & McGuire, M. T. 1987. Mother-infant relationships in vervet monkeys: Response to new adult males. *International Journal of Primatology*, 8, 351–366.

———. 1988. Long-term effects of early mothering behavior on responsiveness to the environment in vervet monkeys. *Developmental Psychobiology*, 21, 711–724.

———. 1993. Maternal protectiveness and response to the unfamiliar in vervet monkeys. *American Journal of Primatology*, 30, 119–129.

———. 1995. Maternal condition and the quality of maternal care in vervet monkeys. *Behaviour*, 132, 733–754.

Fernandez-Duque, E. 2004. High levels of intrasexual competition in sexually monomorphic owl monkeys. *Folia Primatologica*, 75 S1, 260.

Fernandez-Duque, E. & Huntington, C. 2002. Disappearances of individuals from social groups have implications for understanding natal dispersal in monogamous owl monkeys (*Aotus azarai*). *American Journal of Primatology*, 57, 219–225.

Fichtel, C. 2007. Avoiding predators at night: Antipredator strategies in red-tailed sportive lemurs (*Lepilemur ruficaudatus*). *American Journal of Primatology*, 69, 611–624.

———. 2008. Ontogeny of conspecific and heterospecific alarm call recognition in wild Verreaux's sifakas (*Propithecus verreauxi verreauxi*). *American Journal of Primatology*, 70, 127–135.

Fischer, J., Cheney, D. L. & Seyfarth, R. M. 2000. Development of infant baboons' responses to graded bark variants. *Proceedings of the Royal Society B*, 267, 2317–2321.

Fossey, D. 1979. Development of the mountain gorilla (*Gorilla gorilla beringei*): The first thirty-six months. In *The Great Apes* (ed. by D. A. Hamburg & E. R. McCown), 39–186. Menlo Park, CA: Benjamin-Cummings.

Foster, D. L. 1977. Luteinizing hormone and progesterone secretion during sexual maturation of the rhesus monkey: Short luteal phases during the initial menstrual cycles. *Biology of Reproduction*, 17, 584–590.

Fragaszy, D. M. 1986. Time budgets and foraging behavior in wedge-capped capuchins (*Cebus olivaceus*): Age and sex differences. In *Current Perspectives in Primate Social Dynamics* (ed. by D. Taub & F. King), 159–174. New York: Van Nostrand Reinhold.

Fuentes, A. 2000. Hylobatid communities: Changing views on pair bonding and social organization in hominoids. *Yearbook of Physical Anthropology*, 43: 33–60.

Garber, P. & Leigh, S. R. 1997. Ontogenetic variation in small-bodied New World primates: Implications for patterns of reproduction and infant care. *Folia Primatologica*, 86, 1–22.

Glick, B. B. 1979. Testicular size, testosterone level, and body weight in male *Macaca radiata*: Maturational and seasonal effects. *Folia Primatologica*, 32, 268–289.

Godfrey, L. R., Samonds, K. E., Jungers, W. L., Sutherland, M. R. & Irwin, M. T. 2004. Ontogenetic correlates of diet in Malagasy lemurs. *American Journal of Physical Anthropology*, 123, 250–276.

Goodall, J. 1968. The behaviour of free-living chimpanzees in the Gombe Stream Reserve. *Animal Behavior Monographs*, 1, 165–311.

———. 1986. *The Chimpanzees of Gombe: Patterns of Behavior.* Cambridge, MA: Harvard University Press.

Gore, M. A. 1986. Mother-offspring conflict and interference at mother's mating in *Macaca fascicularis*. *Primates*, 27, 205–214.

Gould, L. & Sauther, M. 2007. Lemuriformes. In *Primates in Perspective* (ed. by C. J. Campbell, A. Fuentes, K. C. Mackinnon, M. Panger & S. K. Bearder), 46–72. Oxford University Press.

Gouzoules, H., Gouzoules, S. & Fedigan, L. 1982. Behavioural dominance and reproductive success in female Japanese monkeys (*Macaca fuscata*). *Animal Behaviour*, 30, 1138–1151.

Gunst, N., Boinski, S. & Fragaszy, D. M. 2008. Acquisition of foraging competence in wild brown capuchins (*Cebus apella*), with special reference to conspecifics' foraging artefacts as an indirect social influence. *Behaviour*, 145, 195–229.

Hanby, J. P. & Brown, C. E. 1974. The development of sociosexual behaviours in Japanese macaques *Macaca fuscata*. *Behaviour*, 49, 152–96.

Hane, A. A. & Fox, N. A. 2006. Ordinary variations in maternal care giving influence human infants' stress reactivity. *Psychological Science*, 17, 550–556.

Hanya, G. 2003. Age differences in food intake and dietary selection of wild male Japanese macaques. *Primates*, 44, 333–339.

Harlow, H. F. 1974. *Learning to Love.* New York: Jason Aronson.

Hartman, C. G. 1931. On the relative sterility of the adolescent organism. *Science*, 74, 226–227.

Harvey, P. H. & Clutton-Brock, T. H. 1985. Life history variation in primates. *Evolution*, 39, 559–581.

Hashimoto, C. 1997. Context and development of sexual behavior of wild bonobos (*Pan paniscus*) at Wamba, Zaire. *International Journal of Primatology*, 18, 1–21.

Hauser, M. D. 1987. The behavioral ecology of free-ranging vervet monkeys: Proximate and ultimate levels of explanation. PhD dissertation, University of California, Los Angeles.

Hausfater, G. 1975. Dominance and reproduction in baboons (*Papio cynocephalus*). *Contributions to Primatology*, 7, 1–150.

Hayaki, H. 1985. Social play of juvenile and adolescent chimpanzees in the Mahale Mountains National Park, Tanzania. *Primates*, 26, 342–360.

Henkel, S., Heistermann, M. & Fischer, J. 2010. Infants as costly social tools in male Barbary macaque networks. *Animal Behaviour*, 79, 1199–1204.

Hinde, R. A. & Spencer-Booth, Y. 1967. The behaviour of socially living rhesus monkeys in their first two and a half years. *Animal Behaviour*, 15, 169–196.

Hinde, R. A. 1976. Development of social behavior. In *Behavior of Nonhuman Primates*, Vol. 2 (ed. by A. M. Schrier & F. Stollnitz), 287–334. New York: Academic Press.

Hinde, R. A. & Atkinson, S. 1970. Assessing the roles of social partners in maintaining mutual proximity, as exemplified by mother-infant relations in rhesus monkeys. *Animal Behaviour*, 18, 169–176.

Hiraiwa, M. 1981. Maternal and alloparental care in a troop of free-ranging Japanese monkeys. *Primates*, 22, 309–329.

Hoffman, K. A. 1998. Transition from juvenile to adult stages of development in titi monkeys (*Callicebus moloch*). PhD dissertation, University of California, Davis.

Hohmann, G. 2001. Association and social interactions between strangers and residents in bonobos (*Pan paniscus*). *Primates*, 42, 91–99.

Holmes, S. J. & Robins, L. N. 1988. The role of parental disciplinary practices in the development of depression and alcoholism. *Psychiatry*, 51, 24–35.

Horrocks, J. & Hunte, W. 1983. Maternal rank and offspring rank in vervet monkeys: an appraisal of the mechanisms of rank acquisition. *Animal Behaviour*, 31, 772–782.

Horrocks, J. & Hunte, W. 1993. Interactions between juveniles and adult males in vervets: Implications for adult male turnover. In *Juvenile Primates: Life History, Development, and Behavior* (ed. by M. E. Pereira & L. A. Fairbanks), 228–239. New York: Oxford University Press.

Howell, S., Westergaard, G., Hoos, B., Chavanne, T. J., Shoaf, S. E., Cleveland, A., Snoy, P. J., Suomi, S. J. & Higley, J. D. 2007. Serotonergic influences on life-history outcomes in free-ranging male rhesus macaques. *American Journal of Primatology*, 69, 851–865.

Hrdy, S. 1976. The care and exploitation of nonhuman primate infants by conspecifics other than the mother. In *Advances in the Study of Behavior*, Volume 6 (ed by. J. S. Rosenblatt, R. Hinde, E. Shaw & C. Beer), 101–158. New York: Academic Press.

———. 2009. *Mothers and Others: The Evolutionary Origins of Mutual Understanding.* Cambridge, MA: Harvard University Press.

Humle, T., Snowdown, C. T. & Matsuzawa, T. 2009. Social influences on ant-dipping acquisition in the wild chimpanzees

(*Pan troglodytes verus*) of Bossou, Guinea, West Africa. *Animal Cognition*, 12 Suppl. 1, S37–S48.

Ichise, M., Vines, D. C., Gura, T., Anderson, G. M., Suomi, S.J., Higley, J. D. & Innis, R. B. 2006. Effects of early life stress on [11C] DASB positron emission tomography imaging of serotonin transporters in adolescent peer- and mother-reared rhesus monkeys. *The Journal of Neuroscience*, 26, 4638–4643.

Jack, K. & Fedigan, L. M. 2004. Male dispersal patterns in white-faced capuchins (*Cebus capucinus*). Part 1: Patterns and causes of natal emigration. *Animal Behaviour*, 67, 761–769.

———. 2009. Female dispersal in a female-philopatric species, *Cebus capucinus*. *Behaviour*, 146, 471–497.

Jaeggi, A. V., Dunkel, L. P., van Noordwijk, M. A., Wich, S. A., Sura, A. A. L. & van Schaik, C. P. 2010. Social learning of diet and foraging skills by wild immature Bornean orangutans: Implications for culture. *American Journal of Primatology*, 72, 62–71.

Janson, C. H. & van Schaik, C. P. 1993. Ecological risk aversion in juvenile primates: Slow and steady wins the race. In *Juvenile Primates: Life History, Development, and Behavior* (ed. by M. E. Pereira & L. A. Fairbanks), 57–74. New York: Oxford University Press.

Juárez, C. P., Fernández-Duque, E. & Rotundo, M. 2003. Behavioral sex differences in the socially monogamous night monkeys of the Argentinean Chaco. *Revista de Etología*, 5, 174.

Judge, P. G., Bernstein, I. S. & Ruehlmann, T. E. 1997. Reconciliation and other post-conflict behavior in juvenile rhesus macaques (*Macaca mulatta*). *American Journal of Primatology*, 42, 120.

Kahlenberg, S. & Wrangham, R. 2010. Sex differences in chimpanzees' use of sticks as play objects resemble those of children. *Current Biology*, 20, R1067–R1068.

Kappeler, P. M., & Pereira, M. E. 2002. *Primate Life Histories and Socioecology*. Chicago: University of Chicago Press.

Kappeler, P. M., Pereira, M. E. & van Schaik, C. P. 2003. Primate life histories and socioecology. In *Primate Life History and Socioecology* (ed. by P. M. Kappeler & M. E. Pereira), 1–24. Chicago: University of Chicago Press.

Kempes, M. M., Gulickx, M. M. C., van Daalen, H. J. C., Louwerse, A. L. & Sterck, E. H. M. 2008. Social competence is reduced in socially deprived rhesus monkeys (*Macaca mulatta*). *Journal of Comparative Psychology*, 122, 62–67.

Koford, C. B. 1963. Rank of mothers and sons in bands of rhesus monkeys. *Science*, 141, 356–357.

Kostan, K. M. & Snowdon, C. T. 2002. Attachment and social preferences in cooperatively-reared cotton-top tamarins. *American Journal of Primatology*, 57, 131–139.

Leigh, S. R. 1992. Patterns of variation in the ontogeny of primate body size dimorphism. *Journal of Human Evolution*, 23, 27–50.

———. 1995. Socioecology and the ontogeny of sexual size dimorphism in anthropoid primates. *American Journal of Physical Anthropology*, 97, 339–356.

Leigh, S. R. & Blomquist, G. E. 2007. Life history. In *Primates in Perspective* (ed. by C. Campbell, A. Fuentes, K. C. MacKinnon, M. Panger & S. Bearder), 396–407. New York: Oxford University Press.

Ljungberg, T. & Westlund, K. 2000. Impaired reconciliation in rhesus macaques with a history of early weaning and disturbed socialization. *Primates*, 41, 79–88.

Lonsdorf, E. V. 2005. Sex differences in the development of termite-fishing skills in wild chimpanzees (*Pan troglodytes schweinfurthii*) of Gombe National Park, Tanzania. *Animal Behaviour*, 70, 673–683.

———. 2006. What is the role of the mother in the acquisition of tool-use skills in wild chimpanzees? *Animal Cognition*, 9, 36–46.

Lonsdorf, E. V., Pusey, A. E. & Eberly, L. 2004. Sex differences in learning in chimpanzees. *Nature*, 428, 715–716.

Lovejoy, J. & Wallen, K. 1988. Sexually dimorphic behavior in group-housed rhesus monkeys (*Macaca mulatta*) at 1 year of age. *Psychobiology*, 16, 348–356.

Mackinnon, K. C. 2007. Social beginnings: The tapestry of infant and adult interactions. In *Primates in Perspective* (ed. by C. J. Campbell, A. Fuentes, K. C. Mackinnon, M. Panger, & S. K. Bearder), 571–591. Oxford University Press.

Maestripieri, D. 1994. Infant abuse associated with psychosocial stress in a group-living pigtail macaque (*Macaca nemestrina*) mother. *American Journal of Primatology*, 32, 41–49.

———. 1998. Social and demographic influences on mothering style in pigtail macaques. *Ethology*, 104, 379–385.

Maestripieri, D., Wallen, K. & Carroll, K. A. 1997. Infant abuse runs in families of group-living pigtail macaques. *Child Abuse and Neglect*, 21, 465–471.

Maestripieri, D., Higley, J. D., Lindell, S. G., Newman, T. K., McCormack, K. M. & Sanchez, M. M. 2006. Early maternal rejection affects the development of monoaminergic systems and adult abusive parenting in rhesus macaques (*Macaca mulatta*). *Behavioral Neuroscience*, 120, 1017–1024.

Maggioncalda, A. N., Sapolsky, R. M. & Czekala, N. M. 1999. Reproductive hormone profiles in captive male orangutans: Implications for understanding developmental arrest. *American Journal of Physical Anthropology*, 109, 19–32.

Maggioncalda, A. N., Czekala, N. M. & Sapolsky, R. M. 2002. Male orangutan subadulthood: A new twist on the relationship between chronic stress and developmental arrest. *American Journal of Physical Anthropology*, 118, 25–32.

Mann, D. R., Akinbami, M. A., Gould, K. G., Paul, K. & and Wallen, K. 1998. Sexual maturation in male rhesus monkeys: Importance of neonatal testosterone exposure and social rank. *Journal of Endocrinology*, 156, 493–501.

Manson, J. H. 1993. Sons of low-ranking female rhesus macaques can attain high dominance rank in their natal groups. *Primates*, 34, 285–288.

———. 1996. Male dominance and mount series duration in Cayo Santiago rhesus macaques. *Animal Behaviour*, 51, 1219–1231.

McCowan, B., Franceschini, N. V. & Vicino, G. A. 2001. Age differences and developmental trends in alarm peep responses by squirrel monkeys (*Saimiri sciureus*). *American Journal of Primatology*, 53, 19–31.

Melnick, D. J. & Pearl, M. C. 1987. Cercopithecines in multimale groups: Genetic diversity and population structure. In *Primates Societies* (ed. by B. B. Smuts, D. L. Cheney, R. M. Seyfarth, R. W. Wrangham & T. T. Struhsaker), 121–134. Chicago: University of Chicago Press.

Menard, N., von Segesser, F., Scheffrahn, W., Pastorini, J., Vallet, D., Gaci, B., Martin, R. D. & Gautier-Hion, A. 2001. Is male-infant caretaking related to paternity and/or mating activities in wild Barbary macaques (*Macaca sylvanus*)?

Comptes Rendus de l'Academie des Sciences- Series III- Sciences de la Vie, 324, 601–610.

Mendoza-Granados, D. & Sommer, V. 1995. Play in chimpanzees of the Arnhem Zoo: Self-serving compromises. *Primates*, 36, 57–68.

Michael, R. P. & Wilson, M. 1973. Changes in the sexual behaviour of male rhesus monkeys *(Macaca mulatta)* at puberty: Comparisons with the behavior of adults. *Folia Primatologica*, 19, 384–403.

Mineka, S., Davidson, M., Cook, M. & Keir, R. 1984. Observational conditioning of snake fear in rhesus monkeys. *Journal of Abnormal Psychology*, 93, 355–372.

Missakian, E. A. 1973. The timing of fission among free-ranging rhesus monkeys. *American Journal of Physical Anthropology*, 38, 621–624.

Mitani, J. C. & Watts, D. 1997. The evolution of non-maternal caretaking among anthropoid primates: Do helpers help? *Behavioral Ecology and Sociobiology*, 40, 213–220.

Mitchell, C. L. 1994. Migration alliances and coalitions among adult male South American squirrel monkeys *(Saimiri sciureus)*. *Behaviour*, 130, 169–190.

Moore, J. 1992. Dispersal, nepotism and primate social behavior. *International Journal of Primatology*, 13, 361–378.

Moura, A. C. A. & Lee, P. 2004. Capuchin stone tool use in Caatinga dry forest. *Science*, 306, 1909.

Muris, P., Merckelbach, H., Gadet, B. & Moulaert, V. 2000. Fears, worries, and scary dreams in 4- to 12-year-old children: Their content, developmental pattern, and origins. *Journal of Clinical Child Psychology*, 29, 43–52.

Muroyama, Y. 1995. Developmental changes in mother-offspring grooming in Japanese macaques. *American Journal of Primatology*, 37, 57–64.

Murray, C. M., Eberly, L. E., & Pusey, A. E. 2006. Foraging strategies as a function of season and dominance rank among wild female chimpanzees *(Pan troglodytes schweinfurthii)*. *Behavioral Ecology*, 17, 1020–1028.

Murray, C. M., Mane, S. V. & Pusey, A. E. 2007. Dominance rank influences female chimpanzee *(Pan troglodytes schweinfurthii)* space use: Towards an ideal despotic distribution. *Animal Behaviour*, 74, 1795–1804.

Nash, L. T. 2003. Sex differences in the behavior and the social interactions of immature *Galago senegalensis braccatus*. *Folia Primatologia*, 74, 285–300.

Nekaris, A. & Bearder, S. K. 2007. The lorisiform primates of Asia and mainland Africa. In *Primates in Perspective* (ed. by C. J. Campbell, A. Fuentes, K. C. Mackinnon, M. Panger, S. K. Bearder), 24–45. Oxford: Oxford University Press.

Nguyen, N., van Horn, R. C., Alberts, S. C. & Altmann, J. 2009. "Friendships" between new mothers and adult males: Adaptive benefits and determinants in wild baboons *(Papio cynocephalus)*. *Behavioral Ecology and Sociobiology*, 63, 1331–1344.

Nicholson, N. 1987. Infants, mothers and other females. In *Primate Societies* (ed. by B. B. Smuts, D. L. Cheney, R. M. Seyfarth, R. W. Wrangham & T. T. Struhsaker), 330–342. Chicago: University of Chicago Press.

Nicholson, N. & Demment, M. W. 1982. The transition from suckling to independent feeding in wild baboon infants. *International Journal of Primatology*, 3, 318.

Niemeyer, C. & Anderson, J. 1983. Primate harassment of matings. *Ethology and Sociobiology*, 4, 205–220.

Nieuwenhuijsen, K., Lammers, A. J. J. C., de Neef, K. J. & Slob, A. K. 1985. Reproduction and social rank in female stumptail macaques *(Macaca arctoides)*. *International Journal of Primatology*, 6, 77–99.

Nishida, T. 1988. Development of social grooming between mother and offspring in wild chimpanzees. *Folia Primatologica*, 50, 109–123.

Ohsawa, H., Inoue, M. & Takenaka, O. 1993. Mating strategy and reproductive success of male patas monkeys *(Erythrocebus patas)*. *Primates*, 34, 533–44.

Oluput, W. & Waser, P.M. 2001. Correlates of intergroup transfer in male grey-cheeked mangabeys. *International Journal of Primatology*, 22, 169–187.

Onyango, P. O., Gesquiere, L. R., Wango, E. O., Alberts, S. C. & Altmann, J. 2008. Persistence of maternal effects in baboons: Mother's dominance rank at son's conception predicts stress hormone levels in subadult males. *Hormones and Behavior*, 54, 319–324.

Ottoni, E. B. & Izar, P. 2008. Capuchin monkey tool use: Overview and implications. *Evolutionary Anthropology*, 17, 171–178.

Owens, N. W. 1975. Social play behaviour in free-living baboons, *Papio anubis*. *Animal Behaviour*, 23, 387–408.

Packer, C. 1980. Male care and exploitation of infants in *Papio anubis*. *Animal Behaviour*, 28, 512–520.

Packer, C. & Pusey, A. E. 1979. Female aggression and male membership in troops of Japanese macaques and olive baboons. *Folia Primatologica*, 31, 212–218.

Pagel, M. & Harvey, P. H. 1993. Evolution of the juvenile period in mammals. In *Juvenile Primates: Life History, Development, and Behavior* (ed. by M. Pereira & L. Fairbanks), 28–37. Oxford: Oxford University Press.

Parker, K. J., Buckmaster, C. L., Sundlass, K., Schatzberg, A. F. & Lyons, D. M. 2006. Maternal mediation, stress inoculation, and the development of neuroendocrine stress resistance in primates. *Proceedings of the National Academy of Sciences*, 103, 3000–3005.

Patel, E. R. 2007. Non-maternal infant care in wild Silky sifakas *(Propithecus candidus)*. *Lemur News*, 12, 39–42.

Paul, A. & Thommen, D. 1984. Timing of birth, female reproductive success, and infant sex ratio in semifree-ranging Barbary macaques *(Macaca sylvanus)*. *Folia Primatologica*, 42, 2–16.

Paul, A., Kuester, J. & Arnemann, J. 1996. The sociobiology of male-infant interactions in Barbary macaques, *Macaca sylvanus*. *Animal Behaviour*, 51, 155–170.

Paul, A., Preuschoft, S. & van Schaik, C. 2000. The other side of the coin: Infanticide and the evolution of affiliative male-infant interactions in Old World primates. In *Infanticide by Males and Its Implications* (ed. by C. van Schaik & C. H. Janson), 269–292. Cambridge: Cambridge University Press.

Pellis, S. M. & Iwaniuk, A. N. 1999. The problem of adult play fighting: A comparative analysis of play and courtship in primates. *Ethology*, 105, 783–806.

Pereira, M. E. 1984. Age changes and sex differences in the social behavior of juvenile yellow baboons *(Papio cynocephalus)*. PhD dissertation, University of Chicago.

———. 1988a. Effects of age and sex on intra-group spacing behaviour in juvenile savannah baboons, *Papio cynocephalus cynocephalus*. *Animal Behaviour*, 36, 184–204.

———. 1988b. Agonistic interactions of juvenile savannah ba-boons. I. Fundamental features. *Ethology*, 80, 195–217.

Pereira, M. E. & Altmann, J. 1985. Development of social behavior in free-living nohuman primates. In *Nonhuman Primate Models for Human Growth and Development* (ed. by E. S. Watts), 217–309. New York: Liss.

Pereira, M. E. & Fairbanks, L. 1993. *Juvenile Primates*. New York: Oxford University Press.

Petru, M., Spinka, M., Lhota, S. & Sipek, P. 2008. Head rotations in the play of Hanuman langurs (*Semnopithecus entellus*): Description and analysis of function. *Journal of Comparative Psychology*, 122, 9–18.

Post, D. G. 1984. Is optimization the optimal approach to primate foraging? In *Adaptations for Foraging in Nonhuman Primates* (ed. by P. S. Rodman & J. G. H. Cant), 280–303. New York: Columbia University Press.

Pusey, A. 1990. Behavioural changes at adolescence in chimpanzees. *Behaviour*, 115, 203–246.

Pusey, A. E. & Packer, C. 1987. Dispersal and philopatry. In *Primate Societies* (ed. by B. B. Smuts, D. L. Cheney, R. M. Seyfarth, R. W. Wrangham & T. T. Struhsaker), 250–266. Chicago: University of Chicago Press.

Pusey, A. E., Williams, J. & Goodall, J. 1997. The influence of dominance rank on the reproductive success of female chimpanzees. *Science*, 277, 828–831.

Rapaport, L. G. & Brown, G. R. 2008. Social influences on foraging behavior in young nonhuman primates: Learning what, where and how to eat. *Evolutionary Anthropology*, 17, 189–201.

Repetti, R. L., Taylor, S. E. & Seeman, T. E. 2002. Risky families: Family social environments and the mental and physical health of offspring. *Psychological Bulletin*, 128, 330–366.

Resko, J. A., Goy, R. W., Robinson, J. A. & Norman, R. L. 1982. The pubescent rhesus monkey: Some characteristics of the menstrual cycle. *Biology of Reproduction*, 27, 354–361.

Robbins, M. M. 1995. A demographic analysis of male life history and social structure of mountain gorillas. *Behaviour*, 132, 21–47.

Robinson, J. G. 1986. Seasonal structure in foraging groups of wedge-capped capuchin monkeys *Cebus olivaceus*: Implications for foraging theory. *Smithsonian Contributions to Zoology*, 431, 1–60.

Roney, J. R. & Maestripieri, D. 2003. Social development and affiliation. In *Primate Psychology* (ed. by D. Maestripieri), 171–204. Cambridge: Harvard University Press.

Rosenblum, L. A. & Paully, G. S. 1984. The effects of varying environmental demands on maternal and infant behavior. *Child Development*, 55, 305–314.

Ross, C. 2001. Park or ride? Evolution of infant carrying in primates. *International Journal of Primatology*, 22, 749–771.

———. 2003. Life history, infant care strategies and brain size in primates. In *Primate Life Histories and Socioecology* (ed. by P. M. Kappeler & M. E. Pereira), 266–285. Chicago: University of Chicago Press.

Ross, C. & MacLarnon, A. 2000. The evolution of non-maternal care in anthropoid primates: A test of the hypotheses. *Folia Primatologica*, 71, 93–113.

Ross, S. R., Bloomsmith, M. A., Bettinger, T. M. & Wagner, K. E. 2009. The influence of captive adolescent male chimpanzees on wounding: Management and welfare implications. *Zoo Biology*, 28, 623–634.

Rubenstein, D. I. 1993. On the evolution of juvenile lifestyles in mammals. In *Juvenile Primates: Life History, Development, and Behavior* (ed. by M. E. Pereira & L. A. Fairbanks), 38–57. New York: Oxford University Press.

Sackett, G. P. 1982. Can single processes explain effects of postnatal influences on primate development? In *The Development of Attachment and Affiliative Systems* (ed. by R. N. Emde & R. J. Harmon), 3–12. New York: Plenum.

Sacks, J. J., Smith, J. D., Kaplan, K. M., Lambert, D. A., Sattin, R. W. & Sikes, R. K. 1989. The epidemiology of injuries in Atlanta daycare centers. *Journal of American Medical Association*, 262, 1641–1645.

Sade, D. S. 1976. Population dynamics in relation to social structure on Cayo Santiago. *Yearbook of Physical Anthropology*, 20, 253–262.

Saltzman, W., & Abbott, D.H. 2009. Effects of elevated circulation cortisol concentrations on maternal behavior in common marmoset monkeys (*Callithrix jacchus*). *Psychoneuroendocrinology*, 34, 1222–1234.

Sanchez, M. M., Ladd, C. O. & Plotsky, P. M. 2001. Early adverse experience as a developmental risk factor for later psychopathology: Evidence from rodent and primate model. *Development and Psychopathology*, 13, 419–449.

Schino, G., Rosati, L. & Aureli, F. 1998. Intragroup variation in conciliatory tendency in captive Japanese macaques. *Behaviour*, 135, 897–912.

Schino, G., D'Amato, F. R. & Troisi, A. 1995. Mother-infant relationships in Japanese macaques: Sources of interindividual variation. *Animal Behaviour*, 49, 151–158.

Schino, G., Speranza, L. & Troisi, A. 2001. Early maternal rejection and later social anxiety in juvenile and adult Japanese macaques. *Developmental Psychobiology*, 38, 186–190.

Schneider, M. & Moore, C. 2000. Effects of prenatal stress on development: A non-human primate model. In *The Effects of Early Adversity on Nuero-behavioral development: Minnesota Symposium on Child Psychology*, Vol. 31 (ed. by C. Nelson), 201–244. Mahwah, NJ: Erlbaum.

Schultz, A. H. 1956. Postembryonic age changes. *Primatologia*, 1, 887–964.

Scott, L. M. 1984. Reproductive behavior of adolescent female baboons (*Papio anubis*) in Kenya. In *Female Primates* (ed. by M. F. Small), 77–100. New York: Alan R. Liss.

Setchell, J. M. 2003. Behavioural development in male mandrills (*Mandrillus sphinx*): Puberty to adulthood. *Behaviour*, 140, 1053–1089.

Setchell, J. & Lee, P. C. 2004. Development and sexual selection in primates. In *Sexual Selection in Primates: Causes, Mechanisms and Consequences* (ed. by P. M. Kappeler & C. P. van Schaik), 175–195. Cambridge University Press, Cambridge.

Setchell, J. M. & Kappeler, P. M. 2003. Selection in relation to sex in primates. *Advances in the Study of Behavior*, 33, 97–173.

Seyfarth, R. M. & Cheney, D. L. 1980. The ontogeny of vervet monkey alarm-calling behaviour: A preliminary report. *Zeitschrift für Tierpsychologie*, 54, 37–56.

———. 1986. Vocal development in vervet monkeys. *Animal Behavior*, 34, 1640–1658.

Silk, J. B., Samuels, A. & Rodman, P. S. 1981. The influence of kinship, rank, and sex on affiliation and aggression between adult female and immature bonnet macaques (*Macaca radiata*). *Behaviour*, 78, 111–137.

Smith, D. G. 1982. Inbreeding in three captive groups of rhesus monkeys. *American Journal of Anthropology*, 58, 447–451.

Smith, R. M. & Rubenstein, B. B. 1940. Adolescence of macaques. *Endocrinology*, 26, 667–679.

Smuts, B. B. 1985. *Sex and Friendship in Baboons*. Hawthorne, NY: Aldine.

Snowdon, C. T. & Suomi, S. J. 1982. Paternal behavior in primates. In *Child Nurturance*, Vol. 3: *Studies of Development in Nonhuman Primates* (ed. by H. E. Fitzgerald, J. R. Mullins & P. Gage), 63–108. New York: Plenum.

Snowdon, C. T., Ziegler, T. E. & Widowski, T. M. 1993. Further hormonal suppression of eldest daughter cotton-top tamarins following birth of infants. *American Journal of Primatology*, 31, 11–21.

Soltis, J., Thomsen, R. & Takenaka, O. 2001. The interaction of male and female reproductive strategies and paternity in wild Japanese macaques, *Macaca fuscata*. *Animal Behaviour*, 62, 485–494.

Spinka, M., Newberry, R. C. & Bekoff, M. 2001. Mammalian play: Training for the unexpected. *Quarterly Review of Biology*, 76, 141–168.

Sterck, E. H. 1997. Determinants of female dispersal in Thomas langurs. *American Journal of Primatology*, 42, 179–198.

———. 1998. Female dispersal, social organization, and infanticide in langurs: Are they linked to human disturbance? *American Journal of Primatology*, 44, 235–254.

Stevenson, M. F. & Poole, T. B. 1982. Playful interactions in family groups of the common marmoset (*Callithrix jacchus jacchus*). *Animal Behaviour*, 30, 886–900.

Stewart, K. J. 1981. Social development of wild mountain gorillas. PhD dissertation, University of Cambridge.

Stewart, K. J. & Harcourt, A. H. 1986. Gorillas: Variation in female relationships. In *Primate Societies* (ed. by B. Smuts, D. Cheney, R. M. Seyfarth, R. W. Wrangham & T. T. Struhsaker), 155–164. Chicago: University of Chicago Press.

Stone, A. 2007. Responses of squirrel monkeys to seasonal changes in food availability in an eastern Amazonian forest. *American Journal of Primatology*, 69, 142–157.

Strier, K. B. 2008. The effects of kin on primate life histories. *Annual Review of Anthropology*, 37, 21–36.

Strier, K. B., Boubli, J. P., Possamai, C. B. & Mendes, S. L. 2006. Population demography of northern muriquis (*Brachyteles hypoxanthus*) at the Estacao Biologica de Caratinga/Reserva Particular do Patrimonio Natural-Feliciano Miguel Abdata, Minas Gerais, Brazil. *American Journal of Physical Anthropology*, 130, 227–237.

Struhsaker, T. T. 1973. A recensus of vervet monkeys in the Masai-Amboseli Game Reserve, Kenya. *Ecology*, 54, 930–932.

Sugiyama, Y. 1976. Life history of Japanese macaques. *Advances in the Study of Behavior*, 7, 255–284.

———. 1999. Socioecological factors of male chimpanzee migration at Bossou, Guinea. *Primates*, 40, 61–68.

Suomi, S. J. 1982. Abnormal behavior and primate models of psychopathology. In *Primate Behavior* (ed. by J. Fobes & J. King), 171–215. New York: Academic Press.

———. 1997. Early determinants of behaviour: Evidence from primate studies. *British Medical Bulletin*, 53, 170–184.

———. 1999. Developmental trajectories, early experiences, and community consequences: Lessons from studies with rhesus monkeys. In *Developmental Health and the Wealth of Nations* (ed. by D. Keating & C. Hertzman), 185–200. New York: Guilford.

Symons, D. 1978. *Play and Aggression: A Study of Rhesus Monkeys*. New York: Columbia University Press.

Tanner, J. M. 1962. *Growth at Adolescence*. London: Blackwell Scientific Publications.

Terbough, J. W. 1983. *Five New World Primates: A Study in Comparative Ecology*. Princeton, NJ: Princeton University Press.

Terborgh, J. & Janson, C. H. 1986. The socioecology of primate social groups. *Annual Review of Ecology and Systematics*, 17, 111–136.

Trivers, R. L. 1974. Parent-offspring conflict. *American Zoologist*, 14, 249–264.

Ueno, A. 2005. Development of co-feeding behavior in young wild Japanese macaques (*Macaca fuscata*). *Infant Behavior & Development*, 28, 481–491.

Van Noordwijk, M. A. & van Shaik, C. P. 1985. Male migration and rank acquisition in wild long-tailed macaques (*Macaca fascicularis*). *Animal Behaviour*, 33, 849–861.

Van Schaik, C. P. & van Hoof, J. A. R. A. M. 1983. On the ultimate causes of primate social systems. *Behaviour*, 85, 91–117.

Van Schaik, C. P. & van Noordwijk, M. A. 1989. The special role of male Cebus monkeys in predation avoidance and its effect on group composition. *Behavioral Ecology Sociobiology*, 24, 265–276.

Vasey, N. 2007. The breeding system of wild red ruffed lemurs (*Varecia rubra*): A preliminary report. *Primates*, 48, 41–54.

Visalberghi, E. & Neel, C. 2003. Tufted capuchins (*Cebus apella*) use weight and sound to choose between full and empty nuts. *Ecological Psychology*, 15, 215–228.

Visalberghi, E., Fragaszy, D. M., Izar, P. & Ottoni, E. B. 2005. Terrestriality and tool use. *Science (Letters)*, 308, 951–952.

Visalberghi, E., Addessi, E., Truppa, V., Spagnoletti, N., Ottoni, E., Izar, P. & Fragaszy, D. 2009. Selection of effective stone tools by wild bearded capuchin monkeys. *Current Biology*, 19, 1–5.

Wallen, K. 1996. Nature needs nurture: The interaction of hormonal and social influences on the development of behavioral sex differences in rhesus monkeys. *Hormones and Behavior*, 30, 364–378.

———. 2000. Risky business: Social context and hormonal modulation of primate sexual desire. In *Reproduction in Context* (ed. by K. Wallen & J. Schneider), 289–323. Cambridge, MA: MIT Press.

———. 2001. Sex and context: Hormones and primate sexual motivation. *Hormones and Behavior*, 40, 339–357.

Wang, X., Li, B., Ma, J., Wu, X., Xiao, H., Yang, J. & Liu, Y. 2007. Sex differences in social behavior of juvenile Sichuan snub-nosed monkeys *Rhinopithecus roxellana* at Yuhuangmiao, Mt. Qinling, China. *Acta Zoologica Sinica*, 53, 939–946.

Watts, D. P. 1985. Observations on the ontogeny of feeding behaviour in mountain gorillas (*Gorilla gorilla beringei*). *American Journal of Primatology*, 8, 1–10.

Watts, D. & Pusey, A. 1993. Behavior of juvenile and adolescent great apes. In *Juvenile Primates* (ed. by M. E. Pereira & L. A. Fairbanks), 148–167. New York: Oxford University Press.

Weaver, A. C. & de Waal, F. B. M. 2000. The development of reconciliation in brown capuchins. In *Natural Conflict Resolution* (ed. by F. Aureli & F. B. M. de Waal), 216–218. Berkeley: University of California Press.

Weaver, A., Richardson, R., Worlein, J., de Waal, F. B. M. & Laudenslager, M. 2004. Response to social challenge in young bonnet (*Macaca radiata*) and pigtail (*Macaca nemestrina*) macaques is related to early maternal experiences. *American Journal of Primatology*, 62, 243–259.

Westergaard, G. C., Suomi, S. J., Chavanne, T. J., Houser, L., Hurley, A., Cleveland, A., Snoy, P. J. & Higley, J. D. 2003. Physiological correlates of aggression and impulsivity in free-ranging female primates. *Neuropsychopharmacology*, 28, 1045–1055.

Whitehead, J. M. 1986. Development of feeding selectivity in mantled howling monkeys, *Alouatta palliata*. In *Primate Ontology, Cognition, and Social Behaviour* (ed. by J. G. Else & P. C. Lee), 105–117. Cambridge: Cambridge University Press.

Whiten, A. & Ham, R. 1992. On the nature and evolution of imitation in the animal kingdom: Reappraisal of a century of research. *Advances in the Study of Behavior*, 21, 239–283.

Wich, S. A., Utami-Atmoko, S. S., Mitra Setia, T. Rijksen, H. D., Schurmann, C., van Hooff, J. A. R. A. M., van Schaik, C. P. 2004. Life history of wild Sumatran orangutans (*Pongo abelii*). *Journal of Human Evolution*, 47, 385–398.

Wickings, E. J., Bossi, T. & Dixson, A. F. 1993. Reproductive success in the mandrill, *Mandrillus sphinx*: Correlations of male dominance and mating success with paternity, as determined by DNA fingerprinting. *Journal of the Zoological Society of London*, 231, 563–574.

Widdig, A. 2007. Paternal kin discrimination: The evidence and likely mechanisms. *Biological Reviews*, 82, 319–334.

Wiens, F. & Zitzmann, A. 2003. Social dependence of infant slow lorises to learn diet. *International Journal of Primatology*, 24, 1007–1021.

Wolfe, L. 1978. Age and sexual behavior of Japanese macaques (*Macaca fuscata*). *Archives of Sexual Behavior*, 7, 55–68.

Worlein, J. M. & Sackett, G. P. 1997. Social development in nursery-reared pigtailed macaques (*Macaca nemestrina*). *Americal Journal of Primatology*. 41, 23–35.

Wright, P. C. 1990. Patterns of paternal care in primates. *International Journal of Primatology*, 11, 89–99.

Yamagiwa, J. 1983. Diachronic changes in two eastern lowland gorilla groups (*Gorilla gorilla graueri*) in the Mt. Kahuzi region, Zaire. *Primates*, 24, 174–183.

Young, G. H., Coelho Jr., A. M. & Bramblett, C. A. 1982. The development of grooming, sociosexual behavior, play and aggression in captive baboons in the their first two years. *Primates*, 23, 511–519.

Zinner, D., Alberts, S. C., Nunn, C. L. & Altmann, J. 2002. Evolutionary biology: Significance of primate sexual swellings. *Nature*, 420, 142–143.

Chapter 12 Genetic Consequences of Primate Social Organization

Anthony Di Fiore

OTHER CHAPTERS in this volume reveal that the social systems of nonhuman primates stand out from those of most other mammals in several significant respects. First, primates, more than members of most other mammalian orders, tend to live in stable, bisexual, and sometimes very large social groups, where dispersal is predominantly by individuals of one sex (chapter 9, this volume). Second, because of their long life spans and slow life histories (chapter 10, this volume), it is often possible for closely related individuals (e.g., full or half-siblings) as well as familiar non-kin to associate with one another in the same social groups over long periods of time. Primate groups thus typically comprise multiple, overlapping generations and cohorts of individuals linked to one another and to members of other social groups by complex pedigree relationships. Finally, for many primates, the nature of the social interactions that animals have with group mates and with extragroup individuals can vary dramatically, from highly competitive to indifferent or tolerant to cooperative, resulting in myriad social roles that animals might play with one another, for example, as "dominants," "subordinates," "allies," "friends," or "rivals" (chapter 24, this volume).

Over the past several years, primatologists and other wildlife population ecologists have begun to consider, both theoretically and empirically, how these and other features of animal social systems, including sex-biased dispersal, dominance hierarchies, strong reproductive skew, and patterns of new group formation, influence how individuals within groups and within populations are related

to one another (Chepko-Sade & Halpin 1987; Melnick 1987; Chesser 1991a, b; Melnick & Hoelzer 1992, 1996; Sugg et al. 1996; Storz 1999; Ross 2001; Di Fiore 2003a, 2009; Lukas et al. 2005). It is the aggregate of this set of genetic relationships among the animals in a population, considered in a spatially and socially explicit framework, that constitutes what can be broadly referred to as a population's "genetic structure." The recognition that aspects of primate social systems, underlain as they are by differences in behavior among individuals and between the sexes, can have profound consequences for the genetic structure of primate populations represents a significant departure from the ideas of classical population genetics (fig. 12.1; see also Sugg et al. 1996). Primate population ecologists have likewise come to realize that stochasticity in demographic variables (e.g., variation over time in birth rates, sex ratios at birth, or age- and sex-specific mortality) and historical factors (e.g., disease outbreaks, natural disasters) also influence the genetic structure of primate populations through their indirect effects on primate social organization (e.g., by leading to changes in population density, group composition, or the availability of suitable reproductive partners; fig. 12.2).

The goal of this chapter is to briefly outline how certain aspects of primate social systems are expected to influence primate population genetic structure and patterns of relatedness within and among primate groups and, in turn, how genetic structure within and among groups can influence animals' behavioral decisions about dispersal, social rela-

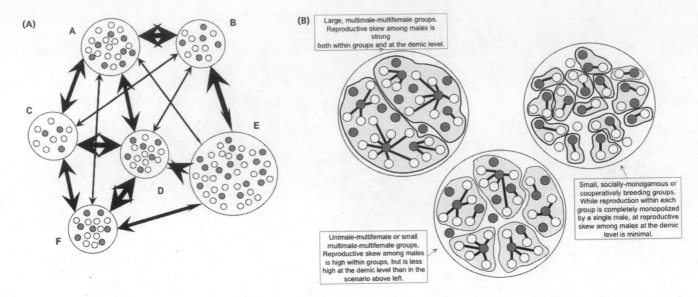

Fig. 12.1. Alternative views of primate population genetic structure.
A. The traditional population genetics approach to characterizing genetic structure imagines that populations of a given taxon are made up of semi-isolated "island" subpopulations within which mating can be modeled as an essentially random process. These subpopulations, or "demes," are connected to one another by gene flow (indicated by arrows), involving the dispersal of some limited number of migrant individuals each generation. Complications (i.e., biological realism) can be incorporated into classical population genetics models by postulating that gene flow might only take place among particular populations, such as those adjacent to one another in space (various "stepping stone" models) or those connected by suitable habitat ("corridor" models), and by allowing unequal rates of migration between demes (e.g., "source-sink" models). In this hypothetical example, the population of interest consists of six subpopulations (demes A through E) of various size and sex composition (dark circles represent males; open circles represent females) which are connected to one another by different rates of gene flow (here, the thickness of the arrows is proportional to the number of migrants per generation). Note that the rate of immigration into deme E differs from the rate of emigration out of it (indicated by the different sizes of the arrowheads on the lines connecting deme E to the other subpopulations) and that some pairs of demes do not exchange migrants (e.g., demes C and E).
B. The "social structure" view of population genetic structure (Sugg et al. 1996) acknowledges that within demes, individual animals are organized into social units (e.g., "groups"), which can impose certain constraints on dispersal and within which mating typically does not take place at random (individuals that mate are indicated by lines connecting males and females). These groups may differ from one another in age and sex composition, and the degree of intra-sexual competition and mating skew seen within groups is related in part to the demographic composition of those units, which can be influenced by a variety of historical factors and demographic stochasticity (see fig. 12.2). Although they are not shown in this figure, social groups can also be organized around sets of related animals (e.g., matrilines in many cercopithecines), which allows for inclusive fitness considerations to influence the dispersal and mating decisions of individual animals (see text for elaboration). Finally, groups can be connected to one another not just by dispersal and immigration (which can be strategic and responsive to demographic conditions), but also by matings between individuals from different groups. Here, three simplified hypothetical "social structures" are suggested for the individuals comprising deme E of fig. 12.1a, which would have very different population genetic consequences. In the figure at upper left, individuals from deme E are divided into two large "multimale-multifemale social groups," within which male reproduction is strongly skewed in favor of a few dominant males (e.g., savanna baboons). In the center figure, individuals from deme E are divided into five smaller social groups, in each of which reproduction is dominated by a single male but occasionally other males have reproductive access to resident females (e.g., Hanuman langurs). In the figure at upper right, individuals from deme E are divided into 14 small social groups, within which reproduction is restricted to a single male-female pair (e.g., "socially monogamous" hylobatids, or cooperatively breeding callitrichines). In each of these three hypothetical scenarios, some groups contain "nonbreeding" individuals of one or both sexes, and the deme also may contain solitary or "floater" individuals that are not associated with a particular social group. Of course, many other variations are also possible. Superimposed on these demic structures is a temporal component that reflects changes in group residency, turnover of breeding individuals, and variation in tenure length of animals at different positions in a social hierarchy.

tionships, and mating. It bears noting from the outset that we expect this interplay between primate social systems and population genetic structure to be dynamic and bidirectional (fig. 12.2). That is, just as we expect there to be predictable genetic consequences associated with certain social structures, so too do we expect individuals' behavioral decisions over dispersal and reproduction to be influenced by the population genetic context in which those decisions are made. For example, individual animals are likely to make very different behavioral choices regarding dispersal if they are living in a group with close same-sexed kin than

in situations in which they are not. Thus, population genetic structure is expected to have just as important consequences for understanding primate social systems as the reverse.

In some ways this chapter is a prospective exercise, since population genetic studies of wild primates are still not common, although fortunately that is changing. The molecular genetic revolution spawned by the development of the polymerase chain reaction (PCR) and the identification of a host of molecular markers (e.g., simple tandem repeats or STRs, single nucleotide polymorphisms or SNPs, and inter-simple sequence repeats or ISSRs) suitable for ex-

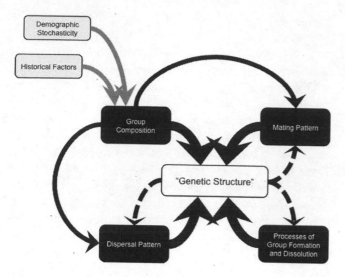

Fig. 12.2. The relationship between key features of social organization and primate population genetic structure. Several aspects of primate social systems—grouping patterns and group composition, dispersal patterns, mating patterns, and patterns of group formation and dissolution (black boxes)—all have direct consequences (thick solid arrows) for how genetic variation is structured within and between primate social groups (loosely referred to as "genetic structure"). Historical factors (e.g., loss of animals due to disease, hunting, natural disasters) and stochasticity in demographic variables (e.g., birth rates, natal sex ratios) also influence primate population genetic structure indirectly, through their effects on group composition (gray arrows). In addition to its direct effect, group composition also affects population genetic structure indirectly through its influence on the mating opportunities and dispersal decisions of individual animals (thin solid black arrows). Finally, individual animals' decisions about mating, dispersal, and group membership (e.g., about which new group to join following a group fission) should also be sensitive to their genetic relatedness to potential social partners. Thus, population genetic structure is likewise expected to have direct effects on these aspects of primate social organization (dotted arrows).

amining within population genetic variation has made it possible for primatologists and other field biologists to look in detail at the genetic structure of wild animal populations, often now using samples that can be collected noninvasively (Di Fiore 2003a; de Ruiter 2004; Di Fiore et al. 2010). But although such studies are becoming increasingly common, data for fully evaluating the influences of many aspects of social organization on genetic structure are still lacking for most primate taxa. Thus, this chapter takes a theoretical tack and reviews what we *expect* to see with respect to population genetic structure and patterns of within group relatedness, given first principles about primate dispersal patterns, patterns of reproductive skew, and patterns of inter- and intragroup social interactions.

In the following, I first discuss why it is important to consider how primate social systems can influence the genetic structure of populations and vice versa. I then give a brief historical introduction to theoretical attempts by behavioral ecologists and population geneticists to explore the links between primate social systems and genetic structure. I then

discuss three major aspects of primate social organization (sensu Di Fiore & Rendall 1994) that are expected to have a significant impact on population genetic structure and on patterns of genetic relatedness within groups and local populations: (1) dispersal patterns, particularly sex-biased dispersal, (2) patterns of reproductive skew, and (3) patterns of group fission, new group formation, and turnover of breeding individuals within groups. For each of these topics, I review some of the relevant molecular studies of nonhuman primates that either support the theoretical link or, more interestingly, have caused primatologists to recognize novelties in the social systems of certain taxa. Following this review, I offer a few comments regarding the limits to our ability to characterize or appreciate some of the links between primate social organization and population genetic structure. I conclude the chapter by outlining some promising directions for future work at the interface of population genetics and primate behavioral ecology, specifically highlighting the need for more extensive theoretical and modeling work.

The Importance of Studying the Interplay between Social Organization and Genetic Structure

An understanding of how population genetic structure and aspects of social systems are related is important because some of the fundamental things that primate biologists would most like to know about our study subjects—the dispersal and reproductive tactics of individual animals, for example—are the very things that are most difficult to observe in the field. Genetic data can also clarify the kinship relationships among individuals, which we presume should have predictive value for understanding variation in social behavior (chapter 21, this volume). Molecular techniques allow us to investigate some of these aspects of behavior and social systems when detailed observational data are lacking (Di Fiore 2003a; de Ruiter 2004), but the step of extrapolating back from genetic data requires an appreciation of how and why social organization is expected to have predictable effects on a taxon's genetic structure.

In addition, understanding the links between social organization and population genetic structure is important for evaluating current models of the evolution of grouping patterns and social behavior in primates. For many years the predominant theoretical paradigm for trying to make sense of patterns of primate sociality and social behavior, particularly affiliative behavior, has been W. D. Hamilton's concept of "kin selection" (Hamilton 1964a, b; West-Eberhard 1975). The idea of kin selection features prominently in models of primate social evolution (e.g., Wrangham 1980;

Isbell 2004) as well as in attempts to explain primate social relationships (Gouzoules & Gouzoules 1987; van Schaik 1989; Sterck et al. 1997; Chapais 2001; Isbell & Young 2002; Silk 2002; chapter 21, this volume). To date, however, there have been few explicit tests of the explanatory power of kin selection in nonhuman primates that actually use genetic data to describe population structure and within-group patterns of relatedness, as these data are lacking for many of even the best-studied taxa (for a review of this problem, see Di Fiore 2003a; Di Fiore 2009). A clear role for kin selection seems evident in several matrilineally-organized cercopithecines, where maternal and (to a lesser degree) paternal kinship have been shown to be important predictors of female social behavior (e.g., yellow baboons, *Papio cynocephalus*, Smith et al. 2003; rhesus monkeys, *Macaca mulatta*, Widdig et al. 2001), as well as in at least one strepsirrhine, the gray mouse lemur (*Microcebus murinus*), where females forage solitarily but form communal nesting associations with female kin (Eberle & Kappeler 2006). However, studies of chimpanzees (*Pan troglodytes*) that have compared behavioral data on social interactions with genetic estimates of kinship suggest that the explanatory value of kinship in shaping primate social behavior may be limited (Goldberg & Wrangham 1997; Mitani et al. 2000; Langergraber et al. 2007a, 2009) Certainly, studies of additional taxa from across the major primate clades are needed.

In retrospect, the idea that kinship may play a limited role in shaping primate social interactions should not be so surprising. Primates are generally long-lived animals with long interbirth intervals, and many taxa show strongly sex-biased dispersal. This sets up a situation in which individual animals may be unlikely to reside in groups where many of the same-sexed animals of similar age (i.e., those animals with whom an individual's personal fitness interests are most likely to overlap, and thus with whom a definable social "relationship" should be most likely to develop) are also kin. If genetic data on additional noncercopithecine taxa, as they become available, continue to suggest a reduced importance for kinship as an explanatory variable of much primate social behavior, a significant paradigm shift may be in order to consider alternative models of primate social interactions (e.g., by-product mutualism, intergroup and interdemic selection, reciprocity and biological markets models: Mesterton-Gibbons & Dugatkin 1992; Noë & Hammerstein 1994; Bradley 1999; Nöe et al. 2001; chapter 22, this volume). Such a shift may already be taking place as increasing emphasis is placed on those factors beyond just kinship that should go into an animal's calculus about how to direct its cooperative and competitive behaviors.

Theoretical and Empirical Background

Beginning in the 1960s and 1970s, anthropological geneticists began to investigate how some aspects of the patterning of human genetic variation in classical molecular markers could be linked to aspects of human social systems in small-scale societies, such as polygyny, patrilocality, and especially the processes of new group formation via social group fission (Neel & Chagnon 1968; Neel 1978; Smouse et al. 1981; reviewed in Hunley et al. 2008). Around the same time, several primate researchers began using such markers to describe aspects of the population genetic structure of a variety of cercopithecine primates (e.g., free-ranging rhesus monkeys, Duggleby 1977, 1978; Cheverud et al. 1978; Chepko-Sade & Olivier 1979; Olivier et al. 1981; Buettner-Janusch et al. 1983; Japanese macaques, *Macaca fuscata*, Nozawa 1972; Aoki & Nozawa 1984; Nozawa et al. 1982; baboons, *Papio* spp., Jolly & Brett 1973, Shotake 1981; geladas, *Theropithecus gelada*, Shotake & Nozawa 1984). However, it was not until the late 1980s that Donald Melnick and collaborators began formally outlining the theoretical links that could be drawn between primate social organization and population genetic structure. In a seminal paper with the same title as this chapter, Melnick (1987) described in detail how the social organization of wild rhesus monkeys and other cercopithecine primates characterized by female philopatry, male-biased dispersal, rank-related differential male mating success, and a tendency for groups to fission along matrilineal lines would be expected to influence the population genetic structure of cercopithecine groups. He then proceeded to test and confirm many of these predictions, while also noting the modulating effects that demographic variation might have in explaining differences in genetic structure between cercopithecine taxa and populations.

Over the next few years, Melnick and Guy Hoelzer (1992, 1996) expanded upon this review of the genetic consequences of cercopithecine social organization. They focused on the fact that many cercopithecine primates live in matrilineally organized social groups, where females interact in affiliative and cooperative ways with their female kin in competitive interactions with females from other matrilines and where females are recruited into the adult population mainly from within their natal social groups. Melnick and Hoelzer (1996) noted that under these conditions mitochondrial gene flow would be severely restricted, as mitochondrial DNA is inherited matrilineally. Thus, they reasoned that mitochondrial DNA diversity at the local level should be very low, while at the regional level, mitochondrial DNA variation should show strong evidence of being structured hierarchically. By contrast, they argued, extensive gene flow brought about by male dispersal should lead to very little genetic structuring to the nuclear genome (Melnick & Hoelzer 1992). Melnick and Hoelzer (1996) compiled data on nuclear and mitochondrial DNA variation for five species of macaques and demonstrated that across all of these species the mitochondrial genome showed much more extensive evidence of population structuring, as predicted by the matrilineally-organized, female-philopatric societies of cercopithecines.

Since these influential papers first laid the groundwork for thinking about the genetic consequences of primate social organization, other researchers have also tackled some other aspects of the expected influence of primate social systems on population genetic structure. In a series of seminal papers, Theresa Pope (1990, 1992, 1998, 2000) examined the population genetic consequences of the mating system and dispersal pattern of ursine howler monkeys (*Alouatta arctoidea*) and discussed how demographic variation influences the links between social organization and population structure in this species. Similarly, Morin et al. (1994) and Di Fiore (2003a, 2009) have examined some of the expected genetic consequences of male philopatry for chimpanzees and South American ateline primates, respectively, and several research groups have used molecular data to investigate the social systems of several solitary and pair-living strepsirrhines, including red-fronted lemurs, *Eulemur rufifrons* (Wimmer & Kappeler 2002), Coquerel's dwarf lemur, *Mirza coquereli* (Kappeler et al. 2002), Lac Alaotra bamboo lemurs, *Hapalemur alaotrensis* (Nievergelt et al. 2002), gray mouse lemurs (Radespiel et al. 2001, 2002; Fredsted et al. 2004, 2005), and western fat-tailed dwarf lemurs, *Cheirogaleus medius* (Fietz et al. 2000; Fredsted et al. 2007).

Outside of primatology there is additional research that explores some of the important links between animal social organization and population genetic structure, though even here the work is not extensive. In an influential edited volume, Chepko-Sade and Halpin (1987) reviewed the influence of dispersal patterns on population genetic structure in mammals. Subsequently, Chesser (1991a, b) began to develop some of the first formal population genetics models to take into account some of the effects of particular features of social organization (e.g., female philopatry) on the organization of genetic variation within populations, while Sugg et al. (1996) and Storz (1999) discussed the impacts of social structure on population genetic structure in animals more generally, focusing on how living in behaviorally-defined breeding groups creates an additional level of population structure or subdivision not often considered in classical population genetic models (Wright 1951, 1965). Ross (2001) has provided an excellent and broad review of the molecular ecology of social behavior, with a strong focus on how various elements of the breeding system (e.g., number of breeders of each sex, genetic relatedness among breeders, patterns of reproductive skew) shape population genetic structure and patterns of relatedness within groups. Below, I discuss some of the key points emerging from this research.

Genetic Consequences of Dispersal Patterns

Effects on Population Genetic Structure

Dispersal—the behavioral mechanism underlying gene flow—is one of the principal factors that influence how a taxon's genetic variation is distributed across space and among social groups. Several aspects of dispersal can influence population genetic structure. First, in many species of primates, as in most other vertebrates, dispersal is strongly sex-biased, with one sex typically dispersing from the natal home range and social group before reproducing while the other sex remains philopatric (Greenwood 1980; Waser & Jones 1983; Melnick & Pearl 1987; Pusey & Packer 1987; Johnson & Gaines 1990). Strongly sex-biased dispersal should result in very clear differences between the sexes in how genetic variation is structured across social groups and populations (Melnick & Hoelzer 1992, 1996; Avise 1994, 2004; Di Fiore 2003a, 2009; Lawson et al. 2007; table 12.1). Specifically, we expect to see much greater population structuring (i.e., genetic differentiation across social groups and across geographic space) in the uniparentally inherited portions of the genome (mitochondrial DNA for females, Y chromosomal DNA for males) that are passed through the philopatric sex relative to the structuring seen in the nuclear genome. This is because both the nuclear and the uniparentally inherited genomes of dispersing individuals are effectively distributed across space, thus homogenizing the distribution of that variation, while the uniparentally inherited genomes of philopatric animals are not reshuffled among social groups. As a result, we expect the mitochondrial genomes of taxa with high levels of female philopatry to show strong evidence of population substructuring, while their autosomal and Y chromosomal markers should reveal little genetic structure. Between the sexes, we would expect females to display even greater evidence of structuring of mitochondrial DNA variation than males. By contrast, in taxa with high levels of male philopatry, Y chromosomal variation is expected to be relatively

Table 12.1. Predicted relationships between dispersal patterns and the patterning of genetic variation and relevant results from molecular studies of wild nonhuman primates

Marker system and measure	Level or comparison	Implied dispersal pattern	Taxa in which relationship has been examined — Common name	Scientific name	References	Notes
Overall mtDNA diversity within social groups and/or local populations	Low	MBD	Rhesus macaque	*Macaca mulatta*	1, 2	
			Long-tailed macaque	*Macaca fascicularis*	1, 2, 3	
			Pig-tailed macaque	*Macaca nemestrina*	1, 2, 3, 4	
			Toque macaque	*Macaca sinica*	1, 2, 5	
			Japanese macaque	*Macaca fuscata*	1, 2	
			Grivet	*Chlorocebus aethiops*	6	
			Yunnan snub-nosed monkey	*Rhinopithecus bieti*	7	
	High	No evident bias or MBD	Ebony leaf monkey	*Trachypithecus auratus*	8	
			Silvered leaf monkey	*Trachypithecus cristatus*	8	
Relative mtDNA diversity among F versus M within social groups and/or local populations	M > F	MBD	Ebony leaf monkey	*Trachypithecus auratus*	8	
	F > M	FBD	Silvered leaf monkey	*Trachypithecus cristatus*	8	
	M and F comparable	No evident bias or MBD	Red woolly monkey	*Lagothrix poeppigii*	9	
Degree of geographic structuring to mtDNA variation	High	MBD	Bonobo	*Pan paniscus*	10, 11	
			Hamadryas baboon	*Papio hamadryas*	12	
			Chimpanzee	*Pan troglodytes*	13, 14, 15	
	Low or none	No evident bias or FBD	Hamadryas baboon	*Papio hamadryas*	12, 16	
Spatial structure and/or genetic differentiation between social groups and/or local populations in mtDNA haplotypes among F versus M	F cluster more than M and/or F differentiation > M	MBD and/or shorter F dispersal distance	Gray mouse lemur	*Microcebus murinus*	17	
	M cluster more than F and/or M differentiation > F	FBD and/or shorter M dispersal distance	Coquerel's dwarf lemur	*Mirza coquereli*	18	
	F and M clustering and/or differentiation comparable	No evident bias in dispersal distances	Siamang	*Symphalangus syndactylus*	19	
Diversity in nonrecombining regions of Y chromosome within social groups and/or local populations among M	High	MBD	Fat-tailed dwarf lemur	*Cheirogaleus medius*	20	
	Low	FBD	Lac Ravelobe mouse lemur	*Microcebus ravelobensis*	21, 22	
Genetic differentiation between groups for F versus M in autosomal markers	F > M	MBD	White-bellied spider monkey	*Ateles belzebuth*	9, 23	
			Hamadryas baboon	*Papio hamadryas*	24	
	M > F	FBD	Red woolly monkey	*Lagothrix poeppigii*	9, 23	
	M and F comparable	No evident bias	Hamadryas baboon	*Papio hamadryas*	24	
Genetic differentiation between groups in nonrecombining regions of Y chromosome among M	High	FBD	Bonobo	*Pan paniscus*	11	
	Low	No evident bias or MBD				

Pattern	Inference	Common name	Scientific name	Reference	Note
Genetic differentiation among groups in uniparentally inherited markers versus autosomes					
mtDNA > autosomes, NRY	MBD	Hamadryas baboon	*Papio hamadryas*	24	1
NRY > autosomes, mtDNA	FBD	Bonobo	*Pan paniscus*	11	
Genetic differentiation between predispersal aged individuals and adults within groups for F versus M					
M > F	MBD				
F > M	FBD				
M and F comparable	No evident bias				
Relatedness among F versus M within groups and/or within local populations					
F > M	MBD	Guereza	*Colobus guereza*	25	
		Yellow baboon	*Papio cynocephalus*	26	
		Long-tailed macaque	*Macaca fascicularis*	27	
		Verreaux's sifaka	*Propithecus v. verreauxi*	28	
		Common marmoset	*Callithrix jacchus*	29	
		Redfronted lemur	*Eulemur fulvus rufus*	30	
		Lac Alaotra bamboo lemur	*Hapalemur alaotrensis*	31	
		Gray mouse lemur	*Microcebus murinus*	32, 33	
M > F	FBD	White-bellied spider monkey	*Ateles belzebuth*	9	2
		Hamadryas baboon	*Papio hamadryas*	24	
		Chimpanzee	*Pan troglodytes*	13	
		Red woolly monkey	*Lagothrix poeppigii*	34	2
M and F comparable	No evident bias	Bonobo	*Pan paniscus*	10	
		Chimpanzee	*Pan troglodytes*	35, 36	3
		White-bellied spider monkey	*Ateles belzebuth*	9	
		Red woolly monkey	*Lagothrix poeppigii*	9, 34	
		Bornean orangutan	*Pongo pygmaeus*	37	

M = male; F = female; mtDNA = mitochondrial DNA; NRY = nonrecombining region of the Y chromosome; MBD = male-biased dispersal; FBD = Female-biased dispersal. Where no taxa are listed in a cell, it indicates that the pattern has not been found or has not yet been examined for wild primates.

Notes

1. Must correct for differences in the mutation rates of markers with different patterns of inheritance (e.g., mtDNA versus NRY versus autosomes) and for the differences in effective population sizes of M versus F.

2. Pattern is not seen in all groups. In some groups, M and F average relatedness was comparable.

3. A pattern of greater average relatedness of M versus F is seen in some groups and/or years analyzed, when adult group size was small.

Sources (see reference list on page XX): 1. Melnick & Hoelzer 1992. 2. Melnick & Hoelzer 1996 3. Perwitasari-Farajallah et al. 1999. 4. Rosenblum et al. 1997b. 5. Hoelzer et al. 1994. 6. Shimada 2000. 7. Liu et al. 2007. 8. Rosenblum et al. 1997a. 9. Di Fiore 2009. 10. Gerloff et al. 1999. 11. Eriksson et al. 2006. 12. Hapke et al. 2001. 13. Vorin et al. 1994. 14. Goldberg & Ruvolo 1997. 15. Mitani et al. 2000. 16. Winney et al. 2004. 17. Fredsted et al. 2004. 18. Kappeler et al. 2002. 19. Lappan 2007. 20. Fredsted et al. 2007. 21. Radespiel et al. 2008. 22. Radespiel et al. 2009. 23. Di Fiore et al. 2009. 24. Hammond et al. 2006. 25. Harris et al. 2009. 26. Altmann et al. 1996. 27. de Ruiter & Geffen 1998. 28. Lawler et al. 2003. 29. Nievergelt et al. 2000. 30. Nievergelt & Kappeler 2002. 31. Nievergelt et al. 2002. 32. Radespiel et al. 2001. 33. Fredsted et al. 2005. 34. Di Fiore & Fleischer 2005. 35. Vigilant et al. 2001. 36. Lukas et al. 2005. 37. Goossens et al. 2006.

more structured than either autosomal or mitochondrial DNA variation.

When dispersal is strongly sex-biased, we would likewise expect the overall level of variation that characterizes the uniparentally inherited genome transmitted through the philopatric sex to be relatively lower than variation in the comparable genome of the dispersing sex. Thus, within social groups of species characterized by female philopatry, there should be much lower diversity in female than in male mitochondrial DNA, because female-mediated gene flow is limited and females are primarily recruited into the adult breeding population from within their social groups (Melnick & Hoelzer 1996; Wallman et al. 1996). Comparably, we expect to see little diversity in the Y chromosomal DNA of males within groups of strongly male-philopatric taxa, while male mitochondrial DNA diversity should be just as high as that among females.

The few genetic studies of primates that have compared the degree of population substructuring in different genomic regions within and between the sexes broadly support these hypothesized genetic consequences of sex-biased dispersal (table 12.1). For example, as discussed above, Melnick and Hoelzer (1992, 1996) found low levels of mitochondrial diversity within groups and documented very clear contrasting patterns of nuclear versus mitochondrial genetic structure among several cercopithecine primates, which are typified by high levels of female philopatry and near universal male dispersal (Melnick & Pearl 1987; Pusey & Packer 1987). Similarly, Shimada et al. (2000) found that the limited variation in mitochondrial DNA of grivets (*Chlorocebus aethiops*) along the Awash River in Ethiopia was more geographically structured for females versus males, also consistent with greater male than female vagility.

In several species where observational studies suggest that female dispersal is common—e.g., chimpanzees, bonobos (*Pan paniscus*), hamadryas baboons (*Papio hamadryas*), woolly monkeys (*Lagothrix* spp.)—mitochondrial DNA diversity within groups and within local populations is relatively high and exceeds that typically observed among cercopithecines (fig. 12.3; Morin et al. 1994; Gerloff et al. 1999; Hapke et al. 2001; Eriksson et al. 2006; Di Fiore 2009), as is predicted when female-meditated gene flow is substantial. Few data exist regarding the diversity and genetic structuring of paternally-inherited Y chromosomes in primates, but among bonobos, Eriksson et al. (2006) found that Y chromosomal microsatellite markers show much greater geographic differentiation than maternally inherited mtDNA, a robust indication that dispersal is strongly female-biased. A similar pattern of greater structuring in Y chromosomal variation compared with mitochondrial diversity has also been noted in hamadryas baboons from the

Fig. 12.3.   A male red woolly monkey (*Lagothrix poeppigii*). Woolly monkeys live in large multimale groups. Males tend to be very tolerant of same-sex groupmates, but can be aggressive to males of other groups. Both observational and genetic data suggest that female dispersal is common in this species, as it is in other members of the same family of New World primates, the Atelidae—but genetic data also suggest substantial levels of male-mediated gene flow. Photo courtesy of Dylan M. Schwindt.

Arabian peninsula (Hammond et al. 2006) and in chimpanzees from Uganda (Langergraber et al. 2007b).

One example highlighting the potential interplay between dispersal behavior, population genetic structure, and social behavior is suggested by a recent study of western lowland gorillas (*Gorilla gorilla gorilla*), which tend to live in small social groups containing a single silverback male and several unrelated females. Bradley et al. (2004) demonstrated that many male silverbacks sampled from one large study population were each closely related to one or more additional silverbacks from other social groups; additionally, they found that silverback males from nearby groups tended to be more closely related than males from more distant groups. This led them to posit a social system that involves networks of male kin dispersed among different social groups. Bradley et al. (2004) suggested that the more frequent and peaceful intergroup interactions that characterize western lowland gorilla groups relative to those of eastern gorillas might be interpreted as an example of nepotistic tolerance among related males in different groups, thus highlighting how population genetic structure might influence patterns of intergroup social dynamics. It is noteworthy that, even in the absence of genetic data, Cheney and Seyfarth (1983) drew similar conclusions from observed patterns of dispersal by male vervet monkeys (*Chlorocebus pygerythrus*).

Effects on Within-Group Relatedness

Strongly sex-biased dispersal also has important theoretical implications for patterns of within-group relatedness for gregarious species. If members predominantly of one sex are recruited as new breeding adults into their natal social groups while members of the opposite sex disperse, then we would expect the average genetic relatedness among post-dispersal-aged group members of the philopatric sex to be greater than that among adults of the dispersing sex (Morin et al. 1994; Goudet et al. 2002; Di Fiore 2003a; Hammond et al. 2006). Thus, among cercopithecine primates, characterized by female philopatry, we would expect mean pairwise relatedness between females within social groups to be greater than that between males. By contrast, for species in which male philopatry is common (e.g., chimpanzees, bonobos, spider monkeys), we would expect males to show greater average levels of relatedness with one another than females.

Several studies of nonhuman primates have now tested these predictions with mixed results (table 12.1). For cercopithecines, Altmann et al. (1996) and de Ruiter and Geffen (1998) found that mean female-female relatedness was significantly greater than mean male-male relatedness in social groups of yellow baboons and long-tailed macaques (*Macaca fascicularis*), respectively. Similar support for this relationship comes from studies of several strepsirrhines in which female philopatry is common (e.g., Verreaux's sifakas, *Propithecus verreauxi*, Lawler et al. 2003; red-fronted lemurs, Wimmer & Kappeler 2002). In contrast, additional studies have failed to find evidence of greater average relatedness among members of the philopatric sex in other primates. Among male philopatric chimpanzees, for example, average relatedness among males within multiple groups did not exceed that among females in most group-years examined (Vigilant et al. 2001; Lukas et al. 2005). Similarly, average male relatedness was not significantly greater than average female relatedness in lowland red woolly monkeys (*Lagothrix poeppigii*), long presumed

Fig. 12.4. A male white-bellied spider monkey (*Ateles belzebuth*). Among spider monkeys, female dispersal and male philopatry are common. Spider monkey males from the same social group cooperate with one another in a variety of contexts, including to conduct patrols of their group's range boundaries and raids into other groups' ranges. Genetic data suggest that sometimes the males in spider monkey groups constitute sets of close kin, but this is not always the case. Photo courtesy of Anthony Di Fiore.

to be a taxon in which male philopatry is common, nor in one of two groups of white-bellied spider monkeys (*Ateles belzebuth*) from the same study sites (fig. 12.4; Di Fiore 2009; Di Fiore et al. 2009). Stochastic demographic factors that affect the genetic composition of groups are one possible reason why observations do not conform to theoretical predictions in some of these cases. For example, in small groups the loss of a single individual can dramatically alter estimates of the average relatedness of subsets of the remaining group members.

Lukas et al. (2005) developed a simulation model to explore the relatedness among males versus that among females under varying scenarios of sex-biased dispersal and differential male and female reproductive success. The results suggest that the expected pattern of greater average relatedness among individuals of the philopatric sex may only hold true in very small social groups with a restricted number of breeding individuals. While this social scenario may be approximated in many primate groups, it is by no means universal. In addition, patterns of average within-group relatedness among males and females can be influenced by factors other than dispersal (e.g., group fission), some of which are reviewed below.

It is important to recognize that for social animals dispersal is not simply a random process, but is rather a product of the behavioral decisions that animals make, and that it can be quite variable within and among populations (fig. 12.5). While classical population genetics theory treats dispersal as an essentially stochastic process that tends to ho-

Fig. 12.5. A young male yellow baboon (*Papio cynocephalus*), near the typical age at natal dispersal, watches another group of baboons in the distance. Dispersal is a fraught process for male baboons, as it often involves spending considerable time as a solitary animal and traveling long distances through unfamiliar areas. Male immigrants also often face aggression from resident males when they try to immigrate into a new group. As a result, males are expected to evaluate their dispersal options carefully. Photo courtesy of Catherine Markham.

individuals frequently should come to be similar to one another genetically and show relatively high mean between-group relatedness while becoming differentiated genetically from other groups.

Additionally, the phenomenon of "parallel dispersal"—in which more than one individual from the same natal social group transfer together into a new group, or in which an individual disperses into a group already containing familiar and/or related conspecifics—is fairly commonplace among primates (chama baboons, *Papio ursinus*, Cheney & Seyfarth 1977; Japanese macaques, Sugiyama 1976; rhesus monkeys, Drickamer & Vessey 1973; Meikle & Vessey 1981; white-faced capuchins, *Cebus capucinus*, Jack & Fedigan 2004; see review in Schoof et al. 2009). For example, Jack and Fedigan (2004) estimated that more than 70% of dispersing male white-faced capuchins either migrate with a familiar conspecific or join groups already containing individuals from their original social groups (chapter 21, this volume; fig 21.3). In this species, parallel-dispersing individuals are likely to be members of the same age cohort, and therefore are likely to be fairly close relatives, given the high degree of mating skew among males and the fact that females are generally philopatric. In joining a new group, then, related parallel dispersers can dramatically and directly influence the genetic structure of that group, particularly if the group's size is small. Still, there is a lack of empirical data for assessing the extent to which patterns of intergroup genetic variation are shaped by either parallel dispersal or biased transfer for any nonhuman primate.

Dispersal, Relatedness, and Population Structure in Solitary Primates

Most taxa of nonhuman primates live in some form of more or less stable, bisexual social groups, but solitary living and ranging characterizes many nocturnal strepsirrhines, including lorisids, galagids, and some nocturnal lemuroids (chapter 2, this volume). Although largely solitary living does not imply that no stable social relationships exist among individuals of a given species, our first cue to those relationships—physical association—is typically absent. Nonetheless, the influence of sex-biased dispersal on population genetic structure and within-population patterns of relatedness is also evident in many solitary species (table 12.1).

Among mouse lemurs and dwarf lemurs, for example, both sexes disperse from their mothers as they mature, but females are presumed to be more philopatric than males, a common pattern among solitary mammals generally (Greenwood 1980; Johnson & Gaines 1990). Genetic studies of Coquerel's dwarf lemurs and gray mouse lemurs indi-

mogenize genetic variation across space and social groups, animals make choices over when and where to disperse, and they can likewise use social tactics to influence the dispersal decisions of others. Additionally, individual animals are not always "free" to disperse anywhere; rather, there are often social or physical constraints on the choice of groups to which a dispersing animal might transfer. The effects of these myriad behavioral choices and constraints on aggregate patterns of spatial genetic structure are not always easily predictable.

Effects of Biased Transfer and Parallel Dispersal

If patterns of individual transfer among social groups are biased—for example, if two groups interchange dispersing individuals more often with one another than with other adjacent groups—then the resulting genetic structure of the population can be very different than if individuals are transferring at random. Over time, groups that exchange

cate that mitochondrial DNA haplotype sharing is common among females and juveniles sampled from within the same local population, while rare haplotypes within the population almost invariably occur in adult males (Kappeler et al. 2002; Wimmer et al. 2002). In Coquerel's dwarf lemurs, estimates of genetic relatedness among females sampled from within a local population were negatively correlated with the geographic distance between those females' centers of activity, while no such association between genetic and geographic distance was found among males (Kappeler et al. 2002). Similarly, among gray mouse lemurs, females with overlapping home ranges and those sharing sleeping sites tended to be more closely related, on average, than females from the population overall (Radespiel et al. 2001; Wimmer et al. 2002; Fredsted et al. 2005). These patterns are consistent with a difference in dispersal between males and females, and they suggest that many solitary primates might be characterized by a similar multigenerational, matrilineal pattern of social organization akin to that seen in many group-living cercopithecines (Clark 1978; Bearder 1987; Radespiel et al. 2001; Kappeler et al. 2002). A recent genetic study of Lac Ravelobe mouse lemurs (*Microcebus ravelobensis*) likewise found evidence of matrilineal social structure (Radespiel et al. 2008, 2009). In this taxon it appears that both males and females delay dispersal and form small sleeping groups with matrilineal kin until sexual maturity, after which dispersal seems somewhat, though not strongly, male-biased. Population genetic studies of other solitary primates, particularly lorisids and galagids, are needed to evaluate whether matrilineal social organization generally characterizes nongregarious primates. Interestingly, a recent study of Bornean orangutans (*Pongo pygmaeus*) characterized by a similar dispersed social system found comparable levels of relatedness among males and among females within the same study area, thus suggesting a similar pattern of dispersal and philopatry with both sexes (Goossens et al. 2006).

Genetic Consequences of Reproductive Skew

One of the robust results of many field studies of nonhuman primates is that male mating behavior tends to be strongly skewed within social groups, with socially dominant individuals generally gaining greater mating access to females than subordinates (Cowlishaw & Dunbar 1991; Kutsukake & Nunn 2006). Altmann's (1962) "priority-of-access" model predicts that the highest-ranking male in a primate group should monopolize matings by controlling access to females during that portion of their reproductive cycle when conception is most likely. Thus, paternity within

groups is expected to be skewed in favor of dominant males. Moreover, because it is difficult for high-ranking males to prevent rivals from mating when multiple females are receptive, the degree of skew is predicted to be negatively related to the average number of females in estrus simultaneously (Altmann 1962), a prediction that has been demonstrated in several recent comparative analyses (Kutsukake & Nunn 2006; Ostner et al. 2008).

Strong reproductive skew among males should have important consequences for primate population genetic structure and patterns of genetic relatedness within groups (Melnick 1987; Pope 1990; Storz 1999). Depending on the predominant dispersal pattern that characterizes the taxon (see above), we might expect somewhat different genetic consequences of strong male reproductive skew. Principally, if a single male or a small number of males is responsible for siring most of the offspring born in a social group over some period of time, then age-cohort mates born during those males' tenure should be more closely related to one another, on average, than they are to other individuals in the group. If male tenure length is long and the same male sires multiple, consecutive cohorts of offspring, the effect of this concentration of paternity would be to increase the average relatedness among group members and reduce effective group size. This effect would be particularly notable in female-philopatric groups, where many females are also matrilineal kin; the offspring from any particular age cohort would thus likely be related both maternally and paternally. Likewise, among female-philopatric groups where male tenure length is long, it is unlikely that dispersing males would soon be able to secure breeding positions in a new group. Thus, the effective rate of gene flow between social groups would be low. Long male tenure length, in combination with strong male reproductive skew and female philopatry, then, should lead to high genetic differentiation among social groups in a local population.

By contrast, average within-group relatedness and the extent of genetic differentiation between social groups should be much lower if male tenure length is short or if dispersal and subsequent mating opportunities for males are more readily available, even if male reproductive success within groups is strongly skewed. In this case, strong paternity skew would lead to subsets of close patrilineal relatives (cohort mates) among the immatures within groups—but unless successive males who sire offspring were themselves close relatives, relatedness across cohorts would be low through the paternal side and would be influenced primarily by the relatedness among breeding females. This, in turn, depends on whether females disperse or are philopatric (table 12.2).

Finally, when male reproductive skew is limited (i.e., when multiple males sire the set of offspring born within

Table 12.2. Expected consequences of dispersal pattern, degree of reproductive skew among males, and male reproductive tenure length for patterns of relatedness within primate social groups

Relatedness category	Reproductive skew among males	Dispersal pattern	
		Females philopatric, males disperse	Males philopatric, females disperse
Cohort mates	High	High, through both maternal and paternal side	Moderate, through paternal side
	Low	Moderate, through maternal line	Low
Members of different cohorts	High	High, through both maternal and paternal side, if male tenure length is long; moderate, if male tenure length is short	Moderate, through paternal side, if male tenure length is long; low, if male tenure length is short
	Low	Moderate, through maternal side	Low
Resident adults			
Male-male	High	Low, unless parallel dispersal by related males is common	High, through both maternal and paternal line
	Low	Low, unless parallel dispersal by related males is common	Moderate to high, through both maternal and paternal line
Female-female	High	High, through both maternal and paternal sides	Low, unless parallel dispersal by related females is common
	Low	Moderate, through maternal side	Low
Male-female	High	Low	Low
	Low	Low	Low

a particular social group over a short period of time), then the relatedness among animals belonging to the same age cohort and the average relatedness between animals from different age cohorts are determined mainly by the relatedness among breeding females, which in turn is dependent on the dispersal pattern.

Overall, the patterns of relatedness that result among both juvenile and adult (breeding-age) individuals within a group, then, depend on a complex calculus that involves the variables discussed above (dispersal pattern, degree of reproductive skew among males, and male reproductive tenure length) as well as other demographic factors (e.g., sex ratio among offspring, sex differences in mortality rates) and the interactions among those factors. Some of the relevant elements in this calculus are summarized in table 12.2.

The discussion above focused on the role of reproductive skew among males because for most primates, and indeed most mammals, males show greater variance in mating and reproductive success than do females (Trivers 1972, 1985). Nonetheless, for some primates we likewise see high skew in female reproductive success, at least over periods of one to a few breeding seasons. This is most prominent among the callitrichines, where reproduction within both sexes is often biased toward a single dominant animal that actively suppresses the reproduction of same-sexed subordinates (French 1997; chapter 3, this volume). Among callitrichines, then, we expect offspring from successive cohorts to be very close relatives—typically full siblings, since both male and female breeders often remain in breeding positions for multiple birth seasons. The callitrichine case is particularly interesting for other reasons as well. First, female callitrichines typically give birth to (fraternal) twin offspring, and the unique morphology of the callitrichine placenta makes it possible for developing twins to directly exchange haemopoetic stem cells that will eventually give rise to blood cells and other cell types produced in bone marrow (Benirschke & Brownhill 1962; Benirschke et al. 1962; Gengozian et al. 1964) and perhaps even germ cells (Haig 1999; Ross et al. 2007; but see Benirschke et al. 1962; Rothe & Koenig 1991). As a result, for particular tissues, callitrichine littermates may be even more "closely related" to one another than they are to either parent or to full siblings from separate pregnancies. Second, in many callitrichine populations, older siblings commonly remain in their natal groups and even assist in the rearing of their younger siblings (Goldizen 1987; Sussman & Garber 1987; Tardif et al. 1993; Tardif 1997), probably due to high population density setting a limit to dispersal and breeding opportunities for animals of both sexes. We would expect overall genetic relatedness within groups to be higher at sites of high versus low population density, because in the latter, maturing offspring would more likely be able to disperse and successfully establish themselves as breeders in their own group.

At the population level, the degree of reproductive skew seen within social groups (or, for solitary taxa, within local populations)—in combination with both demographic factors (e.g., population density) and physical factors (e.g., inhospitable habitat) that limit the dispersal abilities of animals—also influences how genetic variation is structured across space and across the social landscape. In general, greater reproductive skew, in combination with limited dispersal, should result in both a greater rate and greater extent of genetic differentiation between groups or between local populations (Storz 1999). This is because over time, social groups and local populations will accumulate proportionally more copies of the specific genetic variants that the limited number of resident breeders possess, while restricted dispersal minimizes the homogenizing effects of gene flow.

Few empirical data are currently available to evaluate the patterns suggested above. First, genetic studies of paternity in wild primates are still relatively rare, and the kind of longitudinal samples that would permit researchers to investigate the genetic consequences of long versus short tenure length (or, indeed, even what a "typical" male reproductive career may be like) are unavailable for even the best-studied taxa. Returning to the issue of the interplay between genetic structure and social dynamics, it is precisely in cases where reproductive tenure lengths are long that we would expect animals' behavioral decisions to be most influenced by population genetic structure. For example, when male tenure length is long, female offspring may reach sexual maturity in groups where their sire is still resident (and, if that male has enjoyed high reproductive success, where animals of similar age are also paternal relatives). Because of the high costs of inbreeding, these females' behavioral decisions about dispersal and mating are expected to be very sensitive to the demographic and genetic milieu. Indeed, in a comparative analysis across mammals, Clutton-Brock (1989) found female dispersal to be more common among taxa in which the average tenure length for dominant males exceeds the age of female sexual maturity (but see Moore & Ali 1984 for an alternative explanation of this pattern).

Genetic Consequences of Group Dynamics

A third major way in which primate social organization influences population genetic structure, for group-living taxa at least, is through the processes of group formation and dissolution, and through the replacement of breeding individuals. Three decades of fieldwork has taught us that, far from being stagnant, primate social groups are dynamic and ever-changing. The size and composition of established groups can change dramatically with major demographic events like group fusions or fissions, through the loss of a portion of a population due to disease or natural disasters, or through more commonplace demographic events such as births, deaths, emigrations, immigrations, and turnover of breeding individuals.

Group Fissioning

Traditional approaches to considering population genetic structure do not typically take into account dynamics in social group size and composition (Wright 1951, 1965). Human and nonhuman primate population geneticists have long recognized, however, that patterns of new group formation can and do have direct consequences for population genetic structure and for patterns of within-group relatedness (Neel 1978; Cheverud et al. 1978; Smouse et al. 1981; Melnick & Kidd 1983; Duggleby et al. 1984; Melnick 1987; Hunley et al. 2008). Among female-philopatric cercopithecine primates, new groups often form when groups of matrilineally related females fission from their original social group and begin foraging and ranging independently (Missakian 1973; Nash 1976; Cheverud et al. 1978; Chepko-Sade & Sade 1979; Dittus 1988; Ménard & Vallet 1993; Kuester & Paul 1997; Koyama 2003; Widdig et al. 2006). The same pattern of fissioning along female kin lines has also been reported for one platyrrhine with a matrilineal kinship structure similar to that seen in cercopithecines (wedge-capped capuchins, *Cebus olivaceous*, Robinson 1988). Theoretically, such fissioning should result in daughter groups whose females are more closely related to one another, on average, than were females within the original group (Melnick & Kidd 1983; Van Horn et al. 2007). Additionally, average measures of genetic differentiation among groups in the population (e.g., F_{ST}) and the rate of genetic differentiation among social groups are expected to also increase following fissions (Melnick & Kidd 1983; Wade & McCauley 1988; Whitlock & McCauley 1990). An increased relatedness among female group mates following fission has indeed been reported in some studies (e.g., free-ranging rhesus monkeys, Cheverud et al. 1978; Chepko-Sade & Sade 1979; Olivier et al. 1981; some groups of yellow baboons, Van Horn et al. 2007). In some cases, group fissions do not occur only along matrilineal lines (e.g., in the fission of one group of free-ranging rhesus monkeys on Cayo Santiago, Puerto Rico: Widdig et al. 2006). Even in this case, however, the proportion of dyads that consisted of maternal half siblings or of members of the same matriline was greater in daughter groups than in parental groups (Widdig et al. 2006).

Other factors besides matrilineal kinship are known to be important in influencing individuals' decisions about group choice during or following group fission, and these factors likewise have consequences for population genetic structure. Several studies, for example, have examined the relationship between group fissioning and paternal kinship, although a clear, consistent relationship between these variables is lacking. While Kuester and Paul (1997) noted that male members of the same matriline often wind up in different daughter groups following fission of large, free-ranging troops of Barbary macaques (*Macaca sylvanus*), studies of both yellow baboons (Smith 2000; Van Horn et al. 2007) and free-ranging rhesus monkeys (Widdig et al.

2006) revealed a tendency for animals to remain with more patrilineal kin in their social groups following group fissions than would be expected by chance; this effect, however, was not universal and was less pronounced than the maternal kinship effect.

Besides matrilineal and patrilineal kinship, other factors also undoubtedly influence individuals' decisions over group membership following fissioning events, and are thus expected to have consequences for population genetic structure. As discussed above in the context of dispersal, we would expect individual animals, given a choice, to join a daughter group where their personal chances for successful survival and reproduction and their positive effect on the reproduction of relatives is greatest. Thus, an individual's likely dominance rank in a post-fission group, the demographic composition of that group, the number and relatedness of other individuals in its matriline or patriline, and the residual reproductive values of those relatives should all theoretically influence choice of daughter group membership to the extent that such choice is unconstrained by what other animals have chosen. Given these considerations, it is perhaps not surprising that studies of the genetic consequences of group fissioning in cercopithecine primates have found few robust patterns, apart from a trend towards matrilineal relatives remaining in the same group (Widdig et al. 2006; Van Horn et al. 2007). Troop fissioning along matrilines has also been reported for wild ring-tailed lemurs (*Lemur catta*, Ichino 2006; Ichino & Koyama 2006).

Finally, in addition to its effect on patterns of within-group genetic relatedness, group fissioning along matrilineal lines can also influence more landscape-level patterns of genetic variation in cercopithecines and other matrilineally-organized primates (e.g., *Cebus*). Because groups of these taxa tend to show very low mitochondrial DNA diversity, and because immigration by females of these taxa is uncommon, group fissioning coupled with range expansion by daughter groups into new areas can lead to large geographical regions characterized by very similar mitochondrial DNA haplotypes (Melnick & Hoelzer 1996).

Other Patterns of Group Formation and the Turnover of Breeding Individuals

Although other patterns of new group formation in nonhuman primates have not received a lot of attention, for some taxa we know that the process can be very different than what is typical among many cercopithecines. In ursine howler monkeys in Venezuela, for example, new social groups form when a small number of typically unrelated dispersing males and females join together to challenge existing groups and establish a new home range (Crockett

1984; Crockett & Eisenberg 1987; Pope 1990, 2000). In this scenario, within-group relatedness is initially low but increases over time, as reproduction over multiple group-years is dominated by a single adult male (Pope 1990, 1992) and as growing female matrilines compete with one another to expel maturing daughters from other matrilines (Pope 2000). These same conditions should also lead to intergroup genetic differentiation increasing over time, particularly for populations living in saturated environments (Pope 1998).

Still other processes associated with the formation of new groups and the turnover of breeding individuals within established groups can yield different genetic consequences. For example, among socially monogamous primates where facultative dispersal by animals of both sexes is common, several different processes of group formation and turnover have been reported, including (1) the formation of *de novo* pairs by young dispersing animals, who then establish and defend a new territory; (2) the takeover of an existing breeding position by an outside animal, either through displacement or following the death of a resident; and (3) the replacement of a breeder in an established social group by a maturing, natal individual (Leighton 1987; Palombit 1994; Brockelman et al. 1998; Di Fiore et al. 2007; Fernandez-Duque et al. 2008). Each of these processes would have different consequences for population genetic structure and patterns of relatedness within and among groups.

First, if the replacement of breeders comes primarily from within groups (e.g., maturing subadults replace a same-sex parent) and if behavioral mechanisms for inbreeding avoidance are incomplete, then we would expect mean relatedness within groups and homozygosity to increase over time; we would also expect social groups within a local population to gradually diverge from one another genetically. By contrast, if breeders are principally replaced by extra-group animals, mean relatedness within groups should drop after replacement, but the long-term effect on population genetic structure will be contingent on the tenure length of breeders and on the dispersal distance of replacement individuals. Replacement by extragroup animals often results in the formation of "nonnuclear families" (i.e., social groups in which resident juveniles are not closely related to one or both adult breeders), a phenomenon that has now been reported in multiple taxa of socially monogamous primates (e.g., several hylobatids, Palombit 1994; Brockelman et al. 1998; equatorial sakis, *Pithecia aequatorialis*, Di Fiore et al. 2007; Azara's owl monkeys, *Aotus azarae*, Fernandez-Duque et al. 2008).

The establishment of completely new pairs and new territories is probably less common than either of the other two processes of group formation noted above, but at least some cases of new group formation have been reported for hylobatids, wherein young, dispersing animals have successfully set up new territories adjacent to those areas occupied by their parents, sometimes with "parental assistance" (Chivers & Raemaekers 1980; Tilson 1981). A similar process has also been observed among equatorial saki monkeys, where a radio-collared, subadult male left his natal group, joined a parous adult female and her juvenile offspring, and established a territory adjacent to that of his parents for a period of at least six months (Di Fiore & Fernandez-Duque, unpublished data). In such a social scenario, where offspring establish territories near those of their parents or where offspring tend to take over breeding positions in nearby rather than distant territories, genetic differentiation among groups within the population should be correlated with geographic distance. The extent to which this pattern is similar among males and females will depend on sex differences in average dispersal distance; as suggested above, the relationship between genetic and geographic distance should be more pronounced for the more philopatric sex (Lawson Handley & Perrin 2007; Di Fiore 2009).

Challenges for Developing Predictive Models of the Links between Social Organization and Population Genetic Structure

Integrating Historical Contingency and Demographic Variation

Much of the preceding discussion concerns the overall expected genetic consequences of primate dispersal and mating patterns at the population level, but one lesson that emerges from long-term field studies of individually recognized animals is that an individual's decisions about these aspects of behavior can be and are sensitive to the demographic and social context in which they are made (Altmann & Altmann 1979). For example, differences among individuals in choices about dispersal, which are driven in part by differences in the social milieus that animals experience, can obscure some of the expected patterns described above. In general we would expect animals, given a choice, to join or stay in a social group where they predict that their personal opportunities for successful survival and reproduction are greatest—or where they predict that they may best be able to contribute to the survival and reproduction of close kin. Thus, an animal's many decisions about dispersal—whether or not to remain in one's natal group, when to disperse, which of several possible social groups to join, whether to disperse alone or with a peer, and how far to go in search for a new group—are all expected to involve a complex calculus that integrates information about the demographic composition of various groups in the population (e.g., group size, number of potential mates, number and age of same-sex competitors), the residual reproductive values of related individuals in those groups, and the relative risk to personal survival of dispersing versus staying (Altmann & Altmann 1979; Altmann 2000). Such a calculus obviously is unlikely to be the same for all members of a particular class of individuals (e.g., for both males and females) and it is almost certainly different for high- versus low-ranking animals. Moreover, such a calculus is necessarily contingent on animals' assessment of salient cues about their social environment. Assessment processes and their accuracy are determined and constrained by animals' perceptual and cognitive abilities, which in turn may be developmentally contingent. Thus it may be no surprise that deviations from some of the expected, aggregate population genetic patterns discussed above are seen in many wild populations.

Similarly, it is becoming increasingly apparent that historical and demographic circumstances have important roles to play in the expression of primate behavior and, as a result, on elements of primate social structure. In many primate groups, the number of adults of one or the other sex is relatively small, thus limiting options for partner choice for mating or other social interactions. Historical or demographic events that remove even small numbers of individual animals from a social group (e.g., because of infectious disease, natural disasters, predation, and human activities such as hunting) can have a dramatic impact on the set of available partners as constituents in the "biological market," and likewise on patterns of population genetic structure.

Operationalizing "Population Genetic Structure"

A second limitation in our ability to understand or predict the genetic consequences of particular elements of primate social organization concerns the issue of exactly how to operationalize population genetic structure. This is both a practical concern and a philosophical one. Given that we are interested in the genetic consequences associated with particular features of social organization, it becomes necessary for researchers to make some decisions about *at what point in time and over what time frame* population genetic structure is to be characterized, and about *which individuals within a population* are relevant to include in deriving these measures. For some questions we wish to address, this might seem straightforward. For example, if we are interested in the genetic consequences of sex-biased dispersal, then it makes little sense to include predispersal aged individuals in the set of animals for whom we are comparing, say, average female and average male relatedness. The situation is complicated, however, when we acknowledge that in some taxa there can be considerable variation among individuals of "dispersal age," not to mention the fact that sometimes natal individuals of the "dispersing sex" do not, in fact, disperse, while sometimes those of the "philopatric" sex do. Similarly, for some primates the size, composition, and dispersion of individuals within social units might vary seasonally, or over shorter or longer time scales, in response to changing resource availability and changing risk of predation or disease. The within-group fission-fusion association patterns of spider monkeys and chimpanzees and the aggregation of various bands of hamadrayas baboons at protected sleeping sites offer two classic examples. Such variation obviously confuses attempts to partition the genetic variation present in a population hierarchically in the way that most population genetic analyses, including those discussed above, require.

In addition, ambiguity and confusion about the *natural boundaries of primate social units* also make it difficult for us to have confidence in our ability to characterize primate population genetic structure. Among several taxa of South American primates (e.g., bald-headed uacaris, *Cacajao calvus*, Bowler & Bodmer 2009; red woolly monkeys, Di Fiore & Fleischer 2005), it appears that different social units may sometimes coalesce peacefully into larger associations or "supergroups" for variable periods of time. However, the extent to which the composition of constituent social units remains constant or also shifts is unknown. Long-term fieldwork on red woolly monkeys in Ecuador and elsewhere indicates considerable overlap in home range between adjacent social groups (Di Fiore 2003b), extreme spatial dispersion of group members (Peres 1996), and both agonistic and highly tolerant interactions among groups (Di Fiore 1997). As a consequence, presumed social boundaries may be hard to discern and may even be quite permeable (chapter 2, this volume; Di Fiore & Schmitt unpublished data).

Approaching these issues from the perspective of individuals within a population, we need to decide how and when to best incorporate animals that move back and forth between social groups into characterization of the genetic structure of a given population. As one example of this, Di Fiore et al. (2009) have described the case of an adult male red woolly monkey that over the course of two years was sampled in two different social groups, as a member of a bachelor group and as a solitary individual. The question of where genetic data from this individual should be included in a hierarchical analysis of population structure has no obvious answer, although the temporal aspect of the sampling clearly provides some insight into individual dispersal history. The important point to recognize is that just as social organization is flexible and dynamic, so too is what we refer to as population genetic structure.

Developing a "Social Structure" Perspective for Primate Population Genetics

It is important to note that most of the links between primate social organization and population genetic structure discussed above have been proposed primarily on the basis of verbal arguments. Surprisingly, there has been only limited formal modeling of how we might expect the genetic structure of primate populations to relate to particular aspects of primate social organization or of the relative importance or magnitude of the effects that the various factors discussed above may have (but see Laporte & Charlesworth 2002). Two promising, though still incomplete, exceptions are provided by Chesser (1991a, b) and Ross (2001).

Chesser's (1991a, b) models were among the first formal attempts to try to accommodate the influence of particular social organizational features (specifically female philopatry with male dispersal and skewed male mating) into more traditional "demic" population genetic models. These early social structure models reformulate the fixation indices of classic population genetics to accommodate the idea that populations, rather than being divisible into subpopulations or "demes" within which mating can be treated as a random variable, are better considered as being divisible into "lineages" of related, philopatric animals (females), which all then mate with one or a few opposite-sexed dispersers (males) drawn from the larger populations. The theoretical effects of such elements of social organization are akin to those described above: an increase over time in the relatedness of animals within lineages, coupled with an excess in heterozygosity among those animals relative to the population as a whole, though both of these effects are reduced as paternity within female lineages is shared among more males (Chesser 1991a, b).

Ross (2001), too, summarized a simple mathematical framework for exploring the links between animal breeding systems and the kin composition of social groups. Though his framework was principally developed for consideration of colony structure in social insects, it has obvious extensions to other social animals. Ross (2001) notes that the relatedness among the progeny of a single cohort should be a complex function of three principal features of a taxon's breeding system: (1) the number of breeders of each sex contributing to a progeny set, (2) the genetic relatedness between the set of breeding individuals (between each male-female combination that contributes to the progeny set, as well as among the set of same-sex breeders of each sex, if more than one is present), and (3) the degree of reproductive skew among both males and females. All of these features are implicitly highlighted in the verbal models discussed above, but the advantage of a formal framework is to make explicit the effects of each of these aspects of the breeding system. Not surprisingly, the general conclusion that can be drawn from Ross's (2001) formal framework is that the degree of relatedness among progeny in a cohort will be greater when (1) fewer breeders contribute to the progeny set, (2) the relatedness among breeding individuals is high, and (3) both male and female reproduction is strongly skewed.

One key issue that limits the application of these kinds of models to primates and many other social vertebrates is that most primate social groups combine individuals from multiple generations and many cohorts. The models discussed above specifically apply either to the relatedness among individuals within a discrete generation of progeny (Ross 2001) or to the iterated effect of such progeny replacing the adults within their lineage (Chesser 1991a, b). Obviously, primate population geneticists are concerned not just with relatedness among discrete sets of progeny, but rather with characterizing the genetic structure of species that are typically organized into social groups with multiple, overlapping generations, and for which there also exists considerable interindividual variation in such life-history variables as age at first reproduction and interbirth interval. Thus, the dynamic aspect of group structure is, again, not served well by existing models—even those that begin to take a decidedly social-structure perspective.

One promising future direction for exploring the population genetic consequences of primate social organization is, however, through development of explicit "agent-based" computer simulation models that can accommodate differences in individual behavior and can also incorporate other complexities that are difficult to implement in mathematical models, such as continuous reproduction and overlapping generations (Grimm & Railsback 2005; Grimm et al. 2006). In such models, individual computerized "agents" are given attributes (e.g., age, sex, genotype, social group) and a set of rules that governs their behavior, reproduction, and survival (e.g., upon reaching a certain age they must leave a social group with a particular probability, choose a mate unrelated to themselves, and die with a variable probability also related to age). These agents are then allowed to interact with one another in a spatially and socially explicit environment (e.g., a population of a given number of social groups characterized by a specified adult-to-juvenile ratio and a specified sex ratio) while the simulation keeps track of the changing characteristics of individual agents and the population genetic parameters their genotypes imply. Such an approach could be used to effectively explore the influ-

ence of different social organizational variables on population genetic structure (Di Fiore 2010a, b). Agent-based models have been used in some studies of primate behavior and ecology (te Boekhorst & Hogeweg 1994; Hemelrijk 1999, 2002; Pepper & Smuts 2000; te Boekhorst & Hemelrijk 2000; Boyer et al. 2004; Ramos-Fernández et al. 2006; Bryson et al. 2007; reviewed by Dunbar 2002) but have yet to be put to widespread use by primatologists. Among population geneticists, the use of "forward-time" individual-based simulation models (Balloux 2001; Peng & Kimmel 2005; Strand & Niehaus 2007; Neuenschwander et al. 2008) is becoming more common for exploring the effects of various evolutionary forces (e.g., drift, selection, different mutation models), demographic scenarios (e.g., population growth versus stasis), dispersal models, and complex mating behavior on aspects of population genetic structure. To date, most applications of these kinds of models have not incorporated some of the unique features of primate social systems, such as the existence of multiple overlapping generations of individuals within breeding groups, but the approach offers considerable promise (Di Fiore 2010a, b).

Summary and Conclusions

This review has highlighted how several key aspects of primate social organization have dramatic consequences for the structuring of genetic variation within and among primate groups (table 12.2). First, dispersal affects the rate of gene flow among social units and genetic drift influences genetic variation within semi-isolated social units. As a consequence, *sex differences in patterns of philopatry and dispersal* as well as any *barriers to dispersal*—whether imposed by ecological or social factors—exert a very strong influence on the genetic structure of a population, particularly on patterns of relatedness within and between the sexes and on the rate at which breeding groups within a population diverge from one another genetically. Second, *the number of same-sex and opposite-sex animals* present in a group and the *nature of the competitive regime* among these individuals ultimately determines levels of reproductive skew seen among members of each sex, which in turn influences how closely offspring born within a group are related to one another. The degree of relatedness among progeny is also influenced by *how closely breeders are related* to one another. Third, *population density* and *the nature of relationships between group members and extragroup animals* (e.g., members of other groups or floater animals) also affect population genetic structure directly by influencing both the frequency of dispersal between groups and the "permeability" of groups to immigrating dispersers.

Lastly, it is important to reiterate that our understanding of the population genetic consequences of various aspects of primate (and other animal) social systems is still developing. While some generalities concerning the links between social organization and population genetic structure seem robust—and while some significant strides have been made to move from the simplistic "demic" models of classical population genetics to more appropriate "social structure" models (Sugg et al. 1996), which take into account such things as sex differences in mating and dispersal behavior—there is still considerable room for exploring and refining our theoretical models and for considering the consequences of other aspects of primate organization that have not been looked at in detail (e.g., territoriality and the effects of alloparental care). Agent-based simulation models may offer the most promising way forward in this respect. Though they lack the precision of formal mathematical models, they ought to be sufficiently flexible to allow us to incorporate some of the characteristic features of primate social systems that do not lend themselves to analytical solutions, such as nondiscrete generations and individual variation in age at first reproduction, interbirth interval, and other life-history traits.

References

Altmann, J. 2000. Models of outcome and process: Predicting the number of males in primate groups. In *Primate Males: Causes and Consequences of Variation in Group Composition* (ed. by P. M. Kappeler), 236–247. Cambridge: Cambridge University Press.

Altmann, J., Alberts, S. C., Haines, S. A., Dubach, J., Muruthi, P., Coote, T., Geffen, E., Cheesman, D. J., Mututua, R. S., Saiyalel, S. N., Wayne, R. K., Lacy, R. C. & Bruford, M. W. 1996. Behavior predicts genetic structure in a wild primate group. *Proceedings of the National Academy of Sciences, USA*, 93, 5797–5801.

Altmann, S. A. 1962. A field study of the sociobiology of rhesus monkeys, *Macaca mulatta. Annals of the New York Academy of Sciences*, 102, 338–435.

Altmann, S. A. & Altmann, J. A. 1979. Demographic constraints on behavior and social organization. In *Primate Ecology and Human Origins: Ecological Influences on Social Organization* (ed. by I. S. Bernstein & E. O. Smith), 47–63. New York: Garland Press.

Aoki, K. & Nozawa, K. 1984. Average coefficient of relationship within troops of the Japanese monkey and other primate species with reference to the possibility of group selection. *Primates*, 25, 171–184.

Avise, J. C. 1994. *Molecular Markers, Natural History and Evolution.* New York: Chapman and Hall.

———. 2004. *Molecular Markers, Natural History, and Evolution, 2nd Edition.* Sunderland, MA: Sinauer Associates.

Balloux, F. 2001. EASYPOP (version 1.7): A computer program for population genetics simulations. *Journal of Heredity*, 92, 301.

Bearder, S. K. 1987. Lorises, bushbabies, and tarsiers: Diverse societies in solitary foragers. In *Primate Societies* (ed. by B. B. Smuts, D. L. Cheney, R. M. Seyfarth, R. W. Wrangham & T. T. Struhsaker), 11–24. Chicago: University of Chicago Press.

Benirschke, K., Anderson, J. M. & Brownhill, L. E. 1962. Marrow chimerism in marmosets. *Science*, 138, 513–515.

Benirschke, K. & Brownhill, L. E. 1962. Further observations on marrow chimerism in marmosets. *Cytogenetics*, 1, 245–257.

Bowler, M. & Bodmer, R. 2009. Social behavior in fission-fusion groups of red uakari monkeys (*Cacajao calvus ucayalii*). *American Journal of Primatology*, 71, 976–987.

Boyer, D., Miramontes, O., Ramos-Fernández, G., Mateos, J. L. & Cocho, G. 2004. Modeling the searching behavior of social monkeys. *Physica A: Statistical Mechanics and its Applications*, 342, 329–335.

Bradley, B. J. 1999. Levels of selection, altruism, and primate behavior. *Quarterly Review of Biology*, 74, 171–194.

Bradley, B. J., Doran-Sheehy, D. M., Lukas, D., Boesch, C. & Vigilant, L. 2004. Dispersed male networks in western gorillas. *Current Biology*, 14, 510–513.

Brockelman, W. Y., Reichard, U., Treesucon, U. & Raemaekers, J. J. 1998. Dispersal, pair formation and social structure in gibbons (*Hylobates lar*). *Behavioral Ecology and Sociobiology*, 42, 329–339.

Bryson, J. J., Ando, Y. & Lehmann, H. 2007. Agent-based modelling as scientific method: A case study analysing primate social behaviour. *Philosophical Transactions of the Royal Society B: Biological Sciences*, 362, 1685–1698.

Buettner-Janusch, J., Olivier, T. J., Ober, C. L. & Chepko-Sade, B. D. 1983. Models for lineal effects in rhesus group fissions. *American Journal of Physical Anthropology*, 61, 347–353.

Chapais, B. 2001. Primate nepotism: What is the explanatory value of kin selection? *International Journal of Primatology*, 22, 203–229.

Cheney, D. L. & Seyfarth, R. M. 1977. Behavior of adult and immature male baboons during inter-group encounters. *Nature*, 269, 404–406.

———. 1983. Nonrandom dispersal in free-ranging vervet monkeys: Social and genetic consequences. *American Naturalist*, 122, 392–412.

Chepko-Sade, B. D. & Halpin, Z. T. 1987. *Mammalian Dispersal Patterns: The Effects of Social Structure on Population Genetics.* Chicago: University of Chicago Press.

Chepko-Sade, B. D. & Olivier, T. J. 1979. Coefficient of genetic relationship and the probability of intragenealogical fission in *Macaca mulatta. Behavioral Ecology and Sociobiology*, 5, 263–278.

Chepko-Sade, B. & Sade, D. 1979. Patterns of group splitting within matrilineal kinship groups: A study of social group structure in *Macaca mulatta* (Cercopithecidae, Primates). *Behavioral Ecology and Sociobiology*, 5, 67–86.

Chesser, R. K. 1991a. Gene diversity and female philopatry. *Genetics*, 127, 437–447.

———. 1991b. Influence of gene flow and breeding tactics on gene diversity within populations. *Genetics*, 129, 573–583.

Cheverud, J. M., Buettner-Janusch, J. & Sade, D. S. 1978. Social group fission and the origin of intergroup genetic differentiation among the rhesus monkeys of Cayo Santiago. *American Journal of Physical Anthropology*, 49, 449–456.

Chivers, D. J. & Raemaekers, J. J. 1980. Long-term changes in behaviour. In *Malayan Forest Primates: Ten Years' Study in Tropical Rain Forest* (ed. by D. J. Chivers), 209–258. New York: Plenum Press.

Clark, A. B. 1978. Sex ratio and local resource competition in a prosimian primate. *Science*, 201, 163–165.

Clutton-Brock, T. 1989. Mammalian mating systems. *Proceedings of the Royal Society B: Biological Sciences*, 236, 339–372.

Cowlishaw, G. & Dunbar, R. I. M. 1991. Dominance rank and mating success in male primates. *Animal Behaviour*, 41, 1045–1056.

Crockett, C. M. 1984. Emigration by female red howler monkeys and the case for female competition. In *Female Primates: Studies by Women Primatologists* (ed. by M. F. Small), 159–173. New York: Alan R. Liss.

Crockett, C. M. & Eisenberg, J. F. 1987. Howlers: Variations in group size and demography. In *Primate Societies* (ed. by B. B. Smuts, D. L. Cheney, R. M. Seyfarth, R. W. Wrangham & T. T. Struhsaker), 54–68. Chicago: University of Chicago Press.

De Ruiter, J. R. 2004. Genetic markers in primate studies: Elucidating behavior and its function. *International Journal of Primatology*, 25, 1173–1189.

De Ruiter, J. R. & Geffen, E. 1998. Relatedness of matrilines, dispersing males and social groups in long-tailed macaques (*Macaca fascicularis*). *Proceedings of the Royal Society B: Biological Sciences*, 265, 79–87.

Di Fiore, A. 1997. Ecology and behavior of lowland woolly monkeys (*Lagothrix lagotricha poeppigii*, Atelinae) in eastern Ecuador. PhD dissertation, University of California, Davis.

———. 2003a. Molecular genetic approaches to the study of primate behavior, social organization, and reproduction. *Yearbook of Physical Anthropology*, 46, 62–99.

———. 2003b. Ranging behavior and foraging ecology of lowland woolly monkeys (*Lagothrix lagotricha poeppigii*) in Yasuní National Park, Ecuador. *American Journal of Primatology*, 59, 47–66.

———. 2009. Genetic approaches to the study of dispersal and kinship in New World primates. In *South American Primates: Comparative Perspectives in the Study of Behavior, Ecology, and Conservation* (ed. by P. A. Garber, A. Estrada, J. C. Bicca-Marques, E. W. Heymann & K. B. Strier), 211–250. New York: Springer.

———. 2010a. Forward-time, individual-based simulations and their use in primate landscape genetics. In *XXIII Congress of the International Primatological Society*. Kyoto, Japan.

———. 2010b. The influence of social systems on primate population genetic structure: An agent-based modeling approach. In *Social Systems: Demographic and Genetic Issues*. Paimpont, France.

Di Fiore, A., Fernandez-Duque, E. & Hurst, D. 2007. Adult male replacement in socially monogamous equatorial saki monkeys (*Pithecia aequatorialis*). *Folia Primatologica*, 78, 88–98.

Di Fiore, A. & Fleischer, R. C. 2005. Social behavior, reproductive strategies, and population genetic structure of *Lagothrix poeppigii*. *International Journal of Primatology*, 26, 1137–1173.

Di Fiore, A., Lawler, R. L. & Gagneux, P. 2010. Molecular

primatology. In *Primates in Perspective, 2nd Edition* (ed. by C. Campbell, A. Fuentes, K. MacKinnon, S. Bearder & R. Stumpf), 390–416. New York: Oxford University Press.

Di Fiore, A., Link, A., Schmitt, C. A. & Spehar, S. N. 2009. Dispersal patterns in sympatric woolly and spider monkeys: Integrating molecular and observational data. *Behaviour*, 146, 437–470.

Di Fiore, A. & Rendall, D. 1994. Evolution of social organization: A reappraisal for primates by using phylogenetic methods. *Proceedings of the National Academy of Sciences, USA*, 91, 9941–9945.

Dittus, W. P. J. 1988. Group fission among wild toque macaques as a consequence of female resource competition and environmental stress. *Animal Behaviour*, 36, 1626–1645.

Drickamer, L. C. & Vessey, S. H. 1973. Group changing in free-ranging male rhesus monkeys. *Primates*, 14, 359–368.

Duggleby, C. 1977. Blood group antigens and the population genetics of *Macaca mulatta* on Cayo Santiago. II. Effects of social group division. *Yearbook of Physical Anthropology*, 20, 263–271.

Duggleby, C. R. 1978. Blood group antigens and the population genetics of *Macaca mulatta* on Cayo Santiago. I. Genetic differentiation of social groups. *American Journal of Physical Anthropology*, 48, 35–40.

Duggleby, C. R., Sade, D. S., Rawlins, R. G. & Kessler, M. J. 1984. Kin-structured migration and genetic differentiation. *American Journal of Physical Anthropology*, 63, 153.

Dunbar, R. I. M. 2002. Modeling primate behavioral ecology. *International Journal of Primatology*, 24, 785–819.

Eberle, M. & Kappeler, P. M. 2006. Family insurance: Kin selection and cooperative breeding in a solitary primate. *Behavioral Ecology and Sociobiology*, 60, 582–588.

Eriksson, J., Siedel, H., Lukas, D., Kayser, M., Erler, A., Hashimoto, C., Hohmann, G., Boesch, C. & Vigilant, L. 2006. Y-chromosome analysis confirms highly sex biased dispersal and suggests a low effective male population size in bonobos (*Pan paniscus*). *Molecular Ecology*, 15, 939–949.

Fernandez-Duque, E., Juárez, C. P. & Di Fiore, A. 2008. Adult male replacement and subsequent infant care by male and siblings in socially monogamous owl monkeys (*Aotus azarai*). *Primates*, 49, 81–84.

Fietz, J., Zischler, H., Schwiegk, C., Tomiuk, J., Dausmann, K. & Ganzhorn, J. 2000. High rates of extra-pair young in the pair-living fat-tailed dwarf lemur, *Cheirogaleus medius*. *Behavioral Ecology and Sociobiology*, 49, 8–17.

Fredsted, T., Pertoldi, C., Olesen, J. M., Eberle, M. & Kappeler, P. M. 2004. Microgeographic heterogeneity in spatial distribution and mtDNA variability of gray mouse lemurs (*Microcebus murinus*, Primates: Cheirogaleidae). *Behavioral Ecology and Sociobiology*, 56, 1–11.

Fredsted, T., Pertoldi, C., Schierup, M. & Kappeler, P. 2005. Microsatellite analyses reveal fine-scale genetic structure in grey mouse lemurs (*Microcebus murinus*). *Molecular Ecology*, 14, 2363–2372.

Fredsted, T., Schierup, M. H., Groeneveld, L. F. & Kappeler, P. M. 2007. Genetic structure, lack of sex-biased dispersal and behavioral flexibility in the pair-living fat-tailed dwarf lemur, *Cheirogaleus medius*. *Behavioral Ecology and Sociobiology*, 61, 943–954.

French, J. A. 1997. Proximate regulation of singular breeding in callitrichid primates. In *Cooperative Breeding in Mammals* (ed. by N. G. Solomon & J. A. French), 34–75. Cambridge: Cambridge University Press.

Gengozian, N., Batson, J. S. & Eide, P. 1964. Hematologic and cytogenetic evidence for hematopoietic chimerism in the marmoset, *Tamarinus nigricollis*. *Cytogenetics*, 3, 384–393.

Gengozian, N., Brewen, J. G., Preston, R. J. & Batson, J. S. 1980. Presumptive evidence for the absence of functional germ cell chimerism in the marmoset. *Journal of Medical Primatology*, 9, 9–27.

Gerloff, U., Hartung, B., Fruth, B., Hohmann, G. & Tautz, D. 1999. Intracommunity relationships, dispersal pattern, and paternity success in a wild living community of bonobos (*Pan paniscus*) determined from DNA analysis of faecal samples. *Proceedings of the Royal Society B: Biological Sciences*, 266, 1189–1195.

Goldberg, T. & Wrangham, R. W. 1997. Genetic correlates of social behaviour in wild chimpanzees: Evidence from mitochondrial DNA. *Animal Behaviour*, 54, 559–570.

Goldizen, A. 1987. Tamarins and marmosets: Communal care of offspring. In *Primate Societies* (ed. by B. B. Smuts, D. L. Cheney, R. M. Seyfarth, R. W. Wrangham & T. T. Struhsaker), 69–82. Chicago: University of Chicago Press.

Goossens, B., Setchell, J. M., James, S. S., Funk, S. M., Chikhi, L., Abulani, A., Ancrenaz, M., Lackman-Ancrenaz, I. & Bruford, M. W. 2006. Philopatry and reproductive success in Bornean orang-utans (*Pongo pygmaeus*). *Molecular Ecology*, 15, 2577–2588.

Goudet, J., Perrin, N. & Waser, P. 2002. Tests for sex-biased dispersal using bi-parentally inherited genetic markers. *Molecular Ecology*, 11, 1103–1114.

Gouzoules, H. & Gouzoules, S. 1987. Kinship. In *Primate Societies* (ed. by B. B. Smuts, D. L. Cheney, R. M. Seyfarth, R. W. Wrangham & T. T. Struhsaker), 299–305. Chicago: University of Chicago Press.

Greenwood, P. J. 1980. Mating systems, philopatry and dispersal in birds and mammals. *Animal Behaviour*, 28, 1140–1162.

Grimm, V., Berger, U., Bastiansen, F. & Eliassen, S. 2006. A standard protocol for describing individual-based and agent-based models. *Ecological Modelling*, 198, 115–126.

Grimm, V. & Railsback, S. 2005. *Individual-Based Modeling and Ecology*. Princeton, NJ: Princeton University Press.

Haig, D. 1999. What is a marmoset? *American Journal of Primatology*, 49, 285–296.

Hamilton, W. D. 1964a. The genetical evolution of social behavior I. *Journal of Theoretical Biology*, 7, 1–16.

———. 1964b. The genetical evolution of social behavior II. *Journal of Theoretical Biology*, 7, 17–52.

Hammond, R. L., Lawson Handley, L. J., Winney, B. J., Bruford, M. W. & Perrin, N. 2006. Genetic evidence for female-biased dispersal and gene flow in a polygynous primate. *Proceedings of the Royal Society B: Biological Sciences*, 273, 479–484.

Hapke, A., Zinner, D. & Zischler, H. 2001. Mitochondrial DNA variation in Eritrean hamadryas baboons (*Papio hamadryas hamadryas*): Life history influences population genetic structure. *Behavioral Ecology and Sociobiology*, 50, 483–492.

Harris, T. R., Caillaud, D., Chapman, C. A. & Vigilant, L. 2009. Neither genetic nor observational data alone are sufficient for understanding sex-biased dispersal in a social-group-living species. *Molecular Ecology*, 18, 1777–1790.

Hemelrijk, C. K. 1999. An individual-orientated model of the emergence of despotic and egalitarian societies. *Proceedings of the Royal Society B: Biological Sciences*, 266, 361–369.

———. 2002. Self-organizing properties of primate social behavior: A hypothesis for intersexual rank overlap in chimpanzees and bonobos. *Evolutionary Anthropology*, 11, 91–94.

Hunley, K. L., Spence, J. E. & Merriwether, D. A. 2008. The impact of group fissions on genetic structure in native South America and implications for human evolution. *American Journal of Physical Anthropology*, 135, 195–205.

Ichino, S. 2006. Troop fission in wild ring-tailed lemurs (*Lemur catta*) at Berenty, Madagascar. *American Journal of Primatology*.

Ichino, S. & Koyama. 2006. Social changes in a wild population of ringtailed lemurs (*Lemur catta*) at Berenty, Madagascar. In *Ringtailed Lemur Biology: Lemur catta in Madagascar* (ed. by A. Jolly, R. W. Sussman, N. Koyama & H. Rasamimanana), 233–244. New York: Springer.

Isbell, L. A. 2004. Is there no place like home? Ecological bases of female dispersal and philopatry and their consequences for the formation of kin groups. In *Kinship and Behavior in Primates* (ed. by B. Chapais & C. M. Berman), 71–108. Oxford: Oxford University Press.

Isbell, L. A. & Young, T. P. 2002. Ecological models of female social relationships in primates: Similarities, disparities, and some directions for future clarity. *Behaviour*, 139, 177–202.

Jack, K. M. & Fedigan, L. 2004. Male dispersal patterns in white-faced capuchins, *Cebus capucinus*. Part 1: Patterns and causes of natal emigration. *Animal Behaviour*, 67, 761–769.

Johnson, M. L. & Gaines, M. S. 1990. Evolution of dispersal: Theoretical models and empirical tests using birds and mammals. *Annual Review of Ecology and Systematics*, 21, 449–480.

Jolly, C. & Brett, F. 1973. Genetic markers and baboon biology. *Journal of Medical Primatology*, 2, 85–99.

Kappeler, P. M., Wimmer, B., Zinner, D. & Tautz, D. 2002. The hidden matrilineal structure of a solitary lemur: Implications for primate social evolution. *Proceedings of the Royal Society B: Biological Sciences*, 269, 1755–1763.

Koyama, N. F. 2003. Matrilineal cohesion and social networks in *Macaca fuscata*. *International Journal of Primatology*, 24, 797–811.

Kuester, J. & Paul, A. 1997. Group fission in Barbary macaques (*Macaca sylvanus*) at Affenberg Salem. *International Journal of Primatology*, 18, 941–996.

Kutsukake, N. & Nunn, C. L. 2006. Comparative tests of reproductive skew in male primates: The roles of demographic factors and incomplete control. *Behavioral Ecology and Sociobiology*, 60, 695–706.

Langergraber, K. E., Mitani, J. C. & Vigilant, L. 2007a. The limited impact of kinship on cooperation in wild chimpanzees. *Proceedings of the National Academy of Sciences, USA*, 104, 7786–7790.

Langergraber, K. E., Siedel, H., Mitani, J. C., Wrangham, R. W., Reynolds, V., Hunt, K. & Vigilant, L. 2007b. The genetic signature of sex-biased migration in patrilocal chimpanzees and humans. *PLoS One*, 2, e973.

Langergraber, K., Mitani, J. & Vigilant, L. 2009. Kinship and social bonds in female chimpanzees (*Pan troglodytes*). *American Journal of Primatology*, 71, 840–851.

Laporte, V. & Charlesworth, B. 2002. Effective population size and population subdivision in demographically structured populations. *Genetics*, 162, 501–519.

Lawler, R. R., Richard, A. F. & Riley, M. A. 2003. Genetic population structure of the white sifaka (*Propithecus verreauxi verreauxi*) at Beza Mahafaly Special Reserve, southwest Madagascar (1992–2001). *Molecular Ecology*, 12, 2307–2317.

Lawson Handley, L. J. & Perrin, N. 2007. Advances in our understanding of mammalian sex-biased dispersal. *Molecular Ecology*, 16, 1559–1578.

Leighton, D. R. 1987. Gibbons: Territoriality and monogamy. In *Primate Societies* (ed. by B. B. Smuts, D. L. Cheney, R. M. Seyfarth, R. W. Wrangham & T. T. Struhsaker), 135–145. Chicago: University of Chicago Press.

Lukas, D., Reynolds, V., Boesch, C. & Vigilant, L. 2005. To what extent does living in a group mean living with kin? *Molecular Ecology*, 14, 2181–2196.

Meikle, D. B. & Vessey, S. H. 1981. Nepotism among rhesus monkey brothers. *Nature*, 294, 160–161.

Melnick, D. J. 1987. The genetic consequences of primate social organization: A review of macaques, baboons, and vervet monkeys. *Genetica*, 73, 117–135.

Melnick, D. J. & Hoelzer, G. A. 1992. Differences in male and female macaque dispersal lead to contrasting distributions of nuclear and mitochondrial DNA variation. *International Journal of Primatology*, 13, 379–393.

———. 1996. The population genetic consequences of macaque social organization and behaviour. In *Evolution and Ecology of Macaque Societies* (ed. by J. E. Fa & D. G. Lindburg), 413–443. Cambridge: Cambridge University Press.

Melnick, D. J. & Kidd, K. K. 1983. The genetic consequences of social group fission in a wild population of rhesus monkeys (*Macaca mulatta*). *Behavioral Ecology and Sociobiology*, 12, 229–236.

Melnick, D. J. & Pearl, M. C. 1987. Cercopithecines in multimale groups: Genetic diversity and population structure. In *Primate Societies* (ed. by B. B. Smuts, D. L. Cheney, R. M. Seyfarth, R. W. Wrangham & T. T. Struhsaker), 121–134. Chicago: University of Chicago Press.

Ménard, N. & Vallet, D. 1993. Dynamics of fission in a wild Barbary macaque group (*Macaca sylvanus*). *International Journal of Primatology*, 14, 479–500.

Mesterton-Gibbons, M. & Dugatkin, L. A. 1992. Cooperation among unrelated individuals: Evolutionary factors. *Quarterly Review of Biology*, 67, 267–281.

Missakian, E. A. 1973. The timing of fission among free-ranging rhesus monkeys. *American Journal of Physical Anthropology*, 38, 621–624.

Mitani, J. C., Merriwether, D. A. & Zhang, C. 2000. Male affiliation, cooperation and kinship in wild chimpanzees. *Animal Behaviour*, 59, 885–893.

Moore, J. & Ali, R. 1984. Are dispersal and inbreeding avoidance related? *Animal Behaviour*, 32, 94–112.

Morin, P. A., Moore, J. J., Chakraborty, R., Jin, L., Goodall, J. & Woodruff, D. S. 1994. Kin selection, social structure, gene flow, and the evolution of chimpanzees. *Science*, 265, 1193–1201.

Nash, L. T. 1976. Troop fission in free-ranging baboons in the Gombe Stream National Park, Tanzania. *American Journal of Physical Anthropology*, 48, 63–77.

Neel, J. V. 1978. The population structure of an Amerindian tribe, the Yanomama. *Annual Review of Genetics*, 12, 365–413.

Neel, J. V. & Chagnon, N. A. 1968. Demography of two tribes of primitive relatively unacculturated American Indians. *Proceedings of the National Academy of Sciences, USA*, 59, 680–689.

Neuenschwander, S., Hospital, F., Guillaume, F. & Goudet, J. 2008. quantiNemo: An individual-based program to simulate quantitative traits with explicit genetic architecture in a dynamic metapopulation. *Bioinformatics*, 24, 1552–1553.

Nievergelt, C. M., Mutschler, T., Feistner, A. T. C. & Woodruff, D. S. 2002. Social system of the Alaotran gentle lemur (*Hapalemur griseus alaotrensis*): Genetic characterization of group composition and mating system. *American Journal of Primatology*, 57, 157–176.

Noë, R. & Hammerstein, P. 1994. Biological markets: Supply and demand determine the effect of partner choice in cooperation, mututalism, and mating. *Behavioral Ecology and Sociobiology*, 35, 1–11.

Noë, R., van Hooff, J. A. R. A. M. & Hammerstein, P. 2001. *Economics in Nature: Social Dilemmas, Mate Choice, and Biological Markets*. Cambridge: Cambridge University Press.

Nozawa, K. 1972. Population genetics of Japanese monkeys: I. Estimation of the effective troop size. *Primates*, 13, 381–393.

Nozawa, K., Shotake, T., Kawamoto, Y. & Tanabe, Y. 1982. Population genetics of Japanese monkeys: II. Blood protein polymorphisms and population structure. *Primates*, 23, 252–271.

Olivier, T. J., Ober, C. L., Buettner-Janusch, J. & Sade, D. S. 1981. Genetic differentiation among matrilines in social groups of rhesus monkeys. *Behavioral Ecology and Sociobiology*, 8, 279–285.

Ostner, J., Nunn, C. L. & Schülke, O. 2008. Female reproductive synchrony predicts skewed paternity across primates. *Behavioral Ecology*, 19, 1150–1158.

Palombit, R. A. 1994. Dynamic pair bonds in hylobatids: Implications regarding monogamous social systems. *Behaviour*, 128, 65–101.

Peng, B. & Kimmel, M. 2005. SimuPOP: A forward-time population genetics simulation environment. *Bioinformatics*, 21, 3686–3687.

Pepper, J. W. & Smuts, B. B. 2000. The evolution of cooperation in an ecological context: An agent-based model. In: *Dynamics of Human and Primate Societies: Agent-Based Modeling of Social and Spatial Processes* (ed. by T. A. Kohler & G. J. Gumerman), 45–76. Oxford: Oxford University Press.

Peres, C. A. 1996. Use of space, spatial group structure, and foraging group size of gray woolly monkeys (*Lagothrix lagotricha cana*) at Urucu, Brazil. In *Adaptive Radiations of Neotropical Primates* (ed. by M. A. Norconk, A. L. Rosenberger & P. A. Garber), 467–488. New York: Plenum Press.

Pope, T. R. 1990. The reproductive consequences of male cooperation in the red howler monkey: Paternity exclusion in multi-male and single-male troops using genetic markers. *Behavioral Ecology and Sociobiology*, 27, 439–446.

———. 1992. The influence of dispersal patterns and mating systems on genetic differentiation within and between populations of the red howler monkey (*Alouatta seniculus*). *Evolution*, 46, 1112–1128.

———. 1998. Effects of demographic change on group kin structure and gene dynamics of populations of red howling monkeys. *Journal of Mammalogy*, 79, 692–712.

———. 2000. Reproductive success increases with degree of kinship in cooperative coalitions of female red howler monkeys (*Alouatta seniculus*). *Behavioral Ecology and Sociobiology*, 48, 253–267.

Pusey, A. E. & Packer, C. 1987. Dispersal and philopatry. In *Primate Societies* (ed. by B. B. Smuts, D. L. Cheney, R. M. Seyfarth, R. W. Wrangham & T. T. Struhsaker), 250–266. Chicago: University of Chicago Press.

Radespiel, U., dal Secco, V., Drögemüller, C., Braune, P., Labes, E. & Zimmermann, E. 2002. Sexual selection, multiple mating and paternity in grey mouse lemurs, *Microcebus murinus*. *Animal Behaviour*, 63, 259–268.

Radespiel, U., Juric, M. & Zimmermann, E. 2009. Sociogenetic structures, dispersal and the risk of inbreeding in a small nocturnal lemur, the golden-brown mouse lemur (*Microcebus ravelobensis*). *Behaviour*, 146, 607–628.

Radespiel, U., Rakotondravony, R. & Chikhi, L. 2008. Natural and anthropogenic determinants of genetic structure in the largest remaining population of the endangered golden-brown mouse lemur, *Microcebus ravelobensis*. *American Journal of Primatology*, 70, 860–870.

Radespiel, U., Sarikaya, Z., Zimmermann, E. & Bruford, M. W. 2001. Sociogenetic structure in a free-living nocturnal primate population: Sex-specific differences in the grey mouse lemur (*Microcebus murinus*). *Behavioral Ecology and Sociobiology*, 50, 493–502.

Ramos-Fernández, G., Boyer, D. & Gomez, V. P. 2006. A complex social structure with fission-fusion properties can emerge from a simple foraging model. *Behavioral Ecology and Sociobiology*, 60, 536–549.

Robinson, J. G. 1988. Demography and group structure in wedge-capped capuchin monkeys, *Cebus olivaceus*. *Behaviour*, 104, 202–232.

Ross, C. N., French, J. A. & Ortí, G. 2007. Germ-line chimerism and paternal care in marmosets (*Callithrix kuhlii*). *Proceedings of the National Academy of Sciences, USA*, 104, 6278–6282.

Ross, K. G. 2001. Molecular ecology of social behaviour: Analyses of breeding systems and genetic structure. *Molecular Ecology*, 10, 265–284.

Rothe, H. & Koenig, A. 1991. Sex ratio in newborn common marmosets (*Callithrix jacchus*): No indication for a functional germ cell chimerism. *Zeitschrift für Säugetierkunde*, 56, 318–320.

Schoof, V. A. M., Jack, K. M. & Isbell, L. A. 2009. What traits promote male parallel dispersal in primates? *Behaviour*, 146, 701–726.

Shimada, M. K. 2000. Geographic distribution of mitochondrial DNA variations among grivet (*Cercopithecus aethiops aethiops*) populations in central Ethiopia. *International Journal of Primatology*, 21, 113–129.

Shotake, T. 1981. Population genetical study of natural hybridization between *Papio anubis* and *P. hamadryas*. *Primates*, 22, 285–308.

Shotake, T. & Nozawa, K. 1984. Blood protein variations in baboons. II. Genetic variability within and among herds of gelada baboons in the central Ethiopian plateau. *Journal of Human Evolution*, 13, 265–274.

Silk, J. B. 2002. Kin selection in primate groups. *International Journal of Primatology*, 23, 849–875.

Smith, K. 2000. Paternal kin matter: The distribution of social behavior among wild, adult female baboons. PhD dissertation, University of Chicago.

Smith, K., Alberts, S. C. & Altmann, J. 2003. Wild female baboons bias their social behaviour towards paternal half-sisters. *Proceedings of the Royal Society B: Biological Sciences*, 270, 503–510.

Smouse, P. E., Vitzthum, V. J. & Neel, J. V. 1981. The impact of random and lineal fission on the genetic divergence of small human groups: A case study among the Yanomama. *Genetics*, 98, 179–197.

Sterck, E., Watts, D. & van Schaik, C. P. 1997. The evolution of female social relationships in nonhuman primates. *Behavioral Ecology and Sociobiology*, 41, 291–309.

Storz, J. 1999. Genetic consequences of mammalian social structure. *Journal of Mammalogy*, 80, 553–569.

Strand, A. E. & Niehaus, J. M. 2007. KERNELPOP, a spatially explicit population genetic simulation engine. *Molecular Ecology Notes*, 7, 969–973.

Sugg, D. W., Chesser, R. K., Dobson, F. S. & Hoogland, J. L. 1996. Population genetics meets behavioral ecology. *Trends in Ecology & Evolution*, 11, 338–342.

Sugiyama, Y. 1976. Life history of male Japanese monkeys. *Advances in the Study of Behavior*, 7, 255–284.

Sussman, R. W. & Garber, P. A. 1987. A new interpretation of the social organization and mating system of the Callitrichidae. *International Journal of Primatology*, 8, 73–92.

Tardif, S. D. 1997. The bioenergetics of parental behavior and the evolution of alloparental care in marmosets and tamarins. In *Cooperative Breeding in Mammals* (ed. by N. G. Solomon & J. A. French), 11–33. New York: Cambridge University Press.

Tardif, S. D., Harrison, M. L. & Simek, M. A. 1993. Communal infant care in marmosets and tamarins: Relation to energetics, ecology, and social organization. In *Marmosets and Tamarins: Systematics, Behaviour, and Ecology* (ed. by A. B. Rylands), 220–234. Oxford: Oxford University Press.

Te Boekhorst, I. J. A. & Hemelrijk, C. K. 2000. Nonlinear and synthetic models for primate societies. In *Dynamics of Human and Primate Societies: Agent-Based Modeling of Social and Spatial Processes* (ed. by T. A. Kohler & G. J. Gumerman), 19–44. Oxford: Oxford University Press.

Te Boekhorst, I. J. A. & Hogeweg, P. 1994. Self-structuring in artificial "chimps" offers new hypotheses for male grouping in chimpanzees. *Behaviour*, 130, 229–252.

Tilson, R. L. 1981. Family formation strategies of Kloss's gibbons. *Folia Primatologica*, 35, 259–287.

Trivers, R. L. 1972. Parental investment and sexual selection. In *Sexual Selection and the Descent of Man: 1871–1971* (ed. by B. G. Campbell), 136–179. Chicago: Aldine Publishing.

———. 1985. *Social Evolution*. Menlo Park, CA: Benjamin/Cummings Publishing Company, Inc.

Van Horn, R. C., Buchan, J. C., Altmann, J. & Alberts, S. C. 2007. Divided destinies: Group choice by female savannah baboons during social group fission. *Behavioral Ecology and Sociobiology*, 61, 1823–1837.

Van Schaik, C. P. 1989. The ecology of social relationships amongst female primates. In *Comparative Socioecology: The Behavioural Ecology of Humans and Other Mammals* (ed. by V. Standen & R. Foley), 195–218. Oxford: Blackwell Scientific Publications.

Vigilant, L., Hofreiter, M., Siedel, H. & Boesch, C. 2001. Paternity and relatedness in wild chimpanzee communities. *Proceedings of the National Academy of Sciences, USA*, 98, 12890–12895.

Wade, M. J. & McCauley, D. E. 1988. Extinction and recolonization: Their effects on the genetic differentiation of local populations. *Evolution*, 42, 995–1005.

Wallman, J., Hoelzer, G. A. & Melnick, D. J. 1996. The effects of social structure, geographical structure, and population size on the evolution of mitochondrial DNA: I. A simulation model. *Computer Applications in the Biosciences*, 12, 481–489.

Waser, P. M. & Jones, W. T. 1983. Natal philopatry among solitary mammals. *Quarterly Review of Biology*, 58, 355–390.

West-Eberhard, M. J. 1975. The evolution of social behavior by kin selection. *Quarterly Review of Biology*, 50, 1–33.

Whitlock, M. C. & McCauley, D. E. 1990. Some population genetic consequences of colony formation and extinction: Genetic correlations within founding groups. *Evolution*, 44, 1717–1724.

Widdig, A., Nürnberg, P., Bercovitch, B., Trefilov, A., Berard, J. B., Kessler, M. J., Schmidtke, J., Streich, W. J. & Krawczak, M. 2006. Consequences of group fission for the patterns of relatedness among rhesus macaques. *Molecular Ecology*, 15, 3825–3832.

Widdig, A., Nürnberg, P., Krawczak, M., Streich, W. J. & Bercovitch, F. B. 2001. Paternal relatedness and age proximity regulate social relationships among adult female rhesus macaques. *Proceedings of the National Academy of Sciences, USA*, 98, 13769–13773.

Wimmer, B. & Kappeler, P. M. 2002. The effects of sexual selection and life history on the genetic structure of redfronted lemur, *Eulemur fulvus rufus*, groups. *Animal Behaviour*, 64, 557–568.

Wimmer, B., Tautz, D. & Kappeler, P. M. 2002. The genetic population structure of the gray mouse lemur (*Microcebus murinus*), a basal primate from Madagascar. *Behavioral Ecology and Sociobiology*, 52, 166–175.

Wrangham, R. W. 1980. An ecological model of female-bonded primate groups. *Behaviour*, 75, 262–300.

Wright, S. J. 1951. The genetical structure of populations. *Annals of Eugenics*, 15, 323–353.

———. 1965. The interpretation of population structure by F-statistics with special regard to systems of mating. *Evolution*, 19, 395–420.

References for Table 12.1

Altmann, J., Alberts, S. C., Haines, S. A., Dubach, J., Muruthi, P., Coote, T., Geffen, E., Cheesman, D. J., Mututua, R. S., Saiyalel, S. N., et al. 1996. Behavior predicts genetic structure in a wild primate group. *Proceedings of the National Academy of Sciences, USA*, 93, 5797–5801.

De Ruiter, J. R. & Geffen, E. 1998. Relatedness of matrilines, dispersing males and social groups in long-tailed macaques (*Macaca fascicularis*). *Proceedings of the Royal Society B: Biological Sciences*, 265, 79–87.

Di Fiore, A. 2009. Genetic approaches to the study of dispersal and kinship in New World primates. In: *South American Primates: Comparative Perspectives in the Study of Behavior,*

Ecology, and Conservation (ed. by Garber, P.A., Estrada, A., Bicca-Marques, J.C., Heymann, E.W., and Strier, K.B.), 211–250. New York: Springer.

Di Fiore, A. & Fleischer, R. C. 2005. Social behavior, reproductive strategies, and population genetic structure of *Lagothrix poeppigii*. *International Journal of Primatology*, 26, 1137–1173.

Eriksson, J., Siedel, H., Lukas, D., Kayser, M., Erler, A., Hashimoto, C., Hohmann, G., Boesch, C. & Vigilant, L. 2006. Y-chromosome analysis confirms highly sex-biased dispersal and suggests a low effective male population size in bonobos (*Pan paniscus*). *Molecular Ecology*, 15, 939–949.

Fredsted, T., Pertoldi, C., Olesen, J. M., Eberle, M. & Kappeler, P. M. 2004. Microgeographic heterogeneity in spatial distribution and mtDNA variability of gray mouse lemurs (*Microcebus murinus*, Primates: Cheirogaleidae). *Behavioral Ecology and Sociobiology*, 56, 1–11.

Fredsted, T., Pertoldi, C., Schierup, M. & Kappeler, P. 2005. Microsatellite analyses reveal fine-scale genetic structure in grey mouse lemurs (*Microcebus murinus*). *Molecular Ecology*, 14, 2363–2372.

Fredsted, T., Schierup, M. H., Groeneveld, L. F. & Kappeler, P. M. 2007. Genetic structure, lack of sex-biased dispersal and behavioral flexibility in the pair-living fat-tailed dwarf lemur, *Cheirogaleus medius*. *Behavioral Ecology and Sociobiology*, 61, 943–954.

Gerloff, U., Hartung, B., Fruth, B., Hohmann, G. & Tautz, D. 1999. Intracommunity relationships, dispersal pattern, and paternity success in a wild living community of bonobos (*Pan paniscus*) determined from DNA analysis of faecal samples. *Proceedings of the Royal Society B: Biological Sciences*, 266, 1189–1195.

Goldberg, T. & Ruvolo, M. 1997. The geographic apportionment of mitochondrial genetic diversity in east African chimpanzees, *Pan troglodytes schweinfurthii*. *Molecular Biology and Evolution*, 14, 976–984.

Goossens, B., Setchell, J. M., James, S. S., Funk, S. M., Chikhi, L., Abulani, A., Ancrenaz, M., Lackman-Ancrenaz, I. & Bruford, M. W. 2006. Philopatry and reproductive success in Bornean orang-utans (*Pongo pygmaeus*). *Molecular Ecology*, 15, 2577–2588.

Hammond, R. L., Lawson Handley, L. J., Winney, B. J., Bruford, M. W. & Perrin, N. 2006. Genetic evidence for female-biased dispersal and gene flow in a polygynous primate. *Proceedings of the Royal Society B: Biological Sciences*, 273, 479–484.

Hapke, A., Zinner, D. & Zischler, H. 2001. Mitochondrial DNA variation in Eritrean hamadryas baboons (*Papio hamadryas hamadryas*): Life history influences population genetic structure. *Behavioral Ecology and Sociobiology*, 50, 483–492.

Harris, T. R., Caillaud, D., Chapman, C. A. & Vigilant, L. 2009. Neither genetic nor observational data alone are sufficient for understanding sex-biased dispersal in a social-group-living species. *Molecular Ecology*, 18, 1777–1790.

Hoelzer, G., Dittus, W., Ashley, M. & Melnick, D. 1994. The local distribution of highly divergent mitochondrial DNA haplotypes in toque macaques *Macaca sinica* at Polonnaruwa, Sri Lanka. *Molecular Ecology*, 3, 451–458.

Kappeler, P. M., Wimmer, B., Zinner, D. & Tautz, D. 2002. The hidden matrilineal structure of a solitary lemur: Implications for primate social evolution. *Proceedings of the Royal Society B: Biological Sciences*, 269, 1755–1763.

Lappan, S. 2007. Social relationships among males in multimale siamang groups. *International Journal of Primatology*, 28, 369–387.

Lawler, R. R., Richard, A. F. & Riley, M. A. 2003. Genetic population structure of the white sifaka (*Propithecus verreauxi verreauxi*) at Beza Mahafaly Special Reserve, southwest Madagascar (1992–2001). *Molecular Ecology*, 12, 2307–2317.

Liu, Z., Ren, B., Wei, F., Long, Y., Hao, Y. & Li, M. 2007. Phylogeography and population structure of the Yunnan snub-nosed monkey (*Rhinopithecus bieti*) inferred from mitochondrial control region DNA sequence analysis. *Molecular Ecology*, 16, 3334–3349.

Lukas, D., Reynolds, V., Boesch, C. & Vigilant, L. 2005. To what extent does living in a group mean living with kin? *Molecular Ecology*, 14, 2181–2196.

Melnick, D. J. & Hoelzer, G. A. 1992. Differences in male and female macaque dispersal lead to contrasting distributions of nuclear and mitochondrial DNA variation. *International Journal of Primatology*, 13, 379–393.

———. 1996. The population genetic consequences of macaque social organization and behaviour. In *Evolution and Ecology of Macaque Societies* (ed. by Fa, J. E. & Lindburg, D. G.), 413–443. Cambridge: Cambridge University Press.

Mitani, J. C., Merriwether, D. A. & Zhang, C. 2000. Male affiliation, cooperation and kinship in wild chimpanzees. *Animal Behaviour*, 59, 885–893.

Morin, P. A., Moore, J. J., Chakraborty, R., Jin, L., Goodall, J. & Woodruff, D. S. 1994. Kin selection, social structure, gene flow, and the evolution of chimpanzees. *Science*, 265, 1193–1201.

Nievergelt, C., Digby, L., Ramakrishnan, U. & Woodruff, D. 2000. Genetic analysis of group composition and breeding system in a wild common marmoset (*Callithrix jacchus*) population. *International Journal of Primatology*, 21, 1–20.

Nievergelt, C. M., Mutschler, T., Feistner, A. T. C. & Woodruff, D. S. 2002. Social system of the Alaotran gentle lemur (*Hapalemur griseus alaotrensis*): Genetic characterization of group composition and mating system. *American Journal of Primatology*, 57, 157–176.

Perwitasari-Farajallah, D., Kawamoto, Y. & Suryabroto, B. 1999. Variation in blood proteins and mitochondrial DNA within and between local populations of longtail macaques, *Macaca fascicularis* on the island of Java, Indonesia. *Primates*, 40, 581–595.

Radespiel, U., Rakotondravony, R. & Chikhi, L. 2008. Natural and anthropogenic determinants of genetic structure in the largest remaining population of the endangered golden-brown mouse lemur, *Microcebus ravelobensis*. *American Journal of Primatology* 70, 860–870.

Radespiel, U., Sarikaya, Z., Zimmermann, E. & Bruford, M. W. 2001. Sociogenetic structure in a free-living nocturnal primate population: Sex-specific differences in the grey mouse lemur (*Microcebus murinus*). *Behavioral Ecology and Sociobiology*, 50, 493–502.

Rosenblum, L., Supriatna, J., Hasan, M. & Melnick, D. 1997a. High mitochondrial DNA diversity with little structure within and among leaf monkey populations (*Trachypithecus cristatus* and *Trachypithecus auratus*). *International Journal of Primatology*, 18, 1005–1028.

Rosenblum, L., Supriatna, J. & Melnick, D. 1997b. Phylogeographic analysis of pigtail macaque populations (*Macaca nemestrina*) inferred from mitochondrial DNA. *American Journal of Physical Anthropology*, 104, 35–45.

Shimada, M. K. 2000. Geographic distribution of mitochondrial DNA variations among grivet (*Cercopithecus aethiops aethiops*) populations in central Ethiopia. *International Journal of Primatology*, 21, 113–129.

Vigilant, L., Hofreiter, M., Siedel, H. & Boesch, C. 2001. Paternity and relatedness in wild chimpanzee communities. *Proceedings of the National Academy of Sciences, USA*, 98, 12890–12895.

Wimmer, B. & Kappeler, P. M. 2002. The effects of sexual selection and life history on the genetic structure of redfronted lemur, *Eulemur fulvus rufus*, groups. *Animal Behaviour*, 64, 557–568.

Winney, B. J., Hammond, R. L., Macasero, W., Flores, B., Boug, A., Biquand, V., Biquand, S. & Bruford, M. W. 2004. Crossing the Red Sea: Phylogeography of the hamadryas baboon, *Papio hamadryas hamadryas*. *Molecular Ecology*, 13, 2819–2827.

Chapter 13 Human Survival and Life History in Evolutionary Perspective

Michael Gurven

A PRIMARY GOAL of evolutionary anthropology is to determine and quantify differences between humans and other species, especially our primate relatives, and to reconstruct the evolutionary history of our species. As a species of seven billion individuals, humans occupy almost every habitat on the planet. Our exceptionally long lives, encephalized brains, extreme sociality, and penchant for cumulative cultural learning have all likely contributed to the biological success of *Homo sapiens*. Evolutionary anthropology has the goal of describing human universals while also explaining variation among and within human populations in genes, behavior, psychology, and culture.

This chapter focuses on human survival and life history: theoretical approaches and key features of human adaptability across time and space. I focus on small-scale societies of hunter-gatherers, because more than 90% of human history has been spent living in them. Even though domestication is a more recent feature of human societies, I also include forager-horticulturalists because they share many features with hunter-gatherers, such as natural fertility, egalitarianism, kin-based society, high work effort, and similar life spans. The first section summarizes several evolutionary approaches to studying humans. The second section characterizes human subsistence and sociality as important aspects of evolved life history. The third section integrates cultural learning and psychological adaptations related to social cognition and predator avoidance. I conclude by suggesting directions for future research.

Human Evolutionary Behavioral Sciences

Several complementary approaches comprise the evolutionary study of humans: human behavioral ecology, dual gene-culture inheritance theory, and evolutionary psychology. Each has been influential in helping to understand aspects of human nature and the human condition. This three-pronged approach to studying humans is no doubt linked to the fact that the humans studying and being studied are self-reflective, culturally rich, loquacious, and disputatious primates.

Human Behavioral Ecology

Human behavioral ecology (HBE) applies principles of natural selection to explain behavioral and cultural diversity in human populations (Borgerhoff Mulder 1991; Cronk 1991; Smith & Winterhalder 1992). It explores how features of the physical and social environment shape the suite of behaviors or "strategies" of individuals, and applies cost-benefit logic of constrained optimization to design models and to make formal predictions about the conditions that favor particular behaviors. Its roots are in biology (evolutionary biology, animal behavior, population and community ecology, life-history theory), anthropology (cultural ecology, hunter-gatherer studies), and economics (microeconomics of consumer choice). Because of its focus on the adaptive nature of behavior, the HBE tradition studies behavioral and cultural traits likely to have direct or indirect fitness consequences. These include the study of subsistence, mating, parenting, and costly social behaviors (Cronk et al. 2000). In all human societies, people extract resources from their environment; find mates; defend access to resources; protect, feed, and care for offspring; and form and maintain social partners and alliances. People trade off time and energy among these tasks. The reliance on trade-offs, borrowed from life-history theory in biology (chapter 10, this volume), is a fundamental tool in HBE. Fitness payoffs—whether measured directly, in terms of survival and reproductive success, or indirectly, as in foraging return rate, access to more sexual partners, or higher social status—result from optimal allocations made to multiple competing activities. Optima vary according to individual condition, state, and ecological setting; they also may depend on strategies employed by others in a frequency-dependent manner, often requiring game-theoretic modeling (Maynard Smith 1982).

Dual Inheritance Theory

Dual inheritance theory (DIT) or gene-culture coevolution analyzes the interaction of genetic and cultural evolutionary processes (Cavalli-Sforza & Feldman 1981; Boyd & Richerson 1985; Henrich & McElreath 2003). While selection pressures continue to promote changes in gene frequency in human populations (Hawks et al. 2007), particularly with respect to diet, metabolism, and immune function, complex adaptive genetic design appears to remain fairly stable over recent history; the pace of cultural change is extremely rapid in comparison. Exponential population growth, rapid industrialization, and technological change in the last millennium highlight the need to include culture formally in our understanding of human adaptation to a wide range of environments. Given cultural diversity before the post-agricultural population explosion, it further behooves us to take culture seriously. Emphasis has been on the evolution of social learning strategies for obtaining behaviors in moderately noisy environments, such as prestige bias (i.e., imitate the powerful, Henrich & Gil-White 2001) and conformist bias (i.e., imitate the majority, Henrich and Boyd 1998). Reliance on these strategies can lead to both adaptive and maladaptive outcomes. The opportunity to retain skills, transmit them to others, and build on them is a main reason why humans are the only species to show the "ratcheting" of cumulative culture (Boyd & Richerson 1996; Tomasello 1999; chapter 32, this volume). To date, much work in DIT has relied on mathematical models that show the conditions favoring different learning strategies in varied environments and populations. Recent empirical studies have focused, however, on the evolution and maintenance of social norms, ethnic markers, morality, and novel technology (e.g., McElreath 2004; Mesoudi & O'Brien 2008).

Evolutionary Psychology

Evolutionary psychology (EP) studies human nature by mapping the problem-solving design features of psychological adaptations (Tooby & Cosmides 1989; Buss 1995; Crawford & Krebs 1998; Daly & Wilson 1999). The metaphor of the mind is a Swiss Army knife, or a nested network of computer algorithms, with domain-specific adaptations in the brain functionally designed to solve specific recurring problems in our ancestral past. This ancestral past has been referred to as the environment of evolutionary adaptedness (EEA). Examples of these recurring problems include finding mates, evading predators, choosing trustworthy allies, and avoiding toxins and pathogens. Attention has focused on phenotypic cues of mate choice (e.g., facial attractiveness, fluctuating asymmetry, waist-to-hip ratios, indicators of ambition, wealth and social status), sex differences (e.g., spatial abilities, mate preferences, aggression, parental investment), cooperation (e.g., cheater detection, altruistic intent, punitive sentiment), emotions (e.g., jealousy, anger, limerence, guilt) and other topics (e.g., predator-prey cognition, supernatural beliefs, aesthetics; see Buss 1999; Gaulin & McBurney 2004). Because our brains evolved in specialized ways to solve problems, individuals are best viewed as modularized "adaptation executors" rather than general-domain "fitness maximizers" (Tooby & Cosmides 1992). Psychological adaptations are believed to be instantiated as concrete neural circuits common to all members of a species. Cultural variation is "evoked" from the interaction of a context-dependent psychology and environmental cues.

Similarities and Points of Contention: Toward a Synthesis

All three of these approaches place prime importance on the role of natural selection as the sieve that shapes patterns of human variation. While evolution also occurs from founder effects in small populations, random mutation, and gene flow, only natural selection produces complex adaptations. However, alternative goals, methods, disciplinary culture, and mutual misconceptions have fomented disagreements among researchers adopting different approaches. Several recent reviews explore similarities and differences among these three subdisciplines and raise important criticisms (Winterhalder & Smith 2000; Panksepp & Panksepp 2000; Smith et al. 2001; Laland & Brown 2002; Kaplan & Gangestad 2005; Sear et al. 2007). I highlight what I view as key distinctions in five areas.

The primary goal

HBE attempts to explain behavioral variation by testing whether observed behavior matches optimal behavior predicted by an optimality model. It thereby "black boxes" the proximate psychological mechanisms (the fundamental focus of EP) that individuals employ to achieve adaptive outcomes. The reliance on the "phenotypic gambit" (Grafen 1984) in HBE and EP obviates the need to identify the genetic bases of behavior. EP focuses on the cues that minds use as critical inputs when solving specific fitness-relevant problems in the EEA. DIT stresses the role of learning rules that individuals use to choose behaviors and skills, but it neither assumes nor requires domain specificity.

Universals versus variation

Much work in EP has helped to identify human universals rather than explain intra- or interpopulation variation, whereas HBE and DIT explicitly focus on variation. Unfortunately, most EP studies are conducted among members of WEIRD (Western, educated, industrialized, rich, democratic) societies, which represent only a small portion of the tapestry of past and present human experience (Henrich et al. 2010). Debates over the relevant roles of potential genetic difference, innate predisposition, and the timing of cultural conditioning are contentious (Atran et al. 2005; Norenzayan & Heine 2005).

The role of culture

Only DIT models culture explicitly, HBE often assumes rather than explains cultural traits as constraints or parameters of the local environment, or asserts that cultural variation may be the manifestation of local equilibria in different environments. EP mostly ignores culture, as it is largely believed to be "evoked" as a developmental fine-tuning of evolved psychology.

Methods

HBE often tests optimality models using behavioral data collected during long-term fieldwork, often among traditional populations, while EP has principally relied on controlled experiments among student populations. DIT constructs mathematical models underlying the epidemiology, or cause and spread, of ideas and the adaptive or maladaptive consequences.

The role of fitness

EP takes a functional design approach to adaptive problems to characterize the structure of psychological mechanisms. It is less attuned to behavioral outputs and their fitness-maximizing potential, even though it is assumed that psycho-logical adaptations on average result in fitness-maximizing behavior in the trait-specific EEA. HBE focuses explicitly on behavior, and often measures direct fitness or indirect proxies of fitness. According to EP, individuals living in novel environments are not expected to display evidence of fitness-maximizing behavior if their adaptations evolved in a different ecology; this results in "mismatches" between observed behavior and theoretical predictions. However, adaptive behavior is not difficult to find in modern societies (e.g., Nettle & Pollet 2008) despite the common perception that the past few hundred years have severed the link between adaptive design and fitness outcomes. Given the trade-offs associated with allocations of scarce resources to potentially competing modules, selection should act on functional output, which may require compromised design in any single module.

A common goal of evolutionary anthropology is to understand how and why our evolved psychology produces phenotypically plastic responses in different socioecological contexts, and the extent to which psychology and behavior are currently adaptive and maximize fitness. At the population level, an aim is to understand how the cultural landscape of humans jointly alters behavior and gene frequencies (Boyd & Richerson 1985; Laland et al. 2001). Several of the differences noted above are beginning to fade as lines between the three fields become blurry. HBE practitioners often run experiments and study people in developed countries, a growing number of DIT and EP researchers conduct fieldwork and experiments, and life-history theory and the logic of optimization have recently influenced EP while HBE and DIT have increasingly paid more explicit attention to psychology. Practitioners in HBE, and in EP to some extent, have also started to study culture, norms, and historical trends. The number of collaborations among practitioners who use these disparate perspectives is growing (e.g., the Cross Cultural Experimental Games Project, the Culture and the Mind Project, and the Inheritance of Inequality in Pre-Modern Societies Project). Substantial advances in our understanding of human culture, psychology, and behavior will require the methods and theoretical insights of all three approaches.

Human Life History

Table 13.1 summarizes some key life-history differences between humans and our nearest primate relatives, chimpanzees (*Pan troglodytes*). Compared to other mammals, and even other primates of similar body size, humans have long lives, large brains, and bodies that grow and develop

Table 13.1. Comparison of life-history traits among traditional human hunter-gatherers, living in natural fertility conditions, and wild chimpanzees

Trait	Definition	Unit	Humans	Chimpanzees	Difference
Brain volume	Total volume of brain	cm³	1201	400	200%
Juvenile period	Weaning to menarche	Years	12.9	5	158%
Adult life span	Life expectancy at age 15	Years	37.7	14	148%
Maximum life span	Oldest observed individual	Years	121	66[a]	83%
Fertility rate	Inverse of interbirth interval	#/yr	0.29	0.18	71%
Juvenile survival	Probability of living to age 15	Percentage	57	42	36%
Interbirth interval	Time between successive births	Months	41.3	66.7	−71%
Extrinsic mortality rate	Young adult mortality rate	Percentage per year	1.1	3.7	−70%
Neonate mass	Mass at birth	kg	3.4	1.4	143%
Age at menarche	Birth to menstruation	Years	15	10	50%
Age at first reproduction	Birth to reproduction	Years	19.1	13	47%
Age at last reproduction	Birth to last reproduction	Years	39	27.7	41%
Adolescence	Menarche to first reproduction	Years	4.1	3	37%
Fetal growth rate[b]	Conception to birth	g/day	12.6[a]	6.1[a]	107%
Total fertility rate	Total number of live births	Live births	6.1	5	22%
Gestation length	Conception to birth	days	269[a]	228[a]	18%
Body size	Average adult female mass	kg	47	35	34%

[a] Not specific to natural fertility or wild populations
[b] Fetal growth rate = neonate mass (g) /gestation length (days)

References
Brain volume: Aiello & Dean 1990
Juvenile period: Walker et al. 2006
Extrinsic mortality rate, adult lifespan, juvenile survival: Gurven & Kaplan 2007; Hill et al. 2001
Maximum life span: Finch 2007
Neonate mass: Lee et al. 1991
Age at first and last reproduction: Kaplan et al. 2000; Walker et al. 2006
Interbirth interval, total fertility rate: Gurven & Kaplan 2007; Kaplan et al. 2000
Age at menarche, gestation length: Walker et al. 2006; Wood 1994
Adolescence, body size: Walker et al. 2006

slowly (chapter 10, this volume). This evolved human life history has a number of derived or exaggerated features that likely contributed to the success of *Homo sapiens* (Hill & Kaplan 1999; Kaplan et al. 2000; Gurven & Walker 2006): (1) a diet comprising high-quality, nutrient-dense foods that come in large packages; (2) learning-intensive, technology-intensive, and often cooperative, food acquisition techniques; (3) an encephalized brain that facilitates the learning and storage of rich context-dependent information, effective imitation of conspecifics, and development of creative solutions to fitness-relevant problems; (4) a long period of juvenile dependence to support brain development, growth, and learning; (5) low juvenile and adult mortality rates, generating a long productive life span; (6) a three-generational system of downward resource flows from grandparents to parents and children; (7) pair bonds and biparental investment, with men specializing in energetic support and defense, and women combining energetic support with direct care of children; (8) cooperative arrangements among kin and unrelated individuals (including upward resource flows) that reduce

variance in food availability through sharing, divisions of labor, and mutual aid.

One feature of human life history that has received considerable attention is survival beyond the age of last reproduction. Postreproductive longevity is a robust feature of hunter-gatherers and of the human life cycle, but it is rare in mammals, including primates. A hunter-gatherer who survives to age 15 can expect to live an additional 30 to 44 years, and a survivor at age 40 can expect to live an additional 20 to 28 years (fig. 13.1). Much of the variation in age-specific life expectancies among preindustrial populations occurs in the first few years of life, where the force of selection, measured as the fitness elasticity of survival, has the greatest relative impact on the life course (Jones 2009). Age trajectories of adult mortality, however, show a roughly similar profile across the life course. The remarkably detailed demographic studies of !Kung, Hadza, Ache, and Agta hunter-gatherer populations living traditional lifestyles confirm this pattern (Howell 1979; Blurton Jones et al. 1992; Hill & Hurtado 1996; Early & Headland 1998), and their mortality profiles resemble those from ag-

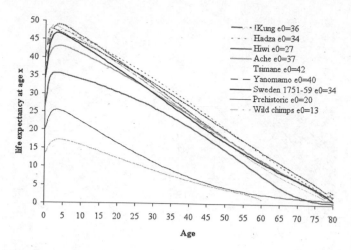

Fig. 13.1. Age-specific life expectancy, e_x. Average number of years of life remaining conditional on reaching age on x-axis. All curves are based on life-table estimates using the Siler method. Populations include hunter-gatherers (!Kung, Hadza, Hiwi, Ache), forager-horticulturalists (Tsimane, Yanomamo), earliest historical European data (Sweden, 1751–59), and a composite of 12 prehistoric populations of mortuary samples. These contrast with a composite of four wild chimpanzee populations. Life expectancies at birth (e_0) are given in the legend. See Gurven and Kaplan (2007) for additional methodological details.

ricultural populations, even in Europe until the nineteenth century (fig. 13.1). The rate of mortality increase in most subsistence populations follows a Gompertz-like pattern, with the mortality rate doubling every seven to nine years. Reports of short life span in early paleodemographic studies are likely due to high rates of contact-related infectious disease and violence, and to methodological problems such as poor age estimates of older individuals, biased preservation of the skeletons of infants and older individuals, and improper use of model life tables (Pennington 1996; O'Connell et al. 2002).

Human adult mortality rates tend to be about 1% to 1.5% per year, while chimpanzee rates are about three times greater (Hill et al. 2001). Chimpanzees, under the most favorable conditions in captivity, show much higher rates of adult mortality and a significantly shorter life span than do foraging humans under the worst conditions (Gurven & Kaplan 2007). This is true in spite of the available evidence, which suggests that members of both species seem to die from relatively similar macro-causes, with the exception of predation (table 13.2). The majority of deaths are due to infections and illness. Despite the cultural importance of dangerous predators, as represented by mythologies, stories, songs, and games, death by predation is rare among extant foragers. Grouping patterns, weapons, warning displays (e.g., fires), and other cultural means of avoiding predators contribute to the reduced impact of predation on human survivorship (Wrangham et al. 2006). In contrast, predation by leopards is a substantial cause of death among Taï

chimpanzees (Boesch & Boesch-Achermann 2000); lions may also be predators of Mahale chimpanzees (Tsukahara 1993), while snakes are a common source of fear and predation among primates more generally (Isbell 2009). Violent death caused by other humans, however, appears to be a common feature of human societies, accounting for 12.5% of 3,328 documented hunter-gatherer and horticultural deaths. Although violent death rates are comparable among human and chimpanzee populations (table 13.2), nonlethal aggression is more common among chimpanzees than among humans (Wrangham et al. 2006) and violent death rates are greater in farming societies than among hunter-gatherers (Keeley 1996). Infanticide is commonly practiced in many small-scale societies. Infants at greatest risk of being abandoned or killed are those who are sickly, unwanted, fathered out of wedlock, and/or female, as well as those viewed as bad omens, such as twins (Milner 2000).

While humans senesce more slowly than chimpanzees, it is still an open question whether the pace of aging has slowed down in recent history. Adult mortality has declined, but this does not mean that the rate of functional, physiological decay has fallen in tandem with it. Aging is

Table 13.2. Causes of death among humans and chimpanzees. Human data ($n = 3,221$) come from seven groups of hunter-gatherers and forager-horticulturalists (see Gurven & Kaplan 2007 for details). Chimpanzee sample ($n = 289$) is based on known reported deaths from Kasekela (Williams et al. 2008), Mahale (Nishida et al. 2003) and Taï (Boesch & Boesch-Achermann 2000) communities. The similar prevalence of violent deaths among humans and chimpanzees mirrors the conclusions of Wrangham et al. (2006), using different samples.

Cause	Humans		Chimpanzees	
	n	% known	n	% known
All illnesses	2333	72.4	128	50.4
Respiratory[a]	292	22.2	35	13.8[c]
Gastrointestinal[a]	239	18.1		
Fever[a]	107	8.1		
Other[a]	317	24.1		
Senescence	306	9.5	28	11.0
Accidents	166	5.2	6	2.4
Violence	354	11.0	35	13.8
Homicide[b]	164	6.0		
Warfare[b]	137	5.0		
Predation			28	11.0
Human-caused			11	4.3
Other	62	1.9	18	7.1
Total	3221	100.0	289	100.0

[a] Illness breakdown does not exist for all human groups. These percentages are based on a risk set of 1,644 individuals, and adjusted to sum to 72.4%.
[b] Information on violence-related deaths does not exist for all human groups. These percentages are based on a risk set of 2,272 individuals, and adjusted to sum to 11.0%.
[c] Respiratory illness accounts for 48% of all illnesses in Gombe, 20% in Mahale, and 0% in Taï.

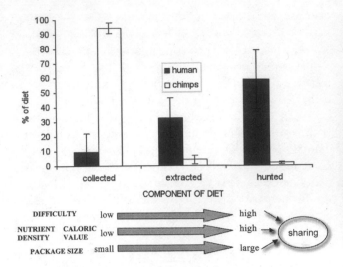

Fig. 13.2. Diets of human foragers and chimpanzees. Adapted from Kaplan et al. 2000, table 3 and figure 4.

Fig. 13.3. A Hadza man tracks an animal he has shot in the bush near Lake Eyasi, northern Tanzania. Successful hunting requires the acquisition of knowledge and skills, a process that starts in childhood but extends well into adulthood. Photo courtesy of Frank Marlowe.

often tricky to define and measure. The crudest but best available method for making inferences about past aging patterns uses historical mortality data to measure age-related changes in mortality. For example, longitudinal analysis of European mortality data suggested that senescence has slowed over the past couple of centuries, where senescence was defined in several different ways (Gurven & Fenelon 2009). This observation is consistent with the notion that reductions in "extrinsic" age-independent mortality (e.g., infectious disease, accidents, and other nondegenerative causes), should lead to greater investments in repair and maintenance, thereby resulting in longer life span, as originally hypothesized by Williams (1957; chapter 10, this volume). It is likely that low extrinsic mortality is a critical factor underlying the life history of long-lived species such as humans, clams on the ocean floor, giant tortoises, and trees on high, dry plains. Whereas other long-lived species with low extrinsic mortality often inhabit microbe-free and predator-free microenvironments, the lower extrinsic mortality of early humans may have come from effective group defense against predation and from the nurturing of sick and injured individuals (Gurven et al. 2000; Sugiyama 2004). Among early humans, low juvenile-adult mortality was likely a prerequisite for further reductions in adult mortality and the further slowing of the life course.

Intraspecific variation in adult mortality rates plays a vital role in shaping life-history traits by altering the valuation of present versus future benefits and costs (chapter 10, this volume). With a shorter expected time horizon, present payoffs (e.g., enjoying a cigarette) are preferred despite future costs (e.g., lung cancer), and future benefits (e.g.,

cardiovascular health) will be devalued if they require present costs (e.g., exercise and dietary restraint, Hill 1993). Life history traits such as physical growth and reproductive maturation are accelerated in populations that experience high mortality (Walker et al. 2006). Psychology and behavior are similarly affected by the same temporal trade-offs. Individuals with uncertain future prospects are more likely to discount the future and adjust behavioral strategies in a facultative manner; in modern societies, they exhibit earlier menarche, younger age at first sexual intercourse, higher reproductive rates, higher frequency of risk-taking and invest less in education and personal health (e.g., Hill et al. 1997; Chisholm 1999; Bereczkei & Csanaky 2001; Brumbach et al. 2009).

Human Ecological Niche

In comparison with non-human primate diets, the vast majority of human hunter-gatherer diets consist of nutrient-dense, calorically rich resources (fig. 13.2). Quantitative data based on behavioral sampling (summarized in Kaplan et al. 2000) and an independent survey of 229 human groups both show that animal foods constitute over 60% of modern hunter-gatherer diets (Cordain et al. 2000; fig. 13.2), whereas the diets of nonhuman primates depend heavily on leaves and fruits (chapter 7, this volume). Ethnographies from different foraging societies suggest that meat acquisition requires a high level of skill and coordination, as hunters navigate over large ranges and integrate extensive cues, signs, and context-specific knowledge concerning animal behavior and ecology (fig. 13.3; see Blurton

Fig. 13.4. Hadza women roast tubers of ekwa, *Vigna frutescens* (Fabaceae). Humans appear biologically adapted to including cooked food in their diet, because without the extra energy provided by cooking, many raw foods are inadequately processed by the relatively small human intestinal system. For example, cooking renders foods so soft that humans have much lower daily chewing times than other primates in relation to body mass. Cooking also increases the range of edible items and the net energy gain, thus allowing high reproductive rates. Photo courtesy of Richard Wrangham.

Jones & Konner 1976; Leibenberg 1990). Other foods have also often been shown to be difficult to acquire and process. For example, roots, tubers, nuts, and palm hearts must first be located and then extracted from a solid substrate.

The reliance on difficult-to-acquire foods that slow-growing children and even adolescents may be ill-equipped to obtain on their own is a critical starting point for several theories attempting to distinguish human life history evolution from the general primate pattern (chapter 10, this volume). The two most prominent of these are the grandmother hypothesis (GH) and the embodied capital model (ECM). GH argues that postreproductive life span evolved because helping daughters reproduce and improving grand-offspring survivorship both yield greater fitness gains than giving birth past age 50. Part of the initial motivation for this idea came from observations of older "hard-working" Hadza women who collected and distributed tubers in the African savanna. ECM argues that contributions of older men and women helped to increase postreproductive life span, and that this period coevolved with the longer developmental (training) period early in life, and with increased brain size (Kaplan et al. 2000; Kaplan & Robson 2002).

The control-of-fire hypothesis complements these models by arguing that human use of fire for cooking helps increase the efficiency of provisioning by promoting food digestibility and energy, and by allowing early weaning through increased availability of weaning foods (Wrangham 2009; Wrangham & Carmody 2010). It also reduces extrinsic mortality by detoxifying certain foods, helping to eliminate food-borne pathogens, and deterring predators (fig. 13.4).

According to GH, body size is the primary determinant of age profiles of productivity. Limited production of juvenile foragers is due to the strength-intensive nature of hunting and foraging activities. Development according to ECM requires additional investments in brain-based capital, due to the learning-intensive nature of the human foraging and social niche. ECM argues that high levels of knowledge, skill, coordination, and strength are required to exploit the resources human foragers consume, and that the attainment of those abilities requires extensive learning. As mentioned in the previous section, while these and other theories may help to explain a substantial slowing of the human life course, an initial lowering of extrinsic juvenile-adult mortality, perhaps due to helping behavior, sharing, and/or cooking, was an important prerequisite to help push humans into the highly cooperative, slow-growing, and long-living life-history niche. Allowances usually made for pregnant and lactating women, who are less able to forage themselves but nonetheless receive ample food, enable a high level of human fertility with short interbirth intervals. The average forager female has about six births (table 13.1), which places a substantial burden on household feeding requirements.

Optimal Foraging Behavior

Food acquisition is a vital activity that has affected selection on primate physiology, behavior, and social organization (chapter 7, this volume). While much has been written about the feeding habits of particular primate species (e.g., Garber 1987), there have been few comprehensive optimality studies of food choice other than tests of ideal-distribution theory, the effects of dominance relationships (e.g., Janson & Chapman 1999; Koenig & Borries 2006), and consideration of feeding trade-offs in the presence of potential predators (chapter 8, this volume).

Studies of human foraging behavior were some of the first applications of evolutionary and ecological theory made by cultural anthropologists. Despite their simplicity, these studies have helped researchers to predict the suite of resources that comprise human diets in different locales, and the optimal choice of resource patches and habitats. Hunter-gatherers have knowledge of hundreds of animal and plant species, yet their diverse diets tend to focus on a much more limited set of species. Optimality models usually assume that a forager's goal is to maximize the rate of caloric intake per unit of time spent foraging (Smith 1981). The most influential optimal foraging models are the "prey choice model" and the "patch choice model" (MacArthur & Pianka 1966; Charnov 1976; Stephens & Krebs 1986).

The prey choice model predicts that a forager should pursue any resource encountered if the expected gain from pursuit outweighs the expected gain of continued search for randomly encountered food items, i.e. the long-term rate of caloric gain. This simple model has been used to predict the suite of resources people target in forest, arctic, desert, and marine environments in ethnographic and archaeological contexts (e.g., Hill & Hawkes 1983; Smith 1991; Gremillion 2002; Thomas 2007a). Adjustments to the models have been made to account for unique characteristics of human foragers (Stephens & Krebs 1986; Giraldeau & Caraco 2000), where search may not be random, encounters may be simultaneous, foragers may be sensitive to risk, and novel techniques or technology are employed. For example, improvements in technology that increase the average caloric return rate—such as shotguns instead of bows and arrows, or trucks and snowmobiles instead of walking and/or dogsleds—have each been demonstrated to decrease the number of resource types (i.e., diet breadth) pursued, as increased efficiency leads to a preference for only highly profitable resources. Domesticated hunting dogs have also been shown to increase prey encounter rates (Koster 2008) and reduce prey handling times (Ikeya 1994). The prey-choice model has also been used to explain dietary transitions. The reduced processing or handling costs of seeds and grains raise the profitability of those resources. Historical declines in the abundance of megafauna in North America have been posited as important catalysts in the subsequent adoption of plant and animal domestication (see Kennett & Winterhalder 2006).

The patch-choice model, based on Charnov's marginal value theorem (Charnov 1976), addresses the question of how long to spend in a resource patch where the rate of caloric gains declines as more time is spent in that patch. Gains might decline during foraging due to prey depletion or resources becoming more difficult to obtain over time. As in the prey choice model, the caloric benefits of remaining in the patch are compared to the foregone benefits of abandoning the patch and moving to the next one. It is often in a forager's best interest to leave patches before they are depleted. While not rigorously tested in humans, most empirical examples instead focus on the amount of time spent foraging in different habitats that vary in their mean profitabilities (see Kaplan & Hill 1992 for review). Case studies show that while foragers spend more time in more profitable patches, less profitable patches are also frequently targeted, but often in ways that make adaptive sense (Smith 1991; Sosis 2002; Thomas 2007b).

The limitations of optimal foraging models help highlight unique aspects of human social organization that require explanation. Optimal foraging models are best applied to animals with herbivorous or carnivorous diets who do not cooperate in production or distribution and who forage by random search. This is because most models maximize a single currency (calories) and do so from only one individual's perspective. Humans are omnivores; they consume a mixture of macro- and micronutrients, they coordinate and cooperate during foraging activities, and they share resources and information. These characteristics have themselves become the focus of much study (see below). Model adjustments emphasize the importance of mixed diets, cognitive limitations of the forager, risk preferences, field processing and transport costs, central-place foraging, information gathering and sharing, and divisions of labor by sex and age (Barlow & Metcalfe 1996; Winterhalder et al. 1999; Stephens et al. 2007). Unfortunately, many of these model adjustments have yet to be rigorously developed or tested in humans. For these reasons, traditional optimal foraging models sometimes make predictions that do not hold most of the time: men should actively gather roots and other plant products, and women should hunt (see below).

Group mobility, group size, and territoriality have also been considered in light of spatial and temporal resource patchiness (Dyson-Hudson & Smith 1978; Cashdan 1983), per-capita production rates over time and space (Beckerman 1983), Fretwell's ideal-free distribution theory, and member-joiner conflict (Smith 1985). Resource predictability and abundance are important factors underlying land tenure regimes and group defense; a recent formalized version of the "economic defensibility model" seems to explain cross-cultural variability in territorial behavior among hunter-gatherers (Baker 2003). The fluid composition of hunter-gatherer groups and of post-marital residence rules has been considered an adaptive response to fluctuating resource availability, demographic stochasticity, mating opportunities, and intergroup raiding (Kelly 1995; Marlowe 2005).

Becoming "Expert": Adult Productivity

The complex feeding and social niche of humans requires substantial learning to achieve adult-level proficiency. Changes in foraging proficiency with age have now been examined among the Ache (Walker et al. 2002), Gidra (Ohtsuka 1989), Hadza (Blurton Jones & Marlowe 2002), Hiwi (Kaplan et al. 2000), Mardu (Bird & Bliege Bird 2005), Machiguenga and Piro (Gurven & Kaplan 2005), Meriam (Bird & Bliege Bird 2002; Bliege Bird & Bird 2002), Mikea (Tucker & Young 2005) and Tsimane (Gurven et al. 2006). Most of these studies show that men's hunting success peaks in the age range of 35 to 50 years, while other foraging and fishing activities peak by about age 20. Several cases of extraction activities show similar delays in productivity (albeit not as extreme), such as shellfish collecting among Gidjingali (Meehan 1982) and mongongo nut processing among Okavango Delta peoples (Bock 2002; fig. 13.5a). Conversely, among the Meriam of Mer Island, increases in children's productivity in several fishing and hunting activities closely tracked changes in physical growth (Bird & Bliege Bird 2002, 2005; Bliege Bird & Bird 2002). Tucker and Young (2005) also found few differences in productivity rates between children and adults in tuber extraction. Tests based on observational, interview, and experimental data collected among Tsimane Amerindians of the Bolivian Amazon suggest that body size alone cannot explain the long delay until peak hunting productivity (fig. 13.5b). Rates of indirect encounter (e.g., smells, sounds, tracks, and scat) and shooting of stationary targets are two components of hunting ability explained primarily by physical

size alone, but more difficult components of hunting, such as direct encounter of important prey items and successful capture, require substantial skill (fig. 13.6). Those skills can take a hunter an additional 10 to 20 years to refine after achieving adult body size (Gurven et al. 2006). Similar conclusions were reached by Walker et al. (2002) based on an analysis of Ache hunting performance.

By the time hunter-gatherer men marry they are well on their way to becoming better hunters, but they still require much experience, or on-the-job training. As pointed out by Blurton Jones and Marlowe (2002) in reference to the Hadza, learning need not be completed before maturation, marriage, or reproduction (cf. Bjorklund 1997). Similarly, learning need not occur in constant increments throughout the entire prereproductive period in order for the long delay to be linked to later production as stipulated by the ECM. A combination of size-dependent learning and stepwise ratcheting up of strength and skill (or "punctuated equilibrium": Bock 2002) may characterize increases in performance during development. Furthermore, developmental milestones occur in an ordered sequence, with each one building cumulatively on prior achievements. Critical learning may occur not continuously but in dispersed stages, as has been found to be the case with physical growth (Lampl et al. 1992). This would suggest that gaps in learning during critical periods may be more detrimental to later performance and require more catch-up time than gaps that occur at other ages. Hunters raised in another culture or restricted from lengthy experience early in life rarely achieve the level of proficiency of hunters immersed in the traditional lifestyle. One experiment in which young Ache men were paid for each animal they killed over a 13-month period showed that while these men spent much more time hunting, there was no net increase in hunting return rate, encounter rates, or likelihood of a kill upon pursuit (Walker et al. 2002). These inexperienced hunters did not become highly proficient hunters, even though they likely benefited from the social transmission of relevant information on game behavior, locations, and successful hunting strategies from other men.

Many of the groups in which substantial delays in hunting performance have been reported are groups of primarily small game hunters. While the source of large game is open to interpretations of hunting and scavenging (e.g., Binford 1981; Blumenschine et al. 1994), small game was probably hunted by early *Homo*. Given that all chimpanzee hunting is directed towards small game such as red colobus monkeys (chapter 8, this volume), the first place to look for expanded hunting among early hominins is in the increased frequency and success in obtaining small game. There is evidence that small game hunting has been an important component of human diets for at least 200,000 years (Stiner 2002).

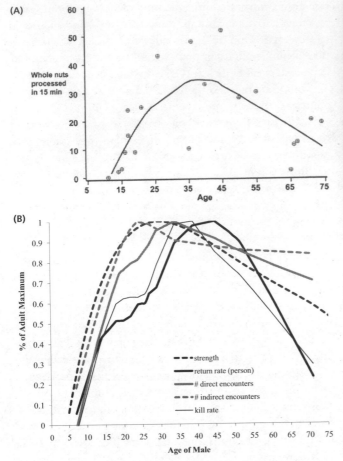

(A)

(B)

Fig. 13.6.   A Tsimane boy and his pet capuchin monkey. Caring for and playing with pets is a common occurrence in hunting societies, even though these same animals are treated as food in a different context. Pets require attention, training, and feeding. The practical value of maintaining pets likely stems from their role as educational "props" in teaching children about animal behavior. Some pets may also detect intruders and clean up scraps. Photo courtesy of Michael Gurven.

Fig. 13.5.   (a) Mongongo nut processing by age. The graph depicts age-specific trends among females from four ethnic groups in the Okavango Delta in the number of nuts processed in 15 minutes. Taken from Bock 2002, figure 10. (b) Hunting performance by age. The graph depicts age-specific trends among Tsimane forager-horticulturalist men, standardized to adult maximum values. Shown in order of proficiency are muscular strength, indirect and direct encounters with animals, kill rate, and caloric return rate. Adapted from Gurven et al. 2006, figure 7. All trend lines are lowess curves.

Even megafaunal "specialists" were likely hunting small and medium-sized game (Byers & Ugan 2005). The earlier practice of small game hunting by *Homo* merits investigation. It is not known whether the cognitive skills required for small versus large game hunting differ, or whether those differences depend on prey and ecology. Often, small game hunters rely on a greater variety of different prey items than do large game hunters, but such differences have yet to be studied systematically. Human hunting is nonetheless unique among primates, due to the expanded day range, the widespread sharing of spoils, and the long-distance transport of kills to a home base for processing and consumption by others (Stiner 1991).

While much attention has been given to skill-intensive aspects of male hunting, arguably many other human economic and cultural skills also take considerable time to master. Furthermore, obtaining sufficient social capital and building reliable support networks may take substantial time and skill, and may be especially helpful during periods of conflict, sickness, and injury (Simmons 1945; Wiessner 1981). Ritual knowledge and crucial roles as leaders, orators, and shamans are held mostly by older adults. Basic skills may be acquired by the time of reproductive maturation, but many more years may be required for an individual to become highly proficient. Among Tsimane forager-farmers, for example, older adults are named as experts in manufacturing activities, such as handbag weaving and bow and arrow making, and in music and storytelling. The average age of male and female experts for a large variety of skills in different domains is consistently over 40 years old. Experiments in the United States confirm that older adults are effective storytellers. Their stories are more memorable when told by them than when told by younger individuals (Mergler et al. 1984); the older adults employ more emotional modulation and show greater attention to detail (Adams et al. 2002).

Even child care abilities may improve with parity. Among several nonhuman primates, primaparous mothers have lower fertility and experience higher infant mortality than do multiparous mothers (Bercovitch et al. 1998; Robbins et al. 2006; chapter 15, this volume). While adolescent subfecundity and higher infant mortality among primaparous human mothers has also been documented, it is unclear whether any of these effects are due to inexperience and lack of proficiency among first-time mothers.

Delayed Maturation

The extension of juvenility through childhood and adolescence is seen as a prominent, recently derived feature of human life history (Bogin & Smith 1996; Bogin 1997; Dean et al. 2001; Bock & Sellen 2002). Although primate growth patterns may be highly variable, the standard primate pattern is to proceed from infancy to juvenility to adulthood (chapter 11, this volume). Four explanations for delayed maturation in primates have been proposed, rooted in concepts of social competition, risk aversion, trade-offs between growth and reproduction, and learning- and skill-based food acquisition strategies (Pagel & Harvey 1993; Leigh 2001; Pereira & Fairbanks 2002; chapters 10 and 11, this volume). Social explanations focus on intragroup competition, in which extra time is needed to develop social competency (Dunbar 1998; Barton 1999). The risk-aversion hypothesis argues that slow growth among group-living primates reduces resource competition and thereby serves to decrease the risk of dying due to fluctuations in the food supply (Janson & van Schaik 1993). A third hypothesis views optimal age at reproductive maturation as a trade-off between increased production from the benefits of growing longer (and hence larger) and the decreased probability of reaching reproductive maturity, because with each additional unit of time invested in growth there is some risk of dying (Charnov 1993). This model has been extended and applied to humans in the form of the grandmother hypothesis. As described earlier, the GH applies Charnov's (1993) model of optimal age at reproduction to explain the juvenile period as an artifact of selection on longer life span due to the indirect fitness benefits that accrue to magnanimous grandmothers (Hawkes et al. 1998).

Finally, learning- and skills-based models focus on the difficult adult foraging niche of many primates, especially humans, where much time early in life is devoted to acquiring the critical coordination, skills, and knowledge necessary for proficient adult foraging (Bogin 1997; Ross & Jones 1999). ECM extends this approach to explain delayed maturation, extended life span, and increased encephalization as a coevolutionary response to the demands of the difficult human foraging niche (Hill & Kaplan 1999; Kaplan et al. 2000; Kaplan & Robson 2002). According to this model, natural selection acts to extend life span when early learning yields high production payoffs over the duration of a longer adult life span. Similarly, longer life span creates selection pressure to grow more slowly early in life, and to spend this time learning when such investments lead to higher production payoffs later in life. The ECM also indirectly incorporates the risk-reduction hypothesis. The gains from learning act to increase the optimal level of investment in mortality reduction at all ages (Kaplan & Robson 2002). As a result, human children should have the lowest mortality of any primate. The ECM links foraging and ultimately reproductive success to brain-based "embodied" capital, which includes the suite of skills, knowledge, and abilities that affect future performance, and to the maintenance and repair mechanisms that act to reduce mortality so that later gains can be realized.

The separation of reproductive and economic maturation is an important feature of human life history. Humans are capable of reproducing before they can fully support their own needs because of the contributions made by others, such as grandmothers, husbands, in-laws, and older children. The "pooled energy budget" from others' caloric contributions not only expands women's energetic budget beyond self-production, but also enables girls to reduce their own activity, thereby permitting more resources to be allocated to growth, maintenance, and reproduction (Kramer et al. 2009).

An interesting paradox is that while poor conditions indicative of high extrinsic mortality might select for earlier maturation and a faster life history, faster physical growth and secular declines in age at menarche due to improved nutrition and health are well documented in many human populations. In a study examining juvenile growth rates in 21 hunter-gatherer and horticultural populations, societies with larger and taller adults (i.e., with better nutrition and health) developed faster and earlier, while higher juvenile mortality rates were associated with earlier menarche and age at first birth (Walker et al. 2006).

Sexual Division of Labor

One ubiquitous feature of human societies is an extensive division of labor among men and women. Apart from breastfeeding and child care responsibilities of females, such a pattern of coordination and collaboration in resource production rarely manifests in nonhuman primate societies.

Five features of hunter-gatherer socioecology are likely responsible for the sexual division of labor among foragers: (1) offspring are dependent during infancy, childhood, and even adolescence; fast fertility rates result in compound dependency of multiple offspring; (2) humans are committed to carrying children and providing high-quality childcare, traits shared throughout the Primate order which are incompatible with hunting; (3) an adequate diet requires macronutrients typically found in mutually exclusive types of foods; (4) peak efficiency in many foraging activities is delayed due to time-dependent learning; and (5) some tasks exhibit sex-differentiated comparative advantage. These conditions are common to most foraging groups, and are together responsible for a discrete sexual division of labor (Murdock & Provost 1973; Kelly 1995; Gurven & Hill 2009).

A central aspect of the division of labor is marriage and family formation. All human groups recognize marriage as a way for men and women to regulate their sexual activity and form cooperative bonds in raising children. Sometimes this bond is defined by monogamy, but mild polygyny may better characterize the spectrum of traditional human societies (Quinlan 2008). Marriage is characterized by intensive food sharing within a family, a division of labor in the organization of other household tasks, and the care of children. Among foragers, the reproductive careers of men and women are usually linked. While divorce is common in many foraging groups, many couples have the majority of their children together, and men often have their last child when their wives reach menopause. The relationship between men and women in foraging societies is arguably the most intense and multifaceted cooperative relationship in which they engage, although the reliance on husbands varies cross-culturally (Jankowiak et al. 2002).

Many individuals other than spouses contribute to the welfare of family members, leading to the claim that humans are "cooperative breeders" (Hrdy 1999, 2009). Divisions of labor among extended family members, as well as among non-kin, are not uncommon. Activities that provide benefits to others' offspring in addition to one's own occupy a relatively large percentage of post-juvenile daily energy expenditure among foragers, consistent with the notion of "pooled energy budgets" described above. While the net flow of resources is generationally downward, resource transfers and helping behavior are bidirectional and occur also among siblings and from young to old.

The Social Brain and Cultural Transmission

Humans as Cultural Animals

Does the human life history require unique cognitive abilities and elaborate culture, or did it instead promote their subsequent evolution? Attempts to answer this question have led to a flurry of studies over the past several decades detailing numerous aspects of primate culture and social cognition (e.g., Boesch & Tomasello 1998; Perry 2006; Lycett et al. 2007; Watson & Caldwell 2009). While cultural "traditions" have been well documented in primates (chapter 31, this volume), especially among chimpanzees, orangutans (*Pongo* spp.), and capuchin monkeys (*Cebus* spp.), human culture is often distinguished by the accumulation of ratchet-like modifications within and among generations, and its reliance on imitation and explicit guidance and instruction. Other aspects of cognition purported to uniquely represent "phylogenetic mind gaps" separating humans and other primates include generative computation, the combining of representations from separate cognitive domains, the ample use of mental symbols, and abstract thought (Hauser 2009). These abilities, combined with other cognitive capacities, particularly intention reading and social learning, have been linked to the development of complex symbolic language, tools, technology, institutions, and belief systems.

Social learning plays a critical role in the intergenerational transmission of knowledge and practices (fig. 13.7; chapters 11, 31, and 32, this volume). Among foragers, social learning probably increases the rate at which human children, adolescents, and adults learn how to hunt and gather efficiently. In these societies, children and adolescents spend many years listening to others tell stories about different foraging experiences before engaging in these activities themselves. Social learning may help foragers use the personal experience of others to improve their own foraging efficiency more rapidly than would be possible without such prior knowledge. Social learning of foraging skills,

Fig. 13.7. Three generations of an American matriline. Intergenerational transfers are a critical component of the human life course, including in developed countries like the United States of America. The grandmother taught her daughters to quilt, and her daughters taught their daughters. The grandmother and her husband traveled to visit their daughters and grandchildren in several US states; they also often paid for travel vacations for the extended family. Although much of the grandmother's savings were needed to cover medical expenses and a care facility because of her severe Alzheimer's disease, she still bequeathed family heirlooms and valuables to her grandchildren late in her life and upon her death. Photo courtesy of Anne Pisor.

however, is not unique to humans. In nonhuman primates, the frequency of social transmission of information strongly predicts variation in brain size, and most of this information pertains to foraging (Reader & Laland 2002).

Among humans, language helps improve the reliability and efficiency of social learning. It involves a specific set of cognitive adaptations (Pinker 1994) and is posited as an important component of the "hominid entry into the cognitive niche" (Barrett et al. 2007). Language lowers the cost of transmitting information by allowing more precise ways of sharing information about the world. It helps to improve coordination among group members and reap the gains of mutualistic cooperation (Alvard & Nolin 2002). Novel solutions to local problems, obtained either directly from personal experience or from others, can be communicated effectively and to a broad audience. In many domains, language allows the communication of much more information than can ever be gained by personal experience in a single lifetime. The rapid accumulation of sequenced skills allows for complex tool manufacture and other cultural adaptations, and accounts for the ratchet-like cumulative nature of human culture (chapter 32, this volume). Culture, leading to the development of tools, clothing, and fire, has influenced selection on genes affecting diet, nutrition and digestion, disease resistance, dentition, and cognition (Durham 1991; Laland et al. 2001).

The ability to imitate and learn from others has cognitive prerequisites that may be uniquely human (Caro & Hauser 1992; Premack & Premack 1996). A variety of learning mechanisms appear to bolster the specialized learning of language, food preferences and aversions, and danger (Gallistel 1990). The decoding of other actors' mental states to infer their intentions (as distinct from outcomes), "theory of mind," and other mental representations are important for effective social learning (Tomasello 1999) and for maintaining long-term cooperative interactions. Even humans' closest relatives, chimpanzees, have limited social cognition and do not achieve full-fledged humanlike belief-desire psychology (Call & Tomasello 2008; chapter 30, this volume). Imitation and emulation therefore go beyond the mimicking of physical movements; behavioral adjustments help an actor achieve an intended goal (chapter 31, this volume).

Much learning in small-scale societies may be observational, or may involve a combination of approval, disapproval, and correction by others. Effective pedagogy may require additional psychological adaptations for purposefully transmitting information to others willing to learn (Csibra & Gergely 2006). Anecdotal reports that overt teaching is absent in these societies may be surprising, but even slight corrections can effectively guide naïve individuals towards target behaviors or skills (Castro & Toro 2004). Much

learning in infancy and childhood is likely transmitted vertically from parents to offspring (Hewlett & Cavalli-Sforza 1986). Among adolescents and adults, horizontal transmission among peers and kin is also common. Preferred models for cultural learning include successful, skillful, and influential group members (Henrich & Gil-White 2001).

Cognitive and Social Niches

Human diets are inherently risky: foraging luck is often highly variable, and food sharing is a fundamental solution. Hunters, in particular, often return to camp empty-handed after a full day's search, especially when pursuing large game (Hawkes et al. 1991). Food sharing among families buffers against the risk of daily food shortfalls associated with hunting large, mobile packages. Cultural norms governing resource distribution likely coevolve with systems of production in ways that help motivate and reward productive effort among group members (Gurven 2006). Among the Lamalera, for example, shares of whale meat are given to participating hunters and to nonhunters who contribute productive capital (e.g., specialist sail and boat makers; Alvard & Nolin 2002). In any group-oriented production system, however, actual decisions and transactions may deviate from the normative patterns (e.g., see Altman 1987, Bailey 1990); there is room for social navigation towards selfish ends by skillful actors. Widespread sharing is vulnerable to exploitation by cheaters who consume resources without providing or sharing them in turn. Indeed, much gossip and conversation around the campfire concerns accusations of repeated stinginess, greed, or laziness, as well as identification and condemnation of possible second-party defections on kin obligations (Wiessner 2005). These conversations often focus on actor intentions, beliefs, and circumstances, in addition to the outcome, in order for the participants to form their opinions.

A growing body of evidence shows that food is not shared equally among all band members in most hunting and gathering societies, except under special circumstances (Gurven 2004; Kaplan & Gurven 2005). People have preferred partners, with whom reciprocal exchange is greatest. Efficient sharing requires the monitoring of contributions made by other group members, and the monitoring requires not just observing others but also inferring the intentions of others. Identifying acts of cheating may be difficult in real-life situations. Thus, the ability to negotiate profitable partnerships requires social intelligence and the ability to understand how one's actions will affect future access to vital resources. Moreover, some food sharing may act as investment in social capital that affects future cooperative interactions. The encephalized primate brain required for navigating a difficult feeding niche must therefore also be a social brain that can strategically share game and other resources (Stanford 1999).

Cognitive substrates for solving economic and social problems may be shared. Selection should act on the total effects of increased abilities, summed over all routes through which those abilities affect fitness. For example, inferences about an animal's behavior, such as its likely escape strategies if it detects the hunter's presence, are critical for hunting success. Other humans and prey are both intentional agents, and so animal "mind reading" and human mind-reading may involve similar cognitive abilities. Hunters often use visual cues, folklore, and observations of tracks and spoor to test their hypotheses about animal behavior. Leibenberg (1990) has argued that such inferences, with empirical verification, form the basis for protoscientific thinking. In a review of the comparative anatomy of primate brains, Rilling (2006) notes that selection uniquely modified the human brain to deviate from the rules of brain design that prevail among other primates. The human brain displays unique modifications in the prefrontal cortex associated with symbolic thinking, knowledge of appropriate social behavior, decision making, planning, cognitive control, and working memory. Bering and Povenilli (2003) propose that the critical divide between the minds of apes and humans is not just a difference of 1,000 cm^3 that enables humans to do the same things much better, but rather a novel ability to think about things that cannot be directly observed by the senses. Humans are often fanatical about thinking about and discussing the hidden world of causation, such as what others are thinking, what their ulterior motives are, how a tool works, or why people get sick (chapter 32, this volume).

In addition to its role in negotiating sharing decisions in foraging societies, the social brain of humans is an ecological brain that helps facilitate coordination and cooperation among group members. Divisions of labor among kin and unrelated group members require coordination and task specialization. Effort is often allocated to different tasks in coordinated ways so as to maximize group production, which must then be distributed among coordinating group members. The chief benefit of such social organization is the formation of synergistic economies of scale in which the sum of the joint production of n actors is greater than the sum of each of the n actors producing alone. Age and sex-based specialization and task complementarity enable human foragers to reap gains from such economies of scale. The evolution of complex human social organization over the past five millennia has often involved increasing returns to scale in production, as in intensive agriculture, irrigation networks, patron-client relations, and armies.

Other Psychological Adaptations

I briefly highlight two additional categories of psychological adaptation to complement the above discussion on life history, subsistence, and cognition. Encounters with predator and prey are common, and salient fitness-relevant experiences among primates (chapter 8, this volume) and human hunter-gatherers are no exceptions. Even though death by animal predators is very low among hunter-gatherers (table 13.2), fear of attack from lions, snakes, and jaguars is common, and much cultural lore emphasizes these animals. Species-typical investments may have led to mortality reduction from predation, thereby contributing to the lower level of extrinsic mortality in human populations as compared with that in primate populations. Evading predators and hunting prey (i.e., avoiding being killed and killing when hungry) is likely to have selected for certain cognitive mechanisms. Fear helps organize one's bodily resources towards seeking escape routes or avoiding predators and dangerous situations altogether (Öhman & Mineka 2001). As described earlier, predator and prey are intentional agents, and so detecting and making decisions about such agents may represent evolved features of the "agency system" (Byrne & Whiten 1988; Barrett 2005). These include monitoring of directional eye gaze, autonomous movement, cause and effect contingency, reasoning about belief and desire, and the types of mind-reading skills outlined earlier (see Barrett 2005 for a summary). These kinds of adaptations, as part of a "predator-prey inference system" (Barrett 1999), are likely present in many mammalian species. However, two distinctions can be made in the human case. First, no evidence exists for evolved perceptual templates for true predators among humans, thus suggesting a diverse set of predators over human history and/or the substantial influence of social learning of fear in response to certain animals, as has been observed among rhesus monkeys (Mineka et al. 1984). Although fear and phobia of snakes and spiders have long been described (Agras et al. 1969; LoBue & DeLoache 2008), snakes, other than constrictors, and spiders do not prey on humans, but rather attack in self-defense. Second, current data suggest that predator-prey inference systems require attention to actor goals and intentions, but not necessarily to beliefs. The latter type of attention may instead require a developed "theory of mind" (Baron-Cohen 1995).

A second area where evidence of functional design is expected comes from the relatively high level of paternal investment among humans. This high level should be reflected in evolved motivational adaptations that help facilitate long-term mateships and high-investment paternal care. Physiological data on male-female and male-offspring bonding mechanisms and hormones that promote such bonds suggest that human males were selected to increase cooperative sentiment with female partners and to help raise highly dependent offspring (Gray et al. 2004). In other primates that show significant paternal care, such as cotton-top tamarins and common marmosets, a male's prolactin levels increase in synchrony with his mate's; an expectant father seems physiologically responsive to his mate's pregnancy and his offspring's imminent birth (Ziegler et al. 2006). A similar response is found among humans, but not among other nonpaternal species (Storey et al. 2000). Male couvade pregnancy symptoms are not uncommon in cultures with high levels of partner intimacy and paternal care (Elwood & Mason 1994). Vasopressin and oxytocin have also been found to help modulate attachment, support, and pair-bonding behavior in male rodents (Heinrichs & Domes 2008). Studies among humans are underway; for example, humans with a certain vasopressin receptor subtype (V1aR) associated with monogamous behavior in rodents were happier in their marriages and felt greater affiliation with their partners (Walum et al. 2008).

Future Directions

I have documented key life-history features of humans and their associated foraging niche, as well as the cooperativeness, social maneuvering, mind reading, and cumulative culture that go along with our encephalized brains. These features evolved in the context of humans living as hunter-gatherers in a world of other hunter-gatherers. There is still an open question regarding the extent to which evolution has continued to shape the genetics underlying human behavior, personality, and cognition, in addition to the well-documented recent adaptations to climate, pathogens, and regional diet. As the size of populations around the world has exploded since the advent of agriculture, the selective sweeps of many favorable alleles have been detected; the future may shed light on their relevance (Hawks et al. 2007). As genome-wide scans become more affordable, the exploration of gene-by-environment interactions will help shed light on both ancient (e.g., Varki & Nelson 2007) and recent (e.g., Bersaglieri et al. 2004) adaptations.

Much of evolutionary behavioral science that is related to "surviving and growing in a difficult and dangerous world" in anthropological populations has emphasized the efficiency of subsistence behavior, cooperative production and sharing, the value of social learning, and life-history variability. Many components of these themes remain to be explored in detail. For example, no systematic study of age-related changes in navigational skills, tracking skills, knowledge of animal behavior, techniques employed, or motivation has been conducted in traditional populations to elucidate the extent of on-the-job learning that is required to attain proficiency in hunting. Systematic research can also help reveal the relative importance of intergenerational social learning while at the same time exploring the functional significance of childhood and adolescence. Why do many human activities require years of learning before peak proficiency is gained? Such activities would include the food-production tasks and their subcomponents listed above, but also unstudied activities like child care and tool manufacture. To clarify the significance of the human social brain, more research is needed on age-specific changes in the accrual of social capital. How and why does it take so long for humans to become adept social adults?

Almost all studies of kin cooperation and human life history among small-scale societies emphasize their indirect effects on fitness (e.g., the presence of a maternal grandmother on her offspring's fertility or her grand-offspring's mortality), or on caloric contributions to the diet, but little attention has been paid to the specific resource transfers that could potentially affect the fitness of kin, especially nonfood contributions. For example, how do older adults acting as leaders, mediators of conflict, coordinators of group activity, and repositories of cultural knowledge affect the fitness of others? Are these benefits directed preferentially towards kin? What about the contributions made by young adults? Here both the EP and HBE approaches can make important contributions. DIT can also contribute by assessing whether cultural transmission across three generations, rather than just two, increases the fitness value of postreproductive individuals in ways not captured by the direct transfer of resources.

The popularity of the GH has led to a focus on grandmothers as caretakers and helpers whose support helps bolster high fertility, slow growth, and long life span, but many other individuals, including even juveniles and adolescents, are also likely to be important contributors (fig. 13.8; Kramer 2005). The "pooled energy budget" concept is essential to all models of human life history in which members of a group contribute to help subsidize slow juvenile growth and female fertility (Hrdy 1999, 2005; Reiches et al. 2009). Future work is needed to clarify who helps, under what conditions, and in what way, and how potential conflicts of interest are resolved. Indeed, the stability of helping behavior among cooperative breeders is not easily reconciled with standard evolutionary models of altruism (Bergmüller et al. 2007). Similarly, the extensive flow of resources and information among non-kin also requires further explanation, as simple models such as reciprocity and

Fig. 13.8. A key adaptation in the evolution of human reproduction and life history is the significant enhancement of alloparental caregiving. In many societies, postreproductive grandmothers are important in reducing the burden on the baby's mother. Thus the mother can work less while recovering from pregnancy and nursing her infant. This, along with subsidies from others, helps the mother return to a fertile, ovulatory state, thereby contributing to the short interbirth intervals characteristic of humans. Photo © Pete Leonard/Corbis.

kin selection cannot explain many of the sharing norms of hunter-gatherers. Complex divisions of labor, group augmentation, and the strategic use of sharing as social insurance are themes that have not been adequately addressed. We also need better studies of multicurrency trades, as well as models of how sharing norms, ownership rights, production schemes, and intergenerational nutrient flows might work in human populations.

To better elucidate the evolution of an extended life span in humans, more attention to the differences in physiology and disease etiology between humans and other primates, such as chimpanzees, will be informative (e.g., Nissi et al.

2009). Also needed is a better understanding of the costs and benefits of investments in innate and adaptive immune response in pathogen-rich and energy-limited environments (McDade 2003). The fields of evolutionary medicine and ecological immunology aim to understand health and disease in ecological and evolutionary contexts, invoking life-history trade-offs, host-parasite dynamics, and other evolutionary principles. Understanding human growth, development, aging, and health in light of the evolutionary history of our species is a major aim of biodemography (Finch & Kirkwood 2000), of evolutionary physiology (Garland and Carter 1994), and of the developmental origins of health and disease (DOHaD) paradigm (Gluckman & Hanson 2006). Each of these areas has benefited, and will continue to benefit, from collaborations with anthropologists working in preindustrial and transitional populations as well as with primatologists who work with wild and captive ape and monkey populations.

Conclusion

Future study of human behavior from an evolutionary perspective will require a synthesis of the three approaches (HBE, DIT, EP) in a diverse range of study populations (Henrich et al. 2010). The functional design of the mind should not be ignored in behavioral ecological studies, just as behavioral outcomes, constrained optimality, and fitness should not be ignored in psychological studies. A better understanding of trait change over time will require a greater appreciation of gene-culture coevolution. Comparative phylogenetic approaches will also be useful for making sense of existing cultural variation (Borgerhoff Mulder 2001; Nunn et al. 2006). The role of initial conditions and both social and physical ecology in shaping the patterned unraveling of history may be best illuminated through further integration of HBE with DIT. Yesterday's decisions and outcomes are today's parameters and constraints. To explain the behavioral diversity that exists within and among human populations is a daunting yet exciting goal. To what extent is such diversity due to reaction norms in variable environments (HBE), to different cultural trajectories in similar or different environments (DIT), to genetic variation, or to a combination and interaction of all three? The future offers hope for further integrating knowledge about humans across the social and biological sciences. Further communication across traditional disciplinary divides can push the biological study of humans to be further consilient with mammalian behavioral biology and to elucidate the origins and maintenance of derived human features.

References

Agras, S., Sylvester, D. & Oliveau, D. 1969. The epidemiology of common fears and phobias. *Comprehensive Psychiatry*, 10, 151–156.

Aiello, L. & Dean, C. 1990. *An Introduction to Human Evolutionary Anatomy*. New York: Academic Press.

Altman, J. C. 1987. *Hunter-Gatherers Today: An Aboriginal Economy of North Australia*. Canberra: Australian Institute of Aboriginal Studies.

Alvard, M. & Nolin, D. 2002. Rousseau's whale hunt? Coordination among big game hunters. *Current Anthropology*, 43, 533–559.

Atran, S., Medin, D. L. & Ross, N. O. 2005. The cultural mind: Environmental decision making and cultural modeling within and across populations. *Psychological Review*, 112, 744–776.

Bailey, R. C. 1990. *The Behavioral Ecology of Efe Pygmy Men in the Ituri Forest, Zaire*. Anthropological Papers. Museum of Anthropology, University of Michigan.

Baker, M. J. 2003. An equilibrium conflict model of land tenure in hunter-gatherer societies. *Journal of Political Economy*, 111, 124–173.

Barlow, K. R. & Metcalfe, D. 1996. Plant utility indices: Two Great Basin examples. *Journal of Archaeological Science*, 23, 351–372.

Baron-Cohen, S. 1995. *Mindblindness: An Essay on Autism and Theory of Mind*. Cambridge, MA: MIT Press.

Barrett, H. C. 1999. Human cognitive adaptations to predators and prey, PhD dissertation, University of California, Santa Barbara.

———. 2005. Adaptations to predators and prey. In *The Handbook of Evolutionary Psychology* (ed. by Buss, D. M.), 200–223. New York: John Wiley & Sons.

Barrett, H. C., Tooby, J. & Cosmides, L. 2007. The hominid entry into the cognitive niche. In *The Evolution of Mind: Fundamental Questions and Controversies* (ed. by Gangestad, S. W. & Simpson, J. A.), 241–248. New York: Guilford Press.

Barton, R. A. 1999. The evolutionary ecology of the primate brain. In *Comparative Primate Socioecology* (ed. by Lee, P. C.), 167–203. Cambridge: Cambridge University Press.

Beckerman, S. 1983. Optimal foraging group size for a human population: The case of Bari fishing. *American Zoologist*, 23, 283–290.

Bercovitch, F. B., Lebron, M. R., Martinez, H. S. & Kessler, M. J. 1998. Primigravidity, body weight, and costs of rearing first offspring in rhesus macaques. *American Journal of Primatology*, 46, 135–144.

Bereczkei, T. & Csanaky, A. 2001. Stressful family environment, mortality, and child socialisation: Life-history strategies among adolescents and adults from unfavourable social circumstances. *International Journal of Behavioral Development*, 25, 501–508.

Bergmüller, R., Johnstone, R. A., Russell, A. F. & Bshary, R. 2007. Integrating cooperative breeding into theoretical concepts of cooperation. *Behavioural Processes*, 76, 61–72.

Bering, J. M. & Povinelli, D. J. 2003. Comparing cognitive development. In *Primate Psychology* (ed. by Maestripieri, D.), 205–233. Cambridge, MA: Harvard University Press.

Bersaglieri, T., Sabeti, P. C., Patterson, N., Vanderploeg, T., Schaffner, S. F., Drake, J. A., Rhodes, M., Reich, D. E. &

Hirschhorn, J. N. 2004. Genetic signatures of strong recent positive selection at the lactase gene. *American Journal of Human Genetics*, 74, 1111–1120.

Binford, L. R. 1981. *Bones: Ancient Men and Modern Myths*. New York: Academic Press.

Bird, D. W. & Bliege Bird, R. 2002. Children on the reef: Slow learning or strategic foraging. *Human Nature*, 13, 269–297.

———. 2005. Mardu children's hunting strategies in the Western Desert, Australia: Foraging and the evolution of human life histories. In *Hunter Gatherer Childhoods* (ed. by Hewlett, B. S. & Lamb, M.E.), 129–146. New York: Aldine de Gruyter.

Bjorklund, D. F. 1997. The role of immaturity in human development. *Psychological Bulletin*, 122, 153–169.

Bliege Bird, R. & Bird, D. 2002. Constraints of knowing or constraints of growing? Fishing and collecting by the children of Mer. *Human Nature*, 13, 239–267.

Blumenschine, R. J., Cavallo, J. A. & Capaldo, S. D. 1994. Competition for carcasses and early hominid behavioral ecology: A case study and a conceptual framework. *Journal of Human Evolution*, 27, 197–213.

Blurton Jones, N. G. & Konner, M. J. 1976. !Kung knowledge of animal behavior. In *Kalahari Hunter-Gatherers* (ed. by Lee, R. B. & DeVore, I.), 325–348. Cambridge: Harvard University Press.

Blurton Jones, N. G. & Marlowe, F. W. 2002. Selection for delayed maturity: Does it take 20 years to learn to hunt and gather? *Human Nature*, 13, 199–238.

Blurton Jones, N., Smith, L., O'Connell, J., K., H. & Samuzora, C. L. 1992. Demography of the Hadza, an increasing and high density population of savanna foragers. *American Journal of Physical Anthropology*, 89, 159–181.

Bock, J. 2002. Learning, life history, and productivity: Children's lives in the Okavango Delta, Botswana. *Human Nature*, 13, 161–198.

Bock, J. & Sellen, D. W. 2002. Childhood and the evolution of the human life course: An introduction. *Human Nature*, 13, 153–161.

Boesch, C. & Boesch-Achermann, H. 2000. *Chimpanzees of the Tai Forest: Behavioural Ecology and Evolution*. Oxford: Oxford University Press.

Boesch, C. & Tomasello, M. 1998. Chimpanzee and human cultures. *Current Anthropology*, 39, 591–614.

Bogin, B. 1997. Evolutionary hypotheses for human childhood. *Yearbook of Physical Anthropology*, 104, 63–90.

Bogin, B. & Smith, B. H. 1996. Evolution of the human life cycle. *American Journal of Human Biology*, 8, 703–716.

Borgerhoff Mulder, M. 1991. Human behavioural ecology. In *Behavioural Ecology: An Evolutionary Approach* (ed. by Krebs, J. R. & Davies, N.), 69–98. Oxford: Blackwell Scientific Publications.

———. 2001. Using phylogenetically based comparative methods in anthropology: More questions than answers. *Evolutionary Anthropology*, 10, 99–111.

Boyd, R. & Richerson, P. J. 1985. *Culture and the Evolutionary Process*. Chicago: University of Chicago Press.

Boyd, R. & Richerson, P. J. 1996. Why culture is common, but cultural evolution is rare. In *Evolution of Social Behaviour Patterns in Primates and Man* (ed. by Runciman, W. G., Maynard Smith, J. & Dunbar, R. I. M.), 77–94. Oxford: Oxford University Press.

Brumbach, B. H., Figueredo, A. J. & Ellis, B. J. 2009. Effects of harsh and unpredictable environments in adolescence on development of life history strategies. *Human Nature*, 20, 25–51.

Buss, D. M. 1995. Evolutionary psychology: A new paradigm for psychological science. *Psychological Inquiry*, 6, 1–30.

———. 1999. *Evolutionary Psychology: The New Science of the Mind*: Allyn & Bacon.

Byers, D. A. & Ugan, A. 2005. Should we expect large game specialization in the late Pleistocene? An optimal foraging perspective on early Paleoindian prey choice. *Journal of Archaeological Science*, 32, 1624–1640.

Call, J. & Tomasello, M. 2008. Does the chimpanzee have a theory of mind? 30 years later. *Trends in Cognitive Science*, 12, 187–192.

Caro, T. M. & Hauser, M. D. 1992. Is there teaching in nonhuman animals? *Quarterly Journal of Biology*, 67, 151–174.

Cashdan, E. 1983. Territoriality among human foragers: Ecological models and an application to four Bushman groups. *Current Anthropology*, 24, 47–66.

Castro, L. & Toro, M. A. 2004. The evolution of culture: From primate social learning to human culture. *Proceedings of the National Academy of Sciences of the United States of America*, 101, 10235–10240.

Cavalli-Sforza, L. & Feldman, M. 1981. *Cultural Transmission and Evolution: A Quantitative Approach*. Princeton: Princeton University Press.

Charnov, E. 1976. Optimal foraging: The marginal value theorem. *Theoretical Population Biology*, 9, 129–136.

Chisholm, J. S. 1999. *Death, Hope and Sex: Steps to an Evolutionary Ecology of Mind and Morality*. Cambridge: Cambridge University Press.

Cordain, L., Brand Miller, J., Eaton, S. B., Mann, N., Holt, S. H. A. & Speth, J. D. 2000. Plant-animal subsistence ratios and macronutrient energy estimations in hunter-gatherer diets. *American Journal of Clinical Nutrition*, 71, 682–692.

Crawford, C. B. & Krebs, D. 1998. *Handbook of Evolutionary Psychology: Ideas, Issues, and Applications* Mahwah, NJ: Lawrence Erlbaum Associates.

Cronk, L. 1991. Human behavioural ecology. *Annual Review of Anthropology*, 20, 25–53.

Cronk, L., Chagnon, N. & Irons, W. 2000. *Adaptation and Human Behavior: An Anthropological Perspective*. New York: Aldine de Gruyter.

Csibra, G. & Gergely, G. 2006. Social learning and social cognition: The case for pedagogy. In *Processes of Change in Brain and Cognitive Development: Attention and Performance, XXI* (ed. by Munakata, Y. & Johnson, M. H.), 249–274. Oxford: Oxford University Press.

Daly, M. & Wilson, M. I. 1999. Human evolutionary psychology and animal behaviour. *Animal Behaviour*, 57, 509–519.

Dean, C., Leakey, M. G., Reid, D., Schrenk, F., Schwartz, G. T., Stringer, C. & Walker, A. 2001. Growth processes in teeth distinguish modern humans from *Homo erectus* and earlier hominins. *Nature*, 414, 628–631.

Dunbar, R. I. M. 1998. The social brain hypothesis. *Evolutionary Anthropology*, 6, 178–190.

Durham, W. H. 1991. *Coevolution: Genes, Culture, and Human Diversity*: Stanford, CA: Stanford University Press.

Dyson-Hudson, R. & Smith, E. A. 1978. Human territoriality:

An ecological reassessment. *American Anthropologist*, 80, 21–41.

Early, J. D. & Headland, T. N. 1998. *Population Dynamics of a Philippine Rain Forest People: The San Ildefonso Agta*. Gainesville: University Press of Florida.

Elwood, R. W. & Mason, C. 1994. The couvade and the onset of paternal care: A biological perspective. *Ethology and Sociobiology*, 15, 145–156.

Finch, C. 2007. *The Biology of Human Longevity*. San Diego: Academic Press.

Finch, C. E. & Kirkwood, T. B. L. 2000. *Chance, Development, and Aging*. Oxford: Oxford University Press.

Gallistel, C. R. 1990. *The Organization of Learning*. Cambridge, MA: MIT Press.

Garber, P. A. 1987. Foraging strategies among living primates. *Annual Review of Anthropology*, 16, 339–364.

Garland, T. & Carter, P. A. 1994. Evolutionary physiology. *Annual Review of Physiology*, 56, 579–621.

Gaulin, S. J. C. & McBurney, D. 2004. *Evolutionary Psychology*. Upper Saddle River, NJ: Pearson Prentice Hall.

Giraldeau, L. & Caraco, T. 2000. *Social Foraging Theory*. Princeton, NJ: Princeton University Press.

Gluckman, P. D. & Hanson, M. A. 2006. *Developmental Origins of Health and Disease*. Cambridge: Cambridge University Press.

Grafen, A. 1984. Natural selection, kin selection and group selection. In *Behavioural Ecology: An Evolutionary Approach, 2nd edition* (ed. by Krebs, J. & Davies, N. B.), 62–84. Oxford: Blackwell Scientific Publications.

Gray, P. B., Chapman, J. F., Burnham, T. C., McIntyre, M. H., Lipson, S. F. & Ellison, P. T. 2004. Human male pair bonding and testosterone. *Human Nature*, 15, 119–131.

Gremillion, K. 2002. Foraging theory and hypothesis testing in archaeology: An Exploration of methodological problems and solutions. *Journal of Anthropological Archaeology*, 21, 142–164.

Gurven, M. 2004. To give or to give not: An evolutionary ecology of human food transfers. *Behavioral and Brain Sciences*, 27, 543–583.

———. 2006. The evolution of contingent cooperation. *Current Anthropology*, 47, 185–192.

Gurven, M., Allen-Arave, W., Hill, K. & Hurtado, M. 2000. "It's a Wonderful Life": signaling generosity among the Ache of Paraguay. *Evolution and Human Behavior*, 21, 263–282.

Gurven, M. & Fenelon, A. 2009. Has the rate of actuarial aging changed over the past 250 years? A comparison of small-scale subsistence populations, and Swedish and English cohorts. *Evolution*, 63, 1017–1035.

Gurven, M. & Hill, K. 2009. Why do men hunt? A re-evaluation of "man the hunter" and the sexual division of labor. *Current Anthropology*, 50, 51–74.

Gurven, M. & Kaplan, H. S. 2005. Determinants of time allocation to production across the lifespan among the Machiguenga and Piro Indians of Peru. *Human Nature*, 17, 1–49.

———. 2007. Longevity among hunter-gatherers: A cross-cultural comparison. *Population and Development Review*, 33, 321–365.

Gurven, M., Kaplan, H. & Gutierrez, M. 2006. How long does it take to become a proficient hunter? Implications for the evo-

lution of delayed growth. *Journal of Human Evolution*, 51, 454–470.

Gurven, M. & Walker, R. 2006. Energetic demand of multiple dependents and the evolution of slow human growth. *Proceedings of the Royal Society of London, Series B: Biological Sciences*, 273, 835–841.

Hauser, M. D. 2009. The possibility of impossible cultures. *Nature*, 460, 190–196.

Hawkes, K. 2003. Grandmothers and the evolution of human longevity. *American Journal of Human Biology*, 15, 380–400.

Hawkes, K., O'Connell, J. F. & Blurton Jones, N. G. 1991. Hunting income patterns among the Hadza: Big game, common goods, foraging goals and the evolution of the human diet. *Philosophical Transactions of the Royal Society of London (B)*, 334, 243–251.

Hawkes, K., O'Connell, J. F., Blurton Jones, N. G., Alvarez, H. & Charnov, E. L. 1998. Grandmothering, menopause, and the evolution of human life histories. *Proceedings of the National Academy of Sciences USA*, 95, 1336–1339.

Hawks, J., Wang, E. T., Cochran, G. M., Harpending, H. C. & Moyzis, R. K. 2007. Recent acceleration of human adaptive evolution. *Proceedings of the National Academy of Sciences, USA*, 104, 20753–20758.

Hewlett, B. S. & Cavalli-Sforza, L. L. 1986. Cultural transmission among Aka pygmies. *American Anthropologist*, 88, 922–934.

Henrich, J. & Boyd, R. 1998. The evolution of conformist transmission and the emergence of between-group differences. *Evolution and Human Behavior*, 19, 215–241.

Henrich, J. & Gil-White, F. 2001. The evolution of prestige: Freely conferred status as a mechanism for enhancing the benefits of cultural transmission. *Evolution and Human Behavior*, 22, 1–32.

Henrich, J., Heine, S. J. & Norenzayan, A. 2010. The weirdest people in the world? *Behavioral and Brain Sciences*, 33, 61–83.

Henrich, J. & McElreath, R. 2003. The evolution of cultural evolution. *Evolutionary Anthropology*, 12, 123–135.

Heinrichs, M. & Domes, G. 2008. Neuropeptides and social behavior: Effects of oxytocin and vasopressin in humans. *Progress in Brain Research*, 170, 337–350.

Hill, E. M., Ross, L. T. & Low, B. S. 1997. The role of future unpredictability in human risk-taking. *Human Nature*, 8, 287–325.

Hill, K. 1993. Life history theory and evolutionary anthropology. *Evolutionary Anthropology*, 2, 78–88.

Hill, K., Boesch, C., Goodall, J., Pusey, A., Williams, J. & Wrangham, R. 2001. Mortality rates among wild chimpanzees. *Journal of Human Evolution*, 40, 437–450.

Hill, K. & Hawkes, K. 1983. Neotropical hunting among the Ache of Eastern Paraguay. In *Adaptive Responses of Native Amazonians* (ed. by Hames, R. & Vickers, W.), 139–188. New York: Academic Press.

Hill, K. & Hurtado, A. M. 1996. *Ache Life History: The Ecology and Demography of a Foraging People*. New York: Aldine de Gruyter.

Hill, K. & Kaplan, H. 1999. Life history traits in humans: Theory and empirical studies. *Annual Review of Anthropology*, 28, 397–430.

Howell, N. 1979. *Demography of the Dobe !Kung*. New York: Academic Press.

Hrdy, S. 1999. *Mother Nature: A History of Mothers, Infants and Natural Selection*. New York: Pantheon.

———. 2005. Comes the child before the man: How cooperative breeding and prolonged post-weaning dependence shaped human potentials. In *Hunter-Gatherer Childhoods* (ed. by Hewlett, B. S. & Lamb, M. E.), 65–91. Piscataway, NJ: Transaction.

———. 2009. *Mothers and Others: The Evolutionary Origins of Mutual Understanding*. Cambridge, MA: Harvard University Press.

Ikeya, K. 1994. Hunting with dogs among the San in the central Kalahari. *African Study Monographs*, 15, 119–134.

Isbell, L. A. 2009. *The Fruit, the Tree, and the Serpent: Why We See So Well*. Boston: Harvard University Press.

Jankowiak, W., Nell, M. D. & Buckmaster, A. 2002. Managing infidelity: A cross-cultural perspective. *Ethnology*, 41, 85–101.

Janson, C. H. & Chapman, C. A. 1999. Resources and the determination of primate community structure. In *Primate Communities* (ed. by Fleagle, J. G., Janson, C. H. & Reed, K. E.), 237–267. Cambridge: Cambridge University Press.

Janson, C. H. & van Schaik, C. P. 1993. Ecological risk aversion in juvenile primates: Slow and steady wins the race. In *Juvenile Primates: Life History, Development and Behavior* (ed. by Pereira, M. & Fairbanks, L.), 57–76. New York: Oxford University Press.

Jones, J. H. 2009. The force of selection on the human life cycle. *Evolution and Human Behavior*, 30, 305–314.

Kaplan, H. S. & Gangestad, S. W. 2005. Life history theory and evolutionary psychology. In *Handbook of Evolutionary Psychology* (ed. by Buss, D. M.), 68–95. New Jersey: John Wiley & Sons.

Kaplan, H. & Gurven, M. 2005. The natural history of human food sharing and cooperation: A review and a new multi-individual approach to the negotiation of norms. In *Moral Sentiments and Material Interests: The Foundations of Cooperation in Economic Life* (ed. by Gintis, H., Bowles, S., Boyd, R. & Fehr, E.). Cambridge, MA: MIT Press.

Kaplan, H. & Hill, K. 1992. The evolutionary ecology of food acquisition. In *Evolutionary Ecology and Human Bheavior* (ed. by Smith, E. A. & Winterhalder, B.), 167–201. New York: Aldine de Gruyter.

Kaplan, H., Hill, K., Lancaster, J. B. & Hurtado, A. M. 2000. A theory of human life history evolution: Diet, intelligence, and longevity. *Evolutionary Anthropology*, 9, 156–185.

Kaplan, H. S. & Robson, A. J. 2002. The emergence of humans: The coevolution of intelligence and longevity with intergenerational transfers. *Proceedings of the National Academy of Sciences*, 99, 10221–10226.

Keeley, L. H. 1996. *War Before Civilization*. New York: Oxford University Press.

Kelly, R. L. 1995. *The Foraging Spectrum: Diversity in Hunter-Gatherer Lifeways*. Washington: Smithsonian Institution Press.

Kennett, D. J. & Winterhalder, B. 2006. *Behavioral Ecology and the Transition to Agriculture*. Berkeley: University of California Press.

Koenig, A. & Borries, C. 2006. The predictive power of socio-ecological models: A reconsideration of resource characteristics, agonism and dominance hierarchies. In *Feeding Ecology in Apes and Other Primates* (ed. by Hohmann, G., Robbins, M. M. & Boesch, C.), 263–284. Cambridge: Cambridge University Press.

Koster, J. M. 2008. Hunting with dogs in Nicaragua: An optimal foraging approach. *Current Anthropology*, 49, 935–944.

Kramer, K. L. 2005. Children's help and the pace of reproduction: Cooperative breeding in humans. *Evolutionary Anthropology*, 14, 224–237.

Kramer, K. L., Greaves, R. D. & Ellison, P. T. 2009. Early reproductive maturity among Pumé foragers: Implications of a pooled energy model to fast life histories. *American Journal of Human Biology*, 21, 430–437.

Laland, K. N. & Brown, G. R. 2002. *Sense and Nonsense: Evolutionary Perspectives on Human Behaviour*. New York: Oxford University Press.

Laland, K. N., Odling-Smee, J. & Feldman, M. W. 2001. Cultural niche construction and human evolution. *Journal of Evolutionary Biology*, 14, 22.

Lampl, M., Veldhuis, J. D. & Johnson, M. L. 1992. Saltation and stasis: A model of human growth. *Science*, 258, 801–803.

Lee, P. C., Majluf, P. & Gordon, I. J. 1991. Growth, weaning and maternal investment from a comparative perspective. *Journal of Zoology*, 225, 99–114.

Leibenberg, L. 1990. *The Art of Tracking: The Origin of Science*. Cape Town: David Phillip.

Leigh, S. R. 2001. Evolution of human growth. *Evolutionary Anthropology*, 10, 223–236.

LoBue, V. & DeLoache, J. S. 2008. Detecting the snake in the grass. *Psychological Science*, 19, 284–289.

Lycett, S. J., Collard, M. & McGrew, W. C. 2007. Phylogenetic analyses of behavior support existence of culture among wild chimpanzees. *Proceedings of the National Academy of Sciences USA*, 104, 17588–17593.

MacArthur, R. H. & Pianka, E. R. 1966. On optimal use of a patchy environment. *American Naturalist*, 100, 603–609.

Marlowe, F. W. 2005. Hunter gatherers and human evolution. *Evolutionary Anthropology*, 14, 54–67.

Maynard Smith, J. 1982. *Evolution and the Theory of Games*. Cambridge: Cambridge University Press.

McDade, T. W. 2003. Life history theory and the immune system: Steps toward a human ecological immunology. *Yearbook of Physical Anthropology*, 46, 100–125.

McElreath, R. 2004. Social learning and the maintenance of cultural variation: An evolutionary model and data from East Africa. *American Anthropologist*, 106, 308–321.

Meehan, B. 1982. *Shell Bed to Shell Midden*. Canberra: Australian Institute of Aboriginal Studies.

Mergler, N. L., Faust, M. & Goldstein, M. D. 1984. Storytelling as an age-dependent skill: Oral recall of orally presented stories. *International Journal of Aging & Human Development*, 20, 205–228.

Mesoudi, A. & O'Brien, M. J. 2008. The cultural transmission of Great Basin projectile-point technology I: An experimental simulation. *American Antiquity*, 73, 3–28.

Milner, L. S. 2000. *Hardness of Heart/Hardness of Life: The Stain of Human Infanticide*. Lanham, MD: University Press of America.

Mineka, S., Davidson, M., Cook, M. & Keir, R. 1984. Observational conditioning of snake fear in rhesus monkeys. *Journal of Abnormal Psychology*, 93, 355–372.

Murdock, G. P. & Provost, C. 1973. Factors in the division of labor by sex: A cross-cultural analysis. *Ethnology*, 12, 203–225.

Nettle, D. & Pollet, T. V. 2008. Natural selection on male wealth in humans. *American Naturalist*, 172, 658–666.

Nishida, T., Corp, N., Hamai, M., Hasegawa, T., Hiraiwa-Hasegawa, M., Hosaka, H., Hunt, K. D., Itoh, N., Kawanaka, K., Matsumoto-Oda, A., Mitani, J. C., Nakamura, M., Norikoshi, K., Sakamaki, T., Turner, L., Uehara, S. & Zamma, K. 2003. Demography, female life history and reproductive profiles among the chimpanzees of Mahale. *American Journal of Primatology*, 59, 99–121.

Nissi, V., Dan, A., James, G. H., Tho, P., Christopher, J. G., Monica, C., James, M., Elizabeth, S., Jo, F., James, G. E. & Ajit, V. 2009. Heart disease is common in humans and chimpanzees, but is caused by different pathological processes. *Evolutionary Applications*, 2, 101–112.

Norenzayan, A. & Heine, S. J. 2005. Psychological universals: What are they and how can we know? *Psychological Bulletin*, 131, 763–784.

Nunn, C. L., Borgerhoff Mulder, M. & Langley, S. 2006. Comparative methods for studying cultural trait evolution: A simulation study. *Cross-Cultural Research*, 40, 177–209.

O'Connell, J. F., Hawkes, K., Lupo, K. D. & Blurton Jones, N. G. 2002. Male strategies and Plio-Pliestocene archaeology. *Journal of Human Evolution*, 43, 831–872.

Öhman, A. & Mineka, S. 2001. Fear, phobias and preparedness: Toward an evolved module of fear and fear learning. *Psychological Review*, 108, 483–522.

Ohtsuka, R. 1989. Hunting activity and aging among the Gidra Papuans: A biobehavioral analysis. *American Journal of Physical Anthropology*, 80, 31–39.

Pagel, M. D. & Harvey, P. H. 1993. Evolution of the juvenile period in mammals. In *Juvenile Primates: Life History, Development, and Behavior* (ed. by Pereira, M. E. & Fairbanks, L.A.), 28–37. New York: Oxford University Press.

Panksepp, J. & Panksepp, J. B. 2000. The seven sins of evolutionary psychology. *Evolution and Cognition*, 6, 108–131.

Pennington, R. L. 1996. Causes of early human population growth. *American Journal of Physical Anthropology*, 99, 259–274.

Pereira, M. E. & Fairbanks, L. A. 2002. *Juvenile Primates: Life History, Development, and Behavior*. Chicago: University of Chicago Press.

Perry, S. E. 2006. What cultural primatology can tell anthropologists about the evolution of culture. *Annual Review of Anthropology*, 35, 171–190.

Pinker, S. 1994. *The Language Instinct: How the Mind Creates Language*. New York: Harper Collins.

Premack, D. & Premack, A. J. 1996. Why animals lack pedagogy and some cultures have more of it than others. In *The Handbook of Education and Human Development* (ed. by Olson, D. R. & Torrance, N.), 302–323. Oxford: Blackwell.

Quinlan, R. J. 2008. Human pair-bonds: Evolutionary functions, ecological variation and adaptive development. *Evolutionary Anthropology*, 17, 227–238.

Reader, S. M. & Laland, K. N. 2002. Social intelligence, innovation, and enhanced brain size in primates. *Proceedings of the National Academy of Sciences USA*, 99, 4436–4441.

Reiches, M. W., Ellison, P. T., Lipson, S. F., Sharrock, K. C., Gardiner, E. & Duncan, L. G. 2009. Pooled energy budget and human life history. *American Journal of Human Biology*, 21, 421–429.

Rilling, J. K. 2006. Human and non-human primate brains: Are they allometrically scaled versions of the same design? *Evolutionary Anthropology*, 15, 65–77.

Robbins, A. M., Robbins, M. M., Gerald-Steklis, N. & Steklis, H. D. 2006. Age-related patterns of reproductive success among female mountain gorillas. *American Journal of Physical Anthropology*, 131, 511–521.

Ross, C. & Jones, K. E. 1999. Socioecology and the evolution of primate reproductive rates. In *Comparative Primate Socioecology* (ed. by Lee, P. C.), 73–110. Cambridge: Cambridge University Press.

Sear, R., Lawson, D. W. & Dickins, T. E. 2007. Synthesis in the human evolutionary behavioural sciences. *Journal of Evolutionary Psychology*, 5, 3–28.

Simmons, L. 1945. *The Role of the Aged in Primitive Society*. New Haven: Yale University Press.

Smith, E. A. 1985. Inuit foraging groups: Some simple models incorporating conflicts of interest, relatedness, and central-place sharing. *Ethology and Sociobiology*, 6, 27–47.

———. 1991. *Inujjuamiut Foraging Strategies: Evolutionary Ecology of an Arctic Hunting Economy*. Hawthorne, NY: Aldine de Gruyter.

Smith, E. A., Borgerhoff-Mulder, M. & Hill, K. 2001. Controversies in the evolutionary social sciences: A guide for the perplexed. *Trends in Ecology & Evolution*, 16, 128–135.

Smith, E. A. & Winterhalder, B. 1992. *Evolutionary Ecology and Human Behavior*. Hawthorne, NY : Aldine De Gruyter.

Sosis, R. 2002. Patch choice decisions among Ifaluk fishers. *American Anthropologist*, 104, 583–598.

Stanford, C. G. 1999. *The Hunting Apes: Meat Eating and the Origins of Human Behavior*. Princeton, NJ: Princeton University Press.

Stephens, D. W., Brown, J. S. & Ydenberg, R. C. 2007. *Foraging: Behavior and Ecology*. Chicago: University of Chicago Press.

Stephens, D. & Krebs, J. R. 1986. *Foraging Theory*. Princeton, NJ: Princeton University Press.

Stiner, M. 1991. An interspecific perspective on the emergence of the modern human predatory niche. In *Human Predators and Prey Mortality* (ed. by Stiner, M.), 149–185. Boulder: Westview Press.

———. 2002. Carnivory, coevolution, and the geographic spread of the genus *Homo*. *Journal of Archaeological Research*, 10, 1–63.

Storey, A. E., Walsh, C. J., Quinton, R. L. & Wynne-Edwards, K. E. 2000. Hormonal correlates of paternal responsiveness in new and expectant fathers. *Evolution and Human Behavior*, 21, 79–95.

Sugiyama, L. S. 2004. Illness, injury, and disability among Shiwiar forager-horticulturalists: Implications for health-risk buffering for the evolution of human life history. *American Journal of Physical Anthropology*, 123, 371–389.

Thomas, F. R. 2007a. The behavioral ecology of shellfish gathering in western Kiribati, Micronesia. 1: Prey Choice. *Human Ecology*, 35, 179–194.

———. 2007b. The behavioral ecology of shellfish gathering in western Kiribati, Micronesia. 2: Patch choice, patch sampling, and risk. *Human Ecology*, 35, 515–526.

Tomasello, M. 1999. *The Cultural Origins of Human Cognition*. Cambridge: Harvard University Press.

Tooby, J. & Cosmides, L. 1989. Evolutionary psychology and the generation of culture, part I. *Ethology and Sociobiology*, 10, 29–49.

———. 1992. The psychological foundations of culture. In *The Adapted Mind: Evolutionary Psychology and the Generation of Culture* (ed. by Barkow, J., Cosmides, L. & Tooby, J.), 19–136. New York: Oxford University Press.

Tsukahara, T. 1993. Lions eat chimpanzees: The first evidence of predation by lions on wild chimpanzees. *American Journal of Primatology*, 29, 1–11.

Tucker, B. & Young, A. G. 2005. Growing up Mikea: Children's time allocation and tuber foraging in southwestern Madagascar. In *Hunter-Gatherer Childhoods* (ed. by Hewlett, B. & Lamb, M.), 147–171. New York: Aldine de Gruyter.

Varki, A. & Nelson, D. L. 2007. Genomic comparisons of humans and chimpanzees. *Annual Review of Anthropology*, 36, 191–209.

Walker, R., Gurven, M., Hill, K., Migliano, H., Chagnon, N., De Souza, R., Djurovic, G., Hames, R., Hurtado, A. M., Kaplan, H., Kramer, K., Oliver, W. J., Valeggia, C. & Yamauchi, T. 2006. Growth rates and life histories in twenty-two small-scale societies. *American Journal of Human Biology*, 18, 295–311.

Walker, R., Hill, K., Kaplan, H. & McMillan, G. 2002. Age-dependency in skill, strength and hunting ability among the Ache of eastern Paraguay. *Journal of Human Evolution*, 42, 639–657.

Walum, H., Westberg, L., Henningsson, S., Neiderhiser, J. M., Reiss, D., Igl, W., Ganiban, J. M., Spotts, E. L., Pederson, N. L., Eriksson, E. & Lichtenstein, P. 2008. Genetic variation in the vasopressin receptor 1a gene (AVPR1A) associates with pair-bonding behavior in humans. *Proceedings of the National Academy of Sciences USA*, 105, 14153–14156.

Watson, C. F. & Caldwell, C. A. 2009. Understanding behavioral traditions in primates: Are current experimental approaches too focused on food? *International Journal of Primatology*, 30, 143–167.

Wiessner, P. 1981. Measuring the impact of social ties on nutritional status among the !Kung San. *Social Science Information*, 20, 641–678.

———. 2005. Norm enforcement among the Ju/'hoansi bushmen: A case of strong reciprocity? *Human Nature*, 16, 115–145.

Williams, G. C. 1957. Pleitropy, natural selection and the evolution of senescence. *Evolution*, 11, 398–411.

Williams, J. M., Lonsdorf, E. V., Wilson, M. L., Schumacher-Stankey, J., Goodall, J. & Pusey, A. E. 2008. Causes of death in the Kasekela chimpanzees of Gombe National Park, Tanzania. *American Journal of Primatology*, 70, 766–777.

Winterhalder, B., Lu, F. & Tucker, B. 1999. Risk-sensitive adaptive tactics: Models and evidence from subsistence studies in biology and anthropology. *Journal of Archaeological Research*, 7, 301–348.

Winterhalder, B. & Smith, E. A. 2000. Analyzing adaptive strategies: Human behavioral ecology at twenty-five. *Evolutionary Anthropology Issues News and Reviews*, 9, 51–72.

Wood, J. W. 1994. *Dynamics of Human Reproduction: Biology, Biometry and Demography*. New York: Aldine de Gruyter.

Wrangham, R. W. 2009. *Catching Fire*. New York: Basic Books.

Wrangham, R. W. & Carmody, R. 2010. Human adaptation to the control of fire. *Evolutionary Anthropology*, 19, 187–199.

Wrangham, R. W., Wilson, M. L. & Muller, M. N. 2006. Comparative rates of violence in chimpanzees and humans. *Primates*, 47, 14–26.

Ziegler, T. E., Prudom, S. L., Schultz-Darken, N. J., Kurian, A. V. & Snowdon, C. T. 2006. Pregnancy weight gain: Marmoset and tamarin dads show it too. *Biology Letters*, 22, 181–183.

Part 3
Mating and Rearing Offspring

REPRODUCTION is the quintessential core of evolution. Part 3 of this book explores the problems and challenges most immediately related to reproduction: the mating and rearing strategies of individuals. Primates are among the most valuable organisms for examining these important processes. This single order manifests an extraordinary diversity of mating systems and strategies, exemplifying the array of patterns found across vertebrates as a whole with only a few exceptions (e.g., leks, parthenogensis, hemaphroditism). Likewise, rearing strategies vary greatly, from exclusive maternal care to biparental care with significant male contributions, and to cooperative and communal breeding. Each of these strategies may include a variety of adaptations in behavior, physiology, and genetics. This diversity reflects not just the great range of solutions, but the catholic nature of the constituent problems. Successful reproduction may require that individuals search for and find mates; differentiate among mates based on their quality; compete with rivals for access to fertilizations; compete with rivals to be chosen by mates; produce viable offspring; invest in offspring differentially based on their reproductive prospects as well as on the availability of mating opportunities; and compete for and accrue the resources, help, and protection needed to successfully rear offspring and ensure their future reproduction.

In *The Descent of Man, and Selection in Relation to Sex*, Charles Darwin established the framework for analyzing such strategies with his theory of sexual selection. Following its logic, part 3 evaluates the reproductive strategies of the sexes separately in light of the reproductive constraints confronted by males and females. The coevolutionary interaction of male and female reproductive strategies, involving both cooperation and conflict, is also critically evaluated in these chapters.

Two chapters provide an organizing nucleus for this section by addressing the magnitude and sources of variation in reproductive performance in females and males. Anne Pusey tackles these issues for female primates in chapter 15, arguing appropriately that longitudinal data are critically needed for this analysis, especially those data encompassing the *entire* reproductive careers of individuals. Although slow life histories constitute a significant methodological hurdle for acquiring such data in primates, the data have recently become available from long-term studies of select species. These studies now reveal that variation in lifetime reproductive success in females is often considerably greater than previously anticipated. The challenge, of course, is to explain this variation. Pusey reviews available data on the effects of age, parity, physical condition, dominance status, and access to food, helpers, allies, or protectors on various components of fertility, such as age at menarche and first reproduction, longevity, reproductive rate, and infant survival. Some interesting surprises have emerged, such as the effect of female dominance rank being negligible in some species with rigid, "despotic" hierarchies, but being relatively strong in some species with weak, "egalitarian" hierarchies. Although most lifetime reproductive data currently derive from Old World species, platyrrhines and strepsirrhines provide important insights, e.g., on the potential effects of allomaternal care and physiological condition.

Susan C. Alberts undertakes the analogous analysis for males in chapter 18. Once again, relevant data are difficult to collect. This is not only because of the slow reproductive rates of primates, but also because of two problems particularly associated with males. First, internal fertilization means that reproductive success has historically been difficult to measure directly, thereby necessitating the occasional use of proxies of variable analytical value, such as mating success. Happily, this methodological hurdle is rapidly being overcome as paternity tests using noninvasively collected DNA are becoming more common. Second, the fact that males are the dispersing sex in many primates limits longitudinal data on individual performance. Nevertheless, data on variance in male reproductive success have been accumulated from various taxa, and Alberts highlights four components of particular importance: when a male begins his reproductive career (age at first reproduction), how long his career lasts (longevity), how effectively he competes for females and allocates his mating effort among them (infant production), and how well he improves the chances that his offspring will live to reproduce (infant survival). This variation can be understood in terms of traditionally recognized variables such as male competitive ability (dominance and fighting ability) and age, but also in terms of more recently acknowledged factors such as maternal rank, affiliation with other males and females, and alternative tactics (e.g., "invasions" by nonresident males). Most data address how males may monopolize access to female reproductive events, but attention recently has begun to focus on how males might improve the reproductive performance of females themselves (e.g., by enhancing the survival of infants through infanticide protection).

Following up on these considerations of variation in reproductive performance, the other chapters in this section focus on particular aspects of reproductive strategies in males and females. In chapter 17, Martin N. Muller and Melissa Emery Thompson evaluate male reproductive strategies by framing their analysis within the fundamental trade-off facing males: the search for additional mating opportunities versus investment in the care and protection of offspring. They argue that primate males have evolved a great variety of solutions to this basic evolutionary conundrum. Partly because of their focus on cercopithecines and great apes, early studies tended to emphasize contest competition for mates, and the attributes that contribute to success in that domain. Muller and Emery Thompson bring a fresh analysis to this question by considering a larger sample of cross-species data on dimorphism, improved assays of intrasexual competition, and integration of additional variables. For example, recent data suggesting that canine dimorphism is constrained by feeding adaptations

help to explain some of the relevant variation. Of special importance is that recent field research has increasingly emphasized postcopulatory mechanisms (e.g., sperm competition) as well as more newly appreciated aspects of precopulatory competition, such as scramble competition and alternative mating tactics. Parental investment in offspring by males has traditionally received little attention in light of the apparent importance of male-male competition, but Muller and Emery Thompson review the accumulating evidence of its strategic importance in certain primates, both monogamous and polygamous.

In chapter 16, Peter M. Kappeler addresses the second pillar of sexual selection theory: female choice. The importance of mate choice was not established until the late 1970s and 1980s, when an explosion of supportive theoretical models and empirical studies of nonprimate animals was published. Primatologists were similarly slow to recognize the significance of female choice, partly because of neglect of the subject, but also partly because the apparent predominance of male-male competition in primate societies appeared to place severe constraints on opportunities for females to exercise mating preferences. The latter debate has continued and expanded, and Kappeler directly addresses the question of why mate choice may be rare in primates. Nevertheless, there is limited empirical evidence and theoretical rationale for female choice in some primates. Kappeler reviews the evidence for direct benefits (e.g., paternal care) and indirect benefits (e.g., good genes) in primates. Of particular interest is whether male ornaments are honest indicators of male genetic quality. This is relatively well established in birds and fish, but less clearly established in primates. Nevertheless, new data suggest that some strepsirrhines may use olfactory cues to choose mates on the basis of genetic compatibility. Most work on mate choice in primates has focused on precopulatory mechanisms (the choosing of sexual partners), but postcopulatory mechanisms based on differential treatment of sperm in the female reproductive tract—cryptic female choice—is suggested by insect studies. There are some limited, indirect data suggesting the possible importance of this mechanism in primates.

The nature of mammalian reproductive physiology dictates that maternal investment strategies are crucially important components of female reproductive strategy. In chapter 14, Maria A. van Noordwijk explores variation in investment provided by primate mothers, beginning with gestation and lactation, and continuing on to care in the form of transportation, protection, provisioning, and learning opportunities. Once again, primates exemplify immense variation in maternal strategies. Van Noordwijk examines how the timing, magnitude, and distribution of the care a female provides her offspring is influenced by the following factors: the predictability of fluctuating food resources, the developmental (growth) trajectory of infants, the sex of offspring (sons versus daughters), the size and composition of groups, and the availability of care provided by allomothers. Conflict between mother and offspring structures some elements of maternal strategy, such as prenatal investment during gestation and the timing of weaning, but it is less relevant for understanding other processes, such as postweaning care. One of the special features of primate maternal strategy is the great importance of postweaning investment, which is sometimes based in lifelong, mutually supportive bonds that benefit both mothers and their offspring.

In chapter 19, Ryne A. Palombit examines infanticide in primates, both as a reproductive strategy of males and as a selective force in the evolution of counterstrategies in females. The possibility that the killing of dependent infants is a sexually selected reproductive strategy of males constitutes one of the earliest ideas in primatology, originating with Sarah Blaffer Hrdy in 1974. As Palombit notes, the now relatively large number of populations generating evidence of infanticide suggests that the behavior is widespread in primates. Female reproductive physiology and slow life histories amplify the potential benefits to male primates of removing infants and thus accelerating their mothers' return to fertilizable condition. As Palombit, Kappeler, and others point out in this volume, infanticide risk is one of the most important constraints on female reproduction in primates. Although Hrdy's sexual selection hypothesis has received considerable empirical support, one of the results of recent research is the explication of alternative benefits of infanticide for males. Today it seems clear that infanticide is not a unitary phenomenon, but has several possible functions. It may be widespread, but it is not necessarily a common behavior, and this directs attention to what has become in recent years one of the most interesting and compelling aspects of these analyses: female counterstrategies to infanticide. Polyandrous mating (promiscuity) and association with male (or female) protectors are relatively better known and supported options, but these counterstrategies comprise myriad mechanisms and may also be supplemented by other kinds of counterstrategies. Collectively, these phenomena constitute important components of female reproductive strategies. Echoing Oliver Schülke and Julia Ostner (chapter 9), Palombit emphasizes how infanticide has likely acted as a selective force in the evolution of primate social systems.

Few topics in anthropology and behavioral biology have engendered as much debate and controversy as the adaptive significance of sexual behavior and reproduction in *Homo sapiens*. Frank Marlowe addresses these issues in the final chapter of part 3. Using forager (hunter-gatherer) societies as a useful reference point, he evaluates the diverse, cross-cultural patterns of human mating and rearing strategies within the broad comparative framework provided by non-human primates, particularly the great apes. Not surprisingly, humans share many features of reproduction with the closely related apes, but they also offer an array of striking distinctions that set them quite apart. First, although mating systems vary considerably, breeding typically occurs within the context of mate bonds embedded in larger social units. Second, the sexual division of labor and food sharing by males is widespread, and it carries important implications for reproduction. Third, alloparental care of young is highly developed and arguably critical for reproductive success. This suite of adaptations has conferred such profound advantages that highly prolific humans came to occupy virtually all suitable habitats long before the advent of agriculture. Unsurprisingly, multiple competing hypotheses account for these phenomena. Marlowe examines these ideas in light of empirical evidence, human variation, and sexual selection theory. He concludes by considering the implications of the demographic transition currently underway in industrialized societies for our understanding of human reproduction.

Prominent in all these chapters is the empirical evidence which has increased not just in numbers, but also in diversity. Virtually all of these authors marshal a variety of behavioral, ecological, demographic, genetic, molecular, hormonal, and in some cases even immunological data to evaluate ideas and test hypotheses. As more data have come in from the traditionally less studied primates—notably the platyrrhines and strepsirrhines—the comparative foundation of these analyses is strengthened along with, of course, the resulting conclusions. As every author in this part points out, there is no question of the great need for more research, and indeed one of the consequences of recent research has been the generation of some exciting new ideas that require careful testing. But there is also no question that our understanding of at least some of many problems and solutions of mating and rearing has also improved over the years.

Chapter 14 From Maternal Investment to Lifetime Maternal Care

Maria A. van Noordwijk

ALL FEMALE mammals provide energetic investment in their offspring in the form of gestation and lactation. Primates stand out among mammals, however, in that the mother-offspring bond rarely ends at weaning. This is a result of their general gregariousness and relatively slow development (chapter 10, this volume). In fact, the potential for prolonged investment in some form exists as long as mother and offspring survive and inhabit the same home range, and especially when they belong to the same social unit. Maternal support may be prolonged throughout the mother's life span and may even be expanded to include grand-offspring (Fairbanks 1988, 2000). Indeed, despite being weaned, juvenile primates, who are still relatively small (initially on average about 30% of maternal weight: Ross 2003), require the "umbrella" care provided by the mother and other members of her social group. Such care is often shared with (half)-siblings of various ages. It provides the offspring not only with antipredator and social protection (thus improving its chances of survival and access to food), but also with numerous opportunities for social learning, and it sometimes includes possibilities to obtain food directly from the mother or other group members. In some species it even includes being carried, and active food sharing by others than the mother. Thus, in addition to the basic direct physical investment shared with all other mammals (e.g., nursing, carrying, thermoregulation), primate mothers also offer their offspring prolonged social care (agonistic and antipredator protection, learning opportunities; see chapter 11, this volume). The long association between mother and offspring, which is even further extended for the philopatric sex, provides great potential for long-term maternal influence on offspring fitness.

Trivers (1972) defined *parental investment* (PI) in terms of the cost to the parent, so that it reflects any parental expenditure of time and energy that reduces the parent's ability to invest in other offspring. The broader term *parental care* includes any parental behavior and expenditure that is likely to increase the offspring's chances of survival and success (Clutton-Brock 1991). Parental care can be depreciable—that is, it contributes exclusively to the offspring concerned because the relevant resource is reduced in availability to others as it is allocated. Examples of this include carrying, food provisioning, and unilateral agonistic support in conflicts. Care can also be nondepreciable—that is, the relevant resource is not reduced in availability when invested, and thus is potentially shareable across multiple offspring at the same time. Examples of this form of care include vigilance and facilitation of opportunities for learning.

In this chapter, I examine variables that affect *maternal investment* and *care* throughout an offspring's life, such as the commencement of reproduction in relation to maternal condition and environmental fluctuations; the interactions between mother and offspring during gestation and lactation; the opportunities for maternal care beyond weaning; care by others; and the possibility of differential investment and care in sons versus daughters. Before examining these issues in detail, however, we must deal with two general issues that characterize all interactions surrounding reproduction. First, reproduction is an arena of evolutionary conflict that occurs not only between mothers and offspring, but also between the two parents and among offspring (Trivers 1974; Parker et al. 2002). Second, because maternal energy expenditure on behalf of the offspring cannot usually be measured directly in wild primates, we must address the indirect measures that are available to field workers.

Conflicts of Interest

As in other mammals, female primates carry the entire burden of energetic investment in offspring through gestation, and the lion's share during lactation (fig. 14.1). How much energy a mother provides to each offspring is likely to affect her own survival chances, as well as those of her offspring, and therefore her own lifetime reproductive success. Because a younger sibling is at best a full sibling to the current offspring, the latter should prefer more and longer investment directed to itself than to its future sibling. The mother, however, is equally related to all her offspring. Thus, throughout development there is a potential conflict between mother and offspring about how much she invests in them and when she should end this investment (Trivers 1974; Clutton-Brock 1991).

Despite the potential for conflict, mother and offspring obviously share an interest in ensuring that maternal investment not only allows the infant to survive and ultimately reproduce, but also (at least in primates) allows the mother to survive to provide important postweaning care. Such care can hardly be enforced by the offspring, since the mother can refuse to provide it if the attendant costs are too high. Thus, in primates the potential for conflict largely ends with weaning, whereas prolonged care is often expected.

Less visible than the behavioral conflict over the timing of termination of exclusive investment is the physiological tug-of-war between mother and offspring in utero (see below). This chemical interaction affects the fetus's physiological settings, and also turns out to influence its lifelong responses to food availability, social stress, and thus susceptibility to disease. Another major conflict hidden in utero is that between the two parents. Paternal copies of genes may be selected to stimulate the developing fetus to maximize its extraction of resources from the mother, whereas the mother's gene copies are expected to limit investment to a lower level that optimizes the mother's lifetime reproductive success (Moore & Haig 1991).

A group-living primate mother typically combines investment in her youngest offspring with supplemental care for older offspring of various ages, which still live in association with her. During the relatively long postweaning immature phase of primates, there is ample opportunity for competition among siblings of different ages. In such conflicts, primate mothers tend to support the younger, weaker offspring against their own stronger siblings, resulting in the "youngest ascendancy dominance rank acquisition" in matrilineal social groups (Kawai 1965, see also chapter 21, this volume).

Measuring Maternal Investment and Its Impact

It is very difficult to estimate maternal investment directly. One approach to measuring it noninvasively is to record a mother's activity pattern during the various phases of reproduction, and to estimate the cost of provided care in comparison with that of nonmothers (or of the same female in the nonreproductive state). However, under natural conditions many other factors vary over time, and reproduction may be timed to take advantage of this variation. Thus, only a few dedicated studies under natural conditions have used this method of estimating the costs of gestation, milk production, carrying, and so on in primates (e.g., baboons, *Papio* spp., Altmann 1980, Dunbar 1988, Barrett et al. 2006a; spectral tarsiers, *Tarsius tarsier*, Gursky 2002; siamangs, *Symphalangus syndactylus*, Lappan 2009).

It is easier to estimate the cost to the mother in terms of time invested exclusively in a particular offspring or litter by measuring interbirth interval (the intervals between the birth of one infant surviving at least to weaning and the next birth). Within a species, gestation length varies very little (Martin 1990), but the duration of the lactation period varies within and among individuals and populations. Similarly, variation is found in the overlap between lactation and gestation of the next offspring, or even the duration of a distinguishable recovery period for the mother to reach a condition threshold for conceiving again (Bercovitch 1987; Mas-Rivera & Bercovitch 2008). In all scenarios, variation in interbirth intervals provides a reasonable estimate of differences among females (or conditions or populations) to acquire enough energy to raise healthy, surviving offspring (e.g., baboons: Altmann & Alberts 2003, 2005; Garcia et al. 2009; long-tailed macaques, *Macaca fascicularis*: van Noordwijk & van Schaik 1999). Similarly, differences in the cost of raising sons or daughters to weaning age can be estimated through the duration of the subsequent birth intervals (see below). Time will therefore be the currency used here to compare maternal *investment* across conditions and species.

Obviously, time cannot be used to estimate the quality of post-weaning *care*. Maternal care is recognized by its value to the offspring, thus it should improve the survival and reproductive success of the offspring. Since many fac-

Fig. 14.1. In almost all monkeys and apes, infants experience relatively long periods of dependence on caregivers, both for nourishment and for locomotion. With a few notable exceptions, it is mothers that bear the burden of most of these costs, such as this female ursine howler monkey (*Alouatta arctoidea*) in Hato Masaguaral, Venezuela. Photo courtesy of Carolyn M. Crockett.

tors could influence the offspring's success in life, one can isolate the influence of maternal care by a comparison of the survival and success of orphans with that of immatures living in association with their surviving mothers (see below).

Maternal Investment and Maternal Condition

Access to Food

Mothers need to find the balance between reproduction and the growth and maintenance needed for their own survival (age-related investment decisions are discussed in chapter 10, this volume). Studies of food-enhanced primate groups show that females with access to more food (enhanced maternal condition) invest more heavily in reproduction. This is shown by their faster-developing offspring and shorter interbirth intervals (e.g., Japanese macaques, *Macaca fuscata*, Watanabe et al. 1992, Kurita et al. 2008; yellow baboons, *Papio cynocephalus*, Altmann & Alberts 2005). Under natural conditions, a mother's social status mediates access to food for herself and her offspring. Indeed, high-ranking females often have larger and/or faster-growing infants, which are weaned after a shorter lactation period (e.g., baboons, Johnson 2003; Altmann & Alberts 2005; Garcia et al. 2009). This results in a shorter interbirth interval after a surviving offspring (e.g., macaques, Sugiyama & Ohsawa 1982, van Noordwijk & van Schaik 1999; mandrills, *Mandrillus sphinx*, Setchell et al. 2002; chacma baboons, *Papio ursinus*, Cheney et al. 2004, 2006), and faster-maturing daughters (lower age at menarche and first reproduction: vervet monkeys, *Chlorocebus pygerythrus*, Cheney et al.

1988; macaques, Bercovitch & Berard 1993, van Noordwijk & van Schaik 1999; mandrills, Setchell et al. 2002; baboons, Bercovitch & Strum 1993, Johnson 2003, Altmann & Alberts 2005). Thus, variability in interbirth interval, in offspring weight or age at weaning, and in age of sexual maturation indicates differences in the rate of a mother's investment in her offspring, and is closely linked to access to food (see also chapter 15, this volume).

Timing of Reproductive Events

Few if any mammals experience a continuous food supply. Fluctuations in food availability and quality often come in rather predictable yearly cycles, but they can also be less predictable, with pronounced interannual variation (van Schaik et al. 1993; chapter 7, this volume). Since female reproduction is largely dependent on access to sufficient food, it is important to understand how females time their reproduction relative to periods of food abundance and (unpredictable) shortage. Primates tend to have a long reproductive cycle, with an especially long lactation period in comparison to other mammals of similar size (chapter 10, this volume). Thus, mothers need to schedule conception so that peak needs (during lactation) will coincide either with a future peak in food availability or with built-up internal (fat) reserves that buffer for low food supply. Either way, limiting the onset of reproduction to a particular season is potentially costly to females, since a reproductive failure would necessarily delay subsequent breeding. Thus, seasonal reproduction must have compensatory benefits in order to be favored by selection, although those benefits do not necessarily reflect the same reproductive strategy in all seasonally breeding primates.

Seasonal breeding is common in most taxonomic groups, but it follows different scenarios in different species. Drent and Daan (1980) identified two distinct reproductive strategies in birds that breed at high latitudes. In some species, the females store fat before the onset of reproduction and adjust the number of eggs they lay to these reserves. Conversely, the females of other species do not store fat in advance, but instead dynamically adjust their clutch size to their food intake rate. A similar distinction in reproductive scenarios was independently proposed for tropical primates, inspired by the high interannual variance in fruit abundance in Southeast Asia as compared to more predictable seasonal variation in food availability in other regions (van Schaik & van Noordwijk 1985). Stearns (1992) incorporated these different investment scenarios into general life-history models to explain variation in the number of offspring produced, the timing of the reproductive cycle, and two extreme maternal strategies of energy acquisition to support reproduction. Mothers following the *income breeding* scenario completely support their reproduction through daily foraging, whereas those following the *capital breeding* scenario store reserves whenever possible, mostly before and during gestation, to support lactation even when conditions are suboptimal.

With few exceptions (e.g., hooded seals, *Cystophora cristata*, which have an extremely short lactation period: Iverson et al. 1995), mammals cannot sustain lactation on built-up fat reserves alone. Therefore, instead of defining a strict dichotomy in breeding scenarios, the labels *income breeding* and *capital* breeding mark the endpoints of a continuum that ranges from species that use only an external trigger for the onset of (seasonal) reproduction irrespective of physical condition to species that require a minimum maternal condition before conception (and additional fat storage whenever possible). This continuum has been helpful in explaining the variation in temporal patterns of maternal investment in many mammals (e.g., Festa-Bianchet et al. 1998; Boyd 2000; Broussard et al. 2005; Wheatley et al. 2008), including primates (e.g., Richard et al. 2002; Brockman & van Schaik 2005; Janson & Verdolin 2005; Lewis & Kappeler 2005). There is little evidence that maternal condition affects litter size in primates since most species give birth to single offspring. However, for species in which twins and triplets are common, females in better condition are more likely to produce triplets, at least in captivity (common marmosets, *Callithrix jacchus*, Tardif & Jaquish 1997, Nievergelt & Martin 1999; ring-tailed lemurs, *Lemur catta*, Pereira et al. 1993, Nunn & Pereira 2000).

Even though detailed data on female fat storage are lacking for most primate species, the available empirical evidence is consistent with the income-capital continuum perspective (table 14.1; see also evidence presented in Brockmann & van Schaik 2005). First, income breeders undergo a very small number of ovarian cycles before conception, whereas that number is variable in capital breeders, which have a high maternal condition threshold.

Second, income breeders tend to have lower frequency of failed pregnancies (abortions), and are instead more vulnerable to high rates of infant mortality (e.g., ring-tailed lemurs, Gould et al. 2003; Japanese macaques, Kurita et al. 2008) or increased rates of infant rejection and abandonment by mothers in poor condition (e.g., vervet monkeys, Fairbanks & McGuire 1995).

Third, income breeders tend to wean infants at a particular absolute age and relative to a temporary food peak, which allows the mother to align her next conception to the following year's food cycle. Thus, we expect to find a less variable interbirth interval or a bimodal distribution. At the capital-breeding end of the continuum, infants are weaned once they have reached a certain condition. Here, length of lactation (age at weaning) and thus interbirth intervals tend to be more variable among females, whereas weight of the weanling is less variable (e.g., baboons, Lycett et al. 1998, Johnson 2003, Altmann & Alberts 2005, Barrett et al. 2006b, Cheney et al. 2006, Garcia et al. 2009; Phayre's leaf monkey, *Trachypithecus obscurus phayrei*, Borries et al. 2008; white-faced capuchins, *Cebus capucinus*, Fedigan et al. 2008; siamangs, Lappan 2008).

Fourth, income breeders using an external trigger continue to reproduce seasonally when translocated or provisioned with a constant diet (e.g., common squirrel monkeys, *Saimiri sciureus*, Schiml et al. 1999; lemurs, Rasmussen 1985). Capital breeders show more variation in seasonality between populations and become less seasonal in their

Table 14.1. Maternal parameters expected under the extreme scenarios of the income-capital continuum (cf. Brockman & van Schaik 2005)

	Income scenario	Capital scenario
Maternal physiology	Little fat storage	Fat storage whenever possible, especially during gestation
Ovarian cycles to conception	Few	Variable
Ovarian activity: Conception	Triggered by exogenous factor (e.g., photoperiod)	Triggered by endogenous factor (physiological condition)
Maternal recuperation period	None	Variable after raising of offspring to weaning
Mating/birth seasonality	Pronounced seasonality	No strict seasonality
Effect of translocation or provisioning on reproduction	Same external trigger, same seasonality	Less seasonality
Timing of gestation	Preceding predictable high food availability	During high food availability
Timing of lactation	Before and during high food availability	During and after high food availability
Timing of weaning	During high food availability	Delayed if food availability insufficient
Birth rate	Annually or semiannually; low interannual variation	High interannual variation depending on food/body condition
Abortion rate	Low	Higher, affected by food acquisition
Infant mortality	Food-dependent	Relatively low
Effect of maternal dominance rank	Little effect on birth rate, some effect on infant survival	Variable effect on birth rate and age of weaning through maternal condition; also possible effect on infant growth rate

reproduction under provisioning or after translocation to a less seasonal environment (e.g., baboons, Barrett et al. 2006; long-tailed macaques, Kavanagh & Laursen 1984).

Finally, the income-capital continuum also helps us to understand differences between populations in the changes that take place in maternal activity budget during pregnancy and lactation. More seasonal and income breeders have to increase their intake (or decrease their expenditure) during lactation, whereas at the capital breeding side of the spectrum, fluctuations in activity budget will be less pronounced (cf. Altmann 1980 vs. Barrett et al. 2006a).

On average, female reproductive physiology is expected to take advantage of predictable changes in food availability. With lengthening of the reproductive cycle and decreasing predictability of the environment, however, a mother needs to provide a buffer to ensure a regular flow of nutrients to her developing offspring. This constraint has led to variation among primates in the timing of reproduction relative to changes in the environment, as well as variation in patterns of maternal physiology and reproductive failure.

Prenatal Investment: Gestation

Mother-Fetus Interaction

During gestation, a mother hosts her offspring inside her own body and allows it to grow by taking nutrients from her. The exchange of nutrients and waste products between mother and offspring takes place through the placenta, an organ formed by the fetus. The anatomical details of the placenta's connection with the uterus vary among mammalian taxa, with three distinct types of fetal invasion of maternal tissue. The least invasive type, epitheliochorial placentation, is found in the strepsirrhines. Through special glands in the uterine wall, the mother releases nutrients for diffuse exchange over a large area of contact between maternal and fetal tissue. All other primates have haemochorial placentation, characterized by fetal tissue entering the uterine wall and connecting directly to maternal blood vessels. The degree of invasiveness varies, however, and it appears to be lowest in the platyrrhines (Carter 2007). Martin (2008) concluded that both these forms of placentation are derived from the intermediate ancestral endotheliochorial form, but that there is no evidence for a difference in efficiency (as estimated by fetal growth rate), and that precocial and large-brained offspring can be produced irrespective of placental anatomy. Thus, the demand of a higher fetal brain growth rate seems not to have been the selective force that resulted in the two extreme forms of placental organization of the two major primate lineages. However, some differences in

vulnerability to pregnancy complications may be related to these anatomical differences.

A mother's offspring is related to her by only 50%, so her body needs to suppress a strong immunological response to reject the implantation of such a foreign body inside her uterus. It is possible that this suppression needs to be even stronger in species with closer contact between maternal and embryonic/fetal tissues (especially in species with invasive haemochorial placentation, such as the Cercopithecoidea and Homoinoidea). However, variation in such a maternal immunological response is still poorly documented (Martin 2008).

After implantation, the intimate mother-fetus relationship may be shaped partly by an evolutionary conflict between the maternal and fetal cells over the quantity and rate of maternal nutrients taken by the fetus (Haig 1993). The fetus is expected to prefer greater access to maternal resources than the mother is willing to allow. Within the fetus, paternally "imprinted" genes are expected to counteract maternal mechanisms that limit investment in the fetus, which ultimately reflects sexual conflict. The interaction between opposing chemical signals from mother and fetus also serves as information exchange, indicating the competence of the fetus and the limits of maternal abilities to provide nutrients (Haig 1993), either from reserves or ongoing food acquisition.

The intimate and prolonged contact between cells with different genetic backgrounds also requires a carefully balanced interaction between the fetal and maternal expression of (species-specific) major histocompatibility complex (MHC) genes in order to avoid severe harm to mother or offspring from a suboptimal exchange of nutrients and information, or even from premature expulsion of the fetus itself (Bainbridge 2000). The haemochorial placenta releases an array of chemical stimuli directly into the mother's bloodstream. These fetal triggers cause the mother to maintain production of progesterone, raise her blood pressure (thereby increasing the volume of blood reaching the fetus), and reduce her control over nutrient flow to the fetus relative to her own tissues (e.g., through placental lactogen, which makes more glucose available to the fetus by counteracting maternal insulin). Pregnancy complications such as pre-eclampsia and gestational diabetes, endangering the life of both mother and fetus, are best known in humans, but there is some evidence for their occurrence in captive great apes and Cercopithecidae (Carter 2007). These problems reflect an extreme outcome of the chemical conflict between fetus and mother. Little is known about the prevalence of such phenomena among wild primates, but among strepsirrhines and plathyrrhines these are expected to be rare, due to their less invasive placentation systems. In general, a mammal fetus causing severe harm to its mother will have a negligible chance of survival if this harm prevents the mother from providing postnatal nursing care. Thus, the prenatal tug-of-war is likely to favor the mother's survival over that of her helpless offspring.

Developmental Plasticity and Prenatal Effects

On the basis of consistent correlations between low birth weight and the incidence of health problems later in life in humans, Hales and Barker (1992) formulated the thrifty phenotype hypothesis, which initially proposed that inadequate early nutrition has major long-term consequences via impaired development of the endocrine pancreas, which greatly increases susceptibility to type 2 diabetes in adults with abundant access to food. Later research expanded the hypothesis to include adjustments of many physiological settings with lifelong, mainly suboptimal, consequences. These strong effects of early development on later physiological functioning indicate sensitive periods in the development of anatomical and physiological features, as well as limitations on adult responsiveness to environmental changes. These constraints emphasize the role of the environment a mother offers to her offspring.

Physiological priming by conditions experienced during fetal development, now known as metabolic imprinting (Metcalfe & Monoghan 2001), has been demonstrated in many species, from invertebrates to vertebrates (Bateson et al. 2004). Some studies even suggest that a female prenatally affected by her mother's condition in turn affects the condition of her own offspring. For example, captive, single-housed female rhesus monkeys (*Macaca mulatta*) born with low birth weight reached average adult size and weight, but nonetheless produced daughters with low birth weight (Price & Coe 2000). This pattern suggests an intergenerational transmission of fetal growth restrictions, known as gestational imprinting. For humans there are similar indications of this phenotypic inertia (e.g., Gluckman et al. 2005; Kuzuwa 2005).

Prenatal adjustments may be adaptive if the conditions experienced during early development have long-term predictive value (e.g., Bateson 2001, 2008; Gluckman et al. 2007, 2008). For this to occur, such conditions need to persist for a long time relative to an individual's life expectancy. In long-lived species, however, individuals are likely to experience a variety of food conditions over their lifetime, and permanent anatomical or physiological adjustments to unusual food conditions made early in life (such as adjustment of the number of kidney cells or of sensitivity to insulin) thus hold little adaptive potential. Instead, they are likely to reflect a trade-off between the mother's own interests and those of the affected offspring. In general, if early conditions are extreme or relatively temporary, physiological priming probably reflects inevitable trade-offs with long-term consequences (Desai & Hales 1997; Wells 2003, 2007; Bateson et al. 2004). When these effects are carried into the next generation, the trade-off interpretation is even more likely to be correct than the hypothesis that early physiological responses are adaptive.

Studies of humans are beginning to reveal a range of both physiological and behavioral correlates of prenatal and early postnatal exposure to poor nutrition, psychosocial stress, and even physical abuse (e.g., Hales & Barker 2001; Mick et al. 2002; Lahti et al. 2006; Phillips 2007; Entringer et al. 2009; McGowan et al. 2009). Such early exposure appears to have an epigenetic effect—for example, by modifying the expression of steroid receptor genes (Champagne & Curley 2009) or causing stable changes in neuronal networks (Cirulli et al. 2009a). In captive rhesus monkeys, the available milk energy to the infant during the early postnatal period was found to correlate with measurements of the infants' confidence, whereas the level of nutrition during peak lactation had no effect (Hinde & Capitanio 2010), thus suggesting an early fixation of physiological settings. For obvious reasons, little is known about prenatal effects on an individual's physiology as a juvenile or adult in wild primates. However, infant yellow baboons with a low growth rate early in life did not show compensatory growth as weaned immatures when conditions became favorable (Altmann & Albert 2005). In addition, for a cohort of female baboons in the same population, Altmann (1991) found a significant correlation between nutrient acquisition (relative protein and energy intake) as yearlings and reproductive success as adults. This suggests that early conditions and ontogenetic experience, which are both directly affected by the mother and her condition, can have long-term effects on metabolic function. These effects are more consistent with the consequences of maternal tradeoffs than with adaptive responses.

Onyango et al. (2008) reported that fecal glucocorticoid levels of five- to six-year-old male yellow baboons correlated with maternal rank at their conception, which remained unchanged during their first year of life. Since these hormones are involved in the stress response, this pattern suggests that these males suffer chronic stress, which may affect their growth, immunocompetence, and reproduction. This result was unexpected because in this population subadult males, prior to their dispersal from the natal group, rank above all adult females in the dominance hierarchy (Pereira 1988). Maternal glucocorticoid (GC) may directly affect the fetus, the infant through lactation, or the maturing youngster via maternal behavior. Larger data sets are needed to clarify the consistency of the suggested connection between maternal behavioral environment and sons' stress levels later in life. A cross-fostering approach can help to separate the effects of maternal behavior from other physiological maternal effects. For example, Maestripieri et al. (2007) showed resemblance between infant rejection behavior by foster mothers and that of foster daughters towards their own first offspring, apparently mediated by early adjustment of neuroendocrine settings.

In summary, a primate mother's physiological and behavioral environment may have profound, but largely unexplored, long-term effects on the lives of her offspring. More work is needed to resolve which of these effects are adaptive responses of offspring or outcomes of maternal compromises, such as reduction of investment in pregnancy and lactation under suboptimal circumstances to increase survival for future reproduction (cf. Wells 2003; Baker et al. 2009).

Post-Natal Investment

Lactation and Carrying

Unlike many other mammals, most primate newborns typically cling to their mother and are carried on her body wherever she goes until they are capable of independent locomotion. Thus, most primate mothers not only nourish their offspring with milk, but also carry its increasing weight during their daily travel and foraging (fig. 14.2). In a few haplorrhines (notably callitrichines) and some "monogamous" species (e.g., owl monkeys, *Aotus* spp., and siamang), males or other helpers take over some of the carrying task (see chapter 17, this volume), but nourishment through lactation remains the responsibility of the mother. In contrast, many strepsirrhine mothers leave their young offspring in a nest or tree hole or pick them up orally to transport and then "park" them in ever-changing hiding places while they forage (Kappeler 1998; Ross 2001; chapter 2, this volume).

Fulfilling all the nutritional requirements of an actively moving and growing immature is the most energy-demanding phase of maternal investment, requiring more energy than gestation. Lactation represents more than 75% of the total maternal energy cost of reproduction (Oftedal 1984). Initially the mother provides all nutrition through milk, but this is followed by a sometimes very long transitional period in which milk supplements the infant's own increasing efforts to find, select, and process solid foods. Compared to that of other mammals, the milk of most primates contains a high level of lactose, but little fat and protein (Tilden & Oftedal 1997). This high-lactose milk, which the infant can easily digest, is thought to enable rapid postnatal growth of the primate brain, even though some growth and certainly differentiation of the brain continues throughout the juvenile phase (Martin 1996; Leigh 2004). Thus, primate mothers provide a buffered food supply during the most nutrient-sensitive phase of their offspring's brain development (chapter 10, this volume).

The gross energy density of milk varies only little between different primate species, including the small callitrichines (Power et al. 2002). However, Tilden and Oftedal (1997) found higher proportions of fat and protein in the milk of those strepsirrhines that "park" their offspring while foraging (e.g., lorises, galagos, and some lemurs) than in those that carry their young and allow them to suckle continuously. In parking species, infant growth tends to be relatively fast and mothers' lactation tends to last shorter than gestation (van Schaik & Kappeler 1997; van Schaik et al. 1999; Gursky 2007; Gould & Sauther 2007). On the other hand, female strepsirrhines that carry their infants resemble haplorrhines in several respects, such as milk composition, longer lactation periods, and slower postnatal growth. Thus, primate milk tends to be diluted and provided over a longer period wherever nursing is furnished "on demand" in frequent, short bouts.

Weaning

A mother faces the challenge of finding the moment in her infant's life at which additional investment through milk no longer significantly affects the offspring's success, and can therefore be withheld, even though continued nursing would still benefit it. Figure 14.3 shows how natural selection can set the optimal timing of this termination in investment. This difference between a mother's own optimum and that of her infant suggests it may be in the mother's interest to minimize investment in her current offspring in order to enhance her own survival and future reproductive success, despite putting the survival of the current infant at risk (Trivers 1974; Fairbanks & McGuire 1995). In most primates, high-yield milk production can hardly be combined with simultaneous gestation, and thus the termination of milk provisioning to one offspring coincides, more or less, with the onset of investment in the next. Therefore, the termination of prolonged milk provisioning, known as weaning, is the most conspicuous period of mother-offspring conflict.

Weaning marks the completion of the transition from milk to a solid food diet. To be weaned successfully, an infant needs to be able to digest solid food, travel independently (unless "helpers" relieve the mother of the cost of carrying), and forage competently in the face of moderate fluctuations in food supply. This last requirement ensures

Fig. 14.2. A six-year-old Bornean orangutan (*Pongo pygmaeus*) suckling, illustrating the long exclusive investment made by orangutan females, which exhibit the longest interbirth intervals of all primates. Photo courtesy of Johanna van Schaik.

that the infant has enough food-handling skills to buffer a temporary energy shortage. Thus, the optimal timing of weaning is affected by the infant's mass, age (physiological maturation), and competence in locomotion (in the absence of helpers; see fig. 14.3), food processing, and foraging (recognizing foods). This is especially needed in species whose diets require either complicated foraging skills (i.e., needing practice before efficiency is reached—e.g., hunting, tool use), or knowledge of a vast diversity of food items. Indeed, Jaeggi et al. (2008) showed that infant Bornean orangutans (*Pongo pygmaeus*), who associate almost exclusively with their mothers, have already acquired full knowledge of their

mother's diet by the time they are fully weaned at five to seven years of age. Similarly, gorilla (*Gorilla gorilla*) weanlings know the adult diet (Nowell & Fletcher 2008), and weaned immature chimpanzees (*Pan troglodytes*) and Sumatran orangutans (*Pongo abelii*) can use tools common to their populations (Lonsdorf 2005; van Noordwijk & van Schaik 2005). However, chimpanzee males can practice communal hunting skills only from adolescence onwards, since their mothers do not usually participate actively in hunts and thus cannot model this behavior for their sons (Boesch & Boesch-Achermann 2000).

The larger the infant grows, the more its being carried

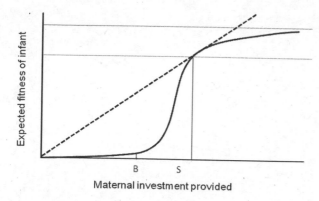

Fig. 14.3. The theoretically optimal moment for the mother to switch (S) her investment from her currently dependent infant to her next offspring, which is equally related to her. For the infant, more investment in itself would contribute more to its expected fitness than the mother's investment into a full- or half-sibling. B indicates the moment of birth.

interferes with the mother's movements, especially during foraging (Altmann 1980). Indeed, several studies have shown that carrying during maternal foraging and resting decreases sooner than during the mother's travel (fig. 14.4), whereas the end of lactation largely coincides with the end of body contact during maternal nonresting activity (e.g., in geladas, *Theropithecus gelada*, Barrett et al. 1995; capuchins, Fragaszy et al. 2004). Carrying the young infant, however, serves not only for transportation but also for thermoregulation. Schino and Troisi (1998) found more attempts to maintain body contact by infant Japanese macaques at lower temperatures. The mothers frequently rejected these attempts, suggesting a rarely noticed mother-offspring conflict over thermoregulation.

There may be a conflict over when the infant is allowed to nurse (Altmann 1980; Barrett et al. 1995), but Bateson (1994) and Barrett et al. (2006) have argued that there is no parent-offspring conflict over the timing of weaning, since the interests of mother and offspring in the infant's survival are the same (see also Maestripieri 2002). In this view, weaning tantrums are honest signals of the offspring's needs, and therefore they reflect an imbalance between those needs and what the mother offers (Barrett & Henzi 2000). This view can be reconciled with classic parent-offspring conflict by considering two different situations. Where immature mortality is largely due to extrinsic factors (predation, disease, infanticide), and thus largely independent of the quality of prolonged maternal investment, infants may be weaned relatively early and sometimes forcefully, which is indicative of conflict (vervets, Hauser 1993; Fairbanks & McGuire 1995; baboons, Lycett et al. 1998). On the other hand, if mortality is strongly affected by care, mothers are expected to invest heavily in each offspring until it can be

completely independent (e.g., Drakensberg chacma baboon population: Barrett et al. 2006b). The validity of this argument depends on the extent to which a particular mortality risk is extrinsic and thus affects individuals irrespective of their care-dependent physical condition.

The degree of seasonality of food availability and reproduction may similarly affect the balance between the interests of infant and mother. If gestation and lactation together take about 12 months in a very seasonal environment, an infant requiring just a few months of extra lactation may cause its mother to skip a full reproductive cycle. In primate populations with a less seasonal reproductive schedule or one that easily fits within a seasonal cycle, females will have more flexibility in adjusting their investment to the needs of their current offspring. Here we expect that fluctuations in food availability (either extrinsic or mediated through group size or dominance) affect the duration of lactation and thus interbirth intervals (e.g., long-tailed macaques, van Noordwijk & van Schaik 1999; chacma baboons, Barrett et al. 2006b; Phayre's leaf monkeys, Borries et al. 2008).

Differences in weaning age between populations of baboons (Lycett et al. 1998; Barrett et al. 2006b), gorillas (Nowell & Fletcher 2007), and orangutans (van Noordwijk & van Schaik 2005; van Noordwijk et al. 2009) have been attributed to differences in care-dependent mortality risk. In most cases, however, it remains an open question whether and how a mother can accurately assess her offspring's needs and survival chances, and thus whether her prolonged nutritional investment affects the offspring's chance of success.

Parent-offspring conflict is over the extent of investment and the timing of a mother's switch from investing in the current offspring towards starting her next reproductive effort. Since in primates this transition does not mark the end of mother-offspring association, the severity of the conflict over the timing of this switch is mainly determined by the degree to which an offspring's survival depends on mother's milk as a continued supplement to its nutrition.

Post-Weaning Maternal Care

One of the most striking features of primate development is that weanlings still rely critically for survival on additional low level, nonexclusive support provided during their prolonged postweaning association with the mother and others. Such support can be in the form of (1) protection against inclement weather, predators, and conspecifics, (2) sharing food or other resources or tolerating scrounging, and (3) providing learning opportunities. Females may even

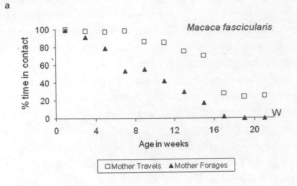

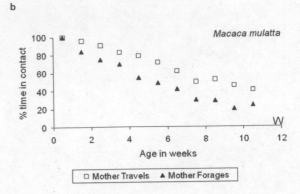

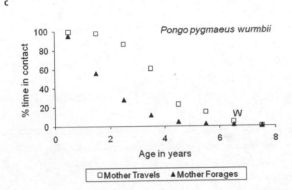

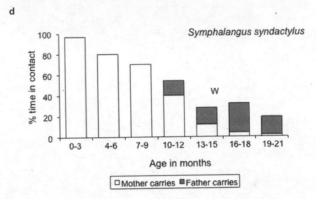

Fig. 14.4. Body contact and/or of an infant at different ages during the mother's travel (true transportation of the infant) and her mostly stationary foraging. Infants tend to reach locomotory independence before weaning (W) in (a) wild, mostly arboreal, long-tailed macaques (*Macaca fascicularis*) in Ketambe (data from Karssemeijer et al. 1990), (b) partly terrestrial rhesus macaques (*M. mulatta*) in a seminatural habitat, Chhatari (data from Johnson 1986), (c) arboreal Bornean orangutans (*Pongo pygmaeus wurmbii*) in Tuanan (data from van Noordwijk et al. unpublished), (d) wild arboreal pair-living siamangs (*Symphalangus syndactylus*) at Way Canguk, in which the father starts to carry the offspring from around the time of weaning (data from Lappan 2008).

provide these forms of care to grand-offspring (Fairbanks 1988, 2000; Pavelka et al. 2009).

In group-living primates, this postweaning association between mother and offspring may last much longer than the preweaning phase, even for youngsters of the dispersing sex. This extended maternal care does not prevent the mother from having subsequent offspring. The resulting simultaneous association of the mother with several immature offspring of different ages—"stacking"—is the rule among diurnal primates, with the notable exception of the orangutan, whose rather solitary lifestyle limits permanent association of a mother to, at most, two immature offspring at a given time (van Noordwijk & van Schaik 2005).

It is difficult to estimate the cost of postweaning care to the mother. Some of the costs potentially are shared with other individuals in the group, in that immatures can follow other group members to food sources and can benefit from their vigilance. The mother may benefit from this arrangement as well, in that she can wean her offspring at

a relatively young age before it can survive independently (van Noordwijk & van Schaik 2005).

The value of this continued care to the offspring appears to be appreciable. Although unweaned primate infants generally do not survive the death of their mother (Thierry & Anderson 1986; Fairbanks 2000; Boesch et al. 2010), the fate of weaned immatures orphaned at different ages, relative to their peers, is much less known. Among wild long-tailed macaques, juveniles that lost their mothers after weaning suffered significantly higher mortality (fig. 14.5). In this wild population, with a generally high survival rate (0.66 from birth to four years), the youngest survivors all had a closely related adult female in the group. Thus, mothers apparently provide valuable services beyond those generally obtained from group living. Protection against intragroup competition (via agonistic support) and, perhaps, against thermal challenges during the night (via huddling), are probably more consistently provided by mothers than by other group members.

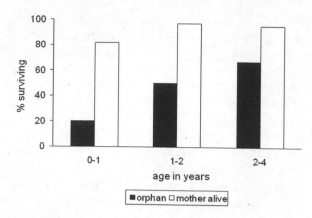

Fig. 14.5. Percentages of orphaned long-tailed macaques that survived to reach the next age cohort after the death of their mothers, compared to immatures with surviving mothers. Survival is lower for orphans whose mothers died after weaning but before they were one year old ($G_{adj}0 = 7.26$, $p < 0.001$), and for those whose mothers died when they were one to two years old ($G_{adj} = 8.36$, $p < 0.01$). But maternal death could affect the survival of three- to four-year-old juveniles as well ($G_{adj} = 2.79$, $p < 0.10$).

Among some primates living in multifemale, unimale groups subject to male takeover events, such as many langurs, weaned immatures may accompany an ousted male (likely their father), when he leaves the natal group and joins an "age-graded-all-male" band, which actually may include immature females (see chapter 19, this volume; fig. 19.7; Hrdy 1977; Rajpurohit & Sommer 1993; Steenbeck et al. 2000). In these species, a physical and social separation between mother and weaned offspring is quite common. Rajpurohit & Sommer (1993) reported that 50% of juvenile Hanuman langurs (*Semnopithecus entellus*) that joined an all-male band died within a year, whether they were living with their putative father or not. Although no formal comparison was presented, this number is likely to be much higher than the mortality rate of mother-associated immatures still living in their natal home range (which may be of higher quality than the range of an all-male band). This comparison supports the notion of the benefits of continued association with the mother.

Certainly, however, survival of juvenile orphans is not uncommon. In such cases (chapter 17, this volume), it could be argued that the prolonged care provided by the mother is not essential. However, even where orphan survival is relatively high, female orphans rarely acquire as high a social status as expected in light of their maternal rank (Hasegawa & Hiraiwa 1980; Johnson 1987; Lee & Johnson 1992; Johnson 2006), although occasionally fraternal support allows them to reach higher rank than expected (chacma baboons: Engh et al. 2009). Lacking competitive support, orphans are also reported to grow at a slower rate (chacma baboons, Johnson 2006). In addition, orphaned

female Japanese macaques have lower success in raising their firstborns than do primiparous females with mothers in the group (Hasegawa & Hiraiwa 1980). The presence of the mother is even reported to enhance mating chances of philopatric bonobo sons (*Pan paniscus*, Surbeck et al. 2010), thus suggesting reduced mating success for male orphans. In other species, much less is known about the fate of sons that survive their mother's death, especially after dispersal from the natal group. Alberts and Altmann (1995a) reported that orphan yellow baboon males disperse at a younger age than their peers, but studies of other cercopithecines did not confirm this effect (Kuester & Paul 1999; Fairbanks 2000; van Noordwijk & van Schaik 2001).

The importance of postweaning support from adults is underscored by another phenomenon. Compared with other species, primate immatures arguably have a stronger tendency to maintain an attachment with a primary caregiver well into juvenility. Usually this primary caregiver is the mother, but in her absence it can be another (usually related) female, an older sibling, or an adult or subadult male with whom the orphan spends as much time in proximity and grooming as it would have done with its mother (Japanese macaques, Hasegawa & Hiraiwa 1980; Nozaki 2009; white-faced capuchins, Perry 2008; chimpanzees, Boesch et al. 2010). Orphans play an active role in maintaining such mother-replacing relationships: they adopt their caregivers more persistently than vice versa, and just as with the mother-offspring relationship, such adopted bonds may last a lifetime, at least among females (Yamada et al. 2005; Nakamichi & Yamada 2007). Even male caretakers frequently accept adoptions (but not always: Boesch et al. 2010). For example, a gorilla orphan tends to be adopted by its putative father, who will share a nest with it (Stewart 2001; Yamagiwa & Kahekwa 2001), and male chimpanzees may do the same (Boesch et al. 2010). If immature adoptees are almost weaned at the time of the mother's death, they seem to have a good chance of survival at least until maturity (chimpanzees, Watts & Pusey 1993, Boesch et al. 2010; ursine howler monkeys, *Aloutta arctoidea*, Pope 2000; chacma baboons, Engh et al. 2001; Japanese macaques, Nozaki 2009). Although the benefits to the young immatures of such relationships seem obvious, no studies have yet shown a significant effect on orphan survival. Even less is known about the benefits, and particularly the costs, to caregivers.

Overall, current data suggest that in primates, offspring with a prolonged mother-offspring association have better chances of survival and reproduction, and thus selection is expected for the mother's survival, which could benefit not only her youngest offspring, but her older ones and other maternal kin as well.

Allomaternal Support: Who Cares, and How?

In dispersed, relatively solitary primates such as orangutans and some nocturnal strepsirrhines, mothers are the sole source of support for their offspring. In many gregarious species, particularly those living in bisexual, multifemale societies, membership in a social group provides mothers and offspring with some shareable benefits, such as enhanced predator detection and access to food. Several recent studies have shown a strong correlation between a female's reproductive parameters (especially offspring survival) and the number of her adult maternal relatives in the group (e.g., ursine howler monkeys, Pope 2000; chacma baboons, Cheney et al. 2006; white-faced capuchin monkeys, Fedigan et al. 2008) or the number of close associates (baboons, Silk et al. 2003; 2009; Japanese macaques, Pavelka et al. 2009). Even though the support provided by these relatives differs between species, these data suggest that primate mothers benefit from a social network to raise their offspring. This effect can be seen as an extension of the maternal effects discussed in the previous section.

Primate mothers and immatures can also receive direct, targeted attention from certain group members in the form of babysitting, carrying, allonursing, and provisioning. When mothers mutually provide allonursing and perhaps other forms of care to each other's offspring, we speak of *communal breeding*. In principle, communal care can be exchanged and more or less fairly reciprocated among mothers. When mothers and offspring receive extensive care from nonmothers—that is, males and nonreproductive adults or subadults—we speak of *cooperative* breeding. To test whether seemingly helpful behavior actually constitutes care, either the mother or the offspring should have enhanced reproduction or survival when more help is received.

In species with a strict dominance hierarchy, mothers are often reluctant to let others, especially non-kin, handle their offspring (Chism 2000). Nevertheless, other adult females and immatures of either sex in particular are attracted to young infants and try to get access to them through grooming the mother (e.g. Muroyama 1994; Henzi & Barret 2002), whereupon they handle the infants and even carry them away from their mothers. This does not occur specifically when such transportation is needed, however, and low-ranking mothers have difficulty retrieving infants from dominant handlers (Chism 2000). Fairbanks (1990) suggested that vervet monkey infant handlers themselves benefit by gaining experience. However, in her study of bonnet macaques (*Macaca radiata*), Silk (1999) found no clear benefits to infant handlers, mothers, or infants, and therefore concluded that infant handling is most likely a by-product

of selection for natal attractivity to enlist appropriate maternal care. Thus, infant carrying by others than the mother in hierarchically structured primate groups probably cannot be considered care, since it does not relieve the mother from providing care (cf. Hrdy 1976).

In some species, temporary aggregations of same-aged infants can be found near just one or a few vigilant females while the other mothers concentrate on foraging without being distracted or hindered by their active infant (e.g., Nilgiri black langur, *Semnopithecus johnii*, Poirier 1968; patas monkeys, vervet, guenons, Chism 2000). However, despite some indications that this crèche-allocare may enable females to achieve a shorter interbirth interval (Fairbanks 1990), it is still unclear whether all mothers take turns and benefit equally from such shared child supervision. In some colobines, juvenile helpers (who tend to be at least paternally related) carry infants during long-distance travel (Kirkpatrick 2007; Xi et al. 2008), but it is not yet clear how valuable this allocarrying is for the mother or her offspring. Thus, in many species the real value of infant carrying for the mother (reduced investment), the infant (increased chances of survival), or the caregiver (improved maternal skills), as well as the degree of reciprocity involved (communal care), are largely unresolved.

Allonursing is a form of allocare with a clear energetic cost (milk) to the provider. This has been reported for strepsirrhines (gray mouse lemurs, *Microcebus murinus*, Eberle & Kappeler 2006; silky sifakas, *Propithecus candidus*, Patel 2007), New World monkeys (ursine howler monkeys, Pope 2000; capuchins, Fragaszy et al. 2004; squirrel monkeys, Soltis et al. 2005), and several Old World monkeys (guenons, patas, talapoin monkeys, Chism 2000; colobines, Kirkpatrick 2007), but rarely among macaques, baboons, and great apes. Since only other lactating mothers can provide allonursing, the potential for reciprocity is clear. Occasionally some allonursing is provided by females who have not given birth (e.g., ring-tailed lemurs, Pereira & Izard 1989) or who have lost their own infants (Bolivian squirrel monkeys, *Saimiri boliviensis*, Milligan et al. 2008).

If mothers are equally likely to provide some allonursing to each other's infants, they don't gain or lose net energy from this form of communal care, but might gain foraging efficiency in not having to stay in permanent close proximity to their infants (Eberle & Kappeler 2006; Perry 2008). However, in populations of different capuchin species, low-ranking mothers nursed infants of high-ranking mothers more often than vice versa, thus suggesting an uneven advantage (O'Brien 1988; Perry 1996; Baldovino & di Bitetti 2006).

True allocare that reduces the amount of energy a mother expends on carrying her relatively heavy offspring is found in species living in single-female groups (e.g., siamangs, Lappan 2008; spectral tarsier, Gursky 2000). A comparison of siamang groups with different levels of male care showed that mothers that received more carrying help from their male partners had shorter interbirth intervals (Lappan 2008, 2009). In a small number of primates that tend to give birth to twins or even triplets, male partners of the mother do most of the carrying and provisioning (Callitrichines: Bales et al. 2002; Fite et al. 2005). Mothers may carry and retrieve their offspring even less when the available helpers are experienced at allocare (Fite et al. 2005).

Females in such cooperative breeding systems compete heavily over access to helpers. This competition may be shown by overt aggression or harassment of other females, thereby limiting group membership, suppressing other females' reproductive physiology, or even killing subordinate females' newborns (see chapter 15, this volume, fig. 15.5; Digby 2000; common marmoset, *Callithrix jacchus*, Saltzman et al. 2009). However, breeding females are also reported to disproportionally groom the females that help them, which can be interpreted as an incentive to keep them in the group as caregivers (common marmoset, Lazaro-Perea et al. 2004). Thus, successful reproduction of females in cooperative breeding primates has become completely dependent on allocare, and recruiting helpers is an essential component of reproduction for these females (fig. 14.6). As

Fig. 14.6.   Allomaternal care is especially pronounced in the callitrichines. For example, callitrichine twin infants are carried most of the time during their first three to four weeks of life, but other group members, such as this adult male moustached tamarin (*Saguinus mystax*) in Peru, relieve mothers of much of the cost of this activity. Photo courtesy of Julia Diegmann.

a consequence, mothers have an extended conflict of interest with their own weaned offspring, who are unable to reproduce in their natal group. Older offspring can either provide allocare for siblings, which is in the mother's interest, or try to find a breeding position in another group, which may be even more costly to them than helping their mother (e.g., saddle-back tamarin, *Saguinus fuscicollis*, Goldizen et al. 1996). Only one daughter can take over the mother's support group upon her death, typically after a reproductive tenure of only a few years. Thus, whereas the outcome of mother-infant conflict over investment depends on the mother's access to sufficient help, her conflict with her sexually mature offspring depends on demographic conditions.

Comparative analyses (Mitani & Watts 1997; Ross & MacLarnon 2000; Ross 2003) have suggested that species with some form of allocare (such as guarding, carrying, or allonursing) have faster infant growth and shorter interbirth intervals than species that lack such help. However, because infants of species with allocare are not larger as weanlings and do not start reproduction earlier than those without allocare, it was concluded that almost all allocare can be interpreted as being beneficial for the mother, but not directly beneficial for individual infants. Yet a direct survival benefit to allocare for infants is suggested by within-species comparisons of the level of received allocare in obligate cooperative breeders. These species show a high risk of maternal abandonment of infants when helpers are lacking (several Callitrichine species, Santos et al. 1997; Bardi et al. 2001). In addition, orphaned gray mouse lemur and white-faced capuchin monkey infants, living in settings with a relatively high level of allonursing, appear to have increased chances of survival through adoption by their allomothers, although sample sizes remain small (Eberle & Kappeler 2006; Perry 2008). Such enhancement of offspring survival would benefit both the mother and the offspring—although, as with all insurance policies, only a few individuals reap the benefits.

In conclusion, in at least some primate species a mother's minimally required investment in an offspring, as well as her reproductive success, depend on the quality and quantity of the allocare her offspring receives, which in turn largely depends on her own social relationships. An infant's chance of survival may improve in the presence of allocaregivers (in cooperative breeders) or allomothers (in communal breeders). More importantly, perhaps, in both communal and cooperative breeders a mother can decrease the amount of care she gives her offspring in response to the level of allocare received, and thus she can save energy for investment in her next offspring—whereas the infant's physical development is hardly affected by allocare. In species with despotic female hierarchies, true allocare is relatively rare due to the potential for very uneven distribution of the benefits.

Maternal Investment in Sons and Daughters

If unequal numbers of females and males are present in a population, mothers producing offspring of the rarer sex benefit by obtaining more grand-offspring, which will drive the secondary sex ratio toward equality (Fisher 1930). Thus, equal investment in sons and daughters is expected. However, many studies on wild mammals have found deviations from an equal sex ratio (Clutton-Brock & Iason 1986; Sheldon & West 2004). The reason is that individual females may achieve greater reproductive success via offspring of one sex because their social position or age differentially affects either the costs of raising sons or daughters or the expected reproductive success of these offspring.

Several hypotheses identify benefits to mothers of biasing their investment according to the sex of their offspring. The Trivers-Willard (1973) hypothesis predicts that females will bias the sex ratio according to differences in their ability to enhance the future reproductive success of sons versus daughters. The local resource competition (LRC) hypothesis posits that mothers will bias the sex ratio of their offspring in favor of the dispersing sex (usually male in primates). This is because adults of the philopatric sex (usually female) compete over food and other resources with their own mothers (as well as their sisters), thereby reducing both their mothers' ability to reproduce and their own (Clark 1978). However, if offspring of the philopatric sex improve their mother's additional reproduction (e.g., directly through allocare in cooperatively breeding species or indirectly through matriline size effects), then the mothers are expected to bias sex ratios toward the philopatric sex as per the local resource enhancement (LRE) hypothesis (Emlen et al. 1986).

Silk and Brown (2008) found that the adult size dimorphism of a species, which is often used as a proxy for the relative energy costs of raising sons versus daughters, was not independently correlated with the average birth sex ratio in that species. In primate species in which the conditions of LRC apply, females give birth to significantly more offspring of the dispersing sex (Johnson 1988; Silk & Brown 2008), whereas in LRE species with cooperative breeding, the birth sex ratio is biased towards the helping sex. Even though these results suggest adaptive deviations from the expected 1:1 sex ratio for the average conditions for species, they may mask further fine-tuning at the population or individual level. For example, van Schaik & Hrdy (1991) pointed out that the severity of local competition may vary both within and between populations (as well as with provisioning or captive conditions), and that this may explain why the costs and benefits of differential investment could vary even within populations. In a meta-analysis, Schino

(2004) indeed found a clear effect of food availability on birth sex ratios in different primate populations. Nunn and Pereira (2000) also found evidence for fine-tuning within a species. Their study of one provisioned population of ring-tailed lemurs suggested both LRE and LRC effects on offspring sex ratio. Females that had just formed a new group (through fission) and young females that received targeted aggression preceding expulsion overproduced daughters that could serve as potential allies (LRE). At the population level, however, females tended to overproduce dispersing sons at larger group sizes (LRC). Thus, multiple factors affect the potential benefits to mothers of differential investment in sons and daughters, and under different conditions, different pressures may prevail (cf. Cockburn et al. 2002).

The facultative sex ratio adjustment hypothesis of Trivers and Willard (1973) assumes not only that the variance in reproductive success is higher for one sex (usually males, in polygynous animals) but also that the mother's condition (including her physical, nutritional, and social status) during the investment period differentially influences the future reproductive success of her sons and daughters. Thus, mothers in good condition are expected to overproduce high-quality sons with high reproductive success prospects, whereas mothers in poor condition would have more grand-offspring if they overproduce daughters whose reproductive success is less affected by their quality.

For primates, we still lack quantitative evidence that maternal condition (or, more likely, dominance rank as a proxy for ability to garner food or protect offspring) during the offspring's early development differentially affects the later reproductive success of sons and daughters. Obviously, this assumption is difficult to test in long-lived species in which there is a long delay between weaning and the onset of reproduction, and in which accurate lifetime reproductive data are typically much harder to collect for members of the dispersing sex than for the philopatric sex. In primates with female philopatry, the expected reproductive future of daughters is similar to that of their mothers under ecologically and socially stable conditions. A mother's influence on her son's reproductive prospects, however, is usually limited because she can effectively support her immature sons only up to a certain size, above which agonistic support from a female no longer meaningfully affects male-male competition outcomes (bonobos may be an exception: Surbeck et al. 2010). In some cercopithecines, the sons of high-ranking mothers mature faster and enjoy better physical condition in early adulthood than do the sons of low-ranking females (Alberts & Altmann 1995a, b; Bercovitch et al. 2000; Setchell et al. 2001; Garcia et al. 2009). Such developmental advantages may well enhance these males' reproductive success after successful dispersal, as has been reported for

long-tailed macaques (van Noordwijk & van Schaik 1999, 2001). However, the same high-ranking females also produce daughters with reproductive advantages (see also Johnson 2003; Altmann & Alberts 2005); thus, the differential payoff of biased investment in sons is still unclear.

Biased sex ratios at birth reflect biased early investment, possibly mediated by sex-biased early abortion under the influence of glucose and hormone levels (Grant 2007; James 2007; Cameron et al. 2008). Some epigenetic consequence of adverse conditions during early development may also affect sons and daughters differently (Price & Coe 2000; Cirulli et al. 2009b). Also, females could bias their postnatal investment in sons and daughters, resulting in different interbirth intervals. For example, the duration and energy costs of maternal care of unweaned offspring are greater for sons than for daughters in chimpanzees (Nishida et al. 2003), mountain gorillas (Robbins et al. 2007), Hanuman langurs, (Ostner et al. 2005), rhesus macaques (Bercovitch et al. 2000), and gray mouse lemurs (Colas 1999). This sex difference in early investment may, however, be counterbalanced by compensatory maternal care later in life in philopatric female offspring, where daughters (and their offspring) receive disproportionately more grooming and agonistic support than do sons (Fairbanks 1988).

In summary, at the species level systematic deviations from an equal sex ratio at birth consistent with the predictions of the LRC and LRE hypotheses have been shown. Fine-tuning of this adjustment at the population or individual level may contribute to seemingly inconsistent results. In most cases, however, it remains an open question how much primate mothers influence their sons' lifetime reproductive success, whereas a mother's long-lasting impact on her daughters' reproduction is well documented, at least in groups with matrilineal hierarchies. This makes it still hard to test Trivers and Willard's hypothesis of maternal condition effects on offspring sex ratio in primates.

From Maternal Investment to Lifetime Maternal Care?

Primate mothers bestow on their offspring lifelong physiological maternal effects, reflecting the outcome of a mother-offspring conflict over investment affected by the mother's condition during the offspring's early development. However, all primate mothers also maintain individualized relationships with their weaned offspring and other members of their social group, or with neighbors in adjacent ranges. On the other hand, there is much variation among primates in the extent of care provided to immatures by mothers and others, whether offspring of both sexes disperse, or whether mothers maintain lifelong associations with adult daughters (in many primates) or sons (in, for example, chimpanzees and bonobos: Stumpf 2007).

Prolonged maternal association may be beneficial to the young offspring, but it is not parental investment in that it does not directly reduce a mother's future reproduction. At least where group living is ecologically beneficial, the mother's costs of extended association with offspring are low—and if the extended association does benefit the offspring, the mothers also derive inclusive fitness benefits. The same would be true for association and care extended to other relatives. Females in species with despotic matrilineal hierarchies that regulate access to food and safety benefit most from strong supportive bonds, and their offspring benefit from receiving prolonged protection against other group members. Here a mother's dominance rank can affect her daughters' early reproductive success, and daughters acquire their own dominance rank with matrilineal support (fig. 14.7). Thus, daughters benefit from a strong bond with their mothers, but the mothers may benefit from associating with their daughters as well, through agonistic support or other supportive social interactions. The benefits for a female's reproductive success of extended association and matrilineal bonds may be so high that, in the absence of maternal kin, a female invests in establishing and maintaining a limited number of strong replacement bonds (Silk et al. 2006a, b), which may even help her to overcome the disadvantages of low dominance rank (Silk et al. 2009). Thus, prolonged mother-offspring care may have given rise to similar relationships among other dyads, using the same

bonding mechanism (grooming exchange and agonistic support) to obtain similar reproductive benefits.

Long life span and individualized social relationships may have facilitated the development of strong, lifelong bonds between mothers and their offspring, with mutual benefits. Similar long-lasting bonds are found in some nonprimate gregarious mammals (e.g., elephants, *Loxodonta africana*, Archie et al. 2006). In most mammals, however, females provide one-sided maternal investment in offspring, which is accompanied by conflict over the level and continuation of investment. Primate mothers tend to live in social settings in which the period of minimally required maternal investment is followed by an extended association, which for the philopatric sex lasts a lifetime. These associations enable lifelong, mutually supportive bonds that benefit both mothers and their offspring. Thus, in at least some primates, the quality of a female's lifelong maternal care may well be an important component of her reproductive success.

Summary and Conclusions

In this chapter I have examined some of the patterns and variables affecting maternal investment and care among primates. For example, variation is found in the timing of reproduction relative to more or less predictable changes in food abundance. With lengthening of the reproductive cycle (conception to weaning) and decreasing predictability in food supply, mothers are more likely to store fat reserves to ensure sufficient nutrient supply to their dependent offspring. Thus, primate species vary in their patterns of maternal physiology, and in the likelihood and timing of reproductive failure. In addition, the value of a mother's continued investment for the survival of the offspring affects the flexibility in the timing of her termination of maternal investment. Furthermore, maternal effects triggered during early development can have lifelong physiological consequences. In the long-lived primates, these seem to reflect a compromise between the mother's physical and social condition and the offspring's demands, and they indicate sensitive developmental periods followed by limits to physiological plasticity later in life.

Finally, with their relatively slow development and tendency to have a gregarious lifestyle, primate mothers not only invest in their offspring through gestation and lactation, but also maintain a long-lasting association with their weaned offspring, during which maternal care is still valuable to the offspring. This bond may still award benefits during young adulthood for offspring of the nondispersing sex.

Fig. 14.7. One of the most significant features of maternal care in primates is the prolonged association with adult offspring, especially daughters, and the resultant opportunities for continued support. Here, a mother-and-daughter pair of white-faced capuchin monkeys (*Cebus capucinus*) at Santa Rosa, Costa Rica, have formed a coalition against a rival female. Photo courtesy of Susan Perry.

References

Alberts, S. & Altmann, J. 1995a. Balancing costs and opportunities: Dispersal in male baboons. *American Naturalist*, 145, 279–306.

———. 1995b. Preparation and activation: determinants of age at reproductive maturity in male baboons. *Behavioral Ecology and Sociobiology*, 36, 397–406.

Altmann, J. 1980. *Baboon Mothers and Infants*. Cambridge, MA: Harvard University Press.

Altmann, J. & Alberts, S. C. 2003. Variability in reproductive success viewed from a life-history perspective in baboons. *American Journal of Human Biology*, 15, 401–409.

———. 2005. Growth rates in a wild primate population: ecological influences and maternal effects. *Behavioral Ecology and Sociobiology*, 57, 490–501.

Altmann, S. A. 1991. Diets of yearling female primates (*Papio cynocephalus*) predict lifetime fitness. *Proceedings of the National Academy of Sciences USA*, 88, 420–423.

Archie, E. A., Moss, C. J. & Alberts, S. C. 2006. The ties that bind: Genetic relatedness predicts the fission and fusion of social groups in wild African elephants. *Proceedings of the Royal Society B*, 273, 513–522.

Bainbridge, D. R. J. 2000. Evolution of mammalian pregnancy in the presence of the maternal immune system. *Reviews of Reproduction*, 5, 67–74.

Baker, J., Hurtado, A. M., Pearson, O. M., Hill, K. R., Jones, T. & Frey, M. A. 2009. Developmental plasticity in fat patterning of Ache children in response to variation in interbirth intervals: A preliminary test of the roles of external environment and maternal reproductive strategies. *American Journal of Human Biology*, 21, 77–83.

Baldovino, M. C. & Di Bitetti, M. S. 2008. Allonursing in tufted capuchin monkeys (*Cebus nigritus*): Milk or pacifier? *Folia Primatologica*, 79, 79–92.

Bales, K., French, J. A. & Dietz, J. M. 2002. Explaining variation in maternal care in a cooperatively breeding mammal. *Animal Behaviour*, 63, 453–461.

Bardi, M., Petto, A. J. & Lee-Parritz, D. E. 2001. Parental failure in captive cotton-top tamarins (*Saguinus oedipus*). *American Journal of Primatology*, 54, 159–169.

Barrett, L., Dunbar, R. I. M. & Dunbar, P. 1995. Mother-infant contact as contingent behaviour in gelada baboons. *Animal Behaviour*, 49, 805–810.

Barrett, L., Halliday, J. & Henzi, S. P. 2006a. The ecology of motherhood: The structuring of lactation costs by chacma baboons. *Journal of Animal Ecology*, 75, 875–886.

Barrett, L. & Henzi, S. P. 2000. Are baboon infants Sir Phillip Sydney's offspring? *Ethology*, 106, 645–658.

Barrett, L., Henzi, S. P. & Lycett, J. E. 2006b. Whose life is it anyway? Maternal investment, developmental trajectories, and life history strategies in baboons. In *Reproduction and Fitness in Baboons: Behavioral, Ecological, and Life History Perspectives* (ed. by L. Sweddell & S. R. Leigh) 199–224. New York: Springer.

Bateson, P. 1994. The dynamics of parent-offspring relationships in mammals. *Trends in Ecology & Evolution*, 9, 399–403.

———. 2001. Fetal experience and good adult design. *International Journal of Epidemiology*, 30, 928–934.

———. 2008. Preparing offspring for future conditions is adaptive. *Trends in Endocrinology and Metabolism*, 19, 111.

Bateson, P., Barker, D., Clutton-Brock, T., Deb, D., D'Udine, B., Foley, R. A., Gluckman, P., Godfrey, K., Kirkwood, T., Mirazon Lahr, M., McNamara, J., Metcalfe, N. B., Monaghan, P., Spencer, H. G. & Sultan, S. E. 2004. Developmental plasticity and human health. *Nature*, 430, 419–421.

Bearder, S. K. 1999. Physical and social diversity among nocturnal primates: A new view based on long term research. *Primates*, 40, 267–282.

Bercovitch, F. B. 1987. Female weight and reproductive condition in a population of olive baboons (*Papio anubis*). *American Journal of Primatology*, 12, 189–195.

Bercovitch, F. B. & Berard, J. D. 1993. Life history costs and consequences of rapid reproductive maturation in female rhesus macaques. *Behavioral Ecology and Sociobiology*, 32, 103–109.

Bercovitch, F. B. & Strum, S. C. 1993. Dominance rank, resource availability, and reproductive maturation in female savanna baboons. *Behavioral Ecology and Sociobiology*, 33, 313–318.

Bercovitch, F. B., Widdig, A. & Nürnberg, P. 2000. Maternal investment in rhesus macaques (*Macaca mulatta*): Reproductive costs and consequences of raising sons. *Behavioral Ecology and Sociobiology*, 48, 1–11.

Boesch, C. & Boesch-Achermann, H. 2000. *The Chimpanzees of the Taï Forest: Behavioural Ecology and Evolution*. Oxford: Oxford University Press.

Boesch, C., Bole, C., Eckhardt, N. & Boesch, H. 2010. Altruism in forest chimpanzees: The case of adoption. *PlosOne*, 5, e8901.

Borries, C., Larney, E., Lu, A., Ossi, K. & Koenig, A. 2008. Costs of group size: Lower developmental and reproductive rates in larger groups of leaf monkeys. *Behavioral Ecology*, 19, 1186–1191.

Boyd, I. L. 2000. State-dependent fertility in pinnipeds: Contrasting capital and income breeders. *Functional Ecology*, 14, 623–630.

Brockman, D. K. & van Schaik, C. P. 2005. Seasonality and reproductive function. In *Seasonality in Primates: Studies of Living and Extinct Human and Non-Human Primates* (ed. by D. K. Brockman & C. P. van Schaik), 269–305. Cambridge: Cambridge University Press.

Broussard, D. R., Dobson, F. S. & Murie, J. O. 2005. The effects of capital on an income breeder: Evidence from female Columbian ground squirrels. *Canadian Journal of Zoology*, 83, 546–552.

Cameron, E. Z., Lemons, P. R., Bateman, P. W. & Bennett, N. C. 2008. Experimental alteration of litter sex ratios in a mammal. *Proceedings of the Royal Society B*, 275, 323–327.

Carter, A. M. 2007. Animal models of human placentation: A review. *Placenta*, 28, Supplement A, Trophoblast Research, 21, S41–S47.

Champagne, F. A. & Curley, J. P. 2009. Epigenetic mechanisms mediating the long-term effects of maternal care on development. *Neuroscience and Biobehavioral Reviews*, 33, 593–600.

Cheney, D. L. & Seyfarth, R. M. 1987. The influence of intergroup competition on the survival and reproduction of female vervet monkeys. *Behavioral Ecology and Sociobiology*, 21, 375–386.

Cheney, D. L., Seyfarth, R. M., Andelman, S. J. & Lee, P. C. 1988. Reproductive success in vervet monkeys. In *Reproductive Success* (ed. by T. H. Clutton-Brock), 384–402. Chicago: University of Chicago Press.

Cheney, D. L., Seyfarth, R. M., Fischer, J., Beehner, J., Bergman, T., Johnson, S. E., Kitchen, D. M., Palombit, R. A., Rendall, D. & Silk, J. B. 2004. Factors affecting reproduction and mortality among baboons in the Okavango delta, Botswana. *International Journal of Primatology*, 25, 401–428.

Cheney, D. L., Seyfarth, R. M., Fischer, J., Beehner, J. C., Bergman, T. J., Johnson, S. E., Kitchen, D. M., Palombit, R. A., Rendall, D. & Silk, J. B. 2006. Reproduction, mortality, and female reproductive success in chacma baboons of the Okavango Delta, Botswana. In *Reproduction and Fitness in Baboons: Behavioral, Ecological, and Life History Perspectives* (ed. by L. Swedell & S. R. Leigh), 147–176. New York: Springer.

Chism, J. 2000. Allocare patterns among cercopithecines. *Folia Primatologica*, 71, 55–66.

Cirulli, F., Francia, N., Berry, A., Aloe, L., Alleva, E. & Suomi, S. J. 2009a. Early life stress as a risk factor for mental health: Role of neurotrophins from rodents to non-human primates. *Neuroscience and Biobehavioral Reviews*, 33, 573–585.

Cirulli, F., Francia, N., Branchi, I., Antonucci, M. T., Aloe, L., Suomi, S. J. & Alleva, E. 2009b. Changes in plasma levels of BDNF and NGF reveal a gender-selective vulnerability to early adversity in rhesus macaques. *Psychoneuroendocrinology*, 34, 172–180.

Clark, A. B. 1978. Sex ratio and local resource competition in a prosimian primate. *Science*, 201, 163–165.

Clutton-Brock, T.H. 1991. *The Evolution of Parental Care*. Princeton, NJ: Princeton University Press.

Clutton-Brock, T. H. & Iason, G. R. 1986. Sex ratio variation in mammals. *Quarterly Review of Biology*, 61, 339–374.

Cockburn, A., Legge, S. & Double, M. C. 2002. Sex ratios in birds and mammals: Can the hypotheses be disentangled? In *Sex Ratios: Concepts and Research Methods* (d. by Hardy, I. C. W.), 266–286. Cambridge: Cambridge University Press.

Colas, S. 1999. Evidence for sex-biased behavioral maternal investment in the gray mouse lemur (*Microcebus murinus*). *International Journal of Primatology*, 20, 911–926.

Desai, M. & Hales, N. 1997. Role of fetal and infant growth in programming metabolism in later life. *Biological Reviews*, 72, 329–248.

Digby, L. 2000. Infanticide by female mammals: Implications for the evolution of social systems. In *Infanticide by Males and its implications* (ed. by C. P. van Schaik & C. H. Janson), 423–446. Cambridge: Cambridge University Press.

Drent, R. H. & Daan, S. 1980. The prudent parent: energetic adjustments in avian breeding. *Ardea*, 68, 225–252.

Dunbar, R. I. M. 1988. *Primate Social Systems*. Ithaca, NY: Cornell University Press.

Eberle, M. & Kappeler, P. M. 2006. Family insurance: Kin selection and cooperative breeding in a solitary primate (*Microcebus murinus*). *Behavioral Ecology and Sociobiology*, 60, 582–588.

Emlen, S. T., Emlen, J. M. & Levin, S. A. 1986. Sex-ratio selection in species with helpers-at-the-nest. *American Naturalist*, 127, 1–8.

Engh, A. L., Hoffmeier, R. R., Seyfarth, R. M. & Cheney, D. L. 2009. O brother, where art thou? The varying influence of older siblings in rank acquisition by female baboons. *Behavioral Ecology and Sociobiology*, 64, 97–104.

Entringer, S., Kumsta, R., Hellhammer, D. H., Wadhwa, P. D. & Wüst, S. 2009. Prenatal exposure to maternal psychosocial stress and HPA axis regulation in young adults. *Hormones & Behavior*, 55, 292–298.

Fairbanks, L. A. 1988. Vervet monkey grandmothers: Interactions with infant grandoffspring. *International Journal of Primatology*, 9, 425–441.

———. 1990. Reciprocal benefits of allomothering for female vervet monkeys. *Animal Behaviour*, 40, 553–562.

———. 2000. Maternal investment throughout the life span in Old World monkeys. In *Old World Monkeys* (ed. by P. F. Whitehead & C. J. Jolly), 341–367. Cambridge: Cambridge University Press.

Fairbanks, L. A. & McGuire, M. T. 1995. Maternal condition and the quality of maternal care in vervet monkeys. *Behaviour*, 132.

Fedigan, L. M., Carnegie, S. D. & Jack, K. M. 2008. Predictors of reproductive success in female white-faced capuchins (*Cebus capucinus*). *American Journal of Physical Anthropology*, 137, 82–90.

Festa-Bianchet, M., Gaillard, J.-M. & Jorgenson, J. T. 1998. Mass-and density-dependent reproductive success and reproductive costs in a capital breeder. *American Naturalist*, 152, 367–379.

Fisher, R. A. 1930. *The Genetical Theory of Natural Selection*. Oxford: Oxford University Press.

Fite, J. E., Patera, K. J., French, J. A., Rukstalis, M., Hopkins, E. C. & Ross, C. N. 2005. Opportunistic mothers: Female marmosets (*Callithrix kuhlii*) reduce their investment in offspring when they have to, and when they can. *Journal of Human Evolution*, 49, 122–142.

Fragaszy, D. M., Fedigan, L. M. & Visalberghi, E. 2004. *The Complete Capuchin: The Biology of the Genus Cebus*. Cambridge: Cambridge University Press.

Garcia, C., Lee, P. C. & Rosetta, L. 2009. Growth in colony living anubis baboon infants and its relationship with maternal activity budgets and reproductive status. *American Journal of Physical Anthropology*, 138, 123–135.

Gluckman, P. D., Hanson, M. A. & Beedle, A. S. 2007. Early life events and their consequences for later disease: A life history and evolutionary perspective. *American Journal of Human Biology*, 19, 1–19.

Gluckman, P. D., Hanson, M. A., Beedle, A. S. & Spencer, H. G. 2008. Predictive adaptive responses in perspective. *Trends in Endocrinology and Metabolism*, 19, 108–110.

Gluckman, P. D., Hanson, M. A., Spencer, H. G. & Bateson, P. 2005. Environmental influences during development and their later consequences for health and disease: Implications for the interpretation of empirical studies. *Proceedings of the Royal Society B*, 272, 671–677.

Goldizen, A. W., Mendelsohn, J., van Vlaardingen, M. & Terborgh, J. 1996. Saddle-back tamarin (*Saguinus fuscicollis*) reproductive strategies: Evidence from a thirteen-year study of a marked population. *American Journal of Primatology*, 38, 57–83.

Gould, L. & Sauther, M. 2007. Lemuriformes. In *Primates in Perspective* (ed. by C. J. Campbell, A. Fuentes, K. C. MacKinnon, M. Panger & S. K. Bearder), 46–72. Oxford: Oxford University Press.

Gould, L., Sussman, R. W. & Sauther, M. L. 2003. Demographic and life-history patterns in a population of ring-tailed lemurs (*Lemur catta*) at Beza Mahafaly Reserve, Madagascar: A 15-year perspective. *American Journal of Physical Anthropology*, 120, 182–194.

Grant, V. J. 2007. Could maternal testosterone levels govern mammalian sex ratio deviations? *Journal of Theoretical Biology*, 246, 708–719.

Gursky, S. 2007. Tarsiiformes. In *Primates in Perspective* (ed. by C. J. Campbell, A. Fuentes, K. C. MacKinnon, M. Panger & S. K. Bearder), 73–85. Oxford: Oxford University Press.

———. 2000. Allocare in a nocturnal primate: Data on the spectral tarsier, *Tarsius spectrum*. *Folia Primatologica*, 71, 39–54.

———. 2002. The behavioral ecology of the spectral tarsier, *Tarsius spectrum*. *Evolutionary Anthropology*, 11, 226–234.

Haig, D. 1993. Genetic conflicts in human pregnancy. *Quarterly Review of Biology*, 68, 495–532.

Hales, C. N. & Barker, D. J. P. 1992. Type 2 (non-insulin-dependent) diabetes mellitus: The thrifty phenotype hypothesis. *Diabetologia*, 35, 595–601.

———. 2001. The thrifty phenotype hypothesis. *British Medical Bulletin*, 60, 5–20.

Hasegawa, T. & Hiraiwa, M. 1980. Social interactions of orphans observed in a free-ranging troop of Japanese monkeys. *Folia Primatologica*, 33, 129–158.

Hauser, M. D. 1993. Do vervet monkey infants cry wolf? *Animal Behaviour*, 45, 1242–1244.

Henzi, S. P. & Barrett, L. 2002. Infants as a commodity in a baboon market. *Animal Behaviour*, 63, 915–921.

Hinde, K. & Capitano, J. P. 2010. Lactational Programming? Mother's milk energy predicts infant behavior and temperament in rhesus macaques (*Macaca mulatta*). *American Journal of Primatology*, 71, 1–8.

Hrdy, S.B. 1976. Care and exploitation of nonhuman primate infants by conspecifics other than the mother. *Advances in the Study of Behavior*, 6, 101–158

———. 1977. *The Langurs of Abu: Female and Male Strategies of Reproduction*. Cambridge, MA: Harvard University Press.

Iverson, S. J., Oftedal, O. T., Bowen, W. D., Boness, D. J. & Sampugna, J. 1995. Prenatal and postnatal transfer of fatty acids from mother to pup in the hooded seal. *Journal of Comparative Physiology B*, 165, 1–12.

Jaeggi, A. V., van Noordwijk, M. A. & van Schaik, C. P. 2008. Begging for information: Mother-offspring food sharing among wild Bornean orangutans. *American Journal of Primatology*, 70, 533–541.

James, W. H. 2008. Some comments on the paper of Grant (2007). *Journal of Theoretical Biology*, 253, 401–404.

Janson, C. & Verdolin, J. 2005. Seasonality of primate births in relation to climate. In *Seasonality in Primates: Studies of Living and Extinct Human and Non-Human Primates* (ed. by D. K. Brockman & C. P. van Schaik), 307–350. Cambridge: Cambridge University Press.

Johnson, C. N. 1988. Dispersal and the sex ratio at birth in primates. *Nature*, 332, 726–728.

Johnson, J. A. 1987. Dominance rank in juvenile olive baboons, *Papio anubis*: The influence of gender, size, maternal rank and orphaning. *Animal Behaviour*, 35, 1694–1708.

Johnson, R. L. 1986. Mother-infant contact and maternal maintenance activities among free-ranging rhesus monkeys. *Primates*, 27, 191–203.

Johnson, S. E. 2003. Life history and the competitive environment: Trajectories of growth, maturation, and reproductive output among chacma baboons. *American Journal of Physical Anthropology*, 120, 83–98.

———. 2006. Maternal characteristics and offspring growth in chacma baboons: A life history perspective. In *Reproduction and Fitness in Baboons: Behavioral, Ecological, and Life History Perspectives* (ed. by L. Swedell, L. & S.R. Leigh), 177–197. Oxford: Oxford University Press.

Kappeler, P. M. 1998. Nests, tree holes, and the evolution of primate life histories. *American Journal of Primatology*, 46, 7–33.

Karssemeijer, G. J., Vos, D. R. & van Hooff, J. A. R. A. M. 1990. The effect of some non-social factors on mother-infant contact in long-tailed macaques (*Macaca fascicularis*). *Behaviour*, 113, 273–291.

Kavanagh, M. & Laursen, E. 1984. Breeding seasonality among long-tailed macaques, *Macaca fascicularis*, in peninsular Malaysia. *International Journal of Primatology*, 5, 17–29.

Kawai, M. 1965. On the system of social ranks in a natural troop of Japanese monkeys, 1: Basic rank and dependent rank. In *Japanese Monkeys* (ed. by K. Imanishi, K. & S.A. Altmann), 66–86. Atlanta: Emory University Press.

Kirkpatrick, R. C. 2007. The Asian colobines. Diversity among leaf-eating monkeys. In *Primates in Perspective* (ed. by C. J. Campbell, A. Fuentes, K. C. MacKinnon, M. Panger & S. K. Bearder), 186–200. Oxford: Oxford University Press.

Kurita, H., Sugiyama, Y., Ohsawa, H., Hamada, Y. & Watanabe, K. 2008. Changes in demographic parameters of *Macaca fuscata* at Takasakiyama in relation to decrease of provisioned foods. *International Journal of Primatology*, 29, 1189–1202.

Kuzuwa, C. W. 2005. Fetal origins of developmental plasticity: Are fetal cues reliable predictors of future nutritional environments? *American Journal of Human Biology*, 17, 5–21.

Laengin, K. M., Norris, D. R., Kyser, T. K., Marra, P. P. & Ratcliffe, L. M. 2006. Capital versus income breeding in a migratory passerine bird: Evidence from stable-carbon isotopes. *Canadian Journal of Zoology*, 84, 947–953.

Lahti, J., Räikkönen, K., Kajantie, E., Heinonen, K., Pesonen, A.-K., Järvenpää, A. L. & Strandberg, T. 2006. Small body size at birth and behavioural symptoms of ADHD in children aged five to six years. *Journal of Child Psychology and Psychiatry*, 47, 1167–1174.

Lappan, S. 2008. Male care of infants in a siamang (*Symphalangus syndactylus*) population including socially monogamous and polyandrous groups. *Behavioral Ecology and Sociobiology*, 62, 1307–1317.

Lappan, S. 2009. The effects of lactation and infant care on adult energy budgets in wild siamangs (*Symphalangus syndactylus*). *American Journal of Physical Anthropology*, 140, 290–301.

Lazaro-Perea, C., De Fatima Arruda, M. & Snowdon, C. T. 2004. Grooming as a reward? Social function of grooming between females in cooperatively breeding marmosets. *Animal Behaviour*, 67, 627–636.

Lee, P. C. & Johnson, J. A. 1992. Sex differences in alliances, and the acquisition and maintenance of dominance status among immature primates. In *Coalitions and Alliances in Humans and Other Animals* (ed. by A. H. Harcourt & F. B. M. de Waal), 391–414. Oxford: Oxford University Press.

Leigh, S. R. 2004. Brain growth, life history, and cognition in primate and human evolution. *American Journal of Primatology*, 62, 139–164.

Lewis, R. J. & Kappeler, P. M. 2005. Are Kirindy sifaka capital or income breeders? It depends. *American Journal of Primatology*, 67, 365–369.

Lonsdorf, E. V. 2005. Sex differences in the development of termite-fishing skills in wild chimpanzees, *Pan troglodytes schweinfurthii*, of Gombe National Park, Tanzania. *Animal Behaviour*, 70, 673–683.

Lycett, J. E., Henzi, S. P. & Barrett, L. 1998. Maternal investment in mountain baboons and the hypothesis of reduced care. *Behavioral Ecology and Sociobiology*, 42, 49–56.

Martin, R. D. 1990. *Primate Origins and Evolution: A Phylogenetic Reconstruction*. London: Chapman & Hall.

———. 1996. Scaling of the mammalian brain: The maternal energy hypothesis. *News in Physiological Science*, 11, 149–156.

———. 2008. Evolution of placentation in primates: Implications of mammalian phylogeny. *Evolutionary Biology*, 35, 125–145.

Mas-Rivera, A. & Bercovitch, F. B. 2008. Postpartum recuperation in primiparous rhesus macaques and development of their infants. *American Journal of Primatology*, 70, 1047–1054.

McGowan, P. O., Sasaki, A., D'Alessio, A. C., Dymov, S., Labonté, B., Szyf, M., Turecki, G. & Meaney, M. J. 2009. Epigenetic regulation of the glucocorticoid receptor in human brain associates with childhood abuse. *Nature Neuroscience*, 12, 342–348.

Metcalfe, N. B. & Monaghan, P. 2001. Compensation for a bad start: Grow now, pay later? *Trends in Ecology and Evolution*, 16, 254–260.

Mick, E., Biederman, J., Prince, J., Fischer, M. J. & Farafone, S. V. 2002. Impact of low birth weight on attention-deficit hyperactivity disorder. *Journal of Developmental & Behavioral Pediatrics*, 23, 16–22.

Milligan, L. A., Gibson, S. V., Williams, L. E. & Power, M. L. 2008. The composition of milk from Bolivian squirrel monkeys (*Saimiri boliviensis boliviensis*). *American Journal of Primatology*, 70, 35–43.

Mitani, J. C. & Watts, D. 1997. The evolution of non-maternal caretaking among anthropoid primates: Do helpers help? *Behavioral Ecology and Sociobiology*, 40, 213–220.

Moore, T. & Haig, D. 1991. Genomic imprinting in mammalian development: A parental tug-of-war. *Trends in Genetics*, 7, 45–49.

Muroyama, Y. 1994. Exchange of grooming for allomothering in female patas monkeys. *Behaviour*, 128, 103–119.

Nakamichi, M. & Yamada, K. 2007. Long-term grooming partnerships between unrelated adult females in a free-ranging group of Japanese monkeys (*Macaca fuscata*). *American Journal of Primatology*, 69, 652–663.

Nievergelt, C. M. & Martin, R. D. 1999. Energy intake during reproduction in captive common marmosets (*Callithrix jacchus*). *Physiology and Behaviour*, 65, 849–854.

Nishida, T., Corp, N., Hamai, M., Hasegawa, T., Hiraiwa-Hasegawa, M., Hosaka, K., Hunt, K. D., Itoh, N., Kawanaka, K., Matsumoto-Oda, A., Mitani, J. C., Nakamura, M., Norikoshi, K., Sakamaki, T., Turner, L., Uehara, S. & Zamma, K. 2003. Demography, female life history, and reproductive

profiles among chimpanzees of Mahale. *American Journal of Primatology*, 59, 99–121.

Nowell, A. A. & Fletcher, A. W. 2008. The development of feeding behaviour in wild western lowland gorillas (*Gorilla gorilla gorilla*). *Behaviour*, 145, 171–193.

Nozaki, M. 2009. Grandmothers care for orphans in a provisioned troop of Japanese macaques (*Macaca fuscata*). *Primates*, 50, 85–88.

Nunn, C. L. & Pereira, M. E. 2000. Group histories and offspring sex ratios in ringtailed lemurs (*Lemur catta*). *Behavioral Ecology and Sociobiology*, 48, 18–28.

O'Brien, T. G. 1988. Parasitic nursing behavior in the wedge-capped capuchin monkey (*Cebus olivaceus*). *American Journal of Primatology*, 16, 341–344.

Oftedal, O. T. 1984. Milk composition, milk yield, and energy output at peak lactation: A comparative review. *Symposia of the Zoological Society of London*, 51, 33–85.

Onyango, P. O., Gesquiere, L. R., Wango, E. O., Alberts, S. C. & Altmann, J. 2008. Persistence of maternal effects in baboons: Mother's dominance rank at son's conception predicts stress hormone levels in subadult males. *Hormones & Behaviour*, 54, 319–324.

Ostner, J., Borries, C., Schuelke, O. & Koenig, A. 2005. Sex allocation in a colobine monkey. *Ethology*, 111, 924–939.

Ottoni, E. B., de Resende, B. D. & Izar, P. 2005. Watching the best nutcrackers: What capuchin monkeys (*Cebus apella*) know about others' tool-using skills. *Animal Cognition*, 8, 215–219.

Parker, G. A., Royle, N. J. & Hartley, I. R. 2002. Intrafamilial conflict and parental investment: A synthesis. *Philosophical Transactions Royal Society London B*, 357, 295–307.

Patel, E. R. 2007. Non-maternal infant care in wild Silky sifakas (*Propithecus candidus*). *Lemur News*, 12, 39–42.

Pavelka, M. S. M., Fedigan, L. M. & Zohar, S. 2009. Availability and adaptive value of reproductive and postreproductive Japanese macaque mothers and grandmothers. *Animal Behaviour*, 64, 407–414.

Pereira, M. E. 1988. Agonistic interactions of juvenile savannah baboons. I. Fundamental features. *Ethology* 79, 195–217.

———. 1993. Seasonal adjustments of growth rate and adult body weight in ringtailed lemurs. In *Lemur Social Systems and their Ecological Basis* (ed. by P. M. Kappeler & J. U. Ganzhorn). 205–221. New York: Plenum.

Perry, S. 1996. Female-female social relationships in wild white-faced capuchin monkeys, *Cebus capucinus*. *American Journal of Primatology*, 40, 167–182.

———. 2008. *Manipulative Monkeys: The Capuchins of Lomas Barbudal*. Cambridge, MA: Harvard University Press.

Phillips, D. I. W. 2007. Programming of the stress response: A fundamental mechanism underlying the long-term effects of the fetal environment? *Journal of Internal Medicine*, 261, 453–460.

Poirier, F. E. 1968. The Nilgiri langur (*Presbytis johnii*) mother-infant dyad. Primates, 9, 45–68.

Pope, T. R. 2000. Reproductive success increase with degree of kinship in cooperative coalitions of female red howler monkeys (*Alouatta seniculus*). *Behavioral Ecology and Sociobiology*, 48, 253–267.

Power, M. L., Oftedal, O. T. & Tardif, S. D. 2002. Does the milk of Callitrichid monkeys differ from that of larger Anthropoids? *American Journal of Primatology*, 56, 117–127.

Price, K. C. & Coe, C. L. 2000. Maternal constraint on fetal growth patterns in the rhesus monkey (*Macaca mulatta*): The intergenerational link between mothers and daughters. *Human Reproduction*, 15, 452–457.

Rahpurohit, L. S. & Sommer, V. 1993. Juvenile male emigration from natal one-male troops of hanuman langurs. In *Juvenile Primates: Life History, Development, and Behavior* (ed. by M. E. Pereira & L.A. Fairbanks), 86–103. Oxford: Oxford University Press.

Rapaport, L. G. & Brown, G. R. 2008. Social influences on foraging behavior in young nonhuman primates: Learning what, where, and how to eat. *Evolutionary Anthropology*, 17, 189–201.

Rasmussen, D. T. 1985. A comparative study of breeding seasonality and litter size in eleven taxa of captive lemurs (*Lemur* and *Varecia*). *International Journal of Primatology*, 6, 501–517.

Richard, A. F., Dewar, R. E., Schwartz, M. & Ratsirarson, J. 2002. Life in the slow lane? Demography and life histories of male and female sifaka (*Propithecus verreauxi verreauxi*). *Journal of Zoology London*, 256, 421–436.

Robbins, A. M., Robbins, M. M. & Fawcett, K. 2007. Maternal investment of the Virunga Mountain Gorillas. *Ethology*, 113, 235–245.

Ross, C. 2001. Park or ride? Evolution of infant carrying in primates. *International Journal of Primatology*, 22, 749–771.

———. 2003. Life history, infant care strategies, and brain size in primates. In *Primate Life Histories and Socioecology* (ed. by P. M. Kappeler & M. E. Pereira), 266–284. Chicago: University of Chicago Press.

Ross, C. & MacLarnon, A. 2000. The evolution of non-maternal care in anthropoid primates: A test of the hypotheses. *Folia Primatologica*, 71, 93–113.

Saltzman, W., Digby, L. J. & Abbott, D. H. 2009. Reproductive skew in female common marmosets: What can proximate mechanisms tell us about ultimate causes? *Proceedings of the Royal Society B*, 276, 389–399.

Santos, C. V., French, J. A. & Otta, E. 1997. Infant carrying behavior in callitrichid primates: *Callithrix* and *Leontopithecus*. *International Journal of Primatology*, 18, 889–907.

Schiml, P. A., Mendoza, S. P., Saltzman, W., Lyons, D. M. & Mason, W. A. 1999. Annual physiological changes in individually housed squirrel monkeys (*Saimiri sciureus*). *American Journal of Primatology*, 47, 93–103.

Schino, G. 2004. Birth sex ratio and social rank: Consistency and variability within and between primate groups. *Behavioral Ecology and Sociobiology*, 15, 850–856.

Schino, G. & Troisi, A. 1998. Mother-infant conflict over behavioral thermoregulation in Japanese macaques. *Behavioral Ecology and Sociobiology*, 43, 81–86.

Setchell, J. M., Lee, P. C., Wickings, E. J. & Dixson, A. F. 2001. Growth and ontogeny of sexual size dimorphism in the mandrill (*Mandrillus sphinx*). *American Journal of Physical Anthropology*, 115, 349–360.

———. 2002. Reproductive parameters and maternal investment in mandrills (*Mandrillus sphinx*). *International Journal of Primatology*, 23, 51–68.

Sheldon, B. C. & West, S. A. 2004. Maternal dominance, maternal condition, and offspring sex ratio in ungulate mammals. *American Naturalist*, 163, 40–54.

Silk, J. B. & Brown, G. R. 2008. Local resource competition and local resource enhancement. *Proceedings of the Royal Society B*, 275, 1761–1765.

Silk, J. B., Alberts, S. C. & Altmann, J. 2003. Social bonds of female baboons enhance infant survival. *Science*, 302, 1231–1234.

Silk, J. B., Willoughby, E. & Brown, G. R. 2005. Maternal rank and local resource competition do not predict birth sex ratios in wild baboons. *Proceedings of the Royal Society B*, 272, 859–864.

Silk, J. B., Altmann, J. & Alberts, S. C. 2006a. Social relationships among adult female baboons (*Papio cynocephalus*) I. Variation in the strength of social bonds. *Behavioral Ecology and Sociobiology*, 61, 183–195.

Silk, J. B., Alberts, S. C. & Altmann, J. 2006b. Social relationships among adult female baboons (*Papio cynocephalus*). II. Variation in the quality and stability of social bonds. *Behavioral Ecology and Sociobiology*, 61, 197–204.

Silk, J. B., Beehner, J. C., Bergman, T. J., Crockford, C., Engh, A. L., Moscovice, L. R., Wittig, R. M., Seyfarth, R. M. & Cheney, D. L. 2009. The benefits of social capital: Close social bonds among female baboons enhance offspring survival. *Proceedings of the Royal Society B*, 276, 3099–3104.

Soltis, J., Wegner, F. H. & Newman, J. D. 2005. Urinary prolactin is correlated with mothering and allo-mothering in squirrel monkeys. *Physiology & Behaviour*, 84, 295–301.

Stearns, S. C. 1992. *The Evolution of Life Histories*. Oxford: Oxford University Press.

Steenbeek, R., Sterck, E. H. M., de Vries, H. & van Hooff, J. A. R. A. M. 2000. Costs and benefits of the one-male, age-graded and all-male phase in wild Thomas's langur groups. In *Primate Males* (ed. by P. M. Kappeler), 130–145. Cambridge: Cambridge University Press.

Stewart, K. 2001. Social relationships of immature gorillas and silverbacks. In *Mountain Gorillas: Three Decades of Research of Karisoke* (ed. by M. M. Robbins, P. Sicotte & K. J. Stewart), 183–213. Cambridge: Cambridge University Press.

Stumpf, R. 2007. Chimpanzees and bonobos: Diversity within and between species. In *Primates in Perspective* (ed. by C. J. Campbell, A. Fuentes, K. C. MacKinnon, M. Panger & S. K. Bearder), 321–344. New York: Oxford University Press.

Sugiyama, Y. & Ohsawa, H. 1982. Population dynamics of Japanese monkeys with special reference to the effect of artificial feeding. *Folia Primatologica*, 39, 238–263.

Surbeck, M., Mundry, R. Hohmann, G. 2010. Mothers matter! Maternal support, dominance status and mating success in male bonobos (*Pan paniscus*). *Proceedings of the Royal Society B*, 278, 590–598.

Tardif, S. D. & Jaquish, C. E. 1997. Number of ovulations in the marmoset monkey (*Callithrix jacchus*): Relation to body weight, age and repeatability. *American Journal of Primatology*, 42, 323–329.

Thierry, B. & Anderson, J. R. 1986. Adoption in anthropoid primates. *International Journal of Primatology*, 7, 191–216.

Tilden, C. D. & Oftedal, O. T. 1997. Milk composition reflects pattern of maternal care in prosimian primates. *American Journal of Primatology*, 41, 195–211.

Trivers, R. L. 1972. Parental investment and sexual selection. In *Sexual Selection and the Descent of Man*. (ed. by B. Campbell), 136–179. Chicago: Aldine.

————. 1974. Parent-offspring conflict. *American Naturalist*, 14, 249–264.

Trivers, R. L. & Willard, D. 1973. Natural selection of parental ability to vary the sex ratio of offspring. *Science*, 179, 90–92.

Van Noordwijk, M. A. & van Schaik, C. P. 1999. The effects of dominance rank and group size on female lifetime reproductive success in wild long-tailed macaques, *Macaca fascicularis*. *Primates*, 40, 105–130.

————. 2001. Career moves: Transfer and rank challenge decisions by male long-tailed macaques. *Behaviour*, 138, 359–395.

————. 2005. Development of ecological competence in Sumatran orangutans. *American Journal of Physical Anthropology*, 127, 79–94.

Van Noordwijk M. A., Sauren S. E. B., Nuzuar, Abulani A., Morrogh-Bernard H. C., Utami Atmoko S. S. & van Schaik C. P. 2009. Development of independence: Sumatran and Bornean orangutans compared. In *Orangutans: Geographic Variation in Behavioral Ecology and Conservation* (ed. by S. A. Wich, S. S. Utami Atmoko, T. Mitra Setia & C. P. van Schaik), 189–203. New York: Oxford University Press.

Van Schaik, C. P. 1983. Why are diurnal primates living in groups? *Behaviour*, 87, 120–144.

————. 1989. The ecology of social relationships amongst female primates. In *Comparative Socioecology* (ed. by V. Standen & R. A. Foley), 195–218. Oxford: Blackwell.

Van Schaik, C. P. & Hrdy, S. B. 1991. Intensity of local resource competition shapes the relationship between maternal rank and sex ratios at birth in cercopithecine primates. *American Naturalist*, 138, 1555–1562.

Van Schaik, C. P. & Kappeler, P. M. 1997. Infanticide risk and the evolution of male-female association in primates. *Proceedings of the Royal Society B*, 264, 1687–1694.

Van Schaik, C. P. & Paul, A. 1996. Male care in primates: Does it ever reflect paternity? *Evolutionary Anthropology*, 5, 152–156.

Van Schaik, C. P., Terborgh, J. W. & Wright, S. J. 1993. The phenology of tropical forests: Adaptive significance and consequences for primary consumers. *Annual Review of Ecology and Systematics*, 24, 353–377.

Van Schaik, C. P. & van Noordwijk, M. A. 1985. Interannual variability in fruit abundance and reproductive seasonality in Sumatran long-tailed macaques (*Macaca fascicularis*). *Journal of Zoology*, 206, 533–549.

Van Schaik, C. P., van Noordwijk, M. A. & Nunn, C. L. 1999. Sex and social evolution in primates. In *Comparative Primate Socioecology* (ed. by P. C. Lee), 204–240. Cambridge: Cambridge University Press.

Warner, D. A., Bonnet, X., Hobson, K. A. & Shine, R. 2008. Lizards combine stored energy and recently acquired nutrients flexibly to fuel reproduction. *Journal of Animal Ecology*, 77, 1242–1249.

Watanabe, K., Mori, A. & Kawai, M. 1992. Characteristic features of the reproduction of Koshima monkeys, *Macaca fuscata fuscata*: A summary of thirty-four years of observation. *Primates*, 33, 1–32.

Watts, D. P. & Pusey, A. E. 1993. Behavior of juvenile and adolescent great apes. In *Juvenile Primates. Life History, Development, and Behavior* (ed. by M. E. Pereira & L. A. Fairbanks), 148–167. Oxford: Oxford University Press.

Wells, J. C. K. 2003. The thrifty phenotype hypothesis: Thrifty offspring or thrifty mother? *Journal of Theoretical Biology*, 221, 143–161.

————. 2007. Flaws in the theory of predictive adaptive responses. *Trends in Endocrinology and Metabolism*, 18, 331–337.

Wheatley, K. E., Bradshaw, C. J. A., Harcourt, R. G. & Hindell, M. A. 2008. Feast or famine: Evidence for mixed capital-income breeding strategies in Weddell seals. *Oecologia*, 155, 11–20.

Wrangham, R. W. 1980. An ecological model of female-bonded primate groups. *Behaviour*, 75, 262–299.

Xi, W. Z., Li, B. G., Zhao, D. P., Hi, W. H. & Zhang, P. 2008. Benefits to female helpers in wild *Rhinopithecus roxellana*. *International Journal of Primatology*, 29, 593–600.

Yamada, K., Nakamichi, M., Shizawa, Y., Yasuda, J., Imakawa, S., Hinobayashi, T. & Minami, T. 2005. Grooming relationships of adolescent orphans in a free-ranging group of Japanese macaques (*Macaca fuscata*) at Katsuyama: A comparison among orphans with sisters, orphans without sisters, and females with a surviving mother. *Primates*, 46, 145–150.

Yamagiwa, J. & Kahekwa, J. 2001. Dispersal patterns, group structure, and reproductive parameters of eastern lowland gorillas at Kahuzi in the absence of infanticide. In *Mountain Gorillas. Three Decades of Research at Karisoke* (ed. by M. M. Robbins, P. Sicotte & K. J. Stewart), 91–122. Cambridge: Cambridge University Press.

Chapter 15 Magnitude and Sources of Variation in Female Reproductive Performance

Anne Pusey

AMONG THE chimpanzees at Gombe National Park in Tanzania, Fifi, the daughter of a high-ranking female, lived to the age of 46 years and gave birth to nine infants, seven of which are still living, while Gilka, a small, low-ranking female, a victim of polio and orphaned at the age of six years, lost all three of her offspring in infancy and died of disease at the age of 18 years. How typical are individual differences in reproductive performance of such magnitude? Can we find general factors that explain these differences? Does the magnitude of difference among females vary across species?

Following Bateman's (1948) classic experiments in *Drosophila* showing that variance in reproductive success was higher in males than in females, and Trivers' (1972) realization that sex differences in investment in gametes and offspring generally generate higher competition for mates among males than among females, the existence and importance of differences in female reproductive success were often downplayed (Clutton-Brock 1988a). But long-term studies of a number of species have revealed that the difference between variance in male and female lifetime reproductive success is often not as great as expected, and that variation in female reproductive success is sometimes considerable (Clutton-Brock 1988b, 2007). Examination of this variation allows us to identify the opportunity for and strength of selection on different components of reproductive performance (Arnold & Wade 1984a, b). Identification of phenotypic differences that correlate with differences in reproductive performance helps us to understand adaptation (Grafen 1988).

Considerable attention has recently been focused on how the intensity and nature of competition among females for the resources necessary for reproduction influences the nature of social behavior and female reproductive skew. For example, among group-living carnivores, reproductive skew is high in spotted hyenas (*Crocuta crocuta*, Holekamp et al. 1996), in which aggressive, high-ranking females monopolize access to meat, and in meerkats (*Suricata suricatta*), in which the dominant female receives help in rearing her offspring from other group members who, if female, rarely breed themselves (Clutton-Brock et al. 2001). On the other hand, skew is low in lionesses (*Panthera leo*), in which reproductive success depends on group territoriality, cub survival is highest when females breed simultaneously, and formidable weaponry renders fighting too costly to support dominance hierarchies (Packer et al. 2001).

In primates, a long-standing interest has been how ecological factors, particularly the distribution of food resources, influence female spacing patterns and the intensity and modes of female competition (chapter 9, this volume). Since early studies of Japanese macaques (*Macaca fuscata*) first identified conspicuous dominance hierarchies among females (Kawai 1958), the importance and significance of rank-related behavior has been the focus of much research. Nevertheless, because of the slow life histories of primates, data on the influence of dominance rank on female reproductive success are still sparse. Besides rank, other attributes of females that may cause differences in reproduction include body size and physical condition. Recent attention has also been focused on the adaptive significance of females' social bonds with other group members by examining how differences in these bonds influence reproductive success (chapters 14, 19, and 24, this volume).

To examine the causes of variation in female reproductive performance, it is important to determine whether there are age effects. If reproductive success varies with age across the life span, this fact must be taken into account when comparing, for example, rank effects on reproduction. Examination of age-specific reproduction in populations of individually identified females of known age is also the only way to produce accurate data on age of first reproduction, longevity, and birth schedules across the life span. These variables are essential for the cross-species comparisons employed in considerations of life-history theory and evolution (chapter 10, this volume) and in addressing closely related questions such as whether females show reproductive senescence or terminal investment. Longitudinal observation of individuals also allows the examination of proximate causes of age differences in reproductive performance, such as the importance of maternal experience.

In this chapter I examine long-term studies of known-aged primates to examine the effects of age on reproduction, as well as the various factors that account for variation in female reproductive performance. Although my focus is on differences among females in the same group, I briefly discuss ecological influences that affect all females in the group and that need to be considered in such analyses. I review data from wild populations, and from captive and free-ranging individuals living in breeding groups. I have attempted to include species from across the primate lineage, but far more information is available from the larger and more easily studied terrestrial species, such as baboons (*Papio* spp.), than from their smaller, more arboreal, or nocturnal cousins, particularly the New World monkeys and strepsirrhines.

Components of Reproductive Performance

Relative lifetime reproductive success is often taken to be the gold standard measure of individual fitness and is useful for examining the adaptive significance of a variety of traits (but see Grafen 1988). Because of the slow life histories of primates, however, very few studies have yet been able to measure this variable. Better progress has been made on measuring several components of reproductive performance that contribute to lifetime reproductive success.

Infant Growth Rates, Age at Sexual Maturation, and Age at First Reproduction

The speed with which offspring grow and reach sexual maturity is an important aspect of a female's reproductive performance. Although across species infants are weaned at a fixed proportion of adult body weight (Lee 1999), and although trade-offs between growth, reproduction, and longevity determine age at first reproduction and other life history traits (chapters 10 and 14, this volume), there is intraspecific variation in growth rate, as well as other life history traits (Lee & Kappeler 2003). In the absence of a negative effect on infant survival, the quicker a female's infant grows and reaches weaning age, the sooner she can have another infant. Infant growth rates have recently been measured in some populations of wild primates.

Offspring age at maturation should influence age at first reproduction, which in turn is likely to correlate with the offspring's reproductive success, provided that age at first reproduction does not influence longevity (Altmann et al. 1988). In several species it is possible to measure age at menarche in females, either from the appearance of sexual swellings and menstruation (e.g., in baboons: see below) or from first mating activity with adult males. In all long-term studies of known-aged individuals, it is possible to measure age at first reproduction for females. Age at sexual maturation in males is much harder to record for a number of reasons, but it has been recorded in a few primate species (chapter 18, this volume).

Reproductive Rate

Several measures of reproductive rate exist in the primate literature. Fertility is often measured as the number of births per female in a certain time interval. However, because many infants die early in life and because many females whose infants die reproduce again sooner than do females whose infants survive (e.g., baboons, Altmann et al. 1978; reviewed in van Schaik 2000), the total birthrate is not necessarily a good measure of a female's reproductive success, and it may even be inversely correlated with the number of surviving offspring. Another common measure is the interbirth interval exhibited by individual females. Because of the effect of early infant death in hastening female reproduction, many studies separate intervals between the birth of a surviving infant and the birth of the next infant from intervals after the previous infant has died, and only report differences in the former. A few studies report differences in both types of intervals or use multivariate statistics to control for the effect of death of the previous infant. Another less frequently used measure of reproductive rate is the rate of production of infants that survive to various ages.

Offspring Survival

Infant survival is a relatively easily measured component of reproductive success that is likely to be influenced by pat-

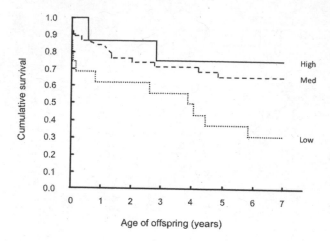

Fig. 15.1. Kaplan-Meier cumulative survival plot of offspring at the ages of zero to seven years for Gombe chimpanzee (*Pan troglodytes*) females of high, middle, and low rank at the birth of their offspring. Solid line indicates high rank, dashed line indicates middle rank, and dotted line indicates low rank. Mortality was highest at early ages. In a Cox–proportional hazards regression with multiple variables, the mother's rank and age at the infant's birth were both significant factors. Infant survival declined with the mother's age. Redrawn from Pusey et al. 1997.

terns of maternal investment and behavior (chapter 14, this volume). In many species, mortality rates are high in the first year or two and then drop to lower rates (e.g., Cheney et al. 1988; Altmann & Alberts 2003; fig. 15.1). Thus, infant survival beyond the period of high mortality is often correlated strongly with survival to sexual maturity. Many studies report survival to one or two years. This has the advantage of keeping sample size large, even in short-term studies. However, variation in offspring survival occurs throughout the period before sexual maturity, and it may continue to be influenced by maternal factors and behavior.

Longevity

In species where females usually produce widely spaced single young, female longevity and reproductive life span are likely to be highly correlated with reproductive success because long-lived females have more time to produce offspring (e.g., see Rhine et al. 2000). Data on female longevity are still scarce, however, because many primates live for decades, and few studies last that long.

Relative Contributions of Different Components to Lifetime Reproductive Success

Several methods exist to estimate the contribution of different components of reproductive performance to lifetime reproductive success or mean fitness. Four primate studies with long-term demographic data have used Brown's (1988) method to estimate the relative contribution of reproductive life span, fertility (births per year), and offspring survival to variance in lifetime reproductive success in adult females (wild vervet monkeys, *Chlorocebus pygerythrus*, Cheney et al. 1988; wild chacma baboons, *Papio ursinus*, Cheney et al. 2004; captive Japanese macaques, Fedigan et al. 1986; captive Barbary macaques, *Macaca sylvanus*, Paul & Kuester 1996). In all of these species except the vervets (which suffer high mortality from predation), reproductive life span contributed most to variance in lifetime reproductive success, and in all four species, offspring survival was considerably more important than fertility. Alberts and Altmann (2003b) used a matrix model to examine the effects of different components of reproductive performance on mean female fitness in wild yellow baboons (*Papio cynocephalus*). They found that changes in offspring survival had much stronger effects than changes in fertility on changes in mean fitness. Thus it appears that at least in haplorrhines, traits that influence offspring survival are under particularly strong selection. In species that can produce litters of more than one young (such as the smaller strepsirrhines or callitrichines) changes in fertility may be more important (Lee & Kappeler 2003).

Ecological and Group-Level Factors Influencing Female Reproductive Performance

In this chapter I am most interested in the magnitude and causes of differences in reproductive performance among females of the same population. To measure these differences it is often necessary to take into account ecological factors that may vary over time and influence all individuals in the groups or populations under study. Female reproduction can be greatly influenced by food availability, predation rates, and disease. Group size and composition also influence female reproduction to varying extents.

Food Availability

The influence of food availability on female reproduction is particularly marked and well known (reviews in Sadlier 1969; Lee & Kappeler 2003). Comparisons of captive and wild primate populations of the same species usually demonstrate large differences in infant growth rates, age at menarche or first birth, and interbirth interval between live young, all of which are attributable at least in part to the higher availability of food in captivity (e.g., chimpanzees, *Pan troglodytes*, Tutin 1994; baboons, Bronikowski et al. 2002, Johnson 2003, Garcia et al. 2006; long-tailed macaques, *Macaca fascicularis*, van Noordwijk & van Schaik

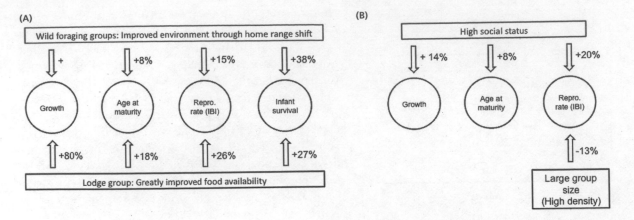

Fig. 15.2.   Effects of (A) foraging environment and (B) social status and group size (density) on components of reproductive success in female baboons at Amboseli, Kenya. Wild foraging groups had no access to human food; Lodge Group had access to the garbage dump of a tourist lodge. Components of reproductive performance are enclosed in circles; "Repro. rate (IBI)" indicates reproductive rate as measured by interbirth interval after surviving offspring. Environmental effects are enclosed in boxes. "+" indicates a positive effect, "–" a negative effect. Arrows indicate demonstrated effects. Modified and redrawn from Altmann & Alberts 2003b.

1999; ring-tailed lemurs, *Lemur catta*, Parga & Lessnau 2005). Among ring-tailed lemurs, differences in food availability in captivity also influence rates of twinning (Pereira 1993). In the wild, access to food from human activities (intentional provisioning, garbage dumps, etc.) also speeds growth and maturation as well as shortening interbirth intervals (e.g., Japanese macaques, Sugiyama & Ohsawa 1982; Mori et al. 1997; baboons; fig. 15.2). Annual differences in food availability—such as between mast years in Asia, when trees synchronously produce huge fruit crops, and nonmast years—also influence female reproduction (long-tailed macaques, van Noordwijk & van Schaik 1999; Bornean orangutans, *Pongo pygmaeus*, Knott 2001). Droughts and hurricanes can also severely influence food availability and reproductive performance (fig. 15.3; e.g., Verreaux's sifakas, *Propithecus verreauxi*, Richard et al. 2002; ring-tailed lemurs, Gould et al. 2003; yellow baboons, Beehner et al. 2006; Central American black howler monkeys, *Alouatta pigra*, Pavelka et al. 2007). In a few studies, long-term changes in habitat and food availability over time have been shown to influence female reproduction (baboons, Bronikowski et al. 2002; Altmann & Alberts 2003a; vervets, Cheney et al. 1988). In yellow baboons, females in troops that shifted their home range from poor to better habitat improved their reproductive rates (fig. 15.2). Finally, some studies have found effects of population density on aspects of female reproductive performance. Female rhesus macaques (*Macaca mulatta*) on Cayo Santiago Island, Puerto Rico, experienced their first births at younger ages when population density was low (Bercovitch & Berard 1993). In the Gombe chimpanzees, interbirth intervals were longer when community range size was smaller and population density higher (Williams et al. 2004).

Predation and Infanticide

The effects of differences in predation pressure on female reproduction are less well known, but interpopulation differences in predation pressure have been linked to reproductive rates and longevity (Hill & Lee 1998; Hill et al. 2000; Bronikowski et al. 2002; Cheney et al. 2004). Changes in adult survival rates and reproductive success over time in the same population have been attributed to changes in predation rates on the yellow baboons of Mikumi National Park, Tanzania (Wasser et al. 2004). When new males enter groups in some species, infant survival declines because of infanticide (chapter 19, this volume), and in the Gombe chimpanzees, infant survival was exceptionally low during a period in which a mother and daughter committed infanticide and cannibalism (Goodall 1977).

Group Size

Intraspecific (as well as interspecific) variation in group size is likely linked to broad ecological differences in food availability and predation risk (Hill & Lee 1998; reviewed in Lee & Kappeler 2003; chapter 9, this volume). Within populations, group size has been found to influence female reproduction in various ways. Often, larger group size is associated with lower rates of reproduction, probably because of increased density and intragroup competition for food. In both yellow baboons (fig. 15.2) and olive baboons (*Papio anubis*, Packer et al. 2000), females living in large groups showed longer interbirth intervals. Similar group size effects were also indicated in long-tailed macaques (van Noordwijk & van Schaik 1999) and Phayre's leaf monkey (*Trachypithecus obscurus phayrei*, Borries et al. 2008). How-

Fig. 15.3.   Relatively frequent droughts in Madagascar challenge lemur mothers like this Verreaux's sifaka at Kirindy Forest. During the drought of 1991–92, only 27% of the adult females in a population at Beza Mahafaly reproduced, compared to the overall average of 48% at that site (Richard et al. 2002). Photo courtesy of Rebecca J. Lewis.

ever, smaller groups had lower birthrates than larger groups in the Japanese macaques of Yakushima Island, Japan (Suzuki et al. 1998), and this was attributed to the superior performance of larger groups in intergroup competition for resources, as suggested by Wrangham (1980).

Age Effects on Reproductive Performance

Several primate studies have examined patterns of fertility over the lifespan. Figure 15.4 shows how age influences birthrates—defined as the number of infants born to females of a particular age class divided by the number of females observed in that age class—in both captive and wild populations of three species of macaques. Very similar patterns are shown by baboons (Packer et al. 1998), chimpanzees (Emery Thompson et al. 2007), mountain gorillas (*Gorilla gorilla beringei*, Robbins et al. 2006), Hanuman langurs (*Semnopithecus entellus*, Borries et al. 1991), Ver-

reaux's sifakas (Richard et al. 2002), and ring-tailed lemurs (Koyama et al. 2001; Gould et al. 2003). In most species, birthrates show an initial increase with age, reach a plateau for a number of years, then decline, often steeply, at the end of life. This general pattern is also common in other species of large mammals.

Lower Reproductive Performance in Young Females

The low birthrates at early ages in the plots in fig. 15.4 are due partly to the fact that individuals vary in the age of first reproduction so that some of the females in the earliest age categories have not started to breed. An additional effect of age on birthrates at the beginning of reproduction is revealed, however, by analysis of age-specific interbirth intervals. In many species, young females breed more slowly (reviewed in Anderson 1986; Silk 1990; Bercovitch et al. 1998). Table 15.1 summarizes studies that have accumulated enough data to examine age effects on components of reproduction.

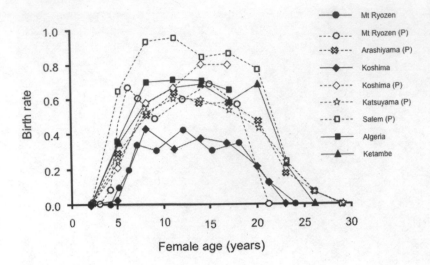

Fig. 15.4. Comparison of age-specific fertility for several populations of macaques of different species, with different levels of nutrition (P = provisioned): Japanese macaques (*Macaca fuscata*) in a period of wild living with no provisioning and a period of provisioning, both at Mount Ryozen, Japan (Sugiyama & Ohsawa, 1982); provisioned at Arashiyama, Japan (Koyama et al. 1992); provisioned and not provisioned at Koshima Island, Japan (Watanabe et al. 1992); and provisioned at Katsuyama, Japan (Itoigawa et al. 1992). Barbary macaques (*Macaca sylvanus*) in a free-living captive population at Salem, Germany (Paul & Kuester 1996); and in a wild population in Algeria (Menard & Vallet 1996). Long-tailed macaques (*Macaca fascicularis*) in a wild population at Ketambe, Indonesia (van Noordwijk & van Schaik 1999). Redrawn from van Noordwijk & van Schaik 1999.

Entries in cells indicate the presence or absence of statistical significance, unless otherwise noted. In almost all studies examining interbirth intervals, primiparous females had significantly longer intervals than multiparous females. In most of these cases, only the intervals following surviving young were measured, but a few studies also analyzed intervals following infants that died. Female mountain gorillas showed significantly longer interbirth intervals after first infants than after second infants, both when the previous infant survived and when it died (Robbins et al. 2006). While no difference was uncovered in the interbirth intervals following surviving young of primiparous and multiparous chimpanzees (Goodall 1983), more recent analysis of interbirth intervals in a larger sample from Gombe, which included both types of intervals in a multivariate analysis, found that the interval after the first infant was longer than the interval after later ones (Jones et al. 2010).

Young females also perform poorly on other reproductive measures. Fifteen of 21 studies of both captive and wild populations from a wide variety of primates found that the survival of first infants or the survival of the infants of young females was lower than that of other females' infants (table 15.1). Some studies found that the age of females at first birth also influenced their infants' survival. In mountain gorillas, for example, while there was no significant difference in infant survival between first- and second-born infants, infants born to primiparous females younger than nine years were significantly less likely to survive than those born to older females (Robbins et al. 2006). Similar significant effects of age on infant survival of primaparas were found in the rhesus monkeys of Cayo Santiago (Sade 1990) and the Barbary macaques of Salem (Paul & Kuester 1996), and a nonsignificant trend in the same direction occurred in blue monkeys (*Cercopithecus mitis*, Cords & Chowdhury 2010). Consistent with these patterns, firstborn infants were found to have lower body mass in yellow baboons at Amboseli, Kenya (Altmann & Alberts 2005). Finally, studies of olive and yellow baboons revealed that rates of conception were low in young females, and increased with age (table 15.1; Packer et al. 1998; Beehner et al. 2006).

The poor performance of young females has several non–mutually exclusive explanations. First, the females themselves are still maturing in body size and/or aspects of their reproductive and neuroendocrine systems (reviewed in Bercovitch et al. 1998; Setchell et al. 2002). Thus, during their initial reproductive efforts, they are likely to be trading off investment in their own continued growth against investment in their offspring.

Second, poor success in young females may derive from inexperience in foraging efficiently or rearing infants effectively. In observational studies, it is difficult to tease apart the effects of experience and learning from maturational effects. Evidence for poor maternal care by primiparas and its improvement with maternal experience comes from captive chimpanzees (Rogers & Davenport 1970). In the wild, infant death was attributed to maternal incompetence in one primiparous female chimpanzee at Gombe (Goodall 1983); lower maternal competence of primiparas has also been described in yellow baboons (Altmann & Alberts 2005) and orphaned primiparas in Japanese macaques (Hiraiwa 1981). In a comprehensive analysis of factors influencing interbirth intervals in captive bonnet macaques (*Macaca radiata*), Silk (1990) found that females with experience in raising surviving infants had shorter interbirth intervals than those without such experience, and she suggested that this difference resulted in part from learned improvements in maternal behavior. Perhaps the best evidence of the effect of maternal experience on infant survival comes from captive vervets, where infant mortality was generally high for primiparas (Fairbanks & McGuire 1984) but significantly lower for those with more experience as allomothers in immaturity (Fairbanks 1990).

Lower Reproductive Performance of Old Females

After an increase with the age of mothers, the birthrates of many species reach a plateau that lasts for many years during which prime-aged females produce infants at similar rates (fig. 15.4). Then, at the oldest ages, birthrate often declines. Analysis of the influence of age on interbirth intervals across the life span in different species generates conflicting results (table 15.1) that may be due partly to the underlying curvilinear relationship of birthrate with age and partly to the different kinds of statistical analyses employed. Some studies suggest diminished reproductive output among older females. For example, in one study of the olive baboons of Gilgil, Kenya, that divided females into several age categories, both young and old females had markedly longer interbirth intervals than prime-aged females (Strum & Western 1982). Similarly, a multivariate analysis that had first birth interval as a factor revealed that interbirth intervals in female chimpanzees at Gombe increased moderately with age (Jones et al. 2010). In another aggregate analysis of several populations of chimpanzees that did not distinguish first interval from later ones, interbirth interval also increased with age (Emery Thompson et al. 2007). Other studies, however, indicate no such negative effect of age. For example, a linear regression analysis of olive baboon data from another group in the Gilgil population (Smuts & Nicolson 1989) and from Gombe (Packer et al. 2000) revealed a significant *decrease* in interbirth intervals (between births of surviving young) with advancing age. Though suggestive of quicker reproduction among older females, these results possibly could have been strongly influenced by long intervals after first birth (as well as by small sample size of older females, in the case of the Gilgil baboons). Finally, many other studies have simply found no significant effect of age on interbirth interval (table 15.1).

Several studies have measured the survival of infants of older females in various ways. In some but not all studies, the survival of infants born to old females was significantly lower (table 15.1). In olive baboons, infant survival drops steeply when females reach the age of 20 years (Packer et al. 1998) and, controlling for higher infant mortality to primiparas and female rank, there is an almost significant decrease in infant survival across the life span (Packer et al. 1995). In chacma baboons, old females also show significantly lower infant survival than prime-aged females (Cheney et al. 2004), as do chimpanzees at Gombe, after

controlling for mother's rank (Pusey et al. 1997). Again, as discussed for interbirth intervals, some of the differences in the significance of age on infant survival in table 15.1 may be due to differences in statistical analyses.

A few studies have measured the effect of the mother's age on infant size and maturation rates. After controlling for maternal rank, age-specific offspring body mass among chacma baboons decreased significantly with the mother's age, especially after the age of 14 years (Johnson 2006). Likewise, there was a nonsignificant tendency for older, parous yellow baboon females to give birth to relatively light infants (Altmann & Alberts 2005). In the Gombe olive baboons, daughters experienced their first birth significantly later as their mother's age increased, once rank and troop identity were taken into account (Packer et al. 1995).

Other signs of decline in reproductive performance in older females include a decrease in conception rate in baboons (Packer et al. 1998; Beehner et al. 2006) and a significant increase in miscarriage rate with age in the olive baboons of Gombe (Packer et al. 1998) but not in the yellow baboons of Amboseli (Beehner et al. 2006) or Mikumi (Wasser et al. 1998). Older captive vervet monkeys also show higher rates of miscarriage (Fairbanks & McGuire 1984), which appear to characterize mountain gorillas as well (Robbins et al. 2006).

In summary, a significant decline in various aspects of the reproductive performance of aging females has been documented in several species or populations of primates. In nonhuman primates (but perhaps not in humans), this decline in reproduction coincides with general bodily senescence (Packer et al. 1998; Emery Thompson et al. 2007). A fascinating specific example of the probable influence of bodily senescence on reproductive performance comes from a study of Milne-Edwards sifakas (*Propithecus edwardsi*, King et al. 2005). Analysis of tooth wear and food particle size (from fecal analyses) demonstrated that females after the age of 18 years have such worn teeth that their ability to reduce food to small particle size is reduced. Such females continue to produce offspring but their infants are significantly less likely to survive, especially in dry years, possibly because the females' lower ability to process food reduces their milk production.

Although life-history theory predicts increased investment in each offspring by aging females of decreasing reproductive value (Williams 1966), this expectation is difficult to test because the reproductive effort required by an older female to maintain an infant's body weight or survival equivalent to that of a younger female's infant may be greater (Clutton-Brock 1984; Fessler et al. 2005). Paul et al. (1993) found that old female Barbary macaques weaned their infants significantly later and had high infant survival.

Table 15.1. The effect of age on female reproductive performance.

	Wild, provisioned, or captive	Sample size	Infant growth	Daughter's age at first birth	Interbirth interval longer after first than after subsequent births	Interbirth interval changed with age	Change in rate of conception	Change in rate of miscarriage	Lower survival of first infants	Lower survival of infants of young females	Lower survival of infants of old females	Reference
Propithecus verreauxi	Wild	426 marked individuals, 16 yrs								Yes	Yes	Richard et al. 2002
Propithecus edwardsi	Wild	41 infants									Yes	King et al. 2005
Lemur catta	Wild	24 females, 10 yrs							Yes	Yes		Koyama et al. 2001
Alouatta palliata	Wild	9 females, 8–11 yrs										Glander 1980; Clarke & Glander 1984
Alouatta seniculus	Wild	83 females, 5 yrs				No			No			Pope 2000
Cebus capucinus	Wild	31 females, 21 yrs			Yes				No		No	Fedigan et al. 2008
Cercopithecis mitis	Wild	66 females, 29 yrs							No			Cords & Chowdhury 2010
Chlorocebus pygerythrus	Wild	33 females, 7 yrs								Yes		Cheney et al. 1988
Chlorocebus pygerythrus	Captive	21 females, 8 yrs						Higher in older females				Fairbanks & McGuire 1984
Macaca fascicularis	Wild	65 females, 12 years				No			No	Yes		Van Noordwijk & van Schaik 1999
Macaca fuscata (Arashiyama East)	Provisioned	200+ females, 30 yrs			Yes	Longer in females aged 20+					No	Koyama et al. 1992
Macaca fuscata (Arashiyama West)	Captive	80 females, 8 yrs				No					No	Fedigan et al.1986
Macaca fuscata (Katsuyama)	Provisioned	156 females, 28 yrs			Yes				Yes		Trend	Itoigawa et al. 1992
Macaca fuscata (Mt. Ryozen)	Wild/provisioned	134 females, 11 yrs							Yes		Yes	Sugiyama & Ohsawa 1982
Macaca fuscata (Koshima Island)	Provisioned	88 females, 34 yrs			Yes					No	No	Watanabe et al. 1992
Macaca mulatta (Cambridge)	Captive	12 females, 2yrs										Gomendio 1989
Macaca mulatta (La Paguera)	Captive	51 females								Yes		Drickamer 1974

Macaca mulatta (Yerkes)	Captive	71 females, 5 yrs		Yes				Yes	Yes	Wilson et al. 1978; Wilson et al. 1983
Macaca radiata	Captive	150+ females,17 yrs		Yes						Silk 1990
Macaca sylvanus	Captive	90–207 females, 4–11 yrs		Yes				Yes		Paul & Thomen 1984; Paul & Kuester1996
Mandrillus sphinx	Captive	26 females, 14 yrs	Older for older mothers	Yes				Yes		Setchell et al. 2002
Papio anubis (Gombe)	Wild	138–167 females, 25–30+ yrs			Decrease	Curvilinear	Higher in older females	Yes	Yes	Packer et al.1995, 1998, 2000
Papio anubis (Gilgil, Eburru cliffs)	Wild	56 females, 4 yrs		Trend	Decrease			Yes		Smuts & Nicolson 1989
Papio anubis (Gilgil, Pump house gang)	Wild	117 births, 10 yrs		Longer in young than prime females	Curvilinear			Yes		Strum & Western 1982
Papio cynocephalus (Amboseli)	Wild	up to 100+ females, 12–30+ yrs	Firstborn infants lighter			Curvilinear	No	Yes		Altmann & Alberts 2005; Beehner et al. 2006; Altmann et al. 1988
Papio cynocephalus (Mikumi)	Wild	106 females, 22 yrs			No		No			Wasser et al. 1998
Papio ursinus	Wild	42 females, 10 yrs	Infants of older females lighter		No				Yes	Yes	Johnson 2006; Cheney et al. 2004
Gorilla gorilla	Wild	66 females , 36 yrs		Yes	No		More common in older females	No		Robbins et al. 2006
Pan troglodytes (Gombe)	Wild (provisioned)	13–42 females, 34–44 yrs		Yes	Increase				Yes	Jones et al. 2010; Pusey et al. 1997
Pan troglodytes (several populations)	Wild (some provisioned)				Increase					Emery Thompson et al. 2007

On the other hand, the lower weight for age of the offspring of older female baboons (Johnson 2006) does not support the hypothesis.

Effects of Competition and Dominance Rank

Efforts to understand the relationships between resource distribution, female distribution, and the types of competitive behavior exhibited by female primates have a long history and are discussed elsewhere in this book (see, e.g., chapter 9, this volume). Here I focus on whether and how modes of female competition, particularly dominance rank, lead to differences in female reproductive performance within populations.

Individuals of most primate species live in permanent social groups. Female group size differs widely across species, and this can affect the kind of competition that ensues. In species where the number of females in the group is typically small, there may be severe competition for group membership. In ursine howlers (*Alouatta arctoidea*), for example, the size of a group does not exceed four adult females, which may aggressively evict young females as well as deter female immigration (Crockett & Pope 1988; see also chapter 19, this volume). In stable, saturated populations, emigrant females appear to have very low chances of breeding, and this probably increases female reproductive skew (Pope 2000). Targeting and evicting of females is also observed in other species in which small group size appears to be selected for, such as some lemurs (Vick & Pereira 1989; reviewed by Erhart & Overdorff 2008).

In callitrichines, not only is female group size small, but usually only one female in a group breeds, and is then assisted by other group members (chapter 3, this volume). As in other cooperative breeders (e.g., meerkats, Clutton-Brock et al. 2001), competition among females is strong and manifests in several ways. First, female marmosets are highly antagonistic towards extragroup females. Second, infanticide by the dominant female may occur if a subordinate female gives birth (fig. 15.5; Saltzman 2003; Bezerra et al. 2007; Digby & Saltzman 2009). The various forms of reproductive suppression exhibited by young subordinate females (Abbott et al. 1998) have likely evolved in response to this potential risk (Saltzman et al. 2009). Third, although long-term data for callitrichine species are scarce, it is likely that reproductive skew in adult females is pronounced, with some individuals retaining the breeding position for several years and producing multiple successful litters while others may never breed at all (Garber et al. 1993; Goldizen et al. 1996). Young females appear to queue for the breeding position held by the older dominant female.

For the majority of diurnal primates living in larger groups, however, all females breed and eviction is rare. In some but not all of these species, dominance hierarchies of varying stability occur (chapter 9, this volume). Among the cercopithecines, macaques, baboons, and vervet monkeys form strong, stable nepotistic hierarchies in which females typically assume a rank just below their mothers and retain that rank for life, yielding matrilines that in turn are ranked linearly. In some guenons (*Cercopithecus* spp.), however, dominance hierarchies are either undetectable or less distinct (chapter 5, this volume). Among colobines, pronounced female hierarchies occur in some populations of Hanuman langurs, but in that case female status is primarily individualistic, with rank being highest in young adults and then falling with advancing age (chapter 4, this volume). Capuchin monkeys (*Cebus* spp.) have pronounced female hierarchies that are not as strongly nepotistic as in some Old World monkeys (Manson et al. 1999). Among species in which females routinely transfer to other groups, female hierarchies are not discernable in northern muriquis (*Brachyteles hypoxanthus*, Strier 1990), but they are in some others. In mantled howlers (*Alouatta palliata*), for example, young, newly immigrant females achieve high rank, which then declines with age as newer females immigrate (Glander 1980). In contrast, female rank rises with age in chimpanzees (Nishida 1989; Wittig & Boesch 2003; Murray et al. 2006) and mountain gorillas (Robbins et al. 2007).

Although the clear expectation is that dominance hierarchies are related to competition for resources, and thus that high rank should lead to higher reproductive performance, early reviews of the primate literature found limited and contradictory support for this prediction (Fedigan 1983; Harcourt 1987; Silk 1987). In the last 20 years, long-term studies have lengthened and increased in number. Here I review these studies to address the following questions: (1) Does rank influence reproductive performance? (2) What are the proximate causes of these rank effects? (3) Are the effects of dominance on reproductive success strongest in those species with the most pronounced hierarchies?

Effects of Dominance Rank on Reproductive Performance

Even now, most of the data available to address this question still come from studies of cercopithecines (table 15.2). The older data originating from studies of captive and free-ranging, but mostly provisioned, macaques have gradually been augmented by new data from additional studies of macaques and baboons from both captivity and the wild, and, most notably, from long-term research on several popula-

Fig. 15.5. In cooperatively breeding callitrichines, competition for breeding positions is often extremely intense among the resident females. This can result in physiological suppression of rivals or direct attacks upon their offspring, as in the case of this fatally wounded, one-month-old infant common marmoset, killed by an adult female in its own group in northeastern Brazil. Photo courtesy of Antonio Souto.

tions of wild baboons. Other taxa are still poorly represented and, for the most part, have smaller sample sizes that provide less power for statistical analysis.

Offspring Size, Growth, and Maturation

Mother's rank positively affects a number of measures of offspring maturation (table 15.2). Infant growth of higher-ranking females was hastened in wild yellow baboons (Altmann & Alberts 2005) and also in the female, but not male, offspring of chacma baboons (Johnson 2003). Consistent with the implications of these higher growth rates, daughters of high-ranking females reached menarche significantly earlier in three of four baboon populations (table 15.2) and sons of high-ranking female yellow baboons experienced testicular enlargement earlier (Altmann & Alberts 2005; Charpentier et al. 2008). In the Gombe chimpanzees, daughters of high-ranking females experienced their first mature sexual cycle (which is strongly correlated with menarche) significantly earlier (Pusey et al. 1997).

Although menarche is strongly correlated with age at

first birth in the yellow baboons of Amboseli (Charpentier et al. 2008), the effect of rank on first birth has not been reported there. In 5 of 13 other populations depicted in table 15.2, high rank was significantly correlated with daughter's early age at first birth, and in the Cayo Santiago rhesus macaques, rank of the matriline significantly influenced age of first reproduction during years of high density (Bloomquist et al. 2009). Two more studies showed a strong trend in the same direction. The studies with no effect of rank on age at first birth tended to have smaller sample sizes (table 15.2).

Birth Rates and Interbirth Intervals

In addition to three studies of tamarins (*Saguinus* spp. and *Leontopithecus* spp.), in which generally only the highest-ranking female breeds, data from 17 populations of species with multiple breeding females show that birthrates, expressed as births per unit time for females of different rank classes, were significantly higher for higher-ranking females in five studies, and in no studies were they significantly lower

Table 15.2. The effect of dominance rank on female reproductive performance

	Wild, provisoned, or captive	Sample size and duration	Infant growth rate positively correlated with mother's rank	Menarche earlier in daughters of high-ranking females	Age at first birth earlier in daughters of high-ranking females	Higher birthrate for high-ranking females	Interbirth interval shorter for high-ranking females	Miscarriage rate	Infant survival higher for high-ranking females	Rate of production of surviving offspring higher for high-ranking females	Longevity greater in high-ranking females	Reference
Lemur catta	Wild	60 group yrs over 13 yrs				No			No	Low-ranking produced fewer than middle-ranking but not high-ranking females		Takahata et al. 2008
Alouatta palliata	Wild	9 females, 8–11 yrs							Not for highest-ranking female			Glander 1980
Leontopithecus rosalia	Wild	100+ females, 8 yrs				Yes						Dietz & Baker 1993
Saguinus fuscicollis	Wild	~50 females, 13 yrs				Yes						Goldizen et al. 1996
Saguinus mystax	Wild/translocated	31 females, 10 yrs				Yes						Garber et al. 1993
Cebus capucinus	Wild	31 females, 21 yrs					No		No	No		Fedigan et al. 2008
Cercopithecis mitis	Wild	16 females, 19 yrs					No					Cords 2002
Chlorocebus pygerythrus (Amboseli)	Wild	33 females, 7 yrs			No	No						Cheney et al. 1988
Chlorocebus pygerythrus	Wild	17 females, 2yrs				Yes, for 1 of 2 groups						Whitten 1983
Chlorocebus pygerythrus	Captive	21 females, 8 yrs				Yes	Yes		No	Yes		Fairbanks & McGuire 1984
Macaca fuscata (Arashiyama East)	Provisioned	200+ females, 30 yrs				No			No			Koyama et al. 1992
Macaca fuscata (Arashiyama West)	Captive	46–80 females, 8 yrs			Yes	Yes, for numerical rank but not categorical rank	No		No		No	Gouzoules et al. 1982; Fedigan et al. 1986

Species (site)	Condition	Sample						Reference
Macaca fuscata (Katsuyama)	Provisioned	156 females, 28 yrs		Yes				Itoigawa et al. 1992
Macaca fuscata (Mt. Ryozen)	Wild/provisioned	134 females, 11 yrs		Yes, during provisioning	Yes, for central rather than peripheral females during provisioning			Sugiyama & Ohsawa 1982
Macaca fuscata (Koshima Island)	Provisioned	88 females, 34 yrs		Yes, for top-ranking females				Watanabe et al. 1992
Macaca fuscata (Yakushima)	Wild	36 females, 19 yrs, but sample for rank data was smaller		No	No			Takahata et al. 1998
Macaca fuscata (Kinkazan)	Wild	< 50 females, 11 yrs	No	No				Takahata et al. 1998
Macaca mulatta (La Paguera)	Captive	437 females	Yes	Yes	Yes			Drickamer 1974
Macaca mulatta (Cayo Santiago)	Captive	700+ females, 30+ yrs	Yes (high-ranking matrilines)				No	Bloomquist 2009; Bercovitch & Berard 1993
Macaca mulatta (Yerkes)	Captive	71 females	Yes	Yes	Yes			Wilson et al. 1978
Macaca radiata	Captive	150+ females, 17 yrs	No	No	Yes			Silk 1990
Macaca sylvanus (Gibraltar)	Provisioned	19 females, 6 yrs	No	Yes, for 1 of 3 groups		No		Kummerli & Martin 2005
Macaca sylvanus	Captive	207 females, 11 yrs	Yes	Nonsignificant trend	No	Yes		Paul & Kuester 1996
Macaca fascicularis	Wild	65 females, 12 yrs	Positive trend	Yes	No	Yes, in large groups	Trend	Van Noordwijk & van Schaik 1999
Mandrillus sphinx	Captive	26 females, 14 yrs	Yes	Yes				Setchell et al. 2002
Papio anubis (Gombe)	Wild	138–167 females, 25–30+ yrs	Yes	Shorter post-partum amenorrhea	Higher in high-ranking females		No	Packer et al. 1995, 1998, 2000
Papio anubis (Gilgil, Pump house gang)	Wild	19 females, 10 yrs	No	Yes				Bercovitch & Strum 1993
Papio anubis (Gilgil, Eburru cliffs)	Wild	56 females, 4 yrs	Yes	Yes	No			Smuts & Nicolson 1989
Papio anubis	Captive	23 females, 2 yrs		Yes				Garcia et al. 2006

(continued)

Table 15.2. (continued)

	Wild, provisoned, or captive	Sample size and duration	Infant growth rate positively correlated with mother's rank	Menarche earlier in daughters of high-ranking females	Age at first birth earlier in daughters of high-ranking females	Higher birthrate for high-ranking females	Interbirth interval shorter for high-ranking females	Miscarriage rate	Infant survival higher for high-ranking females	Rate of production of surviving offspring higher for high-ranking females	Longevity greater in high-ranking females	Reference
Papio cynocephalus (Amboseli)	Wild	Up to 100+ females, 12–30+ yrs	Yes	Yes		Not in early study	Yes	No effect of rank	No			Altmann & Alberts 2003a, 2005; Altmann et al. 1988; Charpentier et al. 2008; Beehner et al. 2006; Silk et al. 2003
Papio cynocephalus (Mikumi)	Wild	75 females, 22 yrs		Yes	Positive trend		Yes	Lower in high-ranking females	Yes for survival to 4 yrs (but not to 2 yrs)		Yes	Wasser et al. 2004
Papio ursinus	Wild	42 females, 10 yrs	Yes for daughters	No			Yes		No		Yes	Johnson 2003; Cheney et al. 2004; Silk et al. 2009; 2010
Semnopithecus entellus	Wild	13 females, 11 yrs				Yes				Yes		Borries et al. 1991
Gorilla gorilla	Wild	66 females, 36 yrs					Yes		No	Yes		Robbins et al. 2007
Pan troglodytes	Wild (provisioned)	13–42 females, 34–44 yrs		Yes (first mating with adult males)			Yes		Yes	Yes (excluding 1 sterile female)	Trend	Pusey et al. 1997; Jones et al. 2010

(table 15.2). In two more studies, high-ranking females had higher birthrates in at least one of the study groups, and in the Japanese monkeys of Mount Ryozen, birthrates were higher for central (high-ranking) than peripheral (low-ranking) females during periods of provisioning, but not during wild feeding (Sugiyama & Ohsawa 1982). In the captive Japanese macaques of Arashiyama West in Texas, there was a significant effect of female numerical rank on the production of infants, although this effect disappeared when females were grouped into high-, middle-, and low-ranking categories (Gouzoules et al. 1982).

Eleven of 17 studies found that high-ranking females had significantly shorter interbirth intervals (generally following the birth of a surviving infant), determined in various ways with or without the control of other factors (table 15.2). In a twelfth study, this time of olive baboons, the length of total interbirth interval was not reported, but one relevant component of this interval, post-partum amenorrhoea, was significantly shorter among females of higher rank (Packer et al. 1995). The strength of the effect of rank sometimes varied with other factors. For example, low-ranking female long-tailed macaques were significantly less likely than their high-ranking rivals to give birth in the year following the birth of a surviving offspring, particularly in bad years of low food availability (van Noordwijk & van Schaik 1999).

Conception Rates and Pregnancy Failure

In the Amboseli baboons, no effect of rank was found on conception rate (conceptions per cycle; Beehner et al. 2006). The effect of rank on miscarriage rate varies among baboon populations (table 15.2). Packer et al. (1995) found a significantly higher rate of miscarriage among high-ranking female olive baboons at Gombe. Among yellow baboons, however, miscarriages were significantly more common in low-ranking females at Mikumi (Wasser et al. 2004) and were unrelated to female rank at Amboseli (Beehner et al. 2006). In captive mandrills (*Mandrillus sphinx*), there was a trend towards high rates of stillbirth in low-ranking females (Setchell et al. 2002).

Infant Survival

Overall, rank has been found to have significant effects on infant survival less frequently (in 5 of 20 studies; table 15.2) than on interbirth interval. Among olive baboons, infant survival in the first year was significantly correlated with mother's rank in both bivariate and multivariate analyses at Gombe (Packer et al. 1995). At Mikumi there was no effect of female rank on infant survival to 18 months, but survival to four years was significantly positively correlated with mother's rank (Wasser et al. 2004). No effects of rank on infant survival to one year were found in the early years

of the Amboseli study (Altmann et al. 1988), or in a more recent multivariate analysis with a larger sample size (Silk et al. 2003). Mother's rank had no significant effects on infant chacma baboon survival to one year and, indeed, infant mortality tended to be higher for high-ranking mothers in this population (Cheney et al. 2004). A later multivariate analysis of factors influencing infant mortality in the same population found no significant effect of rank (Silk et al. 2009). However, infant survival was significantly elevated among high-ranking females in three populations of captive macaques, and in the Japanese macaques of Mount Ryozen, central females had higher infant survival than peripheral females, especially during provisioning (although no statistics are presented; table 15.2). In Gombe chimpanzees, mother's rank was significantly correlated with offspring survival, controlling for female age (Pusey et al. 1997).

Rate of Production of Surviving Young

For high-ranking females, the combination of shorter interbirth intervals and infant survival either higher than or similar to that of low-ranking females should result in higher rates of production of surviving young. This measurement is reported in only nine studies, but in seven of them this expectation is upheld (table 15.2). In captive vervets and the Barbary macaques at Salem, Germany (but not those of Gibraltar), rank had a significantly positive effect on the annual number of infants surviving to one year. In wild Hanuman langurs, rank had significant positive effects on the production of infants surviving to one year and to two years (Borries et al. 1991). Dominant females enjoyed higher rates of infant survival to three years in mountain gorillas (Robbins et al. 2007) and to five years in Gombe chimpanzees (fig. 15.6; Pusey et al. 1997). Sometimes the influence of maternal rank is mediated by group size. In wild long-tailed macaques, for example, high-ranking females produced more infants surviving to one year than did low-ranking females, but this effect was only significant in large groups (van Noordwijk & van Schaik 1999). In wild ring-tailed lemurs in large groups, low-ranking females produced fewer infants surviving to one year than did middle-ranking (but not high-ranking) females (Takahata et al. 2008). In the only study of rank in forest guenons, Cords (2002) found a linear dominance hierarchy in blue monkeys, but no significant effect of rank on the production of infants surviving to one year.

Female Longevity

Few studies have yet reported the effect of rank on longevity. High-ranking female yellow baboons at Mikumi survived significantly longer and also had longer reproductive life spans (Wasser et al. 2004). High-ranking chacma ba-

Fig. 15.6. Female chimpanzee Flo (*left*), estimated to be more than 50 years old, with her five-month-old infant and five-year-old son. This shorter than average interbirth interval is unusual for such an old female, but is consistent with Flo's high dominance rank. Photo courtesy of the Jane Goodall Institute / Hugo van Lawick.

boons also lived longer (Silk et al. 2010), but there was no such correlation between rank and longevity among Gombe olive baboons (Packer et al. 2005). Similarly, dominant females appeared to live longer in wild long-tailed macaques (van Noordwijk & van Schaik 1999), but there was no such effect among the Japanese macaques of Arashiyama West (Fedigan et al. 1986) or the Cayo Santiago rhesus macaques (reviewed in Bercovitch & Berard 1993). In the Gombe chimpanzees there was a nonsignificant tendency for females that ranked high at age 20 to live longer than those who ranked low at that age (Pusey et al. 1997).

Lifetime Reproductive Success

Even among the longest studies, few have yet reported analyses of the effects of female rank on lifetime reproductive success. In the Mikumi baboons, rank had a significant effect on lifetime production of offspring, measured as the number of infants surviving to 18 months and to four years (Wasser et al. 2004). At Gombe, lifetime production of olive baboon infants surviving to two years was uncorrelated with maternal rank overall, but a significant correlation emerged when two females who never gave birth were excluded from the analysis (Packer et al. 1995). Similarly, among the female chimpanzees of Gombe, the lifetime production of infants surviving to five years was positively associated with rank after exclusion of a sterile female from the sample (Pusey et al. 1997). No effect of rank was found on lifetime reproductive success in the rhesus macaques of Cayo Santiago (reviewed in Bercovitch & Berard 1993), or in the Japanese macaques of Arashiyama West (Fedigan et al. 1986).

Summary: The Importance of Rank

In aggregate, these data strongly support the hypothesis that female dominance rank affects reproduction, as exemplified by diverse measures across a variety of primate species. These results provide a clear adaptive explanation for the rank-related agonistic behavior widely observed in primates. In many cases, high rank had significant positive effects, but in no cases did it have a significantly negative effect. From the pattern of significant effects in table 15.2, there is some indication that rank may improve female reproductive success more through its effects on infant growth and reproductive rate than through its effect on infant survival. Thus it appears that while an infant of a low-ranking female grows more slowly, the mother compensates by weaning the infant later, and consequently experiences a longer interbirth interval. In this way she may maximize her infants' survival, which has a stronger effect than fertility on variance in lifetime reproductive success and female fitness (see above).

Proximate Causes of Rank Effects

A major effect of high rank is likely to be greater access to food. Evidence is accumulating that female rank is often correlated with a number of measures of feeding success (reviewed in Silk 1987; Harcourt 1987, 1989; van Noordwijk & van Schaik 1999). For example, high-ranking female chimpanzees gain better access to high-quality food in contest competition with others (Wittig & Boesch 2003), forage more efficiently (Murray et al. 2006), and acquire long-term access to higher-quality core areas (Murray et al. 2006, 2007). Nevertheless, differential access to food is not necessarily the whole story. The effects of rank on reproductive performance also operate in captive settings, where food is usually freely available (table 15.2). For example, in a study of captive olive baboons there was no difference in food intake by females of different rank, yet interbirth intervals were still correlated with rank (Garcia et al. 2006). Possibly the low-ranking females had to expend more energy in avoiding the high-ranking females, thus leading to a more negative energy balance, but it was also suggested that neuroendocrine effects of stress from aggression disrupted reproductive cycles, as has been suggested in other studies (Garcia et al. 2006, and references therein).

Species Differences in the Importance and Effects of Rank

A major hypothesis in primatology is that the form and strength of dominance hierarchies are influenced by the form and strength of competition for food, and notably that species and populations whose food is more patchily distributed and monopolizable have more pronounced despotic hierarchies among individuals (van Schaik 1989; Sterck et al. 1997; Isbell & Young 2002; chapter 9, this volume). We might thus expect that rank has stronger effects on reproductive performance in species with the most

pronounced hierarchies, and that reproductive skew in these species may be more extreme. Data to test this idea are still sparse, however. The majority of studies in table 15.2 are of baboons, macaques, or other cercopithecines whose dominance hierarchies have been classified as despotic. Yet among some of these, such as Japanese macaques, the effects of rank on reproduction are not pronounced. Moreover, significant rank effects are evident in species sometimes classified as egalitarian, such as gorillas, chimpanzees, and langurs. Sample sizes are still too small to compare relative reproductive skew in lifetime reproductive success in different species.

One provocative suggestion to explain a lack of strong rank effect on reproductive performance is that traits that help to enhance or maintain high rank may be subject to stabilizing selection. In the Gombe olive baboons, Packer et al. (1995) found that while female rank influenced many components of reproduction, high-ranking females had significantly higher rates of miscarriage, a few matured late, and some produced few or even no offspring. Consequently, the overall effect of rank on female lifetime reproductive success was not significant. Packer and colleagues argued that this low reproductive performance by some females reflected the adverse effects on reproduction of traits that conferred high competitive ability (including masculinization), and they urged others to investigate the occurrence of such anomalous females in other primate populations. Altmann et al. (1995) countered, however, by raising questions about the Gombe data and by demonstrating the absence of such effects among high-ranking yellow baboons at Amboseli. This further provoked a rebuttal and reinterpretation of the original hormonal data in favor of the hypothesis by Packer (1995). Since then, analysis of a larger data set from Amboseli has found no correlation of miscarriage rate with rank (Beehner et al. 2006), and in the Mikumi baboons, low-ranking females actually had higher rates of miscarriage (Wasser et al. 2004). Nevertheless, high-ranking sterile females were reported in the Barbary macaques of Salem (Paul & Thommen 1984), and continued investigation of the existence and causes of anomalous female reproductive performance in other studies is warranted.

Effects of Individual Condition and Quality

Besides differences in age and rank, it is likely that individual females have intrinsic constitutional differences that affect their reproductive performance. A recent multivariate analysis of data from Gombe chimpanzees found significant effects on interbirth interval of a number of factors, including the survival of the previous infant, the mother's age, and the mother's rank (Jones et al. 2010). However, a large proportion of the variance was accounted for by the effects of individual females, some of whom had longer interbirth intervals than others. A "phenotypic quality effect" for each individual female correlated significantly with a measurement of relative body mass (Pusey et al. 2005), suggesting that larger females had shorter interbirth intervals (Jones et al. 2010). Similar results have been found in nonprimates (bighorn sheep, *Ovis canadensis*, Berube et al. 1999; subantarctic fur seals, *Arctocephalus tropicalis*, Beaupelt & Guinet 2007).

Among Verreaux's sifakas, Richard et al. (2002) also noted considerable consistent variation in fertility among individual females, with some giving birth every year for up to nine years in a row, while others failed to give birth over many years. These reproductive differences were not correlated with individual differences in body weight (or any other factor), however. Nonetheless, the possible relevance of female body size is suggested by data showing that female mass was greatly influenced by environmental variability and that females that were heavier at the time of mating were more likely to give birth the following season (Richard et al. 2000). In a captive baboon group, female body weight was not correlated with rank, but heavier females had shorter interbirth intervals (Garcia et al. 2006).

Effects of health on reproduction are rarely reported, but in a cross-population study of wild chimpanzees, Emery Thompson et al. (2007) found that interbirth interval generally increased with age, but when females over 25 years of age were divided into "healthy" and "unhealthy" groups on the basis of their subsequent survival, reproductive rates declined only in the "unhealthy" group. At Gombe, female chimpanzees infected with SIV had lower fertility and higher infant mortality (Keele et al. 2009).

One of the few primate studies to examine the potential effects of genetic differences among females on reproductive performance was a study of the Amboseli yellow baboons. In these primates, levels of heterozygosity were not correlated with age at menarche in a multivariate model (Charpentier et al. 2008). However, the population contains hybrids of yellow (*P. cynocephalus*) and olive baboons (*P. anubis*), and males with the highest proportion of *anubis* ancestry matured significantly earlier.

Effects of Alloparents and Social Support

Although mothers are the exclusive providers of care to their offspring in many primate species, in some species, other individuals help carry, protect, and even nurse or provision infants that are not their own (chapter 14, this volume). In particular, in the cooperatively breeding callitrichines, adult males and previous offspring perform a great deal of infant care. While cross-taxa correlations between allocare and the speed of maturation of young suggest a causal influence of help from others on female reproductive rate (Mitani & Watts 1997; Ross & Maclarnon 2000), few studies have yet demonstrated that access to alloparents accounts for intraspecific variation in female reproductive performance. Among the callitrichines, several studies have shown that female reproduction is strongly influenced by group composition and the presence of helpers. In a colony of captive cotton-topped tamarins (*Saguinus oedipus*), infant mortality was significantly higher when helpers were not available (Bardi et al. 2001). Reviewing data from free-ranging golden lion tamarins (*Leontopithecus rosalia*) and three species of callitrichines, Bales et al. (2000) found that the number of surviving infants in the group was correlated with the number of resident adult males. Nevertheless, quantitative data on the amount of direct caregiving by males were not available, and the authors suggested that this effect could also be due to other activities of the males, such as territorial defense.

Among noncallitrichines, the best evidence that access to allocare increases mothers' reproductive performance comes from a captive colony of vervet monkeys in which females whose infants were carried often by allomothers (usually juvenile females) had shorter birth intervals (Fairbanks 1990). Less direct evidence of the importance of allomothers comes from two species of New World monkey. In white-faced capuchins (*Cebus capucinus*), interbirth interval (but not infant survival) was significantly negatively correlated with the number of close kin in the group in a multivariate analysis (Fedigan et al. 2008). Communal nursing occurs in this species, and Fedigan and colleagues suggest that allocare from relatives contributes to this outcome. Similarly, in ursine howlers, established groups containing close kin show higher infant survival than new groups composed of non-relatives, which Pope (2000) attributes to possibly higher levels of allomothering by kin.

In baboons, where allocare is uncommon, several studies have nevertheless revealed the importance of close social relationships for infant survival or speed of maturation of infants (chapter 24, this volume). In the yellow baboons of Amboseli, females with closer social bonds showed higher infant survival to one year over their lifetimes, both in a bivariate analysis and a multivariate analysis in which the mother's rank had no significant effect (Silk et al. 2003). Similarly, chacma baboon infants of females that had stronger social bonds with adult females—especially close female kin—survived longer, while rank had no effect (Silk et al.

2009). Finally, at Amboseli, the number of maternal half sisters present was significantly correlated with age at menarche in a multivariate analysis such that the offspring of females with more maternal half sisters reached menarche earlier (Charpentier et al. 2008). In chacma baboons, the presence of close relatives had a more complex effect. The presence of a female's mother or adult sister lowered the age at first reproduction for high and mid-ranking females, but low-ranking females gave birth later when these female kin were present (Cheney et al. 2004). The mechanisms underlying these relationships have not yet been elucidated, but they could include stress reduction and direct benefits such as protection from harassment or improved access to food sources (Silk et al. 2003; chapters 19 and 24, this volume).

Methodological Issues and Future Directions

It is clear from this review that many factors simultaneously influence female reproductive performance, including extrinsic factors, such as ecological conditions and group size, as well as individual attributes of females, such as age, rank, body size, physical condition, and particular social bonds. While it is often possible to make general comparisons between populations and species on the basis of whether particular factors affect components of reproductive performance, it is still difficult to draw general conclusions about differences in the magnitude of these effects. This is partly because studies differ in the questions posed, the sample sizes available, and the statistical analyses employed. While some studies are able to examine factors such as age and rank as continuous numerical variables, others categorize females as young, prime, and old or as high- or low-ranking. When taking account of confounding factors such as group size or habitat quality, some studies divide the data by categories of the confounding variable and perform bivariate statistics on the variables of interest separately within each category; others employ multivariate statistics. Two methodological issues that influence a wide variety of analyses pertinent to this chapter are the treatment of censored data and the treatment of repeated measures.

Several components of reproductive performance involve intervals—for example, the intervals from birth to menarche, between births, or from infant birth to death (infant survival). Most studies have only used completed intervals in their analyses. However, that approach excludes censored intervals, meaning those not yet closed by the event of interest, as in the case of a female with a dependant infant that has not yet given birth again. Such censored intervals often form a considerable percentage of the available data (e.g., Smuts & Nicolson 1989). The exclusive use of completed intervals is likely to create a bias toward short intervals (see Galdikas & Wood 1990). Techniques of survival analysis allow the inclusion of censored intervals and provide a better picture of the distribution of the data. Such techniques have been employed in a few of the studies discussed here (e.g., Smuts & Nicolson 1989; Pusey et al. 1997; Silk et al. 2009) and should be used more generally in the future.

For some components of reproductive performance, the unit of analysis is based on the individual female (e.g., the total number of surviving offspring produced per female), but for many others the unit of analysis is the individual interval or event. In the latter case, individual females may contribute several intervals (repeated measures) to the analysis, such that they are not statistically independent. Techniques exist to address this potential problem, and they have been used increasingly in recent studies (e.g., Robbins et al. 2006, 2007). The study of individual effects can be biologically interesting in itself—as in the example of the Gombe chimpanzees, in which the effect of individual females accounted for much of the variance in interbirth intervals, and in which females who tended toward shorter intervals had higher body mass (Jones et al. 2010).

Long-term studies of primates are immensely challenging to maintain, due not only to the difficulties of obtaining continuous funding, but also to the risks of habitat loss or political unrest in host countries (Clutton-Brock & Sheldon 2010). This makes the data they produce extremely valuable. To answer questions concerning the influence of such factors as dominance rank among species, it will be helpful for future researchers to standardize their methodology and analysis. Reanalysis of older datasets would also be useful.

Summary

Increasingly long-term studies of individual primate species are gradually allowing us to gain a more complete picture of female reproduction and the factors that affect it. Variation in female reproductive performance is often considerable. In general, young females reproduce slowly at first and experience higher infant mortality. They then reach a plateau in reproductive rate before showing a decline in reproductive performance as they age. Dominance rank has pervasive effects on reproductive performance in many species. However, the available data do not clearly support the expectation that these effects are stronger in species with more despotic hierarchies. In some species, reproductive performance is also affected by body weight, health, the presence of allomothers, and the strength of social bonds. Better understanding of the relative influence of these factors on female reproductive performance in different species requires continued long-term study and standardization of methodology and analysis.

References

Abbott, D. H., Saltzman, W., Schultz-Darken, N. J. & Tannenbaum, P. L. 1998. Adaptations to subordinate status in female marmoset monkeys. *Comparative Biochemistry and Physiology, Part C: Pharmacology, Toxicology and Endocrinology*, 119, 261–274.

Alberts, S.C. & Altmann, J. 2003. Matrix models for primate life history analysis. In *Primate Life Histories and Socioecology* (ed. by Kappeler, P. M. & Pereira, M. E.), 66–102. Chicago: University of Chicago Press.

Altmann, J. & Alberts, S. 2003a. Intraspecific variability in fertility and offspring survival in a nonhuman primate: Behavioral control of ecological and social sources. In *Offspring: Human Fertility Behavior in a Biodemographic Perspective*. (ed. by Wachter, K.), 140–169. Washington: National Academy Press.

———. 2003b. Variability in reproductive success viewed from a life-history perspective in baboons. *American Journal of Human Biology*, 15, 401–409.

———. 2005. Growth rates in a wild primate population: Ecological influences and maternal effects. *Behavioral Ecology and Sociobiology*, 57, 490–501.

Altmann, J., Altmann, S. A. & Hausfater, G. 1978. Primate infant's effects on mother's future reproduction. *Science*, 201, 1028–1030.

Altmann, J., Hausfater, G. & Altmann, S. A. 1988. Determinants of reproductive success in savannah baboons, *Papio cynocephalus*. In *Reproductive Success* (ed. by Clutton-Brock, T. H.), 403–418. Chicago: University of Chicago Press.

Altmann, J., Sapolsky, R, & Licht, P. 1995. Baboon fertility and social status. *Nature*, 377, 688–689.

Anderson, C. 1986. Female age: Male preference and reproductive success. *International Journal of Primatology*, 7, 305–326.

Arnold, S. J. & Wade, M. J. 1984a. On the measurement of natural and sexual selection. *Evolution*, 38, 709–719.

———. 1984b. On the measurement of natural and sexual selection: Applications. *Evolution*, 38, 720–734.

Bales, K., Dietz, J., Baker, A., Miller, K. & Tardif, S. D. 2000. Effects of allocare-givers on fitness in infants and parents in callitrhichid primates. *Folia primatologica*, 71, 127–138.

Bardi, M., Petto, A. J. & Lee-Parritz, D. E. 2001. Parental failure in captive cotton-top tamarins (*Saguinus oedipus*). *American Journal of Primatology*, 54, 159–169.

Bateman, A. J. 1948. Intra-sexual selection in *Drosophila*. *Heredity*, 2, 249–268.

Beauplet, G. & Guinet, C. 2007. Phenotypic determinants of individual fitness in female fur seals: Larger is better. *Proceedings of the Royal Society B: Biological Sciences*, 274, 1877–1883.

Beehner, J. C., Onderdonk, D. A., Alberts, S. C. & Altmann, J. 2006. The ecology of conception and pregnancy failure in wild baboons. *Behavioral Ecology*, 17, 741–750.

Bercovitch, F. B. & Berard, J. D. 1993. Life history costs and consequences of rapid reproductive maturation in female rhesus macaques. *Behavioral Ecology and Sociobiology*, 32, 103–109.

Bercovitch, F. B. & Strum, S. C. 1993. Dominance rank, resource availability, and reproductive maturation in female savanna baboons. *Behavioral Ecology and Sociobiology*, 33, 313–318.

Bercovitch, F. B., Lebron, M. R., Martinez, H. S. & Kessler, M. J. 1998. Primigravidity, body weight, and costs of rearing first offspring in rhesus macaques. *American Journal of Primatology*, 46, 135–144.

Berube, C. H., Fest-Bianchet, M. & Jorgenson, J. T. 1999. Individual differences, longevity, and reproductive senescence in bighorn ewes. *Ecology*, 80, 2555–2565.

Bezerra, B. M., Souto, A. D. & Schiel, N. 2007. Infanticide and cannibalism in a free-ranging plurally breeding group of common marmosets (*Callithrix jacchus*). *American Journal of Primatology*, 69, 1–8.

Bloomquist, G. E. 2009. Environmental and genetic causes of maturational differences among rhesus macaque matrilines. *Behavioral Ecology and Sociobiology*, 63, 1345–1352.

Borries, C., Sommer, V. & Srivastava, A. 1991. Dominance, age and reproductive success in free-ranging female hanuman langurs (*Presbytis entellus*). *International Journal of Primatology*, 12, 231–257.

Borries, C., Larney, E., Lu, A., Ossi, K. & Koenig, A. 2008. Costs of group size: Lower developmental and reproductive rates in larger groups of leaf monkeys. *Behavioral Ecology*, 19, 1186–1191.

Bronikowski, A. M., Alberts, S. C., Altmann, J., Packer, C., Carey, K. D. & Tatar, M. 2002. The aging baboon: Comparative demography in a nonhuman primate. *Proceedings of the National Academy of Sciences of the United States of America*, 99, 9591–9595

Brown, D. 1988. Components of lifetime reproductive success In *Reproductive Success* (ed. by Clutton-Brock, T. H.). 439–453. Chicago: University of Chicago Press.

Charpentier, M. J. E., Tung, J., Altmann, J. & Alberts, S. C. 2008. Age at maturity in wild baboons: Genetic, environmental and demographic influences. *Molecular Ecology*, 17, 2026–2040.

Cheney, D. L., Seyfarth, R. M., Andelman, S. J. & Lee, P. C. 1988. Reproductive success in vervet monkeys. In *Reproductive Success* (ed. by Clutton-Brock, T. H.), 384–402. Chicago: University of Chicago Press.

Cheney, D. L., Seyfarth, R. M., Fischer, J., Beehner, J., T., B., Johnson, S. E., Kitchen, D. M., Palombit, R. A., Rendall, D. & Silk, J. B. 2004. Factors affecting reproduction and mortality among baboons in the Okavango Delta, Botswana. *International Journal of Primatology*, 25, 401–428.

Clarke, M. R. & Glander, K. E. 1984. Female reproductive success in a group of free-ranging howling monkeys (*Alouatta palliata*) in Costa Rica. In *Female Primates: Studies by Women Primatologists*, 111–126. New York: Alan R. Liss, Inc.

Clutton-Brock, T. H. 1984. Reproductive effort and terminal investment in iteroparous animals. *American Naturalist*, 123, 212–229.

———. 1988a. Reproductive Success. In *Reproductive Success* (ed. by Clutton-Brock, T. H.). 472–485. Chicago: University of Chicago Press.

———. 1988b. *Reproductive Success*. Chicago: University of Chicago Press.

———. 2007. Sexual selection in males and females. *Science*, 318, 1882–1885.

Clutton-Brock, T. H. & Sheldon, B. C. 2010. The seven ages of *Pan*. *Science*, 327, 1207–1208.

Clutton-Brock, T. H., Brotherton, P. N. M., Russell, A. F., O'Riain, M. J., Gaynor, D., Kansky, R., Griffin, A., Manser, M., Sharpe, L., McIlrath, G. M., Small, T., Moss, A. & Monfort, S. 2001. Cooperation, control, and concession in meerkat groups. *Science*, 291, 478–481.

Cords, M. (2002). Friendship among adult female blue monkeys (*Cercopithecus mitis*). *Behaviour*, 139, 291–314.

Cords, M. & Chowdhury, S. 2010. Life history of *Cercopithecus mitis stuhlmanni* in the Kakamega Forest, Kenya. *International Journal of Primatology*, 31, 433–455.

Crockett, C. M. & Pope, T. R. 1988. Inferring patterns of aggression from red howler monkey injuries. *American Journal of Primatology*, 14, 1–21.

Dietz, J. M. & Baker, A. 1993. Polygyny and female reproductive success in golden lion tamarins, *Leontopithecus rosalia*. *Animal Behaviour*, 46, 1067–1078.

Digby, L. & Saltzman, W. 2009. Balancing cooperation and competition in callitrichine primates: Examining the relative risk of infanticide across species. In *The Smallest Anthropoids: The Callimico/Marmoset Radiation* (ed. by Davis, L. E., Ford, S. M. & Porter L.), 135–154. New York: Springer.

Drickamer, L. C. 1974. A ten-year summary of reproductive data for free-ranging *Macaca mulatta*. *Folia Primatologica*, 21, 61–80.

Emery Thompson, M., Jones, J. H., Pusey, A. E., Brewer-Marsden, S., Goodall, J., Marsden, D., Matsuzawa, T., Nishida, T., Reynolds, V., Sugiyama, Y. & Wrangham, R. W. 2007. Aging and fertility patterns in wild chimpanzees provide insights into the evolution of menopause. *Current Biology*, 17, 2150–2156.

Erhart, E. M. & Overdorff, D. J. 2008. Rates of agonism by diurnal lemuroids: Implications for female social relationships. *International Journal of Primatology*, 29, 1227–1247.

Fairbanks, L. A. 1990. Reciprocal benefits of allomothering for female vervet monkeys. *Animal Behaviour*, 40, 553–562.

Fairbanks, L. A. & McGuire, M. T. 1984. Determinants of fecundity and reproductive success in captive vervet monkeys. *American Journal of Primatology*, 7, 27–38.

Fedigan, L. M. 1983. Dominance and reproductive success in primates. *Yearbook of Physical Anthropology*, 26, 91–129.

Fedigan, L., Fedigan, L., Gouzoules, S., Gouzoules, H. & Koyama, N. 1986. Lifetime reproductive success in female Japanese macaques. *Folia primatologica*, 47, 143–157.

Fedigan, L. M., Carnegie, S. D. & Jack, K. M. 2008. Predictors of reproductive success in female white-faced capuchins (*Cebus capucinus*). *American Journal of Physical Anthropology*, 137, 82–90.

Fessler, D. M. T., Navarette, C. D., Hopkins, W. & Izard, M. K. 2005. Examining the terminal investment hypothesis in humans and chimpanzees: Associations among maternal age, parity and birth weight. *American Journal of Physical Anthropology*, 127, 95–104.

Garber, P. A., Encarnacion, F., Moya, L. & Pruetz, J. D. 1993. Demographic and reproductive patterns in moustached tamarin monkeys (*Saguinus mystax*): Implications for reconstructing platyrrhine mating systems. *American Journal of Primatology* 29, 235–254.

Garcia, C., Lee, P.C. & Rosetta, L. 2006. Dominance and reproductive rates in captive female olive baboons, *Papio anubis*. *American Journal of Physical Anthropology* 131, 64–72.

Glander, K. E. 1980. Reproduction and population growth in free-ranging mantled howling monkeys. *American Journal of Physical Anthropology*, 53, 23–36.

Goldizen, A. W., Mendelson, J., Van Vlaardingen, M. & Terborgh, J. 1996. Saddle-back tamarin (*Saguinus fuscicollis*) reproductive strategies: Evidence from a thirteen-year study of a marked population. *American Journal of Primatology* 38, 57–83.

Gomendio, M. 1989. Differences in fertility and suckling patterns between primiparous and multiparous rhesus mothers (*Macaca mulatta*). *Journal of Reproduction and Fertility*, 87, 529–542.

Goodall, J. 1977. Infant killing and cannibalism in free-living chimpanzees. *Folia Primatologica*, 28, 259–282.

———. 1983. Population dynamics during a 15 year period in one community of free-living chimpanzees in the Gombe National Park, Tanzania. *Zeitschrift für Tierpsychologie*, 61, 1–60.

Gould, L., Sussman, R. W. & Sauther, L. 2003. Demographic and life-history patterns in a population of ring-tailed lemurs (*Lemur catta*) at Beza Mahafaly Reserve, Madagascar: A 15-year perspective. *American Journal of Physical Anthropology*, 120, 182–194.

Gouzoules, H., Gouzoules, S. & Fedigan, L. 1982. Behavioral dominance and reproductive success in female Japanese monkeys (*Macaca fuscata*). *Animal Behaviour*, 30, 1138–1150.

Grafen, A. 1988. On the uses of data on lifetime reproductive success. In *Reproductive Success* (ed. by Clutton-Brock, T. H.). 454–471. Chicago: University of Chicago.

Harcourt, A. H. 1987. Dominance and fertility among female primates. *Journal of Zoology, London*, 213, 471–487.

Hill, R. A. & Lee, P. C. 1998. Predation risk as an influence on group size in cercopithecoid primates: Implications for social structure. *Journal of Zoology, London.*, 245, 447–456.

Hiraiwa, M. 1981. Maternal and alloparental care in a troop of free-ranging Japanese macaques. *Primates*, 22, 309–329.

Holekamp, K. E., Smale, L. & Szykman, J. 1996. Rank and reproduction in the female spotted hyaena. *Journal of Reproduction and Fertility*, 108, 229–237.

Isbell, L. A. & Young, T. P. 2002. Ecological models of female social relationships in primates: Similarities, disparities, and some directions for future clarity. *Behaviour*, 139, 177–202.

Itoigawa, N., T., T., Ukai, N., Fujii, H., Kurokawa, T., Koyama, T., Ando, A., Watanabe, Y. & Imakawa, S. 1992. Demography and reproductive parameters of a free-ranging group of Japanese macaques (*Macaca fuscata*) at Katsuyama. *Primates*, 33, 49–68.

Johnson, S. E. 2003. Life history and the competitive environment: Trajectories of growth, maturation, and reproductive output among Chacma baboons. *American Journal of Physical Anthropology*, 120, 83–98.

———. 2006. Maternal characteristics and offspring growth in chacma baboons: A life history perspective. In *Reproduction and Fitness in Baboons*. (ed. by Swedell, L. & Leigh, S. R.). 177–197. New York: Springer.

Jones, J. H., Wilson, M. L., Murray, C. M. & Pusey, A. E. 2010. Phenotypic quality influences fertility in Gombe chimpanzees. *Journal of Animal Ecology*, 79, 1262–1269.

Kawai, M. 1958. On the system of social ranks in a natural group of Japanese monkeys. *Primates*, 1, 11–48.

Keele, B. F., Jones, J. H., Terio, K. A., Estes, J. D., Rudicell, R. S., Wilson, M. L., Li, Y. Y., Learn, G. H., Beasley, T. M., Schumacher-Stankey, J., Wroblewski, E., Mosser, A., Raphael, J., Kamenya, S., Lonsdorf, E. V., Travis, D. A., Mlengeya, T., Kinsel, M. J., Else, J. G., Silvestri, G., Goodall, J., Sharp, P. M., Shaw, G. M., Pusey, A. E. & Hahn, B. H. 2009. Increased mortality and AIDS-like immunopathology in wild chimpanzees infected with SIVcpz. *Nature*, 460, 515–519.

King, S. J., Arrigo-Nelson, S. J., Pochron, S. T., Semprebon, G. M., Godfrey, L. R., Wright, P. C. & Jernvall, J. 2005. Dental senescence in a long-lived primate links infant survival to rainfall. *Proceedings of the National Academy of Sciences USA*, 102, 16579–16583.

Knott, C. 2001. Female reproductive ecology of the apes. In *Reproductive Ecology and Human Evolution*. (ed. by Ellison, P. T.), 429–463. New York: Aldine de Gruyter.

Koyama, N., Nakamichi, M., Oda, R., Miyamoto, N., Ichino, S. & Takahata, Y. 2001. A ten-year summary of reproductive parameters for ring-tailed lemurs at Berenty, Madagascar. *Primates*, 42, 1–14.

Koyama, N., Takahata, Y., Huffman, M. A. & Norikoshi, K. 1992. Reproductive parameters of female Japanese macques: Thirty years of data from the Arashiyama troops, Japan. *Primates*, 33, 33–47.

Kummerli, R. & Martin, R. D. 2005. Male and female reproductive success in *Macaca sylvanus* in Gibraltar: No evidence for rank dependence. *International Journal of Primatology*, 26, 1229–1249.

Lee, P.C. 1999. Comparative ecology of postnatal growth and weaning among haplorhine primates. In *Comparative Primate Socioecology* (ed. by Lee, P. C.) 111–136. Cambridge: Cambridge University Press.

Lee, P. C. & Kappeler, P. M. 2003. Socioecological correlates of phenotypic plasticity of primate life histories. In *Primate Life Histories and Socioecology* (ed. by Kappeler, P. M. & Pereira, M. E.), 41–65. Chicago: University of Chicago Press.

Manson, J. H., Rose, L., Perry, S. & Gros-Louis, J. 1999. Dynamics of female-female relationships in wild *Cebus capucinus*: Data from two Costa Rican sites. *International Journal of Primatology*, 20, 679–706.

Menard, N. & Vallet, D. 1996. Demography and ecology of Barbary macaques (*Macaca sylvanus*) in two different habitats. In *Evolution and Ecology of Macaque Societies* (ed. by Fa, J. E. & Lindburg, D. G.). 106–131. Cambridge: Cambridge University Press.

Mitani, J. & Watts, D. P. 1997. The evolution of non-maternal caretaking among anthropoid primates: Do helpers help? *Behavioral Ecology and Sociobiology*, 40, 213–220.

Mori, A., Yamaguchi, N., Watanabe, K. & Keiko, S. 1997. Sexual maturation of female Japanese macques under poor nutritional conditions and food enhance perineal swelling in the Koshima troop. *International Journal of Primatology*, 18, 553–579.

Murray, C. M., Eberly, L. E. & Pusey, A. E. 2006. Foraging strategies as a function of season and rank among wild female chimpanzees (*Pan troglodytes*). *Behavioral Ecology*, 17, 1020–1028.

Murray, C. M., Mane, S. V. & Pusey, A. E. 2007. Dominance rank influences female space use in wild chimpanzees, *Pan troglodytes*: Towards an ideal despotic distribution. *Animal Behaviour*, 74, 1795–1804.

Nishida, T. 1989. Social interactions between resident and immigrant female chimpanzees. In *Understanding Chimpanzees* (ed. by Heltne, P. G. & Marquardt L.A.), 68–89. Cambridge, MA: Harvard University Press.

Packer, C. 1995. Baboon fertility and social status: Reply. *Nature*, 377, 689.

Packer, C. R., Collins, D. A., Sindimwo, A. & Goodall, J. 1995. Reproductive constraints on aggressive competition in female baboons. *Nature*, 373, 60–63.

Packer, C. R., Tatar, M. & Collins, D. A. 1998. Reproductive cessation in female mammals. *Nature*, 393, 807–811.

Packer, C. R., Collins, D. A. & Eberly, L. E. 2000. Problems with primate sex ratios. *Philosophical Transactions of the Royal Society of London, Series B*, 355, 1627–1635.

Packer, C. R., Pusey, A. E. & Eberly, L. E. 2001. Egalitarianism in female lions. *Science*, 293, 690–693.

Parga, J. A. & Lessnau, R. G. 2005. Female age-specific reproductive rates, birth seasonality, and infant mortality in ring-tailed lemurs on St. Catherines Island: 17-year reproductive history of a free-ranging colony. *Zoo Biology*, 24, 295–309.

Paul, A. & Kuester, J. 1996. Differential reproduction in male and female Barbary macaques. In *Evolution and Ecology of Macaque Societies* (ed. by Fa, J. E. & Lindburg, D. G.), 293–317. Cambridge: Cambridge University Press.

Paul, A. & Thommen, D. 1984. Timing of birth, female reproductive success and infant sex ratio in semifree ranging Barbary macaques (*Macaca sylvanus*). *Folia primatologica*, 42, 2–16.

Paul, A., Kuester, J. & Podzuweit, D. 1993. Reproductive senescence and terminal investment in female Barbary macaques (*Macaca sylvanus*) at Salem. *International Journal of Primatology*, 14, 105–124.

Pavelka, M. S. M., McGoogan, K. C. & Steffens, T. S. 2007. Population size and characteristics of *Alouatta pigra* before and after a major hurricane. *International Journal of Primatology*, 28, 919–929.

Pereira, M. E. 1993. Seasonal adjustment of growth rate and adult body weight in ringtailed lemurs. In *Lemur Social Systems and their Ecological Basis* (ed. by Kappeler, P. M. & Ganzhorn, J. U.), 205–221. New York: Plenum Press.

Pope, T. R. 2000. Reproductive success increases with degree of kinship in cooperative coalitions of female red howlers (*Alouatta seniculus*). *Behavioral Ecology and Sociobiology*, 48, 253–267.

Pusey, A., Williams, J. M. & Goodall, J. 1997. The influence of dominance rank on the reproductive success of female chimpanzees. *Science*, 277, 828–831.

Rhine, R. J., Norton, G. W. & Wasser, S. K. 2000. Lifetime reproductive success, longevity and reproductive life history of female yellow baboons (*Papio cynocephalus*) of Mikumi National Park, Tanzania. *American Journal of Primatology*, 51, 229–241.

Richard, A. F., Dewar, R. E., Schwartz, M. & Ratsirarson, J. 2000. Mass change, environmental variablity and female fertility in wild *Propithecus verreaxi*. *Journal of Human Evolution*, 39, 381–391.

Richard, A. F., Dewar, R. E., Schwartz, M. & Ratsirarson, J. 2002. Life in the slow lane? Demography and life histories of male and female sifaka (*Propithecus verreauxi verreauxi*). *Journal of Zoology, London.*, 256, 421–436.

Robbins, A. M., Robbins, M. M., Gerald-Steklis, N. & Steklis, H. D. 2006. Age-related patterns of reproductive success among female mountain gorillas. *American Journal of Physical Anthropology*, 131, 511–521.

Robbins, M. M., Robbins, A. M., Gerald-Steklis, N. & Steklis, H. D. 2007. Socioecological influences on the reproductive success of female mountain gorillas (*Gorilla beringei beringei*). *Behavioral Ecology and Sociobiology*, 61, 919–931.

Rogers, C. M. & Davenport, R. K. 1970. Chimpanzee maternal behavior. In *The Chimpanzee* (ed. by Bourne, G. H.), 361–368. Baltimore: University Park Press.

Ross, C. & MacLarnon, A. 2000. The evolution of non-maternal care in anthropoid primates: A test of the hypotheses. *Folia primatologica*, 71, 93–113.

Sade, D. S. 1990. Intrapopulation variation in life history parameters. In *Primate Life History and Evolution* (ed. by DeRousseau, C. J.), 181–194. New York: Wiley-Liss.

Sadlier, R. E. F. S. 1969. *The Ecology of Reproduction in Wild and Domestic Mammals*. London: Methuen.

Saltzman, W. 2003. Reproductive competition among female common marmosets (*Callithrix jacchus*): proximate and ultimate causes. In *Sexual Selection and Reproductive Competition in Primates: New Perspectives and Directions* (ed. by Jones, C. B.), 197–229. Norman, OK: American Society of Primatologists.

Saltzman, W., Digby, L. J. & Abbott, D. H. 2009. Reproductive skew in female common marmosets: What can proximate mechanisms tell us about ultimate causes? *Proceedings of the Royal Society B: Biological Sciences*, 276, 389–399.

Setchell, J. M., Lee, P. C., Wickings, E. J. & Dixson, A. F. 2002. Reproductive parameters and maternal investment in mandrills (*Mandrillus sphinx*). *International Journal of Primatology*, 23, 51–68.

Silk, J. B. 1987. Social behavior in evolutionary perspective. In *Primate Societies* (ed. by Smuts, B. B., Cheney, D. L., Seyfarth, R. M., Wrangham, R. W. & Struhsaker, T. T.), 318–329. Chicago: University of Chicago Press.

———. 1990. Sources of variation in interbirth intervals among captive bonnet macaques (*Macaca radiata*). *American Journal of Physical Anthropology*, 82, 213–230.

Silk, J. B., Alberts, S. C. & Altmann, J. 2003. Social bonds of female baboons enhance infant survival. *Science*, 302, 1231–1234.

Silk, J. B., Beehner, J., Bergman, T. J., Crockford, C., Engh, A. L., Moscovice, L. R., Wittig, R. M., Seyfarth, R. M. & Cheney, D. L. 2009. The benefits of social capital: Close social bonds among female baboons enhance offspring survival. *Proceedings of the Royal Society B: Biological Sciences*, 276, 3099–3104.

Silk, J. B., Beehner, J. C., Bergman, T. J., Crockford, C., Engh, A. L., Moscovice, L. R., Wittig, R. M., Seyfarth, R. M. & Cheney, D. L. 2010. Strong and consistent social bonds enhance the longevity of female baboons. *Current Biology*, 20, 1359–1361.

Smuts, B. B. & Nicolson, N. 1989. Reproduction in wild baboons. *American Journal of Primatology*, 19, 229–246.

Sterck, E. H. M., Watts, D. P. & van Schaik, C. P. 1997. The evolution of female social relationships in nonhuman primates. *Behavioral Ecology and Sociobiology*, 41, 291–309.

Strier, K. B. 1990. New world primates, new frontiers: Insights from the woolly spider monkey, or muriqui (*Brachyteles arachnoides*). *International Journal of Primatology*, 11, 7–19.

Strum, S. C. & Western, J. D. 1982. Variations in fecundity with age and environment in olive baboons (*Papio anubis*). *American Journal of Primatology*, 3, 61–76.

Sugiyama, Y. & Ohsawa, H. 1982. Population dynamics of Japanese monkeys with special reference to the effect of artificial feeding. *Folia Primatologica*, 39, 238–263.

Suzuki, S., Noma, N. & Izawa, K. 1998. Inter-annual variation of reproductive parameters and fruit availability in two populations of Japanese macaques. *Primates*, 39, 313–324.

Takahata, Y., Koyama, N., Ichino, S., Miyamoto, N., Nakamichi, M. & Soma, T. 2008. The relationship between female rank and reproductive parameters of the ringtailed lemur: A preliminary analysis. *Primates*, 49, 135–138.

Takahata, Y., Suzuki, S., Agetsuma, N., Okayasu, N., Sugiura, H., Takahashi, H., Yamagiwa, J., Izawa, K., Furuichi, T., Hill, D., Maruhashi, T., Saito, C., Sato, S., Sprague, D.S. 1998. Reproduction of wild Japanese macaque females of Yakushima and Kinkazan islands: A preliminary report. *Primates*, 39, 339–349.

Trivers, R. L. 1972. Parental investment and sexual selection. In *Sexual Selection and the Descent of Man.* (ed. by Campbell, B.), 136–179. Chicago: Aldine-Atherton.

Tutin, C. E. G. 1994. Reproductive success story: Variability among chimpanzees and comparisons with gorillas. In *Chimpanzee Cultures* (ed. by Wrangham, R. W., McGrew, W. C., de Waal, F. B. M. & Heltne, P. G.), 181–193. Cambridge, MA: Harvard University Press.

Van Noordwijk, M. A. & van Schaik, C. P. 1999. The effects of dominance rank and group size on female lifetime reproductive success in wild long-tailed macaques, *Macaca fascicularis*. *Primates*, 40, 105–130.

Van Schaik, C. P. 1989. The ecology of social relationships amongst female primates. In *Comparative Socioecology* (ed. by Standen, V. & Foley, R. A.), 195–218. Oxford: Blackwell Press.

———. 2000. Infanticide by male primates: The sexual selection hypothesis revisited. In *Infanticide by Males and its Implications* (ed. by van Schaik, C. P. & Janson, C. H.), 27–60. Cambridge: Cambridge University Press.

Vick, L. G. & Pereira, M. E. 1989. Episodic targeted aggression and the life histories of *Lemur* social groups. *Behavioral Ecology and Sociobiology*, 25, 3–12.

Wasser, S. K., Norton, G. W., Rhine, R. J., Klein, N. & Kleindorfer, S. 1998. Ageing and social rank effects on the reproductive system of free-ranging yellow baboons (*Papio cynocephalus*) at Mikumi National Park, Tanzania. *Human Reproduction Update*, 4, 430–438.

Wasser, S. K., Norton, G. W., Kleindorfer, S. & Rhine, R. J. 2004. Population trend alters the effects of maternal dominance rank on lifetime reproductive success in yellow baboons (*Papio cynocephalus*). *Behavioral Ecology and Sociobiology*, 56, 338–345.

Watanabe, K., Mori, A. & Kawai, M. 1992. Characteristic features of the reproduction of Koshima monkeys, *Macaca fuscata fuscata*: A summary of thirty-four years of observation. *Primates*, 33, 1–32.

Whitten, P. L. 1983. Diet and dominance among female vervet

monkeys (*Cercopithecus aethiops*). *American Journal of Primatology*, 5, 139–159.

Williams, G. C. 1966. *Adaptation and Natural Selection.* Princeton, NJ: Princeton University Press.

Williams, J. M., Oehlert, G., Carlis, J. V. & Pusey, A. E. 2004. Why do male chimpanzees defend a group range? *Animal Behaviour*, 68, 523–532.

Wilson, M. E., Gordon, T. P. & Bernstein, I. S. 1978. Timing of births and reproductive success in rhesus monkey social groups. *Journal of Medical Primatology*, 7, 202–212.

Wilson, M. E., Walker, M. L. & Gordon, T. P. 1983. Consequences of first pregnancy in rhesus monkeys. *American Journal of Physical Anthropology*, 61, 103–110.

Wittig, R. M. & Boesch, C. 2003. Food competition and linear dominance hierarchy among female chimpanzees of the Tai National Park. *International Journal of Primatology*, 24, 847–867.

Wrangham, R. W. 1980. An ecological model of female-bonded primate groups. *Behaviour*, 75, 262–299.

Chapter 16 Mate Choice

Peter M. Kappeler

History and Theory

Because potential mates differ in quality and because matings impose costs, primates, like all other animals, should be choosy about whom to mate with, rather than mate randomly. Mate choice can be defined as "the outcome of the inherent propensity of an individual to mate more readily with certain phenotypes of the opposite sex (i.e., a mating preference or bias) and the extent to which an individual engages in mate sampling before deciding to mate (i.e., choosiness)" (Kokko et al. 2006). Mate choice has been studied primarily with the goal of identifying and understanding the basis of preferences for members of the opposite sex, with an historical emphasis on female preferences (Andersson 1994; Ryan 1997). Charles Darwin (1859, 1871) was initially inspired to develop the theoretical foundations for research on mate choice by attempting to reconcile the existence of elaborate secondary sexual traits, such as weapons or colorful body parts in males, with his theory of evolution by natural selection. He suggested that these extravagant traits (or ornaments) exist despite their detrimental effects on survival because they improve success in combat with rivals or increase males' attractiveness to females who happen to prefer a particular ornament. Of particular interest to primatologists, in 1876 Darwin published a paper in *Nature*, entitled "Sexual selection in relation to monkeys," in which he focused explicitly on that subject, but neither his specific interest in primate sexual skins nor his more general theory on mate choice generated much initial interest for empirical studies for at least two reasons. First, Darwin's theory of mate choice lacked a convincing mechanism, as he ascribed a true aesthetic sense to females as the basis of their choice. Second, Alfred Wallace (1891) published an influential rebuttal to Darwin's theory, in which he argued that sexual selection was at most a process secondary to natural selection because, in his view, mate choice was based on traits favored by natural selection.

Important theoretical work in the 20th century eventually generated sufficient interest in mate choice to make it one of the major topics in current behavioral ecology (summarized in Andersson 1994; Danchin & Cézilly 2008). Ronald Fisher (1930) was the first to demonstrate that female mate preferences can indeed evolve. Accordingly, sons of males exhibiting a trait favored by a corresponding female preference enjoy a relative mating advantage: an idea that came to be known as the "sexy son" hypothesis (Weatherhead & Robertson 1979). The notion that male traits and female preferences for them can invade and coevolve in a population through a runaway process, once they are genetically coupled, was later adopted and formally demonstrated by Peter O'Donald (1980) and Russell Lande (1981). This mechanism is based on the assumption that female preferences are essentially arbitrary with respect to the nature of the male trait they focus on, although the focal trait may have been initially advantageous to males. Amotz Zahavi (1975, 1977) proposed more specifically that female preferences evolved for male traits that indicate vigor and good health. Under this type of selection, these male traits become exaggerated and eventually constitute a handicap for survival, thereby providing females with honest information about the bearer's quality. Zahavi argued that females with a preference for male traits that act as handicaps produce offspring with higher than average viability. William D. Hamilton and Marlene Zuk (1982) provided the most influential application of this idea by proposing that male ornaments reliably reveal current health status by reflecting genetic ability to resist parasites and other pathogens. These two processes outlined by Fisher and Zahavi should not be seen as competing alternative hypotheses, but rather as part of a continuum of outcomes of a unitary evolutionary process (Kokko et al. 2002). The recently introduced term "Fisher-Zahavi process" emphasizes the fact that indirect female benefits (see below) can include both improved viability and greater attractiveness of offspring (Kokko & Jennions 2008). With these theoretical developments in the 1970s and 1980s, the study of mate choice became established as a research topic that would dominate sexual selection studies for decades to come. Today the study of mate choice is a highly dynamic field of research with many questions requiring additional theory and new data (Kokko et al. 2006; Jones & Ratterman 2009), but firmly grounded in a theoretical framework that guides and stimulates empirical work in a diversity of taxa (Andersson 1994; Arnqvist & Rowe 2005).

The impetus for studying mate choice in primates, in particular, was largely a paradigm shift in the study of sexual behavior initiated by several female primatologists in the early 1980s. Influential books by Sarah Blaffer Hrdy (1981) and Linda Fedigan (1982) paved the way for new thinking about sex roles in primates (cf. Crook & Gartlan 1966) and helped to generate additional important early writings (Small 1984, 1993; Smuts 1985). The first set of mate choice studies in primates was largely descriptive and was summarized by Small (1989) and Keddy-Hector (1992). Since then, methodological progress, especially in molecular genetics, has facilitated more deductive studies by permitting paternity tests and the quantification of mate quality, for example. Moreover, extensions of the underlying theory have broadened the spectrum of interrelated problems to include questions about male choice and sexual coercion. Finally, studies on mate choice in insects and birds have identified additional mechanisms of mating bias, such as sperm competition and cryptic female choice, and have inspired primatologists to initiate complementary research projects. The resulting increase in information from a growing number of field studies as well as from controlled laboratory experiments resulted in the publication of additional reviews (Paul 2002; Setchell & Kappeler 2003; Manson 2007) and books (e.g., Jones 2003; Kappeler & van Schaik 2004; Dixson 2009; Muller & Wrangham 2009) on sexual selection in primates at ever shorter intervals.

In this chapter, I will not attempt an exhaustive summary; instead I will highlight current concepts and recent primate studies, focusing on the mechanisms and functions of mate choice in both sexes. In particular, I will first explore the general operational scope for mate choice in primates on the basis of fundamental features of their life histories and social systems before focusing on a functionally related but often neglected aspect of mate choice: how to avoid matings with heterospecifics and relatives. In the next section I will summarize theory and evidence related to the potential direct and indirect benefits, as well as the costs, of mate choice before I turn to its underlying behavioral and physiological mechanisms. In the final section I will abandon the paradigm of classical sex roles and review the evidence for the effects of female competition and male mate choice in generating mating bias.

Mate Choice in Primates: How Much Scope Is There?

Why Mate Choice in Primates May Be Rare

Before delving into the details, it is worth considering several theoretical points about primate life histories and social systems that raise doubts about the existence or importance of mate choice in primates. First of all, primate females are notoriously polyandrous. In the majority of species, females copulate repeatedly with several if not all of the males of their group or social neighborhood during a single conceptive cycle. In chimpanzees (*Pan troglodytes*), for example, hundreds of ejaculations from all group males typically precede a single conception (Wrangham 1993), which is hardly indicative of any form of choice (cf. Stumpf & Boesch 2005). Such female polyandry is commonly attributed to the females' dilemma of confusing paternity in view of the high infanticide risk faced by mammals with slow reproductive rates (chapter 19, this volume). Since primates typically exhibit relatively slow reproductive rates, related life-history constraints may severely restrict females' options for choice because of a premium on mating with multiple males.

Second, compared to many birds and fish, primate males generally lack conspicuous visual ornaments. To be sure, males in some taxa (e.g., *Eulemur, Cacajao, Pithecia, Alouatta, Mandrillus, Theropithecus, Hylobates, Pongo*) exhibit colorful ornaments that distinguish them from females (Dixson 1998; Bradley & Mundy 2008), but these taxa clearly represent a minority. The most striking morphological differences between the sexes concern the larger body and canine size of males in most species (chapter 17, this volume). Although size may also act as a visual ornament (Andersson

& Iwasa 1996), such sexual dimorphism clearly seems functionally linked to male combat. Given the ubiquity of male-male contest and the physical inferiority of females, there may not be much room left for females to exercise any choice among consorts. At the very least, the interaction between male-male competition and female choice is difficult to predict (Hunt et al. 2009). This is partly because females may promote the conditions for male-male competition instead of choosing mates directly, thereby increasing the chances that a successful competitor will also be a high-quality sire, a process designated as *indirect* mate choice (Wiley & Poston 1996).

Third, a sizeable share of primates lives in pairs or one-male social units, where members of both sexes theoretically have no choice about with whom to mate, as long as their sexual activities are confined to their social units (Setchell & Kappeler 2003). Thus, if they are trapped in a harem or stuck with just one partner, demographic constraints may limit females' ability to exercise mate choice beyond this initial choice of a social mate.

Finally, because of the physiological requirements of gestation and lactation, primate males have limited opportunities to provide extensive, direct paternal care and therefore are selected to desert their mates after fertilization in favor of pursuing copulations with other mates (Williams 1966). This general strategy to maximize reproductive success may sometimes conflict with female interests. Because their reproductive investment is limited simply to an ejaculate in the majority of species, primate males may have little incentive to be particularly choosy about their sexual partners.

Why Mate Choice in Primates May Be Common

There may be constraints ostensibly limiting the exercise of mate choice in females and males, but there are equally valid reasons to expect it to be common and widespread. First, confusing paternity is not totally incompatible with increasing its probability for particular males. In promiscuous species, females can form temporary consortships or long-term friendships, develop sexual swellings, or produce olfactory signals of imminent ovulation, all of which can skew paternity probabilities to particular sires in spite of promiscuous mating behavior (Nunn 1999; chapter 19, this volume).

Second, most female primates may be smaller than males, but even the biggest males require a minimum of female co-operation for successful mating. Thus, mate choice not only involves selection of a particular mate, but can also be exercised by rejecting unwelcome suitors. Just walking away or not lifting the tail are simple yet effective mechanisms in this context. Accordingly, forced copulations among primates

are rare and have been observed at nontrivial rates only in orangutans (*Pongo* spp., Fox 2002; chapter 6, this volume), in which females actually mediate the risk of assertive matings with nonprime males by associating preferentially with prime flanged males around the time of ovulation (Knott et al. 2010; chapter 17, this volume).

Third, living with one male or in a harem does not necessarily imply a commitment for life. In several pair-living species, extrapair copulations (Palombit 1994a) or paternities (Schülke et al. 2004) have been detected, and mate switching is an option for most species (e.g., Palombit 1994b; Brockelman et al. 1998; Fernandez-Duque & Huntington 2002). Likewise, female dispersal is found in most harem-living species (e.g., Steenbeek et al. 2000; Robbins et al. 2009). Moreover, there may be no conflict of interest between the sexes if males with particular qualities end up defending a harem, because females should prefer exactly these males in order to increase the chances that their sons will also become successful harem holders.

Fourth, females should be selective for two additional selfish reasons. If they mate with either a member of another species or with a close relative, they will have to pay the costs of the reproductive failure stemming from the deleterious consequences of hybridization or inbreeding.

Finally, matings can also be costly for males, which potentially increases the advantages of choosiness in several ways. First, while the production of a single sperm or ejaculate may be physiologically cheap, their constant production and delivery may induce a trade-off in males that favors selectivity (Dixson & Anderson 2004). Moreover, the maximizing of mating opportunities typically intensifies competition with male rivals, and thus the attendant risks of injury or death (Promislow 1992). Even in conditions where scramble competition predominates among males, increases in male mortality rates can result from heightened competition (Kraus et al. 2008). Alternative mating strategies—including enhanced mate choosiness—may have evolved among males in order to reduce these costs and risks (chapter 17, this volume).

Thus, considering only the most general aspects of primate biology, a number of interesting constraints, conflicts of interests, dilemmas, and trade-offs become apparent. In the following sections, I will examine in more detail how females and males deal with these problems.

The Other Side of the Coin: Mate Avoidance

Mate choice plays an important albeit often overlooked role in discriminating *against* potential mates. Because the disproportionately high costs of reproduction reduce the reproductive rates of females, each offspring represents a large share of their total lifetime reproductive success. Consequently, females should be particularly careful to avoid two types of mistakes: mating with a male from a different species and mating with a close relative. In both cases, genetic and/or somatic handicaps result in lost reproductive effort or compromised offspring viability (Coyne & Orr 1998; Keller & Waller 2002), both of which reduce female fitness. This aspect of mate choice has been well captured in a definition by Wiley and Poston (1996): "Mate choice includes all behavior by an individual that restricts membership in its set of potential mates."

Species Recognition

Mating with a member of another species may seem paradoxical at first glance, because species are commonly defined as closed reproductive units, but members of closely related species are often able to produce viable, and more rarely fertile, offspring. In some cases the resulting hybrids may give rise to distinct populations or even new species. East African baboons (*Papio* spp.) provide the best known primate examples of hybridization resulting in animals with intermediate morphological and behavioral traits and, in some cases, new taxa (Sugawara 1988; Phillips-Conroy et al. 1991; Zinner et al. 2009). Typically, however, heterospecific mating leads to developmental failure and termination of the reproductive event at the female's expense. Reinforcement—that is, selection for prezygotic isolation among sympatric species—is therefore expected (Marshall et al. 2002), and female precopulatory mate choice is its most effective mechanism. In primates, the selective pressure on species identification ought to be strongest in solitary species with sympatric congeners, such as bush babies (Galagidae) or mouse lemurs (*Microcebus* spp.), but it should also be pronounced in species that are regular members of mixed-species associations, such as tamarins (*Saguinus* spp.) and guenons (*Cercopithecus* spp.). It should be much weaker in group-living species with female philopatry and in species without sympatric congeners.

Whether and how female primates identify members of their own species has not been explicitly studied, which limits tests of the above prediction for the time being. Circumstantial evidence suggests, however, that species recognition is widespread and is based on stimuli in different modalities. For example, several macaque (*Macaca* spp.) species, including members of the Sulawesi radiation, exhibit a preference for pictures of their own species when they are given the opportunity to press a lever to watch pictures of different species (Fujita 1987; Fujita et al. 1997). Similar studies with macaques and chimpanzees raised under different captive conditions revealed the importance of early social experience in forming such preferences, which thus are unlikely to be innate (e.g., Tanaka 2007). Visual self-recognition tests provide another method for making inferences about potential internal templates for species recognition, but so far only the great apes have passed such tests (Suddendorf & Collier-Baker 2009; chapter 30, this volume), thus suggesting that this ability might be limited to a minority of species. In contrast, discrimination among species based on olfactory cues has been demonstrated experimentally in several species, including brown capuchins (*Cebus apella*, Ueno 1994), bush babies (*Otolemur* spp., Clark 1988) and true lemurs (e.g., brown lemurs, *Eulemur fulvus*, Harrington 1979). Pronounced acoustic differences in species-specific loud calls have also been functionally related to species recognition (e.g., bush babies, Bearder et al. 1995; Bearder 1999), but the necessary playback experiments demonstrating that primates actually discriminate heterospecific calls have only rarely been conducted (tarsiers, *Tarsius* spp., Nietsch & Kopp 1998; macaques, Muroyama & Thierry 1998; gibbons, *Hylobates* spp., Raemaekers & Raemaekers 1985; Mitani 1987). Finally, penises in some strepsirrhines possess elaborate, species-specific spines (Anderson 2000) that could play a role in tactile species recognition during copulation.

Thus, the ability to recognize conspecifics is, as expected, widespread among primates, but more experimental tests with receptive females are desirable, not least because such studies could also inform taxonomic controversies and, hence, conservation issues (see, e.g., Tattersall 2007) by asking the primates themselves which individuals they consider to be conspecifics. Once we have learned more about variation in male traits signaling species identity and the strength of potential female preferences for them, it should also be possible to explore the possibility that female mate choice has been a powerful force in primate speciation events (fish, Haesler & Seehausen 2005; butterflies, Chamberlain et al. 2009).

Inbreeding Avoidance

Mating with a close relative can lead to inbreeding depression, which in turn can reduce fitness significantly (Ralls & Ballou 1982; Charpentier et al. 2007; but see, e.g., Smith 1995). A recent exemplary primate study revealed that inbred ring-tailed lemurs (*Lemur catta*) suffer heavier parasite load, compromised immunocompetency, and shorter life span than more outbred conspecifics (Charpentier et al. 2008a). Given these substantial costs, it is not surprising that three behavioral mechanisms have evolved to reduce the risk of inbreeding: sex-biased dispersal (Pusey 1987; Perrin & Mazalov 1999; Lehmann & Perrin 2003), extragroup matings (Isvaran & Clutton-Brock 2007), and kin discrimination (Pusey & Wolf 1996; Mateo 2009). Research on the ability to discriminate between kin and non-kin has identified several possible mechanisms in primates (reviewed in Widdig 2007; see also Penn & Frommen 2010), but this line of research has been motivated primarily by tests of kin selection (Silk 2002; chapter 21, this volume). Evidence that female and male primates discriminate against mating with kin therefore comes primarily from behavioral and genetic studies rather than specifically designed experiments. Such evidence comes from all major primate radiations, including ring-tailed lemurs (Pereira & Weiss 1991), white-faced capuchins (*Cebus capucinus*, Muniz et al. 2006), Thomas langurs (*Presbytis thomasi*, Sterck et al. 2005), western red colobus (*Procolobus badius*, Starin 2001), macaques (Melnick et al. 1984; Paul & Kuester 1985), baboons (Packer 1979; Alberts & Altmann 1995), mandrills (*Mandrillus sphinx*, Charpentier et al. 2005), golden snub-nosed monkeys (*Rhinopithecus roxellana*, Zhao et al. 2008), chimpanzees (Pusey 1980) and gorillas (*Gorilla gorilla*, Bradley et al. 2007).

However, inbred matings offer a theoretical inclusive fitness benefit to both parents through the disproportionate propagation of genes identical by descent (Kokko & Ots 2007). As a result, there might be an optimal balance between inbreeding and outbreeding, but the available evidence from primates indicates that genetic dissimilarity (of major histocompatibilty complex [MHC] genes) is more important than similarity (mandrills, Setchell et al. 2010), and that genetic similarity or dissimilarity does not affect mate choice (dwarf lemurs, Schwensow et al. 2008; baboons, Huchard et al. 2010a). In some species, males sometimes breed in their natal group (e.g., baboons, Alberts & Altmann 1995), indicating that incest avoidance is not absolute and ubiquitous. In light of the available data, however, avoidance of inbreeding appears to be more important to female primates than avoidance of extreme outbreeding.

What Females Want: Benefits and Costs of Mate Choice

Once a receptive female has limited the pool of potential mates by excluding males of other species and close relatives, she should try to choose a conspecific male of the highest possible quality, assuming that the costs of mate searching are low (which should be a realistic assumption for most primates). The quality of a mate can be assessed according to two criteria. Males vary in traits that affect female reproductive success directly or in genetic traits that

contribute to offspring quality. With respect to the latter, male genetic traits can either confer benefits to male offspring alone (Fisherian selection), or to offspring of both sexes (good genes). The prerequisite for an informed choice is the existence of accessible and honest phenotypic indicators of male quality.

Direct Benefits: Mostly Paternal Care

Direct benefits of mate choice include an increase in a choosy individual's fitness that is not based on the genetic quality of its offspring (Møller & Jennions 2001; Kokko et al. 2006). Fitness in this context is usefully viewed as comprising three components: fertility, fecundity, and offspring survival. First, female fertility describes the proportion of eggs that are actually fertilized, or of young actually weaned. In the context of mate choice, mating with infertile males can compromise female fertility. Mutations on the Y chromosome are the major cause of male infertility in primates (Tyler-Smith 2008), but their prevalence is so rare that it is unlikely to affect mate choice in natural populations. Moreover, most primate females mate polyandrously, so that sperm is not limiting, and there are no known phenotypic indicators of this disability. Second, female fecundity refers to the probability of conception, and there is evidence that certain male traits (such as body size or ornaments) or the quality of nuptial gifts are correlated with female fecundity in oviparous species (e.g., Qvarnström et al. 2000; Cunningham & Russell 2000). Although there is variation in fecundity and fertility among female primates—for example, as a function of social status or resource access (Lee 1987; Altmann et al. 1995; chapter 15, this volume)—there is currently no evidence suggesting that part of this variation is due to sire effects. Moreover, because modal litter size in most primates is one (chapter 10, this volume), there is very little scope for variation in fertility within one reproductive event to act selectively. A recent study reported a positive long-term association between male meat sharing with estrous females and copulation frequency in chimpanzees (Gomes & Boesch 2009), but it remains to be demonstrated that direct nutritional benefits actually impact long-term female fecundity because a recent meta-analysis indicated that systematic meat sharing by male chimpanzees with receptive females in exchange for copulations is rare and exceptional (Gilby et al. 2010).

The choice of a particular male could, however, affect female reproductive success directly if males vary in traits that impact offspring survival. Variation among males in the ability or willingness to invest paternal care is the most important factor in this context. Males vary in this respect because paternal care can be very costly (Trillmich 2010).

Direct paternal care—males guarding, nursing, or carrying offspring (Wright 1990)—is most variable among species and it is found among some lemurs (summarized in Patel 2007), several New World primates (Goldizen 1987; Schradin et al. 2003; Rotundo et al. 2005), and siamangs (*Symphalangus syndactylus*, Lappan 2008; chapter 17, this volume). Because individuals in most of these species live in pairs or polyandrous groups, variation in traits related to paternal care could have consequences for female choice. This would require males to provide females with relevant information—for example, through their infant carrying effort as nonreproductive helpers—but there is mixed evidence for this prediction (Price 1990; Tardiff & Bales 1997). Males can also promote infant survival by protecting offspring from infanticidal attacks or nonlethal harassment—for instance, by forming temporary friendships with individual mothers (Palombit et al. 1997; Lemasson et al. 2008; Nguyen et al. 2009; chapter 19, this volume). More indirect forms of paternal care include infant grooming, playing, or agonistic support of juveniles; these were shown to be differentially directed at offspring in yellow baboons (*Papio cynocephalus*, Buchan et al. 2003) and chimpanzees (Lehmann et al. 2007). Males can also promote offspring survival by antipredator vigilance or by contributing to group defense (Soltis 2002), but such positive effects of group augmentation are probably mostly side effects of males' selfish grouping behavior (Clutton-Brock 2002), and it remains untested whether they affect female choice. Another positive effect of males on offspring through yet unknown mechanisms is indicated by the observation in baboons that the extended presence of fathers accelerates their daughter's maturation, which in turn positively affects their lifetime reproductive success (Charpentier et al. 2008b). Finally, females can reap more proximate direct benefits for themselves by preferring healthy males with low parasite load, because this reduces the probability that pathogens will be transmitted to themselves or their offspring (Able 1996; Nunn 2003). Thus, various potential direct female benefits from choosing a particular male exist in primates with different mating systems, but only some of these effects have been explicitly demonstrated, and more research is required to also demonstrate that these benefits actually influence female mating decisions.

Indirect Benefits: Good and Compatible Genes

Whenever females obtain no direct benefits through the choice of a particular male, but only the sperm necessary for fertilization, females can nevertheless optimize mate choice by selecting a male of superior genetic quality. Because the genetic quality of the offspring can affect offspring growth, fecundity, survival, or sexual attractiveness, such benefits

are considered to be indirect (Kokko et al. 2006). Although (theoretical) studies of possible indirect benefits have dominated the mate choice literature, such benefits appear to enjoy relatively little empirical support, compared to other mechanisms of sexual selection (Kotiaho & Puurtinen 2007), and most existing empirical studies of them are piecemeal (Hettyey et al. 2010). Females can evaluate male quality as the presence of genes in suitors that (1) underlie a male ornament that is generally attractive to females in the population, (2) are good in an absolute sense because they confer adaptively advantageous traits to offspring, or (3) in toto best complement the maternal genotype (Kokko et al. 2002; Mays & Hill 2004; Neff & Pitcher 2005).

Ronald Fisher (1930; see above) was the first to provide a hypothetical explanation for the persistence of elaborate male ornaments through female choice. Accordingly, females that prefer attractive males obtain indirect genetic benefits because the genetic basis of male attractiveness, which could be based on a trait that was initially advantageous for males, is passed on to their sons. Because these "sexy sons" will, in turn, be more attractive than the sons of other females who lack a preference for this male ornament, females with such a preference will be rewarded with more grand-offspring. Genetic coupling of male ornament and female preference generates a self-enforcing runaway process, resulting in ever-increasing ornaments and preferences. Evidence that mate choice is based on the Fisher process has been very limited, even in model systems that are more amenable to the necessary experimentation to test its assumptions and predictions (e.g., stalk-eyed flies, David et al. 2000; guppies, Brooks 2000). There is currently no evidence to suggest that this process has operated in primates. Boinski (1987) could not exclude the possibility that a preference of female Central American squirrel monkeys (*Saimiri oerstedii*) for the biggest males may represent a primate example of runaway selection, but there are equally plausible alternative explanations for these phenomena, and, more important, behavioral data alone cannot conclusively test hypotheses about indirect benefits (see Neff & Pitcher 2005).

"Good genes" or "indicator" models assume that ornaments are honest, condition-dependent signals of male quality that females use to choose males who will confer superior genes to their offspring. After Grafen's (1990) formal theoretical demonstration that Zahavi's handicap principle can indeed lead to the production of costly male ornaments that females can use as reliable indicators of male genetic quality, predictions of this hypothesis were tested and confirmed in numerous taxa (Ryan 1997), but only rarely in primates (see below). An important theoretical objection to this notion, the lek paradox (Kirkpatrick &

Ryan 1991), emphasized the fact that the alleles underlying male quality should become quickly fixed in a population, thereby eliminating the grounds for female choice. Hamilton and Zuk (1982) suggested a solution to this paradox with their insight that parasites and other pathogens are in a continuous evolutionary arms race with their hosts, which maintains additive genetic variance of the male traits subject to female choice. This hypothesis shifted the focus to male immunocompetence as an important fitness determinant. The link to male ornaments was provided by Folstad and Karter's (1992) hypothesis that testosterone and other androgens act as a double-edged sword. On the one hand, testosterone promotes the expression of male secondary sexual traits, including ornaments; on the other hand, it reduces the efficiency of the immune system. Thus, it is predicted that males with the genetically most competent immune system are able to pay the costs of these exaggerated ornaments. The available empirical evidence provides little support for this prediction in nonprimates, however (Roberts et al. 2004).

To demonstrate female choice for good genes convincingly, it is necessary to (1) document variation in males' expression of ornaments, (2) confirm a positive correlation between male quality and ornament expression, (3) show that females perceive variation in male ornaments and exhibit a preference for more elaborate versions, and (4) demonstrate that females who choose ornamented males also produce offspring with higher fitness (see Snowdon 2004). Moreover, active choice needs to be clearly defined, in order to separate its effects from passive attraction or male-male competition, and this can only be achieved by demonstrating the associated costs (Kotiaho & Puurtinen 2007). Most of these steps require controlled experiments that are not feasible with most primates (but see e.g., Craul et al. 2004; Nikitopoulos et al. 2005), especially in the wild. It is therefore not surprising that primatologists have contributed very little evidence—and even that evidence mostly partial and indirect—for female choice based on a "good genes" mechanism.

Indicators of Male Quality

What constitutes a male ornament, and do female primates actually base their mating decisions on such a phenotypic feature? In this context it is important to note that ornaments are not restricted to visual signals, but may exist in any modality a female can perceive. Two general ideas exist about the types and prevalence of male ornaments in primates. First, female primates may have a general preference for male ornaments that potentially serve as viability indicators but, unlike weapons and body size, cannot be used to

coerce females (Pradhan & van Schaik 2009). Second, it has also been argued that, in contrast to animals where the sexes only meet for the purpose of mating, most female primates have information about several aspects of male quality from daily interactions with them (Kappeler & van Schaik 2002). Male ornaments may therefore provide biologically significant information for females only in species in which individuals live solitarily or in very large groups. Indirect evidence for this notion can be construed from the fact that orangutans, as the only diurnal solitary primates, possess several striking male secondary sexual characteristics (Delgado & van Schaik 2000). Moreover, mandrills, who live in hordes of hundreds of individuals (Abernethy et al. 2002), are the most strikingly ornamented male primates, and female mandrills direct proceptive behavior preferentially at males with the brightest red facial coloration (Setchell 2005). Similar importance of male ornaments may exist in drills (*Mandrillus leucophaeus*, Astaras et al. 2008), baldheaded uakaris (*Cacajao calvus*, Barnett & Brandon-Jones 1997) and geladas (*Theropithecus gelada*, Dunbar 1983), but sexually dichromatic *Eulemur* do not support this prediction. Nevertheless, female brown lemurs looked longer at pictures of more colorful males in a study examining the potential signaling function of male coloration (Cooper & Hosey 2003). Similarly, female rhesus monkeys (*Macaca mulatta*) spent more time looking at artificially reddened pictures of male faces (Waitt et al. 2003), thus suggesting that visual male ornaments may play an underappreciated function in a wider range of diurnal primates.

Apart from strikingly colored fur or skin, or conspicuous manes or other body parts (Dixson 1998), body size constitutes a possible visual ornament. There is some evidence from observational studies that some female primates prefer larger males (e.g., Central American squirrel monkeys, Boinski 1987; gray mouse lemurs, *Microcebus murinus*, Eberle & Kappeler 2004a). But the primary function of large male body size is seen in male-male competition (chapter 17, this volume), and experimental evidence for such a preference is preliminary (Craul et al. 2004). Furthermore, fluctuating asymmetry in bilateral traits is used by females in numerous other taxa, including humans (Manning 1995; Scheib et al. 1999), as a phenotypic source of visual information about male genetic quality (Watson & Thornhill 1994). The experiments needed to determine whether females perceive and evaluate this trait have been conducted only with rhesus monkeys so far. In that species, females and males spent more time looking at symmetrical images of opposite-sexed conspecifics (Waitt & Little 2006).

It should be emphasized that male ornaments exist in other modalities, and that these may be more important than visual signals, especially to females of nocturnal species. For example, the amplitude and rate of certain vocal signals have been suggested to signal male quality honestly in birds, anurans, and crickets because their production is energetically costly (Chappell et al. 1995), developmentally stressful (Spencer et al. 2003), and positively associated with longevity and reproductive success (Forstmeier et al. 2002; see also Reby & McComb 2003). Various loud calls of male primates have also been argued to advertise male quality (e.g., Thomas's leaf monkey, *Presbytis thomasi*, Wich et al. 2003; review, Delgado 2006). A female preference for male calls with particular acoustic features—as measured through increased affinity, proceptivity, or receptivity—has been demonstrated in only a few species, including orangutans (Fox 2002), gray-cheeked mangabeys (*Lophocebus albigena*, Arlet et al. 2008), and gray mouse lemurs (Craul et al. 2004). For other taxa, such as macaques, the evidence is mixed (reviewed in Manson 2007). In all these species, however, calling behavior is confounded with other variables, such as age or social status, so that more controlled experiments are required to establish the potential importance of male primate vocal ornaments for female choice (Snowdon 2004).

The same is true for variation in male odors, which have the potential to provide honest information about male quality (Penn & Potts 1998), but which have been studied experimentally in only a few species (e.g., pygmy slow loris, *Nycticebus pygmaeus*, Fisher et al. 2003; ring-tailed lemurs, Knapp et al. 2006; Scordato & Drea 2007; Charpentier et al. 2008c, 2010). The latter experiments on ring-tailed lemurs and similar ones with mandrills (Setchell et al. 2011), in particular, established the important basis of honest cues of genetic constitution by quantifying both genetic heterozygosity and signal content. Study of callitrichines can also be expected to clarify the potential function of odors in female and male mate choice (Heymann 2003; Snowdon et al. 2006), but a comparable level of detailed information is not yet available. Thus, across primates there is variation among males in traits that potentially act as signals of male quality. However, evidence that females use this information is available from only a handful of species, and evidence for positive effects of such indirect benefits on offspring fitness is currently nonexistent.

Trivers (1972) first proposed the idea that selection should favor the choice of mates that are compatible in terms of producing advantageous offspring genotypes. This genetic compatibility hypothesis has received serious consideration only in the last decade, however (Mays & Hill 2004). Its biological significance is provided by the potential advantages of genetic heterozygosity (Hansson & Westerberg 2002). Females may prefer males that are genetically dissimilar to themselves, rather than searching or waiting for a male

with absolutely the best genes—so there should be a set of desirable mates for each individual female, rather than a single best male upon which the preferences of all females converge. The most general biological criterion for genetic compatibility might be the major histocompatibility complex (MHC) genotype of a prospective mate. MHC genes play an important role in immune defense by coding for glycoproteins that expose antigens from parasites or other pathogens to "killer" T-lymphoctye cells that destroy them. By increasing the MHC heterozygosity of their offspring through choice of a congruent mate, females can therefore accrue indirect benefits through enhanced offspring fitness (Kalbe et al. 2009).

To date, four primate studies have examined the role of MHC genes in mate choice. In pair-living western fat-tailed dwarf lemurs (*Cheirogaleus medius*), females preferred to both live and mate (extrapair matings are common: Fietz et al. 2000) with males that had higher MHC diversity and lower overlap with female MHC genes than randomly assigned males from the same population (Schwensow et al. 2008a). In solitary gray mouse lemurs, where females mate with up to seven different males during a single receptive period (fig. 16.1; Eberle & Kappeler 2004b), fathers also had a higher number of MHC supertypes (i.e., alleles with similar antigen motifs), a larger proportion of which differed from those of the mothers than did nonfathers (Schwensow et al. 2008b). In chacma baboons (*Papio ursinus*), males prefer to mate with females sporting attractive sexual swellings that are in good condition (Huchard et al. 2009) and carry particular MHC genotypes (Huchard et al. 2010b), but genetic analyses revealed no evidence of mate choice for MHC dissimilarity, diversity, or rare MHC genotypes (Huchard et al. 2010a). In mandrills, the probability that a given male sired offspring rose with increasing MHC dissimilarity but also with overall genetic dissimilarity, so that the mechanism underlying this mating bias remains obscure (Setchell et al. 2010). Thus, because most of these studies revealed an effect of MHC genotype on mating bias or fertilization probability, it is likely that the genetic compatibility hypothesis will receive more widespread attention in future research (see also Setchell & Huchard 2010).

Costs and Constraints

The reproductive interests of females and males diverge in all species, except those practicing strict, long-term monogamy (Trivers 1972). For example, there is sexual conflict over the mating rate, the level of solicitation necessary for female receptivity, the degree of female fidelity, and the level of maternal investment in current offspring (Holland & Rice 1998; Chapman et al. 2003). Over evolutionary time,

Fig. 16.1. Female gray mouse lemurs (*Microcebus murinus*) are only receptive for a few hours per year. They mate with several males and appear to exercise postcopulatory mate choice based on male MHC genotype. Photo courtesy of Manfred Eberle.

attempting to take food from them (van Noordwijk & van Schaik 2009) or by facultatively associating with less coercive males at times of higher conception risk (Knott et al. 2010). Males may impose costs on females by constraining their grouping, feeding, and ranging behavior (e.g., Slater et al. 2008) or by transmitting sexually transmitted diseases (Nunn et al. 2000).

To separate observed mating outcomes as a result of passive female attraction or male-male competition from active female choice, copulations as a result of the latter ought to be associated with a cost (Kotiaho & Puurtinen 2007). Thus, because females cannot increase their current reproductive success by increasing the number of matings beyond those necessary for fertilization (Bateman 1948), female polyandry requires explanation (Zeh & Zeh 2003; Andersson 2005), particularly if we assume that it is the result of active choice. The benefits of mating with multiple males that more than compensate these costs can also be separated into direct and indirect ones and are strongly influenced by life-history traits. In primates and other mammals, potential direct benefits of polyandry for females include additional paternal investment, reduced costs of sexual harassment, and reduced infanticide risk (Clarke et al. 2009). Indirect benefits of polyandry include increased offspring survival in littering species (e.g., Fisher et al. 2006) and facilitation of cryptic female choice (see below). Thus, to better understand this trade-off between the advantages of polyandry and the costs and constraints of mate choice in primates, many more empirical studies are now needed to test the various theoretical predictions or patterns already documented in other taxa.

sexual conflict is therefore expected to generate significant constraints on mate choice. In primates, the risk of infanticide by males that have not mated with a female imposes the greatest constraint on females (Muller & Wrangham 2009), creating for them a dilemma of both confusing and focusing paternity sufficiently for different classes of males (chapter 19, this volume). Various female counterstrategies against infanticide or other forms of sexual coercion have in turn created opportunities for the development of inter- and intraspecific variation in male mating strategies (chapter 17, this volume), but temporal monopolization ("consortships"), intimidation, harassment, and forced copulations are nevertheless common across species (see also Clutton-Brock & Parker 1995). At the proximate level, male aggressive and coercive behavior can induce immediate costs in terms of injuries and even death, particularly in sexually dimorphic species, where males are bigger and equipped with larger canines (Smuts & Smuts 1993). In chimpanzees, male aggression toward fecund females also induces a physiological stress response (Muller et al. 2007). As a potential counterstrategy, female orangutans may have developed a test for a male's tendency toward violence by

Mechanisms of Mate Choice

Because active female choice, passive female attraction to some male trait, and male-male competition without female choice can all lead to identical direct or indirect female benefits, identification of benefits alone can not be used to infer which particular mechanism has generated a mating bias (Kotiaho & Puurtinen 2007). Thus, the benefits and mechanisms of mate choice need to be separated conceptually. For valid practical reasons, primatologists have so far been able to contribute only little towards a general understanding of the criteria underlying mate choice. However, the social behavior of primates has been studied in more detail in more species than in other taxa, except perhaps birds. These observations include many aspects of mating behavior, so that relatively much is known about proceptive and receptive behavior, as well as about mating decisions of primates. It is important to note, however, that female sexual assertiveness is not equivalent to female choice (Small 1989). Moreover, because promiscuity is so widespread among primates, it is important to distinguish between matings and fertilizations; after all, the typical single egg of most primate females can only be fertilized by one male, no matter how many copulations a female performs. It is therefore useful and necessary to distinguish between pre- and postcopulatory mechanisms of female choice (Birkhead & Pizzari 2002).

At the precopulatory stage, females can exercise mate choice through various behavioral mechanisms. In species with female transfer, females can choose which group to join. For example, Thomas's langur females factor male condition, which is a predictor of infanticide risk, into their dispersal decisions (Sterck et al. 2005). Similarly, gorilla females avoid emigration into groups with related silverbacks (Bradley et al. 2007). In species with less mobile females, they can nevertheless visit neighboring groups temporarily or mate with extragroup males (e.g., olive colobus, *Procolobus verus*, Korstjens & Noë 2004; brown howlers, *Alouatta guariba*, de Souza Fialho & Setz 2007). In pair-living species, females can also solicit extrapair copulations from neighboring or roaming males (e.g., western fat-tailed dwarf lemurs, Fietz et al. 2000). At the more proximate behavioral level, females can exercise proceptive choice by actively soliciting copulations from certain males when conception is most likely (Barbary macaques, *Macaca sylvanus*, Brauch et al. 2008; chimpanzees, Stumpf & Boesch 2005; Pieta 2008; Central American black howler monkeys, *Alouatta pigra*, van Belle et al. 2009). Alternatively, females can exhibit proceptive choice by rejecting the mating advances of certain males through refusal to cooperate in ways necessary for successful copulation, or by actively repelling males (gray-cheeked mangabeys, Arlet et al. 2007; long-tailed macaques, *Macaca fascicularis*, Engelhardt et al. 2006; chimpanzees, Stumpf & Boesch 2006). By synchronizing or lengthening their receptive periods, or by concealing ovulation, females can also decrease the monopolization potential of dominant males and hence increase their opportunities for choice (Hanuman langurs, *Semnopithecus entellus*, Heistermann et al. 2001; mandrills, Charpentier et al. 2005).

Whenever precopulatory choice is costly, females have the theoretical ability to bias the outcome of competition among sperm from two or more males. Postcopulatory choice can rely on behavioral or cryptic mechanisms. Female copulation calls, which occur in some catarrhines, have been suggested to function in postcopulatory choice by encouraging postcopulatory mate guarding by preferred males, thereby reducing the risk of sperm competition from other males (Maestripieri & Roney 2005). A study on Guinea baboons (*Papio papio*) supported the predictions of this hypothesis (Maestripieri et al. 2005), and chimpanzee females also uttered more copulation calls while mating with high-ranking males (Townsend et al. 2008), but it remains unknown how females trade this benefit off against the consequent thwarting of paternity. When mate guarding is ineffective, female copulation calls may attract additional males and incite sperm competition instead (Pradhan et al. 2006). In Barbary macaques, females can manipulate the likelihood of ejaculation by calling, and calls emitted during ejaculatory copulations differ in acoustic structure from those emitted in other copulations (Pfefferle et al. 2008a), thereby providing other males with information about mating outcome (Pfefferle et al. 2008b).

Cryptic female choice has been widely demonstrated in insects, but the underlying genetic and physiological mechanisms have remained largely obscure (Primakoff & Myles 2002; Turner et al. 2008). Given widespread promiscuity, cryptic female choice is expected to be common among primates (Birkhead & Kappeler 2004), but it has rarely been subject to direct study (Dixson 2002; Reeder 2003), perhaps because it is difficult to separate experimentally from potential mechanisms of sperm competition and because the necessary invasive experiments are impossible with primates. Four main mechanisms of cryptic female choice in primates are conceivable, however. Females could bias the fertilization probability of sperm from different males by mating with preferred males at times when ovulation is most likely, thus ensuring that sperm from preferred males is available at the right time. This could also be interpreted, however, as a form of precopulatory choice that affects cryptic choice. Females also have the possibility of ejecting or voiding sperm after copulations with nonpreferred males (see Pizzari & Birkhead 2000). Furthermore, changes in vaginal physiology, immunological reactions with sperm, or egg-sperm interactions may influence sperm from different males differentially. Finally, at least some haplorrhine females could use orgasm to selectively accelerate and facilitate sperm transport after the ejaculations of preferred males (Birkhead & Kappeler 2004). In summary, the mechanisms of precopulatory choice have perhaps been studied in more detail in primates than in most other higher taxa, but the postcopulatory mechanisms remain largely obscure.

Modern Sex Roles

So far, this review has followed traditional sex roles in emphasizing the importance of female mate choice and male competition in shaping sex-specific reproductive strategies. This paradigm of eager males who benefit more from repeated matings than the more discriminating females is based on Darwin's initial choice of examples to distinguish the two processes (Clutton-Brock & McAuliffe 2009), Bateman's (1948) influential experiments with fruit flies, and Trivers's (1972) insights into sex-specific parental investment. Recent empirical and theoretical work has shown and emphasized, however, that decisions about "who mates with whom" are also influenced by competition among females and male mate choice (Gowaty 2004; Dewsbury 2005; Drea 2005; Clutton-Brock 2007, 2009). To provide a comprehensive perspective on additional determinants of mating outcome, I briefly touch upon these two processes as well. Other aspects of male mating strategies are addressed in more detail elsewhere (chapters 17 and 19, this volume).

Female Competition

Traditionally, the study of competition among female primates was relegated to the realm of competition for resources necessary for successful reproduction (Sterck et al. 1997; chapter 9, this volume). In group-living species foraging on defendable resources, females commonly compete for social rank and establish dominance relationships or hierarchies, and female rank is often positively associated with reproductive success (chapter 15, this volume). However, in some mammalian species females also compete for access to mates and exhibit secondary sexual traits that are thought to have a function in this context (Clutton-Brock 2009). Among primates, polyandrous callitrichines provide the best known examples of female-female reproductive competition. The dominant females suppress reproduction in subordinates physiologically; they sometimes are larger than males, and they scent-mark more (Heymann 2003; chapter 15, this volume). Competition is also intense among females in group-living lemurs (chapter 2, this volume), as is evidenced by the expulsion of related females from groups that grow beyond a critical size (Vick & Pereira 1989), high androgen levels (von Engelhardt et al. 2000; Drea 2007), male mimicry of young females (Barthold et al. 2009), and masculinized genitalia (Drea & Weil 2008). The lack of sexual size dimorphism in lemurs may also be ultimately related to female competition (fig. 16.2; Kappeler 1990). Females evict each other from their groups or aggressively resist immigration by new females in several other taxa, including ursine howler monkeys (*Alouatta arctoidea*, Pope 2000), golden lion tamarins (*Leontopithecus rosalia*, French & Inglett 1989), and chimpanzees (Pusey et al. 2008). Furthermore, there is variation across female primates in the relative size of their canines that is related to overall levels of competition (Plavcan et al. 1995), thus demonstrating that selection on weapons is sensitive to variation in female competition as well. More directly rele-

Fig. 16.2.   An adult male (*left*) and adult female (*right*) red-fronted lemur (*Eulemur rufifrons*) sit next to one another at Kirindy, Madagascar. In many lemur species, sexual size dimorphism is negligible, in spite of evidence for male-male competition. Possible reasons for size monomorphism include phylogenetic constraints, the importance of agility over physical strength in male-male competition, and female competition and dominance. Photo courtesy of Peter Kappeler.

vant to mating competition, it has been observed in several species—including rhesus and Barbary macaques, patas monkeys (*Erythrocebus patas*), langurs, and gorillas—that females use aggression and harassment to prevent their rivals from mating (Zumpe & Michael 1985; summarized in Setchell & Kappeler 2003). Reproductive competition among females may also occur postnatally—for example, though differential infant handling (Kleindorfer & Wasser 2004) or even infanticide (Digby 1995; Townsend et al. 2007; see fig. 15.5, this volume). Finally, females may compete among each other for access to particular males and, hence, future paternal care (chacma baboons, Palombit et al. 2001). Thus, reproductive competition among females is apparently widespread among primates, relying on various mechanisms and resulting in various levels of reproductive skew.

Male Mate Choice

Males are generally considered to obey the "copulatory imperative" (Ghiselin 1974), trying to maximize their reproductive success and not disdaining any mating opportunities that may offer themselves. This traditional view of the eager male, however, has neglected potential male costs of mating, such as time and energy expended on mate guarding and risk of injury. First, sperm is not unlimited, and even though individual spermatozoa represent a smaller energy investment than an oocyte, sperm are produced and ejaculated by the millions (Dewsbury 1982). It is therefore conceivable that sperm can become limiting when several females are receptive simultaneously, thus creating for males a trade-off between a current and a potential future mating opportunity. Evidence that male primates can be sperm-limited comes from an experiment with rhesus macaques in which males kept in male-female pairs mated every day, whereas males housed with multiple females limited their matings to each female's fertile period (Wallen 2001).

Second, in solitary species, finding a receptive female is associated with significant search costs in terms of time, energy, and additional risk (Kraus et al. 2008). Whether and how males incorporate these costs into their search strategies remains unknown. Third, and perhaps most importantly, achieving copulations is commonly associated with a significant risk of aggression or injury inflicted by rivals in nonmonogamous species. Depending on his own competitive abilities, a male may therefore either forego potential mating opportunities or focus his investment on particular females or times of their receptivity to minimize these costs. Male yellow baboons and chimpanzees, for example, concentrate their mating efforts on females experiencing conceptive cycles (Alberts et al. 2006; Emery Thompson & Wrangham 2008), and male mandrills focus their mate guarding on high-ranking and parous females (Setchell & Wickings 2006). Fourth, in pair-living species, in which males have relatively high paternity assurance, they are expected to be choosy about female quality, which in turn may vary as a function of a mate's age, health, or genetic constitution.

The fact that male preferences can matter has been demonstrated in experiments with house mice, in which matings with preferred females had striking effects on offspring survival and reproductive success (Gowaty et al. 2003). Males of different species appear to prefer different traits in females; in chimpanzees they prefer mating with older females (Muller et al. 2006), whereas in ringtailed lemurs they prefer females with high reproductive potential (Parga 2006). Finally, the sexual swellings of some catarrhines have also been suggested to act as indicators of individual quality for choosy males (fig. 16.3, Domb & Pagel 2002), but there is little empirical support for this notion (Zinner et al. 2002; Huchard et al. 2009; chapter 18, this volume). Thus, when the costs of reproduction are high or females vary in quality, males are expected to adjust their competitive strategies so that they are most likely to copulate successfully and produce surviving offspring (chapter 18, this volume). How competition and choice by members of both sexes interact to reconcile sex-specific reproductive strategies will have to be revealed by comprehensive future studies of primate reproduction.

Fig. 16.3. Sexual swellings elicit considerable interest from males, as in this chacma baboon, closely inspecting the swelling of a female who has just presented to him. The extent to which sexual swellings form the basis of male mate choice, however, is still little known in this and other baboon species. Photo courtesy of Ryne A. Palombit.

Summary and Conclusions

These are exciting times for the study of mate choice in primates. Compared to some other areas of research on primate behavior, the study of sexual selection, and of mate choice in particular, can rely on a robust and sophisticated theoretical framework. Studies on nonprimate taxa are providing a stimulating comparative perspective, and they continue to highlight the importance of considering taxon-specific life-history constraints in our effort to explain interspecific variation in mate choice. In primates, with their relatively slow life histories, the females of most species are trapped in the trade-off between confusing and concentrating paternity, which constrains their options for exercising active mate choice. Nevertheless, there is evidence that females of all major primate taxa exhibit choosiness among potential mates, and that mating biases exist. Which processes bring these mating biases about and which male traits influence mating decisions remains poorly known. One of the main insights of this review, therefore, is the observation that empirically unequivocal evidence for direct, active mate choice in primates is rare or missing because it is very difficult to separate it from the outcomes of indirect choice or male-male competition in observational studies. Creative ideas for conclusive experiments are therefore required.

At a more specific level, the ability to discriminate against heterospecifics and relatives appears to be widespread, but the underlying mechanisms have been studied in only a few species, and only a small fraction of those studies have focused on mate choice. This is an area with great potential for interesting integrative future research because it touches not only on sexual selection, but also on topics in communication, cooperation, and cognition. This type of investigation may even become relevant for conservation, because females can potentially be used as bio-detectors of conspecific and heterospecific males in taxa where taxonomic status is uncertain or disputed.

Females appear to derive the greatest direct benefit from paternal care. However, more research on inter- and intraspecific variation in paternal care is indicated for several reasons. The diversity of positive paternal effects on offspring in different social systems has apparently not been studied exhaustively, and there is very little evidence demonstrating that females actually evaluate variation in paternal effort in those species where it occurs. Because studies of nonprimate taxa have indicated that direct benefits of mate choice are quantitatively much more important than indirect ones, it seems a worthwhile investment for primatologists to study potential direct benefits in more detail.

With respect to indirect benefits of mate choice, primates provide many opportunities to study potential male ornaments in different modalities, but only a few studies have actually documented the range and correlates of variation between individuals. Several studies have indicated that females are sensitive to this variation—and this is a huge playground for additional future experiments—but the theoretically required link to offspring fitness has not yet been established for any primate species. The genetic compatibility hypothesis deserves additional testing in this context because existing studies indicate that MHC-assortative matings are widespread.

Behavioral components of precopulatory mate choice (i.e., mating behavior) have been studied in more species and in more detail in primates than in any other higher taxon. Nevertheless, a clean separation of active direct mate choice from other mechanisms that also generate mating bias has not yet been accomplished. In particular, clever experimental designs (e.g., Wong et al. 2004) or comprehensive empirical approaches (see Byers & Waits [2006] for an example of a field study of a mammal) are required to incorporate fitness costs of active choice. Given the typical need for invasive research to illuminate mechanisms of cryptic female choice, we will probably remain ignorant about many of these presumably widespread mechanisms, but molecular research on gametes in vitro may provide some insights (e.g., Clark & Swanson 2005; Nascimento et al. 2008).

Finally, male mate choice in primates is a topic that has not been given due attention in the past, mainly for historical reasons. It will be interesting to see what future research reveals about the criteria that underlie male choosiness and which behavioral mechanisms they employ to implement it, in the face of competition from other males. Also, with some of the smaller primate species, breeding experiments may actually be feasible and may reveal whether mate choice by either sex really matters in terms of its effects on offspring fitness.

We have come a long way in the study of mate choice in primates since Darwin's *Nature* paper. Research in the past two decades, in particular, has expanded our understanding of the underlying processes and mechanisms beyond the insights generated by the early studies of macaques and baboons. Much remains to be learned, however, and new molecular and digital methods of experimentation, along with detailed theory, should maintain the study of mate choice as one of the most integrative and active areas of primate behavioral research.

References

Abernethy, K. A., White, L. J. T. & Wickings, E. J. 2002. Hordes of mandrills (*Mandrillus sphinx*): Extreme group size and seasonal male presence. *Journal of Zoology, London*, 258, 131–137.

Able, D. J. 1996. The contagion indicator hypothesis for parasite-mediated sexual selection. *Proceedings of the National Academy of Sciences, USA*, 93, 2229–2233.

Alberts, S. & Altmann, J. 1995. Balancing costs and opportunities: Dispersal in male baboons. *American Naturalist*, 145, 279–306.

Alberts, S. C., Buchan, J. C. & Altmann, J. 2006. Sexual selection in wild baboons: From mating opportunities to paternity success. *Animal Behaviour*, 72, 1177–1196.

Altmann, J., Sapolsky, R. & Licht, P. 1995. Baboon fertility and social status. *Nature*, 377, 688–689.

Anderson, M. J. 2000. Penile morphology and classification of bushbabies (subfamily Galagoninae). *International Journal of Primatology*, 21, 815–836.

Andersson, M. 1994. *Sexual Selection*. Princeton, NJ: Princeton University Press.

———. 2005. Evolution of classical polyandry: Three steps to female emancipation. *Ethology*, 111, 1–23.

Andersson, M. & Iwasa, Y. 1996. Sexual selection. *Trends in Ecology and Evolution*, 11, 53–58.

Arlet, M. E., Molleman, F. & Chapman, C. 2007. Indications for female mate choice in grey-cheeked mangabeys, *Lophocebus albigena johnstoni*, in Kibale National Park, Uganda. *Acta Ethologica*, 10, 89–95.

———. 2008. Mating tactics in male grey-cheeked mangabeys (*Lophocebus albigena*). *Ethology*, 114, 851–862.

Arnqvist, G. & Rowe, L. 2005. *Sexual Conflict*. Princeton, NJ: Princeton University Press.

Astaras, C., Mühlenberg, M. & Waltert, M. 2008. Note on drill (*Mandrillus leucophaeus*) ecology and conservation status in Korup National Park, southwest Cameroon. *American Journal of Primatology*, 70, 306–310.

Barnett, A. A. & Brandon-Jones, D. 1997. The ecology, biogeography and conservation of the uakaris, *Cacajao* (Pitheciinae). *Folia Primatologica*, 68, 223–235.

Barthold, J., Fichtel, C. & Kappeler, P. M. 2009. What is it going to be? Pattern and potential function of natal coat change in sexually dichromatic redfronted lemurs (*Eulemur fulvus rufus*). *American Journal Physical Anthropology*, 138, 1–10.

Bateman, A. J. 1948. Intrasexual selection in *Drosophila. Heredity*, 2, 349–368.

Bearder, S. K. 1999. Physical and social diversity among nocturnal primates: A new view based on long-term research. *Primates*, 40, 267–282.

Bearder, S. K., Honess, P. E. & Ambrose, L. 1995. Species diversity among galagos with special reference to mate recognition. In *Creatures of the Dark: The Nocturnal Prosimians* (ed. by Alterman, L., Doyle, G. A. & Izard, M. K.), 331–352. New York: Plenum Press.

Birkhead, T. R. & Kappeler, P. M. 2004. Post-copulatory sexual selection in birds and primates. In *Sexual Selection in Primates: New and Comparative Perspectives* (ed. by Kappeler, P. M. & van Schaik, C. P.), 151–171. Cambridge: Cambridge University Press.

Birkhead, T. R. & Pizzari, T. 2002. Postcopulatory sexual selection. *Nature Reviews Genetic*, 3, 262–273.

Boinski, S. 1987. Mating patterns in squirrel monkeys (*Saimiri oerstedi*): Implications for seasonal sexual dimorphism. *Behavioral Ecology and Sociobiology*, 21, 13–21.

Boyko, R. H. & Marshall, A. J. 2009. The willing cuckold: Optimal paternity allocation, infanticide and male reproductive strategies in mammals. *Animal Behaviour*, 77, 1397–1407.

Bradley, B. J. & Mundy, N. I. 2008. The primate palette: The evolution of primate coloration. *Evolutionary Anthropology*, 17, 97–111.

Bradley, B. J., Doran-Sheehy, D. M. & Vigilant, L. 2007. Potential for female kin associations in wild western gorillas despite female dispersal. *Proceedings of the Royal Society of London B*, 274, 2179–2185.

Brauch, K., Hodges, J. K., Engelhardt, A., Fuhrmann, K., Shaw, E. & Heistermann, M. 2008. Sex-specific reproductive behaviours and paternity in free-ranging Barbary macaques (*Macaca sylvanus*). *Behavioral Ecology and Sociobiology*, 62, 1453–1466.

Brockelman, W. Y., Reichard, U., Treesucon, U. & Raemaekers, J. J. 1998. Dispersal, pair formation and social structure in gibbons (*Hylobates lar*). *Behavioral Ecology and Sociobiology*, 42, 329–339.

Brooks, R. 2000. Negative genetic correlation between male sexual attractiveness and survival. *Nature*, 406, 67–70.

Buchan, J. C., Alberts, S. C., Silk, J. B. & Altmann, J. 2003. True paternal care in a multi-male primate society. *Nature*, 425, 179–181.

Byers, J. A. & Waits, L. 2006. Good genes sexual selection in nature. *Proceedings of the National Academy of Sciences USA*, 103, 16343–16345.

Chamberlain, N. L., Hill, R. I., Kapan, D. D., Gilbert, L. E. & Kronforst, M. R. 2009. Polymorphic butterfly reveals the missing link in ecological speciation. *Science*, 326, 847–850.

Chapman, T., Arnqvist, G., Bangham, J. & Rowe, L. 2003. Sexual conflict. *Trends in Ecology and Evolution*, 18, 41–47.

Chappell, M. A., Zuk, M., Kwan, T. H. & Johnsen, T. S. 1995. Energy cost of an avian vocal display: Crowing in red junglefowl. *Animal Behaviour*, 49, 255–257.

Charpentier, M. J. E., Peignot, P., Hossaert-McKey, M., Gimenez, O., Setchell, J. M. & Wickings, E. J. 2005. Constraints on control: Factors influencing reproductive success in male mandrills (*Mandrillus sphinx*). *Behavioral Ecology*, 16, 614–623.

Charpentier, M. J. E., Widdig, A. and Alberts, S. C. 2007. Inbreeding depression in non-human primates: A historical review of methods used and empirical data. *American Journal of Primatology*, 69, 1370–1386.

Charpentier, M. J. E., Williams, C. V. & Drea, C. M. 2008a. Inbreeding depression in ring-tailed lemurs (*Lemur catta*): Genetic diversity predicts parasitism, immunocompetence, and survivorship. *Conservation Genetics*, 9, 1605–1615.

Charpentier, M. J. E., Van Horn, R. C., Altmann, J. & Alberts, S. C. 2008b. Paternal effects on offspring fitness in a multimale primate society. *Proceedings of the National Academy of Sciences, USA*, 105, 1988–1992.

Charpentier, M. J. E, Boulet, M. & Drea, C. M. 2008c. Smelling right: The scent of male lemurs advertises genetic quality and relatedness. *Molecular Ecology*, 17, 3225–3233.

Charpentier, M. J. E., Crawford, J. C., Boulet, M. & Drea, C. M. 2010. Message "scent": Lemurs detect the genetic relatedness and quality of conspecifics via olfactory cues. *Animal Behaviour*, 80, 101–108.

Clark, A. B. 1988. Interspecific differences and discrimination of auditory and olfactory signals of *Galago crassicaudatus* and *Galago garnettii. International Journal of Primatology*, 9, 557–571.

Clark, N. L. & Swanson, W. J. 2005. Pervasive adaptive evolution in primate seminal proteins. *PLoS Genetics*, 1, e35.

Clarke, P. M. R., Henzi, S. P. & Barrett, L. 2009. Sexual conflict in chacma baboons, *Papio hamadryas ursinus*: Absent males select for proactive females. *Animal Behaviour*, 77, 1217–1225.

Clutton-Brock, T. H. 2002. Breeding together: Kin selection and mutualism in cooperative vertebrates. *Science*, 296, 69–72.

———. 2007. Sexual selection in males and females. *Science*, 318, 1882–1885.

———. 2009. Sexual selection in females. *Animal Behaviour*, 77, 3–11.

Clutton-Brock, T. H. & McAuliffe, K. 2009. Female mate choice in mammals. *Quarterly Review of Biology*, 84, 3–27.

Clutton-Brock, T. H. & Parker, G. A. 1995. Sexual coercion in animal societies. *Animal Behaviour*, 49, 1345–1365.

Cooper, V. J. & Hosey, G. R. 2003. Sexual dichromatism and female preference in *Eulemur fulvus* subspecies. *International Journal of Primatology*, 24, 1177–1188.

Coyne, J. A. & Orr, H. A. 1998. The evolutionary genetics of speciation. *Philosophical Transactions of the Royal Society of London B*, 353, 287–305.

Craul, M., Zimmermann, E. & Radespiel, U. 2004. First experimental evidence for female mate choice in a nocturnal primate. *Primates*, 45, 271–274.

Crook, J. H. & Gartlan, J. C. 1966. Evolution of primate societies. *Nature*, 210, 1200–1203.

Cunningham, E. J. A. & Russell, A. F. 2000. Egg investment is influenced by male attractiveness in the mallard. *Nature*, 404, 74–77.

Danchin, E. & Cézilly, F. 2008. Sexual selection: Another evolutionary process. In *Behavioural Ecology* (ed. by Danchin, E., Giraldeau, L.-A. & Cézilly, F.), 363–426. Oxford: Oxford University Press.

Darwin, C. 1859. *On the Origin of Species by Means of Natural Selection; or, the Preservation of Favoured Races in the Struggle for Life*. London: John Murray.

———. 1871. *The Descent of Man and Selection in Relation to Sex.* London: John Murray.

———. 1876. Sexual selection in relation to monkeys. *Nature,* 15, 18–19.

David, P., Bjorksten, T., Fowler, K. & Pomiankowski, A. 2000. Condition-dependent signalling of genetic variation in stalk-eyed flies. *Nature,* 406, 186–188.

Delgado, R. A. Jr. 2006. Sexual selection in the loud calls of male primates: Signal content and function. *International Journal of Primatology,* 27, 5–25.

Delgado, R. A. Jr. & van Schaik, C. P. 2000. The behavioral ecology and conservation of the orangutan (*Pongo pygmaeus*): A tale of two islands. *Evolutionary Anthropology,* 9, 201–218.

De Souza Fialho, M. & Setz, E. Z. F. 2007. Extragroup copulations among brown howler monkeys in southern Brazil. *Neotropical Primates,* 14, 28–30.

Dewsbury, D. A. 1982. Ejaculate cost and male choice. *American Naturalist,* 119, 601–610.

———. 2005. The Darwin-Bateman paradigm in historical context. *Integrative and Comparative Biology,* 45, 831–837.

Digby, L. 1995. Infant care, infanticide, and female reproductive strategies in polygynous groups of common marmosets (*Callithrix jacchus*). *Behavioral Ecology and Sociobiology,* 37, 51–61.

Dixson, A. F. 1998. *Primate Sexuality: Comparative Studies of the Prosimians, Monkeys, Apes, and Human Beings.* Oxford: Oxford University Press.

———. 2002. Sexual selection by cryptic female choice and the evolution of primate sexuality. *Evolutionary Anthropology,* 11, 195–199.

———. 2009. *Sexual Selection and the Evolution of Human Mating Systems.* Oxford: Oxford University Press.

Dixson, A. F. & Anderson, M. J. 2004. Sexual behavior, reproductive physiology and sperm competition in male mammals. *Physiology and Behavior,* 83, 361–371.

Domb, L. G. & Pagel, M. 2001. Sexual swellings advertise female quality in wild baboons. *Nature,* 410, 204–206.

Drea, C. M. 2005. Bateman revisited: The reproductive tactics of female primates. *Integrative and Comparative Biology,* 45, 915–923.

———. 2007. Sex and seasonal differences in aggression and steroid secretion in *Lemur catta*: Are socially dominant females hormonally "masculinized"? *Hormones and Behavior,* 51, 555–567.

Drea, C. M. & Weil, A. 2008. External genital morphology of the ring-tailed lemur (*Lemur catta*): Females are naturally "masculinized." *Journal of Morphology,* 269, 451–463.

Dunbar, R. I. M. 1983. Structure of gelada baboon reproductive units. III. The male's relationship with his females. *Animal Behaviour,* 31, 565–575.

Eberle, M. & Kappeler, P. M. 2004a. Sex in the dark: determinants and consequences of mixed male mating tactics in *Microcebus murinus*, a small solitary nocturnal primate. *Behavioral Ecology and Sociobiology,* 57, 77–90.

Eberle, M. & Kappeler, P. M. 2004b. Selected polyandry: Female choice and inter-sexual conflict in a small nocturnal solitary primate (*Microcebus murinus*). *Behavioral Ecology and Sociobiology,* 57, 91–100.

Emery Thompson, M. & Wrangham, R. W. 2008. Male mating interest varies with female fecundity in *Pan troglodytes*

schweinfurthii of Kanyawara, Kibale National Park. *International Journal of Primatology,* 29, 885–905.

Engelhardt, A., Heistermann, M., Hodges, J. K., Nürnberg, P. & Niemitz, C. 2006. Determinants of male reproductive success in wild long-tailed macaques (*Macaca fascicularis*): Male monopolisation, female mate choice or post-copulatory mechanisms? *Behavioral Ecology and Sociobiology,* 59, 740–752.

Fedigan, L. M. 1982. *Primate Paradigms: Sex Roles and Social Bonds.* Montreal: Eden Press.

Fernandez-Duque, E. & Huntington, C. 2002. Disappearances of individuals from social groups have implications for understanding natal dispersal in monogamous owl monkeys (*Aotus azarai*). *American Journal of Primatology,* 57, 219–225.

Fietz, J., Zischler, H., Schwiegk, C., Tomiuk, J., Dausmann, K. H. and Ganzhorn, J. U. 2000. High rates of extra-pair young in the pair-living fat-tailed dwarf lemur, *Cheirogaleus medius. Behavioral Ecology and Sociobiology,* 49, 8–17.

Fisher, R. A. 1930. *The Genetical Theory of Natural Selection.* Oxford: Clarendon Press.

Fisher, H. S., Swaisgood, R. R. & Fitch-Snyder, H. 2003. Countermarking by male pygmy lorises (*Nycticebus pygmaeus*): Do females use odor cues to select mates with high competitive ability? *Behavioral Ecology and Sociobiology,* 53, 123–130.

Fisher, D. O., Double, M. C., Blomberg, S. P., Jennions, M. D. & Cockburn, A. 2006. Post-mating sexual selection increases lifetime fitness of polyandrous females in the wild. *Nature,* 444, 89–92.

Folstad, I. & Karter, A. J. 1992. Parasites, bright males, and the immunocompetence handicap. *American Naturalist,* 139, 603–622.

Forstmeier, W., Kempenaers, B., Meyer, A. & Leisler, B. 2002. A novel song parameter correlates with extra-pair paternity and reflects male longevity. *Proceedings of the Royal Society of London B,* 269, 1479–1485.

Fox, E. A. 2002. Female tactics to reduce sexual harassment in the Sumatran orangutan (*Pongo pygmaeus abelii*). *Behavioral Ecology and Sociobiology,* 52, 93–101.

French, J. A. and Inglett, B. J. 1989. Female-female aggression and male indifference in response to unfamiliar intruders in lion tamarins. *Animal Behaviour,* 37, 487–497.

Fujita, K. 1987. Species recognition by five macaque monkeys. *Primates,* 28, 353–366.

Fujita, K., Watanabe, K., Widarto, T. H. & Suryobroto, B. 1997. Discrimination of macaques by macaques: The case of Sulawesi species. *Primates,* 38, 233–245.

Ghiselin, M. T. 1974. *The Economy of Nature and the Evolution of Sex.* Berkeley: University of California Press.

Gilby, I. C., Emery Thompson, M., Ruane, J. D. & Wrangham, R. W. 2010. No evidence of short-term exchange of meat for sex among chimpanzees. *Journal of Human Evolution,* 59, 44–53.

Goldizen, A. W. 1987. Tamarins and marmosets: Communal care of offspring. In *Primate Societies* (ed. by Smuts, B. B., Cheney, D. L., Seyfarth, R. M., Wrangham, R. W. & Struhsaker, T. T.), 34–43. Chicago: University of Chicago Press.

Gomes, C. M. and Boesch, C. 2009. Wild chimpanzees exchange meat for sex on a long-term basis. *PLoS ONE,* 4, e5116.

Gowaty, P. A. 2004. Sex roles, contests for the control of reproduction, and sexual selection. In *Sexual Selection in Primates:*

New and Comparative Perspectives (ed. by Kappeler, P. M. & van Schaik, C. P.), 37–54. Cambridge: Cambridge University Press.

Gowaty, P. A., Drickamer, L. C. & Schmid-Holmes, S. 2003. Male house mice produce fewer offspring with lower viability and poorer performance when mated with females they do not prefer. *Animal Behaviour*, 65, 95–103.

Grafen, A. 1990. Biological signals as handicaps. *Journal of Theoretical Biology*, 144, 517–546.

Haesler, M. P. & Seehausen, O. 2005. Inheritance of female mating preference in a sympatric sibling species pair of Lake Victoria cichlids: Implications for speciation. *Proceedings of the Royal Society of London B*, 272, 237–245.

Hamilton, W. D. & Zuk, M. 1982. Heritable true fitness and bright birds: A role for parasites? *Science*, 218, 384–387.

Hansson, B. & Westerberg, L. 2002. On the correlation between heterozygosity and fitness in natural populations. *Molecular Ecology*, 11, 2467–2474.

Harrington, J. E. 1979. Responses of *Lemur fulvus* to scents of different subspecies of *Lemur fulvus* and to scents of different species of Lemuriformes. *Zeitschrift für Tierpsychologie*, 49, 1–9.

Heistermann, M., Ziegler, T., van Schaik, C. P., Launhardt, K., Winkler, P. & Hodges, J. K. 2001. Loss of oestrus, concealed ovulation and paternity confusion in free-ranging hanuman langurs. *Proceedings of the Royal Society of London B*, 268, 2445–2451.

Hettyey, A., Hegyi, G., Puurtinen, M., Hoi, H., Török, J. & Penn, D. J. 2010. Mate choice for genetic benefits: Time to put the pieces together. *Ethology*, 116, 1–9.

Heymann, E. W. 2003. Scent marking, paternal care, and sexual selection in Callitrichines. In *Sexual Selection and Reproductive Competition in Primates: New Perspectives and Directions* (ed. by Jones, C. B.), 305–325. Norman, OK: American Society of Primatologists.

Holland, B. & Rice, W. R. 1998. Chase-away sexual selection: Antagonistic seduction versus resistance. *Evolution*, 52, 1–7.

Hrdy, S. B. 1979. Infanticide among animals: A review, classification, and examination of the implications for the reproductive strategies of females. *Ethology and Sociobiology*, 1, 13–40.

———. 1981. *The Woman that Never Evolved*. Cambridge, MA: Harvard University Press.

Huchard, E., Courtiol, A., Benavides, J. A., Knapp, L. A., Raymond, M. & Cowlishaw, G. 2009. Can fertility signals lead to quality signals? Insights from the evolution of primate sexual swellings. *Proceedings of the Royal Society of London B*, 276, 1889–1897.

Huchard, E., Knapp, L. A., Wang, J., Raymond, M. & Cowlishaw, G. 2010a. MHC, mate choice and heterozygote advantage in a wild social primate. *Molecular Ecology*, 19, 2545–2561.

Huchard, E., Raymond, M., Benavides, J., Marshall, H., Knapp, L. & Cowlishaw, G. 2010b. A female signal reflects MHC genotype in a social primate. *BMC Evolutionary Biology*, 10, 96.

Hunt, J., Breuker, C. J., Sadowski, J. A. & Moore, A. J. 2009. Male-male competition, female mate choice and their interaction: Determining total sexual selection. *Journal of Evolutionary Biology*, 22, 13–26.

Isvaran, K. & Clutton-Brock, T. H. 2007. Ecological correlates of extra-group paternity in mammals. *Proceedings of the Royal Society of London B*, 274, 219–224.

Jones, C. B. 2003. *Sexual Selection and Reproductive Competition in Primates: New Perspectives and Directions*. Norman, OK: American Society of Primatologists.

Jones, A. G. & Ratterman, N. L. 2009. Mate choice and sexual selection: What have we learned since Darwin? *Proceedings of the National Academy of Sciences USA*, 106, 10001–10008.

Kalbe, M., Eizaguirre, C., Dankert, I., Reusch, T. B. H., Sommerfeld, R. D., Wegner, K. M. & Milinski, M. 2009. Lifetime reproductive success is maximized with optimal major histocompatibility complex diversity. *Proceedings of the Royal Society of London B*, 276, 925–934.

Kappeler, P. M. 1990. The evolution of sexual size dimorphism in prosimian primates. *American Journal of Primatology*, 21, 201–214.

Kappeler, P. M. & van Schaik, C. P. 2002. Evolution of primate social systems. *International Journal of Primatology*, 23, 707–740.

———. 2004. *Sexual Selection in Primates: New and Comparative Perspectives*. Cambridge: Cambridge University Press.

Keddy-Hector, A. C. 1992. Mate choice in non-human primates. *American Zoologist*, 32, 62–70.

Keller, L. F. & Waller, D. M. 2002. Inbreeding effects in wild populations. *Trends in Ecology and Evolution*, 17, 230–241.

Kirkpatrick, M. & Ryan, M. J. 1991. The evolution of mating preferences and the paradox of the lek. *Nature*, 350, 33–38.

Kleindorfer, S. & Wasser, S. K. 2004. Infant handling and mortality in yellow baboons (*Papio cynocephalus*): Evidence for female reproductive competition? *Behavioral Ecology and Sociobiology*, 56, 328–337.

Knapp, L. A., Robson, J. & Waterhouse, J. S. 2006. Olfactory signals and the MHC: A review and a case study in *Lemur catta*. *American Journal of Primatology*, 68, 568–584.

Knott, C. D., Emery Thompson, M., Stumpf, R. M. & McIntyre, M. H. 2010. Female reproductive strategies in orangutans, evidence for female choice and counterstrategies to infanticide in a species with frequent sexual coercion. *Proceedings of the Royal Society of London B*, 277, 105–113.

Kokko, H. & Jennions, M. D. 2008. Parental investment, sexual selection and sex ratios. *Journal of Evolutionary Biology*, 21, 919–948.

Kokko, H. & Ots, I. 2007. When not to avoid inbreeding. *Evolution*, 60, 467–475.

Kokko, H., Brooks, R., McNamara, J. M. & Houston, A. I. 2002. The sexual selection continuum. *Proceedings of the Royal Society of London B*, 269, 1331–1340.

Kokko, H., Jennions, M. D. & Brooks, R. 2006. Unifying and testing models of sexual selection. *Annual Reviews in Ecology, Evolution and Systematics*, 37, 43–66.

Korstjens, A. H. & Noë, R. 2004. Mating system of an exceptional primate, the olive colobus (*Procolobus verus*). *American Journal of Primatology*, 62, 261–273.

Kotiaho, J. S. & Puurtinen, M. 2007. Mate choice for indirect genetic benefits: Scrutiny of the current paradigm. *Functional Ecology*, 21, 638–644.

Kraus, C., Eberle, M. & Kappeler, P. M. 2008. The costs of risky male behavior: Sex differences in seasonal survival in a small sexually monomorphic primate. *Proceedings of the Royal Society of London B*, 275, 1635–1644.

Lande, R. 1981. Models of speciation by sexual selection on polygenic traits. *Proceedings of the National Academy of Sciences, USA*, 78, 3721–3725.

Lappan, S. 2008. Male care of infants in a siamang (*Symphalangus syndactylus*) population including socially monogamous and polyandrous groups. *Behavioral Ecology and Sociobiology*, 62, 1307–1317.

Lee, P. C. 1987. Nutrition, fertility, and maternal investment in primates. *Journal of Zoology, London*, 213, 409–422.

Lehmann, L. & Perrin, N. 2003. Inbreeding avoidance through kin recognition: Choosy females boost male dispersal. *American Naturalist*, 162, 638–652.

Lehmann, J., Fickenscher, G. & Boesch, C. 2006. Kin biased investment in wild chimpanzees. *Behaviour*, 143, 931–955.

Lemasson, A., Palombit, R. A. & Jubin, R. 2008. Friendships between males and lactating females in a free-ranging group of olive baboons (*Papio hamadryas anubis*): Evidence from playback experiments. *Behavioral Ecology and Sociobiology*, 62, 1027–1035.

Maestripieri, D. & Roney, J. R. 2005. Primate copulation calls and postcopulatory female choice. *Behavioral Ecology*, 16, 106–113.

Maestripieri, D., Leoni, M., Raza, S. S., Hirsch, E. J. & Whitham, J. C. 2005. Female copulation calls in Guinea baboons: Evidence for postcopulatory female choice? *International Journal of Primatology*, 26, 737–758.

Manning, J. T. 1995. Fluctuating asymmetry and body weight in men and women: Implications for sexual selection. *Ethology and Sociobiology*, 16, 145–153.

Manson, J. H. 2007. Mate choice. In *Primates in Perspective* (ed. by Campbell, C. J., Fuentes, A., MacKinnon, K. C., Panger, M. & Bearder, S. K.), 447–463. Oxford: Oxford University Press.

Marshall, J. L., Arnold, M. L. & Howard, D. J. 2002. Reinforcement: The road not taken. *Trends in Ecology and Evolution*, 17, 558–563.

Mateo, J. M. 2009. Kinship signals in animals. In *Encyclopedia of Neuroscience* (ed. by Squire, L. R.), 281–289. Oxford: Academic Press.

Mays, H. L. Jr. and Hill, G. E. 2004. Choosing mates: Good genes versus genes that are a good fit. *Trends in Ecology and Evolution*, 19, 554–559.

Melnick, D. J., Pearl, M. C. & Richard, A. F. 1984. Male migration and inbreeding avoidance in wild rhesus monkeys. *American Journal of Primatology*, 7, 229–243.

Mitani, J. C. 1987. Species discrimination of male song in gibbons. *American Journal of Primatology*, 13, 413–423.

Møller, A. P. & Jennions, M. D. 2001. How important are direct fitness benefits of sexual selection? *Naturwissenschaften*, 88, 401–415.

Muller, M. N. & Wrangham, R. W. 2009. *Sexual Coercion in Primates and Humans: An Evolutionary Perspective on Male Aggression against Females*. Cambridge, MA: Harvard University Press.

Muller, M. N., Thompson, M. E. & Wrangham, R. W. 2006. Male chimpanzees prefer mating with old females. *Current Biology*, 16, 2234–2238.

Muller, M. N., Kahlenberg, S. M., Emery Thompson, M. & Wrangham, R. W. 2007. Male coercion and the costs of promiscuous mating for female chimpanzees. *Proceedings of the Royal Society of London B*, 274, 1009–1014.

Muniz, L., Perry, S., Manson, J. H., Gilkenson, H., Gros-Louis, J. & Vigilant, L. 2006. Father-daughter inbreeding avoidance in a wild primate population. *Current Biology*, 16, R156-R157.

Muroyama, Y. & Thierry, B. 1998. Species differences of male loud calls and their perception in Sulawesi macaques. *Primates*, 39, 115–126.

Nascimento, J. M., Shi, L. Z., Meyers, S., Gagneux, P., Loskutoff, N. M., Botvinick, E. L. & Berns, M. W. 2008. The use of optical tweezers to study sperm competition and motility in primates. *Journal of The Royal Society Interface*, 5, 297–302.

Neff, B. D. & Pitcher, T. E. 2005. Genetic quality and sexual selection: An integrated framework for good genes and compatible genes. *Molecular Ecology*, 14, 19–38.

Nguyen, N., Van Horn, R., Alberts, S. & Altmann, J. 2009. "Friendships" between new mothers and adult males: Adaptive benefits and determinants in wild baboons (*Papio cynocephalus*). *Behavioral Ecology and Sociobiology*, 63, 1331–1344.

Nietsch, A. & Kopp, M.-L. 1998. Role of vocalization in species differentiation of Sulawesi tarsiers. *Folia Primatologica*, 69, 371–378.

Nikitopoulos, E., Heistermann, M., de Vries, H., van Hooff, J. A. R. A. M. & Sterck, E. H. M. 2005. A pair choice test to identify female mating pattern relative to ovulation in long-tailed macaques, *Macaca fascicularis*. *Animal Behaviour*, 70, 1283–1296.

Nunn, C. L. 1999. The evolution of exaggerated sexual swellings in primates and the graded-signal hypothesis. *Animal Behaviour*, 58, 229–246.

———. 2003. Behavioural defences against sexually transmitted diseases in primates. *Animal Behaviour*, 66, 37–48.

Nunn, C. L., Gittleman, J. L. & Antonovics, J. 2000. Promiscuity and the primate immune system. *Science*, 290, 1168–1170.

O'Donald, P. 1980. *Genetic Models of Sexual Selection*. Cambridge: Cambridge University Press.

Packer, C. 1979. Inter-troop transfer and inbreeding avoidance in *Papio anubis*. *Animal Behaviour*, 27, 1–36.

Palombit, R. A. 1994a. Extra-pair copulations in a monogamous ape. *Animal Behaviour*, 47, 721–723.

———. 1994b. Dynamic pair bonds in hylobatids: Implications regarding monogamous social systems. *Behaviour*, 128, 65–101.

Palombit, R. A., Seyfarth, R. M. & Cheney, D. L. 1997. The adaptive value of "friendships" to female baboons: Experimental and observational evidence. *Animal Behaviour*, 54, 599–614.

Palombit, R. A., Cheney, D. L. & Seyfarth, R. M. 2001. Female-female competition for male "friends" in wild chacma baboons, *Papio cynocephalus ursinus*. *Animal Behaviour*, 61, 1159–1171.

Parga, J. A. 2006. Male mate choice in *Lemur catta*. *International Journal of Primatology*, 27, 107–131.

Patel, E. R. 2007. Non-maternal infant care in wild silky sifakas (*Propithecus candidus*). *Lemur News*, 12, 39–42.

Paul, A. 2002. Sexual selection and mate choice. *International Journal of Primatology*, 23, 877–904.

Paul, A. & Kuester, J. 1985. Intergroup transfer and incest avoidance in semifree-ranging Barbary macaques (*Macaca sylvanus*) at Salem (FRG). *American Journal of Primatology*, 8, 317–322.

Penn, D. J. & Frommen, J. G. 2010. Kin recognition: An over-view of conceptual issues, mechanisms and evolutionary theory. In *Animal Behaviour: Evolution and Mechanisms* (ed. by Kappeler, P. M.), 55–85. Heidelberg: Springer.

Penn, D. J. & Potts, W. K. 1998. Chemical signals and parasite-mediated sexual selection. *Trends in Ecology and Evolution*, 13, 391–396.

Pereira, M. E. & Weiss, M. L. 1991. Female mate choice, male migration, and the threat of infanticide in ringtailed lemurs. *Behavioral Ecology and Sociobiology*, 28, 141–152.

Perrin, N. & Mazalov, V. 1999. Dispersal and inbreeding avoidance. *American Naturalist*, 154, 282–292.

Pfefferle, D., Brauch, K., Heistermann, M., Hodges, J. K. & Fischer, J. 2008a. Female Barbary macaque (*Macaca sylvanus*) copulation calls do not reveal the fertile phase but influence mating outcome. *Proceedings of the Royal Society of London B*, 275, 571–578.

Pfefferle, D., Heistermann, M., Hodges, J. K. & Fischer, J. 2008b. Male Barbary macaques eavesdrop on mating outcome: A playback study. *Animal Behaviour*, 75, 1885–1891.

Phillips-Conroy, J. E., Jolly, C. J. & Brett, F. L. 1991. Characteristics of hamadryas-like male baboons living in anubis baboon troops in the Awash hybrid zone, Ethiopia. *American Journal of Physical Anthropology*, 86, 353–368.

Pieta, K. 2008. Female mate preferences among *Pan troglodytes schweinfurthii* of Kanyawara, Kibale National Park, Uganda. *International Journal of Primatology*, 29, 845–864.

Pizzari, T. and Birkhead, T. R. 2000. Female feral fowl eject sperm of subdominant males. *Nature*, 405, 787–789.

Plavcan, J. M. 2004. Sexual selection, measures of sexual selection, and sexual dimorphism in primates. In *Sexual Selection in Primates: New and Comparative Perspectives* (ed. by Kappeler, P. M. & van Schaik, C. P.), 230–252. Cambridge: Cambridge University Press.

Plavcan, J. M., van Schaik, C. P. & Kappeler, P. M. 1995. Competition, coalitions and canine size in primates. *Journal of Human Evolution*, 28, 245–276.

Pope, T. R. 2000. The evolution of male philopatry in neotropical monkeys. In *Primate Males: Causes and Consequences of Variation in Group Composition* (ed. by Kappeler, P. M.), 219–235. Cambridge: Cambridge University Press.

Pradhan, G. R. & van Schaik, C. P. 2009. Why do females find ornaments attractive? The coercion-avoidance hypothesis. *Biological Journal of the Linnean Society*, 96, 372–382.

Pradhan, G. R., Engelhardt, A., van Schaik, C. P. & Maestripieri, D. 2006. The evolution of female copulation calls in primates: A review and a new model. *Behavioral Ecology and Sociobiology*, 59, 333–343.

Price, E. C. 1990. Infant carrying as a courtship strategy of breeding male cotton-top tamarins. *Animal Behaviour*, 40, 784–786.

Primakoff, P. & Myles, D. G. 2002. Penetration, adhesion, and fusion in mammalian sperm-egg interaction. *Science*, 296, 2183–2185.

Promislow, D. E. L. 1992. Costs of sexual selection in natural populations of mammals. *Proceedings of the Royal Society of London B*, 247, 203–210.

Pusey, A. E. 1980. Inbreeding avoidance in chimpanzees. *Animal Behaviour*, 28, 543–552.

———. 1987. Sex-biased dispersal and inbreeding avoidance in birds and mammals. *Trends in Ecology and Evolution*, 2, 295–299.

Pusey, A. E. & Wolf, M. 1996. Inbreeding avoidance in animals. *Trends in Ecology and Evolution*, 11, 201–206.

Pusey, A. E., Murray, C., Wallauer, W., Wilson, M., Wroblewski, E. & Goodall, J. 2008. Severe aggression among female *Pan troglodytes schweinfurthii* at Gombe National Park, Tanzania. *International Journal of Primatology*, 29, 949–973.

Qvarnström, A., Pärt, T. & Sheldon, B. C. 2000. Adaptive plasticity in mate preference linked to differences in reproductive effort. *Nature*, 405, 344–347.

Raemaekers, J. J. & Raemaekers, P. M. 1985. Field playback of loud calls to gibbons (*Hylobates lar*): Territorial, sex-specific and species-specific responses. *Animal Behaviour*, 33, 481–493.

Ralls, K. & Ballou, J. 1982. Effects of inbreeding on infant mortality in captive primates. *International Journal of Primatology*, 3, 491–505.

Reby, D. & McComb, K. 2003. Anatomical constraints generate honesty: Acoustic cues to age and weight in the roars of red deer stags. *Animal Behaviour*, 65, 519–530.

Reeder, D. 2003. The potential for cryptic female choice in primates: Behavioral, anatomical, and physiological considerations. In *Sexual Selection and Reproductive Competition in Primates: New Perspectives and Directions* (ed. by Jones, C. B.), 255–302. Norman, OK: American Society of Primatologists.

Robbins, A. M., Stoinski, T., Fawcett, K. & Robbins, M. M. 2009. Leave or conceive: Natal dispersal and philopatry of female mountain gorillas in the Virunga volcano region. *Animal Behaviour*, 77, 831–838.

Roberts, M. L., Buchanan, K. L. & Evans, M. R. 2004. Testing the immunocompetence handicap hypothesis: A review of the evidence. *Animal Behavior*, 68, 227–239.

Rotundo, M., Fernandez-Duque, E. & Dixson, A. F. 2005. Infant development and parental care in free-ranging *Aotus azarai azarai* in Argentina. *International Journal of Primatology*, 26, 1459–1473.

Ryan, M. J. 1997. Sexual selection and mate choice. In *Behavioural Ecology: An Evolutionary Approach* (ed. by Krebs, J. R. & Davies, N. B.), 179–202. Oxford: Blackwell.

Scheib, J. E., Gangestad, S. W. & Thornhill, R. 1999. Facial attractiveness, symmetry and cues of good genes. *Proceedings of the Royal Society of London B*, 266, 1913–1917.

Schradin, C., Reeder, D. M., Mendoza, S. P. & Anzenberger, G. 2003. Prolactin and paternal care: Comparison of three species of monogamous new world monkeys (*Callicebus cupreus*, *Callithrix jacchus*, and *Callimico goeldii*). *Journal of Comparative Psychology*, 117, 166–175.

Schülke, O., Kappeler, P. M. & Zischler, H. 2004. Small testes size despite high extra-pair paternity in the pair-living nocturnal primate *Phaner furcifer*. *Behavioral Ecology and Sociobiology*, 55, 293–301.

Schwensow, N., Fietz, J., Dausmann, K. & Sommer, S. 2008a. MHC-associated mating strategies and the importance of overall genetic diversity in an obligate pair-living primate. *Evolutionary Ecology*, 22, 617–636.

Schwensow, N., Eberle, M. & Sommer, S. 2008b. Compatibility counts: MHC-associated mate choice in a wild promiscuous primate. *Proceedings of the Royal Society of London B*, 275, 555–564.

Scordato, E. S. & Drea, C. M. 2007. Scents and sensibility: Information content of olfactory signals in the ringtailed lemur, *Lemur catta*. *Animal Behaviour*, 73, 301–314.

Setchell, J. M. 2005. Do female mandrills prefer brightly colored males? *International Journal of Primatology*, 26, 715–735.

Setchell, J. M. & Huchard, E. 2010. The hidden benefits of sex: Evidence for MHC-associated mate choice in primate societies. *BioEssays*, 32, 940–948.

Setchell, J. M. & Kappeler, P. M. 2003. Selection in relation to sex in primates. *Advances in the Study of Behavior*, 33, 87–173.

Setchell, J. M. & Wickings, E. J. 2006. Mate choice in male mandrills (*Mandrillus sphinx*). *Ethology*, 112, 91–99.

Setchell, J. M., Charpentier, M. J. E., Abbott, K. M., Wickings, E. J. & Knapp, L. A. 2010. Opposites attract: MHC-associated mate choice in a polygynous primate. *Journal of Evolutionary Biology*, 23, 136–148.

Setchell, J. M., Vaglio, S., Abbott, K. M., Moggi-Cecchi, J., Boscaro, F., Pieraccini, G. & Knapp, L. A. 2011. Odour signals major histocompatibility complex genotype in an Old World monkey. *Proceedings of the Royal Society B: Biological Sciences*, 278, 274–280.

Silk, J. B. 2002. Kin selection in primate groups. *International Journal of Primatology*, 23, 849–875.

Slater, K. Y., Schaffner, C. M. & Aureli, F. 2008. Female-directed male aggression in wild *Ateles geoffroyi yucatanensis*. *International Journal of Primatology*, 29, 1657–1669.

Small, M. F. 1984. *Female Primates: Studies by Woman Primatologists*. New York: Alan R. Liss.

———. 1989. Female choice in nonhuman primates. *Yearbook of Physical Anthropology*, 32, 103–127.

———. 1993. *Female Choices: Sexual Behavior of Female Primates*. Ithaca, NY: Cornell University Press.

Smith, D. G. 1995. Absence of deleterious effects of inbreeding in rhesus macaques. *International Journal of Primatology*, 16, 855–870.

Smuts, B. B. 1985. *Sex and Friendship in Baboons*. Hawthorne, NY: Aldine.

Smuts, B. B. & Smuts, R. W. 1993. Male aggression and sexual coercion of females in nonhuman primates and other mammals: Evidence and theoretical implications. *Advances in the Study of Behavior*, 22, 1–63.

Snowdon, C. T. 2004. Sexual selection and communication. In *Sexual Selection in Primates: New and Comparative Perspectives* (ed. by Kappeler, P. M. & van Schaik, C. P.), 57–70. Cambridge: Cambridge University Press.

Snowdon, C. T., Ziegler, T. E., Schultz-Darken, N. J. & Ferris, C. F. 2006. Social odours, sexual arousal and pairbonding in primates. *Philosophical Transactions of the Royal Society of London B*, 361, 2079–2089.

Soltis, J. 2002. Do primate females gain nonprocreative benefits by mating with multiple males? Theoretical and empirical considerations. *Evolutionary Anthropology*, 11, 187–197.

Spencer, K. A., Buchanan, K. L., Goldsmith, A. R. & Catchpole, C. K. 2003. Song as an honest signal of developmental stress in the zebra finch (*Taeniopygia guttata*). *Hormones and Behavior*, 44, 132–139.

Starin, E. D. 2001. Patterns of inbreeding avoidance in Temminck's red colobus. *Behaviour*, 138, 453–465.

Steenbeek, R., Sterck, E. H. M., de Vries, H. and van Hooff, J. A. R. A. M. 2000. Costs and benefits of the one-male, age-graded and all-male phases in wild Thomas's langur groups. In *Primate Males: Causes and Consequences of Variation in Group Composition* (ed. by Kappeler, P. M.), 130–145. Cambridge: Cambridge University Press.

Sterck, E. H. M., Watts, D. P. & van Schaik, C. P. 1997. The evolution of female social relationships in nonhuman primates. *Behavioral Ecology and Sociobiology*, 41, 291–309.

Sterck, E. H. M., Willems, E. P., van Hooff, J. A. R. A. M. & Wich, S. A. 2005. Female dispersal, inbreeding avoidance and mate choice in Thomas langurs (*Presbytis thomasi*). *Behaviour*, 142, 845–868.

Stumpf, R. M. & Boesch, C. 2005. Does promiscuous mating preclude female choice? Female sexual strategies in chimpanzees (*Pan troglodytes verus*) of the Taï National Park, Côte d'Ivoire. *Behavioral Ecology and Sociobiology*, 57, 511–524.

———. 2006. The efficacy of female choice in chimpanzees of the Taï Forest, Côte d'Ivoire. *Behavioral Ecology and Sociobiology*, 60, 749–765.

Suddendorf, T. & Collier-Baker, E. 2009. The evolution of primate visual self-recognition: Evidence of absence in lesser apes. *Proceedings of the Royal Society of London B*, 276, 1671–1677.

Sugawara, K. 1988. Ethological study of the social behavior of hybrid baboons between *Papio anubis* and *P. hamadryas* in free-ranging groups. *Primates*, 29, 429–448.

Tanaka, M. 2007. Development of the visual preference of chimpanzees (*Pan troglodytes*) for photographs of primates: Effect of social experience. *Primates*, 48, 303–309.

Tardiff, S. D. & Bales, K. 1997. Is infant-carrying a courtship strategy in callitrichid primates? *Animal Behaviour*, 53, 1001–1007.

Tattersall, I. 2007. Madagascar's lemurs: Cryptic diversity or taxonomic inflation? *Evolutionary Anthropology*, 16, 12–23.

Townsend, S. W., Slocombe, K. E., Emery Thompson, M. & Zuberbühler, K. 2007. Female-led infanticide in wild chimpanzees. *Current Biology*, 17, R355–R356.

Townsend, S. W., Deschner, T. & Zuberbühler, K. 2008. Female chimpanzees use copulation calls flexibly to prevent social competition. *PLoS ONE*, 3, e2431.

Trillmich, F. 2010. Parental care: Adjustments to conflict and cooperation. In *Animal Behaviour: Evolution and Mechanisms* (ed. by Kappeler, P. M.), 267–298. Heidelberg: Springer.

Trivers, R. L. 1972. Parental investment and sexual selection. In *Sexual Selection and the Descent of Man, 1871–1971* (ed. by Campbell, B.), 136–179. Chicago: Aldine.

Turner, L. M., Chuong, E. B. & Hoekstra, H. E. 2008. Comparative analysis of testis protein evolution in rodents. *Genetics*, 179, 2075–2089.

Tyler-Smith, C. 2008. An evolutionary perspective on Y-chromosomal variation and male infertility. *International Journal of Andropology*, 31, 376–382.

Ueno, Y. 1994. Olfactory discrimination of urine odors from five species by tufted capuchin (*Cebus apella*). *Primates*, 35, 311–323.

Van Belle, S., Estrada, A., Ziegler, T. E. & Strier, K. B. 2009. Sexual behavior across ovarian cycles in wild black howler monkeys (*Alouatta pigra*): Male mate guarding and female mate choice. *American Journal of Primatology*, 71, 153–164.

Van Noordwijk, M. A. & van Schaik, C. P. 2009. Intersexual food transfer among orangutans: Do females test males for coercive tendency? *Behavioral Ecology and Sociobiology*, 63, 883–890.

Van Schaik, C. P. & Janson, C. H. 2000. *Infanticide by Males and its Implications*. Cambridge: Cambridge University Press.

Vick, L. G. & Pereira, M. E. 1989. Episodic targeting aggression and the histories of *Lemur* social groups. *Behavioral Ecology and Sociobiology*, 25, 3–12.

Von Engelhardt, N., Kappeler, P. M. & Heistermann, M. 2000. Androgen levels and female social dominance in *Lemur catta*. *Proceedings of the Royal Society of London B*, 267, 1533–1539.

Waitt, C. & Little, A. C. 2006. Preferences for symmetry in conspecific facial shape among *Macaca mulatta*. *International Journal of Primatology*, 27, 133–145.

Waitt, C., Little, A. C., Wolfensohn, S., Honess, P., Brown, A. P., Buchanan-Smith, H. M. & Perrett, D. I. 2003. Evidence from rhesus macaques suggests that male coloration plays a role in female primate mate choice. *Proceedings of the Royal Society of London B*, 270, S144–S146.

Wallace, A. R. 1891. *Natural Selection and Tropical Nature*. London: Macmillan.

Wallen, K. 2001. Sex and context: Hormones and primate sexual motivation. *Hormones and Behavior*, 40, 339–357.

Watson, P. J. & Thornhill, R. 1994. Fluctuating asymmetry and sexual selection. *Trends in Ecology and Evolution*, 9, 21–25.

Weatherhaed, P. J. & Robertson, R. J. 1979. Offspring quality and the polygyny threshold. "The sexy son hypothesis." *American Naturalist*, 113, 201–208.

Wich, S. A., van der Post, D. J., Heistermann, M., Möhle, U., van Hooff, J. A. R. A. M. & Sterck, E. H. M. 2003. Life-phase related changes in male loud call characteristics and testosterone levels in wild Thomas langurs. *International Journal of Primatolgy*, 24, 1251–1265.

Widdig, A. 2007. Paternal kin discrimination: The evidence and likely mechanisms. *Biological Reviews*, 82, 319–334.

Wiley, R. H. & Poston, J. 1996. Indirect mate choice, competition for mates, and coevolution of the sexes. *Evolution*, 50, 1371–1381.

Williams, G. C. 1966. *Adaptation and Natural Selection: A Critique of Some Current Evolutionary Thought*. Princeton, NJ: Princeton University Press.

Wong, B. B. M., Jennions, M. D. & Keogh, J. S. 2004. Sequential male mate choice in a fish, the Pacific blue-eye *Pseudomugil signifer*. *Behavioral Ecology and Sociobiology*, 56, 253–256.

Wrangham, R. W. 1993. The evolution of sexuality in chimpanzees and bonobos. *Human Nature*, 4, 47–79.

Wright, P. C. 1990. Patterns of paternal care in primates. *International Journal of Primatology*, 11, 89–102.

Zahavi, A. 1975. Mate selection: A selection for handicap. *Journal of theoretical Biology*, 53, 205–214.

———. 1977. The cost of honesty (further remarks on the handicap principle). *Journal of Theoretical Biology*, 67, 603–605.

Zeh, J. A. & Zeh, D. W. 2003. Toward a new sexual selection paradigm: Polyandry, conflict and incompatibility. *Ethology*, 109, 929–950.

Zhao, D., Ji, W., Li, B. & Watanabe, K. 2008. Mate competition and reproductive correlates of female dispersal in a polygynous primate species (*Rhinopithecus roxellana*). *Behavioural Processes*, 79, 165–170.

Zinner, D., Alberts, S., Nunn, C. L. & Altmann, J. 2002. Significance of primate sexual swellings. *Nature*, 420, 142–143.

Zinner, D., Groeneveld, L., Keller, C. & Roos, C. 2009. Mitochondrial phylogeography of baboons (*Papio* spp.): Indication for introgressive hybridization? *BMC Evolutionary Biology*, 9, 83.

Zumpe, D. & Michael, R. P. 1985. Mate competition between female rhesus monkeys. *Naturwissenschaften*, 72, 382–384.

Chapter 17 Mating, Parenting, and Male Reproductive Strategies

*Martin N. Muller and
Melissa Emery Thompson*

For primate mothers, the biology of internal gestation and lactation obligates substantial direct parental investment. Because primate fathers are not equally constrained, the question of whether to care for young—and, if so, to what degree—is a fundamental one for male reproductive strategies. Solicitous males can gain reproductive advantage if their investment enhances offspring survival or quality. However, providing care entails a range of energetic expenses and opportunity costs. The most salient of these is the trade-off between caring for offspring and pursuing additional mating opportunities (Trivers 1972; Maynard Smith 1977).

Across primates, diverse compromises have been struck between mating and parenting effort. At one extreme, males are socially monogamous and deeply involved with infant care, transporting, grooming, and, on rare occasions, sharing food with offspring (Azara's night monkeys, *Aotus azarae*, Rotundo et al. 2005). They may even show physiological adaptations for care, such as weight gain during a mate's pregnancy in preparation for postpartum investment (common marmosets, *Callithrix jacchus*; cotton-top tamarins, *Saguinus oedipus*: Ziegler et al. 2006). At the opposite extreme, males are highly polygynous, providing little or no direct care and competing intensely for sexual access to mates (chimpanzees, *Pan troglodytes*, Muller & Wrangham 2004a; gorillas, *Gorilla gorilla*, Harcourt & Stewart 2007). In between lies considerable ambiguity, such as males of some species that occasionally tend unrelated infants, apparently to win favor with the mothers (olive baboons, *Papio anubis*, Smuts & Gubernick 1992). In general, however, the incidence of paternal care among primates is strikingly high compared to that of other mammals (Kleiman & Malcolm 1981; Whitten 1987).

When circumstances favor little or no investment by fathers, operational sex ratios are skewed toward males, males enjoy higher potential reproductive rates than females, and intense competition for mating opportunities can result (Trivers 1972; Emlen & Oring 1977; Clutton-Brock & Parker 1992). Many of the conspicuous features exhibited by male primates, including the thick manes of male baboons (*Papio* spp.), the bright facial coloration of male mandrills (*Mandrillus sphinx*), and the large canines of male gorillas, were interpreted by Darwin (1871) as sexually selected traits that functioned either to attract females or to deter rivals. When to invest in such ornamentation and weaponry, in light of the attendant physiological and behavioral costs, is another major consideration for male reproductive strategies. The relevant trade-off in this context is between immediate mating access and long-term survival.

Again, the range of solutions is catholic. In some species, males compete aggressively to monopolize females, and sexual selection has favored large body size and impressive canines that are employed in precopulatory contests. In others, males forego aggression, investing instead in reproductive physiology devoted to postcopulatory sperm competition (e.g., northern muriquis, *Brachyteles hypoxanthus*, Strier 1990). And in a few species, the costs of agonistic competition are so high that some males adopt alternative reproductive tactics (Setchell 2008), foregoing the development of secondary sexual characteristics and pursuing surreptitious matings (e.g., Bornean orangutan, *Pongo pygmaeus*, Crofoot & Knott in press).

This chapter considers two primary aspects of male reproductive strategies. We first examine the conditions that favor paternal care in primates, elucidating the forms such care takes across species, and the proximate hormonal mechanisms that modulate mating and parenting effort. We then consider the ways in which sexual selection has shaped male mating strategies in the contexts of scramble and contest competition. This chapter focuses primarily on species-level traits. Some intraspecific sources of variation in male reproductive success are explored by Alberts (chapter 18, this volume).

Male Care

Patterns of Paternal Care

Direct male care occurs in fewer than 10% of mammals (Kleiman & Malcolm 1981). The true figure is likely lower, as some species exhibit care in captivity but not in the wild (Woodroffe & Vincent 1994). Such pervasive paternal indifference may be driven by the mammalian arrangement of internal fertilization followed by gestation and lactation, which decreases both paternity certainty and opportunities for male care while increasing the relative benefits of remating (Dawkins & Carlisle 1976).

Among primates, by contrast, direct male care is seen in approximately 40% of genera (Kleiman & Malcolm 1981; Whitten 1987). Why is this taxonomic order so exceptional? The prevalence of paternal care among such long-lived species is ostensibly puzzling, as life-history models suggest that a high probability of future survival, coupled with paternity uncertainty, favors increased mating effort (Mauck et al. 1999). Two additional factors, however, are likely important in selecting for increased parenting. First, the long-term male-female associations found in most primate species increase opportunities for care and the formation of male-infant relationships (Kleiman & Malcolm 1981; van Schaik & Paul 1996). Second, primate infants are relatively costly, and this can increase the value of male care for survival (van Schaik & Paul 1996).

Male-infant interactions in primates take a variety of forms and show substantial interindividual and contextual variation (table 17.1). Here we are concerned primarily with direct care—that is, behaviors that are immediately beneficial for infant welfare. We distinguish this from indirect or incidental care—actions that might benefit offspring but would also be performed in the absence of infants (e.g., territorial defense: Kleiman & Malcolm 1981).

The most obvious forms of direct care are those behaviors, such as carrying and provisioning infants, that directly substitute for maternal care and incur regular and measurable costs (e.g., locomotor expenses). For example, in cotton-top tamarins, caregiving results in weight loss, with males losing more weight when fewer helpers are available (Achenbach & Snowdon 2002). Another prevalent form of direct care is protection from infanticide or predation, a service that entails risks for males, and one that mothers may be less effective at providing (chapters 8 and 19, this volume). Finally, males also provide subtler forms of investment in infants, such as grooming them, playing with them, and tolerating them in proximity, that incur apparently trivial costs yet may have long-term social and energetic benefits for the infants.

Paternity Certainty and Male Care

The willingness of males to bear costs associated with infant care has often been attributed to high levels of paternity certainty, which lead to direct genetic benefits from increased offspring survival (Trivers 1972). This explanation is consistent with the distribution of intense care in species traditionally considered to be monogamous breeders. Socially monogamous male owl monkeys (*Aotus* spp.) and titi monkeys (*Callicebus* spp.), for example, carry infants more than twice as often as mothers do in the first months of life, both in captivity and the wild (Wright 1984; Mendoza & Mason 1986; Robinson et al. 1987; see also chapters 3 and 11, this volume, and figs. 3.3 and 11.2). And in owl monkeys, males provision offspring over three times more frequently than do mothers (Wolovich et al. 2008). Males often provide foods that infants could obtain on their own, apparently because this fosters higher infant growth rates (Garber & Leigh 1997; Wolovich et al. 2008).

It has gradually become apparent that across the Primate order, paternity certainty is neither a necessary nor a sufficient condition for the occurrence of paternal care (Smuts & Gubernick 1992). Among owl monkeys, for example, care may be provided by a male that has been present only after an infant's birth (Azara's night monkeys, Fernandez-Duque et al. 2008). Significant, direct male care occurs in many New World callitrichines, as well as in the siamang (*Symphalangus syndactylus*). These primates live not only in pairs but also in polyandrous units in which both males provide care (Goldizen 1987; Whitten 1987; Lappan 2008). Furthermore, in several species of baboon and macaque, males commonly carry, cuddle, play with, and protect infants, despite promiscuous mating that is expected to reduce paternity certainty (Smuts & Gubernick 1992).

Sophisticated kin recognition mechanisms may promote paternal investment in the absence of paternity monopolization (Widdig 2007; chapter 21, this volume). In humans, for example, males report greater willingness to invest when they perceive that their mates have been faithful or that the offspring resemble themselves (Platek et al. 2003; Anderson et al. 2007). In some cercopithecines, care is biased toward probable offspring, such as those born to mothers with whom the male has frequently mated (yellow baboons, *Papio cynocephalus*, Buchan et al. 2003; chacma baboons, *Papio ursinus*, Busse & Hamilton 1981; Palombit et al. 1997; long-tailed macaque, *Macaca fascicularis*, van Schaik & van Noordwijk 1988). However, it is still unclear how sensitive or reliable these mechanisms are, and similar care is sometimes observed without bias toward probable offspring (Barbary macaque, *Macaca sylvanus*, Paul et al. 1992, 1996; olive baboon, Strum 1984).

In the case of callitrichines, intense investment by males in polyandrous groups may be contingent on some remarkable genetic adaptations. Not only can offspring in a single litter be sired by multiple fathers, but they can be chimeric, exchanging genetic material during early development. In at least one species (Wied's black-tufted-ear marmosets, *Callithrix kuhlii*), chimerism extends to the germ line, such that offspring can actively pass on genes from multiple sires (Ross et al. 2007). Males provide more care to infants with chimeric traits than to those without (Ross et al. 2007).

On the other hand, some primate males with high paternity certainty show relatively little investment in offspring. Leaders of single-male groups, for example, often have near-complete paternity certainty for infants sired during their tenure. Yet while these males do protect offspring from various threats (see below), they make few efforts at active caregiving. Silverback mountain gorillas (*Gorilla gorilla beringei*) live primarily in one-male groups, dominate reproduction when in multimale groups (85% of infants: Bradley et al. 2005), and maintain long tenures. Despite silverbacks being a "spatial focus" for infants and juveniles, "prolonged friendly contact between silverbacks and infants under 2 years old is extremely rare" (Stewart 2001: 186). Hamadryas baboons (*Papio hamadryas*) show such low levels of extrapair mating in the wild that vasectomizing leader males provides effective contraception for an entire group (Biquand et al. 1994). Yet most interactions between adult males and immatures involve not harem leaders but follower males, who may cultivate future mates by kidnapping and caring for them as juveniles (Kummer 1968; Swedell 2006).

Costly Infants and Male Care

Direct male care may be adaptive when it strongly contributes to infant survival or condition. In some birds, for example, females are less successful at raising offspring without male provisioning (Bart & Tornes 1989). On the other hand, across mammalian taxa, male herbivores rarely provide substantive care, perhaps because a relatively ubiquitous food source renders provisioning unnecessary (Kleiman & Malcolm 1981). Folivorous primates show a similar lack of paternal care, with the principal exception of infanticide protection (chapter 19, this volume).

In primates it is hypothesized that high litter weight relative to maternal weight obliges males to provide intensive care in species with infant carrying, because the costs of transporting infants are too high for the mother to bear alone (Kleiman 1977; Wright 1984, 1990; Garber & Leigh 1997). In callitrichines, twin infants total 14% to 25% of the mother's weight at birth, and approximately 50% at weaning (Goldizen 1987). Even in captivity, a lack of rearing assistance can lead mothers to reject infants (Bardi et al. 2001). Adult males share food with infants both actively and passively, invest in carrying more than do adolescent group members, and increase their investment as infants grow larger (fig. 17.1; review: Goldizen 1987; saddle-back tamarins, *Saguinus fuscicollis*: Terborgh & Goldizen 1985, Savage et al. 1996; moustached tamarins, *Saguinus mystax*: Garber 1997; Huck et al. 2004).

In some species, high costs of infant transport are accommodated without alloparental care by "parking" infants at the nest during foraging (Wright 1990). However, this is not feasible when predators are abundant. Consequently, predation risk may be an important factor in promoting paternal care. This is supported by an experiment in captive Goeldi's monkeys (*Callimico goeldii*), in which the introduction of predators (ferrets, to simulate tayra) hastened the onset of male caregiving (Schradin & Anzenberger 2003).

Predation pressure may also favor rapid postnatal growth to reduce the period of extreme vulnerability. This in turn increases the daily costs of lactation, further promoting paternal care to defray maternal energy expenditures (Wright 1990). Species with elevated alloparental care do have rapid growth rates and wean at younger ages than do closely related species (Garber & Leigh 1997; Mitani & Watts 1997; Ross & MacLarnon 2000). However, it is likely that alloparenting has driven adaptations for faster reproduction (e.g., twinning in callitrichines, semiannual birth seasons in Goeldi's monkeys), as well as the reverse.

In both captivity and the wild, individuals carrying infants spend less time traveling (Tardif & Bales 1997), are less vigilant (Savage et al. 1996), and rarely eat or forage (Goldizen 1987; Price 1992). Given these constraints, Goldizen and Terborgh (1989, Terborgh & Goldizen 1985) argued that even mated pairs of saddle-back tamarins could not successfully raise offspring in the wild without assistance. Also, when more caregivers are present, each can invest less time in care (cotton-top tamarin, Savage et al.

Table 17.1. Direct paternal care in selected primate species

Suborder Family *Species*	Modal social group	Female mating pattern	Infant/ maternal weight*	Types of care					Care correlated with paternity?	Care correlated with future mating success?	References
				Carry/ babysit	Provision	Infanticide protection/ agonistic intervention	Predator protection	Triadic interactions			
Strepsirrhini											
• **Cheirogaleidae**											
Cheirogaleus medius	MF pairs	Monogamous (high EPC)	11%	Up to 28%	No	n/a	Yes	No	Inferred, but males care for EP young (genetic)	Inferred	Fietz & Dausmann 2003
Microcebus murinus	Solitary	Promiscuous (high male skew)	15%	No	No	No	No	No	n/a	n/a	Andres et al. 2003
Phaner furcifer	Dispersed MF pairs	Monogamous (high EPC)	?	No	No	No	No	No	n/a	n/a	Schülke 2005
• **Daubentoniidae**											
Daubentonia madagascariensis	Semisolitary	Promiscuous	?	No	No	No	No	No	n/a	n/a	Jolly 1998
• **Lemuridae**											
Eulemur fulvus	MM-MF	Promiscuous (high male skew)	4%	No	No	No	No	No	n/a	n/a	Overdorff 1998; Wimmer & Kappeler 2002
Hapalemur griseus	MF pairs	Monogamous (low EPC)	4%	Up to 20%	No	?	?	No	Inferred	Unknown, but male transfer frequent	Wright 1990; Jolly 1998; Nievergelt et al. 2002
• **Lepilemuridae**											
Propithecus verreauxi	MM-MF	Promiscuous (high male skew)	3%	No	No	No	No	No	n/a	n/a	Kappeler & Schäffler 2008
• **Indriidae**											
Avahi laniger	MF pairs	Monogamous	4%^	No	No	No	No	No	n/a	n/a	Wright 1990, Jolly 1998
Indri indri	MF pairs	Monogamous?	3%	No	No	No	Yes	No	Inferred	Inferred	Wright 1990
• **Lorisidae**											
Nycticebus coucang	MF pairs	Monogamous?	4%	No	No	No	No	No	n/a	n/a	Wiens & Zitzmann 2003
• **Galagidae**											
Galago senegalensis	Semi-solitary	Promiscuous	9%	No	No	No	No	No	n/a	n/a	Wright 1990
Haplorrhini											
• **Tarsiidae**											
Tarsius tarsier	MF pairs	Monogamous?	20–33%	No	No	n/a	Yes	No	n/a	n/a	Gursky 2000
• **Cebidae**											
Callimico goeldii	MF pairs	Monogamous/ limited polyandry	9%	Yes	Yes	?	?	No	Yes (behavioral)	Inferred	Whitten 1987; Jurke et al. 1995
Cebus apella	MM-MF	Promiscuous (high male skew)	12%	Yes	No	Yes	Indirect	No	Yes (behavioral)	Inferred	Robinson & Janson 1987

Species											
Saguinus fuscicollis	1F-MM	Limited polyandry	16%	Up to 96%	Yes	Yes	Yes	No	Inferred, but >1 male cares	Inferred	Goldizen 1987, 1989
Saimiri sciureus	MM-MF	Promiscuous	14–20%	No	No	No	No	No	n/a	n/a	Whitten 1987
• Aotidae											
Aotus azarae	MF pairs	Monogamous?	10% ^	Up to 67%	No	Yes	Yes	No	Inferred, but care provided by nonfather	Inferred	Fernández-Duque et al. 2008, pers. comm...; Wolovich et al. 2008
• Pitheciidae											
Callicebus moloch	MF pairs	Monogamous?	10%	Up to 75%	Yes	Yes	Yes	No	Inferred	Inferred	Wright, 1984; Fragaszy et al. 2005
• Atelidae											
Alouatta pigra	1F-MM	Limited polyandry	8% ^	5%	Yes	Yes	No	No	Inferred	?	Bolin 1981; Kitchen et al. 2004
Brachyteles arachnoides	MM-MF	Promiscuous	7% ^	No	No	No	No	No	n/a	n/a	Guimarães & Strier 2001
• Cercopithecidae											
Macaca nemestrina	MM-MF	Promiscuous	8%	No	No	No	No	No	n/a	n/a	Whitten 1987
Macaca sylvanus	MM-MF	Promiscuous	6–8% ^	No	Yes	n/a?	Yes	Yes	No (genetic)	No (behavioral, genetic)	Kuester & Paul 1986; Paul et al. 1996
Papio cynocephalus/ursinus	MM-MF	Promiscuous	6%	Yes	Yes?	Yes?	Yes	Yes	Yes (behavioral, genetic)	Yes (behavioral)	Bercovitch 1991; Buchan et al. 2003; Busse & Hamilton 1981; Palombit et al. 1997
Semnopithecus entellus	MM-MF	Promiscuous (high male skew)	5–7% ^	No	Yes	No	No	No	Yes (behavioral)	Yes (behavioral)	Borries 1997
• Hylobatidae											
Symphalangus syndactylus	MF pairs, or 1F-MM	Monogamy/ limited polyandry	5%	Up to 27%	No	No	No	No	Inferred, but >1 male cares	Inferred	Chivers 1974, Lappan 2008
Hylobates lar	MF Pairs, or 1F-MM	Monogamy/ limited polyandry	8%	No	No	No	No	No	n/a	n/a	Wright 1990
• Hominidae											
Pongo abelii / pygmaeus	Semisolitary 1M-MF, MM-MF	Promiscuous	5%	No	No	No	No	No	n/a	n/a	Mitani & Watts 1997
Gorilla gorilla	MM-MF	Monogamous/ promiscuous	2%	No	No	Yes	No	No	Inferred	Inferred	Stewart 2001
Pan troglodytes	MM-MF	Promiscuous	6%	No	No	No	No	No	n/a	n/a	Goodall 1986
Homo sapiens	MM-MF	Monogamous (low EPC)	8%	Up to 100%	Yes	Yes	Yes	No	Yes (behavioral)	Yes (behavioral)	Marlowe 2000

This table is not intended to be exhaustive, either in species diversity or in the range of behaviors that might be considered care; it is instead designed to convey the general patterns of paternal behavior in primates. We provide a representative sample of species from different taxa and a range of social systems. We display only direct care behaviors that are exhibited commonly within a species. Play and grooming are omitted because of difficulties in making comparisons across taxa. Species vary in the extent to which these behaviors are part of the social repertoire, and in species for which play and grooming are relevant, they typically occur at some nonzero rate between adult males and infants; in the absence of other forms of care, it is difficult to attribute these interactions to care as opposed to general social behavior. Infanticide and predator protection are indicated as "yes" only when there is evidence that individual males actively protect infants—for example, by intervening in attacks or by changing their vigilance behavior in relation to the presence of vulnerable infants.
*If not provided in the primary reference, infant/maternal weight ratios are from Wright 1990, Harvey & Clutton-Brock 1985, or Garber & Leigh 1997, and they indicate combined weight of a litter where multiple births are common. ^ Birth weights are unavailable, and are instead estimated from closely related species of similar adult body size. "Inferred" indicates that males may expect a high paternity certainty due to social grouping and mating systems; "behavioral" indicates that males providing care were viewed as probable fathers due to their observed mating history with a female, and "genetic" means that genetic studies have linked caregiving with true paternity.

Fig. 17.1. A 2.5-month-old infant saddle-backed tamarin (*Saguinus fuscicollis*) (above) at Estación Biológica Quebrada Blanco, Peru, vocalizes as it begs for food from an adult that is eating an insect it has just caught. A few moments after this photograph was taken, the adult passively shared the food by allowing the youngster to take a piece of it. Photo courtesy of Laurence Culot.

1996; *Callithrix* spp., *Leontopithecus* spp., Santos et al. 1997). Larger groups, and particularly groups with more adult males, have higher infant survival rates, an advantage that may ultimately offset the costs of shared paternity (*Saguinus* spp., Garber 1997; cotton-top tamarins, Savage et al. 1996; pygmy marmosets, *Cebuella pygmaea*, Heymann & Soini 1999).

In siamangs, male carrying behavior generally commences in the second year, when infant weight is relatively high and mothers begin to reject their offspring (Chivers 1974; Lappan 2008). Male caregiving is absent in other gibbon species with smaller body sizes (Smith & Jungers 1997). In contrast to callitrichines (Savage et al. 1996), the amount of care given by siamang males in monogamous pairs is both relatively and absolutely more (and the mother's less) than that of males in polyandrous groupings. Furthermore, infant survival is generally high, suggesting that females have less leverage in trading off paternity certainty for increased care (Lappan 2008).

In humans, birth and weaning weights are a relatively small proportion of maternal weight (Dettwyler 1995). Infants are born exceptionally helpless, however, thus necessitating care that interferes with their mothers' foraging (e.g. Marlowe 2003). Human children also require a substantial period of investment beyond weaning, meaning that concurrent care must normally be provided for multiple dependents (Lancaster & Kaplan 2009; chapter 20, this volume). Hrdy (2009) argues that such constraints make humans obligate cooperative breeders, whether assistance comes from fathers or from other individuals.

By contrast, the more moderate forms of male care found among cercopithecines are not necessary for survival, as not all infants have male caregivers. However, such care likely improves infant quality. For example, baboon males sometimes support infants during agonistic encounters, or tolerate infants in proximity, potentially increasing their access to high-quality food (Altmann 1980). Although these effects are difficult to quantify, one long-term study of yellow baboons revealed that a biological father's presence in the group during the immature period significantly accelerated the timing of maturation in daughters and, in the case of high-ranking males, sons (Charpentier et al. 2008). The possibility that male care, even in small doses, benefits infants is supported by the observation that infants (and/or their mothers) are typically responsible for maintaining proximity with males, and for soliciting care (siamang, Lappan 2008; dusky titi monkeys, *Callicebus moloch*, Mendoza & Mason 1986; Azara's owl monkeys, Wolovich et al. 2008; rhesus macaques, *Macaca mulatta*, Berenstain et al. 1981; mountain gorillas, Stewart 2001; chacma baboons, Palombit et al. 1997). The benefits to males are more difficult to identify, unless one considers that caregiving, rather than interfering with mating effort, actually enhances the probability of future mating.

Male Care as Mating Effort

Although numerous studies show a link between probability of paternity and male-infant affiliation, Smuts & Gubernick (1992) note that these relationships are rarely if ever independent of male-female bonds. Consequently, van Schaik & Paul (1996) questioned whether caregiving by male primates can ever be construed as direct investment in the survival of genetic offspring, as opposed to mating effort. In support of the mating effort hypothesis, a male's relationship with the mother appears to be the key predictor of infant care in many species (fig. 17.2; rhesus macaques, Berenstain et al. 1981; stumptailed macaques, *Macaca arctoides*, Smith & Peffer-Smith 1984; baboons, Smith & Whitten 1988; Strum 1984, 1987).

There are several ways in which male care might function as mating effort, the most obvious being female choice for solicitous males (Wittenberger & Tilson 1980). Care can be an effective form of mating effort if it promotes infant survival, and if females compare and differentially reward male investment (Smuts & Gubernick 1992). This scenario is consistent with the prevalence of active caregiving in promiscuous primates, in which males must continually negotiate mating access, as opposed to harem species, in which males maintain privileged access to females.

Even the intense paternal care exhibited by callitrichines

Fig. 17.2. A male olive baboon in Laikipia, Kenya, grooms an infant as her mother sits nearby. Do males form relationships with infants because they are their fathers or because they are interested in their mothers as potential mates? This is not always clear. Photo courtesy of Ryne A. Palombit.

has been attributed to mating effort, because males actively compete to carry infants (cotton-top tamarins, Price 1991), and the resumption of estrus mere weeks after parturition grants caregivers preferential mating access (Garber 1997). In some species, such as baboons (Seyfarth 1978; Smuts 1985), care has been found to increase a male's relative mating success with particular mothers. In others, caring males gain no such advantage (e.g., common marmoset, cotton-top tamarins, Tardif & Bales 1997; Barbary macaques, Paul et al. 1996).

Male care can also contribute to future reproduction by subsidizing the energy investment of females in current offspring, which increases female reproductive rates. For example, in one siamang population, females who received more care from males had shorter birth intervals (Lappan 2008). Such investment promotes the male's ability to mate with a female sooner, which may be advantageous even if he does not gain exclusive access. This hypothesis is supported by a phylogenetic comparison showing increased birth rates in species with alloparental care (Mitani & Watts 1997).

Males may also use care or tolerance of infants to gain nonreproductive (or indirectly reproductive) benefits, such as acceptance in a new social group, or female coalition-

ary support (Smuts 1985; Palombit et al. 1997). This could explain why some macaque males bias their investment towards the offspring of high-ranking mothers (Smuts & Gubernick 1992). In cercopithecines, males may also use care strategically to negotiate relationships with other males (Paul et al. 1996).

In several primate species, males have one-on-one caring interactions with infants but later use the same infants in a seemingly exploitative manner by carrying them during agonistic encounters with other males (Whitten 1987). Carrying an infant appears to reduce aggression from other males and temporarily increases the immediate status of the carrying male, allowing him to more effectively supplant others (olive baboon, Packer 1980; stump-tailed macaque, Silk & Samuels 1984, but see Barbary macaque, Taub 1980). Infants are used particularly in facilitating interactions with males of higher rank (Barbary macaques, Paul et al. 1996; yellow baboons, Collins 1986; sooty mangabeys, *Cercocebus atys*, Busse & Gordon 1984).

The benefits of carrying infants during male-male interactions are not clear, but they likely vary across species and contexts (Whitten 1987). When aggressively engaging a new immigrant, a male may hold a vulnerable infant in order to protect it (Busse & Hamilton 1981). When challenging males within the group, a male may carry an infant because mutual interest in its welfare or fear of provoking its kin prevents the level of aggression from escalating (Taub 1980; Dunbar 1984a; Stein 1984). There is a general consensus that males are not threatening to harm infants when they seize them during competitive encounters with other males (Whitten 1987). Rather, they may invest in particular infants in order to enhance their own access to them for such use (Packer 1980).

Male Protection

One of the most critical and unique contributions that a primate father can make is to protect his infant from external threats to its survival. Males, who are frequently larger than females and often possess long, sharp canines, can play a major role in group defense, particularly against predators (e.g., *Cebus* spp., van Schaik & van Noordwijk 1989). However, some primate males play a more specific and direct role in protecting their infants, and it is here that we find unambiguous examples of paternal effort.

As noted previously, alpha males in one-male groups or species with extreme male reproductive skew have high paternity certainty, yet invest relatively little in direct, nurturing care, such as carrying and provisioning. In these species, however, infanticide is a major source of mortality, causing more than 30% of infant deaths in some populations (chapter 19, this volume). The ability to protect offspring from infanticide is thus a critical determinant of male reproductive success.

Because infanticide is committed most frequently by new immigrants or by males from outside the group, infant protection is often intertwined with mate defense. However, closer examination suggests the primacy of infant protection. For example, when exposed to recordings of strange male long calls, Central American black howler monkey (*Alouatta pigra*) males were more likely to respond defensively with vocal displays if vulnerable infants were present, even when this risked confrontation with a larger group of males (Kitchen 2004). Chacma baboon males sometimes form "friendships" with female mates. In playback experiments, a male's willingness to respond to his friend's scream was predicted by the presence of her infant and by the pairing of her scream with vocalizations of an infanticidal male (Palombit et al. 1997). These friendships apparently have no bearing on future consort success with the mother (Weingrill 2000), but the importance of the male's actions is reflected by the fact that having a male protector reduces a lactating female's stress response in the presence of an immigrant male (Beehner et al. 2005). In several species, old males, despite being deposed and engaging in little mating activity, remain in the group for some time, actively protecting infants from new males (geladas, *Theropithecus gelada*, Dunbar 1984b; Hanuman langur, *Semnopithecus entellus*, Borries et al. 1999; gorilla, Stewart 2001; see also chapter 19, this volume).

Paternal protection can extend to other arenas. In yellow baboons, for example, males selectively intervene in noninfanticidal conflicts on behalf of their juvenile offspring (Buchan et al. 2003). Silverback mountain gorillas frequently intervene in social conflicts involving immature individuals. Because all juveniles are likely progeny, male interventions favor younger, more vulnerable antagonists (Watts 1997; Stewart 2001). Additional intriguing examples of male care can be found following the orphaning of infants (rhesus macaques, Berman 1982; Japanese macaques, *Macaca fuscata*, Hasegawa & Hiraiwa 1980; stump-tailed macaques, Smith & Peffer-Smith 1984; yellow baboons, Rhine et al. 1980; gorillas, Fossey 1979). While it is not clear how often these cases involve actual fathers, a motivation for paternal effort is likely because mating effort would more appropriately be directed at an infant with a living mother.

Ultimately, isolating a single driving force for male care may be difficult, as similar forms of male care may function alternately, or simultaneously, as mating effort and parenting effort. For example, in a cleverly designed experiment with captive vervet monkeys (*Chlorocebus pygerythrus*), males who were likely fathers gave more care to infants than males who were not. However, the unlikely fathers were also kind to infants when the mother was watching, and mothers rewarded caring males by being less aggressive to them after the interaction (Keddy-Hector et al. 1989). In tamarins, it is hypothesized that new immigrants provide care to ingratiate themselves with breeding females, while the substantially greater care provided by previously-mated males constitutes parenting effort (Garber 1997). In the Hadza, a group of human hunter-gatherers, Marlowe (1999) concludes that significant amounts of direct care given to stepchildren promote mating relationships with their mothers, while a tendency to bias investment towards biological children nonetheless reflects true paternal care (see also chapter 20, this volume).

Finally, male caregiving is conspicuously rare among strepsirrhines and tarsiers (chapter 2, this volume). This is puzzling. Several species have little or no male care despite high paternity certainty, owing either to apparent monogamy (e.g,. indri, *Indri indri*, Wright 1990) or high male reproductive skew (e.g., Verreaux's sifaka, *Propithecus verreauxi*, Kappeler & Schäffler 2008). Others lack paternal care despite large litter sizes (e.g., *Microcebus* spp.), or very high infant-maternal weight ratios (e.g., *Tarsius* spp., Wright 1990). Lemurs often use alternatives to paternal care, such as parking infants in lieu of carrying them (Wright 1990) or exploiting other types of caregivers (Patel 2007). Female dominance might be expected to promote male care in lemurs if females have increased leverage over males. It has been proposed, however, that foraging subordinacy may be a concession by males to females that substitutes for direct male care, particularly as female dominance in lemurs correlates with high reproductive costs, the same condition that predicts male care in anthropoids (Pollock 1979; Young et al. 1990).

The Physiology of Mating and Parenting Effort

The physiological mechanisms mediating male investment in reproductive effort have only recently been studied in detail in the wild. Such work has focused on the endocrine system, which plays a critical role in allocating resources toward the competing demands of growth, maintenance, and reproduction by coordinating morphological, physiological, and behavioral responses to environmental challenges (Ketterson & Nolan 1992; Wingfield et al. 2000). The steroid hormone testosterone, in particular, is thought to promote mating effort at the expense of both parenting effort and survival (Ketterson & Nolan 1992; Hau 2007).

Specifically, testosterone's multiple organizational and activational effects in primates can be understood as manifestations of mating effort. For example, testosterone drives the process of male genital differentiation and supports spermatogenesis (Dixson 1998). It also spurs the development of sexually dimorphic ornaments and armaments that feature in male mating competition (e.g., red faces in mandrills, Setchell et al. 2008; chest patches in Verreaux's sifakas, Lewis 2009), including the dimorphic muscle tissue that males utilize in agonistic encounters (Kemnitz et al. 1988; Bribiescas 2001). Finally, testosterone supports behavioral aspects of mating effort, by promoting libido and facilitating aggressive responses in reproductive contexts (Dixson 1998; Isidori et al. 2005; Archer 2006).

The relationship between testosterone and reproductive aggression has been best studied in birds, which show dramatic interspecific and individual differences in temporal patterns of testosterone production, explicable by variation in the timing and intensity of male mating competition (Wingfield et al. 2000; Hirschenhauser et al. 2003; Goymann et al. 2007). According to the challenge hypothesis (Wingfield et al. 1990), testosterone levels increase when males must respond to threats from conspecifics, particularly during territory formation and mate guarding, and they decrease when males must care for offspring. Experimental manipulations of male birds have shown that in some species, high levels of testosterone suppress paternal behavior in favor of male-male competition (Hegner & Wingfield 1987).

Aspects of the challenge hypothesis have been studied in a range of primates, with generally supportive results. First, in both seasonal and nonseasonal breeders, testosterone levels have been observed to rise during periods of short-term male mating competition (ring-tailed lemurs, *Lemur catta*, Cavigelli & Pereira 2000; Gould & Ziegler 2007; red-fronted lemurs, *Eulemur rufifrons*, Ostner et al. 2008; long-tailed macaques, Girard-Buttoz et al. 2009; white-thighed colobus, *Colobus velerosus*, Teichroeb & Sicotte 2008; mandrills, Setchell et al. 2008; chimpanzees, Muller & Wrangham 2004a). By contrast, species in which mating occurs with little or no direct male aggression do not exhibit predictable increases in testosterone across mating periods (northern muriquis, Strier et al. 1999; mustached tamarins, Huck et al. 2005; brown capuchins, *Cebus apella*, Lynch et al. 2002).

Second, for species in which relative dominance rank determines a male's mating access, testosterone levels predictably rise when rank is being contested. Thus, testosterone increases have been observed during intervals when existing dominance hierarchies grow unstable (mandrills, Setchell et al. 2008; olive baboons, Sapolsky 1983, 1993; chacma baboons, Beehner et al. 2006), as well as when new males attempt to enter established groups (mantled howler monkeys, *Alouatta palliata*, Cristóbal-Azkarate et al. 2006). This suggests that testosterone is positively associated with the acquisition of dominance rank primarily when rank is dependent on aggression (Muller & Wrangham 2004a).

In humans, generally positive but inconsistent associations between testosterone and aggressive behavior have spurred researchers to examine the steroid's effects at the psychological level (Mazur & Booth 1998; Archer 2006). Numerous studies have found that testosterone is associated with the motivation to dominate others (Schultheiss et al. 1999, 2003; Sellers et al. 2007). It also enhances responsiveness to social challenges (van Honk et al. 1999, 2001; Benderlioglu et al. 2004; Hermans et al. 2008), which can result in a lower latency of reactive aggression in high-testosterone men (Kouri et al. 1995). Such mechanisms help to explain the variable associations between aggression and testosterone in humans, as not all high-testosterone men find themselves in environments in which they face persistent social challenges.

Also consistent with the challenge hypothesis, primates with highly solicitous males show a negative correlation between testosterone production and the expression of paternal care. For example, wild golden lion tamarin (*Leontopithecus rosalia*) males maintained higher testosterone levels in the mating season than during the birth/infant care season (Bales et al. 2006). In captive Wied's black-tufted-ear marmosets, significant declines in testosterone were observed three to four weeks following parturition, coinciding with the peak rate of infant carrying by fathers (Nunes et al. 2000). This effect was particularly pronounced in experienced fathers. Furthermore, across individuals, testosterone was negatively correlated with infant carrying (Nunes et al. 2001). Direct contact with infants may be responsible for this pattern, as captive common marmosets showed a drop in serum testosterone levels within 20 minutes of exposure to the scent of their infants (Prudom et al. 2008).

In humans, two longitudinal studies found that men taking prenatal classes had decreased testosterone levels in the weeks surrounding parturition (Storey et al. 2000; Berg & Wynne-Edwards 2001). And a series of cross-sectional studies have variously demonstrated that men in committed long-term relationships, or fathers, or both maintain lower levels of testosterone than unpaired men (Gray & Campbell 2009; Alvergne et al. 2009; Kuzawa et al. 2009). It is not clear whether these associations primarily reflect a lower rate of pair bonding by high-testosterone men or a direct suppressive effect of pair bonding and fatherhood on testosterone levels (e.g., van Anders & Watson 2006). In either case, the magnitude of such effects appears to be dependent on population-specific characteristics, such as level of paternal involvement (Muller et al. 2009a).

Also unclear in human studies is whether pair bonding or paternal involvement has a greater impact on testosterone production. Kuzawa et al. (2009) reported that in a Philippine population, decreased testosterone production was more strongly associated with fatherhood than pair-bonding status, and they suggested that pair-bonded men did not fully commit to reducing their mating effort until they had become fathers. This interpretation is consistent with data from other non-Western populations, in which behavioral measures of male mating effort steadily decrease with a couple's fertility (Winking et al. 2007). Further evidence for a link between androgens and men's mating effort comes from studies showing a positive association between testosterone and interest in extrapair mating (McIntyre et al. 2006), incidence of extramarital affairs (Booth & Dabbs 1993), and number of sexual partners (Bogaert & Fisher 1995; van Anders et al. 2007).

Male Mating Effort

Sexual Selection and Male Mating Effort

Mating effort entails a variety of costs for primate males. For species in which male-male competition is intense, escalated aggression can lead to serious wounding or death (e.g., Drews 1996). At a less dramatic level, males may show chronically elevated energy expenditure related to aggressive displays, territorial aggression, and fighting. In chimpanzees, for example, high-ranking males enjoy preferential access to resources but still exhibit high levels of cortisol (a hormone produced in response to stress) and low levels of C-peptide (a marker of insulin production and, thus, energy balance) relative to those of lower-ranked males (Muller & Wrangham 2004b; Emery Thompson et al. 2009). Perhaps because of a reduction in intrasexual competition, males in caregiving species live longer than those in closely-related species with no paternal care (Allman et al. 1998).

For species that provide little or no paternal care, the fundamental life-history trade-off is between mating effort and long-term survival (Clutton Brock 1988). When and how intensely to invest in mating effort is thus a major consideration for male reproductive strategies. In chapter 18 of this volume, Alberts examines factors that lead to male reproductive success within species. The remainder of this chapter focuses on the ways in which sexual selection has shaped traits related to male-male or intrasexual competition.

Female choice, or intersexual competition, is another potentially important aspect of male mating strategies, but clear evidence for such choice in primates has proven elusive (chapter 16, this volume). It is possible that in species with intense male-male competition, the males that are successful at mating tend to be of high quality, which reduces the benefits of active female choice (Clutton-Brock & McAuliffe 2009). Alternatively, female choice may be just as important for primates as it is for birds, but female preferences are routinely masked by the effects of male-male competition and sexual coercion (Clutton-Brock & McAuliffe 2009; Muller & Wrangham 2009). As is described below, in many primates intense mating competition has selected for the increased body size and large canines that males of some species employ to constrain female mating behavior (Smuts & Smuts 1993; Clutton-Brock & Parker 1995; Muller & Wrangham 2009).

Competition for Mates

In most primate species, investment by fathers is minimal and males invariably enjoy higher potential reproductive rates than females. Intense competition for mating opportunities is the result, including both sperm competition and aggressive competition. The latter takes two general forms. In the long term, males can compete to monopolize or maintain permanent access to a group of females. In the short term (i.e., a single reproductive cycle), males can compete to mate with estrous females. The relative intensity of these forms varies between species, depending on group composition (see chapter 9, this volume, for the ultimate determinants of group composition). Gorilla males, for example, live primarily in one-male groups, so mating competition within any particular reproductive cycle is minimal, but competition to attract females to a group and retain them is pronounced (Watts 1996). Chimpanzees, by contrast, live in multimale-multifemale communities, so competition among males for access to fecund females is intense (Muller & Mitani 2005). However, males also cooperate to defend a territory against males from other groups. In many species with multimale social systems, they also compete within the group for dominance status, which is predictive of success in short-term mating competition (chapter 18, this volume).

Studies of sexual selection frequently employ socionomic sex ratio (the number of adult females to adult males in breeding groups), or related measures such as the operational sex ratio (see below), as proxies for the intensity of male competition. These measures often correlate with levels of sexual dimorphism in size and behavior. However, they do not explain all of the variability in male relationships. Male chimpanzees and male northern muriquis, for example, both live in large multimale-multifemale communities, but only chimpanzees exhibit intense aggressive competition over status (Strier 1990; Muller & Mitani 2005).

Body Size Dimorphism

Primates exhibit a wide range of sexual dimorphism in body size, from the monomorphic lemurs, callitrichines, and gibbons to the massively dimorphic baboons, gorillas, and orangutans in which males can be twice the size of females (fig. 17.3). Darwin (1871) was the first to argue that sexual selection, acting through male-male competition, could explain increased male size across a broad range of taxa. This conclusion was later supported by formal analyses showing that within species, large males outcompete small males, and that across species, increased mating competition, as assayed by socionomic sex ratio, is associated with elevated levels of dimorphism (Clutton-Brock et al. 1977; Clutton-Brock et al. 1988; LeBoeuf & Reiter 1988; Andersson 1994).

Although the first statistical treatment of sexual selection and sexual dimorphism was done with primates (Clutton-Brock et al. 1977), the importance of sexual selection as a cause of dimorphism in the order has long been contested (reviewed in Plavcan 2001, 2004). Early studies noted that dimorphism increases with body size and is not strongly related to socionomic sex ratio when monogamous species are excluded from the analysis (Clutton-Brock et al. 1977). This led some investigators to conclude that dimorphism in body size can be accounted for almost entirely by a combination of allometry and phylogeny (e.g., Leutenegger & Cheverud 1982; Cheverud et al. 1985; Martin et al. 1994).

A consistent shortcoming of such nonfunctional approaches, however, is their failure to explain *why* levels of dimorphism should be greater in larger animals. This is particularly problematic given the substantial costs associated with increasing male size (Promislow 1992; Key & Ross 1999). Nor do these hypotheses explain why in some species, such as the orangutan, large body size in males appears to be a context-dependent strategy rather than an inevitable feature of development (see below). It has also been proposed that dimorphism could arise through selection for smaller female size (Martin et al. 1994), but Lindenfors and Tullberg's (1998) phylogenetic analysis shows that this is unlikely.

More productive approaches to body size dimorphism in primates have focused on testing the sexual selection hypothesis with improved measures of male-male competition. Mitani and colleagues (1996), for example, noted that the socionomic sex ratio does not account for temporal changes in the availability of fecundable females, which varies across species, owing to differences in gestation length, ovarian cycle length, and the number of ovarian cycles per conception. They substituted the operational sex ratio, a measure of reproductively active males to females that ac-

Fig. 17.3. When social groups contain many females, male-male competition favors large body size and canines in males. Adult male gelada baboons in Ethiopia are approximately 1.5 times heavier than adult females. Photo courtesy of Jacinta C. Beehner.

counts for the temporal distribution of females, as a proxy for male-male competition and found a strong, positive relationship with body size dimorphism. In a separate study, Plavcan and van Schaik (1997a) ranked a larger sample of primates in an index of "competition levels" based on the observed frequency and intensity of male-male competition. They found that this measure was also strongly related to sexual dimorphism in body size (for similar results, but with phylogenetic controls, see Plavcan 2004).

An advantage of these approaches that employ sensitive measures of male-male competition is that they can help to explain the observed correlation between body size and dimorphism in functional terms. Operational sex ratios predictably increase with body size because interbirth intervals increase with body size (Mitani et al. 1996). Plavcan & van Schaik's (1997a) competition levels also increase with body size, though the reasons for this are not as clear. Contest competition for mates is likely to be prevalent in larger-bodied species simply because the females in such species can rely on relatively abundant, low-quality food, and thus tolerate living in groups that are potentially monopolizable by males (Jarman 1974; chapter 9, this volume).

A consistent finding in studies of primate body size is that arboreal species maintain lower levels of dimorphism than do terrestrial species (reviewed in Plavcan 2001). This pattern is often cited as support for the hypothesis that terrestrial species suffer from increased predation risk, selecting for large male size to deter predators. The evidence that arboreal primates suffer less predation than terrestrial species, however, is mixed (Isbell 1994; Shultz et al. 2004; cf. chapters 8 and 10, this volume). An alternative hypothesis is that efficient arboreal locomotion places constraints on body size that limit dimorphism.

Kappeler (1990) noted that many lemurs exhibit high levels of male-male competition in the absence of significant body size dimorphism. He suggested that speed and agility, rather than size and strength, have a greater impact on male reproductive success in these species. Lawler et al. (2005) tested this idea with detailed long-term data from the monomorphic Verreaux's sifaka. These sifaka males show high rates of aggressive mating competition, which frequently takes the form of elaborate and prolonged arboreal chases. Lawler and colleagues showed that directional selection acts to increase the size and muscularity of male legs, while stabilizing selection acts to constrain male body mass. They concluded that, for sifaka males at Beza Mahafaly, Madagascar, agility is a critical component of reproductive success. Although comprehensive data of this kind are not available for most species, a similar trade-off between body size and maneuverability may be a consideration for many arboreal primates.

Human populations show varying degrees of dimorphism, with a mean ratio of 1.1 to 1.2 for body weight of adult males compared to that of adult females (reviewed in Dixson 2009). Although Mitani and colleagues (1996) did not consider human data, this places us with other species exhibiting low operational sex ratios, consistent with the large number of mating days per conception in humans (Wrangham et al. 1999). Dixson (2009), however, argues that this figure may be misleading, because humans show relatively high levels of dimorphism in muscularity and strength that are not reflected in overall weight dimorphism, since females maintain larger fat stores than males. If muscle mass, rather than body weight, is compared, human dimorphism increases to 1.53.

Lancaster and Kaplan (2009) argue that dimorphism in human musculature is not likely to be driven by sexual selection, but rather results from the sexual division of labor, with men showing adaptations to sustained and heavy work (see also Wolpoff 1976; Frayer 1980). Darwin (1871) rejected this idea, based on his observation that "the women in all barbarous nations are compelled to work at least as hard as the men." The data necessary to rigorously test this idea are lacking, but Daly and Wilson's (1983) survey of more than 200 human societies suggests that women engage in many forms of sustained work (e.g., carrying water, gathering fuel, planting crops) at equal or greater rates than men. Furthermore, even if dimorphism in musculature were the result of men's work, this would not explain the extreme sexual dimorphism exhibited by humans in other traits, such as facial hair, body hair, pattern baldness, larynx size, and voice pitch (Dixson 2009).

Despite these complexities, most studies conclude that, across primates, sexual dimorphism in body size is well explained by sexual selection acting through male-male contest competition (Plavcan 2001, 2004). Importantly, however, an absence of dimorphism does not necessarily imply the existence of pair bonds or a lack of aggressive male-male competition (Lawler 2009).

Canine Dimorphism

Sexual dimorphism in canine size varies widely across primates, with some species showing complete monomorphism and others maintaining male canine heights greater than four times that of females (Plavcan 2004). Developments in the study of canine dimorphism have largely paralleled those previously discussed for body size dimorphism (reviewed in Plavcan et al. 1995; Plavcan 2004). Canines are frequently employed as formidable weapons, and Darwin (1871) first proposed that sexual selection, acting via intrasexual competition, could account for the exaggerated version in some males. Early investigations supported this view, showing that canine dimorphism is generally associated with mating system (Harvey et al. 1978). Later studies noted that this association was weak or nonexistent when monogamous species were excluded, and proposed that body size and phylogeny were more important factors (e.g., Leutenegger & Cheverud 1982). Finally, studies employing more sensitive measures of male-male competition reiterated the importance of sexual selection (Plavcan & van Schaik 1992; Plavcan et al. 1995).

Canine dimorphism, however, shows interesting patterns that are distinct from those associated with body size dimorphism. Plavcan (2004), for example, reported that canine and body size dimorphism are only modestly correlated across primates. Furthermore, canine dimorphism correlates with Plavcan and van Schaik's (1992) competition levels, but not with Mitani and colleagues' operational sex ratio (Plavcan 2004). What can account for these discrepancies?

First, in some species, such as the monogamous gibbons,

dentition is monomorphic not because males have reduced canines, but because females have impressive, male-like canines. This likely reflects the fact that females find canines just as useful in contest competition over food as males do in contest competition over mates. As one test of this hypothesis, Plavcan (2004) looked at primate species in which the intensity of aggression among females was high, and found a positive correlation between female group size and relative female canine size (ratio of female canine size to body weight). Large body size might also be useful for female competition in these species, but female body size is presumably under tighter energetic constraints than canine size, owing to the demands of reproduction.

Second, species in which coalitions frequently affect the outcome of agonistic interactions appear to have reduced canines in comparison to closely related species that lack coalitionary aggression (Plavcan et al. 1995). This reduction is more pronounced in females than in males, because coalitionary aggression is common among female-bonded cercopithecines. However, male canines are also smaller than expected in species exhibiting coalitionary intergroup interactions, such as chimpanzees, spider monkeys (*Ateles* spp.), and muriquis (Plavcan et al. 1995).

Third, it appears that canine height is constrained by maximal gape because, to be useful as weapons, upper and lower canines cannot overlap when the mouth is fully open (Hylander & Vinyard 2006). Maximal gape, in turn, is constrained by the placement of the jaw muscles. Positioning these muscles anteriorly allows for greater bite force in the posterior dentition, but it limits the opening of the mouth, and thus limits maximal canine size (Hylander & Vinyard 2006; Hylander 2009). This could be an important constraint in species that, for dietary reasons, require significant postcanine bite force, including early hominins (Hylander 2009).

Detailed, long-term data on male canine height and reproductive success are available from only two species: Verreaux's sifakas and mandrills. Lawler and colleagues (2005) showed that, for the monomorphic Verreaux's sifaka, directional selection was not acting on male canine size. In the massively dimorphic mandrill, by contrast, reproductive success was strongly correlated with maximal adult canine size (Leigh et al. 2008). Furthermore, almost all conceptions occurred after males had achieved their maximum canine height, but before significant breakage had occurred (Leigh et al. 2008).

As was the case with body size dimorphism, then, canine dimorphism is largely explained by sexual selection, but acting under a different set of constraints (Plavcan 2004). And once again, the presence of these constraints indicates that a reduction in male canine size, in and of itself, does not imply a lack of aggressive male-male competition. Across haplorrhines, the effects of sexual selection on canine size are stronger than those on body size, suggesting that in this suborder canines may be more important than body size in male-male competition (Thorén et al. 2006).

Sperm Competition

Although contest competition features prominently in the mating strategies of male primates, scramble competition, in the form of post-mating sperm competition, is also widespread (reviewed in Dixson 1998). Sperm competition can occur in the absence of precopulatory contest competition—as in southern muriqui males, who sometimes peacefully queue on branches, waiting to mate with cycling females (Milton 1985)—or alongside aggressive competition, as in chimpanzees. In either case, sexual selection has had multiple effects on male reproductive physiology: some subtle, others marked.

Early work on sperm competition focused on interspecific variation in testes size. Parker (1970) was the first to suggest that, in species where females mate with multiple partners, males who produced more sperm would have an advantage in achieving fertilization. Short (1979) recognized that variation in testes size, and consequently sperm production, was related to mating system in hominoids. Gorillas, he noted, live primarily in one-male groups, mate infrequently, and have a combined testes weight of 30 grams. Chimpanzees, by contrast, live in multimale-multifemale communities in which mating is frequent and promiscuous, and despite their much smaller body size, they have a combined testes weight of more than 120 grams (fig. 17.4).

Fig. 17.4. Owing to sperm competition, primates living in multimale groups maintain large testes for their body size. In some species the effect is dramatic. Pictured is a single chimpanzee testis (below) compared to the same male's brain (above). Photo courtesy of Martin N. Muller.

In the first systematic review of testes size and mating system across primates, Harcourt and colleagues (1981) showed that testes size predictably increases with body size, but that promiscuous species maintain testes that are larger than expected. Males living in pairs or in one-male groups have comparatively small testes. Moreover, primates with relatively large testes have higher rates of sperm production, maintain larger sperm reserves in the epididymis, and produce more sperm per ejaculate than males with smaller testes (Møller 1988), all patterns that have now been documented in a wide range of taxa (Møller 1989; Birkhead & Møller 1998).

More recent studies have focused on additional aspects of male reproductive physiology that are likely to provide an advantage in sperm competition. Increases in sperm midpiece volume, for example, reflect larger numbers and volumes of the mitochondria that power sperm movement, and thus higher sperm motility (Anderson & Dixson 2002). Across both primates and mammals generally, volume of the sperm midpiece is correlated with mating system and relative testes size (Anderson & Dixson 2002; Dixson & Anderson 2004; Anderson et al. 2005). Swimming speed and power have recently been measured directly in large numbers of individual sperm, and both are significantly higher in chimpanzees and macaques, which maintain multimale breeding systems, than in gorillas or humans (Nascimento et al. 2008). Kleven and colleagues (2009) replicated this effect with sperm from 42 bird species, recording faster swimming speeds in species with high rates of extrapair paternity. Finally, in both primates and other mammals, species with multimale mating systems exhibit short vas deferens with a large ratio of muscle to lumen (Anderson et al. 2004). Shortening the vas deferens decreases the total distance over which sperm must be transported during copulation, and thickening the longitudinal muscle layers of the vas deferens appears to increase the force of sperm transport.

Abnormally shaped, or pleiomorphic, sperm occur at varying rates in primates (e.g., Seuanez et al. 1977). Baker and Bellis (1988) noted high levels of pleiomorphic sperm in humans and proposed that this had resulted from sperm competition, with different sperm morphs adapted to different roles. Harcourt (1991) and Dixson (2009) offer detailed and devastating critiques of this hypothesis, showing that across mammals, levels of abnormal sperm actually decrease with the incidence of multimale mating. Additional studies in which ejaculates from different men were mixed and carefully examined for sperm interaction provided experimental refutation of the Baker and Bellis hypothesis (Moore et al. 1999). It now appears that pleiomorphic sperm result from errors in meiosis that are minimized in species experiencing high levels of sperm competition (Dixson 2009).

While there is no evidence for contest competition among primate sperm, males in some species do produce copulatory plugs, formed by proteins from the seminal vesicles (semenogelin 1 and 2) interacting with the prostatic enzyme vesiculase (Dixson 2009). It is not clear whether such plugs function to block the sperm of rival males or simply to retain sperm within the female reproductive tract. However, they are clearly associated with promiscuous mating. Dixson and Anderson (2002) used a four-point scale to rate seminal coagulation in 26 primate genera (1 = no coagulation; 4 = solid plug). They found that this rating, as well as seminal vesicle size, was greater in species with multimale mating systems. Furthermore, SEMG1 and SEMG2, the genes encoding semenogelin 1 and 2, show evidence of strong positive selection in promiscuous species, like the chimpanzee, whereas they appear to be nonfunctional pseudogenes in some species with one-male mating systems, such as gorillas and at least one gibbon (Dorus et al. 2004; Carnahan & Jensen-Seaman 2008).

Evolutionary psychologists have often supported Baker and Bellis's notion that sperm competition played an important role in human evolution (Shackelford & Goetz 2006; Goetz et al. 2007). However, most of this literature is not grounded in a comparative perspective. It has been asserted, for example, that the morphology of the human penis, with its prominent glans and pronounced coronal ridge, is an adaptation for displacing previously deposited sperm from the female reproductive tract (Gallup et al. 2003). Dixson (2009) observes that this claim overlooks the fact that the gorilla penis shows precisely the same morphology, albeit in miniature.

Harcourt and colleagues' (1981) data on testes size are often cited as evidence for human sperm competition, as they indicate that humans are "intermediate" between chimpanzees and gorillas in the ratio of testes weight to body weight (e.g., Shackelford & Goetz 2006). Dixson (2009) discusses complications of this idea, including the fact that very few human males ($n = 4$) were represented in the 1981 study. He assembled data from 14 human populations comprising more than 7,000 subjects, and showed that the range of variation in human testes size is considerable, with some populations being indistinguishable from the gorilla sample. Furthermore, testes weight is only one relatively crude index of sperm competition. A robust test of the hypothesis that sperm competition has played a role in human evolution requires examination of the whole suite of physiological traits discussed above. Dixson's (2009) thorough comparative analysis shows that on almost any conceivable measure, humans cluster near

Table 17.2. Physiological correlates of sperm competition in selected species

	Cebuella pygmaea	Gorilla gorilla	Pan troglodytes	Macaca mulatta	Homo sapiens	Reference
Sperm midpiece volume (µm³)	3.3	6.9	7.8	10.5	3.8	Anderson et al. 2005
Vas deferens muscle:lumen ratio	6.23	—	—	11.21	6.31	Anderson et al. 2004
Testes weight/body weight (%)	—		0.55 (n = 2) W			Muller, unpublished data
		0.02 (n = 2) W	0.27 (n = 3) C	0.56 (n = 26) C	0.06 (n = 4)	Harcourt et al. 1981
					0.029 (n = 100) China*	Dixson 2009
					0.056 (n = 325) India*	Dixson 2009
					0.047 (n = 132) USA*	Dixson 2009
Sperm swimming speed (µm/sec)	—	23.69	85.35	77.54	54.60	Nascimento 2008
Sperm swimming power (mW)	—	2.87	65.52	50.11	7.94	Nascimento 2008
Sperm pleiomorphism (%)	—	29	4.5	—	27	Seuanez et al. 1977
Seminal vesicles size (1–4) (1 = vestigial, 4 = large)	—	2	4	4	3	Dixson 2009
Semen coagulation rating (1–4) (1 = no coagulation, 4 = solid plug)	—	2	4	3	2	Dixson 1998
SEMG2 gene: Ka/Ks ratio†	—	0.61	2.52	1.28	0.91	Dorus et al. 2004

Comparison of humans with two primates that engage in intense sperm competition (*Pan troglodytes* and *Macaca mulatta*) and two species with little or no sperm competition (*Gorilla gorilla* and *Cebuella pygmaea*). *Cebuella pygmaea* is one of the most monogamous marmosets (Goldizen 2003). For testes weights, (C) indicates captive and (W) wild individuals.

*Most human samples report testes weights but not body weights. For the Indian and Chinese samples, an average human male body weight of 63.5 kg (Dixson 2009) was used to calculate a percentage. For the US sample, an average male body weight of 77 kg (Williamson 1993) was used.

†Ratio of the rate of nonsynonymous substitutions (Ka) to the rate of synonymous substitutions (Ks) in a gene. Numbers close to 1 are consistent with neutral selection. Numbers higher than 1 are consistent with positive selection.

gorillas and other pair-living or one-male group species, and not with multimale breeders like chimpanzees or rhesus macaques. Furthermore, the fact that humans are "intermediate" between chimpanzees and gorillas on some traits may simply reflect the fact that humans have a more recent polyandrous ancestor than do gorillas, rather than indicating an intermediate level of sperm competition in our species (e.g., Carnahan & Jensen-Seaman 2008). Table 17.2 summarizes data on physiological traits related to sperm competition. The reader will see that a broad comparative perspective strongly challenges the idea that sperm competition has played a significant role in recent human evolution (see also chapter 20, this volume).

Integrating the Effects of Sexual Selection

Although socionomic sex ratio is a relatively crude measure of male competition, it is instructive to examine the distribution of male sexual traits across different group sizes. Figure 17.5 shows a simplified scheme of the effects of monogamy (one male, one female), polyandry (more than one male, one female), polygyny (one male, more than one female), and polygynandry (more than one male, more than one female) on such traits. Figure 17.6 illustrates these patterns with average values from a large number of primate species in each category. It is apparent that traits related to sperm competition are largely driven by the number of males in

a group and are expressed independently of traits related to contest competition, which are driven by the number of females in a group (Harvey & Harcourt 1984). It is also worth noting that, despite living in multimale-multifemale groups, humans exhibit the distinctive pattern of low canine dimorphism, low body size dimorphism, and smaller than expected testes for body weight that generally characterizes monogamous primates.

Alternative Reproductive Tactics

When mating competition is intense and reproduction is skewed toward a small number of males, selection may favor conditional strategies that allow low-quality males to preserve some reproductive success without investing in the costly morphological and behavioral components of mating effort (reviewed by Gross 1996; Setchell 2008; Taborsky et al. 2008). Such conditional strategies favor a wide range of alternative reproductive tactics. These include, for example, sneaking copulations with females instead of fighting for them, mimicking females to avoid aggression from dominant males, and roving in search of fecund females instead of defending a territory (Gross 1996). Because the range of specific tactics varies widely across taxa, the general term "bourgeois tactics" is sometimes used to refer to all forms of conventional contest competition for access to territories and mates, with alternatives labeled "parasitic

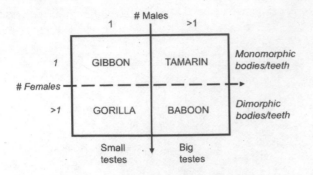

Fig. 17.5. Simplified scheme of male sexual traits and primate group composition, with an example for each combination.

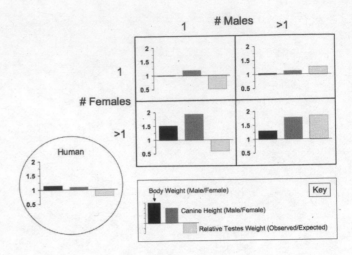

Fig. 17.6. Male sexual traits and primate group composition. Bars show means of generic means. Dimorphism data are from Plavcan (2004). Observed versus expected relative testes weights (including that of *Homo sapiens*) are from Harvey and Harcourt (1984). Sample sizes (N) are listed as #genera (#*species*) for the categories body weight dimorphism (B), canine height (C), and relative testes weight (T). Monogamous sample: $N_B = 4(12)$, $N_C = 4(9)$, $N_T = 2(3)$. Polyandrous sample: $N_B = 3(5)$, $N_C = 3(4)$, $N_T = 2(2)$. Polygynandrous sample: $N_B = 20(43)$, $N_C = 17(37)$, $N_T = 10(16)$. Polygynous sample: $N_B = 14(28)$, $N_C = 13(24)$, $N_T = 7(8)$. Data on human body weight dimorphism are from Dixson (2009). Data on human canine dimorphism are from Plavcan and van Schaik (1997b). Mating system classification follows Plavcan (2004), with all of his species included save *Presbytis melalophos*, *Mico argentatus*, *Leontopithecus rosalia*, and *Saguinus midas*, which had variable mating systems that were difficult to classify. Orangutans (*Pongo* spp.) were also excluded because male bimaturism complicates estimates of sexual dimorphism.

tactics" (Taborsky et al. 2008). To qualify as true "alternative" tactics, these behavioral or morphological variants must show a discontinuous distribution across males.

Alternative reproductive tactics in vertebrates are normally plastic traits, capable of expression by any male under the appropriate conditions (Gross 1996). In some primates, for example, secondary sexual characteristics are plastic traits that can develop permanently ("fixed sequence traits"), as in orangutans, or reversibly ("reverse sequence traits"), as in mandrills (Setchell 2008). In species with plastic traits, shifts from parasitic to bourgeois tactics are frequently driven by androgens, such as testosterone (Oliviera et al. 2008).

The decision to switch from one tactic to another is made by evaluating the potential fitness benefits of each, given the individual's competitive qualities relative to those of other males in the population (Gross 1996). These are influenced by factors such as energetic status, health, and age. The average payoffs for different conditional strategies are not expected to be equal, but each individual should adopt the specific tactic that maximizes his own fitness.

Setchell (2008) discusses numerous candidates for alternative reproductive tactics in primates. Some of these are potential examples of discontinuous traits. For example, in orangutans, and perhaps in some galagos and pottos (*Perodicticus potto*), territoriality is a bourgeois strategy of dominant males, and parasitic nomads attempt to sneak copulations with females in the absence of territory holders. A similar dynamic occurs between bourgeois group members and parasitic intruders, who make brief incursions into groups during breeding periods (Japanese macaques, Hanuman langurs, blue monkeys [*Cercopithecus mitis*], patas monkeys [*Erythrocebus patas*], red-tail monkeys [*Cercopithecus ascanius*]: Setchell 2008).

Setchell also includes the cultivation of female friendships, the maintenance of male coalitions, sexual coercion, and infanticide as potential alternative reproductive tactics.

However, these traits may not all show a clear discontinuous distribution. In some species they may indeed be parasitic strategies, whereas in others they may be practiced by all males, but to different degrees or in slightly different contexts (e.g., sexual coercion: Muller & Wrangham 2009).

One of the most striking examples of an alternative reproductive tactic in primates is the arrested secondary sexual development of orangutan males that generates two discrete morphs (see fig. 6.5, chapter 6, this volume). "Flanged" and "unflanged" males are both sexually mature and show broad overlap in chronological age. However, unflanged males are smaller and lack secondary sexual characteristics such as protruding cheek pads, prominent throat pouches, high testosterone levels, and roaring "long-call" vocalizations (Crofoot & Knott in press). By delaying their maturation, some males likely avoid the costs associated with flanged status, including increased energy requirements, reduced mobility, and dangerous aggression from other flanged males.

The two male orangutan morphs employ markedly different mating tactics. Flanged males are territorial, range over relatively small areas, and are preferentially approached by ovulatory females (Knott & Kahlenberg 2007; Knott et al. 2010). Unflanged males, whose small body size increases their mobility, range widely in search of mates, and frequently employ force to obtain copulations from resistant females (Utami et al. 2002).

Genetic information on reproductive success in unflanged males is still limited (Utami et al. 2002; Goossens et al. 2006), so it is unclear whether unflanged males simply make the best of a bad job, with a decreased probability of siring offspring (Mitani 1985; Galdikas 1985; Schürmann & van Hooff 1986), or whether the tactic enjoys payoffs equal to those enjoyed by flanged males. Maggioncalda and colleagues (1999) proposed that arrested and developed males may achieve equivalent fitness through a balance of the costs and benefits associated with flanging. Others have discussed the potential importance of frequency-dependent selection on maintaining both fertile morphs (Rodman & Mitani 1987; Gross 1996; Utami et al. 2002).

Male mandrills also display bimaturism in secondary sexual characteristics. Like flanged orangutans, "fatted" male mandrills (those with bulkier rumps and brighter coloration) have higher testosterone levels than other males (Dixson et al. 1993) and their physical appearance is indicative of relative dominance (Setchell & Dixson 2002). In mandrills, however, most of the characteristics associated with fatting are reversible, waxing and waning as the male's status fluctuates (Setchell & Dixson 2001). This flexibility in somatic and reproductive investment contrasts markedly with the permanent life history decision faced by orangutan males.

It is notable that, with the exception of humans, orangutans are the only primate species in which forced copulations occur frequently, despite the fact that other forms of sexual aggression and coercive behavior are prevalent in the order (Muller & Wrangham 2009). It is tempting to characterize forced copulation as an alternative reproductive tactic practiced by low-quality, unflanged males, and similar arguments have been made characterizing rape as an alternative tactic for men with poor sexual prospects (Thornhill & Palmer 2000). On the one hand, forced copulation clearly represents a method for obtaining sexual access when conventional mate attraction has failed. On the other hand, this cannot be a complete explanation, because rape is commonly performed by high-status males, both human and orangutan, including those who have had prior consensual sexual interactions with their victims (Knott 2009; Emery Thompson 2009). Thus, forced copulation does not show the clear, discontinuous distribution across males that would be consistent with an alternative tactic.

Why is forced copulation so rare in primates? Two prominent explanations focus on costs to the male. First, females who are commonly alone or separated from protective kin networks (as in humans and orangutans) may be more susceptible to coercion because they lack allies (Smuts & Smuts 1993). However, vulnerability is an incomplete explanation, since forced copulations are absent in many semisolitary species. Second, pronounced differences in body size surely facilitate forcible restraint by flanged orangutan males. However, smaller unflanged males appear just as successful at overcoming female resistance, and extreme body size dimorphism occurs in other species without forced copulation (e.g., gorillas, chacma baboons). A more complete explanation of forced copulation will likely require investigating the costs and benefits of female resistance, rather than focusing on costs to the male (Arnqvist & Rowe 2005). For orangutans, high rates of female resistance imply high costs of mating (Knott 2009). For females that live in multimale mating systems, where the benefits of paternity confusion are substantial, forced copulations may be rare simply because intense resistance is rare. It is striking, for example, that the small number of forced copulations reported in chimpanzees mostly involved close relatives, for whom females are expected to maintain strong aversions owing to the high costs of inbreeding (Muller et al. 2009b).

Summary and Questions for Future Research

Primate males face a fundamental reproductive trade-off between providing care for existing offspring and pursuing additional mating opportunities. Mammalian biology leaves fathers less constrained than mothers in caring for young, favoring male mating effort. However, paternal care is widespread among primates, occurring in at least 40% of genera. Why the strepsirrhines and tarsiers are underrepresented in this sample remains a puzzle for future investigation.

Male solicitude takes a variety of forms. Direct care, such as transporting and provisioning offspring, appears to be most common in species with costly infants and high levels of paternity certainty. Another prevalent form of care is protection from infanticide and predation, which is linked to paternity certainty but observed in species with otherwise low levels of male-infant affiliation. In many primates, a male's relationship with the mother, rather than his degree of paternity certainty, is the main predictor of less intense forms of infant care (holding, grooming, playing with, and tolerating infants in proximity), suggesting that these frequently function as mating effort, making females more likely to mate with the caregiver. Further research is needed, however, to determine how frequently male care functions solely as mating effort rather than as true paternal investment. Ideally, such work would distinguish whether male care increases offspring quality (as opposed to simply survival), or increases a female's reproductive rate.

For species that provide low levels of paternal care, the fundamental life-history trade-off is between mating effort and long-term survival. At the mechanistic level, investment in mating effort appears to be modulated by the steroid hormone testosterone. Additional studies of wild primates are now needed to elucidate the action of other hormones, such as oxytocin and prolactin, that likely play a role in supporting male care.

Across primates, sexual selection has favored traits that increase male effectiveness in mating competition. Sexual dimorphism in body and canine size is well explained by sexual selection acting through male-male contest competition. It is not clear, however, why body size dimorphism should be generally reduced in arboreal species. Additional work is needed to test whether dimorphism in terrestrial species can be partly attributed to defense against predation, or whether arboreal locomotion places constraints on body size or aggression. Future research may also account for the odd human pattern of low canine dimorphism, mild body weight dimorphism, and extreme dimorphism in musculature and strength. Especially useful would be studies evaluating the relative importance of male-male competition and male provisioning in selecting for dimorphic male musculature.

Scramble competition occurs via sperm competition that, in multimale groups, selects for a suite of specialized physiological traits including large testes for body size. These traits are expressed independently of traits related to contest competition, which are driven by the number of females in a group. When mating competition is intense and reproduction is skewed toward a few males, selection may favor conditional strategies that allow low-quality males to maintain some reproductive success without costly investment in male-male competition. In most cases, however, it is not clear whether such strategies enjoy equal payoffs or simply represent males making the "best of a bad job." Future studies are needed to assess the payoffs associated with different mating strategies, and to identify which of these show a clear discontinuous distribution across males.

References

Achenbach, G. G. & Snowdon, C. T. 2002. Costs of caregiving: Weight loss in captive adult male cotton-top tamarins (*Saguinus oedipus*) following the birth of infants. *International Journal of Primatology*, 23, 179–189.

Allman, J., Rosin, A., Kumar, R. & Hasenstaub, A. 1998. Parenting and survival in anthropoid primates: Caretakers live longer. *Proceedings of the National Academy of Science USA*, 95, 6866–6869.

Altmann, J. 1980. *Baboon Mothers and Infants*. Cambridge, MA: Harvard University Press.

Alvergne, A., Faurie, C. & Raymond, M. 2009. Variation in testosterone levels and male reproductive effort: Insight from a polygynous human population. *Hormones and Behavior*, 56, 491–497.

Anderson, K. G., Kaplan, H., & Lancaster, J. B. 2007. Confidence of paternity, divorce, and investment in children by Albuquerque men. *Evolution and Human Behavior*, 28, 1–10.

Anderson, M. J. & Dixson, A. F. 2002. Sperm competition: Motility and the midpiece in primates. *Nature*, 416, 496.

Anderson, M. J., Nyholt, J. & Dixson, A. F. 2004. Sperm competition affects the structure of the mammalian vas deferens. *Journal of Zoology*, 264, 97–103.

———. 2005. Sperm competition and the evolution of sperm midpiece volume in mammals. *Journal of Zoology*, 267, 135–142.

Andersson, M. 1994. *Sexual Selection*. Princeton, NJ: Princeton University Press.

Andres, M., Solignac, M & Perret, M. 2003. Mating system in mouse lemurs: Theories and facts, using analysis of paternity. *Folia Primatologica*, 74, 355–366.

Archer, J. 2006. Testosterone and human aggression: An evaluation of the challenge hypothesis. *Neuroscience and Biobehavioral Reviews*, 30, 319–345.

Arnqvist, G. & Rowe, L. 2005. *Sexual Conflict*. Princeton, NJ: Princeton University Press.

Baker, R. R. & Bellis, M. A. 1988. "Kamikaze" sperm in mammals. *Animal Behaviour*, 36, 936–939.

Bales, K. L., French, J. A., McWilliams, J., Lake, R. A. & Dietz, J. M. 2006. Effects of social status, age, and season on androgen and cortisol levels in wild male golden lion tamarins (*Leontopithecus rosalia*). *Hormones and Behavior*, 49, 88–95.

Bardi, M., Shimizu, K., Fujita, S., Borgognini-Tarli, S. & Huffman, M. 2001. Hormonal correlates of maternal style in captive macaques (*Macaca fuscata* and *M. mulatta*). *International Journal of Primatology*, 647–662.

Bart, J. & Tornes, A. 1989. Importance of monogamous male birds in determining reproductive success. *Behavioral Ecology & Sociobiology*, 24, 109–116.

Beehner, J. C., Bergman, T. J., Cheney, D. L., Seyfarth, R. M. & Whitten, P. L. 2005. The effect of new alpha males on female stress in free-ranging baboons. *Animal Behaviour*, 69, 1211–1221.

———. 2006. Testosterone predicts future dominance rank and mating activity among male chacma baboons. *Behavioral Ecology and Sociobiology*, 59, 469–479.

Benderlioglu, Z., Sciulli, P. W. & Nelson, R. J. 2004. Fluctuating asymmetry predicts human reactive aggression. *American Journal of Human Biology*, 16, 458–469.

Bercovitch, F. B. 1991. Mate selection, consortship formation, and reproductive tactics in adult female savanna baboons. *Primates*, 32, 437–452.

Berenstain, L., Rodman, P. S. & Smith, D. G. 1981. Social relations beween fathers and offspring in a captive group of rhesus monkeys (*Macaca mulatta*). *Animal Behaviour*, 29, 1057–1063.

Berg, S. J. & Wynne-Edwards, K. E. 2001. Changes in testosterone, cortisol, and estradiol levels in men becoming fathers. *Mayo Clinic Proceedings*, 76, 582–592.

Berman, C. M. 1982. The social development of an orphaned rhesus infant on Cayo Santiago: Male care, foster mother-orphan interaction and peer interaction. *American Journal of Primatology*, 3, 131–141.

Biquand, S., Boug, A., Biquand-Guyot, V. & Gautier, J.-P. 1994. Management of commensal baboons in Saudi Arabia. *Revue d'Ecologie*, 49, 213–222.

Birkhead, T. R. & Moller, A. P. 1998. *Sperm Competition and Sexual Selection*. London: Academic Press.

Bogaert, A. F. & Fisher, W. A. 1995. Predictors of university men's number of sexual partners. *The Journal of Sex Research*, 32, 119–130.

Bolin, I. 1981. Male parental behavior in black howler monkeys (*Alouatta palliata pigra*) in Belize and Guatemala. *Primates*, 22, 349–360.

Booth, A. & Dabbs, J. 1993. Testosterone and men's marriages. *Social Forces*, 72, 463–477.

Borries, C. 1997. Infanticide in seasonally breeding multimale groups of Hanuman langurs (*Presbytis entellus*) in Ramnagar (South Nepal). *Behavioral Ecology and Sociobiology*, 41, 139–150.

Borries, C., Launhardt, K., Epplen, C., Epplen, J. T. & Winkler, P. 1999. Males as infant protectors in Hanuman langurs (*Presbytis entellus*) living in multimale groups: Defence pattern, paternity and sexual behaviour. *Behavioral Ecology and Sociobiology*, 46, 350–356.

Bradley, B. J., Robbins, M. M., Williamson, E. A., Steklis, H. D., Steklis, N. G., Eckhardt, N., Boesch, C. & Vigilant, L. 2005. Mountain gorilla tug-of-war: Silverbacks have limited control over reproduction in multimale groups. *Proceedings of the National Academy of Science USA*, 102, 9418–9423.

Bribiescas, R. G. 2001. Reproductive ecology and life history of the human male. *Yearbook of Physical Anthropology*, 44, 148–176.

Buchan, J. C., Alberts, S. C., Silk, J. B. & Altmann, J. 2003. True paternal care in a multi-male primate society. *Nature*, 425, 179–181.

Busse, C. & Hamilton, W. J. 1981. Infant carrying by male chacma baboons. *Science*, 212, 1281–1283.

Busse, C. D. & Gordon, T. P. 1984. Infant carrying by adult male mangabeys (*Cercocebus atys*). *American Journal of Primatology*, 6, 133–141.

Carnahan, S. J. & Jensen-Seaman, M. I. 2008. Hominoid seminal protein evolution and ancestral mating behavior. *American Journal of Primatology*, 70, 939–948.

Cavigelli, S. A. & Pereira, M. E. 2000. Mating season aggression and fecal testosterone levels in male ring-tailed lemurs (*Lemur catta*). *Hormones and Behavior*, 37, 246–255.

Charpentier, M. J. E., van Horn, R. C., Altmann, J. & Alberts, S. C. 2008. Paternal effects on offspring fitness in a multi-male primate society. *Proceedings of the National Academy of Sciences USA*, 105, 1988–1992.

Cheverud, J., Dow, M. & Leutenegger, W. 1985. The quantitative assessment of phylogenetic constraints in comparative analyses: Sexual dimorphism in body weight among primates. *Evolution*, 39, 1335–1351.

Chivers, D. J. 1974. *Contributions to Primatology, Volume 4. The Siamang in Malaya: A Field Study of a Primate in Tropical Rain Forest*. Basel: S. Karger.

Clutton-Brock, T. H. & McAuliffe, K. 2009. Female mate choice in mammals. *Quarterly Review of Biology*, 84, 3–27.

Clutton-Brock, T. H. & Parker, G. A. 1992. Potential reproductive rates and the operation of sexual selection. *Quarterly Review of Biology*, 67, 437–456.

———. 1995. Sexual coercion in animal societies. *Animal Behavior*, 49, 1345–1365.

Clutton-Brock, T. H., Harvey, P. H. & Rudder, B. 1977. Sexual dimorphism, socionomic sex ratio and body weight in primates. *Nature*, 269, 797–800.

Clutton-Brock, T. H., Albon, S. & Guinness, F. 1988. Reproductive success in male and female red deer. In *Reproductive Success* (ed. by Clutton-Brock, T. H.), 335–343. Chicago: University of Chicago Press.

Collins, D. A. 1986. Relationships between adult male and infant baboons. In *Primate Ontogeny, Cognition and Social Behvior* (ed. by Else, J. G. & Lee, P. C.), 205–218. New York: Cambridge University Press.

Cristóbal-Azkarate, J., Chavira, R., Boeck, L., Rodríguez-Luna, E. & Veà, J. J. 2006. Testosterone levels of free-ranging resident mantled howler monkey males in relation to the number and density of solitary males: A test of the challenge hypothesis. *Hormones and Behavior*, 49, 261–267.

Crofoot, M. C. & Knott, C. D. In press. What we do and do not know about orangutan male dimorphism. In *Great and Small Apes of the World* (ed. by Galdikas, B. M. F., Briggs, N., Sheeran, L. K. & Shapiro, G. L.) New York: Kluwer.

Daly, M. & Wilson, M. 1983. *Sex, Evolution, and Behavior*. Belmont, CA: Wadsworth.

Darwin, C. 1871. *The Descent of Man*. London: Murray.

Dawkins, R. & Carlisle, T. R. 1976. Parental investment, mate desertion and a fallacy. *Nature*, 161, 131–133.

Detwyller, K. A. 1995. The hominid blueprint for the natural age at weaning in modern human populations. In *Breastfeeding: Biocultural Perspectives* (ed. By Stuart-Macadam, P. & Dettwyler, K. A.), 39–74. Hawthorne, NY: Aldine de Gruyter.

Dixson, A. F. 1998. *Primate Sexuality: Comparative Studies of the Prosimians, Monkeys, Apes, and Human Beings*. New York: Oxford University Press.

———. 2009. *Sexual Selection and the Origins of Human Mating Systems*. Oxford: Oxford University Press.

Dixson, A. F. & Anderson, M. J. 2002. Sexual selection, seminal coagulation and copulatory plug formation in primates. *Folia Primatologica*, 73, 63–69.

Dixson, A. F., Bossi, T. & Wickings, E. J. 1993. Male dominance and genetically determined reproductive success in the mandrill (*Mandrillus sphinx*). *Primates*, 34, 525–532.

Dorus, S., Evans, P. D., Wyckoff, G. J., Choi, S. S. & Lahn, B. T. 2004. Rate of molecular evolution of the seminal protein gene SEMG2 correlates with levels of female promiscuity. *Nature Genetics*, 36, 1326–1329.

Drews, C. 1996. Contexts and patterns of injuries in free-ranging male baboons (*Papio cynocephalus*). *Behaviour*, 133, 443–474.

Dunbar, R. I. M. 1984a. Infant use by male gelada in agonistic contexts: Agonistic buffering, progeny protection or soliciting support? *Primates*, 25, 28–35.

———. 1984b. *Reproductive Decisions: An Economic Analysis of Gelada Baboon Social Strategies*. Princeton, NJ: Princeton University Press.

Emery Thompson, M. 2009. Human rape: revising evolutionary perspectives. In *Sexual Coercion in Primates: An Evolutionary Perspective on Male Aggression against Females* (ed. by Muller, M. N. & Wrangham, R. W.), 346–376. Cambridge, MA: Harvard University Press.

Emery Thompson, M., Muller, M. N., Wrangham, R. W., Lwanga, J. S. & Potts, K. B. 2009. Urinary C-peptide tracks seasonal and individual variation in energy balance in wild chimpanzees. *Hormones and Behavior*, 55, 299–305.

Emlen, S. T. & Oring, L. W. 1977. Ecology, sexual selection, and the evolution of mating systems. *Science*, 197, 215–223.

Fernandez-Duque, E., Juarez, C. P. & Di Fiori, A. 2008. Adult male replacement and subsequent infant care by male and siblings in socially monogamous owl monkeys (*Aotus azarai*). *Primates*, 49, 81–84.

Fietz, J. & Dausmann, K. H. 2003. Costs and potential benefits of parental care in the nocturnal fat-tailed dwarf lemur (*Cheirogaleus medius*). *Folia Primatologica*, 74, 246–258.

Fossey, D. 1979. Development of the mountain gorilla (*Gorilla gorilla beringei*): The first thirty-six months. In *The Great Apes* (ed. by Hamburg, D. A. & McCown, E. R.), 139–186. Menlo Park, CA: Benjamin/Cummings.

Fragaszy, D. M, Schwarz, S., & Shimosaka, D. 1982. Longitudinal observations of care and development of infant titi monkeys (*Callicebus moloch*). *American Journal of Primatology*, 2, 191–200.

Frayer, D. W. 1980. Sexual dimorphism and cultural evolution in the Late Pleistocene and Holocene of Europe. *Journal of Human Evolution*, 9, 399–415.

Galdikas, B. M. F. 1985. Subadult male orangutan sociality and reproductive behavior at Tanjung Puting. *American Journal of Primatology*, 8, 87–99.

Gallup, G. G. J., Burch, R. L., Zappieri, M. L., Parvez, R. A., Stockwell, M. L. & Davis, J. A. 2003. The human penis as a semen displacement device. *Evolution and Human Behavior*, 24, 277–289.

Garber, P. A. 1997. One for all and breeding for one: Cooperation and competition as a tamarin reproductive strategy. *Evolutionary Anthropology*, 5, 187–199.

Garber, P. A. & Leigh, S. R. 1997. Ontogenetic variation in small-bodied New World primates: Implications for patterns of reproduction and infant care. *Folia Primatologica*, 68, 1–22.

Girard-Buttoz, C., Heistermann, M., Krummel, S. & Engelhardt, A. 2009. Seasonal and social influences on fecal androgen and glucocorticoid excretion in wild male long-tailed macaques (*Macaca fascicularis*). *Physiology and Behavior*, 98, 168–175.

Goetz, A. T., Shackelford, T. K., Platek, S. M., Starratt, V. G. & McKibbin, W. F. 2007. Sperm competition in humans: Implications for male sexual psychology, physiology, anatomy, and behavior. *Annual Review of Sex Research*, 18, 1–22.

Goldizen, A. W. 1987. Tamarins and marmosets: Communal care of offspring. In *Primate Societies* (ed. by Smuts, B., Cheney, D. L., Seyfarth, R. M., Wrangham, R. W. & Struhsaker, T. T.), 34–43. Chicago: University of Chicago Press.

———. 2003. Social monogamy and its variations in callitrichids: Do these relate to the costs of infant care? In *Monogamy: Mating Strategies and Partnerships in Birds, Humans, and Other Mammals* (ed. by Reichard, U. H. & Boesch, C.), 212–247. Cambridge: Cambridge University Press.

Goldizen, A. W. & Terborgh, J. 1989. Demography and dispersal patterns of a tamarin population: Possible causes of delayed breeding. *American Naturalist*, 134, 208–224.

Goodall, J. 1986. *The Chimpanzees of Gombe: Patterns of Behavior*. Cambridge, MA: Belknap Press.

Goossens, B., Setchell, J. M., James, S. S., Funk, S. M., Chikhi, L., Abulani, A., Ancrenaz, M., Lackman-Ancrenaz, I. & Bruford, M. W. 2006. Philopatry and reproductive success in Bornean orang-utans (*Pongo pygmaeus*). *Molecular Ecology*, 15, 2577–2588.

Gould, L. & Ziegler, T. E. 2007. Variation in fecal testosterone levels, inter-male aggression, dominance rank and age during mating and post-mating periods in wild adult male ring-tailed lemurs (*Lemur catta*). *American Journal of Primatology*, 69, 1325–1339.

Goymann, W., Landys, M. M. & Wingfield, J. C. 2007. Distinguishing seasonal androgen responses from male-male androgen responsiveness: Revisiting the Challenge Hypothesis. *Hormones and Behavior*, 51, 463–476.

Gray, P. B. & Campbell, B. C. 2009. Human male testosterone, pair bonding and fatherhood. In *Endocrinology of Social Relationships* (ed. by Ellison, P. T. & Gray, P. B.). Cambridge, MA: Harvard University Press.

Gross, M. R. 1996. Alternative reproductive strategies and tactics: Diversity within sexes. *Trends in Ecology and Evolution*, 11, 92–98.

Guimarães, V. d. A. & Strier, K. 2001. Adult male-infant interactions in wild muriquis (*Brachyteles arachnoides hypoxanthus*). *Primates*, 42, 395–399.

Gursky, S. 2000. Allocare in a nocturnal primate: data on the spectral tarsier, *Tarsius spectrum*. *Folia Primatologica*, 71, 39–54.

Harcourt, A. H. 1991. Sperm competition and the evolution of non-fertilising sperm in mammals. *Evolution*, 45, 314–328.

Harcourt, A. H. & Stewart, K. 2007. *Gorilla Society: Conflict, Compromise & Cooperation between the Sexes*. Chicago: University of Chicago Press.

Harcourt, A. H., Harvey, P. H., Larson, S. G. & Short, R. V. 1981. Testis weight, body weight and breeding system in primates. *Nature*, 293, 55–57.

Harvey, P. H. & Clutton-Brock, T. H. 1985. Life history variation in primates. *Evolution*, 39, 559–581.

Harvey, P. H. & Harcourt, A. H. 1984. Sperm competition, testes size, and breeding systems in primates. In *Sperm Competition and the Evolution of Animal Mating Systems* (ed. by Smith, R. L.), 589–600. New York: Academic Press.

Harvey, P., Kavanagh, M. & Clutton-Brock, T. H. 1978. Sexual dimorphism in primate teeth. *Journal of Zoology*, 186, 474–485.

Hasegawa, T. & Hiraiwa, M. 1980. Social interactions of orphans observed in a free-ranging troop of Japanese monkeys. *Folia Primatologica*, 33, 129–158.

Hau, M. 2007. Regulation of male traits by testosterone: Implications for the evolution of vertebrate life histories. *Bioessays*, 29, 133–144.

Hegner, R. E. & Wingfield, J. C. 1987. Effects of experimental manipulation of testosterone levels on parental investment and breeding success in male house sparrows. *Auk*, 104, 462–469.

Hermans, E. J., Ramsey, N. F. & Van Honk, J. 2008. Exogenous testosterone enhances responsiveness to social threat in the neural circuitry of social aggression in humans. *Biological Psychiatry*, 63, 263–270.

Heymann, E. W. & Soini, P. 1999. Offspring number in pygmy marmosets, *Cebuella pygmaea*, in relation to group size and the number of adult males. *Behavioral Ecology and Sociobiology*, 46, 400–404.

Hirschenhauser, K., Winkler, H. & Oliveira, R. F. 2003. Comparative analysis of male androgen responsiveness to social environment in birds: The effects of mating system and paternal incubation. *Hormones and Behavior*, 43, 508–519.

Hrdy, S. B. 2009. *Mothers and Others: The Evolutionary Origins of Mutual Understanding*. Cambridge, MA: Belknap, Harvard University Press.

Huck, M., Löttker, P. & Heymann, E. 2004. The many faces of helping: Possible costs and benefits of infant carrying and food transfers in wild moustached tamarins (*Saguinus mystax*). *Behaviour*, 141, 915–934.

Huck, M., Löttker, P., Heymann, E. W. & Heistermann, M. 2005. Characterization and social correlates of fecal testosterone and cortisol excretion in wild male *Saguinus mystax*. *International Journal of Primatology*, 26, 159–179.

Hylander, W. L. 2009. The functional significance of canine height reduction in early hominins. *American Journal of Physical Anthropology*, S48, 154.

Hylander, W. L. & Vinyard, C. J. 2006. The evolutionary significance of canine reduction in hominins: Functional links between jaw mechanics and canine size. *American Journal of Physical Anthropology*, S42, 107.

Isbell, L. A. 1994. Predation on primates: Ecological patterns and evolutionary consequences. *Evolutionary Anthropology*, 3, 61–71.

Isidori, A. M., Giannetta, E., Gianfrilli, D., Greco, E. A., Bonifacio, V., Aversa, A., Isidori, A., Fabbri, A. & Lenzi, A. 2005. Effects of testosterone on sexual function in men: Results of a meta-analysis. *Clinical Endocrinology*, 63, 381–394.

Jarman, P. J. 1974. The social organization of antelope in relation to their ecology. *Behaviour*, 48, 215–267.

Jolly, A. 1998. Pair-bonding, female aggression and the evolution of lemur societies. *Folia Primatologica*, 69, 1–13.

Jurke, M. H., Pryce, C. R., Hug-Hodel, A. & Döbeli, M. 1995. An investigation into the socioendocrinology of infant care and postpartum fertility in Goeldi's monkey (*Callimico goeldii*). *International Journal of Primatology*, 16, 453–474.

Kappeler, P. M. 1990. The evolution of sexual size dimorphism in prosimian primates. *American Journal of Primatology*, 21, 201–214.

Kappeler, P. & Schäffler, L. 2008. The lemur syndrome unresolved: Extreme male reproductive skew in sifakas (*Propithecus verreauxi*), a sexually monomorphic primate with female dominance. *Behavioral Ecology & Sociobiology*, 62, 1007–1015.

Keddy-Hector, A. C., Seyfarth, R. M. & Raleigh, M. J. 1989. Male parental care, female choice and the effect of an audience in vervet monkeys. *Animal Behaviour*, 1989, 262–271.

Kemnitz, J. W., Sladky, K. K., Flitsch, T. J., Pomerantz, S. M. & Goy, R. W. 1988. Androgenic influences on body size and composition of adult rhesus monkeys. *American Journal of Physiology: Endocrinology and Metabolism*, 255, E857-E864.

Ketterson, E. D. & Nolan, V. 1992. Hormones and life histories: An integrative approach. *American Naturalist*, 140, S33-S62.

Key, C. & Ross, C. 1999. Sex differences in energy expenditure in non-human primates. *Proceedings of the Royal Society B*, 266, 2479–2485.

Kitchen, D. M. 2004. Alpha male black howler monkey responses to loud calls: Effect of numeric odds, male companion behavior and reproductive investment. *Animal Behaviour*, 67, 125–139.

Kleiman, D. G. 1977. Monogamy in mammals. *Quarterly Review of Biology*, 52, 39–69.

Kleiman, D. G. & Malcolm, J. R. 1981. The evolution of male parental investment in mammals. In *Parental Care in Mammals* (ed. by Gubernick, D. J. & Klopfer, P. H.), 347–387. New York: Plenum Publishing.

Kleven, O., Fossøy, F., Laskemoen, T., Robertson, R. J., Rudolfsen, G. & Lifjeld, J. T. 2009. Comparative evidence for the evolution of sperm swimming speed by sperm competition and female sperm storage duration in passerine birds. *Evolution*, 63, 2466–2473.

Knott, C. D. 2009. Orangutans: Sexual coercion without sexual violence. In *Sexual Coercion in Primates and Humans: An Evolutionary Perspective on Male Aggression against Females* (ed. by Muller, M. N. & Wrangham, R. W.), 81–111. Cambridge, MA: Harvard University Press.

Knott, C. D. & Kahlenberg, S. M. 2007. Orangutans in perspective: Forced copulations and female mating resistance. In *Primates in Perspective* (ed. by Campbell, C. J., Fuentes, A., Mackinnon, K. C., Panger, M. A. & Bearder, S. K.), 290–304. New York: Oxford University Press.

Knott, C., Thompson, M., Stumpf, R. & McIntyre, M. 2010. Female reproductive strategies in orangutans, evidence for female choice and counterstrategies to infanticide in a species with frequent sexual coercion. *Proceedings of the Royal Society B*, 277, 105–113.

Kouri, E. M., Lukas, S. E., Pope, H. G. & Oliva, P. S. 1995. Increased aggressive responding in male volunteers following the administration of gradually increasing doses of testosterone cypionate. *Drug and Alcohol Dependence*, 40, 73–79.

Kuester, J. & Paul, A. 1986. Male-infant relationships in semifree-ranging Barbary macaques (*Macaca sylvanus*) of Affenberg Salem/FRG: Testing the "male care" hypothesis. *American Journal of Primatology*, 10, 315–327.

Kummer, H. 1968. *Social Organization of Hamadryas Baboons*. Chicago: University of Chicago Press.

Kuzawa, C. W., Gettler, L. T., Muller, M. N., McDade, T. W. & Feranil, A. B. 2009. Fatherhood, pairbonding and testosterone in the Philippines. *Hormones and Behavior*, 56, 429–435.

Lancaster, J. B. & Kaplan, H. 2009. The endocrinology of the human adaptive complex. In *Endocrinology of Social Relationships* (ed. by Ellison, P. T. & Gray, P. B.), 95–119. Cambridge, MA: Harvard University Press.

Lappan, S. 2008. Male care of infants in a siamang (*Symphalangus syndactylus*) population including socially monogamous and polyandrous groups. *Behavioral Ecology and Sociobiology*, 62, 1307–1317.

Lawler, R. R. 2009. Monomorphism, male-male competition, and mechanisms of sexual dimorphism. *Journal of Human Evolution*, 57, 321–325.

Lawler, R. R., Richard, A. F. & Riley, M. A. 2005. Intrasexual selection in Verreaux's sifaka (*Propithecus verreauxi verreauxi*). *Journal of Human Evolution*, 48, 259–277.

LeBoeuf, B. & Reiter, J. 1988. Lifetime reproductive success in northern elephant seals. In *Reproductive Success* (ed. by Clutton-Brock, T.), 344–362. Chicago: University of Chicago Press.

Leigh, S. R., Setchell, J. M., Charpentier, M., Knapp, L. A. & Wickings, E. J. 2008. Canine tooth size and fitness in male mandrills (*Mandrillus sphinx*). *Journal of Human Evolution*, 55, 75–85.

Leutenegger, W. & Cheverud, J. 1982. Correlates of sexual dimorphism in primates: Ecological and size variables. *International Journal of Primatology*, 3, 387–402.

Lewis, R. J. 2009. Chest staining variation as a signal of testosterone levels in male Verreaux's sifaka. *Physiology and Behavior*, 96, 586–592.

Lindenfors, P. & Tullberg, B. S. 1998. Phylogenetic analyses of primate size evolution: The consequences of sexual selection. *Biological Journal of the Linnean Society*, 64, 413–447.

Lynch, J. W., Ziegler, T. E. & Strier, K. B. 2002. Individual and seasonal variation in fecal testosterone and cortisol levels of wild male tufted capuchin monkeys, *Cebus apella nigritus*. *Hormones and Behavior*, 41, 275–287.

Maggioncalda, A. N., Sapolsky, R. M. & Czekala, N. M. 1999. Reproductive hormone profiles in captive male orangutans: Implications for understanding developmental arrest. *American Journal of Physical Anthropology*, 109, 19–32.

Marlowe, F. 1999. Male care and mating effort among Hadza foragers. *Behavioral Ecology & Sociobiology*, 46, 57–64.

———. 2000. Parental investment and the human mating system. *Behavioural Processes*, 51, 45–61.

Marlowe, F. W. 2003. A critical period for provisioning by Hadza men: Implications for pair bonding. *Evolution and Human Behavior*, 24, 217–229.

Martin, R. D., Willner, L. A. & Dettling, A. 1994. The evolution of sexual size dimorphism in primates. In *The Differences between the Sexes* (ed. by Short, R. V. & Balaban, E.), 159–202. Cambridge: Cambridge University Press.

Mauck, R. A., Marschall, E. A. & Parker, P. G. 1999. Adult survival and imperfect assessment of parentage: Effects on male parenting decisions. *The American Naturalist*, 154, 99–109.

Maynard Smith, J. 1977. Parental investment: A prospective analysis. *Animal Behaviour*, 25, 1–9.

Mazur, A. & Booth, A. 1998. Testosterone and dominance in men. *Behavioral and Brain Sciences*, 21, 353–397.

McIntyre, M., Gangestad, S. W., Gray, P. B., Flynn Chapman, J., Burnham, T. C., O'Rourke, M. T. & Thornhill, R. 2006. Romantic involvement often reduces men's testosterone levels—but not always: The moderating role of extrapair sexual interest. *Journal of Personality and Social Psychology*, 91, 642–651.

Mendoza, S. P. & Mason, W. A. 1986. Parenting within a monogamous society. In *Primate Ontogeny, Cognition and Social Behaviour* (ed. by Else, J. G. & Lee, P. C.), 255–266. New York: Cambridge University Press.

Milton, K. 1985. Mating patterns of woolly spider monkeys, *Brachyteles arachnoides*: Implications for female choice. *Behavioral Ecology and Sociobiology*, 17, 53–59.

Mitani, J. C. 1985. Mating behaviour of male orangutans in the Kutai Game Reserve, Indonesia. *Animal Behaviour*, 33, 392–402.

Mitani, J.C., Gros-Louis, J. & Richards, A. 1996. Sexual dimorphism, the operational sex ratio, and the intensity of male competition among polygynous primates. *American Naturalist*, 147, 966–980.

Mitani, J. C. & Watts, D. P. 1997. The evolution of non-maternal caretaking among anthropoid primates: Do helpers help? *Behavioral Ecology & Sociobiology*, 40, 213–220.

Møller, A. P. 1988. Ejaculate quality, testes size and sperm production in primates. *Journal of Human Evolution*, 17, 479–488.

———. 1989. Ejaculate quality, testes size and sperm production in mammals. *Functional Ecology*, 3, 91–96.

Moore, H. D. M., Martin, M. & Birkhead, T. R. 1999. No evidence for killer sperm or other selective interactions between human spermatozoa in ejaculates of different males in vitro. *Proceedings of the Royal Society B*, 266, 2343–2350.

Muller, M. N. & Mitani, J. C. 2005. Conflict and cooperation in wild chimpanzees. *Advances in the Study of Behavior*, 35, 275–331.

Muller, M. N. & Wrangham, R. W. 2004a. Dominance, aggression and testosterone in wild chimpanzees: A test of the "Challenge Hypothesis." *Animal Behaviour*, 67, 113–123.

———. 2004b. Dominance, cortisol and stress in wild chimpanzees (*Pan troglodytes schweinfurthii*). *Behavioral Ecology and Sociobiology*, 55, 332–340.

———. 2009. *Sexual Coercion in Primates and Humans.* Cambridge, MA: Harvard University Press.

Muller, M. N., Marlowe, F. W., Bugumba, R. & Ellison, P. T. 2009a. Testosterone and paternal care in East African foragers and pastoralists. *Proceedings of the Royal Society B*, 276, 347–354.

Muller M. N., Kahlenberg, S. & Wrangham, R. W. 2009b. Male aggression against females in chimpanzees. In *Sexual Coercion in Primates and Humans: An Evolutionary Perspective on Male Aggression against Females* (ed. by Muller, M. N. & Wrangham, R. W.), 184–217. Cambridge, MA: Harvard University Press.

Nascimento, J. M. 2008. Analysis of sperm motility and physiology using optical tweezers. PhD dissertation, University of California, San Diego.

Nascimento, J. M., Shi, L. Z., Meyers, S., Gagneux, P., Loskutoff, N. M., Botvinick, E. L. & Berns, M. W. 2008. The use of optical tweezers to study sperm competition and motility in primates. *Journal of the Royal Society Interface*, 5, 297–302.

Nievergelt, C. M., Mutschler, T., Feistner, A. T. C., & Woodruff, D. S. 2002. Social system of the Alaotran gentle lemur (*Hapalemur griseus alaotrensis*): Genetic characterization of group composition and mating system. *American Journal of Primatology*, 57, 157–176.

Nunes, S., Fite, J. E. & French, J. A. 2000. Variation in steroid hormones associated with infant care behaviour and experience in male marmosets (*Callithrix kuhlii*). *Animal Behaviour*, 60, 857–865.

Nunes, S., Fite, J. E., Patera, K. J. & French, J. A. 2001. Interactions among paternal behavior, steroid hormones, and parental experience in male marmosets (*Callithrix kuhlii*). *Hormones and Behavior*, 39, 70–82.

Oliveira, R. F., Canario, A. V. M. & Ros, A. F. H. 2008. Hormones and alternative reproductive tactics in vertebrates. In *Alternative Reproductive Tactics: An Integrative Approach* (ed. by Oliveira, R. F., Taborsky, M. & Brockmann, H. J.), 132–173. Cambridge: Cambridge University Press.

Ostner, J., Kappeler, P. M. & Heistermann, M. 2008. Androgen and glucocorticoid levels reflect seasonally occurring social challenges in male redfronted lemurs (*Eulemur fulvus rufus*). *Behavioral Ecology and Sociobiology*, 62, 627–638.

Overdorff, D. J. 1998. Are *Eulemur* species pair-bonded? Social organization and mating strategies in *Eulemur fulvus rufus* from 1988–1995 in Southeast Madagascar. *American Journal of Physical Anthropology*, 105, 153–166.

Packer, C. 1980. Male care and exploitation of infants in *Papio anubis*. *Animal Behaviour*, 27, 37–45.

Palombit, R. A., Seyfarth, R. M. & Cheney, D. L. 1997. The adaptive value of "friendships" to female baboons: Experimental and observational evidence. *Animal Behaviour*, 54, 599–614.

Parker, G. A. 1970. Sperm competition and its evolutionary consequences in the insects. *Biological Reviews of the Cambridge Philosophical Society*, 45, 525–567.

Patel, E. R. 2007. Non-maternal infant care in wild silky sifakas (*Propithecus candidus*). *Lemur News*, 12, 39–42.

Paul, A., Kuester, J. & Arnemann, J. 1992. DNA fingerprinting reveals that infant care by male Barbary macaques (*Macaca sylvanus*) is not paternal investment. *Folia Primatologica*, 58, 93–98.

———. 1996. The sociobiology of male-infant interactions in Barbary macaques, *Macaca sylvanus*. *Animal Behaviour*, 51, 155–170.

Platek, S. M, Critton, S. R., Burch, R. L., Frederick, D. A., Myers, T. A., & Gallup, G. G., Jr. 2003. How much paternal resemblance is enough? Sex differences in hypothetical investment decisions but not in the detection of resemblance. *Evolution and Human Behavior*, 24, 81–87.

Plavcan, J. M. 2001. Sexual dimorphism in primate evolution. *Yearbook of Physical Anthropology*, 44, 25–53.

———. 2004. Sexual selection, measures of sexual selection, and sexual dimorphism in primates. In *Sexual Selection in Primates: New and Comparative Perspectives* (ed. by Kappeler, P. M. & van Schaik, C. P.), 230–252. Cambridge: Cambridge University Press.

Plavcan, J. M. & van Schaik, C. P. 1992. Intrasexual competition and canine dimorphism in anthropoid primates. *American Journal of Physical Anthropology*, 87, 461–477.

———. 1997a. Intrasexual competition and body size dimorphism in anthropoid primates. *American Journal of Physical Anthropology*, 103, 37–68.

———. 1997b. Interpreting hominid behavior on the basis of sexual dimorphism. *Journal of Human Evolution*, 32, 345–374.

Plavcan, J. M., van Schaik, C. P. & Kappeler, P. M. 1995. Competition, coalitions and canine size in primates. *Journal of Human Evolution*, 28, 245–276.

Pollock, J. I. 1979. Female dominance in *Indri indri*. *Folia Primatologica*, 31, 143–161.

Price, E. C. 1991. Competition to carry infants in captive families of cotton-top tamarins (*Saguinus oedipus*). *Behaviour*, 118, 66–88.

———. 1992. The benefits of helpers: Effects of group and litter size on infant care in tamarins (*Saguinus oedipus*). *American Journal of Primatology*, 26, 179–190.

Promislow, D. E. L. 1992. Costs of sexual selection in natural populations of mammals. *Proceedings of the Royal Society B*, 247, 203–210.

Prudom, S. L., Broz, C. A., Schultz-Darken, N., Ferris, C. T., Snowdon, C. & Ziegler, T. E. 2008. Exposure to infant scent lowers serum testosterone in father common marmosets (*Callithrix jacchus*). *Biology Letters*, 4, 603–605.

Rhine, R. J., Norton, G. W., Roertgen, W. J., & Klein, H. D. 1980. The brief survival of free-ranging baboons infants (*Papio cynocephalus*) after separation from their mothers. *International Journal of Primatology*, 1, 401–409.

Robinson, J. G. & Janson, C. H. 1987. Capuchins, squirrel monkeys, and atelines: Socioecological convergence with Old World primates. In *Primate Societies* (ed. by Smuts, B., Cheney, D. L., Seyfarth, R. M., Wrangham, R. W. & Struhsaker, T. T.), 343–357. Chicago: University of Chicago Press.

Robinson, J. G., Wright, P. C. & Kinzey, W. G. 1987. Monogamous cebids and their relatives: Intergroup calls and spacing. In *Primate Societies* (ed. by Smuts, B., Cheney, D. L., Seyfarth, R. M., Wrangham, R. W. & Struhsaker, T. T.), 44–53. Chicago: University of Chicago Press.

Rodman, P. S. & Mitani, J. C. 1987. Orangutans: Sexual dimorphism in a solitary species. In *Primate Societies* (ed. by Smuts, B., Cheney, D. L., Seyfarth, R. M., Wrangham, R. W. & Struhsaker, T. T.), 146–154. Chicago: University of Chicago Press.

Ross, C. & MacLarnon, A. 2000. The evolution of non-maternal care in anthropoid primates: A test of the hypotheses. *Folia Primatologica*, 71, 93–113.

Ross, C. N., French, J. A. & Orti, G. 2007. Germ-line chimerism and paternal care in marmosets (*Callithrix kuhlii*). *Proceedings of the National Academy of Sciences USA*, 104, 6278–6282.

Rotundo, M., Fernandez-Duque, E. & Dixson, A. F. 2005. Infant development and parental care in free-ranging *Aotus azarai azarai*. *International Journal of Primatology*, 26, 1459–1473.

Santos, C. V., French, J. A. & Otta, E. 1997. Infant carrying behavior in callitrichid primates: *Callithrix* and *Leontopithecus*. *International Journal of Primatology*, 18, 889–907.

Sapolsky, R. M. 1983. Endocrine aspects of social instability in the olive baboon (*Papio anubis*). *American Journal of Primatology*, 5, 365–379.

———. 1993. The physiology of dominance in stable versus unstable social hierarchies. In *Primate Social Conflict* (ed. by Mason, W. A. & Mendoza, S. P.), 171–204. Albany: SUNY Press.

Savage, A., Snowdon, C. T., Giraldo, L. H. & Soto, L. H. 1996. Parental care patterns and vigilance in wild cotton-top tamarins (*Saguinus oedipus*). In *Adaptive Radiations of Neotropical Primates* (ed. by Norconk, M. A., Rosenberger, A. L. & Garber, P. A.), 187–199. New York: Plenum Press.

Schradin, C. & Anzenberger, G. 2003. Mothers, not fathers, determine the delayed onset of male carrying in Goeldi's monkey (*Callimico goeldii*). *Journal of Human Evolution*, 45, 389–399.

Schülke, O. 2005. Evolution of pair-living in *Phaner furcifer*. *International Journal of Primatology*, 26, 903–919.

Schultheiss, O. C., Campbell, K. L. & McClelland, D. C. 1999. Implicit power motivation moderates men's testosterone responses to imagined and real dominance success. *Hormones and Behavior*, 36, 234–241.

Schultheiss, O. C., Dargel, A. & Rohde, W. 2003. Implicit motives and gonadal steroid hormones: Effects of menstrual cycle phase, oral contraceptive use, and relationship status. *Hormones and Behavior*, 43, 293–301.

Shultz, S., Noe, R. N., McGraw, W. S., & Dunbar, R. I. M. 2004. A community-level evaluation of the impact of prey behavioural and ecological characteristics on predator diet composition. *Proceedings of the Royal Society B*, 271, 725–732.

Schürmann, C. L. & van Hooff, J. A. R. A. M. 1986. Reproductive strategies of the orang-utan. *International Journal of Primatology*, 7, 265–287.

Sellers, J. G., Mehl, M. R. & Josephs, R. A. 2007. Hormones and personality: Testosterone as a marker of individual differences. *Journal of Research in Personality*, 41, 126–138.

Setchell, J. M. 2008. Alternative reproductive tactics in primates. In *Alternative Reproductive Tactics: An Integrative Approach* (ed. by Oliveira, R. F., Taborsky, M. & Brockmann, H. J.), 373–398. Cambridge: Cambridge University Press.

Setchell, J. M. & Dixson, A. F. 2001. Changes in the secondary sexual adornments of male mandrills (*Mandrillus sphinx*) are associated with gain and loss of alpha status. *Hormones and Behavior*, 39, 177–184.

———. 2002. Developmental variables and dominance rank in adolescent male mandrills (*Mandrillus sphinx*). *American Journal of Primatology*, 56, 9–25.

Setchell, J. M., Smith, T., Wickings, E. J. & Knapp, L. A. 2008. Social correlates of testosterone and ornamentation in male mandrills. *Hormones and Behavior*, 54, 365–372.

Seuanez, H. N., Carothers, A. C., Martin, D. E. & Short, R. V. 1977. Morphological abnormalities in spermatozoa of man and great apes. *Nature*, 270, 345–347.

Seyfarth, R. M. 1978. Social relationships among adult male and female baboons. II: Behaviour throughout the female reproductive cycle. *Behaviour*, 64, 227–247.

Shackelford, T. K. & Goetz, A. T. 2006. Comparative evolutionary psychology of sperm competition. *Journal of Comparative Psychology*, 120, 139–146.

Short, R. V. 1979. Sexual selection and its component parts, somatic and genital selection, as illustrated by man and the great apes. *Advances in the Study of Behavior*, 9, 131–158.

Silk, J. B. & Samuels, A. 1984. Triadic interactions among *Macaca radiata*: Passports and buffers. *American Journal of Primatology*, 6, 373–376.

Smith, R. J. & Jungers, W. L. 1997. Body mass in comparative primatology. *Journal of Human Evolution*, 32, 523–559.

Smith, E. O. & Peffer-Smith, P. G. 1984. Adult male-immature interactions in captive stumptail macaques (*Macaca arctoides*). In *Primate Paternalism* (ed. by Taub, D. M.), 88–112. New York: Van Nostrand Reinhold.

Smith, E. O. & Whitten, P. L. 1988. Triadic interactions in savanna-dwelling baboons. *International Journal of Primatology*, 9, 409–424.

Smuts, B. 1985. *Sex and Friendship in Baboons*. New York: Aldine de Gruyter.

Smuts, B. & Gubernick, D. J. 1992. Male-infant relationships in nonhuman primates: Paternal investment or mating effort? In *Father-Child Relations: Cultural and Biosocial Contexts* (ed. by Hewlett, B. S.), 1–30. New York: Aldine de Gruyter.

Smuts, B. B. & Smuts, R. W. 1993. Male aggression and sexual coercion of females in nonhuman primates and other mammals: Evidence and theoretical implications. *Advances in the Study of Behavior*, 22, 1–63.

Stein, D. M. 1984. Ontogeny of infant-male relationships during the first year of life for yellow baboons (*Papio cynocephalus*). In *Primate Paternalism* (ed. by Taub, D. M.), 213–243. New York: Van Nostrand Reinhold.

Stewart, K. J. 2001. Social relationships of immature gorillas and silverbacks. In *Mountain Gorillas: Three Decades of Research at Karisoke* (ed. by Robbins, M. M., Sicotte, P. & Stewart, K. J.), 183–214. Cambridge: Cambridge University Press.

Storey, A. E., Walsh, C. J., Quinton, R. L. & Wynne-Edwards, K. E. 2000. Hormonal correlates of paternal responsiveness in new and expectant fathers. *Evolution and Human Behavior*, 21, 79–95.

Strier, K. B. 1990. New World primates, new frontiers: Insights from the woolly spider monkey, or muriqui (*Brachyteles arachnoides*). *International Journal of Primatology*, 11, 7–19.

Strier, K. B., Ziegler, T. E. & Wittwer, D. J. 1999. Seasonal and social correlates of fecal testosterone and cortisol levels in wild male muriquis (*Brachyteles arachnoides*). *Hormones and Behavior*, 35, 125–134.

Strum, S. C. 1984. Why males use infants. In *Primate Paternalism* (ed. by Taub, D. M.), 146–185. New York: Van Nostrand Reinhold.

———. 1987. *Almost Human: A Journey into the World of Baboons*. New York: W.W. Norton.

Swedell, L. 2006. *Strategies of Sex and Survival in Hamadryas Baboons: Through a Female Lens*. Upper Saddle River, NJ: Pearson Prentice Hall.

Taborsky, M., Oliveira, R. F. & Brockmann, H. J. 2008. The evolution of alternative reproductive tactics: Concepts and questions. In *Alternative Reproductive Tactics: An Integrative Approach* (ed. by Oliveira, R. F., Taborsky, M. & Brockmann, H. J.), 1–21. Cambridge: Cambridge University Press.

Tardif, S. D. & Bales, K. 1997. Is infant-carrying a courtship strategy in callitrichid primates. *Animal Behaviour*, 53, 1001–1007.

Taub, D. M. 1980. Testing the "agonistic buffering" hypothesis. *Behavioral Ecology and Sociobiology*, 6, 187–197.

Teichroeb, J. A. & Sicotte, P. 2008. Social correlates of fecal testosterone in male ursine colobus monkeys (*Colobus vellerosus*): The effect of male reproductive competition in aseasonal breeders. *Hormones and Behavior*, 54, 417–423.

Terborgh, J. & Goldizen, A. W. 1985. On the mating system of the cooperatively breeding saddle-backed tamarin (*Saguinus fuscicollis*). *Behavioral Ecology and Sociobiology*, 16, 293–299.

Thorén, S., Lindenfors, P., and Kappeler, P. M. 2006. Phylogenetic analyses of dimorphism in primates: Evidence for stronger selection on canine size than on body size. *American Journal of Physical Anthropology*, 130, 50–59.

Thornhill, R. & Palmer, C. T. 2000. *A Natural History of Rape: Biological Bases of Sexual Coercion*. Cambridge, MA: MIT Press.

Trivers, R. L. 1972. Parental investment and sexual selection. In *Sexual Selection and the Descent of Man 1871–1971* (ed. by Campbell, B.), 136–179. Chicago: Aldine.

Utami, S. S., Goossens, B., Bruford, M. W., de Ruiter, J. R. & van Hooff, J. A. R. A. M. 2002. Male bimaturism and reproductive success in Sumatran orangutans. *Behavioral Ecology*, 13, 643–652.

Van Anders, S. M. & Watson, N. V. 2006. Relationship status and testosterone in North American heterosexual and non-heterosexual men and women: Cross-sectional and longitudinal data. *Psychoneuroendocrinology*, 31, 715–723.

Van Anders, S. M., Hamilton, L. D. & Watson, N. V. 2007. Multiple partners are associated with higher testosterone in North American men and women. *Hormones and Behavior*, 51, 454–459.

Van Honk, J., Tuiten, A., Verbaten, R., Hout, M. v. d., Koppeshaar, H., Thijssen, J. & Haan, E. d. 1999. Correlations among salivary testosterone, mood, and selective attention to threat in humans. *Hormones and Behavior*, 36, 17–24.

Van Honk, J., Tuiten, A., Hermans, E., Putman, P., Koppeschaar, H., Thijssen, J., Verbaten, R. & van Doornen, L. 2001. A single administration of testosterone induces cardiac accelerative responses to angry faces in healthy young women. *Behavioral Neuroscience*, 115, 238–242.

Van Schaik, C. P. & Paul, A. 1996. Male care in primates: Does it ever reflect paternity? *Evolutionary Anthropology*, 5, 152–156.

Van Schaik, C. P. & van Noordwijk, M. A. 1988. Male careers in Sumatran long-tailed macaques (*Macaca fascicularis*). *Behaviour*, 107, 24–43.

———. 1989. The special role of male *Cebus* monkeys in predator vigilance and its effects on group composition. *Behavioral Ecology and Sociobiology*, 24, 265–276.

Watts, D. P. 1996. Comparative socio-ecology of gorillas. In *Great Ape Societies* (ed. by McGrew, W. C., Marchant, L. F. & Nishida, T.), 16–28. Cambridge: Cambridge University Press.

———. 1997. Agonistic interventions in wild mountain gorilla groups. *Behaviour*, 134, 23–57.

Weingrill, T. 2000. Infanticide and the value of male-female relationships in mountain chacma baboons. *Behaviour*, 137, 337–359.

Whitten, P. L. 1987. Infants and adult males. In *Primate Societies* (ed. by Smuts, B., Cheney, D. L., Seyfarth, R. M., Wrangham, R. W. & Struhsaker, T. T.), 343–357. Chicago: University of Chicago Press.

Widdig, A. 2007. Paternal kin discrimination: The evidence and likely mechanisms. *Biological Reviews*, 82, 319–334.

Wiens, F. & Zitzmann, A. 2003. Social dependence of infant slow lorises to learn diet. *International Journal of Primatology*, 24, 1007–1021.

Williamson, D. F. 1993. Descriptive epidemiology of body weight and weight change in U.S. adults. *Annals of Internal Medicine*, 119, 646–649.

Wimmer, B., & Kappeler, P. M. 2002. The effects of sexual selection and life history on the genetic structure of redfronted lemur, *Eulemur fulvus*, groups. *Animal Behaviour*, 64, 557–568.

Wingfield, J. C., Hegner, R. E., Dufty, A. M. & Ball, G. F. 1990. The "challenge hypothesis": Theoretical implications for patterns of testosterone secretion, mating systems, and breeding strategies. *American Naturalist*, 136, 829–846.

Wingfield, J. C., Jacobs, J. D., Tramontin, A. D., Perfito, N., Meddle, S., Maney, D. L. & Soma, K. 2000. Toward an ecological basis of hormone-behavior interactions in reproduction of birds. In *Reproduction in Context* (ed. by Wallen, K. & Schneider, J. E.), 85–128. Cambridge, MA: MIT Press.

Winking, J., Kaplan, H., Gurven, M. & Rucas, S. 2007. Why do men marry and why do they stray? *Proceedings of the Royal Society B*, 274, 1643–1649.

Wittenberger, J. F. & Tilson, R. L. 1980. The evolution of monogamy: Hypotheses and evidence. *Annual Review of Ecology and Systematics*, 11, 197–232.

Wolovich, C. K., Perea-Rodriguez, J. P. & Fernandez-Duque, E. 2008. Food transfers to young and mates in wild owl monkeys. *American Journal of Primatology*, 70, 211–221.

Wolpoff, M. H. 1976. Some aspects of the evolution of hominid sexual dimorphism. *Current Anthropology*, 17, 579–606.

Woodroffe, R. & Vincent, A. 1994. Mother's little helpers: Patterns of male care in mammals. *Trends in Ecology and Evolution*, 9, 294–197.

Wrangham, R. W., Jones, J. H., Laden, G., Pilbeam, D. & Conklin-Brittain, N. L. 1999. The raw and the stolen: Cooking and the ecology of human origins. *Current Anthropology*, 40, 567–594.

Wright, P. C. 1984. Biparental care in *Aotus trivirgatus* and *Callicebus moloch*. In *Female Primates: Studies by Women Primatologists* (ed. by Small, M. E.), 59–75. New York: Alan R. Liss.

———. 1990. Patterns of paternal care in primates. *International Journal of Primatology*, 11, 89–102.

Young, A. L., Richard, A. F. & Aiello, L. C. 1990. Female dominance and maternal investment in strepsirhine primates. *American Naturalist*, 135, 473–488.

Ziegler, T. E., Prudom, S. L., Schultz-Darken, N., Kurian, A. V. & Snowdon, C. T. 2006. Pregnancy weight gain: Marmoset and tamarin dads show it too. *Biology Letters*, 2, 181–183.

Chapter 18 Magnitude and Sources of Variation in Male Reproductive Performance

Susan C. Alberts

THE STUNNING reproductive potential of organisms like salmon—at the opposite end of the slow-versus-fast life-history continuum from primates—illustrates two very important differences between males and females. First, males produce many more gametes than females (male salmon produce on average 200 billion sperm per ejaculate, while females typically produce 3,000 to 8,000 eggs per clutch (Kazakov 1981; Beacham & Murray 1993). Second, male salmon show almost 10 times more variance in zygote production than females, and many males produce no zygotes at all (Fleming & Gross 1994).

This paradox—that males have greater reproductive potential than females, but also a greater probability of failing to reproduce—lies at the heart of two important and closely related evolutionary insights. The first is that females' lower reproductive potential is usually associated with both a greater investment in each zygote and a greater control over the means and circumstances of reproduction (e.g., the size of the nutritional package supplied to the zygote, the location and sturdiness of nests, the extent of post-zygotic care, and often the identity of the father; Gowaty 1997). This in turn leads to selection pressure on males to influence these female reproductive decisions. This is the heart of sexual selection: it is selection pressure on one sex to influence the reproductive decisions of the other sex.

The second important evolutionary insight is that the greater investment by females in each reproductive event usually takes females out of the reproductive population for longer periods of time than males, resulting in an operational sex ratio typically biased towards males (chapter 17,

this volume). The resulting overabundance of males relative to females generally results in greater variance in reproductive success among males than among females (Trivers 1972; Andersson 1994). This means that the pressure on males to influence female decisions will, in general, be stronger than the reverse.

Of course, both sexes experience selection pressure to influence the reproductive decisions of the other sex. But the tendency for females to make more of the reproductive decisions, combined with the typical male-biased operational sex ratio, means that in most species, any trait a male exhibits that successfully influences female reproductive decisions and pushes him to the upper end of the distribution of male reproductive success will tend to have a larger payoff than will a comparable trait in females. The female's trait may push her to the upper end of the female reproductive success distribution, but that distribution will usually (though not always) be narrower and flatter than the male distribution, so that the selection payoff simply won't be as great. She will experience sexual selection on the trait, but the selection will not be as intense as selection on comparable male traits. The result is a world in which males generally (though not always) have more elaborate and exaggerated sexually selected traits than females.

The reproductive performance of an individual is the sine qua non of its Darwinian fitness. Hence, the components and sources of variance in reproductive performance are the subjects of intense scrutiny in evolutionary ecology. The focus here is on the components of reproductive performance—age at maturity, longevity, infant production, and infant survival—in male primates, and the sources of variance in these components.

Measuring Reproductive Performance

Measuring reproductive performance in primates is challenging for three reasons. First, primates live a long time; most studies of reproductive performance will encompass a relatively short part of the life span. Second, one or both sexes disperse in all primate species, so data on dispersers will be constrained to the time period in which the animal remains in the study population. Hence, even if the researcher observes the study population long enough to potentially capture a large fraction of the animal's life, measurements of its reproductive performance will most likely be limited to certain windows of time. Third (and this problem is particular to males), actual production of offspring, the most important measure of reproductive performance, is difficult to measure. This problem is common in all species with internal fertilization; measuring production of offspring by females is straightforward, but measuring offspring production for males requires genetic analyses. Consequently, most studies of male reproductive performance use proxies for offspring production, such as the number of matings that a male obtains. In many cases this proxy is a very good one, but several factors, including cryptic female choice (chapter 16, this volume), difficult-to-detect ovulation, or unobtrusive copulations that are hard to observe, may, in any given population, make mating activity an unreliable index of paternity.

Individual reproductive performance over the lifetime can be understood relatively simply as the product of life-history components. In studies of mammalian life histories, four components are usually recognized, all of which are ultimately derived from measures of age-specific survival and fertility: (1) age at maturity or at first reproduction, (2) fertility, (3) longevity, and (4) survival of the individual's offspring. However, the designation of life-history components can be tailored to accommodate any type of life history. For instance, survival during the larval, pupal, and adult stages in insects can be treated as separate life-history components. Once life-history components have been designated, individual reproductive performance (or success) can be calculated using several different approaches (reviewed in Grafen 1988; Coulson et al. 2006).

Formal analyses of lifetime reproductive success in male primates are rare (e.g., Altmann et al. 1988; Cheney et al. 1988; Struhsaker & Pope 1991; Kuester et al. 1995; Lawler 2007; for a review of studies in females see chapter 15, this volume). Nonetheless, the first step in understanding sources of variation in reproductive performance is to measure its components and to identify the traits and conditions that affect them. A good many studies of male primates provide data on life-history components, and they are reviewed below.

Age at Maturity and First Reproduction

For both males and females, sexual development is a continuous process rather than a discrete event, but for females, a single, distinct physiological indicator of either reproductive potential (menarche) or first reproduction (first birth) is usually observable. In the case of males, detecting the onset of sexual maturity and first reproduction is more challenging. The male equivalent of menarche is testicular development, which has no well-defined moment of attainment and in some species is hard to detect because the scrotal sac is small or difficult to see. Age at first reproduction for males is detectable only through genetic paternity analyses in species in which females mate multiply. For these reasons, age at maturity or first reproduction is rarely reported for male primates, either in captivity or the wild.

Nonetheless, the estimation of age at maturity or first reproduction in male primates is not beyond our reach. Nor is it impossible to identify sources of variance in these important life-history milestones. Indeed, a number of options are open to researchers attempting to estimate age at maturity in males, and such data may already be fairly widely available for primate field studies (table 18.1). The means and variances for these milestones are less commonly reported for males than for females, however, and research on age at sexual maturity and first reproduction in males lags far behind the corresponding research on females.

In the few cases where variance in age at maturity for males has been examined, testicular development is usually the target of analysis. Interspecific comparisons of puberty reveal that male hamadryas baboons (*Papio hamadryas*) experience testicular development at an earlier life stage than do sympatric olive baboons (*Papio anubis*), because hamadryas baboons achieve fully adult testis size a number of months before they achieve full body size or social maturity (Jolly & Phillips-Conroy 2003, 2006). This accelerated pubertal development appears to enable an earlier start to reproductive life in male hamadryas baboons (Jolly & Phillips-Conroy 2006; Zinner et al. 2006). This is potentially advantageous in light of the lower intra-sexual competition for reproductive females that hamadryas males experience relative to males in other baboon social systems, and the resulting greater opportunities for subadult males to begin cultivating relationships with fe-

Table 18.1. Maturational milestones for male primates

Milestone	What it signals	Limitations and considerations	Examples of reported means or analyses of variation
Scrotal development (testicular enlargement)	Puberty/physical maturation	Scrotal sac easy to view on a regular basis (e.g., monthly). Achievement must be scored consistently across individuals.	Yellow baboons, Alberts & Altmann 1995; mandrills, Setchell et al. 2005; chimpanzees, Watts & Pusey 1993
First ejaculatory copulation with an adult female	Reproductive capability or beginning of regular reproductive activity	May represent puberty (beginning of sperm production) in some species, and full attainment of adulthood (regular access to reproductive females) in others. Care must be taken to differentiate between these outcomes if possible.	Northern muriquis, Possamai et al. 2005; capuchin monkeys, Fragaszy et al. 2004; chimpanzees, Watts & Pusey 1993
First paternity	First reproduction	Generally impossible without genetic analysis of paternity, and challenging even then, if males disperse.	Chimpanzees, Wroblewski et al. 2009; mandrills, Setchell et al. 2005; brown capuchin monkeys, Wirz & Rivielo 2008; rhesus macaques, Bercovitch et al. 2003
Attainment of adult dominance rank (multimale species)	Beginning of regular reproductive activity	Not a perfect assay of first reproduction, but in some cases a very good assessment of beginning of adulthood.	Yellow baboons, Alberts & Altmann 1995a; long-tailed macaques, van Noordiwjk & van Schaik 2001
Attainment of primary access to a female group (in unimale species)	Beginning of regular reproductive activity	Not a perfect assay of first reproduction, but in some cases a very good assessment of the beginning of adulthood. May exhibit considerably more variance than attainment of adult dominance rank in multimale species and a very long tail on the right side of distribution.	Mountain gorillas, Robbins 1995
Species-specific signals of adulthood (e.g., silvering of the back in gorillas, attainment of bright coloration in mandrills)	Varies from species to species	As with scrotal development, achievement must be scored consistently, which may be challenging if the process is gradual.	Mandrills, Setchell & Dixson 2002

males at a young age (Jolly & Phillips-Conroy 2006; Zinner et al. 2006).

Several intraspecific analyses of cercopithecines have shown that testicular development is enhanced by high social dominance rank of the male, of his mother, or of both (mandrills, *Mandrillus sphinx*, Wickings & Dixson 1992; Setchell & Wickings 2006; rhesus macaques, *Macaca mulatta*, Bercovitch 1993; Bercovitch & Berard 1993; Dixson & Nevison 1997; yellow baboons, *Papio cynocephalus*, Alberts & Altmann 1995b). In other cases, variance in age at attainment of adult dominance rank has been examined, and again an advantage accrues to males with high-ranking mothers (yellow baboons, Alberts & Altmann 1995b; long-tailed macaques, *Macaca fascicularis*, van Noordwijk & van Schaik 2001). These results raise the question of whether male reproductive success may be enhanced by early puberty, as it is among females in some species (chapter 15, this volume).

This hypothesis has not been directly tested, but Charpentier et al. (2008) showed that male yellow baboons that experienced earlier puberty (testicular enlargement) also attained adult rank earlier, albeit several years after testicular enlargement. In addition, Paul et al. (1992) showed that, among Barbary macaques (*Macaca sylvanus*), sons of

high-ranking mothers experienced first reproduction earlier than other males (i.e., high-born sons were more likely than low-born sons to produce an offspring by six years of age). High maternal rank also confers an advantage on sons by enhancing their growth rates, resulting in an approximate 20% acceleration during maturation, in both yellow baboons (Altmann & Alberts 2005) and rhesus macaques (Bercovitch et al. 2000). Finally, Onyango et al. (2008) found in yellow baboons that sons of high-ranking mothers had significantly lower levels of stress hormones than sons of low-ranking mothers. Cumulatively, these results provide evidence for a general syndrome, at least in male cercopithecines, in which high maternal rank confers persistent potential advantages throughout development to sons as it does to daughters (chapter 14, this volume).

However, the influence of maternal rank on male reproductive performance may be attenuated for male cercopithecines after adulthood is reached, and in this regard males would be quite different from many female primates. For instance, Alberts & Altmann (1995b) showed that the attainment of adult dominance rank by male yellow baboons (which is possible only with the attainment of adult body size) is not sufficient to entirely predict the onset of actual reproductive activity, because demographic factors (the rel-

(A)

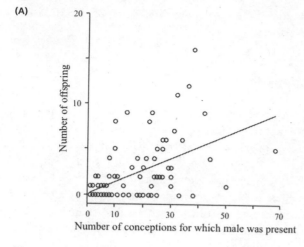

(B)

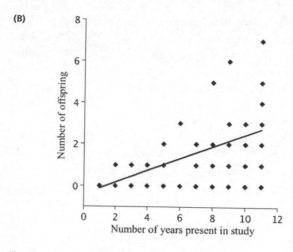

(C)

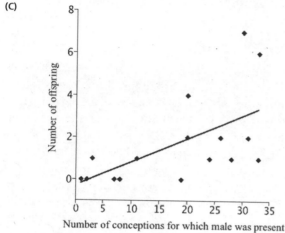

Fig. 18.1. Longevity alone can increase a male's reproductive success. Examples include: (A) baboons (after Alberts et al. 2006), (B) sifakas (after Lawler 2007), and (C) chimpanzees (after Wroblewski et al. 2009).

ative availability of females) intervene. That is, even rapidly maturing males may experience a delay in reproduction if female availability is low at the time of their maturation and if male-male competition is intense. Further, Packer et al. (2000) found that among male olive baboons, maternal rank had no influence on the highest rank sons eventually attained as adults. On the whole, although males may sometimes experience persistent advantages from having high-ranking mothers during their development (e.g., see Meikle et al. 1984; Meikle & Vessey 1988), it is likely that for male primates in most natural populations, both demographic processes and social factors will interfere with and ultimately dilute the influence of maternal rank after development, to a much greater extent than is true for female primates.

Longevity

Again, we know relatively little about variance in this life-history component for male primates; very few data exist

on individual life spans in wild male primates. This gap is attributable to the combined effects of (1) primate longevity in general—researchers must be present for decades to obtain longevity data—and (2) male dispersal in many of the longest-studied primate species, which makes male deaths, and the ages at which they occur, difficult to detect. However, both theoretical and empirical work suggests that longevity is a critical component of reproductive success. In general, fitness will be much more sensitive to changes in adult survival than to changes in adult fertility (e.g., Brault & Caswell 1993; Fisher et al. 2000; Gaillard et al. 2000; Heppell et al. 2000; Altmann & Alberts 2003), thus indicating that traits affecting survival experience strong selection in long-lived mammals. Further, persistence alone—staying active in a population for a relatively long time—can have a significant positive influence on a male's lifetime reproductive output (e.g., yellow baboons, Altmann et al. 1988; Alberts et al. 2006; Verreaux's sifakas, *Propithecus verreauxi*, Lawler 2007; chimpanzees, *Pan troglodytes*, Wroblewski et al. 2009; fig. 18.1).

Male age-specific reproductive patterns will, as a rule, shape age-specific survival patterns very directly because reproductive patterns will determine the force of selection on survival at different ages. For instance, more intense reproductive competition among males may result in faster aging among males than females, because in such circumstances males will generally have shorter reproductive life spans than females (regardless of how long they live), resulting in relaxed selection on longevity in males relative to females. In other words, if reproductive skew is age-biased so that younger, prime males enjoy greater reproductive success than older males, then selection on survival in the older age classes will be relaxed, because traits favoring longer-lived males will have relatively little impact on fitness. Longevity will decrease as a consequence. So, for instance, one would predict that species in which reproduction is heavily skewed towards younger adult males, because they occupy higher rank positions, will show shorter life spans and a more rapid decline in survival with increasing age than will congeners in which mating opportunities are more evenly distributed among age classes or even skewed to older ages. For example, in long-tailed macaques it appears that the highest-ranking males are virtually always young adults (van Noordwijk & van Schaik 2001) who monopolize reproduction (de Ruiter et al. 1994), whereas in Japanese macaques (*Macaca fuscata*), the reproductive success of older males may equal or sometimes even exceed that of younger males (Inoue et al. 2008). The prediction—as yet untested—would be that male Japanese macaques would experience lower relative adult mortality than long-tailed macaques, because they experience greater selection for longevity.

Ample evidence supports the notion that patterns of survival are subject to selection in this way. For instance, Clutton-Brock & Isvaran (2007) compared life spans of polygynous and monogamous species of birds and mammals under conditions of high versus low male-male competition. They found that in polygynous species, males have significantly shorter life spans than females and experience faster increases in mortality with age than females, while in monogamous species males and females have quite similar mortality patterns. Several nonprimate vertebrate species provide dramatic illustrations of the intimate relationship between longevity and reproductive performance. In the long-tailed manakin, male reproductive success is entirely dependent on the formation of long-term duetting partnerships with one or two other males. Males in such partnerships may spend a decade or more "practicing" elaborate duets before having an opportunity to mate. This neotropical bird has an unusually long life span (up to 14 years) for its tiny size (18 grams), and this appears to be an evolutionary consequence of the extremely delayed reproduction experienced by males, which never reproduce before five years of age and do not reach prime reproductive age until 10 years of age (McDonald 1993). African elephant males, too, exhibit an unusual pattern of reproductive performance in which older males have much greater success than younger males; in fact, reproductive performance continues to increase throughout adult life until the last decade, when it experiences a modest decline. This reproductive pattern is probably a consequence of indeterminate growth in this species, in which growth continues throughout most or all of adult life—a very uncommon pattern in mammals—and it is hypothesized to have been one of the selection pressures that produced the great longevity of elephants (Hollister-Smith et al. 2007; Poole et al. 2011).

Producing Offspring

Males are confronted with three different problems when they attempt to produce offspring. The first problem is that of gaining sexual access to fertile females, and the second, a closely related problem, is that of allocating reproductive effort among those females. Both of these problems are discussed here. The third problem, that of achieving success in post-mating competition, is covered in chapter 17. The differences in these problems mean that males have evolved distinct, if sometimes overlapping, solutions to each.

Gaining Access to Mates

Variation in mating success is one of the best studied aspects of primate behavioral biology. The quest to obtain data relevant to this topic has yielded hundreds of papers over the past 50 years. Our knowledge on it is taxonomically limited, however (fig. 18.2). Of a sample of 237 studies of male mating tactics and/or male mating success between 1963 and 2009, most of them are on cercopithecines (161), hominids (39), or Malagasy strepsirrhines (16). Further, the cercopithecine studies are mostly on macaques or baboons (139 of 161). Colobines and New World primates are greatly underrepresented. This taxonomic bias is created by the exigencies of fieldwork and the difficulties of observing arboreal species; the best represented species are either terrestrial or semiterrestrial (macaques, baboons, vervets, and the hominids); the best represented arboreal species (the Malagasy lemurs) tend to be low-canopy dwellers and live in relatively open forest habitat.

In spite of the taxonomic bias, the available data shed considerable light on factors that affect male sexual access to mates. The particular tactics adopted by males may vary across social and mating systems, but for every male primate the problem comes down to three different sources of variance in mating success: (1) direct male-male competition, (2) alternative tactics that circumvent direct competition, and (3) female choice. The first two of these factors are covered in this chapter; the third is covered in chapter 16.

Competitive ability

For most male primates, intrasexual competition (both contest and scramble competition) plays a major role in reproductive performance. Contest competition is more common among male primates, but both these forms of competition can easily blend into each other in animal mating systems,

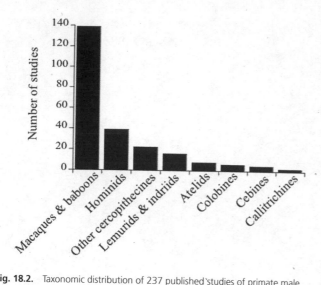

Fig. 18.2. Taxonomic distribution of 237 published studies of primate male mating success

and primates are no exception. Contest competition usually manifests itself as dominance rank, but it can also take the form of physiological suppression, in which subordinate males show less-developed secondary sexual characteristics, reduced circulating testosterone, and little or no sexual behavior (reviewed in Setchell & Kappeler 2003; chapter 17, this volume). In species with spatially dispersed females, scramble competition may be a more important tactic than contest competition. For instance, in the gray mouse lemur (*Microcebus murinus*) a solitary species, successful males were those that began searching early and roamed widely (Eberle & Kappeler 2004). However, even here contest competition played a role, as males often guarded females once they found them, and chased away rivals (Eberle & Kappeler 2004).

For some male primates, tactics other than male-male competition play a greater role in determining mating success. For example, in some lemurs female dominance over males may lead to a degree of female control that can subvert male-male competition (Pereira & Weiss 1991). Even in many of the lemurs, however, male dominance rank, and hence male fighting ability, is a determinant of mating success (fig. 18.3; Nievergelt et al. 2002; Wimmer & Kappeler 2002; Eberle & Kappeler 2004; Kappeler & Port 2008; Kappeler & Schaffler 2008).

Another example is the northern muriqui (*Brachyteles hypoxanthus*), a New World monkey in which relationships among males within groups are egalitarian and tolerant, and in which male-male contest competition for access to females appears to be absent (Strier 1990,r 2000; Strier et al. 2002). Instead, males share sexual access to females with close male associates (Strier et al. 2002). Not surpris-

ingly, male muriquis have large testes relative to their body size, indicating that considerable sperm competition occurs. Interestingly, the males may be less inclined to share access to females when females are conceptive, thus suggesting the possibility of cryptic contest competition in this species (Strier et al. 2002). In addition, aggression between male muriquis of different groups may be intense (Talebi et al. 2009). Nonetheless, the general absence of antagonistic or aggressive behavior among muriqui males that live in the same group suggests that more research on reproductive performance in the rarely studied cebids may yield unusual and surprising behavior patterns.

In spite of these exceptions, the importance of male-male competition (especially contest competition) to male reproductive performance cuts across primate taxa. This includes both direct contests and honest displays of competitive superiority that avoid costly fighting (chapter 17, this volume). The importance of male contest competition has been repeatedly documented in New World monkeys, Old World monkeys, apes, and strepsirrhines (Ellis 1995; Di Fiore 2003). It also cuts across mating systems; researchers have reported an impact of male contest competition on male reproductive performance in all but pair-bonded species (where it is difficult to measure, but certainly plays less of a role than in other systems).

In the early literature on this topic (1970s–1990s), large differences were noted among studies in the impact of male fighting ability and contest competition (i.e., dominance rank) on male mating success (reviewed in Dewsbury 1982; Cowlishaw & Dunbar 1991; Ellis 1995). These discrepancies resulted in considerable uncertainty and discussion about whether dominance rank "really matters" for male primates (Bernstein 1981; Dewsbury 1982; Berenstain & Wade 1983; Fedigan 1983; Bercovitch 1986; Barton & Simpson 1992; Cowlishaw & Dunbar 1992; Dunbar & Cowlishaw 1992a, 1992b; Simpson & Barton 1992; Bercovitch 1992a, 1992b). Several key analyses eventually revealed that the differences among studies stemmed, for the most part, from the interesting fact (not obvious at the time) that in most species the consequences of high dominance rank vary over time (Cowlishaw & Dunbar 1991; Bulger 1993; Alberts et al. 2003; see also Strum 1982 for an early prediction of this sort). Thus, in one population of a given species, dominance rank may almost completely determine male mating success once female availability has been taken into account (e.g., Bulger 1993), while in another population of the same species, or even another time period in the same population, dominance rank may account for little or none of the variance that males experience in gaining access to mates (reviewed in Bulger 1993; Alberts et al. 2003; Alberts et al. 2006; Rodriguez-Llanes et al. 2009).

Fig. 18.3. In many lemur species of Madagascar, males form stable dominance relationships, as in the case of these two male red-fronted lemurs (*Eulemur rufifrons*) from Kirindy, western Madagascar. The importance of rank to male reproductive success in this population is reflected by one study demonstrating that 71% of all infants were sired by dominant males (Kappeler & Port 2008). Other variables, however, such as the relative abundance of females, may be equal if not more important predictors of reproductive skew among males. Photo courtesy of Rebecca Lewis.

For instance, Ellis (1995) found that a majority (63%) of 165 relevant studies reported higher mating success (frequencies of copulation or numbers of sex partners) for high-ranking males than for low-ranking males, with no taxonomic pattern in whether or not the relationship was positive (fig. 18.4). Similarly, in 37 studies of rank and mating success across primate species, Cowlishaw & Dunbar (1991) reported that the correlation between rank and mating success was, on average, positive (mean = 0.37), but varied from $r_s = 1.0$ to $r_s = -0.63$. Alberts et al. (2003) reported similar variability within a single population of baboons over 32 group-years; the correlation for the relationship between dominance rank and mating success was, on average, positive (mean = 0.56 ± 0.04), but varied from $r_s = 1.0$ to $r_s = -0.7$ over these years (Alberts et al. 2003).

It might be argued that mating success is a flawed measure of reproductive performance, and that the best measure is actual offspring production. If we restrict our analysis to genetic studies that involved paternity analysis, however, we see a similar pattern as with mating success. That is, the range of values for the correlation between rank and mating success is similar whether we are comparing across diverse taxa ($r_s = 1.0$ to $r_s = -0.31$; fig. 18.5a), across species within a single genus, *Macaca* (from $r_s = 0.97$ to $r_s = -0.54$; fig. 18.5b), or within a single social group of one species, *M. mulatta* (from $r_s = 0.63$ to $r_s = -0.60$; fig. 18.5c).

These various data sets make two important points. One is that on average, high-rank confers reproductive benefits on male primates, and this explains why males invest so heavily in rank relationships. The relationship is unpredictable, however; any given male may find himself in a situation where, even if he attains high rank, he performs relatively poorly compared to other males. The second point is that the relationship between rank and reproductive performance in male primates shows similar variability at all levels of analysis: across genera, within genera, and within

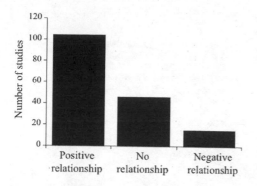

Fig. 18.4. Number of studies showing positive, negative, and no relationship between male dominance rank and mating success (from Ellis 1995)

populations of a single species. Very importantly, this suggests that similar processes are driving variation in the relationship between rank and reproductive performance in most, if not all, primate taxa.

Competitive ability and density dependence. The best-supported source of variation in the relationship between rank and reproductive performance is male density. That is, the number of males in the social group appears to be a primary cause of this variability both across and within species, so that in a number of species, the highest ranking male's ability to monopolize access to females is inversely related to the number of other males he is living with (Cowlishaw & Dunbar 1991; Alberts et al. 2003). Density clearly explains variation in the relationship between rank and reproductive performance, both within and between taxa. High-ranking males gain a greater proportion of fertilizations in smaller groups than in larger ones.

This represents density-dependent sexual selection on fighting ability. It means that the intensity of selection on fighting ability, which is a major component of male contest competition, will be relaxed under conditions in which high rank results in fewer reproductive benefits. Density-dependent sexual selection, in which the efficacy of a sexually selected trait varies with group size or population density, is quite common in animals (e.g., ungulates, Clutton-Brock et al. 1997; Coltman et al. 1999; Yoccoz et al. 2002; fish, Einum et al. 2008; Lehtonen & Lindstrom 2008; Urbach & Cotton 2008; invertebrates, McLain 1982; Conner 1989; Pilastro et al. 1997; Nielsen & Watt 2000; Levitan 2004; Bertin & Cezilly 2005).

In theory, of course, males can circumvent the problem of the density-dependent nature of contest competition by controlling their social density—by living in social groups with only females (i.e., in socially monogamous or socially polygynous groups rather than in polygynandrous groups).

Indeed, in a cross-taxon analysis of how effectively the dominant male monopolizes reproduction in a group, the most important determinant was whether he lived in a single-male social system or a multimale social system (Clutton-Brock & Isvaran 2006). Even males in single-male social systems are not entirely successful in excluding other males from mating, though they are more effective in doing so than males living in groups with other males. However, a male's ability to exclude other males from his group (and hence to keep them away from the females he is interested in) will depend upon female density and dispersion patterns (Emlen & Oring 1977), factors that are beyond the male's control for the most part. With the exception of lemurs (Kappeler 2000), single-male primate groups are virtually always small (1–5 females) and groups containing more than 10 females virtually always contain multiple males who cannot escape ongoing, regular contest competition with other males if they are to reproduce (Altmann 2000). When there are intermediate numbers of females, males strive with varying success to prevent other males from gaining access to them (see fig. 20.1 in Altmann 2000; see also Cords 2000).

In species with male-biased dispersal, males in multimale groups can and do attempt to regulate their density by leaving groups with high male density and moving to groups with lower density (e.g., Alberts & Altmann 1995a; Altmann 2000; Fedigan & Jack 2004; Jack & Fedigan 2004; Whitten & Turner 2004; Clarke et al. 2008; Muniz et al. 2010; see also Pochron & Wright 2003 for a counterexample). Indeed, in some species a male's ability to regulate his reproductive performance by adjusting his demographic environment may be quite nuanced, taking into account not only the availability of females and the number of other males in the group, but his own competitive ability as well. For instance, Alberts & Altmann (1995a) found that adult male yellow baboons that achieved greater than the median level of reproductive success for their age group were less likely to emigrate from a group than were males that were below the median mating success for their age group. Similarly, Clarke et al. (2008) found that male chacma baboons (*Papio ursinus*) in good physical condition tended to remain in groups even during periods when there were many more males than cycling females, while males in poor physical condition did not do so.

Competitive ability and age. In most species, male reproductive performance declines with age after the prime years of adulthood. Typically, this decrease reflects changes in a male's dominance rank over time. Adult males in peak physical condition—usually not the youngest adult males, but those among the younger adults—rank highest and

(A)

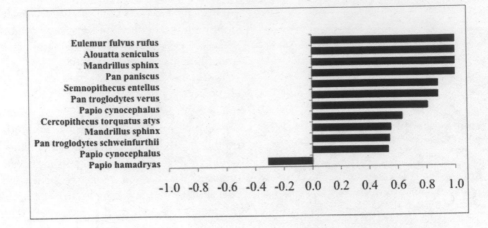

(B)

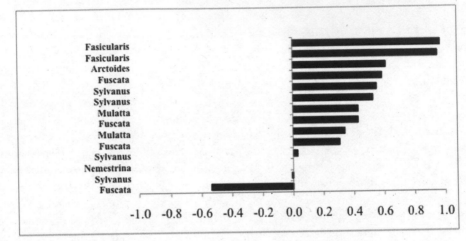

(C)

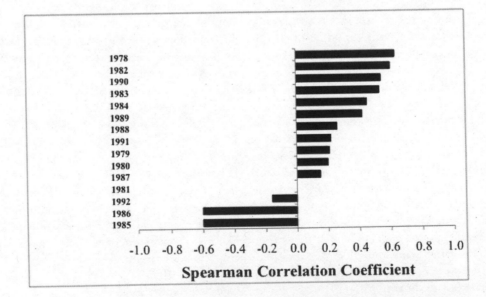

Fig. 18.5. Variation in the relationship between male dominance rank and reproductive success is similar at all levels of analysis: (a) variation across studies of diverse taxa (Alberts et al. 2006; Altmann et al. 1996; Boesch et al. 2006; Dixson et al. 1993; Gerloff et al. 1999; Gust et al. 1998; Launhardt et al. 2001; Pope 1990; Setchell et al. 2005; Smith 1994; Wimmer & Kappeler 2002; Wroblewski et al. 2009), (b) variation within studies of the genus *Macaca* (Rodriguez-Llanes et al. 2009), and (c) variation across years within a group of *M. mulatta* (Smith 1994).

achieve the most reproductive success; both dominance rank and reproductive performance then usually decline steadily with age after the prime years. This pattern has been demonstrated repeatedly in cercopithecines in multi-male groups (e.g., long-tailed macaques, van Noordwijk & van Schaik 2001; yellow baboons, Alberts et al. 2006; mandrills, Setchell et al. 2005; rhesus macaques, Bercov-itch et al. 2003; Widdig et al. 2004; see Lawler 2007 for a counterexample in Verreaux's sifakas).

Among provisioned populations of Japanese macaques and rhesus macaques, the typical relationship between age and dominance rank is reversed. Males in these populations tend to follow a "seniority" rule, in which older males hold the highest-ranking positions, and new immigrant males rank among the lowest. Manson (1998) notes that group sizes in these populations far exceed those typically seen in natural populations, and he proposes that the seniority rule results from the very large numbers of males in these provisioned groups. He points out that the probability of success for any immigrant male primate attempting to wrest the al-pha position from the current alpha declines as the number of males in the group increases. This is simply because in larger groups, which represent larger samples of the popu-lation, it is more likely that the best fighter in the group is better than the immigrant male. Consequently, immigrant males limit the extent to which they challenge high-ranking males, and a seniority rule emerges. Interestingly, in spite of the seniority rule for dominance rank, older males in these populations do not consistently experience higher mating or paternity success than do younger adult males (Bercovitch et al. 2003; Widdig et al. 2004; Inoue & Tak-enaka 2008). This may be driven both by female prefer-ences for immigrant (usually young) males versus long-term (older) residents, and by the use of alternative mating tactics by males (Berard et al. 2004; Inoue & Takenaka 2008).

Alternative Tactics

The greatest threat to the reproductive hegemony of a male is the presence of other males in his proximity (Clutton-Brock & Isvaran 2006). And in most primate species, males live in close proximity to each other. One consequence of this arrangement is pervasive contest competition among males and the frequent employment of mate guarding (close, persistent following of a cycling female and attempts to prevent other males from gaining sexual access to her) as the primary male mating tactic in primates. In most primate species, however, males also employ alternative tactics to mate guarding and direct competition, and these tactics can contribute substantially to variance in male reproductive performance. The alternatives that have been studied fall broadly into four categories: (1) engaging in furtive cop-ulation, (2) using cooperative coalitions among males to undermine the reproductive domination of higher-ranking males, (3) developing affiliative relationships with females that allow competitively inferior males to mate success-fully via female mate choice, and (4) in one-male groups, overwhelming the resident male with simultaneous influxes of multiple nonresident males. These tactics should not be thought of as entirely distinct or mutually exclusive, and some or all of them may also involve some level of contest competition. Each of them is discussed below.

Furtive copulations. This tactic depends on gaining sexual access to a receptive female without being detected by com-petitive males. "Sneaky copulations involve rapid, furtive matings usually under cover of vegetation and frequently on the group periphery out of view of the dominant males." (Berard et al. 1994, p. 194). Furtive copulations, involving either resident or extragroup males, have been reported in a number of primate species (e.g., ring-tailed lemurs, *Lemur catta*, Parga 2009; Barbary macaques, Modolo & Mar-tin 2008; chacma baboons, Crockford et al. 2007; patas monkeys, *Erythrocebus patas*, Ohsawa et al. 1993; Japa-nese macaques, Soltis et al. 2001). In general, however, this tactic is much less understood or documented than others. Because furtive copulations involve stealth, they should be more effective (and hence should occur more frequently) in forest settings or other visually complex environments than in open habitats, and in high-density populations where the sheer number of individuals makes it difficult for competi-tive males to monitor them. This hypothesis has not been quantitatively tested. However, it does seem to be the case that in open-habitat populations, furtive copulations result in relatively little reproductive success, while in crowded or visually complex environments they may be almost as successful as mate guarding. For example, in Amboseli yel-low baboons living in an open habitat, subadult males, who only mated furtively, fathered just 1.5% of offspring (Al-berts et al. 2006). In contrast, among rhesus macaques on the island of Cayo Santiago, off the coast of Puerto Rico (living at high density in a visually complex habitat), males that mated furtively achieved 40% of the reproductive success that consorting males did (Berard et al. 1994). In general, males using furtive copulations perform less well than males that mate guard (e.g., Ohsawa et al. 1993; Be-rard et al. 1994; Soltis et al. 2001; Alberts et al. 2006). Not surprisingly, furtive copulations are often the primary tactic used by subadult males (e.g., Alberts et al. 2006; Modolo & Martin 2008).

Male-male coalitions. In some primate species, males gain access to reproductive females by cooperating with each other. For the most part, this takes the form of cooperatively displacing high-ranking males, either by invading groups together as a semi-permanent coalition (e.g., mantled howlers, *Alouatta palliata*, Dias et al. 2010), or by forming temporary alliances among males within a group that result in simple turnovers of mate-guarding episodes (e.g., baboons, Bercovitch 1988; Noë & Sluijter 1990; Barbary macaques, Kuester & Paul 1992). Such coalitions can be an extremely effective means of redistributing mating opportunities, and in some cases they may result in an almost complete reversal of the typical relationship between dominance rank and mating success (Bercovitch 1988; Noë & Sluijter 1990; Alberts et al. 2003). Male-male coalitions have been most intensively studied in baboons, which show interesting intrageneric diversity in coalition formation. In chacma baboons of southern Africa, male-male coalitions are virtually absent, while in the eastern African olive and yellow baboons, male-male coalitions are common (fig. 18.6; reviewed in Bulger 1993; Henzi et al. 1999; Alberts et al. 2003). The power of male-male coalitions to affect male reproductive performance is highlighted by the observation that among chacma baboons the highest-ranking male nearly always monopolizes mating, while in olive and yellow baboons lower-ranking males may experience substantial mating success (Bercovitch 1986; Bulger 1993; Alberts et al. 2003).

Another type of male-male coalition aimed directly at obtaining access to females (as opposed to those aimed at permanent changes in dominance rank; see Pandit & van Schaik 2003), is one in which pairs or trios of high-ranking males cooperate with each other, preventing other males from mating and sharing sexual access to a cooperatively guarded female. This has been reported in one population of chimpanzees (Watts 1998), and has also been described in another fission-fusion society, bottle-nosed dolphins (Connor et al. 2001), but not in other primates.

Affiliative relationships with females. In a number of primate species, males and females form close social bonds. In theory, the males could pursue such bonds as an alterna-

Fig. 18.6. Two adult male baboons (*Papio cynocephalus*) at Amboseli, Kenya, engage in a male-male coalition against a third male (*center*), who is performing a submissive grimace or teeth-bared display. Photo courtesy of Courtney L. Fitzpatrick.

tive to contest competition for access to females. Indeed, the existence of single-male groups in both monogamous and polygynous species could be construed as evidence that forming affiliative bonds with females circumvents the need for continual contest competition among males. Apart from these single-male social systems, however, there is relatively little support for the hypothesis that males in any species increase their future reproductive success substantially by forming affiliative relationships with females. Smuts (1985) reported that male baboons forming "friendships" with an-estrous females were more likely to mate with those females when they began sexually cycling (fig. 17.2, chapter 17, this volume). However, this result has been difficult to replicate (e.g., Nguyen et al. 2008), and stronger evidence supports the hypothesis that these male-female "friendships" in ba-boons are associated with male care and protection of in-fants rather than future mating between friends (Palombit et al. 1997; Nguyen et al. 2008; Moscovice et al. 2009; chapter 19, this volume).

Invasions by nonresident males. In a number of polygy-nous species, groups of nonresident males temporarily in-vade single-male social groups during periods of high fertile female availability (e.g., Ohsawa et al. 1993; Chism & Rog-ers 1997; Borries 2000; reviewed in Cords 1987, 2000). By doing so, they effectively overwhelm the resident male and take mating opportunities from him. This influx tac-tic can be thought of as an attempt to shift the nature of male-male competition away from contest competition and towards scramble competition. As such, it represents an al-ternative to contest competition. It can be quite effective, and invading males may engage in extensive mating during influxes (Cords 2000). It has only been observed in species with single-male, multifemale groups; perhaps it does not occur in multimale species because resident males in these species can effectively repel influxes through coordinated defense.

Allocating Mating Effort

In contrast to the problem of gaining access to mates, the problem of how males should allocate their mating effort has received relatively little attention in the primate liter-ature. This is probably because most males reproduce in contexts in which receptive females are quite scarce and the competition for them is relatively intense. This would seem to mitigate against any choosiness on the part of males; presumably, the opportunity cost a male experiences by forgoing mating with any receptive female is greater than the gains of being choosy, simply because of the scarcity of

receptive females. However, a number of studies provide evidence that male primates do indeed allocate their mating effort with some care.

For instance, ample evidence supports the hypothesis that males increase their mating effort *within* female cycles as a function of the probability of ovulation (reviewed in Nunn 1999 for species with sexual swellings). In addition, several recent studies of baboons provide evidence that males al-locate their mating effort *between* females as a function of the probability of conception. That is, males prefer to mate with females that are experiencing conceptive, as opposed to nonconceptive, cycles (Bulger 1993; Weingrill et al. 2000; Weingrill et al. 2003; Alberts et al. 2006; Gesquiere et al. 2007). Female fecundability and male interest both probably increase gradually from a female's first postpar-tum cycle through to her conceptive cycle, and male prefer-ences for the penultimate cycle may be quite similar to male preferences for the conceptive cycle (e.g., Bercovitch 1987). Nonetheless, the overall effect of greater male investment in the conceptive cycle is that the highest-ranking male in the group obtains a disproportionate number of conceptive matings (Bulger 1993; Weingrill et al. 2000; Weingrill et al. 2003; Alberts et al. 2006; Gesquiere et al. 2007).

Sexual swellings in primates may have evolved via male mate choice as a "reliable indicator" of female quality (Pa-gel 1994). That is, sexual swelling size might signal endur-ing differences among females in their ability to conceive and raise offspring. If so, males should prefer to mate with females that have larger swellings, because it will enhance their own reproductive success. Several studies have at-tempted to test this hypothesis, but none have supported it. Domb & Pagel (2001) demonstrated that female baboons with larger swellings had higher infant survival and were preferred by males, but this effect disappeared after control-ling for substantial differences in female body size that were associated with different feeding conditions across social groups (Zinner et al. 2002). Setchell et al. (2006) found no effect of sexual swelling size on male reproductive effort in mandrills. Huchard et al. (2009) reported greater male mat-ing effort towards female baboons with larger swellings, but because the authors could not differentiate conceptive from nonconceptive cycles, and because conceptive cycles have larger swellings, they could not differentiate between male choice among females as a simple function of the probabil-ity of conception on that particular cycle, and male choice for females of better quality (see additional discussion of sexual swellings in chapters 16 and 19, this volume, and fig. 16.3, this volume).

Although no evidence supports the idea that males dis-criminate among females on the basis of stable interindivid-ual differences in swelling size, several studies do indicate that male primates often show preferences for two specific female traits—age and dominance rank—and allocate their mating effort accordingly. It has long been known that males discriminate against adolescent, and possibly even young adult, females, almost certainly because these females have lower fertility or infant survival rate than fully adult females (reviewed in Anderson 1986). More recently, Muller et al. (2006) reported that male chimpanzees actively prefer older adult females to younger ones, even among parous females (fig. 18.7). They propose that this might reflect male preferences for more experienced mothers, or male preferences for females with higher viability. As yet, this finding has not been replicated in other species. Male preferences for females of high rank, too, are likely to reflect the higher reproductive performance of those females (chapter 15, this volume; Keddy-Hector 1992; Kuester & Paul 1996; Parga 2006; Setchell & Wickings 2006). However, it is still fairly uncommon for studies to report strong male mating preferences, suggesting that either this is an understudied phenomenon, or that the opportunity costs of acting on mating preferences are indeed greater than the gains associated with choosiness.

Infant Survival

The general expectation for primates, as for all animals, is that males will provide little or no paternal care except when paternity certainty is high. In external fertilizers, such as fish, paternal certainty can be achieved by mate guarding and egg guarding during spawning; in internal fertilizers, paternal care is generally expected only under monogamy (reviewed in Clutton-Brock 1991). Because few primates are monogamous, the common view of male primates is that, for most species, paternal care represents a relatively minor part of their reproductive performance. However, a more detailed and nuanced examination of male primate behavior indicates that paternal contributions to offspring

Fig. 18.7. In spite of their promiscuous mating system, males of the Kanyawara chimpanzee community of Kibale, Uganda, actively prefer to mate with older, not younger, adult females. This is reflected not only in rates of copulation, but also in measures of female attractiveness, such as the proportion of copulations that are preceded by solicitation displays like the one performed by this male (*right*), whose erect penis is a courtship signal. Photo courtesy of Martin N. Muller.

survival and development are often substantial and have played an enormously important role in shaping primate social systems, not only in monogamous species, but also in nonmonogamous species in which paternal certainty is presumed to be low (see also chapters 17 and 19, this volume).

In a few species (all of the callitrichines, one hylobatid—siamangs, *Symphalangus syndactylus*—and two genera of the cebids, *Aotus* and *Callicebus*), males provide extensive paternal care to individual offspring, most prominently in the form of carrying (Wright 1990; Tardif 1994; Snowdon 1996; Lappan 2008; chapter 3, this volume). The best-documented effect of infant carrying (by fathers and/or by other group members) in all these taxa is to increase female reproductive output (Garber & Leigh 1997; Mitani & Watts 1997; Ross & MacLarnon 2000). For instance, in callitrichines, fathers and other group members begin to carry infants soon after their birth, and by freeing the mother of the energy demands of carrying (see Achenbach & Snowdon 2002; see also chapter 14, this volume, fig. 14.6) they apparently allow her to invest more in lactation, thus facilitating rapid growth and development of immatures and therefore hastening the next reproductive event (Goldizen 1987; Snowdon 1996; Garber 1997; Garber & Leigh 1997; Mitani & Watts 1997). In siamangs, carrying by males is variable, but when it does occur, it typically begins much later, near the end of the infant's first year of life. In spite of this relatively late onset, females have shorter interbirth intervals when paired with males that carry their young (Lappan 2008, 2009). Only in the callitrichines is paternal care (and care by other adults in the group) known to directly improve infant survival (reviewed in Snowdon 1996). However, even if the primary role of paternal care in other "carrying" species is to enhance female reproductive output, the male must clearly also play a role in the offspring's well-being, because the mother's reproductive output is only enhanced if the offspring that are cared for by males actually survive.

Even in species in which extensive carrying does not occur, male investment in immatures has left an enduring stamp on multiple aspects of both male and female behavior (reviewed in Fernandez-Duque et al. 2009). In 1997, Kappeler and van Schaik demonstrated that in primates, the evolutionary transition from "infant parking" (stowing the infant in a protected nest or burrow) to full-time carrying by the mother (resulting in an infant that the father could locate and maintain proximity to) was accompanied in lockstep by an evolutionary transition from solitary living or unisexual groups to permanent male-female associations (and hence permanent male-infant associations). Van Schaik and Kappeler (1997) interpreted this as evidence

that male-female associations are the result of selection on males to protect infants, primarily from sexually selected infanticide.

Indeed, the hypothesis that sexually selected infanticide shapes the behavioral strategies of both sexes has spurred compelling research ever since it was first proposed (chapter 19, this volume). In the past 10 years, more and more of this research has produced direct evidence of paternal care in the form of protection against infanticide in a number of primate species. For instance, in chacma baboons, infanticide is relatively common, accounting for at least 38% of infant deaths (and often more) in Moremi, Botswana (Palombit et al. 2000; Cheney et al. 2004). As a consequence, females pursue "friendships" with males, and males—often fathers—act as protectors against potentially infanticidal males (Palombit et al. 2001; Moscovice et al. 2009). In Hanuman langurs (*Semnopithecus entellus*), another species in which infanticide has a substantial impact on infant survival, both true fathers and potential fathers (males that were present in the group when the infant was conceived) actively protect infants against infanticidal males (Borries et al. 1999a, b). Male defense against infanticide occurs in a number of other species (chapter 19, this volume).

When infanticide is not a major source of infant mortality, males in nonmonogamous species may still care for and enhance the fitness of their offspring. In Amboseli baboons, in which sexually selected infanticide is rare in comparison to that in chacma baboons (Janson & van Schaik 2000; Henzi & Barrett 2003; Palombit 2003), adult male baboons engage in paternal care. Specifically, they differentiate their own offspring from the offspring of other males and bias their support in agonistic disputes towards their own offspring (Buchan et al. 2003). In addition, males, often fathers, protect female "friends" with neonates against the stress that can result from other females' attempts to handle neonates (Nguyen et al. 2009). Furthermore, paternal care appears to have long-term consequences; the longer fathers live in the same social group with their offspring, the earlier those offspring reach maturity (Charpentier et al. 2008). Paternal presence may accelerate maturation by enhancing growth, reducing social stress through paternal interventions in agonistic disputes, or both. In any case, it shows an unexpectedly strong effect of fathers in a multimale social system, and raises the possibility that direct, day-to-day paternal care might be more widespread than anticipated, even in multimale systems.

In spite of the clear importance—direct or indirect—of fathers for infant primates, no studies have yet examined the effect of fathers, and in particular of variation in paternal characteristics on infants, in anything like the detail that has been possible for the effect of mothers. In addition, the question of how fathers balance investment in offspring care with investment in offspring production remains largely unknown for primates (see Magrath & Komdeur 2003 for a general review of this issue). For both of these questions, identifying a male's offspring is the initial challenge. Until this has been done for a large number of males and offspring, it will remain beyond our reach to examine the trade-off between paternal care and mating effort, or to measure the impact of paternal dominance rank, paternal age, or paternal presence—common currencies in measuring maternal effects—on infant survival in most primate species.

Summary and Conclusions

In most species, males have greater reproductive potential than females, but also a greater probability of failing to reproduce. The consequence is strong sexual selection on males to influence female reproductive decisions—in other words, strong selection for traits that enhance a male's access to high-quality reproductive opportunities. The relevant traits are those that affect how early in life a male begins his reproductive career (first reproduction), how long his reproductive career lasts (longevity), and how effectively he is able to compete for and allocate effort among females during his reproductive career (fertility). Male primates, as members of a lineage with slow life histories (i.e., long lives and low fertility) should also experience strong selection to influence the survival of the relatively few offspring they will produce in their lifetimes. Good quantitative data are available for sources of variance in fertility, but sources of variance in age at first reproduction, in longevity, and in behaviors that influence infant survival are greatly understudied.

As a final point, it is useful to note that ultimately, variance in male reproductive performance depends entirely upon variance in female fertility. No matter how "good" a male is, he is ultimately constrained by the factors that influence female fertility—environmental effects on *female* reproduction, density effects on *female* reproduction, and individual differences among *females*—because his reproductive performance depends upon theirs. As a consequence, while a male's reproductive performance is dependent upon all of the factors discussed above, it is also dependent upon something much more difficult to quantify: his ability, given a group of females, to increase their reproductive performance by enhancing their resource base.

Almost all work on variance in male reproductive performance and mating success has focused on differences among males in their ability to gain a larger "piece of the reproductive pie." However, a male's ability to increase the reproductive performance of the females he mates with has gained notable attention in recent years. For instance, it is thought to be the ultimate factor driving intercommunity aggression between male chimpanzees (reviewed in Muller & Mitani 2005; see also chapter 6, this volume, fig. 6.8). As yet, however, few data are available to estimate the importance of this factor—a male's ability to influence the fertility of his mates—relative to that of other sources of variance in male reproductive performance. This represents a rich potential source of questions for future studies of many different species.

References

Achenbach, G. G. & Snowdon, C. T. 2002. Costs of caregiving: Weight loss in captive adult male cotton-top tamarins (*Saguinus oedipus*) following the birth of infants. *International Journal of Primatology*, 23, 179–189.

Alberts, S. C. & Altmann, J. 1995a. Balancing costs and opportunities: Dispersal in male baboons. *American Naturalist*, 145, 279–306.

Alberts, S. C. & Altmann, J. 1995b. Preparation and activation: Determinants of age at reproductive maturity in male baboons. *Behavioral Ecology and Sociobiology*, 36, 397–406.

Alberts, S. C., Buchan, J. C. & Altmann, J. 2006. Sexual selection in wild baboons: From mating opportunities to paternity success. *Animal Behaviour*, 72, 1177–1196.

Alberts, S. C., Watts, H. E. & Altmann, J. 2003. Queuing and queue-jumping: Long-term patterns of reproductive skew in male savannah baboons, *Papio cynocephalus. Animal Behaviour*, 65, 821–840.

Altmann, J. 2000. Models of outcome and process: Predicting the number of males in primate groups. In *Primate Males* (ed. by P. Kappeler). Cambridge: Cambridge University Press.

Altmann, J. & Alberts, S. C. 2003. Intraspecific variability in fertility and offspring survival in a nonhuman primate: Behavioral control of ecological and social sources. In *Offspring: Human Fertility Behavior in Biodemographic Perspective* (ed. by Wachter, K. W. & Bulatao, R. A.), 140–169. Washington: National Academies Press.

———. 2005. Growth rates in a wild primate population: Ecological influences and maternal effects. *Behavioral Ecology and Sociobiology*, 57, 490–501.

Altmann, J., Altmann, S. & Hausfater, G. 1988. Determinants of reproductive success in savannah baboons (*Papio cynocephalus*). In *Reproductive Success* (ed. by T. H. Clutton-Brock), 403–418. Chicago: University of Chicago Press.

Anderson, C. M. 1986. Female age: Male preference and reproductive success in primates. *International Journal of Primatology*, 7, 305–326.

Andersson, M. 1994. *Sexual Selection*. Princeton, NJ: Princeton University Press.

Barton, R. A. & Simpson, A. J. 1992. Does the number of males influence the relationship between dominance and mating success in primates? *Animal Behaviour*, 44, 1159–1161.

Beacham, T. D. & Murray, C. B. 1993. Fecundity and egg size

variation in North American Pacific salmon (*Oncorhynchus*). *Journal of Fish Biology*, 42, 485–508.

Berard, J. D., Bercovitch, F. B., Widdig, A., Trefilov, A., Kessler, M. J., Schmidtke, J., Nurnberg, P. & Krawczak, M. 2004. Male reproductive output and alternative male life-history strategies in rhesus macaques. *Folia Primatologica*, 75, 157.

Berard, J. D., Nurnberg, P., Epplen, J. T. & Schmidtke, J. 1994. Alternative reproductive tactics and reproductive success in male rhesus macaques. *Behaviour*, 129, 177–201.

Bercovitch, F. B. 1986. Male rank and reproductive activity in savanna baboons. *International Journal of Primatology*, 7, 533–550.

———. 1987. Reproductive success in male savanna baboons. *Behavioral Ecology and Sociobiology*, 21, 163–172.

———. 1988. Coalitions, cooperation and reproductive tactics among adult male baboons. *Animal Behaviour*, 36, 1198–1209.

———. 1992a. Re-examining the relationship between rank and reproduction in male primates. *Animal Behaviour*, 44, 1168–1170.

———. 1992b. Dominance rank, reproductive success and reproductive tactics in male primates: A reply to Dunbar and Cowlishaw. *Animal Behaviour*, 44, 1174–1182.

———. 1993. Dominance rank and reproductive maturation in male rhesus macaques (*Macaca mulatta*). *Journal of Reproduction and Fertility*, 99, 113–120.

Bercovitch, F. B. & Berard, J. D. 1993. Life history costs and consequences of rapid reproductive maturation in female rhesus macaques. *Behavioral Ecology and Sociobiology*, 32, 103–109.

Bercovitch, F. B., Widdig, A. & Nürnberg, P. 2000. Maternal investment in rhesus macaques (*Macaca mulatta*): Reproductive costs and consequences of raising sons. *Behavioral Ecology and Sociobiology*, 48, 1–11

Bercovitch, F. B., Widdig, A., Trefilov, A., Kessler, M. J., Berard, J. D., Schmidtke, J., Nurnberg, P. & Krawczak, M. 2003. A longitudinal study of age-specific reproductive output and body condition among male rhesus macaques, *Macaca mulatta*. *Naturwissenschaften*, 90, 309–312.

Berenstain, L. & Wade, T. D. 1983. Intrasexual selection and male mating strategies in baboons and macaques. *International Journal of Primatology*, 4, 201–235.

Bernstein, I. S. 1981. Dominance: The baby and the bathwater. *Behavioral and Brain Sciences*, 4, 419–457.

Bertin, A. & Cezilly, F. 2005. Density-dependent influence of male characters on mate-locating efficiency and pairing success in the waterlouse *Asellus aquaticus*: An experimental study. *Journal of Zoology*, 265, 333–338.

Borries, C. 2000. Male dispersal and mating season influxes in hanuman langurs living in multi-male groups. In *Primate Males: Causes and Consequences of Variation in Group Composition* (ed. by P. Kappeler), 146–158. Cambridge: Cambridge University Press.

Borries, C., Launhardt, K., Epplen, C., Epplen, J. T. & Winkler, P. 1999a. DNA analyses support the hypothesis that infanticide is adaptive in langur monkeys. *Proceedings of the Royal Society of London Series B-Biological Sciences*, 266, 901–904.

———. 1999b. Males as infant protectors in Hanuman langurs (*Presbytis entellus*) living in multimale groups: Defence pattern, paternity and sexual behaviour. *Behavioral Ecology and Sociobiology*, 46, 350–356.

Brault, S. & Caswell, H. 1993. Pod-specific demography of killer whales (*Orcinus orca*). *Ecology*, 74, 1444–1454.

Buchan, J. C., Alberts, S. C., Silk, J. B. & Altmann, J. 2003. True paternal care in a multi-male primate society. *Nature*, 425, 179–181.

Bulger, J. B. 1993. Dominance rank and access to estrous females in male savanna baboons. *Behaviour*, 127, 67–103.

Charpentier, M. J. E., Tung, J., Altmann, J. & Alberts, S. C. 2008. Age at maturity in wild baboons: Genetic, environmental and demographic influences. *Molecular Ecology*, 17, 2026–2040.

Cheney, D. L., Seyfarth, R. M., Andelman, S. J. & Lee, P. C. 1988. Reproductive success in vervet monkeys. In *Reproductive Success* (ed. by T. H. Clutton-Brock), 384–402. Chicago: University of Chicago Press.

Cheney, D. L., Seyfarth, R. M., Fischer, J., Beehner, J., Bergman, T., Johnson, S. E., Kitchen, D. M., Palombit, R. A., Rendall, D. & Silk, J. B. 2004. Factors affecting reproduction and mortality among baboons in the Okavango Delta, Botswana. *International Journal of Primatology*, 25, 401–428.

Chism, J. & Rogers, W. 1997. Male competition, mating success and female choice in a seasonally breeding primate (*Erythrocebus patas*). *Ethology*, 103, 109–126.

Clarke, P. M. R., Henzi, S. P., Barrett, L. & Rendall, D. 2008. On the road again: Competitive effects and condition-dependent dispersal in male baboons. *Animal Behaviour*, 76, 55–63.

Clutton-Brock, T. H. & Isvaran, K. 2006. Paternity loss in contrasting mammalian societies. *Biology Letters*, 2, 513–516.

———. 2007. Sex differences in ageing in natural populations of vertebrates. *Proceedings of the Royal Society of London Series B-Biological Sciences*, 274, 3097–3104.

Clutton-Brock, T. H., Rose, K. E. & Guinness, F. E. 1997. Density-related changes in sexual selection in red deer. *Proceedings of the Royal Society of London Series B-Biological Sciences*, 264, 1509–1516.

Coltman, D. W., Smith, J. A., Bancroft, D. R., Pilkington, J., MacColl, A. D. C., Clutton-Brock, T. H. & Pemberton, J. M. 1999. Density-dependent variation in lifetime breeding success and natural and sexual selection in Soay rams. *American Naturalist*, 154, 730–746.

Conner, J. 1989. Density-dependent sexual selection in the fungus beetle, *Bolitotherus cornutus*. *Evolution*, 43, 1378–1386.

Connor, R. C., Heithaus, M. R. & Barre, L. M. 2001. Complex social structure, alliance stability and mating access in a bottlenose dolphin "super-alliance." *Proceedings of the Royal Society of London Series B-Biological Sciences*, 268, 263–267.

Cords, M. 1987. Forest guenons and patas monkeys: Male-male competition in one-male groups. In: *Primate Societies* (ed. by B. B. Smuts, D. L. Cheney, R. M. Seyfarth, R. W. Wrangham & T. T. Struhsaker), 98–111. Chicago: University of Chicago Press.

———. 2000. The number of males in guenon groups. In *Primate Males: Causes and Consequences of Group Composition* (ed. by P. Kappeler), 84–96. Cambridge: Cambridge University Press.

Coulson, T., Benton, T. G., Lundberg, P., Dall, S. R. X., Kendall, B. E. & Gaillard, J. M. 2006. Estimating individual contributions to population growth: Evolutionary fitness in ecological time. *Proceedings of the Royal Society of London Series B-Biological Sciences*, 273, 547–555.

Cowlishaw, G. & Dunbar, R. I. M. 1991. Dominance rank and mating success in male primates. *Animal Behaviour*, 41, 1045–1056.

———. 1992. Dominance and mating success: A reply to Barton and Simpson. *Animal Behaviour*, 44, 1162–1163.

Crockford, C., Wittig, R. M., Seyfarth, R. M. & Cheney, D. L. 2007. Baboons eavesdrop to deduce mating opportunities. *Animal Behaviour*, 73, 885–890.

De Ruiter, J. R., van Hooff, J. A. R. A. M. & Scheffrahn, W. 1994. Social and genetic aspects of paternity in wild long-tailed macaques (*Macaca fascicularis*). *Behaviour*, 129, 203–224.

Dewsbury, D. A. 1982. Dominance rank, copulatory behavior and differential reproduction. *Quarterly Review of Biology*, 57, 135–159.

Di Fiore, A. 2003. Molecular genetic approaches to the study of primate behavior, social organization, and reproduction. In *Yearbook of Physical Anthropology*, 46, 62–99.

Dias, P. A. D., Rangel-Negrín, A., Vea, J. J. & Canales-Espinosa, D. 2010. Coalitions and male-male behavior in *Alouatta palliata*. *Primates*, 51, 91–94.

Dixson, A. F. & Nevison, C. M. 1997. The socioendocrinology of adolescent development in male rhesus monkeys (*Macaca mulatta*). *Hormones and Behavior*, 31, 126–135.

Domb, L. & Pagel, M. 2001. Sexual swellings advertise female quality in wild baboons. *Nature*, 410, 204–206.

Dunbar, R. I. M. & Cowlishaw, G. 1992a. Incest and other artifacts: A reply. *Animal Behaviour*, 44, 1166–1167.

———. 1992b. Mating success in male primates: Dominance rank, sperm competition and alternative strategies. *Animal Behaviour*, 44, 1171–1173.

Eberle, M. & Kappeler, P. M. 2004. Sex in the dark: Determinants and consequences of mixed male mating tactics in *Microcebus murinus*, a small solitary nocturnal primate. *Behavioral Ecology and Sociobiology*, 57, 77–90.

Einum, S., Robertsen, G. & Fleming, I. A. 2008. Adaptive landscapes and density-dependent selection in declining salmonid populations: Going beyond numerical responses to human disturbance. *Evolutionary Applications*, 1, 239–251.

Ellis, L. 1995. Dominance and reproductive success among nonhuman animals: A cross-species comparison. *Ethology and Sociobiology*, 16, 257–333.

Emlen, S. T. & Oring, L. W. 1977. Ecology, sexual selection and the evolution of mating systems. *Science*, 197, 215–223.

Fedigan, L. M. 1983. Dominance and reproductive success in primates. *Yearbook of Physical Anthropology*, 26, 91–129.

Fedigan, L. M. & Jack, K. M. 2004. The demographic and reproductive context of male replacements in *Cebus capucinus*. *Behaviour*, 141, 755–775.

Fernandez-Duque, E., Valeggia, C. R. & Mendoza, S. P. 2009. The biology of paternal care in human and nonhuman primates. *Annual Review of Anthropology*, 38, 115–130.

Fisher, D. O., Hoyle, S. D. & Blomberg, S. P. 2000. Population dynamics and survival of an endangered wallaby: A comparison of four methods. *Ecological Applications*, 10, 901–910.

Fleming, I. A. & Gross, M. R. 1994. Breeding competition in a Pacific salmon (Coho, *Oncorhynchus kisutch*): Measures of natural and sexual selection. *Evolution*, 48, 637–657.

Fragaszy, D. M. 2004. *The Complete Capuchin: The Biology of the Genus Cebus*. Cambridge: Cambridge University Press.

Gaillard, J. M., Festa-Bianchet, M., Yoccoz, N. G., Loison, A. & Toïgo, C. 2000. Temporal variation in fitness components and population dynamics of large herbivores. *Annual Review of Ecology and Systematics*, 31, 367–393.

Garber, P. A. 1997. One for all and breeding for one: Cooperation and competition as a tamarin reproductive strategy. *Evolutionary Anthropology*, 5, 187–199.

Garber, P. A. & Leigh, S. R. 1997. Ontogenetic variation in small-bodied New World primates: Implications for patterns of reproduction and infant care. *Folia Primatologica*, 68, 1–22.

Gesquiere, L. R., Wango, E. O., Alberts, S. C. & Altmann, J. 2007. Mechanisms of sexual selection: Sexual swellings and estrogen concentrations as fertility indicators and cues for male consort decisions in wild baboons. *Hormones and Behavior*, 51, 114–125.

Goldizen, A. W. 1987. Facultative polyandry and the role of infant carrying in wild saddle-back tamarins (*Saguinus fuscicollis*). *Behavioral Ecology and Sociobiology*, 20, 99–109.

Gowaty, P. A. 1997. Sexual dialectics, sexual selection, and variation in reproductive behavior. In *Feminism and Evolutionary Biology: Boundaries, Intersections, and Frontiers* (ed. by P. A. Gowaty), 351–384. New York: Chapman & Hall.

Grafen, A. 1988. On the uses of data on lifetime reproductive success. In *Reproductive Success* (ed. by T. H. Clutton-Brock), 454–471. Chicago: University of Chicago Press.

Henzi, P. & Barrett, L. 2003. Evolutionary ecology, sexual conflict, and behavioral differentiation among baboon populations. *Evolutionary Anthropology*, 12, 217–230.

Henzi, S. P., Weingrill, T. & Barrett, L. 1999. Male behaviour and the evolutionary ecology of chacma baboons. *South African Journal of Science*, 95, 240–242.

Heppell, S. S., Caswell, H. & Crowder, L. B. 2000. Life histories and elasticity patterns: Perturbation analysis for species with minimal demographic data. *Ecology*, 81, 654–665.

Hollister-Smith, J. A., Poole, J. H., Archie, E. A., Vance, E. A., Georgiadis, N. J., Moss, C. J. & Alberts, S. C. 2007. Age, musth and paternity success in wild male African elephants, *Loxodonta africana*. *Animal Behaviour*, 74, 287–296.

Huchard, E., Courtiol, A., Benavides, J. A., Knapp, L. A., Raymond, M. & Cowlishaw, G. 2009. Can fertility signals lead to quality signals? Insights from the evolution of primate sexual swellings. *Proceedings of the Royal Society of London Series B-Biological Sciences*, 276, 1889–1897.

Inoue, E., Inoue-Murayama, M., Vigilant, L., Takenaka, O. & Nishida, T. 2008. Relatedness in wild chimpanzees: Influence of paternity, male philopatry, and demographic factors. *American Journal of Physical Anthropology*, 137, 256–262.

Inoue, E. & Takenaka, O. 2008. The effect of male tenure and female mate choice on paternity in free-ranging Japanese macaques. *American Journal of Primatology*, 70, 62–68.

Jack, K. M. & Fedigan, L. 2004. Male dispersal patterns in white-faced capuchins, *Cebus capucinus*. Part 2: Patterns and causes of secondary dispersal. *Animal Behaviour*, 67, 771–782.

Janson, C. H. & van Schaik, C. P. 2000. The behavioral ecology or infanticide by males. In *Infanticide by Males and Its Implications* (ed. by C. P. van Schaik & C. P. Janson), 469–494. Cambridge: Cambridge University Press.

Jolly, C. J. & Phillips-Conroy, J. E. 2003. Testicular size, mating system, and maturation schedules in wild anubis and hamadryas baboons. *International Journal of Primatology*, 24, 125–142.

———. 2006. Testicular size, developmental trajectories, and

male life history strategies in four baboon taxa. In *Reproduction and Fitness in Baboons: Behavioral, Ecological, and Life History Perspectives* (ed. by L. Swedell & S. R. Leigh). New York: Springer.

Kappeler, P. 2000. Causes and consequences of unusual sex ratios among lemurs. In *Causes and Consequences of Variation in Group Composition* (ed. by P. Kappeler), 55–63. Cambridge: Cambridge University Press.

Kappeler, P. M. & Port, M. 2008. Mutual tolerance or reproductive competition? Patterns of reproductive skew among male redfronted lemurs (*Eulemur fulvus rufus*). *Behavioral Ecology and Sociobiology*, 62, 1477–1488.

Kappeler, P. M. & Schaffler, L. 2008. The lemur syndrome unresolved: Extreme male reproductive skew in sifakas (*Propithecus verreauxi*), a sexually monomorphic primate with female dominance. *Behavioral Ecology and Sociobiology*, 62, 1007–1015.

Kazakov, R. V. 1981. Peculiarities of sperm production by anadromous and parr Atlantic salmon (*Salmo salar* L.) and fish cultural characteristics of such sperm. *Journal of Fish Biology*, 18, 1–7.

Keddy-Hector, A. C. 1992. Mate choice in non-human primates. *American Zoologist*, 32, 62–70.

Kuester, J. & Paul, A. 1992. Influence of male competition and female mate choice on male mating success in Barbary macaques (*Macaca sylvanus*). *Behaviour*, 120, 192–217.

———. 1996. Female-female competition and male mate choice in Barbary macaques (*Macaca sylvanus*). *Behaviour*, 133, 763–790.

Kuester, J., Paul, A. & Arnemann, J. 1995. Age related and individual differences of reproductive success in male and female Barbary macaques (*Macaca sylvanus*). *Primates*, 36, 461–476.

Lappan, S. 2008. Male care of infants in a siamang (*Symphalangus syndactylus*) population including socially monogamous and polyandrous groups. *Behavioral Ecology and Sociobiology*, 62, 1307–1317.

———. 2009. The effects of lactation and infant care on adult energy budgets in wild siamangs (*Symphalangus syndactylus*). *American Journal of Physical Anthropology*, 140, 290–301.

Lawler, R. R. 2007. Fitness and extra-group reproduction in male Verreaux's sifaka: An analysis of reproductive success from 1989–1999. *American Journal of Physical Anthropology*, 132, 267–277.

Lehtonen, T. K. & Lindstrom, K. 2008. Density-dependent sexual selection in the monogamous fish *Archocentrus nigrofasciatus*. *Oikos*, 117, 867–874.

Levitan, D. R. 2004. Density-dependent sexual selection in external fertilizers: Variances in male and female fertilization success along the continuum from sperm limitation to sexual conflict in the sea urchin *Strongylocentrotus franciscanus*. *American Naturalist*, 164, 298–309.

Magrath, M. J. L. & Komdeur, J. 2003. Is male care compromised by additional mating opportunity? *Trends in Ecology & Evolution*, 18, 424–430.

Manson, J. 1998. Evolved psychology in a novel environment: Male macaques and the "seniority rule." *Human Nature: An Interdisciplinary Biosocial Perspective*, 9, 97–117.

McDonald, D. B. 1993. Demographic consequences of sexual selection in the long-tailed manakin. *Behavioral Ecology*, 4, 297–309.

McLain, D. K. 1982. Density dependent sexual selection and positive phenotypic assortative mating in natural-populations of the soldier beetle, *Chauliognathus pennsylvanicus*. *Evolution*, 36, 1227–1235.

Meikle, D. B., Tilford, B. L. & Vessey, S. H. 1984. Dominance rank, secondary sex ratio, and reproduction of offspring in polygynous primates. *American Naturalist*, 124, 173–188.

Meikle, D. B. & Vessey, S. H. 1988. Maternal dominance rank and lifetime survivorship of male and female rhesus monkeys. *Behavioral Ecology and Sociobiology*, 22, 379–383.

Mitani, J. C. & Watts, D. 1997. The evolution of non-maternal caretaking among anthropoid primates: Do helpers help? *Behavioral Ecology and Sociobiology*, 40, 213–220.

Modolo, L. & Martin, R. D. 2008. Reproductive success in relation to dominance rank in the absence of prime-age males in Barbary macaques. *American Journal of Primatology*, 70, 26–34.

Moscovice, L. R., Heesen, M., Di Fiore, A., Seyfarth, R. M. & Cheney, D. L. 2009. Paternity alone does not predict long-term investment in juveniles by male baboons. *Behavioral Ecology and Sociobiology*, 63, 1471–1482.

Muller, M. N. & Mitani, J. C. 2005. Conflict and cooperation in wild chimpanzees. *Advances in the Study of Behavior*, 35, 275–331.

Muller, M. N., Thompson, M. E. & Wrangham, R. W. 2006. Male chimpanzees prefer mating with old females. *Current Biology*, 16, 2234–2238.

Muniz, L., Perry, S., Manson, J. H., Gilkenson, H., Gros-Louis, J. & Vigilant, L. 2010. Male dominance and reproductive success in wild white-faced capuchins (*Cebus capucinus*) at Lomas Barbudal, Costa Rica. *American Journal of Primatology*, 72, 1118–1130.

Nguyen, N., Gesquiere, L. R., Wango, E. O., Alberts, S. C. & Altmann, J. 2008. Late pregnancy glucocorticoid levels predict responsiveness in wild baboon mothers (*Papio cynocephalus*). *Animal Behaviour*, 75, 1747–1756.

Nguyen, N., Van Horn, R. C., Alberts, S. C. & Altmann, J. 2009. "Friendships" between new mothers and adult males: Adaptive benefits and determinants in wild baboons (*Papio cynocephalus*). *Behavioral Ecology and Sociobiology*, 63, 1331–1344.

Nielsen, M. G. & Watt, W. B. 2000. Interference competition and sexual selection promote polymorphism in *Colias* (Lepidoptera, Pieridae). *Functional Ecology*, 14, 718–730.

Nievergelt, C. M., Mutschler, T., Feistner, A. T. C. & Woodruff, D. S. 2002. Social system of the alaotran gentle lemur (*Hapalemur griseus alaotrensis*): Genetic characterization of group composition and mating system. *American Journal of Primatology*, 57, 157–176.

Noë, R. & Sluijter, A. A. 1990. Reproductive tactics of male savanna baboons. *Behaviour*, 113, 117–170.

Nunn, C. L. 1999. The evolution of exaggerated sexual swellings in primates and the graded-signal hypothesis. *Animal Behaviour*, 58, 229–246.

Ohsawa, H., Inoue, M. & Takenaka, O. 1993. Mating strategy and reproductive success of male Patas monkeys (*Erythrocebus patas*). *Primates*, 34, 533–544.

Onyango, P. O., Gesquiere, L. R., Wango, E. O., Alberts, S. C. & Altmann, J. 2008. Persistence of maternal effects in baboons: Mother's dominance rank at son's conception predicts stress hormone levels in subadult males. *Hormones and Behavior*, 54, 319–324.

Packer, C., Collins, D. A. & Eberly, L. E. 2000. Problems with primate sex ratios. *Philosophical Transactions of the Royal Society*, 355, 1627–1635.

Pagel, M. 1994. Evolution of conspicuous estrous advertisement in old world monkeys. *Animal Behaviour*, 47, 1333–1341.

Palombit, R. A. 2003. Male infanticide in wild savanna baboons: Adaptive significance and intraspecific variation. In *Sexual Selection and Reproductive Competition in Primates: New Perspectives and Directions* (ed. by C. B. Jones), 364–411. Norman, OK: American Society of Primatologists.

Palombit, R. A., Cheney, D. L. & Seyfarth, R. M. 2001. Female-female competition for male "friends" in wild chacma baboons, *Papio cynocephalus ursinus*. *Animal Behaviour*, 61, 1159–1171.

Palombit, R. A., Seyfarth, R. M. & Cheney, D. L. 1997. The adaptive value of friendships to female baboons: Experimental and observational evidence. *Animal Behaviour*, 54, 599–614.

Palombit, R. A., van Schaik, C. P. & Janson, C. H. 2000. Infanticide and the evolution of male-female bonds in animals. In *Infanticide by Males and Its Implications* (ed. by C. P. van Schaik & C. H. Janson), 239–268. Cambridge: Cambridge University Press.

———. 2003. A model for leveling coalitions among primate males: Toward a theory of egalitarianism. *Behavioral Ecology and Sociobiology*, 55, 161–168.

Parga, J. A. 2006. Male mate choice in *Lemur catta*. *International Journal of Primatology*, 27, 107–131.

———. 2009. Dominance rank reversals and rank instability among male *Lemur catta*: The effects of female behavior and ejaculation. *American Journal of Physical Anthropology*, 138, 293–305.

Paul, A., Kuester, J. & Arnemann, J. 1992. Maternal rank affects reproductive success of male Barbary macaques (*Macaca sylvanus*): Evidence from DNA fingerprinting. *Behavioral Ecology and Sociobiology*, 30, 337–341.

Pereira, M. E. & Weiss, M. L. 1991. Female mate choice, male migration and the threat of infanticide in ringtailed lemurs. *Behavioral Ecology and Sociobiology*, 28, 141–152.

Pilastro, A., Giacomello, E. & Bisazza, A. 1997. Sexual selection for small size in male mosquitofish (*Gambusia holbrooki*). *Proceedings of the Royal Society of London Series B-Biological Sciences*, 264, 1125–1129.

Pochron, S. T. & Wright, P. C. 2003. Variability in adult group compositions of a prosimian primate. *Behavioral Ecology and Sociobiology*, 54, 285–293.

Poole, J. H., Lee, P. C. & Moss, C. J. 2011. Longevity, competition and musth: A long-term perspective on male reproductive strategies. In *The Amboseli Elephants: A Long-Term Perspective on a Long-Lived Mammal* (ed. by C. J. Moss, H. Croze & P. C. Lee), 272. Chicago: University of Chicago Press.

Possamai, C. B., Young, R. J., de Oliveira, R. C. R., Mendes, S. L. & Strier, K. B. 2005. Age-related variation in copulations of male northern muriquis (*Brachyteles hypoxanthus*). *Folia Primatologica*, 76, 33–36.

Robbins, M. 1995. A demographic analysis of male life history and social structure of mountain gorillas. *Behaviour*, 132, 21–48.

Rodriguez-Llanes, J. M., Verbeke, G. & Finlayson, C. 2009. Reproductive benefits of high social status in male macaques (*Macaca*). *Animal Behaviour*, 78, 643–649.

Ross, C. & MacLarnon, A. 2000. The evolution of non-maternal care in anthropoid primates: A test of the hypotheses. *Folia Primatologica*, 71, 93–113.

Setchell, J. M., Charpentier, M. & Wickings, E. J. 2005. Sexual selection and reproductive careers in mandrills (*Mandrillus sphinx*). *Behavioral Ecology and Sociobiology*, 58, 474–485.

Setchell, J. M., Charpentier, M. J. E., Bedjabaga, I. B., Reed, P., Wickings, E. J. & Knapp, L. A. 2006. Secondary sexual characters and female quality in primates. *Behavioral Ecology and Sociobiology*, 61, 305–315.

Setchell, J. M. & Kappeler, P. M. 2003. Selection in relation to sex in primates. In *Advances in the Study of Behavior, Vol. 33* (ed. by Slater, P., Rosenblatt, J., Snowdon, C., Roper, T. & Naguib, M.), 87–173. San Diego: Academic Press.

Setchell, J. M. & Wickings, E. J. 2006. Mate choice in male mandrills (*Mandrillus sphinx*). *Ethology*, 112, 91–99.

Simpson, A. J. & Barton, R. A. 1992. Dominance and mating success: Avoiding incest and artifact. *Animal Behaviour*, 44, 1164–1165.

Smuts, B. B. 1985. *Sex and Friendship in Baboons*. Hawthorn, NY: Aldine.

Snowdon, C. T. 1996. Infant care in cooperatively breeding species. In *Parental Care: Evolution, Mechanisms, and Adaptive Significance* (ed. by Rosenblatt, J. S.), 643–689. San Diego: Academic Press.

Soltis, J., Thomsen, R. & Takenaka, O. 2001. The interaction of male and female reproductive strategies and paternity in wild Japanese macaques, *Macaca fuscata*. *Animal Behaviour*, 62, 485–494.

Strier, K. B. 1990. New World primates, new frontiers: Insights from the woolly spider monkey, or muriqui (*Brachyteles arachnoides*). *International Journal of Primatology*, 11, 7–19.

———. 2000. From binding brotherhoods to short-term sovereignty: The dilemma of male Cebidae. In *Primate Males* (ed. by P. Kappeler), 72–83. Cambridge: Cambridge University Press.

Strier, K. B., Dib, L. T. & Figueira, J. E. C. 2002. Social dynamics of male muriquis (*Brachyteles arachnoides hypoxanthus*). *Behaviour*, 139, 315–342.

Struhsaker, T. T. & Pope, T. R. 1991. Mating system and reproductive success: A comparison of two African forest monkeys (*Colobus badius* and *Cercopithecus ascanius*). *Behaviour*, 117, 182–205.

Strum, S. C. 1982. Agonistic dominance in male baboons: An alternative view. *International Journal of Primatology*, 3, 175–202.

Talebi, M. G., Beltrao-Mendes, R. & Lee, P. C. 2009. Intra-community coalitionary lethal attack of an adult male southern muriqui (*Brachyteles arachnoides*). *American Journal of Primatology*, 71, 860–867.

Tardif, S. D. 1994. Relative energetic cost of infant care in small-bodied neotropical primates and its relation to infant care patterns. *American Journal of Primatology*, 34, 133–143.

Trivers, R. L. 1972. Parental investment and sexual selection. In *Sexual Selection and the Descent of Man* (ed. by B. Campbell), 136–179. Chicago: Aldine.

Urbach, D. & Cotton, S. 2008. Comment: On the consequences of sexual selection for fisheries-induced evolution. *Evolutionary Applications*, 1, 645–649.

Van Noordwijk, M. A. & van Schaik, C. P. 2001. Career moves: Transfer and rank challenge decisions by male long-tailed macaques. *Behaviour*, 138, 359–395.

Van Schaik, C. P. & Kappeler, P. M. 1997. Infanticide risk and the evolution of male-female association in primates. *Proceedings of the Royal Society of London Series B-Biological Sciences*, 264, 1687–1694.

Watts, D. P. 1998. Coalitionary mate guarding by male chimpanzees at Ngogo, Kibale National Park, Uganda. *Behavioral Ecology and Sociobiology*, 44, 43–55.

Watts, D. P. & Pusey, A. E. 1993. Behavior of juvenile and adolescent great apes. In *Juvenile Primates: Life History, Development, and Behavior* (ed. by M. E. Pereira & L. A. Fairbanks), 148–167. Oxford: Oxford University Press.

Weingrill, T., Lycett, J. E., Barrett, L., Hill, R. A. & Henzi, S. P. 2003. Male consortship behaviour in chacma baboons: The role of demographic factors and female conceptive probabilities. *Behaviour*, 140, 405–427.

Weingrill, T., Lycett, J. E. & Henzi, S. P. 2000. Consortship and mating success in Chacma baboons (*Papio cynocephalus ursinus*). *Ethology*, 106, 1033–1044.

Whitten, P. L. & Turner, T. R. 2004. Male residence and the patterning of serum testosterone in vervet monkeys (*Cercopithecus aethiops*). *Behavioral Ecology and Sociobiology*, 56, 565–578.

Wickings, E. J. & Dixson, A. F. 1992. Testicular function, secondary sexual development, and social status in male mandrills (*Mandrillus sphinx*). *Physiology & Behavior*, 52, 909–916.

Widdig, A., Bercovitch, F. B., Streich, W. J., Sauermann, U., Nurnberg, P. & Krawczak, M. 2004. A longitudinal analysis of reproductive skew in male rhesus macaques. *Proceedings of the Royal Society of London Series B-Biological Sciences*, 271, 819–826.

Wimmer, B. & Kappeler, P. M. 2002. The effects of sexual selection and life history on the genetic structure of redfronted lemur, *Eulemur fulvus rufus*, groups. *Animal Behaviour*, 64, 557–568.

Wirz, A. & Riviello, M. C. 2008. Reproductive parameters of a captive colony of capuchin monkeys (*Cebus apella*) from 1984 to 2006. *Primates*, 49, 265–270.

Wright, P. C. 1990. Patterns of paternal care in primates. *International Journal of Primatology*, 11, 89–102.

Wroblewski, E. E., Murray, C. M., Keele, B. F., Schumacher-Stankey, J. C., Hahn, B. H. & Pusey, A. E. 2009. Male dominance rank and reproductive success in chimpanzees, *Pan troglodytes schweinfurthii*. *Animal Behaviour*, 77, 873–885.

Yoccoz, N. G., Mysterud, A., Langvatn, R. & Stenseth, N. C. 2002. Age- and density-dependent reproductive effort in male red deer. *Proceedings of the Royal Society of London Series B: Biological Sciences*, 269, 1523–1528.

Zinner, D., Alberts, S. C., Nunn, C. L. & Altmann, J. 2002. Evolutionary biology: Significance of primate sexual swellings. *Nature*, 420, 142–143.

Zinner, D., Krebs, E., Schrod, A. & Kaumanns, W. 2006. Early sexual maturity in male Hamadryas baboons (*Papio hamadryas hamadryas*). *American Journal of Physical Anthropology*, 129, 584–590.

Chapter 19 Infanticide: Male Strategies and Female Counterstrategies

Ryne A. Palombit

For the reproduction of the race, there are two instincts needed, the sexual and the parental, and the way these two are organized is, to say the least, curious.
—Charles Galton Darwin

SEXUAL CONFLICT arises when the reproductive strategies of one sex impose fitness costs on the other (Trivers 1972; Arnqvist & Rowe 2005). Sexual conflict assumes many forms in primates (Palombit 2010), but a clear example is the killing of dependent infants by conspecific males. Male infanticide is of special importance in understanding primate biology because it is arguably the most fundamental constraint on the reproduction of females (chapter 16, this volume). As noted previously (chapters 14, 17, 18, this volume), the physiology of reproduction in primates means that females are occupied with gestation and lactation for a significant portion of their lives, during which time they are unavailable for fertilization by males, who are themselves free of such reproductive constraints and potentially capable of producing large numbers of offspring. In such circumstances, a mating strategy that prematurely returned anestrous females to a fertilizable condition would be selected for in males, depending upon the intensity of intrasexual competition. Thus, male infanticide is more common in mammals characterized by prolonged infant dependency (fig. 19.1) and altricial young (van Schaik 2000a; Wolff & MacDonald 2004).

The coevolutionary dynamic of sexual conflict tells us that once such a strategy is established in one sex, members of the other sex will experience selection for strategies that counter this original strategy, depending upon the magnitude of the costs they suffer. There can be little doubt that loss of an infant constitutes a serious fitness cost for females that reproduce at slow rates. Thus, it is likely that some, perhaps many, aspects of female primate biology have evolved as anti-infanticide counterstrategies. In this sense, male infanticide merits analytical scrutiny in the primates. In this chapter I examine evidence that infanticide is a reproductive strategy of males and the selective force behind myriad counterstrategies in females.

Approximately 20 years ago, infanticide had been observed in 10 species, and suspected in three more (Hiraiwa-Hasegawa 1988). Currently, field studies of 56 populations from 35 species provide direct observations of male infanticide (table 19.1). If infant deaths or disappearances accompanied by indirect evidence of male killing are included, infanticide has been observed or inferred in the wild in 80 populations from 51 species, with the following taxonomic distribution: 5 lemur species, 11 platyrrhine species, 11 colobines, 20 cercopithecines, and 4 hominoids. Studies in captivity provide further evidence of infanticide for 10 of these species, as well as for eleven additional species. It has been observed in unimale and multimale groups, fission-fusion systems, and multilevel societies. Being widespread does not make infanticide a unitary phenomenon, however. Hrdy (1979) originally evaluated five proposed causes of infanticide by males, but now almost a dozen adaptive hypotheses, as well as two nonadaptive hypotheses, provide potential explanations (table 19.2).

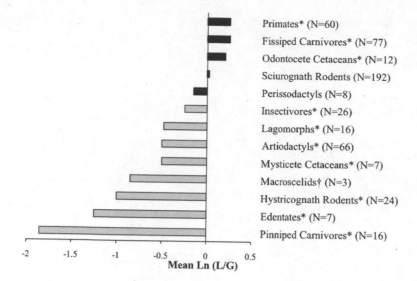

Fig. 19.1. Mean lactation/gestation (L/G) ratios for non-volant mammals (modified from van Schaik 2000a). The log-transformed L/G ratio is different from 1 when indicated by * ($p < 0.05$) or by † ($p < 0.10$). Sample sizes (number of species) are indicated in parentheses. Taxa in which deliberate sexually selected infanticide by males has been recorded (black bars) generally have L/G ratios greater than 1, whereas taxa lacking such observations (grey bars) typically have ratios significantly smaller than 1. Redrawn from van Schaik 2000a.

Infanticide as a Male Strategy

Sexually Selected Male Infanticide

Given the significance of intrasexual mating competition and polygyny in primates (chapter 17, this volume), it is unsurprising that the sexual selection hypothesis (Hrdy 1974) has received the most attention and empirical support. As predicted by this hypothesis, a frequent context of infanticidal behavior is turnover or replacement of the male occupying a position as sole or dominant breeder in the group (table 19.2). Van Schaik's (2000b) conservative analysis of the individual records provides further support by showing that generally, (1) males have a low probability of being related to the infants they attack, often due to their absence from the social unit at the time infants were conceived; (2) loss of a dependent infant significantly accelerates the resumption of ovarian cycling; and (3) on average, infanticide increases the probability of subsequent sexual access to mothers relative to the alternative strategy of not attempting infanticide. Alternative hypotheses that infanticide is a consequence of abnormal social pathology (Curtin & Dolhinow 1978; Boggess 1979) or generalized male aggression (Bartlett et al. 1993) are thus not well supported (van Schaik & Janson 2000).

Although much (but not all) of the controversy originally surrounding the sexual selection hypothesis has abated (Rees 2009), tests of the hypothesis continue, with at least three outcomes of note. First, in the last decade or so, inves-

tigations have exploited new sources of data besides behavioral observations, notably DNA and hormones. Genetic paternity tests verify that males are unrelated to the infants they attack or are the fathers of the subsequent offspring of their victims' mothers (Pereira & Weiss 1991; Borries et al. 1999a; Soltis et al. 2000; Morelli et al. 2009). Field hormonal data clarify infanticide in a population of chacma baboons (*Papio ursinus*) in which recently immigrated males that rise to the top of the dominance hierarchy may attack infants. Following this alpha male turnover, stress hormone levels rise significantly among the group's anestrous females, but not in estrous females, even though the latter are frequent targets of aggressive displays by new immigrant males (see below). Should an infant actually be killed by the new alpha male, stress hormones rise even higher, but again only in anestrous females. Risk of infanticide thus appears to underlie the female stress response (Beehner et al. 2005; Engh et al. 2006; Kitchen et al. 2006). Field hormonal data have also clarified the physiological mechanism underlying resumption of ovarian cycles following the killing of infants (Harris & Montfort 2003), as well as male and female counterstrategies (see below).

A second outcome of recent research is greater appreciation that male infanticide operates in a greater variety of conditions than is implied by the classic Hanuman langur (*Semnopithecus entellus*) model of aseasonally breeding primates living in unimale groups. In particular, it occurs predictably in primates that reproduce seasonally or live in multimale groups.

Table 19.1. Reports of male infanticide in nonhuman primates. Taxa are included if original authors observed or inferred infanticide by males from indirect evidence. Multiple report entries on separate lines for a particular taxon indicate studies of different populations.

Taxon	Reference	Setting[1]	Context[2]	Infanticide observed?[3]
LEMURIFORMES				
Eulemur fulvus	Pereira & Weiss 1991	C	1, 2	Attacks
	Jolly et al. 2000	W	4	Yes
Eulemur macaco	Pereira & Weiss 1991	C	1 or 2	No
Lemur catta	Pereira & Weiss 1991	C	1, 2	Attacks
	Jolly et al. 2000; Ichino 2005	W	1, 2, 4	Yes
Lepilemur edwardsi	Rasoloharijaona et al. 2000	W	1	Yes
Mirza zaza	Stanger et al. 1995; Kappeler, pers. comm.	C	7	No
Propithecus edwardsi	Wright 1995; Erhart & Overdorff 1998; Morelli et al. 2009	W	1, 2	Yes
Propithecus verreauxi	Lewis et al. 2003	W	2	Yes
	Kubzdela pers. comm. to Brockman & Whitten 1996	W	?	?
	Richard et al. (unpublished data, cited by Brockman et al. 2001); Littlefield 2010	W	2	Yes
Varecia variegata	Pereira & Weiss 1991; White 1993	C	1, 2	Attacks
LORISIFORMES				
Otolemur crassicaudatus	Buettner-Janusch 1964	C	7	Yes
TARSIIFORMES				
Cephalopachus bancanus	Roberts 1994	C	7	Attacks
PLATYRRHINI				
Alouatta arctoidea	Agoramoorthy & Rudran 1995; Crockett & Janson 2000	W	2, 3	Yes
Alouatta caraya	Zunino et al. 1986; Peker et al. 2006	W	1	Yes
	Calegaro-Marques & Bicca-Marques 1996	W	1	No
Alouatta guariba	Galetti et al. 1994	W	1	No
Alouatta palliata	Clarke 1983; Clarke & Glander 1984; Clarke et al. 1994	W	1, 3	No
Alouatta pigra	Brockett et al. 1999	W	1	No
	Knopff et al. 2004	W	1?	Yes
	Van Belle et al. 2010	W	1, 2	Yes
Alouatta seniculus	Izawa 1997; Izawa & Lozano 1994; Crockett 2003	W	2	Yes
Ateles belzebuth	Gibson et al. 2008	W	8	Yes
Ateles geoffroyi	Gibson et al. 2008	W	6	Yes
Cebus apella	Ramírez-Llorens et al. 2008	W	2	Yes
Cebus capucinus	Fedigan 2003; Fedigan & Jack 2004; Jack & Fedigan 2009	W	1, 4?	Yes
	Manson et al. 2004; Vogel & Fuentes-Jiménez 2006	W	1, 4?, 6	Yes
Cebus olivaceus	Valderrama et al. 1990	W	3, 7	Yes
COLOBINAE				
Colobus guereza	Oates 1977	W	2	No
	Onderdonk 2000; Harris & Monfort 2003; Chapman & Pavelka 2005	W	2, 4	Yes
Colobus vellerosus	Sicotte et al. 2007; Teichroeb & Sicotte 2008a	W	1, 2	Yes
Nasalis larvatus	Agoramoorthy & Hsu 2005	W	1	Yes
Presbytis thomasi	Steenbeek 2000	W	1, 4	Yes
Procolobus badius	Starin 1994	W	?	No
Procolobus rufomitratus	Marsh 1979; Mowry 1995	W	1, 4	No
	Struhsaker & Leland 1985	W	3	Yes
Pygathrix nemaeus	Angst & Thommen 1977	C	7	Yes
Rhinopithecus bieti	Xiang & Grueter 2007	W	1?	Yes
Rhinopithecus roxellana	Zhang et al. 1999	C	1	Attacks
	Wang et al. 2004	W	1	No
Semnopithecus entellus	Sugiyama 1965; Sugiyama 1966	W	1	Yes
	Hrdy 1974	W	1	Yes
	Newton 1986	W	1	Yes
	Ross 1993	W	1	Yes
	Sommer 1994	W	1	Yes
	Borries 1997	W	2	Yes
	Rajpurohit et al. 2008; Mohnot et al. 2008	W	1	Yes
Semnopithecus vetulus	Rudran 1973a	W	1	Attacks
Trachypithecus villosus	Wolf & Fleagle 1977	W	1	No
CERCOPITHECINAE				
Cercocebus atys	Fruteau et al. 2010	W	2	Attacks
Cercocebus galeritus	Kinnaird 1990	W	1 or 2	Attacks
Cercocebus torquatus	Busse & Gordon 1983; Gust 1994	C	3	Yes
	Böer & Sommer 1992	C	1	Yes

Table 19.1. (continued)

Taxon	Reference	Setting[1]	Context[2]	Infanticide observed?[3]
Cercopithecus ascanius	Struhsaker 1977	W	1	Yes
Cercopithecus campbelli	Galat-Luong & Galat 1979	W	1	No
Cercopithecus mitis	Butynski 1990	W	1	Yes
	Böer & Sommer 1992	C	1	Yes
	Fairgrieve 1995	W	1	Yes
	Cords & Fuller 2010	W	1, 4	Yes
Chlorocebus pygerythrus	Isbell et al. 2002	W	2	No
	Fairbanks & McGuire 1987	C	1	Attacks
Chlorocebus sabaeus	Horrocks 1986; Horrocks & Baulu 1988	W	1, 2	No
Erythrocebus patas	Enstam et al. 2002	W	1	Attacks
Macaca cyclopis	Hsu et al. 2006	W	1 (2?)	?
Macaca fascicularis	Angst & Thommen 1977	C	1, 3	Yes
	Pallaud 1984	C	6	No
	De Ruiter et al. 1994	W	3	Attacks
Macaca fuscata	Soltis et al. 2000	W	3	Yes
	Yamada & Nakamichi 2006	W	4	Yes
Macaca mulatta	Angst & Thommen 1977	C	1	?
	Camperio-Ciani 1984	W	3	Yes
Macaca nemestrina	Kyes et al. 1995	C	2	Yes
	Clarke et al. 2001	C	1	Yes
Macaca radiata	Singh et al. 2006	W	1	Yes
Macaca silenus	Lindburg et al. 1989	C	1 or 2	?
Macaca sinica	Dittus 1988	W	?	?
Macaca sylvanus	Angst & Thommen 1977	C	1	?
Macaca thibetana	Berman et al. 2007	W	3, 6	Yes
Mandrillus leucophaeus	Böer & Sommer 1992	C	1	Attacks
Papio anubis	Collins et al. 1984	W	2, 3, 4	Yes
	Smuts 1985	W	?	Heard
Papio cynocephalus	Shopland 1982; Altmann et al. 2006	W	4	Yes
Papio hamadryas	Zuckerman 1932	C	7	Attacks
	Rijksen 1981	C	7	Yes
	Angst & Thommen 1977	C	1, 7	Yes
	Gomendio & Colmenares 1989	C	1	Yes
	Kaumanns et al. 1989; Zinner et al. 1993	C	1, 7	Yes
	Swedell 2000; Swedell & Tesfaye 2003	W	1	Yes
Papio ursinus	Palombit et al. 2000	W	2, 3	Yes
	Weingrill 2000	W	2	Yes
	Henzi & Barrett 2003	W	2	?
	Huchard et al. 2010	W	3	Yes
Theropithecus gelada	Angst & Thommen 1977	C	1	No
	Moos et al. 1985	C	2	Yes
	Mori et al. 1997; Mori et al. 2003	W	1, 2	Yes
	Beehner & Bergman 2008	W	1	Yes
HOMINOIDEA				
Bunopithecus hoolock	Alfred & Sati 1991	W	7	Yes
Hylobates lar	Borries et al., in press	W	1, 2	No
Gorilla gorilla beringei	Watts 1989; Robbins 1995	W	1, 4, 5	Yes
Gorilla gorilla graueri	Yamagiwa & Kahekwa 2004; Yamagiwa et al. 2009	W	1, 4, 5	Yes
Gorilla gorilla gorilla	Stokes et al. 2003	W	1, 5	No
Pan troglodytes	Goodall 1986; Wilson et al. 2004; Williams et al. 2008; Murray et al. 2007	W	4, 6, 8	Yes
	Watts & Mitani 2000; Sherrow & Amsler 2007	W	4	Yes
	Nishida et al. 2003	W	6	Yes
	Suzuki 1971; Bakuneeta et al. 1993; Newton-Fisher 1999	W	4	Yes
	Arcadi & Wrangham 1999	W	3, 4, 7	Yes
	Boesch et al. 2008	W	4	Yes
	Spijkerman et al. 1990	C	3?, 8	Yes

[1] "W"= wild, free-ranging; "C" = captivity

[2] Contexts: 1. male immigration with replacement or after male disappearance (or, in captivity, after male removal); 2. male immigration without male replacement; 3. rise in rank of resident male; 4. attack by extragroup male(s); 5. following female immigration; 6. attack by high-ranking male(s); 7. attack by likely father; 8. other.

[3] For entries with multiple references, "yes" refers to at least one of the citations in which at least one infanticide was directly observed; "no" refers to all of the citations.

Table 19.2. Hypotheses for male infanticide in primates

Hypothesis	Rationale	Proposed example taxa	Reference
Adaptive hypotheses:			
Sexual selection	When mating opportunities are limited by intrasexual competition, infanticide prematurely ends lactational amenorrhea and returns mothers to ovulatory states sooner for subsequent fertilization by the infanticidal male	*Semnopithecus entellus* and numerous other species (see table 19.3 taxa with context "1" or "2" or "3")	1
Nutritional exploitation	Infants are cannibalized for nutritional benefits (e.g., protein intake)	*Papio cynocephalus, Papio hamadryas*	2, 3
Breeding status acquisition	Infanticide accelerates male acquisition of high rank or breeding status/ immigration	Cercopithecinae, *Alouatta, Semnopithecus entellus*	4, 5, 6, 7
Mate acquisition I & II	In species with female transfer or dispersal, infanticide by an extragroup male promotes female immigration to the male's unit by (I) advertising superior male ability to protect against infanticide, or (II) disrupting mother-subadult daughter bond (thus facilitating daughter's dispersal)	I. *Presbytis thomasi, Colobus polykomos, Gorilla gorilla beringei, Pan troglodytes?* II. *Pan troglodytes*	I. 8, 9, 10, 11 II. 12
Control of female mating	Intragroup infanticide induces females to mate more restrictively with infanticidal male(s) than with other male(s)	*Pan troglodytes*	13
Removal of current competitors' offspring	Infanticide removes genes of rival male(s)	*Erythrocebus patas*	14
Removal of future competitors	Killing of male infants eliminates future competitors and/or undermines rival males' coalitionary power	*Alouatta palliata, Pan troglodytes, Semnopithecus entellus, Colobus vellerosus*	15, 16, 17, 18, 19
Resource competition: Eliminataion of competitors	Infanticide enhances access to limiting resources by the infanticidal male and/or his offspring	*Semnopithecus vetelus, Colobus guereza, Macaca thibetana?, Pan troglodytes*	20, 21, 22, 23
Resource competition: Range expansion	Infanticide by extragroup males induces mothers to withdraw from boundary areas, thereby facilitating expansion of the killers' range (to acquire more resources)	*Pan troglodytes*	23, 24, 25, 26
Avoidance of misdirected paternal investment	Infanticide prevents or terminates male parental investment in offspring sired by another male	*Homo sapiens*	27, 28
Nonadaptive hypotheses:			
Social pathology	Infanticide is abnormal behavior caused by severe overcrowding recently induced by human disturbance	*Semnopithecus entellus*	29, 30, 31, 32
Generalized aggression / side effect	Infanticide is (I) an accidental by-product of aggressive interactions of males acquiring breeding/dominance status, or (II) an incidental side effect of male xenophobic territoriality	I. Primates with male infanticide (e.g., *Macaca thibetana?*) II. *Pan troglodytes*	I. 33, 34, 22 II. 35

Sources: 1. Hrdy 1974. 2. Altmann et al. 2006. 3. Zinner et al. 1993. 4. Angst & Thommen 1977. 5. Crockett 2003. 6. Newton 1987. 7. Rajpurohit et al. 2003. 8. Steenbeek 2000. 9. Korstjens et al. 2005. 10. Watts 1989. 11. Wilson & Wrangham 2003. 12. Goodall 1986. 13. Hamai et al. 1992. 14. Enstam et al. 2002. 15. Clarke 1983. 16. Takahata 1985. 17. Hiraiwa-Hasegawa & Hasegawa 1994. 18. Sommer 1994. 19. Teichroeb & Sicotte 2008a. 20. Rudran 1973a. 21. Harris & Monfort 2003. 22. Berman et al. 2007. 23. Williams et al. 2002. 24. Pusey 2001. 25. Williams et al. 2004. 26. Watts et al. 2002. 27. Daly & Wilson 1984. 28. Daly & Wilson 1994. 29. Dolhinow 1977. 30. Curtin & Dolhinow 1978. 31. Boggess 1979. 32. Boggess 1984. 33. Rijksen 1981. 34. Bartlett et al. 1993. 35. Goodall 1990.

Male Infanticide in Seasonal Breeders

Sexually selected infanticide was originally considered unlikely among seasonal breeders, for good reason: if photoperiodicity triggers female breeding status, then death of an infant would not necessarily accelerate the resumption of cycling and sexual access to fecund females (Hrdy & Hausfater 1984; Sauther & Sussman 1993). Yet infanticide has been recorded in a growing number of seasonally breeding primates—for example, all of the lemurs listed in table 19.1 (except perhaps northern dwarf lemurs, *Mirza zaza*, an apparently aseasonal reproducer) as well as populations of Hanuman langurs, patas monkeys (*Erythrocebus patas*), Japanese macaques (*Macaca fuscata*), and snub-nosed monkeys (*Rhinopithecus bieti, R. roxellana*). Data from most of these taxa are still too limited for rigorous testing of the sexual selection hypothesis, but two adaptive mechanisms have been proposed. First, if the interbirth interval exceeds the period between breeding seasons, then infanticide potentially *does* accelerate female return to estrus for subsequent fertilization by the male, depending upon the probability that he will reside in the group during the next breeding season. This has been suggested for Milne-Edwards' sifaka (*Propithecus edwardsi*, Wright 1995; Morelli et al. 2009; fig. 19.2) and for Nepalese Hanuman langurs (Borries 1997), in which females can breed annually but have interbirth intervals of two years or

Fig. 19.2. Milne-Edwards sifakas (*Propithecus edwardsi*) from Ranomafana National Park, Madagascar. Photo courtesy of Patricia C. Wright.

more. A second alternative is based on the argument that females commencing reproduction early in the breeding season rear infants with greater survival chances, but the capacity to breed early depends upon the mother's condition at the time (Hrdy & Hausfater 1984). Thus, by alleviating the energy costs of lactation and infant carrying, infanticide makes females available for mating earlier in the next breeding season for the siring of more viable offspring. Jolly et al. (2000) invoke this explanation for wild ring-tailed lemurs (*Lemur catta*) in light of the apparently high costs of reproduction for females and the variable timing of births; they note, however, that this advantage of infanticide may be more likely in poor habitats or bad years (see also Pereira & Weiss 1991). Finally, although low body dimorphism in lemurs might seem to suggest low male-male competition, recent evidence that male mating and reproductive skew are pronounced in some lemurs in which infanticide has been observed establishes a rationale

for considering the sexual selection hypothesis (Wimmer & Kappeler 2002; Kappeler & Port 2008; Kappeler & Schäffler 2008).

Infanticide in Multimale Groups

Sexually selected infanticide is expected to be less likely in multimale groups than in unimale groups, since its potential benefits are lower due to multimale mating (Kutsukake & Nunn 2006) and its costs are potentially higher due to the presence of male defenders of the infant (see below). Nevertheless, infanticide is a viable reproductive strategy in multimale groups in at least two scenarios.

First, young, newly dominant males that can monopolize sexual access to receptive females potentially benefit from infanticide (Broom et al. 2004). In chacma baboons, for example, a newly immigrant male that attains alpha status has, on average, only seven months in that position, during which he may have nearly exclusive mating access to periovulatory females (Palombit 2003). Thus, infanticide potentially improves mating opportunities for these new alpha males (Palombit et al. 2000). Similar patterns of infanticide have been attributed to other multimale species (table 19.1). Generally, immigrants that do not attain high rank are less aggressive to infants and elicit less fear from lactating mothers (e.g., Palombit et al. 2000; Teichroeb & Sicotte 2008a). Infanticide apparently accounts for a nearly threefold increase in infant death rates during takeover years over its level in years without male replacement in black capuchin monkeys (*Cebus nigritis*, Ramirez-Llorens et al. 2008). The prediction that dominant males will have exclusive access to females has generally been tested indirectly with observations of mating; more valuable genetic data establish strong reproductive skew in long-tailed macaques (*Macaca fascicularis*, de Ruiter et al. 1994), Japanese macaques (Soltis et al. 2001), and Verreaux's sifaka (*Propithecus verreauxi*, Kappeler & Schäffler 2008). Occasionally, infanticide is perpetrated by longtime resident males that rise to high rank (Busse & Gordon 1983; Palombit et al. 2000), but this appears to be less common than attacks by recent immigrants that become dominant, presumably because the probability of paternity of infants in the group is lower for the latter.

A second scenario for infanticide in the multimale context involves replacement of *all* of the adult males in group by a new set of immigrants. In white-faced capuchin monkeys (*Cebus capucinus*), this turnover occurs in two forms related to parallel dispersal: either when all resident males are aggressively expelled by a coalition of incoming extragroup males, or when residents emigrate on their own and are then replaced (chapters 12 and 21, this volume, see fig. 21.3). In both situations, the new males have been observed to commit infanticide and subsequently mate with the mothers upon resumption of estrus (Fedigan & Jack 2004; Manson et al. 2004; Perry 2008; fig. 19.3). Data are insufficient to assess if the male immigrants vary individually in infanticidal behavior, as predicted, for example, on the basis of rank (Broom et al. 2004). An analysis of 18 years of demographic data from a Costa Rican population reveals that male replacement surpasses maternal rank and resource availability as the greatest predictor of infant survival in the first year of life, which is three times less likely following takeover than in non-takeover periods (Fedigan et al. 2008). Except for ostensibly rare events noted in other species (e.g., Singh et al. 2006), this multimale replacement pattern of infanticide is currently unique to white-faced capuchin monkeys among all primates, though not among mammals: male lions (*Panthera leo*) show similar coalitionary takeovers of prides followed by infanticide (Pusey & Packer 1994).

Alternative Adaptive Hypotheses for Male Infanticide

A third development in recent research is recognition that some benefits of infanticide are not wholly consistent with the sexual selection hypothesis. These advantages form the basis of alternative adaptive hypotheses (table 19.2).

For example, infanticide accounts for one-third of infant mortality in Thomas's langurs (*Presbytis thomasi*) living in unimale groups, but it is not committed by males that have recently immigrated into a group in a takeover. Rather, infanticidal attacks come from *extragroup* males that reside in neighboring bisexual groups or all-male bands, or that are solitary (Sterck 1997; Steenbeek 1999a). The consequence of an infant's death is not just termination of postpartum amenorrhea in its mother, but also her departure from her original group to join the killer male (Steenbeek 2000). Thus, infanticidal males may acquire mates by exposing a rival male's inability to protect the infants in his group and thereby prompting the transfer of the infants' mothers as well as of cycling parous females (Sterck 1997). This hypothesis is supported by evidence that females with infants are often specifically and disproportionately targeted in aggressive interactions by extragroup males (Steenbeek 1999a, 2000).

Infanticide as mate acquisition is more likely to operate in species in which the costs of female transfer are relatively low, which in turn implicates folivores with reduced between-group competition (chapter 9, this volume). Besides Thomas's leaf monkeys, the clearest evidence for this strategy comes from white-thighed colobus (*Colobus vellerosus*), in which infanticide eliminates one-third to half of the infants born in groups that can be unimale or multimale

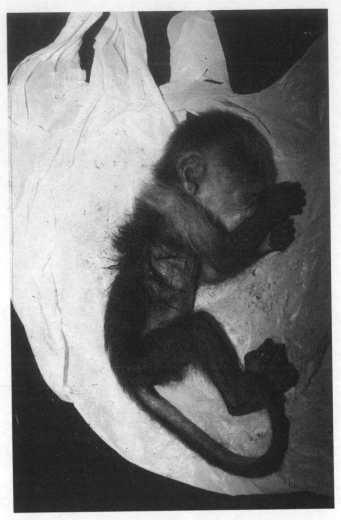

Fig. 19.3. An infant white-faced capuchin monkey (*Cebus capucinus*) killed following the expulsion of adult males from its group by a cohort of takeover males at Lomas Barbudal, Costa Rica (see chapter 8 of Perry 2008). Photo courtesy of Susan Perry.

(Teichroeb & Sicotte 2008a; fig. 19.4). In addition to instigating takeover events, multiple extragroup males also make incursions or "forays" into bisexual groups, attacking mothers with infants in particular, which Saj and Sicotte (2005) interpret as a strategy to procure females through transfer. Similar forays, hostile intergroup relations, and infanticide have been described in one population of guerezas (*Colobus guereza*, Oates 1977a; table 19.1). Fashing (2001) rejects the mate acquisition hypothesis in another west Kenyan population, however, because of the absence of directed aggression to mothers or infants during intergroup interactions. In their study of king colobus (*Colobus polykomos*), Korstjens et al. (2005) also did not observe direct attacks on mothers and infants, but they refrain from rejecting the hypothesis categorically, since male forays clearly targeted groups with in-

Fig. 19.4. A group of white-thighed, or ursine, colobus monkeys (*Colobus vellerosus*) in central Ghana. Infanticide is perpetrated both by new immigrant males, as part of a takeover strategy, and by extragroup males, possibly as a strategy of mate acquisition. Photo courtesy of Lisa MacDonald and Pascale Sicotte.

Fig. 19.5. Infanticide among Gombe chimpanzees (*Pan troglodytes*). Adult male Figan carries the body of an infant who Bygott (1972) had observed being killed and partially cannibalized by other adult males several hours earlier. Photo courtesy of J. David Bygott.

fants significantly more than groups without infants. Finally, female transfer to an extragroup male that has previously killed her infant has also been noted in folivorous mountain gorillas (*Gorilla gorilla beringei*, Watts 1989), but the more common pattern involves females' transfer after loss of the silverback (see below).

An alternative hypothesis closely related to mate acquisition is that infanticide directly helps males achieve breeding or high dominance status in a group. For example, although many of the 68 cases of infanticide in howler monkeys (*Alouatta* spp.) are consistent with the sexual selection takeover model, Crockett (2003, p. 354) notes that infanticide sometimes *precedes* a male's rise to dominance in the group, and thus appears to be part of "a larger strategy to assume the position of breeding male." This function has been suggested for some cercopithecines and colobines as well (table 19.2).

Study of the chimpanzee has generated several additional adaptive hypotheses (table 19.2). Male infanticide is widespread and "relatively common," though rates of occurrence vary greatly across populations and over time (Arcadi & Wrangham 1999, p. 337; fig. 19.5). The patterning of infanticidal behavior in chimpanzees is arguably the most variable of that in any primate, but two contexts are generally recognized: attacks by males in the adult female's own community, and attacks by alien males from a neighboring community. Some observations of intracommunity infanticide are partly consistent with the sexual selection hypothesis in that high-ranking males have killed infants assumed

to be sired by low-ranking or extragroup males, based on observations of mating or prolonged absence from the community range (Nishida & Kawanaka 1985; Takahata 1985; Hamai et al. 1992; Sakamaki et al. 2001). Independent genetic evidence suggests that fertilizations by outside males are rare but possible (Vigilant et al. 2001). Sometimes, however, an attacker is clearly at least a possible father of the victim, and the subsequent mating of the mother is either unreported or not clearly restricted to the killer male alone, thus leaving this form of infanticide difficult to explain (Arcadi & Wrangham 1999).

Intercommunity infanticide offers puzzling features too. Given the pronounced hostility that generally (though variably) characterizes intergroup relations in chimpanzees (Wilson & Wrangham 2003; chapter 6, this volume, fig. 6.8), it is not surprising that these infanticides are often noteworthy for their excessive violence (Goodall 1986). Indeed, an apparent overrepresentation of male infants among the victims of intergroup infanticide at Mahale initially suggested a strategy to eliminate future rivals and undermine the coalitionary power of neighboring communities (Kawanaka 1981; Hiraiwa-Hasegawa 1987); this male bias disappears, however, when overall sex ratios at birth are considered (Nishida et al. 2003). Although males target infants on some occasions (e.g., Newton-Fisher 1999), they more commonly direct these "savage assaults" at the mothers themselves (Goodall 1990, p. 102), in notable contrast to the more general pattern of sexually selected infanticide. Partly because some of these victims are older females, Wolfe and Schulman (1984) suggested that such attacks reflect a conditional male strategy to induce the recruitment of females of high reproductive value and reject those of low reproductive value. But this model is not well supported by evidence that males prefer older females as mates (Muller et al. 2006) and by the paucity of observations of maternal transfer to the killers' communities following infanticide (Arcadi & Wrangham 1999). Thus, Goodall (1990) speculated that intercommunity infanticide is perhaps a nonadaptive consequence of the well-known xenophobic territoriality of males.

Pusey (2001) and Williams et al. (2002, 2004) recast Goodall's idea into an adaptive version based on the resource competition hypothesis, which argues that infanticide removes competitors of the male or his offspring (table 19.2). They argue that infanticide is an integral part of the sometimes lethal territorial defense against *all* outsiders. A proposed indirect benefit of intercommunity infanticide is that the victimized mothers withdraw from dangerous border areas, thus facilitating the expansion of the killers' territory and the subsequent acquisition of more resources for the members (notably females) of their community. This range expansion hypothesis has not been widely tested, but it is supported by observations at Ngogo, Uganda, that infanticide and other intercommunity killings were followed by a 22% increase in territory size at the expense of the victims' range (Watts et al. 2002; Mitani et al. 2010) and by suggestive but somewhat more equivocal patterns at Gombe (Goodall 1986). The need for further study of both the causes and the functions of this form of infanticide is underscored by recent evidence suggesting a surprisingly successful intercommunity transfer of several females *with* their infants (Emery Thompson et al. 2006).

Occurrence of Male Infanticide

Infanticide by males may be widespread and predictable in primates, but it is relatively infrequent for several reasons.

First, its rarity is at least partly an artifact of observational difficulties. Features of methodology—such as short study duration and the inhibitory effect of the presence of human observers on the behavior of unhabituated extragroup or immigrant males—are proposed impediments to direct observation (Borries 1997; Crockett 2003). Additionally, in some species infanticidal attacks are by nature spontaneous, opportunistic, and "over in seconds" (e.g., Jolly et al. 2000, p. 37; Palombit 2003).

Second, a defining feature of infanticide is its variability and apparent conditionality. For example, infanticide is perpetrated by some, but not all, of the immigrant male chacma baboons that attain alpha status, thus prompting Palombit et al. (2000) to suggest that it is a facultative reproductive strategy whose relative benefits vary with conditions such as the relative abundance of estrous females, the ages of infants, the number of resident adult males, or attributes of the new alpha male himself (see also Yamamura et al. 1990). Accordingly, rates of infanticide fluctuated considerably over time in this population, accounting for 75% of infant mortality in one two-year period, but 0% in other periods of similar duration (Cheney et al. 2004). Similarly, in white-thighed colobus, infants were attacked in all groups that received male incursions, but not by all males that made those incursions (Teichroeb & Sicotte 2008a). Within-population variation may be influenced by social, demographic, or ecological conditions affecting the costs and benefits of infanticide. Given evidence of the genetic basis of infanticidal behavior (Perrigo et al. 1993), it may also reflect genetic polymorphism among males (e.g., Palombit et al. 2000; Cords & Fuller 2010).

The limited available data suggest similar trends across primates generally (table 19.3). In some populations, infanticide is a major source—if not *the* major cause—of infant mortality. But even in these populations, if longitudinal data are available, it is usually described as highly variable over time. In other populations of the same or closely related species, infanticide may be virtually absent.

The underlying causes of this variation remain obscure, but some tentative conclusions are possible. Several studies provide evidence that, *within populations*, the rate of breeding male turnover is positively correlated with the rate of infanticide, as predicted by the sexual selection hypothesis (Borries 1997; Sterck 1998; Crockett & Janson 2000; Steenbeek 2000; see also Packer et al. 1988). In a preliminary *interspecific* analysis that controlled for species differences in reproductive rate, Janson & van Schaik (2000) obtain the same result, but only for primates characterized by female philopatry. In species where females disperse, however, the infanticide rate appears to vary *inversely* with the replacement rate. Although the significance of this relationship

Table 19.3. Relative rates of infanticide* for selected primate species with reports of male infanticide

Taxon	Relative rate (%)	Comment	Reference
LEMURIFORMES			
Propithecus edwardsi	33	83% of immigrant males (n = 6) observed or suspected of infanticide	Wright 1995; Morelli et al. 2009
PLATYRRHINI			
Alouatta arctoidea	≥ 44	85% during some periods	Crockett 2003
Alouatta caraya	47		Zunino et al. 1986
Alouatta palliata	32		Clarke 1983; Clarke & Glander 1984
Alouatta seniculus	53		Crockett 2003
Cebus apella	20	62.5% (n = 24) of infants born during takeover years died, cf. 22.5% (n = 8) born in years without male replacements	Janson & Di Bitetti, unpublished data;† Ramírez-Llorens et al. 2008
Cebus capucinus	59	Male replacement in the year following birth is the greatest predictor of infant mortality	Fedigan & Jack 2004
COLOBINAE			
Colobus vellerosus	71		Teichroeb & Sicotte 2008a
Procolobus rufomitratus	30	Value is midrange based on two estimates†	Struhsaker & Leland 1985
Presbytis thomasi	33		Steenbeek 2000
Semnopithecus entellus			
Jodphur	41		Sommer 1994†
Kanha	50		Newton 1987
Ramnagar	33	Range: 31–63; primary cause of infant mortality	Borries & Koenig 2000
CERCOPITHECINAE			
Cercopithecus mitis			
Ngogo	≥ 47		Butynski 1990
Kanyawara	0		Butynski 1990
Kakamega	17		Cords & Fuller 2010
Chlorocebus pygerythrus			
Amboseli	—	3 infanticides from 148 live births	Isbell et al. 2002
Segera	0		Isbell et al. 2002
Macaca fascicularis	20		Van Noordwijk & van Schaik, unpublished data†
Papio anubis			
Gombe	7–15		Palombit 2003
Gilgil / Chololo	≈ 0	1 infanticide	Palombit 2003; Henzi & Barrett 2003
Maasai Mara	0		Sapolsky, pers. comm.; Palombit 2003
Segera	≈ 0	1 infanticide in 11 years	Palombit, unpublished data
Papio cynocephalus			
Amboseli	< 5		Altmann & Alberts, pers. comm.; Palombit 2003
Mikumi	≈ 0	Only 1 possible infanticide of orphan	Rhine et al. 1980; Palombit 2003
Tana River	0		Henzi & Barrett 2003
Papio ursinus			
Okavango	≥ 38	75% during some periods; primary cause of infant mortality	Palombit et al. 2000; Cheney et al. 2004
Drakensberg	75	n = 4 deaths	Henzi & Barrett 2003
De Hoop	44		Henzi & Barrett 2003
Cape Point	100	n = 5 deaths	Henzi & Barrett 2003
Mikuzi	0	n = 4 deaths	Henzi & Barrett 2003
HOMINOIDEA			
Hylobates lar	≥ 50	n = 6 deaths	Borries et al., in press
Gorilla gorilla beringei			
Virungas	≥ 37	Rate lower since mid-1980s (see text)	Watts 1989
Gorilla gorilla graueri	11.5		Yamagiwa, pers. comm.
Pan troglodytes			
Gombe	≥ 6–10%	Value refers to infants under the age of 5 years and excludes infanticide committed by females or unknown individuals; all cases were intercommunity male infanticide	Calculated from fig. 4 in Williams et al. 2008; also Pusey & Wilson, pers. comm.
Mahale	22–35	Value refers to infants < 1 year of age	Hamai et al. 1992
Taï	Very low	1 or 2 infanticides in approximately 25 years of observation	Boesch et al. 2008

* Percentage of infant deaths due to male infanticide.
† See Janson & van Schaik 2000.

disappears when the number of males per group is controlled for, the theoretical paradox it implies merits further study. Overall, however, the variable with the strongest predictive power is the number of males in groups, which is negatively correlated with infanticide rate. A weaker, though significant, dietary association also exists, such that infanticide is less frequent in frugivores than in folivores (see below).

A third possible reason for the rarity of infanticide is that females have evolved effective counterstrategies which, in some cases, reduce their vulnerability.

Female Counterstrategies to Male Infanticide

Hrdy (1979) argued that the female biology should include adaptations that deter the killing of their offspring. Mathematical models also suggested that low levels of infanticide can exert strong selection pressure in such slowly reproducing animals (Chapman & Hausfater 1979; Hausfater et al. 1982; Yamamura et al. 1990). And yet, in an early review of infanticide, Hiraiwa-Hasegawa (1988, p. 140) concluded that "there seems to be no effective female counterstrategy in wild primates," which she attributed to stronger selection on male persistence than on female resistance (sensu Krebs & Dawkins 1979). A more likely explanation of the apparent scarcity of relevant evidence, however, is that the subject of female counterstrategies has been largely neglected until fairly recently, as attention was focused instead on clarifying male infanticide as strategy, not selection pressure (Janson & van Schaik 2000). This situation has changed dramatically in the last two decades. A large array of interrelated female counterstrategies to infanticide have been proposed, studied, and, to varying degrees, clarified empirically (table 19.4).

Sexual Behavior: Promiscuity versus Targeted Mating

Hrdy (1979, 1981) attributed the sexual assertiveness of female haplorrhines largely to a strategy of establishing some nonzero probability of relatedness to their infants in as many potentially infanticidal males as possible. Experimental studies on rodents revealed that ejaculatory copulation and its timing reduces infanticide risk (vom Saal & Howard 1982; vom Saal 1985; Mennella & Moltz 1988; Perrigo et al. 1992). In primates, the majority of infanticides are indeed perpetrated by males known or suspected to lack prior sexual access to the mothers (table 19.3). In a direct test of the hypothesis, Soltis et al. (2000) found that male Japanese macaques attacked (sometimes fatally) the infants of females they had not mated with eight times more frequently than infants whose mothers were their former sexual partners. Moreover, van Noordwijk & van Schaik's (2000) comparative analysis provides indirect evidence that promiscuity is linked to infanticide deterrence: primate species vulnerable to infanticide are characterized by multimale mating more often (62% of 47 species) than less vulnerable species (9% of 11 species).

One unanswered question is whether females can beneficially pursue this strategy with extragroup males, which offers potentially significant advantages in light of the infanticidal tendencies of new immigrants. Observations of copulations with extragroup males have been noted in unimale units (Hrdy 1977; Sommer 1988; Oshawa 1991; Carlson & Isbell 2001; Guo et al. 2010; Cords 2000; Macleod et al. 2002; Swedell & Saunders, 2005; Treves et al. 2003; Agoramoorthy & Rudran 1995), as well as in some multimale groups (Starin 1981; Saito et al. 1998; Cords 2000; Teichroeb et al. 2005; Arlet et al. 2007; Singh et al. 2006). Attempts to mate outside the group are likely to provoke aggressive interference from resident males (see below) and it is unclear how female vulnerability to attack varies with the conditions surrounding extragroup matings. If those costs vary, then extragroup matings may be a feasible counterstrategy for females in some species or under certain circumstances. Indeed, Clarke et al. (1993) suggested that obtaining copulations specifically with extragroup males may be one function of nonconceptive postpregnancy estrus generally (see below).

Given the potential value of male protection in deterring infanticide (see below), Clarke et al. (2009a) revise the paternity confusion hypothesis by arguing that females should follow a mating strategy that concentrates paternity in the best available male defender while at the same time diluting it across as many other males as possible. In principle, such a mixed strategy is possible if males have variable access to the information on which they base paternity estimates (Garcia et al. 2009), which may arise partly through female manipulation of the relevant cues (Clarke et al. 2009a).

Nunn's (1999a) graded signal hypothesis for sexual swellings provides an example of how such a system could operate (see chapter 16, this volume, fig. 16.3). From the perspective of paternity confusion alone, *concealing* ovulation should offer greater protection from infanticide than signaling it (Andelman 1987), but Nunn based his model on evidence that the timing of ovulation is highly variable in species with sexual swellings. Because the probability of ovulation is highest at maximal tumescence, females are likely to be fertilized by a high-quality male that mate guards them at that time and who protects the infants later. But because the probability of ovulation is variable *and* nonzero at other times during the swelling cycle, paternity can be confused among males that copulate at those times.

Heistermann et al. (2001) provide direct evidence for this reproductive system in the Hanuman langur, which lacks exaggerated swellings, but which in the case of this study was characterized by a multimale structure. Their hormonal and genetic data confirmed that the timing of ovulation was extremely variable and occasionally yielded fertilizations by subordinate males, in spite of the dominant male's pronounced sexual monopolization of receptive females (see also Deschner et al. 2004). Also consistent with this view are observations that females in some species mate highly selectively during periovulatory periods and promiscuously at other times (Matsumoto-Oda 1999a; Stumpf & Boesch 2005; Watts 2007; Stumpf et al. 2008), a pattern that may even extend to extragroup males (Arlet et al. 2007; Tobler 2009). Moreover, Garcia et al. (2009) argue that high-ranking male Japanese macaques have disproportionately greater access to the cues of ovulation and that, consequently, ovulation is less concealed from them than it is from low-ranking males. Finally, Pradhan et al. (2006) argue that female copulation calls contribute to the mixed strategy, since their occurrence can intensify either mate guarding by the consort male or sexual interest by nonconsort males. In preliminary support of this hypothesis, they report that such calls are relatively infrequent and are linked to copulations with dominant males in species with effective mate guarding (thus suggesting a paternity concentration role), whereas these calls are frequent and unrelated to the rank of suitors in species with greater female promiscuity (thus suggesting the paternity confusion benefit; see also Maestripieri & Roney 2005).

The mixed strategy's ostensibly delicate balance between confusing and concentrating paternity warrants careful study. For example, because male chimpanzees direct significantly higher rates of courtship *and* copulatory interference toward females who are most likely to conceive, Emery Thompson & Wrangham (2008) conclude that males not only assess the profitability of matings based on swelling status, but use that information to *thwart* the female counterstrategy of paternity confusion through promiscuity. However, since females copulate more *selectively* when fertilization potential is high (see above), sexual interference at this time is more likely to limit the female's attempt to concentrate paternity. Unless the "interfering" males are actually the female's *preferred* partners from a fitness perspective (sensu Cordero & Eberhard 2003, Eberhard 2005), then these patterns suggest that swellings have to some degree "backfired": they provide males with the means to subvert the very benefit they were designed to confer upon females. This is the nature of sexual conflict, of course, but how the benefits of swellings outweigh their costs (assuming that they do) requires clarification. Increasing appreciation that the sexual swelling is a *multicomponent* signal comprising several visual (color, size, shape) and olfactory dimensions (Higham et al. 2008; Clarke et al. 2009b; Huchard et al. 2009), augmented by vocal stimuli, may provide opportunities to elucidate female manipulation of paternity information. Likewise, the proximate mechanisms underlying male paternity assessment are poorly understood, but the possibility that they rely on more than sexual history is raised by observations that temporal proximity to an infant's birth was crucial for forestalling a male's infanticidal tendencies in sooty mangabeys (*Cercocebus atys*, Gust et al. 1995). As Clarke et al. (2009a) point out, one aspect of paternity assessment that must be *absent* for the mixed strategy to fully benefit females is male ability to recognize infants reliably as kin (e.g., Widdig 2007).

An array of ancillary reproductive and sexual characters may further promote matings with multiple males (table 19.4).

Extended Mating Periods

The longer follicular phases of catarrhines with sexual swellings, compared to those of catarrhines without them, suggests a mechanism to increase opportunities to mate with more males (van Schaik et al. 2000). Across mammalian species generally, however, the duration of mating periods is not clearly associated with vulnerability to infanticide (van Noordwijk & van Schaik 2000). On the other hand, evidence that this counterstrategy operates intraspecifically is provided by observations that female Hanuman langurs living in multimale groups have lengthier receptive periods than do females in a unimale setting (Heistermann et al. 2001).

Postconception Mating

In many species of primates, females routinely exhibit sexual receptivity (including production of sexual swellings) and nonconceptive mating at various times during pregnancy and/or lactation (table 19.4; see also van Schaik et al. 1999). Rates of postconception or postpartum copulation vary considerably across species, from 23% to 74% of all observed copulations (Soltis 2002). Several workers note that rates of copulation are virtually identical for anestrous and estrous females (Rowell 1970; Manson et al. 1997; Carnegie et al. 2006), and that in at least one case, sexual swellings were *more* common among gestating females than among cycling females (Shelmidine et al. 2007). Because of its potential to contribute to paternity confusion, postconception mating has been agued to be an anti-infanticide counterstrategy in a variety of species, such as white-faced capuchins (Carnegie et al. 2006), Tana River mangabeys (*Cercocebus galeritus*, Kinnaird 1990), collared mangabeys

Table 19.4. Proposed female counterstrategies to male infanticide

Counterstrategy	Proposed rationale	Proposed example taxa	Reference
SEXUAL BEHAVIOR I: OBSCURE PATERNITY AMONG POTENTIALLY INFANTICIDAL MALES			
Promiscuity	Enhanced sexual receptivity and proceptivity leads to mating with many males, including extragroup males when possible.	Many primates, some other mammals (e.g., rodents)	1, 2, 3, 4
Sexual swellings	Advertising ovulatory status stimulates male-male competition, reduces dominant male's ability to monopolize copulations (except at peak estrus; see below), thus facilitating multimale matings.	Catarrhines in multimale societies, particularly Papionini, Pan troglodytes	5, 6
Supplementary advertisement of ovulatory status and/or copulation	Female vocal and/or visual advertisement of copulation by calls and/or postcopulatory "darts" incites male-male competition and polyandrous mating.	Catarrhines in multimale societies, e.g., Papio anubis, P. cynocephalus, P. ursinus, Cercocebus torquatus, Macaca fascicularis, M. sylvanus, M. tonkeana, Pan troglodytes	7, 8, 9
Lengthened estrous periods	Longer estrus provides more opportunities for multimale mating during fecund cycle.	Catarrhines with sexual swellings, Semnopithecus entellus	10, 11
Concealed ovulation	Ovulation is unaccompanied by signals and/or is variable in timing, thereby obscuring paternity.	Semnopithecus entellus, Chlorocebus pygerythrus	11, 12
Postconception and/or postpartum "estrus"	Copulation and/or development of anovulatory sexual swellings by females occurs regularly during part of pregnancy and/or lactation.	Species in which postconception mating and infanticide have been observed, and/or species in which postconception mating is a proposed counterstrategy to infanticide: Alouatta palliata, Cebus capucinus, Cercocebus agilis, Cercopithecus atys, C. galeritus, C. ascanius, C. mitis, Chlorocebus pygerythrus, Erythrocebus patas, Gorilla gorilla beringei, G. g. graueri, Macaca fascicularis, M. fuscata. M. mulatta, M. nemestrina, M. radiata, M. silenus, M. sylvanus, Mandrillus leucophaeus, Pan troglodytes, P. paniscus, Papio ursinus, P. rufomitratus, Propithecus verreauxi, Rhinopithecus roxellana, Semnopithecus entellus, Trachypithecus villosus	13, 14, 15, 16, 17, 18, 12, 19, 20, 21, 22, 23, 24, 25, 26, 27, 28, 29, 30, 31, 32, 33, 34, 35, 36
Situation-dependent receptivity	Exposure to new (or extragroup) male(s) facultatively causes (I) "deceptive" pseudoestrus and/or sexual swellings among anestrous females, or (II) longer receptive periods among cycling females.	I. Propithecus verreauxi, Cercopithecus mitis, Papio hamadryas, Semnopithecus entellus?, Procolobus badius, Cebus apella II. Cercopithecus mitis, Macaca fuscata	I. 34, 20, 37, 38, 32, 39 II. 20, 19 40
SEXUAL BEHAVIOR II: FOCUS COPULATIONS OR CONCEPTIONS ON PARTICULAR MALE(S)			
Female mate choice I	Females limit periovulatory matings (concentrate paternity) in male likely to be future protector against infanticide.	Papio hamadryas, Papio ursinus?	41,
Female advertisement of ovulatory status and/or copulation	Sexual swellings and/or copulation calls intensify mate guarding, thus reducing polyandrous mating at peak estrus and concentrating paternity in a future male defender of the infant.	Catarrhines in multimale societies, e.g., Papio anubis, P. cynocephalus, P. ursinus, Cercocebus torquatus, Macaca fascicularis, M. sylvanus, M. tonkeana, Pan troglodytes	5, 7, 42, 6
Female mate choice II	If the dominant male is more likely than subordinates to commit infanticide, cycling females preferentially mate with him.	Cebus olivaceus	43
MANIPULATION OF REPRODUCTION			
Superfetation	Conception occurs during pregnancy, with embryonic diapause of fertilized eggs.	European badger	44
Pregnancy block (Bruce effect)	Fertilized egg is aborted and/or fetus is reabsorbed in response to increased infanticide risk following male replacement.	Papio hamadryas, Papio cynocephalus, Semnopithecus entellus, Theropithecus gelada, rodents	45, 46, 47, 48, 49, 50, 51
Abandonment of infant	Mothers may (I) abandon an infant that has been injured by a takeover male but is still alive, or (II) abandon a healthy infant in presence of familiar ousted male (instead of residing with infant in group with takeover male).	I. Colobus vellerosus II. Semnopithecus entellus	I. 52 II. 53
Acceleration of weaning	Following turnover in breeding male, mothers wean older infants sooner, and return to estrus.	Chlorocebus pygerythrus, Papio hamadryas	54, 45
Temporarily reduced fecundity	Immediately following male takeover, females temporarily inhibit their ovulatory cycles until new male(s) are established and are unlikely to be replaced by other males.	Lions, Presbytis thomasi	55, 56
Breeding synchrony	Synchronizing ovulation prevents one male from monopolizing matings and generates multimale groups (which have lower rates of infanticide; see below) or matings with extragroup males.	Lions, Eulemur fulvus rufus, Papio spp.?, Cercopithecus mitis	57, 58, 59, 60

INDIVIDUAL DEFENSE

Category	Prediction/Description	Species examples	Sources
Maternal aggression: intragroup	Mothers individually attack actual or potentially infanticidal male in group (e.g., new immigrant).	Some rodents; most primates, but apparently effective only in few species, e.g., *Lemur catta, Cephalopachus bancanus*	61, 62, 63, 64, 65, 66, 67
Maternal aggression: extragroup	Mothers participate more in aggressive intergroup interactions to deter immigration of extragroup males (and future infanticide). See below for coalitionary aggression in this context.	*Cebus capucinus?*	68
Maternal vigilance and protectiveness	Under conditions of elevated infanticide risk, mothers increase visual monitoring of location of potentially infanticidal male and/or infant, and restrain and/or carry infants more.	*Alouatta pigra, Cercocebus galeritus, Chlorocebus pygerythrus, Erythrocebus patas, Presbytis thomasi, Colobus vellerosus, Papio ursinus, Pan troglodytes*	69, 70, 54, 71, 72, 73, 74, 75, 76, 77, 78, 79
Evasion, sexual segregation	Mothers vulnerable to infanticide (I) increase time spent alone and/or away from areas with heightened risk, and (II) facultatively participate less in intergroup interactions.	I. Lions, bears, *Semnopithecus entellus, Pan troglodytes, Procolobus badius?, Papio ursinus* (see text); II. *Presbytis thomasi, Pan troglodytes*, gregarious primates generally?	I. 80, 81, 82, 53, 83, 84, 85, 86, 33, 87 II. 88, 89, 90

SOCIAL DEFENSE: FEMALE-FEMALE

Category	Prediction/Description	Species examples	Sources
Female aggregation	(I) Females that live in permanent groups with more females experience less infanticide risk through dilution or defense effects. (II) Lactating females in a group aggregate or cluster together temporarily.	I. *Semnopithecus entellus*, many primates? II. *Papio ursinus*	91, 92, 93
Female coalitionary defense	Females collectively attack infanticidal male and/or deter male immigration.	Lions, *Propithecus verreauxi?, Cercopithecus mitis, Procolobus badius temminckii, Pan paniscus?*; see table 5	94, 95, 96, 97, 98, 99
Residency in small groups	If infanticide risk rises with female group size, females maintain small groups via secondary transfer and/or group fission.	Lions, *Alouatta pigra, A. arctoidea, Colobus guereza, C. vellerosus, Presbytis thomasi, Propithecus edwardsi?, Semnopithecus entellus, Gorilla gorilla gorilla*	100, 101, 102, 103, 104, 105, 106, 107, 108
Female aggression to immigrant and/or natal females	If infanticide risk rises with female group size, resident females aggressively deter immigration by extragroup females.	*Alouatta palliata, A. pigra?, A. arctoidea, Colobus vellerosus, Propithecus edwardsi?, Semnopithecus entellus, Gorilla gorilla beringei*	109, 110, 111, 112, 113, 106, 104, 53, 114

SOCIAL DEFENSE: FEMALE-MALE

Category	Prediction/Description	Species examples	Sources
Male defense	Females associate with male defenders in permanent female group or male-female pair, in temporary coalitionary relationships with male(s) within groups (during lactation), or in relatively loose heterosexual associations (in dispersed societies).	*Alouatta arctoidea, Cebus capucinus, Cercocebus atys, C. galeritus, Cheirogaleus major?, Colobus vellerosus, Hylobates spp.?, Papio hamadryas, Papio ursinus, Presbytis thomasi, Procolobus badius, P. rufomitratus, Hylobates lar, Gorilla gorilla beringei, G. g. gorilla?, G. g. graueri, Pongo pygmaeus?*	102, 115, 116, 70, 117, 118, 119, 120, 121, 122, 74, 123, 98, 124, 125, 126, 127, 128, 129, 130
Transfer with evicted male	After male turnover, mothers leave the group to remain with the original evicted male until infant is weaned or passed on to the male.	*Cebus capucinus, Presbytis rubicunda?, Semnopithecus vetulus, S. entellus*	131, 132, 133, 53
Transfer to group with better protector male(s)	Cycling females immigrate to a new group with better protector male(s).	*Cebus capucinus?, Colobus vellerosus, Procolobus badius, P. rufomitratus, Presbytis thomasi, Macaca radiata, Gorilla gorilla beringei, G. g. graueri, G. g. gorilla*	131, 104, 98, 134, 74, 135, 125, 136, 128, 108
Residency in multimale groups	When female transfer is possible, females immigrate to multimale groups. In female-philopatric species, females promote multimale structure of groups.	*Eulemur fulvus rufus, Propithecus verreauxi, Alouatta palliata, Alouatta spp.?, Cebus capucinus, Chlorocebus pygerythrus, Macaca fuscata?, Gorilla gorilla beringei?*	58, 137, 138, 131, 139, 140, 141

Sources: 1. Van Noordwijk & van Schaik 2000. 2. Soltis et al. 2000. 3. Clarke et al. 2009c. 4. Wolff & MacDonald 2004. 5. Nunn 1999a. 6. Deschner et al. 2004. 7. Pradhan et al. 2006. 8. Bercovitch 1995. 9. Smuts 1985. 10. Van Schaik et al. 2000. 11. Heistermann et al. 2001. 12. Andelman 1987. 13. Glander 1980. 14. Carnegie et al. 2006. 15. Walker et a. 2004. 16. Cordon & Busse 1984. 17. Kinnaird 1990. 18. Gust 1994. 19. Cords 1984. 20. Pazol 2003. 21. Loy 1981. 22. Robbins 2003. 23. Yamagiwa & Kahekwa 2001. 24. Hadidian & Bernstein 1979. 25. Soltis et al. 1999. 26. Wilson et al. 1982. 27. Clarke et al. 1993. 28. Möhle et al. 2005. 29. Small 1990. 30. Wallis 1997. 31. Bielert 1986. 32. Struhsaker & Leland 1985. 33. Starin 1988. 34. Brockman & Whitten 1996. 35. Guo et al. 2010. 36. Sommer et al. 1992. 37. Zinner & Deschner 2000. 38. Sommer 1994. 39. Ramirez-Llorens et al. 2008. 40. Takahata et al. 1994. 41. Swedell & Saunders 2006. 42. Maestripieri & Roney 2005. 43. O'Brien 1991. 44. Yamaguchi et al. 2006. 45. Colmenares & Gomendio 1988. 46. Pereira 1983. 47. Agoramoorthy et al. 1988. 48. Mori & Dunbar 1985. 49. Mori et al. 2003; Roberts et al. 2012. 50. Huck 1984. 51. Drickamer 2007. 52. Sicotte et al. 2007. 53. Hrdy 1977. 54. Fairbanks & McGuire 1987. 55. Pusey & Packer 1994. 56. Steenbeek 1999b. 57. Bertram 1975. 58. Ostner & Kappeler 2004. 59. Altmann 1990. 60. Cords 1986. 61. Agrell et al. 1998. 62. Lonstein & Gammie 2002. 63. Gammie et al. 2007. 64. Weber & Olsson 2008. 65. Pereira & Weiss 1991. 66. Ichino 2005. 67. Roberts 1994. 68. Crofoot 2007. 69. Treves et al. 2003. 70. Gust 1994. 71. Fairbanks & McGuire 1993. 72. Enstam et al. 2002. 73. Steenbeek et al. 1999. 74. Sterck 1997. 75. Macintosh & Sicotte 2009. 76. Brent et al. 2008. 77. Busse 1984. 78. Barrett et al. 2006. 79. Pusey & Packer 1983. 81. Wielgus & Bunnell 2000. 82. Ben-David et al. 2004. 83. Wrangham & Smuts 1980. 84. Sakura 1994. 85. Matsumoto-Oda 1999b. 86. Otali & Gilchrist 2006. 87. Palombit et al. 2001. 88. Wich et al. 2004. 89. Goodall 1986. 90. Van Schaik 1996. 91. Treves & Chapman 1996. 92. Brereton 1995. 93. Cowlishaw 1999. 94. Richard et al. 1993. 95. Packer & Pusey 1984. 96. Butynski 1982. 97. Cords & Fuller 2010. 98. Starin 1994. 99. De Waal 1997. 100. Pusey & Packer 1987. 101. Ostro et al. 2001. 102. Crockett & Janson 2000. 103. Chapman & Pavelka 2005. 104. Teichroeb et al. 2009. 105. Sterck et al. 2005. 106. Wright 1995. 107. Koenig & Borries 2001. 108. Stokes et al. 2003. 109. Glander 1992. 110. Knopff & Pavelka 2006. 111. Rudran 1979. 112. Sekulic 1982. 113. Crockett 1984. 114. Watts 1991. 115. Vogel & Fuentes-Jiménez 2006. 116. Fruteau et al. 2010. 117. Kappeler 1997. 118. Teichroeb & Sicotte 2008a. 119. Van Schaik & Dunbar 1990. 120. Swedell 2006. 121. Palombit 2009. 122. Weingrill 2000. 123. Steenbeek 1999a. 124. Borries et al. in press. 125. Watts 1989. 126. Sterck et al. 1997. 127. Stokes 2004. 128. Yamagiwa et al. 2009. 129. Delgado & van Schaik 2000. 130. Mitra Setia & van Schaik 2007. 131. Jack & Fedigan 2009. 132. Davies 1987. 133. Rudran 1973b. 134. Marsh 1979b. 135. Singh et al. 2006. 136. Robbins et al. 2009b. 137. Lewis 2008. 138. Treves 2001. 139. Isbell et al. 2002. 140. Yamada & Nakamichi 2006. 141. Robbins et al. 2009c.

(*C. torquatus*, Gordon et al. 1991), western red colobus (*Procolobus badius*, Starin 1988), and golden snub-nosed monkeys (*R. roxellana*, Guo et al. 2010). Data are insufficient to test the hypothesis directly, but one interspecific pattern is consistent with this proposed function. Assuming that the high rates of postconception mating in noninfanticidal callitrichines is related to their peculiar system of communal breeding, postconception mating in primates is significantly more common in primates that are vulnerable to infanticide than in those that are less vulnerable (van Noordwijk & van Schaik 2000).

Doran-Sheehy et al. (2009) argue that postconception mating cannot contribute to paternity confusion in species with sexual swellings, because changes in the appearance of the sexual skin differ in conceptive versus nonconceptive contexts. Observations that postpartum swellings excite less (or even no) sexual interest by males endorse this view (Kinnaird 1990; Gust 1994). However, the magnitude of difference between conceptive and nonconceptive anogenital swellings varies a great deal across primates, from conspicuous contrasts (e.g., Wallis 1997), to overall similarity (Gordon et al. 1991; Walker et al. 2004), the latter of which may facilitate this counterstrategy.

Situation-Dependent Sexual Receptivity

A related phenomenon is situation-dependent receptivity among pregnant and lactating females, which differs from the postconception receptivity described above in being a *facultative* onset or intensification of nonconceptive estrus following male turnover or exposure to new males. Originally designated "pseudoestrus" (Hrdy 1974, 1977), this pattern has been observed in a number of cercopithecines, colobines, platyrrhines, and strepsirrhines in the wild as well as in captivity (table 19.4). As with postconception mating, situation-dependent receptivity may involve the production of functionally "deceptive" sexual swellings in anestrous females following male takeover (Zinner et al. 2000).

In blue monkeys (*Cercopithecus mitis*), influxes of multiple extragroup males are followed by rapid increases in estrogen levels in pregnant females and their "sudden" and "dramatic" initiation of sexual activity (Pazol 2003, p. 492). As predicted by van Schaik et al. (2000), these receptive periods last longer than those in noninflux contexts, and their duration accurately predicts the number of males mated. Lactating females also copulate with influx males, but only when their infants are more than four months old. Because infants beyond this age are partly independent, Pazol suggests that their mothers could approach stranger males alone. Although the alternative explanation that hormone fluctuations arise incidentally from endocrine maintenance of gestation could not be fully rejected, such data

shed light on a possible physiological mechanism underlying situation-dependent receptivity.

In black capuchin monkeys, male takeover is followed by elevated sexual activity by pregnant and lactating females, but Ramirez-Llorens et al. (2008) describe an interesting variant of this putative counterstrategy. Estrous females also mated with the new males, which were nonconceptive since the matings occurred well outside the mating season. Notably, during the same period, the females in a neighboring group with long-term male residents showed the typical pattern of absence of mating.

Postconception mating and situation-dependent receptivity apparently lower infanticide risk only in some (Struhsaker & Leland 1985; Fairgrieve 1995; Zinner et al. 2000) but not all species (Sommer 1987; Hasegawa 1989; Yamagiwa & Kahekwa 2004). Soltis (2002) suggests two possible reasons why postconception mating apparently works in only some species, and why it is absent in some species with infanticide: (1) the sexes are currently at different points in the coevolutionary arms race, or (2) females in some species rely on other counterstrategies to deter infanticide. Postconception swellings and receptivity could also arise incidentally from hormonal fluctuations (Saayman 1975), although this seems unlikely given the costs females incur (e.g., Sommer 1994). Finally, matings by anestrous females may offer benefits besides infanticide protection, such as feticide deterrence (see below), male support for an increase in dominance rank (Ramirez-Llorens et al. 2008), female competition (Sommer 1994; Doran-Sheehy et al. 2009; Stoinski et al. 2009), or diminished harassment from males (sensu Li et al. 2007).

Manipulation of Reproduction

Among primates, embryo abandonment following male turnover (the Bruce effect) has been observed in a few catarrhines, both in the wild and in captivity (table 19.4). Its adaptive rationale is underscored by the observation by Agoramoorthy et al. (1988) in Hanuman langurs that 70% of the pregnant females that, unlike some other females, did not abort following takeover subsequently lost their offspring to infanticide. Similarly, in geladas (*Theropithecus gelada*), lactating females that abort following replacement of the unit male enjoy shorter interbirth intervals than lactating females that do not terminate pregnancy and attempt to rear their infants (usually unsuccessfully; Roberts et al. 2012). The apparent rarity of the Bruce effect may have several causes, including underreporting (van Noordwijk & van Schaik 2000) or the effectiveness of female promiscuity in obscuring paternity (sensu Pillay & Kinahan 2009).

An alternative, but related, explanation of the Bruce effect is that it reflects a male strategy to hasten a mother's return to fertilizable condition by inducing abortion through stressful aggression. The extreme male aggression that characterizes some reports of primate abortion is consistent with this interpretation (Pereira 1983; Agoramoorthy et al. 1988), as is independent evidence of male-induced disruption of female reproductive physiology (Smith 1981) and of female-induced abortions in rivals (Wasser & Stirling 1988). In summary, available data permit the tentative conclusion that Bruce effects reflect some adaptive form of sexual conflict in primates, as female counterstrategy, as male strategy, or as both.

Several other reproductive patterns have been argued, on the basis of preliminary results, to reduce the costs of infanticide for females (table 19.4). First, the abandonment of infants can potentially cut reproductive losses for a female in a manner analogous to that of pregnancy block, but observed cases are even fewer in number. Second, the capacity for accelerated weaning to forestall sexually selected infanticide is suggested by a report that a nearly weaned mountain gorilla infant who accompanied his mother to a new male's group—the most common context of infanticide in this population—escaped attacks by the silverback long enough to survive (Sicotte 2000). In a study of captive hamadryas baboons (*Papio hamadryas*) in which infanticide had been previously noted, Colmenares and Gomendio (1988) attributed the absence of infanticide following one episode of male replacement to significantly abbreviated amenorrhea by the lactating females. Premature weaning promises greater success with older infants than with younger ones, which may account for observations that vervet monkey (*Chlorocebus pygerythrus*) mothers in groups experiencing male replacements increased their rejections of infants older than six months (Fairbanks & McGuire 1987). A final proposed reproductive counterstrategy is breeding synchrony, which may reduce infanticide risk by promoting the multimale structure of groups (Altmann 1990; Nunn 1999b), which is considered below, or through matings with extragroup males by females in unimale groups (Cords et al. 1986).

Individual Counterstrategies of Defense

Several counterstrategies of individual behavior by mothers have been suggested to deter infanticide (table 19.4).

Maternal aggression
Most direct observations of infanticidal attacks in primates include descriptions of maternal defense, which may be protracted and direct (even to the point of biting the attacker; e.g., Yamagiwa & Kahekwa 2004), but in the end largely ineffective (references in table 19.1; see also Mae-

Fig. 19.6. An adult female ring-tailed lemur (*Lemur catta*) at the Duke Primate Center attacks an adult male that has not signaled submission. Photo courtesy of Peter M. Kappeler.

stripieri 1992). In the majority of cases, the most that maternal defense seems capable of achieving is a postponement of infanticide. A rare exception is provided by ring-tailed lemurs. In a captive population in which infanticidal attacks and infant deaths were recorded, lactating females violently assaulted immigrant males "on sight" for three months, usually with success (fig. 19.6; Pereira & Weiss 1991, p. 145; see also Ichino 2005). Maternal aggression also extended to resident males who had not been previously observed to copulate with the mother.

Notwithstanding this lemur, the reasons for the apparent failure of maternal defense in primates generally remain unclear. Pronounced dimorphism is unlikely to account for this failure, since mothers in some less dimorphic species do not appear more successful defensively (though relevant data are few). Another reason for the failure may simply be that the often opportunistic attacks of infanticidal males are usually launched under conditions that compromise maternal defense.

Crofoot (2007) proposed a variant of this counterstrategy by arguing that female white-faced capuchins, including mothers, increase their participation in aggressive intergroup interactions in order to deter immigration by extragroup males. Her data partly supported this hypothesis, as do observations of coalitionary defense among female western red colobus monkeys (see below).

Maternal vigilance and infant protectiveness
In light of their apparent inability to prevent infanticide physically, mothers may rely instead on early detection and evasion of infanticidal threat. For example, female chacma baboons exhibited higher rates of vigilance when lactating

than at other times, especially during the first four months of their infants' lives (Barrett et al. 2006). Likewise, in Central American black howler monkeys (*Alouatta pigra*), lactating females committed more than six hours per day to visual scanning, 50% more than other females (Treves et al. 2003).

Available data limit the important task of differentiating between infanticide risk and other contingencies possibly targeted by maternal vigilance, such as predation (e.g., Gould 1996; Onishi & Nakamichi 2008; chapter 8, this volume). Treves et al. (2001) argue that because per capita vigilance rates vary systematically with intragroup dispersion of individuals but not with group size in Central American black howlers (and other primates), reduction of predation risk is less likely to be a function than countering of conspecific threat. Such threat includes, but is not limited to, infanticide. In two monkey species in which infanticide is perpetrated primarily by extragroup males, maternal vigilance is patterned accordingly. In white-thighed colobus, vigilance is positively correlated with the encounter rate of extragroup males, and mothers are more watchful in range overlap areas and during intergroup interactions than in contexts with high predation risk (Macintosh & Sicotte 2009). Among female Thomas's langurs, it is *only* mothers with infants that show elevated vigilance in range overlap areas (Steenbeek et al. 1999).

Increased maternal restraint and retrieval of infants following male turnover has been described in several species in which infanticide occurs (Busse 1984; Fairbanks & McGuire 1987; Gust 1994; Brent et al. 2008). An experimental study showed that female Thomas's langurs responded to playbacks of audio recordings simulating the presence of extragroup males by moving closer to their infants (Wich et al. 2004). Sterck (1997) also noted that mothers whose infants usually traveled independently resumed carrying them during intergroup interactions involving extragroup male chases of females. The effectiveness of maternal restraint and vigilance may be higher when infants are young and consistently close to their mothers, and it is likely to decline as infants become more mobile and independent (e.g., Enstam et al. 2002).

Solitary evasion

Mothers may elude detection by infanticidal males by increasing time spent away from them. For gregarious primates, a strategy of cryptic evasion may be expressed in ways more subtle than permanent sexual segregation. For example, Palombit et al. (2001) present preliminary evidence that some lactating female chacma baboons who lost social access to a male defender increased the time they spent away (at least 25 meters) from other individuals in the group, without actually leaving it. Newton (1986, p. 786) observed that approaches of an all-male band to a group of Hanuman langurs caused mothers but not other females to flee to concealed areas such as "subterranean streams," and also that occasionally mothers with infants may even leave groups after male takeover (fig. 19.7).

When infanticide is committed by extragroup males, evasion may be achieved via diminished involvement of mothers in interactions with other groups (van Schaik 1996; but see Fashing 2001) or all-male bands. In their experimental test of this hypothesis, Wich et al. (2004) revealed first that female Thomas's langurs discriminate between audio playbacks of calls of males in all-male bands and of bisexual groups, and also that they respond according to the greater infanticidal threat posed by the former: the females traveled cautiously away from both types of calls, suggesting general avoidance of extragroup males, but they moved sooner and in a more compact group formation upon hearing the calls of a male from an all-male band (see also McComb et al. 1993). Steenbeek (1996) observed that after the sudden loss of the resident male, females avoided other groups but were still attacked by extragroup males.

Increased solitariness as an anti-infanticide strategy is more likely in fission-fusion societies (Aureli et al. 2008). Accordingly, females in several populations of chimpanzees spend more time alone when they are lactating (table 19.4). This pattern may also reflect maternal avoidance of female aggression, but current evidence suggests that females obtain protection from intrasexual harassment through association with adult males (Pusey et al. 2008). Heightened avoidance of community borders by mothers has also been noted, which most likely functions to reduce vulnerability to infanticide by extragroup males (Emery Thompson et al. 2006).

Social Defense: Female-Female Interactions

Female gregariousness in primates has traditionally been interpreted as a solution to the problems posed by feeding ecology and predation (chapter 9, this volume). There is growing evidence, however, that in some taxa infanticide has been an important selection pressure on several aspects of female social life.

Female association and anti-infanticide coalitionary defense

Brereton (1995) suggested that female primates live together to reduce the costs of infanticide and sexual coercion. This hypothesis is supported by limited evidence. For example, under conditions of higher infanticide risk, langur (*Presbytis* spp.) groups with more females had lower rates of immature mortality (Treves & Chapman 1996). In chacma

Fig. 19.7. Two Hanuman langur (*Semnopithecus entellus*) mothers (*far right*) temporarily traveling with an all-male band. One way for mothers to evade infanticidal males is to follow evicted males. A mother may even attempt to leave a nearly weaned infant with its likely father and return alone to her group and matrilineal kin. This putative strategy rarely works, however, as infants old enough to survive without their mothers are also mobile enough to find their way back to them, and back into harm's way. Photo courtesy of Sarah B. Hrdy/Anthro Photo.

baboons with male infanticide, Cowlishaw (1999) reported that female kin cluster together more closely during lactation, as Palombit et al. (2001) also observed among some mothers who lost their male defenders. Hormonal data from this baboon population also suggest that bonds with females may alleviate the stress surrounding infanticide-related turnover in the alpha male (Wittig et al. 2008). What is less available for testing this hypothesis is a clear empirical demonstration of *how* female gregariousness reduces infanticide risk. Possible mechanisms include dilution effects (sensu Hamilton 1971), early detection and alarm-based evasion of infanticidal males or situations (sensu Pulliam 1973), coalitionary aggression to extragroup males to deter takeover (and *future* infanticide; Crofoot 2007), or coalitionary defense against infanticidal males, as in lions (sensu Packer & Pusey 1984).

Few data are available for evaluating these mechanisms, except perhaps the last of them. Infanticidal attacks frequently elicit simultaneous responses from multiple females in the group, thus providing opportunities for mutual defense. As with maternal aggression, however, these communal counterattacks are generally unsuccessful or of unclear effectiveness as a counterstrategy, with the current possible exceptions of those in Temminck's red colobus (*Procolobus badius temminckii*) and in blue monkeys (table 19.5). Infant

killing has not been observed in bonobos (*Pan paniscus*), but de Waal (1997) suggests that the well-developed coalitionary behavior of females may have evolved as a highly effective counterstrategy to infanticide.

Female defense, whether by individual mothers or by coalitions of females, generally seems only to delay infanticide. That outcome could still offer females anti-infanticide benefits if accompanied by accelerated weaning. Data are too few to explore this possibility. Likewise, it is not clear why coalitions capable of effectively deterring infanticide are apparently so rare in primates. Pronounced sexual dimorphism is often invoked, but this explanation is not entirely satisfying. Although the blue monkey and Temminck's red colobus are characterized by relatively low dimorphism, so are other species that lack female alliances against infanticide. One possibly relevant difference between primates and lions is that many primate mothers carry their offspring continuously rather than caching them. Since coalitionary counterattack is likely to involve physical contact with the infanticidal male, carrying of infants may elevate the potential costs of aiding another female under attack to prohibitively high levels. On the other hand, the fact that mothers with infants are virtually never involved in the coalitionary aggression of female Temminck's red colobus monkeys apparently does not limit its effectiveness (Starin 1981).

Table 19.5.   Selected observations of female coalitionary defense against male infanticide in primates

Species	Description	Outcome	Generality as counterstrategy	Reference
RELATIVELY LOW EFFECTIVENESS				
Cercocebus atys	Two to six related and/or unrelated females without infants counterattacked male attacking an infant. Female attacks on adult males were rare, but 20% of them involved protection of other females' infants from infanticide (n = 25).	Largely ineffective	Unlikely	Fruteau et al. 2010
Cercopithecus ascanius	Several females chased attacking male to ground, making physical contact.	Infant killed	Unlikely	Struhsaker 1977
Colobus vellerosus	Females collectively chased intruder male, then initiated and participated in loud calling with resident male.	Infanticide delayed, and prevented in one case where weaning accelerated	Possible but rare	Saj & Sicotte 2005; Sicotte et al. 2007; Teichroeb & Sicotte 2008a
Erythrocebus patas	Female coalitions were considered generally effective.	Infant killed	Unclear	Enstam et al. 2002
Gorilla gorilla beringei	Cooperative defense occurred rarely.	Infants killed	Unlikely	Watts 1989; Harcourt & Stewart 2007
Gorilla gorilla graueri	In response to attack, mother and other females collectively counterattacked against silverback.	Infant killed	Unlikely	Yamagiwa & Kahekwa 2004
Pan troglodytes	An infant's grandmother and her female companion cooperatively defended against an infanticidal attack by two adult males.	Infant saved in this instance	Possible but rare	Sakamaki et al. 2001
Papio ursinus	Females screamed and chased infanticidal male following/during several attacks.	Infants killed or, in some cases, saved due to other reasons	Unlikely	Palombit et al. 2000; Kitchen et al. 2009
Procolobus rufomitratus	At least four females threatened and chased attacking male.	Infant killed	Unclear	Marsh 1979a
Rhinopithecus roxellana	Two females in captivity counterattacked and bit male.	Infant killed	Unlikely	Zhang et al. 1999
Semnopithecus entellus	Females chased and aggressed to attacking male.	Infanticide delayed	Unlikely	Hrdy 1977
RELATIVELY HIGH EFFECTIVENESS				
Propithecus edwardsi	An immigrant male that committed infanticide received "intense aggression" from the group's females; coordination of female actions was unclear.	Infanticidal male evicted from group	Likely?	Morelli et al. 2009
Cercopithecus mitis	Female coalitions were "particularly effective in thwarting male attacks."	Deaths of seven infants prevented in one study	Likely	Butynski 1982, p. 14; Macleod 2000; Cords 2002; Cords & Fuller 2010
Procolobus badius temminckii	Females fiercely attacked alien males attempting to join groups; attacks were joined by resident males, but initiated and led by females.	Immigrants expelled or, in at least two cases, killed	Likely	Starin 1994

Another possible explanation is that females in some taxa, particularly folivores, have evolved alternative infanticidal counterstrategies that constrain the development of female-female alliances (see below).

Residency in small groups: The folivore paradox

Socioecological theory predicts that the reduced within-group feeding competition enjoyed by folivorous primates should facilitate larger groups that confer greater benefits (predator protection) to their members (Wrangham 1980; van Schaik 1983; chapter 9, this volume). Yet female group sizes in leaf-eaters are frequently smaller, or at least not significantly larger, than those in frugivores of similar body size. A proposed reason for this apparent paradox derives from appreciation of their greater vulnerability to infanticide due to the predominance of unimale social systems. Thus, since larger associations of females are more attractive to extragroup males for takeover and subsequent infanticide, females should prefer to live in smaller groups (Janson & Goldsmith 1995; Treves & Chapman 1996; Chapman & Pavelka 2005).

Direct evidence for this hypothesis is limited. In ursine howler monkeys (*Alouatta arctoidea*) of Venezuela, female group size was significantly positively associated with infanticide rate per infant born, but not with infant death due to other causes (Crockett & Janson 2000). Similarly, infant survival was significantly negatively correlated with group size in Hanuman langurs, and infanticide rates were lowest in the smallest group and highest in the largest group (Koenig & Borries 2001).

At least three mechanisms may keep groups small: (1) fissioning of larger groups, (2) transfer of females from larger to smaller groups, and (3) aggression from resident females to deter immigration by other females or encourage emigration by natal females. Female transfer in Thomas's leaf monkeys and white-thighed colobus is consistent with the anti-infanticide hypothesis: the majority of successful immigrations are into smaller groups (Sterck et al. 2005; Teichroeb et al. 2009). In western lowland gorillas (*Gorilla gorilla gorilla*), in which male infanticide is suspected, female immigration rate is inversely correlated with group size, while the converse is true of emigration rate (Stokes et al. 2003). In these species, however, female reproductive success or infant survival has not yet been shown to vary inversely with female group size.

Aggression to extragroup or new immigrant females in many of these species has been interpreted as a mechanism for keeping groups small (table 19.4). In mountain gorillas, for example, such hostility to female outsiders increases in larger groups, even though feeding ecology suggests low within-group feeding competition (Watts 1991). Female ursine howlers and, to a significant extent, white-thighed colobus not only discourage immigration attempts by extragroup females through attacks (and possibly vocal displays), but also evict successful immigrants and even natal nulliparous females (Rudran 1979; Sekulic 1982, 1983; Crockett 1984; Teichroeb et al. 2009). Crockett & Pope (1988) interpret the higher incidence of injuries among subadult and extratroop females (as compared with those among adult or natal female residents) as evidence of the intensity of resistance to female immigration. Antagonism to female outsiders is not universal in these species, however (e.g., Thomas's langurs, Sterck 1997), and it is not necessarily always related to infanticide risk (Estrada & Garber 2008).

Ursine howlers and white-thighed colobus provide the strongest evidence for this anti-infanticide hypothesis, though data for Thomas's leaf monkeys and Hanuman langurs are also suggestive. However, this counterstrategy must be tested against the alternative hypothesis that transfer reduces infanticide risk by improving male protection (see below). More research is also needed to reject other explanations for smaller group sizes in folivores, such as (1) ecological constraints (i.e., lower predation risk [Oates 1977a, b] or greater within-group scramble competition for food [Snaith & Chapman 2005]) and (2) social constraints

(e.g., lower female fertility due to the stress-related effects of extra males [Dunbar 1987]). There is good evidence, however, against inbreeding avoidance being the reason for transfer of *parous* females in these taxa (Sterck & Korstjens 2000).

Finally, this counterstrategy has a potentially important implication. Female transfer to avoid infanticide risk potentially generates groups comprising a relatively small number of females of low relatedness. These social conditions may impede alliance formation among group females (sensu Wrangham 1980). Thus, a strategy to reduce infanticide risk through life in small groups may close off the alternative counterstrategy of female coalitionary defense to some extent. Further research is needed to determine whether this is a reason for the apparent rarity of effective anti-infanticide female coalitions in folivores.

Social Defense: Female-Male Interactions

In light of the apparent limits on maternal aggression, solitary evasion, and female coalitionary defense to reduce infanticide risk in many species, association with male defender(s) may constitute a potentially important counterstrategy in primates. Moreover, van Schaik & Kappeler's (1997) phylogenetic analysis revealed that the evolution of permanent female-male association in primates accompanied a shift in maternal strategy from "parking infants" (caching them in nests; see chapter 2, this volume) to carrying them continuously. The implication, they argue, is that the enduring relations between the sexes that so characterizes primates originated as male strategy to protect infants from infanticide, which is facilitated by the greater ease of maintaining contact with infants that are carried by mothers.

There are several possible manifestations of this counterstrategy depending in part upon the ecological conditions confronting females (fig. 19.8).

Dynamic (transfer-based) female gregariousness with male protector

Wrangham (1979) was among the first to argue that some primate groups are better understood not as cooperative aggregations of females for ecological advantage, but rather as associations of multiple females with a male protector against conspecific threat, especially infanticide. He was referring specifically to the mountain gorilla, whose biology offers several consistencies with this interpretation: (1) its relatively abundant and uniformly distributed foliar foods, which indicate no clear foraging-related benefits of female gregariousness (Watts 1985; Harcourt & Stewart 2007; Robbins 2008); (2) unimale groups organized around the social bonds females individually maintain with the silver-

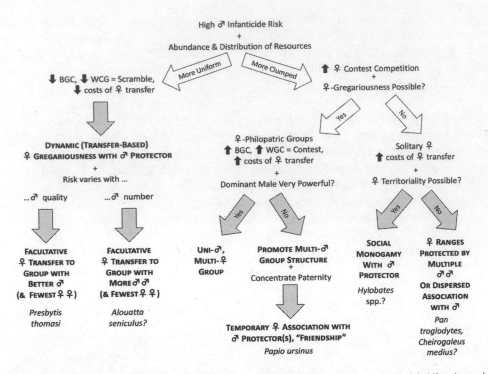

Fig. 19.8. Hypothetical mechanisms for the evolution of male-female association as an anti-infanticide counterstrategy. Each boldfaced entry following a gray arrow represents a counterstrategy that reduces the risk of male infanticide. One example taxon is listed for various versions of the counterstrategy. BCG is between-group competition; WGC is within-group competition. Based upon models by Wrangham (1980), van Schaik and Dunbar (1990), Janson and Goldsmith (1995), Sterck et al. (1997), and Pradhan and van Schaik (2008).

back (Harcourt 1979; Sicotte 1994; Watts 1994, 1996); (3) infanticide as a major source of infant mortality (table 19.3); and (4) a demography and life history which suggests that association with a male protector is much more likely to reduce infanticide risk than a chimpanzee-like strategy of promiscuity and solitary evasion (Harcourt & Greenberg 2001). Thus, there is a consensus that mountain gorilla groups reflect an adaptive female strategy to secure male protection; whether the strategy results from predation (Harcourt & Stewart 2007), from infanticide (Watts 1989), or from both is debated. This explanation has now been suggested for several other folivorous primates in which the costs of female gregariousness are relatively low (table 19.4). But male protection has also been invoked to explain gregariousness in comparatively more frugivorous taxa (Stokes 2004; Jack & Fedigan 2009).

Direct tests of the protective benefits of the resident male are difficult, and therefore rare. In the relevant nonprimate taxa, experimental removal of males has generally resulted in the documented rapid onset of infanticidal attacks by intruder males (reviewed in Palombit 2000). Some support derives from observed sequelae of natural death or disappearance of resident males, which included immediate attacks by extragroup males and the disappearance of one infant in Thomas's leaf monkeys (Steenbeek 1996), and

which is unambiguously the most common context of infanticide in gorillas (Yamagiwa et al. 2009). Sterck et al. (1997) argue that the invariable disintegration of mountain gorilla groups following loss of the silverback indicates greater anti-infanticide benefits of gregariousness than of predation avoidance. In an experimental test of male defense, Wich et al. (2002, 2004) observed that resident male Thomas's leaf monkeys responded more vigorously to audio playbacks of calls by extragroup males more likely to commit infanticide (i.e., members of all-male bands) than to the vocalizations of less threatening males (in bisexual groups).

Female transfer: A better male protector versus more male protectors

Natal dispersal and secondary transfer by females provide a mechanism for improving protection from infanticide through immigration into a group with a better male protector or more males (Sterck & Korstjens 2000). Secondary transfer in several species is virtually always by females that are not lactating or pregnant (Sterck 1997; Teichroeb et al. 2009). Although transfer by lactating females occurs relatively frequently in gorillas, it is usually after loss of the group silverback (Yamagiwa et al. 2009). Despite several potential costs of transfer, several researchers have concluded that relatively low ecological costs make it a viable option for females of more folivorous species (fig. 19.8; reviewed in Sterck & Korstjens 2000). Improvement of male protection is suggested by observations that transfer is triggered by unsuccessful reproduction (e.g., mountain gorillas, Harcourt et al. 1976; Robbins et al. 2009a), or by increased aggression toward mothers by potentially infanticidal extragroup males (e.g., Thomas's langurs, Sterck 1997).

Several studies have suggested that females transfer to males that are better protectors against infanticide (table 19.4). In Thomas's langurs, (1) females move from groups with older, long-tenure males to groups with younger adult males, and (2) females experience fewer potentially infanticidal attacks from extragroup males after transfer (Sterck 1997; Sterck et al. 2005). Sterck (1997) notes, however, that infant survival rates for transferring females are not clearly improved in the new groups, at least in the short term. Observations of mothers transferring to males that have killed their infants (Steenbeek 1999a; Yamagiwa et al. 2009) are also interpreted as supportive evidence, following the (as yet untested) logic of van Schaik (2000b) that the ability to commit infanticide varies in parallel with the ability to defend against it.

The potential benefits of transfer into multimale groups are highlighted by evidence of generally lower infanticide rates in groups with more males (Janson & van Schaik 2000). For example, the presence of additional males is associated with lower rates of takeover infanticide in ursine howler monkeys (Crockett & Janson 2000) and possibly in other *Alouatta* species (Treves 2001), which in turn are attributed to more vigorous resistance to male incursions (Crockett & Janson 2000; Rudran & Fernandez-Duque 2003; see also DeGusta & Milton 1998; Cristóbal-Azkarate et al. 2004). Kitchen (2004) demonstrated experimentally that playback of audio recordings of unfamiliar males' calls elicited stronger defensive responses from male Central American black howlers if vulnerable infants were in their group at the time, even if the playback simulated the presence of potentially more dangerous large cohorts of stranger males. In spite of the apparent protective benefits of males, however, current evidence suggests that transfer in ursine howler monkeys primarily targets groups with fewer females rather than those with more males. Crockett and Janson (2000) propose that the apparent absence of a "small group preference" among female mantled howlers (*Alouatta palliata*) may stem from an alternative strategy to emigrate to multimale groups, which are associated with greater female fertility (Ryan et al. 2008). This strategy may be more likely in this species than in ursine howlers, since multimale groups with up to six males are more common.

Harcourt & Stewart (2007) attribute the absence of in-

fanticide in the Virungas mountain gorillas since the 1980s to the fact that all the groups have been *multimale*, since longitudinal data reveal a clearly lower rate of infanticide in these usually bimale groups than in unimale units (Robbins et al. 2009b). There are two likely reasons: (1) the second male in each group actively counterattacks against infanticidal extragroup males (e.g., Watts 1989), and (2) the loss of one silverback does not automatically expose defenseless mothers to adult males with whom they have no history of copulation. As predicted, females in unimale groups are more likely to emigrate, but the puzzle is that female immigration is *not* biased toward multimale groups and most females (50%–60%) live in unimale groups in this population (Robbins et al. 2009c) as well as in the nearby Bwindi population (Robbins 2001).

In summary, current evidence establishes the potential importance of secondary transfer as infanticide avoidance, but more confirmatory data are needed, particularly on female reproduction (Sterck & Korstjens 2000). Contradictory patterns may reflect an evolutionary disequilibrium (Robbins et al. 2009c). They may also result from currently unclear trade-offs among the (possibly incompatible) anti-infanticide benefits of transfer to groups with (1) fewer females, (2) more powerful male protectors, or (3) more males. In such circumstances, transfer decisions may be influenced more by female group sizes. Transfer benefits unrelated to infanticide protection also need more careful testing. Even in more folivorous species, the quality or number of males in a group may improve female resource acquisition (Ryan et al. 2008), reduce predation risk (Harcourt & Stewart 2007; Breuer et al. 2009; Yamagiwa & Kahekwa 2001), or deter sexual coercion (Stokes 2004).

Association with male defenders in female philopatric groups

In gregarious primates with female philopatry, females may reduce infanticide risk by encouraging multimale group structure through mechanisms such as breeding synchrony, large group size, mating opportunities for subordinate males, or increased relatedness among immigrant males (Andelman 1986; Ridley 1986; Altmann 1990; Nunn 1999; Isbell et al. 2002; Pradhan & van Schaik 2008).

The predominance of multimale group structure has been attributed to this female strategy in some strepsirrhines in which infanticide is known—such as brown lemurs (*Eulemur fulvus*), in which male group size is the main predictor of takeover (Ostner & Kappeler 2004), and Verreaux's sifaka (Lewis 2008), in which multimale protection is argued to account for low frequencies of infanticide (Brockman et al. 2001). The aggressive prevention of intruder male takeover or attacks by new immigrant males has been ob-

served in several multimale species in which infanticide occurs (Brockman et al. 2001; Fruteau et al. 2010; Palombit
et al. 2000; Weingrill 2000; de Ruiter et al. 1994; Borries
1997). In preliminary support of the associated prediction
of paternity concentration, male defense in these species is
exhibited primarily or exclusively by putative sires or, in
rare cases where paternity data are available, by genetic fathers (de Ruiter et al. 1994; Borries et al. 1999a; Huchard
et al. 2010; Moscovice et al. in press).

In some multimale societies with infanticide (e.g., the
chacma baboon), this counterstrategy is based on cohesive associations or "friendships" that commonly develop
between lactating females and certain males (fig. 19.9; reviewed by Palombit 2009). These relationships are maintained more by females than male partners, most likely
because male pursuit of additional fertilizations and protection of current offspring are not easily compatible in
larger groups. Observations of defensive intervention in
potentially infanticidal attacks, as well as responses to audio playback experiments simulating attacks, reveal that a
female chacma baboon's male friend responds disproportionately more strongly to her vulnerability than do other
males, but only while she has an infant (Palombit et al.
1997). The importance of friendship for females is implied
by evidence that they compete for male friends, such that
high-ranking females acquire high-ranking males as friends
and actively displace subordinate females from friendships
with those males. As noted above, endocrine profiles illuminate another dimension of the relationship: compared to
their estrous counterparts, anestrous females maintain high
levels of stress hormones for longer periods following takeover by potentially infanticidal males, and these levels are
even higher for the relatively few mothers who lack a male
friend at the time (Beehner et al. 2005).

It is difficult to assess whether the protective consequences of multimale structure account for the low rates of
infanticide in species such as African baboons, macaques,
and vervets (Smuts & Smuts 1993; Isbell et al. 2002; Palombit 2003). Janson and van Schaik (2000) suggest that the
greater incidence of multimale grouping among more frugivorous primates is the key reason why their lengthier infant dependency periods do not, in fact, generate the predicted *increased* vulnerability to infanticide. This hypothesis
has not been thoroughly tested, but some observations are
suggestive. During periods when male dominance hierarchies were unstable in wild long-tailed macaques, immigration by new males and infant mortality were both elevated
(van Noordwijk & van Schaik 2001). Observed infant killing following the loss of three males, including the alpha
and beta males, led Yamada & Nakamichi (2006) to conclude that reduced male membership in Japanese macaque

Fig. 19.9. A "friendship" in chacma baboons (*Papio ursinus*): A lactating
female grooms her male friend as her infant plays nearby. In chacma baboons,
friendships are maintained primarily by the female partners; the male friend
rarely grooms the female. Photo courtesy of Ryne A. Palombit.

groups increases female vulnerability to infanticide. In his
study of chacma baboons, Cowlishaw (1999) recorded extremely high rates of takeover attempts targeting the only
one of his four study groups that was unimale, even though
that group contained the fewest females. Finally, hormonal
data reveal that testosterone responsiveness of male vervets
increased under conditions of heightened infanticide risk
(Whitten & Turner 2004). Indeed, there is accumulating
evidence from a variety of primates suggesting that males
maintain or facultatively adjust endocrine status in part to
enhance their ability to fend off infanticidal threat aggressively (table 19.6). This pattern has not been observed in
all species in which infanticide has been observed, however
(Gould 2007).

Association with male defenders in other systems

The possibility that this anti-infanticide counterstrategy
may also operate in nongregarious societies deserves further
study. For example, Kappeler (1997) invoked infanticide
avoidance to explain evidence for closer proximity among
western fat-tailed dwarf lemurs (*Cheirogaleus medius*) during the birth season than during the mating season. Similarly, the loose but consistent associations of some female
Sumatran orangutans (*Pongo abelii*) with "flanged" males
has been interpreted in part as an anti-infanticide strategy
(Delgado & van Schaik 2000; Mitra Setia & van Schaik
2007; but see Beaudrot et al. 2009).

One of the most intriguing versions of this hypothesis
is that social monogamy reflects heterosexual bonding
to counter infanticide risk (van Schaik & Dunbar 1990).
Field observations of infanticide in monogamous primates
are extremely rare, but one recent report is consistent with
this proposed function: a lactating female Milne-Edwards

Table 19.6.   Field studies suggesting hormonal bases of male defense against infanticide in wild primates.

Species	Results	Reference
Propithecus verreauxi	Testosterone levels were generally lower in males during breeding season, except for high-ranking males in unstable groups more prone to takeover and infanticide; these males experienced elevated testosterone. Glucocorticoid levels were unrelated to male age, size, rank, or group stability, but were significantly positively correlated with infant presence and number.	Brockman et al. 2001; Brockman et al. 2009
Eulemur fulvus rufus	Testosterone and glucocorticoid levels were unrelated to male rank, but increased significantly during the mating and were highest in the birth season during which infanticide risk is considered significant in the first postpartum month.	Ostner et al. 2002; Ostner et al. 2008
Alouatta palliata	Resident males in multimale groups showed elevated testosterone during periods when solitary, potential immigrant male(s) maintained proximity to the breeding group.	Cristóbal-Azkarate et al. 2006
Colobus vellerosus	Testosterone levels were unrelated to male rank and age, but increased significantly during periods of incursions by extragroup males (during which infanticide as well as copulations may occur), but not during periods of between-group encounters. An overturned adult male that remained in the group showed no increase in testosterone in the month following usurpation by an incoming male, but later showed significantly higher testosterone when the new male commenced attacks on infants and this former alpha male defended them.	Teichroeb & Sicotte 2008b
Chlorocebus pygerythrus	Testosterone responsiveness (1) was significantly positively correlated with infanticide risk (the number of females with unweaned infants in the group), and (2) tended to be higher for males in unimale groups than in multimale groups	Whitten & Turner 2004
Papio ursinus	During periods of male takeover and heightened infanticide risk, higher-ranking males had higher glucocorticoid levels, possibly related to increased need for infant defense as well as destabilized dominance hierarchy.	Bergman et al. 2005

sportive lemur (*Lepilemur edwardsi*) lost her pairmate and gained another one that subsequently killed her infant (Rasoloharijaona et al. 2000). Tests for pair-living strepsirrhines (sensu van Schaik & Kappeler 1993), however, have so far generally rejected the hypothesis (Schülke & Kappeler 2001; Thalmann 2001). The spectral tarsier (*Tarsius tarsier*) has provided partial support in that proximity of pair mates was significantly closer during lactation than during gestation but, somewhat paradoxically, mothers showed increased *avoidance* of their male mates at that time (Gursky 2002). Observations that male replacement in monogamous Azara's owl monkeys (*Aotus azarae*) may promote both paternal care and infant disappearance are thus difficult to interpret (Fernandez-Duque et al. 2008).

The hypothesis has yet to be adequately tested in the taxa that generated it, the gibbons. Arguments against it have included low opportunities for infanticide, the absence of takeover attempts under conditions predicting them, evidence of effective aggression toward males by solitary females without infants, and plausible, though also largely untested, alternative functions for social monogamy (Mitani 1990; Palombit 1999, 2000; Fuentes 2000; Brockelman 2009). It is only recently that intriguing indirect evidence in favor of the hypothesis has emerged from a 29-year record of white-handed gibbon (*Hylobates lar*) demography in Thailand. In the last five years of that period, male immigration rates increased substantially, generating more monogamous or polyandrous groups with new males. In three cases

when males immigrated into groups with an infants, the infants disappeared shortly after the males' arrival, but infant loss was low in other groups not experiencing male replacement (Borries et al. 2011). Although this sample is small and infant deaths were not directly observed, the results are consistent with the predictions of the anti-infanticide hypothesis, and they highlight the great importance of the further study of gibbons.

Conclusions

As more and more data accumulate from field and captive research, there is increasingly greater empirical support for previous arguments that infanticide by males is widespread in primates (Hrdy et al. 1995). Male infanticide thus constitutes a general sexual conflict threat for most female primates, but infanticide itself is not a unitary phenomenon. The sexual selection hypothesis is most supported as a general explanation for male infanticide, but recent studies implicate other adaptive functions under certain conditions and in some taxa. Tests of these hypothetical strategies are only now getting underway.

Likewise, recent research has provided compelling evidence that infanticide has been a selective force behind the evolution of female counteradaptations, of which there are many possible forms involving behavior, reproduction, and physiology. Further research is critically needed, however, along at least two avenues. First, virtually all of the proposed anti-infanticide counterstrategies have alternative explanations that invoke different selective forces (e.g., more than a dozen that involve the functions of sexual swellings) or even nonadaptive causes (e.g., longer follicular phases in species with swelling may simply reflect the longer time needed for the signal to reach maximum size). These competing hypotheses must be carefully tested, which is sometimes complicated by the fact that anti-infanticide hypotheses and alternatives based in feeding competition or predation risk often generate similar predictions (Janson & van Schaik 2000). Second, it seems likely that female counterstrategies may account for variation in rates of infanticide, but the nature and magnitude of this effect remains unclear at the moment. The two general counterstrategies that currently appear to have significant potential to reduce the risk of infanticide are female promiscuity and male protection. The latter highlights an important point: that the ubiquity of sexual conflict does not mean that conflict universally characterizes the *phenotypic* expression of male and female interaction. For example, sexual alliances based on variable degrees of heterosexual cohesion and cooperation appear to be widespread anti-infanticide counterstrategies. Thus, males may be a preeminent evolutionary problem for females, but in different contexts they are also part of the solution. It is perhaps too early to test Nunn and van Schaik's (2000) proposition that infanticide has been a more important selective agent than ecology in primate social evolution, but its significance is well established in a number of taxa.

References

Agoramoorthy, G. & Hsu, M. J. 2005. Occurrence of infanticide among wild proboscis monkeys (*Nasalis larvatus*) in Sabah, Northern Borneo. *Folia Primatologica*, 76, 177–179.

Agoramoorthy, G., Mohnot, S. M., Sommer, V. & Srivastava, A. 1988. Abortions in free-ranging Hanuman langurs (*Presbytis entellus*): A male induced strategy? *Human Evolution*, 3, 297–308.

Agoramoorthy, G. & Rudran, R. 1995. Infanticide by adult and subadult males in free-ranging red howler monkeys, *Alouatta seniculus*, in Venezuela. *Ethology*, 99, 75–88.

Agrell, J., Wolff, J. O. & Ylönen, H. 1998. Counter-strategies to infanticide in mammals: Costs and consequences. *Oikos*, 83, 507–517.

Alfred, J. R. B. & Sati, J. P. 1991. On the first record of infanticide in the hoolock gibbon *Hylobates hoolock* in the wild. *Records of the Zoological Survey of India*, 89, 319–321.

Altmann, J. 1990. Primate males go where the females are. *Animal Behaviour*, 39, 192–195.

Altmann, J., Sayialel, S., Bayes, M., Bruford, M. W. & Alberts, S. C. 2006. Cannibalism in baboons: Sexual selection versus hungry hunters. *International Journal of Primatology*, 27, Supplement 1, 315.

Andelman, S. J. 1986. Ecological and social determinants of cercopithecine mating patterns. In *Ecological Aspects of Social Evolution* (ed. by D. I. Rubenstein & R. W. Wrangham), 201–216. Princeton University Press: Princeton.

———. 1987. Evolution of concealed ovulation in vervet monkeys (*Cercopithecus aethiops*). *American Naturalist*, 129, 785–799.

Angst, W. & Thommen, D. 1977. New data and discussion of infant killing in Old World monkeys and apes. *Folia Primatologica*, 27, 198–229.

Arcadi, C. A. & Wrangham, R. W. 1999. Infanticide in chimpanzees: Review of cases and a new within-group observation from Kanyawara study group in Kibale National Park. *Primates*, 40, 337–351.

Arlet, M. E., Molleman, F. & Chapman, C. 2007. Indications for female mate choice in grey-cheeked mangabeys *Lophocebus albigena johnstoni*. *Acta Ethologica*, 10, 89–95.

Arnqvist, G. & Rowe, L. 2005. *Sexual Conflict*. Princeton, NJ: Princeton University Press.

Aureli, F., Schaffner, C. M., Boesch, C., Bearder, S. K., Call, J., Chapman, C. A., Connor, R. C., Di Fiore, A., Dunbar, R. I. M., Henzi, S. P., Holekamp, K. E., Korstjens, A. H., Layton, R., Lee, P., Lehmann, J., Manson, J. H., Ramos-Fernandez, G., Strier, K. B. & van Schaik, C. P. 2008. Fission-fusion dynamics. *Current Anthropology*, 49, 627–654.

Bakuneeta, C., Inagaki, H. & Reynolds, V. 1993. Identification of wild chimpanzee hair samples from feces by electron microscopy. *Primates*, 34, 233–235.

Barrett, L., Halliday, J. O. & Henzi, S. P. 2006. The ecology of motherhood: The structuring of lactation costs by chacma baboons. *Journal of Animal Ecology*, 75, 875–886.

Bartlett, T. Q., Sussman, R. W. & Cheverud, J. M. 1993. Infant killing in primates: A review of observed cases with specific reference to the sexual selection hypothesis. *American Anthropologist*, 95, 958–990.

Beaudrot, L., Kahlenberg, S. & Marshall, A. 2009. Why male orangutans do not kill infants. *Behavioral Ecology and Sociobiology*, 63, 1549–1562.

Beehner, J. C. & Bergman, T. J. 2008. Infant mortality following male takeovers in wild geladas. *American Journal of Primatology*, 70, 1152–1159.

Beehner, J. S., Bergman, J., Cheney, D. L., Seyfarth, R. M. & Whitten, P. L. 2005. The effect of new alpha males on female stress in free-ranging baboons. *Animal Behaviour*, 69, 1211–1221.

Ben-David, M., Titus, K. & Beier, L. 2004. Consumption of salmon by Alaskan brown bears: A trade-off between nutritional requirements and the risk of infanticide? *Oecologia*, 138, 465–474.

Bercovitch, F. B. 1995. Female cooperation, consortship maintenance, and male mating success in savanna baboons. *Animal Behaviour*, 50, 137–149.

Bergman, T. J., Beehner, J. C., Cheney, D. L., Seyfarth, R. M. & Whitten, P. L. 2005. Correlates of stress in free-ranging male chacma baboons, *Papio hamadryas ursinus*. *Animal Behaviour*, 70, 703–713.

Berman, C., Li, J., Ogawa, H., Ionica, C. & Yin, H. 2007. Primate tourism, range restriction, and infant risk among *Ma-*

caca thibetana at Mt. Huangshan, China. *International Journal of Primatology*, 28, 1123–1141.

Bertram, B. C. 1975. Social factors influencing reproduction in wild lions. *Journal of Zoology*, 177, 463–482.

Bielert, C. 1986. Sexual interactions between captive adult male and female chacma baboons (*Papio ursinus*) as related to the female's menstrual cycle. *Journal of Zoology*, 209, 521–536.

Böer, M. & Sommer, V. 1992. Evidence for sexually selected infanticide in captive *Cercopithecus mitis*, *Cercocebus torquatus*, and *Mandrillus leucophaeus*. *Primates*, 33, 557–563.

Boesch, C., Crockford, C., Herbinger, I., Wittig, R., Moebius, Y. & Normand, E. 2008. Intergroup conflicts among chimpanzees in Taï National Park: Lethal violence and the female perspective. *American Journal of Primatology*, 70, 519–532.

Boggess, J. E. 1979. Troop male membership changes and infant killing in langurs (*Presbytis entellus*). *Folia Primatologica*, 32, 65–107.

———. 1984. Infant killing and male reproductive strategies in langurs (*Presbytis entellus*). In *Infanticide: Comparative and Evolutionary Perspectives* (ed. by G. Hausfater & S. B. Hrdy), 283–310. New York: Aldine.

Borries, C. 1997. Infanticide in seasonally breeding multimale groups of Hanuman langurs (*Presbytis entellus*) in Ramnagar (South Nepal). *Behavioral Ecology and Sociobiology*, 41, 139–150.

Borries, C. & Koenig, A. 2000. Infanticide in hanuman langurs: Social organization, male migration, and weaning age. In *Infanticide by Males and Its Implications* (ed. by C. P. van Schaik & C. H. Janson), 99–122. Cambridge: Cambridge University Press.

Borries, C., Launhardt, K., Epplen, C., Epplen, J. T. & Winkler, P. 1999a. DNA analyses support the hypothesis that infanticide is adaptive in langur monkeys. *Proceedings of the Royal Society of London Series B Biological Sciences*, 266, 901–904.

Borries, C., Savini, T. & Koenig, A. 2011. Social monogamy and the threat of infanticide in larger mammals. *Behavioral Ecology and Sociobiology*, 65, 683–685.

Brent, L. J. N., Teichroeb, J. A. & Sicotte, P. 2008. Preliminary assessment of natal attraction and infant handling in wild *Colobus vellerosus*. *American Journal of Primatology*, 70, 101–105.

Brereton, A. R. 1995. Coercion-defense hypothesis: The evolution of primate sociality. *Folia Primatologica*, 64, 207–214.

Breuer, T., Robbins, A., Olejniczak, C., Parnell, R., Stokes, E. & Robbins, M. 2009. Variance in the male reproductive success of western gorillas: Acquiring females is just the beginning. *Behavioral Ecology and Sociobiology*, 64, 515–528.

Brockelman, W. Y. 2009. Ecology and the social system of gibbons. In *The Gibbons* (ed. by S. Lappan & D. Whittaker), 211–239. New York: Springer.

Brockett, R. C., Horwich, R. H. & Jones, C. B. 1999. Disappearance of infants following male takeovers in the Belizean black howler monkey (*Alouatta pigra*). *Neotropical Primates*, 7, 86–88.

Brockman, D. K. & Whitten, P. L. 1996. Reproduction in free-ranging *Propithecus verreauxi*: Estrus and the relationship between multiple partner matings and fertilization. *American Journal of Physical Anthropology*, 100, 57–69.

Brockman, D. K., Whitten, P. L., Richard, A. F. & Benander, B.

2001. Birth season testosterone levels in male Verreaux's sifaka, *Propithecus verreauxi*: Insights into socio-demographic factors mediating seasonal testicular function. *Behavioral Ecology and Sociobiology*, 49, 117–127.

Brockman, D. K., Cobden, A. K. & Whitten, P. L. 2009. Birth season glucocorticoids are related to the presence of infants in sifaka (*Propithecus verreauxi*). *Proceedings of the Royal Society of London Series B Biological Sciences*, 276, 1855–1863.

Broom, M., Borries, C. & Koenig, A. 2004. Infanticide and infant defence by males: Modelling the conditions in primate multimale groups. *Journal of Theoretical Biology*, 231, 261–270.

Buettner-Janusch, J. 1964. The breeding of galagos in captivity and some notes on their behavior. *Folia Primatologica*, 2, 93–110.

Busse, C. D. 1984. Tail raising by baboon mothers toward immigrant males. *American Journal of Physical Anthropology*, 64, 255–262.

Busse, C. D. & Gordon, T. P. 1983. Attacks on neonates by a male mangabey (*Cercocebus atys*). *American Journal of Primatology*, 5, 345–356.

Butynski, T. M. 1982. Harem-male replacement and infanticide in the blue monkey (*Cercopithecus mitis stuhlmanni*) in the Kibale Forest, Uganda. *American Journal of Primatology*, 3, 1–22.

———. 1990. Comparative ecology of blue monkeys (*Cercopithecus mitis*) in high and low density subpopulations. *Ecological Monographs*, 60, 1–26.

Bygott, J. D. 1972. Cannibalism among wild chimpanzees. *Nature*, 238, 410–411.

Calegaro-Marques, C. & Bicca-Marques, J. C. 1996. Emigration in a black howling monkey group. *International Journal of Primatology*, 17, 229–237.

Camperio-Ciani, A. 1984. A case of infanticide in a free-ranging group of rhesus monkeys (*Macaca mulatta*) in the Jackoo forest, Simla, India. *Primates*, 25, 373–377.

Carlson, A. A. & Isbell, L. A. 2001. Causes and consequences of single-male and multimale mating in free-ranging patas monkeys, *Erythrocebus patas*. *Animal Behaviour*, 62, 1047–1058.

Carnegie, S. D., Fedigan, L. M. & Ziegler, T. E. 2006. Post-conceptive mating in white-faced capuchins, *Cebus capucinus*: Hormonal and sociosexual patterns of cycling, noncycling, and pregnant females. In *New Perspectives in the Study of Mesoamerican Primates* (ed. by A. Estrada, P. A. Garber, M. Pavelka & L. Luecke), 387–409. New York: Springer.

Chapman, C. A. & Pavelka, M. S. M. 2005. Group size in folivorous primates: Ecological constraints and the possible influence of social factors. *Primates*, 46, 1–9.

Cheney, D. L., Seyfarth, R. M., Fischer, J., Beehner, J., Bergman, T., Johnson, S. E., Kitchen, D. M., Palombit, R. A., Rendall, D. & Silk, J. B. 2004. Factors affecting reproduction and mortality among baboons in the Okavango Delta, Botswana. *International Journal of Primatology*, 25, 401–428.

Clarke, A. S., Harvey, N. C. & Lindburg, D. G. 1993. Extended postpregnancy estrous cycles in female lion-tailed macaques. *American Journal of Primatology*, 31, 275–285.

Clarke, M. R. 1983. Infant-killing and infant disappearance following male takeovers in a group of free-ranging howling monkeys (*Alouatta palliata*) in Costa Rica. *American Journal of Primatology*, 5, 241–247.

Clarke, M. R. & Glander, K. E. 1984. Female reproductive success in a group of free-ranging howling monkeys (*Alouatta palliata*) in Costa Rica. In *Female Primates: Studies by Women Primatologists* (ed. by M. F. Small), 111–126. New York: Liss.

Clarke, M. R., Zucker, E. L. & Glander, K. E. 1994. Group take-over by a natal male howling monkey (*Alouatta palliata*) and associated disappearance and injuries of immatures. *Primates*, 35, 435–442.

Clarke, M. R., Blanchard, J. L. & Snyder, J. A. 2001. Infant-killing in pigtailed monkeys: A colony management concern. *Laboratory Primate Newsletter*, 34, 2–5.

Clarke, P. M. R., Pradhan, G. R. & van Schaik, C. P. 2009a. Intersexual conflict in primates: Infanticide, paternity allocation, and the role of coercion. In *Sexual Coercion in Primates and Humans* (ed. by M. N. Muller & R. W. Wrangham), 42–77. Cambridge, MA: Harvard University Press.

Clarke, P. M. R., Barrett, L. & Henzi, S. P. 2009b. What role do olfactory cues play in chacma baboon mating? *American Journal of Primatology*, 71, 493–502.

Clarke, P. M. R., Henzi, S. P. & Barrett, L. 2009c. Sexual conflict in chacma baboons, *Papio hamadryas ursinus*: Absent males select for proactive females. *Animal Behaviour*, 77, 1217–1225.

Collins, D. A., Busse, C. D. & Goodall, J. 1984. Infanticide in two populations of savanna baboons. In *Infanticide: Comparative and Evolutionary Perspectives* (ed. by G. Hausfater & S. B. Hrdy), 193–215. New York: Aldine.

Colmenares, F. & Gomendio, M. 1988. Changes in female reproductive condition following male take-overs in a colony of hamadryas and hybrid baboons. *Folia Primatologica*, 50, 157–174.

Cordero, C. & Eberhard, W. G. 2003. Female choice of sexually antagonistic male adaptations: A critical review of some current research. *Journal of Evolutionary Biology*, 16, 1–6.

Cords, M. 1984. Mating patterns and social structure in redtail monkeys (*Cercopithecus ascanius*). *Zeitschrift für Tierpsychologie*, 64, 313–329.

———. 1986. Promiscuous mating among blue monkeys in the Kakamega Forest, Kenya. *Ethology*, 72, 214–226.

———. 2000. The number of males in guenon groups. In *Primate Males: Causes and Consequences of Variation in Group Composition* (ed. by P. M. Kappeler), 84–96. Cambridge: Cambridge University Press.

———. 2002. Friendship among adult female blue monkeys. *Behaviour*, 139, 291–314.

Cords, M. & Fuller, J. L. 2010. Infanticide in *Cercopithecus mitis stuhlmanni* in the Kakamega Forest, Kenya: Variation in the occurrence of an adaptive behavior. *International Journal of Primatology*, 31, 409–431.

Cowlishaw, G. 1999. Ecological and social determinants of spacing behaviour in desert baboon groups. *Behavioral Ecology and Sociobiology*, 45, 67–77.

Cristóbal-Azkarate, J., Dias, P. & Veà, J. 2004. Causes of intraspecific aggression in *Alouatta palliata mexicana*: Evidence from injuries, demography, and habitat. *International Journal of Primatology*, 25, 939–953.

Cristóbal-Azkarate, J., Chavira, R., Boeck, L., Rodríguez-Luna, E. & Veà, J. J. 2006. Testosterone levels of free-ranging resident mantled howler monkey males in relation to the number and density of solitary males: A test of the challenge hypothesis. *Hormones and Behavior*, 49, 261–267.

Crockett, C. M. 1984. Emigration by female red howler monkeys and the case for female competition. In *Female Primates: Studies by Women Primatologists* (ed. by M. F. Small), 159–173. New York: Liss.

———. 2003. Re-evaluating the sexual selection hypothesis for infanticide by *Alouatta* males. In *Sexual Selection and Reproductive Competition in Primates: New Perspectives and Directions* (ed. by C. B. Jones), 327–365. Norman, OK: American Society of Primatologists.

Crockett, C. M. & Janson, C. H. 2000. Infanticide in red howlers: Female group size, male membership, and a possible link to folivory. In *Infanticide by Males and Its Implications* (ed. by C. P. van Schaik & C. H. Janson), 75–98. Cambridge: Cambridge University Press.

Crockett, C. M. & Pope, T. P. 1988. Inferring patterns of aggression from red howler monkey injuries. *American Journal of Primatology*, 15, 289–308.

Crofoot, M. C. 2007. Mating and feeding competition in white-faced capuchins (*Cebus capucinus*): The importance of short- and long-term strategies. *Behaviour*, 144, 1473–1495.

Curtin, R. & Dolhinow, P. 1978. Primate social behavior in a changing world. *American Scientist*, 66, 468–475.

Daly, M. & Wilson, M. 1984. A sociobiological analysis of human infanticide. In *Infanticide: Comparative and Evolutionary Perspectives* (ed. by G. Hausfater & S. B. Hrdy), 487–502. New York: Aldine.

———. 1994. Stepparenthood and the evolved psychology of discriminative parental solicitude. In *Infanticide and Parental Care* (ed. by S. Parmigiani & F. vom Saal), 121–134. Chur, Switzerland: Harwood Academic Publishing.

Davies, G. 1987. Adult male replacement and group formation in *Presbytis rubicunda*. *Folia Primatologica*, 49, 111–114.

Dawkins, R. & Krebs, J. R. 1979. Arms races between and within species. *Proceedings of the Royal Society of London Series B Biological Sciences*, 205, 489–511.

DeGusta, D. & Milton, K. 1998. Skeletal pathologies in a population of *Alouatta palliata*: Behavioral, ecological, and evolutionary implications. *International Journal of Primatology*, 19, 615–650.

Delgado, R. A. & van Schaik, C. P. 2000. The behavioral ecology and conservation of the orangutan (*Pongo pygmaeus*): A tale of two islands. *Evolutionary Anthropology*, 9, 201–218.

De Ruiter, J. R., van Hooff, J. A. R. A. M. & Scheffrahn, W. 1994. Social and genetic aspects of paternity in wild long-tailed macaques (*Macaca fascicularis*). *Behaviour*, 129, 203–224.

Deschner, T., Heistermann, M., Hodges, K. & Boesch, C. 2004. Female sexual swelling size, timing of ovulation, and male behavior in wild west African chimpanzees. *Hormones and Behavior*, 46, 204–215.

De Waal, F. B. M. 1997. *Bonobo: The Forgotten Ape*. Berkeley: University of California Press.

Dittus, W. P. J. 1988. Group fission among wild toque macaques as a consequence of female resource competition and environmental stress. *Animal Behaviour*, 36, 1626–1645.

Dolhinow, P. 1977. Normal monkeys? *American Scientist*, 65, 266.

Doran-Sheehy, D. M., Fernández, D. & Borries, C. 2009. The strategic use of sex in wild female western gorillas. *American Journal of Primatology*, 71, 1011–1020.

Drickamer, L. C. 2007. Acceleration and delay of reproduction in rodents. In *Rodent Societies: An Ecological and Evolutionary Perspective* (ed. by J. O. Wolff & P. A. Sherman), 106–114. Chicago: University of Chicago Press.

Dunbar, R. I. M. 1987. Habitat quality, population dynamics, and group composition in colobus monkeys (*Colobus guereza*). *International Journal of Primatology*, 8, 299–329.

Eberhard, W. G. 2005. Evolutionary conflicts of interest: Are female sexual decisions different? *American Naturalist*, 165, S19–S25.

Emery Thompson, M., Newton-Fisher, N. & Reynolds, V. 2006. Probable community transfer of parous adult female chimpanzees in the Budongo Forest, Uganda. *International Journal of Primatology*, 27, 1601–1617.

Engh, A. L., Beehner, J. C., Bergman, T. J., Whitten, P. L., Hoffmeier, R. R., Seyfarth, R. M. & Cheney, D. L. 2006. Female hierarchy instability, male immigration and infanticide increase glucocorticoid levels in female chacma baboons. *Animal Behaviour*, 71, 1227–1237.

Enstam, K. L., Isbell, L. A. & de Maar, T. W. 2002. Male demography, female mating behavior, and infanticide in wild patas monkeys (*Erythrocebus patas*). *International Journal of Primatology*, 23, 85–104.

Erhart, E. M. & Overdorff, D. J. 1998. Infanticide in *Propithecus diadema edwardsi*: An evaluation of the sexual selection hypothesis. *International Journal of Primatology*, 19, 73–81.

Estrada, A. & Garber, P. A. 2008. Comparative perspectives in the study of South American primates: Research priorities and conservation implications. In *South American Primates: Comparative Perspectives in the Study of Behavior, Ecology, and Conservation* (ed. by P. A. Garber, A. Estrada, J. C. Bicca-Marques, E. W. Heymann & K. B. Strier). Springer: New York.

Fairbanks, L. A. & McGuire, M. T. 1987. Mother-infant relationships in vervet monkeys: Response to new adult males. *International Journal of Primatology*, 8, 351–366.

———. 1993. Maternal protectiveness and response to the unfamiliar in vervet monkeys. *American Journal of Primatology*, 30, 119–129.

Fairgrieve, C. 1995. Infanticide and infant eating in the blue monkey (*Cercopithecus mitis stuhlmanni*) in the Budongo Forest Reserve, Uganda. *Folia Primatologica*, 64, 69–72.

Fashing, P. J. 2001. Male and female strategies during intergroup encounters in guerezas (*Colobus guereza*): Evidence for resource defense mediated through males and a comparison with other primates. *Behavioral Ecology and Sociobiology*, 50, 219–230.

Fedigan, L. M. 2003. Impact of male takeovers on infant deaths, births and conceptions in *Cebus capucinus* at Santa Rosa, Costa Rica. *International Journal of Primatology*, 24, 723–741.

Fedigan, L. M. & Jack, K. M. 2004. The demographic and reproductive context of male replacements in *Cebus capucinus*. *Ethology*, 141, 755–775.

Fedigan, L. M., Carnegie, S. D. & Jack, K. M. 2008. Predictors of reproductive success in female white-faced capuchins (*Cebus capucinus*). *American Journal of Physical Anthropology*, 137, 82–90.

Fernandez-Duque, E., Juárez, C. P. & di Fiore, A. 2008. Adult male replacement and subsequent infant care by male and siblings in socially monogamous owl monkeys (*Aotus azarai*). *Primates*, 49, 81–84.

Fruteau, C., Range, F. & Noë, R. 2010. Infanticide risk and infant defence in multi-male free-ranging sooty mangabeys, *Cercocebus atys*. *Behavioural Processes*, 83, 113–118.

Fuentes, A. 2000. Hylobatid communities: Changing views on pair bonding and social organization in Hominoids. *Yearbook of Physical Anthropology*, 43, 33–60.

Galat-Luong, A. & Galat, G. 1979. Conséquences comportementales des perturbations sociales repetées sure une troupe de Mones de Lowe, *Cercopithecus campbelli lowei*, de Cote d'Ivoire. *Terre et Vie*, 33, 4–57.

Galetti, M., Pedroni, F. & Paschoal, M. 1994. Infanticide in the brown howler monkey, *Alouatta fusca*. *Neotropical Primates*, 2, 6–7.

Gammie, S. C., Auger, A. P., Jessen, H. M., Vanzo, R. J., Awad, T. A. & Stevenson, S. A. 2007. Altered gene expression in mice selected for high maternal aggression. *Genes, Brain and Behavior*, 6, 432–443.

Garcia, C., Shimizu, K. & Huffman, M. 2009. Relationship between sexual interactions and the timing of the fertile phase in captive female Japanese macaques (*Macaca fuscata*). *American Journal of Primatology*, 71, 1–12.

Gibson, K. N., Vick, L. G., Palma, A. C., Carrasco, F. M., Taub, D. & Ramos-Fernández, G. 2008. Intra-community infanticide and forced copulation in spider monkeys: A multi-site comparison between Cocha Cashu, Peru and Punta Laguna, Mexico. *American Journal of Primatology*, 70, 485–489.

Glander, K. E. 1980. Reproduction and population growth in free-ranging mantled howling monkeys. *American Journal of Physical Anthropology*, 53, 25–36.

———. 1992. Dispersal patterns in Costa Rican mantled howling monkeys. *International Journal of Primatology*, 13, 415–436.

Gomendio, M. & Colmenares, F. 1989. Infant killing and infant adoption following the introduction of new males to an all-female colony of baboons. *Ethology*, 80, 223–244.

Goodall, J. 1986. *The Chimpanzees of Gombe: Patterns of Behavior*. Cambridge, MA: Belknap Press.

———. 1990. *Through a Window*. Boston: Houghton Mifflin.

Gordon, T. P. & Busse, C. D. 1984. Perineal swelling in female mangabeys during pregnancy. *American Journal of Primatology*, 6, 404–405.

Gordon, T. P., Gust, D. A., Busse, C. D. & Wilson, M. E. 1991. Hormones and sexual behavior associated with postconception perineal swelling in the sooty mangabey (*Cercocebus torquatus atys*). *International Journal of Primatology*, 12, 585–597.

Gould, L. 1996. Vigilance behavior during the birth and lactation season in naturally occurring ring-tailed lemurs (*Lemur catta*) at the Beza-Mahafaly Reserve, Madagascar. *International Journal of Primatology*, 17, 331–347.

Guo, S., Ji, W., Li, M., Chang, H. & Li, B. 2010. The mating system of the Sichuan snub-nosed monkey (*Rhinopithecus roxellana*). *American Journal of Primatology*, 72, 25–32.

Gursky, S. L. 2002. Determinants of gregariousness in the spectral tarsier (Prosimian: *Tarsius spectrum*). *Journal of Zoology, London*, 256, 401–410.

Gust, D. 1994. Alpha-male sooty mangabeys differentiate between females' fertile and their postconception maximal swellings. *International Journal of Primatology*, 15, 289–301.

Gust, D. A. 1994. A brief report on the social behavior of the crested mangabey (*Cercocebus galeritus galeritus*) with a comparison to the sooty mangabey (*C. torquatus atys*). *Primates*, 35, 375–383.

Gust, D. A., Gordon, T. P. & Gergits, W. F. 1995. Proximity at birth relates to a sire's tolerance of his offspring among sooty mangabeys. *Animal Behaviour*, 49, 1403–1405.

Hadidian, J. & Bernstein, I. 1979. Female reproductive cycles and birth data from an Old World monkey colony. *Primates*, 20, 429–442.

Hamai, M., Nishida, T., Takasaki, H. & Turner, L. A. 1992. New records of within-group infanticide in wild chimpanzees. *Primates*, 33, 151–162.

Hamilton, W. D. 1971. Geometry for the selfish herd. *Journal of Theoretical Biology*, 31, 295–311.

Harcourt, A. H. 1979. Social relationships between adult male and female mountain gorillas in the wild. *Animal Behaviour*, 27, 325–342.

Harcourt, A. H. & Greenberg, J. 2001. Do gorilla females join males to avoid infanticide? A quantitative model. *Animal Behaviour*, 62, 905–915.

Harcourt, A. H. & Stewart, K. J. 2007. *Gorilla Society: Conflict, Compromise, and Cooperation between the Sexes*. Chicago: University of Chicago Press.

Harcourt, A. H., Stewart, K. J. & Fossey, D. 1976. Male emigration and female transfer in wild mountain gorillas. *Nature*, 263, 226–227.

Harris, T. R. & Monfort, S. L. 2003. Behavioral and endocrine dynamics associated with infanticide in a black and white colobus monkey (*Colobus guereza*). *American Journal of Primatology*, 61, 135–142.

Hasegawa, T. 1989. Sexual behavior of immigrant and resident female chimpanzees at Mahale. In *Understanding Chimpanzees* (ed. by P. G. Heltne & L. A. Marquardt), 99–103. Cambridge, MA: Harvard University Press.

Heistermann, M., Ziegler, T. E., van Schaik, C. P., Launhardt, K., Winkler, P. & Hodges, J. K. 2001. Loss of oestrus, concealed ovulation and paternity confusion in free-ranging Hanuman langurs. *Proceedings of the Royal Society of London Series B Biological Sciences*, 268, 2445–2451.

Henzi, S. P. & Barrett, L. 2003. Evolutionary ecology, sexual conflict, and behavioral differentiation among baboon populations. *Evolutionary Anthropology*, 12, 217–230.

Higham, J. P., MacLarnon, A. M., Ross, C., Heistermann, M. & Semple, S. 2008. Baboon sexual swellings: Information content of size and color. *Hormones and Behavior*, 53, 452–462.

Hiraiwa-Hasegawa, M. 1987. Infanticide in primates and a possible case of male-biased infanticide in chimpanzees. In *Animal Societies: Theories and Facts* (ed. by J. L. Brown & J. Kikkawa), 125–139. Tokyo: Japan Scientific Societies Press.

———. 1988. Adaptive significance of infanticide in primates. *Trends in Ecology and Evolution*, 3, 102–105.

Hiraiwa-Hasegawa, M. & Hasegawa, T. 1994. Infanticide in nonhuman primates: Sexual selection and local resource competition. In *Infanticide and Parental Care* (ed. by S. Parmigiani & F. vom Saal), 137–154. Chur, Switzerland: Harwood Academic Publishing.

Horrocks, J. A. 1986. Life-history characteristics of a wild population of vervet monkeys (*Cercopithecus aethiops*) in Barbados, West Indies. *International Journal of Primatology*, 7, 31–47.

Horrocks, J. A. & Baulu, J. 1988. Effects of trapping on the ver-

vet (*Cercopithecus aethiops*) in Barbados. *American Journal of Primatology*, 15, 223–233.

Hrdy, S. B. 1974. Male-male competition and infanticide among the langurs (*Presbytis entellus*) of Abu, Rajasthan. *Folia Primatologica*, 22, 19–58.

———. 1977. *The Langurs of Abu: Female and Male Strategies of Reproduction*. Cambridge, MA: Harvard University Press.

———. 1979. Infanticide among animals: A review, classification, and examination of the implications for reproductive strategies of females. *Ethology and Sociobiology*, 1, 13–40.

———. 1981. *The Woman That Never Evolved*. Cambridge, MA: Harvard University Press.

Hrdy, S. B. & Hausfater, G. 1984. Comparative and evolutionary perspectives on infanticide: Introduction and overview. In *Infanticide: Comparative and Evolutionary Perspectives* (ed. by G. Hausfater & S. B. Hrdy), xiii–xxxv. New York: Aldine.

Hrdy, S. B., Janson, C. H. & van Schaik, C. P. 1995. Infanticide: Let's not throw out the baby with the bath water. *Evolutionary Anthropology*, 3, 151–154.

Hsu, M. J., Lin, J. F. & Agoramoorthy, G. 2006. Effects of group size of birth rate, infant mortality and social interactions in Formosan macaques at Mt. Longevity, Taiwan. *Ethology, Ecology, and Evolution*, 18, 3–17.

Huchard, E., Benavides, J., Setchell, J., Charpentier, M., Alvergne, A., King, A., Knapp, L., Cowlishaw, G. & Raymond, M. 2009. Studying shape in sexual signals: The case of primate sexual swellings. *Behavioral Ecology and Sociobiology*, 63, 1231–1242.

Huchard, E., Alvergne, A., Féjan, D., Knapp, L., Cowlishaw, G. & Raymond, M. 2010. More than friends? Behavioural and genetic aspects of heterosexual associations in wild chacma baboons. *Behavioral Ecology and Sociobiology*, 64, 769–781.

Huck, U. W. 1984. Infanticide and the evolution of pregnancy block in rodents. In *Infanticide: Comparative and Evolutionary Perspectives* (ed. by G. Hausfater & S. B. Hrdy), 349–365. New York: Aldine.

Ichino, S. 2005. Attacks on a wild infant ring-tailed lemur (*Lemur catta*) by immigrant males at Berenty, Madagascar: Interpreting infanticide by males. *American Journal of Primatology*, 67, 267–272.

Isbell, L. A., Cheney, D. L. & Seyfarth, R. D. 2002. Why vervet monkeys (*Cercopithecus aethiops*) live in multimale groups. In *The Guenons: Diversity and Adaptation in African Monkeys* (ed. by M. E. Glenn & M. Cords), 173–187. New York: Kluwer Academic.

Izawa, K. 1997. Social changes within a group of red howler monkeys (*Alouatta seniculus*). VI. *Field Studies of New World Monkeys, La Macarena, Colombia*, 11, 1–6.

Izawa, K. & Lozano, M. H. 1994. Social changes within a group of red howler monkeys (*Alouatta seniculus*). V. *Field Studies of New World Monkeys, La Macarena, Colombia*, 9, 33–39.

Jack, K. M. & Fedigan, L. M. 2009. Female dispersal in a female-philopatric species, *Cebus capucinus*. *Behaviour*, 146, 471–497.

Janson, C. H. & Goldsmith, M. L. 1995. Predicting group size in primates: Foraging costs and predation risks. *Behavioral Ecology*, 6, 326–336.

Janson, C. H. & van Schaik, C. P. 2000. The behavioral ecology of infanticide by males. In *Infanticide by Males and Its Implications* (ed. by C. P. van Schaik & C. H. Janson), 469–494. Cambridge: Cambridge University Press.

Jolly, A., Caless, A., Cavigelli, S., Gould, L., Pereira, M. E., Pitts, A., Pride, R. E., Rabenandrasana, H. D., Walker, J. D. & Zafison, T. 2000. Infant killing, wounding and predation in *Eulemur* and *Lemur*. *International Journal of Primatology*, 21, 21–40.

Kappeler, P. & Port, M. 2008. Mutual tolerance or reproductive competition? Patterns of reproductive skew among male redfronted lemurs (*Eulemur fulvus rufus*). *Behavioral Ecology and Sociobiology*, 62, 1477–1488.

Kappeler, P. M. 1997. Determinants of primate social organization: Comparative evidence and new insights from Malagasy lemurs. *Biological Reviews of the Cambridge Philosophical Society*, 72, 111–151.

Kappeler, P. M. & Schäffler, L. 2008. The lemur syndrome unresolved: Extreme male reproductive skew in sifakas (*Propithecus verreauxi*), a sexually monomorphic primate with female dominance. *Behavioral Ecology and Sociobiology*, 62, 1007–1015.

Kaumanns, W., Rohrhuber, B. & Zinner, D. 1989. Reproductive parameters in a newly established colony of hamadryas baboons (*Papio hamadryas*). *Primate Report*, 24, 25–33.

Kawanaka, K. 1981. Infanticide and cannibalism in chimpanzees, with special reference to the newly observed case in the Mahale mountains. *Africa Studies Monographs*, 1, 69–91.

Kinnaird, M. 1990. Behavioral and demographic responses to habitat change by the Tana River crested mangabey (Cercocebus galeritus galeritus). PhD dissertation, University of Florida.

Kinnaird, M. F. 1990. Pregnancy, gestation and parturition in free-ranging Tana River crested mangabeys (*Cercocebus galeritus galeritus*). *American Journal of Primatology*, 22, 285–289.

Kitchen, D. M., Beehner, J. C., Bergman, T. J., Cheney, D. L., Engh, A., Palombit, R. A. & Seyfarth, R. M. 2006. Patterns of male aggression towards females among chacma baboons of Botswana (*Papio hamadryas ursinus*). *International Journal of Primatology*, 27, Supplement 1, 78.

Kitchen, D. M., Beehner, J. C., Bergman, T. J., Cheney, D. L., Crockford, C., Engh, A., Fischer, J., Seyfarth, R. M. & Wittig, R. 2009. The causes and consequences of male aggression directed at female chacma baboons. In *Sexual Coercion in Primates and Humans* (ed. by M. N. Muller & R. W. Wrangham), 129–156. Cambridge, MA: Harvard University Press.

Knopff, K. H. & Pavelka, M. S. M. 2006. Feeding competition and group size in *Alouatta pigra*. *International Journal of Primatology*, 27, 1059–1078.

Knopff, K. H., Knopff, A. R. A. & Pavelka, M. S. M. 2004. Observed case of infanticide committed by a resident male Central American black howler monkey (*Alouatta pigra*). *American Journal of Primatology*, 63, 239–244.

Koenig, A. & Borries, C. 2001. Feeding competition and infanticide constrain group size in wild Hanuman langurs. *American Journal of Primatology*, 57, 33–34.

Korstjens, A., Nijssen, E. & Noë, R. 2005. Intergroup relationships in western black-and-white colobus, *Colobus polykomos polykomos*. *International Journal of Primatology*, 26, 1267–1289.

Kutsukake, N. 2006. The context and quality of social relationships affect vigilance behaviour in wild chimpanzees. *Ethology*, 112, 581–591.

Kutsukake, N. & Nunn, C. L. 2006. Comparative tests of reproductive skew in male primates: The roles of demographic factors and incomplete control. *Behavioral Ecology and Sociobiology*, 60, 695–706.

Kyes, R. C., Rumawas, R. E., Sulistiawati, E. & Budiarsa, N. 1995. Infanticide in a captive group of pig-tailed macaques (*Macaca nemestrina*). *American Journal of Primatology*, 36, 135–136.

Lewis, R. J. 2008. Social influences on group membership in *Propithecus verreauxi verreauxi*. *International Journal of Primatology*, 29, 1249–1270.

Lewis, R. J., Razafindrasamba, S. M. & Tolojanahary, J. P. 2003. Observed infanticide in a seasonal breeding prosimian (*Propithecus verreauxi verreauxi*) in Kirindy Forest, Madagascar. *Folia Primatologica*, 74, 101–103.

Li, J., Yin, H. & Zhou, L. 2007. Non-reproductive copulation behavior among Tibetan macaques (*Macaca thibetana*) at Huangshan, China. *Primates*, 48, 64–72.

Lindburg, D. G., Lyles, A. M. & Czekala, N. M. 1989. Status and reproductive potential of lion-tailed macaques in captivity. *Zoo Biology*, 8, 5–16.

Littlefield, B. 2009. Infanticide following male takeover event in Verreaux's sifaka (*Propithecus verreauxi verreauxi*). *Primates*, 51, 83–86.

———. 2010. Infanticide following male takeover event in Verreaux's sifaka (*Propithecus verreauxi verreauxi*). *Primates*, 51, 83–86.

Lonstein, J. S. & Gammie, S. C. 2002. Sensory, hormonal, and neural control of maternal aggression in laboratory rodents. *Neuroscience and Biobehavioral Reviews*, 26, 869–838.

Loy, J. 1981. The reproductive and heterosexual behaviours of adult patas monkeys in captivity. *Animal Behaviour*, 29, 714–726.

MacIntosh, A. J. & Sicotte, P. 2009. Vigilance in ursine black and white colobus monkeys (*Colobus vellerosus*): An examination of the effects of conspecific threat and predation. *American Journal of Primatology*, 71, 919–927.

Macleod, M. 2000. The reproductive strategies of samango monkeys (Cercopithecus mitis erythrarchus). PhD dissertation, University of Surrey Roehampton.

MacLeod, M. C., Ross, C. & Lawes, M. J. 2002. Costs and benefits of alternative mating strategies in Samango monkey males. In *The Guenons: Diversity and Adaptation in African Monkeys* (ed. by M. E. Glenn & M. Cords), 203–216. New York: Kluwer.

Maestripieri, D. 1992. Functional aspects of maternal aggression in mammals. *Canadian Journal of Zoology*, 70, 1069–1077.

Maestripieri, D. & Roney, J. R. 2005. Primate copulation calls and postcopulatory female choice. *Behavioral Ecology*, 16, 106–113.

Manson, J. H., Perry, S. & Parish, A. R. 1997. Nonconceptive sexual behavior in bonobos and capuchins. *International Journal of Primatology*, 18, 767–786.

Manson, J. H., Gros-Louis, J. & Perry, S. 2004. Three apparent cases of infanticide by males in wild white-faced capuchins (*Cebus capucinus*). *Folia Primatologica*, 104–106.

Marsh, C. W. 1979a. Comparative aspects of social organization in the Tana River red colobus, *Colobus badius rufomitratus*. *Zeitschrift für Tierpsychologie*, 46, 337–362.

———. 1979b. Female transference and mate choice among Tana River red colobus. *Nature*, 281, 568–569.

Matsumoto-Oda, A. 1999a. Female choice in the opportunistic mating of wild chimpanzees (*Pan troglodytes schweinfurthii*) at Mahale. *Behavioral Ecology and Sociobiology*, 46, 258–266.

———. 1999b. Mahale chimpanzees: Grouping patterns and cycling females. *American Journal of Primatology*, 47, 197–207.

McComb, K., Pusey, A., Packer, C. & Grinnell, J. 1993. Female lions can identify potentially infanticidal males from their roars. *Proceedings of the Royal Society of London Series B Biological Sciences*, 252, 59–64.

Mennella, J. A. & Moltz, H. 1988. Infanticide in rats: Male strategy and female counter-strategy. *Physiology and Behavior*, 42, 19–28.

Mitani, J. C. 1990. Demography of agile gibbons (*Hylobates agilis*). *International Journal of Primatology*, 11, 411–424.

Mitani, J. C., Watts, D. P. & Amsler, S. J. 2010. Lethal intergroup aggression leads to territorial expansion in wild chimpanzees. *Current Biology*. 20, R507-R508.

Mitra Setia, T. & van Schaik, C. P. 2007. The response of adult orang-utans to flanged male long calls: Inferences about their function. *Folia Primatologica*, 78, 215–226.

Möhle, U., Heistermann, M., Dittami, J., Reinberg, V. & Hodges, J. K. 2005. Patterns of anogenital swelling size and their endocrine correlates during ovulatory cycles and early pregnancy in free-ranging barbary macaques (*Macaca sylvanus*) of Gibraltar. *American Journal of Primatology*, 66, 351–368.

Mohnot, S. M., Winkler, P., Mohnot, U., Sommer, V., Agoramoorthy, G., Rajpurohit, L. S., Srivastava, A., Borries, C., Stephen, D., Chhangani, A. K., Schülke, O., Bhaker, N. R., Little, C., Rajpurohit, R. S., Sharma, G., Rajpurohit, D. S., Devilal, V. P. & Swami, B. 2008. A 30-years history of Kailana-I (K-I or B-19) bisexual troop of Hanuman langurs, *Semnopithecus entellus* at Jodhpur (India). *Primate Report*, 76, 55–61.

Moos, R., Rock, J. & Salzert, W. 1985. Infanticide in gelada baboons (*Theropithecus gelada*). *Primates*, 26, 497–500.

Morelli, T. L., King, S. J., Pochron, S. T. & Wright, P. C. 2009. The rules of disengagement: Takeovers, infanticide, and dispersal in a rainforest lemur, *Propithecus edwardsi*. *Behaviour*, 146, 499–523.

Mori, A., Iwamoto, T. & Bekele, A. 1997. A case of infanticide in a recently found gelada population in Arsi, Ethiopia. *Primates*, 38, 79–88.

Mori, A., Belay, G. & Iwamoto, T. 2003. Changes in unit structures and infanticide observed in Arsi geladas. *Primates*, 44, 217–223.

Mori, U. & Dunbar, R. I. M. 1985. Changes in the reproductive condition of female gelada baboons following takeover of one-male units. *Zeitschrift für Tierpsychologie*, 67, 215–224.

Moscovice, L. R., Di Fiore, A., Crockford, C., Kitchen, D. M., Wittig, R., Seyfarth, R. M. & Cheney, D. L. in press. Hedging their bets? Male and female chacma baboons form friendships based on likelihood of paternity. *Animal Behaviour*.

Mowry, C. B. 1995. Possible infanticidal behaviour by a Tana River red colobus *Procolobus badius rufomitratus*. *African Primates*, 1, 48–51.

Muller, M. N., Thompson, M. E. & Wrangham, R. W. 2006. Male chimpanzees prefer mating with old females. *Current Biology*, 16, 2234–2238.

Murray, C. M., Wroblewski, E. & Pusey, A. E. 2007. New case of intragroup infanticide in the chimpanzees of Gombe National Park. *International Journal of Primatology*, 28, 23–37.

Newton, P. N. 1986. Infanticide in an undisturbed forest population of hanuman langurs, *Presbytis entellus*. *Animal Behaviour*, 34, 785–789.

———. 1987. The social organization of forest hanuman langurs. *International Journal of Primatology*, 8, 199–232.

Newton-Fisher, N. E. 1999. Infant killers of Budongo. *Folia Primatologica*, 70, 167–169.

Nishida, T. & Kawanaka, K. 1985. Within-group cannibalism by adult male chimpanzees. *Primates*, 26, 274–284.

Nishida, T., Corp, N., Hamai, M., Hasegawa, T., Hiraiwa-Hasegawa, M., Hosaka, K., Hunt, K. D., Itoh, N., Kawanaka, K., Matsumoto-Oda, A., Mitani, J. C., Nakamura, M., Norikoshi, K., Sakamaki, T., Turner, L., Uehara, S. & Zamma, K. 2003. Demography, female life history, and reproductive profiles among the chimpanzees of Mahale. *American Journal of Primatology*, 59, 99–121.

Nunn, C. L. 1999a. The evolution of exaggerated sexual swellings in primates and the graded-signal hypothesis. *Animal Behaviour*, 58, 229–246.

———. 1999b. The number of males in primate social groups: A comparative test of the socioecological model. *Behavioral Ecology and Sociobiology*, 46, 1–13.

Nunn, C. L. & van Schaik, C. P. 2000. Social evolution in primates: The relative roles of ecology and intersexual conflict. In *Infanticide by Males and Its Implications* (ed. by C. P. van Schaik & C. H. Janson), 388–420. Cambridge: Cambridge University Press.

O'Brien, T. G. 1991. Female-male social interactions in wedge-capped capuchin monkeys: Benefits and costs of group living. *Animal Behaviour*, 41, 555–567.

Oates, J. F. 1977a. The guereza and its food. In *Primate Ecology: Studies of Feeding and Ranging Behaviour in Lemurs, Monkeys and Apes* (ed. by T. H. Clutton-Brock), 276–321. London: Academic.

———. 1977b. The social life of a black-and-white-colobus monkey. *Zeitschrift für Tierpsychologie*, 45, 1–60.

Onderdonk, D. 2000. Infanticide of a newborn black-and-white colobus monkey (*Colobus guereza*) in Kibale National Park, Uganda. *Primates*, 41, 209–212.

Onishi, K. & Nakamichi, M. 2008. Maternal infant monitoring in a free-ranging group of Japanese macaques (*Macaca fuscata*). *International Journal of Primatology*, 32, 209–222.

Oshawa, H. 1991. Take-over of a harem and subsequent promiscuity in patas monkeys. In *Primatology Today: Proceedings of the XIIIth Congress of the International Primatological Society, Nagoya and Kyoto, 18–12 July 1990* (ed. by A. Ehara, T. Kimura, O. Takenaka & M. Iwamoto), 221–224. Amsterdam: Elsevier Science.

Ostner, J. & Kappeler, P. M. 2004. Male life history and the unusual adult sex ratios of redfronted lemur, *Eulemur fulvus rufus*, groups. *Animal Behaviour*, 67, 249–259.

Ostner, J., Kappeler, P. M. & Heistermann, M. 2002. Seasonal variation and social correlates of androgen excretion in male redfronted lemurs (*Eulemur fulvus rufus*). *Behavioral Ecology and Sociobiology*, 52, 485–495.

Ostner, J., Kappeler, P. & Heistermann, M. 2008. Androgen and glucocorticoid levels reflect seasonally occurring social challenges in male redfronted lemurs (*Eulemur fulvus rufus*). *Behavioral Ecology and Sociobiology*, 62, 627–638.

Ostro, L., Silver, S., Koontz, F., Horwich, R. & Brockett, R. 2001. Shifts in social structure of black howler (*Alouatta pigra*) groups associated with natural and experimental variation in population density. *International Journal of Primatology*, 22, 733–748.

Otali, E. & Gilchrist, J. S. 2006. Why chimpanzee (*Pan troglodytes schweinfurthii*) mothers are less gregarious than nonmothers and males: The infant safety hypothesis. *Behavioral Ecology and Sociobiology*, 59, 561–570.

Packer, C. & Pusey, A. E. 1983. Adaptations of female lions to infanticide by incoming males. *American Naturalist*, 121, 716–728.

———. 1984. Infanticide in carnivores. In *Infanticide: Comparative and Evolutionary Perspectives* (ed. by G. Hausfater & S. B. Hrdy), 31–42. New York: Aldine.

Packer, C., Herbst, L., Pusey, A. E., Bygott, J. D., Hanby, J. P., Cairns, S. J. & Mulder, M. B. 1988. Reproductive success in lions. In *Reproductive Success: Studies of Individual Variation in Contrasting Breeding Systems* (ed. by T. H. Clutton-Brock), 363–383. Chicago: University of Chicago Press.

Pallaud, B. 1984. Conséquences d'un changement de hiérarchie et de territorie dans un group de Macaques crabiers (*Macaca fascicularis*). *Biology of Behaviour*, 9, 89–99.

Palombit, R. A. 1999. Infanticide and the evolution of pair bonds in nonhuman primates. *Evolutionary Anthropology*, 7, 117–129.

———. 2000. Infanticide and the evolution of male-female bonds in animals. In *Male Infanticide and Its Implications* (ed. by C. P. van Schaik & C. H. Janson), 239–268. Cambridge: Cambridge University Press

———. 2003. Male infanticide in wild savanna baboons: Adaptive significance and intraspecific variation. In *Sexual Selection and Reproductive Competition in Primates: New Perspectives and Directions* (ed. by C. B. Jones), 364–411. Norman, OK: American Society of Primatologists.

———. 2009. Friendships with males: A female counterstrategy to infanticide in the Okavango chacma baboons. In *Sexual Coercion in Primates and Humans* (ed. by M. N. Muller & R. W. Wrangham), 377–409. Cambridge, MA: Harvard University Press.

———. 2010. Conflict and bonding between the sexes. In *Mind the Gap: Tracing the Origin of Human Universals* (ed. by P. M. Kappeler & J. B. Silk), 53–84. Berlin: Springer.

Palombit, R. A., Seyfarth, R. M. & Cheney, D. L. 1997. The adaptive value of friendships to female baboons: Experimental and observational evidence. *Animal Behaviour*, 54, 599–614.

Palombit, R. A., Cheney, D. L., Fischer, J., Johnson, S., Rendall, D., Seyfarth, R. M. & Silk, J. B. 2000. Male infanticide and defense of infants in wild chacma baboons. In *Infanticide by Males and Its Implications* (ed. by C. P. van Schaik & C. H. Janson), 123–152. Cambridge: Cambridge University Press.

Palombit, R. A., Cheney, D. L. & Seyfarth, R. M. 2001. Female-female competition for male "friends" in wild chacma baboons (*Papio cynocephalus ursinus*). *Animal Behaviour*, 61, 1159–1171.

Pazol, K. 2003. Mating in the Kakamega forest blue monkeys (*Cercopithecus mitis*): Does female sexual behavior function to manipulate paternity assessment? *Behaviour*, 140, 473–499.

Peker, S., Kowalewski, M. M., Pave, R. & Zunino, G. E. 2006. *Evidencia de infanticidio en Alouatta caraya en el nordeste argentino*. Córdoba: XXII Reunión de Ecología.

Pereira, M. 1983. Abortion following the immigration of an adult male baboon (*Papio cynocephalus*). *American Journal of Primatology*, 4, 93–98.

Pereira, M. E. & Weiss, M. L. 1991. Female mate choice, male migration, and the threat of infanticide in ringtailed lemurs. *Behavioral Ecology and Sociobiology*, 28, 141–152.

Perrigo, G., Belvin, L. & vom Saal, F. 1992. Time and sex in the male mouse: Temporal regulation of infanticide and parental behavior. *Chronobiology International*, 9, 421–433.

Perrigo, G., Belvin, L., Quindry, P., Kadir, T., Becker, J., van Look, C., Niewoehner, J. & vom Saal, F. S. 1993. Genetic mediation of infanticide and parental behavior in male and female domestic and wild stock house mice. *Behavior Genetics*, 23, 525–531.

Perry, S. 2008. *Manipulative Monkeys*. Cambridge, MA: Harvard University Press.

Pillay, N. & Kinahan, A. A. 2009. Mating strategy predicts the occurrence of the Bruce effect in the vlei rat *Otomys irroratus*. *Behaviour*, 146, 139–151.

Pradhan, G. R., Engelhardt, A., van Schaik, C. P. & Maestripieri, D. 2006. The evolution of female copulation calls in primates: A review and a new model. *Behavioral Ecology and Sociobiology*, 59, 333–343.

Pradhan, G. R. & Van Schaik, C. 2008. Infanticide-driven intersexual conflict over matings in primates and its effects on social organization. *Behaviour*, 145, 251–275.

Pulliam, H. R. 1973. On the advantages of flocking. *Journal of Theoretical Biology*, 38, 419–422.

Pusey, A. 2001. Of genes and apes. Chimpanzee social structure and reproduction. In *Tree of Origin* (ed. by F. B. M. de Waal), 613–652. Cambridge, MA: Harvard University Press.

Pusey, A. E. & Packer, C. 1987. The evolution of sex-biased dispersal in lions. *Behaviour*, 101, 275–310.

———. 1994. Infanticide in lions: Consequences and counterstrategies. In *Infanticide and Parental Care* (ed. by S. Parmigiani & F. vom Saal), 277–299. Chur, Switzerland: Harwood Academic Publishing.

Pusey, A., Murray, C., Wallauer, W., Wilson, M., Wroblewski, E. & Goodall, J. 2008. Severe aggression among female *Pan troglodytes schweinfurthii* at Gombe National Park, Tanzania. *International Journal of Primatology*, 29, 949–973.

Rajpurohit, L. S., Chhangani, A. K., Rajpurohit, R. S. & Mohnot, S. M. 2003. Observation of a sudden resident-male replacement in a unimale bisexual troop of Hanuman langurs, *Semnopithecus entellus*, around Jodhpur (India). *Folia Primatologica*, 74, 85–87.

Rajpurohit, I. S., Chhangani, A. K., Rajpurohit, R. S., Bhaker, N. R., Rajpurohit, D. S. & Sharma, G. 2008. Recent observations on resident male change followed by infanticide in Hanuman langurs (*Semnopithecus entellus*) around Jodhpur. *Primate Report*, 75, 33–40.

Ramírez-Llorens, P., Di Bitetti, M. S., Baldovino, M. C. & Janson, C. H. 2008. Infanticide in black capuchin monkeys (*Cebus apella nigritus*) in Iguazú National Park, Argentina. *American Journal of Primatology*, 70, 473–484.

Rasoloharijaona, S., Rakotosamimanana, B. & Zimmermann, E.

2000. Infanticide by a male Milne-Edwards' sportive lemur (*Lepilemur edwardsi*) in Ampijoroa, NW-Madagascar. *International Journal of Primatology*, 21, 41–45.

Rees, A. 2009. *The Infanticide Controversy*. Chicago: University of Chicago Press.

Rhine, R. J., Norton, G. W., Roertgen, W. J. & Klein, H. D. 1980. The brief survival of free-ranging baboon infants (*Papio cynocephalus*) after separation from their mothers. *International Journal of Primatology*, 1, 401–409.

Richard, A. F., Rakotomanga, P. & Schwartz, M. 1993. Dispersal by *Propithecus verreauxi* at Beza Mahafaly, Madagascar: 1984–1991. *American Journal of Primatology*, 30, 1–20.

Ridley, M. 1986. The number of males in a primate troop. *Animal Behaviour*, 34, 1848–1858.

Rijksen, H. D. 1981. Infant killing: A possible consequence of a disputed leader role. *Behaviour*, 78, 138–167.

Robbins, A. M., Stoinski, T. S., Fawcett, K. A. & Robbins, M. M. 2009a. Does dispersal cause reproductive delays in female mountain gorillas? *Behaviour*, 146, 525–549.

Robbins, A. M., Stoinski, T., Fawcett, K. & Robbins, M. M. 2009b. Leave or conceive: Natal dispersal and philopatry of female mountain gorillas in the Virunga volcano region. *Animal Behaviour*, 77, 831–838.

Robbins, A. M., Stoinski, T., Fawcett, K. & Robbins, M. 2009c. Socioecological influences on the dispersal of female mountain gorillas: Evidence of a second folivore paradox. *Behavioral Ecology and Sociobiology*, 63, 477–489.

Robbins, M. M. 1995. A demographic analysis of male life history and social structure of gorillas. *Behaviour*, 132, 21–47.

———. 2001. Variation in the social system of mountain gorillas: The male perspective. In *Mountain Gorillas: Three Decades of Research at Karisoke* (ed. by M. M. Robbins, P. Sicotte & K. J. Stewart), 29–58. Cambridge: Cambridge University Press.

———. 2003. Behavioral aspects of sexual selection in mountain gorillas. In *Sexual Selection and Reproductive Competition in Primates: New Perspectives and Directions* (ed. by C. B. Jones), 477–501. Norman, OK: American Society of Primatologists.

———. 2008. Feeding competition and agonistic relationships among Bwindi *Gorilla beringei*. *International Journal of Primatology*, 29, 999–1018.

Roberts, E. K., Lu, A., Bergman, T. J. & Beehner, J. C. 2012. A Bruce Effect in geladas. *Science*, 335, 1222–1225.

Roberts, M. 1994. Growth, development, and parental care in the western Tarsier (*Tarsius bancanus*) in captivity: Evidence for a "slow" life history and nonmonogamous mating system. *International Journal of Primatology*, 15, 1–28.

Ross, C. 1993. Take-over and infanticide in South Indian Hanuman langurs (*Presbytis entellus*). *American Journal of Primatology*, 30, 75–82.

Rowell, T. E. 1970. Reproductive cycles of two *Cercopithecus* monkeys. *Journal Of Reproduction And Fertility*, 22, 321–338.

Rudran, R. 1973a. Adult male replacement in one-male troups of purple faced langurs (*Presbytis senex senex*) and its effect on population structure. *Folia Primatologica*, 19, 166–192.

———. 1973b. The reproductive cycles of two subspecies of purple-faced langurs (*Presbytis senex*) with relation to enviornmental factors. *Folia Primatologica*, 19, 41–60.

———. 1979. The demography and social mobility of a red howler (*Alouatta seniculus*) population in Venezuela. In *Vertebrate Ecology in the Northern Neotropics* (ed. by J. F. Eisenberg), 107–126. Washington: Smithsonian Institution Press.

Rudran, R. & Fernandez-Duque, E. 2003. Demographic changes over thirty years in a red howler population in Venezuela. *International Journal of Primatology*, 24, 925–947.

Ruehlmann, T. E., Bernstein, I. S., Gordon, T. P. & Balcaen, P. 1988. Wounding patterns in three species of captive macaques. *American Journal of Primatology*, 14, 125–134.

Ryan, S., Starks, P., Milton, K. & Getz, W. 2008. Intersexual conflict and group size in *Alouatta palliata*: A 23-year evaluation. *International Journal of Primatology*, 29, 405–420.

Saayman, G. S. 1975. The influence of hormonal and ecological factors upon sexual behavior and social organization in Old World primates. In *Socioecology and Psychology of Primates* (ed. by R. Tuttle), 181–204. Chicago: Aldine.

Saito, C., Sato, S., Suzuki, S., Sugiura, H., Agetsuma, N., Takahata, Y., Sasaki, C., Takahashi, H., Tanaka, T. & Yamagiwa, J. 1998. Aggressive intergroup encounters in two populations of Japanese macaques (*Macaca fuscata*). *Primates*, 39, 303–312.

Saj, T. L. & Sicotte, P. 2005. Male takeover in *Colobus vellerosus* at Boabeng-Fiema Monkey Sanctuary, central Ghana. *Primates*, 46, 211–214.

Sakamaki, T., Itoh, N. & Nishida, T. 2001. An attempted within-group infanticide in wild chimpanzees. *Primates*, 42, 359–366.

Sakura, O. 1994. Factors affecting party size and composition of chimpanzees (*Pan troglodytes verus*) Bossou, Guinea. *International Journal of Primatology*, 15, 167–183.

Sauther, M. L. & Sussman, R. W. 1993. A new interpretation of the social organization and mating system of the ringtailed lemur (*Lemur catta*). In *Lemur Social Systems and Their Ecological Basis* (ed. by P. M. Kappeler & J. U. Ganzhorn), 111–121. New York: Plenum.

Schülke, O. & Kappeler, P. M. 2001. Protection from infanticide, male resource defense, over-dispersed females and the evolution of pair-living in a nocturnal lemur: *Phaner furcifer*. *Folia Primatologica*, 72, 182–183.

Sekulic, R. 1982. Behavior and ranging patterns of a solitary female red howler (*Alouatta seniculus*). *Folia Primatologica*, 38, 217–232.

———. 1983. The effect of female call on male howling in the red howler monkeys (*Alouatta seniculus*). *International Journal of Primatology*, 4, 291–305.

Shelmidine, N., Borries, C. & Koenig, A. 2007. Genital swellings in silvered langurs: What do they indicate? *American Journal of Primatology*, 69, 519–532.

Sherrow, H. M. & Amsler, S. J. 2007. New intercommunity infanticides by the chimpanzees of Ngogo, Kibale National Park, Uganda. *International Journal of Primatology*, 28, 9–22.

Shopland, J. M. 1982. An intergroup encounter with fatal consequences in yellow baboons (*Papio cynocephalus*). *American Journal of Primatology*, 3, 263–266.

Sicotte, P. 1994. Effect of male competition on male-female relationships in bi-male groups of mountain gorillas. *Ethology*, 97, 47–64.

———. 2000. A case of mother-son transfer in mountain gorillas. *Primates*, 41, 93–101.

Sicotte, P., Teichroeb, J. & Saj, T. 2007. Aspects of male competition in *Colobus vellerosus*: Preliminary data on male and female loud calling, and infant deaths after a takeover. *International Journal of Primatology*, 28, 627–636.

Singh, M., Kumara, H. N., Ananda, K. M., Singh, M. & Cooper, M. 2006. Male influx, infanticide, and female transfer in *Macaca radiata*. *International Journal of Primatology*, 27, 515–528.

Small, M. F. 1990. Promiscuity in barbary macaques (*Macaca sylvanus*). *American Journal of Primatology*, 20, 267–282.

Smith, D. G. 1981. Birth timing and social rank of adult male rhesus monkeys (*Macaca mulatta*). *Journal of Medical Primatology*, 10, 279–283.

Smuts, B. B. 1985. *Sex and Friendship in Baboons*. New York: Aldine.

Smuts, B. B. & Smuts, R. W. 1993. Male aggression and sexual coercion of females in nonhuman primates and other mammals: Evidence and theoretical implications. *Advances in the Study of Behavior*, 22, 1–63.

Snaith, T. & Chapman, C. 2005. Towards an ecological solution to the folivore paradox: Patch depletion as an indicator of within-group scramble competition in red colobus monkeys (*Piliocolobus tephrosceles*). *Behavioral Ecology and Sociobiology*, 59, 185–190.

Soltis, J. 2002. Do primate females gain nonprocreative benefits by mating with multiple males? Theoretical and empirical considerations. *Evolutionary Anthropology*, 11, 187–197.

Soltis, J., Mitsunaga, F., Shimizu, K., Yanagihara, Y. & Nozaki, M. 1999. Female mating strategy in an enclosed group of Japanese macaques. *American Journal of Primatology*, 47, 263–278.

Soltis, J., Thomsen, R., Matusabayshi, K. & Takenaka, O. 2000. Infanticide by resident males and female counter-strategies in wild Japanese macaques (*Macaca fuscata*). *Behavioral Ecology and Sociobiology*, 48, 195–202.

Soltis, J., Thomsen, R. & Takenaka, O. 2001. The interaction of male and female reproductive strategies and paternity in wild Japanese macaques, *Macaca fuscata*. *Animal Behaviour*, 62, 485–494.

Sommer, V. 1987. Infanticide among free-ranging langurs (*Presbytis entellus*) at Jodhpur (Rajasthan/India): Recent observations and a reconsideration of hypotheses. *Primates*, 28, 163–197.

———. 1988. Male competition and coalitions in langurs (*Presbytis entellus*) at Jodhpur, Rajasthan, India. *Human Evolution*, 3, 261–278.

———. 1994. Infanticide among the langurs of Jodhpur: Testing the sexual selection hypothesis with a long-term record. In *Infanticide and Parental Care* (ed. by S. Parmigiani & F. vom Saal), 155–198. Chur, Switzerland: Harwood Academic Publishing.

Sommer, V., Srivastava, A. & Borries, C. 1992. Cycles, sexuality, and conception in free-ranging langurs (*Presbytis entellus*). *American Journal of Primatology*, 28, 1–27.

Spijkerman, R. P., van Hooff, J. A. R. A. M. & Jens, W. 1990. A case of lethal infant abuse in an established group of chimpanzees. *Folia Primatologica*, 55, 41–44.

Stanger, K. F., Coffman, B. S. & Izard, M. K. 1995. Reproduction in Coquerel's dwarf lemur (*Mirza coquereli*). *American Journal of Primatology*, 36, 223–237.

Starin, E. D. 1981. Monkey moves. *Natural History*, 90(1), 36–43.

———. 1988. Gestation and birth-related behaviors in Temminck's red colobus. *Folia Primatologica*, 51, 161–164.

———. 1994. Philopatry and affiliation among red colobus. *Behaviour*, 130, 253–270.

Steenbeek, R. 1996. What a maleless group can tell us about the constraints on female transfer in Thomas's langurs (*Presbytis thomasi*). *Folia Primatologica*, 67, 169–181.

———. 1999a. Tenure related changes in wild Thomas's langurs. I. Between-group interactions. *Behaviour*, 136, 595–625.

———. 1999b. Female choice and male coercion in wild Thomas's langurs. PhD dissertation, Utrecht University.

———. 2000. Infanticide by males and female choice in wild Thomas's langurs. In *Infanticide by Males and Its Implications* (ed. by C. P. van Schaik & C. H. Janson), 153–177. Cambridge: Cambridge University Press.

Steenbeek, R., Piek, R. C., van Buul, M. & van Hooff, J. A. R. A. M. 1999. Vigilance in wild Thomas's langurs (*Presbytis thomasi*): The importance of infanticide risk. *Behavioral Ecology and Sociobiology*, 45, 137–150.

Sterck, E. H. M. 1997. Determinants of female dispersal in Thomas langurs. *American Journal of Primatology*, 42, 179–198.

———. 1998. Variation in langur social organization in relation to the socioecological model, human habitat alteration, and phylogenetic constraints. *Primates*, 40, 199–213.

Sterck, E. H. M. & Korstjens, A. H. 2000. Female dispersal and infanticide avoidance in primates. In *Infanticide by Males and Its Implications* (ed. by C. P. van Schaik & C. H. Janson), 293–321. Cambridge: Cambridge University Press.

Sterck, E. H. M., Watts, D. P. & van Schaik, C. P. 1997. The evolution of female social relationships in nonhuman primates. *Behavioral Ecology and Sociobiology*, 41, 291–309.

Sterck, E. H. M., Willems, E. P., van Hooff, J. A. R. A. M. & Wich, S. A. 2005. Female dispersal, inbreeding avoidance and mate choice in Thomas langurs (*Presbytis thomasi*). *Behaviour*, 142, 845–868.

Stoinski, T. S., Perdue, B. M. & Legg, A. M. 2009. Sexual behavior in female western lowland gorillas (*Gorilla gorilla gorilla*): Evidence for sexual competition. *American Journal of Primatology*, 71, 587–593.

Stokes, E., Parnell, R. J. & Olejniczak, C. 2003. Female dispersal and reproductive success in wild western lowland gorillas. *Behavioral Ecology and Sociobiology*, 54, 329–339.

Stokes, E. J. 2004. Within-group social relationships among females and adult males in wild western lowland gorillas (*Gorilla gorilla gorilla*). *American Journal of Primatology*, 64, 233–246.

Struhsaker, T. T. 1977. Infanticide and social organization in the redtail monkey (*Cercopithecus ascanius schmidti*) in the Kibale forest, Uganda. *Zeitschrift für Tierpsychologie*, 45, 75–84.

Struhsaker, T. T. & Leland, L. 1985. Infanticide in a patrilineal society of red colobus monkeys. *Zeitschrift für Tierpsychologie*, 69, 89–132.

Stumpf, R. & Boesch, C. 2005. Does promiscuous mating preclude female choice? Female sexual strategies in chimpanzees (*Pan troglodytes verus*) of the Taï National Park, Côte d'Ivoire. *Behavioral Ecology and Sociobiology*, 57, 511–524.

Stumpf, R. M., Thompson, M. E. & Knott, C. D. 2008. A comparison of female mating strategies in *Pan troglodytes* and *Pongo* spp. *International Journal of Primatology*, 29, 865–884.

Sugiyama, Y. 1965. On the social change of hanuman langurs (*Presbytis entellus*) in their natural condition. *Primates*, 6, 381–418.

———. 1966. An artificial change in a hanuman langur troop (*Presbytis entellus*). *Primates*, 7, 41–72.

Suzuki, A. 1971. Carnivority and cannibalism observed among forest-living chimpanzees. *Journal of the Anthropological Society of Nippon*, 79, 30–48.

Swedell, L. 2000. Two takeovers in wild hamadryas baboons. *Folia Primatologica*, 71, 169–172.

———. 2006. *Strategies of Sex and Survival in Hamadryas Baboons: Through a Female Lens*. Upper Saddle River, NJ: Pearson Prentice Hall.

Swedell, L. & Saunders, J. 2006. Infant mortality, paternity certainty, and female reproductive strategies in hamadryas baboons. In *Reproduction and Fitness in Baboons: Behavioral, Ecological, and Life History Perspectives* (ed. by L. Swedell & S. R. Leigh), 19–51. New York: Springer.

Swedell, L. & Tesfaye, T. 2003. Infant mortality after takeovers in wild Ethiopian hamadryas baboons. *American Journal of Primatology*, 60, 113–118.

Takahata, Y. 1985. Adult male chimpanzees kill and eat male newborn infant: Newly observed intragroup infanticide and cannibalism in Mahale National Park, Tanzania. *Folia Primatologica*, 44, 161–170.

Takahata, Y., Sprague, D. S., Suzuki, S. & Okayasu, N. 1994. Female competition, co-existence, and the mating structure of wild Japanese macaques on Yakushima Island, Japan. *Physiology and Ecology, Japan*, 29, 163–179.

Teichroeb, J. A., Marteinson, S. & Sicotte, P. 2005. Individuals' behaviors following dye-marking in wild black-and-white colobus (*Colobus vellerosus*). *American Journal of Primatology*, 65, 197–203.

Teichroeb, J. A. & Sicotte, P. 2008a. Infanticide in ursine colobus monkeys (*Colobus vellerosus*) in Ghana: New cases and a test of the existing hypotheses. *Behaviour*, 145, 727–755.

———. 2008b. Social correlates of fecal testosterone in male ursine colobus monkeys (*Colobus vellerosus*): The effect of male reproductive competition in aseasonal breeders. *Hormones and Behavior*, 54, 417–423.

Teichroeb, J. A., Wikberg, E. C. & Sicotte, P. 2009. Female dispersal patterns in six groups of ursine colobus (*Colobus vellerosus*): Infanticide avoidance is important. *Behaviour*, 146, 551–582.

Thalmann, U. 2001. Food resource characteristics in two nocturnal lemurs with different social behavior: *Avahi occidentalis* and *Lepilemur edwardsi*. *International Journal of Primatology*, 22, 287–324.

Tobler, M. 2009. Female reproductive strategies and the ovarian cycle in Hamadryas baboons. Master's thesis, Victoria University of Wellington.

Treves, A. 2001. Reproductive consequences of variation in the composition of howler monkey (*Alouatta* spp.) groups. *Behavioral Ecology and Sociobiology*, 50, 61–71.

Treves, A. & Chapman, C. A. 1996. Conspecific threat, predation avoidance, and resource defense: Implications for grouping in langurs. *Behavioral Ecology and Sociobiology*, 39, 43–53.

Treves, A., Drescher, A. & Ingrisano, N. 2001. Vigilance and aggregation in black howler monkeys (*Alouatta pigra*). *Behavioral Ecology and Sociobiology*, 50, 90–95.

Treves, A., Drescher, A. & Snowdon, C. T. 2003. Maternal watchfulness in black howler monkeys (*Alouatta pigra*). *Ethology*, 109, 135–146.

Trivers, R. L. 1972. Parental investment and sexual selection. In *Sexual Selection and the Descent of Man, 1871–1971* (ed. by B. Campbell), 136–179. Chicago: Aldine.

Valderrama, X., Srikosamatara, S. & Robinson, J. G. 1990. Infanticide in wedge-capped capuchin monkeys, *Cebus olivaceus*. *Folia Primatologica*, 54, 171–176.

Van Belle, S., Kulp, A., Thiessen-Bock, R., Garcia, M. & Estrada, A. 2010. Observed infanticides following a male immigration event in black howler monkeys, *Alouatta pigra*, at Palenque National Park, Mexico. *Primates*, 51, 279–284.

Van Noordwijk, M. A. & van Schaik, C. P. 2000. Reproductive patterns in eutherian mammals: Adaptations against infanticide. In *Infanticide by Males and Its Implications* (ed. by C. P. van Schaik & C. H. Janson), 322–360. Cambridge: Cambridge University Press.

Van Schaik, C. P. 1983. Why are diurnal primates living in groups? *Behaviour*, 87, 120–144.

———. 2000a. Vulnerability to infanticide by males: Patterns among mammals. In *Infanticide by Males and Its Implications* (ed. by C. P. van Schaik & C. H. Janson), 61–71. Cambridge: Cambridge University Press.

———. 2000b. Infanticide by male primates: The sexual selection hypothesis revisited. In *Infanticide by Males and Its Implications* (ed. by C. P. van Schaik & C. H. Janson), 27–60. Cambridge: Cambridge University Press.

Van Schaik, C. P. & Dunbar, R. I. M. 1990. The evolution of monogamy in large primates: A new hypothesis and some crucial tests. *Behaviour*, 115, 30–62. Van Schaik, C. P. & Kappeler, P. M. 1993. Life history, activity period and lemur social systems. In *Lemur Social Systems and Their Ecological Basis* (ed. by P. M. Kappeler & J. U. Ganzhorn), 241–260. New York: Plenum.

Van Schaik, C. P., van Noordwijk, M. A. & Nunn, C. L. 1999. Sex and social evolution in primates. In *Comparative Primate Socioecology* (ed. by P. C. Lee), 204–231. Cambridge: Cambridge University Press.

Van Schaik, C. P., Hodges, J. K. & Nunn, C. L. 2000. Paternity confusion and the ovarian cycles of female primates. In *Infanticide by Males and Its Implications* (ed. by C. P. van Schaik & C. H. Janson), 361–387. Cambridge: Cambridge University Press.

Vigilant, L., Hofreiter, M., Siedel, H. & Boesch, C. 2001. Paternity and relatedness in wild chimpanzee communities. *Proceedings of the National Academy of Sciences of the United States of America*, 98, 12890–12895.

Vogel, E. R. & Fuentes-Jiménez, A. 2006. Rescue behavior in white-faced capuchin monkeys during an intergroup attack: Support for the infanticide avoidance hypothesis. *American Journal of Primatology*, 68, 1–5.

Vom Saal, F. S. 1985. Time-contingent change in infanticide and parental behavior induced by ejaculation in male mice. *Physiology and Behavior*, 34, 7–15.

Vom Saal, F. S. & Howard, L. S. 1982. The regulation of infanticide and parental behavior: Implications for reproductive success in male mice. *Science*, 215, 1270–1272.

Walker, S. E., Strasser, M. E. & Field, L. P. 2004. Reproductive parameters and life-history variables in captive golden-bellied mangabeys (*Cercocebus agilis chrysogaster*). *American Journal of Primatology*, 64, 123–131.

Wallis, J. 1997. A survey of reproductive parameters in the free-ranging chimpanzees of Gombe National Park. *Journal of Reproduction and Fertility*, 109, 297–307.

Wang, H. P., Tan, C. L., Gao, Y. F. & Li, B. G. 2004. A takeover of resident male in the Sichuan snub-nosed monkey (*Rhinopithecus roxellana*) in Qinling Mountains. *Acta Zoologica Sinica*, 50, 859–862.

Wasser, S. K. & Starling, A. K. 1988. Proximate and ultimate causes of reproductive suppression among female yellow baboons at Mikumi National Park, Tanzania. *American Journal of Primatology*, 16, 97–121.

Watts, D. P. 1985. Relations between group size and composition and feeding competition in mountain gorilla groups. *Animal Behaviour*, 33, 72–85.

———. 1989. Infanticide in mountain gorillas: New cases and a reconsideration of the evidence. *Ethology*, 81, 1–18.

———. 1991. Harassment of immigrant female mountain gorillas by resident females. *Ethology*, 89, 135–153.

———. 1994. Social relationships of immigrant and resident female mountain gorillas. II. Relatedness, residence, and relationships between females. *American Journal of Primatology*, 32, 13–30.

———. 1996. Comparative socio-ecology of gorillas. In *Great Ape Societies* (ed. by W. C. McGrew, L. F. Marchant & T. Nishida), 16–28. Cambridge: Cambridge University Press.

Watts, D. P. 2007. Effects of male group size, parity, and cycle stage on female chimpanzee copulation rates at Ngogo, Kibale National Park, Uganda. *Primates*, 48, 222–231.

Watts, D. P. & Mitani, J. C. 2000. Infanticide and cannibalism in male chimpanzees in Ngogo, Kibale National Park, Uganda. *Primates*, 41, 357–365.

Watts, D. P., Mitani, J. C. & Sherrow, H. M. 2002. New cases of inter-community infanticide by male chimpanzees at Ngogo, Kibale National Park, Uganda. *Primates*, 43, 263–270.

Weber, E. M. & Olsson, I. A. S. 2008. Maternal behaviour in *Mus musculus* sp.: An ethological review. *Applied Animal Behaviour Science*, 114, 1–22.

Weingrill, T. 2000. Infanticide and the value of male-female relationships in mountain chacma baboons (*Papio cynocephalus ursinus*). *Behaviour*, 137, 337–359.

White, F. A. 1993. Male transfer in captive ruffed lemurs, *Varecia variegata*. In *Lemur Social Systems and Their Ecological Basis* (ed. by P. M. Kappeler & J. U. Ganzhorn), 41–49. New York: Plenum.

Whitten, P. & Turner, T. 2004. Male residence and the patterning of serum testosterone in vervet monkeys (*Cercopithecus aethiops*). *Behavioral Ecology and Sociobiology*, 56, 565–578.

Wich, S. A., Assink, P. R., Becher, F. & Sterck, E. H. M. 2002. Playbacks of loud calls to wild Thomas langurs (Primates: *Presbytis thomasi*): The effect of familiarity. *Behaviour*, 139, 79–87.

Wich, S. A., Assink, P. R. & Sterck, E. H. M. 2004. Thomas langurs (*Presbytis thomasi*) discriminate between calls of young solitary versus older group-living males: A factor in avoiding infanticide? *Behaviour*, 141, 41–51.

Widdig, A. 2007. Paternal kin discrimination: The evidence and likely mechanisms. *Biological Reviews*, 82, 319–334.

Wielgus, R. B. & Bunnell, F. L. 2000. Possible negative effects of adult male mortality on female grizzly bear reproduction. *Biological Conservation*, 93, 145–154.

Williams, J. M., Pusey, A. E., Carlis, J. V., Farm, B. P. & Goodall, J. 2002. Female competition and male territorial behaviour influence female chimpanzees' ranging patterns. *Animal Behaviour*, 63, 347–360.

Williams, J. M., Oehlert, G. W., Carlis, J. V. & Pusey, A. E. 2004. Why do male chimpanzees defend a group range? *Animal Behaviour*, 68, 523–532.

Williams, J. M., Lonsdorf, E. V., Wilson, M. L., Schumacher-Stankey, J., Goodall, J. & Pusey, A. E. 2008. Causes of death in the Kasekela chimpanzees of Gombe National Park, Tanzania. *American Journal of Primatology*, 70, 766–777.

Wilson, M. E., Gordon, T. P. & Collins, D. C. 1982. Serum 17-β-estradiol and progesterone associated with mating during early pregnancy in female rhesus monkeys. *Hormones and Behavior*, 16, 94–106.

Wilson, M. L. & Wrangham, R. W. 2003. Intergroup relations in chimpanzees. *Annual Review of Anthropology*, 32, 363–393.

Wilson, M. L., Wallauer, W. R. & Pusey, A. E. 2004. New cases of intergroup violence among chimpanzees in Gombe National Park, Tanzania. *International Journal of Primatology*, 25, 523–549.

Wimmer, B. & Kappeler, P. M. 2002. The effects of sexual selection and life history on the genetic structure of redfronted lemur, *Eulemur fulvus rufus*, groups. *Animal Behaviour*, 64, 557–568.

Wittig, R. M., Crockford, C., Lehmann, J., Whitten, P. L., Seyfarth, R. M. & Cheney, D. L. 2008. Focused grooming networks and stress alleviation in wild female baboons. *Hormones and Behavior*, 54, 170–177.

Wolf, K. E. & Fleagle, J. G. 1977. Adult male replacement in a group of silvered leaf-monkeys (*Presbytis cristata*) at Kuala Selangor. *Primates*, 18, 949–955.

Wolfe, K. & Schulman, S. R. 1984. Male response to "stranger" females as a function of female reproductive value among chimpanzees. *American Naturalist*, 123, 163–174.

Wolff, J. O. & MacDonald, D. W. 2004. Promiscuous females protect their offspring. *Trends In Ecology and Evolution*, 19, 127–134.

Wrangham, R. W. 1979. On the evolution of ape social systems. *Social Science Information*, 18, 335–368.

———. 1980. An ecological model of female-bonded primate groups. *Behaviour*, 75, 262–299.

Wrangham, R. W. & Smuts, B. B. 1980. Sex differences in the behavioural ecology of chimpanzees in the Gombe National Park. *Journal of Reproduction and Fertility, Supplement*, 28, 13–31.

Wright, P. C. 1995. Demography and life history of free-ranging *Propithecus diadema edwardsi* in Ranomafana National Park, Madagascar. *International Journal of Primatology*, 16, 835–854.

Xiang, Z.-F. & Grueter, C. C. 2007. First direct evidence of infanticide and cannibalism in wild snub-nosed monkeys (*Rhinopithecus bieti*). *American Journal of Primatology*, 69, 249–254.

Yamada, K. & Nakamichi, M. 2006. A fatal attack on an unweaned infant by a non-resident male in a free-ranging group of Japanese macaques (*Macaca fuscata*) at Katsuyama. *Primates*, 47, 165–169.

Yamagiwa, J. & Kahekwa, J. 2001. Dispersal patterns, group structure, and reproductive parameters of eastern lowland gorillas at Kahuzi in the absence of infanticide. In *Moun-*

tain Gorillas: Three Decades of Research at Karisoke (ed. by
M. M. Robbins, P. Sicotte & K. J. Stewart), 89–122. Cambridge: Cambridge University Press.
———. 2004. First observations of infanticides by a silverback in
Kahuzi-Biega. Gorilla Journal, 29, 6–9.
Yamagiwa, J., Kahekwa, J. & Basabose, A. K. 2009. Infanticide
and social flexibility in the genus Gorilla. Primates, 50,
293–303.
Yamaguchi, N., Dugdale, H. & MacDonald, D. W. 2006. Female
receptivity, embryonic diapause, and superfetation in the European badger (Meles meles): Implications for the reprodutive
tactics of males and females. Quarterly Review Of Biology,
81, 33–48.
Yamamura, N., Hasegawa, T. & Ito, Y. 1990. Why mothers do
not resist infanticide: A cost-benefit genetic model. Evolution,
44, 1346–1357.

Zhang, S. Y., Liang, B. & Wang, L. X. 1999. Infanticide within
captive groups of Sichuan golden snub-nosed monkeys (Rhinopithecus roxellana). Folia Primatologica, 70, 274–276.
Zinner, D. & Deschner, T. 2000. Sexual swellings in female
hamadryas baboons after male take-overs: "Deceptive" swellings as a possible female counter-strategy against infanticide.
American Journal of Primatology, 52, 157–168.
Zinner, D., Kaumanns, W. & Rohrhuber, B. 1993. Infant mortality in captive hamadryas baboons (Papio hamadryas). Primate Report, 36, 97–113.
Zuckerman, S. 1932. The Social Life of Monkeys and Apes. London: K. Paul, Trench, Trubner.
Zunino, G. E., Chalukian, S. C. & Rumiz, D. L. 1986. Infanticidio y desaparición de infantes asociados al reemplaze de ma-
cho en grupos de Alouatta caraya. A Primatologia no Brasil,
2, 185–190.

Chapter 20 The Socioecology of Human Reproduction

Frank W. Marlowe

Tʜɪs ᴄʜᴀᴘᴛᴇʀ focuses on the social and ecological factors that shaped the evolution of reproductive strategies in human foraging societies (hunter-gatherers), the form of subsistence that characterized humans for most of our evolutionary history. It is clear that humans are similar to other primates, particularly the great apes, in many aspects of our reproductive biology and behavior. For example, the same hormonal mechanisms control reproductive activity in both humans and other great apes, and the lifetime fertility rates of women (in natural fertility populations) and great apes are very similar (table 20.1). However, humans also have a number of derived reproductive traits that distinguish us from our closest relatives, chimpanzees (*Pan troglodytes*). These include menopause, concealed ovulation, and continuous sexual receptivity (table 20.2). In addition, there are major differences between humans and other apes in our reproductive systems, the relationship between the sexes, and the patterns of parental investment.

Although there is great diversity across foraging societies, certain common themes emerge. First, pair bonds play an important role in foraging societies, although lifelong pair bonds are not necessarily the rule. Second, the extent of male-male competition is relatively limited and the degree of reproductive skew (table 20.2) in foraging societies is relatively low. Third, the slow growth and development of human children means that mothers are hard pressed to meet their offsprings' needs, and allomaternal care is common. Fourth, humans are characterized by high levels of economic interdependence between men and women.

In this chapter, I focus mainly on the reproductive behavior of contemporary hunter-gatherers (foragers) living in warm habitats (effective temperature ≥ 13° C, below latitudes 40°–50°). I make the distinction between warm and cold climates because humans have occupied arctic climates for less than 30,000 years, and the ability to survive in this habitat has required several highly specialized social and technological adaptations. I mention agricultural and industrialized societies at some points, mainly to illustrate our phenotypic plasticity and the wide range of human variability in contemporary societies.

Social Structures of Warm Climate Foragers

To appreciate the selective forces that shaped the reproductive strategies of men and women, it is useful to have a general idea of the ecological pressures that foragers face. In this section I provide a brief sketch of the ecology of foraging life. As with other primates, human ranging patterns are largely based on the distribution of food resources. Carnivores typically have much larger ranges than herbivores, and human foragers, who rely heavily on hunting, are no exception to this rule. Estimates of the areas used by members of 130 foraging societies range from 22 to 4,500 km² (Marlowe 2005). In comparison, the home ranges of chimpanzees are about 12.5 km² and the ranges of bonobos (*Pan paniscus*) are about 45 km².

The use of such large ranges requires considerable mobility. As a consequence, most foragers do not live in permanent villages or settlements. Instead they reside in tem-

Table 20.1.   Reproductive traits of humans (foragers and *US*) compared to those of the great apes

Trait	Human	Chimpanzee	Bonobo	Gorilla	Orangutan
Gestation (days)	270,[1] 269[2]	225,[1] 228[2]	244,[1] 240[2]	255,[1] 260[2]	260,[1] 250[2]
Menstrual cycle (days)	29.1 (SD = 7.5, n = 2316)[3]	36.1 (n = 53)[3]	41.3 (mean of 3 values S)	31.1 (SD = 7.5, n = 7 captive)[3]	30.5,[3] 28[4]
Age at weaning (years)	2.9 (1.3–4.5, n = 18), 0.25 US[5]	4.5[1]	4.5 (4–5)[6]	2.8[1]	7[1]
Age at menarche (years)	16 (12–17, n = 6), [7] 12.4 US	12	8.5[2]	9, 6.5[2]	12
Age of female at first birth (years)	18.4 (15.9–20, n = 6), 24.8 US[8]	13.3,[1] 14[2]	14.2,[9] 13.5[2]	10, 10.04[2]	15.6, 9.68[2]
Age of female at last birth (years)	45 (menopause),[1] 51.9 US	42 near death[1]	40[10]	? near death	> 41 near death
Potential reproductive span	26–32 years 39 years US	28 years	26 years	42 years	43 years
Interbirth Interval (years)	3.07 (1.8–4, n = 9), 3.69[1]	3.11,[11] 5.46[1]	6.25[1]	3.25,[11] 4.4[1]	7.72,[11] 7[1]
Total fertility Rate	5.14 (.81–8.5, n = 28), 2.1 US	5.39[12]	5.5[13]	5.55[12]	4[14]
Maximum age at death (years)	~85,[1] 100[2]	53.4, 53[2]	50, 40[2]	54, 50[2]	58.7, 57.3[2]
Ovulation signs, estrus (days)	No signs, 24 receptive	Exaggerated swellings, 12.5[11]	Exaggerated swellings, 17 (mean 3 populations)[15]	Behavioral estrus, 2.5[11]	No signs, 31.3,[11] 5[16]
Male-female weight dimorphism	1.19,[7] 1.14[2]	1.19,[11] 1.23[2]	1.36[2]	1.63,[11] 2.12[2]	2.21[11]
Mating system	Monogamy/polygyny	Promiscuity	Promiscuity	Polygyny	Promiscuous polygyny
Social structure	Multimale-multifemale mate bonds	Multimale-multifemale promiscuity	Multimale-multifemale promiscuity	Unimale (or with 2 males) harem polygyny	Solitary; males overlapping female ranges
Mating/paternity	Slight sperm competition	Considerable sperm competition	Considerable sperm competition	Little sperm competition	Little sperm competition?
Paternal investment	Considerable provisioning, with some direct care	Little investment	Little investment	Defense	No investment
Dispersal/philopatry	Multilocal marital residence	Male philopatry	Male philopatry	Dispersal by both sexes, with more by females	Emigration by both sexes
Mate guarding by males	Guarding by both sexes, who spend much time apart	Extra copulations for high-ranking males near time of ovulation	Little to no guarding	Guarding by male, who repels all bachelor males	Some chasing off of small males by large males
Sexual coercion	Occasional rape, frequent subtle coercion	Various forms of male coercion	Little to no coercion	Sex only when females solicit	Frequent rape by small males

Sources: [1]Robson et al. 2006; [2]Lindenfors 2002; [3]Bentley 1999; [4]Knott 2009; [5]Knott 2001; [6]Kuroda 1989; [7]Marlowe & Berbesque 2009a; [8]Johnson & Dye 2005; [9]De Lathouwers & Van Elsacker 2005; [10]Great Apes Survival Project; [11]Mitani et al. 1996; [12]Boesch & Boesch-Achermann 2000; [13]Animal Info; [14]MacDonald 2001; [15]Stanford 1998; [16]Nadler 1995. See also references within these sources.

porary residential "camps," and individuals move from one camp to another several times each year (range = 0–58 moves per year; Marlowe 2005). *Residents of camps do not move en masse*, as some families move in or out of a camp before others do, and so the composition of residential groups is constantly changing. This kind of fluid, multilevel social structure is unusual among primates. The people who occupy the same camps and move between them make up the ethnolinguistic group (or tribe), which has a median population of about 600 (range: 23–11,800) individuals (Marlowe 2005).

Marriage is nearly universal across human societies, and in foraging societies it typically occurs within the ethnolinguistic unit. It is common for newly married couples to live with the wife's parents while the husband performs what anthropologists call "bride service." This means that the new husband labors on behalf of his new wife and her kin, bringing them meat and other foods regularly. This typically lasts for several years. But after that, residence patterns are quite variable, and people have considerable flexibility in their decisions about where to live and with whom. Despite a common impression that men live with their kin more than women do, there is no consistent sex bias in postmarital residence among warm climate foragers

(Marlowe 2004). A couple may live with the wife's kin one month, with the husband's kin the next month, then with the kin of both or neither. The best term for this pattern is multilocal residence.

The great majority of human foragers trace their kinship bilaterally through their mothers and fathers. This extends kinship networks and facilitates peoples' movements between residential camps (Marlowe 2004). With sedentism, agriculture, and wealth accumulation and inheritance, virilocality (living with the husband's kin) and patrilineal descent and inheritance systems become the dominant pattern.

Human Mating Systems

There is great variation in human mating patterns, but in virtually all societies reproduction takes place mainly within marriages. It is important to distinguish the "social mating system" from the "genetic mating system" (table 20.2). The social mating system is what we can observe directly. It might appear to be monogamous, but if females sometimes secretly conceive with an extrapair male, the genetic mating

Table 20.2. Terms and definitions

Attractivity: The female state of being sexually attractive to males.

Estrus: The period in which females are receptive, proceptive, and/or attractive to males.

Mate bonds: Because the term *pair bonds* is so often interpreted as monogamy only, I instead use *mate bonds*, to imply more enduring and possessive relationships than the brief sexual encounters of promiscuity or brief consortship. Mate bonds are assumed to exist primarily for mating; they include monogamy, polyandry, and polygyny.

Mating System: The *social mating system* is what we observe directly: in primates, who mates with whom; in humans, who is married to whom (fig. 20.1). There may be different types of mate bonds between individuals, but a society should be classified by the modal type of mate bond. The *effective mating system* is polygynous if variance in male reproductive success is greater than female reproductive success, effectively polyandrous if female reproductive success is greater, and effectively monogamous if the reproductive success of both sexes is equal. The *genetic mating system* is based on actual parentage as revealed by DNA paternity tests.

Operational sex ratio: The ratio of reproductively active males to females.

Potential reproductive rate: The number of potential offspring with unlimited access to mates.

Proceptivity: The female state of actively initiating sex.

Receptivity: The female state of being receptive to male sexual solicitation.

Reproductive skew: The proportion of matings, or reproductive success, allocated to each member of one sex. High skew occurs when one or a few individuals monopolize mating or reproduction. Low skew occurs when mating or reproduction is more equally distributed.

Reproductive success (RS): The number of surviving offspring controlled for age, or number of offspring surviving to a certain age.

Total fertility rate (TFR): The number of offspring to which a female gives birth in her lifetime, calculated by summing the age-specific fertility across age cohorts.

system would be polyandrous. If males also breed outside their bond it would also be polygynous, and thus polygamous overall. The genetic mating system in humans is rarely measured, because these data can create ethical concerns for researchers and the communities they study. We are therefore constrained to focus more on the social mating system.

The extent of reproductive skew within societies is linked to their mating systems. Like other primates, human females have long periods of gestation and lactation, and this skews the operational sex ratio (table 20.2) in favor of males. In humans, the potential reproductive rate (table 20.2) of males is much higher than that of females. In most animals, this generates higher variance and greater skew in male reproductive success (table 20.2) than female reproductive success, and humans seem to be no exception to this pattern. There is greater genetic polymorphism in human X chromosomes than in autosomes, which indicates that there has long been greater skew in male reproductive success than in female reproductive success, and a smaller effective population of males (Hammer et al. 2008). However, there is considerable variation in the extent of reproductive skew and the variance in reproductive success within and between the sexes across societies. In some societies there is no significant sex difference in the variance in reproductive success, while in others the variance in male reproductive success is four times as high as the variance in female reproductive success (table 20.3).

Mate Choice

The mode of subsistence and the distribution of resources are expected to have an important influence on mate choice. According to the polygyny threshold model (Verner & Willson 1966; Orians 1969), when males defend a territory or resources and there is considerable variation in their holdings, it may be more advantageous for a female to be the second mate of a richer male than the only mate of a poorer male. This is what behavioral ecologists call the polygyny threshold. This model has also been invoked to explain variation in human mating decisions. When men control resources that women need for themselves and their offspring, women may gain greater fitness advantages in polygynous marriages with wealthy men than in monogamous marriages with poor men (Borgerhoff Mulder 1988). In fact, cross-cultural rates of polygny are generally high in pastoral societies, in which men control the herds.

However, rates of polygyny are equally high in horticultural societies, in which women do not rely on men for access to resources. This pattern may seem counterintuitive, but it is not unique to humans. Among birds and mammals,

Table 20.3. Mean reproductive success and reproductive skew in 18 societies

Population	Subsistence	Male reproductive success		Female reproductive success		Ratio of male variance to female variance
		Mean	Variance	Mean	Variance	
Aka	Forager	6.3	8.6	6.2	5.2	1.7
Dobe !Kung	Forager	5.1	8.6	4.7	4.9	1.8
Hadza	Forager	4.3	9.8	3.6	5.1	1.9
Ache	Forager	6.4	15.1	7.8	3.6	4.2
Pitcairn Island	Horticultural/forager	4.6	23.6	4.7	23.2	1
Xavante	Horticultural/forager	3.6	12.1	3.6	3.9	3.1
Dominican	Horticultural	4.4	14.3	5	11.6	1.2
Pimbwe	Horticultural	6	9	6.1	7.3	1.2
Yanomamo	Horticultural	3.7	10.1	3.4	4.4	2.3
Dogon	Horticultural	6.1	10.7	3.2	2.3	4.8
Yomut Turkmen	Pastoral	5.1	8.1	3.9	7.1	1.1
Dazagada	Pastoral	8.6	15,0	6.4	6.5	2.3
Arab	Pastoral	10.3	14.4	8.3	5.1	2.8
Kipsigis	Agricultural/pastoral	10.9	24.4	6.6	5.9	4.2
Finnish	Agricultural/industrial	3.4	6	3.5	7.6	0.8
Norwegian	Agricultural/industrial	4.7	8.5	4.5	8.3	1
Swedish	Agricultural/industrial	2.1	11.5	2.4	9.7	1.2
American (United States)	Industrial	2	2.3	2	1.8	1.3

Where mean reproductive success for males and females is not equal, the data do not come from a closed population.
Source: Brown et al. 2009

males typically provide little paternal investment in species in which polygyny is common. Pair bonding is generally common in species in which males invest in their offspring in the form of provisioning or direct care (Reichard 2003). Among primates, however, paternal care is not restricted to or universal among pair-bonded species (chapter 17, this volume).

This suggests that there must be additional routes to polygyny. When men do not monopolize resources and women have direct access to the resources they need, females may place greater value on physical appearance or other cues of male genetic quality (Marlowe 2003b). If there is consensus among women about the males that are most attractive and women can act on their preferences, polygyny should result. Alternatively, if women acquire most of the resources they need for themselves and their offspring, men may compete for social status in other ways (fighting or other forms of intimidation or prestige), and the men who prevail in these contests may gain greater access to mates (Clutton-Brock 1989). It may benefit a woman to be married to the man with the most social influence over others, even if she must share him with her co-wives. In fact, when men provide very little of the diet, women must feed the men, and under certain circumstances it may be cheaper to share these costs with cowives than to bear the entire cost alone.

When some men do provide substantial food, it should

pay women to trade off genetic quality for provisioning. Figure 20.2 shows how men may switch from direct contest competition to foraging competition if females prefer men who provide the most food. Because of food sharing, hunters can rarely keep all their food, and the polygyny threshold is rarely reached in foraging societies. Among foragers, where males contribute more to the diet the degree of polygyny is lower (Marlowe 2003b), and where male contribution to the diet is greater, female reproductive success is enhanced (Marlowe 2001).

In most foraging societies, the extent of polygyny is quite limited and male reproductive skew is low (Marlowe 2003b). Across foragers, the mean percentage of monogamously married men is 86% (range: 30–100%; $n = 212$) and the mean percentage of married women without a cowife is 79% (range: 10–100%; $n = 51$; Marlowe & Berbesque 2009a). Despite this, there is considerably more reproductive skew in males than in females, mainly because of a sex difference in remarriage rates (table 20.3). Australian aborigines provide a conspicuous exception to these patterns. In these foraging societies, some men have as many as 10 wives while others have none. Australian polygyny is related to age and status; most old men have several wives, while most young men have none (Hart & Pilling 1960; Goodale 1971). Just why the Australians are so polygynous is still something of a mystery. Because marital skew is age-

Fig. 20.1. Although reproduction in humans involves cohesive bonds and jural arrangements between males and females—often labeled "marriage"—the nature of these relationships, and the mating system encompassing them, vary greatly. Most societies sanction polygyny, though its expression is limited, such that even in those cultures monogamous marriage is usually more common. Industrialized cultures are among the few that proscribe polygyny, but there is nevertheless variation—such as the polygyny occurring in the southwestern United States of America, as represented by these two women married to the same man in Herriman, Utah. Although once concealed, polygynous marriage has become increasingly open as fundamentalist Mormons and others push for its decriminalization on religious grounds. Photo © STRINGER/USA/x01447/Reuters/Corbis.

graded, there may be much less skew in lifetime RS than across all adult men.

Polyandry, in which one woman is married to multiple husbands, is common in only a handful of societies, mainly in Tibet, Nepal, and India (Goldstein 1976; Crook & Crook 1988; Levine & Silk 1997; Haddix 2001). However, some societies in South America practice a de facto form of polyandrous mating. Women mate with more than one man and paternity is partitioned among the husband and some number of other men (Beckerman & Valentine 2002). People say there that it takes the semen of several men to create a baby.

Although the previous discussion assumes that men and women exert considerable autonomy over their own mating decisions, arranged marriages are the norm in many traditional societies. In arranged marriages, the parents select their offspring's marriage partners. In some societies the child may have to agree to the partner his or her parents have chosen, but in others children have little say. For example, among the Tiwi foragers of Australia (Hart & Pilling 1960), older men traded daughters and even betrothed them before they were born. When women's husbands are determined for them before they are even born, they obviously have little choice. Parents may negotiate a bride wealth payment for their daughter and then use the resource to their own advantage, or to fund the bride wealth their son needs to acquire a wife. Parents may often choose mates for their children that will enhance their child's reproductive success,

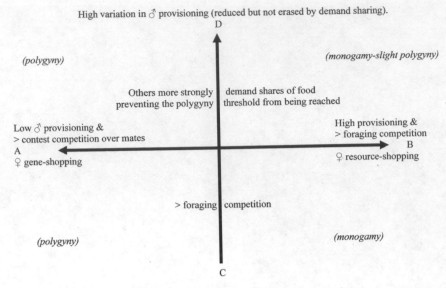

Fig. 20.2.   Forager mating systems. Mearsured along one axis (A-B) is the mean level of male provisioning, and along another axis (C-D) the degree of variation in that level. When males provide most food but there is little variation in the amount they provide, monogamy prevails, because women resource-shop and the polygyny threshold is not reached. Where males provide most food and some provide much more than others, better providers might achieve polygyny (although there is also more demand for them to share food, thus minimizing the variation). Where males provide little food, females gain more from gene-shopping and prefer to mate polygynously, or are indifferent to the outcome of male-male contest competition.

since this will often (but not always) be the best way to enhance their own fitness as well (Buunk et al. 2008).

The amount of parent-offspring conflict over mate choice (Trivers 1974) may depend on residence patterns. For example, if daughters typically move away from their natal households, parents may sell their daughters for their own economic gain, if but if sons-in-law typically come to live with daughters, then parents may have overlapping interests with their daughters.

In many societies, marriage is associated with material transfers between the families of the prospective bride and groom. Bride wealth, money, or livestock that is given by the husband or his parents to the parents of the wife before the marriage is very common across traditional agricultural societies, indicating that females are usually the individuals in reproductive demand. Where there is intense competition among women for desirable husbands, dowry may evolve. Dowry is wealth given to the marrying couple by the wife's family. Its distribution in Eurasia is explained by pronounced social stratification combined with a proscription against polygyny, which together promote intense competition among females to marry the few high-status males (Gaulin & Boster 1990).

Other forms of coercion besides arranged marriages can influence how male-female mate bonds (table 20.2) are forged (Smuts 1995; Thornhill & Palmer 2000). Among the Yanomamo horticulturalists of Brazil, men would some-

times raid a neighboring village. They would often capture a woman and take her back to their own village, where she would be raped by all the men and then become the wife of one of them (Chagnon 1983). In foraging societies with little internal warfare, however, intraethnic rape is rare. There is little opportunity for rape in foraging societies, because there is absolutely no anonymity and the woman's kin could exact revenge (Smith et al. 2001). Rape seems to be an unfortunate consequence of continual estrus, and it may partly explain why forced copulation is also common in orangutans (*Pongo* spp., Fox 2002).

Paternity Certainty and Male Investment

According to theory, males are expected to be reluctant to invest in offspring when they have low confidence in paternity. Across taxa, including primates, relative testes size (testes weight relative to full body weight) is greater in species in which females mate with several males within one fertile period, because sperm competition favors males who produce higher volumes of sperm (Short 1980; Harcourt et al. 1981; Birkhead 2000; chapter 17, this volume). Testes volume is related to number of sperm ejaculated in humans (Simmons et al. 2004). Based on human testes size, men face much less sperm competition than do bonobos and chimpanzees, but more than do gorillas (*Gorilla go-*

rilla) and orangutans (Harcourt 1997). Data on paternity in humans are scarce, and the available data are likely to be biased for various reasons. For example, in a sample of men who pursued paternity cases because they had doubts about their paternity, rates of nonpaternity were about 25%. In contrast, in a sample of men who were seeking genetic counseling, the rate of nonpaternity was only 1.7% (Anderson 2006). The best available data suggest that rates of nonpaternity across a wide variety of different societies are about 4% (Anderson 2006). Another study of Western women in four nations concluded that extrapair copulations occur among 5% to 27% of women under 30 years of age, yielding a nonpaternity estimate of about 2% (Simmons et al. 2004).

In societies with partible paternity (the belief that more than one man can contribute to the formation and development of a fetus), paternity is not monopolized by the husband. Women report different numbers of fathers for each child. Among Ache foragers, secondary fathers include all those who may have had sex with a woman over the whole year prior to giving birth (Hill & Hurtado 1996). Among the Bari horticulturalists of Venezuela, women take additional lovers during pregnancy (Beckerman & Valentine 2002). These multiple paternity societies may have fairly high rates of conception by men other than the recognized husband (Walker et al. 2010).

When women cuckold their husbands by copulating with males of higher quality, the sperm of the two males might compete for fertilization. However, if the benefit of cuckoldry is fertilization by a higher quality male, females should avoid copulating with their husbands during the conceptive window to avoid sperm competition. This sexual tactic may often be difficult for females to execute because within-pair copulation usually occurs about three times a week (every 2.33 days) even after four years of marriage, at least in societies like the United States (Michael et al. 1994). Figure 20.3b shows that the probability of copulation is about 25% per day, or once every four days. Aka foragers report having sex with amazing frequency: three times per night (Hewlett & Hewlett 2008). These frequencies guarantee a husband a high probability of paternity, since the mean longevity of human sperm is about three days (Zinaman et al. 1989).

Female Receptivity and Ovulatory Signaling

Women's ability to manipulate mens' paternity confidence is enhanced by the fact that their ovulation is cryptic and they are receptive throughout their estrous cycles. In contrast, chimpanzees and bonobos have conspicuous genital

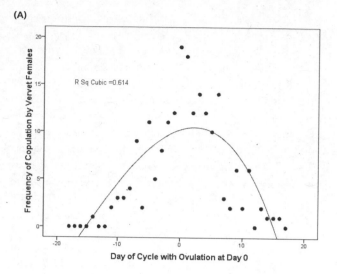

(A)

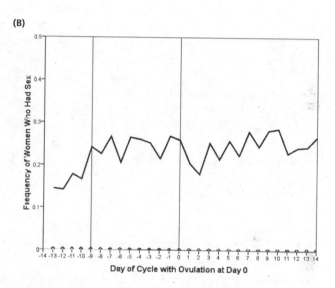

(B)

Fig. 20.3a. Frequency of copulation by day of cycle among female vervet monkeys (*Chlorocebus pygerythrus*) in three groups. Cubic relationship to day of cycle: $r^2 = 0.614$, $p < 0.0005$, $n = 194$ copulations). Redrawn from Andelman 1987, fig. 1.

Fig. 20.3b. Frequency of copulation by day of menstrual cycle, in women from 13 nations, ages 18–40 years. Line marks interpolation to day of cycle. Cubic relationship to day of cycle omits the five days of menses: $r^2 = 0.000$, $p = 0.397$, $n = 14,093$ women. Line in center indicates estimated day of ovulation; line at left indicates end of menses. Data from Brewis & Mayer 2005. Women recruited by demographic and health surveys in Armenia, Benin, Ethiopia, Guinea, Guatemala, Haiti, Khazakstan, Mali, Malawi, Nigeria, Nepal, Peru, and Rwanda. Only women using no chemical contraceptives were surveyed. Each woman was asked whether coitus had occurred within the preceding 24 hours. The independent variable was the woman's current cycle phase, categorized on the basis of the woman's reports of time passed since menstrual bleeding had last begun. The first day of menstrual bleeding was defined as cycle day 1. Cycle days 2 through 5 were classified as menstrual, days 6 through 11 as preovulatory (follicular), days 12 through 17 as ovulatory, and days 18 and up as postovulatory (luteal). Data are converted so that day 0 equals ovulation, day −14 equals the first day of menstrual bleeding, and day +14 and up equals end of luteal phase.

swellings and are sexually receptive only during a limited period within their estrous cycles (chapter 6, this volume).

Ovulation signaling is best viewed as a continuum rather than as a dichotomous trait. On one extreme are species that conspicuously advertise ovulation, though not always honestly; on the other extreme are those with no obvious signs of ovulation within the menstrual cycle. Ovulation can be advertised through morphological changes, such as genital swellings; through physiological changes, such as scent or pheromones; or through behavioral changes. Estrus is the period during which a female is (1) sexually receptive or (2) proceptive (initiating sex), and (3) especially sexually attractive to males (table 20.2). Morphological signaling can vary from exaggerated sexual swellings to no anogenital cues at all, while estrus can vary from a brief period of intense attractiveness or proceptivity around ovulation to little or no change across the cycle. If morphological and behavioral signals are confined to the conceptive period, ovulation is loudly and honestly advertised. When there are no morphological signals and females are receptive and attractive throughout their cycle, ovulation is effectively concealed and estrus is continual.

When swellings and estrus extend over a much longer period than the conception window, ovulation is at least partly obscured. Chimpanzees and bonobos are good examples of this because their swelling last about 14 to 15 days, which is 38% to 40% of their cycle length (Kutsukake & Nunn 2006). However, swellings do vary in size across this period, as does the intensity of male-male competition, thus indicating that ovulation is not totally concealed (Emery & Whitten 2003; Emery Thompson & Wrangham 2008). In chacma baboons (*Papio ursinus*), the period of maximal swelling corresponds to the period of maximum fertility, and females with larger swellings across cycles have a higher probability of conception. Furthermore, males copulate preferentially with females with larger swellings, which therefore appear to advertise ovulation and female genetic quality (Huchard et al. 2009) or current condition.

While exaggerated swellings occur in a minority of primates (about 25 species), behavioral estrus is typical of most species and is usually marked. The fact that males become more interested in females at particular times indicates that they detect changes in females: often a change in scent or pheromones, if there are no morphological changes. But the estrous behavior of females is itself a stimulus for males. Copulations in some species are limited to a few days near ovulation, but in other species they are not so limited (Martin 2007). For example, female vervet monkeys (*Chlorocebus pygerythrus*) may copulate throughout their cycle, and some consider them to have concealed ovulation (Andelman 1987). Yet, as fig. 20.3a shows, copulation frequency clearly peaks near ovulation in vervets.

Across the hominids, only chimpanzees and bonobos have exaggerated sexual swellings (table 20.1). Orangutans and gorillas have extremely slight genital swellings, and humans have no swellings at all. The most parsimonious conclusion is that exaggerated swellings evolved only once in *Pan* after the *Pan-Homo* split (Sillen-Tullberg & Moller 1993). The last common ancestor (LCA) of *Homo* and *Pan* likely had very slight genital swellings or none at all. Orangutans have an ill-defined period of estrus, gorillas a brief estrus, humans continual estrus, and chimpanzees and bonobos an estrous period that overlaps their swellings, with attractivity increasing at maximal swelling (Emery & Whitten 2003; table 20.1). Women have the most effectively concealed ovulation of all haplorrines. Even women who think they know when they are ovulating cannot actually tell (Sievert & Dubois 2005; but see Marinho et al. 1982 on mittelschmerz).

In contrast to many primates, women are sexually receptive throughout their cycles. Copulation rates among married couples show a flat frequency across the cycle, with a drop only during menses (Ford & Beach 1951; Udry & Morris 1968; James 1971; Hrdy & Whitten 1987; Brewis & Meyer 2005). Still, several studies have reported women's mate preferences change across the menstrual cycle, with more masculine males preferred at ovulation and less masculine, presumably more nurturing, males the rest of the time (Penton-Voak et al. 1999; Penton-Voak & Perrett 2001; Roney & Simmons 2008). Gangestad and Thornhill (2008) provide a summary of these findings. In one recent study, however, periovulatory women did not find masculine faces more attractive (Peters et al. 2008), so there is still cause for doubt. There is some evidence that women's libido increases near ovulation; in one study, women reported that they had increased sexual desire and felt more attractive at this time (Roder et al. 2009). Another study found that women attended more to their grooming and clothing around the time of ovulation than during the rest of the cycle (Haselton et al. 2007). One study found that lap dancers earned more in tips during their late follicular (fertile) phase than at other times in their cycle (Miller et al. 2007). This suggests that women who make their living by arousing men may give a more effective performance during the ovulatory phase, presumably because they themselves are more proceptive. This explanation might also explain why the same women earned the least money during their periods of menstruation.

Continual estrus and concealed ovulation presumably evolved because of their benefits for females. Three hypotheses for the evolution of these traits in humans are viable.

1.) Paternity confusion to prevent infanticide. According to this hypothesis, the timing of ovulation was concealed and estrus extended long enough to promote matings with many males. Each male who copulates has some chance of siring that female's subsequent offspring, and therefore refrains from killing it (Hrdy 1981; Smith & Smith 1994; Heistermann et al. 2001). This hypothesis does not seem to be relevant for contemporary humans, because females rarely mate with all or most males in their group, and sexually selected infanticide is uncommon in human societies (Daly & Wilson 1988; Smith & Smith 1994; Hrdy 1999). Of course, conditions could have been quite different when extended estrus first evolved in our human ancestors.

2.) Bonds based on mate guarding and provisioning. Concealment of ovulation meant that males could not limit copulation to the relatively brief period around ovulation, but had to mate guard a female continuously to ensure a fertilization. Females may have benefited from male proximity if males provided direct care of offspring, provisioned them and their offspring, or increased their access to resources (Alexander & Noonan 1979). Females might also have benefited from a male's protection against sexual harassment from other males or predators, or from his defense of food (Wrangham et al. 1999). Additionally, females may have traded sex for food. Strassmann (1981) suggests that it was likely subordinate males who began to offer provisioning or infant care in exchange for sexual access as an alternative mating tactic once estrus was extended (chapter 17, this volume). This is because dominant males already had access to estrous females and monopolized them without such gifts. One drawback to the mate-guarding hypothesis is that extant hunter-gatherer males and females forage separately. Thus, extended estrus and mate bonds would have to have evolved prior to the sexual division of foraging labor.

3.) Strategic extrapair matings. Estrus was extended and ovulation concealed so that a female's long-term mate could not know when to guard her most closely, and this enabled her to pursue extrapair copulations with a different, more desirable male at the time of conception (Benshoof & Thornhill 1979). This may be advantageous to females because in species with pair bonds, both partners should be choosy. Therefore, a female of average mate quality can only acquire a male of average mate quality for a long-term partnership. That same average female can nevertheless interest a highly valued male in a copulation that costs him very little (Symons 1979). The cuckoldry hypothesis outlines a way in which a female can "have it all" reproductively: a good helper or provider of average genetic quality, plus a better sire for her offspring. This hypothesis assumes that mate bonds exist and that females possess a mechanism that modulates sexual behavior with cycle phase. Sufficiently high cuckoldry rates would constitute some evidence to support this hypothesis, though it is not clear how high is high enough.

Sexual Division of Labor

In most primates, males and females forage for the same foods and there is little food sharing. Humans represent a conspicuous exception to this pattern. Almost all human foragers have pronounced sexual division of foraging labor and extensive food sharing (Marlowe 2007). This pattern has long been considered a uniquely human adaptation for household cooperation (Lancaster & Lancaster 1980; chapter 13, this volume). Men tend to target less predictable but high-value foods, like meat and honey, while women tend to target less valuable but more reliable foods like fruit and tubers (fig. 20.4), which guarantee the daily provisioning of their children (Hawkes 1996). The acquisition of these foods is also usually compatible with care of dependent offspring (Brown 1970). When daily food harvest varies greatly across individuals, sharing reduces the chances of an individual going without food on any given day (Winterhalder 1986). Sex-specific foraging specialization also broadens and probably improves the diet's nutritional content (Cordain et al. 2000; Kaplan et al. 2000; Marlowe & Berbesque 2009b). Sex differences in foraging specialization favor sharing, and lead to considerable economic interdependence between males and females.

There is considerable debate about the forces that shape the division of labor between males and females. One hypothesis is that the division of foraging labor is a form of coordinated cooperation between husbands and wives (Hurtado & Hill 1990; Kaplan et al. 2000; Marlowe 2007). Husbands and wives make good trading partners because they have different relative advantages. One mate can nurse and care for dependent offspring while foraging; the other mate has greater size and strength, which may be required for accessing certain foods. The partners also share an interest in their offspring, and may therefore be less likely to defect on their obligations. Good hunters may be desirable mates because they will be reliable providers for their wives and offspring. This hypothesis predicts that women experience higher reproductive success where there is a substantial division of foraging labor, all else being equal. It also predicts that the wives of good hunters will have higher reproductive success than the wives of poor hunters.

Although this hypothesis makes sense, it does not consider the fact that men hunt large-package foods that are shared across households. This means that the wife and children of a poor hunter may eat as well as the wife of a

Fig. 20.4. A Hadza couple gathers berries in northern Tanzania. Although both husband and wife are engaged in the same activity, the husband carries his bow and arrows in case he encounters a hunting opportunity along the way. The wife not only eats berries but fills her sling with them for transport back to the camp. Photo courtesy of Frank Marlowe.

good hunter. An alternative hypothesis proposes that men are less interested in provisioning their households than in gaining reputations as good hunters and generous suppliers of food to everyone. According to this view, hunting serves a signaling function (Hawkes 1990; Bird 1999; Bird et al. 2001; Hawkes et al. 2001b; Hawkes & Bliege Bird 2002; Smith 2004). It is an honest signal of superior phenotypic (possibly genotypic) quality because it is a highly skilled endeavor. Gaining a reputation as a skilled hunter could be useful because to men it means that they would be high-quality mates, beneficial allies, or formidable opponents (fig. 20.5). If hunting has this kind of signaling function, then mens' efforts may not be focused on provisioning their own households. This hypothesis predicts that women will have lower reproductive success when there is more division of foraging labor, at least if food is the most important factor in female reproduction.

It is not yet clear how the extent of sexual division of la-

bor affects women's reproductive success, and considerable debate persists about the kinds of benefits men derive from hunting. There is certainly much sharing of men's foods (see also chapter 13, this volume), but there is no consensus about whether the wives of successful hunters obtain a disproportionate share of their husbands' kills. Among the Hadza, a well-studied foraging group in northwestern Tanzania, for example, some reports indicate that the hunters' household does not obtain any more food than other households (Hawkes et al. 2001), while more recent analyses indicate that the hunter's household eats more of the food he acquires than do other households (Wood 2010). These two hypotheses need not be mutually exclusive; there could be a bit of truth to both. Men may work hard to gain reputations as good hunters to raise their value in the mating market. Even if a good hunter has only one wife, he may be able to marry at a younger age or marry a higher-quality woman than a poor hunter. When women can choose their

Fig. 20.5. A Tsimane hunter of the Bolivian Amazon demonstrates how to shoot a capuchin monkey in the canopy. Though plantains, rice, and sweet manioc comprise up to two-thirds of the calories in the Tsimane diet, hunting is still a highly valued and time-intensive activity. Men in most horticulturalist societies actively hunt, and their hunting ability is a valued trait among mates and social allies. Photo courtesy of Michael Gurven.

mates and insist on exclusive bonds, they may still obtain more food for their households by marrying men who advertise their quality (Marlowe 2003a, 2007). Even if men's foraging is partly mating effort, there is plenty of evidence that men also invest in parenting effort. There is not always a trade-off between the two (Stiver & Alonzo 2009). For example, the children of the most productive Hadza hunters received as much direct care from their fathers as did the children of less proficient hunters (Marlowe 1999). It is also possible that there is variation in the importance of these factors across time and space. For example, in the Arctic, cooperation within the household is essential because men acquire virtually all of the food most of the year while women tend children, make clothes, and process foods (Lee 1968; Kelly 1995; Binford 2001).

Parental Investment and Allomothering

Women give birth to infants that are incapable of holding onto their mothers for several months, which means that they must be carried (Ross 2001). Without a sling to carry her infants, a mother's foraging efficiency is greatly compromised. In addition, the brain of a human infant grows for a longer period than do those of our great ape relatives, which means that hominin infants develop more slowly and must have been more of a burden. Among foragers and in other small-scale societies, mothers do the bulk of infant care (Konner 2005) but also get considerable help from allomothers, especially with the tending of older children (fig. 20.6; Hrdy 2009). Older siblings and even non-kin

Fig. 20.6. Alloparenting is very common among the Mukogodo people of Kenya. Almost all the alloparenting is done by girls and women—usually the child's older sisters or grandmothers, though cousins, nieces, and other female relatives also help. Alloparenting begins very early, with girls as young as five years old helping to care for their younger siblings (Cronk 2004). Photo courtesy of Lee Cronk and Beth L. Leech.

are frequent caretakers (Silk 1990; Ivey Henry et al. 2005; Crittenden & Marlowe 2008). Grandmothers are often a major source of help with direct care and provisioning (Hawkes et al. 1989; Sear et al. 2000). Fathers can also be important sources of care (Hewlett 1991; Marlowe 1999; Gray & Anderson 2010). Forager men spend more time near their young children than do horticultural, pastoral, or agricultural men (Marlowe 2000a). The most involved fathers on record are the Aka foragers. Because men and women often go net hunting together, fathers hold their children a great deal (mean = 22%) of the time (Hewlett 1991). In some industrialized societies, fathers also provide substantial help with child care (Hwang & Lamb 1997; Gray et al. 2002).

The challenge hypothesis proposes that testosterone (T) facilitates male reproductive effort, mainly in the form of mate seeking, acquisition, and guarding, and at the expense of parenting effort (Wingfield et al. 1990; chapter 17, this volume). The data on many avian species fit this pattern (Wingfield et al. 1990; Beletsky et al. 1995; Wingfield et al. 2000). Once a female lays her eggs, her partner's T levels drop as he switches from mating to parenting effort. Lower T has been found in men with young children in some recent studies (Gray et al. 2002, 2006; Gray 2003). Among the largely monogamous Hadza, men provide considerable direct care to their children. Among their pastoralist neighbors, the Datoga, a sizeable percentage of men are polygynous, and men provide little direct care to their children. In accord with the challenge hypothesis, T was lower among Hadza fathers of young children than among those without young children, while there was no such difference between Datoga men (Muller et al. 2009).

When men invest in parenting effort, we should expect them to attend to signs that they are fathers of the children. There is evidence that they invest less in known stepchildren than in their genetic offspring (Anderson et al. 1999; Marlowe 1999), and that they invest more in the children whom they think resemble them (Apicella & Marlowe 2004; Alvergne et al. 2009). As noted above, the rate of cuckoldry in most human societies appears to be fairly low, and male jealousy helps to maintain this. Men exhibit a host of behaviors aimed at increasing paternity confidence (Platek & Shackelford 2006). While infanticide in humans is usually committed by the mother herself or her relatives (Hrdy 1999), the odds of infant homicide and unintentional childhood fatalities occurring are much higher when children live with unrelated adults (Daly & Wilson 1988, 1998; Tooley et al. 2006; Harris et al. 2007; but see Temrin et al. 2000).

In addition to direct allomaternal care, others also provide a substantial part of the diet. Much of this provisioning comes to a woman from her husband (Kaplan et al. 2000; Marlowe 2001; Gurven & Hill 2009; chapter 13, this volume). The widespread sharing of food across households among foragers means that there is not only offspring provisioning and mate provisioning but also provisioning of nonkin, including adults (Hawkes et al. 2001a; Alvard 2004; Gurven 2004). This sharing is partly due to scrounging (Jones 1984), though it appears more voluntary than it does in chimpanzees (Wrangham 1975; Mitani & Watts 2001).

Human foragers wean their infants early (median = 2.5 years, range = 1–4.5, SD = 0.84: Marlowe 2005) compared to chimpanzees (table 20.1). Maternal and allomaternal provisioning of weaning foods is responsible for early human weaning. Some human foods are processed, thus making them easier for young children to consume (Sellen 2007). Mothers continue to provision children after weaning, but are also assisted by others (Draper & Howell 2005; Crittenden & Marlowe 2008; Hill & Hurtado 2009). When forager women seek food and take their infants with them, they have the luxury of leaving their weanlings in camp with allomothers (Marlowe 2006), which helps explain the short (3.5-year) interbirth interval of human foragers (table 20.1). Male contribution to diet predicted female reproductive success among foragers, but it did not predict mortality rate (Marlowe 2001). This suggests that women do as much as possible to reduce mortality themselves, and use male provisioning mainly to speed up their reproduction.

With an early age at weaning and a short interbirth interval, a human mother can have three or four children who are simultaneously young enough to require some provisioning. This, it is often said, is what led to allomaternal provisioning. It makes more sense to reverse the sequence and causality. First, surplus food acquisition became possible (likely via increased tool use), which meant it was possible for some to give food to others. This potential to provision meant that females who managed to get others to help them had shorter interbirth intervals and possibly reduced juvenile mortality. The grandmother hypothesis proposes that grandmothers were crucial helpers, which may explain why only humans among primates have menopause (Hawkes et al. 1998; Shanley et al. 2007; chapter 13, this volume). Other important helpers were presumably long-term mates.

Male-Female Relationships in Contemporary Urban Settings

Although life in wealthy, urban, industrialized democracies is very different from life in foraging societies, many elements of reproductive strategy in those two kinds of societies are more similar to each other than they are to the reproductive strategies of men and women in agricultural societies. Women are generally less oppressed in industrialized nations and foraging societies than they are in agricultural societies, so female mate choice is less constrained. There is bilateral kin reckoning rather than a strong patrilineal bias. The inheritance of material wealth is less important (Borgerhoff Mulder et al. 2009). People can divorce relatively easily, and serial monogamy is common. These similarities are due to the high value placed on individual autonomy and the economic and political independence that makes it possible. However, mating and reproductive patterns in the industrialized democracies also differ in many ways from the patterns in foraging societies. Couples in industrialized societies live neolocally. Living inside dwellings means that food sharing outside the household occurs only on rare occasions. Mothers usually cannot take their infants with them to work, and so they wean them within months of birth, rather than years. Like foragers, they often leave their young children in the care of others while they go to work, but those others are not usually family members or close friends. The sexual division of labor has all but disappeared, at least in the workplace. Social monogamy has become the law (Fortunato & Archetti 2010) and table 20.3 suggests that the reproductive system is effectively monogamous, with similar variation in male and female reproductive success.

Human foragers with cultural practices that included sexual division of foraging labor, mate bonds, allomothers, and food sharing were so successful that they occupied virtually all habitats in every continent except Antarctica. The drop in mortality rates that began in the late 1800s in Europe has resulted in a subsequent drop in birthrates and the onset of the demographic transition. In many societies women are spending more time pursuing an education, thus delaying their first child—the main predictor of lower fertility. The percentage of childless men and women is increasing. In many European nations the birthrate is well below the usual replacement rate of 2.1. In Russia, the total fertility rate (TFR; table 20.2) is only 1.4.

The demographic transition might reflect genetic selection for low stable birthrates (optimal offspring number) that minimize the variance across generations (Chisholm 1993; Borgerhoff Mulder 1998). If so, we should find that those who had a small number of children—two or three, for example—would have more great-grandchildren than those who had a large number of children. So far, there is no evidence that this is true (Kaplan et al. 1995). We should also expect individuals with more resources to have a larger number of offspring than those with fewer resources. That is not the case (Mace 1998). At this time, the demographic transition still poses a challenge to explanation in terms of natural selection.

Summary and Conclusion

Before agriculture, and perhaps throughout the span of modern humans, our ancestors presumably lived in small local groups. Individuals moved between these groups and moved camp frequently within a large area occupied by a larger ethnolinguistic group. There were mate bonds within multimale-multifemale groups. Males and females targeted different foods and shared them upon return to camp. Females often provided an equal or larger share of the diet. Still, the benefits of additional male provisioning meant that females sometimes faced a trade-off between choosing males

on the basis of "good genes" and having help with feeding and caring for offspring. Occasionally, they might cuckold their mates and have both. Concealing ovulation by being attractive and receptive throughout the menstrual cycle (save menses) lowered male-male competition, increased egalitarianism, and gave females greater choice of long-term mates. The effect of investment by kin and mates allowed females to wean their children earlier without increased infant mortality. When males provided a larger part of the diet, mate bonds were more monogamous, though divorce was not uncommon. With plant and animal domestication came greater sedentism, male control of resources, property inheritance, and hierarchy. Despite reduced female choice and independence, fertility increased. In many places where females have more access to education and other resources, fertility has now dropped dramatically.

References

Alexander, R. D. & Noonan, K. M. 1979. Concealment of ovulation, parental care, and human social evolution. In *Evolutionary Biology and Human Social Behavior: An Anthropological Perspective* (ed. by N. Chagnon & W. Irons), 436–453. Belmont, CA.: Duxbury Press.

Alvard, M. 2004. Good hunters keep smaller shares of larger pies. *Behavioral and Brain Sciences*, 27, 560–561.

Alvergne, A., Faurie, C. & Raymond, M. 2009. Father-offspring resemblance predicts paternal investment in humans. *Animal Behaviour*, 78, 61–69.

Andelman, S. J. 1987. Evolution of concealed ovulation in vervet monkeys (*Cercopithecus aethiops*). *American Naturalist*, 129, 785–799.

Anderson, K. G. 2006. How well does paternity confidence match actual paternity? Evidence from worldwide nonpaternity rates. *Current Anthropology*, 47, 513–520.

Anderson, K. G., Kaplan, H., Lam, D. & Lancaster, J. 1999. Paternal care by genetic fathers and stepfathers II: Reports by Xhosa high school students. *Evolution and Human Behavior*, 20, 433–451.

Apicella, C. L. & Marlowe, F. W. 2004. Perceived mate fidelity and paternal resemblance predict men's investment in children. *Evolution and Human Behavior*, 25, 371–378.

Aureli, F., Schaffner, C. M., Boesch, C., Bearder, S. K., Call, J., Chapman, C. A., Connor, R., Di Fiore, A., Dunbar, R. I. M., Henzi, S. P., Holekamp, K., Korstjens, A. H., Layton, R., Lee, P., Lehmann, J., Manson, J. H., Ramos-Fernandez, G., Strier, K. B. & Van Schaik, C. P. 2008. Fission-fusion dynamics new research frameworks. *Current Anthropology*, 49, 627–654.

Bateman, A. J. 1948. Intra-sexual selection in *Drosophila*. *Heredity*, 2, 349–368.

Beckerman, S., Lizarralde, R., Ballew, C., Schroeder, S., Fingelton, C., Garrison, A. & Smith, H. 1998. The Bari partible paternity project: Preliminary results. *Current Anthropology*, 39, 164–167.

Beckerman, S. & Valentine, P. 2002. *Cultures of Multiple Fa-*

thers: *The Theory and Practice of Partible Paternity in Lowland South America*. Gainesville: University Press of Florida.

Beletsky, L. D., Gori, D. F., Freeman, S. & Wingfield, J. C. 1995. Testosterone and polygyny in birds. In *Current ornithology* (ed. by D. M. Power), 1–42. New York: Plenum Press.

Benshoof, L. & Thornhill, R. 1979. The evolution of monogamy and concealed ovulation in humans. *Journal of Social and Biological Structures*, 2, 95–106.

Bentley, G. R. 1999. Aping our ancestors: Comparative aspects of reproductive ecology. *Evolutionary Anthropology*, 7, 175–185.

Bergman, T. J., Ho, L. & Beehner, J. C. 2009. Chest color and social status in male geladas (*Theropithecus gelada*). *International Journal of Primatology*, 30, 791–806.

Betzig, L. 1982. *Despotism and Differential Reproduction: A Darwinian View of History*. Hawthorne, NY: Aldine de Gruyter.

Binford, L. R. 2001. *Constructing Frames of Reference*. Berkeley: University of California Press.

Bird, R. 1999. Cooperation and conflict: The behavioral ecology of the sexual division of labor. *Evolutionary Anthropology*, 8, 65–75.

Bird, R. B., Smith, E. A. & Bird, D. W. 2001. The hunting handicap: Costly signaling in human foraging strategies. *Behavioral Ecology and Sociobiology*, 50, 9–19.

Birkhead, T. 2000. *Promiscuity*. Cambridge, MA: Harvard University Press.

Boesch, C. & Boesch-Achermann, H. 2000. *The Chimpanzees of the Tai Forest: Behavioural Ecology and Evolution*. Oxford: Oxford University Press.

Borgerhoff Mulder, M. 1988. Is the polygyny threshold model relevant to humans? Kipsigis evidence. In *Mating Patterns* (ed. by C. G. N. Mascie-Taylor & A. J. Boyce), 209–230. Cambridge: Cambridge University Press.

———. 1998. The demographic transition: Are we any closer to an evolutionary explanation? *Trends in Ecology and Evolution*, 13, 266–270.

Borgerhoff Mulder, M., Bowles, S., Hertz, T., Bell, A., Beise, J., Clark, G., Fazzio, I., Gurven, M., Hill, K., Hooper, P. L., Irons, W., Kaplan, H., Leonetti, D., Low, B., Marlowe, F. W., McElreath, R., Naidu, S., Nolin, D., Piraino, P., Quinlan, R., Schniter, E., Sear, R., Shenk, M., Smith, E. A., von Rueden, C. & Wiessner, P. 2009. Intergenerational wealth transmission and the dynamics of inequality in small-scale societies. *Science*, 326, 682–688.

Brewis, A. & Meyer, M. 2005. Demographic evidence that human ovulation is undetectable (at least in pair bonds). *Current Anthropology*, 46, 465–471.

Brown, G. R., Laland, K. N. & Borgerhoff Mulder, M. 2009. Bateman's principles and human sex roles. *Trends in Ecology & Evolution*, 24, 297–304.

Brown, J. K. 1970. A note on the division of labor by sex. *American Anthropologist*, 72, 1073–1078.

Bryant, G. A. & Haselton, M. G. 2009. Vocal cues of ovulation in human females. *Biology Letters*, 5, 12–15.

Buunk, A. P., Park, J. H. & Dubbs, S. L. 2008. Parent-offspring conflict in mate preferences. *Review of General Psychology*, 12, 47–62.

Caillaud, D., Levrero, F., Gatti, S., Menard, N. & Raymond, M. 2008. Influence of male morphology on male mating status and behavior during interunit encounters in western lowland gorillas. *American Journal of Physical Anthropology*, 135, 379–388.

Chagnon, N. A. 1983. *Yanomamo: The Fierce People*, Third Edition. New York: Holt, Rinehart and Winston.

Chapais, B. 2008. *Primeval Kinship*. Cambridge, MA.: Harvard University Press.

Chisholm, J. S. 1993. Death, hope, and sex: Life-history theory and the development of reproductive strategies. *Current Anthropology*, 34, 1–24.

Clutton-Brock, T. H. 1989. Mammalian mating systems. *Proceedings of the Royal Society of London*, 236, 339–372.

Clutton-Brock, T. H. & Parker, G. A. 1992. Potential reproductive rates and the operation of sexual selection. *Quarterly Review of Biology*, 67, 437–456.

Cordain, L., Boyd Eaton, S., Miller, J. B., Mann, N. & Hill, K. 2002. The paradoxical nature of hunter-gatherer diets: Meat-based, yet non-atherogenic. *European Journal of Clinical Nutrition*, 56, 42–52.

Cordain, L., Miller, J. B., Eaton, S. B., Mann, N., Holt, S. H. A. & Speth, J. D. 2000. Plant-animal subsistence ratios and macronutrient energy estimations in worldwide hunter-gatherer diets. *American Journal of Clinical Nutrition*, 71, 682–692.

Crittenden, A. N. & Marlowe, F. W. 2008. Allomaternal care among the Hadza of Tanzania. *Human Nature: An Interdisciplinary Biosocial Perspective*, 19, 249–262.

Cronk, Lee. 2004. *From Mukogodo to Maasai: Ethnicity and Cultural Change in Kenya*. Boulder, CO: Westview Press.

Crook, J. H. & Crook, S. J. 1988. Tibetan polyandry: Problems of adaptation and fitness. In *Human Reproductive Behaviour: A Darwinian Perspective* (ed. by L. Betzig, M. B. Mulder & P. Turke), 97–114. Cambridge: Cambridge University Press.

Daly, M. & Wilson, M. 1988. *Homicide*. New York: Aldine de Gruyter.

———. 1998. *The Truth about Cinderella: A Darwinian View of Parental Love*. New Haven: Yale University Press.

Darwin, C. 1871. *Descent of Man and Selection in Relation to Sex*. Princeton, NJ: Princeton University Press.

De Lathouwers, M. & Van Elsacker, L. 2005. Reproductive parameters of female *Pan paniscus* and *P. troglodytes*: Quality versus quantity. *International Journal of Primatology*, 26, 55–71.

Dixson, A. F. 1998. *Primate Sexuality: Comparative Studies of the Prosimians, Monkeys, Apes, and Human Beings*. Oxford: Oxford University Press.

Dixson, A. F. 2009. *Sexual Selection and the Origins of Human Mating Systems*. Oxford: Oxford University Press.

Draper, P. & Howell, N. 2005. The growth and kinship of resources of Ju/'hoansi children. In *Hunter-Gatherer Childhoods: Evolutionary, Developmental and Cultural Perspectives* (ed. by B. S. Hewlett & M. E. Lamb), 262–281. New Brunswick, NJ: Transaction.

Ellison, P. T. 1994. Extinction and descent. *Human Nature: An Interdisciplinary Biosocial Perspective*, 5, 155–165.

———. 2003. Energetics and reproductive effort. *American Journal of Human Biology*, 15, 342–351.

Ember, C. R., Ember, M. & Pasternak, B. 1974. Development of unilineal descent. *Journal of Anthropological Research*, 30, 69–94.

Emery, M. A. & Whitten, P. L. 2003. Size of sexual swellings reflects ovarian function in chimpanzees (*Pan troglodytes*). *Behavioral Ecology and Sociobiology*, 54, 340–351.

Emery Thompson, M. & Wrangham, R. W. 2008. Male mating interest varies with female fecundity in *Pan troglodytes schweinfurthii* of Kanyawara, Kibale National Park. *International Journal of Primatology*, 29, 885–905.

Flinn, M. V. & Low, B. S. 1986. Resource distribution, social competition, and mating patterns in human societies. In *Ecological Aspects of Social Evolution* (ed. by D. I. Rubenstein & R. Wrangham), 217–243. Princeton, NJ: Princeton University Press.

Foley, R. A. 1996. Is reproductive synchrony an evolutionary stable strategy for hunter-gatherers? *Current Anthropology*, 37, 539–545.

Ford, C. S. & Beach, F. A. 1951. *Patterns of Sexual Behavior*. New York: Harper and Row.

Fortunato, L. & Archetti, M. 2010. Evolution of monogamous marriage by maximization of inclusive fitness. *Journal of Evolutionary Biology*, 23, 149–156.

Fox, E. A. 2002. Female tactics to reduce sexual harassment in the Sumatran orangutan (*Pongo pygmaeus abelii*). *Behavioral Ecology and Sociobiology*, 52, 93–101.

Frayser, S. G. 1985. *Varieties of Sexual Experience: An Anthropological Perspective on Human Sexuality*. New Haven: HRAF Press.

Gangestad, S. W. & Thornhill, R. 2008. Human oestrus. *Proceedings of the Royal Society B Biological Sciences*, 275, 991–1000.

Gaulin, S. J. & Boster, J. S. 1990. Dowry as female competition. *American Anthropologist*, 92, 994–1005.

Goldstein, M. C. 1976. Fraternal polyandry and fertility in a high Himalayan valley in northwest Nepal. *Human Ecology*, 4, 223–233.

Goodale, J. C. 1971. *Tiwi Wives: A Study of the Women of Melville Island, North Australia*. Seattle: University of Washington Press.

Gray, P. B. 2003. Marriage, parenting, and testosterone variation among Kenyan Swahili men. *American Journal of Physical Anthropology*, 122, 279–286.

Gray, P. B. & Anderson, K. G. 2010. *Fatherhood: Evolution and Human Paternal Behavior*. Cambridge, MA: Harvard University Press.

Gray, P. B., Kahlenberg, S. M., Barrett, E. S., Lipson, S. F. & Ellison, P. T. 2002. Marriage and fatherhood are associated with lower testosterone in males. *Evolution and Human Behavior*, 23, 193–201.

Gray, P. B., Yang, C. F. J. & Pope, H. G. 2006. Fathers have lower salivary testosterone levels than unmarried men and married non-fathers in Beijing, China. *Proceedings of the Royal Society B Biological Sciences*, 273, 333–339.

Gurven, M. 2004. To give and to give not: The behavioral ecology of human food transfers. *Behavioral and Brain Sciences*, 27, 543–583.

Gurven, M. & Hill, K. 2009. Why do men hunt? A reevaluation of "man the hunter" and the sexual division of labor. *Current Anthropology*, 50, 51–74.

Haddix, K. A. 2001. Leaving your wife and your brothers: When polyandrous marriages fall apart. *Evolution and Human Behavior*, 22, 47–60.

Hammer, M. F., Mendez, F. L., Cox, M. P., Woerner, A. E. & Wall, J. D. 2008. Sex-biased evolutionary forces shape genomic patterns of human diversity. *Plos Genetics*, 4, 8.

Harcourt, A. H. 1997. Sperm competition in primates. *American Naturalist*, 149, 189–194.

Harcourt, A. H., Harvey, P., Larson, S. & Short, R. 1981. Testis weight, body weight and breeding system in primates. *Nature*, 293, 55–57.

Harris, G. T., Hilton, N. Z., Rice, M. E. & Eke, A. W. 2007. Children killed by genetic parents versus stepparents. *Evolution and Human Behavior*, 28, 85–95.

Hart, C. W. & Pilling, A. R. 1960. *The Tiwi of North Australia*. New York: Holt, Rinehart & Winston.

Haselton, M. G., Mortezaie, M., Pillsworth, E. G., Bleske-Rechek, A. & Frederick, D. A. 2007. Ovulatory shifts in human female ornamentation: Near ovulation, women dress to impress. *Hormones and Behavior*, 51, 40–45.

Hawkes, K. 1990. Why do men hunt? Some benefits for risky strategies. In *Risk and Uncertainty in Tribal and Peasant Economies* (ed. by E. Cashdan), 145–166. Boulder: Westview.

———. 1996. Foraging differences between men and women: Behavioural ecology of the sexual division of labour. In *The Archaeology of Human Ancestry: Power, Sex and Tradition* (ed. by S. Shennan & J. Steele), 283–305. London: Routledge.

Hawkes, K. & Bliege Bird, R. 2002. Showing off, handicap signaling, and the evolution of men's work. *Evolutionary Anthropology*, 11, 58–67.

Hawkes, K., O'Connell, J. & Jones, N. G. B. 1989. Hardworking Hadza grandmothers. In *Comparative Socioecology: The Behavioural Ecology of Humans and Other Mammals* (ed. by V. Standen & R. Foley), pp. 341–366. Oxford: Blackwell.

Hawkes, K., O'Connell, J. F. & Jones, N. G. B. 2001a. Hadza meat sharing. *Evolution and Human Behavior*, 22, 113–142.

———. 2001b. Hunting and nuclear families: Some lessons from the Hadza about men's work. *Current Anthropology*, 42, 681–709.

Hawkes, K., O'Connell, J. F., Jones, N. G. B., Alvarez, H. & Charnov, E. L. 1998. Grandmothering, menopause, and the evolution of human life histories. *Proceedings of the National Academy of Sciences of the United States of America*, 95, 1336–1339.

Heistermann, M., Ziegler, T., van Schaik, C. P., Launhardt, K., Winkler, P. & Hodges, J. K. 2001. Loss of oestrus, concealed ovulation and paternity confusion in free-ranging Hanuman langurs. *Proceedings of the Royal Society B Biological Sciences*, 268, 2445–2451.

Hewlett, B. L. & Hewlett, B. S. 2008. A biocultural approach to sex, love, and intimacy in Central African foragers and farmers. In *Intimacies: Love and Sex Across Cultures* (ed. by W. R. Jankowiak), 39–64. New York: Columbia University Press.

Hewlett, B. S. 1991. *Intimate Fathers: The Nature and Context of Aka Pygmy Paternal Infant Care*. Ann Arbor: University of Michigan Press.

Hill, K. & Hurtado, A. M. 1996. *Ache Life History: The Ecology and Demography of a Foraging People*. New York: Aldine de Gruyter.

———. 2009. Cooperative breeding in South American hunter-gatherers. *Proceedings of the Royal Society B Biological Sciences*, 276, 3863–3870.

Hill, K. & Kaplan, H. 1993. On why male foragers hunt and share food. *Current Anthropology*, 34, 701–710.

Hill, K. R., Walker, R., Bozicevic, M., Eder, J., Headland, T., Hewlett, B., Magdalena Hurtado, A., Marlowe, F. W., Wiessner, P. & Wood, B. M. 2011. Coresidence patterns in hunter-gatherer societies show unique human social structure. *Science*, 331, 1286–1289.

Hong, L. K. 1984. Survival of the fastest: On the origin of premature ejaculation. *The Journal of Sex Research*, 20, 109–122.

Hrdy, S. B. 1981. *The Woman That Never Evolved*. Cambridge, MA: Harvard University Press.

———. 1999. *Mother Nature*. New York: Pantheon.

———. 2009. *Mothers and Others: The Evolutionary Origins of Mutual Understanding*. Cambridge, MA.: Belknap Press.

Hrdy, S. B. & Whitten, P. L. 1987. Patterning of sexual activity. In *Primate Societies* (ed. by B. Smuts, D. L. Cheney, R. M. Seyfarth, R. W. Wrangham & T. T. Struhsaker), 370–384. Chicago: University of Chicago Press.

Huchard, E., Courtiol, A., Benavides, J. A., Knapp, L. A., Raymond, M. & Cowlishaw, G. 2009. Can fertility signals lead to quality signals? Insights from the evolution of primate sexual swellings. *Proceedings of the Royal Society B Biological Sciences*, 276, 1889–1897.

Hurtado, A. M. & Hill, K. R. 1990. Seasonality in a foraging society: Variation in diet, work effort, fertility, and sexual division of labor among the Hiwi of Venezuela. *Journal of Anthropological Research*, 46, 293–346.

Hwang, C. P. & Lamb, M. E. 1997. Father involvement in Sweden: A longitudinal study of its stability and correlates. *International Journal of Behavioral Development*, 21, 621–632.

Isaac, G. 1978. The food-sharing behavior of protohuman hominids. *Scientific American*, 238, 90–108.

Ivey Henry, P., Morelli, G. A. & Tronick, E. Z. 2005. Child caretakers among the Efe foragers of the Ituri forest. I: *Hunter-Gatherer Childhoods: Evolutionary, Developmental and Cultural Perspectives* (ed. by B. S. Hewlett & M. E. Lamb), 191–213. New Brunswick, NJ: Transaction.

James, W. H. 1971. Distribution of coitus within human intermenstruum. *Journal of Biosocial Science*, 3, 159–171.

Johnson, T. & Dye, J. 2005. *Indicators of Marriage and Fertility in the United States from the American Community Survey: 2000 to 2003* (ed. by P. Division). Washington: United States Census Bureau. Accessed at www.census.gov/hhs/fertility/data/acs/indicators.abstract.html.

Johnstone, R. A. 2000. Models of reproductive skew: A review and synthesis (invited article). *Ethology*, 106, 5–26.

Johnstone, R. A., Reynolds, J. D. & Deutsch, J. C. 1996. Mutual mate choice and sex differences in choosiness. *Evolution*, 50, 1382–1391.

Jones, N. G. B. 1984. Selfish origin for human food sharing: tolerated theft. *Ethology and Sociobiology*, 5, 1–3.

———. 1987. Tolerated theft: Suggestions about the ecology and evolution of sharing, hoarding, and scrounging. *Social Science Information*, 26, 31–54.

Kaplan, H., Hill, K., Lancaster, J. & Hurtado, A. 2000. A theory of human life history evolution: Diet, intelligence, and longevity. *Evolutionary Anthropology*, 9, 156–185.

Kaplan, H., Lancaster, J. B., Bock, J. A. & Johnson, S. E. 1995. Fertility and fitness among Albuquerque men: A competitive labour market theory. In *Human Reproductive Decisions* (ed. by R. I. M. Dunbar), 96–136. London: St. Martin's Press.

Kelly, R. L. 1995. *The Foraging Spectrum*. Washington: Smithsonian Institution Press.

Knott, C. D. 2001. Female reproductive ecology of the apes: Implications for human evolution. In *Reproductive Ecology and Human Evolution* (ed. by P. T. Ellison), 429–463. Hawthorne, NY: Aldine de Gruyter.

———. 2009. Gunung Palung Orangutan Project. Accessed at people.bu.edu/orang/orangutans.html.

Kokko, H. & Jennions, M. D. 2008. Parental investment, sexual selection and sex ratios. *Journal of Evolutionary Biology*, 21, 919–948.

Kokko, H. & Monaghan, P. 2001. Predicting the direction of sexual selection. *Ecology Letters*, 4, 159–165.

Konner, M. J. 2005. Hunter-gatherer infancy and childhood: The !Kung and others. In *Hunter-Gatherer Childhoods: Evolutionary, Developmental and Cultural Perspectives* (ed. by B. S. Hewlett & M. E. Lamb), 19–64. New Brunswick, NJ: Transaction.

Kuroda, S. 1989. Developmental retardation and behavioral characteristics in the pygmy chimpanzees. In *Understanding Chimpanzees* (ed. by P. G. Heltne & L. Marquardt), 184–193. Cambridge, MA: Harvard Univeristy Press.

Kutsukake, N. & Nunn, C. L. 2006. Comparative tests of reproductive skew in male primates: The roles of demographic factors and incomplete control. *Behavioral Ecology and Sociobiology*, 60, 695–706.

Kvarnemo, C. & Ahnesjo, I. 1996. The dynamics of operational sex ratios and competition for mates. *Trends in Ecology and Evolution*, 11, 404–408.

Lancaster, J. B. & Lancaster, C. S. 1980. The division of labor and the evolution of human sexuality. *Behavioral and Brain Sciences*, 3, 193.

Lee, R. B. 1968. What hunters do for a living; or, how to make out on scarce resources. In *Man the Hunter* (ed. by R. B. Lee & I. DeVore), 30–48. Chicago: Aldine.

———. 1988. Reflections on primitive communism. In *Hunters and Gatherers v.1: History, Evolution and Social Change* (ed. by T. Ingold, D. Riches & J. Woodburn), 252–268. Oxford: Berg.

Levine, N. E. & Silk, J. B. 1997. Why polyandry fails: Sources of instability in polyandrous marriages. *Current Anthropology*, 38, 375–398.

Lindenfors, P. 2002. Sexually antagonistic selection on primate size. *Journal of Evolutionary Biology*, 15, 595–607.

MacDonald, D. 2001. *The Encyclopedia of Mammals*. Abingdon, UK: Barnes & Noble/Andromeda Oxford.

Mace, R. 1998. The coevolution of human fertility and wealth inheritance strategies. *Philosophical Transactions of the Royal Society B Biological Sciences*, 353, 389–397.

Marinho, A. O., Sallam, H. N., Goessens, L., Collins, W. P. & Campbell, S. 1982. Ovulation side and occurrence of mittelschmerz in spontaneous and induced ovarian cycles. *British Medical Journal*, 284, 632–632.

Marlowe, F. W. 1999. Showoffs or providers? The parenting effort of Hadza men. *Evolution and Human Behavior*, 20, 391–404.

———. 2000a. Paternal investment and the human mating system. *Behavioural Processes*, 51, 45–61.

———. 2000b. The patriarch hypothesis. *Human Nature*, 11, 27–42.

———. 2001. Male contribution to diet and female reproductive success among foragers. *Current Anthropology*, 42, 755–760.

———. 2003a. A critical period for provisioning by Hadza men: Implications for pair bonding. *Evolution and Human Behavior*, 24, 217–229.

———. 2003b. The mating system of foragers in the standard cross-cultural sample. *Cross-Cultural Research*, 37, 282–306.

———. 2004. Marital residence among foragers. *Current Anthropology*, 45, 277–284.

———. 2005. Hunter-gatherers and human evolution. *Evolutionary Anthropology*, 14, 54–67.

———. 2006. Central place provisioning: The Hadza as an example. In *Feeding Ecology in Apes and Other Primates* (ed. by G. Hohmann, M. Robbins & C. Boesch), 359–377. Cambridge: Cambridge University Press.

———. 2007. Hunting and gathering: The human sexual division of foraging labor. *Cross-Cultural Research*, 41, 170–195.

Marlowe, F. W. & Berbesque, J. C. 2009a. The operational sex ratio (OSR) among hunter-gatherers: Cause or effect of male-male competition? *American Journal of Physical Anthropology*, 183–183.

———. 2009b. Tubers as fallback foods and their impact on Hadza hunter-gatherers. *American Journal of Physical Anthropology*, 140, 751–758.

Martin, R. D. 2007. The evolution of human reproduction: A primatological perspective. *Yearbook of Physical Anthropology*, 50, 59–84.

McClintock, M. K. 1998. Regulation of ovulation by human pheromones. *Nature*, 392, 177–179.

McWhirter, N. & McWhirter, R. 2010. *Guinness Book of World Records*. New York: Sterling.

Michael, R. T., Gagnon, J. H., Laumann, E. O. & Kolata, G. 1994. *Sex in America: A Definitive Survey*. Boston: Little, Brown and Company.

Miller, G., Tybur, J. M. & Jordan, B. D. 2007. Ovulatory cycle effects on tip earnings by lap dancers: Economic evidence for human estrus? *Evolution and Human Behavior*, 28, 375–381.

Milton, K. 1985. Mating patterns of wooly spider monkeys, *Brachyteles arachnodes*: Implications for female choice. *Behavioral Ecology and Sociobiology*, 17, 53–59.

Mitani, J. C., Gros-Louis, J. & Richards, A. F. 1996. Sexual dimorphism, the operational sex-ratio, and the intensity of male competition in polygynous primates. *American Naturalist*, 147, 966–980.

Mitani, J. C. & Watts, D. P. 2001. Why do chimpanzees hunt and share meat? *Animal Behaviour*, 61, 915–924.

Muller, M. N., Marlowe, F. W., Ellison, P. T. & Bugumba, R. 2009. Testosterone and paternal care in East African foragers and pastoralists. *Proceedings of the Royal Society B Biological Sciences*, 276, 347–354.

Nadler, R. 1995. Sexual behavior of orangutans (*Pongo pygmaeus*): Basic and implied implications. In *The Neglected Ape* (ed. by R. Nadler, B. F. M. Galdikas, L. K. Sheeran & N. Rosen), 223–237. New York: Plenum.

Nunn, C. L. 1999. The evolution of exaggerated sexual swellings in primates and the graded-signal hypothesis. *Animal Behaviour*, 58, 229–246.

Orians, G. H. 1969. On the evolution of mating systems in birds and mammals. *American Naturalist*, 103, 589–603.

Penton-Voak, I. S. & Perrett, D. I. 2001. Male facial attractiveness: Perceived personality and shifting female preferences for male traits across the menstrual cycle. *Advances in the Study of Behavior*, 30, 219–259.

Penton-Voak, I. S., Perrett, D. I., Castles, D. L., Kobayashi, T., Burt, D. M., Murray, L. K. & Minamisawa, R. 1999. Menstrual cycle alters face preference. *Nature*, 399, 741–742.

Peters, M., Rhodes, G. & Simmons, L. W. 2008. Does attractiveness in men provide clues to semen quality? *Journal of Evolutionary Biology*, 21, 572–579.

Platek, S. M. & Shackelford, T. K. 2006. *Female Infidelity and Paternal Uncertainty: Evolutionary Perspectives on Male Anti-Cuckoldry Tactics*. Cambridge: Cambridge University Press.

Pond, C. M. 1997. The biological origins of adipose tissue in humans. In *The Evolving Female: A Life-History Perspective* (ed. by M. E. Morbeck, A. Galloway & A. Zihlman), 147–162. Princeton, NJ: Princeton University Press.

Pusey, A., Williams, J. & Goodall, J. 1997. The influence of dominance rank on the reproductive success of female chimpanzees. *Science*, 277, 828–831.

Reichard, U. H. 2003. Monogamy: Past and present. In *Monogamy: Mating Strategies and Partnerships in Birds, Humans, and Other Mammals* (ed. by U. H. Reichard & C. Boesch), 3–25. Cambridge: Cambridge University Press.

Reynolds, J. D. 1996. Animal breeding systems. *Trends in Ecology and Evolution*, 11, 68–72.

Roberts, S. C., Havlicek, J., Flegr, J., Hruskova, M., Little, A. C., Jones, B. C., Perrett, D. I. & Petrie, M. 2004. Female facial attractiveness increases during the fertile phase of the menstrual cycle. *Proceedings of the Royal Society B Biological Sciences*, 271, S270-S272.

Robson, S. L., van Schaik, C. P. & Hawkes, K. 2006. The derived features of human life history. In *The Evolution of Human Life History* (ed. by K. Hawkes & R. R. Paine), 17–44. Santa Fe: School of American Research.

Roder, S., Brewer, G. & Fink, B. 2009. Menstrual cycle shifts in women's self-perception and motivation: A daily report method. *Personality and Individual Differences*, 47, 616–619.

Rodseth, L., Wrangham, R. W., Harrigan, A. M. & Smuts, B. 1991. The human community as a primate society. *Current Anthropology*, 12, 221–254.

Roney, J. R. & Simmons, Z. L. 2008. Women's estradiol predicts preference for facial cues of men's testosterone. *Hormones and Behavior*, 53, 14–19.

Ross, C. 2001. Park or ride? Evolution of infant carrying in primates. *International Journal of Primatology*, 22, 749–771.

Sear, R., Mace, R. & McGregor, I. A. 2000. Maternal grandmothers improve nutritional status and survival of children in rural Gambia. *Proceedings of the Royal Society B Biological Sciences*, 267, 1641–1647.

Sellen, D. W. 2007. Evolution of infant and young child feeding: Implications for contemporary public health. *Annual Review of Nutrition*, 27, 123–148.

Shanley, D. P., Sear, R., Mace, R. & Kirkwood, T. B. L. 2007. Testing evolutionary theories of menopause. *Proceedings of the Royal Society B Biological Sciences*, 274, 2943–2949.

Short, R. V. 1980. The origins of human sexuality. In *Reproduction in Mammals: Human Sexuality* (ed. by C. R. Austin & R. V. Short), 1–33. Cambridge: Cambridge University Press.

Sievert, L. L. & Dubois, C. A. 2005. Validating signals of ovulation: Do women who think they know, really know? *American Journal of Human Biology*, 17, 310–320.

Silk, J. B. 1990. Human adoption in evolutionary perspective. *Human Nature*, 1, 25–52.

Sillen-Tullberg, B. & Moller, A. P. 1993. The relationship between concealed ovulation and mating systems in anthropoid primates: A phylogenetic analysis. *American Naturalist*, 141, 1–25.

Simmons, L. W., Firman, R. C., Rhodes, G. & Peters, M. 2004. Human sperm competition: Testis size, sperm production and rates of extrapair copulations. *Animal Behaviour*, 68, 297–302.

Singh, D. & Bronstad, P. M. 2001. Female body odor is a potential cue to ovulation. *Proceedings of the Royal Society B Biological Sciences*, 268, 797–801.

Smith, E. A. 2004. Why do good hunters have higher reproductive success? *Human Nature*, 15, 343–364.

Smith, E. A., Mulder, M. B. & Hill, K. 2001. Controversies in the evolutionary social sciences: A guide for the perplexed. *Trends in Ecology & Evolution*, 16, 128–135.

Smith, E. A. & Smith, S. A. 1994. Inuit sex-ratio variation: Population control, ethnographic error, or parental manipulation? *Current Anthropology*, 35, 595–624.

Smith, E. O. & Worthman, C. M. 1995. The evolution of copius menstruation. Unpublished manuscript.

Smuts, B. 1995. The evolutionary origins of patriarchy. *Human Nature: An Interdisciplinary Biosocial Perspective*, 6, 1–32.

Stallmann, R. R. & Harcourt, A. H. 2006. Size matters: The (negative) allometry of copulatory duration in mammals. *Biological Journal of the Linnean Society*, 87, 185–193.

Stammbach, E. 1987. Desert, forest, and montane baboons: Multilevel societies. In *Primate Societies* (ed. by B. Smuts, D. Cheney, R. Seyfarth, R. Wrangham & T. Struhsaker), 112–120. Chicago: University of Chicago Press.

Stanford, C. B. 1998. The social behavior of chimpanzees and bonobos. *Current Anthropology*, 39, 399–420.

Stiver, K. A. & Alonzo, S. H. 2009. Parental and mating effort: Is there necessarily a trade-off? *Ethology*, 115, 1101–1126.

Strassmann, B. I. 1981. Sexual selection, paternal care, and concealed ovulation in humans. *Ethology and Sociobiology*, 2, 31–40.

———. 1996a. The evolution of endometrial cycles and menstruation. *Quarterly Review of Biology*, 71, 181–220.

———. 1996b. Menstrual hut visits by Dogon women: A hormonal test distinguishes deceit from honest signaling. *Behavioral Ecology*, 7, 304–315.

———. 1997. The biology of menstruation in *Homo sapiens*: Total lifetime menses, fecundity, and nonsynchrony in a natural-fertility population. *Current Anthropology*, 38, 123–129.

———. 1999. Menstrual synchrony pheromones: Cause for doubt. *Human Reproduction*, 14, 579–580.

Strier, K. B. 1997. Mate preferences of wild muriqui monkeys (*Brachyteles arachnoides*): Reproductive and social correlates. *Folia Primatologica*, 68, 120–133.

Sugarman, M. & Kendall Tackett, K. A. 1995. Weaning ages in a sample of American women who practice extended breastfeeding. *Clinical Pediatrics*, 34, 642–647.

Symons, D. 1979. *The Evolution of Human Sexuality*. Oxford: Oxford University Press.

Temrin, H., Buchmayer, S. & Enquist, M. 2000. Step-parents and infanticide: New data contradict evolutionary predictions. *Proceedings of the Royal Society B Biological Sciences*, 267, 943–945.

Thornhill, R. & Palmer, C. T. 2000. *A Natural History of Rape: Biological Bases of Sexual Coercion*. Cambridge, MA.: MIT Press.

Tooley, G. A., Karakis, M., Stokes, M. & Ozanne-Smith, J. 2006. Generalising the Cinderella Effect to unintentional childhood fatalities. *Evolution and Human Behavior*, 27, 224–230.

Trevathan, W. R., Burleson, M. H. & Gregory, W. L. 1993. No evidence for menstrual synchrony in lesbian couples. *Psychoneuroendocrinology*, 18, 425–435.

Trivers, R. L. 1972. Parental investment and sexual selection. In *Sexual Selection and the Descent of Man* (ed. by B. Campbell), 136–179. Chicago: Aldine.

———. 1974. Parent-offspring conflict. *American Zoologist*, 14, 249–264.

Trivers, R. L. & Willard, D. E. 1973. Natural selection of parental ability to vary the sex ratio of offspring. *Science*, 179, 90–92.

Udry, J. R. & Morris, N. M. 1968. Distribution of coitus in menstrual cycle. *Nature*, 220, 593–596.

Van den Berghe, P. 1979. *Human Family Systems*. Prospect Heights: Waveland.

Vehrencamp, S. L. 1983. A model for the evolution of despotic versus egalitarian societies. *Animal Behaviour*, 31, 667–682.

Verner, J. & Willson, M. F. 1966. The influence of habitats on mating systems of North American passerine birds. *Ecology*, 47, 143–147.

Vigilant, L., Hofreiter, M., Siedel, H. & Boesch, C. 2001. Paternity and relatedness in wild chimpanzee communities. *Proceedings of the National Academy of Sciences of the United States of America*, 98, 12890–12895.

Walker, R. S., Flinn, M. V. & Hill, K. R. 2010. Evolutionary history of partible paternity in lowland South America. *Proceedings of the National Academy of Sciences of the United States of America*, 107, 19195–19200.

Whyte, M. K. 1980. Cross-cultural codes dealing with the relative status of women. In *Cross-Cultural Samples and Codes* (ed. by H. Barry & A. Schlegel), 335–361. Pittsburgh: Pittsburgh University Press.

Wingfield, J. C., Hegner, R. E., Dufty Jr., A. M. & Ball, G., F. 1990. The "challenge hypothesis": Theoretical implications for patterns of testosterone secretion, mating systems, and breeding strategies. *American Naturalist*, 136, 829–846.

Wingfield, J. C., Jacobs, J. D., Tramontin, A. D., Perfito, N., Meddle, S., Maney, D. L. & Soma, K. 2000. Toward an ecological basis of hormone-behavior interactions in reproduction of birds. In *Reproduction in Context* (ed. by K. Wallen & J. E. Schneider), 85–128. Cambridge, MA: MIT Press.

Winterhalder, B. 1986. Diet choice, risk, and food sharing in a stochastic environment. *Journal of Anthropological Archaeology*, 5, 369–392.

Wolovich, C. K., Feged, A., Evans, S. & Green, S. M. 2006. Social patterns of food sharing in monogamous owl monkeys. *American Journal of Primatology*, 68, 663–674.

Wood, B. M. 2010. Household and kin provisioning by Hadza males. PhD thesis, Harvard University.

Woodburn, J. 1982. Egalitarian societies. *Man*, 17, 431–451.

Wrangham, R. W. 1975. The behavioral ecology of chimpanzees in Gombe National Park, Tanzania. PhD thesis, Cambridge University.

Wrangham, R. W., Jones, J. H., Laden, G., Pilbeam, D. & Conklin-Brittain, N. 1999. The raw and the stolen: Cooking and the ecology of human origins. *Current Anthropology*, 40, 567–594.

Zinaman, M., Drobnis, E. Z., Morales, P., Brazil, C., Kiel, M., Cross, N. L., Hanson, F. W. & Overstreet, J. W. 1989. The physiology of sperm recovered from the human cervix: Acrosomal status and response to inducers of the acrosome reaction. *Biology of Reproduction*, 41, 790–797.

Part 4
Getting Along

OVER THE LAST 25 years, behavioral ecologists have come to appreciate the fact that sociality provides valuable benefits to individuals, such as greater safety from predators, but also imposes important costs, including greater competition for scarce resources. The study of social behavior provides important insights about how natural selection has shaped social strategies to help individuals maximize the benefits they accrue and minimize the costs they incur from living in close proximity to conspecifics. Primatologists have often led the way in these investigations, supported by a long tradition of detailed observations of individually recognized animals and an understanding of the value of long-term studies.

In some species, cooperative strategies have evolved to help individuals cope with competitive pressures. Primatologists have devoted considerable effort to documenting the pattern of cooperative interactions in an effort to understand the evolutionary processes that have shaped them. In chapter 21, Kevin E. Langergraber evaluates the role of kin selection in the evolution of cooperation in the Primate order. He reviews the conceptual foundation of Hamilton's theory of kin selection, a body of new work on the mechanisms that underlie kin recognition in primates, and evidence for biases in favor of maternal and paternal kin in a wide range of primate species. His review underscores the importance of kinship as an organizing force in primate social life, but also points out instances in which cooperation in social groups may have been incorrectly attributed to kin selection.

In chapter 22, Ian C. Gilby examines the evolutionary processes that shape cooperative interactions among unrelated individuals. Gilby uses evolutionary game theory to explore the adaptive logic of cooperation. He evaluates competing claims about the role of contingent reciprocity in regulating cooperation in primate groups, and discusses alternative mechanisms, such as biological markets theory and mutualism, that may facilitate cooperation among unrelated individuals. Gilby discusses the empirical evidence of such cooperation, pointing out the strengths and limitations of claims about the evolutionary processes that shape behavioral exchanges.

For many years there has been a rather well-defined separation of the proximate and ultimate factors underlying behavior. Proximate mechanisms were mainly studied in the laboratory, where carefully controlled experiments could be conducted and biological samples could be obtained on a regular schedule. Ultimate factors shaping behavior were addressed in field studies, where the ecological factors that influence behavior could be identified and the adaptive consequences of behavioral strategies could be assessed under ecologically relevant conditions. Over the last 20 years, however, there has been an active effort to integrate our understanding of the proximate and ultimate mechanisms that influence the nature and patterns of interactions among individuals. Several chapters in this part of the book address the mechanisms that underlie behavior. In chapter 23, Filippo Aureli, Orlaith N. Fraser, Colleen M. Schaffner, and Gabriele Schino review a body of work concerned with the proximate mechanisms that regulate the formation and maintenance of social relationships. They highlight the role of emotions as a device for synthesizing the frequency, content, and value of past interactions with various partners. They also examine the range of behavioral mechanisms, including reconciliation and redirected aggression, that primates use to cope with conflict and preserve valuable social bonds.

The work reviewed in the first three chapters of this section is based on the premise that natural selection has shaped social strategies in adaptive ways. The logical extension of this premise is that variation in sociality ought to produce short-term effects on individuals' welfare and long-term effects on their fitness. In chapter 24, Joan B. Silk addresses this issue. New methods for extracting hormones from feces have enabled field workers to gain new insights about animals' ability to cope with a range of stressors, such as the arrival of potentially infanticidal males and the loss of preferred partners. Data from several long-term studies have also allowed researchers to characterize the nature of social bonds among individuals, and to examine the correlation between the strength of social bonds and important components of lifetime fitness, including infant survival and longevity. This body of work suggests that social bonds have positive effects on the ability to cope with stress and fitness.

In chapter 25, Keith Jensen assesses the psychological processes that underlie behavior, focusing on the motives that shape individuals' decisions about how to interact with others. Jensen explains how experimental methods from behavioral economics have been adapted recently to elucidate the nature of preferences that underlie helpful and harmful behavior in other primates. In these experiments, subjects are offered choices that have different outcomes (payoffs) for themselves and other individuals; the choices they make reveal their preferences. Jensen considers how the extent of primates' knowledge of others' goals, intentions, and desires may interact with their social preferences to generate motivation to help or harm others.

The work Jensen reviews was directly inspired by studies of social preferences in humans, which demonstrate that human cooperation is influenced by a set of social preferences including a concern for the welfare of others, an aversion to inequity, and a range of moral sentiments. This body of work has played an important role in reshaping our ideas about the nature of human societies, the topic of the last chapter in this section of the book. In chapter 26, Michael Alvard outlines the similarities and differences in the societies of humans and other primates, and describes a growing body of evidence which supports the conclusion that group-level cooperation plays a much larger role in human societies than in the societies of other primates. Alvard discusses how cultural evolutionary processes may have facilitated the development of social complexity and group-level cooperation over the course of human history.

Chapter 21 Cooperation among Kin

Kevin E. Langergraber

O VER THE LAST 50 years, evolutionary biologists have accumulated abundant evidence that cooperative behavior is biased in favor of genetic relatives. These kin biases have frequently been attributed to kin selection (Alexander 1974; Wilson 1975; Gouzoules 1984; Gouzoules & Gouzoules 1987; Bernstein 1991; Silk 2001, 2002, 2009). Although kin selection may be a powerful force in some species and contexts, recent empirical and theoretical developments in primates and other taxa suggest that the role of kin selection in primate cooperation may have been overestimated (Clutton-Brock 2002; West et al. 2002). For example, as more studies of cooperatively breeding birds and mammals have accumulated, it has become increasingly clear that cooperative breeding is not limited to species which show high levels of within-group relatedness, and that individuals do not always adjust their cooperation according to their level of relatedness with the recipient (Alcock 1998; Clutton-Brock 2002; West et al. 2002). Instead, many behaviors that have been thought to arise solely or principally through kin selection may actually enhance individual fitness directly. At the same time that the direct benefits that individuals gain from cooperating with relatives have been underestimated, the indirect benefits that individuals gain from cooperating with relatives may have been overestimated, as these benefits may often come at the expense of individuals who are also closely related to the cooperator (West et al. 2001).

In this chapter I assess the evidence for the role of kin selection in cooperation within primate groups in light of these recent developments. I begin with a brief overview of

kin selection theory and expand upon the points raised in the paragraph above. Next I consider the various ways in which primates might recognize their kin. I then assess how these issues affect kin-selected cooperation in species with different social systems. Because whether a sex typically is philopatric or disperses from the natal group has profound effects on the opportunity for cooperation with same-sexed relatives (chapter 12, this volume), I organize this section along these lines, beginning with the effect of kinship and cooperation in philopatric females and ending with the effect of kinship on cooperation in dispersing males.

Hamilton's Equation

Kin selection theory indicates that individuals can gain inclusive fitness benefits indirectly through the reproduction of relatives (indirect fitness) as well as through their own reproduction (direct fitness; Hamilton 1963, 1964a, b; Maynard Smith 1964). The specific conditions under which individuals should bias their behavior towards kin are described mathematically in Hamilton's rule $rb > c$, in which r equals the degree of genetic relatedness between actor and recipient, b equals the direct fitness benefit to the recipient, and c equals the direct fitness cost to the actor. For a particular behavior to be selected, the increase in reproduction to the recipient of the act must be greater than the decrease in reproduction of the individual who performed it, discounted by the degree of relatedness between the two individuals. Kin selection was originally proposed as a solution to phenomena that were puzzling from the perspective of the traditional Darwinian and Fisherian view of natural selection as the process leading to adaptations that maximize individual reproductive success (i.e., direct fitness), the most famous example being the sterile worker castes in eusocial insects. It is perhaps for this historical reason that kin selection is often referred to as something different from or a special case of natural selection, and is considered a term to be invoked only when discussing the evolution of behaviors that have fitness consequences for nondescendant relatives. However, most theoreticians now agree that kin selection is simply an extension of natural selection that accounts for the effects of behavior on the fitness of relatives and allows the fitness effects to be partitioned into their direct and indirect components (West et al. 2007). Thus, parental care, which before Hamilton was always considered an obvious consequence of natural selection for individual fitness, is under kin selection and inclusive fitness theory considered merely a special case of caring for close relatives (Dawkins, 1979). In this chapter I attempt to maintain a balance between the historical usages and theoretical accuracy of these terms by referring to "kin selection" as the process whereby a behavior evolves because of its positive effects on the fitness of relatives, including direct descendants (e.g., offspring, grand-offspring) as well as collateral kin (e.g., siblings, uncles), and "natural selection" as the process whereby a behavior evolves because of its effects on the fitness of the individual who performs it. For example, if a mother gives an alarm call to warn her daughter of a predator, and it increases her daughter's likelihood of survival and reproduction, this is kin selection. In contrast, if alarm calling increases the probability that the mother will produce another offspring in the future, but it does not affect her daughter's reproduction, this is natural selection.

Hamilton originally proposed his theory to explain the existence of "altruistic" behaviors, whose average consequences decrease the actor's direct fitness while increasing the recipient's direct fitness, with fitness being measured over the lifetime of individuals and relative to all other competing individuals in the population (table 21.1). This definition of altruism (which can be quite different from that used colloquially and in the group selection and social science literature; West et al. 2007) has two logical consequences. First, the only possible route to the evolution of altruistic behavior is through kin selection: if a behavior results in direct fitness costs to an actor, these costs must be compensated for by a sufficient increase in the direct fitness of relatives (for a discussion of the exception of "greenbeard" effects, see West et al. 2007). Second, altruism should not be common in nature, as situations under which the increase in the direct fitness of the recipient is greater

Table 21.1. A classification of behaviors according to their effects on the lifetime fitness of actors and recipients

	Effect on recipient's lifetime direct fitness	
	+	−
Effect on actor's lifetime direct fitness		
+	Mutual benefit	Selfishness
−	Altruism	Spite

than the decrease in the direct fitness of the actor, after the former has been discounted by the degree of relatedness between them, may not frequently occur. Thus, Hamilton's equation suggests that when cooperation between relatives is observed in nature, we should expect a priori that individuals are gaining fitness directly as well as through the reproduction of relatives—that is, that it has evolved by a combination of both natural and kin selection. Hamilton's equation is thus not limited only to our understanding of strictly altruistic acts; even if a behavior has positive effects on the fitness of both the actor and the recipient (i.e., the behavior is mutualistic rather than altruistic), actors can preferentially direct that behavior toward recipients that are direct and collateral relatives, thus resulting in indirect fitness benefits contributing positively to its spread among the population. Thus, in this review I adopt a broad definition of cooperation as behavior that increases the lifetime reproductive success of recipients, regardless of whether its effects on the actor's lifetime fitness are negative, neutral, or positive.

Finally, competition between relatives can have dramatic consequences for the profitability of kin-biased cooperation. Theoretical work in this area was spurred by Hamilton's suggestion that altruism could evolve via kin selection in two distinct ways: (1) individuals use kin recognition abilities to direct their altruistic behavior according to their degree of relatedness with the recipient, and (2) limited dispersal (population viscosity) leads to the elevation of relatedness at the local level (e.g., within a social group), such that it pays for individuals to direct altruism indiscriminately to their neighbors (Hamilton 1964a, b). While the first of these two routes is straightforward and uncontroversial, the second has been the subject of a great deal of theoretical debate (see Kummerli et al. 2009 and references therein). Hamilton and other theorists later realized that although limited dispersal brings relatives together to cooperate, it also brings them together to compete (Wilson et al. 1992; Queller 1994; West et al. 2001). If all the competition for resources among the extra offspring created by cooperation occurs on the local scale (e.g., within the social group), then the costs of competition among kin can exactly cancel out the benefits of cooperation. Put more simply, it does not pay to increase the fitness of a relative if those benefits come at the expense of another equally closely related individual. The potential negative effect of kin competition on the evolution of cooperation has consequences not only for our understanding of the effect of patterns of sex-biased dispersal on indiscriminate cooperation (i.e., kin selection theory does not necessarily predict that the higher average relatedness of the philopatric sex should result in higher levels of indiscriminate cooperation than among members of the immigrant sex), but also forces us to carefully consider how the kinship structure of groups, and the ability of individuals to recognize those kin, can affect the indirect fitness payoffs of nepotistic behavior towards relatives. I will return to this latter point in the discussion of the extent of nepotism towards paternal siblings in male versus female philopatric primate species.

Kin Recognition

The preceding discussion suggests that the ability to recognize kin and differentiate them according to degree of relatedness plays a key role in the evolution of cooperation via kin selection. The two mechanisms most commonly proposed as being used by primates to recognize their kin are phenotype matching and familiarity. In phenotype matching, an individual recognizes his kin by assessing some aspect of the phenotype of others, and classifies as kin those individuals whose phenotype is similar to his own (self-referent phenotype matching, or the "armpit effect") or to that of a known relative. In familiarity, an individual recognizes as kin those individuals with whom he has become familiar during some critical stage of his life. These same mechanisms are also presumably involved in the ability of primates to recognize the kin relationships of other individuals (Bergmann et al. 2003).

Phenotype Matching

There are several phenotypic cues that individuals might use to recognize their relatives. A large body of research indicates that odor, which is largely regulated by variation in the highly polymorphic major histocompatability complex (MHC) loci genes, may be the most salient cue for phenotype matching in animals (Mateo & Johnston 2000; Yamazaki et al. 2000; Mateo 2003, 2009; Mateo & Holmes 2004). However, MHC-related odor cues have received relatively little attention in the study of phenotype-

matching–based kin recognition in primates (Charpentier et al. 2008a; Macdonald et al. 2008). Charpentier and colleagues (2008a) found that the chemical profiles of the scent glands of captive ring-tailed lemur (*Lemur catta*) males varied throughout the year and were more similar among related individuals, but only during the mating season. The authors speculated that semiochemical profiles may be costly to maintain, and that they are activated only during the mating season because it is then that related individuals need to identify and avoid competition with relatives. However, they did not investigate whether males actually use odor to identify their kin. Several studies have investigated the role of MHC-determined odor variation in human mate choice. Although the effects have been small and inconsistent, most studies have found that individuals prefer the smell of individuals whose MHC alleles are most dissimilar to their own, perhaps because genetic dissimilarity maximizes offspring MHC heterozygosity (Havlicek & Roberts 2009). These results raise the possibility that an ability which originally evolved in the context of mate choice may also allow primates to identify kin.

Most empirical research on the modality that primates may potentially use to identify kin through phenotype matching has focused on visual similarity. Compared to many other animals, primates are heavily reliant on vision, and studies of brain structure (Barton et al. 1995) and gene expression (Yoav et al. 2004; Gilad et al. 2007) suggest that primates may have faced an evolutionary trade-off between the development of olfactory and visual abilities. In an early study, Parr and de Waal (1999) found that chimpanzees (*Pan troglodytes*) could match photos of unfamiliar chimpanzee mothers and sons at levels moderately above chance, but could not distinguish daughters from unrelated females. This asymmetry was interpreted as a result of the greater selective pressure for mother-son inbreeding avoidance in a species in which males are philopatric. However, a follow-up study using the same set of photos, this time with human (*Homo sapiens*) rather than chimpanzee judges, showed that this asymmetry was likely the result of artifacts of the photo stimuli; human judges were able to identify both mother-son and mother-daughter dyads at levels moderately above chance after these artifacts were removed (Vokey et al. 2004). Human judges have also been shown to have similarly moderate abilities to identify photos of mother-daughter pairs in photos of gorillas (*Gorilla gorilla*) and mandrills (*Mandrillus sphinx*), but not in yellow baboons (*Papio cynocephalus*) (Alvergne et al. 2009). While human judges are also significantly better than chance at identifying full siblings, parent-offspring pairs, and grandparent–grand-offspring pairs belonging to their own species, they cannot differentiate avuncular pairs and cousins from unrelated dyads (Maloney & Dal Martello 2006; Alvergne et al. 2007; Kaminski et al. 2009). In all of these studies, judges were recognizing the kin relations of others rather than their own kin. Thus, the process was more akin to using a relative rather than oneself as a kin template. This is probably how facially based visual phenotype matching would work in nature, as it is difficult to imagine how individuals could gain visual access to their own faces. While the very moderate abilities of judges to identify kin in these studies casts doubt on the efficacy of visual-similarity–based phenotype matching, the relatively impoverished nature of the photo stimuli compared to what would be available in the real world (i.e., complete, moving bodies) suggests that this may not be the case. It is also possible that primates and other animals do not rely on a single sensory modality to assess phenotypic similarity, and that moderate effects in each domain combine to produce adequately reliable kin recognition.

Familiarity

Several lines of evidence suggest that familiarity plays a very important role in primate kin recognition (Gouzoules 1984; Rendall 2004). This process begins in a critical period early after birth during which mothers and offspring use a variety of olfactory, visual, and vocal cues to recognize one another (Rendall 2004; Rendall et al. 2009). In captivity, females will accept strange infants and raise them as their own, thus suggesting that they learn to identify their infants during the first days of life (Klopfer 1970; Steve 1986; Michael & Jacquelyn 1989). The long-term intimate association that occurs between an offspring and its mother in turn facilitates identification of more distant maternal kin (Chapais 2001). In a female philopatric society, for example, maternal sisters will be drawn into proximity because of their common attraction to their mother. Similarly, a female will spend time with her nieces and nephews because she will often be near her sister, their mother. How far this process extends is unknown, as it is difficult to determine whether a lack of nepotism towards a particular kin category results from a lack of recognition or from a failure to satisfy the *b/c* ratio in Hamilton's equation. For example, one study of affiliation among female rhesus macaques (*Macaca mulatta*) indicated that the threshold for nepotistic biases was about $r = 0.125$, the level of relatedness for aunts and nieces (Kapsalis & Berman 1996b). In contrast, an experimental study on the same population that examined responses of individuals to vocalizations of their relatives indicated that differential responses extended further (Rendall et al. 1996). Individuals may also use maternal social networks to identify the maternal kinship relations of third parties, as is suggested by a number of field audio playback experiments and observational studies that indicate that individuals direct both aggression and reconciliation towards the kin of their recent opponents (Cheney & Seyfarth 2004; chapter 28, this volume).

Whatever the limits of maternal social recognition networks, several experimental studies suggest that familiarity plays an important role in their development. Two early studies of pig-tailed (*Macaca nemestrina*) and stump-tailed (*Macaca arctoides*) macaques showed that individuals reared in isolation from their maternal and paternal kin nevertheless showed a slight tendency to preferentially interact with them over unfamiliar individuals when introduced into a new social group (Wu et al. 1980; Mackenzie et al. 1985). These results, however, could not be replicated in studies of pig-tailed macaques (Fredrickson & Sackett 1984; Sackett & Fredrickson 1987), long-tailed macaques (*Macaca fascicularis*, Welker et al. 1987), and savannah baboons (Erhart et al. 1997). However, subjects in only one of these studies were given access to their mothers' phenotypes before the test, which would allow the possibility of both self-referent and kin-template–based phenotype matching for recognizing unfamiliar maternal kin (Erhart et al. 1997). In addition, none of these cross-fostering experiments were conducted on species that rely heavily on olfaction, which, by virtue of its reliance on the highly variable MHC, may be an inherently better sensory modality than vision for assessing phenotypic similarity. Thus, it may be premature to conclude from cross-fostering experiments that phenotype matching plays no role in primate kin recognition before further studies are conducted using a wider variety of rearing conditions and species that rely more heavily on olfaction, such as many nocturnal species.

In most multimale primate groups, females mate with multiple males around the time of conception (Chapais 2001; Rendall 2004). Most researchers assumed that paternal kin recognition would therefore depend on the existence of some kind of phenotype matching. However, in much the same way that primates use regularities in their environment to recognize and bias their behavior towards their maternal kin, they may use similar kinds of cues to identify their paternal kin. In many multimale primate species there is a high degree of reproductive skew, with only a few males siring the majority of the offspring at any given time. If high skew is combined with a consistent turnover in the males that monopolize mating opportunities (e.g., one male sires all the offspring in a given year, and a different male sires all the offspring in the next year, and so on), then individuals who are similar in age will more likely be paternal siblings than individuals who are different in age (Altmann 1979). Familiarity among peers could thus be the basis of recognition of paternal siblings. Primates may also

Cooperation among Kin 495

use familiarity to recognize other paternal kin. An adult male, for example, could use his own mating history with a female to determine the likelihood that he is the father of her offspring, while an immature individual could identify her father based on the bond he shares with her mother (see Widdig 2007 for a discussion of several other possibilities).

In sum, primates have a variety of ways to identify several different types of maternal and paternal kin, with most evidence pointing towards familiarity as the most important kin recognition mechanism. In the following sections we will explore the evidence for how primates use these mechanisms to discriminate among their kin in various demographic scenarios and dispersal systems.

Fig. 21.1. The influence of maternal kinship on patterns of grooming and other forms of cooperation has been particularly well studied in primate species that form matrilineal dominance hierarchies, such as the yellow baboons (*Papio cynocephalus*) depicted here, in Amboseli, Kenya. Photo courtesy of Susan C. Alberts.

Kinship and Cooperation among the Philopatric Sex

Maternal Kinship and Female Cooperation in Philopatric Females

Species with matrilineal dominance hierarchies

The vast majority of research on the effect of maternal kinship on primate cooperation has been conducted in taxa in which females are philopatric and males disperse at sexual maturity (fig. 21.1). The effect of maternal kinship on cooperation has been particularly well studied among female vervets (*Chlorocebus pygerythrus*), baboons, and rhesus and Japanese (*Macaca fuscata*) macaques (for reviews, see Gouzoules 1984, Gouzoules & Gouzoules 1987; Silk 1987, 2001, 2002, 2006; Walters 1987; Chapais 2001; Chapais & Bélisle 2004; Kapsalis 2004). These were the first primate species in which long-term observations of individually identified animals took place, allowing researchers to build multigenerational pedigrees of maternal kin relationships (Kawai 1958; Kawamura 1958; Sade 1965; Hausfater 1975; Cheney 1980; Seyfarth 1980; Seyfarth & Cheney 1984). In these and other species with classical matrilineal dominance hierarchies (e.g., Thibetan macaques, *Macaca thibetana*, Berman et al. 2004; Berman et al. 2008), dominance ranks are almost entirely predictable from maternal kinship. A daughter outranks all females that her mother outranks, and her mother outranks all of her own daughters for most or all of her life. Females usually outrank their older sisters (the "youngest sister ascendancy rule"), and thus the daughters of younger sisters also outrank the daughters of older sisters (Kawamura 1958; chapter 5, this volume; but see Hill and Okayasu 1995). Finally, all members of a given matriline rank above or below all members of the other matrilines within the social group.

Extensive observational and experimental data show that coalitionary support from maternal kin plays a key role in the genesis of classical matrilineal dominance hierarchies. For the first few years of their lives, young females have undecided dominance relations with females that rank below their own mothers. Coalitionary support in agonistic conflicts allows a young female to dominate and eventually outrank all females that her mother can outrank. During the rank acquisition stage, support in conflicts comes almost entirely from maternal kin. Experiments with Japanese macaques, for example, show that a juvenile female who acts submissively when isolated with dominant peers can outrank them when her mother ($r = 0.5$), sister ($r = 0.25$), grandmother ($r = 0.25$), or great-grandmother ($r = 0.125$) is present to provide her with coalitionary support. In contrast, they are not able to intimidate dominant peers in the presence of aunts ($r = 0.125$), grand-aunts ($r = 0.0625$), and cousins ($r = 0.0625$) because these relatives provide coalitionary support less consistently (Chapais et al. 1997, 2001). Coalitionary support among adults follows a similar pattern of decreasing frequency with decreasing maternal relatedness, and it may play a role in the maintenance of dominance ranks and the long-term stability of matrilineal dominance hierarchies (Chapais 1983; Datta 1983a; Chapais et al. 1991; Kapsalis and Berman 1996a, b; Silk et al. 2004). Probably because of a combination of increased success in within-group competition for food, decreased levels of aggression from conspecifics, and lower rates of predation, females with higher dominance rank often have higher fertility and offspring survival, longer reproductive careers, and earlier reproducing daughters, all of which ultimately translates into higher estimated lifetime direct fitness (Harcourt 1987; Silk 1987; Altmann & Alberts 2003; Charpentier et al. 2008c; chapters 15 and 24, this volume).

For many years the links between (1) coalitionary support according to degree of relatedness, (2) coalitionary support and dominance rank, and (3) dominance rank and reproductive success were cited as evidence that cooperation in primates evolved because of its effects on the fitness of relatives. Recently, however, it has been suggested that patterns of cooperation in species with classical matrilineal dominance hierarchies do not arise solely, or perhaps even predominantly, as a result of individuals striving to increase the reproductive success of their kin (Chapais 2001, 2004, 2006). These critiques stem from two major characteristics of coalitionary behavior in species with matrilineal dominance hierarchies: the tendency of individuals to form coalitions against individuals who rank lower than themselves, and the tendency of individuals to intervene on behalf of the higher-ranking of two individuals. These so-called "conservative" coalitions are probably frequent because they represent a low-cost way to gain immediate access to a resource or to reinforce the rank of an aggressor over that of the target. Whatever the reason for their existence, the presence of conservative coalitions has consequences for our understanding of the maintenance of classical matrilineal dominance hierarchies and the role of kin selection in these interactions.

First, the effect of conservative coalitions and interventions among non-kin is to reinforce the matrilineal dominance hierarchy that was initially established by support from maternal kin. Thus, classic nepotistic matrilineal dominance hierarchies are not entirely nepotistic, and conservative coalitions among kin as well as non-kin may play a major role in their long-term stability. For example, an experimental study of Japanese macaques showed that a female of the highest-ranking matriline, A, was unable to maintain her dominance rank when isolated with members of the lowest-ranking matriline, C, but instead ranked between C and the mid-ranking matriline B when the latter were also present, as members of matriline B supported her in conflicts with members of matriline C (Chapais et al. 1991). A similar experiment showed that artificially created subgroups consisting mostly of unrelated and distantly related females from the two highest ranking matrilines could often use coalitionary support to maintain their ranks over members of the lowest-ranking matriline (Chapais et al. 1991). In both of these experiments, however, females were isolated from the group for relatively short periods of time (six and three days, respectively), and it is not clear whether these non-kin and distant kin coalitions would be effective in maintaining matrilineal dominance ranks over long time periods. Determining the relative importance of kin and non-kin support in the long-term stability of matrilineal

dominance hierarchies will ultimately require the acquisition of more long-term data on the effects of kin loss on dominance rank (Silk 2001).

Other research is consistent with the hypothesis that interspecific variation in the behavior of females can play a key role in determining interspecific differences in the structure of matrilineal dominance hierarchies. It has been shown in several species with matrilineal dominance hierarchies that young females are consistently favored by their mothers and other close kin in disputes with their older sisters, presumably because residual reproductive value is negatively correlated with age (Schulman & Chapais 1980; Horrocks & Hunte 1983) or because their size-related vulnerability means that they benefit more from protection (Hill & Okayasu 1995). In other species, however, dominant non-kin do not consistently intervene on behalf of young females over their older sisters, and youngest ascendancy is more rarely observed (Prudhomme & Chapais 1993a, b).

Individuals also may enhance their own fitness, and that of their relatives, when they support those relatives in conflicts with other individuals. Thus, high rates of coalitionary support among maternal kin may arise partly from the fact that maternal kin tend to hold adjacent ranks, and therefore dominate the same set of unrelated individuals (Chapais 2001, 2006; Chapais & Bélisle 2004). Thus, like support from non-kin, support from maternal kin also serves to reinforce the matrilineal dominance hierarchy for reasons that have little to do with the effect of coalitionary support on the fitness of recipients. It has also been argued that in some cases, support for relatives does not even result in them receiving a benefit as a by-product of the intervener maximizing her own fitness for selfish reasons, because the intervener intervenes only after her relative has won her dispute, and thus she does not receive an immediate benefit from her relative's intervention (Chapais 1983; Datta 1983a; Chapais et al. 1991; Kapsalis & Berman 1996a, b; Silk et al. 2004). In these cases it is difficult to exclude the possibility that the intervention does not result in the recipient obtaining a longer-term benefit in the form of lower levels of received aggression in the future. In other cases, the maternal relative clearly would benefit from the actor's support, but not enough so that the actor refrains from forming a coalition against her. For example, experimental and observational data show that while juvenile females typically side with their close maternal kin in disputes with unrelated individuals, they often side with an unrelated individual against a maternal sister with whom they are competing for rank (Chapais et al. 1994).

Although primates frequently gain personal fitness benefits when they form coalitions with maternal kin, not all cases of coalition formation are mutualistic. Some of the best evidence for altruistic behavior among primates comes from coalitions in which individuals support others in conflicts with individuals who rank higher than themselves. In these cases, the coalitionary supporter is unlikely to reverse ranks with the target even if the coalition is successful, and may risk severe retaliation. As predicted by Hamilton's rule, the vast majority of costly coalitions are performed by close maternal kin (Kurland 1977; Watanabe 1979; Walters 1980; Silk 1982; Chapais 1983; Cheney 1983; Datta 1983b; Netto & van Hoof 1986; Hunte & Horrocks 1987; Pereira 1988, 1989; Chapais et al. 1991).

Rank reversals between mothers and adult daughters in baboons provide another example of behavior that fits predictions derived from Hamilton's rule. Drawing on parent-offspring conflict theory (Trivers 1974), Combes and Altmann (2001) predicted that a mother should consent to being outranked by her daughter when the mother's residual reproductive value is lower than that of her daughter, devalued by their degree of relatedness. At this point there is no conflict of interest between mother and daughter, as a rank reversal will maximize the inclusive fitness of both individuals. Alternatively, old females may involuntarily decline in rank because they can no longer win contests with their daughters due to physical deterioration. To test these alternatives, Combes and Altmann compared the rank trajectories of older females with and without mature daughters. As would be predicted if rank reversals were a product of kin selection rather than physical deterioration, older females were much more likely to experience rank reversals with females immediately below them in the hierarchy if those females were their daughters.

In sum, the indirect fitness benefits, derived from increasing the fitness of relatives by supporting them in coalitions to increase their rank, play a key role, but not the only role, in the generation of matrilineal dominance hierarchies. However, there are other ways in which females living in matrilineal dominance hierarchies can increase the fitness of their relatives besides increasing their dominance rank through coalitionary support. Considerable research suggests that spatial proximity, grooming, tolerated cofeeding, and reconciliation are preferentially directed towards maternal kin in species with matrilineal dominance hierarchies (Gouzoules 1984; Gouzoules & Gouzoules 1987; Silk 1987, 2001, 2002, 2006; Walters 1987; Chapais 2001; Schino 2001; Chapais & Bélisle 2004; Kapsalis 2004). The long-standing assumption that these behaviors have positive effects on the fitness of their recipients has recently been supported by the demonstration that in two separate baboon populations, females who spend more time in close spatial proximity to each other and groom other individuals have higher reproductive success, and that this effect occurs independently of the effects of dominance rank (Silk

et al. 2003, 2009). Like coalition formation, rank similarity probably contributes to the high frequency of affiliative behavior among maternal kin. Seyfarth, for example, argued that if individuals can trade grooming for coalitionary support, and high-ranking individuals are the most effective coalition partners, then time constraints and competition for partners will result in grooming occurring most often between individuals of similar rank (Seyfarth 1977, 1980). While the hypothesis that primates trade grooming for coalitionary support remains controversial, recent metanalyses show that such trade may occur, and that females may increase their own fitness and that of their close maternal relatives when they groom closely ranked individuals (Henzi & Barrett 1999; Schino 2001, 2007; Stevens et al. 2005; Schino & Aureli 2008). The benefits that females gain directly from grooming individuals of high rank, however, appear to be lower than the benefits they gain from grooming close kin. Female baboons, for example, groom their adjacently ranked close kin more frequently than they do their adjacently ranked non-kin (Silk et al. 1999, 2006a), while multivariate analyses show that the effect of close maternal kinship on proximity, grooming, and cofeeding is stronger than the effect of rank distance (de Waal 1991; Kapsalis & Berman 1996a, b).

Philopatric females in other species

We know less about the effect of maternal kinship on cooperative behavior among females in female philopatric species outside of the well-studied rhesus and Japanese macaques, vervets, and baboons. Methodological differences between studies further complicate efforts to determine species differences in the extent of maternal kin biases in cooperative behavior. Despite these difficulties, there is a growing body of evidence to suggest that female philopatric species can differ quite substantially in the extent of maternal kin biases in cooperation, and that this variation is associated with variation in the dominance styles of the species (de Waal & Luttrell 1989; Aureli et al. 1997; Sterck et al. 1997; Boinski et al. 2001, 2005; Chapais 2004; Kapsalis 2004; Thierry 2004, 2007; Thierry et al. 2008). Although this interspecific variation has been hypothesized to result from phylogenetic constraints (Di Fiore & Rendall 1994; Thierry 2007) or self-organization (Hemelrijk 1999), most empirical and theoretical research has focused on how feeding competition influences kinship and female social relationships (Wrangham 1980; van Schaik 1989; Sterck et al. 1997; chapter 9, this volume).

Paternal Kinship and Cooperation in Philopatric Females

Uncertainty about the paternity of infants born in primate groups has limited our ability to investigate the effects of paternal kinship on cooperation among philopatric females. However, several studies have now used genetic data to assess the influence of paternal kinship on the pattern of interactions among females. Studying 34 adult females in a group of free-ranging rhesus macaques, Widdig and colleagues (2001) showed that a composite affiliation index of approaches, close spatial proximity, and grooming was significantly higher among paternal siblings than among unrelated dyads. Affiliation was higher among members of the same age cohort than among members of different age cohorts, and members of the same age cohort were more than 3.3 times more likely than members of different age cohorts to be paternal siblings (Widdig, personal communication), thus suggesting that age proximity might underlie recognition of paternal siblings. However, females preferred paternal sisters who were members of their own age cohorts over unrelated females who belonged to their own age cohort, and they also preferred paternal sisters who were not members of their own age cohorts over unrelated females who were not members of their own age cohort. These results suggest that age proximity is not the only cue that females use to recognize paternal siblings. Additional analyses suggested that elevated affiliation among paternal siblings could not be explained as a result of attraction to individuals who are similar in rank, or as a side effect of high rates of affiliation among mothers whose infants were fathered by the same male. Based on these findings, Widdig and her colleagues suggested that phenotype matching might play some role in paternal kin recognition.

The levels of affiliation among paternal siblings were considerably lower than among maternal siblings. In a follow-up study that looked at the three affiliative behaviors individually, as well as at physical and nonphysical aggression, only 2 of 10 comparisons (5 behaviors × 2 categories of dyad, same age cohort and different age cohort), showed a significant difference between paternal siblings and unrelated dyads (Widdig et al. 2002). In contrast, all comparisons between maternal siblings and unrelated dyads were statistically significant. Additional investigations into coalitionary behavior revealed further ambiguous results (Widdig et al. 2006b). Females targeted their paternal sisters significantly less often than they did unrelated individuals when the victim ranked lower than themselves. However, females did not intervene on behalf of their paternal sisters more frequently than they did for unrelated individuals when the victim ranked higher than themselves, nor did they form coalitions with their paternal sisters more frequently than with unrelated individuals, regardless of whether the victim ranked lower or higher than themselves. In contrast, females showed clear biases in favor of their maternal sisters in all three of these situations. After group

fissions in this population, members of the same maternal family (mother-daughter and maternal sister dyads) and extended matrilines tended to remain together, but paternal sibling dyads did not (Widdig et al. 2006a). Similarly, after group fissions in semi–free-ranging Barbary macaques (*Macaca sylvanus*), paternal relatives were distributed in groups together no more often than would be expected by chance (Kuester & Paul 1997).

Investigations of paternal sibling biases in female baboons have produced similarly conflicting results over the magnitude of paternal kin biases. In a study of 29 adult female baboons, Smith and colleagues (Smith et al. 2003) found that a composite affiliation index of measures of close spatial proximity and grooming was significantly higher among paternal siblings than unrelated dyads. Rates of affiliation among paternal siblings were substantially higher than among unrelated dyads, and similar to the level observed among maternal siblings. This may be explained by the fact that age similarity was a much better cue for paternal sibship in baboons than in rhesus macaques. Members of the same age cohort were over 13.5 times more likely than members of different age cohorts to be paternal siblings, and female affiliation decreased with age distance. Phenotype matching was invoked to explain the finding that levels of affiliation among females belonging to the same and different age cohorts did not differ, and that a set of paternal sibling dyads matched in age with a set of non-kin dyads had a nonsignificant tendency towards higher affiliation.

Another study of the same population of baboons using a much more extensive long-term data set of 118 females found that affiliation among paternal sisters was much lower than among maternal sisters (Silk et al. 2006). Affiliation among paternal sisters was not significantly higher than among a sample of dyads that were definitively known to be unrelated from parentage analyses. However, affiliation among paternal sisters was slightly but significantly higher than in a much larger sample of dyads that probably contained some unidentified paternal sisters and other paternal relatives. As in the earlier study on this baboon population, affiliation decreased with age difference. However, age similarity was a much less reliable predictor of paternal sibship in this sample, as members of the same age cohort were only 5.4 times more likely than members of different age cohorts to be paternal siblings. There was no evidence that phenotype matching was used as a kin recognition mechanism, as paternal sisters who were members of the same age cohort did not affiliate more frequently than pairs of unrelated females who were members of the same cohort, and paternal sisters who were members of different age cohorts were not preferred over pairs of unrelated females who were members of different age cohorts. Another study of the same baboon population found that females had a much stronger tendency to stay with their maternal kin than with their paternal kin following group fission events, but paternal kin and age-mates were no more likely to end up in the same group than would be expected by chance (Van Horn et al. 2007). These data suggest that the maternal indirect fitness benefits derived from staying with maternal relatives, who are necessarily dissimilar in age, outweighed the paternal indirect fitness benefits derived from remaining with paternal sisters or age-mates.

A study of 10 adult female white-faced capuchin monkeys (*Cebus capucinus*) found that maternal siblings and other close kin preferentially maintained close spatial proximity and groomed more frequently than did unrelated dyads, while paternal siblings did not (Perry et al. 2008). Although this study did not report the percentage of paternal siblings that belonged to the same versus different age cohorts, additional research shows that alpha males in this species sire a disproportionate number of offspring and retain their rank for exceptionally long periods (Muniz et al. 2006). Thus, in this species paternal siblings should be common, but many of them will differ substantially in age, thus reducing the usefulness of age similarity as a mechanism for recognizing paternal siblings.

Finally, two studies of female philopatric primate species have examined the effect of paternal siblingship on patterns of affiliation, but they did not distinguish the sex combination of dyads, and included juveniles and adolescents. In a small group consisting of four female and six male semi–free-ranging sun-tailed monkeys (*Cercopithecus solatus*), close maternal relatives (i.e., $0.25 \leq r \leq 0.5$) had higher levels of grooming and lower levels of aggression than more distantly related dyads, but paternal siblings and age-mates did not (Charpentier et al. 2008b). In contrast, in a study of semi–free-ranging mandrills 11 male and 11 female juveniles affiliated with their older paternal sisters as often as they did with their older maternal sisters, and significantly more often than they did with other adolescent and adult females (Charpentier et al. 2007). While this study did not examine males and females separately, additional analyses showed that juvenile females affiliated much more frequently with adult females than juvenile males did, so females were probably driving this effect. In contrast, when rates of affiliation among juveniles were compared, maternal siblings were preferred over unrelated juveniles, but there was no bias in favor of paternal siblings.

Charpentier and her colleagues wanted to know whether phenotype matching contributed to the mandrills' elevated rates of affiliation with older paternal sisters. They reasoned that if individuals use phenotype matching to recognize kin, then individuals with the same degree of relatedness, regardless of its origin, should have an equivalent number of phenotypic cues in common. Thus they compared the affiliation index between juveniles and older females who were paternal sisters ($r = 0.25$) and dyads that were related by 0.25 but did not share a common maternal ancestor or the same father. The results showed that paternal siblings had significantly higher affiliation levels than the other dyads. These findings suggest that juveniles' preferences for their older paternal sisters are not based on phenotype matching.

In sum, there is limited evidence that female philopatric primates gain indirect fitness benefits by cooperating with their paternal sisters, and age proximity seems to be a more important factor in paternal kin recognition than phenotype matching. In general, paternal kin biases seem to be stronger when levels of affiliation among age-mates are high and when age similarity is a more reliable predictor of paternal sibship status. One important factor that requires further examination is the role of breeding seasonality in determining whether age proximity is a good cue for paternal sibship. On the one hand, when breeding shows low seasonality, male reproductive skew over a given time period should be high, as it is easier for a single male to monopolize multiple females (Ostner et al. 2008), thus increasing the accuracy of age proximity as a cue for paternal sibship status (Altmann 1979). On the other hand, when breeding shows low seasonality, births show low seasonality as well, thus reducing the extent to which individuals will grow up alongside a well-defined set of individuals with whom they have been familiar from birth (van Hoof 2000).

To determine how these opposing forces balance themselves out, and for a more general understanding of paternal kinship and cooperation in primate societies, we will need more data on paternal kinship and behavior from primate species with a wider variety of mating systems. This will be a challenging task, as one key issue common to all of the studies reviewed above is that they usually had low power to detect small but biologically meaningful effects. This makes it difficult to interpret the meaning of negative results, both in terms of the overall effect of paternal kinship on patterns of behavior and of the specific kin recognition mechanisms involved.

Maternal Kinship and Cooperation in Philopatric Males

We would expect a priori that the effect of maternal kinship on same-sex cooperation would be lower in philopatric males than in philopatric females because fewer types of readily recognizable maternal kin are available to philopatric males (i.e., only maternal brothers) than to philopatric females (i.e., mothers, daughters, maternal sisters, grandmothers, granddaughters, etc.). Strong social bonds and male philopatry occur in several primate species, including bonobos (*Pan paniscus*, Kano 1992), chimpanzees (Mitani et al. 2002a), northern muriquis (*Brachyteles hypoxanthus*, Strier et al. 2002), Geoffroy's spider monkeys (*Ateles geoffroyi*, Ahumada 1992), Central American squirrel monkeys (*Saimiri oerstedii*, Boinski 1994; Boinski and Mitchell 1994), and eastern red colobus monkeys (*Procolobus rufomitratus*, Struhsaker 2000). However, we know very little about how kinship affects the pattern of affiliation and cooperation in any of these species except chimpanzees.

Male chimpanzees are similar to resident-nepotistic female primate species (see chapter 9, this volume) in several ways: they establish strongly linear dominance hierarchies, have formal signals of submission, and form highly differentiated social relationships that are thought to be important in the acquisition of high dominance rank, which is in turn positively correlated with reproductive success (Constable et al. 2001; Vigilant et al. 2001; Muller & Mitani 2005; Boesch et al. 2006; Wroblewski et al. 2009). Males engage in a variety of affiliative and cooperative behaviors, including preferential association in fission-fusion parties, maintenance of close spatial proximity, grooming, coalition formation, and meat sharing (Muller & Mitani 2005). Research relying on mitochondrial DNA haplotype sharing information initially suggested that maternal kinship was a poor predictor of who cooperated with whom (Goldberg & Wrangham 1997; Mitani et al. 2000; Mitani et al. 2002b). However, more recent research, employing a larger suite of loci with sex-specific modes of inheritance that allowed maternal sibship to be accurately determined, showed that all five of these behaviors occur significantly more often among maternal siblings than among unrelated dyads (fig. 21.2; Langergraber et al. 2007). Maternal siblings also groom one another more equitably and have longer lasting social bonds than unrelated individuals (Mitani 2009). Despite these clear and strong biases in favor of maternal siblings, the overall impact of kinship on patterns of cooperation among males within groups is limited. Unlike among female baboons (Silk et al. 2006), among male chimpanzees the vast majority of dyads that form strong affiliative and cooperative social bonds are unrelated. This is likely to be due to the males' inability to recognize some of their maternal kin. Assuming that immigrant females are unrelated, the only same-sex maternal relatives that male chimpanzees can readily recognize are their maternal siblings. Many males have no maternal brothers. Without an extended network of various types of close kin, male chimpanzee societies cannot create the corporate matrilineal dominance hierarchies observed in some female philopatric primate species.

Like male chimpanzees, male bonobos also form linear dominance hierarchies, but they lack formal signals of submission, may have lower reproductive skew (Gerloff et al. 1999), and do not show such highly differentiated patterns of affiliation and cooperation. The few bonobo maternal brothers studied thus far show a moderate tendency to associate and groom with one another more often than unrelated males do (Furuichi & Ihobe 1994; Hohmann et al. 1999). Finally, in the unusually egalitarian northern muriqui, dominance hierarchies are not detectable, aggression is rare, and few maternal brothers who have been identified thus far do not seem to preferentially maintain close spatial proximity or embrace one another more frequently than do unrelated males (Strier et al. 2002). Thus, the pattern of nepotism among philopatric males seems to resemble the pattern among philopatric females: the extent of nepotism decreases with the extent of despotism.

Maternal kinship may also influence cooperation in the hierarchically organized social structure of the hamadryas baboon (*Papio hamadryas*, Colmenares 2004). In this species, the smallest unit of grouping is the one-male unit, which consists of a single breeding male, a variable number of females, and sometimes one or two follower males (Kummer 1995; chapter 5, this volume). One-male units group together to form clans, clans group together to form bands, and bands group together to form troops. Studies conducted in captivity show that, among leader-follower dyads within one-male units, maternal brothers are much more likely to maintain close spatial proximity to one another than are unrelated males (Colmenares 1992, 2004). It is unclear, however, whether the relatedness between leaders and followers influences the timing or severity of rank reversals, which always involve a period of severe and prolonged fighting (Colmenares 2004). Strong bonds among leaders of different one-male units within a clan play a key role in the clan's cohesiveness. Although all males of a clan are related, strong bonds are formed predominantly between maternal siblings (Colmenares 1992, 2004). Leader males whose one-male unit belonged to a clan were less likely to be successfully challenged for ownership of females, and less likely to be supplanted from food resources by one-male units from outside their clan than were leader males who did not belong to a clan. Together, these results suggest that males gain kin-selected benefits through clan formation, through both the maternal and paternal sides of their genomes, but that maternal kinship plays the key role in mediating this process.

Paternal Kinship and Cooperation in Philopatric Males

Age proximity may provide a much less useful cue for identifying paternal siblings for philopatric males than it does

Fig. 21.2. An adult male chimpanzee of the Ngogo community in Kibale National Park, Uganda, grooms another adult male. Genetic data confirm that grooming and other affiliative behaviors are exchanged at higher rates between males who are maternal siblings than among non-kin. These kin preferences explain little of the overall variation in male-male cooperation, however, since most close long-term social bonds among males involve unrelated individuals, as is the case with these two males. Photo courtesy of Kevin Langergraber.

for philopatric females. To see why, consider a situation in which sex-biased dispersal is complete, all immigrants are unrelated, and age proximity is a perfect cue for paternal sibship (i.e., all paternal siblings are members of the same age cohort). In female philopatric species, females live alongside a suite of different relatives whom they should be able to readily recognize via familiarity, including their mothers, grandmothers, granddaughters, maternal sisters, great-grandmothers, great grand-daughters, and maternal aunts and nieces. They will also live alongside a group of relatives whom they should not so easily recognize, such as their maternal and paternal cousins and more distant maternal kin. Imagine that a female (A, for ally) witnesses a dispute between an individual member of the same age cohort as herself (P, for peer) and a female member of a different age cohort from her own (N, for nonpeer). Assuming that the effects of both relatives on her direct fitness are equal, A should support P against N if she is more closely related to P than to N. A is related to P by 0.25, because all

peers are paternal siblings. So she should not support P if N is one of her maternal siblings, her grandmother, or her granddaughter—all of whom are related to her by 0.25. She should support P against N if N is her aunt or niece ($r = 0.125$) or her cousin ($r = 0.0625$). Only in a minority of cases will she not know whether N is a relative (i.e., a maternal or paternal cousin, or more distant maternal kin), and in these cases N is anyway less closely related to her than is P. In contrast, when a philopatric male is placed in the same situation, there is more ambiguity as to whether P is more closely related to him than N. A male would know whether N is a maternal sibling, but may not know whether N is a father, son, uncle, nephew, grandfather, or grandson. Thus, kin competition may explain why males in male-philopatric species are less likely to show nepotistic biases towards their paternal siblings than are females in female-philopatric species: even when individuals can identify their paternal brothers with perfect certainty, the indirect benefits they gain by acting nepotistically towards them will be reduced by the extent to which those gains come at the expense of other (unrecognizable) relatives.

The influence of paternal kinship on male cooperation in male-philopatric species has only been investigated directly in chimpanzees. Even though male chimpanzees preferentially cooperate with individuals who are similar in age (Mitani et al. 2002b), paternal siblings do not cooperate more frequently than unrelated males (Langergraber et al. 2007). Age proximity was a poorer indicator of paternal siblingship in this group of chimpanzees than in any of the other primate species where this has been investigated. The average degree of genetic relatedness among members of the same age cohort (five years difference or less) was also very low, and not significantly higher than that of dyads that belonged to different age cohorts. These results suggest that males would not increase the fitness of their relatives by preferentially cooperating with individuals of a similar age. The absence of paternal sibling biases has also been found in a study of the play behavior of immature chimpanzees (of both sexes) conducted in a different chimpanzee community (Lehmann et al. 2006).

While male chimpanzees do not appear to cooperate preferentially with their paternal siblings, there is a limited amount of evidence that suggests that they may favor their immature offspring. Fathers were less aggressive to mothers who had sired their offspring 6 to 12 months previously, and they played longer with their own offspring than with unrelated infants (Lehmann et al. 2006). However, these effects were not significant across all immature age classes, and were not evident in other measures of parental investment (i.e., playing frequency, grooming duration and frequency, and aggression towards mothers when their infant was zero to six months old). The mechanism that males could use to recognize their offspring are unclear. Mothers and sires did not associate more frequently than other male-female pairs, either at the time surrounding conception or after the birth of their infant, thus suggesting that males did not use familiarity with the mother to recognize their offspring. Males also did not copulate more frequently with the mother of their offspring than with other females, thus suggesting that they could not use their prior mating behavior as a recognition cue.

Well before techniques for assessing genetic relatedness were available, observers of wild hamadryas baboons hypothesized that close paternal kinship was responsible for the close social bonds between leader and follower males within a one-male unit, and between leaders of different one-male units within the same clan (Abbeglen 1984; Kummer 1995). In these studies, paternal kinship was inferred from morphological resemblance and age differences. More recent results from captive studies suggest that maternal kinship is much more important than paternal kinship in the development of male social bonds: strong bonds among fathers and sons are much less frequent than strong bonds among maternal brothers, and bonds among paternal brothers actually occur less often than would be expected by chance (Colmenares 1992). Future research on wild populations employing molecular genetic techniques to resolve kin relations will be required to resolve this issue. Similarly, genetic analyses of Central American squirrel monkeys will be required to determine whether males who are similar in age and who maintain close spatial proximity much more frequently than do dyads of other age-sex combinations are paternal siblings (Boinski & Mitchell 1994). In this species, levels of mating skew within breeding seasons and high levels of turnover between breeding seasons in the identity of the male that mates most frequently may make age similarity a relatively reliable cue for paternal siblingship.

Kinship and Within-Group Cooperation in the Dispersing Sex

Primates have the opportunity to influence the reproductive success of their relatives even if they disperse from their natal group. Two of the main contexts in which this occurs among primates are parallel dispersal with kin (kin-biased dispersal) and male parental care.

Parallel dispersal occurs when two or more individuals disperse from one group and emigrate to the same new group. This process, in both its kin-biased and non–kin-biased forms, has important fitness consequences in a variety of animal species (Packer et al. 1991; McNutt 1996; Koenig et al. 2000; Yaber & Rabenold 2002; Williams & Rabenold 2005; Marker et al. 2008; Sharp et al. 2008). There are three main ways in which parallel dispersal operates in primates. First, in individuals with well-defined breeding and birth seasons, individuals from the same birth cohort may disperse together. In such cases, parallel dispersers will not be maternal siblings, who necessarily belong to different birth cohorts, but may often be paternal siblings or more distant maternal or paternal kin (fig. 21.3). Second, individuals from the same group may disperse together at the same time following group dissolution or fission to join or form a new group. Here, parallel dispersers can be any type of maternal or paternal relative of any age combination that has occurred in the original group. Third, individuals can preferentially disperse into groups that contain previously dispersed individuals from their own source group. Under this form of parallel dispersal, individuals may join a variety of types of maternal and paternal kin who will usually be around the same age or older than themselves.

Fig. 21.3. One form of parallel dispersal is the joint transfer of paternal half siblings, as in the case of these two white-faced capuchins (*Cebus capucinus*) of Costa Rica, who have adopted the "overlord" posture used in coalition-ary aggression against rivals. These brothers, subjects in the Lomas Barbudal Monkey Project, left their natal group together and successfully took over the group they immigrated into, ousting the former resident males. Photo courtesy of Wiebke Lammers.

Kinship and Cooperation in Dispersing Females

Among species in which females disperse, the effects of kin-biased dispersal are best documented in mountain go-rillas (*Gorilla gorilla beringei*). Female mountain gorillas show both natal and secondary transfer, the latter often occurring after the death of the silverback male and the dis-solution of the group (Robbins et al. 2008, 2009). Despite routine dispersal, females mountain gorillas live with at least one close maternal or paternal relative for the majority of their lives (Watts 1996). Female mountain gorillas display nepotistic biases in spatial proximity, the tolerance of close approaches, grooming, and coalitionary support. They are also less aggressive to close kin than to others. In contrast to what has typically been found among female philopatric primates, levels of these affiliative and cooperative behav-iors are approximately half as high in paternal kin as in

maternal kin, while being only slightly higher than those in unrelated dyads. This difference in the extent of nepo-tism among paternal kin may occur because most mountain gorilla groups have only one male, and even in multimale groups reproductive skew is high and often involves related males (Bradley et al. 2005). Thus, being born in the same natal group may be a better cue of paternal relatedness sta-tus in dispersing female mountain gorillas than is being a member of the same age cohort in philopatric female rhe-sus macaques and baboons. One limitation in the available studies on nepotism in mountain gorillas, however, is that paternal siblingship was not determined directly through genetic analyses, but instead was assumed when individu-als were born in the same natal group containing the same sexually mature male(s) (Watts 1994, 1996). Whether kin-biased social bonds are an important component of gorilla female fitness is also unclear, as even related females inter-act with each other much less frequently than they do with dominant adult males (Watts 1994).

In contrast to female mountain gorillas, adult female chimpanzees can have bonds as strong as those among males. Rates of affiliation and cooperation are lower among female dyads than among male dyads on average (Halperin 1979; Wrangham & Smuts 1980; Kawanaka 1984; Hasegawa 1990; Williams et al. 2002), but at some study sites the strongest social bonds are formed between females (Gilby & Wrangham 2008; Lehmann and Boesch 2008; Langergraber et al. 2009). The function of strong social bonds among females is unknown, but they may en-hance females' ability to gain access to areas of the group's territory that have high-quality food resources, which is in turn positively related to reproductive success (Murray et al. 2006, 2007, 2008; Emery Thompson et al. 2007; Townsend et al. 2007; Kahlenberg et al. 2008). Despite the potential fitness consequences of strong social bonds, there is little evidence that kin-biased dispersal allows females to form them with relatives. Although it has long been known that nondispersing females will often form strong bonds with their mothers (Williams et al. 2002; Gilby & Wrangham 2008), genetic analyses show that few females coreside with close female relatives, and that the vast majority of females who form close social bonds are not close maternal or paternal kin (99%, *n* = 126 dyads, Langergraber et al. 2009). Kin recognition difficulties may explain why female chimpanzees, and probably also female bonobos (Hashi-moto et al. 1996; Hohmann et al. 1999), do not disperse to groups that contain their close kin. With a 50:50 sex ratio at birth, high infant mortality, an average interbirth interval of five to six years, and an average dispersal age of 11 years (Boesch & Boesch-Achermann 2000; Nishida et al. 2003; Sugiyama 2004), dispersing maternal siblings will rarely be able to recognize one another by virtue of the bond they shared with their mother in the natal group. Similarly, the low accuracy of age similarity as a paternal sibling recognition mechanism, combined with the rarity of two or more females reaching dispersal age at the same time, may explain why paternal siblings do not frequently emigrate together into the same communities. Thus, in the absence of a more direct kin recognition mechanism such as phenotype matching, demographic conditions can set limits on behaviors that would otherwise have positive effects on inclusive fitness.

In contrast, life-history characteristics may promote parallel migration of paternal sisters in western red colobus females (*Procolobus badius*, Starin 1994). In this species, both breeding and births are highly seasonal, and the majority of dispersing females leave their group in the company of age-mates to join the same new group. Females who have originated from the same natal group show much lower levels of aggression and much higher rates of grooming than do females who have come from different natal groups. Despite high breeding seasonality, a single breeding male is responsible for the vast majority of copulations in a group, and the identity of this male almost always changes from year to year (Starin 2001), thus suggesting that dispersing age-mates are frequently paternal siblings. This interpretation is supported by comparisons with eastern red colobus, in which birth seasonality is lower, male mating skew within a breeding season is lower, males can achieve high mating success across successive breeding seasons, and females do not form differentiated social relationships (Struhsaker & Leland 1987; Struhsaker & Pope 1991).

Kinship and Cooperation in Dispersing Males

Parallel dispersal by males occurs in a wide variety of primate species, with one study reporting that it occurred in 25 of 57 species that were deemed to be sufficiently well studied to determine its presence or absence (Schoof et al. 2008). In some species, parallel dispersal among males may play a key role in successful integration into new social groups (e.g., ring-tailed lemurs, Sussman 1992; vervet monkeys, Cheney and Seyfarth 1983; long-tailed macaques, van Noordwijk & van Schaik 1985, 2001; Bolivian squirrel monkeys, *Saimiri boliviensis*, Mitchell 1994; white-faced capuchins, Jack & Fedigan 2004b, a) or to successfully establish and defend a new group (e.g., ursine howlers, *Alouatta arctoidea*, Pope 1990).

Parallel dispersal of maternal brothers has been reported in several primate species (white-faced capuchins, Jack & Fedigan 2004a, b; vervets, Cheney & Seyfarth 1983; rhesus macaques, Meikle & Vessey 1981). One early study of rhesus macaques showed that immigrant maternal brothers

spend more time in close proximity, form coalitions more frequently, and interrupt each other's mating attempts less frequently than do unrelated males (Meikle & Vessey 1981). Residence length within a group is also longer for males who live alongside their maternal brothers. As length of residence is correlated with dominance rank and probably reproductive success in this species (Widdig et al. 2004), this further suggests that co-migration of maternal brothers also may have resulted in increased direct and indirect fitness benefits.

Parallel dispersal may actually be more common among paternal kin than among maternal kin. Many species show high levels of male skew combined with dispersal of age-mates, who may be paternal siblings (e.g., Bolivian squirrel monkeys, Mitchell 1994; long-tailed macaques, van Noordwijk & van Schaik 1985, 2001; red-fronted lemurs, *Eulemur rufifrons*, Ostner & Kappeler 2004; Kappeler & Port 2008). While a recent comparative analysis showed that male parallel dispersal is not more common in species thought to have high levels of reproductive skew (i.e., seasonally breeding species and species in which the most successful male sires more than 50 % of the offspring: Schoof et al. 2008), very few studies have actually used molecular genetic analyses to directly examine the paternal kin relationships of parallel dispersing males.

In a rare exception, Pope (1990) found that in ursine howler monkeys, coalitions of closely related males (i.e., fathers and sons, full siblings, and maternal and paternal half siblings) lasted significantly longer and were more stable (i.e., with fewer rank changes between males) than coalitions of unrelated males. In both types of coalitions, the dominant coalition member monopolized reproduction. In coalitions of related males, the deposed male would usually remain in the group, even though he was no longer reproducing. In contrast, coalitions of unrelated males typically ended after rank reversals. Together, these results suggest that subordinate males are willing to incur direct fitness costs to aid their relatives' reproduction through coalitionary group formation and defense. One shortcoming of these analyses is that the determination of the kinship relationships of a substantial fraction of males was based on whether those males had originated from the same group, not on genetic analyses.

Care of Offspring by Males

In species in which males disperse, males may gain inclusive fitness benefits by providing care for their offspring and those of close relatives. Offspring care by males improves offspring survival in a number of primate species (Wright 1990; Dietz 2004; chapters 17 and 18, this volume). Comparative analyses show, however, that nonmaternal care, including care from actual fathers, tends to benefit mothers by increasing their reproductive rate more than it benefits offspring by increasing their survival (Mitani & Watts 1997; Ross & MacLarnon 2000). Indeed, in a wide variety of primate species, males sometimes care for offspring that are not their own or to whom they are unrelated, suggesting that both natural selection and kin selection may shape males' interactions with immatures. Caring for an infant may enhance a male's prospects of mating with the mother in the future (Smuts & Gubernick 1992) or inheriting the breeding position in the group. Unfortunately, for most primate species providing offspring care, we lack the necessary pedigree and observational data to determine the relative importance of these factors.

The behavior of males in pair-bonded primate species suggest that both natural selection and kin selection may shape patterns of male care of offspring. For example, in the socially monogamous Azara's night monkey (*Aotus azarae*), the single adult male in the group frequently carries the group's infant, who is presumably his own (Fernandez-Duque et al. 2008; chapter 3, this volume, fig. 3.3). Recent observations, however, suggest that when the adult male is evicted from the group, the infant is initially cared for mostly by older group members who presumably are siblings, and later primarily by the new resident male (Fernandez-Duque et al. 2008). While these findings suggest that male infant care arises due to a combination of (1) inclusive fitness benefits from increased survival of related offspring and (2) increased future mating success, the relative importance of these selective forces cannot yet be determined. Siamang (*Symphalangus syndactylus*) groups usually consist of a single adult male and an unrelated adult female, but in one population many groups have two adult males that are unrelated to the resident female (Lappan 2008). Males in one-male groups carry infants substantially more often than do males in two-male groups (Lappan 2008). As males of infants in one-male groups presumably are the fathers of the infants they care for, this suggests that increased survival of related offspring contributes to the evolution of this behavior. Reduced care by males in two-male groups may reflect uncertainty about paternity which reduces male investment, may be a means of enhancing future mating opportunities with the resident female, or may increase each male's prospects for gaining a single breeding position within the group. Again, more research is needed to assess the the relative importance of offspring versus mating benefits in the evolution of male care for offspring in siamangs.

Baboons provide some of the best evidence for the evolution of paternal care via kin selection (chapters 17, 18, 19, this volume). Although the kin recognition mechanism involved is unknown, male yellow baboons at Amboseli selectively support their own juvenile offspring in conflicts (Buchan et al. 2003), and having a father present in the group accelerates offspring maturation (Charpentier et al. 2008d). In both yellow and chacma (*Papio ursinus*) baboons, males preferentially care for the offspring of females with whom they form affiliative "friendships" (Moscovice et al. 2009; Nguyen et al. 2009). Males are often, but not always, the fathers of their female friends' offspring, and they do not enjoy higher future mating success with those female friends in the future (Moscovice et al. 2009; Nguyen et al. 2009). Males may frequently provide care to offspring who are not their own because it is a relatively low-cost activity, and there is always some chance that a male is the father of the offspring if he has mated even only once with the mother. While this paternity confusion hypothesis accords well with data from chacma baboons (Moscovice et al. 2009), yellow baboon males frequently care for infants whose mothers they were never observed mating (Nguyen et al. 2009). In these cases, however, we cannot exclude the possibility of matings when observers were not present. Elucidating the myriad benefits that males gain from offspring care is an important area of future research.

While it has long been known that in captivity males of various cooperatively breeding callitrichine species do not provide care in proportion to their level of kinship with the infant, there are few data from the wild to assess the role of kin selection in the evolution of this social and mating system. Genetic studies carried out on wild callitrichine populations thus far have typically employed a small number of autosomal microsatellite loci (Nievergelt et al. 2000; Huck et al. 2004; Huck et al. 2005). While this is sufficient for determining the identity of the father when the mother is known from behavioral observations and has been genotyped herself, assigning paternity in the absence of a known genotyped parent typically requires many more autosomal loci. Identifying more distant kin relations, such as maternal and paternal half siblings, typically requires even more autosomal loci and the combination of marker systems with sex-specific modes of inheritance, such as mtDNA and the Y chromosome (Csillery et al. 2006; Langergraber et al. 2007; Van Horn et al. 2008). Studies of kinship and caring in two groups of mustached tamarins (*Saguinus mystax*) have been able to demonstrate, however, that while single dominant males likely sired all of the offspring within their groups, they do not consistently provide the highest level of offspring care (Huck et al. 2004, 2005). Similarly, male reproductive skew is probably high in common marmosets (*Callithrix jacchus*), but the kin relationships between other adult male caregivers and the offspring are unknown.

Recent laboratory research on germ-line chimerism in

marmosets suggests that identifying kinship through traditional means may be inappropriate from the point of view of kin selection theory. Callitrichine fraternal twins often exchange embryonic stem cells through shared blood cells in the same placenta, which results in an individual having more than one genetically distinct population of cells that originated from more than one zygote (Haig 1999). In effect, the individuals are genetic chimeras whose cells, to varying extents, are a mixture of their own and their siblings' genotypes. Thus, fraternal twins will, on average, be more closely related to one another than the expected value of $r = 0.5$ for fraternal twins. While previous research had suggested that chimerism was limited to blood-derived somatic tissues, recent research on captive Wied's black-tufted-ear marmosets (*Callithrix kuhlii*) has shown that this process also extends to germ-line cells. Ross and colleagues (2007) showed that four of seven individuals' sperm cells were genetically chimerical, and that individuals often pass sibling alleles acquired in utero from their twins down to their own offspring. In these cases, males were, in a sense, siring their fraternal twin's offspring.

Ross and her colleagues discovered that there was substantial within-individual variation in the amount of tested tissue types that were chimerical. This raises the question of how individuals should recognize offspring who have different r values for different tissue types. Ross and her colleagues hypothesized that epithelial skin cells would be most likely to regulate phenotype-matching-based kin recognition, and they examined the effects of chimerism in epithelial tissue on parenting behavior. They found that fathers with chimerical offspring ($n = 10$) carried them significantly more often than did fathers with nonchimerical offspring ($n = 20$). The authors speculated that chimerical offspring may have matched their fathers at more kin-recognition alleles than did nonchimerical offspring to their fathers, thus elevating the chimerical father's perceived relatedness to his offspring. However, as mothers showed exactly the opposite pattern, this result deserves replication with a larger sample, as well as with a variety of different tissue types that could conceivably be involved in phenotype matching. If different tissue types are involved in kin recognition, and fathers and their chimerical offspring differ in the proportion of alleles they share across tissue types, this could result in conflicting messages about phenotypic similarity.

Additional research is needed to determine if germ-line chimerism extends to all callitrichine species, and to fully understand the role of kin selection in the evolution of cooperative breeding in these species. Previous studies of the genetics of wild callitrichine genetics relied on DNA derived from epithelial cells shed from the digestive tract and present in feces, but they did not report multiple alleles per locus

(Nievergelt et al. 2000; Huck et al. 2004; Huck et al. 2005). In contrast, Ross and colleagues (2007) found multiple chimerical individuals when they examined epithelial cells from saliva and feces. Whether these differences represent true species differences or methodological errors will need to be determined by further study.

Summary and Conclusions

A large body of evidence suggests that kin selection contributes to the evolution of cooperative behavior in primate species with many different types of social systems. Although there are some examples of altruistic cooperation in primates, in most cases primates increase their own fitness directly as well as the fitness of their relatives when they cooperate, and thus most cooperation involves the complementary effects of natural selection and kin selection. The extent to which cooperators can supplement the direct fitness benefits of cooperation with indirect fitness benefits depends on the ability to direct their cooperation preferentially towards relatives. Considerable research suggests that in many primate species, individuals become familiar with their close maternal kin during their development, and bias their fitness-increasing behaviors towards them throughout their lives. Primates also often bias their positive behaviors towards individuals with whom they are familiar due to similarity in age, and in such cases they may also gain indirect fitness benefits through the paternal line, depending on the particulars of the mating system. Finally, the indirect fitness benefits that individuals gain by cooperating to increase the fitness of relatives may be negated by the extent to which those fitness gains come at the expense of other relatives; kin recognition abilities thus may interact with the nature and scale of competition for fitness-limiting resources to determine the patterns of cooperation within primate groups.

Determining the relative importance of direct versus indirect fitness benefits in any specific case is made difficult by the problems involved in estimating all three parameters of Hamilton's rule in animals as long-lived as primates. Future developments in molecular genetics for obtaining massive quantities of DNA sequence data from low-quality template material, currently being pioneered in studies of ancient DNA, should soon make it easier for fieldworkers to use noninvasively collected source materials (e.g., feces and hair) to more accurately and precisely determine the kin relations of their study subjects. Genetic analyses are particularly important for determining whether and under what conditions individuals can recognize and cooperate with their paternal kin, and how paternal kin recognition abilities affect the extent of nepotism among other types of relatives.

Measurements of individual benefits and costs to cooperators are more difficult to determine, as these benefits and costs properly should be measured over the lifetimes of individuals. Researchers should also consider effects on the fitness of other individuals who may not be immediately involved in the cooperative act. It is comforting to see, however, that some of the long-held assumptions about the positive fitness effects of cooperative behaviors are receiving at least some level of empirical support from long-term measurements of fitness (Silk et al. 2003, 2009). As long-term field studies accumulate more data on the complete life histories of individuals, we will be in a better position to get reasonable estimates of the long-term costs and benefits of cooperative acts, and the kinship structure of groups (chapter 12, this volume). Together, these data will allow a more precise understanding of the complicated ways in which kin cooperation and competition interact. But for many species we do not have even a basic knowledge of the patterns of nepotism and kinship structure. These will also be required if we are to achieve what might be called the "holy grail" of research on kinship and cooperation in primates: an understanding of the role of kin selection in generating the great intra- and interspecific variations in primate social systems.

References

Abbeglen, J. J. 1984. *On Socialization in Hamadyas Baboons.* Cranbury, NJ: Associated University Press.

Ahumada, J. A. 1992. Grooming behavior of spider monkeys (*Ateles geoffroyi*) on Barro Colorado Island, Panama. *International Journal of Primatology*, 13, 33–49.

Alexander, R. D. 1974. The evolution of social behavior. *Annual Review of Ecology and Systematics*, 5, 325–385.

Altmann, J. 1979. Age cohorts as paternal sibships. *Behavioral Ecology and Sociobiology*, 6, 161–164.

Altmann, J. & Alberts, S. C. 2003. Variability in reproductive success viewed from a life-history perspective in baboons. *American Journal of Human Biology*, 15, 401–409.

Alvergne, A., Faurie, C. & Raymond, M. 2007. Differential facial resemblance of young children to their parents: Who do children look like more? *Evolution and Human Behavior*, 28, 135–144.

Alvergne, A., Huchard, E., Caillaud, D., Charpentier, M. J. E., Setchell, J. M., Ruppli, C., Fejan, D., Martinez, L., Cowlishaw, G. & Raymond, M. 2009. Human ability to recognize kin visually within primates. *International Journal of Primatology*, 30, 199–210.

Aureli, F., Das, M. & Veenema, H. C. 1997. Differential kinship effect on reconciliation in three species of macaques (*Macaca fascicularis*, *M. fuscata*, and *M. sylvanus*). *Journal of Comparative Psychology*, 111, 91–99.

Barton, R. A., Purvis, A. & Harvey, P. H. 1995. Evolutionary radiation of visual and olfactory brain systems in primates, bats and insectivores. *Philosophical Transactions of the Royal Society of London Series B Biological Sciences*, 348, 381–392.

Bergman, T. J., Beehner, J. C., Cheney, D. L. & Seyfarth, R. M. 2003. Hierarchical classification by rank and kinship in baboons. *Science*, 302, 1234–1236.

Berman, C. M., Ionica, C. S. & Li, J. H. 2004. Dominance style among *Macaca thibetana* on Mt. Huangshan, China. *International Journal of Primatology*, 25, 1283–1312.

Berman, C. M., Ogawa, H., Ionica, C., Yin, H. B. & Li, J. H. 2008. Variation in kin bias over time in a group of Tibetan macaques at Huangshan, China: Contest competition, time constraints or risk response? *Behaviour*, 145, 863–896.

Bernstein, I. S. 1991. The correlation between kinship and behavior in non-human primates. In *Kin Recognition in Animals* (ed. by Hepper, P.), 6–29. Cambridge: University of Cambridge Press.

Boesch, C. & Boesch-Achermann, H. 2000. *The Chimpanzees of the Taï Forest.* Oxford: Oxford University Press.

Boesch, C., Kohou, G., Nene, H. & Vigilant, L. 2006. Male competition and paternity in wild chimpanzees of the Taï forest. *American Journal of Physical Anthropology*, 130, 103–115.

Boinski, S. 1994. Affiliation patterns among male Costa Rican squirrel-monkeys. *Behaviour*, 130, 191–209.

Boinski, S., Kauffman, L., Ehmke, E., Schet, S. & Vreedzam, A. 2005. Dispersal patterns among three species of squirrel monkeys (*Saimiri oerstedii*, *S. boliviensis* and *S. sciureus*): I. Divergent costs and benefits *Behaviour*, 142, 525–632.

Boinski, S. & Mitchell, C. L. 1994. Male residence and association patterns in Costa Rican squirrel-monkeys (*Saimiri oerstedii*). *American Journal of Primatology*, 34, 157–169.

Boinski, S., Sughrue, K., Selvaggi, L., Quatrone, R., Henry, M. & Cropp, S. 2002. An expanded test of the ecological model of primate social evolution: Competitive regimes and female bonding in three species of squirrel monkeys (*Saimiri oerstedii*, *S. boliviensis*, and *S. sciureus*). *Behaviour*, 139, 227–261.

Bradley, B. J., Robbins, M. M., Williamson, E. A., Steklis, H. D., Steklis, N. G., Eckhardt, N., Boesch, C. & Vigilant, L. 2005. Mountain gorilla tug-of-war: Silverbacks have limited control over reproduction in multimale groups. *Proceedings of the National Academy of Sciences of the United States of America*, 102, 9418–9423.

Buchan, J. C., Alberts, S. C., Silk, J. B. & Altmann, J. 2003. True paternal care in a multi-male primate society. *Nature*, 425, 179–181.

Chapais, B. 1983. Dominance, relatedness, and the structure of female relationships in rhesus monkeys. In *Primate Social Relationships: An Integrated Approach* (ed. by Hinde, R. A.), 208–217. Oxford: Blackwell.

———. 2001. Primate nepotism: What is the explanatory value of kin selection? *International Journal of Primatology*, 22, 203–229.

———. 2004. How kinship generates dominance structures: A comparative perspective. In *Macaque Societies: A Model for the Study of Social Organization* (ed. by Thierry, B., Sing, V. & Kaumanns, W.), 186–204. Cambridge: Cambridge University Press.

———. 2006. Kinship, competence and cooperation in primates. In *Cooperation in Primates and Humans: Mechanisms and*

Evolution (ed. by Kappeler, P. M. & Van Schaik, C. P.), 47–64. Heidelberg: Springer-Verlag.

Chapais, B. & Bélisle, P. 2004. Constraints on kin selection in primate groups. In *Kinship and Behavior in Primates* (ed. by Chapais, B. & Berman, C. M.), 365–386. New York: Oxford University Press.

Chapais, B., Gauthier, C., Prudhomme, J. & Vasey, P. 1997. Relatedness threshold for nepotism in Japanese macaques. *Animal Behaviour*, 53, 1089–1101.

Chapais, B., Girard, M. & Primi, G. 1991. Non-kin alliances, and the stability of matrilineal dominance relations in Japanese macaques. *Animal Behaviour*, 41, 481–491.

Chapais, B., Prudhomme, J. & Teijeiro, S. 1994. Dominance competition among siblings in Japanese macaques: Constraints on nepotism. *Animal Behaviour*, 48, 1335–1347.

Chapais, B., Savard, L. & Gauthier, C. 2001. Kin selection and the distribution of altruism in relation to degree of kinship in Japanese macaques (*Macaca fuscata*). *Behavioral Ecology and Sociobiology*, 49, 493–502.

Charpentier, M. J. E., Boulet, M. & Drea, C. M. 2008a. Smelling right: The scent of male lemurs advertises genetic quality and relatedness. *Molecular Ecology*, 17, 3225–3233.

Charpentier, M. J. E., Deubel, D. & Peignot, P. 2008b. Relatedness and social behaviors in *Cercopithecus solatus*. *International Journal of Primatology*, 29, 487–495.

Charpentier, M. J. E., Peignot, P., Hossaert-Mckey, M. & Wickings, E. J. 2007. Kin discrimination in juvenile mandrills, *Mandrillus sphinx*. *Animal Behaviour*, 73, 37–45.

Charpentier, M. J. E., Tung, J., Altmann, J. & Alberts, S. C. 2008c. Age at maturity in wild baboons: Genetic, environmental and demographic influences. *Molecular Ecology*, 17, 2026–2040.

Charpentier, M. J. E., Van Horn, R. C., Altmann, J. & Alberts, S. C. 2008d. Paternal effects on offspring fitness in a multimale primate society. *Proceedings of the National Academy of Sciences of the United States of America*, 105, 1988–1992.

Cheney, D. 1983. Extrafamilial alliances among vervet monkeys. In *Primate Social Relationships: An Integrated Approach* (ed. by Hinde, R. A.), 278–286. Oxford: Blackwell.

Cheney, D. & Seyfarth, R. 1983. Nonrandom dispersal in free-ranging vervet monkeys: Social and genetic consequences. *American Naturalist*, 122, 392–412.

Cheney, D. L. 1980. Vocal recognition in free-ranging vervet monkeys. *Animal Behaviour*, 28, 362–367.

Cheney, D. L. & Seyfarth, R. 2004. The recognition of other individuals' kinship relationships. In *Kinship and Behavior in Primates* (ed. by Chapais, B. & Berman, C. M.), 347–364. Oxford: Oxford University Press.

Colmenares, F. 1992. Clans and harems in a colony of hamadryas and hybrid baboons: Male kinship, familiarity and the formation of brother-teams. *Behaviour*, 121, 61–94.

———. 2004. Kinship structure and its impact on behavior in multilevel societies. In *Kinship and Behavior in Primates* (ed. by Chapais, B. & Berman, C. M.), 242–270. New York: Oxford University Press.

Combes, S. L. & Altmann, J. 2001. Status change during adulthood: Life-history by-product or kin selection based on reproductive value? *Proceedings of the Royal Society of London Series B Biological Sciences*, 268, 1367–1373.

Constable, J. L., Ashley, M. V., Goodall, J. & Pusey, A. E. 2001.

Noninvasive paternity assignment in Gombe chimpanzees. *Molecular Ecology*, 10, 1279–1300.

Csillery, K., Johnson, T., Beraldi, D., Clutton-Brock, T., Coltman, D., Hansson, B., Spong, G. & Pemberton, J. M. 2006. Performance of marker-based relatedness estimators in natural populations of outbred vertebrates. *Genetics*, 173, 2091–2101.

Datta, S. 1983a. Relative power and the maintenance of rank. In *Primate Social Relationships: An Integrated Approach* (ed. by Hinde, R. A.), 103–112. Oxford: Blackwell.

Datta, S. B. 1983b. Patterns of agonistic interference. In *Primate Social Relationships: An Integrated Approach* (ed. by Hinde, R. A.), 289–297. Oxford: Blackwell.

Dawkins, R. 1979. Twelve misunderstandings of kin selection. *Zeitschrift für Tierpsychologie*, 51, 184–200.

De Waal, F. B. M. 1991. Rank distance as a central feature of rhesus monkey social organization: a sociometric analysis. *Animal Behaviour*, 41, 383–395.

De Waal, F. B. M. & Luttrell, L. M. 1989. Toward a comparative socioecology of the genus *Macaca*: Different dominance styles in rhesus and stumptail macaques. *American Journal of Primatology*, 19, 83–109.

Di Fiore, A. & Rendall, D. 1994. Evolution of social organization: A reappraisal for primates by using phylogenetic methods. *Proceedings of the National Academy of Sciences of the United States of America*, 91, 9941–9945.

Dietz, J. 2004. Kinship structure and reproductive skew in cooperatively breeding primates. In *Kinship and Behavior in Primates* (ed. by Chapais, B. & Berman, C. M.), 223–241. New York: Oxford University Press.

Emery Thompson, M., Kahlenberg, S. M., Gilby, I. C. & Wrangham, R. W. 2007. Core area quality is associated with variance in reproductive success among female chimpanzees at Kibale National Park. *Animal Behaviour*, 73, 501–512.

Erhart, E. M., Coelho, A. M. & Bramblett, C. A. 1997. Kin recognition by paternal half-siblings in captive *Papio cynocephalus*. *American Journal of Primatology*, 43, 147–157.

Fernandez-Duque, E., Juarez, C. P. & Di Fiore, A. 2008. Adult male replacement and subsequent infant care by male and siblings in socially monogamous owl monkeys (*Aotus azarai*). *Primates*, 49, 81–84.

Fredrickson, W. T. & Sackett, G. P. 1984. Kin preferences in primates (*Macaca nemestrina*): Relatedness or familiarity. *Journal of Comparative Psychology*, 98, 29–34.

Furuichi, T. & Ihobe, H. 1994. Variation in male relationships in bonobos and chimpanzees. *Behaviour*, 130, 211–228.

Gerloff, U., Hartung, B., Fruth, B., Hohmann, G. & Tautz, D. 1999. Intracommunity relationships, dispersal pattern and paternity success in a wild living community of bonobos (*Pan paniscus*) determined from DNA analysis of faecal samples. *Proceedings of the Royal Society of London Series B Biological Sciences*, 266, 1189–95.

Gilad, Y., Wiebe, V., Przeworski, M., Lancet, D. & Paabo, S. 2004. Loss of olfactory receptor genes coincides with the acquisition of full trichromatic vision in primates. *Plos Biology*, 2, e5.

———. 2007. Correction: Loss of olfactory receptor genes coincides with the acquisition of full trichromatic vision in primates (vol. 2, pg. 120, 2004). *Plos Biology*, 5, 1383.

Gilby, I. C. & Wrangham, R. 2008. Association patterns among wild chimpanzees (*Pan troglodytes schweinfurthii*) reflect sex differences in cooperation. *Behavioral Ecology and Sociobiology*, 62, 1831–1842.

Goldberg, T. & Wrangham, R. W. 1997. Genetic correlates of social behavior in wild chimpanzees: Evidence from mitochondrial DNA. *Animal Behaviour*, 54, 559–570.

Gouzoules, S. 1984. Primate mating systems, kin associations, and cooperative behavior: Evidence for kin recognition. *Yearbook of Physical Anthropology*, 27, 99–134.

Gouzoules, S. & Gouzoules, H. 1987. Kinship. In *Primate Societies* (ed. by Smuts, B., Cheney, D. L., Seyfarth, R. M., Wrangham, R. & Struhsaker, T. T.), 299–305. Chicago: University of Chicago Press.

Haig, D. 1999. What is a marmoset? *American Journal of Primatology*, 49, 285–296.

Halperin, S. 1979. Temporary association patterns in free ranging chimpanzees: An assessment of individual grouping preferences. In *The Great Apes* (ed. by Hamburg, D. & McCown, E.), 491–499. Menlo Park, CA: Benjamin/Cummings.

Hamilton, W. D. 1963. The evolution of altruistic behavior. *American Naturalist*, 97, 354–356.

———. 1964a. The genetical evolution of social behaviour. I. *Journal of Theoretical Biology*, 7, 1–16.

———. 1964b. The genetical evolution of social behaviour. II. *Journal of Theoretical Biology*, 7, 17–52.

Harcourt, A. H. 1987. Dominance and fertility among female primates. *Journal of Zoology*, 213, 471–487.

Hasegawa, T. 1990. Sex differences in ranging patterns. In *The Chimpanzees of the Mahale Mountains: Sexual and Life History Strategies* (ed. by Nishida, T.), 99–114. Tokyo: University of Tokyo Press.

Hashimoto, C., Furuichi, T. & Takenaka, O. 1996. Matrilineal kin relationship and social behavior of wild bonobos: Sequencing the D-loop region of mitochondrial DNA. *Primates*, 37, 305–318.

Hausfater, G. 1975. Dominance and reproduction in baboons (*Papio cynocephalus*): Quantitative analysis. *Contributions to Primatology*, 7, 2–150.

Havlicek, J. & Roberts, S. C. 2009. MHC-correlated mate choice in humans: A review. *Psychoneuroendocrinology*, 34, 497–512.

Hemelrijk, C. K. 1999. An individual-orientated model of the emergence of despotic and egalitarian societies. *Proceedings of the Royal Society of London Series B Biological Sciences*, 266, 361–369.

Henzi, S. P. & Barrett, L. 1999. The value of grooming to female primates. *Primates*, 40, 47–59.

Hill, D. A. & Okayasu, N. 1995. Absence of youngest ascendancy in the dominance relations of sisters in wild Japanese macaques (*Macaca fuscata yakui*). *Behaviour*, 132, 367–379.

Hohmann, G., Gerloff, U., Tautz, D. & Fruth, B. 1999. Social bonds and genetic ties: Kinship association and affiliation in a community of bonobos (*Pan paniscus*). *Behaviour*, 136, 1219–1235.

Horrocks, J. A. & Hunte, W. 1983. Rank relations in vervet sisters: A critique of the role of reproductive value. *American Naturalist*, 122, 417–421.

Huck, M., Lottker, P., Bohle, U. & Heymann, E. W. 2005. Paternity and kinship patterns in polyandrous moustached tamarins (*Saguinus mystax*). *American Journal of Physical Anthropology*, 127, 449–464.

Huck, M., Lottker, P. & Heymann, E. W. 2004. The many faces of helping: Possible costs and benefits of infant carrying and food transfer in wild moustached tamarins (*Saguinus mystax*). *Behaviour*, 141, 915–934.

Hunte, W. & Horrocks, J. A. 1987. Kin and nonkin interventions in the aggressive disputes of vervet monkeys. *Behavioral Ecology and Sociobiology*, 20, 257–263.

Jack, K. M. & Fedigan, L. M. 2004a. Male dispersal patterns in white-faced capuchins, *Cebus capucinus*. Part 1: Patterns and causes of natal emigration. *Animal Behaviour*, 67, 761–769.

———. 2004b. Male dispersal patterns in white-faced capuchins, *Cebus capucinus*. Part 2: Patterns and causes of secondary dispersal. *Animal Behaviour*, 67, 771–782.

Kahlenberg, S. M., Emery Thompson, M. & Wrangham, R. W. 2008. Female competition over core areas in *Pan troglodytes schweinfurthii*, Kibale National Park, Uganda. *International Journal of Primatology*, 29, 931–947.

Kaminski, G., Dridi, S., Graff, C. & Gentaz, E. 2009. Human ability to detect kinship in strangers' faces: Effects of the degree of relatedness. *Proceedings of the Royal Society of London Series B-Biological Sciences*, 276, 3193–3200.

Kano, T. 1992. *The Last Ape*. Stanford, CA: Stanford University Press.

Kappeler, P. M. & Port, M. 2008. Mutual tolerance or reproductive competition? Patterns of reproductive skew among male redfronted lemurs (*Eulemur fulvus rufus*). *Behavioral Ecology and Sociobiology*, 62, 1477–1488.

Kapsalis, E. 2004. Matrilineal kinship and primate behavior. In *Kinship and Behavior in Primates* (ed. by Chapais, B. & Berman, C. M.), 153–176. Oxford: Oxford University Press.

Kapsalis, E. & Berman, C. M. 1996a. Models of affiliative relationships among free-ranging rhesus monkeys (*Macaca mulatta*). 2. Testing predictions for three hypothesized organizing principles. *Behaviour*, 133, 1235–1263.

———. 1996b. Models of affiliative relationships among free-ranging rhesus monkeys (*Macaca mulatta*). 1. Criteria for kinship. *Behaviour*, 133, 1209–1234.

Kawai, M. 1958. On the system of social ranks in a natural troop of Japanese monkey. I. Basic rank and dependent rank. *Primates*, 1, 111–130.

Kawamura, S. 1958. Matriarchal social ranks and in the Minoo-B troop: A study of the rank system of Japanese monkeys. *Primates*, 1–2, 149–156.

Kawanaka, K. 1984. Association, ranging, and the social unit in chimpanzees of the Mahale mountains, Tanzania. *International Journal of Primatology*, 5, 411–434.

Klopfer, P. H. 1970. Discrimination of young in galagos. *Folia Primatologica*, 13, 137–143.

Koenig, W. D., Hooge, P. N., Stanback, M. T. & Haydock, J. 2000. Natal dispersal in the cooperatively breeding acorn woodpecker. *Condor*, 102, 492–502.

Kuester, J. & Paul, A. 1997. Group fission in Barbary macaques (*Macaca sylvanus*) at Affenberg Salem. *International Journal of Primatology*, 18, 941–966.

Kummer, H. 1995. *In Quest of the Sacred Baboon*. Princeton, NJ: Princeton University Press.

Kummerli, R., Gardner, A., West, S. A. & Griffin, A. S. 2009.

Limited dispersal, budding dispersal, and cooperation: An experimental study. *Evolution*, 63, 939–949.

Kurland, J. A. 1977. *Kin Selection in the Japanese Monkey*. Basel: Karger.

Langergraber, K., Mitani, J. & Vigilant, L. 2009. Kinship and social bonds in female chimpanzees (*Pan troglodytes*). *American Journal of Primatology*, 71, 840–851.

Langergraber, K. E., Mitani, J. C. & Vigilant, L. 2007. The limited impact of kinship on cooperation in wild chimpanzees. *Proceedings of the National Academy of Sciences of the United States of America*, 104, 7786–7790.

Lappan, S. 2008. Male care of infants in a siamang (*Symphalangus syndactylus*) population including socially monogamous and polyandrous groups. *Behavioral Ecology and Sociobiology*, 62, 1307–1317.

Lehmann, J. & Boesch, C. 2008. Sexual differences in chimpanzee sociality. *International Journal of Primatology*, 29, 65–81.

Lehmann, J., Fickenscher, G. & Boesch, C. 2006. Kin biased investment in wild chimpanzees. *Behaviour*, 143, 931–955.

Macdonald, E. A., Fernandez-Duque, E., Evans, S. & Hagey, L. R. 2008. Sex, age, and family differences in the chemical composition of owl monkey (*Aotus nancvmaae*) subcaudal scent secretions. *American Journal of Primatology*, 70, 12–18.

Mackenzie, M. M., McGrew, W. C. & Chamove, A. S. 1985. Social preferences in stump-tailed macaques (*Macaca arctoides*): Effects of companionship, kinship, and rearing. *Developmental Psychobiology*, 18, 115–123.

Maloney, L. T. & Dal Martello, M. F. 2006. Kin recognition and the perceived facial similarity of children. *Journal of Vision*, 6, 1047–1056.

Marker, L. L., Wilkerson, A. J. P., Sarno, R. J., Martenson, J., Breitenmoser-Wuersten, C., O'Brien, S. J. & Johnson, W. E. 2008. Molecular genetic insights on cheetah (*Acinonyx jubatus*) ecology and conservation in Namibia. *Journal of Heredity*, 99, 2–13.

Mateo, J. M. 2003. Kin recognition in ground squirrels and other rodents. *Journal of Mammalogy*, 84, 1163–1181.

———. 2009. The causal role of odours in the development of recognition templates and social preferences. *Animal Behaviour*, 77, 115–121.

Mateo, J. M. & Holmes, W. G. 2004. Cross-fostering as a means to study kin recognition. *Animal Behaviour*, 68, 1451–1459.

Mateo, J. M. & Johnston, R. E. 2000. Kin recognition and the "armpit effect": Evidence of self-referent phenotype matching. *Proceedings of the Royal Society of London Series B Biological Sciences*, 267, 695–700.

Maynard Smith, J. 1964. Group selection and kin selection. *Nature*, 201, 145–147

McNutt, J. W. 1996. Sex-biased dispersal in African wild dogs, *Lycaon pictus*. *Animal Behaviour*, 52, 1067–1077.

Meikle, D. B. & Vessey, S. H. 1981. Nepotism among rhesus monkey brothers. *Nature*, 29, 160–161.

Michael, J. O. & Jacquelyn, A. D. 1989. Infant cross-fostering between Japanese (*Macaca fuscata*) and rhesus macaques (*M. mulatta*). *American Journal of Primatology*, 18, 245–250.

Mitani, J. C. 2009. Male chimpanzees form enduring and equitable social bonds. *Animal Behaviour*, 77, 633–640.

Mitani, J. C., Merriwether, D. A. & Zhang, C. B. 2000. Male affiliation, cooperation and kinship in wild chimpanzees. *Animal Behaviour*, 59, 885–893.

Mitani, J. C. & Watts, D. P. 1997. The evolution of non-maternal caretaking among anthropoid primates: Do helpers help? *Behavioral Ecology and Sociobiology*, 40, 213–220.

Mitani, J. C., Watts, D. P. & Muller, M. N. 2002a. Recent developments in the study of wild chimpanzee behavior. *Evolutionary Anthropology*, 11, 9–25.

Mitani, J. C., Watts, D. P., Pepper, J. W. & Merriwether, D. A. 2002b. Demographic and social constraints on male chimpanzee behaviour. *Animal Behaviour*, 64, 727–737.

Mitchell, C. L. 1994. Migration alliances and coalitions among adult male South American squirrel monkeys (*Saimiri sciureus*). *Behaviour*, 130, 169–190.

Moscovice, L. R., Heesen, M., Di Fiore, A., Seyfarth, R. M. & Cheney, D. L. 2009. Paternity alone does not predict long-term investment in juveniles by male baboons. *Behavioral Ecology and Sociobiology*, 63, 1471–1482.

Muller, M. N. & Mitani, J. C. 2005. Conflict and cooperation in wild chimpanzees. *Advances in the Study of Behavior*, 35, 275–331.

Muniz, L., Perry, S., Manson, J. H., Gilkenson, H., Gros-Louis, J. & Vigilant, L. 2006. Father-daughter inbreeding avoidance in a wild primate population. *Current Biology*, 16, R156–R157.

Murray, C. M., Eberly, L. E. & Pusey, A. E. 2006. Foraging strategies as a function of season and rank among wild female chimpanzees (*Pan troglodytes*). *Behavioral Ecology*, 17, 1020–1028.

Murray, C. M., Mane, S. V. & Pusey, A. E. 2007. Dominance rank influences female space use in wild chimpanzees, *Pan troglodytes*: Towards an ideal despotic distribution. *Animal Behaviour*, 74, 1795–1804.

Murray, C. M., Wroblewski, E. & Pusey, A. 2008. New case of intragroup infanticide in the chimpanzees of Gombe National Park. *International Journal of Primatology*, 28, 23–37.

Netto, W. J. & van Hoof, J. A. R. A. M. 1986. Conflict interference and the development of dominance relationships in immature *Macaca fascicularis*. In *Primate Ontogeny, Cognition, and Social Behaviour* (ed. by Else, J. G. & Lee, P. V.), 291–300. Cambridge: Cambridge University Press.

Nguyen, N., Van Horn, R. C., Alberts, S. C. & Altmann, J. 2009. "Friendships" between new mothers and adult males: Adaptive benefits and determinants in wild baboons (*Papio cynocephalus*). *Behavioral Ecology and Sociobiology*, 63, 1331–1344.

Nievergelt, C. M., Digby, L. J., Ramakrishnan, U. & Woodruff, D. S. 2000. Genetic analysis of group composition and breeding system in a wild common marmoset (*Callithrix jacchus*) population. *International Journal of Primatology*, 21, 1–20.

Nishida, T., Corp, N., Hamai, M., Hasegawa, T., Hiraiwa-Hasegawa, M., Hosaka, K., Hunt, K. D., Itoh, N., Kawanaka, K., Matsumoto-Oda, A., Mitani, J. C., Nakamura, M., Norikoshi, K., Sakamaki, T., Turner, L., Uehara, S. & Zamma, K. 2003. Demography, female life history, and reproductive profiles among the chimpanzees of Mahale. *American Journal of Primatology*, 59, 99–121.

Ostner, J. & Kappeler, P. M. 2004. Male life history and the unusual sex ratios of redfronted lemur (*Eulemur fulvus rufus*) groups. *Animal Behaviour*, 67, 249–259.

Ostner, J., Nunn, C. L. & Schulke, O. 2008. Female reproductive synchrony predicts skewed paternity across primates. *Behavioral Ecology*, 19, 1150–1158.

Packer, C., Gilbert, D. A., Pusey, A. E. & O'Brien, S. J. 1991. A molecular genetic analysis of kinship and cooperation in African lions. *Nature*, 351, 562–565.

Parr, L. A. & de Waal, F. B. M. 1999. Visual kin recognition in chimpanzees. *Nature*, 399, 647–648.

Pereira, M. E. 1988. Agonistic interactions of juvenile female savanna baboons .1. Fundamental features. *Ethology*, 79, 195–217.

———. 1989. Agonistic interactions of juvenile savanna baboons .2. Agonistic support and rank acquisition. *Ethology*, 80, 152–171.

Perry, S., Manson, J. H., Muniz, L., Gros-Louis, J. & Vigilant, L. 2008. Kin-biased social behaviour in wild adult female white-faced capuchins, *Cebus capucinus*. *Animal Behaviour*, 76, 187–199.

Pope, T. R. 1990. The reproductive consequences of male cooperation in the red howler monkey: Paternity exclusion in multi-male and single-male troops using genetic markers. *Behavioral Ecology and Sociobiology*, 27, 439–446.

Prudhomme, J. & Chapais, B. 1993a. Aggressive interventions and matrilineal dominace relations in semifree-ranging barbary macaques (*Macaca sylvanus*). *Primates*, 34, 271–283.

———. 1993b. Rank relations among sisters in semi-free-ranging barbary macques (*Macaca sylvanus*). *International Journal of Primatology*, 14, 405–420.

Queller, D. C. 1994. Genetic relatedness in viscous populations. *Evolutionary Ecology*, 8, 70–73.

Rendall, D. 2004. "Recognizing kin": Mechanisms, media, minds, modules, and muddles. In *Kinship and Behavior in Primates* (ed. by Chapais, B. & Berman, C. M.), 295–316. Oxford: Oxford University Press.

Rendall, D., Notman, H. & Owren, M. J. 2009. Asymmetries in the individual distinctiveness and maternal recognition of infant contact calls and distress screams in baboons. *Journal of the Acoustical Society of America*, 125, 1792–1805.

Rendall, D., Rodman, P. S. & Emond, R. E. 1996. Vocal recognition of individuals and kin in free-ranging rhesus monkeys. *Animal Behaviour*, 51, 1007–1015.

Robbins, A. M., Stoinski, T., Fawcett, K. & Robbins, M. M. 2009. Leave or conceive: Natal dispersal and philopatry of female mountain gorillas in the Virunga volcano region. *Animal Behaviour*, 77, 831–838.

Robbins, A. M., Stoinski, T. S., Fawcett, K. A. & Robbins, M. M. 2008. Does dispersal cause reproductive delays in female mountain gorillas? *Behaviour*, 146, 525–549.

Ross, C. & MacLarnon, A. 2000. The evolution of non-maternal care in anthropoid primates: A test of the hypotheses. *Folia Primatologica*, 71, 93–113.

Sackett, G. P. & Fredrickson, W. T. 1987. Social preferences by pigtailed macaques: Familiarity versus degree and type of kinship. *Animal Behaviour*, 35, 603–606.

Sade, D. S. 1965. Some aspects of parent-offspring and sibling relations in a group of rhesus monkeys, with a discussion of grooming. *American Journal of Physical Anthropology*, 23, 1–17.

Schino, G. 2001. Grooming, competition and social rank among female primates: A meta-analysis. *Animal Behaviour*, 62, 265–271.

———. 2007. Grooming and agonistic support: A meta-analysis of primate reciprocal altruism. *Behavioral Ecology*, 18, 115–120.

Schino, G. & Aureli, F. 2008. Grooming reciprocation among female primates: A meta-analysis. *Biology Letters*, 4, 9–11.

Schoof, V. A. M., Jack, K. M. & Isbell, L. A. 2009. What traits promote male parallel dispersal in primates? *Behaviour*, 146, 701–726.

Schulman, S. R. & Chapais, B. 1980. Reproductive value and rank relations among macaque sisters. *American Naturalist*, 115, 580–593.

Seyfarth, R. 1977. A model of social grooming among adult female monkeys. *Journal of Theoretical Biology*, 65, 671–698.

Seyfarth, R. M. 1980. The distribution of grooming and related *Behaviours* among adult female vervet monkeys. *Animal Behaviour*, 28, 798–813.

Seyfarth, R. M. & Cheney, D. L. 1984. The natural vocalizations of non-human primates. *Trends in Neurosciences*, 66–73.

Sharp, S. P., Simeoni, M. & Hatchwell, B. J. 2008. Dispersal of sibling coalitions promotes helping among immigrants in a cooperatively breeding bird. *Proceedings of the Royal Society of London Series B Biological Sciences*, 275, 2125–2130.

Silk, J. B. 1982. Altruism among female *Macaca radiata*: Explanations and analysis of patterns of grooming and coalition-formation. *Behaviour*, 79, 162–188.

———. 1987. Social behavior in evolutionary perspective. In *Primate Societies* (ed. by Smuts, B. B., Cheney, D. L., Seyfarth, R. M., Wrangham, R. W. & Struhsaker, T. T.), 318–329. Chicago: University of Chicago Press.

———. 2001. Ties that bond: The role of kinship in primate societies. In *New Directions in Anthropological Kinship* (ed. by Stone, L.), 71–92. Boulder: Rowman and Littlefield.

———. 2002. Kin selection in primate groups. *International Journal of Primatology*, 23, 849–875.

———. 2006. Practicing Hamilton's rule: Kin selection in primate groups. In *Cooperation in Primates and Humans* (ed. by Kappeler, P. M. & van Schaik, C. P.), 25–46. Berlin: Springer.

———. 2009. Nepotistic cooperation in non-human primate groups. *Philosophical Transactions of the Royal Societyof London Series B-Biological Sciences*, 364, 3243–3254.

Silk, J. B., Alberts, S. C. & Altmann, J. 2003. Social bonds of female baboons enhance infant survival. *Science*, 302, 1231–1234.

———. 2004. Patterns of coalition formation by adult female baboons in Amboseli, Kenya. *Animal Behaviour*, 67, 573–582.

Silk, J. B., Altmann, J. & Alberts, S. C. 2006. Social relationships among adult female baboons (*Papio cynocephalus*) I. Variation in the strength of social bonds. *Behavioral Ecology and Sociobiology*, 61, 183–195.

Silk, J. B., Beehner, J. C., Bergman, T. J., Crockford, C., Engh, A. L., Moscovice, L. R., Wittig, R. M., Seyfarth, R. M. & Cheney, D. L. 2009. The benefits of social capital: Close social bonds among female baboons enhance offspring survival. *Proceedings of the Royal Society of London Series B Biological Sciences*, 276, 3099–3104.

Smith, K., Alberts, S. C. & Altmann, J. 2003. Wild female baboons bias their social behaviour towards paternal half-sisters. *Proceedings of the Royal Society of London Series B-Biological Sciences*, 270, 503–510.

Smuts, B. B. & Gubernick, D. J. 1992. Male-infant relationships in nonhuman primates: Paternal investment or mating effort?

In *Father-Child Relations: Cultural and Biosocial Contexts.* (ed. by Hewlett, B. S.), 1–30. New York: Aldine de Gruyter.

Starin, E. D. 1994. Philopatry and affiliation among red colobus. *Behaviour,* 130, 253–270.

———. 2001. Patterns of inbreeding avoidance in Temminck's red colobus. *Behaviour,* 138, 453–465.

Sterck, E. H. M., Watts, D. P. & van Schaik, C. P. 1997. The evolution of female social relationships in nonhuman primates. *Behavioral Ecology and Sociobiology,* 41, 291–309.

Steve, S. 1986. Infant cross-fostering in rhesus monkeys (*Macaca mulatta*): A procedure for the long-term management of captive populations. *American Journal of Primatology,* 11, 229–237.

Stevens, J. R., Cushman, F. A. & Hauser, M. D. 2005. Evolving the psychological mechanisms for cooperation. *Annual Review of Ecology and Systematics,* 36, 499–518.

Strier, K. B., Dib, L. T. & Figueira, J. E. C. 2002. Social dynamics of male muriquis (*Brachyteles arachnoides hypoxanthus*). *Behaviour,* 139, 315–342.

Struhsaker, T. T. 2000. Variation in adult sex ratios of red colobus monkey groups: Implications for interspecific comparisons. In *Primate Males* (ed. by Kappeler, P. M.), 108–119. Cambridge University Press: Cambridge.

Struhsaker, T. T. & Leland, L. 1987. Colobines: Infanticide by adult males. In *Primate Societies* (ed. by Smuts, B. B., Cheney, D. L., Seyfarth, R. M., Wrangham, R. W. & Struhsaker, T. T.), 83–97. Chicago: University of Chicago Press.

Struhsaker, T. T. & Pope, T. R. 1991. Mating system and reproductive success: A comparison of 2 African forest monkeys (*Colobus badius* and *Cercopithecus ascanius*). *Behaviour,* 117, 182–205.

Sugiyama, Y. 2004. Demographic parameters and life history of chimpanzees at Bossou, Guinea. *American Journal of Physical Anthropology,* 124, 154–165.

Sussman, R. W. 1992. Male life history and intergroup mobility among ring-tailed lemurs (*Lemur catta*). *International Journal of Primatology,* 13, 395–413.

Thierry, B. 2004. Social epigenesis. In *Macaque Societies: A Model for the Study of Social Organization* (ed. by Thierry, B., Sing, V. & Kaumanns, W.), 267–290. Cambridge: Cambridge University Press.

———. 2007. Unity in diversity: Lessons from macaque societies. *Evolutionary Anthropology,* 16, 224–238.

Thierry, B., Aureli, F., Nunn, C. L., Petit, O., Abegg, C. & De Waal, F. B. M. 2008. A comparative study of conflict resolution in macaques: Insights into the nature of trait covariation. *Animal Behaviour,* 75, 847–860.

Townsend, S. W., Slocombe, K. E., Emery Thompson, M. & Zuberbuhler, K. 2007. Female-led infanticide in wild chimpanzees. *Current Biology,* 17, R355-R356.

Trivers, R. 1974. Parent-offspring conflict. *American Zoologist,* 14, 249–264.

Van Hoof, J. A. R. A. M. 2000. Relationships among nonhuman primate males: A deductive framework. In *Primate Males* (ed. by Kappeler, P.), 183–191. Cambridge: Cambridge University Press.

Van Horn, R. C., Altmann, J. & Alberts, S. C. 2008. Can't get there from here: Inferring kinship from pairwise genetic relatedness. *Animal Behaviour,* 75, 1173–1180.

Van Horn, R. C., Buchan, J. C., Altmann, J. & Alberts, S. C.

2007. Divided destinies: Group choice by female savannah baboons during social group fission. *Behavioral Ecology and Sociobiology,* 61, 1823–1837.

Van Noordwijk, M. A. & van Schaik, C. P. 1985. Male migration and rank acquisition in wild long-tailed macaques (*Macaca fascicularis*). *Animal Behaviour,* 33, 849–861.

———. 2001. Career moves: Transfer and rank challenge decisions by male long-tailed macaques. *Behaviour,* 138, 359–395.

Van Schaik, C. P. 1989. The ecology of social relationships amongst female primates. In: *Comparative Socioecology: The Behavioural Ecology of Humans and Other Mammals* (ed. by Standon, V. & Foley, R. A.), 195–218. Oxford: Blackwell Scientific Publications.

Vigilant, L., Hofreiter, M., Siedel, H. & Boesch, C. 2001. Paternity and relatedness in wild chimpanzee communities. *Proceedings of the National Academy of Sciences of the United States of America,* 98, 12890–12895.

Vokey, J. R., Rendall, D., Tangen, J. M., Parr, L. A. & de Waal, F. B. M. 2004. Visual kin recognition and family resemblance in chimpanzees. *Journal of Comparative Psychology,* 118, 194–199.

Walters, J. 1980. Interventions in the development of dominance relations in female baboons. *Folia Primatologica,* 34, 61–89.

———. 1987. Kin recognition in non-human primates. In *Kin Recognition in Animals* (ed. by Fletcher, D. J. C. & Michener, C. D.), 359–394. Chichester, UK: Wiley.

Watanabe, K. 1979. Alliance formation in a free-ranging troop of Japanese macaques. *Primates,* 20, 459–474.

Watts, D. P. 1994. Social relationships of immigrant and resident female mountain gorillas. 2. Relatedness, residence, and relationships between females. *American Journal of Primatology,* 32, 13–30.

———. 1996. Comparative socioecology of gorillas. In *Great Ape Societies* (ed. by McGrew, W. C., Marchant, L. & Nishida, T.), 16–28. Cambridge: Cambridge University Press.

Welker, C., Schwibbe, M. H., Schaferwitt, C. & Visalberghi, E. 1987. Failure of kin recognition in *Macaca fascicularis*. *Folia Primatologica,* 49, 216–221.

West, S. A., Griffin, A. S. & Gardner, A. 2007. Social semantics: Altruism, cooperation, mutualism, strong reciprocity and group selection. *Journal of Evolutionary Biology,* 20, 415–432.

West, S. A., Murray, M. G., Machado, C. A., Griffin, A. S. & Herre, E. A. 2001. Testing Hamilton's rule with competition between relatives. *Nature,* 409, 510–513.

Widdig, A. 2007. Paternal kin discrimination: The evidence and likely mechanisms. *Biological Reviews,* 82, 319–334.

Widdig, A., Bercovitch, B., Streich, W. J., Suaermann, U., Nürnberg, P. & Krawczak, M. A longitudinal oanalysis of reproductive skew in male rhesus macaques. *Proceedings of the Royal Society of London Series B Biological Sciences,* 271, 819–826.

Widdig, A., Nürnberg, P., Bercovitch, B., Trefilov, A., Berard, J. B., Kessler, M. J., Schmidtke, J., Streich, W. J. & Krawczak, M. 2006a. Consequences of group fission for the patterns of relatedness among rhesus macaques. *Molecular Ecology,* 15, 3825–3832.

Widdig, A., Nürnberg, P., Krawczak, M., Streich, W. J. & Bercovitch, F. 2002. Affiliation and aggression among adult female rhesus macaques: A genetic analysis of paternal cohorts. *Behaviour,* 139, 371–391.

———. 2001. Paternal relatedness and age proximity regulate social relationships among adult female rhesus macaques. *Proceedings of the National Academy of Sciences of the United States of America*, 98, 13769–13773.

Widdig, A., Streich, W. J., Nurnberg, P., Croucher, P. J. P., Bercovitch, F. B. & Krawczak, M. 2006b. Paternal kin bias in the agonistic interventions of adult female rhesus macaques (*Macaca mulatta*). *Behavioral Ecology and Sociobiology*, 61, 205–214.

Williams, D. A. & Rabenold, K. N. 2005. Male-biased dispersal, female philopatry, and routes to fitness in a social corvid. *Journal of Animal Ecology*, 74, 150–159.

Williams, J. M., Liu, H. Y. & Pusey, A. E. 2002. Costs and benefits of grouping for female chimpanzees at Gombe. In *Behavioural Diversity in Chimpanzees and Bonobos* (ed. by Boesch, C., Hohmann, G. & Marchant, L. F.), 192–203. Cambridge: Cambridge University Press.

Wilson, D. S., Pollock, G. B. & Dugatkin, L. A. 1992. Can altruism evolve in purely viscous populations? *Evolutionary Ecology*, 6, 331–341.

Wilson, E. O. 1975. *Sociobiology*. Cambridge, MA: Belknap Press of Harvard University Press.

Wrangham, R. & Smuts, B. B. 1980. Sex differences in the behavioural ecology of chimpanzees in the Gombe National Park, Tanzania. *Journal of Reproductive Fertility Supplement*, 28, 13–31.

Wrangham, R. W. 1980. An ecological model of female-bonded primate groups. *Behaviour*, 75, 262–300.

Wright, P. C. 1990. Patterns of paternal care in primates. *International Journal of Primatology*, 11, 89–102.

Wroblewski, E. E., Murray, C. M., Keele, B. F., Schumacher-Stankey, J. C., Hahn, B. H. & Pusey, A. E. 2009. Male dominance rank and reproductive success in chimpanzees, *Pan troglodytes schweinfurthii*. *Animal Behaviour*, 77, 873–885.

Wu, H. M. H., Holmes, W. G., Medina, S. R. & Sackett, G. P. 1980. Kin preference in infant *Macaca nemestrina*. *Nature*, 285, 225–227.

Yaber, M. C. & Rabenold, K. N. 2002. Effects of sociality on short-distance, female-biased dispersal in tropical wrens. *Journal of Animal Ecology*, 71, 1042–1055.

Yamazaki, K., Beauchamp, G. K., Curran, M., Bard, J. & Boyse, E. A. 2000. Parent-progeny recognition as a function of MHC odortype identity. *Proceedings of the National Academy of Sciences of the United States of America*, 97, 10500–10502.

Chapter 22 Cooperation among Non-kin: Reciprocity, Markets, and Mutualism

Ian C. Gilby

COOPERATION AMONG non-kin has puzzled behavioral ecologists, primatologists, and anthropologists for decades. The core question is deceptively simple. If animals act in their own self interest (Darwin 1859), why do they sometimes provide aid to unrelated individuals? The quick answer is that non-kin cooperation evolves when helping others increases the helper's own direct fitness. This is reflected in the broad definition of cooperation I adopt in this chapter—"joint action for mutual benefit" (Mesterton-Gibbons & Dugatkin 1992; Clements & Stephens 1995). However, exactly *how* and *when* these benefits accrue is fiercely debated, with primates at the center of the controversy.

Unraveling the causes and consequences of cooperation is essential for understanding other quintessential features of primate societies, including social bonding (e.g., Gilby & Wrangham 2008; chapters 23 and 24, this volume;) and intergroup dynamics (e.g., Mitani & Watts 2005; Crofoot et al. 2008). Cooperation is also central to debates over the nature of empathy and social preferences in nonhuman primates (Silk et al. 2005; Warneken et al. 2007; chapter 25, this volume;). Ultimately, understanding the cooperative behavior of nonhuman primates provides insight into our own species, for which cooperation is particularly important (chapters 13, 26, and 32, this volume).

Why do primates cooperate with non-kin? My approach to this rather complex problem starts with a simple question. When an animal is faced with the opportunity to cooperate (e.g., to join a group hunt), I ask whether cooperating would yield a higher immediate payoff for that individual than not cooperating. If so, then some form of mutualism will most likely explain the evolution of cooperation in this context. Alternatively, if defection (*failing to cooperate*) generates a greater immediate net benefit for that individual, cooperation can only evolve via a form of contingent exchange in which cooperation is conditional upon previous or future cooperation or the market value of a commodity that a potential partner can offer.

This distinction between immediately mutualistic and exchange-based cooperation has important consequences for understanding the nature of primate intelligence. It is often assumed that contingent exchange requires cognitive abilities that are beyond the capacities of most primates. In some cases, this assumption has been stated explicitly (Stevens & Cushman 2004; Stevens & Gilby 2004; Stevens & Hauser 2004). However, others argue that such exchanges only require simple "emotionally based bookkeeping" (Schino & Aureli 2009), rather than advanced cognitive abilities. Whether contingent exchange is cognitively or emotionally based is a matter that must be resolved empirically on a case-by-case basis. Understanding the distribution of contingent exchanges across primate taxa is a critical first step.

Assessing Benefits and Costs

Evolutionary game theory enables behavioral ecologists to analyze animal decision-making when the consequences of an individual's decisions depend on the behavior of others

(Maynard Smith 1982). Ultimately, the consequences of a given behavior should be measured in terms of a change in an individual's relative fitness, but this is unrealistic under most circumstances. Instead, behavioral ecologists make logical assumptions about the effect of a single action. For example, we assume that it is costly for females to relinquish food items to competitors because nutritional status is generally correlated with female fertility.

As an example of simple evolutionary game theory applied to cooperation, imagine a scenario in which two individuals (A and B) simultaneously encounter a potential prey item. Each individual can either hunt (cooperate, C) or do nothing (defect, D). Fig. 22.1 illustrates the four potential payoffs for individual A, whose optimal decision depends upon the actions of B. If both animals hunt, then A receives a net payoff in amount of R (the reward for mutual cooperation). If B hunts and A defects, then A receives a payoff of T (the temptation to defect). If both players defect, then A receives P (the punishment for mutual defection). Finally, if A cooperates and B defects, A receives S (the sucker's payoff). (Note that the labels R, T, P, and S are based on the outcomes of a specific payoff scenario, the prisoner's dilemma, described below). Careful inspection of the payoff matrix allows us to identify whether and when cooperation is an "evolutionarily stable strategy" (ESS; Maynard Smith 1972, 1982). A strategy is an ESS if, when adopted by most members of a population, it cannot be invaded by the spread of any rare alternative strategy. I begin by discussing scenarios in which cooperation is an ESS.

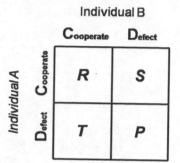

Individual B

Fig. 22.1. Game theoretical payoff matrix. R, S, T, and P represent payoffs to individual A, based on its behavior and that of individual B. For example, individual A receives a payoff of R when both A and B cooperate.

When Defection Yields No Immediate Benefit

Cooperation will be an ESS when there is no incentive to take advantage of others' cooperation. There are several ways in which this might occur. For example, imagine that two predators are faced with a single large prey item that is almost impossible for a single hunter to capture. The probability of capture increases considerably if both individuals hunt together. The carcass is shared, regardless of who kills it, because it is more than sufficient to satiate them both. The payoffs of this game are shown in fig. 22.2a. Should A hunt or defect? If B hunts and A defects, there is only a very slim chance that A will benefit, because it is so hard for a single individual to succeed. Therefore, if B hunts, A will receive the greatest payoff if she hunts as well. Now, what is A's best strategy if B defects? A will not obtain any meat by defecting, but there is a small chance that A might succeed by herself. Therefore, if B defects, the best strategy for A is to hunt. As a result, A should hunt, regardless of B's behavior. This simple scenario is an example of a *true mutualism*, in which cooperation is a pure ESS (R > T and

S > P): hunting always yields a higher payoff than defecting, regardless of the behavior of other individuals.

A *by-product mutualism* is a form of mutualism in which an individual's selfish actions incidentally benefit others (Brown 1983; Dugatkin 1997). This occurs when R > S > T ≈ P (fig. 22.2b). To illustrate a by-product mutualism, imagine sharks hunting small schooling fish. When a single shark attacks a school, it can capture several small fish. If it doesn't hunt, it gets nothing. However, when the schooling fish detect the shark, they disperse in panic. This makes it easier for other sharks to capture prey. The more hunters, the greater the net benefit for all, even though each hunter can successfully hunt on its own. Again, cooperation is a pure ESS, because there is never any reason to defect.

A third situation in which cooperation is a pure ESS is a *manipulative mutualism* (Stevens & Stephens 2002; Stevens 2004). Here, individual B affects the payoff structure such that cooperation is immediately beneficial to individual A. For example, harassment may change the payoff structure so that an individual that possesses a desirable food item may be better off sharing rather than risking the costs of defending it (Stevens & Stephens 2002). In this case it is in the possessor's immediate best interests to share. Critically, it is the behavior of other individuals rather than extrinsic ecological factors that make sharing beneficial to the food possessor.

There are also situations in which cooperation is part of a "mixed" ESS. This is illustrated by a stag hunt (Skyrms 2004; Silk 2007a), in which individuals choose between hunting small prey (e.g., hare) or large prey (e.g., stag). A lone hunter has a high chance of killing a hare and obtaining a small net payoff. A stag yields a greater per capita payoff, but it can only be captured by two individuals working together. In this case, the ESS depends upon the initial proportion of cooperators and defectors in the population. At one extreme, if the population is composed only of hare

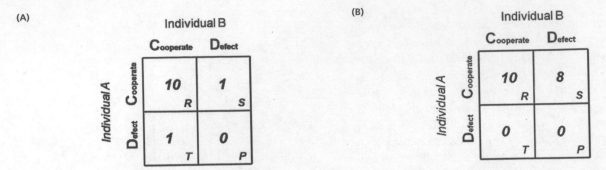

Fig. 22.2. (A) A true mutualism, in which the strategy "cooperate" is a pure ESS (R > T and S > P). (B) A by-product mutualism.

hunters, stag hunting will not spread, because at least one other cooperator is required to kill a stag (fig. 22.3). However, if the population is comprised solely of stag hunters, there will be no incentive to begin hunting hare, and cooperation will persist. The system may also reach a stable equilibrium at which a certain proportion of individuals hunt (or each individual has a certain probability of hunting) stag. One important feature of the stag hunt is that once a hunt (of either prey type) has begun, neither party gains an advantage from cheating by hanging back during the chase or abandoning the hunt altogether.

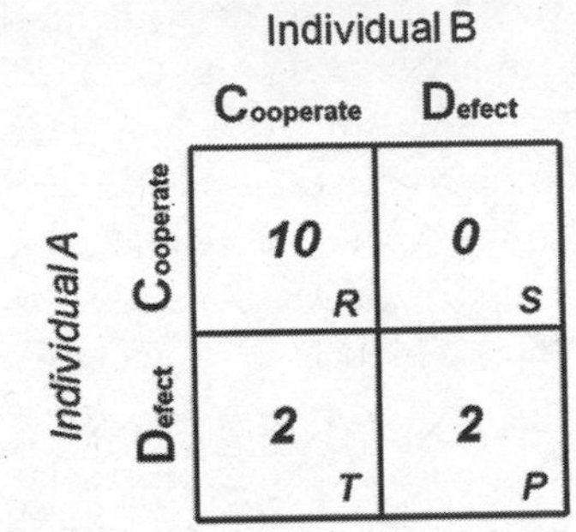

When Defection Yields Immediate Benefits

In the scenarios described above, individuals obtain a greater net benefit by cooperating than by defecting—and in these cases cooperation isn't much of an evolutionary puzzle. However, what if cooperation is costly, and an individual could gain an immediate net benefit by defecting? This situation is often represented by the prisoner's dilemma (PD), which is depicted by a payoff structure in which $T > R > P > S$ (e.g., fig. 22.4). Here, the payoff for mutual cooperation (R) is relatively high, but an individual can obtain an even greater payoff by exploiting the cooperative tendencies of others ($T > R$). To revisit our hunting example, imagine that two predators encounter a single prey item that is larger than a single predator can consume. One individual is able to bring down the prey, but at substantial cost (risk of injury, energy expenditure, etc.). If both individuals hunt, then the individual costs are considerably lower. Therefore, cooperation yields a higher payoff than hunting alone. However, if individual B hunts and A defects, A obtains the benefits of meat without paying the costs (T), a payoff that is greater than what she would receive by cooperating. Also, if B defects, then A should also defect, since the payoff for exploited cooperators (S) is lower than the payoff for

Fig. 22.3. A stag hunt

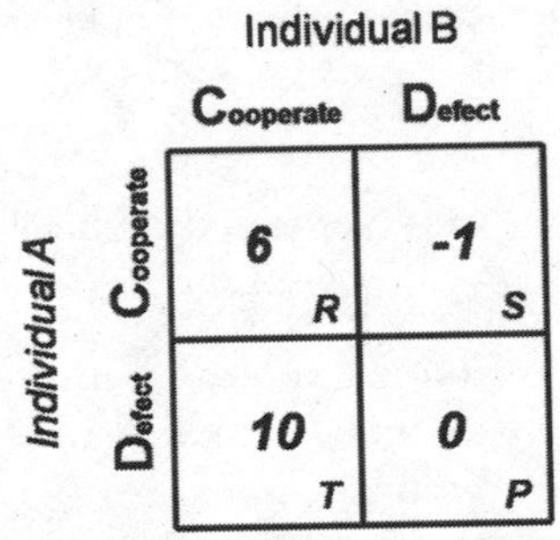

Fig. 22.4. A prisoner's dilemma

mutual defection (P). Therefore, the ESS in a PD is to defect, even though this yields a lower combined payoff than if both individuals had cooperated.

If this is the case, then how can we explain the existence of cooperation in circumstances where there is an apparent cost to cooperation?

Contingent reciprocity

The PD models a single interaction, and assumes that the two players have no history of past interactions and will not meet again. For individuals that live in a social group, this is clearly an unrealistic assumption. This prompted Trivers (1971) to propose *reciprocal altruism* (also known as *reciprocity*) as a mechanism that might promote cooperation in circumstances where there is an immediate temptation to defect. The logic of reciprocal altruism is very simple: the tendency to perform a costly act benefiting another may evolve if the recipient returns the favor in the future. Provided that the benefit to the recipient outweighs the cost to the actor in each iteration, cooperation yields a net benefit for both participants (Trivers 1971). Therefore, the immediate benefit of defection is outweighed by a larger benefit that accrues when the act is reciprocated. Reciprocity should be most likely to evolve in social species in which individuals can recognize each other and can exchange roles (Trivers 1971). Primates are excellent candidates because their sociality, long life spans, and low dispersal rates promote frequent interactions. Additionally, the non-linear dominance hierarchies and social tolerance exhibited by many species place more individuals in a position to reciprocate (van Schaik & Kappeler 2006).

Starting with a famous game theory computer tournament (Axelrod & Hamilton 1981) theoreticians have developed countless detailed mathematical models exploring the nuances of reciprocity within the iterated prisoner's dilemma (IPD) paradigm (reviewed in Dugatkin 1997; Trivers 2006). The majority of these theoretical exercises address some form of *contingent reciprocity* by which individuals keep track of the occurrence of past cooperation and defection and respond in kind. In Axelrod and Hamilton's (1981) tournament, the simple "tit-for-tat" strategy (which is equivalent to contingent reciprocity) was one of a number of strategies that was able to persist in the IPD paradigm.

Biological markets

Biological markets theory (or market-based exchange) proposes that cooperation is maintained via a system in which animals trade inalienable commodities, which by definition cannot be taken by force (Noë et al. 1991; Noë & Hammerstein 1994, 1995). Individuals are classified according to the commodity they can provide, and they exchange that commodity with those in another trading class. Both classes benefit from receiving the commodity that the other class controls. For example, high-ranking individuals can offer tolerance to low-ranking individuals at a monopolizable feeding site, and they may trade tolerance for grooming (e.g., Barrett et al. 1999). In this respect, reciprocal altruism and biological markets theory make the same prediction: that subordinate animals may groom dominants in return for access to valued foods (fig. 22.5).

However, the critical difference between contingent reciprocity and biological markets theory is that in the latter, animals choose with whom to interact. A key assumption of reciprocity theory is that individuals interact randomly but dispense altruism nonrandomly in light of prior interactions. Animals may therefore find themselves forced into a series of noncooperative, mutually detrimental interactions. By contrast, in a biological market, trading partners are chosen from a range of alternatives, "via a mechanism of outbidding competition, in such a way that profit is maximized" (Barrett & Henzi 2006). In theory, this allows an individual to interact only with partners that maximize her own payoff.

Proponents of biological market theory argue that because of partner choice, defection is not the primary obstacle to cooperation, as it is with contingent reciprocity. However, it seems that in most putative cases of market-based exchange in primates, the immediate benefits of defecting are in fact greater than zero. For example, in the grooming-for-feeding access example used above, after the subordinate grooms the dominant for the appropriate amount of time, the dominant individual could defect, thus denying the subordinate access to the feeding site. The dominant would have received the benefits of being groomed without paying any costs. Therefore, I argue that for cooperation to persist in a primate biological market, individuals must be able to overcome the temptation to defect (as with contingent reciprocity). In contrast, market exchanges in nonprimate taxa, such as cleaner/client fish (Bshary 2001), typically involve immediate benefits to both participants.

A critical prediction arising from biological markets theory is that the relative value of exchangeable resources should be determined by the laws of supply and demand (Barrett & Henzi 2001, 2006). For example, if grooming is exchanged for feeding tolerance, a high-ranking individual in a population with highly monopolizable resources should receive relatively more grooming in return from subordinates than in a population in which food is not monopolizable (Barrett et al. 1999). Relative commodity value should also vary within a population as supply of and demand for the relevant commodities fluctuate according to ecology and demography. Therefore, the decision to cooperate depends upon the immediate value of the goods and services a potential trading partner can offer, not upon the probability of future cooperation. Some proponents of biological markets theory in primates argue that such short-term transactions provide a better explanation for observed patterns of cooperation than do reciprocal exchanges over the long term (Henzi & Barrett 1999; Barrett & Henzi 2002, 2006,

Fig. 22.5. Grooming, as in these vervet monkeys, is a critical currency in primate biological market theory. Photo courtesy of Ian C. Gilby.

2007; Henzi & Barrett 2007; but see Fruteau et al. 2009; Schino et al. 2009).

Implications

The preceding exercise enabled us to conclude that under conditions in which there is no benefit to defecting, mutualism suffices to explain the evolution of cooperation. However, when an individual could benefit immediately by defecting, cooperation can only evolve via a form of exchange that is contingent upon previous cooperation (contingent reciprocity) or the value of the resource that a potential partner can offer (biological markets). Several researchers have argued that for such cooperation to be maintained, certain cognitive abilities are required. First, in deciding whether or not to cooperate, a contingent reciprocator must be capable of recalling the history of interactions (or, at least, the outcome of the most recent interaction) with the potential recipient (Visalberghi 1997; de Waal 2000; Barrett & Henzi 2002; Stevens & Cushman 2004; Stevens & Gilby 2004; Stevens & Hauser 2004; Barrett & Henzi

2006). Otherwise, there would be selective pressure for an individual to exploit his partner's generosity by providing slightly less in the next interaction, eventually leading to the dissolution of cooperation. Second, because many species, including primates, value small immediate rewards over larger, delayed rewards (Stevens et al. 2005), contingent reciprocators (and often traders in a biological market) must be able to overcome the tendency to discount the value of future benefits—the longer the anticipated delay, the greater the temptation to defect. Schino and Aureli (2009) point out, however, that partner choice models hinge upon the outcomes of past (rather than future) interactions, thus alleviating the problems associated with temporal discounting: individuals simply do not interact with previous defectors. The complications associated with discounting of delayed rewards may also be somewhat alleviated if costs and benefits are measured in different currencies, such as grooming, food sharing, or coalitionary support. In a service economy (de Waal 1997) in which animals exchange several different commodities, the potential delay between reciprocated favors is reduced. However, this leads to another putative cognitive constraint: the ability to assess the relative value of different currencies (Stevens & Gilby 2004; Stevens & Hauser 2004).

As an alternative to cognitively demanding processes, some argue that loose accounting mechanisms may suffice to maintain exchange-based cooperation. Schino and Aureli (2009) propose that emotions are just such a mechanism. Rather than keeping track of the magnitude of favors given and received, exchanges may be maintained via a system by which individuals make social decisions based on emotionally mediated rules of thumb (chapter 23, this volume). This is similar to the concept of attitudinal reciprocity (de Waal 2000), in which animals mirror each others' general attitudes rather than engaging in strict scorekeeping. While these are intriguing and influential ideas, these mechanisms would seem to be rather unstable, as individuals would be vulnerable to exploitation by those that take slightly more than they give (Silk 2005).

Mutualism is likely to be the most common explanation for non-kin cooperation in most animal taxa (Dugatkin 1997). In contrast, contingent reciprocity and market-based exchanges are common in all human societies. When did this latter tendency arise? Are such exchanges unique to our lineage, or can the roots be traced to our primate ancestors? Identifying the extent to which contingent reciprocity, market-based exchanges, and mutualism explain cooperation in nonhuman primates will allow us to more fully understand the evolutionary history of cooperation in our own species.

Empirical Studies of Cooperation among Nonrelatives

Primates groom one another, form coalitions, share food, hunt together, and collectively defend their territories. In these contexts, individuals exchange goods or services—but it is problematic to make any generalizations about the relative costs and benefits of cooperating (or defecting) in these interactions. Consider food sharing, for example. When donors transfer food to others, do they always incur costs, or are there circumstances in which sharing confers an immediate benefit on the donor? Despite this uncertainty, it is often assumed (implicitly or explicitly) that these behaviors carry an immediate cost to actors, thus resulting in a prisoner's dilemma. As discussed above, contingent reciprocity and market-based exchange provide a way to escape the PD.

Contingent Reciprocity

Testing for contingent reciprocity in wild populations is challenging. Observational studies typically ask whether dyadic rates of aid given and received are positively correlated. However, while such correlations are consistent with contingent reciprocity, they do not allow us to confidently reach the conclusion that individual A cooperates because of earlier cooperation by B. Challenges include choosing (and being able to test) the appropriate time frame during which reciprocation might occur, as well as identifying and controlling for confounding effects of additional variables (e.g., dominance rank, age, and association frequency) that could also generate a reciprocal pattern. For example, if low-ranking individuals tend to associate (e.g., if they are excluded from prime feeding sites by dominants), then these subordinates would have more opportunities to groom each other. Under these circumstances, and if grooming were distributed randomly among available partners, then at the troop level, dyadic rates of grooming given and received would be positively correlated, even though individuals did not have preferential grooming partners. To rule out such *symmetry-based reciprocity* (de Waal & Luttrell 1986, 1988; Brosnan & de Waal 2002), it is important to identify and then control for variables such as age and association frequency before concluding that individuals are engaging in contingent reciprocity. However, one must use caution when doing so. Some individuals may choose to associate *in order to* groom. In this case, controlling for association frequency would mask evidence of contingent reciprocity. Therefore, when designing their analyses, researchers must ask whether there is evidence that individuals tend to associate for reasons other than cooperation.

Despite the disadvantages of correlational data, it is important to determine the extent to which cooperation in wild populations is consistent with contingent reciprocity. Several studies have found a reciprocal pattern of cooperation after controlling for appropriate variables. For example, among male chimpanzees (*Pan troglodytes*) at Ngogo in Kibale National Park, Uganda, there were strong positive correlations in dyadic rates of exchange within and across several currencies, including grooming, coalitionary support, and meat sharing (Mitani & Watts 1999, 2001; Mitani et al. 2000, 2002; Watts 2000, 2002; Mitani 2006). These studies used partial matrix correlation techniques to control for several variables including association, age, rank, and kinship. Similarly, at Taï National Park, Côte d'Ivoire, grooming given was positively correlated with grooming received after controlling for association, rank (Boesch & Boesch-Achermann 2000; Gomes et al. 2009), age and sex (Gomes et al. 2009). Also at Taï, meat sharing was correlated with long-term copulation rates (Gomes & Boesch 2009; but see Gilby et al. 2010). Symmetry-based reciprocity was also ruled out in studies of food sharing by captive chimpanzees (de Waal 1989) and agonistic interventions in captive chimpanzees, rhesus monkeys (*Macaca mulatta*), and stumptail macaques (*Macaca arctoides*, de Waal & Luttrell 1988).

Two meta-analyses of previously published data found positive correlations between grooming given and received in 22 primate species (Schino & Aureli 2008) and between grooming given and support received in 14 species (Schino 2007). While these analyses are consistent with contingent reciprocity, several of the studies included in the meta-analyses did not control for the possible effects of symmetry-based reciprocity (e.g., chimpanzees, Nishida & Hosaka 1996; Arnold & Whiten 2003; white-faced capuchins, *Cebus capucinus*, Manson et al. 1999; Japanese macaques, *Macaca fuscata*, Ventura et al. 2006; and hamadryas baboons, *Papio hamadryas*, Leinfelder et al. 2001). Therefore, the results should be interpreted with a degree of caution.

Finally, several studies failed to find a significant correlation between help given and received, either within or between currencies. For example, meat sharing among male chimpanzees at Gombe National Park, Tanzania, was not correlated with dyadic grooming rates (Gilby 2006). Similarly, chimpanzees did not trade food for short-term mating in captivity (Hemelrijk et al. 1992, 1999) or in the wild (Mitani & Watts 2001; Gilby 2006; Gomes & Boesch 2009; Gilby et al. 2010). Moreover, given the reduced likelihood of publishing nonsignificant results (Møller & Jennions 2001), there may be unpublished studies that failed to find a correlation between help given and received. Nevertheless, it is important to acknowledge the difficulty of assessing the benefits and costs of social behavior. For example, we generally assume that sharing is costly to the donor and beneficial to the recipient. However, Moore (1984) points out that by sharing, a male may actually obtain immediate social benefits by being in a "control role." Begging for and receiving food from another may reinforce an individual's relatively low social status (Moore 1984; Zahavi 1990). Similarly, does grooming benefit the groomer as well as the recipient? Such considerations may explain cases in which "help" given is not correlated with "help" received.

Nevertheless, as illustrated above, correlations between help given and received are common in primate studies. While some of these examples may be due to symmetry-based reciprocity or an alternative mechanism (e.g., harassment; see below), they are consistent with contingent reciprocity. However, such correlational studies are ultimately inconclusive with regard to causality, although the absence of unambigouous evidence of contingent reciprocity in wild populations does not necessarily constitute evidence of absence.

Studies of captive animals are key to determining whether primates engage (or are capable of engaging) in contingent reciprocity. Several such studies found evidence that cooperation by A depended upon recent cooperation by B. In a colony of chimpanzees at the Chester Zoo in the United Kingdom, Koyama et al. (2006) found that individuals were more likely to provide agonistic support to those who had recently groomed them. Specifically, there was more grooming on days before conflicts with support than on days before conflicts without support. Similarly, de Waal (1997) found that food sharing among captive chimpanzees was more likely to occur if the recipient had groomed the donor within the past 90 minutes than if no grooming had occurred. Critically, these exchanges were partner-specific, and there was some evidence of turn-taking, for grooming increased subsequent sharing by the groomee, but not by the groomer. However, the effect was rather small, and the contested resources in this experiment were bundles of leaves, which are neither highly valued nor shared by chimpanzees in the wild. In fact, leaves were chosen because "high-energy foods . . . provoke physical violence, whereas branches and foliage rarely do so" (de Waal 1989).

Melis et al. (2008) tested whether chimpanzees have the capacity for contingent reciprocation in a controlled experiment at the Ngamba Island chimpanzee sanctuary in Uganda. In a joint pulling task, subjects recruited previous cooperators more often than they recruited previous defectors. The authors conclude that the results provide "some support for the hypothesis that chimpanzees are capable of contingent reciprocity," but they acknowledge that the effect was relatively weak, and "suggest that models of immediate reciprocation and detailed accounts of recent exchanges . . . may not play a large role in guiding the social decisions of chimpanzees" (Melis et al. 2008, p. 951).

In one particularly ingenious experiment, Hemelrijk (1994) manipulated grooming rates among long-tailed macaques (*Macaca fascicularis*) by applying a sticky solution of syrup and seeds to the backs of certain individuals (thus encouraging others to groom them), and then incited conflicts within the group. The results showed that females were more likely to support another macaque after being groomed by that individual in the recent past than without such prior grooming. This study is similar to the famous experiment by Seyfarth and Cheney (1984) in which wild vervet monkeys (*Chlorocebus pygerythrus*) were more interested in audio broadcasts (playbacks) of distress calls by those who had recently groomed them than in recordings of those who had not. These results were replicated in a study of chacma baboons (*Papio ursinus*, Cheney et al. 2010).

Contingent reciprocity is often thought to be a critical component of a close social bond between individuals (Mitani 2006). According to this view, closely bonded dyads "invest" in each other through repeated acts of cooperation that are reciprocated across time. Well-differentiated, long-lasting bonds have been described in many primate taxa, including apes (Gilby & Wrangham 2008; Mitani 2009), Old World monkeys (Silk et al. 2006), and New World monkeys (Boinski & Mitchell 1994), and they may have positive fitness consequences (Silk et al. 2003a, 2010; Silk 2007b, c). In support of this idea, several studies have suggested that reciprocation may take place over a longer time frame than bouts or days. For example, Schino et al. (2009) found that female brown capuchins (*Cebus apella*) preferentially groomed those individuals that groomed them the most, even when instances of immediate reciprocation (within seven minutes) were excluded from the analysis. Among the Taï chimpanzees, grooming interactions were significantly more balanced when lumped over a two-year period than within bouts or days (Gomes et al. 2009). Similarly, free-ranging olive baboons (*Papio anubis)* in Laikipia, Kenya, demonstrated more evenly balanced grooming across multiple bouts than within single bouts (Frank & Silk 2009a). These data suggest that long-term social relationships are important in some monkeys and apes.

It is important to note that many of the examples of contingent reciprocity described above appear to entail very low costs. What are the costs to a healthy captive chimpanzee of sharing a small amount of leaves? How great are the energy and opportunity costs of grooming? Schino and Aureli (2009) argue that most cases of reciprocity among animals "consist of innumerable episodes of (almost) costless behaviors." As a result, there may be benefits to defecting, but they are so low that there is very little incentive to defect. Schino and Aureli (2009) propose, "Since individual episodes of failed reciprocation have negligible or no cost, it is more important to choose partners on the basis of their overall willingness to return benefits than to avoid being cheated." However, while it is true that each transaction may appear to be insignificant, the logic of the PD depends on the relative rather than absolute magnitude of the payoffs. As long as the long-term benefits of cooperating outweigh the benefits of defection, then there should be incentive to cooperate. If the difference between the payoffs is so low that animals cannot discern a difference between them, then there is no PD, and cooperation is not a mystery.

Biological Markets

There are several primate studies that support biological markets theory. In a widely cited study, Barrett et al. (1999) found that among chacma baboons, the distribution of grooming was more balanced when food resources could not be monopolized by dominant individuals. They argued

that in certain populations, and under certain conditions within the same population, dominant individuals could control access to a resource—an inalienable commodity for which lower-ranking individuals could trade grooming. Since grooming could be exchanged for feeding privileges, the rates of grooming given and received were not tightly correlated. When resources were not monopolizable, however, grooming was exchanged in kind, resulting in a more equitable distribution and a tighter correlation between giving and receiving.

Next, the Barrett and Henzi team tested the prediction that the value of tradable commodities should vary according to the principles of supply and demand, again using data on free-ranging chacma baboons. They took advantage of the curious phenomenon of infant handling in baboons, in which females touch, hold, and sometimes carry other females' infants (fig. 22.6). Mothers never initiate infant handling, which suggests that they do not benefit from these interactions. It is unclear exactly why nonmothers are attracted to infants in this way (Silk 1999; Silk et al. 2003b), but their obvious eagerness to handle infants suggests that they obtain some benefit. For example, handling infants may enhance one's future parenting skills (Silk et al. 2003b). Nevertheless, the situation creates the potential for a biological market because one class of traders (mothers) possesses a commodity (infants) for which others can trade another commodity (grooming). Henzi and Barrett (2002) found that females devoted more time to grooming mothers when the supply of infants was low than when there were more infants available. Thus, supply and demand seemed to affect the "price" that females were willing to pay for access to infants. Note, however, that Frank and Silk (2009b) failed to replicate these results in olive baboons.

Another example of the effect of supply and demand on exchanges among primates comes from an experimental study of wild vervet monkeys. When Fruteau and colleagues (2009) enabled a single low-ranking female to provide food to the rest of the troop by triggering the opening of a container, that female received significantly more grooming than she had before she was able to provide food. Then, when the experimenters enabled a second low-ranking female to provide food, thereby doubling the supply of food providers, the amount of grooming that the first provider received from other troop members decreased, but still remained higher than pre-experimental levels.

There is also evidence of mating markets in some primate groups. Male long-tailed macaques sometimes groom receptive females before mating with them, and in one study, the longer a male groomed a female, the more likely it was that she mated with him (Gumert 2007). Also, the amount of time that a male groomed a female during grooming-

Fig. 22.6. Infant handling is a common behavior among baboons, such as these olive baboons in Laikipia, Kenya. An adult female picks up, examines, and manipulates a 19-day old infant as it clings to its mother. Photo courtesy of Ryne A. Palombit.

mating interchanges was related to the number of other females in the immediate vicinity, suggesting that supply and demand influenced the value of grooming.

Barrett and Henzi (2006) downplay the importance of contingent reciprocity for explaining primate cooperation. In particular, Henzi and Barrett (2007, p. 73) argue that "'relationships' . . . need not, and probably do not, take the long-term, temporally consistent form that has been attributed to them, and that . . . short-term contingent response to current need . . . may provide a more satisfactory evolutionary account of female coexistence." Instead, they suggest that biological markets theory should be considered a theoretical framework distinct from contingent reciprocity (Barrett & Henzi 2006). As illustrated above, both mechanisms have received varying amounts of support and criticism. As is often the case with this complex and diverse order, it is likely that both mechanisms are important, even within the same species or population. Therefore, I suggest

that rather than treating biological markets theory as an alternative to reciprocity, future studies may benefit by considering the two together. The emphasis that market-based exchange places on partner choice is important, but it is also clear that long-term relationships matter. For example, long-term reciprocal relationships may maintain low-cost cooperation among non-kin, but market forces could alter the short-term dynamics of these relationships. Closely bonded individuals may reciprocate grooming, but the economics of the exchange will depend upon relative rank and the availability of other tradable commodities.

Mutualism

Mutualism is considered to be an important mechanism for cooperation in other taxa (Dugatkin 1997). For example, group hunting in African wild dogs (*Lycaon pictus*, Creel and Creel 1995) and African lions (*Panthera leo*, Scheel and

Packer 1991), food calling in passerine birds (reviewed in Dugatkin 1997), and territory defense in pied wagtails (*Motacilla alba*, Davies and Houston 1981) are well documented examples of mutualism. Despite its prevalence in other taxa, mutualism has received relatively little attention in primates. Below, I discuss several examples of cooperative behavior in primates that may be the product of mutualism.

Meat sharing among chimpanzees

Meat sharing occurs in all chimpanzee populations that hunt vertebrate prey (for reviews, see Uehara 1997; Muller and Mitani 2005; Gilby 2006). Hunting is most commonly an adult male activity, although females often make kills, particularly by opportunistic captures of concealed prey (Goodall 1986; Pruetz & Bertolani 2007; see also chapters 4 and 6, this volume, and fig. 4.5). If adult males are present, however, females rarely control the carcass. Therefore, most research has focused on meat sharing by adult male chimpanzees. When a male captures a prey item (usually a red colobus monkey, *Procolobus* spp.), he is typically surrounded by several other chimpanzees who beg for meat in a variety of ways (fig. 22.7). They may simply sit and stare, pull on the carcass, or place their hands over the meat

possessor's mouth (Teleki 1973; Wrangham 1975; Goodall 1986; Stanford 1998; Gilby 2006). Possessors share either passively (by allowing others to take pieces) or actively (by dropping a piece into an outstretched hand or relinquishing a mouthful).

Males might share meat with reciprocating partners, and as discussed earlier, there is evidence that sharing is reciprocated in some populations (e.g., Ngogo: Mitani & Watts 2001; Mitani 2006). However, it is also evident that some sharing is the product of coercion or harassment. Thus, males might give up meat because the costs of monopolizing it are greater than the costs of losing small amounts of it. If so, sharing is actually mutualistic. This idea has been described as sharing under pressure (Wrangham 1975), tolerated theft (Blurton Jones 1984, 1987), or harassment (Stevens & Stephens 2002). Moore (1984) suggested that a donor benefits both by avoiding conflict and by gaining social status. In an experimental study, Stevens (2004) manipulated the potential for harassment by varying the presence or absence of a mesh partition between a food possessor and a beggar. In this paradigm, captive chimpanzees and Bolivian squirrel monkeys (*Saimiri boliviensis*) rarely shared unless they were forced to do so.

Fig. 22.7. An adult female chimpanzee (*right*) grimaces and begs persistently for a scrap of meat from several male hunters who have killed a red colobus monkey at Gombe National Park, Tanzania. Photo courtesy of David Bygott.

Several lines of evidence suggest that harassment also influences the pattern of meat sharing among free-ranging chimpanzees at Gombe. The feeding efficiency of males who possessed meat was negatively correlated with the number of beggars, indicating that there was a cost to resisting harassment (Gilby 2006). Second, the duration of harassment (in the form of pulling on the carcass or reaching hand-to-mouth) increased the probability of sharing. Individuals that did not harass the possessor during a begging bout rarely received meat. Third, beggars typically stopped begging after a sharing event, and this allowed the possessor to eat more efficiently. Together, these results support the notion that meat sharing at Gombe is a manipulative mutualism in which the behavior of the beggars changes the payoff structure, making sharing the best immediate option for the meat possessor.

At Gombe, there is considerable variation in begging behavior, both within and among individuals (Gilby 2006). If harassment is a successful strategy for getting meat, why don't all individuals harass with equal intensity? One likely explanation is that begging is an honest signal of need—individuals who rarely get meat may be more willing to risk being the target of a display or attack by the meat possessor. It is also possible that long-term cooperative partners beg more persistently than others, either because they are less likely to be attacked or because they expect reciprocation. If so, then a reciprocal sharing pattern would emerge as a complicated combination of mutualism and reciprocity. Harassment would provide the proximate motivation for sharing, and it would be in the meat possessor's immediate self-interest to share when the costs of hoarding meat exceed the costs of relinquishing a portion. But if the beggar's persistence or the possessor's responses to harassment are based on expectations derived from previous interactions, then contingent reciprocity would be the ultimate force regulating the outcome of these interactions.

Third-party interventions

Third-party interventions, or coalitions, occur when one individual joins an ongoing conflict between two other individuals (de Waal & Harcourt 1992; fig. 22.8). The most common interpretation of such behavior among non-kin is that the intervening individual is helping one of the combatants, and the behavior is maintained by some form of reciprocity. Indeed, several studies have demonstrated a positive correlation between coalitionary support given and received (e.g., chimpanzees, reviewed in Muller & Mitani 2005). However, there is also evidence that individuals may derive immediate benefits when they participate in coalitions. For example, in East Africa, male olive baboons form coalitions to break up high-ranking males' consortships

with receptive females. Packer (1977) showed that males at Gombe most often solicited the help of males who most often solicited them, and he interpreted these coalitions as a form of reciprocal altruism. However, a later study on a different population showed that the intervening male was likely to benefit immediately by mating opportunistically with the female (Bercovitch 1988). In this case, third-party interventions seem to be a form of mutualism.

Male chimpanzees often intervene in conflicts among other males, and these interventions may have important effects on male dominance hierarchies. De Waal (1984) and Nishida (1983) have argued that males intervene opportunistically and may turn against former coalition partners when it is beneficial to do so. In some situations, by-product mutualism may be a more important factor in males' decisions about when to intervene than contingent reciprocity.

Group-level cooperation

A collective action problem arises when one individual can potentially profit from the joint action of other individuals. When one can avoid the costs of participation but still reap the rewards, the best strategy is to defect, and this makes group-level cooperation inherently unstable. Territorial defense provides one situation in which collective action problems are likely to arise. For example, in one study, white-faced capuchins benefited from a home field advantage in territorial disputes; they were more likely to win encounters that occurred near the center of their home ranges than encounters that occurred toward the edges (Crofoot et al. 2008). Crofoot et al. (2008) proposed that this was likely due to increasing rates of defection as parties moved away from the center of their home range. Indeed, subsequent playback experiments with this population demonstrated that focal subjects were more likely to run away from simulated conflicts that occurred at the edge of their range than from those that occurred in the center (Crofoot & Gilby 2012). The costs of losing an encounter near the center of the home range are extremely high for individuals, so it is mutually beneficial for everyone to participate in intergroup encounters. Interestingly, this does not seem to be the case for chimpanzees, who have not shown a location-dependent response to playbacks of intruders (Wilson et al. 2001). The behavior of specific individuals during aggressive intergroup interactions has not been well studied. In their review of the topic, Kitchen and Beehner (2007) found that within-sex variation in participation in intergroup aggression was often correlated with dominance rank and past reproductive access. For example, during encounters between vervet monkey troops in Amboseli National Park, Kenya, high-ranking females were more aggressive than low-ranking females (Cheney et al. 1981). Similarly, in intergroup con-

Fig. 22.8. A male-male coalition against a rival in yellow baboons of Amboseli, Kenya. Photo courtesy of Joan B. Silk.

tests, high-ranking male chacma baboons were more likely than low-ranking males to give loud calls, and they did so at a faster rate (Kitchen et al. 2004). At Ngogo, the overall frequency with which individual male chimpanzees participated in boundary patrols was positively correlated with variation in mating success, even after controlling for dominance rank (Watts & Mitani 2001). These results support the notion that those who stand to profit the most should be most likely to participate in cooperative territory defense (Cheney 1987; Nunn 2000). Therefore, those who refrain from participating in cooperative territory defense may not actually be defecting, since they do not share equally in the benefits of intergroup dominance.

Hunting of red colobus monkeys by chimpanzees is probably the best-studied example of group-level cooperation among primates. At all sites where the two species are sympatric, hunting probability increases with chimpanzee party size (Gombe, Gilby et al. 2006; Kanyawara, Gilby et al. 2008; Gilby & Wrangham 2007; Ngogo, Mitani & Watts 2001; Mahale, Hosaka et al. 2001), or hunts typically involve multiple individuals (Taï, Boesch & Boesch-

Achermann 2000). This suggests that individuals benefit by hunting in groups. Several social hypotheses have been proposed to explain this phenomenon, including the notion that males are motivated to obtain meat they can trade with females for sex (Stanford et al. 1994) or use to reward other males for political support (Mitani & Watts 2001). Once again, it is assumed that sharing entails immediate costs which are offset by future opportunities for exchange. However, work from Gombe (Gilby et al. 2006) and Kanyawara (Gilby & Wrangham 2007; Gilby et al. 2008) casts doubt on this idea, providing evidence that communal hunting by chimpanzees is best described as a form of mutualism. At Taï, males who hunt together obtain more calories per capita than males who hunt alone or in smaller parties. Males are more likely to share meat with males who have participated in the hunt than with nonparticipants, thereby helping to maintain the mutualistic benefits of joint action (Boesch 1994, 2002, 2006; Boesch & Boesch-Achermann 2000). Chimpanzee hunting behavior at Taï may therefore be described as a form of stag hunt (Silk 2007a). In contrast, there is no per capita caloric benefit to hunting in groups at

Gombe (Boesch 1994; Gilby et al. 2006), Ngogo (Mitani & Watts 2001), or Kanyawara (Gilby et al. 2008). However, meat provides a concentrated source of valuable micronutrients, as well as calories, and it is likely to be valuable in small quantities. If hunting in groups increases the chances that an individual obtains even a small scrap of meat, then hunting at these sites may also be mutualistic (Tennie et al. 2009). Indeed, the probability that a given male obtains meat (in any amount) increases with the number of hunters at Gombe (Tennie et al. 2009), Kanyawara (Gilby et al. 2008), and Ngogo (Watts & Mitani 2002).

Whether the benefits of communal hunting are measured in calories or micronutrients, the mechanism that yields these mutual benefits is unclear. At Taï, Boesch and colleagues describe "drivers," "ambushers," "blockers," and "chasers" who coordinate their behavior in space and time (Boesch & Boesch 1989; Boesch & Boesch-Achermann 2000; Boesch 1994, 2002). In these descriptions it is im-

plied that the decision to adopt a certain role is based on increasing the chances of group success. In contrast, Gilby and Connor (2010) argue that by-product mutualism suffices to explain group hunting in chimpanzees: each individual hunts for himself, but experiences an increased chance of success when others hunt. When several individuals hunt simultaneously, prey defenses are diluted, thus increasing the probability of a kill for *each* participant. In some cases this strategy may result in differentiated roles, but it does not represent a true team task (Anderson & Franks 2001) in which hunters act to increase the probability of a group kill at a cost to their own chances of success.

If individual success rates increase with the number of other hunters, how are hunts initiated? Certain males may act as catalysts by being the first to hunt (fig. 22.9). Once they begin hunting, they engage red colobus defenders, thereby reducing the number of defenders available to attack other hunters. Other males then take advantage of

Fig. 22.9. A keen and successful hunter from a young age, the adult male Frodo has influenced chimpanzee hunting behavior at Gombe for more than 20 years. Photo courtesy of Ian C. Gilby.

the reduced costs of hunting. Data from Kanyawara support this idea. There, a hunt was significantly more likely to occur if at least one of two particular males was present when a party encountered red colobus than if they were both absent (Gilby et al. 2008). Parties without these "impact" males almost never hunted, and other males in parties that included one or both of these males almost never hunted unless one or both of the impact males hunted. This phenomenon may help to explain differences in hunting frequency within and between sites. At Taï, for example, the maturation of a particularly persistent hunter significantly changed hunting frequency in the community (Boesch & Boesch 1989). However, the factors responsible for inter-individual variation in hunting behavior remain unknown.

Summary and Conclusions

The study of non-kin primate cooperation is at an exciting stage. It is becoming clear that no single mechanism can explain the observed patterns of cooperative behavior within or between populations. Therefore, in the next research phase, we should adopt an approach in which we rigorously consider multiple hypotheses, including those that invoke immediately mutualistic mechanisms. A model study would begin by testing simple explanations before invoking contingent reciprocity and biological markets. Rather than striving to identify a single mechanism, a more constructive approach might be to identify the amount of variation that each hypothesis can explain. With regard to food sharing, for example, it may be that simple mechanisms such as harassment may explain food sharing among some classes, but that a form of contingent exchange may explain sharing among closely bonded dyads. It is also critical to recognize the flexibility that is inherent in the system: individuals may adopt variable strategies and respond differently, depending on ecological or demographic conditions.

This approach has at least two critical advantages. First, it provides a way to compare cooperation across taxa, thereby identifying ways in which nonhuman primates may be different from social carnivores, birds, insects, and humans. Second, it alleviates the potential bias toward tests of hypotheses that invoke cognitively complex mechanisms. Ultimately, a pluralistic approach will be more illuminating.

References

Anderson, C. & Franks, N. R. 2001. Teams in animal societies. *Behavioral Ecology*, 12, 534–540.

Arnold, K. & Whiten, A. 2003. Grooming interactions among the chimpanzees of the Budongo Forest, Uganda: Tests of five explanatory models. *Behaviour*, 140, 519–552.

Axelrod, R. & Hamilton, W. D. 1981. The evolution of cooperation. *Science*, 211, 1390–1396.

Barrett, L. & Henzi, S. P. 2001. The utility of grooming in baboon troops. In *Economics in Nature* (ed. by Noë, R., van Hooff, J. A. R. A. M. & Hammerstein, P.), 119–145. Cambridge: Cambridge University Press.

———. 2002. Constraints on relationship formation among female primates. *Behaviour*, 139, 263–289.

———. 2006. Monkeys, markets, and minds: Biological markets and primate sociality. In *Cooperation in Primates and Humans: Mechanisms and Evolution* (ed. by Kappeler, P. M. & van Schaik, C. P.), 209–232. Berlin: Springer-Verlag.

———. 2007. Social brains, simple minds: does social complexity really require cognitive complexity? *Philosophical Transactions of the Royal Society*, 362, 561–575.

Barrett, L., Henzi, S. P., Weingrill, T., Lycett, J. E. & Hill, R. A. 1999. Market forces predict grooming reciprocity in female baboons. *Proceedings of the Royal Society of London Series B Biological Sciences*, 266, 665–670.

Bercovitch, F. B. 1988. Coalitions, cooperation and reproductive tactics among adult male baboons. *Animal Behaviour*, 36, 1198–1209.

Blurton Jones, N. G. 1984. A selfish origin for human food sharing: Tolerated theft. *Ethology and Sociobiology*, 5, 1–3.

———. 1987. Tolerated theft: Suggestions about the ecology and evolution of sharing, hoarding and scrounging. *Social Science Information*, 26, 31–54.

Boesch, C. 1994. Cooperative hunting in wild chimpanzees. *Animal Behaviour*, 48, 653–667.

———. 2002. Cooperative hunting roles among Taï chimpanzees. *Human Nature: An Interdisciplinary Biosocial Perspective*, 13, 27–46.

———. 2006. Cooperative hunting in chimpanzees: Kinship or mutualism? In *Cooperation in Primates and Humans: Mechanisms and Evolution* (ed. by Kappeler, P. M. & van Schaik, C. P.), 139–150. Berlin: Springer-Verlag.

Boesch, C. & Boesch, H. 1989. Hunting behavior of wild chimpanzees in the Taï National Park. *American Journal of Physical Anthropology*, 78, 547–573.

Boesch, C. & Boesch-Achermann, H. 2000. *The Chimpanzees of the Taï Forest: Behavioural Ecology and Evolution*. Oxford: Oxford University Press.

Boinski, S. & Mitchell, C. L. 1994. Male residence and association patterns in Costa-Rican squirrel monkeys (*Saimiri oerstedi*). *American Journal of Primatology*, 34, 157–169.

Brosnan, S. F. & de Waal, F. B. M. 2002. A proximate perspective on reciprocal altruism. *Human Nature*, 13, 129–152.

Brown, J. L. 1983. Cooperation: A biologist's dilemma. In *Advances in Behaviour* (ed. by Rosenblatt, J. S.), 1–37. New York: Academic Press.

Bshary, R. 2001. The cleaner fish market. In *Economics in Nature: Social Dilemmas, Mate choice and Biological Markets* (ed. by Noe, R., Hammerstein, P. & van Hooff, J. A. R. A. M.), 146–172. Cambridge: Cambridge University Press.

Cheney, D. L. 1987. Interactions and relationships between groups. In *Primate Societies* (ed. by Smuts, B. B., Cheney, D. L., Seyfarth, R. M., Wrangham, R. W. & Struhsaker, T. T.), 267–281. Chicago: University of Chicago Press.

Cheney, D. L., Lee, P. C. & Seyfarth, R. M. 1981. Behavioral correlates of non-random mortality among free-ranging female vervet monkeys. *Behavioral Ecology and Sociobiology*, 9, 153–161.

Cheney, D. L., Moscovice, L. R., Heesen, M., Mundry, R. & Seyfarth, R. M. Contingent cooperation between wild female baboons. *Proceedings of the National Academy of Sciences of the United States of America*, 107, 9562–9566.

Clements, K. C. & Stephens, D. W. 1995. Testing non-kin co-operation: Mutualism and the Prisoner's Dilemma. *Animal Behaviour*, 50, 527–535.

Creel, S. & Creel, N. M. 1995. Communal hunting and pack size in African wild dogs, *Lycaon pictus*. *Animal Behaviour*, 50, 1325–1339.

Crofoot, M. C. & Gilby, I. C, 2012. Cheating monkeys undermine group strength in enemy territory. *Proceedings of the National Academy of Sciences (USA)*, 109, 501–505.

Crofoot, M. C., Gilby, I. C., Wikelski, M. C. & Kays, R. W. 2008. Interaction location outweighs the competitive advantage of numerical superiority in *Cebus capucinus* intergroup contests. *Proceedings of the National Academy of Sciences (USA)*, 105, 577–581.

Darwin, C. 1859. *On the Origin of Species*. Cambridge, MA: Harvard University Press.

Davies, N. B. & Houston, A. I. 1981. Owners and satellites: The economics of territory defence in the pied wagtail *Motacilla alba*. *Journal of Animal Ecology*, 50, 157–180.

De Waal, F. B. M. 1984. Sex differences in the formation of co-alitions among chimpanzees. *Ethology and Sociobiology*, 5, 239–255.

———. 1989. Food sharing and reciprocal obligations among chimpanzees. *Journal of Human Evolution*, 18, 433–459.

———. 1997. The chimpanzee's service economy: Food for grooming. *Evolution and Human Behavior*, 18, 375–386.

———. 2000. Attitudinal reciprocity in food sharing among brown capuchin monkeys. *Animal Behaviour*, 60, 253–261.

De Waal, F. B. M. & Harcourt, A. H. 1992. Coalitions and alliances: a history of ethological research. In *Coalitions and Alliances in Humans and Other Animals* (ed. by Harcourt, A. H. & de Waal, F. B. M.), 1–19. Oxford: Oxford University Press.

De Waal, F. B. M. & Luttrell, L. M. 1986. The similarity principle underlying social bonding among female rhesus monkeys. *Folia Primatologica*, 46, 215–234.

———. 1988. Mechanisms of social reciprocity in three primate species: Symmetrical relationship characteristics or cognition? *Ethology and Sociobiology*, 9, 101–118.

Dugatkin, L. A. 1997. *Cooperation among Animals:. An Evolutionary Perspective*. Oxford: Oxford University Press.

Frank, R. E. & Silk, J. B. 2009a. Impatient traders or contingent reciprocators? Evidence for the extended time course of grooming exchanges in baboons. *Behaviour*, 146, 1123–1135.

———. 2009b. Grooming exchange between mothers and non-mothers: The price of natal attraction in wild baboons (*Papio anubis*). *Behaviour*, 146, 889–906.

Fruteau, C., Voelkl, B., van Damme, E. & Noë, R. 2009. Supply and demand determine the market value of food providers in wild vervet monkeys. *Proceedings of the National Academy of Sciences of the United States of America*, 106, 12007–12012.

Gilby, I. C. 2006. Meat sharing among the Gombe chimpanzees:

Harassment and reciprocal exchange. *Animal Behaviour*, 71, 953–963.

Gilby, I. C. & Connor, R. C. 2010. The role of intelligence in group hunting: Are chimpanzees different from other social predators? In *The Mind of the Chimpanzee: Ecological and Experimental Perspectives* (ed. by Lonsdorf, E. V., Ross, S. R. & Matsuzawa, T.), 220–233. Chicago: University of Chicago Press.

Gilby, I. C. & Wrangham, R. W. 2007. Risk-prone hunting by chimpanzees (*Pan troglodytes schweinfurthii*) increases during periods of high diet quality. *Behavioral Ecology and Sociobiology*, 61, 1771–1779.

———. 2008. Association patterns among wild chimpanzees (*Pan troglodytes schweinfurthii*) reflect sex differences in co-operation. *Behavioral Ecology and Sociobiology*, 62, 1831–1842.

Gilby, I. C., Eberly, L. E., Pintea, L. & Pusey, A. E. 2006. Ecological and social influences on the hunting behaviour of wild chimpanzees (*Pan troglodytes schweinfurthii*). *Animal Behaviour*, 72, 169–180.

Gilby, I. C., Eberly, L. E. & Wrangham, R. W. 2008. Economic profitability of social predation among wild chimpanzees: Individual variation promotes cooperation. *Animal Behaviour*, 75, 351–360.

Gilby, I. C., Emery Thompson, M., Ruane, J. & Wrangham, R. W. 2010. No evidence of short-term exchange of meat for sex among chimpanzees. *Journal of Human Evolution*, 59, 44–53.

Gomes, C. M. & Boesch, C. 2009. Wild chimpanzees exchange meat for sex on a long-term basis. *PLoS ONE*, 4, e5116.

Gomes, C. M., Mundry, R. & Boesch, C. 2009. Long-term reciprocation of grooming in wild West African chimpanzees. *Proceedings of the Royal Society B-Biological Sciences*, 276, 699–706.

Goodall, J. 1986. *The Chimpanzees of Gombe: Patterns of Behavior*. Cambridge, MA.: Harvard University Press.

Gumert, M. D. 2007. Payment for sex in a macaque mating market. *Animal Behaviour*, 74, 1655–1667.

Hemelrijk, C. K. 1994. Support for being groomed in long-tailed macaques, *Macaca fascicularis*. *Animal Behaviour*, 48, 479–481.

Hemelrijk, C. K., Meier, C. & Martin, R. D. 1999. "Friendship" for fitness in chimpanzees? *Animal Behaviour*, 58, 1223–1229.

Hemelrijk, C. K., van Laere, G. J. & van Hooff, J. A. R. A. M. 1992. Sexual exchange relationships in captive chimpanzees? *Behavioral Ecology and Sociobiology*, 30, 269–275.

Henzi, S. P. & Barrett, L. 1999. The value of grooming to female primates. *Primates*, 40.

———. 2002. Infants as a commodity in a baboon market. *Animal Behaviour*, 63, 915–921.

———. 2007. Coexistence in female-bonded primate groups. *Advances in the Study of Behavior*, 37, 43–81.

Hosaka, K., Nishida, T., Hamai, M., Matsumoto-Oda, A. & Uehara, S. 2001. Predation of mammals by the chimpanzees of the Mahale Mountains, Tanzania. In *All Apes Great and Small, Volume I. African Apes* (ed. by Galdikas, B., Briggs, N., Sheeran, L., Shapiro, G. & Goodall, J.), 107–130. New York: Klewer Academic Publishers.

Kitchen, D. M. & Beehner, J. C. 2007. Factors affecting individual participation in group-level aggression among non-human primates. *Behaviour*, 144, 1551–1581.

Kitchen, D. M., Cheney, D. L. & Seyfarth, R. M. 2004. Factors mediating inter-group encounters in savannah baboons (*Papio cynocephalus ursinus*). *Behaviour*, 141, 197–218.

Koyama, N. F., Caws, C. & Aureli, F. 2006. Interchange of grooming and agonistic support in chimpanzees. *International Journal of Primatology*, 27, 1293–1309.

Leinfelder, I., De Vries, H., Deleu, R. & Nelissen, M. 2001. Rank and grooming reciprocity among females in a mixed-sex group of captive hamadryas baboons. *American Journal of Primatology*, 55, 25–42.

Manson, J. H., Rose, L. M., Perry, S. & Gros-Louis, J. 1999. Dynamics of female-female relationships in wild *Cebus capucinus*: Data from two Costa Rican sites. *International Journal of Primatology*, 20, 679–706.

Maynard Smith, J. 1972. *On Evolution*. Edinburgh: Edinburgh University Press.

———. 1982. *Evolution and the Theory of Games*. Cambridge: Cambridge University Press.

Melis, A. P., Hare, B. & Tomasello, M. 2008. Do chimpanzees reciprocate received favours? *Animal Behaviour*, 76, 951–962.

Mesterton-Gibbons, M. & Dugatkin, L. A. 1992. Cooperation among unrelated individuals: Evolutionary factors. *Quarterly Review of Biology*, 67, 267–281.

Mitani, J. C. 2006. Reciprocal exchange in chimpanzees and other primates. In *Cooperation in Primates: Mechanisms and Evolution*. (ed. by Kappeler, P. M. & van Schaik, C. P.), 101–113. Heidelberg: Springer-Verlag.

———. 2009. Male chimpanzees form enduring and equitable social bonds. *Animal Behaviour*, 633–640.

Mitani, J. C., Merriwether, D. A. & Zhang, C. 2000. Male affiliation, cooperation and kinship in wild chimpanzees. *Animal Behaviour*, 59, 885–893.

Mitani, J. C. & Watts, D. P. 1999. Demographic influences on the hunting behavior of chimpanzees. *American Journal of Physical Anthropology*, 109, 439–454.

———. 2001. Why do chimpanzees hunt and share meat? *Animal Behaviour*, 61, 915–924.

———. 2005. Correlates of territorial boundary patrol behaviour in wild chimpanzees. *Animal Behaviour*, 70, 1079–1086.

Mitani, J. C., Watts, D. P., Pepper, J. W. & Merriwether, D. A. 2002. Demographic and social constraints on male chimpanzee behaviour. *Animal Behaviour*, 64, 727–737.

Møller, A. P. & Jennions, M. 2001. Testing and adjusting for publication bias. *Trends in Ecology and Evolution*, 16, 580–586.

Moore, J. 1984. The evolution of reciprocal sharing. *Ethology and Sociobiology*, 5, 5–14.

Muller, M. N. & Mitani, J. C. 2005. Conflict and cooperation in wild chimpanzees. *Advances in the Study of Behavior*, 35, 275–331.

Nishida, T. 1983. Alpha status and agonistic alliance in wild chimpanzees (*Pan troglodytes schweinfurthii*). *Primates*, 24, 318–336.

Nishida, T. & Hosaka, K. 1996. Coalition strategies among adult male chimpanzees of the Mahale Mountains, Tanzania. In *Great Ape Societies* (ed. by McGrew, W. C., Marchant, L. & Nishida, T.), 114–134. Cambridge: Cambridge University Press.

Noë, R. & Hammerstein, P. 1994. Biological markets: Supply and demand determine the effect of partner choice in cooperation, mutualism and mating. *Behavioral Ecology and Sociobiology*, 35, 1–11.

———. 1995. Biological markets. *Trends in Ecology and Evolution*, 10, 336–339.

Noë, R., van Schaik, C. P. & van Hooff, J. A. R. A. M. 1991. The market effect: An explanation for pay-off assymmetries among collaborating animals. *Ethology*, 87, 97–118.

Nunn, C. L. 2000. Collective benefits, free-riders, and male extra-group conflict. In *Primate Males* (ed. by Kappeler, P. M.), 192–204. Cambridge: Cambridge University Press.

Packer, C. 1977. Reciprocal altruism in *Papio anubis*. *Nature*, 265, 441–443.

Pruetz, J. D. & Bertolani, P. 2007. Savanna chimpanzees, *Pan troglodytes verus*, hunt with tools. *Current Biology*, 17, 412–417.

Scheel, D. & Packer, C. 1991. Group hunting behavior of lions: A search for cooperation. *Animal Behaviour*, 41, 697–709.

Schino, G. 2007. Grooming and agonistic support: A meta-analysis of primate reciprocal altruism. *Behavioral Ecology*, 18, 115–120.

Schino, G. & Aureli, F. 2008. Grooming reciprocation among female primates: A meta-analysis. *Biology Letters*, 4, 9–11.

———. 2009. Reciprocal altruism in primates: Partner choice, cognition and emotion. *Advances in the Study of Behavior*, 39, 45–69.

Schino, G., Di Giuseppe, F. & Visalberghi, E. 2009. The time frame of partner choice in the grooming reciprocation of *Cebus apella*. *Ethology*, 115, 70–76.

Seyfarth, R. M. & Cheney, D. L. 1984. Grooming, alliances and reciprocal altruism in vervet monkeys. *Nature*, 308, 541–543.

Silk, J. B. 1999. Why are infants so attractive to others? The form and function of infant handling in bonnet macaques. *Animal Behaviour*, 57, 1021–1032.

———. 2005. The evolution of cooperation in primate groups. In *Moral Sentiments and Material Interests: On the Foundations of Cooperation in Economic Life* (ed. by Gintis, H., Bowles, S., Boyd, R. & Fehr, E.), 43–73. Cambridge, MA: MIT Press.

———. 2007a. The strategic dynamics of cooperation in primate groups. In *Advances in the Study of Behavior*, 37, 1–41.

———. 2007b. The adaptive value of sociality in mammalian groups. *Philosophical Transactions of the Royal Society B Biological Sciences*, 362, 539–559.

———. 2007c. Social components of fitness in primate groups. *Science*, 317, 1347–1351.

Silk, J. B., Alberts, S. C. & Altmann, J. 2003a. Social bonds of female baboons enhance infant survival. *Science*, 302, 1231–1234.

Silk, J. B., Rendall, D., Cheney, D. L. & Seyfarth, R. M. 2003b. Natal attraction in adult female baboons (*Papio cynocephalus ursinus*) in the Moremi Reserve, Botswana. *Ethology*, 109, 627–644.

Silk, J. B., Alberts, S. C. & Altmann, J. 2006. Social relationships among adult female baboons (*Papio cynocephalus*). II. Variation in the quality and stability of social bonds. *Behavioral Ecology and Sociobiology*, 61, 197–204.

Silk, J. B., Beehner, J. C., Bergman, T. J., Crockford, C., Engh, A. L., Moscovice, L. R., Wittig, R. M., Seyfarth, R. M. & Cheney, D. L. 2010. Strong and consistent social bonds enhance the longevity of female baboons. *Current Biology*, 20, 1359–1361.

Silk, J. B., Brosnan, S. F., Vonk, J., Henrich, J., Povinelli, D. J., Richardson, A. S., Lambeth, S. P., Mascaro, J. & Schapiro, S. J. 2005. Chimpanzees are indifferent to the welfare of unrelated group members. *Nature*, 437, 1357–1359.

Skyrms, B. 2004. *The Stag Hunt and the Evolution of Social Structure*. Cambridge, UK: Cambridge University Press.

Stanford, C. B. 1998. *Chimpanzee and Red Colobus*. Cambridge, MA: Harvard University Press.

Stanford, C. B., Wallis, J., Mpongo, E. & Goodall, J. 1994. Hunting decisions in wild chimpanzees. *Behaviour*, 131, 1–18.

Stevens, J. R. 2004. The selfish nature of generosity: Harassment and food sharing in primates. *Proceedings of the Royal Society of London Series B Biological Sciences*, 271, 451–456.

Stevens, J. R. & Cushman, F. A. 2004. Cognitive constraints on reciprocity and tolerated scrounging. *Behavioral and Brain Sciences*, 27, 569–570.

Stevens, J. R. & Gilby, I. C. 2004. A conceptual framework for nonkin food sharing: Timing and currency of benefits. *Animal Behaviour*, 67, 603–614.

Stevens, J. R., Hallinan, E. V. & Hauser, M. D. 2005. The ecology and evolution of patience in two New World monkeys. *Biology Letters*, 1, 223–226.

Stevens, J. R. & Hauser, M. D. 2004. Why be nice? Psychological constraints on the evolution of cooperation. *Trends in Cognitive Sciences*, 8, 60–65.

Stevens, J. R. & Stephens, D. W. 2002. Food sharing: A model of manipulation by harassment. *Behavioral Ecology*, 13, 393–400.

Teleki, G. 1973. *The Predatory Behavior of Wild Chimpanzees*. Lewisburg, PA: Bucknell University Press.

Tennie, C., Gilby, I. C. & Mundry, R. 2009. The meat-scrap hypothesis: Small quantities of meat may promote cooperation in wild chimpanzees (*Pan troglodytes*). *Behavioral Ecology and Sociobiology*, 63, 421–431.

Trivers, R. L. 1971. The evolution of reciprocal altruism. *Quarterly Review of Biology*, 46, 35–57.

———. 2006. Reciprocal altruism: 30 years later. In *Cooperation in Primates and Humans: Mechanisms and Evolution* (ed. by Kappeler, P. M. & van Schaik, C. P.), 68–83. Berlin, New York: Springer.

Uehara, S. 1997. Predation on mammals by the chimpanzee (*Pan troglodytes*). *Primates*, 38, 193–214.

Van Schaik, C. P. & Kappeler, P. M. 2006. Cooperation in primates and humans: Closing the gap. In *Cooperation in Primates and Humans: Mechanisms and Evolution* (ed. by Kappeler, P. M. & van Schaik, C. P.), 3–21. Berlin: Springer-Verlag.

Ventura, R., Majolo, B., Koyama, N. F., Hardie, S. & Schino, G. 2006. Reciprocation and interchange in wild Japanese macaques: Grooming, cofeeding, and agonistic support. *American Journal of Primatology*, 68, 1138–1149.

Visalberghi, E. 1997. Success and understanding in cognitive tasks: A comparison between *Cebus apella* and *Pan troglodytes*. *International Journal of Primatology*, 18, 811–830.

Warneken, F., Hare, B., Melis, A. P., Hanus, D. & Tomasello, M. 2007. Spontaneous altruism by chimpanzees and young children. *PLoS Biology*, 5, 1414–1420.

Watts, D. P. 2000. Grooming between male chimpanzees at Ngogo, Kibale National Park, I: partner number and diversity and grooming reciprocity. *International Journal of Primatology*, 21, 189–210.

———. 2002. Reciprocity and interchange in the social relationships of wild male chimpanzees. *Behaviour*, 139, 343–370.

Watts, D. P. & Mitani, J. C. 2001. Boundary patrols and intergroup encounters in wild chimpanzees. *Behaviour*, 138, 299–327.

———. 2002. Hunting behavior of chimpanzees at Ngogo, Kibale National Park, Uganda. *International Journal of Primatology*, 23, 1–28.

Wilson, M. L., Hauser, M. D. & Wrangham, R. W. 2001. Does participation in intergroup conflict depend on numerical assessment, range location or rank for wild chimpanzees? *Animal Behaviour*, 61, 1203–1216.

Wrangham, R. W. 1975. The behavioural ecology of chimpanzees in Gombe National Park, Tanzania. PhD dissertation, Cambridge University.

Zahavi, A. 1990. Arabian babblers: The quest for social status in a cooperative breeder. In *Cooperative Breeding in Birds: Long-Term Studies of Ecology and Behavior* (ed. by Stacey, P. B. & Koenig, W. D.), 103–130. Cambridge: Cambridge University Press.

Chapter 23 The Regulation of Social Relationships

Filippo Aureli, Orlaith N. Fraser,
Colleen M. Schaffner, and Gabriele Schino

PRIMATES ESTABLISH and maintain social relationships with other group members throughout their lives. In this chapter we first briefly present a framework for conceptualizing primate social relationships, and then review ways to systematically characterize variation in their quality. The main part of the chapter focuses on the social mechanisms that establish and maintain social relationships. We review the role of aggression and affiliation, as well as specific mechanisms that prevent aggressive escalation and mitigate its negative consequences. Most emphasis is on the behavior of the two partners in each relationship, but the role of third parties in regulating social relationships is also considered. Our review focuses on studies of nonhuman primates living in different social systems, but comparisons with findings on nonprimate species are made when possible.

A Framework for Social Relationships

In his chapter in *Primate Societies*, Robert Hinde used his influential framework about social relationships as an example of how studies on nonhuman primates have been helpful in identifying basic principles relevant to human behavior (Hinde 1987). According to this framework, a social relationship is a concept that links the observable social interactions among group members to the inferred group social structure (Hinde 1976, 1979, 1983). When the occurrence and outcome of any single interaction between two individuals affects the occurrence and outcome of their subsequent interactions, we can say these two individuals have a social relationship. Thus, the unique history of interactions between two individuals constitutes their social relationship. According to this view, social relationships are emergent properties derived from the patterning of interactions between individuals across time (see chapters 24 and 28, this volume).

Kummer (1978) added a functional dimension to Hinde's framework, and proposed that social relationships can be regarded as investments that benefit the individuals involved because they increase predictability. Moreover, an individual's interactions with a partner may influence the likelihood that the partner behaves in a beneficial way, thus maximizing the benefits (Kummer 1978) or minimizing the losses (Cords 1997). As a consequence, individuals are expected to invest more effort in establishing and maintaining relationships with valued partners: those with whom they have the most profitable social relationships (Cords 1997). Accordingly, partner preferences are expected to be based on individual qualities, behavioral tendencies, and availability.

This view of the nature of social relationships has been criticized by Henzi and Barrett (2007; Barrett & Henzi 2002), who have argued that the abstraction of social relationships from a series of interactions introduces a temporal component that reflects the way we think about our own relationships but is not relevant to the animals themselves. Henzi and Barrett argue that the temporal component introduces a prospective element in which individuals are motivated to cultivate social relationships for future benefits.

They claim that the Hinde/Kummer perspective conflates an evolutionary account of relationships in terms of their fitness-enhancing properties with the proximate mechanisms that motivate animals to interact. They conclude that social relationships do not correspond to any real-world entity for the animals and cannot provide the proximate motivation for the behaviors nonhuman primates display. Their alternative view explains the coexistence of nonhuman primates in social groups essentially as spatial arrangements based on short-term contingent responses to current needs.

Contrary to Henzi and Barrett's view, the Hinde/Kummer perspective of social relationships does not imply that animals are motivated by anticipated benefits or are conscious of the long-term outcome of their actions. Instead, it implies that animals are influenced by the history of their interactions with their partners (see Seyfarth & Cheney 2012 for a detailed discussion). Being able to predict the actions and responses of their partners with reasonable accuracy based on past interactions (i.e., on their social relationship) clearly facilitates social intercourse. Some of the benefits of well developed and differentiated social relationships among primates include selective tolerance around resources, cooperative hunting, food sharing, mating privileges, agonistic support, and protection against external threats (Cords 1997; van Schaik & Aureli 2000), which in turn provide fitness benefits (Silk et al. 2003, 2009; chapter 24, this volume).

Individuals need to modify their behavior according to the quality of their relationships with potential partners in a particular context. How do animals assess the quality of their relationships with others? We suggest that primates base these assessments on the information contained in the various interactions previously exchanged with their partners (Aureli & Schaffner 2002). Somehow they must integrate information about the frequency, duration, quality, and consequences of different types of interactions with each partner, convert this information into some kind of common currency, and update it across time. Such bookkeeping requires memory and computational skills that may exceed the cognitive capacities of nonhuman animals, including primates.

However, primates may be able to accomplish this task in a less cognitively demanding way (Aureli & Schaffner 2002). Emotions can provide individuals with a timely relationship assessment to guide social decisions (Aureli & Schaffner 2002; Aureli & Whiten 2003; Aureli & Schino 2004; Schino & Aureli 2009). This can be achieved by considering the mediating role of emotion implicit in the concept of intervening variables (Hinde 1972; Aureli & Whiten 2003). A particular emotional state (without implying subjective feeling) would lead an individual to take a particular motivational stance, which constrains decision-making and influences behavior in a way that is appropriate to the situation (Aureli & Whiten 2003; cf. Fessler & Gervais 2010). An individual's emotional experience is certainly affected by the frequency and quality of previous interactions with other group members (Aureli & Schaffner 2002; Aureli & Whiten 2003; Aureli & Schino 2004; cf. Schino & Aureli 2009; Seyfarth & Cheney 2012). Emotional states may then express a crucial integration of the information contained in the various interactions between two partners, and they may change over time, depending on the nature of those interactions. Such partner-dependent emotional experience can therefore be functionally equivalent to the processes of bookkeeping, frequency/duration computation, and quality conversion of the social interactions needed for continuous relationship assessment. For example, the anxiety that former opponents experience in the aftermath of a conflict is strongly dependent on the quality of their relationship, not on the differential likelihood of renewed aggression (Aureli 1997; Kutsukake & Castles 2001; Cooper et al. 2007; Koski et al. 2007). Postconflict anxiety can then function as the proximate mechanism for reconciliation (i.e., the affiliative reunion of former opponents), which is more likely to occur between individuals with stronger relationships (see below). Thus, emotional differences can be at the core of the observed variation in social interactions, reflecting the variation in relationship quality across partners (Aureli & Schaffner 2002; Aureli & Whiten 2003; Aureli & Schino 2004; fig. 23.1).

In order to establish and maintain their social relationships, individuals need to recognize and interact with other group members over the course of relatively long periods and remember their previous interactions (Cords 1997), possibly using emotional mediation. It is unlikely that these abilities are unique to primates. When *Primate Societies* was published, the social behavior of species other than primates was poorly studied (Wrangham 1983; Harcourt 1988; Rowell 2000), but in recent years there has been a dramatic increase in knowledge about a variety of group-living nonprimate species (e.g., spotted hyenas, Holekamp et al. 2007; elephants, Moss et al. 2011; corvids, Emery et al. 2007; Heinrich & Bugnyar 2007; dolphins, Connor 2007; Lusseau 2007; goats, Schino 1998; coatis, Romero & Aureli 2008; fish, Bshary et al. 2002). These species share various similarities with primates, including reconciliation, coalitionary support, and the establishment of well-differentiated relationships. Although the patterns of these kinds of interactions may be similar across taxa, it is possible that there are differences between primate and nonprimate species in the mechanisms that regulate so-

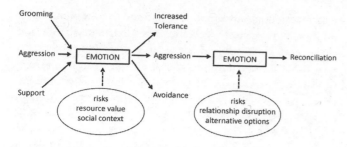

Fig. 23.1. Schematic example of emotional mediation. The patterns of social interactions exchanged (e.g., grooming, aggression, support) between two partners (i.e., social relationship, sensu Hinde) affect each individual's emotion. This partner-dependent emotional experience influenced by contingent factors (marked by broken lines) then mediates the relative occurrence of mechanisms of conflict management (e.g., tolerance, aggression, avoidance, and possible reconciliation).

cial relationships: how they are established, changed, and maintained (Cords 1997; Schino 2000; Bshary et al. 2002). Before reviewing these mechanisms, we briefly consider how variation in the quality of social relationships can be measured.

Measuring Relationship Quality

Relationship quality is difficult to measure with precision (Dunbar & Shultz 2010), but primatologists have largely relied on two different approaches. First, characteristics such as the frequency of particular behaviors (e.g., grooming, aggression, agonistic support, and food sharing) have been used as indicators (e.g., Aureli et al. 1989; Kudo & Dunbar 2001; Cooper et al. 2005; Majolo et al. 2005; Wittig & Boesch 2005). Second, relationship quality has been inferred from the likely fitness advantages provided by relationships with relatives (e.g., matrilineal kin in despotic species of macaques: Aureli et al. 1997; Chapais et al. 2001), same-sex partners (e.g., male-male dyads in chimpanzees, *Pan troglodytes*, Mitani et al. 2000; Koski et al. 2007), or from other characteristics of the individuals involved.

Relationship quality is unlikely to consist of a single dimension, but is likely to comprise several aspects. Cords & Aureli (2000) proposed three components: value, compatibility, and security. Value means the benefits afforded by the relationship. Compatibility is defined as the degree of tolerance within a dyad. Security means the predictability or consistency of interactions between partners over time. Whereas the measures outlined above can approximate the differences in the quality of relationships within a group, choosing the best behavior to represent each component can be difficult and runs the risk of reflecting the investigator's assumptions more than the animals' perspectives.

Broad categories of relationship quality based on kinship and age-sex combinations may mask the effects of individual and dyadic variability within categories (Fraser et al. 2008a). Reliance on a single behavior may limit the interpretability of the results (Silk 2002b), and yet combining multiple variables into a single, meaningful indicator of relationship quality can be difficult.

Statistical methods, such as principal components analysis (PCA), provide a way to systematically assess the dimensions of social relationships. This methodology was pioneered in studies of mother-infant relationships (Simpson & Howe 1980; Tanaka 1980; Fairbanks & McGuire 1987; Schino et al. 1995) to reduce many independent variables to a few behavioral dimensions that describe mother-infant relationships (see Soltis 1999 for male-female mating relationships). Adapting these methods to the study of adult social relationships in chimpanzees, Fraser et al. (2008a) entered nine behavioral variables into a PCA and extracted three components that appeared to match the three dimensions outlined by Cords and Aureli (2000). A similar approach was successful in obtaining comparable components in Japanese macaques (*Macaca fuscata*, Majolo et al. 2010) as well as ravens (*Corvus corax*, Fraser & Bugnyar 2010a). Somewhat different components were obtained in spider monkeys (Rebecchini et al. 2011), and when the asymmetric nature of social relationships in Japanese macaques was taken into account (Majolo et al. 2010). The advantage of using such extracted components is that they offer comprehensive, conceptually coherent measures and provide an objective assessment of the quality of relationships relevant to a particular study group.

Formation of Social Relationships

Lonsdorf and Ross (chapter 11, this volume) address the development of social relationships while an individual grows up. Here, we focus on the establishment of relationships during adulthood. When two individuals meet for the first time, they do not have a history of past interactions that can inform each of them about the likely behavior of their partner. We know very little about how social relationships develop among unfamiliar adult primates in the wild. This is likely due to the difficulty of observing rare events, of following new immigrants that are not habituated to the presence of observers, and of knowing whether immigrants are actually strangers to the residents. Therefore, our understanding of these processes is based on a handful of captive studies in which pairs of strangers were introduced to one another.

Introductions tend to conform to a "species-typical etiquette" as interactions follow a predictable sequence, from aggression to brief contacts to grooming (Kummer 1975, 1995; Maxim 1976; Welker et al. 1980; Mendoza 1993; Baker & Aureli 2000). However, the composition of the dyad may influence the sequence of elements. In geladas (*Theropithecus gelada*), for example, the first interaction in a male-male dyad is typically aggressive, and is followed by presenting, mounting, and grooming, whereas in female-female and male-female dyads aggression is largely bypassed (Kummer 1975). In chimpanzees, interactions within female-female and female-male dyads initially involve dominance and agonism, and are followed by brief friendly touches and grooming (Baker & Aureli 2000). De Waal (1986) hypothesized that the establishment of clearcut dominance relations in primates is a prerequisite for the exchange of affiliative interactions. Schino et al. (1990) reported that during two-hour experimental encounters in long-tailed macaques (*Macaca fascicularis*), unfamiliar pairs that did not establish dominance relationships exchanged little grooming and exhibited progressively more frequent behavioral indicators of anxiety. Conversely, unfamiliar pairs that quickly established dominance relationships did not differ from familiar pairs in their grooming or anxiety levels. These results indicate that clear and mutually acknowledged status differences allow for the development of friendly relationships in some species, particularly those with well-established dominance hierarchies.

In species that form pair bonds or live in small family groups, the establishment of dominance relations may not be a necessary prerequisite for friendly interactions. In marmosets and tamarins, arguably the strongest social bonds exist between the breeding individuals (Kleiman 1977). Relationships between breeding adults from newly formed and well established pairs differed significantly in the quality and quantity of social and sexual interactions (golden lion tamarins, *Leontopithecus rosalia*, Ruiz 1990; cotton-top tamarins, *Saguinus oedipus*, Savage et al. 1988; common marmosets, *Callithrix jacchus*, Evans & Poole 1983). Newly formed pairs had higher rates of huddling, grooming, and sniffing than established pairs. During the first weeks of pair formation, Wied's black-tufted-ear marmosets (*Callithrix kuhlii*) showed higher rates of sexual behavior, but rates of grooming and grooming solicitation steadily increased across the first 80 days of pair formation (Schaffner et al. 1995). The exchange of sexual behavior, but not dominance displays, may be an important initial step to establishing social relationships in opposite-sexed pairs of adult marmosets and tamarins. Different behavioral patterns might be important in developing social relationships in same-sex dyads.

Much less is known about how relationships develop when individuals are embedded in a social context in which multiple partners are available. This is particularly relevant since, under natural conditions (e.g., immigration into a new group), the simultaneous formation of multiple relationships is presumably the norm. Studies of experimental group formation do not generally report information in enough detail to allow evaluation of the dynamics of multiple relationship formation. A notable exception is the experimental work of Kummer (1975) on group formation in geladas. He reported that in the first hours after group formation, dominant females actively prevented subordinate females from interacting with the alpha male, and males interfered with females' interactions with other males. These observations suggest that the geladas were actively involved in monitoring each other's interactions and tried to steer the development of third-party relationships to their own advantage.

Another study assessed the impact of an additional breeding male on the development of social relationships by comparing newly formed breeding trios and breeding pairs of Wied's black-tufted-ear marmosets (Schaffner & French 2004). Two key differences emerged in these contexts. The males from polyandrous trios copulated with females more frequently than did the males in breeding pairs, but sat in contact with those females less than half as often. In addition, rates of grooming and grooming solicitation in trios did not increase across the 80-day observation period, as did the corresponding rates for newly formed breeding pairs. These findings suggest that the presence of an additional male slows the development of social relationships between breeding males and females.

The manner in which relationships develop in captivity likely reflects how relationships develop in the wild, although the challenges faced by new immigrants are inevitably more complicated because they must form social relationships with multiple individuals that already possess highly differentiated relationships among themselves. More research is needed about how relationships are formed in the wild, but the limited evidence suggests that some of the same patterns occur. Aggression is often a first step in developing social relationships in captivity, and in the wild new immigrants frequently receive high rates of aggression, particularly from same-sex conspecifics (Pusey & Packer 1987). As in captive experiments, social integration may be facilitated by developing social relationships with opposite-sex partners, thus bypassing the risk of aggression (e.g., Kummer 1975; Baker & Aureli 2000). Male chimpanzees pave the way for immigrant females by defending them from attacks by resident females, particularly when they are in estrus (Pusey et al. 2008). Female vervet monkeys (*Chlorocebus pygerythrus*) exhibit friendly behavior toward new immigrant males (Henzi & Lucas 1980), and new immigrant female Geoffroy's spider monkeys (*Ateles geoffroyi*), who are the recipients of aggression from resident females (Asensio et al. 2008), are better tolerated in mixed-sex subgroups than in same-sex subgroups, presumably due to the presence of resident males. Finally, the timing of the development of social relationships among immigrants and residents in wild settings is likely to be prolonged, based on the finding that the addition of one further individual slowed the development of social relationships in captive Wied's black-tufted-ear marmosets (Schaffner & French 2004). The fact that relationships change over time and are influenced by the social context highlights the importance of mechanisms for maintaining social relationships. We turn to these mechanisms in the next section.

Maintaining Social Relationships

Affiliation as Negotiation

Various neuropeptides are associated with affiliative behaviors and social bonding in several mammalian species (for recent reviews see Dunbar 2010; Massen et al. 2010; Seyfarth & Cheney 2012). As a consequence affiliative behaviors, particularly grooming, have been interpreted as tools to establish and maintain valuable social relationships (Dunbar 2010). Although it is often plagued by "prospective-like" language (Fairbanks 1993), this interpretation is consistent with evidence that grooming emerges in the initial phases of relationship formation and is exchanged for rank-related benefits (Seyfarth 1980; Silk 1992; Hemelrijk 1994; Barrett & Henzi 2002; Ventura et al. 2006; Schino 2007; chapter 22, this volume).

Grooming may also be used to regulate aggression and tolerance (fig. 23.2). Grooming has a calming influence on the recipient, reducing behavioral and physiological correlates of anxiety and stress (Schino et al. 1988; Aureli et al. 1999), and it facilitates access to valued commodities (infant handling, Henzi & Barrett 2002; Gumert 2007a; mating, Gumert 2007b). In some cases, primates devote more time to grooming those that pose the greatest threat (Perry 1996; Schino et al. 2005), but we know next to nothing about the underlying processes that mediate these interactions. A recent experiment in meerkats (*Suricata suricatta*, Madden & Clutton-Brock 2009) showed that reducing grooming through the pharmacological elimination of external parasites did not lead to an increase in aggression, suggesting that more complex or indirect causal relations between grooming and aggression are involved. Similar experiments with primates would be extremely useful.

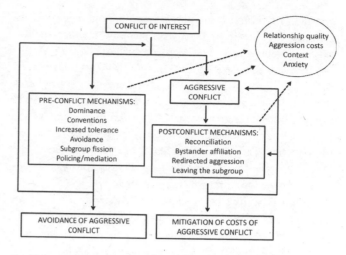

Fig. 23.2. Heuristic model of the options for conflict management. The temporal sequence of events is reported from top to bottom. Arrows that go from bottom to top represent the reinitiation of the temporal sequence when pre- or postconflict mechanisms are only partially successful. Broken lines mark the possible influence of selected factors. Based on de Waal's (1996) fig. 12.1, Aureli & Smucny's (2000) fig. 10.4, and Wittig & Boesch's (2003) fig. 5.

On a shorter time scale, grunts and embraces can signal benign intent and facilitate friendly interactions in contexts in which a reduction in tension and ambiguity is advantageous, such as in the aftermath of a fight and before attempting to handle another female's infant (Cheney et al. 1995; Silk et al. 2000; Kutsukake et al. 2006; Slater et al. 2007).

Aggression as Negotiation

Traditional views suggested that aggression is an individual expression bearing little relation to the behavior of other group members (including the recipients of aggression or bystanders), or that the occurrence of aggressive conflict is dictated purely by the resource value and the associated physical risks (reviewed in de Waal 1996). Aggression, however, is by its very definition a social behavior and is thus likely to be influenced by the relationship between those involved. Aggression may in fact function as a negotiation tool, used as a means to modify a partner's behavior (de Waal 1996). Within a relationship, partners are expected to behave in predictable ways. Violation of such expectations may lead to a protest and punishment in the form of aggression (Clutton-Brock & Parker 1995). Aggression may allow the boundaries of the relationship to be assessed, so that both partners can establish whether the relationship has changed substantively, or whether a conflict of interest or violation of expected behavior merely reflects a temporary "hiccup" in an unchanged relationship (van Schaik & Aureli 2000).

Aggression, however, is not an inevitable outcome of

conflicts of interest. In fact, potential opponents have a number of possible options for managing conflict, including increasing tolerance or active avoidance (fig. 23.2). According to the relational model (de Waal 1996, 2000a), the value of the source causing the conflict and the value of the relationship between competitors contribute to the process of deciding whether to escalate a contest. As the value of the resource or commodity in question increases, the benefits of aggression increase; but as the value of the relationship between competitors increases, the overall benefit of aggression decreases as relationship damage becomes more costly. However, if damage to the relationship can be repaired through postconflict mechanisms (see below), the benefits of aggressive escalation may outweigh the costs even between highly valuable partners. Thus, the value of the commodity, the quality of the relationship between partners, the possibility of relationship repair, and the risk of injury all influence whether aggression escalates (de Waal 2000a).

In support of the relational model, male wild chimpanzees were most likely to fight in social contexts, such as conflicts over females, whereas females fought over food, thus suggesting that both sexes were most likely to fight over resources that have the largest impact on fitness (Wittig & Boesch 2003a). Furthermore, aggression increases with the intensity of competition in many primate species (Janson 1988; Barton & Whiten 1993; Sterck & Steenbeek 1997; Wittig & Boesch 2003b), although this may simply be a result of an increase in the frequency of conflicts of interest, and not an increase in an individual's aggressive tendencies per conflict of interest (Wittig & Boesch 2003b). The relational model, however, does not take into account the relative fighting abilities of potential opponents, and thus the likelihood of winning the conflict, or the conflict duration, which may influence the energetic costs of aggression. Therefore, Wittig and Boesch (2003a) extended the relational model to incorporate these factors, and tested it on wild male chimpanzees. They found that aggressive conflicts initiated by dominant individuals were likely to be longer and more intense than conflicts initiated by subordinates, but that males avoided initiating conflicts with valuable partners. Conversely, subordinates initiated conflicts with lower physical costs, but did not avoid conflicts with valuable partners, which could be mitigated through subsequent reconciliation. These results suggest that chimpanzees pursue different strategies that confer net benefits depending on the likelihood of winning the conflict (Wittig & Boesch 2003a).

Pre-Conflict Mechanisms

The most effective strategy for reducing the costs of a conflict of interest is to prevent aggressive escalation. Indeed, it would be ideal to reduce the likelihood of a conflict of interest in the first place. Primates seem to use conventions, such as dominance (Preuschoft & van Schaik 2000; see below), prior ownership (Kummer & Cords 1991), and adherence to routines (Menzel 1993) to regulate access to resources (fig. 23.2). Greetings and other rituals may increase tolerance and decrease aggressive interactions in potentially competitive contexts (Colmenares et al. 2000; Kuester & Paul 2000; Whitham & Maestripieri 2003; Kutsukake et al. 2006). Thus, during periods of tension, when aggression is more likely, appeasing and reassuring behavior might prevent such conflicts from occurring (Judge 2000).

One situation in which conflict avoidance in captive groups can be easily studied is the period preceding regularly scheduled feeding times. During the first few minutes after food is delivered, aggression rates are often high as individuals compete for access to favored items. If they can anticipate this situation, they are expected to behave in ways that reduce tension and promote tolerance in an effort to reduce the likelihood of conflict (Koyama 2000). Individuals in a number of primate species accomplish this feat. For example, stump-tailed macaques (*Macaca arctoides*) selectively increased their grooming of the alpha male, who was most likely to initiate aggression before feeding times (Mayagoitia et al. 1993). Conversely, chimpanzees increased their grooming of kin and usual grooming partners, possibly to increase tolerance of those most likely to share food with them (Koyama & Dunbar 1996). Brown capuchin monkeys (*Cebus apella*) engaged in more grooming before scheduled feeding, which reduced the risk of aggressive escalation and increased co-feeding during the subsequent feeding period (Polizzi di Sorrentino et al. 2010). However, they did not select their pre-feeding grooming partners on the basis of the benefits expected in the upcoming competitive situation (e.g., individuals that pose the highest risk). An increase in kissing and embracing just prior to feeding has been shown to reduce food-related aggression in chimpanzees (de Waal 1992). Bonobos (*Pan paniscus*) increase their rates of social play prior to feeding times, seemingly to reduce tension and increase tolerance, as playing among adults is correlated with rates of co-feeding (Palagi et al. 2006). During feeding, however, bonobos increase the rate of sociosexual behaviors, possibly as a mechanism for reassurance and appeasement (de Waal 1992; Hohmann & Fruth 2000; Palagi et al. 2006).

Species showing high degrees of fission-fusion dynamics may reduce intragroup competition (Kummer 1971; Aureli et al. 2008), and thus the likelihood of aggressive escalation, by fissioning into temporary subgroups according to local resource availability (Anderson et al. 2002; Asensio et al. 2008; spotted hyenas, *Crocuta crocuta*, Smith et al. 2008; fig. 23.2). This mechanism can also reduce conflicts over travel direction and activities by generating more homo-

Fig. 23.3.   Fusion in Geoffroy's spider monkeys in Santa Rosa National Park, Costa Rica. The monkeys in the background are from a subgroup that has just joined another subgroup of monkeys in the foreground. Photograph courtesy of Filippo Aureli.

geneous subgroups (Conradt & Roper 2000). In addition, victims of aggressive conflicts may leave their subgroups to reduce the likelihood of renewed aggression (Slater et al. 2008; spotted hyenas, Smith et al. 2008; cf. Kutsukake & Clutton-Brock 2008 for postconflict avoidance of former aggressors in meerkats; fig. 23.2). Fission may therefore function as a conflict management mechanism. Conversely, fusion of subgroups may promote aggressive conflict (fig. 23.3; Muller 2002; Aureli & Schaffner 2007). Under such circumstances, affiliative interactions between members of the fusing subgroups may reduce tension and the likelihood of aggressive escalation. For example, Geoffroy's spider monkeys embrace members of the joining subgroup upon fusion (Aureli & Schaffner 2007; see Schaffner & Aureli 2005 for supporting evidence in a captive group). Moreover, embraces at fusion reduce the probability of subsequent aggression (Aureli & Schaffner 2007), possibly because the exposing of vulnerable body parts inherent in this interaction increases its reliability as a signal of benign intent (Schaffner & Aureli 2005). Chimpanzees and mantled howler monkeys (*Alouatta palliata*) also increase their rate of affiliative interactions after periods of separation, thus suggesting that these behaviors are similarly linked to tension reduction and aggression avoidance (Nishida et al. 1999; Okamoto et al. 2001; Dias et al. 2008; spotted hyenas, East et al. 1993).

Submissive signals, such as pant-grunt greetings in chimpanzees (Noë et al. 1980; de Waal 1982; Goodall 1986) and silent bared teeth in rhesus macaques (*Macaca mulatta*, de Waal & Luttrell 1985), which are consistently directed up the hierarchy, may allow subordinate individuals to signal their competitive inferiority and thus gain tolerance from dominant partners without risking attack (de Waal 1986). Dominance signals directed down the hierarchy, such as mock bites in stump-tailed macaques (Maestripieri 2005), may allow dominant individuals to preempt potentially disruptive conflicts with subordinate partners without wasting energy on aggressive conflict (Preuschoft & van Schaik 2000). Thus, dominance may function as a conflict management strategy by conventionalizing priority of access and avoiding aggressive escalation over competitive resources (Preuschoft & van Schaik 2000).

Third parties may also play a role in maintaining relationships by policing and mediating conflicts (fig. 23.2). A third party may intervene impartially in an ongoing conflict between two group members and stop aggressive escalation through aggressive or peaceful means (Petit & Thierry 2000). The experimental removal of disproportionately powerful group members, who would normally intervene in aggressive conflicts between others, increased levels of aggression and decreased sociopositive interactions within a whole group of southern pig-tailed macaques (*Macaca nemestrina*, Flack et al. 2005).

Postconflict Mechanisms

Despite the range of preconflict mechanisms for avoiding escalation, aggressive conflict nevertheless occurs. The potential for the benefits of aggressive conflict to outweigh its costs can be increased by the existence of postconflict mechanisms (table 23.1; fig. 23.2), which mitigate some costs of aggressive escalation, such as recurrence of hostility, intolerance around resources, increased uncertainty, and stress (Aureli et al. 2002; Arnold and Aureli 2007; Arnold et al. 2010).

Table 23.1. Functions of postconflict interactions

Postconflict interaction	Participants and direction	Function(s) (and beneficiarys)[1]	Relationship quality patterns[1, 2]
Reconciliation	Directed by victim or aggressor to former opponent	Relationship repair	Valuable, compatible relationship between opponents
		Stress reduction (aggressor, victim)	
Postconflict bystander affiliation	Directed by victim to bystander	*Stress reduction (victim)*	*Valuable relationship between victim and bystander*
		Self-protection (victim)	*Low compatibility between victim and bystander*
	Directed by bystander to victim	Stress reduction (victim)	Valuable relationship between victim and bystander
		Opponent relationship repair (opponents, bystander)	Valuable relationship between aggressor and bystander
		Self-protection (bystander)	*Low compatibility between victim and bystander*
	Directed by bystander to aggressor	*Stress reduction (aggressor)*	*Valuable relationship between aggressor and bystander*
		Self-protection (bystander)	*Low compatibility between aggressor and bystander*
		Support (aggressor, possibly bystander)	*Valuable relationship between aggressor and bystander*
	Directed by aggressor to bystander	Self-protection (aggressor)	*Low compatibility between aggressor and bystander*
Postconflict aggressive interactions	Directed by victim to aggressor	Reversal of conflict outcome (victim)	*Low compatibility between participants*
	Directed by aggressor to victim	*Reinforcement of winner effect (aggressor)*	*Low compatibility between participants*
	Directed by victim to bystander	Stress reduction (victim)	*Low compatibility between participants*
		Self-protection (victim)	*Low compatibility between participants*
		Reversal of loser effect (victim)	*Low compatibility between participants*
		Opponent relationship repair, if aggressor supports victim against bystander (aggressor, victim)	*Low compatibility between participants*
		Self-protection: Revenge (victim)	*Bystander is the aggressor's vulnerable kin or close associate*
	Directed by aggressor to bystander	*Reinforcement of winner effect*	*Low compatibility between participants*

[1] Text in italics refers to aspects that are predicted but not yet demonstrated.
[2] The functions are more likely when participants have these relationship qualities.

Reconciliation

Reconciliation, the postconflict affiliative reunion of former opponents, was first reported by de Waal and van Roosmalen (1979) in chimpanzees. Since then, it has been demonstrated in more than 30 primate species and a number of nonprimate species (Aureli & de Waal 2000; de Waal 2000b; Schino 2000; Aureli et al. 2002; Arnold et al. 2010), including recent additions such as spotted hyenas (Wahaj et al. 2001), dogs (*Canis familiaris*, Cools et al. 2007), wolves (*Canis lupus*, Cordoni & Palagi 2008), ravens (Fraser & Bugnyar 2011), and horses (*Equus caballus*, Cozzi et al. 2010; table 23.2). Interspecific reconciliation even occurs between cleaner wrasse (*Labroides dimidiatus*) and their client reef fish (Bshary & Würth 2001). Although many of the studies were carried out in captive settings, reconciliation patterns are not an artifact of captivity (Aureli et al. 2002; Colmenares 2006).

"Reconciliation" is a heuristic label, and a relationship repair function is implied by the term. Whether reconciliation does repair the opponents' relationship, however, is still a matter for debate. Silk (1997, 2000, 2002a) proposed that rather than repairing relationships, reconciliation functions as a signal of benign intent, an honest signal of the end of a conflict that facilitates the occurrence of affiliative interactions between the opponents (i.e., the benign intent hypothesis). The relationship repair hypothesis proposes that reconciliation restores valuable relationships, repairing damage caused by preceding aggressive conflict (Kappeler & van Schaik 1992; de Waal & Aureli 1997; Cords & Aureli 2000; Aureli et al. 2002). The difference between the two models is subtle, and indeed the benign intent hypothesis has been proposed to be simply a proximate explanation for how opponents renew contact with each other, whereas the relationship repair hypothesis offers an ultimate explanation for why they seek to reconcile (Cords & Aureli 1996). Thus, the two models may not be incompatible. The difficulty in refuting one model in favor of the other is that the same data can often be interpreted to support either argument. The key to understanding the function of reconciliation may be the duration of its effects. Most studies have focused on the short-term effects of reconciliation in restoring tolerance, reducing the likelihood of renewed aggression, and decreasing stress and anxiety (e.g., Cords 1992; Cheney et al. 1995; Aureli et al. 2002; Wittig & Boesch 2005; Butovskaya 2008; table 23.1). These short-term effects fit Hinde's framework of social relationships. If reconciliation changes the interaction patterns between former opponents by removing the negative consequences of conflict, it should be interpreted as the first sign of at least partial restoration of the usual pattern of interaction (Cords & Aureli 1996). Koyama (2001) has shown that the

negative consequences of aggressive conflict in wild Japanese macaques last for at least ten days, and that reconciliation effectively reduces this negative impact, supporting a relationship-repair function. Silk et al. (1996) did not detect long-term effects of reconciliation in chacma baboons (*Papio ursinus*), but their analysis did not use comparisons with baseline levels. Another piece of evidence that is difficult to explain with the benign intent hypothesis is that former opponents exhibit higher anxiety levels after conflicts with more valuable partners (Aureli 1997; Kutsukake & Castles 2001; Cooper et al. 2007; Koski et al. 2007), with whom they are more likely to reconcile (references below).

Expanding on the relationship repair hypothesis, Aureli et al. (2002) proposed a framework in which the occurrence of reconciliation is predicted by the potential loss of benefits resulting from aggressive conflict and the need for relationship repair. Thus, reconciliation should be possible in any species in which there are individualized relationships, intragroup aggression, and the potential for aggression to disrupt relationships. If the relationships are of sufficient value, the benefits of relationship repair should outweigh the risks of renewed attack, thus making reconciliation worthwhile (Aureli et al. 2002). Reconciliation is predicted to be affected not only by the value of the relationship but also by its compatibility and security (Cords & Aureli 2000). As the risks of renewed aggression are lower for opponents with a compatible relationship, those partners should be more likely to reconcile. Partners with valuable but highly secure relationships may not reconcile, however, given that the security buffers their relationships from the disruptions of aggressive conflict.

In support of the predictive framework, kin and other valuable partners, such as those who support each other in aggressive conflicts, are more likely to reconcile than those with less valuable relationships (e.g., Aureli et al. 1989; Manson et al. 2004; Majolo et al. 2005; Watts 2006; Fraser et al. 2010). Also, reconciliation rates are higher between partners who affiliate with each other at higher rates (e.g., Call et al. 1989; Arnold & Barton 2001a; Preuschoft et al. 2002; Palagi et al. 2004). Perhaps, though, it is the absence of reconciliation that is most illuminating. Red-bellied tamarins (*Saguinus labiatus*) are one of the very few primate species in which reconciliation could not be demonstrated. Cooperation and interdependency in offspring care and predator protection may be so crucial for all members of tamarin groups that their relationships cannot afford to be damaged. Aggression is rare, mild, and possibly inconsequential for social relationships (Schaffner & Caine 2000). In fact, no evidence for a postconflict reduction in tolerance between former opponents was found, even when aggression flared

Table 23.2. Occurrence of reconciliation across species and settings

Species	No. of studies (wild/captive)	Reconciliation demonstrated (CCT range)†	Other demonstrated postconflict interactions
Primates			
Callithrix jacchus[1]	0/1	Yes (31.4)	
Cebus apella[2–4]	0/3	Yes (12.6–50.0)	
Cebus capucinus[5, 6]	1/1	Yes (18.5–27)	
Cercocebus torquatus[7]	0/1	Yes	
Chlorocebus pygerythrus[8]	1/0	Yes	Redirected aggression, bystander affiliation[a]
Colobus guereza[9]	0/1	Yes (45.1)	
Erythrocebus patas[10]	0/1	Yes	
Eulemur fulvus rufus[11, 12]	0/2	Yes (6.4–26.5)	
Eulemur macaco[12]	0/1	No	
Gorilla gorilla[13–15]	1/2	Yes (14.4–29.0)	Bystander affiliation[b, c]
Lemur catta[11, 16, 17]	0/3	Yes (41.5)	
Macaca arctoides[18–21]	0/3	Yes (34.9)	Bystander affiliation[b]
Macaca assamensis[22]	1/0	Yes (11.2)	
Macaca fascicularis[23–25]	1/2	Yes (13.9)	Redirected aggression, bystander affiliation[d]
Macaca fuscata[26–33]	3/5	Yes (7.9–30.3)	Redirected aggression
Macaca maurus[34]	1/0	Yes (40.0)	
Macaca mulatta[33–37]	0/3	Yes (7.1–8.1)	
Macaca nemestrina[38, 39]	0/1	Yes (20.4–41.9)	Bystander affiliation[e]
Macaca nigra[40]	0/1	Yes (40.4)	
Macaca radiata[41]	1/0	Yes (29.3)	
Macaca silensus[42]	0/1	Yes	
Macaca sylvanus[43, 44]	0/2	Yes	
Macaca thibetana[45]	1/0	Yes (7.9)	Quadratic affiliation[f]
Macaca tonkeana[37, 46]	0/2	Yes	Bystander affiliation[b, c]
Pan paniscus[47–49]	1/2	Yes (35.6)	Bystander affiliation[b, c]
Pan troglodytes[50–59]	3/6	Yes (12.3–47.5)	Bystander affiliation[b]
Papio anubis[60, 61]	1/1	Yes (15.6–27.3)	Bystander affiliation,[g] quadratic affiliation[f]
Papio hamadryas[62, 63]	0/2	Yes (19.1)	Bystander affiliation[c]
Papio papio[64]	0/1	Yes (27.0)	
Papio ursinus[54]	1/0	Yes	
Propithecus verreauxi[66]	1/0	Yes (38.0–48.8)	
Rhinopithecus bieti[67]	0/1	Yes (54.5)	
Rhinopithecus roxellanae[68, 69]	1/1	Yes (43.4–58.0)	
Saimiri sciureus[70]	0/1	Yes	
Saguinus labiatus[71]	0/1	No	
Saguinus oedipus[72]	0/1	Yes (37.2)	
Semnopithecus entellus[73]	1/0	Females only	
Theropithecus gelada[74]	0/1	Yes (29.8)	
Trachypithecus obscurus[75, 76]	0/1	Yes (41.3–51.3)	Bystander affiliation[c]
Nonprimates			
Capra hircus[77]	0/1	Yes (16.5)	
Canis familiaris[78]	0/1	Yes	Bystander affiliation[b, c]
Canis lupus[79, 80]	0/1	Yes (53.2)	Bystander affiliation[b, c]
Corvus corax[81, 82]	0/2	Yes (16.0)	Bystander affiliation[b, c]
Corvus frugilegus[83]	0/1	No	Bystander affiliation[b, c]
Crocuta crocuta[84]	1/0	Yes (11.3)	
Equus caballus[85]	1/0	Yes (26.5)	Bystander affiliation[h]
Suricata suricatta[86]	1/0	No	
Tursiops truncatus[87–89]	0/3	Yes (44.0)	

†If available. A single CCT value is provided when CCTs are available for only one group. When CCTs are available for multiple groups, values for groups with the lowest and highest CCT values are provided. CCT values were calculated at either the individual or the group level, and thus caution must be taken when comparing CCT values for different studies.
[a]between bystander and recipient of aggression, initiator unknown; [b]directed by a bystander to the recipient of aggression; [c]directed by the recipient of aggression to a bystander; [d]between aggressor and bystander, initiator unknown; [e]directed by bystander to conflict opponent, with aggressor and recipient not distinguished; [f]between two bystanders; [g]directed by aggressor to a bystander; [h]between bystander and conflict opponent, with aggressor and recipient not distinguished, and initiator unknown.

References: [1]Westlund et al. 2000; [2]Verbeek & de Waal 1997; [3]Weaver & de Waal 2003; [4]Daniel et al. 2009; [5]Leca et al. 2002; [6]Manson et al. 2005; [7]Gust & Gordon 1993; [8]Cheney & Seyfarth 1989; [9]Björnsdotter et al. 2000; [10]York & Rowell 1988; [11]Kappeler 1993; [12]Roeder et al. 2002; [13]Watts 1995; [14]Cordoni et al. 2006; [15]Mallavarapu et al. 2006; [16]Rolland & Roeder 2000; [17]Palagi et al. 2005; [18]de Waal & Ren 1988; [19]Perez-Ruis & Mondragon-Ceballos 1994; [20]Call et al. 1999; [21]Call et al. 2002; [22]Cooper & Bernstein 2002; [23]Cords & Killen 1998; [24]Aureli et al. 1989; [25]Aureli et al. 1992; [26]Aureli et al. 1993; [27]Chaffin et al. 1995; [28]Petit et al. 1997; [29]Schino et al. 1998; [30]Koyama 2001; [31]Kutsukake & Castles 2001; [32]Abegg et al. 2003; [33]Majolo et al. 2005; [34]Matsumura 1996; [35]de Waal & Yoshihara 1983; [36]Call et al. 1996; [37]Demaria & Thierry 2001; [38]Judge 1991; [39]Castles et al. 1996; [40]Petit et al. 1997; [41]Cooper et al. 2007; [42]Abegg et al. 1996; [43]Aureli et al. 1994; [44]Patzelt et al. 2009; [45]Berman et al. 2006; [46]De Marco et al. 2010; [47]de Waal 1987; [48]Hohmann & Fruth 2000; [49]Palagi et al. 2004; [50]de Waal & van Roosmalen 1979; [51]Arnold & Whiten 2001; [52]Preuschoft et al. 2002; [53]Fuentes et al. 2002; [54]Wittig & Boesch 2003a; [55]Kutsukake & Castles 2004; [56]Palagi et al. 2006; [57]Koski et al. 2007; [58]Koski & Sterck 2007; [59]Fraser & Aureli 2008; [60]Castles & Whiten 1998; [61]Meishvili et al. 2005; [62]Judge & Mullen 2005; [63]Romero et al. 2008; [64]Petit & Thierry 1994; [65]Silk et al. 1996; [66]Palagi et al. 2008; [67]Grüter 2004; [68]Ren et al. 1991; [69]Zhang et al. 2010; [70]Pereira et al. 2000; [71]Schaffner & Caine 2000; [72]Peñate et al. 2009; [73]Sommer et al. 2002; [74]Swedell et al. 1997; [75]Arnold & Barton 2001a; [76]Arnold & Barton 2001b; [77]Schino 1998; [78]Cools et al. 2008; [79]Cordoni & Palagi 2008; [80]Palagi & Cordoni 2009; [81]Fraser & Bugnyar 2010b; [82]Fraser & Bugnyar 2011; [83]Seed et al. 2007; [84]Wahaj et al. 2001; [85]Cozzi et al. 2010; [86]Kutsukake & Clutton-Brock 2008; [87]Samuels & Flaherty 2000; [88]Weaver 2003; [89]Holobinko & Waring 2009.

up again (Schaffner et al. 2005). After aggressive conflicts, the tamarins were just as likely to return to their preconflict activities as during control periods, thus suggesting that the highly secure and valuable nature of their relationships protected those relationships from damage. Reconciliation did not take place, as there was no need for repair.

A large body of evidence supports variation in relationship quality as a key predictor of the relative occurrence of reconciliation across dyads in all studied species (Cords & Aureli 2000; de Waal 2000b; Aureli et al. 2002; Watts 2006; Arnold et al. 2010; see above; fig. 23.2). It is, however, rare that all conflicts between two partners are reconciled. It is possible that some conflicts do not disturb the relationship, and that reconciliation is thus unnecessary, regardless of the relationship quality. For example, conflicts that occur over food are less likely to be followed by reconciliation than conflicts in other contexts, possibly because they have the obvious purpose of obtaining a resource, and thus may not affect the opponents' relationship (Aureli 1992; Arnold & Aureli 2007). Similarly, recent research has indicated that conflicts initiated with a bluff display in chimpanzees are less likely to be followed by reconciliation, possibly because the unfocused nature of the attack does not disrupt the opponents' relationships (Fraser et al. 2010). As reconciliation requires the participation of both parties, withholding it from an opponent may also be used as a form of punishment or a means of further negotiation. Asymmetry in the value of a relationship, and thus a differential need for repair, may enable the more valuable partner, who has less to gain from reconciliation, to use consent for reconciliation as a bargaining tool in negotiations over future interactions (de Waal 1996; Aureli & Schaffner 2006).

Reconciliation can be described as being *implicit* or *explicit*, depending on whether the behavior used to reconcile is regularly used in other contexts (de Waal 1993). For example, rhesus macaques reconcile implicitly using behaviors that are used often in other contexts, such as grooming. In contrast, explicit forms of reconciliation involve behaviors that are rarely expressed in other contexts, such as the "hold-bottom" behavior of stump-tailed macaques or the kissing of chimpanzees (fig. 23.4; de Waal & van Roosmalen 1979; de Waal & Ren 1988; Fraser & Aureli 2008). Explicit styles tend to characterize species with high levels of reconciliation and a relaxed dominance style, whereas for implicit styles of reconciliation the reverse is true (de Waal & Ren 1988; Fraser & Aureli 2008; Thierry et al. 2008). Variation in the average quality of social relationships among species may account for the differences in reconciliation style. Individuals in social systems with high levels of tolerance and affiliation may need to use behaviors that are relatively specific to a particular context, such as reconciliation, to make their intentions explicit. In more

Fig. 23.4. "Kiss" in chimpanzees, which is often used for reconciliation. Photo courtesy of Frans B. M. de Waal.

despotic social systems, in which the use of explicit behavior might be too risky, even proximity may be enough to repair the opponents' relationships (Arnold et al. 2010). Although relative spatial position after a conflict may influence the occurrence of reconciliation (Call 1999; Puga-Gonzalez et al. 2009), variation in reconciliation levels cannot simply be attributed to the levels of general proximity or affiliation, because the standardized measure for conciliatory tendency fully corrects for baseline levels (Veenema et al. 1994; see Arnold et al. 2010 for further discussion). Even within species that display an explicit style of reconciliation, explicit gestures are used only for some conciliatory contacts. Partners with more valuable and compatible relationships may require more explicit behaviors for reconciliation than those for whom the rarity of affiliative interactions would lend more meaning and more risk to affiliative postconflict contacts, and for whom implicit forms of reconciliation may thus suffice (Fraser & Aureli 2008; Arnold et al. 2010).

Postconflict Bystander Affiliation

Affiliative postconflict interactions may also occur with bystanders, or third parties uninvolved in the previous conflict (fig. 23.2). For example, after an aggressive conflict between two chimpanzees, another member of the group who has not been involved in the aggression may approach and embrace or groom the former victim. The functions of such interactions are likely to depend upon their direction, upon the quality of the relationships between the opponents and between each opponent and the bystander, and upon the behaviors used (table 23.1). De Waal and van Roosmalen (1979) reported the occurrence of affiliative postconflict interaction between recipients of aggression and bystanders in chimpanzees and labeled it "consolation," but the implied calming function was not tested. Since then, postconflict bystander affiliation has been found in several primate species (Das 2000; Watts et al. 2000; Fraser et al. 2009), and it has been recently reported in rooks (*Corvus frugilegus*, Seed et al., 2007), ravens (Fraser & Bugnyar 2010b), dogs (Cools et al. 2008), wolves (Palagi et al. 2009), and horses (Cozzi et al. 2010). As the cognitive requirements and motivations for a bystander to offer unsolicited postconflict affiliation are likely to differ from those for postconflict affiliation solicited by former opponents (Verbeek & de Waal 1997), most studies distinguished the direction of the interaction when investigating its patterns and functions.

The first study to investigate the stress-reduction function of postconflict bystander affiliation directed towards the recipient of aggression used relevant behavioral indicators such as self-scratching and self-grooming (see Maestripieri et al. 1992; Schino et al. 1996; Baker & Aureli 1997; Troisi 2002 for evidence), and found no evidence for such a function in chimpanzees (Koski & Sterck 2007). A subsequent study of chimpanzees, however, showed that postconflict affiliation from bystanders does reduce behavioral indicators of stress in recipients of aggression (Fraser et al. 2008b), thus supporting the conclusion that it serves a stress-reduction and consoling function. Furthermore, bystander affiliation was provided by valuable partners, who are expected to be more responsive to each others' stress, even after controlling for baseline levels of affiliation. The disparity in results obtained in the two studies on the same species suggests that bystander affiliation may not always have the same function. Indeed, a number of alternative hypotheses account for the behavior (Fraser et al. 2009; table 23.1).

Affiliation between bystanders and conflict participants might repair the relationship between former opponents, and thus function as a substitute for reconciliation (the relationship-repair function, Fraser et al. 2009). This type of bystander affiliation has been previously labeled "triadic reconciliation" (Judge 1991) and "kin-mediated reconciliation" (Wittig et al. 2007), as the conflict participant affiliates with the opponent's kin as proxy, thus potentially repairing the relationship between opponents without the risks of direct contact between them. Unfortunately, many studies that found an increase in postconflict affiliation between conflict participants and their opponents' kin did not distinguish between solicited and unsolicited contacts (Judge 1991), or between the relative roles of the interacting individuals (York & Rowell 1988; Cheney & Seyfarth 1989; Castles & Whiten 1998), thus limiting functional interpretation of the results. Experimental findings from chacma baboons, however, have shown that recipients of aggression displayed increased tolerance of the aggressor's presence after hearing affiliative postconflict vocalizations from the aggressor's kin (Wittig et al. 2007), suggesting that such vocalizations have a relationship-repair function. Such a function could also be fulfilled by postconflict affiliation from any bystanders who have valuable relationships with the aggressor, in which case the bystander-aggressor relationship is predicted to be more valuable than the bystander-recipient relationship (Fraser et al. 2009). In support of the relationship-repair function, wild chimpanzees who provided bystander affiliation were found to have more valuable relationships with the aggressor than with the recipient (Wittig 2010). Conversely, in another population of chimpanzees (Romero & de Waal 2010) and in ravens (Fraser & Bugnyar 2010b), bystander affiliation was most likely to be provided by those who had a more valuable relationship with the recipient than with the aggressor.

Another hypothesis about the function of postconflict bystander affiliation proposes that it protects the bystander (the self-protection function, Fraser et al. 2009). As recipients of aggression may redirect that aggression towards others (see below), bystanders may be able to reduce their risk of becoming targets by affiliating with the original recipient (Watts et al. 2000; Call et al. 2002; Koski & Sterck 2009). This self-protection hypothesis may also apply to bystander affiliation with aggressors, as the latter may also direct further aggression toward bystanders after a conflict (Das 2000). Bystander affiliation with a self-protection function is thus predicted to be initiated by those most at risk of being attacked. In chimpanzees, bystanders who initiated postconflict affiliative interactions with either the aggressor or the recipient of aggression reduced the likelihood of receiving aggression from that opponent (Koski & Sterck 2009), thus supporting the hypothesis of a self-protection function. In lowland gorillas (*Gorilla gorilla gorilla*), group members received further hostility from the former aggressor less often if the former aggressor had received bystander affiliation from them (Palagi et al. 2008). Conversely, in chimpanzees (Koski & Sterck 2009) and in long-tailed macaques (Das et al. 1998), receiving postconflict affiliation from a bystander did not influence the likelihood that an individual would in turn attack other bystanders. Aggressors may also protect themselves from aggression from other group members by affiliating with bystanders, as in hamadryas baboons (*Papio hamadryas*). Such aggressor-initiated bystander affiliation, however, did not reduce the aggressor's stress levels (Romero et al. 2009).

It is also possible that postconflict affiliation directed by bystanders to aggressors serves to signal support for the aggressor (the support function, Das 2000). In doing so, it may signal the strength of the alliance to other group members. Evidence for this function can be found in long-tailed macaques. Rather than reducing the likelihood of further aggression by the aggressor, postconflict affiliation from bystanders actually increased it, thus suggesting that bystanders' support encourages the aggressor to continue behaving aggressively (Das 2000).

The function of bystander affiliation is not necessarily uniform across species, or even within species or groups (table 23.1). The quality of the relationship between partners is likely to be a key factor in determining the function of bystander affiliation in specific interactions (Fraser et al. 2009). The initiation of bystander affiliation by valuable partners may have a stress-reduction function when it is directed towards a recipient of aggression, a support function when it is directed towards an aggressor, and a relationship-repair function when it occurs between an aggressor's valuable partner and the recipient of the aggression. A self-protection function may be more likely if the bystander and the opponent share a relationship with low compatibility. Future work on postconflict bystander affiliation should thus consider its function according to the dyad type. The relationship between opponents, which is likely to affect the probability of their reconciliation, may also play a role in bystander affiliation. When the functions of reconciliation and bystander affiliation overlap (such as when bystander affiliation is predicted to reduce the opponents' stress levels or repair their relationship), reconciliation is likely to be the priority for the conflict participants, and thus bystander affiliation is likely to occur only if the costs of reconciliation outweigh its benefits. Accordingly, the occurrence of reconciliation and bystander affiliation was found to be interdependent when a stress-reduction (Fraser et al. 2008b) or relationship-repair (Wittig & Boesch 2003b) function was likely.

Redirection

Not all postconflict mechanisms are friendly. Recipients of aggression may attack third parties in the aftermath of an aggressive conflict (Bastock et al. 1953). In long-tailed macaques, redirecting aggression toward others reduced behavioral indicators of anxiety in original recipients of aggression (Aureli & van Schaik 1991b; fig. 23.2). Thus, redirection may enable recipients of aggression to alleviate their postconflict stress when reconciliation is too risky

(table 23.1). In some cases, the resulting alleviation of stress may even facilitate subsequent reconciliation. Although redirected aggression is unlikely to occur after reconciliation (since it is no longer necessary), reconciliation is more likely to occur after redirection than when redirection has not occurred (Aureli & van Schaik 1991a; table 23.1).

Recipients of aggression may redirect it toward a suitable target to signal to the original aggressor and other group members that they have not been compromised by a previous defeat (Kazem & Aureli 2005). Redirection may therefore reverse the negative effects of losing a conflict. Indeed, recipients of aggression in long-tailed and rhesus macaques are less likely to receive further aggression from other group members over the next 10 minutes if they do redirect aggression than if they do not (Aureli & van Schaik 1991a; Kazem & Aureli 2005). In those cases the target of redirection may not be important because redirected aggression may serve primarily as a signal of the perpetrator's condition to third parties, and thus may usually be directed toward vulnerable targets such as lower-ranking and younger individuals. In some cases, redirected aggression is selectively directed toward close associates of former opponents. Vervet monkeys and Japanese macaques redirect aggression toward their former aggressor's vulnerable kin (Cheney & Seyfarth 1986; Aureli et al. 1992; spotted hyenas, Engh et al. 2005). This form of partner-specific redirection may enable individuals to obtain "revenge" against former aggressors when direct retaliation is too costly. It is possible that by redirecting aggression towards the aggressors' vulnerable kin, former recipients of aggression may deter aggressors from future attacks (Aureli et al. 1992; table 23.1).

Conclusions: Unanswered Questions and Future Research

We believe that our proposal for social relationships based on the integration of Hinde's (1979, 1976, 1983) descriptive framework, Kummer's (1978) functional view, and the proximate mechanisms of emotional mediation (Aureli & Schaffner 2002) is most consistent with the available evidence (see also Seyfarth & Cheney 2012). While further research on the details of social relationships and the mechanisms that animals use to establish and maintain social relationships is welcome, we believe that it does not necessarily need to focus on the stability of affiliative or cooperative relationships over time (Silk et al. 2006a; Silk et al. 2006b; Gilby & Wrangham 2008; Mitani 2009; Henzi et al. 2009). This is because social relationships are dynamic and can change (Hinde 1983; Dunbar 1988; Cords 1997; Cords

& Aureli 2000). What is important is to develop empirical tools to characterize the quality of social relationships and monitor them over time. This implies the use of more integrative methods for assessing relationship quality (cf. Dunbar & Shultz 2010). The use of principal components analysis or similar methods is recommended for future studies involving assessment of the quality of social relationships (Fraser et al. 2008a). Where possible, such analyses should be conducted taking into account asymmetries in relationship quality within dyads (Fraser and Bugnyar 2010a; Majolo et al. 2010).

We know little about the processes that shape the initial development of social relationships among adults. Effort should be made to gather data on the formation of relationships by new immigrants and to determine whether relationship formation follows the same sequence observed in experimental dyad formation (Kummer 1975; Maxim 1976; Welker et al. 1980; Mendoza 1993; Baker & Aureli 2000). In wild groups, immigrants must establish multiple social relationships simultaneously, but this problem is rarely addressed even in experimental work (though see Stammbach 1978; Kummer 1975). Social relationships do not exist in isolation; they are likely influenced by other group members in the social network (Croft et al. 2008; Whitehead 2008). It may be profitable to consider how primates monitor social relationships among members of a group, and to determine whether they actively interfere with those relationships (e.g., de Waal 1982; Dunbar 1988; Mondragón-Ceballos 2001). The study of the regulation of social relationships will certainly benefit from more integration of proximate mechanisms and functions. In particular, investigation of the factors affecting the patterns of different mechanisms for the regulation of social relationships should also test their functions. Multivariate statistical approaches have greatly enhanced analysis of the complex factors affecting the quality and maintenance of social relationships (Schino et al. 1998; Wittig & Boesch 2005; Fraser et al. 2008b; Majolo et al. 2009). Researchers need to use these methods to identify relevant predictor variables and appropriate measures of functional outcomes (see above) to test predictions derived from solid theoretical foundations. At the same time, the evaluation of the cognitive processes involved in the regulation of social relationships should be more widely undertaken using naturalistic observations and experiments (Cheney & Seyfarth 1986; chapter 28, this volume).

Our review has emphasized that in the last few decades most research focused on postconflict mechanisms for the regulation of social relationships. We encourage a similar effort on preconflict mechanisms and other tools for relationship negotiation in the years to come. At the same time, research should focus on a wide range of primate and nonprimate species living in different social systems to achieve a broad comparative perspective, allowing for phylogenetically controlled comparative analyses (Nunn & Barton 2001) and refined predictive frameworks (see Aureli et al. 2002 for an attempt for reconciliation, the best known mechanism), possibly including a social network approach (Croft et al. 2008; Whitehead 2008). Scientists will be then in a position to develop and quantitatively test integrated models of the options and conditions for the regulation of social relationships (fig. 23.2).

References

Abegg, C., Petit, O. & Thierry, B. 2003. Variability in behavior frequencies and consistency in transactions across seasons in captive Japanese macaques (*Macaca fuscata*). *Aggressive Behavior*, 29, 81–93.

Abegg, C., Thierry, B. & Kaumanns, W. 1996. Reconciliation in three groups of lion-tailed macaques. *International Journal of Primatology*, 17, 803–816.

Anderson, D. P., Nordheim, E. V., Boesch, C. & Moermod, T. C. 2002. Factors influencing fission-fusion grouping in chimpanzees in the Tai National Park, Côte d'Ivoire. In *Behavioural Diversity in Chimpanzees and Bonobos* (ed. by Boesch, C., Hohmann, G. & Marchant, L. F.), 90–101. Cambridge: Cambridge University Press.

Arnold, K. & Aureli, F. 2007. Postconflict reconciliation. In *Primates in Perspective, 1st Edition* (ed. by Campbell, C.K., Fuentes, A., MacKinnon, K., Panger, M. & Bearder, S.), 592–608. Oxford: Oxford University Press.

Arnold, K. & Barton, R. 2001a. Postconflict behavior of spectacled leaf monkeys (*Trachypithecus obscurus*). I. Reconciliation. *International Journal of Primatology*, 22, 243–266.

———. 2001b. Postconflict behavior of spectacled leaf monkeys (*Trachypithecus obscurus*). II. Contact with third parties. *International Journal of Primatology*, 22, 267–286.

Arnold, K., Fraser, O. N. & Aureli, F. 2011. Postconflict reconciliation. In *Primates in Perspective, 2nd Edition* (ed. by Campbell, C.K., Fuentes, A., MacKinnon, K., Stumpf, R. M. and Bearder, S.), 608–625. Oxford: Oxford University Press.

Arnold, K. & Whiten, A. 2001. Post-conflict behaviour of wild chimpanzees (*Pan troglodytes schweinfurthii*) in the Budongo forest, Uganda. *Behaviour*, 138, 649–690.

Asensio, N., Korstjens, A. H., Schaffner, C. M. & Aureli, F. 2008. Intragroup aggression, fission-fusion dynamics and feeding competition in spider monkeys. *Behaviour*, 145, 983–1001.

Aureli, F. 1992. Post-conflict behavior among wild long-tailed macaques (*Macaca fascicularis*). *Behavioral Ecology and Sociobiology*, 31, 329–337.

———. 1997. Post-conflict anxiety in nonhuman primates: The mediating role of emotion in conflict resolution. *Aggressive Behavior*, 23, 315–328.

Aureli, F., Cords, M. & van Schaik, C. P. 2002. Conflict resolution following aggression in gregarious animals: A predictive framework. *Animal Behaviour*, 64, 325–343.

Aureli, F., Cozzolino, R., Cordischi, C. & Scucchi, S. 1992. Kin-oriented redirection among Japanese macaques: An expression of a revenge system. *Animal Behaviour*, 44, 283–291.

Aureli, F., Das, M., van Eck, C. J. V., Veenema, H. C., Verleur, D. & van Hooff, J. A. R. A. M. 1993. Post-conflict affiliative interactions among Japanese and Barbary macaques. *Aggressive Behavior*, 19, 49–50.

Aureli, F., Das, M., Verleur, D. & van Hooff, J. A. R. A. M. 1994. Postconflict social interactions among Barbary macaques (*Macaca sylvanus*). *International Journal of Primatology*, 15, 471–485.

Aureli, F., Das, M. & Veenema, H. C. 1997. Differential kinship effect on reconciliation in three species of macaques (*Macaca fascicularis*, *M. fuscata*, and *M. sylvanus*). *Journal of Comparative Psychology*, 111, 91–99.

Aureli, F. & de Waal, F. B. M. 2000. *Natural Conflict Resolution*. Berkeley: University of California Press.

Aureli, F., Preston, S. D. & de Waal, F. B. M. 1999. Heart rate responses to social interactions in free-moving rhesus macaques (*Macaca mulatta*): a pilot study. *Journal of Comparative Psychology*, 113, 59–65.

Aureli, F. & Schaffner, C. 2002. Relationship assessment through emotional mediation. *Behaviour*, 139, 393–420.

———. 2007. Aggression and conflict management at fusion in spider monkeys. *Biology Letters*, 3, 147–149.

Aureli, F. & Schaffner, C. M. 2006. Causes, consequences and mechanisms of reconciliation: the role of cooperation. In *Cooperation in Primates and Humans* (ed. by Kappeler, P. M. & van Schaik, C. P.), 121–136. Berlin: Springer.

Aureli, F., Schaffner, C. M., Boesch, C., Bearder, S. K., Call, J., Chapman, C. A., Connor, R., Di Fiore, A., Dunbar, R. I. M., Henzi, S. P., Holekamp, K., Korstjens, A. H., Layton, R., Lee, P., Lehmann, J., Manson, J. H., Ramos Fernandez, G., Strier, K. B. & van Schaik, C. P. 2008. Fission-fusion dynamics: New research frameworks. *Current Anthropology*, 49, 627–654.

Aureli, F. & Schino, G. 2004. The role of emotions in social relationships. In *Macaque Societies: A Model for the Study of Social Organization* (ed. by Thierry, B., Singh, M. & Kaumanns, W.), 38–60. Cambridge: Cambridge University Press.

Aureli, F. & van Schaik, C. P. 1991a. Post-conflict behaviour in long-tailed macaques (*Macaca fascicularis*): I. The social events. *Ethology*, 89, 89–100.

———. 1991b. Post-conflict behaviour in long-tailed macaques (*Macaca fascicularis*): II. Coping with uncertainty. *Ethology*, 89, 101–114.

Aureli, F., van Schaik, C. P. & van Hooff, J. A. R. A. M. 1989. Functional aspects of reconciliation among captive long-tailed macaques (*Macaca fascicularis*). *American Journal of Primatology*, 19, 39–51.

Aureli, F. & Whiten, A. 2003. Emotions and behavioural flexibility. In *Primate Psychology: The Mind and Behaviour of Human and Nonhuman Primates* (ed. by Maestripieri, D.), 289–323. Cambridge, MA: Harvard University Press.

Baker, K. C. & Aureli, F. 2000. Coping with conflict during initial encounters in chimpanzees. *Ethology*, 106, 527–541.

———. 1997. Behavioural indicators of anxiety: An empirical test in chimpanzees. *Behaviour*, 134, 1031–1050.

Barrett, L. & Henzi, S. P. 2002. Constraints on relationship formation among female primates. *Behaviour*, 139, 263–289.

Barton, R. A., Whiten, A. 1993. Feeding competition among female olive baboons, *Papio anubis*. *Animal Behaviour*, 46, 777–789.

Bastock, M., Morris, D. & Moynihan, M. 1953. Some comments on conflict and frustration in animals. *Behaviour*, 6, 66–84.

Berman, C. M., Ionica, C. S., Dorner, M. & Li, J. H. 2006. Postconflict affiliation between former opponents in *Macaca thibetana* on Mt. Huangshan, China. *International Journal of Primatology*, 27, 827–854.

Björnsdotter, M., Larsson, L. & Ljungberg, T. 2000. Postconflict affiliation in two captive groups of black-and-white guereza *Colobus guereza*. *Ethology*, 196, 289–300.

Bshary, R. & Würth, M. 2001. Cleaner fish *Labroides dimidiatus* manipulate client reefer fish providing tactile stimulation. *Proceedings of the Royal Society of London, Series B: Biological Sciences*, 268, 1495–1501.

Bshary, R., Wickler, W. & Fricke, H. 2002. Fish cognition: A primate's eye view. *Animal Cognition*, 5, 1–13.

Butovskaya, M. 2008. Reconciliation, dominance and cortisol levels in children and adolescents (7–15-year-old boys). *Behaviour*, 145, 1557–1576.

Call, J., Aureli, F. & de Waal, F. B. M. 2002. Postconflict third party affiliation in stumptailed macaques. *Animal Behaviour*, 63, 209–216.

Call, J., Judge, P. G. & de Waal, F. B. M. 1996. Influence of kinship and spatial density on reconciliation and grooming in rhesus monkeys. *American Journal of Primatology*, 39, 35–45.

Castles, D. L. & Whiten, A. 1998. Post-conflict behaviour of wild olive baboons I. Reconciliation, redirection and consolation. *Ethology*, 104, 126–147.

Chapais, B., Savard, L. & Gauthier, C. 2001. Kin selection and the distribution of altruism in relation to degree of kinship in Japanese macaques (*Macaca fuscata*). *Behavioral Ecology and Sociobiology*, 49, 493–502.

Cheney, D. L. & Seyfarth, R. M. 1986. The recognition of social alliances by vervet monkeys. *Animal Behaviour*, 34, 1722–1731.

———. 1989. Redirected aggression and reconciliation among vervet monkeys, *Cercopithecus aethiops*. *Behaviour*, 110, 258–275.

Cheney, D. L., Seyfarth, R. M. & Silk, J. B. 1995. The role of grunts in reconciling opponents and facilitating interactions among adult female baboons. *Animal Behaviour*, 50, 249–257.

Clutton-Brock, T. & Parker, G. 1995. Punishment in animal societies. *Nature*, 373, 209–216.

Colmenares, F. 2006. Is postconflict affiliation in captive nonhuman primates an artifact of captivity? *International Journal Primatology* 27, 1311–1335.

Colmenares, F., Hofer, H. & East, M. L. 2000. Greeting ceremonies in baboons and hyenas. In *Natural Conflict Resolution* (ed. by Aureli, F. & de Waal, F. B. M.), 94–96. Berkeley: University of California Press.

Connor, R. C. 2007. Dolphin social intelligence: Complex alliance relationships in bottlenose dolphins and a consideration of selective environments for extreme brain size evolution in mammals. *Philosophical Transactions of the Royal Society B: Biological Sciences*, 362, 587–602.

Conradt, L. and Roper, T. J. 2000. Activity synchrony and social cohesion: a fission-fusion model. *Proceedings of the Royal Society B: Biological Sciences*, 267, 2213–2218.

Cools, A. K. A., van Hout, A. J. M. & Nelissen, M. H. J. 2008. Canine reconciliation and third-party-initiated postconflict affiliation: Do peacemaking social mechanisms in dogs rival those of higher primates? *Ethology*, 114, 53–63.

Cooper, M., Aureli, F. & Singh, M. 2007. Sex differences in reconciliation and post-conflict anxiety in bonnet macaques. *Ethology*, 113, 26–38.

Cooper, M. A. & Bernstein, I. S. 2002. Counter aggression and reconciliation in Assamese macaques (*Macaca assamensis*). *American Journal of Primatology*, 56, 215–230.

Cooper, M. A., Bernstein, I. S. & Hemelrijk, C. K. 2005. Reconciliation and relationship quality in Assamese macaques (*Macaca assamensis*). *American Journal of Primatology*, 65, 269–282.

Cordoni, G. & Palagi, E. 2008. Reconciliation in wolves (*Canis lupus*): New evidence for a comparative perspective. *Ethology*, 114, 298–308.

Cordoni, G., Palagi, E. & Borgognini Tarli, S. 2006. Reconciliation and consolation in captive western gorillas. *International Journal of Primatology*, 27, 1365–1382.

Cords, M. 1992. Post-conflict reunions and reconciliation in long-tailed macaques. *Animal Behaviour*, 44, 57–61.

———. 1997. Friendships, alliances, reciprocity and repair. In *Machiavellian Intelligence II* (ed. by Whiten, A. & Byrne, R. W.), 24–49. Cambridge: Cambridge University Press.

Cords, M. & Aureli, F. 1996. Reasons for reconciling. *Evolutionary Anthropology*, 5, 42–45.

———. 2000. Reconciliation and relationship qualities. In *Natural Conflict Resolution* (ed. by Aureli, F. & de Waal, F. B. M.), 177–198. Berkeley: University of California Press.

Cords, M. & Killen, M. 1998. Conflict resolution in human and nonhuman primates. In *Piaget, Evolution, and Development* (ed. by Langer, J. & Killen, M.), 193–219. Mahwah, NJ: Lawrence Earlbaum Associates.

Cozzi, A., Sighieri, C., Gazzano, A., Nicol, C. & Baragli, P. 2010. Post-conflict friendly reunion in a permanent group of horses (*Equus caballus*). *Behavioural Processes*, 85, 185–90.

Croft, D. P., James, R. & Krause, J. 2008. *Exploring Animal Social Networks*. Princeton, NJ: Princeton University Press.

Daniel, J. R., Santos, A. J. & Cruz, M. G. 2009. Postconflict behaviour in brown capuchin monkeys (*Cebus apella*). *Folia Primatologica*, 80, 329–340.

Das, M. 2000. Conflict management via third parties: Post-conflict affiliation of the aggressor. In *Natural Conflict Resolution* (ed. by Aureli, F. & de Waal, F. B. M.), 263–280. Berkeley: University of California Press.

Das, M., Penke, Z. & Van Hooff, J. A. R. A. M. 1998. Postconflict affiliation and stress-related behavior of long-tailed macaque aggressors. *International Journal of Primatology*, 19, 53–71.

De Marco, A., Cozzolino R., Dessì-Fulgheri F. & Thierry B. 2010. Conflicts induce affiliative interactions among bystanders in a tolerant species of macaque (*Macaca tonkeana*). *Animal Behaviour*, 80, 197–203.

Demaria, C. & Thierry, B. 2001. A comparative study of reconciliation in rhesus and Tonkean macaques. *Behaviour*, 138, 397–410.

De Waal, F. B. M. 1982. *Chimpanzee Politics*. London: Jonathon Cape.

———. 1986. The integration of dominance and social bonding in primates. *The Quarterly Review of Biology*, 61, 459–479.

———. 1987. Tension regulation and nonreproductive functions of sex in captive bonobos (*Pan paniscus*). *National Geographic Research*, 3, 318–335.

———. 1992. Appeasement, celebration, and food sharing in the two *Pan* species. In *Topics in Primatology* (ed. by Nishida, T., McGrew, W. C., Marler, P. R., Piekford, M. & de Waal, F. B. M.), 37–50. Tokyo: University of Tokyo Press.

———. 1993. Reconciliation among primates: A review of empirical evidence and unresolved issues. In *Primate Social Conflict* (ed. by Mason, W. A. & Mendoza, S. P.), 111–144. Albany: State University of New York Press.

———. 1996. Conflict as negotiation. In *Great Ape Societies* (ed. by McGrew, W. C., Marchant, L. F. & Nishida, T.), 159–172. Cambridge: Cambridge University Press.

———. 2000a. The first kiss: Foundations of conflict resolution research in animals. In *Natural Conflict Resolution* (ed. by Aureli, F. & de Waal, F. B. M.), 15–33. Berkeley: University of California Press.

———. 2000b. Primates: A natural heritage of conflict resolution. *Science*, 289, 586–590.

De Waal, F. B. M. & Aureli, F. 1997. Conflict resolution and distress alleviation in monkeys and apes. In *The Integrative Neurobiology of Affiliation* (ed. by Carter, C. S., Kirkpatrick, B. & Lenderhendler, I.), 317–328. New York: Annals of the New York Academy of Sciences.

De Waal, F. B. M. & Luttrell, L. M. 1985. The formal hierarchy of rhesus monkeys: An investigation of the bared-teeth display. *American Journal of Primatology*, 9, 73–85.

De Waal, F. B. M. & Ren, R. 1988. Comparison of the reconciliation behavior of stumptail and rhesus macaques. *Ethology*, 78, 129–142.

De Waal, F. B. M. & van Roosmalen, A. 1979. Reconciliation and consolation among chimpanzees. *Behavioral Ecology and Sociobiology*, 5, 55–66.

De Waal, F. B. M. & Yoshihara, D. 1983. Reconciliation and redirected affection in rhesus monkeys. *Behaviour*, 85, 224–241.

Dias P. A. D., Rodríguez-Luna E. & Canales-Espinosa D. 2008. The functions of the "greeting ceremony" among male mantled howlers (*Alouatta palliata*) on Agaltepec Island, Mexico. *American Journal of Primatology*, 70, 621–628.

Dunbar, R. I. M. 1988. *Primate Social Systems*. London: Chapman & Hall.

———. 2010. The social role of touch in humans and primates: Behavioural function and neurobiological mechanisms. *Neuroscience and Biobehavioral Reviews*, 34, 260–268.

Dunbar, R. I. M. & Shultz, S. 2010. Bondedness and sociality. *Behaviour*, 147, 775–803.

East, M. L., Hofer, H. & Wickler, W. 1993. The erect "penis" is a flag of submission in a female-dominated society: Greetings in Serengeti spotted hyenas. *Behavioral Ecology and Sociobiology*, 33, 355–370.

Emery, N. J., Seed, A. M., von Bayern, A. M. & Clayton, N. S. 2007. Cognitive adaptations of social bonding in birds. *Philosophical Transactions of the Royal Society B: Biological Sciences*, 362, 489–505.

Engh, A. L., Siebert, E. R., Greenberg, D. A. & Holekamp, K. E. 2005. Patterns of alliance formation and post-conflict aggression indicate spotted hyaenas recognize third party relationships. *Animal Behaviour*, 69.

Evans, S. & Poole, T. B. 1983. Pair-bond formation and breeding success in the common marmoset *Callithrix jacchus jacchus*. *International Journal of Primatology*, 4, 83–97.

Fairbanks, L. A. 1993. Juvenile vervet monkeys: Establishing relationships and practising skills for the future. In *Juvenile Primates: Life History, Development and Behavior* (ed. by Pereira, M. E. & Fairbanks, L. A.), 211–227. Oxford: Oxford University Press.

Fairbanks, L. A. & McGuire, M. T. 1987. Mother-infant relationships in vervet monkeys: Response to new adult males. *International Journal of Primatology*, 8, 351–366.

Fessler, D.M.T. & Gervais, M. 2010. From whence the captains of our lives: Ultimate and phylogenetic perspectives on emotions in humans and other primates. In *Mind the Gap: The Origins of Human Universals* (ed. By Kappeler, P.M. & Silk, J.B.), 261–280. Berlin: Springer.

Flack, J. C., Krakauer, D. C. & de Waal, F. B. M. 2005. Robustness mechanisms in primate societies: A perturbation study. *Proceedings of the Royal Society of London B Biological Sciences*, 272, 1091–1099.

Fraser, O. N. & Aureli, F. 2008. Reconciliation, consolation and postconflict behavioral specificity in chimpanzees. *American Journal of Primatology*, 70, 1114–1123.

Fraser, O. N. & Bugnyar, T. 2010a. The quality of social relationships in ravens. *Animal Behaviour*, 79, 927–933.

———. 2010b Do ravens show consolation? Responses to distressed others. *PLoS ONE*, 5, e10605.

———. 2011. Ravens reconcile after aggressive conflicts with valuable partners. *PLoS ONE*, 6, e18118.

Fraser, O. N., Koski, S. E., Wittig, R. M. & Aureli, F. 2009. Why are bystanders friendly to recipients of aggression? *Communicative and Integrative Biology*, 2, 285–291.

Fraser, O. N., Schino, G. & Aureli, F. 2008a. Components of relationship quality in chimpanzees. *Ethology*, 114, 834–843.

Fraser, O. N., Stahl, D. & Aureli, F. 2008b. Stress reduction through consolation in chimpanzees. *Proceedings of the Proceedings of the National Academy of Sciences of the United States of America*, 105, 8557–8562.

———. 2010. Function and determinants of reconciliation in *Pan troglodytes*. *International Journal of Primatology* 31: 39–57.

Fuentes, A., Malone, N., Sanz, C., Matheson, M. & Vaughan, L. 2002. Conflict and post-conflict behaviour in a small group of chimpanzees. *Primates*, 43, 223–235.

Gilby, I. C. & Wrangham, R. W. 2008. Association patterns among wild chimpanzees (*Pan troglodytes schweinfurthii*) reflect sex differences in cooperation. *Behavioral Ecology and Sociobiology*, 62, 1831–1842.

Goodall, J. 1986. *The Chimpanzees of Gombe: Patterns of Behavior*. Cambridge, MA: Harvard University Press.

Gruter, C. C. 2004. Conflict and postconflict behaviour in captive black and white snub-nosed monkeys (*Rhinopithecus bieti*). *Primates*, 45, 197–200.

Gumert, M. D. 2007a. Grooming and infant handling interchange in *Macaca fascicularis*: the relationship between infant supply and grooming payment. *International Journal of Primatology*, 28, 1059–1074.

———. 2007b. Payment for sex in a macaque mating market. *Animal Behaviour*, 74, 1655.

Gust, D. A. & Gordon, T. P. 1993. Conflict resolution in sooty mangabeys. *Animal Behaviour*, 46, 685–694.

Harcourt, A. H. 1988. Alliances in contests and social intelligence. In *Machiavellian Intelligence* (ed. by Byrne, R. W. & Whiten, A.), 132–152. Oxford: Oxford University Press.

Heinrich, B. & Bugnyar, T. 2007. Just how smart are ravens? *Scientific American*, 296, 64–71.

Hemelrijk, C. K. 1994. Support for being groomed in long-tailed macaques, *Macaca fascicularis*. *Animal Behaviour*, 48, 479–481.

Henzi, S. P. & Barrett, L. 2002. Infants as a commodity in a baboon market. *Animal Behaviour*, 63, 915–921.

———. 2007. Coexistence in female-bonded primate groups. *Advances in the Study of Behavior*, 37, 43–81.

Henzi, S. P. & Lucas, J. W. 1980. Observations on the intertroop movement of adult vervet monkeys (*Cercopithecus aethiops*). *Folia Primatologica*, 33, 220–235.

Henzi, S.P. Lusseau, D., Weingrill, T., van Schaik, C.P. & Barrett, L. 2009. Cyclicity in the structure of female baboon social networks. *Behavioural Ecology and Sociobiology* 63: 1015–1021.

Hinde, R. A. 1972. Concepts of emotion. *Ciba Foundation Symposia*, 8, 3–13.

———. 1976. Interactions, relationships and social structure. *Man*, 11, 1–17.

———. 1979. *Towards Understanding Relationships*. London: Academic Press.

———. 1983. A conceptual framework. In *Primate Social Relationships: An Integrated Approach* (ed. by Hinde, R. A.), 1–7. Oxford, U.K.: Blackwell Scientific Publications.

———. 1987. Can nonhuman primates help us understand human behavior? In *Primate Societies* (ed. by Smuts, B. B., Cheney, D. L., Seyfarth, R., Wrangham, R. W. & Struhsaker, T. T.), 413–420. Chicago: University of Chicago Press.

Hohmann, G. & Fruth, B. 2000. Use and function of genital contacts among female bonobos. *Animal Behaviour*, 60, 107–120.

Holekamp, K. E., Sakai, S. T. & Lundrigan, B. L. 2007. Social intelligence in the spotted hyena (*Crocuta crocuta*). *Philosophical Transactions of the Royal Society B: Biological Sciences*, 362, 523–38.

Holobinko, A. & Waring, G. H. 2009. Conflict and reconciliation behavior trends of the bottlenose dolphin (*Tursiops truncatus*). *Zoo Biology*, 28, 1–19.

Janson C. 1988. Food competition in brown capuchin monkeys (*Cebus apella*): Quantitative effects of group size and tree productivity. *Behaviour*, 105, 53–76.

Judge, P. G. 1991. Dyadic and triadic reconciliation in pigtail macaques (*Macaca nemestrina*). *American Journal of Primatology*, 23, 225–237.

———. 2000. Coping with crowded conditions. In *Natural Conflict Resolution* (ed. by Aureli, F. & de Waal, F. B. M.), 129–154. Berkeley: University of California Press.

Judge, P. G. & Mullen, S. H. 2005. Quadratic postconflict affiliation among bystanders in a hamadryas baboon group. *Animal Behaviour*, 69, 1345–1355.

Kappeler, P. M. 1993. Reconciliation and post-conflict behaviour in ring-tailed lemurs, *Lemur catta*, and redfronted lemur, *Eulemur fulvus rufus*. *Animal Behaviour*, 45, 905–915.

Kappeler, P. M. & van Schaik, C. P. 1992. Methodological and evolutionary aspects of reconciliation among primates. *Ethology*, 92, 51–69.

Kazem, A. J. N. & Aureli, F. 2005. Redirection of aggression: Multiparty signalling within a network? In: *Animal Communication Networks* (ed. by McGregor, P. K.), 191–218. Cambridge: Cambridge University Press.

Kleiman, D. G. 1977. Monogamy in mammals. *Quarterly Review of Biology*, 52, 39–69.

Koski, S. E., Koops, K. & Sterck, E. H. M. 2007. Reconciliation, relationship quality, and postconflict anxiety: Testing the integrated hypothesis in captive chimpanzees. *American Journal of Primatology*, 69, 158–172.

Koski, S. E. & Sterck, E. H. 2009. Post-conflict third-party affiliation in chimpanzees: What's in it for the third party? *American Journal of Primatology*.71, 409–418.

Koski, S. E. & Sterck, E. H. M. 2007. Triadic postconflict affiliation in captive chimpanzees: Does consolation console? *Animal Behaviour*, 73, 133–142.

Koyama, N. F. 2000. Conflict prevention before feeding. In *Natural Conflict Resolution* (ed. by Aureli, F. & de Waal, F. B. M.), 130–132. Berkeley: University of California Press.

———. 2001. The long-term effects of reconciliation in Japanese macaques *Macaca fuscata*. *Ethology*, 107, 975–987.

Koyama, N. F. & Dunbar, R. I. M. 1996. Anticipation of conflict by chimpanzees. *Primates*, 37, 79–86.

Kudo, H. & Dunbar, R. I. M. 2001. Neocortex size and social network size in primates. *Animal Behaviour*, 62, 711–722.

Kuester, J. & Paul, A. 2000. The use of infants to buffer male aggression. In *Natural Conflict Resolution* (ed. by Aureli, F. & de Waal, F. B. M.), 91–93. Berkeley: University of California Press.

Kummer, H. 1971. *Primate Societies: Group Techniques of Ecological Adaptation*. Arlington Heights, IL: AHM Publishing Corporation.

———. 1975. Rules of dyad and group formation among captive baboons (*Theropithecus gelada*). In *Proceedings of the 5th Congress of the International Primatological Society*, 129–160. Basel: Karger.

———. 1978. On the value of social relationships to nonhuman primates: A heuristic scheme. *Social Science Information*, 17, 697–705.

———. 1995. *In Quest of the Sacred Baboon*. Princeton, NJ: Princeton University Press.

Kummer, H. & Cords, M. 1991. Cues of ownership in long-tailed macaques, *Macaca fascicularis*. *Animal Behaviour*, 42, 529–549.

Kutsukake, N. & Castles, D. L. 2001. Reconciliation and variation in post-conflict stress in Japanese macaques (*Macaca fuscata fuscata*): testing the integrated hypothesis. *Animal Cognition*, 4, 259–268.

Kutsukake, N. & Castles, D. L. 2004. Reconciliation and postconflict third-party affiliation among wild chimpanzees in the Mahale Mountains, Tanzania. *Primates*, 45, 157–165.

Kutsukake, N. & Clutton-Brock, T. H. 2008. Do meerkats engage in conflict management following aggression? Reconciliation, submission and avoidance. *Animal Behaviour*, 75, 1441–1453.

Kutsukake, N., Suetsugu, N. & Hasegawa, T. 2006. Pattern, distribution, and function of greeting behaviour among black-and-white colobus. *International Journal of Primatology*, 27, 1271–1291.

Leca, J. B., Fornasieri, I. & Petit, O. 2002. Aggression and reconciliation in *Cebus capucinus*. *International Journal of Primatology*, 23, 979–998.

Lusseau, D. 2007. Why are male social relationships complex in the Doubtful Sound bottlenose dolphin population? *PLoS ONE*, 2, e348.

Madden, J. R. & Clutton-Brock, T. H. 2009. Manipulating grooming by decreasing ectoparasite load causes unpredicted changes in antagonism. *Proceedings of the Royal Society B: Biological Sciences*, 276, 1263–1268.

Maestripieri, D. 2005. Gestural communication in three species of macaques (*Macaca mulatta*, *M. nemestrina*, *M. arctoides*): Use of signals in relation to dominance and social context. *Gesture*, 5, 57–73.

Maestripieri, D., Schino, G., Aureli, F. & Troisi, A. 1992. A modest proposal: Displacement activities as an indicator of emotions in primates. *Animal Behaviour*, 44, 967–979.

Majolo, B., Ventura, R. & Koyama, N. 2009. A statistical modelling approach to the occurrence and timing of reconciliation in wild Japanese macaques. *Ethology*, 115, 152–166.

Majolo, B., Ventura, R. & Koyama, N. F. 2005. Postconflict behaviour among male Japanese macaques. *International Journal of Primatology*, 26, 321–336.

Majolo, B., Ventura, R. & Schino, G. 2010. Asymmetry and dimensions of relationship quality in the Japanese macaque (*Macaca fuscata yakui*). *International Journal of Primatology*, 31, 736–750.

Mallavarapu, S., Stoinski, T. S., Bloomsmith, M. A. & Maple, T. L. 2006. Postconflict behavior in captive western lowland gorillas (*Gorilla gorilla gorilla*). *American Journal of Primatology*, 68, 789–801.

Manson, J. H., Perry, S. & Stahl, D. 2005. Reconciliation in wild white-faced capuchins (*Cebus capucinus*). *American Journal of Primatology*, 65, 205–219.

Massen, J. J. M., Sterck, E. H. M. & de Vos H. 2010. Close social associations in animals and humans: Functions and mechanisms of friendship. *Behaviour*, 147, 1379–1412.

Matsumura, S. 1996. Post-conflict contacts between former opponents among wild moor macaques (*Macaca maurus*). *American Journal of Primatology*, 38, 211–219.

Maxim, P. E. 1976. An interval scale for studying and quantifying social relations in pairs of rhesus monkeys. *Journal of Experimental Psychology*, 195, 123–147.

Mayagoitia, L., Santillan-Doherty, A. M., Lopez-Vergara, L. & Mondragon-Ceballos, R. 1993. Affiliation tactics prior to a period of competition in captive groups of stumptail macaques. *Ethology, Ecology and Evolution*, 5, 435–446.

Meishvili, N., Chalyan, V. & Butovskaya, M. 2005. Studies of reconciliation in anubis baboons. *Neuroscience and Behavioral Physiology*, 35, 913–916.

Mendoza, S. P. 1993. Social conflict on first encounters. In *Primate Social Conflict* (ed. by Mason, W. A. & Mendoza, S. P.), 85–110. Albany: State University of New York Press.

Mitani, J. C. 2009. Male chimpanzees form enduring and equitable social bonds. *Animal Behaviour*, 77, 633–640.

Mitani, J. C., Merriwether, D. A. & Zhang, C. 2000. Male affiliation, cooperation and kinship in wild chimpanzees. *Animal Behaviour*, 59, 885–893.

Mondragón-Ceballos, R. 2001. Interfering in affiliations: Sabotaging by stumptailed macaques, *Macaca arctoides*. *Animal Behaviour*, 62, 1179–1187.

Moss, C. J. & Croze, H. & Lee, P. C. 2011. *The Amboseli Elephant: A Long-Term Perspective on a Long-Lived Species*. Chicago: University of Chicago Press.

Muller, M. N. 2002. Agonistic relations among Kanyawara chimpanzees. In *Behavioral Diversity in Chimpanzees and Bonobos* (ed. by Boesch, C., Hohmann, G. & Marchant, L.), 112–124. Cambridge: Cambridge University Press.

Nishida, T., Kano, T., Goodall, J., McGrew, W. C. & Nakamura, M. 1999. Ethogram and ethnography of Mahale chimpanzees. *Anthropological Science*, 107, 141–188.

Noë, R., de Waal, F. B. M. & van Hooff, J. A. R. A. M. 1980. Types of dominance in a chimpanzee colony. *Folia Primatologica*, 34, 90–110.

Nunn, C. L. & Barton, R. A. 2001. Comparative methods for studying primate adaptation. *Evolutionary Anthropology*, 10, 81–98.

Okamoto, K., Agetsuma, N. & Kojima, S. 2001. Greeting behaviour during party encounters in captive chimpanzees. *Primates*, 42, 161–165.

Palagi, E., Chiarugi, E. & Cordoni, G. 2008. Peaceful postconflict interactions between aggressors and bystanders in captive lowland gorillas (*Gorilla gorilla gorilla*). *American Journal of Primatology*, 70, 949–55.

Palagi, E. & Cordoni, G. 2009. Postconflict third-party affiliation in *Canis lupus*: Do wolves share similarities with the great apes? *Animal Behaviour*, 78, 979–986.

Palagi, E., Cordoni, G. & Borgognini Tarli, S. 2006. Possible roles of consolation in captive chimpanzees (*Pan troglodytes*). *American Journal of Physical Anthropology*, 129, 105–111.

Palagi, E., Paoli, T. & Borgognini Tarli, S. M. 2004. Reconciliation and consolation in captive bonobos (*Pan paniscus*). *American Journal of Primatology*, 62, 15–30.

———. 2005. Aggression and reconciliation in two captive groups of *Lemur catta*. *International Journal of Primatology*, 26, 279–294.

Palagi, E., Paoli, T. & Borgognini Tarli, S. 2006. Short-term benefits of play behavior and conflict prevention in *Pan paniscus*. *International Journal of Primatology*, 27, 1257–1270.

Peñate, L., Peláez, F. & Sánchez, S. 2009. Reconciliation in captive cotton-top tamarins (*Saguinus oedipus*), a cooperative breeding primate. *American Journal of Primatology*, 71, 895–900.

Pereira, M., Schill, J. & Charles, E. 2000. Reconciliation in captive Guyanese squirrel monkeys (*Saimiri sciureus*). *American Journal of Primatology*, 50, 159–167.

Perez-Ruiz, A. L. & Mondragon-Ceballos, R. 1994. Rates of reconciliatory behaviors in stumptail macaques: effects of age, sex, rank and kinship. In *Current Primatology. Vol. II Social Development, Learning and Behaviour* (ed. by Roeder, J. J., Thierry, B., Anderson, J. R. & Herrenschmidt, N.), 147–155. Strasbourg: Presses de l' Universite Louis Pasteur.

Perry, S. 1996. Female-female social relationships in wild white-faced capuchin monkeys, *Cebus capucinus*. *American Journal of Primatology*, 40, 167–182.

Petit, O., Abegg, C. & Thierry, B. 1997. A comparative study of aggression and conciliation in three cercopithecine monkeys (*Macaca fuscata, Macaca nigra, Papio papio*). *Behaviour*, 134, 415–432.

Petit, O. & Thierry, B. 1994. Reconciliation in a group of Guinea baboons. In *Current Primatology. Vol. II Social Development, Learning and Behaviour* (ed. by Roeder, J. J., Thierry, B., Anderson, J. R. & Herrenschmidt, N.), 137–145. Strasbourg: Presses de l' Universite Louis Pasteur.

———. 2000. Do impartial interventions in conflicts occur in monkeys and apes? In *Natural Conflict Resolution* (ed. by Aureli, F. & de Waal, F. B. M.), 267–269. Berkeley: University of California Press.

Polizzi di Sorrentino E., Schino G., Visalberghi E., Aureli F. 2010. What time is it? Coping with expected feeding time in capuchin monkeys. *Animal Behaviour*, 80, 117–1123.

Preuschoft, S. & van Schaik, C. P. 2000. Dominance and communication: Conflict management in various social settings. In *Natural Conflict Resolution* (ed. by Aureli, F. & de Waal, F. B. M.), 77–105. Berkeley: University of California Press.

Preuschoft, S., Wang, X., F., A. & de Waal, F. B. M. 2002. Reconciliation in captive chimpanzees: A reevaluation with controlled methods. *International Journal of Primatology*, 23, 29–50.

Puga-Gonzalez, I., Hildenbrandt, H. & Hemelrijk C. K. 2009. Emergent patterns of social affiliation in primates, a model. *PLoS Computational Biology* 5(12): e1000630.

Pusey, A., Murray, C., Wallauer, W., Wilson, M., Wroblewski, E. & Goodall, J. 2008. Severe aggression among female *Pan troglodytes schweinfurthii* at Gombe National Park, Tanzania. *International Journal of Primatology*, 29, 949–973.

Pusey, A. E. & Packer, C. 1987. Dispersal and philopatry. In *Primate Societies* (ed. by Smuts, B. B., Cheney, D. L., Seyfarth, R. M., Wrangham, R. W. & Struhsaker, T. T.), 250–266. Chicago: Chicago University Press.

Rebecchini L., Schaffner C. M. & Aureli F. 2011. Risk is a component of social relationships in spider monkeys. *Ethology*, 117, 691–699.

Ren, R., Yan, K., Su, Y., Qi, H., Liang, B., Bao, W. & de Waal, F. B. M. 1991. The reconciliation behavior of golden monkeys (*Rhinopithecus roxellanae roxellanae*) in small breeding groups. *Primates*, 32, 321–327.

Roeder, J. J., Fornasieri, I. & Gosset, D. 2002. Conflict and postconflict behaviour in two lemur species with different social organizations (*Eulemur fulvus* and *Eulemur macaco*): A study on captive groups. *Aggressive Behavior*, 28, 62–74.

Rolland, N. & Roeder, J. J. 2000. Do ringtailed lemurs (*Lemur catta*) reconcile in the hour post-conflict? A pilot study. *Primates*, 41, 223–227.

Romero, T. & Aureli, F. 2008. Reciprocity of support in coatis (*Nasua nasua*). *Journal of Comparative Psychology*, 122, 19–25.

Romero, T., Colmenares, F. & Aureli, F. 2008. Postconflict affiliation of aggressors in *Papio hamadryas*. *International Journal of Primatology*, 29, 1951–1606.

———. 2009. Testing the function of reconciliation and third-party affiliation for aggressors in hamadryas baboons (*Papio hamadryas hamadryas*). *American Journal of Primatology*, 71, 60–69.

Romero, T. & de Waal, F. B. M. 2010. Chimpanzee (*Pan troglodytes*) consolation: Third party identity as a window on possible function. *Journal of Comparative Psychology*, 124, 278–286.

Rowell, T. E. 2000. The ethological approach precluded recognition of reconciliation. In *Natural Conflict Resolution* (ed. by Aureli, F. & de Waal, F. B. M.), 227–228. Berkeley: University of California Press.

Ruiz, J. C. 1990. Comparison of affiliative behaviors between old and recently established pairs of golden lion tamarin, *Leontopithecus rosalia*. *Primates*, 31, 197–204.

Samuels, A. & Flaherty, C. 2000. Peaceful conflict resolution in the sea. In *Natural Conflict Resolution* (ed. by Aureli, F. & de Waal, F. B. M.), 229–231. Berkeley: University of California Press.

Savage, A., Ziegler, T. E. & Snowdon, C. T. 1988. Sociosexual development, pair bond formation, and mechanisms of fertility suppression in female cotton-top tamarins (*Saguinus oedipus oedipus*). *American Journal of Primatology*, 14, 345–359.

Schaffner, C. M. & Aureli, F. 2005. Embraces and grooming in captive spider monkeys. *International Journal of Primatology*, 26, 1093–1106.

Schaffner, C. M., Aureli, F. & Caine, N. G. 2005. Following the rules: Why small groups of tamarins do not reconcile conflicts. *Folia Primatologica*, 76, 67–76.

Schaffner, C. M. & Caine, N. G. 2000. The peacefulness of co-operatively breeding primates. In *Natural Conflict Resolution* (ed. by Aureli, F. & de Waal, F. B. M.), 155–169. Berkeley: University of California Press.

Schaffner, C. M. & French, J. A. 2004. Behavioral and endocrine responses in male marmosets to the establishment of multimale breeding groups: Evidence for non-monopolizing facultative polyandry. *International Journal of Primatology*, 25, 709–732.

Schaffner, C. M., Shepherd, R. E., Santos, C. V. & French, J. A. 1995. Development of heterosexual relationships in Wied's black tufted-ear marmosets (*Callithrix kuhli*). *American Journal of Primatology*, 36, 185–200.

Schino, G. 1998. Reconciliation in domestic goats. *Behaviour*, 135, 343–356.

———. 2000. Beyond the primates: Expanding the reconciliation horizon. In *Natural Conflict Resolution* (ed. by Aureli, F. & de Waal, F. B. M.), 225–242. Berkeley: University of California Press.

———. 2007. Grooming and agonistic support: A meta-analysis of primate reciprocal altruism. *Behavioral Ecology*, 18, 115–120.

Schino, G. & Aureli, F. 2009. Reciprocal altruism in primates: Partner choice, cognition and emotions. *Advances in the Study of Behavior*, 39, 45–69.

Schino, G., D'Amato, F. R. & Troisi, A. 1995. Mother-infant relationships in Japanese macaques: Sources of interindividual variation. *Animal Behaviour*, 49, 151–158.

Schino, G., Maestripieri, D., Scucchi, S. & Turilazzi, P. G. 1990. Social tension in familiar and unfamiliar pairs of long-tailed macaques. *Behaviour*, 113, 264–272.

Schino, G., Perretta, G., Taglioni, A., Monaco, V. & Troisi, A. 1996. Primate displacement activities as an ethopharmacological model of anxiety. *Anxiety*, 2, 186–191.

Schino, G., Rosati, L. & Aureli, F. 1998. Intragroup variation in conciliatory tendencies in captive Japanese macaques. *Behaviour*, 135, 897–912.

Schino, G., Scucchi, S., Maestripieri, D. & Turillazzi, P. G. 1988. Allogrooming as a tension-reduction mechanism: A behavioral approach. *American Journal of Primatology*, 16, 43–50.

Schino, G., Ventura, R. & Troisi, A. 2003. Grooming among female Japanese macaques: Distinguishing between reciprocation and interchange. *Behavioral Ecology*, 14, 887–891.

Seed, A. M., Clayton, N. S. & Emery, N. J. 2007. Postconflict third-party affiliation in rooks, *Corvus frugilegus*. *Current Biology*, 17, 152–158.

Seyfarth, R. M. 1980. The distribution of grooming and related behaviours among adult female vervet monkeys. *Animal Behaviour*, 28, 798–813.

Seyfarth, R. M. & Cheney D. L. 2012. The evolutionary origin of friendship. *Annual Review of Psychology*, 63, 153–177.

Silk, J. B. 1997. The function of peaceful post-conflict contacts among primates. *Primates*, 38, 265–279.

———. 1992. The patterning of intervention among male bonnet macaques: Reciprocity, revenge, and loyalty. *Current Anthropology*, 33, 318–325.

———. 2000. The function of peaceful post-conflict interactions: An alternate view. In *Natural Conflict Resolution* (ed. by Aureli, F. & de Waal, F. B. M.), 179–181. Berkeley: University of California Press.

———. 2002a. The form and function of reconciliation in primates. *Annual Review of Anthropology*, 31, 21–44.

———. 2002b. Using the "F"-word in primatology. *Behaviour*, 139, 421–446.

Silk, J. B., Alberts, S. C. & Altmann, J. 2003. Social bonds of female baboons enhance infant survival. *Science*, 302, 1231–4.

———. 2006a. Social relationships among adult female baboons (*Papio cynocephalus*). II. Variation in the quality and stability of social bonds. *Behavioral Ecology and Sociobiology*, 61, 197–204.

Silk, J. B., Altmann, J. & Alberts, S. C. 2006b. Social relationships among adult female baboons (*Papio cynocephalus*). I. Variation in the strength of social bonds. *Behavioral Ecology and Sociobiology*, 61, 183–195.

Silk, J.B., Beehner, J.C., Bergman, T.J., Crockford, C., Engh, A.L., Moscovice, L.R., Wittig, R.M., Seyfarth, R.M. & Cheney, D.L. 2009. The benefits of social capital: Close social bonds among female baboons enhance offspring survival. *Proceedings of the Royal Society of London, Series B*, 276, 3099–3104.

Silk, J., Cheney, D. & Seyfarth, R. 1996. The form and function of post-conflict interactions between female baboons. *Animal Behaviour*, 52, 259–268.

Silk, J. B., Kaldor, E. & Boyd, R. 2000. Cheap talk when interests conflict. *Animal Behaviour*, 59, 423–432.

Simpson, M. J. A. & Howe, S. 1980. The interpretation of individual differences in the behaviour of rhesus monkey infants. *Behaviour*, 72, 127–155.

Slater, K. Y., Schaffner, C. M. & Aureli, F. 2007. Embraces for infant handling in spider monkeys: Evidence for a biological market? *Animal Behaviour*, 74, 455–461.

Slater, K. Y., Schaffner, C. M. & Aureli, F. 2008. Female-directed male aggression in wild spider monkeys (*Ateles geoffroyi*). *International Journal of Primatology*, 29, 1657–1669.

Smith, J. E., Kolowski, J. M., Graham, K. E., Dawes, S. E. & Holekamp, K. E. 2008. Social and ecological determinants of fission-fusion dynamics in the spotted hyaena. *Animal Behaviour*, 76, 619–636.

Soltis, J. 1999. Measuring male-female relationships during the mating season in wild Japanese macaques (*Macaca fuscata yakui*). *Primates*, 40, 453–467.

Sommer, V., Denham, A. & Little, K. 2002. Postconflict behaviour of wild Indian langur monkeys: Avoidance of opponents but rarely affinity. *Animal Behaviour*, 63, 637.

Stammbach, E. 1978. On social differentiation in groups of captive female hamadryas baboons. *Behaviour*, 67, 322–338.

Sterck, E. H. M. & Steenbeek, R. 1997. Female dominance relationships and food competition in the sympatric Thomas langur and long-tailed macaque. *Behaviour*, 134, 769–774.

Swedell, L. 1997. Patterns of reconciliation among captive gelada baboons (*Theropithecus gelada*): A brief report. *Primates*, 38, 325–330.

Tanaka, I. 1980. Variability in the development of mother-infant relationships among free-ranging Japanese macaques. *Primates*, 30, 477–491.

Thierry, B., Aureli, F., Nunn, C. L., Petit, O., Abegg, C. & de Waal, F. B. M. 2008. A comparative study of conflict resolution in macaques: Insights into the nature of trait co-variation. *Animal Behaviour*, 75, 847–860.

Troisi, A. 2002. Displacement activities as a behavioural measure of stress in nonhuman primates and human subjects. *Stress*, 5, 47–54.

Van Schaik, C. P. & Aureli, F. 2000. The natural history of valuable relationships in primates. In *Natural Conflict Resolution* (ed. by Aureli, F. & de Waal, F. B. M.), 307–333. Berkeley: University of California Press.

Veenema, H. C., Das, M. & Aureli, F. 1994. Methodological improvements for the study of reconciliation. *Behavioural Processes*, 31, 29–38.

Ventura, R., Majolo, B., Koyama, N., Hardie, S. & Schino, G. 2006. Reciprocation and interchange in wild Japanese macaques: Grooming, cofeeding, and agonistic support. *American Journal of Primatology*, 68, 1138–1149.

Verbeek, P. & de Waal, F. B. M. 1997. Postconflict behavior of captive brown capuchins in the presence and absence of attractive food. *International Journal of Primatology*, 18, 703–725.

Wahaj, S. A., Guse, K. R. & Holekamp, K. E. 2001. Reconciliation in the spotted hyena (*Crocuta crocuta*). *Ethology*, 107, 1057–1074.

Watts, D. P. 1995. Post-conflict social events in wild mountain gorillas (Mammalia, Hominoidea). I. Social interactions between opponents. *Ethology*, 100, 139–157.

———. 2006. Conflict resolution in chimpanzees and the valuable-relationships hypothesis. *International Journal of Primatology*, 27, 1337–1364.

Watts, D. P., Colmenares, F. & Arnold, K. 2000. Redirection, consolation and male policing: How targets of aggression interact with bystanders. In *Natural Conflict Resolution* (ed. by Aureli, F. & de Waal, F. B. M.), 281–301. Berkeley: University of California Press.

Weaver, A. 2003. Conflict and reconciliation in captive bottlenose dolphins, *Tursiops truncatus*. *Marine Mammal Science*, 19, 836–846.

Weaver, A. C. & de Waal, F. B. M. 2003. The mother-offspring relationship as a template in social development: Reconciliation in captive brown capuchins (*Cebus apella*). *Journal of Comparative Psychology*, 117, 101–110.

Welker, C., Lührmann, B. & Meinel, W. 1980. Behavioural sequences and strategies of female crab-eating monkeys, *Macaca fascicularis* during group formation studies. *Behaviour*, 68, 108–126.

Westlund, K., Ljungberg, T., Borefelt, U. & Abrahamsson, C. 2000. Post-conflict affiliation in common marmosets (*Callithrix jacchus jacchus*). *American Journal of Primatology*, 52, 31–46.

Whitehead, H. 2008. *Analyzing Animal Societies: Quantitative Methods for Vertebrate Social Analysis*. Chicago: University of Chicago Press.

Whitham, J. C. & Maestripieri, D. 2003. Primate rituals: The function of greetings between male Guinea baboons. *Ethology*, 109, 847–859.

Wittig, R. M. 2010. Function and cognitive underpinning of post-conflict affiliation in wild chimpanzees. In *The Mind of the Chimpanzee: Ecological and Experimental Perspectives* (ed. by Lonsdorf, A. V., Ross, S. R. & Matsuzawa, T.), 208–219. Chicago: University of Chicago Press.

Wittig, R. M. & Boesch, C. 2003a. "Decision-making" in conflicts of wild chimpanzees (*Pan troglodytes*): An extension of the Relational Model. *Behavioral Ecology and Sociobiology*, 54, 491–504.

———. 2003b. The choice of post-conflict interactions in wild chimpanzees (*Pan troglodytes*). *Behaviour*, 140, 1527–1559.

———. 2005. How to repair relationships: Reconciliation in wild chimpanzees (*Pan troglodytes*). *Ethology*, 111, 736–763.

Wittig, R. M., Crockford, C., Wikberg, E., Seyfarth, R. M. & Cheney, D. L. 2007. Kin-mediated reconciliation substitutes for direct reconciliation in female baboons. *Proceedings of the Royal Society B: Biological Sciences*, 274, 1109–1115.

Wrangham, R. W. 1983. Social relationships in comparative perspective. In *Primate Social Relationships: An Integrated Approach* (ed. by Hinde, R. A.), 325–334. Oxford: Blackwell Scientific Publications.

York, A. D. & Rowell, T. E. 1988. Reconciliation following aggression in patas monkeys, *Erythrocebus patas*. *Animal Behaviour*, 36, 502.

Zhang, J., Dapeng, Z. & Baoguo, L. 2010. Postconflict behavior among female *Rhinopithecus roxellana* within one-male units in the Qinling Mountains, China. *Current Zoology*, 56, 222–226.

Chapter 24 The Adaptive Value of Sociality

Joan B. Silk

PRIMATES ARE social creatures. All of the diurnal primates, except orangutans (*Pongo* spp.), live in stable social groups and interact regularly with familiar conspecifics. Over evolutionary time, sociality has left its mark on primate morphology and behavior. For example, the extent of sexual dimorphism, the intensity of male-male competition, and the risk of infanticide are more pronounced in species that form one-male groups than in pair-bonded species (chapters 17 and 19, this volume). Comparative analyses also indicate that the size of social groups is positively correlated with brain size, the size of grooming cliques, the prevalence of deception, and the incidence of innovation and social learning (chapter 28, this volume). These correlations suggest that selection has shaped the behavior and morphology of individuals in response to the challenges and opportunities presented by life in social groups. These correlations also imply that sociality has had important adaptive consequences for individuals.

We generally operate on the assumption that variation in social behavior has fitness consequences. For example, when we invoke kin selection to explain why strong kin biases emerge among members of the philopatric sex (chapter 21, this volume), we rely on the premise that social interactions have adaptive consequences for participants. This implies that the short-term benefits that individuals gain each time that they are groomed, embraced, supported in conflicts, or warned that a predator is near incrementally improve their health or their survival and this enhances their lifetime fitness. Similarly, the short-term costs that individuals incur when their grooming solicitations are ignored, they are sup-

planted from a feeding site, or they are defeated in an aggressive conflict will incrementally weaken their health and survival and therefore decrease their lifetime fitness.

Some kinds of social interactions, such as winning a decisive fight over access to a group of females, have dramatic and immediate effects on mating success. We know that the ability to establish high-ranking dominance positions within their groups enhances the reproductive success of males (chapter 18, this volume) and females (chapter 15, this volume) in a wide range of primate species. Thus, we can make a reasonably convincing argument that evolution has favored behavioral strategies for establishing and maintaining high-ranking positions in the dominance hierarchy in many primate species because high rank enhances fitness.

It is much more difficult to determine the impact of other kinds of social interactions or social relationships on individual fitness. For example, when a female monkey is groomed she is cleansed of dirt and debris, her wounds are cleaned, her heart rate may drop, and beta-endorphins may be released (chapter 23, this volume). All of these effects seem to be beneficial for the recipient, but females participate in thousands of grooming bouts with dozens of partners over the course of their lifetimes (fig. 24.1). It is very hard to translate the short-term benefits that individuals gain from being groomed, or from engaging in other kinds of affiliative interactions, into long-term differences in fitness. These difficulties originally prompted Cheney et al. (1986) to write, "Although many studies of insects, birds, and mammals have documented the functional significance

Fig. 24.1. One adult female baboon grooms another at the Moremi Reserve in the Okavango Delta of Botswana. Photo courtesy of Joan B. Silk.

of single interactions such as fights, the reproductive benefits of long-term social bonds are less immediately obvious."

Now, 25 years after that comment was made, we are beginning to gain some traction on this problem. The goal of this chapter is to review what we know about the nature of social bonds and the fitness consequences of sociality. I begin by examining the effects of group size on female reproductive success, and then consider how variation in the composition of social groups affects the fitness of individuals. Next, I examine the nature of dyadic social relationships using data derived from several long-term studies. Finally, I review evidence which suggests that sociality has short-term effects on the welfare of individuals and long-term effects on their reproductive success.

Effects of Group Size

Clutton-Brock and his colleagues (2001) observed that female fitness is generally positively related to group size in species, such as meerkats and lions, in which females rely on helpers to rear their young, but is negatively related to group size in species, like red deer and badgers, in which females rear their young with little or no help from others. These correlations reflect the fact that the size of social groups reflects a compromise between the costs and benefits of grouping (chapter 9, this volume). Living in larger groups may enhance success in intergroup encounters, reduce vulnerability to predators, or increase the availability of helpers. On the other hand, the costs of competition for access to food, sleeping sites, or mates may increase as groups

grow in size. In some primate species there is substantial intragroup competition over access to food. Large groups may also be more attractive targets for male takeover attempts; for species in which male takeover attempts are associated with infanticide, large group size may increase infant mortality rates (Crockett & Janson 2000).

Although all social animals face these conflicting pressures, their responses are influenced by their feeding ecology, level of predation pressure, reproductive systems, and dispersal strategies. Some species respond to competition by dividing into temporary foraging parties and forming fission-fusion groups; others respond by adjusting their activity budgets or altering their food preferences to compensate for the increased costs of intra-group competition (Steenbeek & van Schaik 2001; Borries et al. 2008). If predation pressure is high, then it may be risky for animals to break into small foraging parties. In these and many other species, as groups grow in size, females must travel farther each day because larger groups deplete food patches more rapidly or require larger areas in which to search for food (Janson & Goldsmith 1995; Chapman & Chapman 2000). Comparative studies have shown that as group size increases, daily travel distance also increases (Isbell 1991; Wrangham et al. 1993). Thus, females who live in large groups expend more energy on travel than females who live in smaller groups, and this reduces the amount of energy they can devote to reproduction. Females may also experience higher levels of competition over access to resources or experience higher levels of stress when groups are large (Altmann & Alberts 2003a, b), and this may depress their fertility. Female ring-tailed lemurs (*Lemur catta*), which normally live in multimale-multifemale groups that range in size from about 5 to 25 individuals, had higher cortisol levels when they lived in groups that were larger or smaller than the mean size of groups in the population (Pride 2005). At the same time, however, females in large groups may be more successful in protecting themselves and their infants from male harassment. Female Hanuman langurs (*Semnopithecus entellus*) living in large groups were better able to cope with the threat from extragroup males that took over groups, killed infants, and expelled juveniles from groups (Treves & Chapman 1996).

The complexity of responses to ecological pressures is reflected in the variability of the relationship between group size and reproductive parameters across primates. In a mixed population of yellow (*Papio cynocephalus*) and olive baboons (*Papio anubis*) in the Amboseli basin, the length of interbirth intervals increased as group size increased, and this depressed female fitness (Altmann & Alberts 2003a, b). In this population, groups tended to divide when they became too large, and females living in the smaller daughter groups reproduced more successfully than they did in the original group. Similarly, female long-tailed macaques (*Macaca fascicularis*) living in smaller groups had higher lifetime fitness than females living in larger groups in the same habitat (van Noordwijk & van Schaik 1999). Females in larger groups of Phayre's leaf monkeys (*Trachypithecus obscurus phayrei*) weaned their infants later and had longer interbirth intervals than females in smaller groups (Borries et al. 2008). Negative relationships between group size and female fitness have also reported in several species of howler monkeys (*Alouatta* spp., Treves 2001; mantled howlers, *Alouatta palliata*, Ryan et al. 2008; Central American black howler monkey, *Alouatta pigra*, Van Belle & Estrada 2008, and lion-tailed macaques, *Macaca silenus*, Kumar 1995). However, there was no consistent relationship between group size and female reproductive success in gorillas (*Gorilla gorilla*, Stokes et al. 2003; Robbins et al. 2007), vervets (*Clorocebus pygerthrus*, Cheney & Seyfarth 1987), wedge-capped capuchins (*Cebus olivaceus*, Robinson 1988), or ring-tailed lemurs (Takahata et al. 2006). Female Thomas's leaf monkeys (*Presbytis thomasi*) experienced scramble competition, but females in larger groups still had more surviving infants per year than females in smaller groups (Steenbeek & van Schaik 2001).

Effects of Group Composition

The characteristics of other group members may influence the benefits that individuals derive from living in social groups. For example, in some cooperatively breeding species, such as alpine marmots (*Marmota marmota*) and Seychelles warblers (*Acrocephalus sechellensis*), members of one sex are more helpful than members of the other sex (Komdeur 1996; Allainé et al 2000; Griffin et al. 2005). In such cases, the sex ratio within groups influences the reproductive performance of the breeding pair. Female philopatry is expected to occur when females gain fitness benefits from associating with their female relatives (Isbell 2004; chapter 9, this volume). If that is the case, then the presence of close kin or the size of matrilines may affect females' reproductive success. The presence of kin enhances the reproductive success of females in a number of mammalian taxa. For example, house mice (*Mus domesticus*) typically form groups that include a male and several breeding females. If two females give birth around the same time, they share a nest and rear their offspring communally. Females preferentially nest with related females (Rusu & Krackow 1994; Dobson et al. 2000), and laboratory studies show that when females are allowed to nest with sisters, they produce significantly more offspring than when they are housed with unrelated females

or housed alone (Konig 1994a). The reproductive benefits of living among relatives have been documented in a variety of additional taxa, including several species of microtine rodents, grey seals (*Helichoerus grypus*), and African elephants (*Loxodonta africana*, reviewed in Silk 2007).

Cooperative breeding occurs in several genera of one primate subfamily, the Callitrichinae. Cooperatively breeding callitrichine groups typically contain only one breeding pair, along with same-sexed siblings of the breeding pair, and mature offspring from previous years (French 1997; Tardiff 1997; Dietz 2004). In golden lion tamarins (*Leontopithecus rosalia*), the species for which we have the most complete data in the wild, coresident adult males are generally close kin but only one male sires offspring (Dietz 2004). Chimerism, which increases the degree of relatedness among twins, may promote the evolution of cooperative breeding in this family of primates (chapter 21, this volume). Breeding females typically give birth to fraternal twins and can produce two litters per year. The energy costs of reproduction for callitrichine females are considerably higher than in other primate species (Harvey et al. 1987), and females are only able to sustain these reproductive costs because they get substantial help from other group members. After females give birth, other group members help carry, provision, and protect infants.

For breeding females, who depend on helpers to rear their offspring, the composition of social groups can have important effects on their fitness. In golden lion tamarins, groups with two adult males raised more surviving offspring than groups with only one adult male (Dietz & Baker 1993). In mustached tamarins (*Saguinus mystax*) and common marmosets (*Callithrix jacchus*), the number of adult males was positively associated with the number of surviving infants that the breeding female produced, but not with overall group size (Garber et al. 1984; Koenig 1995). These data suggest that males may be more reliable helpers than females, as is the case in some other cooperatively breeding species (Griffin et al. 2005), or that the presence of potential rivals for the female's breeding position reduces female reproductive success, as seems to be the case in African wild dogs (*Lycaon pictus*, McNutt & Silk 2008).

The composition of social groups also influences the reproductive success of howler monkeys. In ursine howlers (*Alouatta arctoidea*), new groups are formed when unrelated migrants from different groups meet, establish a territory, attract a male, and begin to reproduce. When daughters mature, they may remain in their natal groups or disperse and form new groups. Natal females often remain in their groups if there are only one or two adult female residents, but typically disperse and form new groups when there are three or four females present (Crockett & Pope 1993; Pope

1998). Genetic analyses indicate that recruitment opportunities are skewed in favor of the daughters of a single female (Pope 2000a), and over time the average degree of relatedness within groups increases. In well-established groups, females represent a single matriline and the average degree of relatedness within groups approaches the value for mothers and daughters and full siblings. Variation in the degree of relatedness among females within groups is correlated with their reproductive success, as females in newly formed groups have fewer surviving infants per year than females in well-established groups do (Pope 2000b). These data suggest that the presence of kin may enhance females' reproductive success, but the proximate mechanisms that underlie these effects are not known.

In species with female philopatry, the presence of close maternal kin might be expected to influence females' reproductive performance because kin are likely to be reliable allies. In a captive colony of vervet monkeys, young adult females were harassed less often and were more likely to be defended when their mothers were present than when their mothers were absent, and the presence of grandmothers enhanced the survivorship of their daughters' firstborn infants (Fairbanks & McGuire 1986). Similarly, a mother's presence reduced the age at first birth and the length of interbirth intervals in semi–free-ranging Japanese macaques (*Macaca fuscata*, Pavelka et al. 2002). However, among free-ranging olive baboons at Gombe and chacma baboons (*Papio ursinus*) at the Moremi Reserve in the Okavango Delta of Botswana, the presence or absence of mothers and sisters had no consistent effect on females' reproductive success (Packer et al. 1995; Cheney et al. 2004).

As information about paternal kinship becomes more widely available, it is becoming possible to evaluate the effects of fathers on their offspring's welfare. In the Amboseli baboon population, adult males selectively support their own immature offspring in agonistic disputes (Buchan et al. 2003). The extent of coresidence between fathers and their offspring influences the age at which their offspring reach important developmental milestones (Charpentier et al. 2008). Females whose fathers were present longer while they were infants and juveniles reached menarche at earlier ages than females whose fathers were present for shorter parts of their immature period. In this population, age at sexual maturity is an important source of variation in females' lifetime reproductive success, so the presence of fathers may enhance their daughters' fitness. Fathers had a similar but less consistent effect on their sons' maturation. The extent of coresidence between fathers and sons reduced the age of the sons' testicular enlargement, but this effect was limited to the sons of high-ranking males. Males that reached the age of testicular enlargement sooner also began to consort with females sooner, but it is not clear whether those developmental milestones are linked to variation in lifetime fitness.

The Structure and Consequences of Social Strategies

If the extent and form of sociality reflects a compromise between the benefits and costs of living in social groups, then we would expect natural selection to favor behavioral strategies that enable individuals to increase the benefits and reduce the costs of living in social groups. Thus, natural selection might favor more efficient foraging techniques and morphological traits that enhance the ability of individuals to exclude rivals from access to resources. Natural selection might also favor the capacity to develop social strategies that enable individuals to enhance their benefit-cost ratios. This idea is central to socioecological models of primate social organization, which connect food distribution patterns to female dispersal strategies and the nature of social bonds among females (chapter 9, this volume). It is also a central component of the social intelligence hypothesis, which posits that the challenges of sociality favored the evolution of larger brains and more sophisticated cognitive capacities in the anthropoid primates (chapters 28 and 30, this volume). If this logic is correct, then variation in social strategies should be linked to short-term variation in well-being and long-term variation in fitness outcomes.

Assessing the Nature of Social Relationships

Social relationships are abstractions that represent the history of interactions between two individuals (Hinde 1983; chapter 23, this volume). The behavioral data we collect tell us who does what to whom and how often. When we tabulate these data, we can identify certain regularities in the patterning of associations and social interactions. For example, there are strong kin biases in behavior in many taxa (chapter 21, this volume). We can also use these data to test functional predictions about the patterning of behavior. However, the data do not tell us whether the animals themselves are aware that these patterns exist, or whether the behavioral decisions they make on a daily basis are influenced by their memory of past interactions or their expectations about what might happen in the future (Barrett & Henzi 2002).

We can use information about the frequency of interactions among individuals to quantify the strength and quality of social relationships among pairs of individuals along a number of different dimensions (Silk 2002). They may rarely interact or interact often; they may interact in a lim-

Dimensions of Social Interactions

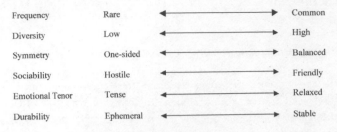

Frequency	Rare	←————→	Common
Diversity	Low	←————→	High
Symmetry	One-sided	←————→	Balanced
Sociability	Hostile	←————→	Friendly
Emotional Tenor	Tense	←————→	Relaxed
Durability	Ephemeral	←————→	Stable

Fig. 24.2. Schematic characterization of social relationships

ited number of behavioral contexts or in a diverse range of situations; their affiliative interactions may be highly asymmetric or well balanced; the tenor of their interactions can range from mostly hostile to mostly friendly; they may be tense or relaxed when together; and they may interact at high rates for short periods or they may interact at consistently high rates over long periods of time. Relationships that consistently fall toward the right side of the continuums in fig. 24.2 can reasonably be categorized as close social bonds.

It is important to emphasize that close social bonds are not necessarily the same as human friendships. The qualities that define close social bonds are consistent with our commonsense notions of friendship, but they leave out the emotional and psychological dimensions of relationships. Human friendship is based on companionship, affection, intimacy, trust, loyalty, acceptance, sympathy, and concern for the other's welfare (Silk 2002). Behavioral data provide us with relatively little access to how our subjects feel about their partners, although we are beginning to gain some insight about this.

The Structure of Social Bonds

Analyses of data collected in several groups of yellow baboons observed over a 16-year period in the Amboseli basin of Kenya and another group of chacma baboons over 16 years in the Moremi Reserve of Botswana demonstrate that females form relationships that fit the operational definition of close social bonds. In each case, the analyses are based on a composite measure of sociality, which includes information about grooming and proximity (Silk et al. 2006a, 2010a). At both sites, females interacted with most other females at low rates, but nearly all interacted with a few partners at substantially higher rates. As in many other species with female philopatry (chapter 21, this volume), the females groomed, associated, and supported their close maternal kin at higher rates than more distantly related females

or non-kin (Silk et al. 2004). Female baboons in Amboseli and Moremi also showed strong social preferences for age mates, perhaps because peers are likely to be paternal half siblings in species, like baboons, with relatively high reproductive skew (Alberts et al. 2006; cf. chapter 21, this volume). In the Amboseli population, paternal half siblings were preferred over true non-kin (Smith et al. 2003; Silk et al. 2006a). Although the availability of close maternal kin had strong effects on the number of close social bonds that females formed, virtually all females in the population formed at least one close social bond even if they had no close maternal relatives in the group.

There was considerable variability in the distribution of grooming within dyads. That is, in some dyads grooming was well-balanced, but in other dyads one female groomed her partner much more than she was groomed in return. It is possible for pairs of females to have high rates of grooming, but for one partner to contribute considerably more grooming than she receives in return. In fact, pairs that formed strong social bonds groomed more equitably than pairs that formed weaker social bonds (Silk et al. 2006b; figure 24.3). This result could be an artifact of maternal kinship, because related partners groomed more equitably than unrelated partners and related partners also formed stronger social bonds than unrelated partners. However, the relationship between grooming equity and the strength of social bonds held when the analysis was limited to unrelated females, so this relationship seemed to be at least partly independent of kinship. The same patterns hold for female chacma baboons in Moremi.

If females have consistent preferences for certain categories of partners, such as close kin, peers, or reciprocating partners, then it seems logical that there would be considerable temporal consistency in the quality of females' relationships across time. That is, the preferences that emerge in cross-sectional analyses ought to generate stable preferences in longitudinal analyses as well. However, the availability of partners varies across time as new partners mature and old partners die or are separated after group fissions. These kinds of demographic forces impose natural limits on the duration of social ties. In addition, not all preferences will generate long-lasting relationships. For example, female monkeys' attraction to newborn infants generates a strong, but short-lived, attraction to their mothers (Henzi & Barrett 2002; Frank & Silk 2009).

To assess the stability of females' *preferences* for favored partners, my colleagues and I evaluated the likelihood that females would have the same top partners in successive years. In our calculations we took into account the fact that the availability of partners changed from year to year, as some females matured and others died or disappeared. The

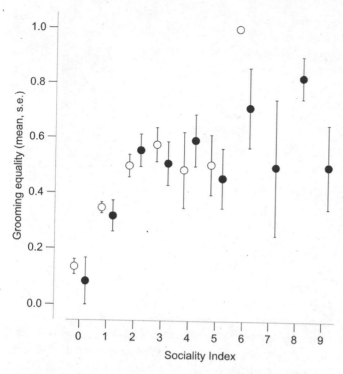

Fig. 24.3. The relationship between the strength of social bonds and the extent of grooming equality in female chacma baboons. The sociality index is based on the relative frequency of grooming and associations; low values represent dyads with weak social bonds and high values represent dyads with strong social bonds. Grooming equality measures the distribution of grooming within dyads. When grooming is evenly balanced within the dyad, grooming equality equals 1; when it is completely one-sided, grooming quality equals 0. Open circles = nonkin; closed circles = kin. From Silk et al. 2006b, fig. 2.

results of these analyses showed that female baboons in Amboseli and Moremi had consistent preferences for their favored partners (Silk et al. 2006b, 2009). For example, mothers and daughters were quite likely to remain top partners as long as they lived together in the group. Not surprisingly, the same factors that affected the strength of social bonds among females also influenced the consistency of females' preferences across time. At the same time, relationships with unrelated top partners were typically ephemeral, existing for one year but not lasting until the next.

Although females had consistent preferences for certain categories of favored partners, the actual length of their relationships with favored partners varied considerably. In Amboseli, the maximum period of coresidence for any pair of adult females was 14 years, but the average period was approximately 3.5 years (Silk et al. 2006b). This set upper limits on the duration of close social bonds. Some pairs of females did form long-lasting close social bonds. For example, 14 pairs of females formed close social bonds that lasted at least five years; five of these were mothers and daughters, five were maternal sisters, and three were known to be paternal sisters. The remaining pair of females

could not be tested for paternity, but were very close in age, making them likely to have been paternal sisters. However, many pairs of females did not form social bonds that lasted very long, and not every female formed an enduring bond with another. Further work is needed to assess how and why females' relationships are disrupted.

This picture of social bonds among these female baboons contrasts with the patterns reported for female chacma baboons in South Africa. Barrett and Henzi (2002; Henzi et al. 2009) found little consistency in the relationships among female chacma baboons in the Drakensburg Mountains of South Africa over a four-year period, or among females at De Hoop near the Cape of Good Hope. They argue that baboon females do not "sustain relationships with a constant and circumscribed set of individuals over time, but instead form only short-term companionships with an array of different partners in response to local ecological contingencies" (Henzi et al. 2009, p. 1019). It is not yet clear whether these differences are the result of differences in the behavior of females across populations or of differences in the methods used to collect and analyze data across sites (Silk et al. 2010a).

The relationships among male chimpanzees parallel relationships among female baboons in Amboseli and Moremi in a number of ways. In chimpanzees, males are the philopatric sex, and they participate in a variety of affiliative and cooperative behaviors including grooming, coalitionary support, cooperative hunting, joint mate guarding, and collective territorial defense (Muller & Mitani 2005). Nepotistic biases among male chimpanzees are not as pronounced as nepotistic biases in baboons and other female-bonded primates, but male chimpanzees do interact at higher rates with maternal kin than with unrelated males (Langergraber et al. 2007).

Mitani (2009) used data on the frequency of association among males over a 10-year period at Ngogo in the Kibale National Park of Uganda to characterize the strength of males' dyadic social relationships, and used information about the distribution of grooming within dyads to assess the quality of males' relationships. He found that maternal brothers groomed more equitably than pairs of unrelated males, and males in the same rank category groomed more equitably than pairs of males from different rank categories. Pairs of chimpanzees who had strong social bonds groomed more equitably than males who had weak social bonds, and this relationship was independent of the effects of kinship or male dominance rank. Males' partner preferences in one year were correlated with their partner preferences in successive years, and these correlations remained significant when the effects of kinship were controlled. Of the 28 males who were present for at least five years, 26 had at least

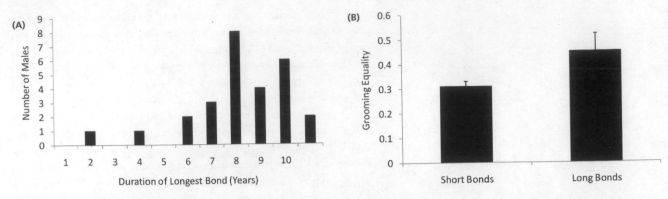

Fig. 24.4 (a) This graph shows the duration of the longest close social bond formed by each of the male chimpanzees that was observed for at least five years. (b) This graph compares the distribution of grooming within dyads that formed long bonds (≥ 5 years) and within more short-lived social bonds. The same procedure was used to assess grooming equality as is depicted in fig. 24.3. Redrawn from Mitani 2009, fig. 6.

one close relationship that lasted at least five years. As in female baboons, pairs of male chimpanzees who formed these long-lasting relationships groomed more equitably than pairs of males who had less enduring social relationships (fig. 24.4).

Short-Term Impacts of Close Social Bonds

In order to find out whether social strategies have short-term impacts on animals' welfare, we need to identify biologically meaningful measures of well-being. Physiological data, such as heart rate and glucocorticoid levels can provide this kind of information, but it is not easy to get it from animals, particularly free-ranging animals, without affecting their behavior. However, there are two different methods that can be used to assess animals' stress levels without handling them directly. Captive studies showed that self-directed behaviors (SDB) were correlated with pharmacological agents that increase anxiety in macaques and marmosets (Aureli & Whiten 2003). Thus, self-directed behaviors provide a useful behavioral proxy for anxiety. In addition, it is now possible to extract glucocorticoids (GC) from feces (Beehner & Whitten 2004), and it is relatively easy (if not particularly glamorous) to collect samples of feces from individuals on a regular basis.

When animals experience stressful events, particularly unpredictable and uncontrollable events such as attacks by predators or by other group members, their levels of glucocorticoids rise. This enables them to mobilize energy reserves for critical activities, such as flight, and to divert energy from other less immediately essential metabolic processes, such as growth and maintenance (Sapolsky 2004). Although stress responses are advantageous short-term responses to acute threats, sustained elevation of glucocorticoid levels has deleterious effects on health and reproduction.

There is considerable variation in the patterning of stress responses within and between species. When dominance hierarchies are maintained through frequent attacks on subordinates, high-ranking individuals have higher stress levels than subordinates. In contrast, when dominance hierarchies are maintained through intimidation and less frequent attacks, dominants tend to have lower stress levels than subordinates (Sapolsky 2004). However, when the hierarchy is unstable and there is great uncertainty about the source and outcome of aggressive interactions, there may be no consistent differences in the stress levels of high- and low-ranking individuals (Sapolsky 2004). Comparative analyses suggest that subordinates have high GC levels in relation to dominants in species with high levels of harassment by dominants and low levels of social support (primates, Abbott et al. 2003; vertebrates, Goymann & Wingfield 2004). Abbott and his colleagues also found that social support and access to kin mediate the effects of low rank on subordinates across species.

There is considerable evidence that everyday events can be stressful for individuals in primate groups. When rhesus monkeys (*Macaca mulatta*) are approached by unrelated higher-ranking members of their groups, their heart rates rise, while approaches by kin and unrelated lower-ranking monkeys produce little change in resting heart rates (Aureli et al. 1999). Similarly, when female olive baboons are in close proximity to higher-ranking females, their rates of self-directed behaviors are elevated (Castles et al. 1999). The immigration of potentially infanticidal males and instability in the male dominance hierarchy also creates sustained anxiety for anestrous female chacma baboons (Beehner et al. 2005; Engh et al. 2006; Wittig et al. 2008).

Social support seems to contribute to variation in GC levels among individuals within groups. These kinds of associations were first explored by Robert Sapolsky, who pioneered endocrinological studies of free-ranging primates. He and his colleagues showed that dominant male olive baboons generally have lower GC levels than subordinate males, but dominant males who rarely groom females or interact with infants have elevated GC levels (Sapolsky & Ray 1992; Virgin & Sapolsky 1997). Similarly, in Amboseli, male yellow baboons who are more socially integrated into their groups have significantly lower GC levels than males who are more isolated (Sapolsky et al. 1997). Recent studies of wild Assamese macaques (*Macaca assamensis*) in Thailand suggest that close social bonds among high-ranking adult males may mitigate the costs of maintaining high rank and contribute to the negative relationship between male dominance rank and GC levels (Ostner et al. 2008).

More recently, a series of studies have examined the role of social support in mediating stress among females in wild baboon populations. In the Moremi Reserve of the Okavango Delta of Botswana, predation rates are high, and female chacma baboons who had suddenly lost close kin experienced significant increases in fecal glucocorticoid (fGC) levels (Engh et al. 2006a). This might not seem surprising, since predatory attacks are the kinds of events that stress responses are designed by natural selection to contend with. However, females who had lost close kin had higher fGC levels than other females who were present in the group but had not suffered personal losses. These data suggest that the females' responses were exaggerated because their own social networks were disrupted (Engh et al. 2006). However, Henzi and Barrett (2007) have argued that this interpretation does not consider the fact that close kin tend to rest close together, and close proximity at the time of the predatory attack may have generated a more intense response.

The threat of infanticide is another potential source of stress for females in Moremi, where infanticide is a major source of mortality for infants (Cheney et al. 2004). Females, especially females with young infants, react fearfully to immigrant males, and male-female "friendships" seem to be designed to protect females and their infants from infanticidal threats (Palombit et al. 1997; chapter 19, this volume). The presence of high-ranking immigrant males produced sustained elevations of the fGC levels of lactating females, whose infants were most vulnerable to infanticidal attacks (Beehner et al. 2005; Engh et al. 2006b). But lactating females who had established "friendships" with adult males had significantly lower fGC levels than females without male "friends" (Engh et al. 2006b; see chapter 19, this volume).

Instability at the top of the adult male dominance hierarchy is another potent stressor for females, presumably because changes in the identity of the top-ranking male are often followed by infanticide. Females' relationships

with one another seemed to provide a means for them to cope with this source of stress during a five-month period of instability in Moremi (Wittig et al. 2008). Females who focused their grooming on a small number of partners (low grooming diversity) before the dominance hierarchy was disrupted experienced smaller increases in fGC than females with more diffuse grooming networks. Moreover, while all females reduced their grooming diversity in the early weeks of the unstable period, females who reduced their grooming diversity the most also reduced their fGC levels the most. Wittig and his colleagues conclude that close social bonds with other adult females allow females "to alleviate the stress associated with social instability."

Moremi females show strong and consistent responses to events, such as predation and infanticide, that have major effects on their lives. But they also show more subtle responses to everyday events. For example, chacma baboons give grunts that function as predictive signals of benign intent (Cheney et al. 1995; Silk et al. 1996; Wittig et al. 2007) and are effective in reconciling aggression (Cheney & Seyfarth 1997; Cheney et al. 1995; Wittig et al 2007). Cycling and pregnant baboons that receive grunts at high rates from higher-ranking adult females have lower fGC levels than females that receive grunts at lower rates. Lactating females do not show the same effect, perhaps because they are the targets of a considerable amount of unwanted attention from other females that want to handle their infants (Crockford et al. 2008; fig. 24.5; see also chapter 22 this volume, fig. 22.6). The structure of females' social networks is also

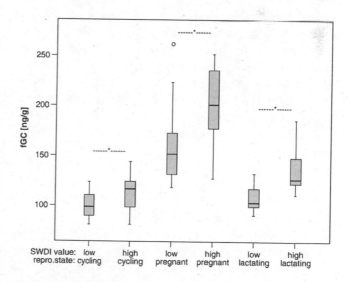

Fig. 24.5. The fecal glucocorticoid (fGC) levels of cycling, pregnant, and lactating females with high or low levels of grooming diversity are compared in this graph. Low values of the Shannon Weaver diversity index (SWDI) represent females that focused their grooming on a small number of partners, and high values of grooming diversity indicate females with more diffuse grooming networks. From Crockford et al. 2008, fig. 3.

related to their fGC levels. Females had lower fGC levels during months in which they concentrated their grooming on a small number of partners than during months in which they distributed their grooming more evenly among potential partners (Crockford et al. 2008). Moreover, the structure of females' grooming networks had a bigger impact on their fGC levels than the overall frequency with which females groomed or were groomed by others.

The Long-Term Impact of Close Social Bonds

If close social bonds provide a coping mechanism for dealing with stress, or if they facilitate the formation of alliances, then they might have positive long-term effects on reproductive performance. This seems to be the case for at least one primate species, *Homo sapiens*. In humans, the disruption of social ties due to death, divorce, or separation is a major source of stress. Social isolation and feelings of loneliness increase the risk of disease, a range of mental disorders, accidents, and mortality (Cacioppo et al. 2000). In contrast, social support is associated with reduced mortality and better physical and mental health, particularly for women (Thorsteinsson & James 1999; Taylor et al. 2000; Kendler et al. 2005).

To determine whether variation in sociality is correlated with fitness outcomes in nonhuman primate groups, we need to match detailed information about the social behavior of individuals with information about their reproductive histories. Few studies can provide these kinds of data on sizable numbers of individuals. However, long-term studies of free-ranging baboons provide the kinds of data that we need to assess the short-term and long-term effects of social relationships. Although we do not know whether the findings of these studies are representative of other taxa, they provide what philosophers call "an existence proof"—they show that it is possible to tackle this question empirically and provide a template for future analyses in other species.

Sociality is associated with fitness outcomes among female yellow baboons in Amboseli. In these analyses we used a composite measure of female sociality based on the amount of time females spent grooming or in close proximity to other adult group members (Silk et al. 2003). We matched these data with information about the proportion of infants that survived to one year of age—an important component of variation in the lifetime reproductive success of females in the Amboseli baboon population. The results of these analyses showed that females that were most fully socially integrated into the group reproduced more successfully than females that were less socially integrated, and this effect was independent of the effects of female dominance rank, the presence of kin, or changes in environmental conditions that influenced female sociality.

These findings from Amboseli are vulnerable to two different criticisms. First, the analysis included interactions with adult males, leaving open the possibility that females' associations with males, not their associations with other females, might have driven the results. This could be important because new mothers often form "friendships" with adult males (see chapter 19, this volume), and those relationships may have long-term effects on offspring survival. Second, female baboons are strongly attracted to other females' infants (Altmann 1980; Bentley-Condit & Smith 1999; Silk et al. 2003b), so the relationship between female sociality and infant survival might reflect elevated levels of sociality for mothers of surviving infants rather than benefits derived from sociality (Henzi & Barrett 2007). However, analyses that were designed to avoid both of these confounds replicated the original results. Among chacma baboons in the Moremi Reserve of Botswana, the strength of dyadic social bonds among adult females when they did not have young infants was a significant predictor of offspring survival (Silk et al. 2009). Moreover, the effects of maternal sociality extended well past the age of weaning.

Variation in breeding life span accounts for nearly half of the variation in lifetime fitness among female baboons (Altmann & Alberts 2003; Cheney et al. 2004), but very little is known about the sources of variation in longevity among primates or other taxa (Clutton-Brock 1988; Newton 1988). Recent work on the Moremi baboons indicates that sociality may enhance longevity. Females that had strong and stable relationships with other females lived significantly longer than females who had weaker and less stable relationships (Silk et al. 2010b). This effect was independent of the effects of dominance, which was also associated with enhanced longevity among Moremi females.

We have not identified the mechanisms that underlie the relationship between female sociality, offspring survival, and life span in Moremi or Amboseli. It is possible that females with stronger and more secure social bonds may be less spatially isolated, making them and their offspring safer from predation. Alternatively, females with strong social bonds may be better shielded from social conflict and able to feed more efficiently. It is also possible that females that are more fully integrated into their groups may be able to cope with stress better than other females (Crockford et al. 2008; Wittig et al. 2008), and this may enhance their reproductive performance. Females' ability to cope with stress may also confer important advantages on their offspring because maternal exposure to environmental and social stressors can be detrimental to their offspring's growth rates, longevity, and behavior, as well as to the organization of their hypothalamic-pituitary-adrenal pathway (Bernardo 1996; Sanchez 2006; Onyango et al. 2008; Weinstock 2008).

Recent data suggests that males may also gain fitness

advantages from forming strong social bonds in some settings. Like chimpanzees, male Assamese macaques rely on support from other males to attain and maintain high rank. Males form strong social bonds with selected partners, and males that form the strongest social bonds also sire the most offspring (Schülke et al. 2010).

Summary

Despite the difficulties of evaluating the relationship between sociality and fitness outcomes in animals such as primates, which tend to have long lives and live in complex social groups, a growing body of evidence suggests that sociality has important impacts on the fitness of individuals. While females may have relatively limited control over some of the factors that influence their fitness—such as the size of the group in which they live, or the presence of close relatives—they may have more control over the nature of their relationships with the other members of their groups. This may be important because the availability of social support and the quality of social bonds affects individuals' ability to cope with stressful events in their lives. The quality of social bonds that females form has long-term impacts on fitness in two populations of baboons. The ability to form close, balanced, and enduring social bonds may have important selective value for individuals. Surprisingly, the benefits of social bonds may also extend to males in some species. Assamese macaques form strong, stable, and supportive social bonds whose quality contributes to variation in reproductive performance. It will be interesting to find out whether social bonds have similar effects among chimpanzees, whose social bonds are similar to those formed by female baboons.

It is important to document the nature of social bonds in a wider range of primate species. We need to develop a consensus about how to quantify the multiple dimensions of social relationships and how to assess their consistency across time. Such standardization will greatly enhance the power of comparative analyses. In addition, it would be useful to explore the sources of variation in the extent of sociality among individuals. Although there is some evidence that the availability of preferred categories of partners contributes to variation in the structure of social networks among female baboons, much of that variation has not been explained. It would be interesting to know, for example, whether individual differences in temperament contribute to variation in sociality, and thereby influence fitness outcomes.

There seem to be strong parallels in humans and other primates between the effects of social support and social bonds on health and survival. In humans, social integra-

tion is associated with lower mortality and better physical and mental health, particularly for women (Thorsteinsson & James 1999; Taylor et al. 2000; Kendler et al. 2005). The strength and quality of these bonds has a more pronounced impact on health outcomes than the size of the social network (Seeman 1996; Hill & Dunbar 2003).

It seems likely that these similarities are the result of our shared ancestry over millions of years. However, we do not yet know how widespread these patterns are among primates, and more data are clearly needed. We need to collect information about the pattern of social behavior among individuals over extended periods of time, to monitor the size and composition of social groups across time, and to keep track of the reproductive history of individuals over the course of their lives. Then we need to integrate these data with information about how individuals respond to potentially stressful events in their lives. Long-term studies are essential to answer these kinds of questions.

References

Abbott, D. H., Keverne, E. B., Bercovitch, F. B., Shiveley, C. A., Mendoza, S. P, Saltzman, W., Snowdon, C. T., Ziegler, T. E., Banjeic, M., Garland, T. Jr. & Sapolsky, R. M. 2003. Are subordinates always stressed? A comparative analysis of rank differences in cortisol levels among primates. *Hormones and Behavior*, 43, 67–82.

Alberts, S. C., Buchan, J. C. & Altmann, J. 2006. Sexual selection in wild baboons: From mating opportunities to paternity success. *Animal Behaviour*, 72, 1177–1196.

Allainé, D. 2004. Sex ratio variation in the cooperatively breeding alpine marmot *Marmota marmota*. *Behavioral Ecology*, 15, 997–1002.

Altmann, J. & Alberts, S. C. 2003a. Variability in reproductive success viewed from a life-history perspective in baboons. *American Journal of Human Biology*, 15, 401–409.

———. 2003b. Intraspecific variability in fertility and offspring survival in a nonhuman primate: behavioral control of ecological and social sources. In *Offspring: Human Fertility Behavior in Biodemographic Perspective* (ed. by Wachter, K. W. & Bulatao, R. A.), 140–169. Washington: National Academies Press.

Aureli, F. & Whiten, A. 2003. Emotions and behavioral flexibility. In *Primate Psychology* (ed. by Maestripieri, D.), 289–323. Cambridge, MA: Harvard University Press.

Aureli, F., Preston, S. D. & de Waal, F. B. M. 1999. Heart rate responses to social interactions in free-moving rhesus macaques (*Macaca mulatta*): A pilot study. *Journal of Comparative Psychology*, 113, 59–65.

Barrett, L. & Henzi, S. P. 2002. Constraints on relationship formation among female primates. *Behaviour*, 139, 263–289.

———. 2005. Monkeys, markets and minds: Biological markets and primate sociality. In *Cooperation in Primates and Humans* (ed. by Kappeler, P. M. & van Schaik, C. P.), 209–232. Berlin: Springer.

Barrett, L. H., Henzi, S. P. & Rendall, D. 2007. Social brains, simple minds: Does social complexity really require cognitive complexity? *Philosophical Transactions of the Royal Society London*, 266, 665–670.

Beehner, J. C. & Whitten, P. L. 2004. Modifications of a field method for faecal steroid analysis in baboons. *Physiology and Behavior*, 82, 269–277.

Beehner, J. C., Bergman, T. J., Cheney, D. L., Seyfarth, R. M. & Whitten, P. L. 2005. The effect of new alpha males on female stress in free-ranging baboons. *Animal Behaviour*, 69, 1211–1221.

Bernardo, J. 1996. Maternal effects in animal ecology. *American Zoologist*, 36, 83–105.

Borries, C., Larney, E., Lu, A., Ossi, K. & Koenig, A. 2008. Costs of group size: Lower developmental rates in larger groups of leaf monkeys. *Behavioral Ecology*, 19, 1181–1186.

Buchan, J. C., Alberts, S. C., Silk, J. B. & Altmann, J. 2003. True paternal care in a multi-male primate society. *Nature*, 425,179–181.

Cacioppo, J. T., Ernst, J. M., Burleson, M. H., McClintock, M. K., Malarkey, W. B., Hawkley, L. C., Kowalewski, R. B., Paulsen, A., Hobson, J. A., Hugdahl, K., Spieger, D. & Berntson, G. G. 2000. Lonely traits and concomitant physiological processes: The MacArthur social neuroscience studies. *International Journal of Psychophysiology*, 35, 143–154.

Castles, D. L., Whiten, A. & Aureli, F. 1999. Social anxiety, relationships and self-directed behaviour among wild female olive baboons. *Animal Behaviour*, 6, 1207–1215.

Charpentier, M. J. E., Van Horn, R. C., Altmann, J. & Alberts, S. C. 2008. Paternal effects on offspring fitness in a multi-male primate society. *Proceedings of the National Academy of Sciences USA*, 105, 1988–1992.

Chapman, C. A. & Chapman, L. J. 2000. Constraints on group size in red colobus and red-tailed guenons: Examining the generality of the ecological constraints model. *International Journal of Primatology*, 21, 565–585.

Cheney, D. L. and Seyfarth, R. M. 1987. The influence of inter-group competition on the survival and reproduction of female vervet monkeys. *Behavioral Ecology and Sociobiology*, 21, 375–386.

———. 1997. Reconciliatory grunts by dominant female baboons influence victim's behavior. *Animal Behaviour*, 54, 409–418.

Cheney, D., Seyfarth, R. & Smuts, B. 1986. Social relationships and social cognition in nonhuman primates. *Science*, 234, 1361–1366.

Cheney, D. L., Seyfarth, R. M., & Silk, J. B. 1995. The role of grunts in reconciling opponents and facilitating interactions among adult female baboons. *Animal Behaviour*, 50, 249–257.

Cheney, D. L., Seyfarth, R. M., Fischer, J., Beehner, J., Bergman, T., Johnson, S. E., Kitchen, D. M., Palombit, R. A. & Silk, J. B. 2004. Factors affecting reproduction and mortality among baboons in the Okavango Delta, Botswana. *International Journal of Primatology*, 25, 401–428.

Clutton-Brock, T. H. 1988. *Reproductive Success*. Chicago: University of Chicago Press.

Clutton-Brock, T. H., Russell, A. F., Sharpe, L. L., Brotherton, P. N. M., McIlrath, G. M., White, S. & Cameron, E. Z. 2001. Effects of helpers on juvenile development and survival in meerkats. *Science*, 293, 2446–2449.

Crockett, C.M. & Janson, C.H. 2000. Infanticide in red howlers: Female group size, male membership, and a possible link for folivory. In *Infanticide by Males and its Implications* (ed. by van Schaik, C. P. & Janson, C. H.), 75–98. Cambridge: Cambridge University Press.

Crockett, C. M. & Pope, T. R. 1993. Consequences for sex difference in dispersal for juvenile red howler monkeys. In *Juvenile Primates: Life History, Development, and Behavior* (ed. by Pereira, M.E. & Fairbanks, L.A.), 104–118. Oxford: Oxford University Press.

Crockford, C., Wittig, R. M., Whitten, P. L., Seyfarth, R. M. & Cheney, D. L. 2008. Social stressors and coping mechanisms in wild female baboons (*Papio hamadryas ursinus*). *Hormones and Behavior*, 53, 254–265.

Dietz, J. M. 2004. Kinship structure and reproductive skew in cooperatively breeding primates. In *Kinship and Behavior in Primates* (ed. by Chapais, B. & Berman, C.), 223–241. Oxford: Oxford University Press.

Dietz, J. M. & Baker, A. J. 1993 Polygyny and female reproductive success in golden lion tamarins, *Leontopithecus rosalia*. *Animal Behaviour*, 46, 1067–1078.

Dobson, F. S., Jacquot, J. & Baudoin, C. 2000. An experimental test of kin association in the house mouse. *Canadian Journal of Zoology*, 78, 1806–1812.

Engh, A. L., Beehner, J. C., Bergman, T. J., Whitten, P. L., Hoffmeier, R. R., Seyfarth, R. M. & Cheney, D. L. 2006a. Female hierarchy instability, male immigration, and infanticide increase glucocorticoid levels in female chacma baboons. *Animal Behaviour*, 71, 1227–1237.

Engh, A. L., Beehner, J. C., Bergman, T. J., Whitten, P. L., Hoffmeier, R. R., Seyfarth, R. M. & Cheney, D. L. 2006b. Behavioural and hormonal responses to predation in female chacma baboons. *Proceedings of the Royal Society, London B*, 273, 707–712.

Fairbanks, L. A. & McGuire, M. T. 1986. Age, reproductive value, and dominance-related behaviour in vervet monkey females: Cross-generational influences on social relationships and reproduction. *Animal Behaviour*, 34, 1710–1721.

Frank, R. & Silk, J. B. Grooming exchange between mothers and non-mothers: The price of natal attraction in wild baboons (*Papio anubis*). *Behaviour*, 136, 889–906.

French, J. A. 1997. Proximate regulation of singular breeding in callitrichid primates. In *Cooperative Breeding in Mammals* (ed. by Solomon, N.G. & French, J. A.), 34–75. Cambridge: Cambridge University Press.

Garber, P. A., Moya, L. & Malaga, C. 1984. A preliminary field study of the moustached tamarin monkeys (*Saguinus mystax*) in northeastern Peru: Questions concerned with the evolution of a communal breeding system. *Folia Primatologica*, 42, 17–32.

Goymann, W. & Wingfield, J. 2004. Allostatic load, social status and stress hormones: The costs of social status matter. *Animal Behaviour*, 67, 591–602.

Griffin, A. S., Sheldon, B. C. & West, S. A. 2005. Cooperative breeders adjust offspring sex ratios to produce helpful helpers. *The American Naturalist*, 166, 628–632

Harvey, P. H., Martin, R. D. & Clutton-Brock, T. H. 1987. Life histories in comparative perspective. In *Primate Societies* (ed. by Smuts, B. B., Cheney, D. L., Seyfarth, R. M., Wrangham, R. W. & Struhsaker, T. T.), 181–196. Chicago: University of Chicago Press.

Henzi, S. P. & Barrett, L. 1999. The value of grooming to female primates. *Primates*, 40, 47–59.

———. 2007. Coexistence in female-bonded primate groups. *Advances in the Study of Behavior* 37, 43–81.

Henzi, S. P., Lusseau, D., Weingrill, T., van Schaik, C. P., & Barrett, L. 2009. Cyclicity in the structure of female baboon social networks. *Behavioral Ecology and Sociobiology*, 63, 1015–1021.

Hill, R. A. & Dunbar, R. I. M. 2003. Social network size in humans. *Human Nature*, 14, 53–72.

Hinde, R. A. 1983. *Primate Social Relationships: An Integreated Approach*. Oxford: Blackwell Scientific.

Isbell, L. A. 1991. Contest and scramble competition: Patterns of female aggression and ranging behavior among primates. *Behavioral Ecology*, 2, 143–155.

———. 2004. Is there no place like home? Ecological bases of female dispersal and philopatry and their consequences for the formation of kin groups. In *Kinship and Behavior in Primates* (ed. by Chapais, B. & Berman, C. M.), 71–108. Oxford: Oxford University Press.

Janson, C. H. & Goldsmith, M. L. 1995. Predicting group size in primates: Foraging costs and predation risks. *Behavioral Ecology*, 6, 326–336.

Kendler, K. S., Myers, J., & Prescott, C. A. 2005 Sex differences in the relationship between social support and risk for major depression: A longitudinal study of opposite-sex twin pairs. *American Journal of Psychology*, 162, 250–256.

Koenig, A. 1995. Group size, composition, and reproductive success in wild common marmosets (*Callithrix jacchus*). *American Journal of Primatology*, 35, 311–317.

Komdeur, J. 1996. Facultative sex ratio bias in the offspring of Seychelles warblers. *Proceedings of the Royal Society, London B*, 263, 661–666.

König, B. 1994a. Fitness effects of communal rearing in house mice: The role of relatedness versus familiarity. *Animal Behavior*, 48, 1449–1457.

———. 1994b. Components of lifetime reproductive success in ommunally and solitarily nursing house mice: A laboratory study. *Behavioral Ecology and Sociobiology*, 34, 275–283.

Kumar, A. 1995. Birth rate and survival in relation to group size in the lion-tailed macaque, *Macaca silenus*. *Primates*, 30, 1–9.

Langergraber, K. E., Mitani, J. C., & Vigilant, L. 2007. The limited impact of kinship on cooperation in wild chimpanzees. *Proceedings of the National Academy of Science*, 104, 7786–7790.

McNutt, J. W. & Silk, J. B. 2008. Pup production, sex ratios, and survivorship in African wild dogs, *Lycaon pictus*. *Behavioral Ecology and Sociobiology*, 62, 1061–1067.

Mitani, J. C. 2009. Male chimpanzees form enduring and equitable social bonds. *Animal Behaviour*.

Muller, M. N. & Mitani, J. C. 2005. Conflict and cooperation in wild chimpanzees. *Advances in the Study of Behavior*, 35, 275–331.

Newton, I. 1989. *Lifetime Reproduction in Birds*. London: Academic Press.

Onyango, P. O., Gesquiere, L. R., Wango, E. O., Alberts, S. C. & Altmann, J. 2008. Persistence of maternal effects in baboons: Mother's dominance rank at son's conception predicts stress hormone levels in sub adult males. *Hormones and Behavior*, 54, 319–324.

Ostner, J., Heistermann, M., & Schülke, O. 2008. Dominance, aggression, and physiological stress in wild male Assamese macaques (*Macaca assamensis*). *Hormones and Behavior*, 54, 613–619.

Palombit, R. A., Seyfarth, R. M. & Cheney, D. L. 1997. The adaptive value of "friendship" to female baboons: Experimental and observational evidence. *Animal Behaviour*, 54, 599–614.

Pavelka, M. S. M., Fedigan, L. M. & Zohar, S. 2002. Availability and adaptive value of reproductive and postreproducive Japanese macaque mothers and grandmothers. *Animal Behaviour*, 64, 407–414.

Packer, C., Collins, D. A., Sindimwo, A. & Goodall, J. 1995. Reproductive constraints on aggressive competition in female baboons. *Nature*, 373, 60–63.

Pope, T. R. 1998. Effects of demographic change on group kin structure and gene dynamics of populations of red howling monkeys. *Journal of Mammalogy*, 79, 692–712.

———. 2000a. Reproductive success increases with degree of kinship in cooperative coalitions of female red howler monkeys (*Alouatta seniculus*). *Behavioral Ecology and Sociobiology*, 48, 253–267.

———. 2000b. The evolution of male philopatry in neotropical monkeys. In *Primate Males* (ed. by Kappeler, P. M.), 219–235. Cambridge: Cambridge University Press.

Pride, E. R. 2005. Optimal group size and seasonal stress in ring-tailed lemurs (*Lemur catta*). *Behavioral Ecology*, 16, 550–560.

Pusey, A. E., Oehlert, G. W., Williams, J. M. & Goodall, J. 2005. Influence of ecological and social factors on body mass of wild chimpanzees. *International Journal of Primatology*, 26, 3–31.

Robbins, M. M., Robbins, A. M., Gerald-Steklis, N. & Steklis, H. D. 2007. Socioecological influences on the reproductive success of female mountain gorillas (*Gorilla beringei beringei*). *Behavioral Ecology and Sociobiology*, 61, 919–931.

Robinson, J. G. 1988. Group size in wedge-capped capuchins *Cebus olivaceus* and the reproductive success of males and females. *Behavioral Ecology and Sociobiology*, 23, 187–197.

Rusu, A. S. & Krackow, S. 2004. Kin-preferential cooperation, dominance-dependent reproductive skew and competition for mates in communally nesting female house mice. *Behavioral Ecology and Sociobiology*, 56, 298–305.

Ryan, S. J., Starks, P. T., Milton, K., & Getz, W. M. 2008. Intersexual conflict and group size in *Alouatta palliata*: A 23-year evalution. *International Journal of Primatology*, 29, 405–420.

Sanchez, M. M. 2006. The impact of early adverse care on HPA axis development: Non-human primate models. *Hormones and Behavior*, 50, 623–631.

Sapolsky, R. & Ray J. 1989. Styles of dominance and their physiological correlates among wild baboons. *American Journal of Primatology*, 18, 1–13.

Sapolsky, R. M. 2004. Social status and health in humans and other animals. *Annual Review of Anthropology*, 33, 393–418.

Sapolsky, R. M., Alberts, S. C. & Altmann, J. 1997. Hypercortisolism associated with social subordinance or social isolation among wild baboons. *Archives of General Psychiatry*, 54, 1137–1143.

Schülke, O., Bhagavatula, J., Vigilant, L., & Ostner, J. 2010. Social bonds enhance reproductive in male macaques. *Current Biology*, 20, 2207–2210.

Seeman, T. A. 1996 Social ties and health: The benefits of social integration. *Annals of Epidemiology*, 6, 442–445.

———. 2002. Using the "F" word in primatology. *Behavior*, 139, 421–446.

———. 2007. The adaptive value of sociality in mammalian groups. *Philosophical Transactions of the Royal Society*, 362, 539–559.

Silk, J. B., Cheney, D. L., & Seyfarth, R. M. 1996. The form and function of post-conflict interactions among female baboons. *Animal Behaviour*, 52, 259–268.

Silk, J. B., Altmann, J., & Alberts, S. C. 2006a. Social relationships among adult female baboons (*Papio cynocephalus*) I. Variation in the strength of social bonds. *Behavioral Ecology and Sociobiology*, 61, 183–195.

Silk, J. B., Alberts, S. C., & Altmann, J. 2006b. Social relationships among adult female baboons (*Papio cynocephalus*) II: Variation in the quality and stability of social bonds. *Behavioral Ecology and Sociobiology*, 61,197–204.

Silk, J. B., Beehner, J. C., Berman, T. J., Crockford, C., Engh, A. L., Moscovice, L. R., Wittig, R. M., Seyfarth, R. M., & Cheney, D. L. 2009. The benefits of social capital: Close social bonds among female baboons enhance offspring survival. *Proceedings of the Royal Society London, Series B.* 276, 3099–3104.

———. 2010a. Female chacma baboons form strong, equitable, and enduring social bonds. *Behavioral Ecology and Sociobiology*, 60, 197–204.

———. 2010b. Strong and consistent social bonds enhance the longevity of female baboons. *Current Biology*, 20, 1359–1361.

Smith, K., Alberts, S. C., & Altmann, J. 2003. Wild female baboons bias their social behaviour towards paternal half sisters. *Proceedings of the Royal Society, London B*, 270, 503–510.

Steenbeek, R., & van Schaik, C. P. 2001. Competition and group size in Thomas' langurs (*Presbytis thomasi*): The folivore paradox revisited. *Behavioral Ecology and Sociobiology*, 49, 100–110.

Stokes, E. J., Parnell, R. J., Olejniczak, C. 2003. Female dispersal and reproductive success in wild western lowland gorillas (*Gorilla gorilla gorilla*). *Behavioral Ecology and Sociobiology*, 54, 329–339.

Takahata, Y., Koyama, N., Ichino, S., & Nakamichi, M. 2006. Influence of group size on reproductive success of female ringtailed lemurs: Distinguishing between IGFC and PFC hypotheses. *Primates*, 46, 383–387.

Tardiff, S. 1997 The bioenergetics of parental behavior and the evolution of alloparental care in marmosets and tamarins. In *Cooperative Breeding in Mammals* (ed. by N. G. Solomon & J. A. French), 11–33. Cambridge: Cambridge University Press.

Taylor, S. E., Cousino, K. L., Lewis, B. P., Gruenewald, T. L., Gurung, R. A. & Updegraff, J. A. 2000. Biobehavioral responses to stress in females: Tend-and-befriend not fight-or-flight. *Pyschological Review*, 107, 411–429.

Thorsteinsson, E. B. & James, J. E. 1999. A meta-analysis of the effects of experimental manipulations of social support during laboratory stress. *Psychology and Health*, 14, 869–886.

Treves, A. 2001. Reproductive consequences of variation in the composition of howler monkey (*Alouatta* spp.) groups. *Behavioral Ecology and Sociobiology*, 50, 61–71.

Treves, A. & Chapman, C. A. 1996 Conspecific threat, predation avoidance, and resource defense: Implications for grouping in langurs. *Behavioral Ecology and Sociobiology*, 39, 43–53.

Van Belle, S. & Estrada, A. 2008. Group size and composition influence male and female reproductive success in black howler monkeys (*Alouatta pigra*). *American Journal of Primatology*, 6, 613–618.

Van Noordwijk, M. A. & van Schaik, C. P. 1999. The effects of dominance rank and group size on female lifetime reproductive success in wild long-tailed macaques, *Macaca fasicularis*. *Primates*, 40, 105–130.

Virgin C. E., & Sapolsky R. l997. Styles of male social behavior and their endocrine correlates among low-ranking baboons. *American Journal of Primatology*, 42, 25–39.

Weinstock, M. 2008 The long-term behavioural consequences of prenatal stress. *Neuroscience and Biobehavioral Reviews*, 32, 1073–1086.

Williams, J. M., Pusey, A. E., Carlis, J. V., Farms, B. P.& Goodall, J. 2002. Female competition and male territorial behaviour influence female chimpanzees' ranging patterns. *Animal Behaviour*, 63, 347–360.

Williams, J. M., Oehlert, G. W., Carlis, J. V., & Pusey, A. E. 2004. Why do male chimpanzees defend a group range? *Animal Behaviour*, 68, 523–532.

Wittig, R. M., Crockford, C., Wikberg, E., Seyfarth, R. M., Cheney, D. L. 2007. Kin-mediated reconciliation substitutes for direct reconciliation in female baboons. *Proceedings of the Royal Society of London B*, 274, 1109–1115.

Wittig, R. M., Crockford, C., Lehmann, J., Whitten, P. L., Seyfarth, R. M., Cheney, D. L. 2008. Focused grooming networks and stress alleviation in wild female baboons. *Hormones and Behavior*, 54, 170–177.

Wrangham, R. W., Gittleman, J. L. & Chapman, C. A. 1993. Constraints on group size in primates and carnivores: Population density and day range as assays of exploitation competition. *Behavioral Ecology and Sociobiology*, 32, 199–209.

Chapter 25 Social Regard: Evolving a Psychology of Cooperation

Keith Jensen

PRIMATES ENGAGE in a variety of acts that provide immediate and delayed benefits to others, such as participating in hunts and territorial defense, sharing food, giving alarm calls, mobbing predators, grooming partners, and supporting others in conflicts. They also act in ways that are harmful to others—from relatively minor acts, such as avoiding social partners and taking food by stealing or begging, to injurious and even lethal acts of physical violence, including infanticide (e.g., Kappeler & van Schaik 2006; chapters 19, 21–24, this volume). Behavioral ecologists have devoted considerable effort to understanding the evolution of these kinds of behaviors, but we are just beginning to consider the psychological mechanisms that drive them. A consideration of these mechanisms is relevant to discussions of the evolution of sociality because natural selection is expected to favor traits that lead to adaptive behaviors. Insights into psychology can inform debates about the evolution of sociality. Bacteria, social insects, primates, and humans cooperate, often in superficially similar ways, but differing selective pressures, phylogenetic histories, and biological constraints will lead to different solutions to the challenges of cooperation. These solutions, at least in the case of humans and other primates, will involve psychological mechanisms of some sort, and perhaps similar ones.

Interest in these mechanisms has arisen as social scientists have begun to probe the nature of human cooperation (e.g., Fehr & Fischbacher 2003; Richerson & Boyd 2005; Tomasello, et al. 2005a; chapter 26, this volume). Division of labor, activities coordinated with strangers, and support of public goods typify the way in which humans cooperate; furthermore, people ostracize free-riders, chastise noncom-

formists, and impose costly punishments on transgressors even when they are not directly affected themselves. Converging evidence suggests that people are motivated to help other people out of a concern for their welfare (e.g., Batson 1991; Herrmann et al. 2008). In addition, moral sentiments, such as a sense of justice and a concern for fairness, motivate people to punish transgressions. Where do these social concerns come from? How do they work? Do we see their antecedents in our living primate relatives, or are they derived traits that evolved sometime in the last five to seven million years of human evolution?

Here, I will address these questions by summarizing the emerging literature that delves into the psychological underpinnings of primate social behavior. First, I will discuss the relationship between the ultimate and proximate levels of social analysis, and establish the conceptual foundation for identifying social regard. Then I will explore the psychological bases for helping and harming behavior, namely social concerns (sensitivity to the welfare of others), social intentions (the ability to infer the goals of others), and social preferences (motivations for outcomes affecting others). For these social motivations, I will evaluate the evidence for both their positive (prosocial) and negative (antisocial) effects. Simply put, do primates other than humans know how others feel? Do they know what they want? Do they care?

The Adaptive Basis of Social Behavior

Hamilton (1964) provided a useful classification of social behaviors based on fitness effects resulting from an individual's behavior (see chapter 21, table 21.1). The individual's behavior can produce fitness gains or losses for both the actor and the recipient, resulting in four categories of interactions. These effects are considered prosocial if the recipient accrues positive fitness gains, or antisocial if the recipient suffers fitness losses. The prosocial behaviors can be subdivided into mutualism, which is beneficial to both the actor and the recipient, and altruism, which is costly to the actor but beneficial to the recipient. Similarly, antisocial behaviors can be divided into selfishness, if the actor benefits at the recipient's expense, and spite, which is costly to both. Typically, it is very difficult to show the lifetime fitness effects of behaviors, especially in long-lived species like primates, so most research focuses on the immediate costs and benefits of single acts (Silk 2005; West et al. 2007a, b; Bshary & Bergmüller 2008; Clutton-Brock 2009). I will focus on the proximate psychological mechanisms that impel the actor to act prosocially or antisocially; furthermore, I will emphasize the immediate consequences for the recipient.

The fact that evolutionary biologists share the same terms with social scientists—notably altruism and spite—creates considerable confusion (West et al. 2007b; Bshary & Bergmüller 2008). Since the emphasis of this chapter is on proximate mechanisms, "evolutionary altruism" and "evolutionary spite" will be used to refer to the ultimate level; "functional" will preface altruism and spite, and "helping," "sharing," "vengeance," "punishment," and other commonly used terms will also be used for functional descriptions of behavior; and "psychological altruism" and "psychological spite," as well as "altruistic motives," "spitefulness," and so on, will refer to the psychological (proximate) mechanisms (Jensen 2010).

Proximate explanations are not incompatible with ultimate explanations; fitness effects are mediated through proximate mechanisms in the form of rewards and punishments, and these govern behaviors in the present. The important point that has been raised in the past is that proximate and ultimate causes are not opposing causes (Mayr 1961; Tinbergen 1963). Proximate causes are themselves subject to further levels of analysis (Seed et al. 2009). Kin selection and reciprocity, which are discussed in chapters 21 and 22, explain how cooperative behaviors can evolve; they neither support nor preclude different psychological mechanisms. Just as natural selection favors the evolution of behaviors that confer direct or indirect fitness benefits to individuals, it is also expected to shape the proximate mechanisms that lead animals to perform adaptive behavior. For example, animals that avoid predators will live longer and leave behind more offspring than animals that seek out predators, and proximate mechanisms such as fearful responses in the presence of dangerous stimuli

will be favored. However, there is not always a direct correspondence between proximate mechanisms and their adaptive outcomes, just as not all behaviors are adaptive. To borrow an example from West et al. (2007a), it is unlikely that elephants were under selective pressure to defecate, despite the benefit for dung beetles. Similarly, selfishly motivated acts can lead to benefits for the recipient as a by-product, and socially motivated acts can incidentally result in benefits for the actor. For example, one person might help a blind man cross the street to impress his girlfriend (selfish motive, socially beneficial outcome), while someone else might provide assistance because he is concerned about the blind man's safety, but inadvertently lead him to the wrong place (prosocial motive, socially harmful outcome).

The Adaptive Basis of Prosocial Behavior

Cooperative behaviors are those that have been selected to provide a benefit for the recipient; the acts can be altruistic (evolutionarily) in that the recipient benefits at a cost to the actor, or mutually beneficial in that both actor and recipient benefit. Both functional and evolutionary altruism are of particular interest to primatologists because acts that are costly for an actor and beneficial to the recipient are not expected to evolve by natural selection. The challenge has been to find adaptive mechanisms to explain behaviors that benefit others. The emphasis of this line of research has been on inclusive fitness (Hamilton 1964), an example of indirect fitness benefits, and reciprocity (Trivers 1971), an example of direct fitness benefits (West et al. 2007b; chapters 21 and 22, this volume). In mutualistic interactions, both the actor and recipient derive fitness benefits, but individuals may opt to free-ride on the efforts of others instead of contributing. For example, they might lag behind in risky encounters (Kitchen & Beehner 2007). Mutually beneficial acts are therefore also challenging to explain from an evolutionary perspective because the temptation to free-ride can derail cooperative outcomes.

The Adaptive Basis of Antisocial Behavior

Antisocial behaviors can readily evolve by natural selection. For example, traits that enhance fighting ability can be adaptive because they improve the individual's success in competitive encounters. Competitive interactions are evolutionarily selfish because they provide immediate benefits to the actor, such as access to mates and food, while imposing fitness costs on the losing recipient. Similarly, functional punishment can benefit the actor by achieving and maintaining social dominance, deterring cheaters and parasites, disciplining offspring, and coercing sexual partners (Clutton-Brock & Parker 1995a). What distinguishes functional punishment from other adaptively selfish behaviors is that the benefits to the actor are delayed. For example, male hamadryas baboons (*Papio hamadryas*) sometimes threaten or attack females that stray from the harem (Kummer 1968; Clutton-Brock & Parker 1995b). The delay in benefits to the actor distinguishes functional punishment from other evolutionarily selfish behaviors because the act is costly at the time it is performed, making it a form of delayed-benefits (functional) spite (Clutton-Brock & Parker 1995a). However, it is difficult to distinguish behaviors that were selected because they changed the subsequent behavior of others from simple acts of aggression that provide immediate benefits (Silk 1992, 2002; Jensen & Tomasello 2010).

Evolutionary spite—suffering a fitness cost to impose a fitness cost on others—is more problematic. It can only evolve when the actor and the target are less related to each other than to the population as a whole (Hamilton 1970). Functionally spiteful behaviors that reduce the fitness of others can indirectly benefit more closely related individuals (Hamilton 1970; Wilson 1975; Gardner & West 2006). It is not clear whether the conditions for evolutionary spite are fulfilled in any primate species, and all previously reported examples of functional spite in primates have been reinterpreted as evolutionarily selfish behaviors (Foster et al. 2001). However, functionally spiteful acts are very different than functionally selfish and mutualistic acts. For example, moving to another tree to eat fruit when competitors are around (scramble competition), chasing others away from the tree (contest competition), and destroying fruit in a rival's tree so that it will starve (spiteful competition) are all adaptively selfish behaviors, yet they are very different phenomena (Jensen 2010).

The Nature of Social Motives

The psychological mechanisms that drive individuals to act are intentions and motives. Psychologists define an intention as a plan of action to achieve a goal (Tomasello & Rakoczy 2003; Tomasello et al 2005a). Motivations are internal states or external influences (incentives) that lead to goal-oriented actions; they can be thought of as the driving force behind intentions. For instance, a hungry individual is motivated to eat. The sight of food alone might motivate consumption. Motivations and incentives can lead to the intention to get the food. If the path to the food is blocked, the animal can change its plan and find a different path, a different means to get the food, or it can search for a different food. Basic, nonsocial motivations and intentions behind activities such as obtaining food are straightforward. The motivations and goals are selfish—getting food to satisfy hunger. There is no motivation or goal that is *for* the fruit.

Social intentions and social motivations are more complex. They lead to actions to achieve goals that are chosen *for their effect* on other individuals. It is the effect on the other individual that is the motivator, not the effect on the actor. Adam Smith referred to the motivations that cause us to act for others as "moral sentiments" (Smith 1759/2005). However, socially motivated acts have effects on the actor as well as another individual, thus making it difficult to determine for a given act whether the actor is motivated by the effects on himself or by the effects for the recipient. While an individual can benefit from his or her actions (for instance, feeling good about helping someone, or receiving delayed benefits), this is not the reason for the action. The effect on the other individual is the ultimate goal—the end in itself—and not an ulterior means to some other end (Batson 1991). Altruists help others for the benefit of the recipients; any personal gains (such as feeling good) are unintended by-products. To take a more concrete example, consider what happens when one female monkey grooms another. The groomer may be motivated out of a desire to help the recipient relax or to clean her wounds; if so, she is motivated by the positive effects that grooming has on the recipient. Alternatively, the groomer may be motivated to groom because the act reduces her own levels of tension, or because it provides her with nutritious ectoparasites to eat; the consequences for the recipient might have little or no bearing on the motivations of the actor. In other words, "altruistic" benefits might just be unintended by-products of selfish motives. The fact that a behavior has a social consequence does not necessarily mean that it has a social motivation any more than eating a fruit has a fruit-motivation. The effects can be unintended by-products of selfish motivations. However, if the individual has nothing to gain from his or her actions, particularly if he or she persists at trying to achieve a goal for the other when his or her initial attempts are blocked, then it is plausible that a social motivation is at work.

Social Concern

One of the ways in which an individual can be motivated for others is out of concern for their welfare. Social concerns are emotions that are influenced by the welfare of others, and these can lead to the intention to perform socially motivated behaviors. Welfare can be the emotions of others (e.g., happiness or sadness), or the conditions that would ultimately lead to such emotions (e.g., success or fail-

Table 25.1. Social concern matrix (from Ortony et al. 1988). As in the Hamilton payoff matrix, payoffs are to the actor and to the recipient. In this matrix, however, instead of fitness, benefits and costs are measured as positive and negative feelings.

	Individual B Positive feelings	Individual B Negative feelings
Individual A Positive feelings	Symhedonia (+ , +)	Schadenfreude (+ , –)
Individual A Negative feelings	Jealousy (–, +)	Empathy (–, –)

ure at achieving a goal). Emotions of an actor that are attuned to the welfare of others are called fortunes-of-others emotions (Ortony et al. 1988). One can categorize various forms of social concern based on their positive and negative effects in much the same way that Hamilton categorized fitness payoffs for actors and recipients (Ortony et al. 1988; table 25.1). In this classification, sadness and suffering are negative; happiness and pleasure are positive. Representative examples of the four cells of the social concern matrix are symhedonia (vicarious joy), feeling happy for someone else's good fortunes; jealousy, feeling angry at the fortunes of others; schadenfreude, feeling pleasure at the misfortunes of others; and empathy, feeling unhappy at the misfortunes of others.

Positive Social Concern

Empathy, the prototypical example of social concern, is having the emotions appropriate to the circumstances of another individual (Hoffman 1982; Eisenberg et al. 1991; Eisenberg 2002). The original German term from which "empathy" is derived—*einfühlung*, which translates literally as "feeling into"—captures this meaning (Wispé 1987). There is an important difference between emotional contagion and empathy. Emotional contagion is essentially "catching" another's emotions. While this can lead to empathy or sympathy, it can also lead to personal distress and emotions or behaviors that are self-oriented. Empathy, on the other hand, involves affective perspective-taking, and it can motivate functionally altruistic behavior (Batson 1991). To what degree theory of mind, namely an understanding of the thoughts, beliefs, and desires of others, is important to affective perspective-taking in humans is unclear (Harwood & Farrar 2006), but some form of putting oneself in the position of others is a necessary part of empathy. For instance, one-and-a-half-year-old children who see an adult experience a distressing situation provide help, even though the adult has not exhibited emotional distress (Vaish et al. 2009). Recent work suggests that chimpanzees (*Pan troglodytes*) have some perspective-taking abilities, though not

a humanlike understanding of beliefs and desires of others (Call & Tomasello 2008; chapter 30, this volume).

The practical difficulty with identifying social concerns is that they do not necessarily lead to actions, and helpful acts do not necessarily require social concern (e.g., Krebs 1970, 1971; Koski & Sterck 2010). Individuals may empathize with others but not help them. For instance, a parent watching a child struggle with a difficult task might resist helping so that the child solves the problem herself. In addition, individuals might provide assistance out of a sense of duty or a concern for their reputation, but have little concern for the other's welfare.

It is therefore very difficult to infer whether nonhuman animals feel social concern. Despite this, the inference has often been made. When other animals perform behaviors that are similar to human actions, like reaching out to a distressed infant or putting an arm around a sick partner, it is very tempting to assume that the same motives are at work. The temptation to assume common motives is especially great with species that are closely related to humans, like chimpanzees, and animals that are intimately associated with humans, such as dogs. Chimpanzees have been inferred to experience grief at the loss of a close associate (Brown 1879), sympathy and pity for a sick conspecific (Yerkes & Yerkes 1929), concern for others who have been in conflicts (de Waal & van Roosmalen 1979), and empathy for injured group members (Boesch 1992). Some primatologists are firmly convinced that other primates are empathic (O'Connell 1995; Preston & de Waal 2002; de Waal 2008).

These claims are very hard to verify. Anecdotal accounts, which are often used as supporting evidence, are especially problematic for at least two reasons. First, it is very difficult to test alternative interpretations of one-time events. Second, the absence of helping, when empathic accounts would predict it, is likely to be underreported (Silk 2007a). Unfortunately, even systematic observations of functionally altruistic acts are unlikely to allow us to tease apart empathy and personal distress. For example, third-party affiliation after conflicts, often called consolation, is sometimes interpreted as being driven by empathy (de Waal & van Roosmalen 1979; Preston & de Waal 2002). According to this view, conflict creates distress, and observers are motivated to alleviate the victim's suffering. However, conflicts are sometimes followed by aggression toward previously uninvolved bystanders, and it is possible that "consolation" may reduce the observers' own concerns about becoming the target of redirected aggression (Koski & Sterck 2009). Even if the consoled individual benefits from reduced stress (Fraser et al. 2008), which is not always the case (Koski et al. 2007), this might be a by-product of the observers' motivation to reduce their own distress. The same may be

Fig. 25.1.   Consolation in chimpanzees may involve tactile contact (A) and tandem walking (B), and may happen more for the sake of the consoler than for the victim. Photos courtesy of the Max Planck Institute for Evolutionary Anthropology.

true of grooming, because groomers may also experience reduced stress (e.g., Shutt et al. 2007; Aureli & Yates 2009); benefits to the groomee could be a by-product of a self-concerned act (fig. 25.1).

It not possible to draw strong conclusions about the psychological motivations or emotions that underlie naturally occurring behaviors, and most research on empathy in humans relies on experiments (cf. Batson 1991). Only a few experiments on empathy in nonhuman animals have been conducted. Nearly 50 years ago, researchers discovered that rats and rhesus monkeys avoided actions that caused pain and distress to conspecifics (Church 1959; Miller et al. 1959; Masserman et al. 1964; see also Langford et al. 2006 for mice). It is not clear from these experiments whether the animals were reacting to their own distress or were motivated out of concern for the others' welfare (Parr 2002; Silk 2007a). In a more recent experiment, chimpanzees showed physiological responses of fearfulness to images of aversive stimuli (hypodermic needles, dartguns) as well as to images of conspecifics being darted and injected (Parr 2001). Because the chimpanzees were as upset by images of needles and dart guns as they were by images of conspecifics being injected, it is difficult to know whether they were aroused by personal distress or empathy (Parr 2002).

Negative Social Concern

Negative social concern occurs when there is a conflict between the positive well-being of the actor and another individual. Feelings such as jealousy, envy, schadenfreude, and gloating are examples of negative social concern. All of these sentiments can be regarded as social because they are driven by a sensitivity to the welfare of others, albeit to their misfortune. They cannot be considered positive in the social sense, even though schadenfreude and gloating

might feel good. For instance, men showed an increased activation in the reward circuit of the brain and a decrease in activation in a region associated with empathy when they saw a cheater in a prisoner's dilemma game receive a painful electric shock (Singer et al. 2006).

Just as empathy motivates functional altruism, in which the primary goal is to improve another individual's welfare, negative social concerns motivate functionally spiteful behavior, in which the primary goal is to diminish another's welfare. Negative social concern can lead to beneficial outcomes such as deterring noncooperative behavior in others. Of course, deterrence of noncooperators also may arise out of other, more moralistic, sentiments (Trivers 1971; Hoffman 2000; Price et al. 2002).

Darwin (1872) imputed jealousy to various animals, including insects, dogs, and orangutans (*Pongo* spp.). Primates sometimes do things that seem costly to themselves and the recipients. For example, stump-tailed macaques (*Macaca artoides*) interfere with copulating pairs (Brereton 1994), and vervet monkeys (*Chlorocebus pygerythrus*) destroy food without consuming it more often near the home ranges of other groups (Horrocks & Hunte 1981). Though these do not satisfy the criteria for evolutionary spite because of potential delayed benefits (Foster et al. 2001), they might be motivated by negative social concerns like jealousy. As with empathy, however, it is very difficult to infer whether antisocial sentiments motivate these behaviors.

Social Intentions

Seeking to understand social motivations by inferring social concerns basically asks the question: do primates care about the welfare of others? Another possibly more tractable approach is to ask: Do they know what others want? This approach attempts to see whether primates recognize the intentions of others and use their intention-reading ability to act in ways that are motivated by the goals of others. There is some suggestive evidence that primates other than humans interpret the behavior of others psychologically (Tomasello et al. 2003), at least to some degree, rather than only behaviorally (Povinelli & Vonk 2003; chapter 30, this volume). Being able to infer the goals of others would allow for more sophisticated social behavior such as intentional deception and coordinated cooperative activities. If primates recognize others as having intentions, then they might use their intention reading to help others achieve their goals if they are prosocially motivated, or to thwart those goals if they are antisocially motivated.

Positive Social Intentions

It is difficult to assess the social motives of individuals who are cooperating. In some experimental settings, for instance, pairs of individuals can learn to work together to achieve a mutually beneficial outcome (reviewed in Silk 2007b). In mutualistic tasks like these, the individuals might be motivated to help their partner achieve her goals, or to achieve the collective, shared goal—what Tomasello and colleagues (2005a) refer to as joint intentions. However, it is conceivable that each individual has the goal of getting the food, recognizes that the partner is essential to achieving that goal (Tomasello et al. 2005b), and uses him as a social tool to achieve his own ends. Although chimpanzees prefer partners with whom they can successfully cooperate (Melis et al. 2006a), benefits to those partners may be incidental by-products of the selfish goal of achieving rewards for themselves.

Recent studies of instrumental helping in chimpanzees have provided more compelling evidence for prosocial motives based on an understanding of others' goals (Warneken & Tomasello 2009). For a brief summary of these studies, see table 25.2. In these experiments, subjects witness a situation in which another individual is unable to achieve her goal, such as retrieving an object that is out of her reach. Contrary to results from studies of reciprocity and cooperation, the actor derives no immediate material benefits from helping. One research paradigm had experimenters reach for or request an out-of-reach object; chimpanzees were willing to return objects even in the absence of food rewards (Warneken & Tomasello 2006; Warneken et al. 2007). They also were more likely to retrieve objects when they were dropped accidentally than when they were thrown out of reach, thus supporting the suggestion of intention reading (Warneken & Tomasello 2006). The subjects also invested some effort to retrieve objects from relatively distant locations and helped both familiar and unfamiliar humans (Warneken et al. 2007). The chimpanzees did not respond to requests for help in any other tasks, perhaps because they did not understand the needs of the experimenters (Warneken & Tomasello 2006). Were these chimpanzees motivated by the social consequences of their actions? The fact that individuals helped a human achieve a goal does not guarantee that they were motivated by the effect their actions had on the human. Subjects in all of the reaching studies had prior training in handing objects when requested to their caregivers, and had been rewarded in the past for doing so. Handing objects to others when this is requested may be compliant, but it does not mean that it is motivated *for* the other. Dogs also fetch objects, but there is no reason to believe that they have the stick thrower's goal in mind, or that they are motivated out of a concern for the stick thrower's welfare.

More convincing evidence of prosocial motivations comes from an experiment in which chimpanzees could open a door for another member of their group (Warneken et al. 2007). In this study, the subject watched another chimpanzee attempt to open a door to gain access to an adjacent enclosure. The other chimpanzee's motivation was to get a piece of food; the food was hidden from the subject's view to reduce competitive motivations that might interfere with helping. The subject had previously learned that by releasing a peg, the chain holding the door shut would slacken, allowing the partner to enter the enclosure. Subjects opened that door more often when the partner was trying to get into the adjacent enclosure than in a control condition in which the other chimpanzee was trying to get into an enclosure that the subject could not open (see fig. 25.2). This is a novel behavior that would not have been reinforced in the past, and it involved a conspecific rather than a human. However, it is not clear whether the subjects were opening the door in response to a noisy demand, or were more attracted to the peg holding the door when it was shaken (stimulus enhancement). In other words, subjects might have been motivated to release the peg because it was an interesting thing to do, particularly when doing so had an interesting outcome (they had to be given another rope as a distracter when preliminary data showed that they released the peg at an overly high rate when there was no apparent reason for doing so). While the finding is intriguing, the conclusions regarding prosocial motivations and goal understanding are equivocal.

To address the role of goal understanding in eliciting help, chimpanzees in another study were given tools that were useful to their partners, but not to themselves (Yamamoto et al. 2009). One subject needed a straw to drink juice from a box attached to the wall, and another needed a cane to hook and pull closer a juice box that was out of reach. Each subject had the other's tool, which they consistently gave to their partner, but only when the partner requested it by trying to capture their attention and actively requesting the tool. Watching a partner struggle to achieve a goal did not induce either chimpanzee to help. This finding contradicts the notion that chimpanzees recognize the goals of others and are spontaneously motivated to help. They comply with low-cost requests, but do not anticipate them.

No study has yet given subjects a choice between an outcome that would help the partner and one that would not help, to determine whether primates actually recognize the goals of other individuals or simply treat all requests as signals to do something. This may be an important point to consider in chimpanzees and other species where begging is used to induce food sharing (Stevens 2004; Gilby 2006; Burkart et al. 2007) and attention getters are used to recruit play partners, beg for food, and so on (fig. 25.3; Tomasello 2008).

Negative Social Intentions

Intention reading can allow an actor to gauge whether another individual intends harm, or has a competitive goal or any other goal that the actor will be motivated to thwart. For instance, the Machiavellian intelligence hypothesis suggests that competition strengthens cognitive skills and favors the evolution of larger brains (Humphrey 1976; Byrne & Whiten 1988; Dunbar 1998; chapter 30, this volume). Animals use tactics such as punishment to deter noncooperators and coerce others (de Waal 1982; de Waal & Luttrell 1988; Clutton-Brock & Parker 1995a), but it is not clear that these antisocial behaviors are based on an understanding of the intentions of others. It is even difficult to determine whether the goal of individuals who punish or coerce others is to change their behavior. For example, the suggestion that monkeys functionally punish individuals for a noncooperative behavior such as failing to give a food call (Hauser 1992) may be explained more simply as aggression arising over conflict for food when ownership has not been signaled (Gros-Louis 2004). Any effects on the subsequent behavior of the target of aggression may be unintended by-products of the "punisher" trying to obtain food or some other resource such as dominance status or mates (fig. 25.4). It is even more challenging to infer what the punisher knows of the intentions of others (see also chapter 30, this volume).

My colleagues and I designed an experiment to determine how chimpanzees would respond to intentional transgressions directed against them (Jensen et al. 2007a; table 25.2). This study is similar to a money-burning game in economics, in which an individual can pay a cost to destroy another person's money (Zizzo & Oswald 2001). In the chimpanzee study, food was placed on a sliding tray that stood on a table between two enclosures. The subject had access to the food on the tray, but after a short time another chimpanzee, the "thief," came into the adjacent enclosure. The thief could appropriate the food by pulling a rope that slid the tray away from the subject. The subject could not get the food back, but she could "punish" the thief by collapsing the table supporting the food tray, scattering food to the floor, beyond the reach of either chimpanzee. Subjects collapsed the table more often when their food was stolen than in two other conditions in which they lost their food when the experimenter moved it away, either in front of an empty enclosure or toward another chimpanzee. In the latter condition, the chimpanzees were not functionally spiteful, that is, they were no more frustrated at seeing another chimpanzee eat their food than they were at simply losing it. Chimpanzees appeared insensitive to unfair outcomes (more on this later). Emotions may have played an important role, because the subjects were more likely to collapse the table when they showed signs of anger (displays and tantrums). This is consistent with the experimental economics literature on humans that shows that anger motivates punishment (e.g., Frank 1988; Pillutla & Murnighan 1996). The fact that chimpanzees responded more strongly to theft than other forms of food loss suggests that their punitive or vengeful responses were motivated by the behavior of other chimpanzees. However, it is still not clear what motivated the subjects. Did they recognize the harmful intent of their rivals and attempt to thwart their goals? Did they attempt to modify the behavior of the thieves (despite the finding that theft increased over time while punishment decreased)? Or were they simply reacting vengefully in response to a personal harm with no regard for the consequences for the target?

Social Preferences

Preferences are motivational choices between outcomes. For example, an individual might prefer bananas over apples. Given a choice, she would eat bananas before she would eat apples. As with simple motivations, the effects on the apples and bananas are not likely to influence the individual's choice. In contrast, social (other-regarding) preferences are motivated by their effect on the target of the action. Individuals who systematically prefer outcomes that benefit others have prosocial (positive other-regarding) preferences, while

Table 25.2. Summary of experimental studies of primate social regard. (The many interesting studies of mutualistic cooperation and reciprocity are not included since these are not designed to distinguish between self-regarding and other-regarding preferences.) Social outcome studies are distinguished as instrumental helping, food sharing (dictator/prosocial games), functionally spiteful outcomes (money burning games), fairness preferences (ultimatum games), and disadvantageous inequity aversion (impunity games); see main text for descriptions. Payoffs are shown as pluses (benefits to actor/partner), minuses (costs) and zeroes (no outcome). Symbols before a slash represent payoffs to actors; partner payoffs (if any) are after the slash. Whether there are potential consequences for actor and partner are shown in separate columns, with additional columns for the potential benefits (+), harm (−) or absence of effect (0); outcome disparities are shown by additional pluses or minuses (e.g. +/+++, −/−−). If tested, prosocial and antisocial regard are described as absent (No), present (Yes) or not tested (not applicable: N/A). Note that these are the interpretations of the authors, not necessarily mine. See main text for further details. Studies are broken down by species, with number of subjects tested. Some papers have multiple studies, with each listed separately, though the same subjects are usually tested.

Social outcome	Personal effect	Social effect	Benefit or harm To actor	Benefit or harm To recipient	Prosocial regard	Antisocial regard	Species	n	Source
Instrumental helping									
Object returned to experimenter helpful (0/+)	No	Yes	0	+	No	N/A	*Cebus apella*	7	Barnes et al. 2008 (study 1)
Object returned to experimenter helpful (0/+)	No	Yes	0	+	No	N/A	*C. apella*	5	Barnes et al. 2008 (study 2)
Object returned to experimenter helpful (0/+)	No	Yes	0	+	Yes	N/A	*Pan troglodytes*	3	Warneken & Tomasello 2006
Object returned to experimenter helpful (0/+)	No	Yes	0	+	Yes	N/A	*P. troglodytes*	36	Warneken et al. 2007 (study 1)
Object returned to experimenter helpful (−/+)	No	Yes	−	+	Yes	N/A	*P. troglodytes*	18	Warneken et al. 2007 (study 2)
Door opened for partner helpful (0/+)	No	Yes	0	+	Yes	N/A	*P. troglodytes*	9	Warneken et al. 2007 (study 3)
Tool handed to partner helpful (0/+)	No	Yes	0	+	Yes	N/A	*P. troglodytes*	6	Yamamoto et al. 2009 (study 1)
Tool handed to partner helpful (0/+)	No	Yes	0	+	Yes	N/A	*P. troglodytes*	6	Yamamoto et al. 2009 (study 2)
Food-sharing (dictator/prosocial games)									
Food outcome for partner null vs. altruistic (0/0 vs. 0/+)	No	Yes	0	+ or 0	Yes	N/A	*Callithrix jacchus*	15	Burkart et al. 2007
Food outcome for partner altruistic (0/+)	No	Yes	0	+	Yes	N/A	*Saguinus oedipus*	5	Hauser et al. 2003 (study 1)
Food outcome for partner null vs. altruistic (0/0 vs. 0/+)	No	Yes	0	+ or 0	No	N/A	*S. oedipus*	14	Cronin et al. 2009 (study 2)
Food outcome for self and partner selfish vs. altruistic (+/0 vs. +/+)	Yes	Yes	+	+ or 0	No	No	*S. oedipus*	14	Cronin et al. 2009 (study 1)
Food outcome for self and partner selfish vs. altruistic (+/0 vs. +/+)	No	Yes	0	+	Yes	N/A	*S. oedipus*	13	Cronin et al. 2010
Food outcome for self and partner selfish vs. generous (+/0 vs. +/+++)	Yes	Yes	+	+ or 0	Yes	N/A	*S. oedipus*	5	Stevens 2010
Food outcome for self and partner null vs. altruistic (0/0 vs. 0/+++)	No	Yes	0	+ or 0	Yes	N/A	*S. oedipus*	5	Stevens 2010
Food outcome for self and partner selfish vs. altruistic (+/0 vs. +/+)	Yes	Yes	+	+ or 0	Yes	No	*C. apella*	8	de Waal et al. 2008 (prosocial conditions)
Food outcome for self and partner mutualistic vs. generous (+/+ vs. +/++)	Yes	Yes	+	+ or ++	Yes	No	*C. apella*	8	de Waal et al. 2008 (inequity conditions)
Food outcome for self and partner mutualistic vs. generous (+/+ vs. +/++)	Yes	Yes	+	+ or ++	No	No	*C. apella*	8	de Waal et al. 2008 (partner visibility control)

Comparison							Species	N	Source
Food outcome for self and partner selfish vs. mutualistic (+/0 vs. +/+)	Yes	Yes	+	– or 0	Yes	No	C. apella	7	Lakshminarayanan & Santos 2008 (prosocial conditions)
Food outcome for self and partner mutualistic vs. generous (++/+ vs. +/++)	Yes	Yes	+	+ or ++	Yes	No	C. apella	7	Lakshminarayanan & Santos 2008 (inequity conditions)
Food outcome for self and partner mutualistic vs. generous (++/+ vs. +/++)	Yes	Yes	+	+ or ++	No	Yes	C. apella	8	Fletcher 2008
Food outcome for self and partner selfish vs. generous (+++/+ vs. +/+++)	Yes	Yes	++	+ or +++	Yes	No	C. apella	6	Takimoto et al. 2010 (middle value food condition; studies 1 & 2)
Food outcome for self and partner mutualistic vs. generous (+++/+ vs. +++/+++)	Yes	Yes	+	+ or +++	Yes	No	C. apella	6	Takimoto et al. 2010 (high value food condition; studies 1 & 2)
Food outcome for self and partner selfish vs. generous (++/+ vs. +/+++)	Yes	Yes	++	+ or +++	No	Yes	C. apella	6	Takimoto et al. 2010 (middle value food condition; study 3)
Food outcome for self and partner mutualistic vs. generous (+++/+ vs. +++/+++)	Yes	Yes	+	+ or +++	No	No	C. apella	6	Takimoto et al. 2010 (high value food condition; study 3)
Food outcome for partner altruistic (0/+)	No	Yes	0	+	No	N/A	Macaca mulatta	2	Wolfle & Wolfle 1939
Food outcome for partner altruistic (0/+)	No	Yes	0	+	No	N/A	Macaca nemestrina	3	Wolfle & Wolfle 1939
Food outcome for partner altruistic (0/+)	No	Yes	0	+	No	N/A	Papio hamadryas	1	Wolfle & Wolfle 1939
Food outcome for self and partner selfish vs. mutualistic (+/0 vs. +/+)	Yes	Yes	+	+ or 0	Yes	Selfish	M. mulatta	2	Colman et al. 1969
Food outcome for self and partner selfish vs. mutualistic (+/0 vs. +/+)	Yes	Yes	+	+ or 0	No	No	Macaca arctoides	2	Colman et al. 1969
Food outcome for self and partner selfish vs. altruistic (+/0 vs. 0/+)	Yes	Yes	+ or 0	+ or 0	No	Spiteful	Macaca fascicularis	11	Schaub 1996
Food outcome for self and partner selfish vs. mutualistic (+/0 vs. +/+)	Yes	Yes	+	+ or 0	Yes	Yes	M. fascicularis	20	Massen et al. 2010
Food outcome for self and partner selfish vs. mutualistic (+/0 vs. +/+)	Yes	Yes	+	+ or 0	No	N/A	P. troglodytes	7	Vonk et al. 2008 (study 1)
Food outcome for self and partner selfish vs. mutualistic (+/0 vs. +/+)	Yes	Yes	+	+ or 0	No	N/A	P. troglodytes	11	Vonk et al. 2008 (study 2)
Food outcome for self and partner selfish vs. mutualistic (+/0 vs. +/+)	Yes	Yes	+	+ or 0	No	No	P. troglodytes	7	Silk et al. 2005 (study 1)
Food outcome for self and partner selfish vs. mutualistic (+/0 vs. +/+)	Yes	Yes	+	+ or 0	No	No	P. troglodytes	11	Silk et al. 2005 (study 2)
Food outcome for self and partner selfish vs. mutualistic (+/0 vs. +/+)	Yes	Yes	+	+ or 0	No	No	P. troglodytes	11	Jensen et al. 2006 (study 1)
Food outcome for partner altruistic vs. weakly spiteful (0/– vs. 0/+)	No	Yes	0	+ or –	No	No	P. troglodytes	11	Jensen et al. 2006 (study 2)
Food outcome for partner altruistic vs. spiteful (0/– vs. 0/+)	No	Yes	0	+ or –	No	No	P. troglodytes	11	Jensen et al. 2006 (study 3)
Spiteful outcomes (money-burning games) Negative food outcome for partner spiteful (0/–)	No	Yes	0	–	N/A	No	P. troglodytes	13	Jensen et al. 2007a (study 1)
Negative food outcome for partner spiteful (0/–)	No	Yes	0	–	N/A	No	P. troglodytes	11	Jensen et al. 2007a (study 2)

(continued)

Table 25.2. (continued)

Social outcome	Personal effect	Social effect	Benefit or harm		Prosocial regard	Antisocial regard	Species	n	Source
			To actor	To recipient					
Fairness preferences (ultimatum games)									
Rejection of unfair food offer by partner spiteful (+/++ vs. –/– –)	Yes	Yes	+ or–	++ or – –	No	No	P. troglodytes	11	Jensen et al. 2007b
Disadvantageous inequity aversion (impunity games)									
Rejection of low-quality food from experimenter disadvantageous inequity aversion (–/0)	Yes	No	–	0	N/A	Yes	S. oedipus	11	Neiworth et al. 2009
Rejection of low-quality food from experimenter disadvantageous inequity aversion (–/0)	Yes	No	–	0	N/A	Yes	C. apella	5	Brosnan & de Waal 2003
Rejection of low-quality food from experimenter disadvantageous inequity aversion (–/0)	Yes	No	–	0	N/A	No	C. apella	6	Dubreuil et al. 2006
Rejection of low-quality food from experimenter disadvantageous inequity aversion (–/0)	Yes	No	–	0	N/A	No	C. apella	8	Roma et al. 2006
Rejection of low-quality food from experimenter disadvantageous inequity aversion (–/0)	Yes	No	–	0	N/A	No	C. apella	12	Dindo & de Waal 2007
Rejection of low-quality food from experimenter disadvantageous inequity aversion (–/0)	Yes	No	–	0	N/A	No	C. apella	5	Fontenot et al. 2007 (study 1)
Rejection of low-quality food from experimenter disadvantageous inequity aversion (–/0)	Yes	No	–	0	N/A	No	C. apella	5	Fontenot et al. 2007 (study 2)
Rejection of low-quality food from experimenter disadvantageous inequity aversion (–/0)	Yes	No	–	0	N/A	Yes	C. apella	13	van Wolkenten et al. 2007
Rejection of low-quality food from experimenter disadvantageous inequity aversion (–/0)	Yes	No	–	0	N/A	No	C. apella	7	Silberberg et al. 2009
Rejection of low-quality food from experimenter disadvantageous inequity aversion (–/0)	Yes	No	–	0	N/A	No	Pongo abelii	7	Bräuer et al. 2006
Rejection of low-quality food from experimenter disadvantageous inequity aversion (–/0)	Yes	No	–	0	N/A	No	P. abelii	4	Bräuer et al. 2009
Rejection of low-quality food from experimenter disadvantageous inequity aversion (–/0)	Yes	No	–	0	N/A	No	Gorilla gorilla	6	Bräuer et al. 2006
Rejection of low-quality food from experimenter disadvantageous inequity aversion (–/0)	Yes	No	–	0	N/A	No	Pan paniscus	4	Bräuer et al. 2006
Rejection of low-quality food from experimenter disadvantageous inequity aversion (–/0)	Yes	No	–	0	N/A	No	P. paniscus	5	Bräuer et al. 2009
Rejection of low-quality food from experimenter disadvantageous inequity aversion (–/0)	Yes	No	–	0	N/A	Yes	P. troglodytes	20	Brosnan et al. 2005
Rejection of low-quality food from experimenter disadvantageous inequity aversion (–/0)	Yes	No	–	0	Yes	Yes	P. troglodytes	16	Erosnan et al. 2010
Rejection of low-quality food from experimenter disadvantageous inequity aversion (–/0)	Yes	No	–	0	N/A	No	P. troglodytes	13	Bräuer et al. 2006
Rejection of low-quality food from experimenter disadvantageous inequity aversion (–/0)	Yes	No	–	0	N/A	No	P. troglodytes	6	Bräuer et al. 2009

Fig. 25.2. The door-opening (helping) study of Warneken et al. (2007, experiment 3) had a helper chimpanzee (shown at top) and a conspecific partner. The partner's goal was to get into the enclosure at the right because there was a banana inside. (The banana was concealed from the helper's view to reduce any competitive motives.) The helper could open the door by releasing a peg that held it shut, thereby allowing the partner through. To account for a chimpanzee's possible attraction to releasing the peg for no reason in particular, there was a control condition in which the banana was in the enclosure to the left, but the helper had no access to the peg holding it shut. Redrawn from Warneken et al. (2007).

those who systematically prefer outcomes that harm others have antisocial (negative other regarding) preferences. Self-regarding preferences, regardless of their effects on other individuals, are not social preferences. Tests of social preferences present individuals with a conflict between selfish outcomes and outcomes that affect others (table 25.2). Social preferences are what remains after selfish preferences have been overcome. (For reviews on social preferences in primates, see Silk 2007a, 2009.)

Fairness preferences are of particular interest to economists and psychologists because people compare themselves to others when assessing their own well-being (Festinger 1954), and a sensitivity to inequity may be an important element in human cooperation (Fehr & Fischbacher 2003). Preferences for equity can be prosocial when they motivate an individual to give resources to others who are less well off (advantageous inequity aversion). Preferences for equity can be antisocial when having less than others is aversive, motivating individuals to decrease the gains of others or increase their own gains relative to others (disadvantageous inequity aversion). Inequity aversion appears to be sensitive to both outcomes (gains and losses relative to others) as well as the intentions of others (Falk & Fischbacher 2006).

Positive Social Preferences

Several experiments have been designed to give subjects a choice between selfish outcomes and outcomes that benefit

Fig. 25.3. Active food transfer or sharing is not commonly observed in chimpanzees. Even between mother and offspring, food is typically yielded only in response to persistent begging or harassment (A) or "shared" passively by tolerating a grab (B). In some cases food is actively withheld (C). Photos courtesy of the Max Planck Institute for Evolutionary Anthropology.

others (table 25.2). The basic premise is similar to that of the dictator game used by economists, in which one player, the allocator, is given a sum of money he can then divide as he wishes with another player. The obvious expectation of rational self-interest is that allocators should give nothing and keep everything for themselves. However, people routinely share money, typically around 20% (Kahneman et al. 1986; Camerer 2003). The assumption is that people do so

Fig. 25.4. Displays and threats involving piloerection and ritualized movements (A) are part of the standard repertoire of aggression in male chimpanzees. These may serve to punish others—for instance, for behaving noncooperatively—but they may also serve to assert dominance, as in the case of this male on his rise to the alpha position, showing a fear grimace (B). Photo courtesy of the Max Planck Institute for Evolutionary Anthropology.

out of a concern for the outcomes of the other individuals, as there are no opportunities for reciprocity or reputation formation in these anonymous one-shot games. The dictator game is a useful tool for detecting positive social preferences, because a self-interested individual should not share at all. Any sharing, in principle, should be due to a concern for the positive well-being of another individual (though see Hagen & Hammerstein 2006). Early studies on cercopithecids gave individuals opportunities to deliver food to other individuals, something they rarely did (Wolfle and Wolfle 1939; Colman et al. 1969; Schaub 1996; but see Massen et al. 2010). To determine whether primates have preferences for outcomes that benefit others, two recent independent studies on chimpanzees made sharing as cost-free as possible (Silk et al. 2005; Jensen et al. 2006). Subjects could choose between a mutually beneficial (prosocial) option that delivered food to themselves and another chimpanzee (1/1 outcome), or a selfish option that resulted in food only for themselves (1/0 outcome; fig. 25.5). (The number before the slash represents the benefit or cost to the subject; the number after the slash is the partner's payoff.) To ensure that the chimpanzees did not simply prefer the prosocial option over the selfish one because the total amount of food visible was greater, each study had a control condition in which a similar piece of food went to an empty enclosure. In these experiments, actors incurred no material cost for choosing the 1/1 option over the 1/0 option. Even so, their behavior was not influenced by the presence of a recipient who could benefit from their choice. These results have been replicated in two other studies (Vonk et al. 2008; Yamamoto & Tanaka 2009). In all of these studies, the chimpanzees appeared indifferent to the effect of their choices on their partners. It has been suggested that the prospect of

getting food, however remote, distracts the subjects, making them oblivious to the needs of others (Warneken & Tomasello 2009; Yamamoto & Tanaka 2009a). It is true that chimpanzees have a very difficult time inhibiting their responses in the presence of food (Boysen & Berntson 1995). However, the same patterns were obtained in experiments in which the actors were unable to obtain rewards for themselves (Jensen et al. 2006, studies 2 and 3; Vonk et al. 2008).

One suggestion is that chimpanzees did not exhibit prosocial preferences because they compete for food in the wild and do not share freely (Stevens 2004; Warneken & Tomasello 2009). Therefore, it is possible that prosocial preferences are more likely to be found in species that share food more readily. For instance, bonobos (*Pan paniscus*), which are regarded as being more socially tolerant than chimpanzees (de Waal 1997), prefer to eat with a conspecific rather than alone (Hare & Kwetuenda 2010). This does not necessarily indicate a preference for benefits to the partner, but only that bonobos value companionship. Among New World monkeys, capuchin monkeys (*Cebus* spp.) are relatively tolerant; they allow conspecifics to sit close to them while eating and permit others to take food from them, though active, unsolicited food donations are uncommon (de Waal et al. 1993). Two studies found evidence for prosocial preferences in tufted or brown capuchin monkeys (*Cebus apella*, Lakshminarayanan & Santos 2008; de Waal et al. 2008). Lakshminarayanan and Santos (2008) gave tufted capuchin monkeys a choice between a high-quality food (marshmallow) for both themselves and a partner (1/1) and a high-quality food for themselves and a low-quality one for the partner (1/1−). In another condition, they could choose between low-quality food for themselves and either the same food or a high-quality food for the partner (1/1

Fig. 25.5. In one of the dictator-style sharing preference studies of (Silk et al. 2005), one chimpanzee (actor, on the left) could choose one of two trays. One tray held a piece of food for both chimpanzees (1/1, mutualistic outcome), and the other held food only for the actor (1/0, selfish outcome). The conspecific partner had no control over the outcome. Redrawn from Silk et al. (2005).

vs. 1/1+). The capuchin monkeys had a small but significant preference for pulling a tray that resulted in the partner getting food of either the same or higher quality, as opposed to food of a lower quality. Perhaps surprisingly (see discussion on inequity aversion), they actually had a preference for choosing the larger reward for the partner rather than choosing two equally low-quality rewards. Another test of prosocial preferences in brown capuchin monkeys used a token-exchange paradigm (de Waal et al. 2008). Each subject was offered a choice between two tokens: one token resulted in the experimenter handing a piece of food only to the subject, and the other resulted in the experimenter giving food to both the subject and a conspecific partner. The monkeys demonstrated a preference for tokens resulting in prosocial outcomes over selfish outcomes, except when the partner was not visible at the time of choice. While, as the authors suggest, it may be the case that "giving is self-rewarding" in this species, it is puzzling that the capuchin monkeys showed a strong selfish tendency when the partner was in the next cage but not visible at the time of choice; furthermore, there was no demonstration that the capuchin monkeys tracked the effects of the tokens, which changed regularly.

Another hypothesis suggests that cooperative breeding is a route to prosociality; species like common marmosets (*Callithrix jacchus*) and cotton-top tamarins (*Saguinus oe-*

dipus) provide food to their offspring, and therefore may have evolved social preferences similar to those of humans (Burkart et al. 2007). In one study, common marmosets were more likely to choose a 0/1 option over a 0/0 option when another monkey was in the adjoining enclosure than when they were alone (Burkart et al. 2007). On the other hand, cotton-top tamarins, which also breed cooperatively, did not show prosocial preferences toward their long-term mates when presented with a choice between 1/1 and 1/0 or 0/1 and 0/0 (Cronin et al. 2009); the same was true when partners could receive more food than subjects (1/3 vs. 1/0 and 0/3 vs. 0/0; Stevens 2010).

Negative Social Preferences

The competitive drive that makes food sharing difficult for chimpanzees and many other nonhuman primates would make aversion to disadvantageous inequity an adaptive social preference. If nonhuman primates are sensitive to situations in which others have more than they do, then they should attempt to increase their own welfare or decrease the welfare of others. One approach to this question uses a variation of the impunity game (Bolton & Zwick 1995). In this game, one player can choose between a fair and an unfair outcome, as in the dictator game, and the second player can accept this offer or reject it. Acceptance results in both players getting the proposed distribution; rejection has no effect on the first player and causes the second player to gain nothing. It can be expected that a rational second player should never reject any offer, and as a result, first players are more selfish in their choices. (Contrast this with the ultimatum game, discussed below.) In the first experiment to attempt to test inequity aversion in nonhuman primates, female brown capuchin monkeys were trained to exchange tokens (small rocks) with an experimenter (Brosnan & de Waal 2003; table 25.2). They did this for either highly preferred food (grapes) or for less desirable food (cucumber pieces). There were four treatments: an equality test in which the subject and a partner in the adjacent enclosure exchanged tokens for cucumber, an inequality test in which the exchanges resulted in the subject getting cucumber while the partner got a grape, an effort control in which the partner got better food and did not have to exchange a token, and a food control in which the grape was put in an empty enclosure while the subject exchanged tokens for cucumbers. The primary finding was that subjects exchanged tokens less and rejected food more (sometimes even throwing the cucumber at the experimenter) in the effort control than in all of the other conditions. They also rejected the outcome more in the inequality test than in the equality test. However, they were equally likely to reject the outcome in the food control condition as to reject it in the inequality test. Brosnan and de Waal suggested that capuchin monkeys evaluate rewards and effort in relative terms, and that this sensitivity to inequality is a precursor to human fairness (see also Brosnan & de Waal 2004; Brosnan 2006).

However, a number of concerns have been raised about this conclusion. Henrich (2004) pointed out that when the subjects rejected the cucumber, they were actually increasing inequity. It is possible that the monkeys rejected the lower-quality food because they developed an expectation for a better one when they saw the grapes (food expectation hypothesis, Wynne 2004; Bräuer et al. 2006); this conclusion cannot be ruled out because of the similar rejection rates in the inequality test and food control. Alternatively, they might have rejected the cucumber because they had previously received grapes (Dubreuil et al. 2006; Roma et al. 2006): this is known as the frustration effect (Tinklepaugh 1928). To address these alternatives, further studies have been conducted on capuchin monkeys and great apes. The results have been mixed, with some finding support for inequity aversion (Brosnan et al. 2005; van Wolkenten et al. 2007; Brosnan et al. 2010) and others not (Bräuer et al. 2006, 2009; Dubreuil et al. 2006; Roma et al. 2006; Dindo & de Waal 2007; Fontenot et al. 2007; Neiworth et al. 2009; Silberberg et al. 2009). The inconsistent findings and subtle differences in experimental procedures make it difficult to evaluate the importance of very minimal effort (token exchanges with experimenters) and outcome disparity on inequity aversion. Disadvantageous inequity aversion, if it exists in nonhuman primates, is not a robust phenomenon.

It is questionable whether the rejection paradigm is the best tool for exploring disadvantageous inequity aversion. In the studies by Brosnan and de Waal and in those that followed, rejection of inequitable offers had no effect on the other individual (Henrich 2004). Nor could those studies reveal anything about understanding of intentions, because the partners were not responsible for the outcomes. It is not even clear that this method elicits inequity aversion in humans (Hachiga et al. 2009). A more direct approach is to have the subjects choose between payoffs that affect themselves as well as others, as in the prosocial preference studies described earlier. If nonhuman primates are averse to disadvantageous inequity, they should show a consistent preference for equitable outcomes, or an aversion to personally disadvantageous outcomes (e.g., 1/1 over 0/1). Furthermore, if they are functionally spiteful (which might be expected of very competitive individuals), they could even show a preference for outcomes that are personally advantageous for themselves or disadvantageous for others (e.g., preferring 1/0 over 1/1, or 0/0 over 0/1). This was not the case for chimpanzees. In a series of experiments described

earlier (Jensen et al. 2006), we gave chimpanzees choices between functionally selfish and functionally mutualistic outcomes (1/0 vs. 1/1), functionally altruistic and functionally weak spiteful outcomes (0/1 vs. 0/0; the functionally spiteful outcome was weak in that doing nothing also resulted in no food for the partner), and functionally altruistic and functionally spiteful outcomes (0/1 vs. 0/0; pulling the 0/0 option resulted in the partner getting nothing). The choices the chimpanzees made did not reveal any social preferences at all. Results for capuchin monkeys were mixed: in one study they showed a preference for generous (personally disadvantageous) offers (Lakshminarayanan & Santos 2008); economists sometimes refer to this as hyperfairness. Two studies did find a preference for equitable over generous outcomes in brown capuchin monkeys (Fletcher 2008; Takimoto et al. 2010), though there was no clear evidence that the monkeys understood the consequences of their choices in these studies (i.e., a nonsocial probe in which the subject would get more food by choosing the generous [1/2] over the fair [1/1] option).

Melis et al. (2009) gave pairs of chimpanzees a simultaneous choice between inequitable and equitable outcomes. The subjects had to work together to pull in a tray of food. One tray was baited equitably (5/5) and the other was baited disproportionately (roughly 8/2). Each chimpanzee needed the other in order to get food, and on most trials the chimpanzees were able to "negotiate" so that both got food. The negotiation amounted to a game of chicken in which each individual sat in front of his preferred amount of food and hoped that the partner caved in first; each tried to maximize his personal payoff, but did not have to be sensitive to inequity or the consequences of his actions for the partner.

The most widely used test for fairness sensitivity in humans is the ultimatum game (Güth et al. 1982; chapter 26, this volume). One individual (the proposer) is given an endowment (say, $50), which he can then divide with another individual (the responder). If the responder accepts this offer, they both keep the allocated amounts; if he rejects, both get nothing. Rejections are functionally spiteful in that both individuals suffer a loss. People routinely reject offers of 20%, and proposers typically offer around 50% (Camerer 2003), though there is considerable variation across human cultures (e.g., Henrich et al. 2005). This is an interesting finding that has generated a large body of research because standard economic theory of rational self-interest predicts that responders should accept any offer, since something is better than nothing, and that as a result, proposers will offer as little as possible.

My colleagues and I presented chimpanzees with a reduced version of the ultimatum game (Jensen et al. 2007b, based on Falk et al. 2003; fig. 25.6; table 25.2). The pro-

Fig. 25.6. In the mini-ultimatum game study by Jensen et al. (2007b), the first chimpanzee, the proposer (on the left), could pull one of two ropes. Pulling resulted in the attached tray coming closer while moving the other rope away. Here, the top tray is baited fairly (5/5) with raisins and the bottom tray unfairly (8/2). After the first chimpanzee makes her choice, the tray is still out of both subjects' reach. Only if the responder (on the right) pulls a rod that is now within his reach will both chimpanzees get the food. If he rejects it by doing nothing, the food will be removed by the experimenter after one minute, and neither subject will get any food. Redrawn from Jensen et al. (2007b).

poser chimpanzee could pull one of two trays. One tray always had an unfair division of food, with eight raisins for herself and two for her partner (8/2). The alternatives were equitable (5/5), equally unfair (8/2), generous (2/8), and even more unfair (10/0). The responder chimpanzee could accept the offer by completing the choice, pulling the chosen tray within reach, or reject the offer by not pulling the tray, thus resulting in both subjects getting nothing. Proposers routinely chose outcomes that were personally favorable regardless of the outcomes for the responders. Responders accepted all non-zero offers; they only reliably rejected offers of nothing (10/0). Unlike humans, chimpanzees did not appear to be willing to pay a cost to punish unfair offers; they showed no sensitivity to the unfair intentions of the proposer or to the resulting unfair outcome (see Falk & Fischbacher 2006). Chimpanzees, then, do not appear to be functionally spiteful (see also Jensen et al. 2007a).

Discussion

Research on the cognitive mechanisms that underlie the prosocial and antisocial behavior of primates helps us understand how cooperative behaviors evolved, how coopera-

tion is sustained, and how both human and nonhuman primate societies work. Inferring cognitive mechanisms from observable behavior is challenging, but recent experiments have provided some insights (table 25.2). While there is no clear evidence to distinguish between empathy and personal distress, future studies can provide nonhuman primates with conflicts, namely between escaping a distressing situation and helping. Also, subjects can witness a distressing situation without seeing expressions of distress. Chimpanzees and possibly other species appear to recognize something of the intentions of others, and can use this understanding to achieve their own objectives. However, they do not appear to have a strong concern about how their own behavior affects the outcomes of others (Call & Jensen 2007). Future work on social intentions could disentangle the various motivations that can lead to social and antisocial outcomes, such as play, habitual responses to previously reinforced actions, compliance, reputation, attraction to stimuli, and so on. As for social preferences, there is considerable debate about whether primates compare their outcomes relative to others and are motivated to rectify inequity. There is very little consistency in the findings for advantageous and disadvantageous inequity aversion, both within and across species. Future work on social preferences will likely focus on experiments in which the subjects create and react to inequity and other unfair outcomes with other members of their species. Adaptations of economic experiments, such as the ultimatum game, will be useful tools for this endeavor.

Social regard, which motivates functionally altruistic, punitive, and spiteful acts in humans, seems to be considerably more limited in nonhuman primates. Chimpanzees and possibly other primates seem to have some understanding of the intentions of others. Contradictory results in experiments on prosocial behavior might reflect differences in goal understanding and prosocial motivations between helping, sharing, informing (Warneken & Tomasello 2009), and comforting (Zahn-Waxler et al. 1992). However, it is not evident why helping and other prosocial motives should be built upon different psychological substrates, or whether food competition is sufficient to blunt prosocial motives in sharing contexts. Given the small number of studies, it is too early to say with any confidence whether helpful outcomes are generated by helpful motivations. Studies of social preferences in which individuals are torn between selfish and social outcomes will likely continue to produce results that will inform the debates. Importantly, tests in which the subjects are not directly affected by the outcomes but can influence the outcomes for others will help separate self-regard from social preferences. Adapting third-party punishment, trust, public goods, and other games used in experimental economics will allow us to illuminate the social preferences of primates more clearly.

Fig. 25.7. Indifference to the welfare of others may typify the social concerns of chimpanzees, despite their social sophistication. Photo courtesy of the Max Planck Institute for Evolutionary Anthropology.

Speculation about whether and why some species and not others may or may not have social concern, social preferences, and the capacity for intention reading is premature at this point because relatively few studies have been conducted on a limited number of species. Even if the bulk of the evidence comes to suggest that the social behavior of primates is not driven for a concern for the welfare of others, that does not mean that the primates are not influenced by their social environment; it is very clear that they are. Primates form social bonds and experience distress when partners and offspring die or when they are isolated from their group; they acquire new skills through observational learning, and so on. While the social environment is important to primates, they may not be aware of or concerned about the influence of their own actions on their social milieu (fig. 25.7). If so, then we can ask how evolution has shaped cooperation in the absence of social regard, and how social regard arose in humans and shaped

the development of our sociality (chapter 26, this volume). Learning more about the cognitive mechanisms that govern social behavior and the affective mechanisms that influence social concern will contribute to answering long-standing questions about the evolution of cooperation.

Summary

There is a rapidly emerging literature into the cognitive processes underlying cooperation in primates. As is to be expected, there is considerable debate, with some studies suggesting that primates share some degree of social concern such as empathy and aversion to inequity. Exploring the psychology underlying cooperative and competitive acts in primates will allow us to understand how their social systems work and have evolved, and will provide insights about how human sociality has evolved. Future research can inform the current debates and confusion in the literature by adapting game-theoretical and experimental economic approaches to confront individuals with conflicts between personal and social outcomes. Proper controls will need to be ensured in all studies; currently, the wide variety of methods used and the varying numbers of controls, such as knowledge probes, make direct comparisons of studies difficult. Collaborative work across research groups using agreed-upon methods would help. Additional measures such as indicators of stress and emotions would greatly enhance the value of the reported findings. Finally, and most obviously, more individuals and more species need to be tested against the various theories and conflicting results.

References

Aureli, F. & Yates, K. 2009. Distress prevention by grooming others in crested black macaques. *Biology Letters*, 6, 27–29.

Barnes, J. L., Hill, T., Langer, M., Martinez, M. & Santos, L. R. 2008. Helping behaviour and regard for others in capuchin monkeys (*Cebus apella*). *Biology Letters*, 4, 638–640.

Batson, C. D. 1991. *The Altruism Question: Toward a Social-Psychological Answer*. Hillsdale, NJ: Lawrence Erlbaum.

Boesch, C. 1992. New elements of a theory of mind in wild chimpanzees. *Behavioral and Brain Sciences*, 15, 149–150.

Bolton, G. E., & Zwick, R. 1995. Anonymity versus punishment in ultimatum bargaining. *Games and Economic Behavior*, 10, 95–121.

Boren, J. J. 1966. An experimental social relation between two monkeys. *Journal of the Experimental Analysis of Behavior*, 9, 691–700.

Boyd, R. & Richerson, P. J. 1992. Punishment allows the evolution of cooperation (or anything else) in sizable groups. *Ethology & Sociobiology*, 13, 171–195.

Boysen, S. T. & Berntson, G. G. 1995. Responses to quantity: Perceptual versus cognitive mechanisms in chimpanzees (*Pan troglodytes*). *Journal of Experimental Psychology: Animal Behavior Processes*, 21, 82–86.

Bräuer, J., Call, J. & Tomasello, M. 2006. Are apes really inequity averse? *Proceedings of the Royal Society of London B Biological Sciences*, 273, 3123–3128.

———. 2009. Are apes inequity averse? New data on the token-exchange paradigm. *American Journal of Primatology*, 71, 175–181.

Brereton, A. R. 1994. Return-benefit spite hypothesis: An explanation for sexual interference in stumptail macaques. *Primates*, 35, 123–136.

Brosnan, S. F. 2006. Nonhuman species' reactions to inequity and their implications for fairness. *Social Justice Research*, 19, 153–185.

Brosnan, S. F. & de Waal, F. B. M. 2003. Monkeys reject unequal pay. *Nature*, 425, 297–299.

———. 2004. Fair refusal by capuchin monkeys. *Nature*, 428, 140.

Brosnan, S. F., Schiff, H. C. & de Waal, F. B. M. 2005. Tolerance for inequity may increase with social closeness in chimpanzees. *Proceedings of the Royal Society of London B Biological Sciences*, 272, 253–258.

Brosnan, S., Silk, J., Henrich, J., Mareno, M., Lambeth, S. & Schapiro, S. 2009. Chimpanzees (*Pan troglodytes*) do not develop contingent reciprocity in an experimental task. *Animal Cognition*, 12, 587–597.

Brosnan, S. F., Talbot, C., Ahlgren, M., Lambeth, S. P. & Schapiro, S. J. 2010. Mechanisms underlying responses to inequitable outcomes in chimpanzees, *Pan troglodytes*. *Animal Behaviour*, 79, 1229–1237.

Brown, A. E. 1879. Grief in the chimpanzee. *The American Naturalist*, 13, 173–175.

Bshary, R. & Bergmüller, R. 2008. Distinguishing four fundamental approaches to the evolution of helping. *Journal of Evolutionary Biology*, 21, 405–420.

Burkart, J. M., Fehr, E., Efferson, C. & van Schaik, C. P. 2007. Other-regarding preferences in a non-human primate: Common marmosets provision food altruistically. *Proceedings of the National Academy of Sciences of the United States of America*, 104, 19762–19766.

Byrne, R. & Whiten, A. 1988. Machiavellian Intelligence. Oxford: Oxford University Press.

Call, J. & Jensen, K. 2007. Chimpanzees may recognize motives and goals, but may not reckon on them. In *Empathy and Fairness* (ed. by Bock, G. & Goode, J.), 56–65. Chichester, UK: John Wiley & Sons.

Call, J. & Tomasello, M. 2008. Does the chimpanzee have a theory of mind? 30 years later. *Trends in Cognitive Sciences*, 12, 187–192.

Camerer, C. 2003. *Behavioral Game Theory*. Princeton, NJ: Princeton University Press.

Charness, G. & Rabin, M. 2002. Understanding social preferences with simple tests. *Quarterly Journal of Economics*, 117, 817–869.

Church, R. M. 1959. Emotional reactions of rats to the pain of others. *Journal of Comparative and Physiological Psychology*, 52, 132–134.

Clutton-Brock, T. 2009. Cooperation between non-kin in animal societies. *Nature*, 462, 51–57.

Clutton-Brock, T. H. & Parker, G. A. 1995a. Punishment in animal societies. *Nature*, 373, 209–216.

———. 1995b. Sexual coercion in animal societies. *Animal Behaviour*, 49, 1345–1365.

Colman, A. D., Liebold, K. E. & Boren, J. J. 1969. A method for studying altruism in monkeys. *Psychological Record*, 19, 401–405.

Cronin, K. A., Schroeder, K. K. E., Rothwell, E. S., Silk, J. B. & Snowdon, C. T. 2009. Cooperatively breeding cottontop tamarins (*Saguinus oedipus*) do not donate rewards to their long-term mates. *Journal of Comparative Psychology*, 123, 231–241.

Cronin, K. A., Schroeder, K. K. E. & Snowdon, C. T. 2010. Prosocial behaviour emerges independent of reciprocity in cottontop tamarins. *Proceedings of the Royal Society B: Biological Sciences*, 277, 3845–3851.

Darwin, C. 1872. *The Expression of Emotion in Man and Animals*, 1st ed. London: John Murray.

De Waal, F. B. M. 1982. *Chimpanzee Politics: Power and Sex Among the Apes*. New York: Johns Hopkins University Press.

———. 1997. *Bonobo: The Forgotten Ape*. Berkely, CA: University of California Press.

———. 2008. Putting the altruism back into altruism: The evolution of empathy. *Annual Review of Psychology*, 59, 279–300.

De Waal, F. B. M., Leimgruber, K. & Greenberg, A. R. 2008. Giving is self-rewarding for monkeys. *Proceedings of the National Academy of Sciences of the United States of America*, 105, 13685–13689.

De Waal, F. B. M. & Luttrell, L. M. 1988. Mechanisms of social reciprocity in three primate species: Symmetrical relationship characteristics or cognition? *Ethology & Sociobiology*, 9, 101–118.

De Waal, F. B. M., Luttrell, L. M. & Canfield, M. E. 1993. Preliminary data on voluntary food sharing in brown capuchin monkeys. *American Journal of Primatology*, 29, 73–78.

De Waal, F. B. M. & van Roosmalen, A. 1979. Reconciliation and consolation among chimpanzees. *Behavioral Ecology and Sociobiology*, 5, 55–66.

Dindo, M. & de Waal, F. B. M. 2007. Partner effects on food consumption in brown capuchin monkeys. *American Journal of Primatology*, 69, 448–456.

Dubreuil, D., Gentile, M. S. & Visalberghi, E. 2006. Are capuchin monkeys (*Cebus apella*) inequity averse? *Proceedings of the Royal Society of London B Biological Sciences*, 273, 1223–1228.

Dufour, V., Pelé, M., Neumann, M., Thierry, B. & Call, J. 2009. Calculated reciprocity after all: Computation behind token transfers in orang-utans. *Biology Letters*, 5, 172–175.

Dunbar, R. I. M. 1998. The social brain hypothesis. *Evolutionary Anthropology*, 6, 178–190.

Eisenberg, N. 2002. Distinctions among various modes of empathy-related reactions: A matter of importance in humans. *Behavioral and Brain Sciences*, 25, 33–34.

Eisenberg, N., Shea, C. L., Carlo, G. & Knight, G. P. 1991. Empathy-related responding and cognition: A "chicken and the egg" dilemma. In *Handbook of Moral Behavior and Development, Volume 2: Research* (ed. by Kurtines, W. M. & Gewirtz, J. L.), 63–88. Hillsdale, NJ: Lawrence Erlbaum.

Falk, A., Fehr, E. & Fischbacher, U. 2003. On the nature of fair behavior. *Economic Inquiry*, 41, 20–26.

———. 2003. The nature of human altruism. *Nature*, 425, 785–791.

———. 2004. Third-party punishment and social norms. *Evolution and Human Behavior*, 25, 63–87.

———. 2006. A theory of reciprocity. *Games and Economic Behavior*, 54, 293–315.

Fehr, E. & Gächter, S. 2002. Altruistic punishment in humans. *Nature*, 415, 137–140.

Festinger, L. 1954. A theory of social comparison processes. *Human Relations*, 7, 117–140.

Fletcher, G. E. 2008. Attending to the outcome of others: Disadvantageous inequity aversion in male capuchin monkeys (*Cebus apella*). *American Journal of Primatology*, 70, 901–905.

Fontenot, M. B., Watson, S. L., Roberts, K. A. & Miller, R. W. 2007. Effects of food preferences on token exchange and behavioural responses to inequality in tufted capuchin monkeys, *Cebus apella*. *Animal Behaviour*, 74, 487–496.

Foster, K. R., Wenseleers, T. & Ratnieks, F. L. W. 2001. Spite: Hamilton's unproven theory. *Annales Zoologici Fennici*, 38, 229–238.

Frank, R. H. 1988. *Passions Within Reason: The Strategic Role of the Emotions*. New York: W. W. Norton.

Fraser, O. N., Stahl, D. & Aureli, F. 2008. Stress reduction through consolation in chimpanzees. *Proceedings of the National Academy of Sciences of the United States of America*, 105, 8557–8562.

Gardner, A. & West, S. A. 2006. Spite. *Current Biology*, 16, R662–R664.

Gilby, I. C. 2006. Meat sharing among the Gombe chimpanzees: Harassment and reciprocal exchange. *Animal Behaviour*, 71, 953–963.

Gintis, H. 2000. Strong reciprocity and human sociality. *Journal of Theoretical Biology*, 206, 169–179.

Gros-Louis, J. 2004. The function of food-associated calls in white-faced capuchin monkeys, *Cebus capucinus*, from the perspective of the signaller. *Animal Behaviour*, 67, 431–440.

Güth, W., Schmittberger, R. & Schwarze, B. 1982. An experimental analysis of ultimatum bargaining. *Journal of Economic Behavior & Organization*, 3, 367–388.

Hachiga, Y., Silberberg, A., Parker, S. & Sakagami, T. 2009. Humans (*Homo sapiens*) fail to show an inequity effect in an "up-linkage" analog of the monkey inequity test. *Animal Cognition*, 12, 359–367.

Hagen, E. H. & Hammerstein, P. 2006. Game theory and human evolution: A critique of some recent interpretations of experimental games. *Theoretical Population Biology*, 69, 339–348.

Hamilton, W. D. 1964. The genetical evolution of social behaviour. I & II. *Journal of Theoretical Biology*, 7, 1–52.

———. 1970. Selfish and spiteful behaviour in an evolutionary model. *Nature*, 228, 1218–1220.

Hammerstein, P. 2003. Why is reciprocity so rare in social animals? A protestant appeal. In *Genetic and Cultural Evolution of Cooperation* (ed. by Hammerstein, P.), 83–93. Cambridge, MA: MIT Press.

Hammerstein, P. & Hagen, E. H. 2005. The second wave of evolutionary economics in biology. *Trends in Ecology & Evolution*, 20, 604–609.

Hare, B. & Kwetuenda, S. 2010. Bonobos voluntarily share their own food with others. *Current Biology*, 20, R230–R231.

Harwood, M. D. & Farrar, M. J. 2006. Conflicting emotions: The connection between affective perspective taking and theory of mind. *British Journal of Developmental Psychology*, 24, 401–418.

Hauser, M. D. 1992. Costs of deception: Cheaters are punished in rhesus monkeys (*Macaca mulatta*). *Proceedings of the National Academy of Sciences*, 89, 12137–12139.

Hauser, M. D., Chen, M. K., Chen, F. & Chuang, E. 2003. Give unto others: Genetically unrelated cotton-top tamarin monkeys preferentially give food to those who altruistically give food back. *Proceedings of the Royal Society of London B Biological Sciences*, 270, 2363–2370.

Henrich, J. 2004. Inequity aversion in capuchins? *Nature*, 428, 139.

Henrich, J., Boyd, R., Bowles, S., Camerer, C., Fehr, E., Gintis, H., et al. 2005. "Economic man" in cross-cultural perspective: Behavioral experiments in 15 small-scale societies. *Behavioral and Brain Sciences*, 28, 795–815.

Herrmann, B., Thöni, C. & Gächter, S. 2008. Antisocial punishment across societies. *Science*, 319, 1362–1367.

Hoffman, M. L. 1982. Development of prosocial motivation: Empathy and guilt. In *The Development of Prosocial Behavior* (ed. by Eisenberg, N.), 281–338. New York: Academic Press.

———. 2000. *Empathy and Moral Development: Implications for Caring and Justice*. New York: Cambridge University Press.

Horrocks, J. & Hunte, W. 1981. "Spite": A constraint on optimal foraging in the vervet monkey, *Cercopithecus aethiops sabaues*, in Barbados. *American Zoologist*, 21, 939.

Humphrey, N. K. 1976. The social function of the intellect. In *Growing Points in Ethology* (ed. by Bateson, P. P. G. & Hinde, R. A.), 303–317. Cambridge: Cambridge University Press.

Jensen, K. 2010. Punishment and spite, the dark side of cooperation. *Philosophical Transactions of the Royal Society B: Biological Sciences*, 365, 2635–2650.

Jensen, K., Call, J. & Tomasello, M. 2007a. Chimpanzees are vengeful but not spiteful. *Proceedings of the National Academy of Sciences of the United States of America*, 104, 13046–13050.

———. 2007b. Chimpanzees are rational maximizers in an ultimatum game. *Science*, 318, 107–109.

Jensen, K., Hare, B., Call, J. & Tomasello, M. 2006. What's in it for me? Self-regard precludes altruism and spite in chimpanzees. *Proceedings of the Royal Society of London B Biological Sciences*, 273, 1013–1021.

Jensen, K. & Tomasello, M. 2010. Punishment. In *Encyclopedia of Animal Behavior, Volume 2* (ed. by Breed, M. D. & Wood, J.), 800–805. Oxford: Academic Press.

Kahneman, D., Knetsch, J. L. & Thaler, R. 1986. Fairness as a constraint on profit seeking: Entitlements in the market. *American Economic Review*, 76, 728–741.

Kappeler, P. M. & van Schaik, C. P. 2006. *Cooperation in Primates and Humans: Mechanisms and Evolution* (1st ed.). Berlin: Springer-Verlag.

Kitchen, D. M. & Beehner, J. C. 2007. Factors affecting individual participation in group-level aggression among non-human primates. *Behaviour*, 144, 1551–1581.

Koski, S. E. & Sterck, E. H. M. 2007. Triadic postconflict affiliation in captive chimpanzees: Does consolation console? *Animal Behaviour*, 73, 133–142.

———. 2009. Post-conflict third-party affiliation in chimpanzees: What's in it for the third party? *American Journal of Primatology*, 71, 409–418.

———. 2010. Empathic chimpanzees: A proposal of the levels of emotional and cognitive processing in chimpanzee empathy. *European Journal of Developmental Psychology*, 7, 38–66.

Krebs, D. L. 1970. Altruism: An examination of the concept and a review of the literature. *Psychological Bulletin*, 73, 258–302.

———. 1971. Infrahuman altruism. *Psychological Bulletin*, 76, 411–414.

Kummer, H. 1968. *Social Organization of Hamadryas Baboons: A Field Study*. Basel: Karger.

Lakshminarayanan, V. R. & Santos, L. R. 2008. Capuchin monkeys are sensitive to others' welfare. *Current Biology*, 18, 999–1000.

Langford, D. J., Crager, S. E., Shehzad, Z., Smith, S. B., Sotocinal, S. G., Levenstadt, J. S., et al. 2006. Social modulation of pain as evidence for empathy in mice. *Science*, 312, 1967–1970.

Massen, J. J. M., van den Berg, L. M., Spruijt, B. M. & Sterck, E. H. M. 2010. Generous leaders and selfish underdogs: Prosociality in despotic macaques. *PLoS ONE*, 5, e9734.

Masserman, J. H., Wechkin, S. & Terris, W. 1964. "Altruistic" behavior in rhesus monkeys. *American Journal of Psychiatry*, 121, 584–585.

Mayr, E. 1961. Cause and effect in biology. *Science*, 134, 1501–1506.

Melis, A. P., Hare, B. & Tomasello, M. 2006a. Chimpanzees recruit the best collaborators. *Science*, 311, 1297–1300.

———. 2009. Chimpanzees coordinate in a negotiation game. *Evolution and Human Behavior*, 30, 381–392.

Miller, R. E., Murphy, J. V. & Mirsky, I. 1959. Relevance of facial expression and posture as cues in communication of affect between monkeys. *Archives of General Psychiatry*, 1, 480–488.

Neiworth, J. J., Johnson, E. T., Whillock, K., Greenberg, J. & Brown, V. 2009. Is a sense of inequity an ancestral primate trait? Testing social inequity in cotton top tamarins (*Saguinus oedipus*). *Journal of Comparative Psychology*, 123, 10–17.

O'Connell, S. M. 1995. Empathy in chimpanzees: Evidence for theory of mind? *Primates*, 36, 397–410.

Ortony, A., Clore, G. L. & Collins, A. 1988. *The Cognitive Structure of Emotions*. Cambridge: Cambridge University Press.

Panchanathan, K. & Boyd, R. 2004. Indirect reciprocity can stabilize cooperation without the second-order free rider problem. *Nature*, 432, 499–502.

Parr, L. A. 2001. Cognitive and physiological markers of emotional awareness in chimpanzees (*Pan troglodytes*). *Animal Cognition*, 4, 223–229.

———. 2002. Understanding other's emotions: From affective resonance to empathic action. *Behavioral and Brain Sciences*, 25, 44–45.

Pillutla, M. M. & Murnighan, J. K. 1996. Unfairness, anger, and spite: Emotional rejections of ultimatum offers. *Organizational Behavior and Human Decision Processes*, 68, 208–224.

Povinelli, D. J. & Vonk, J. 2003. Chimpanzee minds: Suspiciously human? *Trends in Cognitive Sciences*, 7, 157–160.

Preston, S. D. & de Waal, F. B. M. 2002. Empathy: Its ultimate and proximate bases. *Behavioral and Brain Sciences*, 25, 1–20.

Price, M. E., Cosmides, L. & Tooby, J. 2002. Punitive sentiment as an anti-free rider psychological device. *Evolution and Human Behavior*, 23, 203–231.

Richerson, P. J. & Boyd, R. 2005. *Not by Genes Alone: How Culture Transformed Human Evolution*. Chicago: University of Chicago Press.

Roma, P. G., Silberberg, A., Ruggiero, A. M. & Suomi, S. J. 2006. Capuchin monkeys, inequity aversion, and the frustration effect. *Journal of Comparative Psychology*, 120, 67–73.

Schaub, H. 1996. Testing kin altruism in long-tailed macaques (*Macaca fascicularis*) in a food-sharing experiment. *International Journal of Primatology*, 17, 445–467.

Seed, A., Emery, N. & Clayton, N. 2009. Intelligence in corvids and apes: A case of convergent evolution? *Ethology*, 115, 401–420.

Shutt, K., MacLarnon, A., Heistermann, M. & Semple, S. 2007. Grooming in Barbary macaques: Better to give than to receive? *Biology Letters*, 3, 231–233.

Silberberg, A., Crescimbene, L., Addessi, E., Anderson, J. & Visalberghi, E. 2009. Does inequity aversion depend on a frustration effect? A test with capuchin monkeys (*Cebus apella*). *Animal Cognition*, 12, 505–509.

Silk, J. B. 1992. The patterning of intervention among male bonnet macaques: Reciprocity, revenge, and loyalty. *Current Anthropology*, 33, 318–324.

———. 2002. Practice random acts of aggression and senseless acts of intimidation: The logic of status contests in social groups. *Evolutionary Anthropology*, 11, 221–225.

———. 2005. The evolution of cooperation in primate groups.

In *Moral Sentiments and Material Interests* (ed. by Gintis, H., Bowles, S., Boyd, R. & Fehr, E.), 43–73. Cambridge, MA: MIT Press.

———. 2007a. Empathy, sympathy, and prosocial preferences in primates. In *The Oxford Handbook of Evolutionary Psychology* (ed. by Dunbar, R. I. M. & Barrett, L.), 115–126. Oxford: Oxford University Press.

———. 2007b. The strategic dynamics of cooperation in primate groups. *Advances in the Study of Behavior*, 37, 1–41.

———. 2009. Social preferences in primates. In *Neuroeconomics: Decision Making and the Brain* (ed. by Glimcher, P. W. Camerer, C. F. Fehr, E. & Poldrack, R. A.), 269–284. London: Academic Press.

Silk, J. B., Brosnan, S. F., Vonk, J., Henrich, J., Povinelli, D. J., Richardson, A. S., et al. 2005. Chimpanzees are indifferent to the welfare of unrelated group members. *Nature*, 437, 1357–1359.

Singer, T., Seymour, B., O'Doherty, J. P., Stephan, K. E., Dolan, R. J. & Frith, C. D. 2006. Empathic neural responses are modulated by the perceived fairness of others. *Nature*, 439, 466–469.

Smith, A. 1759/2005. *The Theory of Moral Sentiments*: Meta-Libri.

———. 1776/2005. *An Inquiry into the Nature and Causes of the Wealth of Nations*. Pennsylvania: Pennsylvania State University.

Stevens, J. R. 2004. The selfish nature of generosity: Harassment and food sharing in primates. *Proceedings of the Royal Society of London B Biological Sciences*, 271, 451–456.

Stevens, J. R. 2010. Donor payoffs and other-regarding preferences in cotton-top tamarins (*Saguinus oedipus*). *Animal Cognition*, 13, 663–670.

Stevens, J. R. & Hauser, M. D. 2004. Why be nice? Psychological constraints on the evolution of cooperation. *Trends in Cognitive Sciences*, 8, 60–65.

Takimoto, A., Kuroshima, H. & Fujita, K. 2010. Capuchin monkeys (*Cebus apella*) are sensitive to others' reward: An experimental analysis of food-choice for conspecifics. *Animal Cognition*, 13, 249–261.

Tinbergen, N. 1963. On aims and methods of ethology. *Zeitschrift fuer Tierpsychologie*, 20, 410–433.

Tinklepaugh, O. L. 1928. An experimental study of representative factors in monkeys. *Journal of Comparative Psychology*, 8, 197–236.

Tomasello, M. 2008. *Origins of Human Communication*. Cambridge, MA: MIT Press.

Tomasello, M., Call, J. & Hare, B. 2003. Chimpanzees understand psychological states: The question is which ones and to what extent. *Trends in Cognitive Sciences*, 7, 153–156.

Tomasello, M., Carpenter, M., Call, J., Behne, T. & Moll, H. 2005a. Understanding and sharing intentions: The origins of cultural cognition. *Behavioral and Brain Sciences*, 28, 675–735.

———. 2005b. In search of the uniquely human. *Behavioral and Brain Sciences*, 28, 721–727.

Tomasello, M. & Rakoczy, H. 2003. What makes human cognition unique? From individual to shared to collective intentionality. *Mind and Language*, 18, 121–147.

Trivers, R. L. 1971. The evolution of reciprocal altruism. *Quarterly Review of Biology*, 46, 35–57.

Vaish, A., Carpenter, M. & Tomasello, M. 2009. Sympathy through affective perspective taking and its relation to prosocial behavior in toddlers. *Developmental Psychology*, 45, 534–543.

Van Wolkenten, M., Brosnan, S. F. & de Waal, F. B. M. 2007. Inequity responses of monkeys modified by effort. *Proceedings of the National Academy of Sciences of the United States of America*, 104, 18854–18859.

Vonk, J., Brosnan, S. F., Silk, J. B., Henrich, J., Richardson, A. S., Lambeth, S. P., et al. 2008. Chimpanzees do not take advantage of very low cost opportunities to deliver food to unrelated group members. *Animal Behaviour*, 75, 1757–1770.

Warneken, F., Hare, B., Melis, A. P., Hanus, D. & Tomasello, M. 2007. Spontaneous altruism by chimpanzees and young children. *PLoS Biology*, 5, 1414–1420.

Warneken, F. & Tomasello, M. 2006. Altruistic helping in human infants and young chimpanzees. *Science*, 311, 1301–1303.

———. 2009. Varieties of altruism in children and chimpanzees *Trends in Cognitive Sciences*, 13, 397–402.

West, S. A., Griffin, A. S. & Gardner, A. 2007a. Evolutionary explanations for cooperation. *Current Biology*, 17, R661–R672.

———. 2007b. Social semantics: Altruism, cooperation, mutualism, strong reciprocity and group selection. *Journal of Evolutionary Biology*, 20, 415–432.

Wilson, E. O. 1975. *Sociobiology: The New Synthesis*. Cambridge, MA: Harvard University Press.

Wispé, L. 1987. History of the concept of empathy. In *Empathy and Its Development* (ed. by Eisenberg, N.), 17–37. Cambridge: Cambridge University Press.

Wolfle, D. L. & Wolfle, H. M. 1939. The development of cooperative behavior in monkeys and young children. *Journal of Genetic Psychology*, 55, 137–175.

Wynne, C. D. L. 2004. Fair refusal by capuchin monkeys. *Nature*, 428, 140.

Yamamoto, S., Humle, T. & Tanaka, M. 2009. Chimpanzees help each other upon request. *PLoS One*, 4, e7416, 7411–7417.

Yamamoto, S. & Tanaka, M. 2009a. How did altruism and reciprocity evolve in humans? Perspectives from experiments on chimpanzees (*Pan troglodytes*). *Interaction Studies*, 10, 150–182.

———. 2009b. Selfish strategies develop in social problem situations in chimpanzee (*Pan troglodytes*) mother-infant pairs. *Animal Cognition*, 12 (Suppl 1), S27–S36.

Yerkes, R. M. & Yerkes, A. W. 1929. *The Great Apes: A Study of Anthropoid Life*. New Haven: Yale University Press.

Zahn-Waxler, C., Radke-Yarrow, M., Wagner, E. & Chapman, M. 1992. Development of concern for others. *Developmental Psychology*, 28, 126–136.

Zizzo, D. J. & Oswald, A. J. 2001. Are people willing to pay to reduce others' incomes? *Annales d'économie et de statistique*, 63–64, 39–65.

Chapter 26 Human Sociality

Michael Alvard

UMANS, MORE THAN most primates, are torn between the selfish motivations of "mine" and the public good of "ours." Although we are individuals with clear agency and concern for our own self-interest, we are also embedded within a complex social structure which includes large, hierarchically organized groups such as tribes, cults, castes, nations, teams, and political parties. We are members of various kinds of groups, which provide us with much that we cannot obtain on our own and to which we often subordinate ourselves (Johnson & Earle 2000). People, more than most other primates, cooperate—we make sacrifices to help others in ways that can be as trivial as conforming to group norms that proscribe how high the grass should be in the front lawn or as brutal as participating in suicide attacks on enemy forces.

While human sociality is different from that of other primates in its complexity and scale, it has roots in our common mammalian and primate heritage. We engage in a variety of social behaviors that have homologues with the behavior of other mammal species, such as extensive maternal investment (Hrdy 2000; chapter 14, this volume) and strong nepotistic biases (Chapais & Berman 2004; chapter 21, this volume). We also share a certain propensity for reciprocity (Silk 2007b; chapter 22, this volume). Chapais (2008) identifies a number of building blocks of human sociality that are present in living nonhuman primates, though not all of them are found in any one species. For example, stable pair bonds, a key component of human social structure, are also found in some nonhuman primate species, such as gibbons (Fuentes 2002). Multimale-multifemale groups are characteristic of human society and are also found in some,

but not all, nonhuman primates (Terborgh & Janson 1986; chapter 9, this volume).

Humans are clearly different, however. The cultural mechanisms that are present in recognizable form in nonhuman primates and other animals (Whiten et al. 2007; chapter 31, this volume) work in humans to increase social complexity beyond the level that genetic kinship generally provides in these other creatures. While genetic kinship remains important in human societies, mechanisms of cultural kinship and other forms of group identity are able to create larger, hierarchically structured societies (Rodseth et al. 1991; Wiessner 1998; Jones 2000; Alvard 2003a; Rodseth & Wrangham 2004; Chapais 2008). Social structure is the content, quality, and patterning of the relationships between individuals and is one way to think about social complexity (Hinde 1976; chapter 23, this volume). Structure provides the nonrandom assortative interactions fundamental to the evolution of cooperation via natural selection at multiple levels for both genetic and cultural systems (Sober & Wilson 1998; Boyd & Richerson 2002; Traulsen & Nowak 2006). Genetic relationships, for example, provide kinship structure that assorts individuals and facilitates nepotism in humans and other social species. In the other highly social organisms, particularly the eusocial insects, genetic similarities provide the structure for ultrasociality to evolve via the process of kin selection (Ratnieks 1988; Ratnieks & Wenseleer 2005). But assorting according to degree of genetic relatedness is not the only mechanism that can bring altruists together (Hamilton 1975). While accepting the essential nature of Darwinian theory, it is increasingly clear that culture plays the key role in structuring human social behavior beyond the level of the family (Boyd & Richerson 1985; Durham 1991; Henrich & McElreath 2003; Richerson & Boyd 2005b).

Tribal structure is often discussed as a level of complexity that is uniquely human (Richerson & Boyd 1999a, 2005a; Chapais 2008). Tribal society is often hierarchically segmented, in the sense that subgroups are embedded within groups, and this is one important characteristic that distinguishes human from chimpanzee (*Pan troglodytes*) society (Rodseth & Wrangham 2004). Segmentation alone, however, does not differentiate human society from that of other primates and mammals. Primate species like hamadryas baboons (*Papio hamadryas*) and geladas (*Theropithecus gelada*) and nonprimate species like orcas and elephants have societies that are clearly segmented (Silk 2007a; chapter 5, this volume). Richerson & Boyd (2005) define a tribe "as a unit of social organization that organizes people of low genetic relatedness into a common social system without the reliance of central authority." They go on to say that tribes are large (over 500 members), can consist of people that do not live in the same community, do not have clearly defined leaders, and instead rely on institutional norms often organized around an ideology, language, and ethnic identity. Hierarchical tribal structure can bring appropriately large numbers of people into coordinated action like big game cooperative hunting, managing large herds of animals, and intergroup conflict (Kelly 2000). Tribal identities, and the institutions and social preferences they encompass, are one manifestation of cultural kinship: affinity based on socially constructed identity. Over the course of human history, many societies have moved beyond tribal-level social organization to construct larger and more complex societal forms (fig. 26.1; Richerson & Boyd 1999b; Bar-Yosef 2001).

Since the first edition of *Primate Societies*, there have been a number of important theoretical advances in evolutionary approaches to understanding how humans can produce this degree of complex sociality (Campbell 1983; Richerson & Boyd 1998). While it is true that sociality evolves to the extent that benefits outweigh the costs (Krause & Ruxton 2002), the key questions that are now being reevaluated are *For whom are the costs and benefits tallied* and *Over what time frame*? While the canonical answer is that individual interests are maximized over the short-term, this view has been challenged, particularly among students of human social behavior. There is an emerging consensus that human sociality emerges from the evolutionary dynamic between the interests of individuals and those of the groups to which they belong. Selection occurs at multiple levels and sociality is favored, in part, because of benefits that accrue to groups over the span of many generations (Wilson & Sober 1994; Wilson & Wilson 2007).

Many of these new directions have been only slowly appreciated by students of evolutionary approaches to human behavior, perhaps because of the foundational assumptions that focus their attention on the individual and away from sociostructural context. With few exceptions, social behavior is conceived as a series of dyadic interactions between individuals in reciprocal or nepotistic relationships (Trivers 2006). The food sharing literature provides a good example (Gurven 2004). This perspective may appear to suffice for many hunting and gathering groups, but it removes from consideration the possibility, for instance, that ownership of real property, like a carcass, rests with a group or corporate entity (Alvard 2002). The study of mate choice is another example. Both the human evolutionary psychologists and the behavioral ecologists have focused on mate choice from the point of view of the individual, propelling the analysis in certain directions at the expense of others (Gangestad & Simpson 2000; chapter 20, this volume). Female mate choice has received special interest because females often appear to choose mates that are not in their best personal interest (see Low 2005 for a review). While there are good hypotheses to explain aspects of mate choice—for example, how criteria may vary depending on whether the partner will be short or long-term (Schmitt 2005)—what is sometimes ignored is that marriage is often a decision arranged with group interests in mind and often involves potential intergroup coalitionary strategies (Chisholm & Burbank 1991; Beckerman 2000; Shenk 2005; Apostolou 2007; Buunk et al. 2008). This conclusion exists in spite of the work of early human behavioral ecologists like Chagnon (1979a), who emphasized the importance of marriage as a mechanism of coalition formation. The conceptual error misses the fact that pair bonds often form the links that tie together the large social networks that typify human sociality (Chapais 2008).

The primary thesis of this chapter is that group selection on culturally maintained variation is the primary evolutionary force that produced and maintains complex human sociality (Wilson 1975; Campbell 1983; Richerson & Boyd 1998). In the first two sections I will argue that the standard theoretical tools in the primatologist's tool kit are inadequate to fully explain the nature of human sociality. Kin selection theory cannot explain cooperation among unrelated individuals, and neither kin selection nor reciprocity can account for cooperation in large groups. The third section focuses on recent work in the field of experimental game theory and discusses how the limitations of simple models like the prisoner's dilemma game have led theoreticians to think about alternative paths to cooperation exemplified by games of coordination. The next section introduces culture as a key evolutionary force for the production of human

Fig. 26.1. "Yell Practice" by students at Texas A&M University, United States of America, 1951. Yell Practice is a group activity whose purpose is to generate fervor among participants in anticipation of a football game and to rally support for the team. The students in the photo are "humping it," a position taken by students when giving a yell. The highly distinctive yells "are vocalized in perfect unison and accompanied by highly coordinated gestures" (Smith 2007, p.188). It is this sort of stereotyped, coordinated behavior in large groups of unrelated people that characterizes human sociality (Source: Cushing Memorial Library and Archives, Texas A&M University).

sociality. Culture is shown to potentiate group selection by homogenizing behavioral variance within groups while maintaining variance between groups; the nature of cultural group selection is discussed in detail in the penultimate section. This chapter's conclusion reviews the communal imperative of human reproduction and speculates that the social complexity that supports our exceptional life history was produced, at least in part, by cultural group selection.

The Limits of Nepotistic Explanations of Human Sociality

Kin selection has been useful for explaining certain kinds of cooperative relationships in human and nonhuman species (Hamilton 1964), but the extent of social behavior explained by kin selection, especially for humans, is clearly limited. Data show that there are clear nepotistic biases, especially among mothers and offspring and siblings, in many primate groups (Silk 2002, 2006; chapter 21, this volume). For humans, there is much evidence that altruism is also often preferentially directed at close relatives (Dunbar 2008). For example, relatives are more likely than non-kin to provide child care in a variety of foraging societies, including Efe pygmies living in the rain forests of the Congo (Ivey 2000) and the Hadza of Tanzania (Crittenden & Marlowe 2008), although in each case fully a third of the allocare is provided by individuals unrelated to the infant. Food sharing in subsistence economies often has a nepotistic bias. For example, among the indigenous Dolgan of northern Russia, the presence of a genetic kinship link between nuclear families was the strongest predictor of inter-household food sharing (Ziker & Schnegg 2005). Among the Ache foragers of Paraguay, however, no kin bias in meat distribution was found during extended hunting trips (Kaplan & Hill 1985). More recent analysis of sharing in Ache settlements concludes that reciprocal altruism, rather than kin selection, maintains food transfers (Allen-Arave et al. 2008). These results are interesting in light of new data that indicate that although chimpanzees have kin biases, the majority of strong social bonds involve unrelated individuals (Langergraber et al. 2007).

For humans, and perhaps for most animals, kin selection provides the foundation for understanding the distribution of parental care and interactions among close kin—what might be called family-level complexity—but it has significant limitations for explaining the more complex social institutions like tribes, chiefdoms, or states that are important elements of human sociality (Richerson & Boyd 1999a). There are two reasons why it is difficult for kin selection to produce the large cooperative groups seen in humans (see detailed discussion in Alvard 2003a). First, in the absence of extreme reproductive skew and high levels of endogamy, it is difficult for large groups of closely related individuals to form. For most species, as groups become larger the average degree of relatedness within them drops rapidly (Campbell 1983; Richerson & Boyd 1999a; Aviles et al. 2004; Lukas et al. 2005). Second, the egocentric nature of genetic kinship produces conflicts of interest among group members. Only full siblings share identical kinship networks; the genetic kinship networks of other relatives overlap, but are not isomorphic. In an outbred population, ego's cousins need not be related genetically to each other at all, and if cooperative commitment is a function of the coefficient of relatedness, their commitment to each other will not be equivalent to their commitment to ego (Van den Berghe 1979).

In some species of nonhuman primates—rhesus monkeys (*Macaca mulatta*), for example—when groups become large, affiliative behavior becomes limited to subgroups of close kin (Kapsalis & Berman 1996; Berman et al. 1997; Chapais 2001; Hill 2004). It does not appear that large group size necessarily "short circuits" kin selection in people, as subgroups of close kin or families are almost always embedded within large human societies (Davis & Daly 1997). Nonetheless, a distinctive feature of human sociality is the ability to transcend the limitations of genetic kinship, organize large groups, and solve the collective action problems that are inherent therein. For example, the Lamalerans of Lembata, Indonesia, are complex subsistence foragers who cooperatively hunt large marine resources like sperm whale and manta ray. Average relatedness within Lamaleran patrilineages varies widely, and much of the variation is explained in terms of group size. There is a significant negative logarithmic relationship between the total number of lineage members and mean within-lineage relatedness (Alvard 2003a; fig. 26.2). However, it is easier for big lineages to organize whaling operations because of the larger labor pool from which they can recruit (Alvard & Nolin 2002). If kin selection were the only mechanism operating, the Lamalerans would have difficulty organizing themselves into sufficiently large groups to hunt whales. Indeed, genetic kinship explained little of the patterns of affiliation among hunters during hunts, independent of their patrilineage identity (Alvard 2003a). This result suggests that genetic kinship may be less important as a principle for organizing cooperation in sizable human groups than previously thought, and that socially constructed and culturally transmitted identities like patrilineages may play a larger role than is generally appreciated.

Reciprocity, Coordination, and Behavioral Economics

Reciprocity is often modeled with a game that most readers are familiar with, called the prisoner's dilemma (PD) (Poundstone 1992). The game is a dyadic interaction in which each individual has the choice to cooperate or defect with their partner. In the single-round game, defecting provides a better payoff than cooperating in response to a partner's move of either defection or cooperation, even though the average payoff of a cooperating pair is greater than that of a mixed pair or a pair of defectors. The essence of a PD is that mutual cooperation is best for the group (of two, in this case) (Buchanan 1962), but defection is better for the individual and mutual defection is the stable equilibrium (chapter 22, this volume). Much of the theoretical work on human sociality tries to reconcile this solution with the fact that cooperation among unrelated individuals is ubiquitous.

One solution to the problem of how cooperation could evolve in this context was derived from Trivers's (1971) theory of reciprocal altruism. In its simplest form, reciprocity involves helping an individual who has helped you in the past. Axelrod & Hamilton (1981; Axelrod 1984) modeled reciprocity as an *iterated* prisoner's dilemma, wherein play occurs repeatedly between randomly chosen partners who share some probability that they will play with each other again during the next turn. Their well-known computer tournament resulted in the winning solution called "tit-for-tat" (TFT), in which an individual cooperates on the first move and then copies or reciprocates the move of his or her partner. The key to the TFT strategy is that it discriminates against defectors by defecting against them, while it cooperates with other TFT in a type of behavioral assortment (McElreath & Boyd 2007). If the probability of interacting again is sufficiently great, TFT types can enjoy

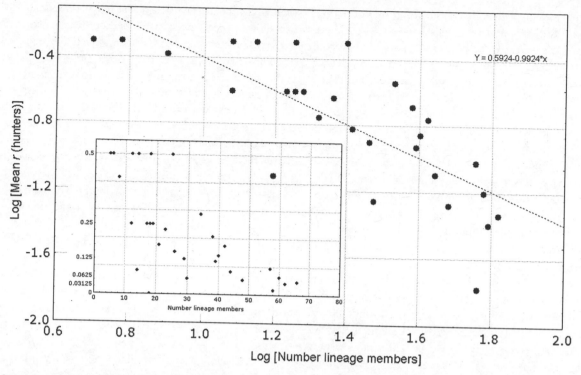

Fig. 26.2. Among the Lamalera whale hunters of Indonesia, mean within-lineage relatedness declines with lineage size. The large plot has a logarithmic scale; the small plot gives the same data on a linear scale (redrawn from Alvard 2003a, fig. 4).

the mutual benefits of cooperation not enjoyed by defectors, and reciprocity can be a stable equilibrium. The success of cooperation in the iterated PD depends on the probability of future play with other partners that play with the same strategy.

While it is not particularly common among nonhuman animals, some evidence exists for reciprocity in grooming and coalition support for females in some cercopithecine primates (Schino 2007; chapter 22, this volume). The extensive ethnographic research on the patterns of food sharing in traditional societies often fits predictions derived from the theory of reciprocal altruism (Gurven 2004). As described earlier, recent work shows that food transfers among households were fairly well balanced among the Ache (Allen-Arave et al. 2008); similar results were obtained with the Ye'kwana horticulturalists of Venezuela (Hames & McCabe 2007) and the Lamalera whale hunters (Nolin 2008).

Reciprocity works well in dyadic situations where partners are easy to monitor and their defections can be targeted directly, but it becomes less tenable in the sizable groups that make human sociality so interesting (Joshi 1987; Boyd & Richerson 1988). In an *n*-person prisoner's dilemma, the decision to cooperate does not depend on the move of one other partner, but rather on 10 or 20 or 100 others. In

larger groups, the response of TFT to defect in the face of one group member's defection punishes the rest and causes cooperation to collapse, while tolerating defectors provides them with benefits that allow them to persevere and spread (Boyd & Richerson 1988). Trivers (2006) finds fault with how Boyd & Richerson (1988) assume that payoffs are allocated. He argues that most interactions within large groups are dyadic, and that it is unrealistic to assume that other members suffer costs or reap benefits from the cooperation or defection of the pair. While this is clearly sometimes the case, Trivers's limited view of sociality ignores a vast suite of situations involving public goods that cannot be understood in terms of serial dyadic PD games.

The *n*-person games closely match the group dynamics common to much human sociality (Olson 1965), and both empirical and experimental work show that cooperation is notoriously difficult to maintain in these sorts of situations. Experimental public goods and trust games are designed to examine the social dilemmas that emerge from *n*-person nondyadic prisoner's dilemma games, and experimental results identify some of the difficulties (Ostrom et al. 1992). A typical public goods game is set up as follows. A group of 10 players is formed, and each player is endowed with $100. Each player is given the option of contributing a fraction of his or her endowment to a common pot, while keeping the balance. The value of the common pot is then doubled and redistributed in even shares back to the 10 players. In this situation, the best group strategy is for all to contribute the maximum amount; if all players do this, the pot grows to $1,000, is doubled to $2,000, and $200 is redistributed to each player. The equilibrium outcome, however, is for each player to contribute nothing. This is because a cheater who contributes nothing to the common pool still receives his share of the pot, $180, and gets to keep his initial endowment, which amounts to $280, while the cooperators receive only their share of the pot, $180. In the absence of additional incentives or threats of punishment, there is an immense temptation for players to limit their contributions to the common pool. This risk motivates others, who may be inclined to retaliate by lowering their own contributions. In anonymous settings in which players do not know one another and do not know what actions other players have taken, people often contribute nontrivial amounts (30–70%) of their endowment in one-shot games, but cooperation quickly decays when the game is iterated (Isaac & Walker 1988; Ledyard 1995). The results closely match what is referred to as the "tragedy of the commons" (Hardin 1968), where self-interested individuals eventually destroy a common, shared resource even though it is in the group's long-term interest to maintain it.

In the field of economics, such self-interest is an a priori assumption (Cropanzano et al. 2005; Gintis et al. 2005). However, this cornerstone of economics and its model of *Homo economicus* has been shaken by empirical results emerging from what until recently was considered a rogue branch of economics. The practioners of behavioral economics claim that people do not always behave in ways that conform to the predictions of the standard models that form the foundation of economic theory—people do not always behave like rational, self interested maximizers (Simon 1982; Thaler 1992; Kahneman 2003; Venkatachalam 2008). Instead, they hypothesize, people sometimes demonstrate concern for the welfare of others and act in ways that appear to be "prosocial" and "other-regarding" (Gintis 2003). Much of the initial insight for this hypothesis was gained from more closely examining anomalies that previously had been ignored by mainstream neoclassical economics. For example, neoclassical economics can not explain why initial contributions in the public goods game are so high, or why people donate to public radio (Thaler 1992).

This approach has been substantially enhanced by data from the relatively new field of experimental economics (Kagel & Roth 1995; Gintis 2000; Camerer et al. 2004). In games such as the ultimatum game (UG), people regularly behave in ways that confound standard economic assumptions, and that suggest that people care about how their decisions affect others. The UG is interesting because it creates a different payoff schedule than a prisoner's dilemma (Guth et al. 1982). The standard UG involves two anonymous players. One player, the proposer, is given an endowment and instructions that she may give any amount of the endowment to the other player, the responder. The responder has two options. He or she can accept the offer, in which case the distribution is made. Alternatively, the responder may reject the offer, in which case neither player receives anything. Both players know the size of the original endowment and are aware of all of the rules. While canonical economic and game theory predicts the lowest possible offers allowed by the rules and no rejections, people playing the game with real money routinely offer significant amounts, and may even split the endowment. Moreover, when low offers are made, they are routinely rejected. Behavioral economists present these results as evidence that people do not always behave in ways consistent with their own self-interest. The rejections also suggest that some people are willing to pay nontrivial amounts to punish others who are perceived as being unfair (Fehr & Schmidt 1999). Interestingly, chimpanzees never rejected any nonzero offers in a modified form of the UG (Jensen et al. 2007; chapter 25, this volume).

Subsequent work by anthropologists has shown that the story is even more interesting. Henrich (2000) found that the Machiguenga Indians, slash-and-burn horticulturalists living in the southeastern Peruvian Amazon, play in ways that more closely match the canonical predictions for the UG: they often make and accept very low offers. This result prompted a broader cross-cultural study of behavior in the UG in 15 small-scale societies (Henrich et al. 2005, 2006). Within these groups, mean offers ranged from 25% of the pot among the Quichua horticulturalists of Ecuador to 58% among the Lamalera whale hunters. In no society did people play in a way that was fully consistent with the selfishness axiom. A substantial amount of the variation in the size of offers and the patterning of rejections across these 15 cultures was related to differences in the patterns of interaction in everyday life. Both market integration and the payoffs to cooperation, which measured the degree to which economic life depended on cooperation with folks outside the immediate family, enhanced offers in the UG (fig. 26.3). For example, the Machiguenga and the Quichua, both swidden horticulturalists, ranked lowest in payoffs to cooperation because they are almost entirely economically independent at the family level. Conversely, the Lamalera whale hunters, with the highest offers, depend on the cooperation of large groups of people from outside the close family. Individual-level economic and demographic variables did not consistently explain behavior

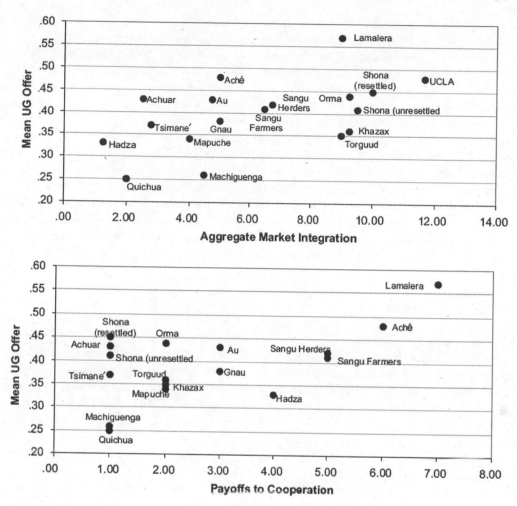

Fig. 26.3. Among 15 societies that played the ultimatum game, mean offers increased as both a function of aggregate market integration (AMI) and payoffs to cooperation (PC) in the everyday life of the people in each society. PC is a relative measure of the degree to which economic life depends on cooperation with non-immediate kin. AMI is a relative measure of how frequently people engage in market exchange. Both PC and AMI are the relative subjective ranks of each group with respect to the others as determined by the ethnographers (reprinted with permission from Henrich et al. 2005, fig. 5).

within or across groups. These results suggest that cultural variation has an important impact on how individuals play the game.

Some evolutionary psychologists argue, alternatively, that the results obtained from these sorts of experiments indicate a maladaptive response by the subjects. Price (2008), for example, argues that human prosociality evolved in ancestral environments characterized by groups of close kin and long-term social interaction: the so-called environment of evolutionary adaptedness or EEA (Bowlby 1969, 1973). He argues that the cooperative behavior observed during the games' short-term, anonymous encounters are mistakes made by minds that evolved to expect ancestral conditions and are now unable to tell the difference. Gintis et al. (2008, p. 251) counter by pointing to evidence (e.g., trade networks) that early modern humans lived in a so-

cial milieu that surely included interactions with strangers at fitness-critical times, arguing that "those who failed to distinguish between long-term and short-term or one-shot interactions would be at a significant fitness disadvantage as a result." Henrich (2005) also notes that people everywhere can distinguish kin, so they should be able to understand when they are playing with strangers.

Results from the UG have motivated researchers to think about the role of both culture and punishment in the maintenance of human sociality. Fehr & Gachter (2002) find experimentally that the altruistic punishment of defectors can maintain high levels of cooperation. In a series of anonymous, iterated public goods games, they first replicated previous results that showed moderate levels of initial contributions with subsequent cooperative decay. They added a second treatment that allowed players to punish other players each round, after all players' initial moves were disclosed. The punishment was altruistic in the sense that it was costly to the actor, reduced the payoff of his target, and benefited the group as a whole. Play was anonymous to rule out reputation effects, and group composition was changed every round to rule out reciprocity. Under this new treatment, high contributors punished low contributors who subsequently increased their contributions, and cooperation was nearly universal at the end of the game. Not only were individuals willing to punish others at a cost to themselves, but the risk of punishment motivated others to maintain high contributions to the public good. Everyday experience and ethnographic data suggest that this sort of altruistic punishment is common in the real world, and that it ranges from gossip, ridicule, ostracism, and threats of supernatural intervention to more severe sanctions like fines, incarceration, and executions (Lancaster 1986; Boehm 1999; Johnson 2005; Masclet 2003; Wiessner 2005; Kerr and Levine 2008).

While the propensity to punish rule breakers varies across cultures (Henrich et al. 2006), people in all populations demonstrate some willingness to administer costly punishment in the face of unfair behavior. There is a catch, however. Such punishment has the ability to bring *any norm, whether cooperative or arbitrary*, to fixation. Known as the "folk theorem," game theory shows that while punishment can produce cooperative equilibria, it can also fix *any* sort of behavior, cooperative or not (Rubinstein 1979; Boyd & Richerson 1992; Boyd 2006). Punishment can maintain cooperation, but it can also maintain genital mutilation, ritual male homosexuality, foot binding, even necktie wearing (see Edgerton 1992). In other words, punishment may be necessary for cooperation to emerge in large groups, but it is not sufficient. Another mechanism is required to create the ubiquity of human cooperation.

One important way out of this conundrum begins with understanding that the prisoner's dilemma may not be the best way of thinking about certain cases of cooperation. Coordination (also referred to as mutualism; chapter 22, this volume) describes situations in which individuals have a common interest (Clements & Stephens 1995; Dugatkin 1997; Winterhalder 1997; Alvard 2001). In contrast to the single-round prisoner's dilemma, in a coordination game the best move for ego depends critically on what the other players do, and people are better off when they behave in concert rather than at odds with one another. In the prototypical coordination game, driving on the same side of the road avoids costly collisions. In the road game, there are two equilibria with equally good payoffs (all to the left or all to the right). An important feature of these games is that there are as many stable solutions as there are behavioral

options around which to coordinate, but not all solutions necessarily generate equal payoffs.

Jean-Jacques Rousseau's parable of the stag hunt is often used to illustrate the nature of coordination games (Skyrms 2004; chapter 22, this volume). Big game hunting requires collective action, while hunting hare can be done successfully alone. Per capita returns from stag hunting are greater than those from hare hunting, and of course killing a hare is better than obtaining nothing, which is what a stag hunter gets if his partner defects and opts for the hare he spies during the stag hunt.

Theoreticians have recently noted that iterating a prisoner's dilemma over time with an evolutionary dynamic transforms the game into a coordination problem with multiple equilibria (Bergstrom 2002; Skyrms 2004; Boyd 2006; Taylor & Nowak 2007; McAdams 2008). If the probability of interacting again is sufficiently great, the advantage that a defector reaps in the one-round-game PD is lost as an otherwise cooperating TFT partner defects, thus reducing the return of a defector in subsequent rounds (See equation 3.1 in Taylor and Nowak 2007). This is important because, as mentioned above, in contrast to single round PD, coordination games have multiple stable equilibria, some of which may be better than others. It is important to keep in mind that a socioecological context that matches a coordination game allows for a cooperative solution, but does not guarantee one: recall the folk theorem. In the stag hunt game, both all stags (cooperation) and all hare (defection) are possible stable solutions. The stag payoff is greater, but it is also more risky because it requires the action of others, which may or may not be forthcoming. Convergence on the cooperative solutions to coordination problems is not a trivial problem, and experimental games often show that people have a difficult time coordinating because of the risk that others will not make the same choice (Van Huyck et al. 1997; Battalio et al. 2001). Nonetheless, and in contrast to nonhuman primates, in human societies people are adept at finding cooperative solutions to coordination problems. Alvard & Nolin (2002) describe the collective action involved in cooperative whale hunting in Lamalera as a coordination game. There are two stable equilibria, with collective action providing the greater payoff. The payoff achieved from cooperating to hunt whales is greater than what could be achieved from solitary pursuits, but it requires coordination among large numbers of people, each of whom may be tempted to follow a strategy equivalent to hunting hare if there are any uncertainties about the commitments of their fellows. To understand how societies like the Lamalerans become common, a mechanism is required that selects between the multiple stable equilibria that arise from coordination. Before moving on to that discussion, however,

I will first examine how such a multiplicity of behavioral options arises and is maintained in the first place.

Culture and Institutions

In genetic inheritance, meiosis generally leads to the unbiased transmission of alternative alleles (Maynard Smith & Szathmary 1995). In contrast, cultural inheritance can produce biased transmission processes, and individuals may preferentially adopt certain cultural variants rather than others from nonparental sources. Biased transmission is a cultural force of evolutionary change that is independent of natural selection on genes. There are various ways in which inheritance might be biased. Prestige bias involves selective imitation of others according to their perceived attributes (Henrich & Gil-White 2001). Conformist transmission occurs when people preferentially adopt the cultural variants that are most frequent in the population (Henrich & Boyd 1998). Conformist biases are adaptive in spatially and temporally variable environments in which trial-and-error learning is time-consuming, error-prone, and costly (Boyd & Richerson 1985). Another kind of biased transmission might be called a lineal bias, in which identity is transmitted in a way that is analogous to genomic imprinting (Ember et al. 1974; Alvard 2003a). Patrilineal identity is culturally transmitted only from the male parent, for example. Genomic imprinting is a non-Mendelian inheritance process in which genes from only one parent or the other are expressed (Reik & Walter 2001).

While aspects of cultural transmission are apparent in nonhuman primates (Chapter 31, this volume), they are not associated with the development of social institutions, as is the case with humans. Institutions are culturally transmitted, and therefore shared, rules of behavior that structure social relations, constrain behavior, and serve collective function (Richerson & Boyd 2001; Hodgson 2006). Tomasello et al. (2005) note, "When individuals in complex social groups share intentions with one another repeatedly in particular interactive contexts, the result is habitual social practices and beliefs that sometimes create what Searle (1995) calls social or institutional facts: such things as marriage, money, and government, which only exist due to the shared practices and beliefs of a group." Nonhuman primates seem to lack the "shared intentionality" that reifies the institutions of humans to the point that they seem to exist and possess interests independent of their individual members. A key benefit to institutions is that they reduce transaction costs during coordination games and drive groups toward equilibrium points (North 1990; Ensminger 1997; Alvard & Nolin 2002; Wiessner 2002; Husain & Bhattacharya

2004). Transaction costs are the costs of establishing and maintaining property rights, and they include such activities as inspection, enforcing agreements, policing, and measurement (Allen 1991).

The idea that culture produces institutional solutions to coordination problems developed from the work of Schelling in his book *The Strategy of Conflict* (1960). In some of the very first experimental games, Schelling asked students to name a time and place to meet a friend from whom they had separated in New York City. A majority identified Grand Central Terminal, and almost everyone selected noon—in spite of nearly an infinite number of equally good possible times and places. Such shared "focal points," Schelling argued, have salience because they develop out of players' shared histories, a quality that anthropologists quickly recognize as shared cultural information. Cultural mechanisms such as language provide people with the ability to infer each other's mental states and form shared notions that enhance the ability to solve simple yet common coordination games. Even in contexts where direct communication is not possible, symbolic cultural markers identify partners who can be trusted to share common solutions (Riolo et al. 2001; Alvard & Nolin 2002; McElreath et al. 2003).

For example, even for a cooperative hunt in which hunters agree that shares will be equitable, the number of ways to butcher and distribute a kill is nearly infinite. Hunters should be indifferent to which particular norm is used, but all must settle on the same one. By sharing the institutional norms that define the payoffs, hunters can proceed with the assurance that their fellows will not lay claim to shares that they themselves believe they own. Imagine the transaction costs for resolving claims if norms of distribution did not exist for the whalers at Lamalera (fig. 26.4). Without the trust associated with shared norms of behavior, like the rules of meat distribution common to human hunters, it is likely that cooperative behavior would collapse in response to high transaction costs. In the case of the Lamaleran whale hunters, certain rules of distribution (and many other norms) are inherited culturally, along with lineage identity from fathers (Alvard 2003a). Such a process forms the basis of the trust people often share with others who are not personally known. In the United States, the anonymous driver bearing down on us on the dark road at night is trusted to pass on our left. Lamalera hunters trust that each of the hundreds of other hunters, even those who are not well known, follow the same set of rules. By understanding that others belong to an identifiable group, one can make predictions about their behavior. The rules for a variety of everyday behaviors involve complex coordination involving large numbers of people.

Fig. 26.4. Lamalerans cooperate to butcher a sperm whale. Photo by Allen Gillespie.

One mechanism neglected in this treatment of human sociality is indirect reciprocity (Alexander 1987; Nowak & Sigmund 2005), which, unlike direct reciprocity, is not a dyadic process. In indirect reciprocity an individual is seen to help someone, develops a reputation as a helper, and thereby attracts the help of other individuals in the future. Information about the behavior of others is key, and one's reputation is enhanced by being seen and/or identified as a cooperator. It is interesting to note that indirect reciprocity involves a process of discrimination and behavioral assortment. Assortment among cooperators is maintained by monitoring the reputation of others in a process of social learning (Henrich & Henrich 2006).

In these ways, culturally biased transmission increases positive assortment, which can facilitate coordination (Wilson & Dugatkin 1997; Alvard 2003b; McElreath et al. 2003; Fletcher & Doebeli 2006; Henrich 2004; Chiang 2008). To understand how large groups with socially beneficial institutions become common, a process is required that selects between the multiple stable equilibria that arise from coordination (van Huyck et al. 1997; Bornstein et al. 2002; Boyd & Richerson 2002; Gurerk et al. 2006).

Cultural Group Selection in Human Societies

It is clear that individuals within groups can have conflicting interests, but the idea that groups can have collective interests has met resistance in a variety of fields. In behavioral ecology, individual self-interest has been the key principle underlying the evolutionary approach to social behavior (Krebs and Dawkins 1984), and human-behavioral ecologists and others interested in human social behavior adopted the same perspective (Alexander 1979, 1987; Chagnon & Irons 1979; Smith & Winterhalder 1992; Winterhalder & Smith 1992). For functionality at the group level, individuals must behave in ways that help one another. People share their meat and pay their taxes; they risk and sometimes give their lives defending their communities; they follow the rules and submit to the harsh yoke and subtle comfort of institutions. The evolutionary obstacle for this kind of behavior is that following the rules can be costly for individuals; while groups that contain more altruists do better in competition against groups that contain fewer, these sorts of behaviors are not usually fitness-maximizing within the group. Overall, it would be better to give the meat to your own children, circumvent payment of your taxes, and avoid being killed far from home.

In response to new theory, data, and methods, multi-level selection perspectives have earned a new credibility (Wilson & Wilson 2007; Wilson 2008). This is especially the case when human sociality and its cultural roots are considered (Henrich 2004; Richerson & Boyd 2005b). A key conceptual breakthrough for understanding multilevel selection was made by George Price in two papers that were published in the 1970s but remained somewhat neglected for many years (Price 1970, 1972). In these papers, Price develops an alternative set of accounting rules for measuring fitness costs and benefits. Price's equation describes change in the mean value of a trait from one generation to the next.

In Price's model, the force of natural selection is partitioned into components that operate at different levels. Selection *within* groups works to reduce the frequency of altruistic traits within each group, because individuals who express altruistic traits incur costs while other group members accrue the resulting benefits. However, if altruists enhance the success of the group as a whole, then selection *among* groups may favor the traits. This will happen when there is a positive correlation between the frequency of altruists and the mean fitness of the group. The Price equation shows that the strength of selection at each level is proportional to the relative amount of variation at each level, and the outcome depends on the relative strength of selection working at each level.

The process of group level selection is made clearer if it is examined in the context of social structure. Following a simple model from Sober & Wilson (1998), imagine a population consisting of altruists and egoists. Altruistic individuals pay a fitness cost to generate a fitness benefit to the other altruists and egoists in the population. Egoists pay no cost, but they benefit from the behavior of the altruists. It is clear that over time, natural selection working at the individual level will remove altruists from the population; but if structure is added to the population, the outcome changes. Now imagine the same population with equal numbers of altruists and egoists, now divided into two groups that vary in the proportions of each. Group A has 80% altruists while group B has 80% egoists. Within each group, selection still works against altruists via individual selection, and they will decline in frequency over time. But if the benefits of being in a group with many altruists are sufficiently great, the frequency of altruists in the entire population can increase in spite of the fact that their frequency declines in each individual group.

A common criticism is that ultimately, actors are acting in their own self-interest if by acting as altruists they increase their group's representation among groups and thus increase their own selfish representation in the global population (West et al. 2007). Thus, paying taxes and contributing to the public good is in my selfish best interest because it makes the nation strong. Sober & Wilson (1998) have the most cogent reply; they refer to this problem as the "averaging fallacy" (see also Okasha 2004). Fitness can be measured absolutely, without considering the structure of the entire population, or relatively, within the local groups. Depending on how fitness is measured, a behavior can appear to be altruistic or selfish. Individuals who engage in costly behaviors, such as sharing meat, suffer a relative disadvantage within groups in comparison to free-riders, but they can absolutely increase in frequency in the population when the effect is averaged across all groups. When structure is ignored and the effects are averaged across groups, the process of group selection is masked. The altruistic behavior is seen as selfish, and is presumed to evolve via selection at the individual level.

One relationship that provides this sort of structure is genetic kinship. Preferential association among kin increases the average relatedness between actors and their partners (Hamilton 1975; Pepper 2000). From the point of view of multilevel selection, such positive assortment reduces genetic variance within groups and increases it between groups. In this way, kin selection is understood not as an alternative, but rather as a special case of group selection (Hamilton 1975; Kerr & Godfrey-Smith 2002; Wilson & Wilson 2007; Wilson 2008). The utility of group selection as a tool for understanding social behavior has been questioned, however, by claims that it provides answers more easily obtained via the kin selection approach (West et al. 2007). While this may or may not turn out to be true for most nonhuman examples, claims that "kin selection models have shown that punishment or cooperation [among humans] are only favored if they provide a direct fitness benefit to the actor, or if limited dispersal leads to an indirect fitness benefit to relatives" (West et al. 2007, p. 378) ignore the fact that the central question for humans is the widespread cooperation among *non-kin*. Granted that kin selection is group selection, not all group selection is kin selection. As will be discussed below, many of the adaptations that characterize human sociality are hypothesized to have arisen as a result of cultural group selection (Henrich 2004).

The key problem with genetic group selection is that migration and other forms of mixing work to erase between-group variation (Williams 1966). Cultural group selection does not suffer from the same problem. Recall that one of the key adaptive features of culture is that it works to reduce conflict within groups by homogenizing behavioral variance within them—often via culturally transmitted ideologies, like religions, for example (Wilson 2002; Norenzayan & Shariff 2008). Cultural forces, such as conformist and prestige-biased transmission, potentiate group selection by reducing the amount of behavioral variation within groups and by increasing the amount of variation among them (Richerson & Boyd 2005b). They can help to maintain cultural boundaries even in the face of substantial migration (McElreath 2004). The outcome is that people assort behaviorally with others who have similar or complementary intentions or capabilities, and, perhaps more importantly, with people who share the norms of behavior that allow cooperative intentions to be realized.

Conclusions

Humans display a number of life-history traits that can be maintained only in conjunction with cooperative social adaptations. Although post-weaning investment is distinctively substantial in primates generally (chapter 14, this volume), human children show much greater nutritional dependence for a long period after they are weaned (Kaplan et al. 2000). To accommodate the burden of multiple dependent young, human hunter-gatherer mothers divert time and effort away from foraging, thus creating a subsistence deficit that must be filled (Lancaster et al. 2000). Although there is currently much debate over the details, there is growing consensus that humans are communal breeders in the sense that mothers must have help from others to raise their children (Bogin 1997; Hrdy 2000, 2009; Hill & Hurtado 2009). In a meta-study of 45 historical and contemporary, mostly natural fertility populations, Sear & Mace (2008) found that help from kin is a universal feature of human child rearing, but that the relative importance of help from fathers, grandmothers, siblings, and others depends on ecological conditions.

There are two points to be drawn from this discussion. The first is that the exceptional life history of *Homo sapiens* requires a degree of bioeconomic social complexity far beyond the chimpanzee pattern. In a human social system, helping occurs between categories of kin that are not usually recognizable among chimpanzees: fathers, cousins, and grandmothers, for example. In addition, the importance of nongenetic kin is underappreciated. One key relationship that is noticeably absent in evolutionary discussions of human sociality is the role of *affinal* kin. The role of lineage mates, pantribal sodalities, and other institutional sources of support is also important, but is rarely examined (see, however, Alvard 2003a; Quinlan & Flinn 2005). Second, dyadic reciprocal altruism is an unlikely mechanism to explain the help required to support human life-history patterns. The conditions that are required for reciprocal exchange—namely, a high probability that ego's partner will repay in the future—are simply not present. "The human life course could not have evolved without long-term imbalances in food transfers within and among families" (Kaplan & Gurven 2005). Data from hunter-gatherers show that net caloric demand increases with the increasing size of older families at a rate that outstrips the families' production (fig. 26.5). Older adults with large families who are most in deficit are not able to pay back, before they grow old and die, the younger families who must subsidize them (Kaplan & Robson 2002; Robson & Kaplan 2003).

A common argument is that predator defense is the primary benefit of living in groups, while the chief cost is competition with other group members for resources (Silk 2007a; chapter 9, this volume). Primates, however, excel at turning what is a cost in many other species into a benefit by leveraging sociality into collective action. Wrangham (1980), in what appears to be a group selection model despite his claims to the contrary, understood this when he explained sociality in female-bonded primates; larger, more cooperative groups of related females have a competitive edge over smaller ones in exploiting patchy food resources. Sociality is clearly a key primate adaptation; the difference between human and nonhuman primates is not collective action per se but rather the fact that human ultrasociality is principally organized around cultural rather than genetic kinship relationships.

One possible path from nonhuman primate kinship to human kinship is described by Chapais (2008) who reconciles Levi-Strauss's (1969) idea of reciprocal exogamy—the exchange of women between kin groups—with what Chapais calls the "exogamy configuration." Like many before him, Chapais points to the pair bond as the key to human sociality (Washburn & Lancaster 1968; Lovejoy 1981). He argues that the structural consequence of pair bonding is the transformation of kinship as we know it in the modern human sense: cultural kinship rather than strictly genetic kinship. One key feature of human sociality is that relationships define other relationships. Enduring pair bonds between adult males and females provide a far greater range of opportunity for beneficial social interactions among parents and offspring, siblings, and fathers' relatives. Importantly, exogamy or dispersal in the context of enduring pair bonds creates links between siblings in *different* groups and their mates, creating bilateral recognition of affines (relatives through marriage). Individuals who may not be related at all geneaologically—two males, for instance—recognize a common interest in the well-being of a third party: one as a brother or a father, the other as a mate. These affinal relationships provide links between local groups and they produce the order of complexity not seen in other primates (Rodseth et al. 1991a). Local groups that would be mortal

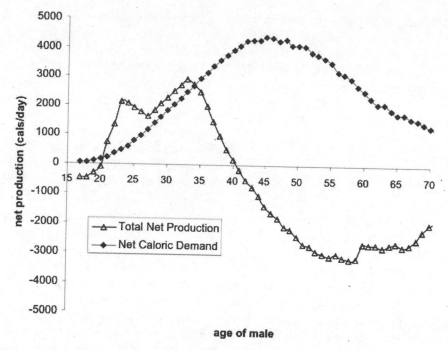

Fig. 26.5. Among Ache hunter-gatherers of Paraguay, net demand on parents increases with age and additional children as indicated by the diamonds. Total net production, indicated by the triangles, does not keep up (reprinted with permission from Kaplan & Gurven 2005, fig. 3.3).

enemies in a chimpanzee social system are often coalitionary partners in human society. This expanded cooperative network may provide support for human communal breeding, and it sets the stage for the creation of tribal-level complexity discussed above.

Cooperative hunting of large, prime adult game (Stiner 2002) appears to provide the resources to fuel the life-history shifts documented by Kaplan and his colleagues (Kaplan et al. 2000), but there are other advantages of large groups that become starkly apparent in the context of direct intergroup competition. In chimpanzees, numerical superiority is essential to success in such competition (Wilson et al. 2001, 2002). The same is true for other mammals, like lions (McComb et al. 1994) and elephants (Archie et al. 2006). Numerical superiority is also key to contemporary human military action (Rotte & Schmidt 2003). One argument for the evolution of institutions like unilineal lineage systems via cultural group selection is that they are larger than kin groups, and this provides great numerical advantages when groups are in conflict (Kelly 2000; Bowles 2009). Classic examples of this are the segmental systems of east Africa, in which lower-order segments organized by unilineal descent principles are easily combined into higher-order segments when needed. Such systems allowed the Nuer to expand at the expense of the Dinka, for example (Evans-Pritchard 1940; Kelly 1985). In large armies, men who are otherwise strangers form kin-like bonds and trust men who share symbolic tags of affinity and institutionalized norms of conduct (Richerson & Boyd 1999b).

Societies vary in their ability to bring together people who are willing to subordinate their own interests to those of the group. Cultural evolution has favored increased social complexity over the last 100,000 years, and especially over the last 10,000 years (Henrich 2004; Richerson & Boyd 2005b). People in these more complex societies come together in the tens of millions for coordinated action. Historically, there is some evidence that simple acephalous tribal institutions, like the segmentary lineages systems among the Lamalera, are replaced by even more inclusive organizing principles. One type of structure is a hierarchy formed around charismatic leaders who can facilitate coordination and punishment in effective ways (Kessler & Cohrs 2008; Van Vugt et al. 2008). For example, in 508 BCE, Cleisthenes reformed the ancient clan-based requirement for Athenian citizenship (*phratries*) and replaced it with one based on locality (*demes*). Interestingly, he was also probably responsible for introducing *ostracism*, a practice in which a vote from more than 6,000 would exile a citizen for 10 years (Lambert 1993). A similar process occurred early in the formation of the Mongol empire, when Genghis Khan dissolved his followers' tribal identities and replaced them with a Mongol imperial identity (Weatherford 2004). The Hebrew tribes were united under the rule of Solomon (Miller and Hayes 2006). A crucial turning point in Zulu history occurred during the reign of Shaka, who united numerous related clans into a single powerful tribe (Morris 1966). In each case, societies that had moved beyond kin-based social organization to form lineage type systems developed social structures that created even larger, more inclusive groups, and these societies were able to dominate other, smaller groups in a cultural group selection process.

Summary

Human sociality has a legacy in its mammalian and primate past. The processes of kin selection and reciprocity remain important in human societies: close familial structure can be produced by the former, and dyadic relationships like friendship by the latter. However, people also live in large, hierarchically, and complexly structured societies and they commonly cooperate with non-kin—features that the aforementioned theoretical tools are inadequate to fully explain (Richerson & Boyd 1999a). The primary thesis of this chapter is that group selection on culturally maintained variation is the primary evolutionary force that originally produced and now maintains human social complexity (Richerson & Boyd 1998).

Results from experimental game theory point to the importance of both culture and punishment in the maintenance of human sociality (Fehr & Gachter 2002; Henrich et al. 2005). Punishment can lead to cooperative outcomes in ultimatum and *n*-person iterated prisoner's dilemma games, but the Folk Theorem notes that punishment can produce any sort of behavior—cooperative or not (Boyd 2006). Perhaps the prisoner's dilemma is not the best way to conceptualize cooperation (Skyrms 2004). As an alternative, coordination games describe situations in which individuals have common interest but multiple stable equilibria also exist—only some of which may be cooperative. Group selection is a process that can select between the equilibria that arise from such coordination.

Cultural forces like conformist and prestige-biased transmission potentiate group selection by reducing the amount of behavioral variation within groups and increasing the amount of variation among them (Henrich 2004). The standard arguments against genetic group selection do not apply to cultural systems. Migration, for example, does not reduce variance between groups in the way it does for genetic systems. Culturally transmitted information can produce shared rules of behavior that structure social relations to produce institutions that can serve collective functions. Nonhuman primates seem to lack the "shared intentionality" that reifies institutions among humans to the point that the institutions seem to possess interests independent of those possessed by their individual members.

Humans are communal breeders, and mothers must have help from others to raise their children (Kaplan et al. 2000). The exceptional life history of humans requires a degree of bioeconomic cooperation that reciprocity and kin selection cannot provide. The conditions required for reciprocal exchange are not present when resources flow from older generations of producers to younger generations of consumers. Sociality is clearly a key primate adaptation; the difference between nonhuman and human primates is not collective action per se, but rather the organization of human ultrasociality around cultural rather than genetic kinship relationships. Structural consequences of pair bonding create new categories of cultural kinship, thereby setting the stage for the creation of higher-order social collectives, like tribes (Chapais 2008).

The communal imperative of our exceptional life history suggests that social complexity was produced, at least in part, by cultural group selection. Societies with the ability to form larger groups have important advantages as group selection is played out in intergroup competition (Bowles 2009). Understanding these higher-order mechanisms and the contexts that favor them is a major challenge for evolutionary anthropologists.

References

Alexander, R. D. 1979. *Darwinism and Human Affairs*. Seattle: University of Washington Press.

———. 1987. *The Biology of Moral Systems*. Hawthorne, NY: A. de Gruyter.

Allen-Arave, W., Gurven, M. & Hill, K. 2008. Reciprocal altruism, rather than kin selection, maintains nepotistic food transfers on an Ache reservation. *Evolution and Human Behavior*, 29, 305–318.

Allen, D. 1991. What are transaction costs? *Research in Law and Economics*, 14, 1–18.

Alvard, M. 2001. Mutualistic hunting. In *The Early Human Diet: The Role of Meat* (ed. by Stanford, C. & Bunn, H.), 261–278. New York: Oxford University Press.

———. 2002. Carcass ownership and meat distribution by big-game cooperative hunters. *Research in Economic Anthropology*, 21, 99–132.

———. 2003a. Kinship, lineage, and an evolutionary perspective on cooperative hunting groups in Indonesia. *Human Nature*, 14, 129–163.

———. 2003b. The adaptive nature of culture. *Evolutionary Anthropology*, 12, 136–149.

Alvard, M. & Nolin, D. 2002. Rousseau's whale hunt? Coordination among big-game hunters. *Current Anthropology*, 43, 533–559.

Apostolou, M. 2007. Sexual selection under parental choice: The role of parents in the evolution of human mating. *Evolution and Human Behavior*, 286, 403–409.

Archie, E. A., Moss, C. J. & Alberts, S. C. 2006. The ties that bind: Genetic relatedness predicts the fission and fusion of social groups in wild African elephants. *Proceedings of the Royal Society B: Biological Sciences*, 273, 513–522.

Aviles, L., Fletcher, J. A. & Cutter, A. D. 2004. The kin composition of social groups: Trading group size for degree of altruism. *American Naturalist*, 164, 132–144.

Axelrod, R. 1984. *The Evolution of Cooperation*. New York: Basic Books.

Axelrod, R. & Hamilton, W. D. 1981. The evolution of cooperation. *Science*, 211, 1390.

Bar-Yosef, O. 2001. From sedentary foragers to village hierarchies: The emergence of social institutions. In *The Origin of Human Social Institutions* (ed. by Runciman, W.), 1–38. Oxford: Oxford University Press.

Battalio, R., Samuelson, L. & Van Huyck, J. 2001. Optimization incentives and coordination failure in laboratory stag hunt games. *Econometrica*, 69, 749–764.

Beckerman, S. 2000. Mating and marriage, husbands and lovers. *Behavioral and Brain Sciences*, 23, 590–591.

Bergstrom, T. C. 2002. Evolution of social behavior: Individual and group selection. *Journal of Economic Perspectives*, 16, 67–88.

Berman, C. M., Rasmussen, K. L. R. & Suomi, S. J. 1997. Group size, infant development and social networks in free-ranging rhesus monkeys. *Animal Behaviour*, 53, 405–421.

Boehm, C. 1999. *Hierarchy in the Forest: The Evolution of Egalitarian Behavior*. Cambridge, MA: Harvard University Press.

Bogin, B. 1997. Evolutionary hypotheses for human childhood. *Yearbook of Physical Anthropology*, 40, 63–89.

Bornstein, G., Gneezy, U. & Nagel, R. 2002. The effect of intergroup competition on group coordination: An experimental study. *Games and Economic Behavior*, 41, 1–25.

Bowlby, J. 1969. *Attachment and Loss*, Vol. I. Attachment. New York: Basic Books.

———. 1973. *Attachment and Loss*, Vol. II. Separation: Anxiety and Anger. New York: Basic Books.

Boyd, R. 2006. Reciprocity: You have to think different. *Journal of Evolutionary Biology*, 19, 1380–1382.

Boyd, R. & Richerson, P. 1985. *Culture and the Evolutionary Process*. Chicago: University of Chicago Press.

———. 1988. The evolution of reciprocity in sizable groups. *Journal of Theoretical Biology*, 132, 337–356.

———. 1992. Punishment allows the evolution of cooperation (or anything else) in sizable groups. *Ethology and Sociobiology*, 13, 171–195.

———. 2002. Group beneficial norms can spread rapidly in a structured population. *Journal of Theoretical Biology*, 215, 287–296.

Bowles, S. 2009. Did warfare among ancestral hunter-gatherers affect the evolution of human social behaviors? *Science*, 324, 1293–1298.

Buchanan, J. M. 1962. The relevance of pareto optimality. *Journal of Conflict Resolution*, 6, 341–354.

Buunk, A. P., Park, J. H. & Dubbs, S. L. 2008. Parent-offspring conflict in mate preferences. *Review of General Psychology*, 12, 47–62.

Camerer, C., Loewenstein, G. & Rabin, M. 2004. *Advances in Behavioral Economics*. Princeton, NJ: Princeton University Press.

Campbell, D. 1983. The two distinct routes beyond kin selection to ultrasociality: Implications for the humanities and social sciences. In *Nature of Prosocial Development: Theories and Strategies* (ed. by Bridgeman, D.), 71–81. New York: Academic Press.

Chagnon, N. 1975. Genealogy, solidarity, and relatedness: Limits to local group size and patterns of fissioning in an expanding population. *Yearbook of Physical Anthropology*, 19, 95–110.

Chagnon, N. & Irons, W. 1979. *Evolutionary Biology and Human Social Behavior: An Anthropological Perspective*. North Scituate, MA: Duxbury Press.

Chapais, B. 2001. Primate nepotism: What is the explanatory value of kin selection? *International Journal of Primatology*, 22, 203–229.

———. 2008. *Primeval Kinship: How Pair-Bonding Gave Birth to Human Society*. Cambridge, MA: Harvard University Press.

Chapais, B. & Berman, C. 2004. *Kinship and Behavior in Primates*. New York: Oxford University Press.

Chiang, Y. S. 2008. A path toward fairness: Preferential association and the evolution of strategies in the ultimatum game. *Rationality and Society*, 20, 173–201.

Chisholm, J. S. & Burbank, V. K. 1991. Monogamy and polygyny in southeast arnhem-land: Male coercion and female choice. *Ethology and Sociobiology*, 12, 291–313.

Crittenden, A. N. & Marlowe, F. W. 2008. Allomaternal care among the Hadza of Tanzania. *Human Nature*, 19, 249–262.

Cropanzano, R., Goldman, B. & Folger, R. 2005. Self-interest: Defining and understanding a human motive. *Journal of Organizational Behavior*, 26, 985–991.

Davis, J. N. & Daly, M. 1997. Evolutionary theory and the human family. *Quarterly Review of Biology*, 72, 407–435.

Dawkins, R. 1976. *The Selfish Gene*. New York: Oxford University Press.

Dugatkin, L. 1997. *Cooperation Among Animals: An Evolutionary Perspective*. New York: Oxford University Press.

Dunbar, R. 2008. Kinship in biological perspective. In *Early Human Kinship: From Sex to Social Reproduction* (ed. by Allen, N., Callan, H., Dunbar, R. & James, W.), 131–150. Oxford: Blackwell Pub.

Durham, W. 1991. *Coevolution: Genes, Culture, and Human Diversity*. Stanford, CA: Stanford University Press.

Edgerton, R. 1992. *Sick Societies: Challenging the Myth Of Primitive Harmony*. New York: Free Press.

Ember, C. R., Ember, M. & Pasternak, B. 1974. On the development of unilineal descent. *Journal of Anthropological Research*, 30, 69–94.

Ensminger, J. 1997. Transaction costs and Islam: Explaining conversion in Africa. *Journal of Institutional and Theoretical Economics-Zeitschrift Fur Die Gesamte Staatswissenschaft*, 153, 4–29.

Fehr, E. & Gachter, S. 2002. Altruistic punishment in humans. *Nature*, 415, 137–140.

Fehr, E. & Schmidt, K. M. 1999. A theory of fairness, competition, and cooperation. *Quarterly Journal of Economics*, 114, 817–868.

Fletcher, J. A. & Doebeli, M. 2006. How altruism evolves: Assortment and synergy. *Journal of Evolutionary Biology*, 19, 1389–1393.

Fuentes, A. 2002. Patterns and trends in primate pair bonds. *International Journal of Primatology*, 23, 953–978.

Gangestad, S. W. & Simpson, J. A. 2000. The evolution of human mating: Trade-offs and strategic pluralism. *Behavioral and Brain Sciences*, 23, 573–587.

Gintis, H. 2000. Beyond *Homo economicus*: Evidence from experimental economics. *Ecological Economics*, 35, 311–322.

———. 2003. Solving the puzzle of prosociality. *Rationality and Society*, 15, 155–187.

Gintis, H., Bowles, S., Boyd, R. & Fehr, E. 2005. *Moral Sentiments and Material Interests: The Foundations of Cooperation in Economic Life*. Cambridge, MA: MIT Press.

Gintis, H., Henrich, J., Bowles, S., Boyd, R. & Fehr E. 2008. Strong reciprocity and the roots of human morality. *Social Justice Research* 21, 241–253.

Gurerk, O., Irlenbusch, B. & Rockenbach, B. 2006. The competitive advantage of sanctioning institutions. *Science*, 312, 108–111.

Gurven, M. 2004. To give and to give not: The behavioral ecology of human food transfers. *Behavioral and Brain Sciences*, 27, 543–560.

———. 2006. The evolution of contingent cooperation. *Current Anthropology*, 47, 185–192.

Guth, W., Schmittberger, R. & Schwarze, B. 1982. An experimental analysis of ultimatum bargaining. *Journal of Economic Behavior & Organization*, 3, 367–388.

Hames, R. & McCabe, C. 2007. Meal sharing among the Ye'kwana. *Human Nature* 18, 1–21.

Hamilton, W. D. 1964a. Genetical evolution of social behaviour 2. *Journal of Theoretical Biology*, 7, 17–52.

———. 1964b. Genetical evolution of social behaviour 1. *Journal of Theoretical Biology*, 7, 1–16.

———. 1975. Innate social aptitudes of man: An approach from evolutionary genetics. In *Biosocial Anthropology* (ed. by Fox, R.), 133–155. London: Malaby Press.

Hardin, G. 1968. Tragedy of the commons. *Science*, 162, 1243–1248.

Hawkes, K., O'Connell, J. F., Jones, N. G. B., Alvarez, H. & Charnov, E. L. 1998. Grandmothering, menopause, and the evolution of human life histories. *Proceedings of the National Academy of Sciences of the United States of America*, 95, 1336–1339.

Henrich, J. 2000. Does culture matter in economic behavior? Ultimatum game bargaining among the Machiguenga of the Peruvian Amazon. *American Economic Review*, 90, 973–979.

———. 2004. Cultural group selection, coevolutionary processes and large-scale cooperation. *Journal of Economic Behavior & Organization*, 53, 3–35.

Henrich, J. & Boyd, R. 1998. The evolution of conformist transmission and the emergence of between-group differences. *Evolution and Human Behavior*, 19, 215–241.

Henrich, J., Boyd, R., Bowles, S., Camerer, C., Fehr, E., Gintis, H., McElreath, R., Alvard, M., Barr, A., Ensminger, J., Henrich, N. S., Hill, K., Gil-White, F., Gurven, M., Marlowe, F. W., Patton, J. Q. & Tracer, D. 2005. "Economic man" in cross-cultural perspective: Behavioral experiments in 15 small-scale societies. *Behavioral and Brain Sciences*, 28, 795–855.

Henrich, J. & Gil-White, F. J. 2001. The evolution of prestige: Freely conferred deference as a mechanism for enhancing the benefits of cultural transmission. *Evolution and Human Behavior*, 22, 165–196.

Henrich, J., & Henrich, N. 2006. Culture, evolution and the puzzle of human cooperation. *Cognitive Systems Research*, 7, 220–245.

Henrich, J. & McElreath, R. 2003. The evolution of cultural evolution. *Evolutionary Anthropology*, 12, 123–135.

Henrich, J., McElreath, R., Barr, A., Ensminger, J., Barrett, C., Bolyanatz, A., Cardaroas, J. C., Gurven, M., Gwako, E., Henrich, N., Lesoronol, C., Marlowe, F., Tracer, D. & Ziker, J. 2006. Costly punishment across human societies. *Science*, 312, 1767–1770.

Hill, D. 2004. The effects of demographic variation on kinship structure and behavior in cercopithecines. In *Kinship and Behavior in Primates* (ed. by Chapais, B. & Berman, C.), 132–150. New York: Oxford University Press.

Hill, K., & Hurtado, A. 2009. Cooperative breeding in South American hunter-gatherers. *Proceedings of the Royal Society B: Biological Sciences*, 276, 3863–3870.

Hinde, R. A. 1976. Interactions, relationships and social structure. *Man*, 11, 1–17.

Hodgson, G. M. 2006. What are institutions? *Journal of Economic Issues*, 40, 1–25.

Hrdy, S. B. 2000. *Mother Nature: Maternal Instincts and How They Shape the Human Species*, New York: Ballantine Books.

———. 2009. *Mothers and Others: The Evolutionary Origins of Mutual Understanding*. Cambridge, MA: Belknap Press of Harvard University Press.

Husain, Z. & Bhattacharya, R. N. 2004. Common pool resources and contextual factors: Evolution of a fishermen's cooperative in Calcutta. *Ecological Economics*, 50, 201–217.

Isaac, R. M. & Walker, J. M. 1988. Group-size effects in public-goods provision: The voluntary contributions mechanism. *Quarterly Journal of Economics*, 103, 179–199.

Ivey, P. K. 2000. Cooperative reproduction in Ituri forest hunter-gatherers: Who cares for Efe infants. *Current Anthropology*, 41, 856–866.

Jensen, K., Call, J., & Tomasello, M. 2007. Chimpanzees are rational maximizers in an ultimatum game. *Science*, 318, 107–109.

Johnson, A. W. & Earle, T. K. 2000. *The Evolution of Human Societies: From Foraging Group to Agrarian State*, 2nd edition. Stanford, CA: Stanford University Press.

Johnson, D. 2005. God's punishment and public goods: A test of the supernatural punishment hypothesis in 186 world cultures. *Human Nature: An Interdisciplinary Biosocial Perspective*, 16, 410–446.

Jones, D. 2000. Group nepotism and human kinship. *Current Anthropology*, 41, 779–809.

Joshi, N. V. 1987. Evolution of cooperation by reciprocation within structured demes. *Journal of Genetics*, 66, 69–84.

Kagel, J. H. & Roth, A. E. 1995. *The Handbook of Experimental Economics*. Princeton, NJ: Princeton University Press.

Kahneman, D. 2003. Maps of bounded rationality: Psychology for behavioral economics. *American Economic Review*, 93, 1449–1475.

Kaplan, H. & Gurven, M. 2005. The natural history of human food sharing and cooperation: A review and a new multi-individual approach to the negotiation of norms. In *Moral Sentiments and Material Interests: The Foundations of Cooperation in Economic Life* (ed. by Gintis, H., Bowles, S., Boyd, R. & Fehr, E.), 76–113. Cambridge, MA: MIT Press.

Kaplan, H. & Hill, K. 1985. Food sharing among Ache foragers: Tests of explanatory hypotheses. *Current Anthropology*, 26, 223–246.

Kaplan, H., Hill, K., Lancaster, J. & Hurtado, A. M. 2000. A theory of human life history evolution: Diet, intelligence, and longevity. *Evolutionary Anthropology*, 9, 156–185.

Kaplan, H. S. & Robson, A. J. 2002. The emergence of humans: The coevolution of intelligence and longevity with intergenerational transfers. *Proceedings of the National Academy of Sciences of the United States of America*, 99, 10221–10226.

Kapsalis, E. & Berman, C. M. 1996. Models of affiliative relationships among free-ranging rhesus monkeys (*Macaca mulatta*). I. Criteria for kinship. *Behaviour*, 133, 1209–1234.

Kelly, R. C. 2000. *Warless Societies and the Origin of War*. Ann Arbor: University of Michigan Press.

Kerr, B. & Godfrey-Smith, P. 2002. On Price's equation and average fitness. *Biology and Philosophy*, 17, 551–565.

Kerr, N. L. & Levine, J. M. 2008. The detection of social exclusion: Evolution and beyond. *Group Dynamics*, 12, 39–52.

Kessler, T. & Cohrs, J. C. 2008. The evolution of authoritarian processes: Fostering cooperation in large-scale groups. *Group Dynamics*, 12, 73–84.

Kramer, K. L. 2005. Children's help and the pace of reproduction: Cooperative breeding in humans. *Evolutionary Anthropology*, 14, 224–237.

Krause, J. & Ruxton, G. D. 2002. *Living in Groups*. New York: Oxford University Press.

Krebs, J. R., & Dawkins, R. 1984. Animal signals: Mind reading and manipulation. In *Behavioral Ecology: An Evolutionary Approach*, Second Edition (ed. by Krebs, J. & Davies, N.), 380-402. Oxford: Blackwell.

Lambert, S. D. 1993. *The Phratries of Attica*. Ann Arbor: University of Michigan Press.

Lancaster, J., Kaplan, H., Hill, K. & Hurtado, A. 2000. The evolution of life history, intelligence, and diet among chimpanzees and human foragers. In *Perspectives in Ethology: Evolution, Culture and Behavior* (ed. by Tonneau, F. & Thompson, N.), 47–72. New York: Plenum Press.

Lancaster, J. B. 1986. Primate social behavior and ostracism. *Ethology and Sociobiology*, 7, 215–225.

Langergraber, K. E., Mitani, J. C. & Vigilant, L. 2007. The limited impact of kinship on cooperation in wild chimpanzees. *Proceedings of the National Academy of Sciences of the United States of America*, 104, 7786-7790.

Ledyard, J. 1995. Public goods: A survey of experimental research. In *The Handbook of Experimental Economics* (ed. by Kagel, J. & Roth, A.), 111–194. Princeton, NJ: Princeton University Press.

Lévi-Strauss, C. 1969. *The Elementary Structures of Kinship*. Boston: Beacon Press.

Lovejoy, C. O. 1981. The origin of man. *Science*, 211, 341–350.

Low, B. S. 2005. Women's lives there, here, then, now: A review of women's ecological and demographic constraints cross-culturally. *Evolution and Human Behavior*, 26, 64–87.

Lukas, D., Reynolds, V., Boesch, C. & Vigilant, L. 2005. To what extent does living in a group mean living with kin? *Molecular Ecology*, 14, 2181–2196.

Masclet, D. 2003. Ostracism in work teams: A public good experiment. *International Journal of Manpower*, 24, 867–887.

Maynard Smith, J. 1982. *Evolution and the Theory of Games*. Cambridge: Cambridge University Press.

Maynard Smith, J. & Szathmary, E. 1995. *The Major Transitions in Evolution*. Oxford: W. H. Freeman Spektrum.

McAdams, R. 2008. Beyond the prisoners' dilemma: Coordination, game theory, and law. *University of Chicago Law and Economics, Olin Working Paper No. 437*.

McComb, K., Packer, C. & Pusey, A. 1994. Roaring and numerical assessment in contests between groups of female lions, *Panthera leo*. *Animal Behaviour*, 47, 379–387.

McElreath, R. 2004. Social learning and the maintenance of cultural variation: An evolutionary model and data from east Africa. *American Anthropologist*, 106, 308–321.

McElreath, R. & Boyd, R. 2007. *Mathematical Models of Social Evolution: A Guide For the Perplexed*. Chicago: University of Chicago Press.

McElreath, R., Boyd, R. & Richerson, P. J. 2003. Shared norms and the evolution of ethnic markers. *Current Anthropology*, 44, 122–129.

Miller, J. M. & Hayes, J. H. 2006. *A History of Ancient Israel and Judah*, 2nd edition. Louisville, KY: Westminster John Knox Press.

Mitani, J. C., Merriwether, D. A. & Zhang, C. B. 2000. Male affiliation, cooperation and kinship in wild chimpanzees. *Animal Behaviour*, 59, 885–893.

Morris, D. R. 1966. *The Washing of the Spears: A History of the Rise of the Zulu Nation Under Shaka and its Fall in the Zulu War of 1879*. London: Cape.

Nolin, D. 2008. Food-sharing networks in Lamalera, Indonesia: Tests of adaptive hypotheses. PhD dissertation, University of Washington.

Norenzayan, A. & Shariff, A. F. 2008. The origin and evolution of religious prosociality. *Science*, 322, 58–62.

North, D. C. 1990. *Institutions, Institutional Change, and Economic Performance*. Cambridge: Cambridge University Press.

Nowak, M. 2006. Five rules for the evolution of cooperation. *Science*, 314, 1560–1563.

Nowak, M. & Sigmund, K. 2005. Evolution of indirect reciprocity. *Nature* 437:1291–1298.

Olson, M. 1965. *The Logic of Collective Action: Public Goods and the Theory of Groups*. Cambridge, MA: Harvard University Press

Ostrom, E., Walker, J. & Gardner, R. 1992. Covenants with and without a sword: Self-governance is possible. *The American Political Science Review*, 86, 404–417.

Poundstone, W. 1992. *Prisoner's Dilemma*. New York: Doubleday.

Price, G. R. 1970. Selection and covariance. *Nature*, 227, 520–521.

———. 1972. Extension of covariance selection mathematics. *Annals of Human Genetics*, 35, 485-&.

Price, M. 2008. The resurrection of group selection as a theory of human cooperation. A review of *Foundations of Human Sociality: Economic Experiments and Ethnographic Evidence from Fifteen Small-Scale Societies* (ed. by Henrich, J., Boyd, R., Bowles, S., Camerer, C., Fehr, E. & Gintis, H.) and *Moral Sentiments and Material Interests: The Foundations of Cooperation in Economic Life* (ed. by Gintis, H., Bowles, S. & Boyd, R.). *Social Justice Research*, 21, 1–13.

Quinlan, R. J. & Flinn, M. V. 2005. Kinship, sex, and fitness in a caribbean community. *Human Nature: An Interdisciplinary Biosocial Perspective*, 16, 32–57.

Ratnieks, F. 1988. Reproductive harmony via mutual policing by workers in eusocial hymenoptera. *American Naturalist*, 132, 217—236

Ratnieks, F. & Wenseleer, T. 2005. Policing insect societies. *Science*, 307, 54–56.

Reik, W. & Walter, J. 2001. Genomic imprinting: Parental influence on the genome. *Nature Reviews Genetics*, 2, 21–32.

Richerson, P. & Boyd, R. 1998. The evolution of human ultrasociality. In *Indoctrinability, Ideology, and Warfare: Evolution-*

ary Perspectives (ed. by Eibl-Eibesfeldt, I. & Salter, F.), 71–95. New York: Berghahn.

———. 2005a. Solving the puzzle of human cooperation. In *Evolution and Culture* (ed. by Jaisson, S. L. a. P.), 105–132. Cambridge, MA: MIT Press.

———. 1999b. Complex societies: The evolutionary origins of a crude superorganism. *Human Nature*, 10, 253–289.

———. 2001. Institutional evolution in the Holocene: The rise of complex societies. In *The Origin of Human Social Institutions* (ed. by Runciman, W.), 197–235. Oxford: Oxford University Press.

———. 2005b. *Not by Genes Alone: How Culture Transformed Human Evolution*. Chicago: University of Chicago Press.

Riolo, R. L., Cohen, M. D. & Axelrod, R. 2001. Evolution of cooperation without reciprocity. *Nature*, 414, 441–443.

Robson, A. J. & Kaplan, H. S. 2003. The evolution of human life expectancy and intelligence in hunter-gatherer economies. *American Economic Review*, 93, 150–169.

Rodseth, L., Smuts, B. B., Harrigan, A. M. & Wrangham, R. W. 1991a. On the human community as a primate society: Reply. *Current Anthropology*, 32, 429–433.

Rodseth, L. & Wrangham, R. 2004. Human kinship: A continuation of politics by other means? In *Kinship and Behavior in Primates* (ed. by Chapais, B. & Berman, C.), 389–419. Oxford: Oxford University Press.

Rodseth, L., Wrangham, R. W., Harrigan, A. M. & Smuts, B. B. 1991b. The human community as a primate society. *Current Anthropology*, 32, 221–254.

Rotte, R. & Schmidt, C. M. 2003. On the production of victory: Empirical determinants of battlefield success in modern war. *Defence and Peace Economics*, 14, 175–192.

Rubinstein, A. 1979. Equilibrium in supergames with the overtaking criterion. *Journal of Economic Theory*, 21, 1–9.

Schino, G. 2007. Grooming and agonistic support: A meta-analysis of primate reciprocal altruism. *Behavioral Ecology*, 18, 115–120.

Séar, R. & Mace, R. 2008. Who keeps children alive? A review of the effects of kin on child survival. *Evolution and Human Behavior*, 29, 1–18.

Shenk, M. 2005. Kin investment in wage-labor economies: Effects on child and marriage market outcomes. *Human Nature*, 16, 81–114.

Silk, J. 1987. Social behavior in evolutionary perspective. In *Primate Societies* (ed. by Smuts, B., Cheney, D., Seyfarth, R., Wrangham, R. & Struhsaker, T.), 318–329. Chicago: University of Chicago Press.

———. 2002. Kin selection in primate groups. *International Journal of Primatology*, 23, 849–875.

———. 2006. Practicing Hamilton's rule: Kin selection in primate groups. In *Cooperation in Primates and Humans* (ed. by Kappeler, P. & van Schaik, C.), 25–46. New York: Springer.

———. 2007a. The adaptive value of sociality in mammalian groups. *Philosophical Transactions of the Royal Society B: Biological Sciences*, 362, 539–559.

———. 2007b. The strategic dynamics of cooperation in primate groups. In *Advances in the Study of Behavior* (ed. by Brockmann, H., Roper, T., Naguib, M., Wynne-Edwards, K., Barnard, C. & Mitani, J.), 1–41. Burlington, MA: Academic Press.

Simon, H. A. 1982. *Models of Bounded Rationality*. Cambridge, M: MIT Press.

Skyrms, B. 2004. *The Stag Hunt and the Evolution of Social Structure*. Cambridge: Cambridge University Press.

Smith, E. A. & Winterhalder, B. 1992. *Evolutionary Ecology and Human Behavior*. New York: Aldine de Gruyter.

Smith, J.M. 2007. The Texas Aggie bonfire: A conservative reading of regional narratives, traditional practices, and a paradoxical place. *Annals of the Association of American Geographers*, 97, 182–201.

Smuts, B., Cheney, D., Seyfarth, R., Wrangham, R. & Struhsaker, T. 1987. *Primate Societies*. Chicago: University of Chicago Press.

Sober, E. & Wilson, D. 1998. *Unto Others: The Evolution and Psychology of Unselfish Behavior*. Cambridge, MA: Harvard University Press.

Stiner, M. C. 2002. Carnivory, coevolution, and the geographic spread of the genus *Homo*. *Journal of Archaeological Research*, 10, 1–63.

Taylor, C. & Nowak, M. A. 2007. Transforming the dilemma. *Evolution*, 61, 2281–2292.

Terborgh, J. & Janson, C. H. 1986. The socioecology of primate groups. *Annual Review of Ecology and Systematics*, 17, 111–136.

Thaler, R. H. 1992. *The Winner's Curse: Paradoxes and Anomalies of Economic Life*. Princeton, NJ: Princeton University Press.

Traulsen, A. & Nowak, M. A. 2006. Evolution of cooperation by multilevel selection. *Proceedings of the National Academy of Sciences of the United States of America*, 103, 10952–10955.

Trivers, R. 1971. The evolution of reciprocal altruism. *Quarterly Review of Biology*, 46, 35–57.

———. 1974. Parent-offspring conflict. *American Zoologist*, 14, 249–264.

———. 2006. Reciprocal altruism: 30 years later. In *Cooperation in Primates and Humans* (ed. by Kappeler, P. & van Schaik, C.), 67–83. Berlin: Springer.

Van den Berghe, P. L. 1979. *Human Family Systems: An Evolutionary View*. New York: Elsevier.

Van Huyck, J. B., Battalio, R. C. & Rankin, F. W. 1997. On the origin of convention: Evidence from coordination games. *Economic Journal*, 107, 576–596.

Van Vugt, M., Hogan, R. & Kaiser, R. B. 2008. Leadership, followership, and evolution: Some lessons from the past. *American Psychologist*, 63, 182–196.

Venkatachalam, L. 2008. Behavioral economics for environmental policy. *Ecological Economics*, 67, 640–645.

Washburn, S. & Lancaster, C. 1968. The evolution of hunting. In *Man the Hunter* (ed. by Lee, R. & DeVore, I.), 293–303. Chicago: Aldine.

Weatherford, J. M. 2004. *Genghis Khan and the Making of the Modern World*, 1st edition. New York: Crown.

Whiten, A., Spiteri, A., Horner, V., Bonnie, K. E., Lambeth, S. P., Schapiro, S. J. & de Waal, F. B. M. 2007. Transmission of multiple traditions within and between chimpanzee groups. *Current Biology*, 17, 1038–1043.

Wiessner, P. 1998. Indoctrinability and the evolution of socially defined kinship. In *Indoctrinability, Ideology, and Warfare: Evolutionary Perspectives* (ed. by Eibl-Eibesfeldt, I. & Slater, F.), 133–250. New York: Berghahn Books.

———. 2002. The vines of complexity: Egalitarian structures and the institutionalization of inequality among the Enga. *Current Anthropology*, 43, 233–269.

———. 2005. Norm enforcement among the Ju/'hoansi bushmen: A case of strong reciprocity? *Human Nature: An Interdisciplinary Biosocial Perspective*, 16, 115–145.

Williams, G. C. 1966. *Adaptation and Natural Selection: A Critique of Some Current Evolutionary Thought*. Princeton, NJ: Princeton University Press.

Wilson, D. S. 2002. *Darwin's Cathedral : Evolution, Religion, and the Nature of Society*. Chicago: University of Chicago Press.

Wilson, D. S. & Dugatkin, L. A. 1997. Group selection and assortative interactions. *American Naturalist*, 149, 336–351.

Wilson, D. S. & Sober, E. 1994. Reintroducing group selection to the human behavioral sciences. *Behavioral and Brain Sciences*, 17, 585–608.

Wilson, D. S. & Wilson, E. O. 2007. Rethinking the theoretical foundation of sociobiology. *Quarterly Review of Biology*, 82, 327–348.

Wilson, E. O. 1975. *Sociobiology: The New Synthesis*. Cambridge, MA: Belknap Press of Harvard University Press.

———. 2008. One giant leap: How insects achieved altruism and colonial life. *BioScience*, 58, 17–25.

Wilson, M. L., Britton, N. F. & Franks, N. R. 2002. Chimpanzees and the mathematics of battle. *Proceedings of the Royal Society B*, 269, 1107–1112.

Wilson, M. L., Hauser, M. D. & Wrangham, R. W. 2001. Does participation in intergroup conflict depend on numerical assessment, range location, or rank for wild chimpanzees? *Animal Behaviour*, 61, 1203–1216.

Winterhalder, B. 1997. Gifts given, gifts taken: The behavioral ecology of nonmarket, intragroup exchange. *Journal of Archaeological Research*, 5, 121–168.

Winterhalder, B. & Smith, E. 1992. Evolutionary ecology and the social sciences. In *Evolutionary Ecology and Human Behavior* (ed. by Smith, E. & Winterhalder, B.), 1–23. New York: Aldine de Gruyter.

Wrangham, R. W. 1980. An ecological model of female-bonded primate groups. *Behaviour*, 75, 262–300.

Wynne-Edwards, V. C. 1962. *Animal Dispersion in Relation to Social Behaviour*. New York: Hafner.

Ziker, J. & Schnegg, M. 2005. Food sharing at meals: Kinship, reciprocity, and clustering in the Taimyr autonomous okrug, northern Russia. *Human Nature*, 16, 178–211.

Part 5

Cognitive Strategies for
Coping with Life's Challenges

PRIMATES HAVE traditionally played a key role in research on comparative cognition. However, the last two decades have seen an unprecedented growth in studies of primate cognition. This body of work is relevant not only to other primatologists, but also to scholars from other disciplines such as economics, artificial intelligence, linguistics, and human developmental psychology.

This growth has been sustained by three main developments. First, primate cognition, like some other areas in psychology, has embraced an evolutionary framework. The emphasis has shifted from ordering and classifying species based on their abilities to seeking the evolutionary forces that have shaped the cognitive mechanisms used by primates to solve ecological and social problems. All the chapters included in this part offer clear testimony to this shift in focus. Second, primate cognition has left its traditional laboratory setting and moved into the field. The work of pioneers like Hans Kummer, who aptly combined field and captive work, and Dorothy L. Cheney and Robert M. Seyfarth, who began conducting carefully controlled experiments in the field, is not an exception anymore; numerous researchers either work in both settings or at least consider research in the field and in the laboratory as complementary approaches. This development is well reflected in the studies conducted by authors in this part and in their chapters.

Third, and perhaps most important for a book devoted to primate societies, primate cognition has broadened its scope beyond the traditional emphasis on ecological problems (i.e., locating, identifying, and extracting food) to social problems. The impact of the change that has occurred in the last 25 years cannot be underestimated. One way to appreciate it is to examine the chapters devoted to primate cognition in *Primate Societies*, published 25 years ago. *Primate Societies* included a section entitled "Communication and Intelligence," composed of four chapters. One chapter was devoted entirely to cognition, but social cognition only occupied a few pages of text. There was also a chapter on social learning and cultural transmission, and two additional chapters were devoted to communication. Of these latter three chapters, only one, which dealt with vocal communication and the origins of language, had a cognitive orientation. Obviously, this is not meant as a criticism. It is simply a clear reflection on the state of the field 25 years ago.

Since then the picture has changed substantially, and all the chapters included in this part have a cognitive orientation from beginning to end. Chapter 27, by Charles Menzel, provides a mixture of research on ecological and social problems, thus creating a bridge between these two branches of primate cognition research. As in *Primate Societies*, there are chapters devoted to social learning and communication. In chapter 31, Andrew Whiten reviews the evidence for social learning mechanisms and discusses the evidence for culture or behavioral traditions among nonhuman primates. Whiten's chapter presents a valuable mixture of observational data accumulated over the years with experimental approaches to the question of the social transmission of knowledge among nonhuman primates. In chapter 29, Klaus Zuberbühler covers olfactory, gestural, and vocal communication, placing a particular emphasis on issues of acquisition, flexibility, and comprehension of signals. Zuberbühler also discusses some implications for the evolution of language. Two chapters are devoted to the social knowledge of primates from two different perspectives. Chapter 28, by Seyfarth and Cheney, explores what primates know about their social groups in terms of the social relations between individuals. Chapter 39, by Josep Call and Laurie R. Santos, explores social knowledge from a more individualistic perspective and investigates what primates know about the minds of other individuals. As in other parts of this book, the final chapter in this part focuses on the similarities and differences between humans and other primates. Here, Esther Herrmann and Michael Tomasello compare and contrast the cognition of human and nonhuman primates and discuss the implications for the evolution of culture.

Chapter 27 Solving Ecological Problems

Charles Menzel

All men by nature desire to know. An indication of this is the delight we take in our senses . . . . we prefer seeing (one might say) to everything else. The reason is that this, most of all the senses, makes us know and brings to light many differences between things. By nature animals are born with the faculty of sensation, and from sensation memory is produced in some of them, though not in others. . . . from memory experience is produced in men; for the several memories of the same thing produce finally the capacity for a single experience. . . . Now art [theoretical knowledge] arises when from many notions gained by experience one universal judgment about a class of objects is produced. . . . Wisdom is knowledge about certain principles and causes.
—Aristotle, *Metaphysics*

HISTORICALLY AND PHILOSOPHICALLY, most of the problems to be discussed in this chapter hark back directly to Aristotle, who is widely credited with the first systematic psychology (Robinson 1989). His treatise *On the Movement of Animals* anticipates the classical ethology of Lorenz (1971) and Tinbergen (1976), as well as some current trends in ecology (Nathan et al. 2008), and his search for a "final cause" for motion remains an active issue in the life sciences (Mayr 1988). Aristotle begins his *Metaphysics* with a commonsense, naturalistic statement of how knowledge advances—in varying ways in different kinds of organisms—from sensory knowledge to memory and connected experience, and to understanding of the principles governing a general class of objects or events. He attributes memory and intelligence to some nonhuman animals but reserves theoretical, systematic forms of knowledge to humans.

To paraphrase Aristotle, in this chapter I will discuss both "physics" (animals' movements) and "metaphysics" (their knowledge). Because the latter subject has been relatively neglected in primate field research and also in some laboratories, at least until recently, I will begin with it. All primates, if not all animals, are motivated to learn about their environments. An indication of this is the way in which they use their perceptual systems, particularly vision, to explore their surroundings. Lemurs, monkeys, and apes turn their eyes, heads, and bodies to examine novel objects and perceive them clearly, and they learn the visual layouts of new pathways, surfaces, objects, and resources as they move about. Primates discriminate a multitude of classes of organisms and objects by sight. They detect slight variations in the quantity (Beran et al. 2008), shape (Smith et al. 2009), and location of environmental features, and they retain such information in memory, although the extent to which they can recall experiences and distant environmental features surely varies across species. Aside from memory, some nonhuman primates also show signs of understanding the underlying causes of events and—in some species and in some situations—signs of forming an overall purpose, goal, or plan that guides their actions (Putney 1985, 2007; Mulcahy & Call 2007; Osvath & Osvath 2008). On this account, then, primates have an immediate and potent drive to see clearly what is present in the environment (e.g., Woodworth 1958, p. 192; Butler 1960; Menzel & Davenport 1961; Menzel 1980) and to explore and register the layout of new areas and objects (Tolman 1932; Menzel &

Menzel 1979; Gibson 1979). Their forms of "knowing" vary in elaboration across species and individuals, from sensory knowledge (e.g., Dominy et al. 2001; Laska et al. 2007) to memory and connected, cumulative experience (e.g., Harlow 1949; Menzel 1964), and to more sophisticated understanding of causes and effects.

What is the empirical basis for this sketch? How do learning and memory capabilities support the everyday activity patterns of primates and help them solve the unique ecological problems that they face, particularly in the context of foraging (chapter 7, this volume)?

The knowledge base underlying primate ranging and foraging remains almost an undiscovered country compared to what is known about birds and insects (e.g., Holldobler & Wilson 2009). In his pioneering full-year study of mountain gorillas (*Gorilla gorilla beringei*), Schaller (1964, p. 191) stated: "Often the group wanders so erratically that I suspect a lack of plan and purpose in . . . [the leader's] actions." That view of nonhuman primates was probably the consensus opinion among field workers, or at least Western field workers, in the 1960s. The focus of attention was not on "metaphysics" (primates' knowledge) but on "physics" (their movement patterns, especially at the intra-individual level of analysis). An example of this focus was the description of species-specific action patterns and the development of ethograms. There were some maps generated of primate travel paths, but what motivated the travels was unknown, because no one had conducted experiments to evaluate the process. Few researchers had recorded in detail the features of the environment that might be furnishing the stimuli for travel. Kummer (1968) and Altmann (1974) were among the first field workers to discuss the potential ecological and perceptual determinants of foraging pathways. Emil Menzel's studies of chimpanzees' (*Pan troglodytes*) movement in an open field (Menzel 1969a, 1971a, 1971b, 1973a, 1973b, 1974) provided a detailed account of the operation of nonhuman primate learning and memory capabilities in a naturalistic setting. Nevertheless, 10 years after the 1987 publication of *Primate Societies*, only a modest number of studies had analyzed primate learning and memory capabilities experimentally in relatively large-scale space. There still were few experimental studies of any sort on wild primates, and even fewer on the operation of learning and memory capabilities in the animals' natural habitat in the context of foraging (reviews: Cheney & Seyfarth 1990; C. Menzel 1997; Tomasello & Call 1997). In the last 10 years, however, researchers have produced a considerable body of experimental and observational work on primate navigation, spatial learning, and memory (reviews: Byrne 2000; Cunningham & Janson 2007a; Janson & Byrne 2007).

The aim of "movement ecology" studies (Boyer et al. 2006; Nathan et al. 2008; Beisner & Isbell 2009) is to characterize organismal movements and to determine the immediate causes and ecological consequences of those patterns. These studies have been facilitated by a number of technological advances—such as miniaturized radio transmitters and satellite networks—that permit researchers to collect high-resolution spatiotemporal movement data (e.g., Pochron 2001; Noser & Byrne 2007b; Markham & Altmann 2008). Movement ecology studies also are based on advances in geographic databases, in computational power, and in analytical methods for characterizing the movements of organisms of any size and in any spatial setting. These technological advances improve our ability to characterize what animals do. In principle, they will also improve our ability to predict where a lemur, monkey, or ape will go next, and to answer questions about how well the animals learn and remember the structure and changes of their environment. Accordingly, the technological improvements will improve our understanding of how cognitive capabilities contribute to the solution of ecologically relevant tasks.

The movements of primates are intrinsically interesting, whether in a forest or cyberspace, and the analysis of paths of movement can shed light on psychological processes. For example, studies of barrier-and-detour tasks have a long history in primate psychology (James 1890; Köhler 1925; Tolman 1932; Lorenz 1971; Menzel & Menzel 2007). The path of travel that a macaque, chimpanzee, adult human, or computer algorithm generates as a solution to a novel detour problem can be characterized graphically and quantitatively using purely objective physical criteria. Physical attributes of paths include, for example, a quantitative measure of how much each discrete movement of a traveler reduces the traveler's remaining distance to the goal, a reasonable measure of efficiency. Other measures include the percentage of movements that involve a change of direction. Needless to say, the paths produced by different animals in the same task differ markedly. A comparison of paths produced in simulated, computer-presented barrier-and-detour tasks raises questions about how primates resemble and differ from one another in psychological dimensions such as (a) their perceptual criteria for what constitutes a "best" or a "least-distance" path; (b) their ability to plan and take into account features of the situation before responding; (c) their ability to navigate by short-term memory, without an immediate view of the barrier; and (d) their ability to profit from general practice with the joystick system (Menzel & Menzel 2007). In sum, this analytical approach for the study of detour behavior uses geometrical and mechanical questions and measurements to classify movement patterns. The findings, in turn, aid the study of psychological processes that inherently have less well defined boundaries. Experimental studies of detour and homing behavior in golden lion tamarin (*Leontopithecus rosalia*) social groups in three-dimensional arboreal environments (Menzel & Beck 2000) and studies of chimpanzee group movement in an outdoor field (E. Menzel 1974, 1987) address closely related conceptual problems, and these experimental approaches complement observational approaches to movement ecology.

The speed of an animal's movement preceding encounters with resources is a topic of increasing interest to field primatologists (e.g., Pochron 2001; Janmaat et al. 2006a; Noser & Byrne 2007b; Normand & Boesch 2009). Speed of movement can be conceived as an aspect of what is learned (Logan 1972). Mammals learn not only what to do but also how fast to do it. A rat typically runs faster when it is given repeated experiences in a maze. On the one hand, this may be an example of motor efficiency increasing with practice. On the other hand, under typical conditions a faster running speed results in earlier access to food. Thus, under typical test conditions there is a positive correlation between a dimension of the reward and a quantitative dimension of the response pattern. It is possible to arrange other types of correlations experimentally, and an interesting finding is that rats learn to run slowly if doing so provides a more immediate or larger reward. The rat runs fast to the extent that it pays to do so (Logan 1972, p. 1055). Variations in the speed of an animal's movement and in the consistency of its direction ("purposive behavior," Tolman 1932; Boden 1972) are of particular importance to social foragers because such movements can provide a sign that allows others to discriminate what types of resources are available. For example, the speed and accelerations of a primate's locomotion in a naturalistic setting can provide group members with a basis for judging the relative incentive value of hidden goals; and a primate that knows the sign value of its own behavior for its competitors may withhold response to a preferred object as long as a dominant animal is watching (Menzel & Halperin 1975).

From this point, I will review briefly several field descriptions and tests of monkey, lemur, and ape travel movements that suggest memory. Then I will review some experimental studies on macaque foraging and chimpanzee memory in more detail. These are mostly my own studies. Finally, to end this chapter, I will offer some general conclusions and mention possible future directions for research.

Naturalistic Background

Thirty years ago, few reputable students of primate learning took field research seriously. Thorndike's (1898) negative opinions regarding studies conducted outside the laboratory were still accepted by some as definitive, and even one

of the most prominent students of primate learning went so far as to claim, "Our own data show without question that the monkeys we raise in the laboratory are . . . brighter . . . than any monkeys ever raised in a wild or feral state" (Harlow & Mears 1979, p. 5). Today we can be more open-minded regarding the relevance of field research to primate learning, especially insofar as we can document our claims more rigorously and quantitatively than was the custom in the past.

The "field" is relevant to primate learning in part because it represents an approximation of the circumstances under which cognitive capabilities evolved. A working hypothesis is that primates, like many other animals, are capable of learning the nature and relative positions of a large number of objects in their habitat, and that they use this information, together with current cues, to find food, to discriminate food from nonfood objects, to treat items as food or nonfood based on the abundance of profitable resources (fig. 27.1; Sayers et al. 2010), and to travel efficiently and safely. Gains in the efficiency of foraging can improve an animal's lifetime reproductive success (Stephens & Krebs 1986; Ritchie 1990; Altmann 1998).

Over the past 20 years, detailed observational field studies have indicated that primates can move toward resource locations from large distances, and long before they can see them (Cunningham & Janson 2007b). Tamarins move along fairly straight lines from one ripe-fruit-bearing tree to another tree of the same species, even though the forests are dense and visibility is quite limited (Garber 1989). Many of the food trees the tamarins use are members of species that all produce ripe fruit at the same time of year, and more than 30% of the time tamarins move from one of these trees to another of the same species. The second tree is typically the closest available unvisited tree of that species, rather than a more distant one (Garber 1989). If by chance all of the trees of that species were clumped together in space, then it would be difficult to rule out the possibility that the monkeys were moving in this way largely by accident. However, some of the nearest-neighbor trees they have been observed to move to have been more than 100 meters away, and some of those trees were approached repeatedly from varying angles. Wild black capuchin monkeys (*Cebus nigritus*) revisited experimental food platforms using travel paths that were shorter than predicted by a random model of search (fig. 27.2; Janson 1998). The random search model assumed a visual detection field of 90 meters, based on the empirically measured detection field of 82 meters. These findings strongly suggest that the animals used spatial memory to relocate the platforms. The capuchins' movements were influenced both by distance and by expected food quantity, but the animals seldom bypassed a smaller nearby quantity in favor of a larger distant quantity unless the difference in

Fig. 27.1. An adult female Himalayan gray langur (*Semnopithecus entellus*) at Langtang National Park, Nepal, prepares to consume a food item she has just extracted from the soil. When profitable resources increase in abundance, she will ignore low-quality, nonseasonal foods—including certain types of roots, bark, and evergreen mature leaves—in favor of better alternatives. Photo courtesy of Ken Sayers.

quantity was substantial (Janson 1998; see also Gomes & Bicca-Marques 2012).

The long-distance travel of baboons and mangabeys also can be oriented toward resources that lie out of view. In the dry season in Limpopo province, South Africa, chacma baboons (*Papio ursinus*) leave their sleeping site and travel rapidly and quite directly over long distances to sparse, depletable fruit sources. Along the way, the baboons bypass visible trees with abundant seeds, and they return to these trees later in the day and feed on the seeds just before climbing into their sleeping site for the night. Noser and Byrne (2007b, p. 264) note that "bypassing a potential type of food in favour of a more distant alternative is not consistent with the notion of merely responding to salient stimuli." Yellow baboons (*Papio cynocephalus*) at Ruaha National Park, Tanzania, usually encounter baobab fruit and *Combretum obovatum* after a long bout of direct, rapid travel. These major components of their diet provide a large number of grams per minute of handling time (high profitability) and are predictable in space (Pochron 2001). The baboons' fast, direct (and thus economical) travel leading up to encounters with profitable, spatially predictable foods differs from their slow, meandering travel prior to encounters with grass seeds and other foods that are less profitable and less predictable in space and time (Pochron 2001). In the dry Erer region of Ethiopia, hamadryas baboons (*Papio hamadryas*) move faster before reaching a food patch, provided that it is a location in which they will stay longer than 15 minutes. Their acceleration begins before the location is visible (Sigg & Stolba 1981; Kummer 1995; the assumption here is that a food patch is larger if the animals stay longer than 15 minutes). Sooty mangabeys (*Cercocebus atys*) and gray-cheeked mangabeys (*Lophocebus albigena*) in African forests approach distant trees that contain fruit

Fig. 27.2. The use of feeding platforms has allowed researchers to clarify foraging decisions by varying systematically the location, productivity, and renewal schedule of foods. This experimental approach has revealed the importance of proximity and also of spatial memory in guiding these decisions in black capuchin monkeys of Iguazu National Park, Argentina. Photos courtesy of Brandon Wheeler.

more frequently, and more quickly, than they approach trees that lack fruits (Janmaat et al. 2006a). The travel movements of the mangabeys suggest that they can orient themselves toward trees that are not visible, from distances of more than 100 meters and, based on their experience in the area, can discriminate between trees that have contained fruit in preceding days and trees that have not contained fruit during that period. The findings on travel speed in baboons and mangabeys are consistent with the fact that these monkeys are social foragers and that the finder's share is likely larger than those of stragglers or scroungers (di Bitetti & Janson 2001).

Travel to distant feeding locations is not limited to frugivorous primates. Guerezas (*Colobus guereza*) are among the least active primates, but social groups at the Kakamega and Kibale forests periodically travel almost twice their normal daily path length to rare food types located outside their normal home range. The resources visited include swamp plants and the bark of *Eucalyptus* trees, both of which have extremely high concentrations of sodium compared to available leaves, and which therefore probably constitute an important supplementary source of this mineral (Fashing et al. 2007).

Much less information is available on the operation of spatial memory in wild strepsirrhines than in haplorrhine primates, but Milne-Edwards sifakas (*Propithecus edwardsi*) and red-fronted lemurs (*Eulemur rufifrons*) appear to use information about the availability and distribution of fruit to improve the efficiency of their foraging routes. Erhart and Overdorff (2008) examined the movements of two social groups of Milne-Edwards sifakas and one group of red-fronted lemurs during a time of year when the animals fed primarily on one or two species of ripe fruit. When a group initiated a travel progression, its next feeding location was usually the closest available patch of a given fruit species, rather than a more distant patch of the same species. Distances between food patches frequently exceeded 50 meters, which was more than the spatial limits of visibility in the forest at the height at which the animals traveled.

During their nightly rounds, gray mouse lemurs (*Microcebus murinus*) revisit stationary food sources, such as gum trees, more frequently than they visit trees where they have previously encountered mobile insect prey (Joly et al. 2007). When wild gray mouse lemurs detect an experimentally introduced nectar source in their home range, they revisit the location on subsequent nights (Joly et al. 2008; Lührs et al 2009). They sometimes even shift their sleeping site closer to the new resource (Lührs et al 2009). Lührs et al. (2009) moved artificial nectar sources from familiar "old" locations to sites that had not contained food during the previous three weeks. The mouse lemurs initially made the mistake of returning to the old locations. This suggested that they put more weight on the locations of established, reliable nectar finds than on stimuli associated with the nectar itself. Within several nights the mouse lemurs detected and began to visit the new locations. Five of these animals were trapped, carried to the edge of their respective home ranges, and released. Three of them moved to an experimental site where they had previously encountered nectar, using an efficient path. The remaining two animals, both males, had other priorities: they wandered widely outside their normal home range in search of receptive females (Lührs et al. 2009).

There are many different types of nonhuman primate diets (chapter 7, this volume), and not all of them permit purposive encounters with food items. A tarsier or galago that subsists largely on insects and other mobile prey may rarely have information on exact prey locations prior to its search. Most of its encounters with mobile food therefore will have a random component. Behaviors employed in finding prey include moving, stopping, and visual and auditory scanning. Even in the case of immobile resources, primates are likely to encounter some resources purposefully and others randomly. Some areas of the home range may be visited regularly, whereas others may be crossed only a few times each year. The strength of a memory trace about food patches surely depends on many factors. These factors might include the amount of time that has elapsed since the animal last visited the patch, the quality of the food found there in the past, the quantity of food remaining in the patch at the time of the animal's last departure, the food's ease of procurement, and the abundance or rarity of that patch type. For many patches of low to moderate quality, the principal retrieval cue for memory may simply be the sight of the patch itself. That is, the animal might "recognize" the patch only when it happens upon it. In contrast, for high-quality patches the animal might "recall" the resource from a distance, prior to visual contact with the area. Experimental evidence for recall capacities in at least some primates will be described below.

A long-standing aim of behavioral science is to identify the particular environmental cues that animals follow when they search for food. In some studies it has been possible to identify specific environmental features that influence where wild primates travel and search. If a baboon discovers a small amount of ripe fruit in its home range, it sometimes moves to other locations that have contained the same type of fruit in past seasons (e.g., *Lycium* "europaeum," Altmann 1998). Gray-cheeked mangabeys revisit fruit trees more frequently after warm and sunny days than after cold and overcast days; higher temperatures and higher levels of

Fig. 27.3. After finding experimentally introduced ripe akebi fruit (*left*), an adult female Japanese macaque stares upward (*right*). Photos courtesy of Jane Gagne and Charles Menzel.

solar radiation speed the maturation of the fruits and of the insect larvae inside (Janmaat et al. 2006b).

Field workers often assume that travel is goal-directed if the animal's path follows a relatively straight line, and if the rate of travel accelerates at certain points along that path. The grounds for that assumption are not always spelled out, but a common—if ancient—premise is that an animal becomes more motivated as it comes closer and closer to its path's natural endpoint. Speed and motivation also are expected to depend on the perceived social and environmental conditions; a primate might accelerate if it has the lead on its competitors and if the expected amount of food at the endpoint is limited. It might decelerate if a snake, a predator, or a higher ranking competitor is expected near the endpoint (Menzel 1971a; Normand & Boesch 2009). Variations in the rate of movement toward goals can be modified on the basis of the outcomes they generate, and are not simply innately determined strengths of a response (Logan 1972). A related question is: How does one objectively know the "start point" and "endpoint" of a path? Lab workers assume that the start point is defined by the procedure and that the food or other incentive they provide is the animal's endpoint or goal, but obviously that is not always the case. The animal might be satiated and uninterested in the food; it might reach the endpoint because it has followed a familiar path or social partner. Such challenges to definition are much greater in the field than in the lab. A typical solution has been to identify a well-organized event that contrasts with random motion. For example, it is usually plausible to argue that a primate sees food when it gives a startle or a sudden change in behavior and moves to the food in a straight line. A similar argument can be made for memory and knowledge more generally, even if the animal engages in some amount of random travel prior to the goal.

If one can define the start point and goal prior to the animal's movement, it provides a more powerful test than if the start point and goal are unknown to the researcher. Encounters between Japanese macaques (*Macaca fuscata*) and ripe fruits have been arranged experimentally in the field, thus permitting a study of where macaques go after finding an initial food item. In one set of experiments I placed a piece of food on the ground along one of the monkeys' normal travel paths (C. Menzel 1991). Animals that found ripe akebi fruit, which is one of their normal food items, stared upwards and then entered trees containing akebi vines (fig. 27.3). They showed this behavior even at times of year when the naturally occurring akebi fruits were not ripe. In contrast, animals that found chocolate, a favored food with no established connection to the test environment, later returned to the location of the find and restricted much of their visual and manual searching to a small area of ground. Animals that found nothing neither entered the akebi vines nor searched near their own starting locations, and were not simply moving to akebi vines already. The Japanese macaques' behavior suggested long-term memory of specific types of food (C. Menzel 1991). Monkeys seem to know that akebi fruit grows on vines in trees but chocolate does not. Comparable performances have been reported in captive Sumatran orangutans (*Pongo abelii*). Notably, orangutans that encountered pieces of banana in an isolated location in a 1,900-square-meter outdoor enclosure soon moved to and inspected other places that had contained bananas on previous days, rather than places that had previously contained grapes (Scheumann & Call 2006).

Contrasting Viewpoints Regarding Space

A detailed look at how individual primates travel to and find food in their particular ecological contexts raises questions about spatial perception and memory. The amount of space used, and the structures used as paths and reference points during travel and feeding, can vary sharply across species or among individuals of the same species (Hediger 1968; E. Menzel 1969b; C. Menzel 1986). The environmental features that serve as choice points, goals, and landmarks are not always within range of sight or hearing at the time of response (Janson 1998; Dominy et al. 2001). Animals often appear to fill in gaps in the information provided by direct visual perception.

To illustrate the role of learning and memory capabilities in foraging, consider where a long-tailed macaque (*Macaca fascicularis*) goes after it encounters initial food items. Like Japanese macaques, long-tailed macaques have diverse diets (Aldrich-Blake 1980; Wheatley 1980; Lucas & Corlett 1991). Many of the types of food they use are rapidly depleted and are hard to detect from a distance (van Schaik et al. 1983). To forage efficiently, macaques should be capable of forming new search strategies immediately for new occurrences of food. That is, they should be innovative (Kawamura 1959; Itani 1965). If a macaque finds one or a few initial food items, how does it find more items of the same food type? Where does it go and what locations does it inspect?

It is possible that macaques (and other animals) organize their searches partly by the spatial proximity of food items to environmental structures. I call this the "structure-guided" search method. The class of structure the animal inspects might vary depending on where the animal has discovered an initial food item. For example, if a macaque finds a piece of highly preferred food within a streambed, it might restrict its subsequent search to locations within the same streambed. Alternatively, if it finds a new type of highly preferred food next to a fern, then it might inspect other ferns in preference to rocks, trees, or logs. In the natural habitat of macaques, foods can be associated spatially with visible borders such as stream edges or forest edges, or with discrete structures such as trees of a particular species (Wheatley 1980).

Alternatively, macaques and other primates might organize their searches by distance from the food location. According to what I have labeled the "spatial gradients" search method, available space is homogeneous and unstructured. It is an undifferentiated area around a reinforcer. The animal allocates its search effort to concentric rings around a food item according to some decreasing function of the Euclidean distance from the item. After finding food, the animal might slow down, turn more quickly, and inspect any location that lies within a reasonable distance of an initial find, regardless of the position of visible environmental structures. For example, after a crow encounters an eggshell on a beach, it will search in a meandering walk up to

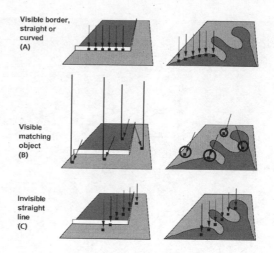

Fig. 27.4. Sample layouts of hidden food. (A) Visible environmental border: food (black squares) lies along a low cement wall between a grass field and a sand field (*left*) or along the border between a stone field and a grass field (*right*). (B) Visible matching objects: food lies next to bases of vertical poles (*left*) or next to wooden crates (*right*). (C) Invisible line: the line cuts across grass and sand fields (*left*) or across stone and grass fields (*right*). Drawings based on photographs in C. R. Menzel 1996b.

about two meters from the site of its original find (Tinbergen et al. 1967). Other birds, fish, and insects also show increased turning and area-restricted search after finding a resource (Bell 1991). Formal analyses of two-dimensional spatial gradients in learning and navigation can be found in Hull's (1952) classic study and also in Reid and Staddon (1998).

To test whether macaques use the structure-guided method, I studied the searching behavior of captive-born long-tailed macaques in an 880-square-meter outdoor enclosure at the Bokengut field station, University of Zürich (Menzel 1996a, b). The enclosure contained tree trunks, grass fields, rock fields, elevated walkways, and visual barriers. Food was hidden in the macaques' enclosure according to one of several different types of rules (fig. 27.4 shows sample layouts). If macaques used the structure-guided method, they were expected to extrapolate their search within dimensions of environmental structure that in their natural habitat would normally be correlated with the distribution of food, such as along visible borders, within visible surface areas, within particular height levels, and within discrete visible objects of the same type.

I found that if a macaque detected a single visible pile of food next to a continuous visible border, such as a border between a grass field and a sand field or between a sand field and a stone field, the animal restricted much of its subsequent manual inspections to other locations along the same border. Significantly, the animals searched along visible borders and found hidden food from the very first trial in the experiment, thus indicating that they were not relying

Fig. 27.5. A sample pairing of food (banana) and a discrete environmental feature (yellow stone) in the Bokengut outdoor enclosure. Photo courtesy of Charles Menzel.

at regular intervals of one meter or three meters along an arbitrarily oriented, invisible straight line (layouts in fig. 27.4; see also Hemmi and Menzel 1995). Further tests showed that the macaques extended their search within visible surface areas, within specific height levels, and in directions associated with a change in food visibility. These results are in strong agreement with the structure-guided hypothesis.

In other tests (C. Menzel 1996a), the long-tailed macaques made far more visits to locations where they had obtained food on the previous day than to locations they had merely inspected on the previous day without finding food. Social factors also were important in their search behavior. The macaques inspected locations where another animal had searched and obtained food more often than they inspected locations where another animal had searched but not found food. They appeared to detect the outcome of another animal's search visually from a distance of several meters.

Taken together, the experimental findings on wild Japanese macaques and captive-born long-tailed macaques indicate that macaques are influenced by many different aspects of a food's distribution in space. They show a degree of innovation in their searching, and they show excellent memory for locations that have contained food in the past (C. Menzel 1997). The immediate use of matching visual cues in trial-unique problems by experimentally naive macaques (C. Menzel 1996a) is far more efficient than would be expected from many prior laboratory studies, but consistent with the foraging patterns of wild long-tailed macaques (van Schaik et al. 1983). The findings on structure-guided foraging suggest that for macaques, and likely for most animals, "space" is not perceived or remembered as an empty container. Instead, it is defined largely by the relative positions of objects (Jammer 1960). These findings can be generalized to other primates. Neotropical primates live only in a narrow band of canopy heights, and this influences the space in which foraging occurs and patches or food types that are encountered or remembered. Many arboreal primates search along tree branches for mushrooms or insects; terrestrial baboons and Himalayan langurs might lift stones along a curving streambed while searching for invertebrates, or move from one small bush to another while foraging for berries. For macaques, pure spatial gradients are disrupted by the presence of visible objects. Specifically, the close spatial proximity of an initial food item to a visible environmental feature was an important determinant of where the macaques searched. The dimensions of environmental structure that macaques and other primates follow include discrete visible objects, visible borderlines, surface areas (Beisner & Isbell 2009), height levels, and directions associated with change in food visibility.

Thus, the experimental findings raise the question of

on slow, gradual learning within the experiment to solve novel problems of this general type. If an animal found food next to a single discrete visible object (e.g., a log, a stone, or a vertical pole) it might walk several meters to another object of the same general type and inspect it manually. Again, the monkeys extended their search from a single baited object to other, similar-looking objects from the very first trial in the experiment. Moreover, if the macaques found a favored food, banana, next to a single object of type A (e.g., a yellow stone; fig. 27.5), and found a less favored food, carrot, next to a single object of type B (e.g., a low post), they inspected other locations of type A more often than they inspected other locations of type B. Macaques also found hidden food items much more quickly when the items were hidden according to a "natural" rule, along a visible border or next to visible matching objects, than when they were hidden according to a relatively "unnatural" rule—that is,

what more there is that primates take into account or anticipate during their foraging. For example, how well can a lemur, macaque, tamarin, or ape recall events, environmental features, and social partners located completely out of sight or hearing of the general search area? In most animal memory studies, an animal's retrieval is tested within the same spatial situation in which learning originally occurred. In this respect, most animal studies address recognition memory rather than recall memory (Shettleworth 1998). With chimpanzees who have been reared with exposure to an artificial language, one can study the types of information that chimpanzees can both recall *and* report about areas outside the immediate sensory environment (C. Menzel 1999, 2005; Schwartz & Evans 2001; Schwartz 2005). The ability to recall objects and locations outside the area in which they were originally encountered has traditionally been viewed as a central factor in the emergence of human thinking, planning, and communication (Lorenz 1971), and recall capabilities would appear to be necessary for animals to evaluate food patches and waterholes that are outside the range of perception. Until recently, however, there have been few systematic data on recall memory capabilities in nonverbal animals.

Chimpanzees

The operation of memory in outdoor environments has been studied in more detail in chimpanzees than in other nonhuman primate species. I will describe some of these studies, beginning with a description of the naturalistic and broader experimental context, and leading up to studies of recall. Chimpanzees show a combination of traits that is rare among mammals: they are large-bodied animals and specialize on a rare and ephemeral type of food (Ghiglieri 1984; see also Gaulin 1979; chapter 6, this volume). Chimpanzees at Kibale National Park, Uganda, devote approximately three times as much of their feeding time to ripe fruit as sympatric monkeys do, and many of the fruit species they eat are rare (Wrangham et al.1998). The chimpanzees show a seasonal improvement in the quality of their diet when ripe fruit becomes available, increasing their intake of free simple sugars and reducing their intake of fiber. Three species of monkeys studied in the same forest do not show this pattern. Chimpanzees maintain a much higher-quality diet than would be expected for a mammal of their body size (Conklin-Brittain et al. 1998).

Chimpanzees often move rapidly in straight lines over long distances to ripe fruit trees, and field observers frequently speculate that the chimpanzees know where they are going and the types of food they are likely to find there (Wrangham 1975, 1977; Ghiglieri 1984; Goodall 1986). It would be interesting, as well as valuable in analyzing goal-directed travel, if field workers would report the level of accuracy of their on-the-spot predictions about where their animals are headed next, and discuss the specific cues they are using to make these judgments. Although experimental data are lacking from the field, the spatial memory capabilities of chimpanzees and their ability to organize efficient itineraries among food sources can be expected to improve their foraging efficiency substantially (Wrangham 1977). Recent detailed observational studies (Normand et al. 2009) on forest chimpanzee travel movements, combined with global positioning system data on more than 12,000 food trees of 17 species, show that chimpanzees visit rare species of trees more frequently and by shorter approach paths than would be predicted by chance. Furthermore, the chimpanzees' initial movement to a distant food source (294 meters away on average) corresponds closely to the direction needed to reach that location, thus suggesting that the animals may have information from their starting point on the direction of a substantial number of food trees without necessarily having a view of most of the environmental features or "landmarks" that will be encountered later along the way (Normand & Boesch 2009). The latter data are also consistent with searches made in straight-line paths for good heuristic reasons (Janson & Byrne 2007). Chimpanzee stone tool transport has also been discussed in light of what appears to be path efficiency (Boesch & Boesch 1984; Boesch & Boesch-Achermann 2000; see also Visalberghi et al. 2009). Ghiglieri (1984) suggested that spatiotemporal memory capabilities enable chimpanzees to obtain sufficient energy from rare, ephemeral, scattered fruit sources under conditions of feeding competition in the Kibale forest. For each chimpanzee there are some 200 monkeys, as well as birds, squirrels, and other chimpanzees, foraging on the same fruit. Based on paleoenvironmental data, Potts (2004) argued that specialization on rare, ephemeral ripe fruit in tropical forests was a central organizing factor in the emergence and subsequent evolutionary development of chimpanzee cognitive capabilities, including their ability to assign attributes to distant locations.

Emil Menzel's experimental studies of captive chimpanzees indicate that chimpanzees not only can remember the locations of a large number of food items (Tinklepaugh 1932), but also that they retain information about the relative positions of those food items (E. Menzel reviews: 1978, 1984). In one study (E. Menzel 1973a), one chimpanzee from a group of six juveniles was carried around the one-acre experimental field and shown a piece of fruit in each of 18 randomly selected sectors of the field. During the cue giving, the chimpanzee was not allowed to do anything other than cling to his carrier and watch as the food was hidden. This eliminated locomotor practice and primary reinforcement from the cue-giving phase of a trial. The field included trees, areas of relatively tall grass, small hills, and other visual barriers. Not all food locations were visible to the chimpanzee from a single vantage point. After the experimenters showed the chimpanzee the food, they returned him to the group, and then released the entire group into the field. The question was how the chimpanzee subject organized his route while collecting the 18 hidden food items. Four different chimpanzees were used as test animals in this way.

The chimpanzee who had been shown the food obtained most of the hidden items. In 16 trials, the test animal found 200 of 288 hidden foods, and the five control animals who had not been shown the food found a total of just 17. The test animals used very efficient routes during their collection of food, and these routes bore no relation to the ones on which they had been carried during cue giving. The test animals clearly took into account more than one or two food locations at a time, because from the outset of the trial they usually headed toward larger clusters of food rather than smaller ones. In another experiment, they typically went toward whichever half of the field contained three food locations rather than two. In other tests, the experimenters hid a piece of preferred food (fruit) in nine locations and a piece of nonpreferred food (vegetable) in nine other locations. The test chimpanzees went first to the fruit locations and rarely reinspected locations they had already emptied.

Other experiments in the series (E. Menzel 1971a, 1973b) showed that group members could pick up information from another, informed, animal about an object's quality (fruit versus vegetable), its relative desirability (e.g., food versus snake), the quantity of food present, and its location. The speed at which a chimpanzee approached a distant, hidden goal provided sufficient information for discriminating whether the item was food or a novel object and who knew its location best (Menzel & Halperin 1975). The group of chimpanzees searched very differently after an operationally designated "leader" had been shown a real snake than they did after the leader had been shown a piece of food that was later removed. When the leader had been shown a snake, group members showed piloerection as they emerged from the release cage, and they moved together slowly toward the location. Not only would they use sticks to probe the location where the snake had last been seen, but they also would broaden their search area, as if taking into account the possibility that the snake had moved. When the leader had been shown food, however, the group typically raced for the location where the food had been seen, and restricted its search to that vicinity.

In summary, these studies showed not only that chimpanzees could remember many food locations in a given trial, but also that they remembered the spatial relationships among those locations (E. Menzel 1973a; Gallistel 1990). The studies also demonstrated transmission of information about the nature and location of hidden objects from one chimpanzee to another at an unexpectedly high level of efficiency (E. Menzel 1973b). Recent studies of geographic orientation and foraging by Cunningham and Janson (2007b), Garber (1989), Janmaat and colleagues (2006a), Normand and Boesch (2009), Noser and Byrne (2007b), and others expand these theoretical and methodological concerns to wild primates. This extension will be important in discovering how well principles of performance derived from relatively controlled situations hold up in uncontrolled real-life situations—and in determining what has been left out of the picture. The ape studies described next can be viewed as lab simulations of orientation and foraging problems in the same spirit.

Studies of Recall Memory in *Pan*

The studies reviewed above assess an animal's knowledge of resource locations primarily from its path of travel and from the locations it inspects. Such analyses potentially underestimate knowledge (Marler & Terrace 1984). In principle, an animal might recall the presence, type, relative quantity, and location of a resource, but not move to it for several days. A current limitation in studies of how well nonverbal animals know their habitat is that there are fewer data on recall memory than on recognition. The distinction between recall and recognition concerns the extent to which information is detached from immediate sensory stimulation. In tests of recall, fewer retrieval cues are typically available than in tests of recognition. As Shettleworth (1998, p. 238) notes:

With humans, either *recognition* or *recall* of past input can be tested. That is, a person can be presented with a stimulus and asked "Have you experienced this before?" (recognition) or simply instructed to "Tell me what you remember" (recall). Performance on recognition tasks is typically better than on comparable recall tasks . . . . Most tests of animal memory are tests of recognition.

To what extent can a nonverbal animal recall and compare distant locations that lie well beyond its immediate range of sight, smell, and hearing (Byrne 2000)? Which attributes of past events are recalled? Can the animal perform recall in different situations and in different ways (Tulving 1983)? What are all the types of information a primate can retrieve and convey regarding past events and distant locations, from a single location in space?

An extension of classic delayed-response methods allows assessment of which event features an animal remembers and can convey to other beings (Herman & Forestell 1985; E. Menzel 1973b; Savage-Rumbaugh et al. 1986; Tulving 1984, p. 224). Symbol-competent chimpanzees, for example, provide a powerful opportunity for investigating the "time in which the chimpanzee lives" (Köhler 1925, p. 272) and the nature of the information they possess and convey about the structure of their environment and past experiences. The symbol system allows one to ask questions regarding memory that would be difficult to pose otherwise; the apes can convey information regarding specific events and environmental features, and can be tested outside the spatial and temporal contexts in which they have encountered them (Savage-Rumbaugh et al. 1983, 1986, 1993).

A Bonobo's Use of Road Signs in a 20-Hectare Forest

The ability to find a direct route between any two locations, such that either location can serve as the starting point and the other as the goal, is customarily regarded as one of the three crucial kinds of evidence for generalized mapping skills in animals (Muller et al. 1996; Collett 2002). There are few direct tests of this ability in outdoor environments in mammals.

Menzel, Savage-Rumbaugh, & Menzel (2002) tested a four-year old male bonobo (*Pan paniscus*), Kanzi, in a heavily wooded area of 20 hectares (55 acres). The forest included 3.5 miles of trails. Kanzi had traveled with human companions in the forest on a daily basis, and he had learned many lexigrams (arbitrarily designated visuographic forms that stand for foods, tools, places, and actions) and the locations of all 17 of the food stations in the forest. Each station in the forest had a different type of food reliably present.

Kanzi's lexigram competence provided a tool for studying his spatial memory organization. In the first experiment, we showed him a "road sign" just outside his indoor sleeping area. The sign indicated, by lexigrams, the location where food was hidden. Only 2 of the 15 locations were visible from the spot where the sign was located; some were as much as 170 meters from the sign. In 99 of 127 test trials, Kanzi went to the location designated by the sign on his first move. In a second experiment, we presented him with the road sign at varied points in the forest rather than at the original fixed place. In these trials the goal was a preferred toy. Kanzi's human companions were never informed about the location of the goal, and distances ranged up to 650 meters. In all 12 trials Kanzi led his companions to the designated location using an efficient path (fig. 27.6).

He appeared to be able to navigate, based on the information provided by a lexigram, from almost any arbitrarily designated starting location in his 20-hectare environment to any one of the numerous goal locations. The ability to cope with unique combinations of starting points and endpoints—to move to any one of many locations in a forest from any given starting point—could benefit animals that range widely and are not "central place" foragers. Stated otherwise, if an animal cannot use any single environmental feature, such as its home base or the North Star, as a fixed reference point, it might rely on more generalized mapping skills. To explore this issue further in the field or in large-scale captive environments, it would be useful to have a method by which the experimenter can define the starting point and endpoint in advance.

Unprompted Recall and Reporting of Hidden Objects by a Chimpanzee

Recall of resources and locations that are beyond immediate perception could be an important part of an animal's foraging strategy, and it is important to clarify specifically what organisms know about their broader surroundings. I will next describe in some detail my first study on memory in a lexigram-competent female chimpanzee named Panzee. The aim of this study (C. Menzel 1999) was to identify the types of information that Panzee could recall and transmit about hidden objects. Panzee was 11 years old at the outset of the experiment, and she had already learned more than 120 lexigrams (fig. 27.7; Beran et al. 1998; see Brakke & Savage-Rumbaugh 1995, 1996 and Rumbaugh & Washburn 2003 for details on Panzee's rearing history).

During this experiment Panzee was housed in an indoor-outdoor facility called the Lanson building (fig. 27.8). When Panzee was indoors she could not see the outdoor cage or the forest. The outdoor cage was bordered by forest, and an area of forest adjacent to the cage, measuring about 160 square meters, served as a test area for the introduction of objects. The indoor and outdoor cages each contained an array of 256 lexigrams (termed a "keyboard") that Panzee could touch. To obtain something from outside the outdoor cage, Panzee had to recruit the assistance of a person and in effect tell them where to go.

This initial study consisted of 34 trial-unique experimental tests, each consisting of a cue-giving phase, a delay phase, and a response phase. The test objects were 26 food types and 7 types of nonfood objects. During cue giving, the experimenter stood outside Panzee's outdoor enclosure, held up an object while Panzee was watching, then walked to a predetermined hiding place in the forest and placed the object on the ground. Each trial used a different location,

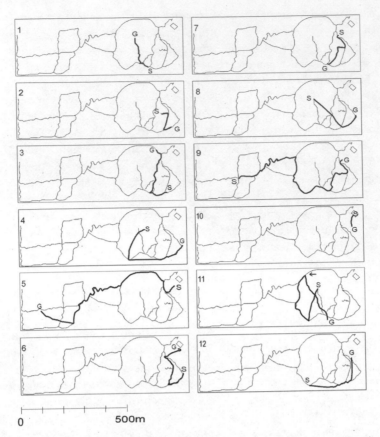

Fig. 27.6. Kanzi's travel routes in the 20-hectare forest. S marks the start point at which he was given the cue. G marks the goal location. Solid line indicates Kanzi's path of travel. Dotted lines indicate human-made trails. The rectangle in the upper right is the laboratory building. The arrow in trial 11 marks the location at which Kanzi was given the cue a second time. Redrawn based on a figure in C. R. Menzel et al. 2002.

and distances from the enclosure to the object ranged up to eight meters. On 32 of the 34 trials, the experimenter placed the object under leaf litter so that it was concealed from view. On the remaining two trials the object was not concealed and Panzee could see it from her outdoor cage. After placing the object, the experimenter left the area, and this began the delay phase of the trial. On 10 trials the experimenter imposed an overnight delay between the cue-giving phase and the response phase by showing Panzee the object outdoors after all caretakers had left for the day. Panzee then did not have a chance to interact with any person until the next morning. On the other 24 trials, Panzee could interact with a person indoors on the same day, typically within a few minutes after seeing the object. In any given opportunity, Panzee might or might not choose to recruit the person's assistance in recovering the hidden object. The delay phase lasted until Panzee initiated recruitment of a person (details below). Thus, she determined the exact time of recruitment and the total length of the delay.

In the response phase of a trial, Panzee interacted with a person who was naive to the locations or identity of hidden items inside the Lanson building. Panzee could use any means necessary (e.g., lexigrams, gesture, vocalization) to get the uninformed person outside to search for the test objects. Fig. 27.8 shows a map of a sample trial; Panzee recruited the assistance of her favorite caregiver and directed him to a prized type of hidden food item (fig. 27.9).

The uninformed persons found all 34 objects that were used in these tests as the result of Panzee's behavior. Panzee almost never (i.e., on only 3 of 268 days) recruited the uninformed person and caused him or her to search when there was no test object in the woods. When a test object was present outdoors, Panzee never pointed to a location that had contained an object in a previous trial.

From trial 1, Panzee was highly effective in attracting the uninformed person's attention, in conveying the type of item that was hidden, and in directing the person to its location. The lexigrams she touched on her keyboard when she recruited a person typically corresponded to the type of object she had seen. Considering just the 33 lexigrams corresponding to the 33 types of objects used in the experiment to be the relevant available choices, 84% of Panzee's

Fig. 27.7. The chimpanzee, Panzee, with her portable lexigram keyboard in her outdoor enclosure (*left*) and using it inside the Lanson building (*right*). Photo courtesy of Carolyn Richardson and the Georgia State University Language Research Center.

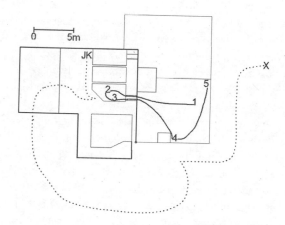

Fig. 27.8. Map of Panzee's recruitment of her favorite caregiver (JK) to find a prized food, M&M candies (location indicated by X), in the woods. 1. Panzee's location in her outdoor enclosure during cue giving. From here she watched an experimenter hide the candies. 2. Panzee's position at her keyboard inside the Lanson building. Here, she held her finger on the "M&M" lexigram and vocalized to attract the attention of her caregiver (JK). 3. Panzee's position while gesturing with an extended arm toward the outdoor enclosure. 4. Panzee's position at outdoor keyboard where she touched the "M&M" lexigram. 5. Panzee's position while pointing toward the hidden M&M candies. The dark line indicates Panzee's approximate path of travel; the dotted line indicates JK's path. Redrawn based on a figure in C. R. Menzel 2005.

Fig. 27.9. After touching the lexigram on her keyboard (visible in background) that corresponds to the type of object that is hidden, Panzee points toward the object's location. Photo courtesy of Charles Menzel.

76 lexigram touches corresponded to the specific type of experimental object presented, rather than to one of the other 32 types of experimental objects. It is obvious that the observed frequency of correct lexigram touches far exceeded the frequency expected by chance. Furthermore, when Panzee recruited a person, she touched lexigrams that she rarely touched otherwise. On 25 of 34 trials, the lexigram corresponding to the object hidden was one that accounted for 1% or less of her routine lexigram use. Panzee used the correct lexigram on 21 of these 25 trials, and on the 4 remaining trials she relied effectively on gesture, including manual pointing, to direct the person to the hidden item.

Several aspects of Panzee's performance suggested that her memory involved a fairly rich information retrieval process. First, she, rather than the experimenters, determined the exact time of reporting, and she reported items after extended delays. Second, she selected from among 256 lexigrams on a keyboard rather than from a small set of alternatives, as in traditional primate matching-to-sample tasks. Third, she selected the lexigrams indoors, without an immediate view of the area in which the object was hidden. Fourth, she did not simply touch lexigrams; she pointed toward the outdoor area from the vicinity of the indoor keyboard, and she persisted in the interaction using a variety of response outputs until the person found the object outdoors. The findings strongly suggest that Panzee could recall which of several dozen types of objects she had been shown from a distance (and had not been allowed to navigate to or to touch) after overnight periods of at least 16 hours. Although in some test situations captive chimpanzees, like other great apes and preverbal human infants, put more weight on positional cues than on visual cues when they locate hidden objects (Haun et al. 2006), it is clear that Panzee retained and transmitted both types of information: spatial locations and their previously seen contents.

If we return to the general issues of ecological cognition and spatial representation, and attempt to incorporate an improved psychological concept of recall memory in animals, here are just a few of the details we must include, assuming that Panzee is representative.

1. Memory can, to a significant extent, be independent from sensory feedback from the original environment in question.
2. Recall can be performed in different ways and in different situations.
3. Specific types of objects can be recalled from a distance after hours or days, not merely after a few seconds or minutes.
4. A variety of indirect signs can serve as inputs for long-term memory.
5. A wide range of distracting events can occur during the retention interval without causing an animal to forget hidden items.
6. A wide range of response outputs can be used to convey a given message.
7. High accuracy of memory for object types and locations can be sustained, across multiple years of study in an outdoor area, and despite hundreds of trials and the possibility of proactive interference.

There is no a priori reason why these conclusions could not hypothetically apply to myriad other primates, and indeed other animals, in particular habitats and ecological situations. The ability to spontaneously and accurately recall information such as the type and quality of food patches, their relative position in space, and the presumptive location of competitors would presumably provide a considerable fitness advantage to its possessor.

Current opinions differ among field workers as to how flexible and generalized nonhuman primate spatial mapping skills are. Normand and Boesch (2009) argue that wild chimpanzees retain precise, accurate information about the distance and direction of an exceptionally large number of food trees and move directly to particular trees from many different starting locations, even within a dense and homogeneous forest that in the authors' estimation includes few distinctive topographical features. The chimpanzees' path of travel is as direct in the periphery of the home range as it is in its center, which might indicate a capacity for using novel shortcuts. Normand and Boesch (2009) argue that, for the chimpanzees to move as efficiently as they do, they must possess a flexible type of spatial knowledge, which the authors compare to a highly veridical Euclidean map. They suggest that chimpanzees are less reliant than monkeys are on the sight of topographical features, pathways, and landmarks, particularly when they navigate on large spatial scales. In contrast, Noser and Byrne (2007a) argue that wild chacma baboons and many other primates frequently adhere to a network of familiar, memorized routes when they visit food trees, and they note that a relatively simple movement rule of this sort can provide a satisfactory yield in the baboon's habitat and can also offer the advantages of speed and ease of use. Noser and Byrne (2007a) argue that the baboon data do not imply knowledge of the precise distances and directions to many locations in space at once, but better fit the predictions of a topologically accurate, remembered network of paths, searchable from any point. On the one hand, our experimental studies of captive *Pan* indicate generalized mapping skills (Menzel et al. 2002) and flexibility in the recall of information about distant objects (Menzel 1999). It is at least conceivable that captive monkeys would be capable of similar feats if they had comparable opportunity, training, motivation, and communication abilities. On the other hand, our studies of macaques suggest that they do not perceive or conceive of space as an empty container, but as something distorted by objects and environmental features. The same is presumably true for apes, even if their representation of distance and direction is more precise and accurate (i.e., more abstract and Euclidean) than is now widely appreciated. Of course, the question is not whether a chimpanzee's notion of large-scale space ultimately differs from a mathematician's conception of physical space (Jammer 1960), but how the spatial information is organized. From an experimental standpoint, we still know little about the differences in the operation of ape and monkey memory in large-scale spaces. Both experimental and observational studies will be required to achieve a synthesis in this exciting area.

Conclusions and Future Directions

A primate's knowledge of distant, remembered portions of its environment is no doubt much less acute than its visual and tactile perception of nearby objects (Dominy et al. 2001). Nevertheless, the past few decades have furnished compelling evidence against the ancient belief that nonhuman animals are totally lacking in true symbolic capabilities and thereby are limited entirely to the "here and now." It seems clear that tamarins, baboons, mangabeys, macaques, and apes sometimes travel as if discriminating locations and objects that are beyond touching, seeing, or hearing. The extent to which memory is independent of feedback from the immediate environment can be studied in a variety of ways, but the use of apes trained in an artificial language provides an especially powerful approach. Some apes at the Language Research Center behave as if distant areas are real, worth communicating about, and structured in terms of objects. The distal environment, as we currently conceive it for chimpanzees (genus *Pan*), clearly must include the types and locations of food and nonfood objects, signs, trails, landmarks, goals, and temporal properties of objects. The big question is, of course, how representative such research subjects are of their species as a whole. I doubt very much that they are brighter than any chimpanzees ever reared in the natural habitat. Instead, I suggest that they serve to show how preliminary our understanding is of the upper limits of primate cognitive capabilities in both captivity and the field. There is much to be learned about how the distant environment is represented in both these animals and in nonprimates, and about how and to what degree the animals use the information.

A promising area for synthesis of field and laboratory approaches to "ecological problem solving" is the experimental study of how well primates estimate and remember the variables that appear, from field studies and foraging theory, to be important for foraging efficiency and reproductive success. For example, there are few systematic evaluations of how well primates estimate and retain information over long time spans regarding the profitability of food patches (Sayers et al. 2010), the abundance of specific categories of food, the maturation rates of fruit resources (Janmaat et al. 2006b), or the variance (risk) associated with particular types of resources. Similarly, there is surprisingly little experimental data on primates' discrimination of distances in outdoor spaces. Many current field reports and models of optimal movement assume that animals can estimate distances in outdoor spaces at a high level of accuracy (chapter 7, this volume).

A second area for future research concerns the use of human observers as a source of prediction. It would be intresting and helpful in analyzing goal-oriented travel if field workers would (a) record their predictions about where their animals are headed next and where they are most likely to find the animals if they are out of sight, (b) analyze the accuracy of their predictions, and (c) identify the principal cues, including activity patterns and situational factors, that allow them to predict their animals' endpoints. Field workers acquire a wealth of tacit knowledge regarding their animals' movements, and their intuitive judgments about the directionality of behavior can draw attention to the types of information that animals possess. By analogy, caregivers' inferences from a captive chimpanzee's behavior regarding what, where, and when objects are hidden have proven to be highly accurate, and from this it can be inferred that the chimpanzee has information about the same ecological variables.

A second promising area for future research is what I have called the "rank ordering" problem in primate foraging. This problem can be described thus: Given multiple resources that vary in quality, quantity, distance and other attributes relevant to foraging efficiency, prioritize ("rank order") the locations of resources according to their overall expected value, and visit those locations according to their rank in priority. The problem amounts to a psychometric scaling of objects and environmental properties. The difference between this type of rank ordering and the "rank ordering" of prey items addressed by the classic prey model in foraging theory (Stephens & Krebs 1986) is that only in the former case does the organism compare multiple items and generate the ordered list. In contrast, in the classic prey model it is the researcher who selects a reasonable nutritional currency and ranks the items; the organism is assumed to encounter prey items individually and sequentially and does not evaluate two or more of them at a time;

the organism also decides, based on its recognition of item types, whether to include each item in the diet. The decision of whether to include each item can be based on an unknown mechanism, potentially a very simple one. This rank ordering problem in foraging can be illustrated by a study we recently conducted at the Language Research Center. Adult chimpanzees were tested individually and watched as an experimenter hid multiple food containers in unique locations in a forest. Later, the chimpanzees recruited otherwise uninformed caregivers and directed them from a tower to each container. The sequence in which containers were recovered was highly orderly and far more efficient than would be expected by chance: the chimpanzees remembered and prioritized the locations by food quantity, type, and proximity, consistent with a strategy of instantaneous rate maximization (iteratively selecting the best available remaining item). Numerous field researchers postulate that primates possess recall and the ability to evaluate resources that are out of view (e.g., Boyer et al. 2006). Studies of recall and rank-ordering capabilities in both primates and nonprimates may provide a systematic approach to some long-standing biological, anthropological, and psychological problems in knowing.

"The occurrence of goal-directed [teleonomic] processes is perhaps the most characteristic feature of the world of living organisms," and activities associated with food getting are an obvious example of goal-directed behavior in the broadest sense of this word (Mayr 1988, p. 45). This chapter has described recent studies of the traveling and foraging patterns of primates and some of the differences between species. To paraphrase the long-standing challenge posed by Aristotle's *On the Movement of Animals*, there remains a need for an investigation of the common ground for all varieties of primate movement, including the following of habitual pathways or topographical features, moving with a random component, and traveling in straight lines. A characterization of the learning, memory, and planning capabilities that underlie traveling and foraging in primates and other animals will remain a significant and exciting component of this endeavor.

References

Aldrich-Blake, F. P. G. 1980. Long-tailed macaques. In *Malayan Forest Primates* (ed. by Chivers, D. J.), 147–165. New York: Plenum Press.

Altmann, S. A. 1998. *Foraging for Survival*. Chicago: University of Chicago Press.

———. 1974. Baboons, space, time and energy. *American Zoologist*, 14, 221–248.

Bell, W. J. 1991. *Searching Behaviour*. New York: Chapman and Hall.

Beisner, B. A. & Isbell, L. A. (2009). Movement ecology in a captive environment: The effects of ground substrate on movement paths of captive rhesus macaques, *Macaca mulatta*. *Animal Behaviour*, 78, 1269–1277.

Beran, M. J., Evans, T. A. & Harris, E. H. 2008. Perception of food amounts by chimpanzees based on the number, size, contour length and visibility of items. *Animal Behaviour*, 75, 1793–1802.

Beran, M. J., Savage-Rumbaugh, E. S., Brakke, K. E., Kelley, J. W. & Rumbaugh, D. M. 1998. Symbol comprehension and learning: A "vocabulary" test of three chimpanzees (*Pan troglodytes*). *Evolution of Communication*, 2, 171–188.

Boden, M. A. 1972. *Purposive Explanation in Psychology*. Cambridge, MA: Harvard University Press.

Boesch, C. & Boesch, H. 1984. Mental map in wild chimpanzees: An analysis of hammer transports for nut cracking. *Primates*, 25, 160–170.

Boesch, C. & Boesch-Achermann, H. 2000. *The Chimpanzees of the Taï Forest: Behavioural Ecology and Evolution*. New York: Oxford University Press.

Boyer, D., Ramos-Fernández, G., Miramontes, A., Mateos, J. L., Concho, G., Larraldo, H., Ramos, H. & Rojas, F. 2006. Scale-free foraging by primates emerges from their interaction with a complex environment. *Proceedings of the Royal Society B*, 273, 1743–1750.

Brakke, K. E. & Savage-Rumbaugh, E. S. 1995. The development of language skills in bonobo and chimpanzee. I. Comprehension. *Language and Communication*, 15, 121–148.

———. 1996. The development of language skills in *Pan*. II. Production. *Language and Communication*, 16, 361–380.

Butler, R. A. 1960. Acquired drives and the curiosity-investigative motives. In *Principles of Comparative Psychology* (ed. by Waters, R. H., Rethlingshafer, D. A. & Caldwell, W. E.), 144–176. New York: McGraw-Hill.

Byrne, R. W. 2000. How monkeys find their way: Leadership, coordination, and cognitive maps of African baboons. In *On the Move: How and Why Animals Travel in Groups* (ed. by Boinski, S. & Garber, P.), 491–518. Chicago: University of Chicago Press.

Cheney, D. L. & Seyfarth, R. M. 1990. *How Monkeys See the World: Inside the Mind of Another Species*. Chicago: University of Chicago Press.

Collett, T. S. (2002). Spatial learning. In *Steven's Handbook of Experimental Psychology, Vol. 3, Learning, Motivation and Emotion, 3rd Edition* (ed. by Gallistel, R., & Pashler, H.), 301–364. New York: John Wiley & Sons.

Conklin-Brittain, N. L., Wrangham, R. W. & Hunt, K. D. 1998. Dietary response of chimpanzees and cercopithecines to seasonal variation in fruit abundance. II. Macronutrients. *International Journal of Primatology*, 19, 971–998.

Cunningham, E. & Janson, C. 2007a. A sociological perspective on primate cognition, past and present. *Animal Cognition*, 10, 273–281.

———. 2007b. Integrating information about location and value of resources by white-faced saki monkeys (*Pithecia pithecia*). *Animal Cognition*, 10, 293–304.

Di Bitetti, M. S., and Janson, C. H. 2001. Social foraging and the finder's share in capuchin monkeys, *Cebus apella*. *Animal Behaviour*, 62, 47–56.

Dominy, N. J., Lucas, P. W., Osorio, D. & Yamashita, N. 2001. The sensory ecology of primate food perception. *Evolutionary Anthropology*, 10, 171–186.

Erhart, E. M. & Overdorff, D. J. 2008. Spatial memory during foraging in prosimian primates: *Propithecus edwardsi* and *Eulemur fulvus rufus*. *Folia Primatologica*, 79, 185–196.

Fashing, P. J., Dierenfeld, E. S. & Mowry, C. B. 2007. Influence of plant and soil chemistry on food selection, ranging patterns, and biomass of *Colobus guereza* in Kakamega forest, Kenya. *International Journal of Primatology*, 28, 673–703.

Gallistel, C. R. 1990. *The Organization of Learning*. Cambridge, MA: MIT Press.

Garber, P. 1989. Role of spatial memory in primate foraging patterns: *Saguinus mystax* and *Saguinus fuscicollis*. *American Journal of Primatology*, 19, 203–216.

Gaulin, S. J. C. 1979. A Jarman/Bell model of primate feeding niches. *Human Ecology*, 7, 1–20.

Ghiglieri, M. P. 1984. *The Chimpanzees of Kibale Forest: A Field Study of Ecology and Social Structure*. New York: Columbia University Press.

Gibson, J. J. 1979. *The Ecological Approach to Visual Perception*. Boston: Houghton Mifflin.

Gomes, D. F. & Bicca-Marques, J. C. 2012. Capuchin monkeys (*Cebus nigritus*) use spatial and visual information during within-patch foraging. *American Journal of Primatology*, 74, 58–67.

Goodall, J. 1986. *The Chimpanzees of Gombe*. Cambridge, MA: Belknap Press of Harvard University Press.

Harlow, H. F. 1949. The formation of learning sets. *Psychological Review*, 56, 51–65.

Harlow, H.F. & Mears, C. 1979. *The Human Model: Primate Perspectives*. Washington: V. H. Winston; New York: Wiley.

Haun, D. B. M., Call, J., Janzen, G. & Levinson, S. C. 2006. Evolutionary psychology of spatial representations in the hominidae. *Current Biology*, 16, 1736–1740.

Hediger, H. 1968. *The Psychology and Behaviour of Animals in Zoos and Circuses*. New York: Dover Publications, Inc.

Hemmi, J. & Menzel, C. R. 1995. Foraging strategies of long-tailed macaques *Macaca fascicularis*: Directional extrapolation. *Animal Behaviour*, 49, 457–464.

Herman, L. M. & Forestell, P. H. 1985. Reporting presence or absence of named objects by a language-trained dolphin. *Neuroscience and Biobehavioral Reviews*, 9, 667–681.

Hölldobler, B. & Wilson, E. O. 2009. *The Super-organism*. New York: W. W. Norton.

Hull, C. L. 1952. *A Behavior System*. New Haven: Yale University Press.

Itani, J. 1965. On the acquisition and propagation of a new food habit in the troop of Japanese monkeys at Takasakiyama. In *Japanese Monkeys: A Collection of Translations* (ed. by Imanishi, K. & Altmann, S. A.), 52–65. Published privately by S. A. Altmann.

James, W. 1890. *The Principles of Psychology*, Vol. 1. New York: Holt.

Jammer, M. 1960. *Concepts of Space*. New York: Harper & Brothers.

Janmaat, K. R. L., Byrne, R. W. & Zuberbühler, K. 2006a. Evidence for a spatial memory of fruiting states of rainforest trees in wild mangabeys. *Animal Behaviour*, 72, 797–807.

———. 2006b. Primates take weather into account when searching for fruits. *Current Biology*, 16, 1232–1237.

Janson, C. H. 1998. Experimental evidence for spatial memory in foraging wild capuchin monkeys, *Cebus apella*. *Animal Behaviour*, 55, 1229–1243.

Janson, C. H. & Byrne, R. 2007. What wild primates know about resources: Opening up the black box. *Animal Cognition*, 10, 357–367.

Joly, M., Scheumann, M. & Zimmermann, E. 2008. Wild mouse lemurs revisit artificial feeding platforms: Implications for field experiments on sensory and cognitive abilities in small primates. *American Journal of Primatology*, 70, 892–896.

Joly, M. & Zimmermann, E. 2007. First evidence for relocation of stationary food resources during foraging in a strepsirhine primate (*Microcebs murinus*). *American Journal of Primatology*, 69, 1045–1052.

Kawamura, S. 1959. The process of sub-culture propagation among Japanese macaques. *Primates*, 2, 43–60.

Köhler, W. 1925. *The Mentality of Apes*. New York: Liveright.

Kummer, H. 1968. *Social Organization of Hamadryas Baboons*. Chicago: University of Chicago Press.

———. 1995. *In Quest of the Sacred Baboon: A Scientist's Journey*. Princeton, NJ: Princeton University Press.

Logan, F.A. 1972. The experimental psychology of animal learning and now. *American Psychologist*, 27, 1055–1062.

Laska, M., Freist, P. & Krause, S. 2007. Which senses play a role in nonhuman primate food selection? A comparison between squirrel monkeys and spider monkeys. *American Journal of Primatology*, 69, 282–294.

Lorenz, K. 1971. *Studies in Animal and Human Behaviour*, Vol. 2. Cambridge, MA: Harvard University Press.

Lucas, P. W. & Corlett, R. T. 1991. Relationship between the diet of *Macaca fascicularis* and forest phenology. *Folia Primatologica*, 57, 201–215.

Lührs, M., Dammhahn, M., Kappeler, P. M. & Fichtel, C. 2009. Spatial memory in the grey mouse lemur (*Microcebus murinus*). *Animal Cognition*, 12, 599–609.

Markham, A. C. & Altmann, J. 2008. Remote monitoring of primates using automated GPS technology in open habitats. *American Journal of Primatology*, 70, 495–499.

Marler, P. & Terrace, H. S. 1984. Introduction to *The Biology of Learning* (ed. by Marler, P. & Terrace, H. S.), 1–13. New York: Springer-Verlag.

Mayr, E. (1988). *Toward a New Philosophy of Biology: Observations of an Evolutionist*. Cambridge, MA: Harvard University Press.

Menzel, C. R. 1980. Head-cocking and visual perception in primates. *Animal Behaviour*, 28, 151–159.

———. 1986. Structural aspects of arboreality in titi monkeys (*Callicebus moloch*). *American Journal of Physical Anthropology*, 70, 167–176.

———. 1991. Cognitive aspects of foraging in Japanese monkeys. *Animal Behaviour*, 41, 397–402.

———. 1996a. Spontaneous use of matching visual cues during foraging by long-tailed macaques (*Macaca fascicularis*). *Journal of Comparative Psychology*, 110, 370–376.

———. 1996b. Structure-guided foraging in long-tailed macaques. *American Journal of Primatology*, 38, 117–132.

———. 1997. Primates' knowledge of their natural habitat: As indicated in foraging. In *Machiavellian Intelligence II: Extensions and Evaluations* (ed. by Whiten, A. and Byrne, R. W.), 207–239. Cambridge: Cambridge University Press.

———. 1999. Unprompted recall and reporting of hidden objects by a chimpanzee (*Pan troglodytes*) after extended delays. *Journal of Comparative Psychology*, 113, 426–434.

———. 2005. Progress in the study of chimpanzee recall and episodic memory. In *The Missing Link in Cognition: Origins of Self-Reflective Consciousness* (ed. by Terrace, H. and Metcalfe, J.), 188–224. New York: Oxford University Press.

Menzel, C. R. & Beck, B. B. 2000. Homing and detour behavior in golden lion tamarin groups. In: *On the Move: How and Why Animals Travel in Groups* (ed. by Boinski, S. and Garber, P.), 299–326. Chicago: Chicago University Press.

Menzel, C. R., Savage-Rumbaugh, E. S., & Menzel, E. W., Jr. 2002. Bonobo (*Pan paniscus*) spatial memory and communication in a 20-hectare forest. *International Journal of Primatology*, 23, 601–619.

Menzel, C. R., Menzel, E. W. Jr., Kelley, J. W., Chan, B., Evans, T. A. & Stone, B. W. 2008. The rank ordering problem in primate foraging. *American Journal of Primatology*, 70, supplement 1, 64.

Menzel, E. W., Jr. 1964. The effects of cumulative experience on responses to novel objects in young isolation-reared chimpanzees. *Behaviour*, 21, 1–12.

———. 1969a. Chimpanzee utilization of space and responsiveness to objects: Age differences and comparison with macaques. *Proceedings of the International Congress of Primatology*, 1, 72–80.

———. 1969b. Naturalistic and experimental approaches to primate behavior. In *Naturalistic Viewpoints in Psychological Research* (ed. by Willems, E. & Rausch, H.), 78–121. New York: Holt, Rinehart & Winston.

———. 1971a. Communication about the environment in a group of young chimpanzees. *Folia Primatologica*, 15, 220–232.

———. 1971b. Group behavior in young chimpanzees: Responses to cumulative novel changes in a large outdoor enclosure. *Journal of Comparative and Physiological Psychology*, 74, 46–51.

———. 1973a. Chimpanzee spatial memory organization. *Science*, 182, 943–945.

———. 1973b. Leadership and communication in young chimpanzees. In *Precultural Primate Behavior* (ed. by Menzel, E.), 192–225. Basel: Karger.

———. 1974. A group of young chimpanzees in a 1-acre field. In *Behavior of Nonhuman Primates*, Vol. 5 (ed. by Schrier, A. M. & Stollnitz, F.), 83–153. New York: Academic Press.

———. 1978. Cognitive mapping in chimpanzees. In *Cognitive Processes in Animal Behavior* (ed. by Hulse, S. H., Fowler, H. & Honig, W. K.), 375–422. Hillsdale, NJ: Lawrence Erlbaum.

———. 1984. Spatial cognition and memory in captive chimpanzees. In *The Biology of Learning* (ed. by Marler, P. & Terrace, H. S.), 509–531. New York: Springer-Verlag.

———. 1987. Behavior as a locationist views it. In *Cognitive Processes and Spatial Orientation in Animal and Man* (ed. by Ellen, P. & Thinus Blanc, C.), 55–72. Dordrecht: Marinus Nijhoff

Menzel, E. W., Jr. & Davenport, R. K., Jr. 1961. Preference for clear versus distorted viewing in the chimpanzee. *Science*, 134, 1531.

Menzel, E. W., Jr. & Halperin, S. 1975. Purposive behavior as a basis for objective communication between chimpanzees. *Science*, 189, 652–654.

Menzel, E. W., Jr. & Menzel, C. R. 1979. Cognitive, developmental and social aspects of responsiveness to novel objects in a family group of marmosets (*Saguinus fuscicollis*). *Behaviour*, 70, 251–279.

———. 2007. Do primates plan routes? Simple detour problems reconsidered. In *Primate Perspectives on Behavior and Cognition* (ed. by D. A. Washburn), 175–206. Washington: American Psychological Association.

Mulcahy, N. J. & Call, J. 2006. Apes save tools for future use. *Science*, 312, 1038–1040.

Muller, R. U., Stead, M. & Pach, J. 1996. The hippocampus as a cognitive graph. *Journal of General Physiology*, 107, 663–694.

Nathan, R., Getz, W. M., Revilla, E., Holyoak, M., Kadmon, R., Saltz, D. & Smouse, P. E. 2008. A movement ecology paradigm for unifying organismal movement research. *Proceedings of the National Academy of Sciences*, 105, 19052–19059.

Normand, E., Ban, S. D. & Boesch, C. 2009. Forest chimpanzees (*Pan troglodytes verus*) remember the location of numerous fruit trees. *Animal Cognition*, 12, 797–807.

Normand, E. & Boesch, C. 2009. Sophisticated Euclidean maps in forest chimpanzees. *Animal Behaviour*, 77, 1195–1201.

Noser, R. & Byrne, R. W. 2007a. Mental maps in chacma baboons (*Papio ursinus*): Using inter-group encounters as a natural experiment. *Animal Cognition*, 10, 331–340.

———. 2007b. Travel routes and planning of visits to out-of-sight resources in wild chacma baboons, *Papio ursinus*. *Animal Behaviour*, 73, 257–266.

Osvath M. & Osvath H. 2008. Chimpanzee (*Pan troglodytes*) and orangutan (*Pongo abelii*) forethought: Self-control and pre-experience in the face of future tool use. *Animal Cognition*, 11, 661–674.

Pochron, S. T. 2001. Can concurrent speed and directness of travel indicate purposeful encounter in the yellow baboons (*Papio hamadryas cynocephalus*) of Ruaha National Park, Tanzania? *International Journal of Primatology*, 22, 773–785.

Potts, R. 2004. Paleoenvironmental basis of cognitive evolution in great apes. *American Journal of Primatology*, 62, 209–228.

Putney, R. T. 1985. Do willful apes know what they are aiming at? *Psychological Record*, 35, 49–62.

———. 2007. Willful apes revisited: The concept of prospective control. In *Primate Perspectives on Behavior and Cognition*, ed. by Washburn, D. A., 207–219. Washington: American Psychological Association.

Reid, A. K. & Staddon, J. E. R. 1998. A dynamic route finder for the cognitive map. *Psychological Review*, 105, 585–601.

Ritchie, M. E. 1990. Optimal foraging and fitness in Columbian ground squirrels. *Oecologia*, 82, 56–67.

Robinson, D. N. 1989. *Aristotle's Psychology*. New York: Columbia University Press.

Rumbaugh, D. M. & Washburn, D. A. 2003. *Intelligence of Apes and Other Rational Beings*. New Haven: Yale University Press.

Savage-Rumbaugh, E. S., McDonald, K., Sevcik, R. A., Hopkins, W. D. & Rubert, E. 1986. Spontaneous symbol acquisition and communicative use by pygmy chimpanzees (*Pan paniscus*). *Journal of Experimental Psychology: General*, 115, 211–235.

Savage-Rumbaugh, E. S., Murphy, J., Sevcik, R., Brakke, K. E., Williams, S. L. & Rumbaugh, D. M. 1993. Language comprehension in ape and child. *Monographs of the Society for Research on Child Development*, 58, 3–4, serial no. 233.

Savage-Rumbaugh, E. S., Pate, J. L., Lawson, J., Smith, S. T. & Rosenbaum, S. 1983. Can a chimpanzee make a statement? *Journal of Experimental Psychology: General*, 112, 457–492.

Sayers, K., Norconk, M. A. & Conklin-Brittain, N. L. 2010. Optimal foraging on the roof of the world: Himalayan langurs and the classical prey model. *American Journal of Physical Anthropology*, 141, 337–357.

Schaller, G. B. 1964. *The Year of the Gorilla*. Chicago: University of Chicago Press.

Scheumann, M. & Call, J. 2006. Sumatran orangutans and a yellow-cheeked crested gibbon know what is where. *International Journal of Primatology*, 27, 575–602.

Schwartz, B. L. 2005. Do nonhuman primates have episodic memory? In *The Missing Link in Cognition: Origins of Self-Reflective Consciousness* (ed. by Terrace, H. & Metcalfe, J.), 225–241. New York: Oxford University Press.

Schwartz, B. L., & Evans, S. 2001. Episodic memory in primates. *American Journal of Primatology*, 55, 71–85.

Shettleworth, S. J. 1998. *Cognition, Evolution and Behavior*. New York: Oxford University Press.

Sigg, H. & Stolba, A. 1981. Home range and daily march in a Hamadryas baboon troop. *Folia Primatologica*, 36, 40–75.

Smith, J. D., Redford, J. S. & Haas, S. M. 2009. The comparative psychophysics of complex shape perception. *Animal Cognition*, 12, 809–821.

Stephens, D. W. & Krebs, J. R. 1986. *Foraging Theory*. Princeton, NJ: Princeton University Press.

Thorndike, E. L. 1898. Animal intelligence: An experimental study of the associative processes in animals. *Psychological Monographs*, 2(8).

Tinbergen, N. 1976. *The Study of Instinct*. New York: Oxford University Press (originally published 1951; new introduction added 1969).

Tinbergen, N., Impekoven, M. & Franck, D. 1967. An experiment on spacing-out as a defence against predation. *Behaviour*, 28, 307–321.

Tinklepaugh, O. L. 1932. Multiple delayed reaction with chimpanzees and monkeys. *Journal of Comparative Psychology*, 13, 207–243.

Tolman, E. C. 1932. *Purposive Behavior in Animals and Men*. New York: Century.

Tulving, E. 1983. *Elements of Episodic Memory*. Oxford: Clarendon Press.

———. 1984. Precis of elements of episodic memory. *Behavioral and Brain Sciences*, 7, 223–268.

Tomasello, M. & Call, J. 1997. *Primate Cognition*. New York: Oxford University Press.

Van Schaik, C. P., van Noordwijk, M. A, de Boer, R. J. & den Tonkelaar, I. 1983. The effect of group size on time budgets and social behaviour in wild long-tailed macaques (*Macaca fascicularis*). *Behavioral Ecology and Sociobiology*, 13, 173–181.

Visalberghi, E., Addessi, E., Truppa, V., Spagnoletti, N., Ottoni, E., Izar, P. & Fragaszy, D. 2009. Selection of effective stone tools by wild bearded capuchin monkeys. *Current Biology*, 19, 213–217.

Wheatley, B. P. 1980. Feeding and ranging of East Bornean *Macaca fascicularis*. In *The Macaques* (ed. by D. G. Lindburg), 215–246. New York: Van Nostrand Reinhold.

Woodworth, R. S. 1958. *Dynamics of Behavior*. New York: Holt.

Wrangham, R. W. 1975. The behavioural ecology of chimpanzees in Gombe National Park, Tanzania. PhD dissertation, Cambridge University.

———. 1977. Feeding behaviour of chimpanzees in Gombe National Park, Tanzania. In *Primate Ecology* (ed. by T. H. Clutton-Brock), 503–538. London: Academic Press.

Wrangham, R. W., Conklin-Brittain, N. L. & Hunt, K. D. 1998. Dietary response of chimpanzees and cercopithecines to seasonal variation in fruit abundance. I. Antifeedants. *International Journal of Primatology*, 19, 949–970.

Chapter 28 Knowledge of Social Relations

Robert M. Seyfarth and Dorothy L. Cheney

The vervet monkeys had moved out of their sleeping trees to forage on the ground. While the adults fed, the juveniles played in a nearby bush. Macauley, the rambunctious son of a low-ranking female, wrestled Carlyle, the daughter of the highest-ranking female, to the ground. Carlyle screamed, chased Macauley away, and then went to forage next to her mother. Apparently the fight had been noticed by others, because a few minutes later Shelley, Carlyle's sister, approached Austen, Macauley's sister, and bit her on the tail.

REGARDLESS OF whether they are strepsirrhines, New World monkeys, Old World monkeys, or apes, nonhuman primates do not interact at random with the members of their own groups. Instead, as reviewed by Aureli and colleagues in chapter 23 of this volume, they interact in different ways with different individuals, in some cases forming close, enduring social bonds that may persist for months or even years (Seyfarth & Cheney 2012 for review). These differentiated relationships constitute the animals' social environment. What do they know about it? Do they recognize the relationships that exist among other individuals? In this chapter we examine primates' social knowledge and consider what it may reveal about the evolution of cognition in human and nonhuman species.

Research on primate social cognition is of general interest for several reasons. First, it addresses central issues in the study of animal learning. A vervet monkey (*Chlorocebus pygerythrus*) is not born knowing about the relation between Shelley and Carlyle; instead, she must learn about other animals' relationships from what she sees and hears around her. This knowledge, it turns out, can be surprisingly complex, not least because it may involve as many as 100 different individuals. Can we explain it using the same principles of classical and operant conditioning that guide our understanding of learning in other species?

Second, data on primate social cognition are central to the social intelligence hypothesis (Jolly 1966; Humphrey 1976), which begins with the observation that primates have larger brains for their body size than other vertebrates (Martin 1990). This difference has arisen, the hypothesis argues, because primate groups are larger and their social relationships are more complex than those in other taxa (e.g., Dunbar 2000, 2003). Their large brains have evolved both to manage their own relationships and to recognize and "track" the relations among others. But are nonhuman primate relationships really different from those in other species? Is their social knowledge more extensive?

Third, if sophisticated social cognition is a pervasive feature of life in primate groups, how did this trait evolve? What, precisely, is the link between social intelligence and increased survival and reproduction?

Over the past 20 years scientists have made considerable progress in addressing these questions. Some are now firmly settled; answers to others are only beginning to emerge.

Brain Specializations for Social Stimuli

Regardless of their brain size or the complexity of their social behavior, all animals respond selectively to stimuli produced by members of their own species. Solitary frogs respond more strongly to their own species' calls than to calls of another species (Gerhardt & Bee 2006); monogamous birds respond more strongly to their own species' song (Benney & Braaten 2000). Primates are no different. Rhesus macaques (*Macaca mulatta*), for example, have "face cells" in the temporal cortex that respond at least twice as vigorously to faces or components of faces (like eyes or mouths) than to other complex visual stimuli (Tsao 2003, 2006). Face cells are surprisingly specialized. Those in the inferior temporal cortex (IT) seem most important for processing facial identity, whereas those in the superior temporal sulcus (STS) seem most important for processing facial expressions. IT and STS are extensively interconnected and probably share face-specific information (see Ghazanfar & Santos 2004 for review). Face cells in STS respond not only to facial expressions but also to the direction of an individual's head orientation and gaze. Their response is greatest when head orientation and gaze direction are congruent, and less strong when they are incongruent (Perrett et al. 1992; Emery & Perrett 2000; Jellema et al. 2000; Eifuku et al. 2004). The STS of rhesus macaques also includes neurons that fire when the monkey observes an individual walking, turning his head, bending, or extending his arm (Perrett et al. 1990). Particularly intriguing are "mirror neurons" in the inferior parietal lobule that show elevated activity both when the subject monkey executes a specific grasping action and when the monkey observes a human or other monkey executing a more or less similar grasp (Rizzolatti & Craighero 2004).

Rhesus macaques also process their own species' vocalizations in ways that are measurably different from the way they process other auditory stimuli. Like humans, they have brain areas that appear to be specialized for the recognition of different speakers (Belin & Zattore 2003; Petkov et al. 2008). Rhesus and Japanese macaques (*Macaca fuscata*) also display a left-brain, right-ear advantage when processing their own species' vocalizations but not when processing other sounds (Petersen et al. 1978; Hefner & Hefner 1984; Weiss et al. 2002; Poremba et al. 2004).

Specialized brain mechanisms for dealing with social stimuli do not, by themselves, confirm the social intelligence hypothesis—particularly because we find such specializations in many species despite wide variation in brain size and apparent social complexity. But the special responsiveness of the primate brain to conspecific voices, faces, gaze direction, and actions are just what we would expect if natural selection had acted to favor individuals skilled in solving social problems. Particularly intriguing is evidence that the STS and mirror neurons are also highly sensitive to intentional, goal-directed behavior, which indicates that they may help animals assess other individuals' intentions.

Recognition of Stable, Long-Term Relations in New and Old World Monkeys

Some features of primate social relationships are stable and predictable over long periods of time. For example, the adult females in many New and Old World monkey species form stable, long-term dominance hierarchies in which individual members of the same matriline—mothers, daughters, and sisters—maintain close relationships characterized by high rates of grooming, solicitations for grooming, mutual tolerance at feeding sites, and aggressive coalitions. Such bonds may last for 5, 10, or even 20 years (baboons, genus *Papio*, Silk et al. 2006a, 2006b; Cheney & Seyfarth 2007; rhesus macaques, Sade 1972; Japanese macaques, Kawai 1958; Koyama 1967; wedge-capped capuchins, *Cebus olivaceus*, O'Brien & Robinson 1993; white-faced capuchins, *Cebus capucinus*, Perry et al. 2008). Members of the same matriline occupy adjacent dominance ranks, and matrilines rank above one another in a "hierarchy of families." Both within and between families, ranks are linear and transitive (A dominates B, B dominates C, C dominates D, and so on). Typically, ranks remain stable for years at a time. Under these conditions, where highly predictable relations are so clear to a human observer, it is reasonable to ask whether the animals themselves recognize the kin and rank relations that exist among their companions. Evidence that they do so comes from both observational and experimental studies.

Recognition of Other Animals' Matrilineal Kin Relations

In many species, natural patterns of aggression and reconciliation suggest that animals have some knowledge of the close relations that exist among matrilineal kin. (Note that when we refer to monkeys' ability to recognize other individuals' kin, we mean only that monkeys recognize the close bonds that exist among kin. There is as yet no evidence that they distinguish the close bond that exists between two sisters, for instance, from the bond that exists between a mother and her adult daughter). Often, for example, an individual that has just been involved in an aggressive interaction with another will redirect aggression by attacking a third, previously uninvolved animal. Judge (1982, 1991) was the first to note that redirected aggression does not occur at random. He found that rhesus macaques do not simply threaten the nearest lower-ranking individual; instead, they target a close matrilineal relative of their opponent. Similar kin-biased redirected aggression occurs in Japanese macaques (Aureli et al. 1992) and vervet monkeys (Cheney & Seyfarth 1986, 1989). In a study of captive stump-tailed macaques (*Macaca arctoides*), Call et al. (2002) found that following aggression recipients directed an increased amount of sociosexual behavior toward all others in their group except their opponent's matrilineal kin, thus providing indirect evidence for their recognition of others' close relations.

When two white-faced capuchin monkeys are involved in an aggressive interaction, one or both of the combatants may attempt to recruit an ally. The monkeys' pattern of recruitment attempts also demonstrates their knowledge of other animals' close relations. Over a two-year period, Perry et al. (2004, 2008) found that individuals preferentially solicited allies who (1) ranked higher than their opponents and (2) had a social relationship with the solicitor that was closer (as measured by the ratio of past affiliative to aggressive interactions) than their relationship with the opponent. Although preferential solicitation of higher-ranking individuals could have occurred because individuals simply followed one of two simple rules ("Solicit an animal that ranks higher than yourself" or "Solicit the highest-ranking individual available"), the preferential solicitation of more closely bonded individuals could be explained only by assuming that solicitors were somehow comparing the bond between the ally and themselves with the bond between the ally and their opponent (Perry et al. 2004, 2008).

Supplementing these observational studies, field audio playback experiments provide independent evidence that individual monkeys recognize other animals' kin relations. In one study, a recording of a juvenile vervet monkey's distress scream was played to a group of three females, one of whom was the juvenile's mother. As expected, the mother looked toward the loudspeaker for longer durations of time than did the two control females. In addition, however, the control females responded by looking toward the mother, often reacting before the mother herself had responded. They behaved as if they associated the call with a specific juvenile, and that juvenile with a specific adult female (Cheney & Seyfarth 1980).

In another study, two unrelated adult female chacma baboons (*Papio ursinus*) served as subjects. On three different occasions, a pair of subjects heard a sequence of calls consisting of two other individuals' threat-grunts and screams. The calls were designed to mimic a common aggressive interaction in which a higher-ranking baboon gives threat-grunts to a lower-ranking animal, and the lower-ranking animal screams. In the first control condition, both of the apparent combatants were unrelated to the subjects. In the second, one of the combatants was a close relative of the dominant subject, while the other was unrelated to either female. In the third test condition, one of the combatants was a close relative of the dominant subject and the other was a close relative of the subordinate subject. When females heard a sequence that mimicked a dispute between

two individuals unrelated to them, they showed little or no reaction. When they heard a sequence that involved a relative of the dominant subject, the subordinate looked at the dominant but the dominant rarely looked at the subordinate. But when they heard a sequence that involved their relatives, both females looked at each other. Equally striking was that the dominant subject was more likely to seek out the subordinate subject and supplant her in the half-hour that followed these trials than in the half-hour that followed the two control sequences. Subjects behaved as if they recognized that a conflict between their families had occurred and had temporarily disrupted their relationship (Cheney & Seyfarth 1999).

Many primates "reconcile" with an opponent by touching, hugging, or behaving in a friendly way towards the opponent after aggression (Cords 1992; de Waal 1996; chapter 23, this volume). In baboons, reconciliation often takes the form of a grunt given by the aggressor to her former victim (Silk et al. 1996). Audio playback experiments confirm that grunts given by the aggressor after a fight change the victim's behavior, thus increasing the likelihood that the victim will interact in a friendly or tolerant way toward the aggressor—hence the description of this behavior as "reconciliation" (Cheney et al. 1995b; Cheney & Seyfarth 1997). Further experiments have shown that the grunts of a close relative of the aggressor can also function as a proxy to reconcile opponents. Specifically, victims of aggression are more likely to tolerate their opponent's proximity in the hour after the dispute if they have heard the grunt of their opponent's relative than if they have heard the grunt of a more dominant individual belonging to a different matriline (Wittig et al. 2007a). Such kin-mediated reconciliation succeeds only when the victim recognizes the relationships that exist among other group members (see also Das 2000; Judge & Mullen 2005). Conversely, if the victim hears the threat-grunt of her opponent's relative shortly after aggression, she is more likely to avoid her opponent and other members of her opponent's matriline (Cheney & Seyfarth 2007; Wittig et al. 2007b).

The experiments on baboons' responses to kin-mediated reconciliation and vocal alliances support the view that baboons recognize other females' matrilineal kinship relations. This is not to say, however, that baboons treat all the members of a matriline as equivalent. Although they recognize that close kin can serve as proxies for each other, they nonetheless distinguish among the different members of a matriline. Hearing a "reconciliatory" grunt from an opponent's relative changes females' behavior toward the opponent and that relative, but not toward other members of the opponent's matriline (Cheney & Seyfarth 2007; see Rendall et al. 1996 for experiments demonstrating the recognition of individuals and of close relations among matrilineal kin in rhesus macaques).

What learning mechanisms underlie this behavior? Considered in isolation, the recognition of other animals' close kin relations would seem to require no special skills beyond those well documented in laboratory studies of learning. All a young primate needs to do is observe and memorize who interacts with whom, and note differences in the quality of their interactions. The recognition of matrilineal kin, however, does not occur in isolation: matrilineal kin relations are embedded in a network of short- and long-term bonds that vary among individuals according to age, rank, reproductive state, and many other variables. Whether social knowledge under natural conditions, in all its simultaneous manifestations, can be explained by simple theories of association remains an open issue. We discuss some competing hypotheses below.

Recognition of Other Animals' Dominance Ranks

Like the bonds among matrilineal kin, linear, transitive dominance relations are a pervasive feature of behavior in many primate groups (chapter 9, this volume). A linear rank hierarchy might emerge because a female simply takes note of who is dominant and subordinate to herself—an egocentric view of the world, but one that nonetheless would result in a linear, transitive rank order. Alternatively, the female might also distinguish among the relative ranks of others. If rank were determined by a physical attribute like size, recognizing other individuals' ranks would be easy. Among most female monkeys, however, there is no relation between rank and size, condition, age, or any other obvious feature. As a result, the problem is considerably more challenging.

There are hints from their behavior that monkeys do recognize other individuals' relative dominance ranks. In a meta-analysis of 14 different species, including New and Old World monkeys, Schino (2001) found consistent evidence that high-ranking females received more grooming and were groomed by more different individuals than lower-ranking females. These data suggest a general preference for grooming high-ranking individuals, but they fall short of showing that each animal recognizes the relative ranks of others. In subsequent papers, Schino used a within-subject regression analysis to test the hypothesis that each individual distributed grooming among others in direct relation to their relative rank. He found a significant rank effect in Japanese macaques (Schino et al. 2007) but not in tufted capuchins (*Cebus apella*, Schino et al. 2009).

In vervet monkeys, females solicit grooming from others by presenting a part of their body to them. The solicited individual may or may not accept the invitation. In general,

females are most likely to accept the solicitations of the highest-ranking female, second most likely to accept those of the second-ranking female, third most likely to accept those of the third-ranking female, and so on (Seyfarth 1980).

Knowledge of other individuals' rank relations also appears in patterns of recruitment and coalition formation. When two female vervets, macaques, or baboons are involved in an aggressive interaction and a third female, higher-ranking than both, joins them to form a coalition, the third female almost always supports the higher-ranking individual (vervets, Seyfarth 1980; Cheney 1983; rhesus macaques, de Waal 1991; Japanese macaques, Chapais 2001; but for contrary data on chimpanzees [*Pan troglodytes*], gorillas [*Gorilla gorilla*], and white-faced capuchins, see de Waal & van Hooff 1981; Watts 1991; and Perry 2003).

Similarly, when recruiting alliance partners, monkeys often appear to be assessing not only their own rank relative to a potential ally but also the rank relation between the prospective ally and their opponent. As already noted, white-faced capuchin monkeys consistently recruit allies who outrank their opponents, but this pattern of recruitment could result from individuals following the simple rule "Solicit an ally who ranks higher than yourself" or "Solicit the highest-ranking animal available." By contrast, in Silk's (1993, 1999) study of captive male bonnet macaques (*Macaca radiata*), the pattern of solicitation could only be explained by assuming that males tried preferentially to recruit allies that outranked both themselves and their opponents. Furthermore, the males' choice of alliance partner varied depending on their opponent's rank. If a male was involved in a fight with the seventh-ranking male, he would attempt to solicit the aid of, say, the fifth-ranking male. But if his opponent was the twelfth-ranking male, he would attempt to recruit the ninth-ranking one. If the male dominance hierarchy had remained stable, memorizing each male's rank might not have been a difficult task, but each month roughly half of the 16 males changed rank. The males' behavior suggests that they carefully monitored all aggressive interactions and constantly updated their list.

These observations are supported by field experiments. In one study, for example, chacma baboons heard a sequence of vocalizations mimicking an interaction that violated the female dominance hierarchy. The sequence consisted of a series of grunts originally recorded from a lower-ranking female (say, the fifth-ranking female) combined with a series of fear barks originally recorded from a higher-ranking female (say, the third-ranking female). This sequence violated the female dominance hierarchy because, whereas baboons routinely grunt both to lower- and higher-ranking individuals, they give fear-barks only to individuals that rank above them. In control trials, subjects heard the same anomalous sequence of calls, but with the grunts of a third female that normally outranked both of the other individuals (say, the second-ranking female). This sequence was consistent with the female dominance hierarchy because the third-ranking female could be giving fear-barks to the second-ranking female rather than the fifth-ranking one. Supporting the view that baboons recognize other individuals' dominance ranks, subjects looked in the direction of the loudspeaker for significantly longer durations when they heard the sequence that violated the dominance hierarchy (Cheney et al. 1995a).

The recognition of other individuals' dominance ranks has also been documented among adult male chacma baboons, whose aggressive contests often involve loud "wahoo" calls given by two or more males as they race through the group, climb trees, and leap from branch to branch (Kitchen et al. 2003; Fischer et al. 2004). In this experiment, individual males heard wahoo sequences that mimicked a contest between either adjacently ranked or disparately ranked males. To control for the fact that wahoo contests involving adjacently ranked males are more common than those involving males of disparate ranks, only the calls of adjacently and disparately ranked males who had interacted at the same rate during the past six months were used as stimuli. High-ranking male subjects responded significantly more strongly to the audio playback of a wahoo contest between males of disparate ranks than to the playback of a contest between males of adjacent ranks (Kitchen et al. 2005). This result might have arisen because adjacently ranked males compete with one another in many different contexts, whereas males of disparate ranks tend to compete only when the resource at stake is highly valued: meat, a sexually receptive female, or an infant vulnerable to infanticide. Whatever the explanation, results suggested that males could assess the rank distance between any two males (Kitchen et al. 2005). The result is particularly striking because, like the ranks of male bonnet macaques described earlier, the ranks of male baboons change often.

Like the recognition of other animals' kin relations, the recognition of other animals' rank relations may have been favored by natural selection because it allows individuals to groom, recruit allies, and form coalitions most effectively. Some scientists have argued that rank recognition should be most well developed in species where dominance hierarchies are most pronounced. Thierry (2008) and Schino and Aureli (2008) review evidence to support this view. For example, the ability to recognize other individuals' ranks implies the ability to make transitive inferences: to recognize that, if A is greater than B and B is greater than C, then A must be greater than C. In an experimental test of transitive inference, Maclean et al. (2008) found that ring-tailed lemurs (*Lemur catta*), which live in large groups with linear dominance hierarchies, performed better than mongoose lemurs (*Eulemur mongoz*), which live in small, monogamous pairs. After extensive training, however, the two species performed equally well.

The ability to engage in transitive inference seems to have evolved independently in many species with linear dominance hierarchies. When forming a coalition, spotted hyenas (*Crocuta crocuta*) consistently join the higher-ranking of two competitors, regardless of which one is winning at the time (Engh et al. 2005). Pinyon jays (*Gymnorhinus cyanocephalus*)—often called "avian baboons"—live in stable flocks of 50 to 500 individuals, each containing individuals that are linked by kinship and arranged in a linear dominance hierarchy. Elegant experiments by Paz-y-Miño et al. (2004) have shown that jays use transitive inference to calculate their own dominance status relative to that of a stranger they have observed interacting with their group mates. Grosenick et al. (2007) performed a similar experiment on fish (*Astatotilapia burtoni*), with similar results.

Regardless of how it evolved, the recognition of other individuals' rank relations, like the recognition of other animals' matrilineal kin relations, requires by itself no special skills in learning and intelligence beyond those well documented in laboratory studies of classical conditioning. Individuals simply need to observe the behavior of others, remember what has happened in the past, and update their knowledge as new information becomes available. In nature, however, recognition of other animals' ranks does not occur on its own; it must necessarily be integrated into a complex matrix of other social relations. We are only beginning to understand how this is achieved.

Integrating Knowledge of Kin and Rank Relations

Having found that chacma baboons recognize the close bonds among matrilineal kin and individual dominance ranks, Bergman et al. (2003) tested whether individuals integrated their knowledge of other individuals' kinship and rank to recognize that the female dominance hierarchy is in fact composed of a hierarchy of families (that is, subgroups of closely bonded females). As background, recall that rank relations among adult female baboons are generally very stable over time, with few rank reversals occurring either within or between families. When rare reversals do occur, their consequences differ significantly depending on who is involved. If, for example, the third-ranking female in matriline B (B_3) rises in rank above her second-ranking sister (B_2), the reversal affects only the two individuals involved; the family's rank relative to other families remains unchanged. However, a rare rank reversal between two females from different matrilines (for example, C_1 rising in rank above B_3) is potentially much more momentous because it can affect entire families, with all the members of one matriline (in this case, the C matriline) rising in rank above all the members of another.

Bergman et al. (2003) played sequences of calls mimicking rank reversals to subjects in paired trials. In one set of trials, the subjects heard an apparent rank reversal involving two members of the same matriline: for example, female B_3 giving threat-grunts while female B_2 screamed. In the other set, the same subjects heard an apparent rank reversal involving the members of two different matrilines: for example, female C_1 giving threat-grunts while female B_3 screamed. As a control, the subjects also heard a fight sequence that was consistent with the female dominance hierarchy. To control for the rank distance separating the subjects and the individuals whose calls were being played, each subject heard a rank reversal (either within- or between-family) that involved the matriline one step above her own (cf. Penn et al. 2008). Within this constraint, the rank distance separating apparent opponents within and between families was systematically varied.

As before, listeners responded with apparent surprise to sequences of calls that appear to violate the existing dominance hierarchy. Moreover, between-family rank reversals elicited a consistently stronger response than did within-family rank reversals (Bergman et al. 2003). The subjects acted as if they classified individuals simultaneously according to both kinship and rank. The classification of individuals simultaneously according to two different criteria has also been documented in Japanese macaques (Schino et al. 2006).

Recognition of More Transient Social Relations

Bonds among matrilineal kin and a linear, transitive female dominance hierarchy are components of monkey social structure that typically remain stable for many years. It is perhaps not surprising, therefore, that primate social cognition has been best documented in these two domains. There is growing evidence, however, that primates also recognize and monitor more transient social bonds.

Hamadryas baboons (*Papio hamadrayas*) in Ethiopia are organized into one-male units, each containing a fully adult male and two to nine adult females (Kummer 1968; Stammbach 1987; chapter 5, this volume). One-male units frequently come into contact with single, unattached males who may attempt to challenge the unit leader in an attempt to take over his females. In the first experimental test of individuals' ability to recognize other animals' relations, Bachmann and Kummer (1980) found that the willingness of a male to challenge a unit leader depended not on the

challenger's dominance rank relative to that of the leader, but on the challenger's perception of the strength of the bond between the leader and his females. Noting that social bonds between adult males and females can change often, Bachmann and Kummer suggested that challengers continually monitor one-male units to assess whether the bonds between a male and his females have weakened.

Just this kind of monitoring seems to occur in multi-male groups of baboons, where males form sexual consortships with an adult female during the week when she is most likely to ovulate (chapter 18, this volume). Sexual consortships constitute a form of mate guarding, and typically involve the highest-ranking male. When a consortship has been formed, lower-ranking males can nonetheless gain mating opportunities by taking advantage of temporary separations between a female and her consort to mate "sneakily" (chapter 18, this volume). To test whether subordinate males monitor sexual consortships for such opportunities in chacma baboons, Crockford et al. (2007) used a two-speaker playback experiment to simulate a temporary separation between the consort pair. One speaker played the consort male's grunt to signal his location. The other speaker, located approximately 40 meters away, played the female's copulation call to signal that she was mating with another male and that further mating opportunities might be available. The subordinate males responded immediately to the apparent separation between the female and her consort by approaching the speaker playing the female's call. By contrast, when the same playback was repeated a few hours after the consortship had ended, the subordinate males showed no interest. Apparently, they already knew that the consortship had ended, and the information was therefore redundant. Thus, males appear to monitor the status of these transient consort relationships very closely, even though they typically last for only a few days (see Smuts 1985 for similar data on animals' recognition of the "friendships" between males and lactating females in olive baboons, *Papio anubis*).

The Recognition of Social Relationships in Fission-Fusion Societies

The fission-fusion societies found in many animals, including spider monkeys (*Ateles* spp., chapter 3, this volume) and chimpanzees (chapter 6, this volume), may present animals with challenges and selective pressures not found in species where all individuals in a group travel, interact, and sleep as a unit (Aureli et al. 2008). Thus far, the only data on social knowledge in a primate with a fission-fusion society come from studies of chimpanzees, so they are the focus of this section. Recognizing other animals' relations in chimpanzee society presents a challenge, for several reasons.

First, whereas the close social bonds found in monkeys involve a relatively small number of behaviors like grooming, coalitions, and tolerance at food sources, the long-term bonds formed by chimpanzees include behaviors as diverse as mate guarding (Watts 1998), coalitions in male-male competition (Nishida & Hosaka 1996), cooperative defense of territories (Mitani et al. 2010), grooming, and meat sharing (Wittig & Boesch 2003; Mitani 2006, 2009). To monitor and distinguish the different relationships within its group, therefore, a chimpanzee must keep track of many more behaviors.

Second, as in many species of New and Old World monkeys (see above), but perhaps to an even greater extent, the interactions that characterize close bonds in chimpanzees are often widely separated in time. Among male chimpanzees at Ngogo, for example, the pairs who groom most often also have the highest rates of coalition formation and participation in border patrols, yet these behaviors do not necessarily occur together in time (Mitani 2006, 2009). Days may pass between a grooming bout and the formation of a coalition or a border patrol. Further complicating matters, there may be striking short-term "imbalances" in cooperative behavior within a close relationship. Pairs with the strongest grooming relations and the strongest bonds overall may have grooming that is highly imbalanced within a bout (one partner does most of the grooming) but highly equitable over longer periods of time (Gomes et al. 2009; Mitani 2009). The most closely bonded chimpanzees, like the most closely bonded baboons, seem "tolerant of temporary imbalances" in their relationship (Silk et al. 2010a, p. 1743; Seyfarth & Cheney 2012). As a result, an observer attempting to distinguish bonds of different strength cannot do this by observing a single grooming session.

Third, the correlations among behaviors that distinguish long-term bonds in chimpanzees can be complex. In the Taï forest, for example, male-male, male-female, and female-female dyads with the strongest, most enduring bonds as measured by meat sharing and the rate of coalition formation also had the highest rates of grooming, but not the highest rates of spatial association. Rates of meat sharing and coalition formation, but not rates of grooming, predicted the rate at which dyads reconciled following aggression (Wittig & Boesch 2005; Wittig 2010). An observer attempting to distinguish closely bonded pairs from others must therefore take note of and remember many different behaviors.

Despite the complexity of their behavior and society, chimpanzees not only maintain close, long-term bonds with specific partners (Mitani 2009; Wittig 2010; chapter 6, this volume), but also recognize and distinguish the different relationships that exist among others (see Seyfarth & Cheney 2012 for review). For example, sometimes after an aggressive interaction between two individuals a previously uninvolved bystander will direct friendly behavior toward the victim. In one study, the probability that a bystander would engage in such behavior depended primarily on the strength of the bond between the bystander and the victim: the stronger their bond, the more likely that such "consolation" would occur (Kutsukake & Castles 2004). In another study, however, the bystander was most likely to direct friendly behavior toward the victim if the bystander had a strong bond with the aggressor and a weak bond with the victim (Wittig 2010). Moreover, if the bystander had a strong bond with the aggressor, this increased the likelihood that the aggressor and victim would tolerate each other's proximity in the near future; but if the bystander had a weak bond with the aggressor, this effect disappeared (Wittig & Boesch 2010). Victims acted as if they recognized the close bond (or lack of it) between the bystander and the aggressor. As a result, they treated the bystander's friendly behavior as "reconciliation by proxy" only if the bystander was a close associate of the aggressor.

Chimpanzees often scream when involved in aggressive disputes. Slocombe and Zuberbühler (2005) found that victims produce acoustically different screams according to the severity of aggression they are receiving. In playback experiments, listeners responded differently to the different scream types (Slocombe et al. 2009). In cases of severe aggression, victims' screams sometimes seemed to exaggerate the severity of the attack, but victims only gave exaggerated screams if their foraging party included at least one listener whose dominance rank was equal to or higher than that of their aggressor (Slocombe & Zuberbühler 2007). Victims seemed to alter their screams depending upon the relationships between their opponent and their potential allies.

The Recognition of Intentions and Motives

Although it now seems clear that many animals recognize other group members' relationships and dominance ranks (cf. Henzi & Barrett 2007), we still know little about whether they attribute to these relationships a particular set of emotions and motives, as humans do. In the more than 30 years since Premack and Woodruff (1978) posed the question "Does the ape have a theory of mind?," much progress has been made in the study of mental state attribution in animals. Many questions, however, remain unresolved (chapters 30 and 32, this volume).

Several lines of evidence suggest that primates routinely attribute simple mental states, like intentions and motives, to others. In the wild, this ability is particularly evident in their response to vocalizations, when individuals must make inferences about the intended recipient of another animal's calls. This is not surprising: primate groups are noisy, tumultuous societies, and an individual could never manage her social interactions if she assumed that every vocalization was directed at her. Inferences about the directedness of vocalizations are probably often mediated by the direction of the caller's gaze. Even in the absence of visual signals, however, monkeys seem to make inferences about the intended recipient of a call based on their knowledge of a signaler's identity and the nature of recent interactions.

Recall, for example, that a female baboon responded strongly when she was played a recording of her aggressor's grunt within minutes after being threatened, but showed little response if she heard the grunt of another dominant female unrelated to her aggressor. In other words, the female responded as if the aggressor's grunt was directed *to her*, but the other individual's grunt was directed to someone else. Moreover, the female's responses to the aggressor's grunt (she approached the aggressor and/or tolerated the aggressor's approach) indicated that she treated the call as a "reconciliatory" signal of benign intent. In other words, she attributed specific motives to the aggressor.

These results were replicated in one test where female baboons heard the "reconciliatory" grunt of their aggressor's kin and likewise treated it as a reconciliatory signal (see above). They were further replicated in a second test where subjects heard an aggressive threat-grunt from an individual after they had either exchanged aggression or groomed with that same individual (Engh et al. 2006c). Subjects who heard a female's threat-grunt shortly after grooming with her ignored the call—they acted as if they assumed that the female was threatening another individual. By contrast, the same subjects responded strongly when they heard the same call after receiving aggression; they acted as if the call was directed *at them*.

In sum, while nonhuman primates may lack a full-blown "theory of mind" like that found in young children (chapters 30 and 32, this volume), they do appear to attribute simple emotions and motivational states to others. Their knowledge of social relationships may well be closely linked to these attributions (de Waal 2008; Schino & Aureli 2009; Seyfarth & Cheney 2012).

Testing the Social Intelligence Hypothesis

Can primate social knowledge be explained by traditional theories of learning? Are the societies of nonhuman primates more complex than those of other species? Is their social intelligence more impressive? Finally, regardless of whether primate social intelligence exceeds that found in other species, what role has social intelligence played in primate evolution? How does it translate into improved reproductive success? These questions, by no means resolved, are central to tests of the social intelligence hypothesis. We consider them in turn.

Are Traditional Theories of Learning Sufficient?

Can a few simple rules explain the complexity of primates' social knowledge? Some learning psychologists believe that they can, and have argued that monkeys' apparent recognition of other individuals' kin (to cite one example) is simply an impressive form of associative learning and conditioning.

It is well known that laboratory animals like rats and pigeons can be taught to group even very different looking stimuli together if they are all associated with the same reward or outcome. In one series of tests, Schusterman and Kastak (1993, 1998) taught Rio, a California sea lion (*Zalophus califonianus*), to group arbitrary symbols into "equivalence classes." Each group consisted of three cards depicting a symbol: for example, a pipe (A_1), a fish (A_2), and a star (A_3). The experimenters arranged the symbols into equivalence classes by displaying one group of cards (say, A_1, A_2, and A_3) next to each other on one side of Rio's enclosure and another group of cards (say, B_1, B_2, and B_3) next to each other on the other side. After a few days' exposure, Rio was presented with one card from the A class and one card from the B class (A_1 and B_1). As soon as she prodded one of the cards with her nose, she was rewarded with food. Assuming that Rio chose A_1 rather than B_1, she then received repeated presentations of the same cards, with A_1 always rewarded and B_1 not rewarded, until she achieved a 90% success rate. Then Rio was tested, first with symbols A_2 and B_2 (transfer test 1) and next with symbols A_3 and B_3 (transfer test 2), to determine whether she had learned to treat all A stimuli as equivalent to each other and all B stimuli as equivalent to each other, at least insofar as they followed the rule "If A_1 is greater than B_1, then A_n is greater than B_n." Rio performed correctly on 28 of 30 transfer tests.

Schusterman and Kastak argue that these relatively simple equivalence judgments constitute a general learning process that underlies much of the social behavior of animals, including the recognition of social relationships by monkeys and apes (for similar arguments see Heyes 1994; Wasserman and Astley 1994; Thompson 1995). This argument has much validity. Indeed, it is hard to imagine how a monkey could learn that two other individuals were members of the same matriline except by grouping them together by virtue of their high rates of association. At the same time, however, the "equivalence classes" found in nonhuman primate groups exhibit complexities not present in laboratory experiments.

First, consider the magnitude of the problem. The sea lion Rio was confronted with a total of 180 dyadic comparisons. This is roughly equivalent to the number of different dyads that confront a monkey in a group of 18 individuals. But the number of possible dyads increases rapidly as group size increases. Baboons, for example, often live in groups of 80 individuals, which contain 3,160 different dyads and 82,160 different triads. As a result, individuals face problems in learning and memory that are not just quantitatively but also qualitatively different from those presented in a typical laboratory experiment. This is important, because it is large numbers that may force primates to develop rules to classify their group mates.

Second, in primate groups no single metric specifies the associations between individuals. It is, of course, a truism that animals can learn which other individuals share a close social relationship by watching them interact. But no single behavioral measure is either necessary or sufficient to recognize such associations. Aggression often occurs at the same rate within and between families, and different family members may groom and associate with each other at widely different rates (e.g., Silk et al. 2010a). Spatial proximity is not a defining characteristic of close bonds among female baboons, but it is invariably present in the bond between a male and his sexual consort (Crockford et al. 2007). In sum, there is no threshold or simple defining criterion for a "close" social bond among primates. By contrast, in Schusterman and Kastak's experiments the spatial and temporal juxtaposition of stimuli provided an easy, one-dimensional method for the formation of equivalence classes.

Third, class members in monkey groups are sometimes mutually substitutable, and sometimes not. Consider, for example, the early experiment on vervet monkeys, in which females who heard a juvenile's screams then looked toward the juvenile's mother. Schusterman and Kastak (1998) argue that this occurs because the scream, the juvenile, and the juvenile's mother form a three-member equivalence class in the females' minds in which any one of the stimuli can be substituted for another. But in fact the call, the juvenile, and the mother are not interchangeable. A female who has a close bond with the juvenile's mother, for example, may interact very little with the juvenile himself. The call is linked primarily to the juvenile, and only secondarily to the mother. Indeed, audio playback experiments on rhesus macaques have shown that, although monkeys do group calls given by members of the same matriline into the same category, they also distinguish among the calls given by different individuals within that matriline (Rendall et al. 1996; see also the "reconciliatory" grunt experiments discussed above).

Fourth, some social relationships are transitive, but others are not. In baboon society, if an infant and a juvenile both associate at high rates with the same adult female, it is usually correct to infer that the two are siblings and will also associate at high rates. Similarly, if a lactating female associates at a high rate with a particular adult male friend, it is probably correct to assume that the male is also closely allied to the female's infant. It would be incorrect, though, to make the same assumption about the juvenile, because males seldom interact at high rates with their friends' older offspring. In fact, the juvenile is more likely to associate with a different male—the male who was his mother's friend when he was an infant.

Further complicating matters, individuals can belong to many different classes simultaneously. An adult female baboon, for instance, belongs to a matrilineal kin group, associates with one or more adult males, holds a particular dominance rank, and may be weakly or strongly linked to other females outside her matriline. Here again, the natural situation is considerably more complex than the situation in most laboratory settings.

Finally, some types of class membership change often. While the rank and kin relations among female monkeys are often relatively stable, other social relationships change often and unpredictably. Yet individuals are able to monitor these changes and make the appropriate adjustments.

In sum, there is no doubt that associative processes and contingency-based learning provide powerful ways to assess the relationships that exist among others. However, in order to conclude that all primate social knowledge results from simple learning mechanisms, we need proof that these mechanisms can account for behavior as complex as that which occurs in free-ranging primate groups. Suppose that the sea lion Rio were trained with an array of 80 items (the approximate size of a baboon group), each of which associated at varying rates with all 79 other items, but at high rates with a subset of the items. Item A, for example, might associate at a high rate with items B, C, D, and E. Item B might also associate with these items, but at a different rate than A. Item B would also associate with some items with which A rarely associated. To complicate matters further, there would also be brief, transient associations of varying duration between pairs of items that cut across the links formed between items that associated at a high rate. Under these circumstances, could Rio learn to group the items that associate at high rates into "kin" classes while simultaneously keeping track of the transient pairings that cut across classes? Perhaps Rio would rise to the occasion; perhaps not.

Laboratory experiments designed to explain complex behavior using the simplest explanation possible have limited external validity if they leave out the very complexity they hope to explain, or depend on extensive training and reinforcement. Primates, after all, derive no immediate and predictable rewards from their knowledge of other individuals' social relationships—unless we assume that they find social spectatorship inherently rewarding. But if we accept this view and assume that primates are motivated by the inherent value of acquiring knowledge about others, our concepts of reward and reinforcement must become considerably broader and more open-ended than they are in most laboratory studies of learning.

Are Primates Special?

Across the animal kingdom, brain size increases with body size. Despite this common scaling principle, however, brain size-to-body-weight ratios differ from one taxonomic group to another. Among mammals, primates have brains that are, on average, larger than the brains of similar-sized nonprimate mammals. They have a higher "index of cranial capacity" (ICC, Martin 1990).

The social intelligence hypothesis (Jolly 1966; Humphrey 1976) attempts to explain this difference. It proposes that all group-living animals confront a multitude of social problems, but that the problems facing primates are more daunting than those facing other species because primate groups are larger, their social relationships more complex, and their frequent formation of coalitions requires a more sophisticated knowledge of other animals' relations (Harcourt 1988; Dunbar 2000). The data reviewed in this chapter demonstrate that primates have an impressive knowledge of their companions' social relationships. But are primate societies really more complex, and is their social intelligence really superior to that found in other species? The issue is currently unresolved.

Some comparative tests of captive apes, monkeys, pigeons, and other animals suggest that primates are more adept than nonprimates at classifying items according to their relative relations (for reviews see Tomasello & Call 1997; Cheney & Seyfarth 2007; Shettleworth 2010). However, there is also good evidence that social complexity and large brains have coevolved in nonprimate species as well as in monkeys and apes. Large group size is positively correlated with larger neocortex size not only in primates (Barton & Dunbar 1997), but also in carnivores (Barton & Dunbar 1997), toothed whales (Connor et al. 1998), and ungulates (Perez-Barberia & Gordon 2005).

Moreover, many nonprimate species display examples of social cognition that rival those found in monkeys and apes.

When competing over access to females, male dolphins (*Tursiops truncatus*) form dyadic and triadic alliances with specific other males, and allies with the greatest degree of partner fidelity are most successful in acquiring access to females (Connor 2007). Field observations suggest that opponents may recognize the bonds that exist among others and selectively retreat when they encounter rivals with a long history of cooperation. Like many primates, lions (Packer & Pusey 1982), horses (Feh 1999), spotted hyenas (Engh et al. 2005), and dolphins (Connor & Mann 2006) intervene selectively on behalf of the higher-ranking animal when forming a coalition (but see Jennings 2009 for the opposite result among deer). Of course, selective intervention could occur simply because individuals intervene on behalf of winners. Arguing against this view, however, hyenas sometimes redirect aggression toward other, previously uninvolved individuals after a fight. When this occurs, they are more likely to attack a relative of their former opponent (Engh et al. 2005). Hyenas also seem to make transitive inferences about other individuals' dominance ranks (Engh et al. 2005). As already noted, pinyon jays recognize other individuals' dominance ranks and use that information to make transitive inferences about the ranks of unfamiliar individuals (Bond et al. 2003; Paz y Miño et al. 2004). Striking examples of social cognition have even been found in monogamous birds (e.g., Peake et al. 2002) and fish (Oliveira et al. 1998; Grosenick et al. 2007), where individuals "eavesdrop" on competitive interactions and remember the identities of winners and losers. These data refute the conclusion that there is a simple causal relation between large group size and knowledge of other animals' relations.

To summarize, while nonhuman primates have a higher ICC than other vertebrates, we cannot yet conclude that this difference has evolved as a result of greater social complexity. Primate societies are complex, and primates' knowledge of each other's social relationships is increasingly well documented, but we cannot yet demonstrate that either their societies or their social knowledge is more complex than that of other species.

The Role of Social Cognition in Primate Evolution

The social intelligence hypothesis argues that perception and cognition in primates have been shaped by the demands of social life—in other words, that primates recognize other animals' relations because such knowledge is essential to their survival and reproduction. This argument rests on two assumptions: first, that the formation of social relationships has a positive effect on reproductive success; and second, that in order to form the most adaptive social bonds, individuals must know about other animals' relations. The first assumption is now well supported; the second is more speculative.

As Silk demonstrates in chapter 24 of this volume, we now have good evidence from studies of baboons and rhesus macaques that the formation of close, enduring social relationships in females reduces stress (Beehner et al. 2005; Engh et al. 2006a, b; Crockford et al. 2008; Wittig et al. 2008; Brent et al. 2011), increases infant survival (Silk et al. 2003, 2009), and increases females' longevity (Silk et al. 2010b). Similar benefits have been documented in horses (Cameron et al. 2009) and dolphins (Frere et al. 2010). Among males, the formation of close, enduring social relationships increases reproductive success in chimpanzees (Nishida & Hosaka 1996; Constable et al. 2001; Boesch 2009), Assam macaques (*Macaca assamensis*, Shülke et al. 2010) and dolphins (Kopps et al. 2010). All of these data support the assumption that close social relationships lead to improved reproductive success.

With regard to social knowledge, it is now clear that individuals in many species observe other animals' interactions and use the information they acquire about others to adjust their own behavior (see Cheney 2011 for review). Is such voyeurism adaptive? We do not yet know, because thus far no study has documented individual differences in social knowledge and linked them to differences in survival or reproduction. At this point, we can only propose that individuals must know as much as possible about other animals' relations—that is, they must have a sophisticated understanding of the individuals in their group, their long-term associations, their short-term bonds, and the motivations that underlie them—if they are to form the social relationships that return the greatest benefit. Social cognition, we suggest, is therefore essential to survival and reproduction (Cheney & Seyfarth 2007).

Summary and Conclusions

Primates live in groups characterized by differentiated social relationships: animals do not interact at random (chapter 23, this volume). Their social environment is thus characterized by predictable patterns of interaction—statistical regularities that an individual must recognize if she is to predict other animals' behavior. Studies of social knowledge demonstrate that individuals do recognize the relationships that exist among others. They recognize other animals' dominance ranks and the close, enduring bonds formed among kin. They also combine their knowledge of kinship and rank, grouping animals simultaneously along two dimensions. Field experiments indicate that baboons recognize when a vocalization is directed at them, thus

suggesting that they attribute motives to others. Further, when baboons hear an aggressive vocalization from one individual and a submissive vocalization from another, they respond as if they assume that one call has caused the other. Their social knowledge thus includes some recognition of causality.

It remains unclear whether primates' social knowledge can be explained as simply the "scaling up" of simple conditioning mechanisms like those found in laboratory studies. Nor do we do know whether primate social knowledge is qualitatively different from that found in other species; although nonhuman primates have a higher index of cranial capacity than other vertebrates, we cannot yet conclude that this difference has evolved as a result of greater social complexity. Finally, there is at present no clear, explicit link between social knowledge and reproductive success. We can only propose that individuals must know as much as possible about other individuals' relations if they are to form the social relationships that return the greatest benefit.

References

Aureli, F., Cozzolino, R., Cordischi, C., & Scucchi, S. 1992. Kin-oriented redirection among Japanese macaques: An expression of a revenge system? *Animal Behaviour* 44, 283–291.

Aureli, F., Fraser, O. N., Schaffner, C. M. & Schino, G., this volume. The regulation of social relationships.

Aureli, F., Schaffner, C. M., Boesch, C., Bearder, S. K., Call, J. Chapman, C. A., Connor, R., Di Fiore, A. Dunbar, R. I. M., Henzi, S. P., Holekamp, K. Korstjens, A. H., Layton, R., Lee, P., Lehman, J., Manson, J. H., Ramos-Fernandez, G., Strier, K. B. & van Schaik, C. P. 2008. Fission-fusion dynamics. *Current Anthropology* 49, 627–654.

Bachmann, C. & Kummer, H. 1980. Male assessment of female choice in hamadryas baboons. *Behavioral Ecology and Sociobiology* 6, 315–321.

Barton, R. A., & Dunbar, R. 1997. Evolution of the social brain. In *Machiavellian Intelligence II: Extensions and Evaluations* (ed. by Whiten, A. & Byrne, R. W.), 240–263. Cambridge: Cambridge University Press.

Beehner, J. C., Bergman, T. J., Cheney, D. L., Seyfarth, R. M. & Whitten, P. L. 2005. The effect of new alpha males on female stress in free-ranging baboons. *Animal Behaviour* 69, 1211–1221.

Belin, P. & Zattore, R. 2003. Adaptation to speaker's voice in right anterior temporal lobe. *NeuroReport* 14, 2105–2109.

Benney, K. S. & Braaten, R. F. 2000. Auditory scene analysis in estrildid finches (*Taeniopygia guttata* and *Lonchura striata domestica*): A species advantage for detection of conspecific song. *Journal of Comparative Psychology*, 114, 174–182.

Bergman, T. J., Beehner, J. C., Cheney, D. L., & Seyfarth, R. M. 2003. Hierarchical classification by rank and kinship in baboons. *Science* 302, 1234–1236.

Boesch, C. 2009. *The Real Chimpanzee: Sex Strategies in the Forest*. Cambridge: Cambridge University Press.

Bond, A. B., Kamil, A. C., & Balda, R. P. 2003. Social complexity and transitive inference in corvids. *Animal Behaviour* 65, 479–487.

Brent, L. J. N., Semple, S., Dubuc, C., Heistermann, M. & MacLarnon, A. 2011. Social capital and physiological stress levels in free-ranging adult female rhesus macaques. *Physiology and Behavior* 102, 76–83.

Call, J., Aureli, F. & de Waal, F. B. M. 2002. Post-conflict third-party affiliation in stumptailed macaques. *Animal Behaviour* 63, 209–216.

Cameron, E. Z., Setsaas, T. H., & Linklater, W. L. 2009. Social bonds between unrelated females increase reproductive success in feral horses. *Proceedings of the National Academy of Sciences USA* 106, 13850–13853.

Chapais, B. 2001. Primate nepotism: What is the explanatory value of kin selection? *International Journal of Primatology* 22, 203–229.

Cheney, D. L. 1983. Extra-familial alliances among vervet monkeys. In *Primate Social Relationships* (ed. by Hinde, R. A.), 278–286. Oxford: Blackwell Scientific.

———. 2011. The extent and limit of cooperation in animals. *Proceedings of the National Academy of Sciences, USA* 108, 10902–10909.

Cheney, D. L. & Seyfarth, R. M. 1980. Vocal recognition in free-ranging vervet monkeys. *Animal Behaviour* 28, 362–367.

———. 1986. The recognition of social alliances among vervet monkeys. *Animal Behaviour* 34, 1722–1731.

———. 1989. Reconciliation and redirected aggression in vervet monkeys. *Behaviour* 110, 258–275.

———. 1990. *How Monkeys See the World*. Chicago: University of Chicago Press.

———. 1997. Reconciliatory grunts by dominant female baboons influence victims' behaviour. *Animal Behaviour* 54, 409–418.

———. 1999. Recognition of other individuals' social relationships by female baboons. *Animal Behaviour* 58, 67–75.

———. 2005. Social complexity and the information acquired during eavesdropping by primates and other animals. In *Animal Communication Networks* (ed. by McGregor, P. K.), 583–603. Cambridge: Cambridge University Press.

———. 2007. *Baboon Metaphysics*. Chicago: University of Chicago Press.

Cheney, D. L., Seyfarth, R. M. & Silk, J. B. 1995a. The responses of female baboons to anomalous social interactions: Evidence for causal reasoning? *Journal of Comparative Psychology* 109, 134–141.

———. The role of grunts in reconciling opponents and facilitating interactions among adult female baboons. *Animal Behaviour* 50, 249–257.

Connor, R. C. 2007. Complex alliance relationships in bottlenose dolphins and a consideration of selective environments for extreme brain size evolution in mammals. *Philosophical Transactions of the Royal Society B* 362, 587–602.

Connor, R. C. & Mann, J. 2006. Social cognition in the wild: Machiavellian dolphins? In *Rational Animals* (ed. by Hurley, S. & Nudds, M.), 329–370. Oxford: Oxford University Press.

Connor, R. C., Mann, J., Tyack, P. L. & Whitehead, H. 1998. Social evolution in toothed whales. *Trends in Ecology and Evolution* 13, 228–232.

Constable, J., Ashley, M., Goodall, J., Pusey, A. E. 2001. Noninvasive paternity assignment in Gombe chimpanzees. *Molecular Ecology* 10, 1279–1300.

Cords, M. 1992. Post-conflict reunions and reconciliation in long-tailed macaques. *Animal Behaviour* 44, 57–61.

Crockford, C., Wittig, R. M., Seyfarth, R. M. & Cheney, D. L. 2007. Baboons eavesdrop to deduce mating opportunities. *Animal Behaviour* 73, 885–890.

Crockford, C., Wittig, R. M., Whitten, P., Seyfarth, R. M. & Cheney, D. L. 2008. Social stressors and coping mechanisms in wild female baboons (*Papio hamadryas ursinus*). *Hormones and Behavior* 53, 254–265.

Das, M. 2000. Conflict management via third parties. In *Natural Conflict Resolution* (ed. by Aureli, F. & de Waal, F. B. M.), 263–280. Berkeley: University of California Press.

Dunbar, R. 2000. Causal reasoning, mental rehearsal, and the evolution of primate cognition. In *The Evolution of Cognition. Vienna Series in Theoretical Biology* (ed. by Heyes, C. & Huber, L.), 205–219. Cambridge, MA: MIT Press.

———. 2003. Why are apes so smart? In *Primate Life Histories and Socioecology* (ed. by Kappeler, P. M. & Pereira, M. E.), 285–298. Chicago: University of Chicago Press.

Eifuku, S., De Souza, W. C., Tamura, R., Nishijo, H. & Ono, T. 2004. Neuronal correlates of face identification in the monkey anterior temporal cortical areas. *Journal of Neurophysiology* 91, 358–371.

Emery, N. J. & Perrett, D. I. 2000. How can studies of the monkey brain help us understand "theory of mind" and autism in humans? In *Understanding Other Minds: Perspectives from Developmental Cognitive Neuroscience*, 2nd ed. (ed. by Baron-Cohen, S. Tager-Flusberg, H. & Cohen, D.), 279–310. Oxford: Oxford University Press.

Engh, A. L., Siebert, E. R., Greenberg, D. A. & Holekamp, K. 2005. Patterns of alliance formation and postconflict aggression indicate spotted hyenas recognize third-party relationships. *Animal Behaviour* 69, 209–217.

Engh, A. L., Beehner, J. C., Bergman, T. J., Whitten, P. L., Hoffmeier, R. R., Seyfarth, R. M. & Cheney, D. L. 2006a. Behavioural and hormonal responses to predation in female chacma baboons (*Papio hamadryas ursinus*). *Proceedings of the Royal Society of London B* 273, 707–712.

———. 2006b. Behavioral and hormonal responses to predation in female chacma baboons (*Papio hamadryas ursinus*). *Proceedings of the Royal Society London B* 273, 707–712.

Engh, A. L., Hoffmeier, R. R., Cheney, D. L. & Seyfarth, R. M. 2006c. Who, me? Can baboons infer the target of vocalizations? *Animal Behaviour* 71, 381–387.

Engh, A. L., Beehner, J. C., Bergman, T. J., Whitten, P. L., Hoffmeier, R. R., Seyfarth, R. M. & Cheney, D. L. 2006d. Female hierarchy instability, male immigration, and infanticide increase glucocorticoid levels in female chacma baboons. *Animal Behaviour* 71, 1227–1237.

Feh, C. 1999. Alliances and reproductive success in Camargue stallions. *Animal Behaviour* 55, 705–713.

Fischer, J., Kitchen, D. M., Seyfarth, R. M. & Cheney, D. L. 2004. Baboon loud calls advertise male quality: Acoustic features and their relation to rank, age, and exhaustion. *Behavioral Ecology and Sociobiology* 56, 140–148.

Frere, C. H., Krützen, M., Mann, J., Connor, R. C., Bejder, L., Sherwin, W. B. 2010. Social and genetic interactions drive fitness variation in a free-living dolphin population. *Proceedings of the National Academy of Sciences USA* 107, 19949–19954.

Gerhardt, H. C. & Bee, M. A. 2006. Recognition and localization of acoustic signals. In *Hearing and Sound Communication in Amphibians*, Vol. 28 (ed. by Narins, P. M., Feng, A. S., Fay, R. R. & Popper, A. N.), 128–146. New York: Springer-Verlag.

Ghazanfar, A. A. & Santos, L. R. 2004. Primate brains in the wild: The sensory bases for social interactions. *Nature Review Neurosciences* 5, 603–616.

Gomes C. M., Mundry, R. & Boesch, C. 2009. Long-term reciprocation of grooming in wild West African chimpanzees. *Proceedings of the Royal Society London B* 276, 699–706.

Grosenick, L., Clement, T. S. & Fernald, R. 2007. Fish can infer social rank by observation alone. *Nature* 446, 102–104.

Harcourt, A. H. 1988. Alliances in contests and social intelligence, In *Machiavellian Intelligence: Social Expertise and the Evolution of Intellect in Monkeys, Apes, and Humans* (ed. by Byrne, R. W. & Whiten, A. A.). Oxford: Oxford University Press.

Hefner, H. E. & Hefner, R. S. 1984. Temporal lobe lesions and perception of species-specific vocalizations by macaques. *Science* 226, 75–76.

Henzi, S. P. and Barrett, L. 2007. Coexistence in female-bonded primate groups. *Advances in the Study of Behavior* 37, 43–81.

Heyes, C. M. 1994. Social cognition in primates. In *Animal Learning and Cognition* (ed. by Macintosh, N. J.), 281–305. New York: Academic Press.

Humphrey, N. K. 1976. The social function of intellect. In *Growing Points in Ethology* (ed. by Bateson, P. & Hinde, R. A.), 303–318. Cambridge: Cambridge University Press.

Jellema, T., Baker, C. I., Wicker B. & Perrett, D. I. 2000. Neural representation for the perception of the intentionality of actions. *Brain and Cognition* 44, 280–302.

Jennings, D., Carlin, C. M. & Gammell, M. P. 2009. A winner effect supports third-party intervenion during fallow deer, *Dama dama*, fights. *Animal Behaviour* 77, 343–348.

Jolly, A. 1966. Lemur social behavior and primate intelligence. *Science* 153, 501–506.

Judge, P. 1982. Redirection of aggression based on kinship in a captive group of pigtail macaques. *International Journal of Primatology* 3, 301.

———. 1991. Dyadic and triadic reconciliation in pigtailed macaques (*Macaca nemestrina*). *American Journal of Primatology* 23, 225–237.

Judge, P. & Mullen, S. H. 2005. Quadratic post-conflict affiliation among bystanders in a hamadryas baboon group. *Animal Behaviour* 69, 1345–1355.

Kawai, M. 1958. On the system of social ranks in a natural group of Japanese monkeys. *Primates* 1, 11–48.

Kitchen, D. M., Seyfarth, R. M., Fischer, J. & Cheney, D. L. 2003. Loud calls as an indicator of dominance in male baboons (*Papio cynocephalus ursinus*). *Behavioral Ecology and Sociobiology* 53, 374–384.

Kitchen, D. M., Cheney, D. L. & Seyfarth, R. M. 2005. Male chacma baboons (*Papio hamadryas ursinus*) discriminate loud call contests between rivals of different relative ranks. *Animal Cognition* 8, 1–6.

Kopps, A. M., Connor, R. C., Sherwin, W. B., Krützen, M. 2010. Direct and indirect fitness benefits of alliance formation in male bottlenose dolphins. Poster for annual meeting of the International Society for Behavioural Ecology.

Koyama, N. 1967. On dominance rank and kinship of a wild Japanese monkey troop in Arashiyama. *Primates* 8, 189–216.

Kummer, H. 1968. *Social Organization of Hamadryas Baboons*. Chicago: University of Chicago Press.

Kutsukake, N. & Castles, D. L. 2004. Reconciliation and post-conflict third-party affiliation among wild chimpanzees in the Mahale Mountains, Tanzania. *Primates* 45, 147–165.

Maclean, E. L., Merrit, D. J. & Brannon, E. 2008. Social complexity predicts transitive reasoning in prosimians primates. *Animal Behaviour* 76, 479–486.

Martin, R. D. 1990. *Primate Origins and Evolution: A Phylogenetic Reconstruction*. Princeton, NJ: Princeton University Press.

Mitani, J. C. 2006. Reciprocal exchanges in chimpanzees and other primates. In *Cooperation in Primates and Humans* (ed. by Kappeler, P. M. & van Schaik, C.), 107–119. Berlin: Springer-Verlag.

———. 2009. Male chimpanzees form enduring and equitable social bonds. *Animal Behaviour* 77, 633–40.

Mitani, J. C., Watts, D. P. & Amsler, S. J. 2010. Lethal intergroup aggression leads to territorial expansion in wild chimpanzees. *Current Biology* 20, R507–508.

Muller, M. & Mitani, J. C. 2005. Conflict and cooperation in wild chimpanzees. *Advances in the Study of Behavior* 35, 275–331.

Nishida, T. & Hosaka, K. 1996. Coalition strategies among adult male chimpanzees of the Mahale Mountains, Tanzania. In *Great Ape Societies* (ed. by McGrew, W. C., Marchant, L. & Nishida, T.), 114–134. Cambridge: Cambridge University Press.

O'Brien, T. G. & Robinson, J. 1993. Stability of social relationships in female wedge-capped capuchin monkeys. In *Juvenile Primates* (ed. by Pereira, M. E. & Fairbanks, L. A.), 197–210. Oxford: Oxford University Press.

Oliveira, R. F., McGregor, P. K. & Latruffe, C. 1998. Know thine enemy: Fighting fish gather information from observing conspecific interactions. *Proceedings of the Royal Society, London B* 265, 1045–1049.

Packer, C. & Pusey, A. E. 1982. Cooperation and competition within coalitions of male lions: Kin selection or game theory? *Nature* 296, 740–742.

Paz-y-Miño, G., Bond, A. B., Kamil, A. C. & Balda, R. P. 2004. Pinyon jays use transitive inference to predict social dominance. *Nature* 430, 778–782.

Peake, T. M., Terry, A. M. R., McGregor, P. K. & Dabelsteen, T. 2002. Do great tits assess rivals by combining direct experience with information gathered by eavesdropping? *Proceedings of the Royal Society, London B* 269, 1925–1929.

Penn, D. C., Holyoak, K. & Povinelli, D. J. 2008. Darwin's mistake: Explaining the discontinuity between human and non-human minds. *Behavioral & Brain Sciences* 31, 109–178.

Percz-Barberia, F. J. & Gordon, I. J. 2005 Gregariousness increases brain size in ungulates. *Oecologia* 145, 41–52.

Perrett, D. I., Harries, M. H, Mistlin, A. J. & Hietanen, J. K. 1990. Social signals analyzed at the single cell level: Someone is looking at me, something touched me, something moved. *International Journal of Comparative Psychology* 4, 25–55.

Perrett, D. I., Hietanen, J. K., Oram, M. W. & Benson, P. J.

1992. Organization and function of cells responsive to faces in the temporal cortex. *Philosophical Transactions of the Royal Society of London B* 335, 23–30.

Perry, S. 2003. Coalitionary aggression in white-faced capuchins: In *Animal Social Complexity* (ed. by de Waal, F. B. M. & Tyack, P.), 111–114. Cambridge, MA: Harvard University Press.

Perry, S., Barrett, H. C. & Manson, J. 2004. White-faced capuchin monkeys show triadic awareness in their choice of allies. *Animal Behaviour* 67, 165–170.

Perry, S., Manson, J., Muniz, L., Gros-Louis, J. & Vigilant, L. 2008. Kin-biased social behaviour in wild adult female white-faced capuchins, *Cebus capucinus. Animal Behaviour* 76, 187–99.

Petersen, M. R., Beecher, M. D., Zoloth, S. R., Moody, D. B. & Stebbins, W. C. 1978. Neural lateralization of species-specific vocalizations by Japanese macaques (*Macaca fuscata). Science* 202, 324–327.

Petkov, C. I., Kayser, C., Steudel, T., Whittingstall, K., Augath, M. & Logothetis, N. 2008. A voice region in the monkey brain. *Nature Neuroscience* 11, 367–374.

Poremba, A., Malloy, M., Saunders, R. C., Carson, R. E., Herskovitch, P. & Mishkin, M. 2004. Species-specific calls evoke asymmetric activity in the monkey's temporal poles. *Nature* 427, 448–451.

Premack, D. & Woodruff, G. 1978. Does the chimpanzee have a theory of mind? *Behavioral and Brain Sciences* 4, 515–526.

Rendall, D., Rodman, P. S., & Emond, R. E. 1996. Vocal recognition of individuals and kin in free-ranging rhesus monkeys. *Animal Behaviour* 51, 1007–1015.

Rizzolatti, G. & Craighero, L. 2004. The mirror-neuron system. *Annual Review of Neuroscience* 27, 169–192.

Sade, D. S. 1972. Sociometrics of *Macaca mulatta*. I. Linkages and cliques in grooming matrices. *Folia Primatologica* 18, 196–223.

Schino, G. 2001. Grooming, competition and social rank among female primates: A meta-analysis. *Animal Behaviour* 62, 265–271.

Schino, G., Tiddi, B. & Polizzi di Sorrentino, E. 2006. Simultaneous classification by rank and kinship in Japanese macaques. *Animal Behaviour* 71, 1069–1074.

Schino, G., Polizzi di Sorrentino, E. & Tiddi, B. 2007. Grooming and coalitions in Japanese macaques (*Macaca fuscata*): Partner choice and the time frame of reciprocation. *Journal of Comparative Psychology* 121, 181–188.

Schino, G., Di Giuseppe, F. & Visalberghi, E. 2009. Grooming, rank, and agonistic support in tufted capuchin monkeys. *American Journal of Primatology* 71, 101–105.

Schino, G. & Aureli, F. 2008. Trade-offs in primate grooming reciprocation: Testing behavioral flexibility and correlated evolution. *Biological Journal of the Linnean Society of London* 95, 439–446.

Schino, G. & Aureli, F. 2009. Reciprocal altruism in primates: Partner choice, cognition, and emotions. *Advances in the Study of Behavior* 39, 45–69.

Schusterman, R. J. & Kastak, D. A. 1993. A California sea lion (*Zalophus californianus*) is capable of forming equivalence relations. *Psychological Record* 43, 823–839.

———. 1998. Functional equivalence in a California sea lion: Relevance to animal social and communicative interactions. *Animal Behaviour* 55, 1087–1095.

Schülke, O., Bhagavatula, J., Vigilant, L. & Ostner, J. 2010. Social bonds enhance reproductive success in male macaques. *Current Biology* 20, 1–4.

Seyfarth, R. M. 1980. The distribution of grooming and related behaviors among adult female vervet monkeys. *Animal Behaviour* 28, 798–813.

Seyfarth, R. M. and Cheney, D. L. 2012. The evolutionary origins of friendship. *Annual Review of Psychology* 63, 153–177.

Shettleworth, S. 2010. *Cognition, Evolution, and Behavior*, 2nd edition. Oxford: Oxford University Press.

Silk, J. B. 1993. Does participation in coalitions influence dominance relationships among male bonnet macaques? *Behaviour* 126, 171–189.

Silk, J. B., Cheney, D. L. & Seyfarth, R. M. 1996. The form and function of post-conflict interactions between female baboons. *Animal Behaviour* 52, 259–268.

Silk, J. B., Alberts, S. C. & Altmann, J. 2003. Social bonds of female baboons enhance infant survival. *Science* 302, 1231–1234.

Silk, J. B., Altmann, J. & Alberts, S. C. 2006a. Social relationships among adult female baboons (*Papio cynocephalus*). I. Variation in the strength of social bonds. *Behavioral Ecology and Sociobiology* 61, 183–195.

Silk, J. B., Altmann, J. & Alberts, S. C. 2006b. Social relationships among adult female baboons (*Papio cynocephalus*). II: Variation in the quality and stability of social bonds. *Behavioral Ecology and Sociobiology* 61, 197–204.

Silk, J. B., Beehner, J. C., Bergman, T., Crockford, C., Engh, A. L., Moscovice, L., Wittig, R. M., Seyfarth, R. M. & Cheney, D. L. 2009. The benefits of social capital: Close bonds among female baboons enhance offspring survival. *Proceedings of the Royal Society London B* 276, 3099–3104.

Silk, J. B., Beehner, J. C., Bergman, T., Crockford, C., Engh, A. L., Moscovice, L., Wittig R. M., Seyfarth, R. M. & Cheney, D. L. 2010a. Female chacma baboons form strong, equitable, and enduring social bonds. *Behavioral Ecology and Sociobiology* 64, 1733–47.

———. 2010b. Strong and consistent social bonds enhance the longevity of female baboons. *Current Biology* 20, 1359–61.

Slocombe, K., Townsend, S. & Zuberbuhler, K. 2009. Wild chimpanzees (*Pan troglodytes*) distinguish between different scream types: Evidence from a playback study. *Animal Cognition* 12, 441–449.

Slocombe, K. & Zuberbuhler, K. 2005. Agonistic screams in wild chimpanzees (*Pan troglodytes*) vary as a function of social role. *Journal of Comparative Psychology* 119, 67–77.

———. 2007. Chimpanzees modify recruitment screams as a function of audience composition. *Proceedings of the National Academy of Sciences, USA* 104, 17228–17233.

Smuts, B. 1985. *Sex and Friendship in Baboons*. New York: Aldine.

Stammbach, E. 1987. Desert, forest, and montane baboons: Multilevel societies. In *Primate Societies* (ed. by Smuts, B. B. Cheney, D. L. Seyfarth, R. M. Wrangham, R. W. & Struhsaker, T. T.), 112–120. Chicago: University of Chicago Press.

Thierry, B. 2008. Primate socioecology, the lost dream of ecological determinism. *Evolutionary Anthropology* 17, 93–96.

Thompson, R. K. R. 1995. Natural and relational concepts in animals. In *Comparative Approaches to Cognitive Science* (ed.

by Roitblat, H. & Myers, J. A.), 175–224. Cambridge, MA: MIT Press.

Tomasello, M. & Call, J. 1997. *Primate Cognition*. Oxford: Oxford University Press.

Tsao, G. Y., Friewald, W. A., Knutsen, T. A., Mandeville, J. B. & Tootell, R. B. 2003. Faces and objects in macaque cerebral cortex. *Nature Neuroscience* 6, 989–995.

Tsao, G. Y., Friewald, W. A., Tootell, R. B. & Livingston, M. S. 2006. A cortical region consisting entirely of face-selective cells. *Science* 311, 670–674.

DeWaal, F. 1991. Rank distance as a central feature of rhesus monkey social organization: A sociometric analysis. *Animal Behaviour* 41, 383–395.

———. 1996. Conflict as negotiation. In *Great Ape Societies* (ed. by McGrew, W. C., Marchant L. & Nishida, T.), 159–172. Cambridge: Cambridge University Press.

———. 2008. Putting the altruism back into altruism: The evolution of empathy. *Annual Review of Psychology* 59, 279–300.

DeWaal, F. & Aureli, F. 1996. Consolation, reconciliation, and a possible difference between macaques and chimpanzees. In *Reaching into Thought: The Minds of the Great Apes* (ed. by Russon, A. E., Bard, K. A. & Parker, S. T.), 80–110. Cambridge: Cambridge University Press.

DeWaal, F. and Van Hoof, J. 1981. Side-directed communication and agonistic interactions in chimpanzees. *Behaviour* 77, 164–198.

Wasserman, E. A. & Astley, S. L. 1994. A behavioral analysis of concepts: Application to pigeons and children. In *Psychology of Learning and Motivation*, Vol. 31 (ed. by Medin, D. L.), 73–132. New York: Academic Press.

Watts, D. 1991. Harrassment of immigrant female mountain gorillas by resident females. *Ethology* 89, 135–153.

———. 1998. Coalitionary mate-guarding by wild chimpanzees at Ngogo, Kibale National Park, Uganda. *Behavioral Ecology and Sociobiology* 44, 43–55.

Wittig, R. M. 2010. The function and cognitive underpinnings of post-conflict affiliation in wild chimpanzees. In *The Mind of the Chimpanzee* (ed. by Lonsdorf, E. V., Ross, S. R. & Matsuzawa, T.), 208–219. Chicago: University of Chicago Press.

Wittig, R. M. & Boesch, C. 2003. The choice of post-conflict interactions in wild chimpanzees (*Pan troglodytes*). *Behaviour* 140, 1527–1559.

———. 2005. How to repair relationships: Reconciliation in wild chimpanzees (*Pan troglodytes*). *Ethology* 111, 736–763.

———. 2010. Receiving post-conflict affiliation from the enemy's friend reconciles former opponents. *PLoS One* 5, e13995.

Wittig, R. M., Crockford, C., Wikberg, E., Seyfarth, R. M. & Cheney, D. L. 2007a. Kin-mediated reconciliation substitutes for direct reconciliation in female baboons. *Proceedings of the Royal Society London B* 274, 1109–1115.

Wittig, R. M., Crockford, C., Seyfarth, R. M. & Cheney, D. L. 2007b. Vocal alliances in chacma baboons, *Papio hamadryas ursinus*. *Behavioral Ecology and Sociobiology* 61, 899–909.

Wittig, R. M., Crockford, C., Lehmann, J. Whiten, P. L., Seyfarth, R. M. & Cheney, D. L. 2008. Focused grooming networks and stress alleviation in wild female baboons. *Hormones and Behavior* 54, 170–177.

Chapter 29 Communication Strategies

Klaus Zuberbühler

THE AIM OF this chapter is to review current evidence of signaling behavior in a wide range of primate species within the three major communication channels: olfaction, vision, and sound. A range of methodologies is discussed, as are the current theories, major hypotheses, ongoing controversies, and directions of future research. In selecting case studies, the bias has been towards more recent examples, especially ones published after *Primate Societies* (Smuts et al. 1987). Material on human communication is also presented, though mainly opportunistically when it is relevant for comparisons, or because it is associated with key scientific advances.

Communication is central to a number of scientific disciplines; definitions of communication vary accordingly, and continue to cause controversy (Rendall et al. 2009). At the broadest level, communication is defined as the process of transferring information from one source to another. Strictly, this definition also encompasses instances in which individuals provide information not by design or intention, but as accidental by-products of other activities. For example, the movement sounds produced by a fleeing forest antelope can be very meaningful to nearby monkeys as an indicator of predator presence (chapter 8, this volume). Although the monkeys learn something important from this acoustic stimulus, it does not involve a specifically designed signal or an intention to inform. A stricter definition of communication thus requires the signaler to "encode" a message into a specific modality with a purpose of making it available to a receiver. The receiver then "decodes" the message, and this can change its inner state and behavior. In

practice, it is not always easy to determine whether design or intention is involved—that is, whether the interaction qualifies as communication in the stricter sense or as mere accidental provisioning of information (Hauser 1996). Eavesdropping—for example, cases of individuals responding to signals of another species—illustrates the difficulties with these definitions (McGregor 2005). In this chapter I adopt a broad, inclusive definition of communication, even if the role of the signaler as an active provider of information is uncertain.

Olfactory Communication

Among the different modalities, chemical communication by olfaction and gustation has received relatively scant empirical attention (Heymann 2006). Although olfactory communication is typically associated with strepsirrhine primates, which have specialized glands for scent marking (Zeller 1987; chapter 2, this volume), it is likely to play a role in all groups of primates (Matsumoto-Oda et al. 2007). For example, common squirrel monkeys (*Saimiri sciureus*) tested with odorants representing different chemical classes showed olfactory sensitivity comparable to that of dogs (Laska et al. 2000), and Southern pig-tailed macaques (*Macaca nemestrina*) have displayed an equally high olfactory sensitivity (Hübener & Laska 2001). Although its status as a signal is not always clear, olfactory information plays an important role in a variety of contexts including foraging, sexual interactions, territorial defense, individual recognition, mother-offspring bonding, and cooperative behaviors (Zeller 1987). For example, primates often smell the mouths of more experienced group members and adjust their foraging behavior accordingly (Chauvin & Thierry 2005; Laidre 2009), or signal ownership of resources by scent marking (e.g., Kappeler 1998).

The main and accessory olfactory bulbs are the two sensory systems involved in transforming chemicals present in the environment into neural activity. The former detects volatile chemical compounds, while the latter responds to fluid-phase chemicals. Across primates, the relative size of the main olfactory bulb correlates predominantly with activity period and diet, while the size of the accessory olfactory bulb correlates with group size and mating systems (Barton 2006). In terrestrial vertebrates, the accessory bulb receives input from the vomeronasal organ, which in most primates is present only in rudimentary form. In contrast, strepsirrhines and some New World monkeys have well-developed vomeronasal organs, which suggests that for those animals chemical signals play a more central role, especially during social communication (Smith et al. 2001). There is genetic evidence that a progressively greater number of olfactory genes have become pseudogenes in Old World monkeys, apes, and humans, a deterioration process probably related to the advent of trichromatic vision (Gilad et al. 2004).

Active Scent Marking

Active scent marking provides good evidence for olfactory communication, especially if it triggers specific responses, such as investigation or countermarking (Kappeler 1998). Scent marking is relatively common in strepsirrhines and callitrichines, and individuals often possess specialized glands on the head, sternum, abdomen, forelegs, and anogenital region (Zeller 1987). In some primates, urine appears to function as a carrier of olfactory signals, as is evidenced by special glands secreting into the urethra or anal canal. Scent marking can be part of various biological functions. For example, in male Milne-Edwards sifakas (*Propithecus edwardsi*), scent marking appears to serve in territorial defense and advertisement of social status (Pochron et al. 2005). In Verreaux's sifakas (*Propithecus verreauxi*), females scent mark resources, while males scent mark during interactions with neighboring groups and as part of mate guarding (Lewis 2005). In ring-tailed lemurs (*Lemur catta*), scents vary individually and seasonally (Scordato et al. 2007), and these differences are discriminated by other group members (Palagi & Dapporto 2007). In particular, males attend to scents of estrous females and to familiar, dominant animals while female responses depend on their own reproductive state (Scordato & Drea 2007). In New World monkeys, golden lion tamarins (*Leontopithe-*

cus rosalia) use scent marks in relation to food resources, to assert their social position, and in response to neighboring individuals (Miller et al. 2003). In common marmosets (*Callithrix jacchus*), scent marking is common during foraging with peaks during twilight hours (de Souza et al. 2006); in this species, scents are also individually discriminated, due to a unique ratio of highly volatile chemicals (Smith 2006). In Nancy Ma's owl monkeys (*Aotus nancymaae*), a socially monogamous nocturnal New World primate, scent samples also differ with the individual's gender, age, and family membership (Macdonald et al. 2008). In white-faced sakis (*Pithecia pithecia*), scents are deposited during specific throat and chest rubbing, a mainly male behavior that peaks during breeding periods and courtship (Setz & Gaspar 1997). In common woolly monkeys (*Lagothrix lagotricha*), scent marking is associated with male sexual activity and intergroup encounters (Di Fiore et al. 2006). In Old World monkeys, apes, and humans, scent-marking is less commonly reported, although scent glands have also been described, for instance, in male mandrills (*Mandrillus sphinx*, Setchell & Dixson 2001).

Self-Anointment

Further evidence for olfactory communication comes from self-anointing behavior, the application of scent-bearing material onto the body. Free-ranging male Geoffroy's spider monkeys (*Ateles geoffroyi*) rub a mix of saliva and plant material on their sternal and axillary body regions. This behavior seems to function not in repelling insects or mitigating skin infections, but in social communication (Laska et al. 2007). White-faced capuchins (*Cebus capucinus*) also rub urine onto their bodies, a behavior that is particularly common in high-status males (Campos et al. 2007). In humans, self-anointing occurs in the form of applying fragrances. In one study, human females rated male faces as more attractive in the presence of pleasant odors than in the presence of unpleasant ones (Dematte et al. 2007). Furthermore, people who are similar in their MHC genotypes (a genomic region involved in immune function) express similar preferences for perfume ingredients, suggesting that self-anointing behavior functions by revealing immunogenetic information (Milinski & Wedekind 2001) or by mimicking MHC-related odor cues (Milinski 2006).

Pheromones

Pheromones are usually defined as externally secreted chemical substances that influence the physiology or behavior of conspecifics, with the underlying assumption that the substance has evolved specifically to function as an olfactory communication signal. The classic example of a pheromone is Bombykol, an unsaturated straight-chain alcohol excreted by female silkworm moths (*Bombyx mori*) to attract males over large distances (Butenandt et al. 1959). Although primates perceive and respond to a range of chemicals emitted by conspecifics, it is often difficult to determine whether the substance has been released as part of an active strategy to influence a receiver or as a by-product of another metabolic process. The issue is particularly contentious for humans. In one early study, female college students were reported to synchronize their menstrual cycles when living in the same dormitory, with unknown olfactory signals suspected to play a key role (McClintock et al. 1971). Support for this idea was later provided by a study reporting that odorless compounds from the armpits of women shortened other women's menstrual cycles if collected prior to the most fertile days, whereas compounds collected later during the menstrual cycle had the opposite effect (Stern & McClintock 1998). However, subsequent studies failed to replicate some of these results, and there is a lack of evidence for menstrual synchrony in nonhuman primates (Schank 2001; Matsumoto-Oda & Kasuya 2005).

Despite these controversies, there is good evidence that humans and nonhuman primates are sensitive to sex steroids released into the environment by conspecifics (Laska et al. 2005; Snowdon et al. 2006). A particularly dramatic effect has been noted in gray mouse lemurs (*Microcebus murinus*). If females are kept in groups, they tend to overproduce sons, a tendency which appears to be driven by urinary odors from other females (Perret 1996). At the same time, dominant males suppress the sexual activity of other males, which also seems to be driven by chemicals present in the urine (Perret 1992). In stump-tailed macaques (*Macaca arctoides*), males exposed to female vaginal secretions collected at different times of the menstrual cycle showed differences in genital exploration, copulation, and coercive behavior (Cerda-Molina et al. 2006). In humans, some sex steroids are known to influence mood (Jacob & McClintock 2000), and women exposed to them rated men as more attractive than others in control conditions (Saxton et al. 2008). Women in the fertile phase of their cycle prefer the body odors of males who score high on social dominance, facial attractiveness, and body symmetry (Rikowski & Grammer 1999; Havlicek et al. 2005; Garver-Apgar et al. 2008). Similarly, men rate odors from women in their fertile phase as more attractive than odors collected at other times, thus suggesting that they can perceive ovulation via sex steroids (Singh & Bronstad 2001; Havlicek et al. 2006), a finding that contrasts with the more classic notion that ovulation is concealed in humans (see also chapter 20, this volume). Sex-specific responses to steroids are also reported from nonhuman primates, including southern pig-tailed macaques and squirrel monkeys (Laska et al. 2006), but the overall patterns of preferences are less clear-cut.

The Major Histocompatibility Complex

A considerable body of research has shown that humans differ in terms of individual preferences for body odors. As a general pattern, the body odors of individuals who are genetically dissimilar in their major histocompatibility complex (MHC) are preferred (Wedekind et al. 1995). This appears to be an adaptive strategy, as MHC heterozygosity is related not only to immunocompetence but also to attractiveness (Lie et al. 2008). Interestingly, however, MHC-related assortative mating may be restricted to individuals of European descent, and has not been found in African couples (Chaix et al. 2008).

The theory that fitness is positively related to diversity in immune genes has also been tested in nonhuman primates, yet so far only with limited success (see also chapter 16, this volume). In ring-tailed lemurs, males produce individually distinct odor profiles, and these apparently tell receivers something about the male's heterozygosity at the major histocompatibility complex (Knapp et al. 2006; Charpentier et al. 2008). Perhaps the best evidence for MHC-based olfactory communication comes from monogamous strepsirrhines: the pair-living Western fat-tailed dwarf lemurs (*Cheirogaleus medius*) and the gray mouse lemurs (Schwensow et al. 2008a, b). In the former, females prefer males that overlap less in terms of their own MHC supertypes than do randomly assigned males (MHC supertypes are groups of MHC alleles with functional similarities at their antigen binding sites). In addition, extrapair matings were observed preferentially in females that were paired with males with similar MHC supertypes, thus suggesting that females seek genetic diversity even after having established pair bonds (Schwensow et al. 2008a). In gray mouse lemurs, breeding pairs are more dissimilar in their MHC supertypes than randomly assigned pairs (Schwensow et al. 2008b). In rhesus monkeys (*Macaca mulatta*), MHC heterozygosity was the strongest predictor of male reproductive success (Widdig et al. 2004), but it was not clear whether this was the result of mate choice. As a general pattern, heterozygous MHC individuals are generally preferred as mating partners, while at the dyadic level MHC-dissimilar individuals are most attractive.

Gestural Communication

Muscle actions integral to basic biological functions sometimes function in visual communication. A well-known example is the bared-teeth display, characterized by marked upper lip retraction by the signaler, as described for long-tailed macaques (*Macaca fascicularis*, Angst 1974). In many primates, facial displays provide information on how the signaler evaluates an ongoing event—for example, by displaying disgust, aggressive intent, pleasure, or fear (Darwin 1872; Andrew 1962; van Hooff 1962; Eibl-Eibesfeldt 1972; Chevalier-Skolnikoff 1973). Receivers benefit from these signals as honest indicators of how the signalers assess the environment (palatability, danger) or the social interactions they anticipate (aggression, play, sex). The benefits to the signalers are less obvious, but must exist in terms of facilitating valuable social interactions or preventing costly ones such as escalated fights.

However, primates use not only facial muscle movements but often other parts of their bodies to generate gestural signals. For example, chimpanzees (*Pan troglodytes*) sometimes hunch their shoulders, and this is nearly always accompanied by hair (or pilo-) erection in contexts of aggression, courtship, and greeting (Nishida et al. 1999). Despite their conspicuousness, it is often surprisingly difficult to decide which elements of an individual's behavioral repertoire have signal character and thus qualify as gestures, a problem already encountered in the olfactory domain. Pika (2007a), for example, defines gesture largely in mentalistic and functional terms as "an expressive movement of limbs or head and body postures that appears to transfer a communicative message, such a request and/or a desired action/event, is directed to a recipient, and is accompanied by the following criteria; gazing at the recipient and/or waiting after the signal has been produced." Genty et al. (2009) offer a somewhat broader definition, including any behavior whose physical force is not sufficient to obtain a desired outcome, provided it is perceivable by a conspecific and shows signs of goal-directedness. But these criteria—physical force, desired outcome, perceptibility, and goal-directedness—require a fair amount of interpretation by the researcher, which can make comparisons between studies difficult.

Acquisition and Development

Facial displays appear to develop under strong genetic control in the different species. In the current literature they are thus often dismissed as uninteresting because the assumption is that individuals have comparably little control over their production, although this has not been tested formally. In bonobos (*Pan paniscus*), for example, the play face is used in highly flexible ways (Palagi 2008), suggesting that individuals may have considerable control over some facial signals (Genty et al. 2009). Interesting also is that newborn chimpanzees and rhesus macaques readily imitate some basic facial displays, such as lip smacking

Table 29.1.   Gestural repertoires in some primate species

Species	Gestures			
	Facial	Tactile	Visual	Auditory
Chimpanzee	10	15	19	4
Gorilla	*	11	16	6
Bonobo	*	8	11	1
Orangutan	3	14	15	0
Siamang	Graded	12	8	0
Macaque	13	12	10	2

* Not reported
Source: Call & Tomasello 2007

and tongue protrusion (Myowa 1996; Ferrari et al. 2006). Body gestures, in contrast, which are generally thought of as highly flexible, show some developmental features that can be oddly rigid. For instance, Pika (2007b) reported that in gorillas (*Gorilla gorilla*), most tactile and visual gestures are already fully developed in very young infants. Call and Tomasello (2007) have recently reviewed some of the gestural literature in which a number of unexpected patterns emerge (table 29.1). First, the gestural repertoire of some monkeys can fall within the apes' range (Hesler & Fischer 2007). Second, there are phylogenetic effects. For example, the Asian apes perform no auditory gestures, such as chest beating or buttress drumming, but relatively many tactile ones. Phylogenetic history could also explain differences in the contexts of production, with Asian apes gesturing mainly in agonistic and affiliative contexts and African apes doing so mainly during play, travel, or feeding (Call & Tomasello 2007).

The process by which gestures are acquired is not well understood. Tomasello (1996) has argued that apes acquire gestures primarily through a process termed "ontogenetic ritualization," in which two individuals shape each other's behaviors during repeated interactions. The receiver learns to anticipate the signaler's forthcoming behavior, which in turn allows the signaler to abbreviate it, and as a result the abbreviated behavior (or "intention movement"; Heinroth 1911) obtains signal character. An example is a chimpanzee mother's abbreviated lowering of her back to which the infant responds by clambering onto her, instead of having to be physically pulled to the right position. Interestingly, imitation and other social learning seem not to play an important role in gestural communication (Tomasello et al. 1994, 1997, see also chapter 31, this volume). Nevertheless, gestures have been of special interest to the debate about primate culture. At a number of sites, chimpanzees "leaf clip" by pulling a leaf repeatedly between their lips or teeth with one hand, thereby producing a conspicuous sound that attracts the attention of others (Nishida et al. 1999). This behavior is used in different contexts at different sites, and may thus be an example of cultural variation (Nishida 1980; de Waal & Seres 1997; Whiten et al. 1999). However, much of the observed variation has to do with the context in which the behavior is produced, not with the transmission process of the gesture itself.

Operant conditioning could also be involved in the acquisition of gestures. Although the operant conditioning hypothesis has not been explored specifically, it remains a distinct possibility that needs to be addressed. In operant conditioning an arbitrary behavior consistently triggers a favorable response in receivers, and will thus become reinforced (e.g., Tanner & Byrne 1993). As this is usually part of a social process, one might be prepared to accept it as a case of social learning. Another relevant finding is that in apes, at any given time, individuals often differ in their gestural repertoire. In the extreme, idiosyncratic gestures have been reported that are sometimes interpreted as individual "inventions" (Fossey 1983; Goodall 1986), but it is possible that conditioning has led to the emergence of such idiosyncratic gestures, generated by keen receivers and extinguished by reluctant ones, or that the gestural repertoire of individuals simply changes in unknown ways throughout their lifetime.

Flexibility

A key reason for studying gestural communication is to obtain insights into a signaler's underlying cognitive architecture. Gestural signals are well suited for this enterprise because individuals appear to have considerable voluntary control over some of them, in contrast to olfaction or even vocalizations. As it is difficult to operationalize volition, the focus has been on flexibility, assessed in terms of sensitivity to the audience, ability to produce sequences or combinations of gestures, or release from basic biological functions ("means-end dissociation," Bruner 1981; Call & Tomasello 2007).

Apes show some awareness of a communication process because they can modify their gestures to enhance their effectiveness. For instance, they use visual gestures primarily when a recipient is looking at them (see chapter 30, this volume), but use auditory and tactile gestures when the recipient is looking away (e.g., Russell et al. 2005; Poss et al. 2006). Some apes actively position themselves in the receiver's visual field before producing the gesture, thus suggesting that they understand how to obtain a desired response (Call & Tomasello 2007). Captive chimpanzees modify their gestures to direct a nonattentive experimenter to an object they want (Povinelli et al. 2003).

Gesture combinations are another focus of study. The ability to combine different kinds of semiotic elements is a fundamental development during language acquisition in humans (Greenfield et al. 2008). Apes occasionally produce gestures in sequences, although they often repeat the same signal multiple times (Tomasello et al. 1994), and when different gestures are assembled into a sequence they tend to belong to the same contextual class (Liebal et al. 2004). The current consensus, therefore, is that gesture sequences increase the likelihood of receiver responses but are not used to generate or convey novel meanings.

Another hallmark of flexibility is the use of the same signal in different contexts or vice versa. This is particularly characteristic of play, which can include a particularly diverse range of gestures, but it has also been seen in other contexts (Cartmill & Byrne 2008). This type of "means-end dissociation" is unusual in animal communication, where signals usually have strong links to very narrow biological functions or contexts (Pollick & de Waal 2007).

Finally, there are some records in the literature which indicate that ape gestures can refer to external objects or events in the signaler's environment (Savage-Rumbaugh et al. 1986; Pika & Mitani 2006). In one example, chimpanzees pointed towards desirable food, but only if the experimenter was present, thus suggesting that they possessed some understanding of the gesture's referential function (Leavens et al. 2004; see also chapter 27, this volume, fig. 27.9). In a recent laboratory study, Sumatran orangutans (*Pongo abelii*) were observed pointing for other apes to request out-of-reach objects, and these gestures often induced the partners to comply and hand over the requested object (Pele et al. 2009). In the wild, evidence for referential gesturing is weak. In one anecdotal report, a young male bonobo sitting in a tree was observed to produce calls while stretching his right arm "with a pointing index and ring finger" in the direction of two groups of human observers who sat camouflaged in the undergrowth 30 meters apart from each other, while turning his head toward his fellow group members. Three minutes later the male repeated this pointing and calling behavior, and it triggered approach by the others, who then looked towards the human observers (Vea & Sabater-Pi 1998). Of course these observations can be interpreted in other ways, but they are at least suggestive. In another more controlled field study with chimpanzees, the receiver of grooming was observed indicating a desired grooming location by "making a relatively loud and exaggerated scratching movement on a part of his body, which could be seen by his grooming partner" (Pika & Mitani 2006). This type of self-scratching was interpreted as a deictic gestural signal because of its seemingly exaggerated nature and its immediate effects on the partner's behavior.

The existence of "iconic" gestures is even more contentious than the previous cases of seemingly "indexical" gestures. Following Peirce (1932), an iconic gesture is characterized by its resemblance to the object or event it seeks to denote; it is a pantomime. This is in contrast to the previously discussed examples of potentially "indexical" gestures, which are characterized by their direct connection to the things they seek to denote. In symbolic gestures the link is completely arbitrary, determined by mere habit or rule. In captive bonobos and gorillas, individuals have been observed to seemingly depict an action a signaler wanted from them, for example by pantomiming it (Savage-Rumbaugh et al. 1977; Tanner & Byrne 1996). Another interesting case is that of male chimpanzees stretching their arms towards a desired female and then making sweeping movements toward themselves (van Hooff 1973). Although these reports are interesting, it is usually not clear whether the signals are the product of an intention to inform, or an attempt to directly reposition the other body.

Generally, examples of pointing and other indexical or iconic gestures are uncommon in apes, especially in the wild, and they are restricted to a narrow range of contexts. A theoretically interesting point is the degree to which they are communicative as opposed to being attempts to regulate executive behavior (Gomez 2007). Some observations suggest that apes do not fully appreciate the communicative nature of their gestures. For example, chimpanzees readily produce auditory gestures (banging, drumming) but, oddly, do not use them as attention-getters for subsequent visual gestures, which suggests that they do not understand these signals' strategic value (Call & Tomasello 2007).

Comprehension and Meaning

With sufficient training, nonhuman primates and other animals can learn a large number of arbitrary signals to communicate with humans symbolically (Premack & Premack 1972; Pepperberg 1999; Herman et al. 2001). For example, Kanzi, a male bonobo, can use more than 200 arbitrary lexigram symbols to communicate about objects, places, and activities (Savage-Rumbaugh & Lewin 1994; see also chapter 27, fig. 27.7). Whether this ability has any equivalents in the natural communication of apes is still largely unknown, partly because there is virtually no information on how primates comprehend naturally produced gestures by conspecifics. In human infants, two-year-olds can already interpret gestures as representations of objects or typical actions associated with these objects. For example, infants chose objects (hammer, brush, baby bottle, book) that an experimenter indicated by performing associated actions, such as hammering the floor with her fist (Tomasello et al. 1999; Striano et al. 2003).

Pointing has already been mentioned. This gesture is universal in humans, but to what degree nonhuman primates comprehend its meaning is part of an ongoing debate. One typical finding is that apes are often surprisingly unable to use human pointing to find hidden food items in object choice tasks (Tomasello 2008), something that domestic dogs have little difficulty with (Bräuer et al. 2006). This is remarkable also because many primates are good at gaze following, and take each other's gaze into account when making foraging decisions (Hare et al. 2000; Byrint 2004; Hauser et al. 2007). One explanation for this apparent lack of understanding has attributed it to a deeper inability to interpret signals of cooperative acts that involve someone wanting to share information (Tomasello 2008).

However, the inability of primates to understand pointing gestures is still far from accepted. Some additional studies suggest that apes do understand pointing, even without additional verbal or gaze cues (Peignot & Anderson 1999; Kumashiro et al. 2003; Pele et al. 2009). In a recent study, three species of apes were tested for their understanding of human-given cues in an object choice task. In contrast to previous studies, the spatial arrangements were altered to reduce the ambiguities of the human cues as well as the perceptual influence of the containers hiding the food. These minor alterations turned out to be highly effective in that the apes followed humans' pointing more than one would expect on the basis of chance (Mulcahy & Call 2009).

What then is the communicative function of primate gestures? One theory is that they carry virtually no meaning. They are not produced to initiate new interactions, but serve as catalysts of already ongoing interactions (Call & Tomasello 2007). This catalyst function was confirmed in a recent experimental study on intergroup encounters in free-ranging chimpanzees (Herbinger et al. 2009). In the experiment, the presence of a neighboring group was simulated with playbacks of audio recordings, a powerful experience for chimpanzees that leads to interesting and coordinated behavior (Mitani & Watts 2005). A range of gestural signals was triggered, but there were no indications that they were produced to refer to the event itself. Instead the gestures were used in bonding and to provide mutual reassurance, a pattern also seen in other contexts (fig. 29.1). If gestures are not given in response to specific external events, receivers cannot form associations between signals and events and will require pragmatic cues to infer the communicative goal.

Vocal Communication

The study of primate vocal communication has a long history. Basic biological functions such as detecting a predator, finding food, or being lost often trigger acoustically spe-

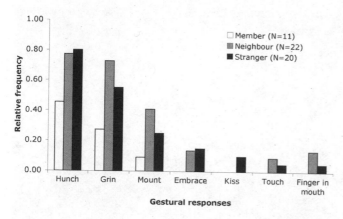

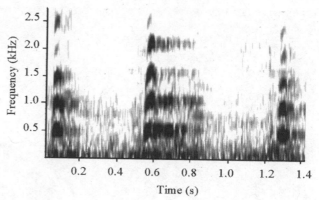

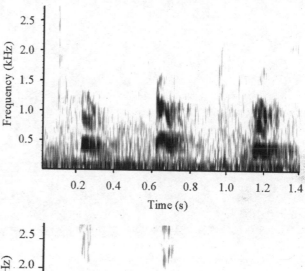

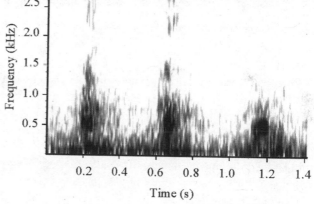

Fig. 29.1. Gestural responses by three different chimpanzee communities in playback experiments simulating the unexpected presence of a familiar group member, a neighbor, or an unfamiliar stranger in the periphery of their home range in the Taï Forest, Ivory Coast (reprinted with permission from Herbinger et al. 2009).

cific call types that differ from species to species. An early characterization of primate vocal repertoires has referred to their discrete versus graded nature (e.g., Byrne 1982). Some species, such as Diana monkeys (*Cercopithecus diana*), produce a set of acoustically distinct call types that can be discriminated with relative ease. Other species living in the same habitat, such as Western red colobus (*Procolobus badius*), produce a few basic call types with considerable acoustic gradation. However, even in species with typically discrete systems, some call types can vary acoustically. For example, vervet monkeys (*Chlorocebus pygerythrus*) produce at least four acoustic variants of grunts, which are given in specific contexts. For example, one variant is used when animals move into an unfamiliar area, while another is given when they approach a more dominant group member. Audio playback experiments have shown that these variants are meaningful to receivers (Cheney & Seyfarth 1982). Similar results have been found for grunts in baboons and chimpanzees (Cheney & Seyfarth 2007; Slocombe & Zuberbühler 2006; Laporte & Zuberbühler 2010; figs. 29.2 and 29.3). These acoustically graded signals are particularly interesting because human vocal production also is acoustically highly graded, while its perception is categorical. One methodological challenge in the study of primate vocal behavior is, therefore, to isolate the psychologically relevant units from the stream of utterances.

Acquisition and Development

The development of the basic vocal repertoire is under relatively strong genetic control in both nonhuman primates and humans. In their first year of life, the vocal behavior of human infants does not differ in obvious ways from other primates (Hammerschmidt & Fischer 2008; Ross et al.

Fig. 29.2. Chimpanzee grunts given in response to different types of foods. Receivers are able to discriminate between these grunt variants and adjust their own foraging behavior accordingly (reprinted with permission from Slocombe & Zuberbühler 2005, 2006).

2009). Infants produce a range of call types in context-specific ways and, according to theory, without much social learning. Nevertheless, the basic frequency contour of crying in newborn babies can differ significantly between cultures, consistent with the intonation pattern of the mother's language (Mampe et al. 2009). Some call types remain in the repertoire and continue to play important communication functions in adulthood during specific types of social interactions (laughter, crying, screams, copulation calls).

Fig. 29.3. Chimpanzees give acoustically graded calls in response to different foods they encounter. Photograph courtesy of Florian Möllers.

Compared to other primates, human infants are intensely vocal and, with the onset of babbling, also produce sounds in a playful way, something not normally seen in primates. An interesting exception is pygmy marmosets (*Cebuella pygmaea*). In this species the vocal behavior of infants resembles the babbling of human infants in a number of ways (Elowson et al. 1998b). Like humans, pygmy marmosets are cooperative breeders in which both parents and older siblings participate in rearing the young. Interestingly, infant babbling appears to stimulate interactions with them (Elowson et al. 1998a). In human infants, the onset of babbling coincides with gaining increasing control over the orofacial musculature, particularly the tongue. This ability develops gradually, allowing infants to modify their acoustic products into more complex syllables, and those syllables into longer sequences. In their second year, infants show signs of vocal imitation when exposed to adults' vowels by producing vocalizations that resemble them (Kuhl & Meltzoff 1996). Very little is known about how nonhuman primate infants learn to produce the acoustic variants that are sometimes observed in adult primates (e.g., Crockford et al. 2004). In vervet monkeys, the production of grunts,

their use in appropriate circumstances, and the response by others to them emerges gradually during the first four years. In both human and nonhuman primates, comprehension develops faster than production, and is driven largely by social learning (see also chapter 31, this volume). For example, infant vervet monkeys monitor the behavior of adults when responding to vocalizations, which suggests that call meaning is learned socially by attending to others' behavioral responses (Seyfarth & Cheney 1986).

Although nonhuman primates may have little control over the basic call morphology, there are occasional reports of dialects and spontaneous idiosyncratic calls (Ouattara et al. 2009a). In wild pygmy marmosets, a species in which infants produce babbling-like vocal behavior, vocal dialects have been reported on a broad scale (de la Torre & Snowdon 2009). The study found significant differences in the structure of two vocalizations in five populations of pygmy marmosets of the same subspecies in northeastern Ecuador, 14 groups in total. In chimpanzees, the "pant-hoot" vocalizations of males housed in two different facilities revealed group-specific acoustic differences (Marshall et al. 1999), something also found in the wild (Mitani et al.

1992; Crockford et al. 2004). Captive chimpanzees have also been reported to produce species-atypical utterances to attract the attention of their caregivers (Hopkins et al. 2007). In orangutans, one captive hybrid individual has been reported to copy human whistling spontaneously and without reinforcement (Wich et al. 2009). Whether these behaviors are indicators of a generative capacity is difficult to assess, mainly because the systematic study of ape vocal behavior in the wild is still in its infancy.

The ontogenetic mechanisms underlying these phenomena are equally poorly understood. In humans, the ability to share attention provides an essential platform for acquiring communication skills, including vocal production. It is possible that joint attention also plays a role in nonhuman primates (Carpenter et al. 1995; Kumashiro et al. 2003), but there are other possibilities. Primates are capable of controlling their vocal output in conditioning paradigms, particularly if social stimuli are used as reinforcers (Pierce 1985). Evidence for variability at the level of call morphology, however, is more contested. Although rhesus monkeys can be conditioned to modify the structure of their calls in response to arbitrary stimuli, the effects are minor and unimpressive (Sutton et al. 1973). More recently, Japanese macaques (*Macaca fuscata*), conditioned to produce coo calls to request food or tools, spontaneously varied the acoustic structure of their calls in subtle but consistent ways depending on the request (Hihara et al. 2003). In a seminal experiment, Japanese and rhesus macaques were cross-fostered for the first years of their lives in an attempt to study the impact of auditory experience on their vocal development. A first study reported significant effects at the level of the fundamental frequency of food calls (Masataka & Fujita 1989), but these results could not be replicated (Owren et al. 1992). Although the second study also found that the calls of cross-fostered Japanese macaques were more similar to those of rhesus macaques than those of their own species, the effects were small and, crucially, within the natural range of Japanese macaque calls.

Overall, these examples demonstrate that nonhuman primates can modify some of their calls, but the degree of control is very limited and the effects minor, compared to those in humans. These studies are nevertheless important for questions of evolutionary origins, especially in understanding how nonhuman primates alter their call morphology and how much control they have over their vocal tracts (Riede et al. 2005).

Flexibility

Considering key indicators of flexibility in vocal signals, such as audience effects, combinations, and means-end dissociations, the following patterns emerge. First, nonhu-

man primates are sensitive to their audience, particularly bystanders (Tomasello & Zuberbühler 2002). Captive chimpanzees are more likely to produce vocalizations as their first communication signals if a human experimenter is looking away from them, but are more likely to use gestures initially if the experimenter is looking towards them (Hostetter et al. 2001). There is not much systematic evidence for such intentional use in the wild, but audience effects have nevertheless been detected there. For instance, female chimpanzees are less likely to produce copulation calls if other high-ranking females are nearby (Townsend et al. 2008). This effect was particularly strong if the female was mating with a high-ranking male. In another field study, chimpanzees were found to "exaggerate" the acoustic structure of their victim screams if their audience included a high-ranking individual who was able to displace the attacker (Slocombe & Zuberbühler 2007). Female bonobos produce copulation calls when mating with males or other females, and their call production is largely determined by social variables (Clay et al. 2011; fig. 29.4). In Thomas's leaf monkeys (*Presbytis thomasi*), males continue to make alarm calls to a predator until every group member has responded with an alarm call, which suggests that males monitor the behavior of their audience (Wich & de Vries 2006). Finally, blue monkey males (*Cercopithecus mitis*) produce more alarm calls if other members of their group are close to a suspected crowned eagle than when they are further away from the eagle, regardless of their own position (Papworth et al. 2008).

Second, studies of forest guenons have revealed some interesting cases of meaningful call combinations (Zuberbühler 2002). Male greater spot-nosed monkeys (*Cercopithecus nictitans*), for example, produce two basic alarm call types, "pyows" and "hacks," which they give in sequences. Eagles tend to elicit a series of hacks; leopards and some other disturbances tend to elicit a series of pyows. In addition, males sometimes produce pyow-hack combinations, which reliably trigger group movements by group members (Arnold & Zuberbühler 2006, 2008; fig. 29.5). In Campbell's mona monkeys (*Cercopithecus campbelli*), adult males produce six different loud call types in response to a range of disturbances (Ouattara et al. 2009b; fig. 29.6). Callers regularly combine the calls into nine structurally unique sequences in highly context-specific ways, such as before travel, after the falling of trees, in response to neighboring groups, toward other animals, in response to danger, and in the presence of predatory eagles and leopards (Ouattara et al. 2009c). As the different call sequences were so tightly linked to specific external events, the Campbell's mona monkey call system provides an unusual case of complexity in animal communication.

How primates acquire such combinatorial calling behav-

Fig. 29.4. Female bonobos produce copulation calls during sexual interactions with both males and other females. Call production is sensitive to caller rank, with females calling more when involved with high-ranking sexual partners, regardless of their sex. Photograph courtesy of Zanna Clay.

ior is largely unknown. In Diana monkeys there is some evidence that learning is involved in the acquisition of sequential signaling (Stephan & Zuberbühler 2008). Diana monkeys at Taï forest (Ivory Coast) regularly interact with two major predators, crowned eagles and leopards, while Diana monkeys at Tiwai Island (Sierra Leone) are hunted only by crowned eagles (fig. 29.7). Field experiments have demonstrated that monkeys at both sites produced the same two basic call types, one made to eagles and the other to leopards and general disturbances. The calls are usually given within repetitive sequences, which differ in the number of their component calls (Zuberbühler et al. 1997). The numbers of calls given in sequences to eagles and general disturbances were identical at both sites, whereas the sequences of calls given to leopards differed, thus suggesting that ontogenetic experience with predators was required for assembling calls into meaningful sequences (Stephan & Zuberbühler 2008).

There are a number of additional examples of meaning at the level of the call sequence. Lar gibbons (*Hylobates lar*) produce duet songs as part of their territorial behavior, but

the same individuals also produce songs when encountering a predator, the clouded leopard. These predator songs consist of the same basic song units as the duet songs, but the units are assembled in different ways (Clarke et al. 2006). In one observational study with chimpanzees in West Africa, about half of all calls were given in combination with other calls or drumming, and there were some context-specific patterns (Crockford & Boesch 2005). Bonobos give five acoustically distinct call types when they find food: barks, peeps, peep-yelps, yelps, and grunts. These calls are almost always given in combination, with the value of the food source determining the composition of the sequence (Clay & Zuberbühler 2009, 2011).

The issue of means-end dissociation has not been addressed much in the literature on primate vocal communication. The general assumption is that vocal signals have strong links with a narrow range of biological functions or contexts, and that callers have little flexibility in this domain—but the empirical basis for these claims is weak. In Barbary macaques (*Macaca sylvanus*), it has been reported that most call types occur in several contexts, and that dif-

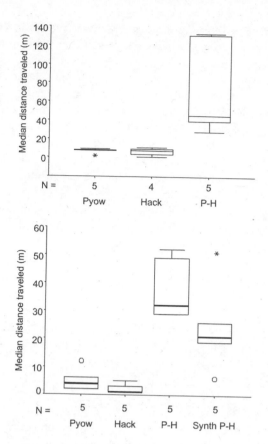

Fig. 29.6. Campbell's monkeys combine some of their calls into longer vocal sequences that are meaningful to others. Photograph courtesy of Eugen Zuberbühler.

Fig. 29.7. Diana monkey alarm calls refer to representations of different predators types. Photograph courtesy of Florian Möllers.

Fig. 29.5. Nigerian male putty-nosed monkeys (*Cercopithecus nictitans martini*) produce two main types of loud calls to external disturbances: "pyows" and "hacks." Crowned eagles typically elicit series of "hacks"; leopards and other terrestrial disturbances tend to elicit series of "pyows" (Arnold et al. 2008). Males sometimes combine pyows and hacks into specific pyow-hack sequences, which reliably trigger group progression both naturally (top figure) and in response to playbacks (bottom figure; reprinted with permission from Arnold & Zuberbühler 2008).

ferent call types are sometimes used within a particular context (Hesler & Fischer 2007), but it is not clear how generalizable this finding is. Other criteria of flexibility, such as response waiting or direction of gaze during vocalization, are also poorly studied despite their relevance for evolutionary and cognitive questions. Systematic research is urgently needed.

Comprehension and Meaning

There is generally good evidence that primate vocalizations can be meaningful to receivers, especially for calls given during encounters with predators or food. The classic case is the vervet monkey alarm call system that encodes acoustically distinct calls for a number of different predator types, such as eagles, leopards, pythons, or baboons (Seyfarth et al. 1980). Predator-specific alarm calls have also been found in many other primate species (e.g., Zuberbühler

et al. 1997; Zuberbühler 2001; Fichtel & Kappeler 2002; Schel et al. 2009; chapter 8, this volume). Calls produced in the context of feeding and during aggressive interactions provide other interesting examples (e.g., Gouzoules et al. 1984; Hauser 1998; Slocombe & Zuberbühler 2005; Slocombe et al. 2009), but there is still much debate about what exactly is encoded by calls about external events or objects, an issue addressed briefly in the next section. It also needs to be pointed out that the ability to extract meaning from vocal signals is by no means a uniquely primate ability, but possibly is a widespread feature of animal communication (e.g., Rainey et al. 2004; Evans & Evans 2007).

The Evolution of Primate Communication

Evolutionary approaches to communication rely much on theoretical modeling, and are typically tested on simple systems (e.g., Ryan 1985). Primates thus tend to play a minor role, although most likely the same principles apply. Key questions in the evolutionary origins of communication concern why signals have evolved, how they maintain their reliability, how their morphology has evolved, and what functions they serve. The following section briefly reviews some important points.

Why Produce Signals?

Primates are very attentive to each other but, as outlined earlier, it is often difficult to determine whether or not this is due to active signaling. For a receiver, it may be irrelevant whether it responds to a signal or to an inadvertent cue ("*Intentionsbewegung*," Lorenz 1939), provided that the receiver can offer a reliable prediction of the event. The situation is different for senders, who may not benefit by providing information, as this is always associated with costs that require compensating personal benefits (Guilford & Dawkins 1991). In practice it is rarely possible to quantify the relevant costs and benefits of communication, so much of the argument remains theoretical. With very few exceptions, animal signals are used during social interactions with conspecifics. Species living in more complex social systems are thus more challenged than solitary or pair-living species. Perhaps it is therefore not so surprising that a positive relationship has been found between vocal repertoire size, group size, and extent of social bonding in nonhuman primates (McComb & Semple 2005). In the vocal domain, social complexity appears to be the main selection force driving the evolution from simple respiratory behavior to controlled vocal signals.

The Evolution of Signal Morphology

Why do long-tailed macaques open their mouths and raise their eyebrows during agonistic interactions (Angst 1974)? Why do vervet monkeys produce acoustically different alarm calls (Seyfarth et al. 1980)? One hypothesis is that animal signals have evolved as abbreviated versions of a more ancestral full behavioral sequence. If receivers respond to the initial segments of such sequences, the so-called intention movements (Heinroth 1911), then the groundwork for evolutionary ritualization has been laid (Bradbury & Vehrencamp 1998). Another hypothesis locates the main driving force with existing predispositions in the receiver. For instance, like many other primates, chimpanzees show piloerection during agonistic interactions (van Hooff 1973). This causes an individual to appear bigger (and probably more dangerous), which may have advantageous consequences during a fight due to a generalized predisposition of animals to respond to body size. In the acoustic domain, primate vocalizations are derivatives of respiration. Receiver biases may have been the main driving force as well, an argument made forcefully by Morton (1977) and followers (e.g. Owren & Rendall 1997). The basic idea is that the various sound structures, which can be produced by an individual's vocal tract, differ in their suitability in eliciting specific behavioral responses or enabling learning opportunities in receivers. The morphology of the calls is evolutionarily determined by their main physiological effects on receivers, and the behavioral outcomes the signalers are interested in. Perhaps for this reason, it is not uncommon to find disparate taxa producing calls that are acoustically remarkably similar, provided their functions are similar. One example is the loud eagle alarm calls of guerezas (*Colobus guereza*) and Diana monkeys (Zuberbühler et al. 1997; Schel et al. 2009).

What Keeps Signals Reliable?

Once signals have evolved a related question concerns the maintenance of their reliability. In other words, how are animals kept from producing signals in situations that are not indicative of the current state or the event witnessed even though receiver responses are beneficial to them (e.g., by crying wolf)? Zahavi (1975) argued that reliability is possible only if the signal is costly or risky to produce. Since receivers will eventually stop responding to inaccurately produced signals, this essentially acts as a selection pressure since costly signals are particularly hard for low-quality or poorly motivated individuals to produce. For instance, in aggressive interactions, adult male chacma baboons (*Papio ursinus*) give loud "wahoo" calls whose acoustic features reveal something about their competitive ability. Changes in a caller's rank are soon followed by corresponding acoustic changes, providing accurate information for others to use in assessing competitive ability (Fischer et al. 2004).

However, nonhuman primates also produce a range of low-amplitude calls frequently and effortlessly during social interactions, and it is difficult to think that this behavior is particularly costly. A number of models have sought to explain the evolution of such low-cost signals. For instance, Maynard Smith (1991) argued that costly signals would only evolve if there were a conflict of interest between signalers and receivers, as during male-male competition. However, if signalers and receivers share common interests, and gain nothing from deceiving each other, then reliable signals need not be costly. A potential example is the low-cost grunts given by baboons during friendly interactions, which appear to provide honest information about callers' intentions. Low-cost signals can even evolve if interests conflict, provided both interactants work towards a common goal (Farrell & Rabin 1996). An example may be the subtle "voting" system used by hamadryas baboons (*Papio hamadryas*) during their daily deliberations in deciding on the direction of group travel (Sigg & Stolba 1981). A third factor that may keep signals from becoming costly has to do with the reputation of the signaler (Silk et al. 2000). As this requires repeated interactions with known individuals, the default case for most primates, the prediction is that costly signals should only evolve in primates in interactions with strangers or members of other species. The male loud calls produced by many forest monkeys appear to fit these predictions.

Symbolic Communication and Language

Mental Representations and Cooperative Motivation

For obvious reasons, research on the evolutionary origins of human language has been particularly interested in primate communication. Questions of the origins of semantics and syntax are of central concern (Hurford 2007), although they have typically been addressed psychologically—that is, by investigating the cognitive processes associated with semantic- and syntax-like phenomena in primate communication.

In humans, speech acts are symbolic, in the sense that acoustic utterances (words, sentences) refer to both material and mental entities. Human infants understand from an early age that symbols (words) and indexical signals (manual gestures) complement each other in their capacity to refer to external objects or events, provided that both kinds of symbols are given by the same signaler (Gliga & Csibra 2009). A key challenge in primate communication research is thus to understand how and why this capacity for symbolic reference has emerged during evolution. In captivity, primates and other animals can be trained to learn a considerable number of symbols in almost completely arbitrary ways, so some of the cognitive capacities may be fairly basic (Premack 1970; Savage-Rumbaugh et al. 1986; Kaminski et al. 2004). Strikingly, however, there is very little evidence that non-human primates use these skills in creative ways and with each other. Instead, artificial symbol systems are typically used in the context in which they have been learned—that is, to request food or entertainment from caregivers.

In the wild, there are numerous examples of how signals naturally produced by animals appear to function in symbolic ways; the dance language of the honeybee is one prominent example (von Frisch 1950). However, the symbolic nature of a communication event may only be appreciated by the researcher, while the animal operates with a less complicated mechanism. Primates and other animals can certainly maintain a variety of mental concepts (Herrnstein et al. 1976), but how similar these concepts are to ours is often almost impossible to decide. In one study, captive long-tailed macaques were trained to respond to pictures of familiar group members (mothers and their offspring). In transfer tests using different sets of animals, the monkeys continued to pick out other mother-offspring pairs, suggesting that they possessed a social concept similar to ours (Dasser 1988). Exploring mental concepts in free-ranging primates is even more challenging. One paradigm, the habituation-dishabituation method, originally developed for pre-linguistic human infants (Eimas 1971), has been used with some success. In one version of this method, an individual hears a first audio playback (e.g., of an eagle alarm call) to activate a putative mental representation (e.g., a predatory eagle). A few minutes later, the same individual hears a second playback stimulus, the probe, which is acoustically different but semantically similar to the first one (e.g., an eagle shriek). If the subject animal is able to generate mental representations, this should affect how it responds to the probe, regardless of the stimulus' physical features. In contrast, if the animal only responds to the physical features of the two playback stimuli, then it should treat them in different ways. Fig. 29.8 shows the responses of Diana monkeys to eagle shrieks as a function of having been primed with either eagle shrieks, semantically similar eagle alarm calls, or semantically different leopard alarm calls five minutes earlier. Priming with eagle shrieks or Diana monkey eagle alarm calls had largely identical effects on how monkeys responded to subsequent eagle shrieks, despite the very different acoustic structure of these stimuli. Priming with Diana monkey leopard alarm calls, however, had a very different effect, although these calls were acoustically very similar to eagle alarm calls (Zuberbühler et al. 1999).

Humans interpret communicative signals mainly in terms of a signaler's presumed intentions (see also chapter 30, this volume). Much of this is based on cultural conventions, which determine how different intentions are related to behavioral outcomes. There is some evidence for similar processes in apes. Captive orangutans can alter their signal production if they are interacting with an unresponsive receiver (Cartmill & Byrne 2007), and bonobos start gesturing if confronted by a reluctant partner during a social game (Pika & Zuberbühler 2008). In the vocal domain, captive chimpanzees have been seen to produce sounds that are not part of their natural vocal repertoire to get the attention of

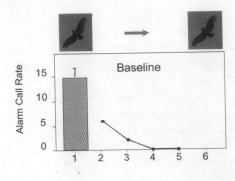

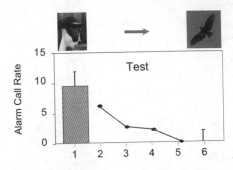

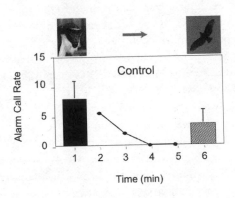

Fig. 29.8. A field experiment to investigate the cognitive processes underlying primate alarm calling behavior. In the baseline condition, a group of Diana monkeys hears a sequence of two playback stimuli, separated by a period of five minutes. The two stimuli are identical in terms of their acoustic and semantic features. In the test condition, the monkeys first hear a conspecific's alarm call to a predator, followed by typical vocalizations of the same predator. The two stimuli are identical in terms of their semantic content, but different in terms of their acoustic features. In the control condition, the monkeys hear a predator alarm call, followed by non-corresponding predator vocalizations, such that the two stimuli are different both in terms of their acoustic and semantic features. In the experiment, the Diana monkeys behaved as if they attended to the semantic features of the stimuli (modified after Zuberbühler et al. 1999).

human caretakers (Hopkins et al. 2007). Humans seem to have gone a step further, however, as both signalers and receivers are able to share intentions with each other (Tomasello 2008; chapter 30, this volume). As a result, human communication is typically bi-directional and based on conventions as to how intentions are communicated (Tomasello & Camaioni 1997). The symbolic nature of human language largely rests on this type of social cognition:

receivers are perceived as intentional agents, with whom experiences can be shared. Research on whether primates are able to perceive and share each other's intentions is still in an early stage (see also chapter 32, this volume).

Gestural and Vocal Theories of Language Evolution

One particularly contentious issue is that of whether human language has originated from gestural or vocal communication. Although it seems more parsimonious to assume that spoken language has evolved directly from other kinds of vocal communication, this route is discounted by many scientists (Hewes 1992; Armstrong & Wilcox 2007). Key arguments are as follows. First, primate vocal behavior is thought to be more emotional than primate gestural signals or human speech. Second, primate manual gestures appear to be produced in more flexible ways than vocalizations, and are more detached from basic biological functions. Finally, gesturing may be cognitively harder because it requires signalers to take the attention of the receiver into account—something that is less necessary in other communication domains (Tomasello & Call 2007; Pollick & de Waal 2007).

Some proponents of a gestural origin of language have identified the missing link in the mirror neuron system, the fact that some manual actions of nonhuman primates share the same neurological substrate as their execution (Rizzolatti and Arbib 1998; Arbib 2005). They also point out that manual gestures show a right-hand bias as compared to noncommunicative manual actions, suggesting that the brain asymmetry involved in human speech may have more ancient evolutionary roots in primate gestural communication, although similar claims have also been made for the vocal domain (Meguerditchian & Vauclair 2009).

One major difficulty in all gestural theories is to explain the change from auditory signals to visual signals and back to auditory signals again, a process that would require two major evolutionary transitions (Seyfarth 2005). One way to address this problem is to argue that speech is actually a gestural behavior that received its final oral form as a result of chance mutations some 100,000 years ago (Corballis 2004; Gentilucci & Corballis 2006). The FoxP2 gene plays a prominent role in these scenarios because two functioning copies of this gene appear to be required for the development of normal speech (Fisher et al. 1998). A comparison of the protein structure of FoxP2 for humans and other primates, including chimpanzees, suggests that in recent human history, two amino-acid changes have occurred which have become fixed in modern humans only very recently (Enard et al. 2002). As intriguing as it is, this scenario puts the emphasis on facial gestures, over which primates appear to have relatively little control, and which may be driven by other neural substrates. Another point is that primates use gestures not in a symbolic referential way, as some use vocal signals, but as a means to coordinate social interactions. Humans use language for both functions.

Future Directions

Olfactory Communication

Olfactory cues linked to MHC alleles play an important role in mate choice and kin recognition in several animal taxa, including humans, but almost nothing is known about how these cues influence behavior in nonhuman primates. Although some primates possess olfactory glands that are used in scent marking, very little is known about the perceptual, developmental, and cognitive processes involved in olfactory communication. How do primates acquire the ability to control olfactory signals, and how do they come to respond to them in appropriate ways? How much awareness is involved when primates perceive an olfactory cue, and what type of learning is required to make an appropriate behavioral decision in response to a chemical stimulus? A more general issue with olfactory stimuli is that, in the absence of active scent marking, it is often difficult to decide whether odors, such MHC-related scents are emitted as part of active communication or as by-products of general metabolic processes that have been left unaltered by natural selection.

Gestural Communication

The systematic study of gestural communication in primates is relatively young, and some methodological problems have yet to be overcome. What counts as a gesture, and how can we distinguish between voluntary and automatic signaling acts? Much of the existing theory about gestural communication is based on captive studies in which individuals live in relatively small areas, which can be spatially and socially demanding. To what degree is the gestural behavior seen in captive apes an ontogenetic adaptation to this special habitat? Systematic field studies are needed to determine how primates, especially apes, use gestures during natural communication. In humans, gestural signals often convey intentional information that requires the receiver to take into account pragmatic cues before understanding the meaning of a signal. Are similar processes taking place in nonhuman primates? Are the patterns of neural activation that occur when watching facial or hand gestures similar in humans and nonhuman primates? Finally, gestures are usually part of a broader communication attempt, which often involves vocalizations. In humans, gestures and vocalizations influence each other, but not much is known about such interactions in nonhuman primates.

Vocal Communication

Human communication is largely bi-directional, and one largely unaddressed problem in primate vocal behavior research concerns the discrepancy between signalers and receivers. Why is it that the comprehension abilities of primates are so advanced, but the production abilities are not? Given that primates do have some control over the acoustic features of certain call types, why are there not more reports of inventions of new calls? Why is there so little evidence of intentional and goal-directed use of vocal signals? More generally, if primates are capable of flexibility as signalers in the visual domain, why does this flexibility not translate to vocal signals? Our knowledge of primate vocalizations is based mainly on studies of various strepsirrhine and monkey species, but the vocal communication of great apes has not been extensively studied. So far, very few experimental audio playback studies, which provide important insights about the function of vocal signals, have been conducted with great apes. Historically, there has been resistance to investigating wild apes with such experimental methods, but recent work has demonstrated that careful playback studies can produce meaningful results (e.g., Slocombe et al. 2009). Little is still known about the ontogeny of vocal competence in signalers and receivers and the social learning mechanisms needed to acquire it. How do primates ensure that they address the right individuals with their calls?

Human Language

Although much is known about the various communication strategies employed by the world's primates, a vast gulf remains between our means of communication and those of other primates. Although interesting analogies have been found at various levels and in a variety of species, it is still difficult to generate a coherent evolutionary scenario to explain the emergence of human language from its primate roots. Nevertheless, research on the communication skills of our living relatives, along with use of the comparative method more generally, is likely to produce important progress in one of the fundamental problems in science: how humans evolved the capacity to communicate in such vastly complex, creative and flexible ways.

Summary

Olfactory cues appear to be relatively ubiquitous as regulators of primate behavior, but outside of the strepsirrhines there are relatively few examples of the strategic, flexible use of olfactory signaling. In many cases it is not clear whether scents have been released actively or whether receivers simply "eavesdrop" on chemical substances excreted as part of general metabolic processes. Scent marking and self-anointment provide some interesting exceptions to this uncertainty, with good evidence that signalers adjust the amplitude and location of these signals to ensure successful communication. In mating, MHC-related odor cues appear to serve as olfactory displays to enhance reproductive success.

Primates can communicate with gestures, produced by parts of the body over which signalers normally have good voluntary control, which makes them interesting as expressions of underlying cognitive processes. For instance, when producing visual gestures, apes are sensitive to the visual attention of the receiver. The degree to which primates learn gestures socially from each other is not well understood. Phylogenetic history and socially influenced conditioning probably play some role during development, which could explain the occasional occurrence of idiosyncratic signals. Although primates can be trained to produce and understand symbolic human gesture systems, there is almost no evidence that their own gestures are meaningful to them in the same way. Instead, gestures appear to facilitate ongoing social interactions. Finally, apart from Plooij (1978), studies of gestural signaling have almost exclusively been done with captive groups, where the sociospatial conditions are often unusual. For instance, wild chimpanzees have not been seen to use explicit gestures during post-conflict interactions, although this is regularly reported in captivity (Arnold & Whiten 2001; de Waal 1982). Systematic studies on gesturing behavior in wild primates are urgently needed.

Vocalizations provide the main channel of communication for many primates. Vocal repertoires are species-specific, even for closely related species, and call development is genetically constrained. Some call types are acoustically more flexible than others; both call use and, to a lesser degree, call morphology can be influenced by experience. Idiosyncratic calls, dialects, and vocal convergence are one manifestation of this, although these phenomena tend to be rare and the exact learning mechanisms remain largely unknown. In contrast to gestures, many vocal signals seem to address a wider audience and as such are less directed to particular receivers. Because many calls are given under relatively specific situations, they are often highly meaningful to receivers, and in some cases semantic content is encoded at the level of the call sequence.

References

Andrew, R. J. 1963. The origin and evolution of the calls and facial expressions of the primates. *Behaviour*, 20, 1–109.

Angst, W. 1974. *Das Ausdrucksverhalten des Javaneraffen Macaca fascicularis raffles*. Berlin: Verlag Paul Parey.

Arbib, M. A. 2005. From monkey-like action recognition to human language: An evolutionary framework for neurolinguistics. *Behavioral and Brain Sciences*, 28, 105–124.

Arcadi, A. C. 2003. Is gestural communication more sophisticated than vocal communication in wild chimpanzees? *Behavioral and Brain Sciences*, 26, 210–211.

Armstrong, D. F. & Wilcox, S. E. 2007. *The Gestural Origin of Language*. Oxford: Oxford University Press.

Arnold, K. & Whiten, A. 2001. Post-conflict behaviour of wild chimpanzees (*Pan troglodytes schweinfurthii*) in the Budongo Forest, Uganda. *Behaviour*, 138, 649–690.

Arnold, K. & Zuberbühler, K. 2006. Language evolution: Semantic combinations in primate calls. *Nature*, 441, 303.

———. 2008. Meaningful call combinations in a nonhuman primate. *Current Biology*, 18, R202–203.

Arnold, K., Pohlner, Y. & Zuberbühler, K. 2008. A forest monkey's alarm call series to predator models. *Behavioral Ecology and Sociobiology*, 62, 549–559.

Barton, R. A. 2006. Olfactory evolution and behavioral ecology in primates. *American Journal of Primatology*, 68, 545–558.

Bonnie, K. E., Horner, V., Whiten, A. & de Waal, F. B. M. 2007. Spread of arbitrary conventions among chimpanzees: A controlled experiment. *Proceedings of the Royal Society B: Biological Sciences*, 274, 367–372.

Bradbury, J. W. & Vehrencamp, S. L. 1998. *Principles of Animal Communication*. Sunderland, MA: Sinauer.

Bräuer, J., Kaminski, J., Riedel, J., Call, J. & Tomasello, M. 2006. Making inferences about the location of hidden food: Social dog, causal ape. *Journal of Comparative Psychology*, 120, 38–47.

Bruner, J. 1981. Intention in the structure of action and interaction. In *Advances in Infancy Research*, Vol. 1 (ed. by Lipsitt, L.), 41–45. Norwood: Ablex.

Butenandt, A., Beckmann, R., Stamm, D. & Hecker, E. 1959. Ueber den Sexuallockstoff des Seidenspinners Bombyx mori: Reindarstellung und Konstitution. *Z. Naturforsch., Teil B*, 14, 283–384.

Byrnit, J. T. 2004. Nonenculturated orangutans' (*Pongo pygmaeus*) use of experimenter-given manual and facial cues in an object-choice task. *Journal of Comparative Psychology*, 118, 309–315.

Byrne, R. W. 1982. Primate vocalizations: Structural and functional approaches to understanding. *Behaviour*, 80, 241–258.

Call, J. & Tomasello, M., eds. 2007. *The Gestural Communication of Apes and Monkeys*. Mahwah, NJ: Lawrence Erlbaum Associates.

Campos, F., Manson, J. H. & Perry, S. 2007. Urine washing and sniffing in wild white-faced capuchins (*Cebus capucinus*): Testing functional hypotheses. *International Journal of Primatology*, 28, 55–72.

Carpenter, M., Tomasello, M. & Savage-Rumbaugh, S. 1995. Joint attention and imitative learning in children, chimpanzees, and enculturated chimpanzees. *Social Development*, 4, 217–237.

Cartmill, E. A. & Byrne, R. W. 2007. Orangutans modify their gestural signaling according to their audience's comprehension. *Current Biology*, 17, 1345–1348.

Cerda-Molina, A. L., Hernandez-Lopez, L., Rojas-Maya, S., Murcia-Mejia, C. & Mondragon-Ceballos, R. 2006. Male-induced sociosexual behavior by vaginal secretions in *Macaca arctoides*. *International Journal of Primatology*, 27, 791–807.

Chaix, R., Cao, C. & Donnelly, P. 2008. Is mate choice in humans MHC-dependent? *Plos Genetics*, 4, e1000184.

Charpentier, M. J. E., Boulet, M. & Drea, C. M. 2008. Smelling right: The scent of male lemurs advertises genetic quality and relatedness. *Molecular Ecology*, 17, 3225–3233.

Chauvin, C. & Thierry, B. 2005. Tonkean macaques orient their food search from olfactory cues conveyed by conspecifics. *Ethology*, 111, 301–310.

Cheney, D. L., & Seyfarth, R. M. 1982. How vervet monkeys perceive their grunts: Field playback experiments. *Animal Behaviour*, 30, 739–751.

———. 2007. *Baboon Metaphysics*. Chicago: University of Chicago Press.

Chevalier-Skolnikoff, S. 1973. Facial expression of emotion in nonhuman primates. In *Darwin and Facial Expression* (ed. by Ekman, P.), 11–90. New York: Academic Press.

Clarke, E., Reichard, U., & Zuberbühler, K. 2006. The syntax and meaning of wild gibbon songs. *PLoS One*, 1, e73.

Clay, Z., Pika, S., Gruber, T. & Zuberbühler, K. 2011. Female bonobos use copulation calls as social signals. *Biology Letters*, 7, 513–516.

Clay, Z. & Zuberbühler, K. 2009. Food-associated calling sequences in bonobos. *Animal Behaviour*, 77, 1387–1396.

———. 2011. Bonobos extract meaning from call sequences. *PLoS One*, 6, e18786.

Corballis, M. C. 2004. FOXP2 and the mirror system. *Trends in Cognitive Sciences*, 8, 95–96.

Crockford, C. & Boesch, C. 2005. Call combinations in wild chimpanzees. *Behaviour*, 142, 397–421.

Crockford, C., Herbinger, I., Vigilant, L. & Boesch, C. 2004. Wild chimpanzees produce group-specific calls: A case for vocal learning? *Ethology*, 110, 221–243.

Darwin, C. 1872/1999 *The Expression of Emotions in Man and Animals* (3 ed.). London: Harper Collins.

Dasser, V. 1988. A social concept in Java monkeys. *Animal Behaviour*, 36, 225–230.

De la Torre, S., & Snowdon, C. T. 2009. Dialects in pygmy marmosets? Population variation in call structure. *American Journal of Primatology*, 71, 333–342.

Dematte, M. L., Osterbauer, R. & Spence, C. 2007. Olfactory cues modulate facial attractiveness. *Chemical Senses*, 32, 603–610.

De Sousa, M. B. C., Moura, S. L. N. & Menezes, A. A. D. 2006. Circadian variation with a diurnal bimodal profile on scent-marking behavior in captive common marmosets (*Callithrix jacchus*). *International Journal of Primatology*, 27, 263–272.

De Waal, F. B. M. 1982. *Chimpanzee Politics*. New York: Harper.

De Waal, F. B. M. & Seres, M. 1997. Propagation of handclasp grooming among captive chimpanzees. *American Journal of Primatology*, 43, 339–346.

Di Fiore, A., Link, A., & Stevenson, P. R. 2006. Scent marking in two western Amazonian populations of woolly monkeys (*Lagothrix lagotricha*). *American Journal of Primatology*, 68, 637–649.

East, M. L., Hofer, H., & Wickler, W. 1993. The erect penis is a flag of submission in a female-dominated society: Greetings in Serengeti spotted hyenas. *Behavioral Ecology and Sociobiology*, 33, 355–370.

Eibl-Eibesfeldt, I. 1972. Similarities and differences between cultures in expressive movements. In *Non-verbal Communica-tion* (ed. by Hinde, R. A.), 297–311. Cambridge: Cambridge University Press.

Eimas, P. D., Siqueland, P., Jusczyk, P. & Vigorito, J. 1971. Speech perception in infants. *Science*, 212, 303–306.

Elowson, A. M., Snowdon, C. T. & Lazaro-Perea, C. 1998a. "Babbling" and social context in infant monkeys: Parallels to human infants. *Trends in Cognitive Sciences*, 2, 31–37.

———. 1998b. Infant "babbling" in a nonhuman primate: Complex vocal sequences with repeated call types. *Behaviour*, 135, 643–664.

Enard, W., Przeworski, M., Fisher, S. E., Lai, C. S. L., Wiebe, V., Kitano, T., Monaco, A. P. & Paabo, S. 2002. Molecular evolution of FOXP2, a gene involved in speech and language. *Nature*, 418, 869–872.

Evans, C. S. & Evans, L. 2007. Representational signalling in birds. *Biology Letters*, 3, 8–11.

Farrell, J. & Rabin, M. 1996. Cheap talk. *Journal of Economic Perspectives*, 10, 103–118.

Ferrari, P. F., Visalberghi, E., Paukner, A., Fogassi, L., Ruggiero, A. & Suomi, S. J. 2006. Neonatal imitation in rhesus macaques. *Plos Biology*, 4, 1501–1508.

Fichtel, C. & Kappeler, P. M. 2002. Anti-predator behavior of group-living Malagasy primates: Mixed evidence for a referential alarm call system. *Behavioral Ecology and Sociobiology*, 51, 262–275.

Fischer, J., Kitchen, D. M., Seyfarth, R. M. & Cheney, D. L. 2004. Baboon loud calls advertise male quality: Acoustic features and their relation to rank, age, and exhaustion. *Behavioral Ecology and Sociobiology*, 56, 140–148.

Fisher, S. E., Vargha-Khadem, F., Watkins, K. E., Monaco, A. P. & Pembrey, M. E. 1998. Localisation of a gene implicated in a severe speech and language disorder. *Nature Genetics*, 18, 168–170.

Fossey, D. 1983. *Gorillas in the Mist*. Boston: Houghton Mifflin.

Garver-Apgar, C. E., Gangestad, S. W. & Thornhill, R. 2008. Hormonal correlates of women's mid-cycle preference for the scent of symmetry. *Evolution and Human Behavior*, 29, 223–232.

Gentilucci, M., & Corballis, M. C. 2006. From manual gesture to speech: A gradual transition. *Neuroscience and Biobehavioral Reviews*, 30, 949–960.

Gilad, Y., Wiebel, V., Przeworski, M., Lancet, D. & Paabo, S. 2004. Loss of olfactory receptor genes coincides with the acquisition of full trichromatic vision in primates. *Plos Biology*, 2, 120–125.

Gliga, T. & Csibra, G. 2009. One-year-old infants appreciate the referential nature of deictic gestures and words. *Psychological Science*, 20, 347–353.

Gomez, J. C. 2007. Pointing behaviors in apes and human infants: A balanced interpretation. *Child Development*, 78, 729–734.

Goodall, J. 1986. *The Chimpanzees of Gombe: Patterns of Behavior*. Cambridge: Harvard University Press.

Gouzoules, S., Gouzoules, H. & Marler, P. 1984. Rhesus monkey (*Macaca mulatta*) screams: Representational signalling in the recruitment of agonistic aid. *Animal Behaviour*, 32, 182–193.

Guilford, T. & Dawkins, M. S. 1991. Receiver psychology and the evolution of animal signals. *Animal Behaviour*, 42, 1–14.

Greenfield, P. M., Lyn, H., & Savage-Rumbaugh, E. S. 2008.

Protolanguage in ontogeny and phylogeny: Combining deixis and representation. *Interaction Studies*, 9, 34–50.

Hammerschmidt, K., & Fischer, J. 2008. Constraints in primate vocal production. In *Evolution of Communicative Flexibility: Complexity, Creativity, and Adaptability in Human and Animal Communication* (ed. by Oller, D. K. & Griebel, U.). Cambridge, MA: MIT Press.

Hare, B., Call, J., Agnetta, B., & Tomasello, M. 2000. Chimpanzees know what conspecifics do and do not see. *Animal Behaviour*, 59, 771–785.

Hauser, M. D. 1996. *The Evolution of Communication*. Cambridge, MA: MIT Press.

———. 1998. Functional referents and acoustic similarity: field playback experiments with rhesus monkeys. *Animal Behaviour*, 56, 1647–1658.

Hauser, M. D., Glynn, D., & Wood, J. 2007. Rhesus monkeys correctly read the goal-relevant gestures of a human agent. *Proceedings of the Royal Society B-Biological Sciences*, 274, 1913–1918.

Havlicek, J., Dvorakova, R., Bartos, L. & Flegr, J. 2006. Non-advertised does not mean concealed: Body odour changes across the human menstrual cycle. *Ethology*, 112, 81–90.

Havlicek, J., Roberts, S. C. & Flegr, J. 2005. Women's preference for dominant male odour: Effects of menstrual cycle and relationship status. *Biology Letters*, 1, 256–259.

Heinroth. 1911. Beitrage zur Biologie der Anatiden. In *Verhandlungen des V. Internationalen Ornithologen Kongresses*. Berlin.

Herbinger, I., Papworth, S., Boesch, C. & Zuberbühler, K. 2009. Vocal, gestural and locomotor responses of wild chimpanzees to familiar and unfamiliar intruders: a playback study. *Animal Behaviour*, 78, 1389–1396.

Herman, L. M., Matus, D. S., Herman, E. Y. K., Ivancic, M. & Pack, A. A. 2001. The bottlenosed dolphin's (*Tursiops truncatus*) understanding of gestures as symbolic representations of its body parts. *Animal Learning and Behavior*, 29, 250–264.

Herrnstein, R. J., Loveland, D. H. & Cable, C. 1976. Natural concepts in pigeons. *Journal of Experimental Psychology: Animal Behavior Processes*, 2, 285–302.

Hesler, N. & Fischer, J. 2007. Gestural communication in barbary macaques (*Macaca sylvanus*): An overview. In *The Gestural Communication of Apes and Monkeys* (ed. by Call, J. & Tomasello, M.), 159–196. Mahwah: Lawrence Erlbaum.

Hewes, G. W. 1992. Primate communication and the gestural origin of language. Special Issue: Celebration of the 50th anniversary of the Wenner-Gren Foundation for Anthropological Research. *Current Anthropology*, 33(suppl), 65–84.

Heymann, E. W. 2001. Interspecific variation of scent-marking behaviour in wild tamarins, *Saguinus mystax* and *Saguinus fuscicollis*. *Folia Primatologica*, 72, 253–267.

———. 2006. The neglected sense-olfaction in primate behavior, ecology, and evolution. *American Journal of Primatology*, 68, 519–524.

Hihara, S., Yamada, H., Iriki, A., & Okanoya, K. 2003. Spontaneous vocal differentiation of coo-calls for tools and food in Japanese monkeys. *Neuroscience Research*, 45, 383–389.

Hopkins, W. D., Taglialatela, J. P. & Leavens, D. A. 2007. Chimpanzees differentially produce novel vocalizations to capture the attention of a human. *Animal Behaviour*, 73, 281–286.

Hostetter, A. B., Cantero, M. & Hopkins, W. D. 2001. Differential use of vocal and gestural communication by chimpanzees (*Pan troglodytes*) in response to the attentional status of a human (*Homo sapiens*). *Journal of Comparative Psychology*, 115, 337–343.

Hubener, F. & Laska, M. 2001. A two-choice discrimination method to assess olfactory performance in pigtailed macaques, *Macaca nemestrina*. *Physiology and Behavior*, 72, 511–519.

Hurford, J. R. 2007. *The Origins of Meaning*. Oxford: Oxford University Press.

Jacob, S. & McClintock, M. K. 2000. Psychological state and mood effects of steroidal chemosignals in women and men. *Hormones and Behavior*, 37, 57–78.

Kaminski, J., Call, J. & Fischer, J. 2004. Word learning in a domestic dog: Evidence for "fast mapping." *Science*, 304, 1682–1683.

Kappeler, P. M. 1998. To whom it may concern: The transmission and function of chemical signals in *Lemur catta*. *Behavioral Ecology and Sociobiology*, 42, 411–421.

Knapp, L. A., Robson, J. & Waterhouse, J. S. 2006. Olfactory signals and the MHC: A review and a case study in *Lemur catta*. *American Journal of Primatology*, 68, 568–584.

Kuhl, P. K. & Meltzoff, A. N. 1996. Infant vocalizations in response to speech: Vocal imitation and developmental change. *Journal of the Acoustical Society of America*, 100, 2425–2438.

Kumashiro, M., Ishibashi, H., Uchiyama, Y., Itakura, S., Murata, A. & Iriki, A. 2003. Natural imitation induced by joint attention in Japanese monkeys. *International Journal of Psychophysiology*, 50, 81–99.

Laidre, M. E. 2009. Informative breath: Olfactory cues sought during social foraging among Old World monkeys (*Mandrillus sphinx, M. leucophaeus*, and *Papio anubis*). *Journal of Comparative Psychology*, 123, 34–44.

Laporte, M. N. C. & Zuberbühler, K. 2010. Vocal greeting behaviour in wild chimpanzee females. *Animal Behaviour*, 80, 467–473.

Laska, M., Bauer, V. & Salazar, L. T. H. 2007. Self-anointing behavior in free-ranging spider monkeys (*Ateles geoffroyi*) in Mexico. *Primates*, 48, 160–163.

Laska, M., Seibt, A. & Weber, A. 2000. "Microsmatic" primates revisited: Olfactory sensitivity in the squirrel monkey. *Chemical Senses*, 25, 47–53.

Laska, M., Wieser, A. & Salazar, L. T. H. 2005. Olfactory responsiveness to two odorous steroids in three species of nonhuman primates. *Chemical Senses*, 30, 505–511.

———. 2006. Sex-specific differences in olfactory sensitivity for putative human pheromones in nonhuman primates. *Journal of Comparative Psychology*, 120, 106–112.

Leavens, D. A., Hopkins, W. D. & Thomas, R. K. 2004. Referential communication by chimpanzees (*Pan troglodytes*). *Journal of Comparative Psychology*, 118, 48–57.

Lewis, R. J. 2005. Sex differences in scent-marking in sifaka: Mating conflict or male services? *American Journal of Physical Anthropology*, 128, 389–398.

Lie, H. C., Rhodes, G. & Simmons, L. W. 2008. Genetic diversity revealed in human faces. *Evolution*, 62, 2473–2486.

Liebal, K., Call, J. & Tomasello, M. 2004. Use of gesture sequences in chimpanzees. *American Journal of Primatology*, 64, 377–396.

Lorenz, K. 1939. Vergleichende Verhaltensforschung. *Zoologischer Anzeiger*, 12, 69–102.

Macdonald, E. A., Fernandez-Duque, E., Evans, S. & Hagey, L. R. 2008. Sex, age, and family differences in the chemical composition of owl monkey (*Aotus nancymaae*) subcaudal scent secretions. *American Journal of Primatology*, 70, 12–18.

Marshall, A., Wrangham, R., & Clark Arcadi, A. 1999. Does learning affect the structure of vocalizations in chimpanzees? *Animal Behaviour*, 58, 825–830.

Matsumoto-Oda, A. & Kasuya, E. 2005. Proximity and estrous synchrony in Mahale chimpanzees. *American Journal of Primatology*, 66, 159–166.

Matsumoto-Oda, A., Kutsukake, N., Hosaka, K. & Matsusaka, T. 2007. Sniffing behaviors in Mahale chimpanzees. *Primates*, 48, 81–85.

Maynard Smith, J. 1991. Honest signalling: The Philip Sidney game. *Animal Behaviour*, 42, 1034–1035.

McClintock, M. K. 1971. Menstrual synchrony and suppression. *Nature*, 229, 244–245.

McComb, K. & Semple, S. 2005. Coevolution of vocal communication and sociality in primates. *Biology Letters*, 1, 381–385.

McGregor, P. 2005. *Animal Communication Networks*. Cambridge: Cambridge University Press.

Meguerditchian, A. & Vauclair, J. 2009. Contrast of hand preferences between communicative gestures and non-communicative actions in baboons: Implications for the origins of hemispheric specialization for language. *Brain and Language*, 108, 167–174.

Milinski, M. 2006. The major histocompatibility complex, sexual selection, and mate choice. *Annual Review of Ecology Evolution and Systematics*, 37, 159–186.

Milinski, M. & Wedekind, C. 2001. Evidence for MHC-correlated perfume preferences in humans. *Behavioral Ecology*, 12, 140–149.

Miller, K. E., Laszlo, K., & Dietz, J. M. 2003. The role of scent marking in the social communication of wild golden lion tamarins, *Leontopithecus rosalia*. *Animal Behaviour*, 65, 795–803.

Mitani, J. C., Hasegawa, T., Gros-Louis, J., Marler, P. & Byrne, R. 1992. Dialects in wild chimpanzees? *American Journal of Primatology*, 27, 233–243.

Mitani, J. C. & Watts, D. P. 2005. Boundary patrols and inter-group encounters in wild chimpanzees. *Behaviour*, 138, 299–327.

Montgomery, K. J. & Haxby, J. V. 2008. Mirror neuron system differentially activated by facial expressions and social hand gestures: A functional magnetic resonance imaging study. *Journal of Cognitive Neuroscience*, 20, 1866–1877.

Morton, E. S. 1977. On the occurrence and significance of motivation: Structural rules in some bird and mammal sounds. *American Naturalist*, 111, 855–869.

Mulcahy, N. J. & Call, J. 2009. The performance of bonobos (*Pan paniscus*), chimpanzees (*Pan troglodytes*), and orangutans (*Pongo pygmaeus*) in two versions of an object-choice task. *Journal of Comparative Psychology*, 123, 304–309.

Myowa, M. 1996. Imitation of facial gestures by an infant chimpanzee. *Primates*, 37, 207–213.

Nishida, T. 1980. The leaf-clipping display: A newly discovered expressive gesture in wild chimpanzees. *Journal of Human Evolution*, 9, 117–128.

Nishida, T., Kano, T., Goodall, J., McGrew, W. C. & Nakamura, M. 1999. Ethogram and ethnography of Mahale chimpanzees. *Anthropological Science*, 107, 141–188.

Ouattara, K., Zuberbühler, K., N'goran, E. K., Gobert, J.-E. & Lemasson, A. 2009a. The alarm call system of female Campbell's monkeys. *Animal Behaviour*, 78, 35–44.

Ouattara, K., Lemasson, A. & Zuberbühler, K. 2009b. Campbell's monkeys use affixation to alter call meaning. *PLoS One*, 4, e7808.

———. 2009. Wild Campbell's monkeys concatenate vocalizations into context-specific call sequences. *Proceedings of the National Academy of Sciences of the United States of America*, 106, 22026–22031.

Owren, M. J., Dieter, J. A., Seyfarth, R. M. & Cheney, D. L. 1992. Food calls produced by adult female rhesus (*Macaca mulatta*) and Japanese (*M. fuscata*) macaques, their normally-raised offspring, and offspring cross-fostered between species. *Behaviour*, 120, 218–231.

Owren, M. J. & Rendall, D. 1997. An affect-conditioning model of nonhuman primate vocal signaling. In *Perspectives in Ethology* (ed. by Owings, D. H., Beecher, M. D. & Thompson, N. S.), 299–346. New York: Plenum Press.

Palagi, E. 2008. Sharing the motivation to play: The use of signals in adult bonobos. *Animal Behaviour*, 75, 887–896.

Palagi, E. & Dapporto, L. 2007. Females do it better: Individual recognition experiments reveal sexual dimorphism in *Lemur catta* (Linnaeus 1758) olfactory motivation and territorial defence. *Journal of Experimental Biology*, 210, 2700–2705.

Papworth, S., Böse, A.-S., Barker, J. & Zuberbühler, K. 2008. Male blue monkeys alarm call in response to danger experienced by others. *Biology Letters*, 4, 472–475.

Peignot, P. & Anderson, J. R. 1999. Use of experimenter-given manual and facial cues by gorillas (*Gorilla gorilla*) in an object-choice task. *Journal of Comparative Psychology*, 113, 253–260.

Pele, M., Dufour, V., Thierry, B. & Call, J. 2009. Token transfers among great apes (*Gorilla gorilla*, *Pongo pygmaeus*, *Pan paniscus*, and *Pan troglodytes*): Species differences, gestural requests, and reciprocal exchange. *Journal of Comparative Psychology*, 123, 375–384.

Pepperberg, I. M. 1999. *The Alex Studies*. Cambridge, MA: Harvard University Press.

Perret, M. 1992. Environmental and social determinants of sexual function in the male lesser mouse lemur (*Microcebus murinus*). *Folia Primatologica*, 59, 1–25.

———. 1996. Manipulation of sex ratio at birth by urinary cues in a prosimian primate. *Behavioral Ecology and Sociobiology*, 38, 259–266.

Pierce, J. D. 1985. A review of attempts to condition operantly alloprimate vocalizations. *Primates*, 26, 202–213.

Peirce, C. S. 1932. *Collected Papers: Elements of Logic*. Cambridge, MA: Harvard University Press.

Pika, S. 2007a. Gestures in subadult bonobos (*Pan paniscus*). In *The Gestural Communication of Apes and Monkeys* (ed. by Call, J. & Tomasello, M.), 41–68. Mahwah: Lawrence Erlbaum.

———. 2007b. Gestures in subadult gorillas (*Gorilla gorilla*). In *The Gestural Communication of Apes and Monkeys* (ed. by Call, J. & Tomasello, M.), 99–130). Mahwah: Lawrence Erlbaum.

Pika, S. & Mitani, J. 2006. Referential gestural communication in wild chimpanzees (*Pan troglodytes*). *Current Biology*, 16, R191–192.

Pika, S. & Zuberbühler, K. 2008. Social games between bonobos and humans: Evidence for shared intentionality? *American Journal of Primatology*, 70, 207–210.

Plooij, F. X. 1978. Some basic traits of language in wild chimpanzees? In *Action, Gesture and Symbol* (ed. by Lock, A.), 111–131. London: Academic Press.

Pollick, A. S. & de Waal, F. B. M. 2007. Ape gestures and language evolution. *Proceedings of the National Academy of Sciences of the United States of America* 104, 8184–8189.

Poss, S. R., Kuhar, C., Stoinski, T. S. & Hopkins, W. D. 2006. Differential use of attentional and visual communicative signaling by orangutans (*Pongo pygmaeus*) and gorillas (*Gorilla gorilla*) in response to the attentional status of a human. *American Journal of Primatology*, 68, 978–992.

Povinelli, D. J., Theall, L. A., Reaux, J. E. & Dunphy-Lelii, S. 2003. Chimpanzees spontaneously alter the location of their gestures to match the attentional orientation of others. *Animal Behaviour*, 66, 71–79.

Premack, A. J. & Premack, D. 1972. Teaching language to an ape. *Scientific American*, 227(4), 92–99.

Premack, D. 1970. A functional analysis of language. *Journal of the Experimental Analysis of Behavior*, 14, 107–125.

Rainey, H. J., Zuberbuhler, K. & Slater, P. J. B. 2004. Hornbills can distinguish between primate alarm calls. *Proceedings of the Royal Society of London Series B: Biological Sciences*, 271, 755–759.

Rako, S., & Friebely, J. 2004. Pheromonal influences on sociosexual behavior in postmenopausal women. *Journal of Sex Research*, 41, 372–380.

Rendall, D., Owren, M. J. & Ryan, M. J. 2009. What do animal signals mean? *Animal Behaviour*, 78, 233–240.

Riede, T., Bronson, E., Hatzikirou, H. & Zuberbühler, K. 2005. Vocal production mechanisms in a nonhuman primate: Morphological data and a model. *Journal of Human Evolution*, 48, 85–96.

Rikowski, A. & Grammer, K. 1999. Human body odour, symmetry and attractiveness. *Proceedings of the Royal Society of London Series B: Biological Sciences*, 266, 869–874.

Rizzolatti, G. & Arbib, M. A. 1998. Language within our grasp. *Trends in Neurosciences*, 21, 188–194.

Ross, M. D., Owren, M. J. & Zimmermann, E. 2009. Reconstructing the evolution of laughter in great apes and humans. *Current Biology*, 19, 1106–1111.

Russell, J. L., Braccini, S., Buehler, N., Kachin, M. J., Schapiro, S. J. & Hopkins, W. D. 2005. Chimpanzee (*Pan troglodytes*) intentional communication is not contingent upon food. *Animal Cognition*, 8, 263–272.

Ryan, M. J. 1985. *The Tungara Frog: A Study in Sexual Selection and Communication*. Chicago: University of Chicago Press.

Savage-Rumbaugh, E. S., Rumbaugh, D. M. & Boysen, S. 1978. Symbolic communication between two chimpanzees (*Pan troglodytes*). *Science*, 201, 641–644.

Savage-Rumbaugh, S. & Lewin, R. 1994. *Kanzi: The Ape at the Brink of the Human Mind*. New York: Wiley.

Savage-Rumbaugh, S., McDonald, K., Sevcik, R. A., Hopkins, W. D. & Rupert, E. 1986. Spontaneous symbol acquisition and communicative use by pygmy chimpanzees (*Pan paniscus*). *Journal of Experimental Psychology: General*, 115, 211–235.

Saxton, T. K., Lyndon, A., Little, A. C. & Roberts, C. 2008. Evidence that androstadienone, a putative human chemosignal, modulates women's attributions of men's attractiveness. *Hormones and Behavior*, 54, 597–601.

Schank, J. C. 2001. Measurement and cycle variability: Reexamining the case for ovarian-cycle synchrony in primates. *Behavioural Processes*, 56, 131–146.

Schel, A. M., Tranquilli, S. & Zuberbühler, K. 2009. The alarm call system of black-and-white colobus monkeys. *Journal of Comparative Psychology*, 123, 136–150.

Schwensow, N., Eberle, M. & Sommer, S. 2008. Compatibility counts: MHC-associated mate choice in a wild promiscuous primate. *Proceedings of the Royal Society B: Biological Sciences*, 275, 555–564.

Schwensow, N., Fietz, J., Dausmann, K. & Sommer, S. 2008. MHC-associated mating strategies and the importance of overall genetic diversity in an obligate pair-living primate. *Evolutionary Ecology*, 22, 617–636.

Scordato, E. S. & Drea, C. M. 2007. Scents and sensibility: Information content of olfactory signals in the ringtailed lemur, *Lemur catta*. *Animal Behaviour*, 73, 301–314.

Scordato, E. S., Dubay, G. & Drea, C. M. 2007. Chemical composition of scent marks in the ringtailed lemur (*Lemur catta*): Glandular differences, seasonal variation, and individual signatures. *Chemical Senses*, 32, 493–504.

Setchell, J. M. & Dixson, A. F. 2001. Circannual changes in the secondary sexual adornments of semifree-ranging male and female mandrills (*Mandrillus sphinx*). *American Journal of Primatology*, 53, 109–121.

Setz, E. Z. F. & Gaspar, D. D. 1997. Scent-marking behaviour in free-ranging golden-faced saki monkeys, *Pithecia pithecia chrysocephala*: Sex differences and context. *Journal of Zoology*, 241, 603–611.

Seyfarth, R. M. 2005. Continuities in vocal communication argue against a gestural origin of language. *Behavioral and Brain Sciences*, 28, 144–145.

Seyfarth, R. M. & Cheney, D. L. 1986. Vocal development in vervet monkeys. *Animal Behaviour*, 34, 1640–1658.

Seyfarth, R. M., Cheney, D. L. & Marler, P. 1980. Vervet monkey alarm calls: Semantic communication in a free-ranging primate. *Animal Behaviour*, 28, 1070–1094.

Sigg, H. & Stolba, A. 1981. Home range and daily march in a hamadryas baboon troop. *Folia primatologica*, 36, 40–75.

Silk, J. B., Kaldor, E. & Boyd, R. 2000. Cheap talk when interests conflict. *Animal Behaviour*, 59, 423–432.

Singh, D. & Bronstad, P. M. 2001. Female body odour is a potential cue to ovulation. *Proceedings of the Royal Society B: Biological Sciences*, 268, 797–801.

Slocombe, K. E., Townsend, S. W. & Zuberbühler, K. 2009. Wild chimpanzees (*Pan troglodytes schweinfurthii*) distinguish between different scream types: Evidence from a playback study. *Animal Cognition*, 12, 441–449.

Slocombe, K. E. & Zuberbühler, K. 2005. Agonistic screams in wild chimpanzees (*Pan troglodytes schweinfurthii*) vary as a function of social role. *Journal of Comparative Psychology*, 119, 67–77.

———. 2005. Functionally referential communication in a chimpanzee. *Current Biology*, 15, 1779–1784.

———. 2006. Food-associated calls in chimpanzees: Responses to food types or food preferences? *Animal Behaviour*, 72, 989–999.

———. 2007. Chimpanzees modify recruitment screams as a function of audience composition. *Proceedings of the National Academy of Sciences of the United States of America*, 104, 17228–17233.

Smith, T. 2006. Individual olfactory signatures in common marmosets (*Callithrix jacchus*). *American Journal of Primatology*, 68, 585–604.

Smith, T. D., Siegel, M. I., & Bhatnagar, K. P. 2001. Reappraisal of the vomeronasal system of catarrhine primates: Ontogeny, morphology, functionality, and persisting questions. *Anatomical Record*, 265, 176–192.

Smuts, B. B., Cheney, D. L., Seyfarth, R. M., Wrangham, R. M., & Struhsaker, T. T., eds. 1987. *Primate Societies*. Chicago: University of Chicago Press.

Snowdon, C. T., Ziegler, T. E., Schultz-Darken, N. J. & Ferris, C. F. 2006. Social odours, sexual arousal and pairbonding in primates. *Philosophical Transactions of the Royal Society B: Biological Sciences*, 361, 2079–2089.

Stephan, C. & Zuberbühler, K. 2008. Predation increases acoustic complexity in primate alarm calls. *Biology Letters*, 4, 641–644.

Stern, K. & McClintock, M. K. 1998. Regulation of ovulation by human pheromones. *Nature*, 392, 177–179.

Striano, T., Rochat, P. & Legerstee, M. 2003. The role of modelling and request type on symbolic comprehension of objects and gestures in young children. *Journal of Child Language*, 30, 27–45.

Sutton, D., Larson, C., Taylor, E. M. & Lindeman, R. C. 1973. Vocalization in rhesus monkey: Conditionability. *Brain Research*, 52, 225–231.

Tanner, J. E. & Byrne, R. W. 1993. Concealing facial evidence of mood: Perspective taking in a captive gorilla. *Primates*, 34, 451–457.

———. 1996. Representation of action through iconic gesture in a captive lowland gorilla. *Current Anthropology*, 37, 162–173.

Tomasello, M. 1996. Do apes ape? In *Social Learning in Animals: The Roots of Culture* (ed. by Galef, B. & Heyes, C.), 319–346. San Diego: Academic Press.

Tomasello, M. 2008. *Origins of Human Communication*. Cambridge, MA: MIT Press.

Tomasello, M. & Call, J. 2007. Ape gestures and the origins of language. In *The Gestural Communication of Apes and Monkeys* (ed. by Call, J. & Tomasello, M.), 17–40. Mahwah: Lawrence Erlbaum.

Tomasello, M., Call, J., Nagell, K., Olguin, R. & Carpenter, M. 1994. The learning and use of gestural signals by young chimpanzees: A trans-generational study. *Primates*, 35, 137–154.

Tomasello, M. & Camaioni, L. 1997. A comparison of the gestural communication of apes and human infants. *Human Development*, 40, 7–24.

Tomasello, M., Striano, T. & Rochat, P. 1999. Do young children use objects as symbols? *British Journal of Developmental Psychology*, 17, 563–584.

Tomasello, M. & Zuberbühler, K. 2002. Primate vocal and gestural communication. In *The Cognitive Animal* (ed. by Bekoff, M., Allen, C. & Burghardt, G. M.), 293–299. Cambridge, MA: MIT Press.

Townsend, S. W., Deschner, T. & Zuberbühler, K. 2008. Female chimpanzees use copulation calls flexibly to prevent social competition. *PLoS One*, 3, e2431.

Van Hooff, J. 1962. Facial expressions in higher primates. *Symposia of the Zoological Society of London*, 8, 97–125.

———. 1973. A structural analysis of the social behavior of a semi-captive group of chimpanzees. In *Expressive Movement and Non-verbal Communication* (ed. by Cranach, M. V. & Vine, I.), 75–162. London: Academic Press.

Vea, J. J. & Sabater Pi, J. 1998. Spontaneous pointing behaviour in the wild pygmy chimpanzee (*Pan paniscus*). *Folia Primatologica*, 69, 289–290.

Von Frisch, K. 1950. *Bees: Their Vision, Chemical Sense, and Language*. Ithaca, NY: Cornell University Press.

Wedekind, C., Seebeck, T., Bettens, F. & Paepke, A. J. 1995. Mhc-dependent mate preferences in humans. *Proceedings of the Royal Society of London Series B: Biological Sciences*, 260, 245–249.

Whiten, A., Goodall, J., McGrew, W. C., Nishida, T., Reynolds, V., Sugiyama, Y., Tutin, C. E. G., Wrangham, R. W. & Boesch, C. 1999. Cultures in chimpanzees. *Nature*, 399, 682–685.

Wich, S. A. & de Vries, H. 2006. Male monkeys remember which group members have given alarm calls. *Proceedings of the Royal Society B: Biological Sciences*, 273, 735–740.

Wich, S. A., Swartz, K. B., Hardus, M. E., Lameira, A. R., Stromberg, E. & Shumaker, R. W. 2009. A case of spontaneous acquisition of a human sound by an orangutan. *Primates*, 50, 56–64.

Widdig, A., Bercovitch, F. B., Streich, W. J., Sauermann, U., Nurnberg, P. & Krawczak, M. 2004. A longitudinal analysis of reproductive skew in male rhesus macaques. *Proceedings of the Royal Society of London Series B: Biological Sciences*, 271, 819–826.

Zahavi, A. 1975. Mate selection: A selection for a handicap. *Journal of Theoretical Biology*, 53, 205–214.

Zeller, A. C. 1987. Communication by sight and smell. In *Primate societies* (ed. by Smuts, B. B., Cheney, D. L., Seyfarth, R. M., Wrangham, R. W. & Struhsaker, T. T.), 433–439. Chicago: University of Chicago Press.

Zuberbühler, K. 2000. Interspecific semantic communication in two forest monkeys. *Proceedings of the Royal Society of London Series B: Biological Sciences*, 267, 713–718.

———. 2001. Predator-specific alarm calls in Campbell's guenons. *Behavioral Ecology and Sociobiology*, 50, 414–422.

———. 2002. A syntactic rule in forest monkey communication. *Animal Behaviour*, 63, 293–299.

Zuberbühler, K., Noë, R. & Seyfarth, R. M. 1997. Diana monkey long-distance calls: Messages for conspecifics and predators. *Animal Behaviour*, 53, 589–604.

Zuberbühler, K., Cheney, D. L. & Seyfarth, R. M. 1999. Conceptual semantics in a nonhuman primate. *Journal of Comparative Psychology*, 113, 33–42.

Chapter 30 Understanding Other Minds

Josep Call and Laurie R. Santos

The unit was resting. An adult female spent 20 min gradually shifting in a seated position over a distance of about 2m to a place behind a rock about 50cm high where she began to groom the subadult male follower of the unit—an interaction often not tolerated by the adult male. As I was observing from a cliff slightly above the unit, I could judge that the adult male leader could, from his resting position, see the tail, back and crown of the female's head, but not her front, arms and face; the subadult male sat in a bent position while being groomed and was also invisible to the leader. The leader could thus see that she was present, but probably not that she groomed. The only aspect that made me doubt that the arrangement was accidental was the exceptionally slow, inch by inch shifting of the female. This had in fact caused me to focus on her behaviour so long before she had reached the final position [fig. 30.1].
—Hans Kummer (in Byrne & Whiten 1990)

Two Meanings of Social Knowledge

Kummer's observation of the sneaky baboon is by no means unique. Indeed, virtually every primatologist has a story about the crafty social tactics used by one of their subjects (for review, see Whiten & Byrne 1988). Such observations have been instrumental in raising researchers' awareness about the kinds of social tactics that primates employ, but they also give the false sense that researchers have come to some consensus about cognitive strategies that primates

use for social reasoning. Indeed, a student new to the field of primate cognition might assume that studies of social cognition have always been at the core of this research area. Such a conclusion, however, would be unwarranted. A few decades ago, observations like those of Kummer cited above constituted the main sources of knowledge regarding what primates knew about the social world. Thankfully, the situation has now changed dramatically. Over the past few decades, much careful empirical work has been devoted to the question of what primates know about social agents and situations.

Current studies of primate social knowledge derive from at least two rather separate traditions (Call 2007). The first is the ethological tradition. Historically, ethologists have tended to focus on social behavior and social dynamics in groups of animals. From this perspective, social cognition is grounded on knowledge about social interactions and relations in the group, such as third-party relations, kin networks, and dominance relationships. In this way the ethological perspective has long focused on social knowledge as it relates to group processes—or what Kummer (1982) famously referred to as the "social field." This focus on group-level processes led to empirical traditions investigating phenomena such as reciprocity and revenge, cooperation, alliance formation, third-party interactions, and kin biases in behavior (see chapters 21, 22, 23, and 28, this volume).

The second tradition that led to the current study of primate social knowledge is a psychological one, in which

Fig. 30.1. Adult female hamadryas baboon grooms a subadult male while concealed from the adult male. Drawing by Leonard Erlbruch, based on an observation by Hans Kummer.

researchers were primarily concerned with the kinds of information-processing mechanisms that primates employed. Psychologists' interest in primate social knowledge has historically centered on the ways in which individuals reasoned about the mental states of others (e.g., Premack and Woodruff 1978). In contrast to the ethological tradition that focused on group processes, the psychological perspective considered social cognition as the set of cognitive strategies utilized by an individual organism. Although this is an eminently psychological tradition, it is also reflected in Kummer's writings (e.g., Kummer 1982) and those of other pioneers such as Emil Menzel, the author of a classic study of deceptive tactics and communication in young chimpanzees (*Pan troglodytes*, Menzel 1974).

That both kinds of approaches are important became increasingly apparent in the decades that followed the initial emergence of these two traditions. The early work of both Kummer and Menzel already contained the seeds of the ethological and psychological approaches that were to flourish in the years to come. The late 1970s in particular brought two key developments for the field. First, Humphrey (1976; see also Jolly 1966) published an influential paper on the evolution of primate social intelligence, arguing that primate intelligence was largely shaped by the social dilemmas primates face—the same dynamics studied in the ethological tradition. In this way, Humphrey's paper was one of the first to directly incorporate ethological insights into the evolution of information-processing capacities. Second, Premack and Woodruff (1978) published a seminal paper investigating whether a chimpanzee could attribute intentions to others. They highlighted the importance of a

representational approach to studying primate social cognition, one that considered not just primates' behaviors but also the cognitive processes they used to represent social problems. The decade that followed these two publications witnessed several influential works on primate social intelligence (de Waal 1982; Whiten & Byrne 1988). Crucially, Richard Byrne and Andrew Whiten paved the way for an interdisciplinary study of primate social knowledge with their prominent volumes that brought together not only ethologists and psychologists but also philosophers, developmental researchers, and others interested in the nature of primate social knowledge (Byrne & Whiten 1988; Whiten 1991).

The last 20 years have witnessed an unprecedented transfer of ideas between researchers from the ethological and psychological traditions. The goal of this chapter is to survey the state of the art in this rapidly expanding field. We have organized our review along some classic empirical dividing lines in the field: those concerning methodologies and topic areas. What will emerge, we hope, are both spots of theoretical consensus and areas where future work is needed to resolve ongoing debates.

Settings and Research Tools for the Study of Social Minds

Before beginning, however, we want to distinguish between the different kinds of methodological approaches that researchers have used to study this area. Each approach has its own strengths and weaknesses, but when taken together, these typically dichotomous approaches can be highly complementary, providing a more complete window into the nature of primate social knowledge.

Observations and Experiments

The first distinction within primate research is between studies that involve observation and experimentation. The observational approach tends to focus on how animals solve social problems in their everyday lives. Observational approaches can involve labor-intensive data collection, and they typically incorporate much larger numbers of subjects than experimental studies. Observational studies also allow researchers to be sure that the skills and limitations they observe are not due merely to task demands or unnaturalistic experimental situations. A weakness of this approach, however, is that it is often difficult to obtain adequate control over the variables of interest. In addition, it is sometimes difficult to test causal hypotheses about social cognition without manipulating something about an animal's social situation. This is where the experimental approach comes in. The experimental approach allows researchers to specifically manipulate particular variables in order to see how they affect primates' responses or problem solving. Although the experimental approach sometimes results in situations that can negatively (or sometimes positively) affect primates' performance, it is still one of the most powerful ways of making inferences about which variables affect primates' cognitive skills.

Although many researchers tend to rely on one of the two methods, it is the research question of interest that should dictate which of these methodologies is best suited to answer it. For instance, if we are interested in knowing the ontogeny of chimpanzees' communicative gestures, observational studies are essential to understanding how these behaviors naturally change over the life of a chimpanzee. On the other hand, if we want to know which stimuli control chimpanzees' communicative gesturing, experimentation may be a more suitable approach. For some questions, though, researchers have productively used both observation and experimentation to learn about primate cognition. Students new to the field may want to consider the relationship between observation and experimentation as that of two partially intersecting circles—there is a common ground that both can cover, but there is also ground that is served by each of them exclusively.

Field and Laboratory

The image of the intersecting circles also illustrates what an ideal relationship between field and laboratory work might look like. As with the observation-versus-experimentation issue discussed above, research in primate social knowledge has involved studies performed on primates living in both naturalistic settings and more controlled environments (e.g., zoos and captive colonies). Although both methodologies have provided tremendous insight into primate social cognition, researchers using these different methods have often sported a relation of mutual indifference sprinkled with an occasional peek over the shoulder. One reason for this indifference concerns the fact that the two methods sometimes provide answers to different kinds of questions. Laboratory settings provide a much more controlled environment, which can allow researchers to manipulate subtle variables to explore the factors that drive certain cognitive skills. Because of this, laboratory research has typically provided richer insight into more specific questions regarding information-processing mechanisms. Field research, on the other hand, is less plagued by issues of sample size than laboratory work, and typically offers a richer environment

in which to explore primates' social cognitive strategies. More recently, many researchers have tried to blur the line between fieldwork and laboratory studies, with some researchers doing "laboratory" work in primate research stations and sanctuaries located in naturalistic settings (e.g., rhesus macaque (*Macaca mulatta*) studies at Cayo Santiago, see Flombaum and Santos 2005; chimpanzee studies at Ngamba Island, Melis et al. 2006b).

Social Problems and Solutions

In this section we review four prominent social problems and the solutions reported for a number of primate species. Although we have tried to integrate findings from all of the settings and research tools described above, our review will focus a bit more on the experimental results used in laboratory settings, as this work is likely to be less familiar to readers in the field of primatology who are interested in the evolution of primate societies. Note also that we plan to leave the interpretations of these data aside, since these will be tackled head-on in the next section of this chapter. This review of the literature is organized around social problems, not around specific psychological mechanisms. It is purposefully agnostic as to whether solving each of these social problems involves the attribution of various mental states, the use of sophisticated behavior reading, or perhaps even more basic associative learning mechanisms. Readers interested in seeing the information presented in this section organized according to the specific kinds of mechanisms that may be involved are directed to other sources (e.g., Suddendorf & Whiten 2001; Call & Tomasello 2005, 2008; Rosati et al. in press). We return to the issue of mechanism at the end of this chapter when discussing some of the recent controversies surrounding this body of work.

Gathering Information about External Entities

One of the useful things about interacting with other social agents is that they may have information about some aspects of the external world that you don't currently have. A central problem for social organisms, then, is detecting the cues produced by other agents that are relevant for learning about the external world. A number of lines of work have explored the kinds of information that primates can obtain by watching others. These sorts of information include learning about which foods are good to eat, how to use a tool, or where to search for hidden food (see chapters 27 and 31, this volume).

One of the most important social cues that primates use

to learn about the external world is where other individuals are looking. Individuals tend to direct their gaze toward interesting and often evolutionarily relevant phenomena, such as the location of food, unusual events, or significant social interactions. Recognizing where others are looking, then, can provide important new information about salient objects and events in the external world. Many primates make use of this tactic, and look in the direction where others are looking. E. Menzel (1974) was the first to systematically describe how chimpanzees used the body orientation, gaze direction, and travel direction of others to locate hidden food in a large outdoor enclosure, but more systematic research on this question was not undertaken until the 1990s. This newer work revealed that lemurs, monkeys, and apes follow others' gaze (see table 30.1, fig. 30.2). Indeed, such skills are not limited to primates; there is evidence that dogs, goats, dolphins, fur seals, and ravens can also follow human gaze (Hare et al. 1998; Miklosi et al. 1998; Tschudin et al. 2001; Bungyar et al. 2004; Scheumann & Call 2004; Kaminski et al. 2005).

More recent work has explored the mechanisms governing primates' gaze-following behavior. One possibility is that gaze following reflects only an automatic orientation response, either hardwired or learned, to turn in a certain direction depending on the orientation of one's partner. A richer psychological interpretation of an orientation response is that subjects follow gaze because they expect to detect something they cannot see from their original visual position. In line with this second interpretation, Povinelli and Eddy (1996a) reported that when a human looked toward a location situated behind an opaque barrier, chimpanzees tried to look around the barrier at the same spot where the human was focusing her attention. These results have been confirmed in great apes (Tomasello et al. 1999; Bräuer et al. 2005) but not in marmosets (Burkart and Heschl 2007). Additionally, Call et al. (1998) found that when chimpanzees follow the gaze of a human experimenter to the ceiling and find nothing of apparent interest, they look back to the face of the experimenter and then back up again. Bräuer et al. (2005) confirmed the existence of such double looks in chimpanzees and other great apes. However, only individuals older than six years of age displayed double looks, even though younger individuals followed gaze perfectly well. Diana monkeys (*Cercopithecus diana*) and pileated gibbons (*Hylobates pileatus*) also engaged in double looks when the informant was not looking at the spot where a target object was located (Scerif et al. 2004; Horton & Caldwell 2006). When exposed to a human looking repeatedly to a location above them where nothing could be found, adult rhesus monkeys and adult chimpanzees decreased their responses, but infants of both

Table 30.1. Primates' use of the gaze of others to look into distant space (gaze following) and to select a baited cup (object choice). Numbers in cells correspond to the references for each study.

Species	Gaze following		Object choice	
	Yes	No	Yes	No
Lemur catta	23			
Eulemur fulvus			21	21
Eulemur macaco		2	21	21
Ateles geoffroyi	1			
Callithrix jacchus	7		7	
Cebus apella	1			3,25
Saguinus oedipus		19	19	
Cercopithecus diana	22			
Cercocebus torquatus	24			
Macaca arctoides	2,24			
Macaca fascicularis	12			
Macaca mulatta	10,24			4
Macaca nemestrina	11,24			
Papio anubis			26	
Gorilla gorilla	6		27	
Hylobates lar			14	
Hylobates pileatus	13			
Pan paniscus	6		18	18
Pan troglodytes	6,8,20,24		5,17,18	5,8,9,16,18
Pongo abelii	6		17	18

Sources: 1. Amici et al. 2009; 2. Anderson & Mitchell 1999; 3. Anderson et al. 1995; 4. Anderson et al. 1996; 5. Barth et al. 2005; 6. Bräuer et al. 2005; 7. Burkart & Heschl 2006; 8. Call et al. 1998; 9. Call et al. 2000; 10. Emery et al. 1997; 11. Ferrari et al. 2000; 12. Goosens et al. 2008; 13. Horton & Caldwell 2006; 14. Inoue et al. 2004; 15. Itakura 1996; 16. Itakura et al. 1999; 17. Itakura & Tanaka 1998; 18. Mulcahy & Call 2009; 19. Neiworth et al. 2002; 20. Povinelli & Eddy 1996a; 21. Ruiz et al. 2009; 22. Scerif et al. 2004; 23. Shepherd & Platt 2008; 24. Tomasello et al. 1998; 25. Vick & Anderson 2000; 26. Vick & Anderson 2003; 27. Peignot & Anderson 1999.

species continued to look (Tomasello et al. 2001). This pattern of performance hints at an interesting developmental transition in the mechanisms that underlie gaze-following behavior: perhaps infants follow gaze as an orientation response, whereas older individuals use more sophisticated psychological mechanisms to regulate this response (e.g., the expectation of a target).

Although many primates follow the gaze of others, there is more controversy about whether they can use others' gaze cues to deduce the location of hidden objects (see Rosati et al., in press, for a review). Typically, an experimenter indicates where a piece of food is located by looking at and gesturing towards one of several possible hiding places. Although many primates naturally follow the gaze of others, few nonhuman primates succeed in this task without training (see table 30.1). Some individual apes consistently succeed at this task, but most do not. Monkeys and lemurs are even less adept at it, often requiring even more training. Recently, however, researchers have argued that primates' mediocre performance in these choice studies has less to do

Fig. 30.2. Following the gaze of others to distant locations is widespread among primates. Courtesy of Nature Neuroscience.

with their abilities to use gaze cues than with the communicative nature of the task (see Hare 2001; Hare and Tomasello 2004; Herrmann and Tomasello 2006). Primates do not naturally cue each other to the location of valued pieces of food, and therefore it makes some sense that primates might not naturally use such cues in this context. In line with this idea, Hare and Tomasello (2004) observed that chimpanzees presented with a competitive situation used gaze cues to locate hidden food. Mulcahy and Call (in press) showed that the performance of great apes in the object choice task improved substantially when the experimenter gave more salient cues. Furthermore, Ruiz et al. (2009) have convincingly shown that the problem may be that subjects do not consistently pay attention to the experimenter giving the cues. Taken together, these results suggest that primates can follow the direction of others' gaze to detect the location of hidden food, but that they tend to do so only in certain situations.

In addition to learning where to look and search, there is evidence that primates can learn how to act on novel objects from watching other social agents (see chapter 31, this volume). Recent work has explored which aspects of other agents' behavior primates attend to in social learning tasks (table 30.1). Are they blindly copying the behaviors of others, or do they take into account psychologically relevant cues in deciding how to act on novel objects? One psychologically relevant cue is whether or not an agent has completed his goal-directed action. If primates use other agents' goals when learning about how to behave, then they should ignore others' unfulfilled efforts when learning about novel objects or foraging locations. Tomasello and Carpenter (2005; see also Myowa-Yamakoshi & Matsuzawa 2000) examined this question, allowing chimpanzees to watch as an experimenter acted on an object. In some cases the experimenter failed to complete her action

(e.g., her hands slipped off the object). After watching these events, chimpanzees executed the action "intended" by the experimenter, and not the action actually demonstrated (see also Meltzoff 1995). Moreover, chimpanzees that saw a human perform one accidental and one intentional action preferentially copied the intentional action (Tomasello & Carpenter 2005). Finally, human-raised chimpanzees, like human infants (Gergely et al. 2002), selectively imitated actions depending on whether they appeared to be freely chosen by the demonstrator or forced by situational constraints (Buttelmann, Carpenter et al. 2007). Taken together, this work suggests that at least one primate species uses the behavior of others to determine how to act on novel objects in a relatively sophisticated way: rather than merely copying other agents, chimpanzees selectively take into account the cues that are psychologically relevant to the agents' intentions and goals.

Predicting the Impending Behavior of Others

Social agents are not just sources of information; as other chapters in this volume show, they are also potential competitors, mates, allies, and grooming partners. It is extremely valuable for individuals to be able to make accurate predictions about the future actions of others. But what information do primates really use to make such predictions? Premack and Woodruff (1978) were among the first to explore whether primates make predictions about others' future actions based on their goals. In their study they investigated whether a chimpanzee, Sarah, was capable of completing actions seen in video sequences in which humans were trying to solve problems. For instance, if Sarah saw a video of a human looking up at a banana hanging from the ceiling, she was given a choice of photographs depicting various kinds of actions that were either appropriate (climbing on a box) or inappropriate (opening a door) for solving the problem. Sarah was able to pick the appropriate photograph in a number of different scenarios, and this led Premack and Woodruff to argue that she "recognized the videotape as representing a problem, understood the actor's purpose, and chose alternatives compatible with that purpose" (p. 515).

Since this seminal work, several other studies have shown that primates seem to make predictions about behavior in a way that is consistent with an understanding of goal-directed action (table 30.1). An important example is the distinction between accidental and intentional acts. Humans distinguish between intended and accidental acts, using the intentional but not the accidental acts to make generalizations about an agent's future behavior. A growing number of researchers have begun exploring whether other primates make a similar distinction (Tomasello & Carpenter 2005; Buttelmann et al. 2007). Call and Tomasello (1998), for example, trained chimpanzees and Sumatran orangutans (*Pongo abelii*) to use a landmark placed on top of one of three opaque containers as an indicator for the location of hidden food. During training, the apes never saw the human actually placing the marker on the container; the marker was already on top of one of the containers when they were presented to the apes. During the test trials, however, an experimenter placed the marker on one of the containers intentionally, but either before or after this he let the marker fall "accidentally" onto one of the other containers. Apes as a group chose the container that the experimenter "intended" to mark, suggesting that they distinguished between actions that were goal-directed and those that were not (see Riedel et al. [in press] for dogs' inability to perform a comparable task).

Other studies, however, have offered alternative explanations for primates' performance in these kinds of tasks. Savage-Rumbaugh et al. (1978), for example, criticized the interpretations of Premack and Woodruff (1978) by arguing that Sarah chose the correct answer using straightforward associative connections between stimuli and responses. Savage-Rumbaugh et al. reported that two language-trained chimpanzees were capable of solving the problems presented to Sarah using a matching-to-sample paradigm without training. In line with this critique, Premack (1988) reported that Sarah failed to discriminate between videotaped sequences that depicted intentional and unintentional actions. Similarly, Povinelli et al. (1998) presented juvenile chimpanzees with one experimenter who feigned clumsiness, "accidentally" spilling a cup of juice she was about to give to the chimpanzee, and a second experimenter who deliberately poured the juice on the floor. When six chimpanzees were later asked to choose between these two ex-

perimenters, none showed a preference for the "clumsy" over the "mean" experimenter. Povinelli and colleagues interpreted this result in terms of the chimpanzees' failure to understand the intentions of the two actors. However, since the chimpanzees did not actually receive any juice from either experimenter, they had little reason to differentiate between them.

The ambiguity in these results has inspired further experiments. Call et al. (2004) tested whether chimpanzees can distinguish between a human who is unwilling to provide food and one who is unable to do so. In the first condition of this experiment, a human experimenter took some grapes lying on a platform and passed a few of them to the subject, but then took another grape from the platform and did not pass it to the subject. In some cases he feigned clumsiness so that the grape fell out of his hand or would not fit through a small delivery hole. In the second condition, the experimenter acted as if he were unwilling to offer the grape (e.g., placing the grape close to the ape and then pulling it away, or leaving the grape on the platform and staring at the ape for no apparent reason). Although chimpanzees did not receive the grape in either condition of this experiment, they exhibited more behaviors, such as knocking on the glass, and left the testing station earlier when faced with an experimenter who was "unwilling" to give them a grape than when faced with one who was "unable" but apparently willing. Phillips et al. (2009) found a similar pattern of performance in brown capuchin monkeys (*Cebus apella*), with capuchins vacating an enclosure sooner when an experimenter acted unwilling to give food than when she acted unable to do so. Phillips and colleagues observed that this response was triggered only by human action. When the same actions were performed by inanimate rods instead of human hands, the capuchins did not differentiate between the two kinds of action. Taken together, the results of these experiments suggest that some primates can use information about experimenters' intentional actions to predict how they will likely interact with them in the future.

Outwitting Competitors

Predicting the actions of others can be particularly useful when competing over access to food and mates—something that occurs often in the lives of almost all group-living primates. It is therefore no coincidence that most examples of tactical deception compiled by Byrne and Whiten (1990) fell into the category of competition. More systematic explorations have confirmed these anecdotal observations. For example, subordinate chimpanzees, spider monkeys, capuchins, and collared mangabeys (*Cercocebus torquatus*) refrain from taking hidden food in the presence of dominant animals (Menzel 1974; Coussi-Korbel 1994; Hirata & Matsuzawa 2001; Amici et al. 2009). These observations demonstrate that primates can successfully react to the presence of competitors, but they also raise questions about exactly how primates represent these competitive social situations. More specifically, these studies are silent on whether subordinate animals react to the presence of the dominant animal, the dominant's current behavior, or the dominant's visual access to the food.

Recently, the question of how subordinate primates actually reason in competitive situations has received much empirical attention. In a landmark paper, Hare et al. (2000) placed a subordinate and a dominant chimpanzee outside a room containing two pieces of food (fig. 30.3). The subordinate chimpanzee could see both pieces of food, but the dominant chimpanzee could see only one (the other was blocked by a small barrier). Subordinate chimpanzees were more likely to approach the piece of food that the dominant individual could not see. Karin-D'Arcy and Povinelli (2002) failed to replicate these results with another group of chimpanzees, but Bräuer et al. (2007) suggested that this failure may have been caused by the shorter distances between the two food pieces that Karin-D'Arcy and Povinelli (2002) used, compared to the distances between pieces of food in the experiment by Hare et al. (2000). Indeed, Bräuer et al. (2007) showed that manipulating the distances affected the likelihood of obtaining positive results, and in doing so, they successfully replicated the original results of Hare et al. (2000). Several control conditions ruled out the possibility that the subordinate chimpanzees were reacting to the gaze, orientation, or initial movements of the dominant animal before deciding which piece of food to go after. Thus, the subordinate chimpanzees seem to have been able to anticipate that the dominant could see only one of the two pieces of food, and thus selectively approached the piece of food that the dominant could not see. Brown capuchins also succeeded in this task, but it was not possible to rule out the possibility that they were responding to the dominants' behavioral cues (Hare et al. 2003).

In a follow-up study, Hare et al. (2001) investigated whether chimpanzees were also able to take into account what their competitors had seen in the past. In this experiment, the setup consisted of two barriers with a piece of food hidden behind one. In experimental trials, dominants either did not see the food being hidden, or food that they saw being hidden was later moved while they were not watching. Subordinates preferentially headed toward the food that dominants had not seen being hidden, and toward sites to which food had been moved, which suggests that they were sensitive to what the dominants had or had not seen a few moments earlier. Since in both conditions

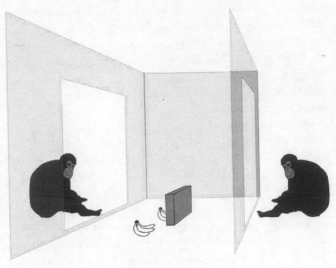

Fig. 30.3. Basic setup of the Hare et al. 2000, 2001 competitive paradigm. Illustrated by Marike Schreiber.

the food items had been hidden behind barriers, this study also helped to rule out the possibility that subordinate individuals simply preferred to forage near protective barriers (Karin-D'Arcy and Povinelli 2002).

Chimpanzees also take into account what other *individuals* have seen. Hare et al. (2001) allowed one dominant individual to witness the baiting procedure, but then exchanged this individual for a different dominant individual who had not seen the baiting. Subordinate chimpanzees were more likely to take food from ignorant dominants than to take it from knowledgeable dominants, thus suggesting that chimpanzees attribute knowledge to particular individuals who have had visual access to certain events. Kaminski, Call, and Tomasello (2008) extended these findings by showing that chimpanzees could capitalize on their ability to predict what other individuals would do on the basis what those other individuals had witnessed. They presented a pair of chimpanzees with a situation in which they took turns choosing food hidden in two of three opaque cups. One subject had seen where both pieces of food were located, and also knew that the competitor had only seen one piece. In one set of experiments the competitor was allowed to make the first choice. The subordinate could not see which cup the competitor chose. Then the subordinate was allowed to choose. Subordinates preferentially selected the food that the competitor had not seen being hidden, thus indicating that they could infer that the competitor would have chosen the cup in which they had seen food hidden, rather than the other cup which also contained food. In contrast, when the subordinates were allowed to choose before their competitors, they did not show any preference, presumably because they knew that both rewards were still available.

Although these studies show that chimpanzees take into account something about competitors' behavior in these contexts, it is unclear whether chimpanzees recognize that the face or the eyes play a privileged role in gathering visual information about the world. When faced with a choice between a human experimenter who is looking toward them and another experimenter who is looking away from them, chimpanzees and rhesus macaques preferentially attempt to steal food from the experimenter who is looking away from them (Flombaum & Santos 2005; Hare et al. 2006). Flombaum and Santos (2005) expanded this work to show that rhesus macaques are sensitive to the state of humans' eyes (open, closed, or averted) when trying to outwit them.

Primates also appear to recognize how auditory cues can affect a competitor's behavior. When interacting with a human competitor, rhesus monkeys selectively steal food from silent containers rather than noisy ones (Santos et al. 2006; fig. 30.4). Significantly, they attend to this auditory cue only when the experimenter cannot see them; they choose between noisy and silent containers at chance levels when an experimenter is already watching them. Similarly, chimpanzees will open a silent trap door, as opposed to a noisy one, to steal a piece of food without alerting a distracted human (Melis et al. 2006). These results demonstrate that chimpanzees and macaques are sensitive not only to what others can or cannot see, but also to what others may hear—a finding suggested earlier by some observational studies (e.g., Byrne & Whiten 1988; Hauser 1992; Boesch & Boesch-Achermann 2000). Taken together, this body of work suggests that primates are adept at attending to cues relevant to outwitting competitors, including those related to what competitors see, hear, have seen previously, and might know in the future.

Enlisting the Participation of Others

Much of the time, individuals' interest in predicting the behavior of others serves purely competitive ends. But primates also need to manipulate others' behavior for more cooperative ends. For instance, primates routinely request the participation of others in a number of activities including travel, food transfer, agonistic support, and mating. Communicative signals play a key role in regulating such exchanges. The importance of communicative signals in this regard raises a question regarding primates' understanding of such signals (see chapter 29, this volume). Do primates understand the nature of their own communicative acts? Do they understand which communicative acts will be effective and which won't?

As was the case with following gaze to external entities, researchers have often made isolated observations about the

Fig. 30.4. Basic setup of the Santos et al. (2006) study demonstrating that rhesus monkeys selectively steal food from silent rather than noisy containers.

role that visual attention plays in communicative exchanges. For instance, de Waal (1982) observed a chimpanzee using his hand to hide his fear grin from opponents during agonistic interactions. Although such observations have great heuristic value, they cannot clearly demonstrate the underlying cognitive representations at work. To investigate this issue systematically, Tomasello et al. (1994) studied whether primates alter their communicative gestures depending on the orientation of the receiver. They found that chimpanzees systematically used visual gestures when others were oriented towards them and could see them, but did not take body orientation into account when they used tactile gestures. Since then, these results have been replicated both in apes (Tomasello et al. 1997; Liebal et al. 2004; Liebal 2007a, b; Pika 2007a, b) and barbary macaques (*Macaca sylvanus*, Hesler & Fischer 2007). Likewise, Call and Tomasello (1994) showed that orangutans gestured less frequently to someone turned away from them than to someone facing them. Other work suggests that chimpanzees and other apes spontaneously beg food from a human who is facing them significantly more often than they do from a human facing away from them (Kaminski et al. 2004).

Such results suggest that primates recognize the importance of others' visual attention when communicating, but other results indicate that this understanding may be rather fragile. Using a begging task, Povinelli and Eddy (1996b) observed that chimpanzees failed to spontaneously discriminate between one experimenter with her face visible and another with a bucket over her head, between one with her eyes visible and a blindfold over her mouth and another with a blindfold over her eyes, and between one with her back turned but looking over her shoulder and another facing forward who was looking away. Although Hattori et al. (2007)

found that brown capuchin monkeys looked longer at a human who looked at them, they did not gesture differently to her than to another human who looked away. In contrast, Kaminski et al. (2004) found that apes begging from a human were sensitive to her face orientation, but only when her body was also oriented toward them. Kaminski et al. (2004) hypothesized that the apes focused on both face and body orientation and extracted different information from each stimulus. Chimpanzees seem to use face orientation to determine whether someone can see them, and body orientation to determine a person's disposition to provide food. Someone facing away from them would not be giving them food unless she first turned around, and so begging would not likely be effective.

Another subject that has received some research attention is the question of whether individual primates can manipulate the attention of others when they want to communicate with them (see Gómez 1991; Hostetter et al. 2001; Liebal et al. 2004a, b). Liebal et al. (2004a) found no evidence that chimpanzees used auditory gestures to call others' attention before using purely visual gestures. Instead, they used tactile gestures to convey their message, or walked around an inattentive recipient to face her and then use a visual gesture. Liebal et al. (2004b) tested these observations experimentally by confronting apes with a human holding food with her back turned. They then examined whether apes would use an auditory gesture to call the human's attention or walk around to face her before begging for food. Results showed that the apes did not use auditory gestures to get the experimenter's attention when she had her back turned. Instead, all species moved around the experimenter when she had the food, thus corroborating the observational findings. Note that this does not mean that apes cannot use auditory gestures or vocalizations to call the attention of others when their preferred options are unavailable. Hostetter et al. (2001) found that chimpanzees uttered vocalizations faster, and more likely as their first communicative behavior, when a human holding food faced away from them. In contrast, they used manual gestures faster and more frequently when the human holding the food was facing toward them. Gómez (1991) also found that chimpanzees established attention with inattentive recipients by touching them before using visual gestures.

Three Key Debates in Primate Social Cognition

In the previous section we noted that our aim was merely to present the empirical results of the relevant studies on primate social problem-solving, and to hold off on any controversial claims regarding the interpretation of these findings.

Here we return to the larger debates that have dominated this line of work, and attempt to find some common ground.

Competition and Cooperation

One of the major breakthroughs in the field of primate cognition involved testing subjects' social knowledge in more ecologically valid situations. Traditionally, many researchers in primate cognition borrowed their empirical questions and experimental methodologies from developmental psychologists interested in children's cognitive development. Sharing methods in this way allowed primate researchers to easily compare their results with those observed in developing humans, and thus it was essential in establishing interspecific comparisons. Unfortunately, however, the use of developmentally inspired tasks also led to a problem: experimental tasks that work well for human children will not necessarily be as effective for testing adult primates. Indeed, many early tasks employed by primate researchers used situations rather foreign to primates, such as another agent intentionally indicating the location of a contested piece of food (e.g., Povinelli et al. 1990; Call et al. 1998; Call & Tomasello 1999). Not surprisingly, primates on the whole performed rather poorly on these sorts of tasks.

The results changed dramatically with the introduction of competitive paradigms. In a pioneering set of studies, Hare and colleagues developed the first competitive perspective-taking tasks for chimpanzees (see Hare et al. 2000, 2001; Hare 2001), with remarkable results. In contrast to what earlier studies might have suggested, chimpanzees tested in these competitive tasks behaved as though they had a complex understanding of cues relevant for psychological states like seeing (Hare et al. 2000) and knowing (Hare et al. 2001). Similar capacities were evidenced when the tasks were used with monkeys (e.g., Flombaum & Santos 2005; Santos et al. 2006).

The advent of competitive methodologies radically changed the way in which comparative researchers think about primate social competence, but these new methods have also presented primate researchers with a theoretical challenge: namely, how should we make sense of primates' performance when tested in different social contexts? This question still remains unresolved. Most researchers agree that a satisfactory theoretical account of primates' social competence must adequately explain both their successes in sociocognitive tasks and their limitations. Nevertheless, researchers have come up with very different explanations to account for the pattern. One view put forth by Tomasello and colleagues (see Tomasello et al. 2005) is that primates' failures on earlier developmentally-inspired social tasks resulted from motivational problems. Under this view, primates possess sociocognitive skills but fail to exhibit them in certain tasks because they lack the motivation to cooperatively share intentions with others. Although this account fits with the available data, it is difficult to test empirically; it is neither clear what representational predictions the motivational account makes, nor easy to devise tasks that selectively manipulate motivation across tasks (see Lyons et al. 2005 for a discussion of this issue). A different explanation, initially championed by Santos and her colleagues, was that primates may show context-dependent performance because their sophisticated sociocognitive abilities may be confined to competitive situations (see Lyons & Santos 2006; Santos et al. 2006). The logic of this view was that primates' social reasoning may be restricted to those domains in which their reasoning capacities initially evolved: presumably those involving deception and competition. However, this view has been challenged by recent empirical work suggesting that the skills of primates are not always so restricted. Although primates *often* perform best in competitive situations, recent studies suggest that they can also exhibit social competence in some noncompetitive situations. To take one example, both chimpanzees and brown capuchins appear to attend to cues related to a person's goals in an instrumental helping task (Barnes et al. 2009; Warneken et al. 2007; chapter 25, this volume), thus suggesting that these species can represent some psychologically relevant cues in cooperative tasks as well—something that should stimulate future research.

Behavior Reading versus Mind Reading

Another still unresolved but perhaps more intensely debated question concerns the mechanisms involved in solving the tasks described earlier in this chapter. The data presented there have generated, and continue to generate, passionate debates about whether primates' performance on sociocognitive tasks requires that they attribute mental states to others. Indeed, this question has dominated studies of primate social knowledge since their inception (e.g., Premack & Woodruff 1978). Thirty years later, the battle lines are still drawn, with two central kinds of hypotheses emerging. Some authors have strongly argued that the available data suggest that primates can attribute certain mental states to others in some situations (Tomasello et al. 2003; Santos et al. 2006). This kind of view has often been referred to as a "*mind*-reading" hypothesis because of its focus on representing the mental states of others. Other authors see nothing in the available data that would grant this conclusion, arguing instead that primates solve social problems by making generalizations about observable behaviors (Heyes 1993, 1998; Povinelli 2004; Povinelli & Vonk 2004; Po-

vinelli and Barth 2005). We will refer to this idea as the "*behavior*-reading" hypothesis.

Let's use an example to illustrate how these two alternatives might play out. A chimpanzee wants to initiate a play bout with a conspecific who is facing away from her. She moves around the conspecific, and once they are facing each other, she uses a visual gesture and a playful chase ensues. According to the mind-reading explanation, the chimpanzee infers that her potential partner cannot *see* her, and so she moves around to alter her perceptual state. According to the behavior-reading explanation, the chimpanzee simply makes generalizations about the observable behavioral cues available, such as body orientation, to determine whether her visual gesture will be effective. Note that such generalizations could result either from an innate predisposition or from learning through repeated interactions.

As this example illustrates, both the mind-reading and behavior-reading ideas postulate that primates can make (often quite sophisticated) generalizations and predictions about others' actions. The difference between the two views involves which *inputs* are used to make those generalizations and predictions. Under the mind-reading view, primates are making generalizations based on others' mental representations; for this reason, mind-reading accounts are often referred to as "representational," since they require that primates represent the representations of others. In contrast, behavior-reading accounts can be considered "nonrepresentational" in that all generalizations are performed over observable behaviors and not unseen mental representations. Thus, although the nonrepresentational and representational accounts both postulate internal representations, they differ on their nature (see also Whiten 2000 for discussion).

Although we have tried to distinguish between two very different classes of mechanistic views, there is considerably more nuance in the way researchers have articulated these two positions. In order to obtain a sense of the range of views, table 30.2 presents the six main positions discussed in the literature today. The first is Heyes's (1993, 1998) view, perhaps one of the first and most discussed nonrepresentational accounts. Heyes argued that there is no convincing evidence that nonhuman animals attribute mental states to solve social problems. Instead, subjects' responses are determined by detecting certain key stimuli. Heyes (1993, 1998) argued that even the most impressive cases of social problem solving could be explained by invoking generalizations from stimuli that have been acquired in the past. According to Heyes (1998), this account would be considerably weakened if subjects were able to use their own experiences with certain stimuli to predict what others would do in that same situation (see "Heyes's challenge" below).

Table 30.2. A survey of different researchers' views of the mechanisms underlying primates' success on social cognition tasks, broken down into representational and nonrepresentational accounts

Nonrepresentational accounts	Representational accounts
Heyes (Heyes 1993, 1998)	Tomasello & Call (Tomasello & Call 1997; Call 2001; Tomasello et al. 2003; Call & Tomasello 2008)
Povinelli and colleagues (Povinelli 2004; Povinelli & Vonk, 2003; Povinelli & Barth 2005)	Santos and colleagues (Lyons & Santos 2006; Santos et al. 2006)
Gómez (Gómez 1991, 2005)	Whiten and Byrne (Whiten & Byrne 1991; Byrne 1995; Suddendorf & Whiten 2001)

The second account, that of Povinelli and colleagues, also falls into the nonrepresentational camp, although there is some variation in their articulation of this view over time (see Povinelli & Vonk 2004; Povinelli & Barth 2005; Penn & Povinelli 2007). Povinelli and colleagues have suggested that primates are incapable of reasoning about any unobservable entities, a category of constructs that would include social representations like mental states, but also physical representations such as gravity or force. Under their view, all reasoning (including that in the social domain) involves making generalizations and forming concepts over observable features. Povinelli and Vonk (2003) concede that some version of the "goggles test" (see below) is one way to move forward, since subjects passing that test would show that they construe their partners as seeing them or not seeing them.

The third account, that of Gómez (1991, 2004), can also be classified as nonrepresentational. Unlike in the two other accounts just described, however, Gómez has placed a strong emphasis on the idea that individuals establish a *causal* link between the gaze direction and the target of attention. Under his view, primates "perceive that there is a link between the behavioral manifestations of attention and the actions of people, although they do not perceive nor represent this link" (Gómez 1991, p. 204). Gómez therefore postulates that primates can recognize that an individual's gaze is *about* an object in the absence of representing that individual's mental state. Gómez does argue, however, that primates represent causal links between stimuli and responses.

The fourth position—that of Tomasello, Call, Hare, and colleagues—involves a representational rather than nonrepresenational account (see Call 2001; Tomasello et al. 2003; Tomasello et al. 2005). Tomasello and Call (1997) were earlier champions of a nonrepresentational position, which they have since adjusted to account for new empiri-cal evidence. Under their new view, they reject the idea that primates merely learn links between stimuli and possible responses. Instead, they argue that primates' attribution of knowledge to others should be understood as their knowing what other individuals have seen in the past, or what kind of information they possess. However, these authors have argued that there is no evidence to date that primates possess a *meta*-representational level of cognition (see also Whiten 2000). For instance, they have argued that there is no evidence that any nonhuman primate is capable of false belief attribution.

The fifth position, that of Santos and colleagues, is a representational account very similar to that of Tomasello and Call (e.g., Santos et al. 2006). Santos and colleagues have argued that primates possess a representational theory of mind that in many ways is qualitatively similar to that of humans. The one difference between humans and nonhumans, under their view, involves representations relating to the beliefs of others. In this way, they agree with Tomasello and Call's position that a major difference between human and nonhuman primate sociocognitive capacities involves the capacity to represent others' beliefs (see Rosati et al. in press).

The sixth position, that of Whiten and Byrne (1991), also falls squarely into the representational camp. In some of their writings, Whiten and Byrne have hypothesized that metarepresentation (or what some psychologists refer to as "secondary representations"; see Perner 1991) may provide a mechanism to explain the joint emergence of pretense, mind reading, and imitation in chimpanzees (see also Suddendorf & Whiten 2001). In the same vein, Byrne (1995) has argued against viewing learning by trial and error as an alternative to mental state attribution. More recently, Byrne (2002) seems to have revised his position to some extent, although it still has a rather representational flavor. Under his new view, mental state representations have become more nodelike and implicit (perhaps with little access to or from other systems), and are created via the detection of statistical regularities of observable behavior. In this regard, Byrne's current position comes closer to that of Gómez, in which linkages rather than the mental states themselves play a pivotal role.

As the above six positions indicate, the debate still rages regarding whether primates' performance on social tasks can be construed as evidence of a representational understanding of mind. We favor a representational account: one that allows primates to *reason* about social dilemmas through the use of abstract representations involving others' representational states of mind. But this interpretation remains quite controversial, and although we believe that other accounts are less plausible, we also admit that they are still viable. Interested readers can consult these debates in greater detail in a number of publications (Heyes 1998; Povinelli & Vonk 2003; Tomasello et al. 2003; Tomasello & Call 2006; Santos et al. 2006; Penn & Povinelli 2007).

Social Cognition across the Primate Order: Monkeys versus Apes?

A final area of continued controversy in the field concerns how sociocognitive skills vary across the Primate order. Historically, researchers have assumed that sociocognitive skills would be richer and more sophisticated in species more closely related to humans. Under this view, apes—who are the "close cousins" of cognitively sophisticated humans—will possess complex sociocognitive skills, whereas monkeys and strepsirrhine primates likely won't. This view gained some support in early accounts of primate social cognition, including the early compendiums of primate deception (e.g., Byrne & Whiten 1988, 1990). Byrne and Whiten (1990) analyzed the available anecdotes about tactical deception using a set of strict criteria. Their analysis revealed that most of the reported deceptive acts involved great apes, and a minority came from cercopithecine monkeys. Indeed, the strepsirrhine primates were completely absent from the available record. This led Byrne and Whiten to propose a breakdown of sociocognitive skills that would shape the field for decades to come: apes, they argued, were sociocognitively sophisticated, but monkeys and strepsirrhine primates were not.

Although incredibly influential, the Byrne and Whiten analysis of social cognition across primates had two critical problems. First, their analysis was based on anecdotal evidence. Second, it was vulnerable to observer bias, as researchers working with monkeys may have been less prone than those investigating apes to recall and interpret certain episodes in a Machiavellian light. Such problems led Tomasello and Call (1997) to make another attempt at analyzing sociocognitive skills across the Primate order. At the time of their 1997 book, new experimental work had just begun systematically exploring social-cognitive skills in primates using cooperative tasks, and the emerging picture was rather grim both for monkeys and apes. As reviewed in the previous section, before the use of competitive experimental methodologies, apes and monkeys seemed to know very little about other social agents. Tomasello and Call (1997) concluded from this evidence that neither apes nor monkeys possessed sophisticated sociocognitive abilities.

A decade later, the evidence supports a different picture. In particular, the advent of competitive methodologies has provided new evidence about primates' sociocognitive skills. This work has led to a new appreciation of their sociocognitive sophistication, but it has also led to a new view of how such sophistication emerged across the Primate order. Nearly all of the social cognitive capacities observed in apes in the last decade—both their sophisticated abilities and their limitations—appear to be shared in at least some monkey species. Both apes and monkeys appear to detect cues relevant to what another individual sees (e.g., see Hare et al. 2000; Flombaum & Santos 2005), hears (e.g., see Melis et al. 2006; Santos et al. 2006), knows (e.g., Hare et al. 2001; Santos et al. 2006), and intends to do (e.g., Call et al. 2004; Phillips et al. 2009). In addition, both apes and monkeys show hints of instrumental helping (e.g., Warneken & Tomasello 2007; Barnes et al. 2009; see also chapter 25, this volume), inequity aversion (e.g., Brosnan & de Waal 2003, 2005), and social learning (e.g., Bonnie & de Waal 2007; Bonnie et al. 2007; see also chapter 31, this volume). We feel that taken together, the present evidence suggests that both apes and monkeys possess social cognitive sophistication. Indeed, the present work suggests little reason to suspect that apes differ qualitatively from monkeys in their social knowledge.

That said, despite new advances and evidence in monkeys and apes, we are at present forced to remain relatively agnostic regarding the sociocognitive skills of strepsirrhine primates. Though the past decade has seen an explosion of work in primate social cognition, the strepsirrhine primates have been woefully underrepresented in this empirical fervor. Thankfully, this appears to be changing with more and more empirical studies focused on social skills in one group of strepsirrhines, the lemurs (e.g., Shepard & Platt 2008; Ruiz et al. 2009; chapter 2, this volume). At present, unfortunately, it is still difficult to know whether this group shares the sociocognitive skills of the monkeys and apes. Although some researchers place their bets on a qualitative separation between strepsirrhines and haplorrhines—with some even referring to the former as "socially unsophisticated" (Byrne 2000, p. 560), we prefer to remain agnostic until more empirical findings weigh in on this issue.

Future Directions

We end our review with a look to the future of primate social cognition research. Specifically, we review areas where we hope new work will bring new insight in the years ahead.

Heyes' Challenge: From Self-Knowledge to Others' Knowledge

There is still considerable controversy about whether primates' performance is indicative of mental state attribution or instead results from nonmentalistic generalizations over observable behaviors. Although some researchers have argued that distinguishing between these accounts is, in principle, impossible (see Povinelli & Vonk 2003; Penn & Povinelli 2007), we think that some experimental procedures could potentially distinguish between these two kinds of accounts. Heyes (1998) described one elegant way to do so. Her idea rests on two features of true mental state attribution. The first is that mental state attribution involves the ability to make inferences from one's own knowledge acquisition processes to those of others. The second is that it should be able to incorporate new information about novel attentional barriers in the environment. Using this logic, Heyes suggested introducing primate subjects to a novel object (e.g., unusual goggles) that control visual access in unexpected ways. For example, imagine a pair of goggles that appear to be opaque but are actually more transparent. The subject would be able to learn about this "transparency" by looking through the goggles. The question of interest is whether subjects would be able to predict another individual's ability to see through the special goggles using only information they have gathered from their own experience with them. If so, Heyes argues, their performance would provide convincing evidence that an account based on observables alone cannot explain the full pattern of primates' performance on mind-reading tasks. To date, no empirical study has taken Heyes's challenge head-on, but we agree that such an empirical test would provide an effective method for negotiating between these two views. That said, we also caution researchers not to fall into the "either-or" trap that so often plagues primate social cognition research (e.g., the idea that *only* a false-belief test can serve as the standard for understanding theory of mind). Although exploring this aspect of primate social cognition is important, we would not argue that researchers must immediately abandon the representational account if primates were to perform poorly on such a task.

Mind Reading and Cooperation

Another area where we anticipate exciting developments in the next few years is at the interface between mind reading and cooperation (see chapter 25, this volume). As reviewed earlier, there has been considerable debate concerning why primates appear to demonstrate such context specificity in their mind reading performance. Although researchers have not yet fully resolved why primates perform so differently on different kinds of social tasks, the empirical stage is now set for future research to look more closely at these issues. Researchers are beginning to learn more about how effectively to design cooperative tasks that are suitable for different primates (e.g., Hare et al. 2007; Melis et al. 2006; chapter 25, this volume). With these new methodological insights, they are now well positioned to explore the kinds of sociocognitive skills primates employ when cooperating rather than competing with others. For example, do primates selectively choose cooperative partners with positive intentions over those with negative ones? Do they recognize the difference between a cooperator who is unable to help them versus one who is unwilling to help them (see Call et al. 2004 for a similar logic)? Similarly, do primates perform actions in order to help a cooperative partner see the problem at hand? Do they recognize the psychological cues relevant to a cooperator knowing about the cooperative task? Although these questions are still untested, we expect that research in the next decade will shed light on this still unresolved theoretical question.

Metacognition and Mind Reading

Thus far in this chapter, we have focused on what primates know about the minds and behaviors of social agents. This focus on reasoning about the minds of *others* leaves open the question of how primates represent their own minds, and whether they have the capacity to make predictions based on their own knowledge and beliefs. Indeed, a number of researchers have begun to explore whether primates have the capacity to think about their own thoughts and beliefs in the context of uncertainty-monitoring tasks (Call & Carpenter 2001; Hampton 2001; Smith et al. 2003; Beran et al. 2006; Kornell et al. 2007). Although this work suggests that primates can represent cues relevant to their own knowledge state in some tasks, there is still some controversy regarding how these "metarepresentational" processes connect with the kinds of processes we have reviewed above. For example, do primates use the same psychologically relevant cues when reasoning about the behavior of themselves and others? Are the limitations observed in primates' social reasoning also observed when they process and predict their own behavior? These are important open questions that we hope future studies can address empirically.

Linking Primate Mind Reading to Human Mind Reading

Traditionally, the field of primate cognition has gained important insights from the field of human cognitive development (see chapter 32, this volume), but more recently other fields such as social psychology, experimental economics, and cognitive neuroscience have also contributed to our understanding of primate cognition in significant ways (see Santos et al. 2006; Brosnan et al. 2009). Primate researchers have recently begun capitalizing on experimental economic tasks to develop new measures of primate cooperation and negotiation (e.g., Silk et al. 2005; Jensen et al. 2007; chapter 25, this volume). Similarly, they have begun employing some of the insights from sociocognitive neuroscience to make sense of primates' performance on mind reading tasks (e.g., Santos et al. 2006).

We see at least two areas where studies of adult humans could provide new ideas for studies of primate mind reading. The first involves insights from "person perception" studies in social psychology. Rather than narrowly focus on "theory of mind" capacities, social psychologists have typically taken a broader view of the processes people use to make sense of others' behavior (e.g., Mitchell 2008). To social psychologists, mind reading capacities go far beyond reading others' intentions and beliefs. They also encompass more varied processes, including attributing personality traits, making social category judgments, and using situational variables to explain behavior. We still know very little about these processes in primates, but some studies have broken some ground. Subiaul et al. (2008) found some evidence that chimpanzees attributed to other chimpanzees reputations for being generous and stingy. Additionally, Russell et al. (2008) found that chimpanzees took third-party interactions into account in deciding which one of two strangers to approach to obtain food.

Cognitive neuroscience is a second area that could enrich studies of primate social cognition. Recently, neuroscientists have begun to dissociate human mind reading at the neural level (see Saxe et al. 2004; Saxe 2006). Such dissociations have produced new insights regarding the component processes that make up human social processing, such as the importance of low-level biological motion cues like animate motion and eye gaze (e.g., Jellema et al. 2000; Pelphrey et al. 2003, 2004) and the fact that belief reasoning may be separable from other aspects of mind-reading (e.g., Saxe & Wexler 2005). For example, Saxe (2006) has argued that primates might lack the sociocognitive processing mediated by the human temporo-parietal-junction, an area thought to be selective for processing beliefs separately from other kinds of mental states. This would explain why primates may succeed in representing cues related to knowledge and ignorance, but fail false-belief tasks (e.g., Kaminski et al. 2008).

Conclusion

Living together creates challenges and opportunities. Faced with limited resources, individuals attempt to outwit competitors by anticipating their behavior. Often two or more individuals cooperate in outwitting a third party, and again the success of partners in achieving a common goal requires their making predictions about each other's behavior. Living together also offers opportunities for primates to gather information from conspecifics by paying attention not only to what they do, but also to where their attention is focused. Much work in the last two decades has been devoted to exploring the social representations underlying the strategic behavior that individuals deploy during social interactions. Whereas some authors argue that those representations are about observable behavior, others have interpreted the evidence as an indication that individuals process social information at a more abstract level that goes beyond observable stimuli. Although traditionally apes, particularly chimpanzees, have been thought to possess more sophisticated sociocognitive skills than monkeys, the evidence currently available suggests that some monkey species are, at the very least, comparable to chimpanzees in a number of social cognition abilities. Moreover, data on some lemur species has begun to trickle in, and even though it is still premature to draw any firm conclusions, lemurs may possess some of the same sociocognitive skills found in monkeys and apes.

References

Amici, F., Call, J. & Aureli, F. 2009. Variation in withholding of information in three monkey species. *Proceedings of the Royal Society of London B*, 276, 3311–3318.

Anderson, J. R. & Mitchell, R. W. 1999. Macaques but not lemurs co-orient visually with humans. *Folia Primatologica*, 70, 17–22.

Anderson, J. R., Montant, M. & Schmitt, D. 1996. Rhesus monkeys fail to use gaze direction as an experimenter-given cue in an object-choice task. *Behavioural Processes*, 37, 47–55.

Anderson, J. R., Sallabery, P. & Barbier, H. 1995. Use of experimenter-given cues during object-choice tasks by capuchin monkeys. *Animal Behaviour*, 49, 201–208.

Barnes, J. L., Martinez, M., Langer, M., Hill, T. & Santos, L. R. 2008. Helping behaviour and regard for others in capuchin monkeys (*Cebus apella*): An evolutionary perspective on altruism. *Biology Letters*, 4, 638–40.

Barth, J., Reaux, J. E. & Povinelli, D. J. 2005. Chimpanzees' (*Pan troglodytes*) use of gaze cues in object-choice tasks: Different methods yield different results. *Animal Cognition*, 8, 84–92.

Behne, T., Carpenter, M., Call, J. & Tomasello, M. 2005. Unwilling versus unable: Infants' understanding of intentional action. *Developmental Psychology*, 41, 328–337.

Beran, M., Smith, J., Redford, J. & Washburn, D. 2006. Rhesus macaques (*Macaca mulatta*) monitor uncertainty during numerosity judgments. *Journal of Experimental Psychology: Animal Behavior Processes*, 32, 111–119.

Boesch, C. 1994. Cooperative hunting in wild chimpanzees. *Animal Behaviour*, 48, 653–667.

Boesch, C. & Boesch-Achermann, H. 2000. *The Chimpanzees of the Taï Forest: Behavioural Ecology and Evolution*. Oxford: Oxford University Press.

Bonnie, K. E. & de Waal, F.B.M. 2007. Copying without rewards: Socially influenced foraging decisions among brown capuchin monkeys. *Animal Cognition*, 10, 283–292.

Bonnie, K. E., Horner, V., Whiten, A. & de Waal, F.B.M. 2007. Spread of arbitrary conventions among chimpanzees: A controlled experiment. *Proceedings of the Royal Society of London B*, 274, 367–372.

Bräuer, J., Call, J. & Tomasello, M. 2004. Visual perspective taking in dogs (*Canis familiaris*) in the presence of barriers. *Applied Animal Behaviour Science*, 88, 299–317.

———. 2005. All great ape species follow gaze to distant locations and around barriers. *Journal of Comparative Psychology*, 119, 145–154.

———. 2007. Chimpanzees really know what others can see in a competitive situation. *Animal Cognition*, 10, 439–448.

Brosnan, S. F. 2008. Inequity responses in nonhuman primates. In *Neuroeconomics* (ed. by Glimcher, P., Fehr, E., Camerer, E. & Poldrack, R.), 285–302. New York: Elsevier.

Brosnan, S. F., Freeman, C. & de Waal, F. B. M. 2006. Partner's behavior, not reward distribution, determines success in an unequal cooperative task in capuchin monkeys. *American Journal of Primatology*, 68, 713–724.

Brosnan, S. F., Newton-Fisher, N. E. & van Vugt, M. 2009. A melding of the minds: When primatology meets personality and social psychology. *Personality and Social Psychology Review*, 13, 129–147.

Bugnyar, T., Stowe, M. & Heinrich, B. 2004. Ravens, *Corvus corax*, follow gaze direction of humans around obstacles. *Proceedings from the Royal Society of London B Biological Sciences*, 271, 1331–1336.

Burkart, J. M. & Heschl, A. 2006. Geometrical gaze following in common marmosets (*Callithrix jacchus*). Perspective taking or behaviour reading? *Journal of Comparative Psychology*, 120, 120–130.

———. 2007. Understanding visual access in common marmosets, *Callithrix jacchus*: Perspective taking or behaviour reading? *Animal Behaviour*, 73, 457–469.

Buttelmann, D., Carpenter, M., Call, J. & Tomasello, M. 2007. Enculturated chimpanzees imitate rationally. *Developmental Science*, 10, F31–F38.

Byrne, R. 1995. *The Thinking Ape: Evolutionary Origins of Intelligence*. New York: Oxford University Press.

Byrne, R. & Whiten, A. 1987. The thinking primate's guide to deception. *New Scientist*, Dec 3. 54–57.

Byrne, R.W. & Whiten, A. 1988. *Machiavellian Intelligence: Social Expertise and the Evolution of Intellect in Monkeys, Apes and Humans*. Oxford: Clarendon Press.

———. 1990. Tactical deception in primates: The 1990 database. *Primate Reports*, 27, 1–101.

Caldwell, C. A. & Whiten, A. 2004. Testing for social learning and imitation in common marmosets, *Callithrix jacchus*, using an "artificial fruit." *Animal Cognition*, 7, 77–85.

Call, J. 2001. Chimpanzee social cognition. *Trends in Cognitive Sciences*, 5, 369–405.

———. 2007. Social knowledge in primates. In *Handbook of Evolutionary Psychology* (ed. by Dunbar, R. I. M. & Barrett, L.), 1–81. Oxford: Oxford University Press.

Call, J., Agnetta, B. & Tomasello, M. 2000. Cues that chimpanzees do and do not use to find hidden objects. *Animal Cognition*, 3, 23–34.

Call, J., Bräuer, J., Kaminski, J. & Tomasello, M. 2003. Domes-

tic dogs are sensitive to the attentional state of humans. *Journal of Comparative Psychology*, 117, 257–263.

Call, J. & Carpenter, M. 2001. Do apes and children know what they have seen? *Animal Cognition*, 4, 207–220.

Call, J., Hare, B. H., Carpenter, M. & Tomasello, M. 2004. "Unwilling" versus "Unable": Chimpanzees' understanding of human intentional action? *Developmental Science*, 7, 488–498.

Call, J., Hare, B. & Tomasello, M. 1998. Chimpanzee gaze following in an object-choice task. *Animal Cognition*, 1, 89–99.

Call, J. & Tomasello, M. 1994. Production and comprehension of referential pointing by orangutans (*Pongo pygmaeus*). *Journal of Comparative Psychology*, 108, 307–317.

———. 1998. Distinguishing intentional from accidental actions in orangutans (*Pongo pygmaeus*), chimpanzees (*Pan troglodytes*), and human children (*Homo sapiens*). *Journal of Comparative Psychology*, 112, 192–206.

———. 1999. A nonverbal theory of mind test. The performance of children and apes. *Child Development*, 70, 381–395.

———. 2005. What chimpanzees know about seeing revisited: An explanation of the third kind. In *Joint Attention: Communication and Other Minds* (ed. by Eilan, N., Hoerl, C., McCormack, T. & Roessler, J.), 45–64. Oxford: Oxford University Press.

———. 2008. Does the chimpanzee have a theory of mind? 30 years later. *Trends in Cognitive Sciences*, 12, 187–192.

Carpenter, M., Akhtar, N. & Tomasello, M. 1998. Fourteen- to 18-month-old infants differentially imitate intentional and accidental actions. *Infant Behavior and Development*, 21, 315–330.

Carruthers, P. 2009. How we know our own minds: The relationship between mindreading and metacognition. *Behavioral and Brain Sciences*, 32, 121–182.

Coussi-Korbel, S. 1994. Learning to outwit a competitor in mangabeys, *Cercocebus t. torquatus*. *Journal of Comparative Psychology*, 108, 164–171.

Custance, D. M., Whiten, A. & Fredman, T. 1999. Social learning of "artifical fruit" processing in capuchin monkeys (*Cebus apella*). *Journal of Comparative Psychology*, 113, 13–23.

Custance, D. M., Whiten, A., Sambrook, T. & Galdikas, B. 2001. Testing for social learning in the "artificial fruit" processing of wildborn orangutans (*Pongo pygmaeus*), Tanjung Puting, Indonesia. *Animal Cognition*, 4, 305–313.

De Waal, F. B. M. 1982. *Chimpanzee Politics: Power and Sex among Apes*. London: Jonathan Cape.

———. 2000. Primates: A natural heritage of conflict resolution. *Science*, 289, 586–590.

———. 2000. Attitudinal reciprocity in food sharing among brown capuchin monkeys. *Animal Behaviour*, 60, 253–261.

Ferrari, P. F., Kohler, E., Fogassi, L. & Gallese, V. 2000. The ability to follow eye gaze and its emergence during development in macaque monkeys. *Proceedings from the National Academy of Sciences USA*, 97. 13997–14002.

Flombaum, J. I. & Santos, L. R. 2005. Rhesus monkeys attribute perceptions to others. *Current Biology*, 15, 447–452.

Fragaszy, D. & Visalberghi, E. 2004. Socially-biased learning in monkeys. *Learning & Behavior*, 32, 24–35.

Ghazanfar, A. A. & Santos, L. R. 2004. Primate brains in the wild: The sensory bases for social interactions. *Nature Reviews Neuroscience*, 5, 603–616.

Gilby, I. C. 2006. Meat sharing among the Gombe chimpanzees:

Harassment and reciprocal exchange. *Animal Behaviour*, 71, 953–963.

Gómez, J. C. 1991. Visual behavior as a window for reading the mind of others in primates. In *Natural Theories of Mind*, (ed. by Whiten, A.), 195–208. Oxford: Blackwell.

———. 2004. *Apes, Monkeys, Children and the Growth of Mind*. Cambridge, MA: Harvard University Press.

Goossens, B. M. A., Dekleva, M., Reader, S.M., Sterck, E. H. M. & Bolhuis, J. J. 2008. Gaze following in monkeys is modulated by observed facial expressions. *Animal Behaviour*, 75, 1673–1681.

Hampton, R. 2001. Rhesus monkeys know when they remember. *Proceedings of the National Academy of Sciences*, 98, 5359–5362.

Hare, B. 2001. Can competitive paradigms increase the validity of social cognitive experiments on primates? *Animal Cognition*, 4, 269–280.

Hare, B., Addessi, E., Call, J., Tomasello, M. & Visalberghi, E. 2003. Do capuchin monkeys (*Cebus apella*) know what conspecifics do and do not see? *Animal Behaviour*, 65, 131–142.

Hare, B., Call, J., Agnetta, B. & Tomasello, M. 2000. Chimpanzees know what conspecifics do and do not see. *Animal Behaviour*, 59, 771–786

Hare, B., Call, J. & Tomasello, M. 1998. Communication of food location between human and dog (*Canis familiaris*). *Evolution of Communication*, 2, 137–159.

———. 2001. Do chimpanzees know what conspecifics know? *Animal Behaviour*, 61, 139–151.

———. 2006. Chimpanzees deceive a human competitor by hiding. *Cognition*, 101, 495–514.

Hare, B., Melis, A. P., Woods, V., Hastings, S. & Wrangham, R. 2007. Tolerance allows bonobos to outperform chimpanzees on a cooperative task. *Current Biology*, 17, 619–623.

Hare, B. & Tomasello, M. 1999. Domestic dogs (*Canis familiaris*) use human and conspecific social cues to locate hidden food. *Journal of Comparative Psychology*, 113, 1–5.

———. 2004. Chimpanzees are more skillful in competitive than in cooperative cognitive tasks. *Animal Behaviour*, 68, 571–581.

Hattori, Y., Kuroshima, H. & Fujita, K. 2007. I know you are not looking at me: Capuchin monkeys' (*Cebus apella*) sensitivity to human attentional states. *Animal Cognition*, 10, 141–148.

Hauser, M. D. 1992. Costs of deception: Cheaters are punished in rhesus monkeys. *Proceedings of National Academy of Sciences, U.S.A.*, 89, 12137–12139.

Henrich, J., McElreath, R., Barr, A., Ensimger, J., Barrett, C., Bolyanatz, A., Cardenas, J. C., Gurven, M., Gwako, E., Henrich, N., Lesorogol, C., Marlowe, F., Tracer, D. & Ziker, J. 2006. Costly punishment across human societies. *Science*, 312, 1767–1770.

Herrmann, E. & Tomasello, M. 2006. Apes' and children's understanding of cooperative and competitive motives in a communicative situation. *Developmental Science*, 9, 518–529.

Herrmann, E., Melis, A. P., & Tomasello, M. 2006. Apes' use of iconic cues in the object-choice task. *Animal Cognition*, 9, 118–130.

Hesler, N. & Fischer, J. 2007. Gestures in Barbary macaques. In *Gestural Communication in Monkeys and Apes* (ed. by Tomasello, M., and Call, J.), 159–195. Erlbaum: New Jersey.

Heyes, C. M. 1993. Anecdotes, training, trapping and triangulating: Can animals attribute mental states? *Animal Behaviour*, 46, 177–188

———. 1998. Theory of mind in nonhuman primates. *Behavioral & Brain Sciences*, 21, 101–134.

Hirata, S. & Matsuzawa, T. 2001. Tactics to obtain a hidden food item in chimpanzee pairs (*Pan troglodytes*). *Animal Cognition*, 4, 285–295.

Hirata, S. & Morimura, N. 2000. Naive chimpanzees' (*Pan troglodytes*) observation of experienced conspecifics in a tool-using task. *Journal of Comparative Psychology*, 114, 291–296.

Horner, V. & Whiten, A. 2005. Causal knowledge and imitation/emulation switching in chimpanzees (*Pan troglodytes*) and children (*Homo sapiens*). *Animal Cognition*, 8, 164–181.

Horner, V., Whiten, A., Flynn, E. & de Waal, F. B. M. 2006. Faithful replication of foraging techniques along cultural transmission chains by chimpanzees and children. *Proceedings of National Academy of Sciences, U.S.A.*, 13878–13883.

Horton, K. E., Caldwell, C. A. 2006. Visual co-orientation and expectations about attentional orientation in pileated gibbons (*Hylobates pileatus*). *Behavioral Processes*, 72, 65–73.

Hostetter, A. B., Cantero, M. & Hopkins, W. D. 2001. Differential use of vocal and gestural communication by chimpanzees (*Pan troglodytes*) in response to the attentional status of a human (*Homo sapiens*). *Journal of Comparative Psychology*, 115, 337–343.

Humphrey, N. K. 1976. The social function of intellect. In *Growing Points in Ethology* (ed. by Bateson, P. P. G. & Hinde, R.A.), 303–317. Cambridge: Cambridge University Press.

Inoue, Y., Inoue, E., & Itakura, I. 2004. Use of experimenter-given directional cues by a young white-handed gibbon (*Hylobytes lar*). *Japanese Psychological Research*, 46, 262–267.

Itakura, S. 1996. An exploratory study of gaze-monitoring in non-human primates. *Japanese Psychological Research*, 38, 174–180.

Itakura, S., Agnetta, B., Hare, B., & Tomasello, M. 1999. Chimpanzees' use of human and conspecific social cues to locate hidden food. *Developmental Science*, 2, 448–456.

Jellema, T., Baker, C. I., Wicker, B., & Perrett, D. I. 2000. Neural representation for the perception of the intentionality of actions. *Brain and Cognition*, 44, 280–302.

Jensen, K., Call, J., & Tomasello, M. 2007. Chimpanzees are rational maximizers in an ultimatum game. *Science*, 318, 107–109.

Jolly, A. 1966. Lemur social behavior and primate intelligence. *Science*, 153, 501–506.

Kaminski, J., Call, J. & Tomasello, M. 2004. Body orientation and face orientation: Two factors controlling apes' begging behavior from humans. *Animal Cognition*, 7, 216–223.

———. 2008. Chimpanzees know what others know but not what they believe. *Cognition*, 109, 224–234.

Kaminski, J., Riedel, J., Call, J., & Tomasello, M. 2005. Domestic goats, *Capra hircus*, follow gaze direction and use social cues in an object choice task. *Animal Behaviour*, 69, 11–18.

Karin-D'Arcy, R., & Povinelli, D. J. 2002. Do chimpanzees know what each other see? A closer look. *International Journal of Comparative Psychology*, 15, 21–54.

Kornell, N., Son, L., & Terrace, H. 2007. Transfer of metacognitive skills and hint seeking in monkeys. *Psychological Science*, 18, 64–71.

Kummer, H. 1982. Social knowledge in free-ranging primates.

In *Animal Mind-Human Mind*, ed. by D. R. Griffin. Berlin: Springer-Verlag.

Liebal, K., Call, J., & Tomasello, M. 2004. Use of gesture sequences in chimpanzees. *American Journal of Primatology*, 64, 377–396.

Liebal, K., Mueller, C., Pika, S. 2007. *Gestural Communication in Nonhuman and Human Primates*. Amsterdam: John Benjamins Publishing Company.

Liebal, K., Pika, S., Call, J., & Tomasello, M. 2004. To move or not to move: How apes adjust to the attentional state of others. *Interaction Studies*, 5, 199–219.

Lyons, D. E. & Santos, L. R. 2006. Ecology, domain specificity, and the evolution of theory of mind: Is competition the catalyst? *Philosophy Compass*, 1, 481–492.

Lyons, D. E., Phillips, W. & Santos, L. R. 2005. Motivation is not enough: A commentary on Tomasello et al.,'s "Understanding and sharing intentions: The origins of cultural cognition. *Behavioural and Brain Sciences*, 28, 708.

Melis, A. P., Call, J. & Tomasello, M. 2006a. Chimpanzees (*Pan troglodytes*) conceal visual and auditory information from others. *Journal of Comparative Psychology*, 120, 154–162.

Melis, A. P., Hare, B. & Tomasello, M. 2006b. Engineering cooperation in chimpanzees: Tolerance constraints on cooperation. *Animal Behaviour*, 72, 275–286.

Meltzoff, A. N. 1995. Understanding the intentions of others: Re-enactment of intended acts by 18-month-old children. *Developmental Psychology*, 31, 838–850.

Menzel, E. W. 1973. Leadership and communication in young chimpanzees. In *Precultural Primate Behaviour* (ed. by E. W. Menzel), 192–225. Basel: Karger.

Miklósi, A., Polgárdi, R., Topál, J. & Csányi V. 1998. Use of experimenter-given cues in dogs. *Animal Cognition*, 1, 113–122.

Mitchell, J. P. 2008. Contributions of functional neuroimaging to the study of social cognition. *Current Directions in Psychological Science*, 17, 142–146.

Mulcahy, N. & Call, J. 2009. The performance of bonobos (*Pan paniscus*), chimpanzees (*Pan troglodytes*) and orangutans (*Pongo pygmaeus*) in two versions of an object choice task. *Journal of Comparative Psychology*, 123, 304–309.

Myowa-Yamakoshi, M. & Matsuzawa, T. 2000. Imitation of intentional manipulatory actions in chimpanzees (*Pan troglodytes*). *Journal of Comparative Psychology*, 114, 381–391.

Neiworth, J. J., Burman, M. A., Basile, B. M. & Lickteig, M. T. 2002. Use of experimenter-given cues in visual co-orienting and in an object-choice task by a New World monkey species, cotton top tamarins (*Saguinus oedipus*). *Journal of Comparative Psychology*, 116, 3–11.

Peignot, P. & Anderson, J. R. 1999. Use of experimenter-given manual and facial cues by gorillas (*Gorilla gorilla*) in an object-choice task. *Journal of Comparative Psychology*, 113, 253–260.

Pelphrey, K. A., Singerman, J. D., Allison, T., & McCarthy, G. 2003. Brain activation evoked by perception of gaze shifts: The influence of context. *Neuropsychologia*, 41, 156–170.

Pelphrey, K.A., Viola, R. J., & McCarthy, G. 2004. When strangers pass: Processing of mutual and averted social gaze in the superior temporal sulcus. *Psychological Science*, 15, 598–603.

Penn, D.C. & Povinelli, D.J. 2007. On the lack of evidence that chimpanzees possess anything remotely resembling a "theory of mind." *Philosophical Transactions of the Royal Society B*, 362, 731–744.

Phillips, W., Barnes, J. L., Mahajan, N., Yamaguchi, M. & Santos, L. R. 2009. "Unwilling" versus "unable": Capuchin monkeys' (*Cebus apella*) understanding of human intentional action. *Developmental Science*, 12(6), 938–945.

Pika, S. 2007. Gestures in sub-adult gorillas. In *The Gestural Communication of Apes and Monkeys* (ed. by J. Call & M. Tomasello), 99–130. New York: Erlbaum.

Povinelli, D. J. 2004. Behind the ape's appearance: Escaping anthropomorhism in the study of other minds. *Daedalus: Journal of the American Academy of Arts and Sciences*, Winter, 29–41.

Povinelli, D. J. & Barth, J. 2005. Reinterpreting behavior: A human specialization? Commentary on Tomasello et al. Understanding and sharing intentions: The origins of cultural cognition. *Behavioral and Brain Sciences*, 28, 712–713.

Povinelli, D. J. & Eddy, T. J. 1996a. Chimpanzees: Joint visual attention. *Psychological Science*, 7, 129–135.

———. 1996b. What young chimpanzees know about seeing. *Monographs of the Society for Research in Child Development*, 61, 1–152.

Povinelli, D. J., Nelson, K. E. & Boysen, S.T. 1990. Inferences about guessing and knowing by chimpanzees (*Pan troglodytes*). *Journal of Comparative Psychology*, 104, 203–210.

Povinelli, D., Perilloux, H., Reaux, J., & Bierschwale, D. 1998. Young chimpanzees' reactions to intentional versus accidental and inadvertent actions. *Behavioral Processes*, 42, 205–218.

Povinelli, D. J. & Vonk, J. 2004. We don't need a microscope to explore the chimpanzee's mind. *Mind and Language*, 19, 1–28.

Premack, D. 1988. 'Does the chimpanzee have a theory of mind?' revisited. In *Machiavellian Intelligence: Social Expertise and the Evolution of Intellect in Monkeys, Apes and Humans*, (ed. by Byrne, W. & Whiten, A.). Oxford: Oxford University Press.

Premack, D. & Woodruff, G. 1978. Does the chimpanzee have a theory of mind? *Behavioral and Brain Sciences*, 1, 515–526.

Riedel, J., Buttlemann, D., Call, J. & Tomasello, M. 2006. Domestic dogs (*Canis familiaris*) use a physical marker to locate hidden food. *Animal Cognition*, 9, 27–35.

Rosati, A. G., Santos, L. R. & Hare, B. 2009. Primate social cognition: Thirty years after Premack and Woodruff. In *Primate Neuroethology* (ed. by Platt, M. & Ghazanfar A. A.), 117–143. New York: Oxford University Press.

Ruiz, A., Gómez, J. C., Roeder, J. J., & Byrne, R. W. 2009. Gaze following and gaze priming in lemurs. *Animal Cognition*, 12, 427–434.

Russell, Y. I., Call, J. & Dunbar, R. I. M. 2008. Image scoring in great apes. *Behavioural Processes*, 78, 108–111.

Santos, L. R., Flombaum, J. I. & Phillips, W. 2006. The evolution of human mind reading. In *Evolutionary Cognitive Neuroscience* (ed. by Platek, S., Keenan, J. P. & Shackelford, T. K.), 433–456. Cambridge: MIT Press.

Santos, L. R., Nissen, A. G. & Ferrugia, J. 2006. Rhesus monkeys (*Macaca mulatta*) know what others can and cannot hear. *Animal Behaviour*, 71, 1175–1181.

Savage-Rumbaugh, E. S., Rumbaugh, D. & Boysen, S. T. 1978. Sarah's problems in comprehension. *Behavioral and Brain Sciences*, 1, 555–557.

Saxe, R., Carey, S., & Kanwisher, N. 2004. Understanding other minds: Linking developmental psychology and functional neuroimaging. *Annual Review of Psychology*, 55, 87–124.

Saxe, R. & Wexler, A. 2005. Making sense of another mind: The role of the right temporo-parietal junction. *Neuropsychologia*, 43, 1391–1399.

Scerif, G., Gómez, J. C.,& Byrne, R. W. 2004. What do Diana monkeys know about the focus of attention of a conspecific? *Animal Behaviour*, 68, 1239–1247.

Scheumann, M. & Call, J. 2004. The use of experimenter-given cues by South African fur seals (*Arctocephalus pusillus*). *Animal Cognition*, 7, 224–230.

Shepherd, S. V. & Platt, M. L. 2008. Spontaneous social orienting and gaze following in ringtailed lemurs (*Lemur catta*). *Animal Cognition*, 11, 13–20.

Smith, J., Shields, W. & Washburn, D. 2003. The comparative psychology of uncertainty monitoring and metacognition. *Behavioral and Brain Sciences*, 26, 317–373.

Stoinski, T. S., Wrate, J. L., Ure, N. & Whiten, A. 2001. Imitative learning by captive western lowland gorillas (*Gorilla gorilla gorilla*) in a simulated food-processing task. *Journal of Comparative Psychology*, 115, 272–281.

Stoinski, T. S. & Whiten, A. 2003. Social learning by orangutans in a simulated food-processing task. *Journal of Comparative Psychology*, 117, 272–282.

Subiaul. F., Vonk, J., Barth, J., & Okamoto-Barth, S. 2008. Chimpanzees learn the reputation of strangers by observation. *Animal Cognition*, 11, 611–623.

Suddendorf, T. & Whiten, A. 2001. Mental evolution and development: Evidence for secondary representation in children, great apes and other animals. *Psychological Bulletin*, 127, 629–650.

Tomasello, M. & Call, J. 1997. *Primate Cognition*. Oxford: Oxford University Press.

Tomasello, M., Call, J., & Gluckman, A. 1997. The comprehension of novel communicative signs by apes and human children. *Child Development*, 68, 1067–1081.

Tomasello, M., Call, J., & Hare, B. 1998. Five primate species follow the visual gaze of conspecifics. *Animal Behaviour*, 55, 1063–1069.

———. 2003. Chimpanzees understand psychological states: The question is which ones and to what extent. *Trends in Cognitive Science*, 7, 153–156.

Tomasello, M., Call, J., Warren, J., Frost, T., Carpenter, M., & Nagell, K. 1997. The ontogeny of chimpanzee gestural signals: A comparison across groups and generations. *Evolution of Communication*, 1, 223–253.

Tomasello, M., & Carpenter, M. 2005. The emergence of social cognition in three young chimpanzees. *Monographs of the Society for Research in Child Development*, 70 (1, serial no. 279).

Tomasello, M., Carpenter, M., Call, J., Behne, T., & Moll, H. 2005. Understanding and sharing intentions: The origins of cultural cognition. *Behavioral and Brain Sciences*, 28, 675–691.

Tomasello, M., Hare, B., & Agnetta, B. 1999. Chimpanzees, *Pan troglodytes*, follow gaze direction geometrically. *Animal Behaviour*, 58, 769–777.

Tomasello, M., Hare, B., & Fogleman, T. 2001. The ontogeny of gaze following in chimpanzees, *Pan troglodytes*, and rhesus macaques, *Macaca mulatta*. *Animal Behaviour*, 61, 335–343

Tschudin, A., Call, J., Dunbar, R. I. M., Harris, G., & van der Elst, C. 2001. Comprehension of signs by dolphins (*Tursiops truncatus*). *Journal of Comparative Psychology*, 115, 100–105.

Vick, S. J., & Anderson, J. R. 2000. Learning and limits of use of eye gaze by capuchin monkeys (*Cebus apella*) in an object-choice task. *Journal of Comparative Psychology*, 114, 200–207.

———. 2003. Use of human visual attention cues by olive baboons (*Papio anubis*) in a competitive task. *Journal of Comparative Psychology*, 117, 209–216.

Visalberghi, E. & Addessi, E. 2003. Food for thoughts: Social learning and the feeding behavior in capuchin monkeys. Insights from the laboratory. In *Traditions in Non-Human Animals: Models and Evidence* (ed. by Fragaszy, D. & Perry, S.), 187–212. Cambridge: Cambridge University Press.

Voelkl, B. & Huber, L. 2007 Imitation as faithful copying of a novel technique in marmoset monkeys. *PLOS One*, 2, e611.

Von Bayern, A. M. P. & Emery, N. J. 2009. Jackdaws respond to human attentional states and social cues in different contexts. *Current Biology*, 19, 602–606.

Vonk, J., Brosnan, S., Silk, J. B., Henrich, J., Schapiro, S., Richardson, A., Lambeth, S. P., & Povinelli, D. J. 2008. Chimpanzees do not take advantage of very low cost opportunities to deliver food to unrelated group members. *Animal Behavior*, 75, 1757–1770.

Warneken, F., Hare, B., Melis, A.P., Hanus, D. & Tomasello, M. 2007. Spontaneous altruism by chimpanzees and young children. *PLoS Biology*, 5, e184.

Warneken, F. & Tomasello, M. 2006. Altruistic helping in human infants and young chimpanzees. *Science*, 311, 1301–1303.

Whiten, A. 1994. Grades of mindreading. In *Children's Early Understanding of Mind* (ed. by Lewis, C. & Mitchell, P.), 47–70. Hillsdale, NJ: Lawrence Erlbaum Associates.

———. 1998. Imitation of the sequential structure of actions by chimpanzees (*Pan troglodytes*). *Journal of Comparative Psychology*, 112, 270–281.

———. 2000. Chimpanzee cognition and the question of mental re-representation. In *Meta-Representations: A Multidisciplinary Perspective* (ed. by Sperber, D.), 139–167. Oxford: Oxford University Press.

Whiten, A., & Byrne, R.W. 1991. The emergence of metarepresentation in human ontogeny and primate phylogeny. In *Natural Theories of Mind* (ed. by Whiten, A.), 267–281. Oxford: Blackwell.

Whiten, A., Custance, D., Gómez, J. C., Teixidor, P., & Bard, K.A. 1996. Imitative learning of artificial fruit processing in children (*Homo sapiens*) and chimpanzees (*Pan troglodytes*). *Journal of Comparative Psychology*, 110, 3–14.

Whiten, A., Goodall, J., McGrew, W., Nishida, T., Reynolds, V., Yugiyama, Y., Tutin, C., Wrangham, R. & Boesch, C. 1999. Cultures in chimpanzees. *Nature*, 399, 682–685.

Whiten, A., Horner, V. & de Waal, F. B. M. 2005. Conformity to cultural norms of tool use in chimpanzees. *Nature*, 437, 737–740.

Chapter 31 Social Learning, Traditions, and Culture

Andrew Whiten

HUMANITY'S CAPACITY for culture is one of the defining characteristics of our species. It has allowed humans to exploit and dominate the planet, and has created forms and levels of behavioral diversity to which no other species comes close. Accordingly, it's easy to think that human culture somehow divorces us from the biological world, releasing us to a large extent from the influence of Darwinian evolutionary forces that govern the rest of life on earth, and separating us qualitatively from even our closest relatives among the primates.

There is some essential truth in this. You need only look around you at the rich and multifarious manifestations of your culture to acknowledge the gulf between us and other primates. Nevertheless, one of the most exciting areas of primatology has increasingly suggested an equal truth: that our cultural nature did not make a sudden recent appearance, but instead evolved from origins about which compelling inferences can be made through comparative primatology. In the last decade or so, the literature describing these discoveries has mushroomed to proportions that no single scientist can encompass—a situation at once exhilarating and daunting. Surveying the high points of this body of work and supplying productive entry points to its deeper exploration are the central aims of this chapter.

To put this review in context, it is important to note that primatology does not stand alone in charting the evolutionary history of the phenomena of interest. Much has been discovered about the evolutionary roots of social learning and tradition through studies of broader swaths of the animal kingdom (Fragaszy & Perry 2003; Whiten et al. 2011),

especially fish (Brown & Laland 2003; Laland & Hoppitt 2003), birds (Lefebvre & Bouchard 2003; Zentall 2004) and nonprimate mammals (Laland & Galef 2009), as well as invertebrates such as insects (Leadbeater & Chittka 2007). However, studies of primates have been particularly influential in shaping our present understanding of social learning and culture.

Another reason not to divorce culture from the rest of biology is that once it is in place, culture can give rise to a new level of evolution: cultural evolution, in which traditions diversify progressively in ways that parallel Darwinan biogenetic evolution. A growing literature describes this, with an inevitable focus on the complexities of our own relatively recent past (Cavalli-Sforza & Feldman 1981; Boyd & Richerson 1985; Richerson & Boyd 2005; Mesoudi et al. 2006; Whiten et al. 2011, 2012). One can think of the evolutionary picture deriving from primatology as sitting, influentially, between the broader animal perspective and the narrower hominin-focused perspective, each of which has its own blossoming literature (Whiten et al. 2012).

This chapter is necessarily selective in its portrayal of this rapidly expanding field. Readers are urged to consult recent complementary works—including those of Subiaul (2007), Rapaport and Brown (2008), Laland and Galef (2009), and Caldwell and Whiten (2010)—which cover important topics that are absent or mentioned only briefly here.

Key Concepts and Terms

I begin by outlining the key "top-level" terms and concepts in this field. More finely distinguished terms are dealt with in relevant sections below.

Heyes (1994) defined social learning as "learning that is influenced by observation of, or interaction with, another animal (typically a conspecific) or its products" (p. 207). This can include such manifestations as copying others' actions (imitation), but it also captures such things as learning from discarded tools or half-processed food items. Social learning is thus differentiated from individual learning, in which an animal learns through its own efforts alone. In social learning we always need to consider at least two individuals: the learner and the learned-from. The latter may possess skills or knowledge that the learner can benefit from assimilating, even if imperfectly. On the other hand, a social learner is vulnerable to picking up maladaptive behavior from others. It is important to recognize that "social learning" covers an enormous gamut of processes, from the entire content of a university education, in the case of some primates, to the experience of having one's learning merely "socially biased" (Fragaszy & Visalberghi 2004) by, for example, following one's mother and developing preferences for certain routes or food trees.

Social learning may be a relatively transitory phenomenon, as when a monkey learns from others' foraging behavior that tree X is a good one to visit—information that is applicable only for a limited period. By contrast, social learning about more durable phenomena can give rise to traditions (fig. 31.1), defined by Fragaszy and Perry (2003, p. xiii) as "a distinctive behavior pattern shared by two or more individuals in a social unit, which persists over time and that new practitioners acquire in part through socially aided learning." We shall look at some of the most interesting discoveries about primate traditions further below.

Defining "culture" is more contentious. Many researchers studying animal traditions treat culture and tradition (as defined above) as synonyms. Others, including many sociocultural anthropologists, note that human cultures represent the products of complex systems of knowledge, belief, and values, and involve more than just the transmission of a set of traditions, so that the terms "culture" and "traditions" should not be conflated. Others worry that use of the term "culture" might be taken to imply that "culture" in animals is assumed to be *homologous* with human culture (sharing common evolutionary roots), as opposed to its possibly being only *analogous* (similar in some respects, but with no direct, ancestral evolutionary linkage; Galef 1992, 2009). Homology may exist, but it cannot be assumed a priori. Galef thus argues that the use of the term "culture" should

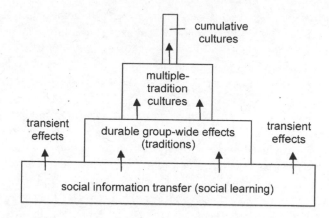

Fig. 31.1. Culture pyramid. The large base layer represents processes of social information transfer, which are increasingly demonstrated to be widespread among vertebrates and possibly invertebrates too (Danchin et al. 2004). Many of these have only transient effects such as focusing attention on a currently productive food source, but others may give rise to the sub-set of consequences that are traditions (level 2). "Cultures" have been distinguished from the existence of a tradition in several ways (see text), one of which focuses on the richness of phenomena associated with multiple traditions, so this is a third layer. Finally, a subset of cultures exhibit cumulative evolutionary changes, most notably in humans. The relative sizes of layers are notional. After Whiten & van Schaik 2007.

hinge on demonstrating humanlike forms of social transmission, such as imitation and teaching. Other researchers adopt different, principled criteria for describing some primate traditions as cultures, and we will look at these more closely later in this chapter (see also fig. 31.1).

A Century of Research

Inspired by the revelations of Darwin and Wallace, comparative psychology and primatology began to establish a fascinating scientific literature concerning social learning, based on experiments with captive animals, around the beginning of the twentieth century. The field has progressed ever since, creating a formidable reading list for anyone who aspires to master this field. Comprehensive listings and selective reviews of the achievements of the twentieth century are provided by Whiten and Ham (1992), Tomasello and Call (1997), and Subiaul (2007).

One can sympathize with the budding scholar of today who, in struggling to assimilate the burgeoning literature of the last decade or two, ignores the older literature in the process. However, this risks neglecting many wonderful pockets of early originality and discovery. To give just a single recent illustration of this, the new literature on culture-diffusion experiments (reviewed below; Whiten & Mesoudi 2008) for many years neglected to cite the earliest pioneering study on the topic, a gem "years before its time," by E. Menzel et al. (1972).[1] Menzel and colleagues exposed three young chimpanzees (*Pan troglodytes*) to novel objects, and then through repeated replacement of the most experienced chimp in the trio with a naïve one, demonstrated the buildup of consistent cultural attitudes of bravura and exploration of objects, an attitude that did not exist in the original trio.

Serious field research relating to social learning and culture began only later in the twentieth century. First came the now-famous studies of the spread of innovations such as sweet-potato washing among Japanese macaques (*Macaca fuscata*) by Imanishi, Itani, Kawai, Kawamura, and their colleagues in the 1950s, well reviewed by Nishida (1987) and updated by Hirata et al. (2001). However, despite such behavioral innovations becoming routinely cited in textbooks as examples of imitation and tradition, the basis for concluding that their spread relied on social learning has more recently been debated. Some critics have noted that spread of the new habits was too slow to be consistent with imitation. Others have expressed worries that the behavior was shaped by the selective rewards given by caretaking staff. These controversies can be followed further through the pages of Galef (1991), de Waal (2001), Hirata et al. (2001), and Caldwell and Whiten (2010). Below I discuss newer, and I believe more compelling, findings from Japanese monkey studies.

The next major development came in the 1960s, when serious ape field research began. Following her discoveries about different forms of tool use, it was not long before Goodall (1973) was writing about the signs of "cultural elements in a chimpanzee community." It is here that we begin to examine the subject in more depth.

Field Studies of Primate Cultures

A Regional "Cross-cultural" Perspective

As chimpanzee field studies proliferated across Africa, researchers began to realize that behavior patterns varied across the different sites, and some such differences were inferred to be likely local traditions (Nishida et al. 1983; Goodall 1986). Evidence from experimental studies with captive subjects has shown that apes readily learn different forms of tool use, so that consistent intersite variations are unlikely to be genetically based, and instead represent local traditions. The accumulating evidence was assembled by McGrew (1992) in a landmark volume boldly titled *Chimpanzee Material Culture*. Charts of variations across sites were later extended to include social and other forms of behavior as more studies accumulated (Boesch & Tomasello 1998).

However, these pioneering studies were based largely on what had been published for each site, so the growing cultural picture was likely incomplete. Primatologists do not necessarily publish full lists of the behavior patterns at their sites, and are particularly unlikely to publish accounts of those that are never seen locally, but are known elsewhere. To achieve a more complete picture, the leaders of the nine longest-term study sites pooled their data, which spanned a total of more than 150 years, extracting from it the behavior patterns that were common for at least one chimpanzee community yet absent at others, with no obvious genetic or environmental explanation for the difference (for example, absence of a given behavior was not of interest where the materials needed for it, such as nuts in the case of nut cracking with natural hammers, were unavailable). This procedure identified as many as 39 chimpanzee traditions across Africa, spanning a variety of types of behavior including foraging techniques, tool use, grooming, social styles, and courtship styles (Whiten et al. 1999, 2001; Lycett et al. 2009; see also chapter 11, this volume, fig. 11.4). For example, in "pestle pounding" a chimpanzee climbs into the top of a palm tree and uses a large frond to pound into the growing point and extract the nutritional pulp below—a striking behavior pattern customary at Bossou in West Africa yet absent a few hundred kilometers away in the Taï Forest. The existence of similar palm trees and the same subspecies of chimpanzee at these sites led the authors to exclude environmental or genetic explanations of the difference, and to identify the behavior as a local tradition at Bossou.

Identifying 39 such variants was remarkable because most reports of animal traditions identify only a single cultural variant, such as birdsong dialects or pine-cone stripping by black rats (Heyes & Galef 1996). The methodology used in the chimpanzee study was soon applied to orangutans (*Pongo* spp.), producing a remarkably similar picture, with two dozen traditions covering a variety of technical and social domains including food processing techniques, tool use, and communication patterns (fig. 31.2; van Schaik et al. 2003). The existence of such multiple-tradition cultures offers a further way in which the concept of "culture" can be interestingly differentiated from the case of a species displaying but a single specific tradition, such as the dialect of a songbird (fig. 31.1).

A certain degree of evolutionary convergence on such cultural complexity has since been identified in some New World primates. Researchers studying white-faced capuchins (*Cebus capucinus*) have pooled their data and identified putative intersite cultural variations in foraging behavior, such as in following army ants to catch the prey they flush out, which is habitual at one site but not at others

Fig. 31.2. Stick tools used to dislodge highly nutritious seeds from the dehiscent woody fruits of *Neesia* while avoiding contact with the stinging hairs lining the open valves. Shown are *Neesia* fruits, one with a tool still inserted, several used tools, and the remainder of a twig used to make a tool. These tools were used by an adult male Sumatran orangutan (*Pongo abelii*), but all individuals (except young infants) in the Suaq Balimbing population are known to use such tools (van Schaik 2004). Photo courtesy of Ellen Meulman.

(Panger et al. 2002; see also Ottoni et al. 2005; Perry 2009), as well as in interactions with other species (Rose et al. 2003). A groundbreaking analysis concerned variation in such strange "social conventions" as inserting fingers into the mouth, nostrils, and even eyes of group mates (fig. 31.3), and various "games" in which small objects such as hairs are put in one monkey's mouth and extracted by another (Perry et al. 2003). These provide compelling evidence for traditions because (1) as the actions were purely social, environmental explanation was unlikely, and (2) in addition to identifying intersite differences, both the spread and the eventual decline of several of these behavior patterns were documented as existing in ways difficult to explain by processes other than social learning. The authors suspect that these peculiar conventions function to test social bonds.

An intriguing Old World counterpart to these discoveries comes from studies of the curious behavior of "stone handling" in Japanese macaques. This involves picking up a few stones and manipulating them in various ways, such as clacking them together or cuddling them. This has become a compelling example of a primate tradition, because its spread has been tracked from its earliest manifestation in just one or two animals to its customary performance across a group (Huffman 1996). Neither genetic nor environmental factors seem to offer plausible explanations for the spread of this apparently functionless activity. Most recently it has been discovered that each of 10 widely dispersed groups that developed this behavior show multiple variants that differentiate them culturally (Leca et al.

2007), echoing the patterning described above for the apes, but here restricted to just the one narrow domain of stone handling (fig. 31.4).

Sapolsky and Share (2004) have described a quite different kind of social convention spreading among olive baboons (*Papio anubis*). The origin of this lay in the deaths through food poisoning of the boldest and most aggressive members of a baboon troop, following their exploitation of human food waste at a tourist lodge. Lacking these males, the troop shifted to a more affiliative balance of social interactions, an "ethos" that was maintained over years when new males entered and which conformed to what Sapolsky and Share described as the new "pacific culture" of the troop.

It is difficult to directly compare these various monkey and ape studies, because only the chimpanzee and orangutan studies have attempted to quantify the variety of traditions in the same way. At present these chimpanzee and orangutan cultures, now described as displaying more than 40 and 30 different social and technical traditions respectively (van Schaik 2009; Whiten 2010) appear to remain unmatched in their scope and extent. The convergence on this pattern of multiple differences across sites in the case of Japanese macaque stone handling, by contrast, concerns variation in just one particular behavior pattern. The closest convergence with the ape picture is the one in capuchins, but here again the range of behavior patterns is relatively restricted, with most of the foraging variants listed by Panger et al. (2002) involving application of the same techniques, such as rubbing and pounding, to different target foodstuffs. Accordingly, Whiten and van Schaik (2007) tentatively concluded that the great apes display a distinctive cultural complexity, possibly associated with their relatively large brain size (the "cultural intelligence hypothesis") and implying that the common great ape ancestor of around 14 million years ago is likely to have begun showing this complexity, continued and elaborated in different ways by its descendants.

Turning an Observational Spotlight onto Social Learning

The relatively "macro" levels of field study outlined above are complemented by "micro" studies examining the development of putative cultural behaviors. For example, Lonsdorf (2005; Lonsdorf et al. 2003) found that young female chimpanzees, who spent significantly more time than young males watching and participating alongside their termite-fishing mothers, mastered the technique more than a year earlier than their male peers, and that unlike the males they also tended to follow their mothers' technique in the length of their fishing tools and depths of their probing—behavior consistent with learning by observation. Similar kinds of

Fig. 31.3. Social conventions in capuchins. Here a male, "'Fonz" has his finger in the eye of a female, "Rumor," while she has her finger in his nostril. For further explanation, see text. Photo courtesy of Susan Perry and the Lomas Barbudal Monkey Project.

correlations consistent with social learning have been identified by Agostini and Visalberghi (2005) in black-horned capuchin monkeys (*Cebus nigritus*), where the profile of foraging techniques and preferences of young males corresponded significantly with that of the adult males they most closely associated with.

At the broadest levels of social learning, young primates' common association first with their mother and then with a wider group, over the long period of immaturity that characterizes the order, provides many opportunities to learn such basic but crucial things as "what, where, and how to eat" (Rapaport & Brown 2008). In many primates, immatures begin their foraging career by taking scraps from adults who are feeding, and then progressing to "cofeeding" simultaneously with them on the same food source. Both of these behaviors are likely to shape the developing individual's knowledge of what to eat and where to find

it, as is shown in the Agostini and Visalberghi study noted above. Figs. 31.4c and 31.4d illustrate this kind of context in the parallel case of stone handling. One step beyond this is provisioning by adults in response to juveniles' begging. This has been commonly documented in the great apes and in New World callitrichines (marmosets and tamarins) and capuchins, but has been little described in Old World monkeys (Rapaport & Brown 2008, table 4). Where such provisioning does occur, the interactions often focus on items that the juvenile needs to learn about, like resources that are hard to find and process (Silk 1978). That this is likely to function in social information transfer as well as straightforward nutrition is supported by a number of studies that show how the behavior becomes more common for items relatively novel to the young, who appear to be "begging for knowledge," in the title of one such study concerning Bornean orangutans (*Pongo pygmaeus*, Jaeggi et al. 2008).

e

Category	SH pattern	Captive troops				Free-ranging troops					
		Ara.A	Wak.A	Takh.	JMC	Kosh.	Ara.E	Sho.A	Sho.B	Tak.B	Tak.C
Investigative activities	Lick										
	Move inside mouth							–	–		
	Pick								–	–	
	Put in mouth										
Locomotion activities	Carry										
	Carry in mouth										
	Move and push/pull										
	Toss walk										
Collection activities	Pick and drop										
	Pick up small stone(s)							–			
Percussive or rubbing sound producing activities	Clack										
	Combine with object									–	
	Flint										
	Flint in mouth										
	Pound on surface									–	
	Rub in mouth										
	Rub stones together										
	Rub with mouth							–	–		–
	Shake in hands							–	–		–
	Slap										
	Swipe										
	Tap in mouth							–			–
Other complex manipulative activities	Flip										
	Put in water										
	Rub/put on fur								–		
	Spin							–			
	Stone groom										
	Throw										
	Throw and jump							–			–
	Throw and run										
	Throw and sway							–			–
	Wash							–	–		–
	Wrap in leaf							–	–	–	–
No. pattern occurrence		5	20	32	19	4	20	11	10	15	19

Fig. 31.4. Stone handling (SH) traditions among Japanese macaques. (a) cuddle; (b) roll in hands; (c) infant beginning to manipulate mother's stones; (d) infant manipulating "own" stones; (e) chart showing frequency of occurrence of the 33 SH patterns that are absent in some studied troops. Black: customary (exhibited by at least 90% of the sampled individuals in at least one age class, or at least 70% of the sampled individuals in at least two age classes). Hatched: habitual (observed at least three times in several individuals, consistent with some degree of social transmission). Spotted: present (observed at least once). White: absent (not observed despite at least 90 hours of observation). White with dash: unknown. Two troops at each of Shodoshima (Sho.A and Sho.B) and Takasakiyama (Tak.B and Tak.C) are neighboring troops that share the same feeding site.

Adults of some species go further, calling juveniles to discovered food items or even actively offering them the food (see also chapter 29, this volume). The latter again appears to be common in great apes and in callitrichines (Rapaport & Brown 2008), two distantly related groups at opposite ends of a spectrum of body size and encephalization. In a study of wild golden lion tamarins (*Leontopithecus rosalia*), adults were observed to avoid eating the insect prey they had discovered and to instead make food-offer calls, after which youngsters approached, investigated, and tackled the prey themselves. This echoes the parental behavior of other predatory mammals such as cats (Caro & Hauser 1980; Thornton & Raihani 2008), thus raising the prospect of a functional level of "'teaching"—an issue we return to later.

One of the greatest challenges faced by the field studies reviewed above is to provide robust evidence that the documented phenomena truly rely on social learning. Social learning can be identified unambiguously with the simplest kind of experiment, contrasting a group of subjects who see a model perform a novel behavior pattern and a control group who see no such model. Unfortunately, such an experiment is very difficult to engineer in the wild, as is experimentally translocating a primate skilled in some behavior from their natal group into a naïve one to discover whether their skill spreads (this might also be considered unethical). At the time of writing, field experiments have been completed for only a handful of nonprimate species (reviewed by Whiten & Mesoudi 2008), such as meerkats (Thornton & Malapert 2009), and have only recently begun for primates (Gruber et al. 2009;[2] van de Waal et al. 2010). Given the dearth of field experiments, debates have erupted about just how strong the evidence is for social learning in the field studies (Galef 1991; Laland & Galef 2009). For example, it is difficult or impossible to be sure that some subtle, unrecognized environmental factor has not shaped a local behavior variant through individual learning.

A case study that illustrates this, but also shows how follow-up research can nevertheless spiral up to more sophisticated and compelling conclusions, concerns ant dipping in chimpanzees. Whiten et al. (1999) had noted regional differences in which chimpanzees in some communities used only a short stick to harvest several ants and transfer them to the mouth, whereas those in other communities used a long wand to gather a ball of ants, and swept them into the mouth using a more complex bimanual swiping movement. The difference was inferred to be cultural because similar raw materials were available at both locations. However, both of these techniques were later found to occur also at a third site, where the long-wand technique was shown to be more common for a species of ant that had a particularly vicious bite and was more likely to be found near the nest hole (Humle & Matsuzawa 2002). Thus, it was possible that the difference in behavior across sites was due to differences in the prey. These findings led to a suite of impressively detailed quantitative studies of both chimpanzees and ants at several sites, permitting a more rigorous and sophisticated analysis that identified both individually and socially learned elements (Humle 2006; Möbius et al. 2008; Shöning et al. 2008). These authors concluded that in addition to the aspects of tool use that have been shaped environmentally by the distribution of different ants, there are cultural differences in whether ants are eaten at different sites, and in some of the ways chimpanzees eat them, including the contrast between the nibble-from-short-stick method used at Taï and the more complex techniques employed elsewhere.

Experimental Tests of Cultural Transmission: Diffusion Experiments

Contrasting with the dearth of social learning experiments in the wild, scores of such experiments with captive primates have been conducted through the past century (Tomasello & Call 1997; Subiaul 2007). These have much illuminated the mechanisms whereby one individual learns from another: the subject of the next section. In such studies, a single observer typically watches a single model. This neat dyadic scenario is useful for distinguishing between learning processes, but is less well suited for investigating traditions and cultures, which by their nature go beyond a pair of individuals. One cannot simply extrapolate from dyadic tests to gauge the capacity of animals to sustain the multiple transmissions required to sustain traditions, so a different design is needed.

A "diffusion experiment" fits this bill. In the simplest design, a novel behavior pattern is "seeded" by training an initial individual, then the extent to which this pattern will spread across a population of conspecifics is examined, in contrast to a control group lacking a model. Ideally this will be done with multiple experimental and control groups. Over the last 30 years a corpus of such studies has built up, but quite haltingly until recent times, with just 33 experiments, covering a variety of animals from fish to apes, identified and reviewed by Whiten and Mesoudi (2008).[3] Aside from the Menzel et al. (1972) study mentioned earlier, this approach has been applied to primates with adequate controls only in the last few years, all such studies being conducted in captivity.

The first of these (Whiten et al. 2005) involved a three-group design. A high-ranking female from each of two separate groups of chimpanzees was taken aside and trained to use a stick tool to extract a grape from an artificial foraging problem (the "panpipes") outside her enclosure mesh, using just one of two different techniques. In one technique, "lift," the stick was hooked through the top of the obstacle and used to lift it up so the food would roll toward the chimp. In the second technique, "poke," the stick was poked through a hidden flap so that it pushed the obstacle backwards until the grape fell and rolled toward the chimp through a lower pipe. Each newly created "expert" was then reunited with her group, where she displayed her new skill. Each of the two techniques was found to spread preferentially in the seeded group, thus providing the first experimental evidence that chimpanzees can sustain such traditions. However, fidelity of transmission was not high for all the chimpanzees, with some in one group discovering and adopting the technique more common in the other group (the "poke"—perhaps the more natural approach for chimpanzees). Nevertheless, the overall difference in traditions remained in place at two- and nine-month retests. A third, control, group that saw no seeded model failed to discover either technique, confirming these were novel routines for the chimps in the experimental conditions.

Such experiments as these and the field studies reviewed in the earlier section thus provide complementary evidence defining the scope of cultural phenomena in the species of interest. The field studies crucially map the scope of putative traditions in the wild and their functional significance, yet are limited in their capacity to show that social learning is definitely responsible, rather than genetics or individual learning. Diffusion experiments in captivity can rigorously test the capacity of the species to learn such inferred traditions.

To date, only a small number of such experiments have been published for primates. These include six with chimpanzees (collated in Whiten et al. 2007), one with guerezas (Colobus guereza, Price and Caldwell 2007), one with macaques (stump-tailed macaques, Macaca arctoides, and rhesus macaques, Macaca mulatta, deWaal and Johanowicz 1993) and two with tufted capuchin monkeys (Cebus apella, Dindo et al. 2008, 2009). The latter study and one on chimpanzees (Horner et al. 2006; incorporating human child subjects also) are the first to apply to primates a specialized "transmission chain" design. In this, just one naive primate watches the initial model; upon later mastering the task, that individual becomes the model for another, and so on along a potential chain of transmission. Here one knows exactly who is learning what, and from whom, at each step in the spread of a tradition. For both the chimpanzees and capuchins, each of two different ways of opening an "artificial fruit" (fig. 31.5) were transmitted with high fidelity along chains of individuals, simulating the passing of traditions through a series of generations. For these species, at least, the capacity to sustain different cultural variants in these ways has at last been clearly demonstrated (figs. 31.6 and 31.7). Consistency between such results and those from the wild is beginning to build a fuller and more compelling picture of the scope of traditions in primates than was possible even a decade ago.

The Mechanisms of Social Learning

The term "social learning" covers a very broad gamut of processes, some relatively simple and widely distributed, others more cognitively sophisticated and, it appears, more restricted in their distribution. As social learning research has developed through over a century of work, more and more distinctions have been made, and full taxonomies of social learning processes have become elaborate (see Whiten et al. 2004). The picture is made more complex because many different authors have offered what they see as improved distinctions, thus promoting either new concepts or new definitions of old ones. "Imitation" offers an illustration of this. Whiten and Ham (1992) defined imitation simply as observationally "learning the form of an act," but dozens of other definitions abound in the literature, typically adding various other criteria, such as the imitator also understanding and copying the goals of the act (Tomasello et al. 1993). Some authors insist on high fidelity matching to the model, others require only that some matching is detectable. Still others split imitation into different kinds. All in all, the field has become notorious for being somewhat treacherous and often bewildering to the uninitiated reader. The more positive side of this is that the rich variety of kinds of social learning is at last being grasped scientifically. In evaluating any one study, the reader is advised to scrutinize the author's definition of the terms at stake and strive to appreciate how they may relate to the wider literature.

Unlike language acquisition in humans, vocal learning in nonhuman primates plays a less important role (Snowdon & Hausberger 1997) than observational learning in the visual mode. Here, given space limitations, we focus on three main classes of observational learning that are most frequently distinguished. More wide-ranging analyses can be consulted elsewhere (Call & Carpenter 2002; Whiten et al. 2004; Hurley & Chater 2005; Hoppitt & Laland 2008; Heyes et al. 2009).

The simplest of the three main processes is *stimulus enhancement*, in which an observer's attention is merely focused on some stimulus through the actions of another individual. This process can include the results of the latter's actions, such as leaving a termite-fishing tool at a termite

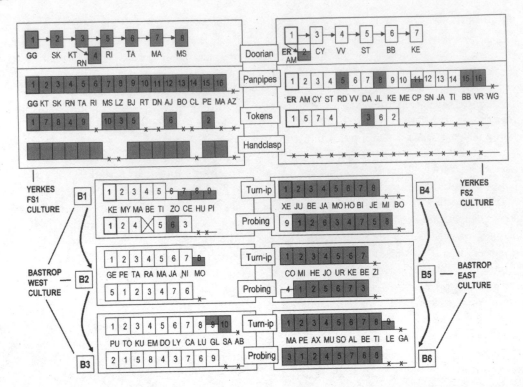

Fig. 31.6.: The results of diffusion experiments in captive chimpanzees. Each rectangle represents a chimpanzee, with a two-character ID code. Different tasks, named in the center, were available in both local populations named on either side, but different techniques (cf. fig. 31.4), coded here as dark versus light, were seeded in one individual, marked here as no. 1, in each population. The "Doorian" experiment, in which one model slid open a door and the other model opened a hatch set into the slide, was run as a "transmission chain" in which each learner later became the model for the next chimpanzee in the chain, as indicated by the arrows; all other experiments involved "open diffusion," with no predetermination of potential order of transmission. At Bastrop, transmission extended from groups B1 to B2 and B3, and from B4 to B5 and B6. Handclasp grooming spread spontaneously in the FS1 population only. Numbers represent order of acquisition for each task. For further explanation, see text and Whiten et al, (2007). These studies demonstrate the capacity of chimpanzees to sustain multiple-traditions cultures, consistent with the interpretation of regional variations among wild chimpanzees.

mound. Learning by *imitation*, by contrast, involves copying some novel aspects of another individual's actions, as explained above. The third process is *emulation*, in which the observer learns from the environmental results of what the other individual does, rather than from imitating their actions. Emulation goes beyond stimulus enhancement to learning such information as the properties or functional significance of what is being manipulated, such that the observer is then more likely to succeed in the task. This might involve the observer choosing a quite different approach (thus contrasting markedly with imitation), but alternatively it might result in an action similar to the original, which might thus look superficially like imitation. A closer look reveals several quite different types of emulation (Whiten et al. 2004).

Observational Learning in Apes

In the study of social learning in apes, particularly chimpanzees, a prime focus has been on whether they truly "ape" in the sense of imitating, or whether this supposition has been

Fig. 31.5. Recording a young chimpanzee working on an "artificial fruit," Ngamba Island, Uganda, 2001. Artificial fruits have been designed to study the observational learning of foraging techniques. They are designed to incorporate "defenses" that need twisting, poking, peeling and the like, simulating the routines required of challenging natural food resources. Once these are removed, the participant gains access to the edible fraction inside. Defenses are typically designed so that they can be dealt with in either of two quite different ways (e.g., "pulling and twisting" versus "poking through"); participants witness only one of these, so the information they acquire can later be precisely measured. See Whiten (1998) and text for more details. Photo courtesy Andrew Whiten.

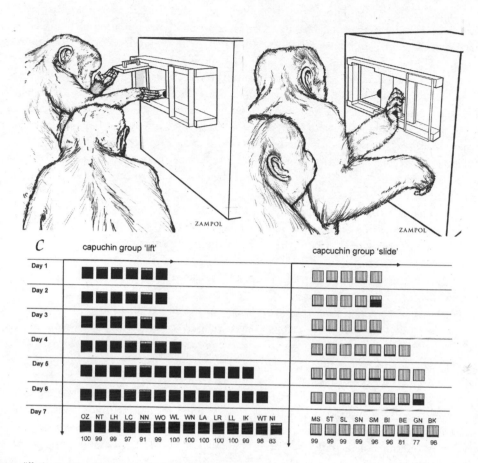

Fig. 31.7. An "open diffusion" study with capuchin monkeys: (a) capuchin performing "lift" technique on artificial doorian fruit; (b) capuchin performing alternative "slide" technique to gain reward; (c) spread of each technique in groups where one male was first taught to use either the lift or slide technique (after Dindo et al. 2009). Each symbol represents the proportion of each technique performed by each individual on consecutive days. Dark = slide, light = lift. Numbers indicate the percentage of actions performed on day 7, corresponding to that seeded in the group on day 1.

based on a failure to recognize alternatives, particularly emulation. The concept of emulation came to prominence following a study in which Tomasello et al. (1987) found chimpanzees failing to copy the trick used by one chimpanzee who used a stick tool to extract food from a foraging device. Since control chimps who saw no model nevertheless handled the stick (even though not copying what the model did with it), the authors concluded that something more than stimulus enhancement was involved. The observers appeared to be learning something about the raking functions afforded by the stick—and this kind of learning has been distinguished as "emulation" (Tomasello 1990).

So are chimpanzees emulators or imitators? If we first ask, "*Can* apes imitate?," then a positive, if qualified "yes" is offered by "do-as-I-do" experiments, in which chimpanzees (Custance et al. 1995) and a Sumatran orangutan (*Pongo abelii*, Call 2001) were first exposed to a training set of actions (e.g., "touch chin") meant to convey the idea that when asked to "do this," they would attempt it. The apes were then tested with a battery of 48 different and, as far as possible, novel kinds of facial, postural, and manual actions (e.g. "touch nose"). In the cases of both the orangutan and the chimpanzees, independent judges viewing videotapes of the experiments could identify matches significantly over chance expectations. The orangutan performed best, fully imitating 58% and partially imitating 36% of the 48 target actions. The fidelity of their imitation was, however, still poor compared to what a child is typically capable of.

Thus, apes *can* achieve bodily imitation, if only at crude levels by human standards. Whether they typically *do* imitate is another question. Answers to this question have been pursued principally by allowing experimental subjects to observe a model solving a novel foraging task, such as opening an "artificial fruit" (fig. 31.5), and comparing their later tendency to duplicate the model's performance with that of a control sample of chimpanzees who have seen no model, and/or who have witnessed a model using a different technique.

The working hypothesis of Tomasello et al. (1993), in a much cited classic overview of the field, is that apes typically emulate rather than imitate. This was supported by a series of experiments, reviewed by Tomasello and Call (1997) and Call et al. (2005), that showed chimpanzees and orangutans failing to imitate in ways that children in the same tests did, and instead acting more in line with emulation. In the latter study, for example, where a conspecific model extracted a reward from a tube in either of two ways (cracking it open or pulling caps off the ends), children typically copied the technique they had seen, even when the model acted merely as if *trying* a technique without success. By contrast, chimpanzees did not show significant matching and, unlike the children, tended to try a method *different* from the one they had seen being attempted without success.

These chimpanzees might be seen as acting more adaptively than the children in this situation, opting for an alternative to the approach they had seen fail. Indeed, more recent research has suggested that chimpanzees may be more *selective* than children in their approach to social learning. In a study by Horner and Whiten (2005), some subjects watched a familiar model first vigorously poke a stick tool into the top of an opaque box, then use it to extract a reward from a lower hole. Other subjects saw the same thing, but with a transparent box where one could now see that the operation in the top hole actually had no causal connection with the lower one. Chimpanzees in this study behaved selectively, tending toward emulation with the transparent box, by omitting the action at the top; but they swung toward a more imitative response in that they ran off the whole sequence, starting at the top, with the opaque box where causal irrelevance could not be detected. Human children, by contrast, tended to copy everything, even with the transparent box—a startling level of conformity I shall return to later.

This developing image of chimpanzees, not so much as "either/or" emulators or imitators but as selective social learners, was reinforced in a different context by Buttelmann et al. (2007), who found that chimpanzees were more likely to copy the body part used to perform an act (e.g., butting with the head) when the model's hands were free, than when the hands were occupied. This evidenced a power of discrimination that developmental psychologists have described as "rational imitation"—the chimpanzee chooses whether or not to copy according to whether the model is rationally choosing to act as they do, as opposed to merely having to act in that way (e.g., using one's head because the hands are unavailable; see also chapter 30, this volume).

The characterization of apes primarily as emulators is also complicated by recent experiments taking a quite different approach. In "ghost conditions," the movements of objects normally manipulated by a model are instead made to happen with no model present (for example, by surreptitiously moving them using fine fishing line), thus offering only the information that emulation focuses on (Tennie et al. 2006; Hopper et al. 2007, 2008). The latter two studies contrasted a complex and challenging task (the "panpipes" described above) with a simpler one (sliding a door to one side to gain access to food behind it). Only in the latter task was there initial evidence of subjects matching their behavior to what they had seen happen, and even then it was fleeting. With the panpipes the chimpanzees failed to learn from the ghost condition, whereas the diffusion studies had shown that they learn much from the normal full demonstration in which a live model performs the actions (Whiten et al. 2005), consistent with "imitation" in the broad sense.[4]

Byrne and Russon (1998) suggested that apes might indeed copy actions, but that they do so at the level of the overarching structure of a complex routine ("program-level-imitation"), without necessarily imitating all the details—in which case their copying might neglect picky details, yet count as imitation in an interesting way. Byrne and Russon offered examples from their studies of wild gorillas (*Gorilla gorilla*) and sanctuary-living Bornean orangutans (Russon & Galdikas 1993). These case studies were not generally viewed as compelling in the peer commentary on Byrne and Russon's (1998) article, and indeed the first attempts to experimentally investigate one of the key gorilla behaviors (manual processing of nettles before ingestion), failed to yield supportive evidence (Tennie et al. 2008). However, if we restrict the idea to action *sequence* learning rather than to the hierarchical organization Byrne and Russon emphasized (Whiten et al. 2006), the Horner and Whiten (2005) study outlined above fits the bill, for chimpanzees showed that they could split the sequence in the case of the transparent box, yet copied both elements in sequence with the opaque box (probing in the top hole, and then in the lower hole). Note that alternative *sequences* of quite different actions were also transmitted with high fidelity in the groups they were seeded in, in diffusion experiments (Whiten et al. 2007).

Apes reared intimately with humans, rather than with conspecifics, have been described as "enculturated," particularly when the relationship converges on that which normally exists between a human child and a parent. Some of the most impressive instances of imitation have been recorded for such apes, stimulating debates about whether these outcomes reflect latent capacities in the apes or whether their minds have been fundamentally reshaped and enhanced by their experiences. Readers are encouraged to consult Bering (2004) and Tomasello and Call (2004) to engage with these debates, which have occupied students of apes more than those researching other species (but see Fredman & Whiten 2008).

Observational Learning in Monkeys

Just as apes have long been assumed to "ape," the expression "monkey see, monkey do" has suggested similar expectations for monkeys that were seemingly reinforced by the studies of Japanese monkey traditions cited earlier.

The idea that monkeys would go beyond learning by stimulus enhancement to exhibiting imitation was endorsed by several of the first wave of experimental studies in the first half of the twentieth century (Whiten & Ham 1992). However, critical scrutiny of these studies later stressed that the experimental designs were weak in discriminating imitation from stimulus enhancement (Galef 1988; Whiten & Ham 1992). Moreover, an extensive series of well controlled experiments by Visalberghi and Fragaszy on tufted capuchins, for which expectations were high because of these monkeys' high brain-to-body ratio and busy manipulativeness, was evaluated in the authors' influential review paper "Do monkeys ape?" (1990) with an essentially negative verdict. For example, capuchins did not copy the actions of a conspecific that used a stick to push food items out of a tube; indeed, they showed better success at this task after additional individual experience with the objects than through opportunities for social learning. After a further decade of studies performed by themselves and others on capuchins and other species of monkeys, Visalberghi and Fragaszy (2002) held essentially to this verdict.

Additional studies have increasingly suggested relatively sophisticated processes of social learning, including imitation. These include studies by other workers on tufted capuchins. Custance et al. (1999) found evidence consistent with emulation or low-fidelity imitation in capuchins' matching of whichever of two ways of opening an artificial fruit they had witnessed. Fredman and Whiten (2008) found evidence of their imitating either of two forms of tool use (prying a lid off a foraging device versus stabbing through it), and Dindo et al. (2008, 2009) demonstrated marked fidelity of matching in diffusion experiments of the kind described earlier in this chapter (fig. 31.7).

At the other end of the scale of brain-to-body ratios, common marmosets (*Callthrix jacchus*) have shown evidence of bodily imitation in matching a model's use of either mouth or hand to open a box (Voelkl & Huber 2000; see also Bugynar & Huber 1997). These workers have also pio-

neered a very precise measurement system to show a very close match between model and observer, consistent with bodily imitation, in the bodily trajectory used in a foraging task (Voelkl & Huber 2007). Kumsashiro et al. (2009) claimed to have "trained" Japanese monkeys to imitate, such that the learned ability transferred to a novel context.

From a quite different perspective, Subiaul et al. (2004) described what they dubbed "cognitive imitation" in rhesus macaques. Subiaul et al. allowed one naive monkey to watch another who was skilled in pressing, in the correct order, four images that lit up in different places in a 4 × 4 grid—a task that had minimal manipulative content yet tested the subjects' grasp of sequencing. The observer monkey mastered the task faster after this experience, implying assimilation of the sequence required in the task. The term "cognitive imitation" is perhaps unfortunate insofar as it suggests that other forms of imitation are not cognitive, but the phenomenon is impressive and it suggests that monkeys are capable of learning much more by observation than was thought just a decade ago.

The discovery of "mirror neurons" in monkeys has raised the prospect of understanding the neural underpinnings of action imitation. Mirror neurons are distinctive in firing both when a monkey performs a certain action (such as picking up a nut) and when it merely sees another individual do the same thing (Rizzolatti 2005). Earlier doubts about the imitative prowess of monkeys, however, led such researchers to attribute other functions to them, such as understanding goal-directed actions in others. The identification of imitation-related mirror neuron activity in humans (Iacoboni 2005), however, coupled with the newer evidence of imitation in monkeys, suggests that a potential linkage between mirror neurons and social learning in non-human primates may yet repay further study (Ferrari et al. 2009).

Teaching

When Galef (1992) argued that the term "culture" should be applied only to cases in which the transmission mechanisms in animals are akin to those of humans, he made reference to teaching as well as imitation. Is teaching part of the social learning process in primates?

If intentional teaching requires some level of "theory of mind," in which the teacher recognizes a state of ignorance in the learner that can be rectified by intervening, one might expect to see teaching in the great apes, for whom the most evidence for such insights has accumulated (Whiten 1999, and see chapter 30, this volume). However, such teaching has been claimed only in the case of the particularly challenging task of nut cracking using natural hammers by chimpanzees (Boesch 1991). Just two cases of demonstration were described, the bulk of the mother's contribution being at the level of stimulating and facilitating her offspring's efforts by tolerating theft of nuts and hammers and sometimes actively providing them. Ape mothers have also occasionally been observed to confiscate nonfood plants from their offspring and cast such items aside (Rapaport & Brown 2008).

Aside from such uncommon cases as these, there is scant evidence of teaching in apes and Old World monkeys. By contrast, recent reviews document more extensive evidence of teaching in other animal taxa, particularly in animals that have to make the transition from weaning to foraging on difficult foods, the most obvious cases of which concern predation (Hoppitt et al. 2008; Thornton & Raihani 2008). Such analyses rest on a conception of teaching that does not require intent but is instead focused on functional criteria, notably supporting such effects as skill development at personal cost (Caro & Hauser 1992). Teaching in this sense has been well demonstrated by experimental studies in meerkats, which provide disabled prey to their young and otherwise adjust their behavior to facilitate the development of hunting skill (Thornton & McAuliffe 2006). Rapaport and Brown (2008) suggest that there is no similar pressure for this kind of social learning in most primates, who can instead negotiate a more gradual acquisition of the adult diet (in which animal prey constitutes only a minor part), relying on the kinds of observational learning outlined above during a typically long, drawn-out period of immaturity and parental dependence.

Perhaps the primate behavior closest to the kind of teaching seen in meerkats occurs in callitrichines. In these species it has been shown that parents are more disposed to offer their young novel foods than to offer them foods they are already familiar with, a possible case of "opportunity teaching" in that parents shape what the young may then learn by their own efforts (Rapaport 1999). Further, as noted earlier, golden lion tamarins have been observed to give food calls upon finding large insect prey, but then inhibit their own capture of the prey so that the young are attracted to it and then deal with it themselves (Rapaport & Ruiz-Miranda 2002).

Interdisciplinary Fertilization and Applications

Before moving to a concluding overview of what this corpus of studies of primate social learning has to tell us about the evolution of culture within primate societies, we should pause for a moment to note the links between this body of work and its sister disciplines. These links often represent two-way streets in which primatology offers information of great interest to other disciplines and also learns from them (Claidière & Whiten 2012).

One long-standing link is with developmental psychology. Reference has already been made to studies that systematically compare social learning in apes and children, to which we return below. There is much more scope for fruitful exchange of methodologies and concepts between the disciplines. For example, in reviewing the child social learning literature, Want and Harris (2002) borrowed from the conceptual framework and methodologies developed in comparative psychology, as did Horowitz (2003) in extending similar studies to adults. Other cross-fertilizations extend to the study of social learning in autism (Whiten 2006). The reciprocal flow from developmental to comparative psychology is also considerable, and is nicely illustrated by the adoption of concepts and methods from child psychology to great effect in the studies of Buttelmann et al. (2007, 2008), described earlier.

A further link is with robotics, where workers in artificial intelligence are striving to create imitative robots and are keen to learn from our growing understanding of social learning in a variety of organisms (Dautenhahn & Nehaniv 2002). It is to be hoped that knowledge transfer will become a two-way street here also.

A quite different link is with the growing areas of phylogenetic and paleoanthropological study of human cultural evolution, which increasingly use sophisticated numerical methods to reconstruct cumulative cultural diversification (Mace & Holden 2005; Whiten et al. 2012). Lycett et al. (2009) have now extended this approach to apes and paleoanthropologists are incorporating primatological discoveries into models of the past (Whiten et al. 2009).

Among more practical applications, two areas can be highlighted. The first is conservation, in which a limited role has been established for the reintroduction of primates into selected habitats. As might be predicted from the work reviewed in this chapter, many primates need to acquire much of the information they need to survive and breed through social learning, whether it is provided by knowledgeable conspecifics or by human caretakers simulating that role (Custance et al. 2002; Stoinski et al. 2003).

A second practical area concerns welfare and management of captive primates, where care staff often informally report that animals have learned such things as being compliant in receiving injections, mothering infants, or mating by watching experienced conspecifics live or on video. There appears to be only minimal systematic testing of such procedures (Lambeth et al 2008), but this seems to be an area in which it would be extremely worthwhile to do so.

Conclusions: The Evolution of Social Learning in Primate Societies

Have we learned enough yet to attempt to construct an evolutionary scenario for the phenomena discussed in this chapter? Barely. The evidence for traditions and cultures in the wild comes from long-term studies of multiple groups, and so it still remains to be reported for most primates, with just a handful of species focused on to date. There is a dearth of such information even for such a well-studied group as the baboons. Even where species are well studied, as is the case with chimpanzees, controversy still surrounds the true nature and scope of their cultural repertoire. However, we may have reached an opportune point to start building a set of working hypotheses. This may help clarify where gaps in our knowledge exist and what is most needed in the next phase of research in this exciting area.

A first hypothesis is that the most basic social learning processes of stimulus enhancement and observational conditioning are available to all primates. Such processes have been identified in much broader groups of animals, extending to both vertebrates and invertebrates (e.g., Curio et al. 1978; Galef & Whiskin 2008). In conjunction with what we know about specific primate life histories and societies reviewed elsewhere in this volume, we may predict that young primates, during their typically long periods of immaturity, will learn to discriminate among such crucial things as foods, foraging sites, and predators, most commonly first through the mother, and later from whatever wider society they are exposed to (chapters, 7, 8, 10, and 11, this volume). The latter process may be moderated significantly by the levels of social tolerance that prevail (van Schaik et al. 1999).

In Old World monkeys there is surprisingly little compelling evidence for more structured modes of social learning (Tomasello and Call 1997, table 9.2: but see Subiaul et al. 2004; Subiaul 2007). By contrast, there is more to social learning in the relatively distantly related New World monkeys and apes (great apes, at least; we still know all too little of social learning in gibbons).

Consider New World monkeys first. As detailed earlier, there is much evidence across several taxa for both provisioning, in response to begging and active giving—in both cases, with a focus on foods that are difficult to harvest and/or process. In the callitrichines there is even evidence of dedicated food calls used in ways that are consistent with "teaching" (encouraging the skills associated with locating and dealing with invertebrate prey). The prevalence of these interactions concerning novel items suggests that the func-tion is not restricted to nutrition directly, but extends to information transfer (Rapaport & Brown 2008). There is also evidence of both marmosets (Voelkl & Huber 2000, 2007) and capuchins (Dindo et al. 2008, 2009) copying the foraging techniques of others. Finally, in capuchins there is evidence of regional variation in multiple traditions involving social conventions, foraging techniques, and possibly tool use (e.g., Perry et al. 2003).

Turning to apes, we see several of these features too: provisioning in response to begging, active food offering, confiscation of problem items, and facility in copying the techniques of others (Whiten et al. 2004; Rapaport & Brown 2008). The copying is relatively sophisticated in extending to the sequential and perhaps hierarchical structure of actions, and the very nature of imitation is understood by apes well enough for them to learn the "do-as-I-do" game and actively test others who appear to be imitating them (Nielsen 2005; Haun & Call 2008), thus going beyond monkeys' recognition of being imitated (Paukner et al. 2005, 2009). Consistent with these capacities, great apes (orangutans and chimpanzees, at least) appear to exhibit the richest multiple-tradition cultures among nonhuman animals, incorporating a diversity of behavior types including social behavior, courtship gambits, foraging techniques, and tool use. The fact that all these features are shared with humans promotes the inference that they also would have been shared with our common ancestor of around 14 million years ago (Whiten 2011).

A thorough analysis of the evolutionary changes involved in the giant leap human culture took from the platform provided by those ancestral features is a substantial task (see Klein & Edgar 2002; Boyd & Silk 2006; chapter 32, this volume). Here we can only note that there have been changes in three principal aspects of culture (Whiten 2005, 2009a). The first concerns the large scale spatiotemporal patterning of culture. As we have seen, ape culture can extend to multiple regional differences in traditions, as occurs in humans. But human culture goes further to display extensive evolutionary change in its own right: the culture of each generation builds on what went before, often leading to what Tomasello (1999) has described as a progressive "ratcheting" of cultural diversity and achievement. In this, culture echoes Darwinian biological evolution (Mesoudi et al. 2006). Examples are to be found in every aspect of human culture, from language (Gray et al. 2009) to religion and technology (Shennan 2002). The second set of changes concerns the contents of culture: the types of behavior that are culturally transmitted. Some such contents are shared with apes, such as tool use, but major new ones have emerged, particularly in the social domain, on the scale of such examples as spoken language and religion (Whiten 2009b).

The third set of changes is in transmission mechanisms. In some societies these mechanisms have become as elaborate as those involved in formal education extending into adulthood, and such conduits as books and the Internet. However, just what forms transmission takes in hunter-gatherer societies, which are more representative of much of our history over the last million years or so, remains frustratingly unclear. Some reports appear to imply that teaching is rare and that observational learning predominates—a situation much more akin to what we see in nonhuman primates (Whiten et al. 2003). But even observational learning takes different forms. There is debate over the nature of such differences (Herrmann et al. 2007; de Waal et al. 2008; Tennie et al. 2009; Whiten et al. 2009), with some researchers concluding that true imitation is limited to human children and is not shared by other apes, and other researchers disagreeing with the contention that the contrast between children and apes is so stark, as has been discussed above. Nevertheless, there is much consensus that children attain unique levels of imitative fidelity. Indeed, recent work has focused on the discovery of what has been dubbed "overimitation," in which children are so ready to copy that they become unable to inhibit themselves from blindly imitating the actions of others even when perceptual cues should tell them the actions are wildly ineffective (Lyons et al. 2007; Whiten et al. 2009; see Nagell et al. 1993 for an early example).

Whatever the features of social learning and culture that separate humans from other primates, it has become clear that extensive foundations for these behaviors must have existed in the various ancestors we share most recently with other apes and further in the past with other primate taxa. Such inferences appear increasingly substantial as aspects of social learning, traditions, and cultures are identified in other primates that exhibit manifest commonalities with human counterparts.

However, this human-centered perspective, while reflecting a common motivation for interest in the topics of this chapter, is but one among many. Social learning and traditions have evolved into their present forms across all of the hundreds of living species "tips" of the primate family tree. A comprehensive analysis of the evolution of these cultural phenomena across the whole primate order is a perfectly valid, nonanthropocentric aspiration for the discipline. At present, the kinds of data required for this have been collected for only a small subset of primates, particularly those highlighted in this chapter—but it is to be hoped that these data will provide a future inspiration for other primatologists to help fill in the gaps.

Notes

1. I am grateful to Josep Call for earlier drawing my attention to the pioneering diffusion experiment of Menzel et al. (1972).

2. Gruber et al. (2009) have shown that wild chimpanzees presented with the same novel honey-dipping problem at one site, where they habitually use sticks for probing, and at a second, where instead leaf sponges are used, go on to apply their local technique, which the authors attribute to their "cultural knowledge." This suggests the feasibility of related diffusion experiments in the field.

3. Whiten and Mesoudi (2008) review 33 diffusion experiments conducted with fish, birds, and mammals.

4. Tennie et al. (2010) further enrich the picture by using a different approach, showing evidence of emulation in chimpanzees when the context does not allow imitation of a technique used by a model.

References

Agostini, I. & Visalberghi, E. 2005. Social influences on the acquisition of sex-typical foraging patterns by juveniles in a group of wild tufted capuchin monkeys (*Cebus nigritus*). *American Journal of Primatology*, 65, 335–351.

Bering, J. 2004. A critical review of the "enculturation hypothesis": The effects of human rearing on great ape social cognition. *Animal Cognition*, 7, 201–212.

Boesch, C. 1991. Teaching among wild chimpanzees. *Animal Behaviour*, 41, 531–532.

Boesch, C. & Tomasello, M. 1998. Chimpanzee and human cultures. *Current Anthropology*, 39, 591–614.

Boyd, R. & Richerson, P. J. 1985. *Culture and the Evolutionary Process*. Chicago: University of Chicago Press.

Boyd, R. & Silk, J. B. 2009. *How Humans Evolved*. 5th edition. New York: W. W. Norton.

Brown, C. & Laland, K. N. 2003. Social learning in fishes: A review. *Fish and Fisheries*, 4, 280–228.

Bugnyar, T. & Huber, L. 1997. Push or pull: An experimental study on imitation in marmosets. *Animal Behaviour*, 54, 817–31.

Buttelmann, D., Carpenter, M., Call, J. & Tomasello, M. 2007. Enculturated chimpanzees imitate rationally. *Developmental Science*, 10, 31–38.

———. 2008. Rational tool use and tool choice in human infants and great apes. *Child Development*, 79, 609–626.

Byrne, R. W. & Russon, A. E. 1998. Learning by imitation: A hierarchical approach. *Behavioral and Brain Sciences*, 21, 667–721.

Caldwell, C. A. & Whiten, A. 2010. Social learning in monkeys and apes: Cultural animals? In *Primates in Perspective*, 2nd edition (ed. by Campbell, C. J., Fuentes, A., MacKinnon, K., Bearder, S. & Stumpf, R.), 652–662. Oxford: Oxford University Press.

Call, J. 2001. Body imitation in an enculturated orangutan (*Pongo pygmaeus*). *Cybernetics and Systems*, 32, 97–119.

Call, J. & Carpenter, M. 2002. Three sources of information in social learning. In *Imitation in Animals and Artifacts* (ed. by Dautenhahn, K. & Nehaniv, C. L.), 211–228. Cambridge, MA: MIT Press.

Call, J., Carpenter, M. & Tomasello, M. 2005. Copying results and copying actions in the process of social learning: Chimpanzees (*Pan troglodytes*) and human children (*Homo sapiens*). *Animal Cognition*, 8, 151–163.

Caro, T. M. & Hauser, M. D. 1992. Is there teaching in non-human animals? *Quarterly Review of Biology*, 67, 151–174.

Cavalli-Sforza, L. L. & Feldman, M. W. 1981. *Cultural Transmission and Evolution: A Quantitative Approach*. Princeton, NJ: Princeton University Press.

Claidière, N. & Whiten, A. 2012. Integrating the study of conformity and culture in humans and non-human animals. *Psychological Bulletin*, 138, 126–145.

Curio, E., Ulrich, E. & Vieth, W. 1978. Cultural transmission of enemy recognition: One function of avian mobbing. *Science*, 202, 899–901.

Custance, D. M., Whiten, A., & Bard, K. A. 1995. Can young chimpanzees (*Pan troglodytes*) imitate arbitrary actions? Hayes and Hayes 1952 revisited. *Behaviour*, 132, 837–859.

Custance, D. M., Whiten, A., & Fredman, T. 1999. Social learning of artificial fruit processing in capuchin monkeys (*Cebus apella*). *Journal of Comparative Psychology*, 113, 13–23.

———. 2002. Social learning and primate reintroduction. *International Journal of Primatology*, 23, 479–99.

Dautenhahn, K. & Nehaniv, C. L. 2002. *Imitation in Animals and Artifacts*. Cambridge, MA: MIT Press.

De Waal, F.B.M. 2001. *The Ape and the Sushi Master: Cultural Reflections of a Primatologist*. New York: Basic Books.

De Waal, F. B. M. & Johanowicz, D. L. 1993. Modification of reconciliation behavior through social experience: An experiment with two macaque species. *Child Development*, 64, 897–908.

De Waal, F.B.M., Boesch, C., Horner, V. & Whiten, A. 2008. Comparing social skills of children and apes. *Science*, 319, 569.

Dindo, M., Thierry, B. & Whiten, A. 2008. Social diffusion of novel foraging methods in brown capuchin monkeys (*Cebus apella*). *Proceedings of the Royal Society B*, 275, 187–193.

Dindo, M., de Waal, F.B.M. & Whiten, A. 2009 In-group conformity sustains different foraging traditions in capuchin monkeys (*Cebus apella*). *PLoS One*, 4, e7858. http://dx.plos.org/10.1371/journal.pone.0007858.

Ferrari, P. F., Bonini, L. & Fogassi, L. 2009. From monkey mirror neurons to primate behaviors: Possible "direct" and "indirect" pathways. *Philosophical Transactions of the Royal Society B*, 364, 2311–2323.

Fragaszy, D. M. & Perry, S. 2003. *The Biology of Traditions: Models and Evidence*. Cambridge: Cambridge University Press.

Fragaszy, D., & Visalberghi, E. 2004. Socially biased learning in monkeys. *Learning and Behavior*, 32, 24–35.

Fredman, T. & Whiten, A. 2008. Observational learning from tool using models by human-reared and mother-reared capuchins monkeys (*Cebus apella*). *Animal Cognition*, 11, 295–309.

Galef, B. G. 1988. Imitation in animals: History, definitions, and interpretation of data from the psychological laboratory. In *Social Learning: Psychological and Biological Perspectives* (ed. by Zentall, T. R. & Galef, B. G., Jr.), 3–28. Hillsdale, NJ: Erlbaum.

———. 1991. Tradition in animals: Field observations and laboratory analyses. In *Interpretation and Explanation in the Study of Behavior, Vol. 1: Interpretation, Intentionality and Communication* (ed. by Bekoff, M. & Jamieson, D.), 74–95. Boulder, CO: Westview Press.

———. 1992. The question of animal culture. *Human Nature*, 3, 157–178.

———. 2009. Culture in animals? In *The Question of Animal Culture* (ed. by Laland, K. N. & Galef, B. G.), 222–246. Cambridge, MA: Harvard University Press.

Galef, B. G. & Whiskin, E. E. 2008. "Conformity" in Norway rats? *Animal Behaviour*, 75, 2035–2039.

Goodall, J. 1986. *The Chimpanzees of Gombe*. Cambridge, MA: Harvard University Press.

Gray, R. D., Drummond, A. J. & Greenhill, S. J. 2009. Language phylogenies reveal expansion pulses and pauses in pacific settlement. *Science*, 323, 479–483.

Gruber, T., Muller, M. N., Strimling, P., Wrangham, R. & Zuberbuhler, K. 2009. Wild chimpanzees rely on cultural knowledge to solve an experimental honey acquisition task. *Current Biology*, 19, 1846–1852.

Haun, D. B. M. & Call, J. 2008. Imitation recognition in great apes. *Current Biology*, 18, R288–290.

Herrmann, E., Call, J., Hernandez-Lloreda, M. V., Hare, B. & Tomasello, M. 2007. Humans have evolved specialized skills of social cognition: The cultural intelligence hypothesis. *Science*, 317, 1360–1365.

Heyes, C. M. 1994. Social learning in animals: Categories and mechanisms. *Biological Reviews*, 69, 207–231.

Heyes, C. M. & Galef, B. G. 1996. *Social Learning in Animals: The Roots of Culture*. San Diego: Academic Press.

Heyes, C. M., Huber, L., Gergely, G. & Brass, M., eds. 2009. Evolution, development and control of imitation. Theme issue, *Philosophical Transactions of the Royal Society B*, 364, whole issue no. 1528, 2291–2443.

Hirata, S., Watanabe, S. & Kawai, M. 2001. "Sweet-potato washing" revisited. In *Primate Origins of Human Behavior and Cognition* (ed. by T. Matsuzawa), 487–508. Tokyo: Springer-Verlag.

Hopper, L. M., Spiteri, A., Lambeth, S. P., Schapiro, S. J., Horner, V. & Whiten, A. 2007. Experimental studies of traditions and underlying transmission processes in chimpanzees. *Animal Behaviour*, 73, 1021–1032.

Hopper, L. M., Lambeth, S. P., Schapiro, S. J. & Whiten, A. 2008. Observational learning in chimpanzees and children studied through "ghost" conditions. *Proceedings of the Royal Society B*, 275, 835–840.

Hoppitt, W. J. E., Brown, G. E., Kendal, R., Rendell, L., Thornton, A., Webster, M. M. & Laland, K. N. 2008. Lessons from animal teaching. *Trends in Ecology and Evolution*, 23, 486–493.

Hoppitt, W. and Laland, K. N. 2008. Social processes influencing learning in animals: A review of the evidence. *Advances in the Study of Behavior*, 38, 105–165.

Horowitz, A. C. 2003. Do humans ape? Or do apes human? Imitation and intention in humans (*Homo sapiens*) and other animals. *Journal of Comparative Psychology*, 117, 325–336.

Horner, V. K. & Whiten, A. 2005. Causal knowledge and imitation/emulation switching in chimpanzees (*Pan troglodytes*) and children. *Animal Cognition*, 8, 164–181.

Horner, V., Whiten, A., Flynn, E. & de Waal, F. B. M. 2006. Faithful replication of foraging techniques along cultural transmission chains by chimpanzees and children. *Proceed-*

ings of the National Academy of Sciences, USA, 103, 13878–13883.

Huffman, M. A. 1996. Acquisition of innovative cultural behaviours in nonhuman primates: A case study of stone handling, a socially transmitted behavior in Japanese macaques. In *Social Learning in Animals: The Roots of Culture* (ed. by Heyes, C. M. & Galef, B. G., Jr.) 267–289. London: Academic Press.

Humle, T. 2006. Ant-dipping in chimpanzees: An example of how microecological variables, tool use, and culture reflect the cognitive abilities of chimpanzees. In *Cognitive Development in Chimpanzees* (ed. by Matsuzawa, T., Tomonaga, M. & Tanaka, M.), 452–75. Tokyo: Springer-Verlag.

Humle, T. and Matsuzawa, T. 2002. Ant-dipping among the chimpanzees of Bossou, Guinea, and some comparisons with other sites. *American Journal of Primatology*, 58, 133–148.

Hurley, S. & Chater, N., eds. 2005. *Perspectives on Imitation: From Neuroscience to Social Science*. Cambridge, MA: MIT Press.

Iacoboni, M. 2005. Understanding others: Imitation, language and empathy. In *Perspectives on Imitation: From Neuroscience to Social Science* (ed. by Hurley, S. & Chater, N.), 77–99. Cambridge, MA: MIT Press.

Jaeggi, A. V., van Noordwijk, M. A. & van Schaik, C. P. 2008. Begging for information: Mother-offspring food sharing among wild Bornean orangutans. *American Journal of Primatology*, 70, 533–541.

Klein, R. G. & Edgar, B. 2002. *The Dawn of Human Culture*. New York: John Wiley and Sons.

Kumashiro, M., Yokoyama, O. & Ishibashi, H. 2008. Imitation of body movements facilitated by joint attention through eye contact and pointing in Japanese monkey. *Public Library of Science One*, 3, e3074, doi:10.1371/journal.pone.0003704.

Laland, K. N. & Galef, B. G., eds. 2009. *The Question of Animal Culture*. Cambridge, MA: Harvard University Press.

Laland, K. N. & Hoppitt, W. 2003. Do animals have culture? *Evolutionary Anthropology*, 12, 150–159.

Lambeth, S. P., Perlman, J. E. & Schapiro, S. J. 2000. Positive reinforcement training paired with videotape exposure decreases training time investment for a complicated task in female chimpanzees. *American Journal of Primatology*, 51 (suppl. 1), 79–80.

Leadbeater, E. & Chittka, L. 2007. Social learning in insects: From miniature brains to concensus building. *Current Biology*, 17, R703–713.

Leca, J.-B., Gunst, N. & Huffman, M. A. 2007. Japanese macaque cultures: Inter- and intra-troop behavioral variability of stone-handling patterns across 10 groups. *Behaviour*, 144, 251–81.

Lefebvre, L. & Bouchard, J. 2003. Social learning about food in birds. In *The Biology of Traditions: Models and Evidence* (ed. by Fragaszy, D. M. & Perry, S.), 94–126. Cambridge: Cambridge University Press.

Lonsdorf, E. V. 2006. What is the role of mothers in the acquisition of termite-fishing behaviors in wild chimpanzees (*Pan troglodytes schweinfurthii*)? *Animal Cognition*, 9, 36–46.

Lonsdorf, E. V., Pusey, E. A. & Eberly, L. 2004. Sex differences in learning in chimpanzees. *Nature*, 428, 715–716.

Lycett, S. J., Collard, M., & McGrew, W. C. 2009. Cladistic analyses of behavioral variation in wild *Pan troglodytes*: Exploring the chimpanzee culture hypothesis. *Journal of Human Evolution*, 57, 337–349.

Lyons, D. E., Young, A. G. & Keil, F. C. 2007. The hidden structure of overimitation. *Proceedings of the National Academy of Sciences, USA*, 104, 19751–19756.

Mace, R. & Holden, C. J. 2005. A phylogenetic approach to cultural evolution. *Trends in Ecology and Evolution*, 20, 116–121.

McGrew, W. C. 1992. *Chimpanzee Material Culture: Implications for Human Evolution*. Cambridge: Cambridge University Press.

Menzel, E. W., Davenport, R. K. & Rogers, C. M. 1972. Protocultural aspects of chimpanzees' responsiveness to novel objects. *Folia Primatologica*, 17, 161–170.

Mesoudi, A., Whiten, A. & Laland, K. N. 2006. Towards a unified science of cultural evolution. *Behavioral and Brain Sciences*, 29, 329–383.

Möbius, Y., Boesch, C., Koops, K., Matsuzawa, T. & Humle, T. 2008. Cultural differences in army ant predation by West African chimpanzees? A comparative study of microecological variables. *Animal Behaviour*, 76, 37–45.

Nagell, K., Olguin, R. S. & Tomasello, M. 1993. Processes of social learning in the tool use of chimpanzees (*Pan troglodytes*) and human children (*Homo sapiens*). *Journal of Comparative Psychology*, 107, 174–186.

Nielsen, M., Collier-Baker, E., Davis, J. M. & Suddendorf, T. 2005. Imitation recognition in a captive chimpanzee (*Pan troglodytes*). *Animal Cognition*, 8, 31–36.

Nishida, T. 1987. Local traditions and cultural transmission. In *Primate Societies* (ed. by Smuts, B.B., Cheney, D. L., Seyfarth, R. M., Wrangham, R. W. & Struhsaker, T T.), 462–474. Chicago: University of Chicago Press.

Nishida, T., Wrangham, R. W., Goodall, J. & Uehara, S. 1983. Local differences in plant-feeding habits of chimpanzees between the Mahale mountains and Gombe National Park. *Journal of Human Evolution*, 12, 467–480.

Ottoni, E. B., Resende, B. D. & Izar, P. 2005. Watching the best nutcrackers: What capuchin monkeys (*Cebus apella*) know about each others' tool-using skills. *Animal Cognition*, 24, 215–219

Panger, M. A., Perry, S., Rose, L., Gros-Luis, J., Vogel, E., Mackinnon, K. C., & Baker, M. 2002. Cross-site differences in foraging behavior of white-faced capuchins (*Cebus capuchinus*). *American Journal of Physical Anthropology*, 119, 52–56.

Paukner, A., Anderson, J. R., Borelli, E., Visalberghi, E. & Ferrari, P. F. 2005. Macaques (*Macaca nemestrina*) recognise when they are being imitated. *Biology Letters*, 1, 219–221.

Paukner, A., Suomi, S. J., Visalberghi, E. & Ferrari, P. F. 2009. Capuchin monkeys display affiliation towards humans who imitate them. *Science*, 325, 880–883.

Perry, S. 2009. Conformism in the food-processing techniques of white-faced capuchin monkeys (*Cebus capucinus*). *Animal Cognition*, 12, 705–716.

Perry, S., Baker, M., Fedigan, L., Gros-Louis, J., Jack, K., Mackinnon, K. C., Manson, J., Panger, M., Pyle, K. & Rose, L. M. 2003. Social conventions in white-face capuchins monkeys: Evidence for behavioral traditions in a neotropical primate. *Current Anthropology*, 44, 241–268.

Price, E., & Caldwell, C. A. 2007. Artificially generated cultural variation between two groups of captive monkeys, *Colobus guereza kikuyuensis*. *Behavioural Processes*, 74, 13–20.

Rapaport, L. G. 1999. Provisioning of young in in golden lion tamarins (*Callitrichidae, Leontopithecus rosalia*): A test of the information hypothesis. *Ethology*, 105, 619–636.

Rapaport, L. G. & Brown, G. R. 2008. Social influences on foraging behavior in young nonhuman primates: Learning what, where, and how to eat. *Evolutionary Anthropology*, 17, 189–201.

Rapaport, L. G. & Ruiz-Miranda, C. R. 2002. Tutoring in wild golden lion tamarins. *International Journal of Primatology*, 23, 1063–1070.

Richerson, P. J. & Boyd, R. 2005. *Not by Genes Alone*. Chicago: University of Chicago Press.

Rizzolatti, G. 2005. The mirror neuron system and imitation. In *Perspectives on Imitation: From Neuroscience to Social Science* (ed. by Hurley, S. & Chater, N.), 55–76. Cambridge, MA: MIT Press.

Rose, L. M., Perry, S., Panger, M., Jack, K., Manson, J., Gros-Louis, J., Mackinnon, K. C. & Vogel, E. 2003. Interspecific interactions between *Cebus capuchinus* and other species in Costa Rican sites. *International Journal of Primatolology*, 24, 759–796.

Russon, A. E. & Galdikas, B. M. F. 1993. Imitation in free-ranging rehabilitant orangutans (*Pongo pygmaeus*). *Journal of Comparative Psychology*, 107, 147–161.

Sapolsky, R. M. & Share, L. J. 2004. A pacific culture among wild baboons: Its emergence and transmission. *Public Library of Science, Biology*, 2, 294–304.

Shennan, S. 2002. *Genes, Memes and Human History: Darwinian Archaeology and Cultural Evolution*. London: Thames and Hudson.

Shöning, C., Humle, T., Möbius, Y. & McGrew, W. C. 2008. The nature of culture: Technological variation in chimpanzee predation on army ants revisited. *Journal of Human Evolution*, 55, 48–59.

Silk, J. B. 1978. Patterns of foodsharing among mother and infant chimpanzees at Gombe National Park, Tanzania. *Folia Primatologica*, 29, 129–141.

Snowdon, C. T. & Hausberger, M. 1997. *Social Influences on Vocal Development*. Cambridge: Cambridge University Press.

Stoinski, T. S., Beck, B. B., Bloomsmith, M. A. & Maple, T. I. 2003. A behavioral comparison of captive born, reintroduced golden lion tamarins and their wild born offspring. *Behaviour*, 140, 137–160.

Subiaul, F. 2007. The imitation faculty in monkeys: Evaluating its features, distribution and evolution. *Journal of Anthropological Science*, 85, 35–62.

Subiaul, F., Cantlon, J. F., Holloway, R. L. & Terrace, H. S. 2004. Cognitive imitation in rhesus macaques. *Science*, 305, 407–410.

Tennie, C., Call, J. & Tomasello, M. 2006. Push or pull: Imitation versus emulation in human children and great apes. *Ethology*, 112, 1159–1169.

———. 2009. Ratcheting up the ratchet: On the evolution of cumulative culture. *Philosophical Transactions of the Royal Society B*, 364, 2405–2415.

———. 2011. Evidence for emulation in chimpanzees in social settings using the floating peanut task. *PLoS One* 5, e10544. DOI: 10.1371/journal.pone.0010544.

Tennie, C., Hedwig, D., Call J. & Tomasello, M. 2008. An experimental study of nettle-feeding in captive gorillas. *American Journal of Primatology*, 70, 584–593.

Thornton, A. & Malapert, A. 2009. The rise and fall of an arbitrary tradition: An experiment with wild meerkats. *Proceedings of the Royal Society B* 276, 1269–1276.

Thornton, A. & McAuliffe, K. 2006. Teaching in wild meerkats. *Science*, 313, 227–229.

Thornton, A. & Raihani, N. J. 2008. The evolution of teaching. *Animal Behaviour*, 75, 1823–1836.

Tomasello, M. 1990. Cultural transmission in the tool use and communicatory signaling of chimpanzees? In *"Language" and Intelligence in Monkeys and Apes: Comparative Developmental Perspectives* (ed. by Parker, S. T. & Gibson, K.), 274–311. Cambridge: Cambridge University Press.

———. 1999. *The Cultural Origins of Human Cognition*. Cambridge, MA: Harvard University Press.

Tomasello, M. & Call, J. 1997. *Primate Cognition*. Oxford: Oxford University Press.

———. 2004. The role of humans in the cognitive development of apes revisited. *Animal Cognition*, 7, 213–215.

Tomasello, M., Davis-Dasilva, M., Camak, L. & Bard, K. 1987. Observational learning of tool use by young chimpanzees. *Human Evolution*, 2, 175–83.

Tomasello, M., Kruger, A. E. & Ratner, H. 1993. Cultural learning. *Behavioral and Brain Sciences*, 16, 595–652.

Van de Waal, E., Renevey, N., Favre, C. M., & Bshary, R. 2010. Selective attention to philopatric models cause directed social learning in wild vervet monkeys. *Proceedings of the Royal Society B* 277, 2105–2111.

Van Lawick-Goodall, J. 1973. Cultural elements in a chimpanzee community. In *Precultural Primate Behavior* (ed. by Menzel, E. W.), 144–84. Basel: Karger.

Van Schaik, C. P. 2004. *Among Orangutans: Red Apes and the Rise of Human Culture*. Cambridge, MA: Belknap Press.

———. 2009. Geographic variation in the behavior of wild great apes: Is it really cultural? In *The Question of Animal Culture* (ed by Laland, K. N. & Galef, B. G.), 70–98. Cambridge, MA: Harvard University Press.

Van Schaik, C. P., Deaner, R. O. & Merrill, M. Y. 1999. The conditions for tool use in primates: Implications for the evolution of material culture. *Journal of Human Evolution*, 36, 719–741.

Van Schaik, C. P., Ancrenaz, M., Borgen, G., Galdikas, B., Knott, C. D., Singleton, I., Suzuki, A., Utami, S. S. & Merrill, M. 2003. Orangutan cultures and the evolution of material culture. *Science*, 299, 102–105.

Visalberghi, E., & Fragaszy, D. 1990. Do monkeys ape? In *"Language" and Intelligence in Monkeys and Apes: Comparative Developmental Perspectives* (ed. by Parker, S. & Gibson, K.), 247–273. Cambridge: Cambridge University Press.

Visalberghi, E., & Fragaszy, D. M. 2002. Do monkeys ape?— Ten years after. In *Imitation in Animals and artifacts*, (ed. by Dautenhahn, K. & Nehaniv, C. L.), 471–499. Cambridge, MA: MIT Press.

Voelkl, B., & Huber, L. 2000. True imitation in marmosets. *Animal Behaviour*, 60, 195–202.

———. 2007. Imitation as faithful copying of a novel technique in marmoset monkeys. *Public Library of Science One*, 2, e611. doi:10.1371/journal.pone.0000611.

Want, S. & Harris, P.L. 2002. How do children ape? Applying concepts from the study of non-human primates to the developmental study of "imitation" in children. *Developmental Science*, 5, 1–13.

Whiten, A. 1998. Imitation of the sequential structure of actions by chimpanzees (*Pan troglodytes*). *Journal of Comparative Psychology*, 112, 270–281.

———. 1999. Parental encouragement in *Gorilla* in comparative perspective: Implications for social cognition. In *The Mentality of Gorillas and Orangutans* (ed. by. Parker, S. T., Miles, H. L. & Mitchell, R. W.), 342–366. Cambridge: Cambridge University Press.

———. 2005. The second inheritance system of chimpanzees and humans. *Nature*, 437, 52–55.

———. 2006. The dissection of imitation and its "cognitive kin" in comparative and developmental psychology. In *Imitation and the Social Mind: Autism and Typical Development* (ed. by Rogers, S. & Williams, J. H. G.), 227–250. New York: Guilford Press.

———. 2009a. The identification of culture in chimpanzees and other animals: From natural history to diffusion experiments. In *The Question of Animal Culture* (ed. by Laland, K. N. & Galef, B. G.), 99–124. Cambridge, MA: Harvard University Press.

———. 2009b. Ape behavior and the origins of human culture. In *Mind the Gap: Tracing the Origins of Human Universals* (ed. by Kappeler, P. & Silk, J. B.), 429–450. Berlin: Springer-Verlag.

———. 2010. A coming of age for cultural panthropology. In *The Mind of the Chimpanzee* (ed. by Lonsdorf, E., Ross, S. & Matsuzawa, T.), 87–100. Chicago: University of Chicago Press.

———. 2011. The scope of culture in chimpanzees, humans and ancestral apes. *Philosophical Transactions of the Royal Society B*, 366, 997–1007.

Whiten, A., Flynn, E., Brown, K. & Lee, T. 2006 Imitation of hierarchical action structure by young children. *Developmental Science*, 9, 575–583.

Whiten, A., Goodall, J., McGrew, W. C., Nishida, T., Reynolds, V., Sugiyama, Y., Tutin, C. E. G., Wrangham, R. W. & Boesch, C. 1999. Cultures in chimpanzees. *Nature*, 399, 682–685.

———. 2001. Charting cultural variation in chimpanzees. *Behaviour*, 138, 1481–1516.

Whiten, A. & Ham, R. 1992. On the nature and evolution of imitation in the animal kingdom: Reappraisal of a centuary of research. *Advances in the Study of Behavior*, 11, 239–283.

Whiten, A., Hinde, R. A., Laland, K. N. & Stringer, C. B. 2011. Culture evolves. *Philosophical Transactions of the Royal Society B* 366, 938–948.

Whiten, A., Hinde, R. A., Stringer, C. B. & Laland, K. N., eds. 2012. *Culture Evolves*. Oxford: Oxford University Press.

Whiten, A., Horner, V. & de Waal, F.B.M. 2005. Conformity to cultural norms of tool use in chimpanzees. *Nature*, 437, 737–740.

Whiten, A., Horner, V., Litchfield, C.A. & Marshall-Pescini, S. 2004. How do apes ape? *Learning and Behavior*, 32, 36–52.

Whiten, A., Horner, V., & Marshall-Pescini, S. R. J. 2003. Cultural Panthropology. *Evolutionary Anthropology*, 12, 92–105.

Whiten, A., McGuigan, H., Hopper, L. M. & Marshall-Pescini, S. 2009. Imitation, over-imitation, emulation and the scope of culture for child and chimpanzee. *Philosophical Transactions of the Royal Society B*, 364, 2417–2428.

Whiten, A. & Mesoudi, A. 2008. Establishing an experimental science of culture: Animal social diffusion experiments. *Philosophical Transactions of the Royal Society B*, 363, 3477–3488.

Whiten, A., Schick, K. & Toth, N. 2009. The evolution and cultural transmission of percussive technology: Integrating evidence from paleoanthropology and primatology. *Journal of Human Evolution*, 57, 420–435.

Whiten, A., Spiteri, A., Horner, V., Bonnie, K. E., Lambeth, S. P., Schapiro, S. J. & de Waal, F. B. M. 2007. Transmission of multiple traditions within and between chimpanzee groups. *Current Biology*, 17, 1038–1043.

Whiten, A. & van Schaik, C. P 2007. The evolution of animal "cultures" and social intelligence. *Philosophical Transactions of the Royal Society B*, 362, 603–620.

Zentall, T. R. 2004. Action imitation in birds. *Learning and Behavior*, 32, 15–23.

Chapter 32 Human Cultural Cognition

Esther Herrmann and Michael Tomasello

THE ANCESTORS of modern humans began traveling down their own evolutionary pathway about six million years ago. About two million years ago, the various hominin species started becoming extreme cognitive specialists. This is evidenced physiologically by the fact that the brains of modern humans are approximately three times larger than those of their nearest great ape relatives (Jerison 1973). It is evidenced behaviorally by the many cognitive practices and products of modern humans, from complex technologies and linguistic and mathematical symbols to rule-based social institutions, that are seemingly unique to the species as well. What happened? And what is the precise nature of the cognitive skills that distinguish humans from other primates? There are three readily observable aspects of uniquely human cognitive practices and products that provide important clues. First, virtually all of humans' most complex cognitive practices and products—everything from kayak building to written language to governments—are created not by single individuals but by individuals collaborating in cultural groups. Human cognition is *collaborative* (Vygotsky 1978). Second, virtually all of humans' most complex practices and products have arisen not instantaneously but rather gradually across generations, accumulating modifications as individuals work to improve on what they have inherited from their forebears to meet current needs. Human cognition is *cumulative* (Tomasello et al. 1993). And third, virtually all of humans' most complex practices and products have a normative and/or institutional dimension based in socially agreed-upon rules

for "how we do things"—with sanctions for doing them "wrong." Human cognition is *normative* (Searle 1995). In general, we may refer to culturally specific practices and products displaying these three characteristics as "cultural cognition."

Despite these readily observable characteristics of human cultural cognition, which mark it as unique in the primate world, identifying the underlying cognitive processes turns out to be more than a little difficult. Nevertheless, in this chapter we attempt to identify the species-universal cognitive processes that enable human cultural cognition and its culturally specific practices and products. We argue that humans do not differ much from other great apes in their basic cognitive skills for dealing with the physical world of objects and their spatial-temporal-causal relations (i.e., without any of the various cultural prostheses, from compasses to symbol systems, that humans use for assistance). In contrast, humans differ significantly from other great apes in their sociocognitive skills for putting their heads together with others in acts of collaboration, communication, and social learning. Humans possess a kind of species-specific "cultural intelligence" based on species-unique skills and motivations to participate with others in collaborative activities with shared intentions and attention: so-called shared intentionality (Bratman 1992). These underlying sociocognitive skills of shared intentionality are what enable particular cultural groups to construct locally adapted cognitive practices and products with collaborative, cumulative, and normative dimensions.

Ape and Human Cognition

Humans (*Homo sapiens*) are a member of the great ape clade, along with chimpanzees (*Pan troglodytes*), bonobos (*Pan paniscus*), gorillas (*Gorilla gorilla*), and orangutans (*Pongo* spp.). As evolutionary background for our account of human cognition, we very briefly characterize great ape cognitive abilities for dealing with both the physical and the social world. We then attempt to situate human cognition in this evolutionary context.

Great Ape Cognition

In thinking about the evolution of primate cognition it is useful to make a distinction between physical cognition and social cognition (Tomasello & Call 1997). Physical cognition deals with inanimate objects and their spatial-temporal-causal relations (see chapter 27, this volume), whereas social cognition deals with other animate beings and their intentional actions, perceptions, and knowledge (see chapters 28 and 30, this volume).

Primate cognition of the physical world presumably evolved mainly in the context of navigation and foraging. To locate food and water, primates need cognitive skills for dealing with space; to choose wisely among multiple food sources they need cognitive skills for dealing with quantities; and to extract food from difficult places they need cognitive skills for understanding causality. Many different studies suggest that great apes (and perhaps other nonhuman primates) understand and interact with the physical world in ways very similar to those of humans. Like humans, other great apes live most basically in a world of permanent objects (and categories and quantities of objects) existing in a mentally represented space. They understand much about various kinds of events in the world, and how these events relate to one another causally (see Tomasello & Call 1997; Call & Tomasello 2005; chapter 27, this volume).

In addition to perceiving and understanding many things in their environments, great apes also recall things they have perceived in the past and anticipate and plan for things that might happen in the future. For example, in a recent experimental study, bonobos and orangutans used a tool to retrieve food, and when the food was gone they dropped the tool and left. Later, when they returned, more food was there but the necessary tool was not. After only a few such trials, the apes began to take the tool with them before using it, planning for the next trial when they would need it again (Mulcahy & Call 2006). Great apes can also make inferences about what one perceived state or event implies about another. For example, in an experiment apes were faced with two cups, and learned that only one of the cups contained food. Then they watched as a human experimenter shook one of the cups. They were able to infer that a cup that clattered when it was shaken contained food, and they were also able to figure out that if one cup made no noise when shaken, then the other cup must contain food. This kind of reasoning by exclusion (since it is not in this one, it must be in the other one) is analogous to disjunctive syllogism in formal logic (Call 2004). Finally, in an experiment with chimpanzees, a banana was hidden inside one of two cups attached to opposite ends of a balance beam, causing the beam to tilt toward one end. Subjects preferentially searched for the banana in the lower cup, successfully inferring the causal effect of its weight on the beam (Hanus & Call 2008). In many different ways, therefore, great ape cognition goes beyond the here and now of the directly perceivable environment.

Primate cognition of the social world evolved mainly in the context of cooperation and competition between group members: to manipulate the behavior of others, primates need communication skills (chapter 29, this volume); to learn things from others, they need social learning skills (chapter 31, this volume); and to compete effectively or solicit the cooperation of others they need cognitive skills for understanding psychological states such as goals and perceptions (chapter 28, this volume). Many studies suggest that great apes understand their social worlds, like their physical worlds, in basically the same way that humans do. Like humans, apes live in a world of identifiable individuals with whom they form various kinds of social relationships and they recognize the third-party social relationships that other individuals have with one another (see Tomasello & Call 1997; chapter 28, this volume).

Great apes' understanding of social behavior and relationships rests on an understanding of the goals and perceptions of others (see Call & Tomasello 2008, chapter 30, this volume, for reviews). For instance, in one experiment chimpanzees were faced with two situations in which a human experimenter was either unwilling (by refusing) or unable (caused by a simulated accident) to provide food. Subjects clearly distinguished between the unwilling and unable experimenters, staying longer and gesturing less when the human was unable than when he was unwilling—even if the human's behavior was very similar in both (Call et al. 2004). This suggests that apes have at least some understanding of the goals of others. In another experiment, when an actor failed in a behavioral attempt or had an accident, human-raised chimpanzees inferred the goal the actor was attempting to achieve (Tomasello & Carpenter 2005). Great apes can even reason about the behavioral decision-making of other individuals. For instance, in a recent study human-raised chimpanzees observed a human successfully solving a problem in a particular way. They sometimes adopted the same method for solving the problem, and sometimes used a different method. Their choice was based on whether the obstacles to solving the problem were the same or different than those that faced the human demonstrator. They seemingly reasoned about why the human had chosen the behavioral means she had (Buttelmann et al. 2007).

Great apes also understand what others can and cannot see, and then act strategically based on this knowledge. In a series of studies a chimpanzee competed with a dominant individual over food. Some pieces of food were visible to both chimpanzees, and some were only visible to the subordinate chimpanzee. Subordinates preferentially pursued the food that was hidden from the dominant's view, and even took into account whether or not the dominant had seen the hiding process in the recent past (Hare et al. 2000, 2001; Bräuer et al. 2007; Kaminski et al. 2008). In a second set of food-competition experiments, chimpanzees demonstrated an understanding of visual and auditory perception by actively concealing their own approach to food either by hiding behind a barrier or by choosing a silent option to obtain a piece of food (Hare et al. 2006; Melis et al. 2006).

If great apes live in basically the same mentally represented physical and social worlds as humans, and if they reason about these worlds in many similar ways as well, then what is the difference in cognition among humans and other apes? What is the difference that leads to all of the complexities of human societies and their many material, symbolic, and institutional artifacts?

An Overall Comparison of Ape and Human Cognition

One proposal to explain uniquely human cognition is that humans' extra-large brains enable them to perform all kinds of cognitive operations more efficiently than other species: greater working memory, faster learning, more robust inferences, longer-range planning, and so on. This is what we may call the general intelligence hypothesis. A variant of this hypothesis suggests that the main difference between human and nonhuman cognition is based on a uniquely human ability to reason about relations of all kinds across both physical and social domains (Penn et al. 2008).

But neither of these variants of the general intelligence hypothesis is supported by the evidence. Herrmann et al. (2007) administered a comprehensive battery of cognitive tests to large numbers of chimpanzees ($n = 106$), orangutans ($n = 32$), and 2.5-year-old human children ($n = 105$; children of this age use some language but are still years away from reading, counting, or formal schooling). The test battery consisted of 16 different nonverbal tasks assessing all kinds

of cognitive skills involving both physical and social problems relevant to primates in their natural environment. The tests relating to the physical world consisted of problems concerning space, quantities, tools, and causality. The tests relating to the social world consisted of problems requiring subjects to imitate another's solution to a problem, communicate nonverbally with others, and read the intentions of others from their behavior. If the difference between human and ape cognition is based on cognitive skills that apply in all domains, such as relational reasoning across the board, then the children in this study should have differed from the apes uniformly across all the different tasks. But this was not the case. In fact, the children and apes had very similar levels of success in tasks related to the physical world; but the children had considerably more sophisticated cognitive skills than either ape species for dealing with the social world (fig. 32.1). It should be noted that children's better performance in the social tasks was not based on their being generally more comfortable or interested in the test situation, given that a human was the experimenter (see de Waal et al. 2008; Herrmann et al. 2008). The comfort level of the children and apes was directly measured, and the children were actually shyer and less interested in interacting with novel materials and an unfamiliar human than were the apes; performance on that measure did not correlate with performance on the social tasks.

Even if group differences in the cognitive abilities of human children and other apes are most apparent in the social domain, it is possible that the individual differences *within* each species are related to a general intelligence factor that applies across the board. For example, it could be that, despite absolute levels of performance, chimpanzees who excel at physical tasks also excel, relative to other chimpanzees, at social tasks. To investigate this possibility, Herrmann et al. (2010) employed factor analysis to examine separately the correlational structure of individual differences in the performance of children and chimpanzees on these same cognitive tasks. For neither species was a single general intelligence factor found; that is, individual differences within neither species could be explained by one underlying general factor (i.e., individual performance on one task did not predict performance on other tasks across the board). Instead, what was found was this. For both species there was a factor of spatial cognition: individuals who were good at one kind of spatial task were also good at other kinds of spatial tasks. Beyond this commonality, the structure of individual differences in cognitive performance between the two species varied. Specifically, chimpanzees' performance in various physical and sociocognitive tasks intercorrelated indiscriminately in one factor, whereas human children's performance showed two distinct and

A) Physical Domain

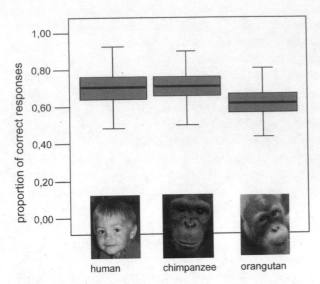

B) Social Domain

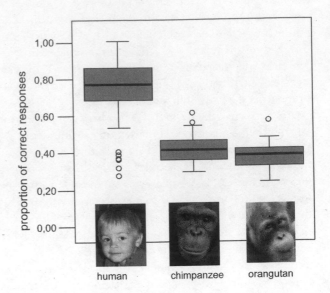

Fig. 32.1: Physical (A) and social (B) cognitive performance of human children, chimpanzees, and orangutans in Herrmann et al. (2007).

separate factors explaining their individual differences: one for physical cognition and one for social cognition (fig. 32.2).

Together, these findings suggest that the main differences between human and great ape cognition are not based on humans being generally more intelligent. Instead, they suggest that humans share many cognitive skills with their closest living relatives, especially for dealing with the physical world, but they have also evolved some specialized and more integrated sociocognitive skills. These skills emerge early in ontogeny and enable developing humans to create all kinds of cognitive skills for dealing with both their physical and social environments by collaborating, communicating, and learning from those around them. Humans have evolved a kind of species-specific cultural intelligence for living and exchanging information in cultural groups.

Human Cultural Cognition

Human cognition is thus one form that great ape cognition can take. Humans share with other great apes all of their most basic cognitive skills; they have just evolved a few new sociocognitive skills. But these few new skills led in human evolution to a fundamental change in process. In parallel to processes of biological evolution, but happening at a much faster time scale, processes of cultural evolution began taking place in human groups (Richerson & Boyd 2005). This led to a plethora of species-unique cognitive practices and products, as individuals putting their heads together in col-

laboration, symbolic communication, and instructed learning—often in the context of rule-based, symbolically based social institutions—were able to do much more cognitively than single individuals alone.

The Herrmann et al. (2007) study identified some of the uniquely human sociocognitive skills leading to processes of cultural evolution in a very general way. Other, more narrowly focused comparative and developmental studies have helped to specify them in more detail. The basic idea is that humans have evolved the ability to participate with one another in acts of shared intentionality in which they form joint goals and joint plans, maintain joint attention, share mutual knowledge, and are motivated to help one another (for general theoretical accounts of shared intentionality see Bratman 1992; Searle 1995; Tomasello et al. 2005). That is to say, uniquely human cognition derives from humans' unique ways of cooperating with one another.

Collaboration, Perspective, and Symbolic Representation

All social animals are, by definition, cooperative in the sense of living together relatively peacefully in social groups. Most social species forage in groups in one way or another, mostly as a defense against predation, and individuals in addition form specific relationships with others, leading to such things as coalitions and alliances in intragroup competition for food and mates (Harcourt & de Waal 1992). In many mammalian species, intergroup conflicts and defense against predators are group activities as well, and some mammals (e.g., social carnivores and some primates) even

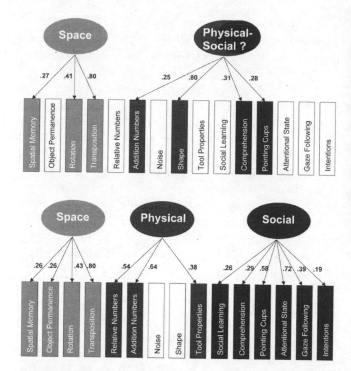

Fig. 32.3. Group foraging in the Hadza. Photograph courtesy of Frank Marlowe.

Fig. 32.2. Correlational structure for chimpanzees (a) and human children (b) in Herrmann et al. 2010. The arrows pointing from a factor to the observed variables (tasks) indicate the causal effect of the factor on the observed variables. The numbers alongside the arrows are standardized coefficients that represent the correlation between the factor and each task. Task descriptions: *Spatial Memory, Object Permanence, Rotation,* and *Transposition*: Locating a reward or tracking a reward after its location changes. *Relative Numbers* and *Addition Numbers*: Discriminating quantities. *Noise, Shape,* and *Tool Properties*: Understanding causal relationships. *Social Learning*: Solving a problem by observing a demonstrated solution. *Comprehension, Pointing Cups,* and *Attentional State*: Understanding communicative cues and producing communicative gestures. *Gaze Following*: Following an actor's gaze direction. *Intentions*: Understanding an actor's intention.

engage in group hunting (Boesch & Boesch-Achermann 2000). Great apes considered together do more or less all of these group things (for reviews see Muller & Mitani 2005; chapter 6, this volume), and so our question is how these are similar to and different from human forms of collaboration.

Compared with other primates, humans engage in an extremely wide array of collaborative activities, and are able to coordinate joint activities involving large numbers of unrelated individuals. These kinds of collaborative efforts are often organized in the context of symbolic structures and formal institutions (Richerson & Boyd 2005). And different cultural groups collaborate in different activities: some in hunting, some in fishing, some in gathering, some in house building, some in playing music, some in governing, and so on, which testifies to the flexibility of the underlying cognitive skills involved (fig. 32.3). In all human cultures there is the collective expectation that members participate in

many different kinds of collaborative activities, and these group expectations (norms) are backed by the threat of real sanctions against those who violate those norms. In addition, there are cultural practices and institutions whose existence is due totally to the collective agreement of group members. As just one example, while nonhuman primates have some understanding of familial relatedness (chapter 28, this volume), humans assign social roles such as "wife," "husband," and "parent" that they all recognize, and which carry ethical obligations to behave in specified ways.

According to shared intentionality theorists, fully cooperative activities have two key characteristics: (1) the participants have a joint goal to which they are jointly committed, and (2) the participants coordinate their interdependent roles, plans, and subplans of action, and even help each other in their roles if needed (e.g. Gilbert 1989; Bratman 1992). We may also add, from a more motivational perspective, that (3) collaborators tend to share the spoils of their efforts among themselves (and perhaps others) in mutually satisfactory ways (i.e., "fairly"). For example, many hunter-gatherer groups forage for certain kinds of foods that are not easily captured by single individuals (e.g., large game, some fish, underground plants, etc.; see Hill & Hurtado 1996 for a review). Typically in these activities, individuals commit themselves to the joint goal of capturing a certain prey or extracting a certain plant, and then they plan together their various roles and how they should be coordinated ahead of time—or else those roles are already common knowledge based on a common history of the practice. The participants share the spoils not only among themselves, but also with others in the group, typically not at the site of capture but rather back at some home base; they are typically under strict social norms about how the food should be allocated, and those who do not share are harshly sanctioned (Gurven 2004).

Male chimpanzees in some groups hunt small animals, but there is controversy over whether this is similar to human cooperative hunting in having joint goals, plans for coordination, and a socially agreed upon division of spoils (compare, e.g., Tomasello et al. 2005 and Boesch 2005). Experiments help to settle the issue. For example, in Crawford's (1937, 1941) classic studies, pairs of chimpanzees were presented with a heavy box containing food, which they could bring within reach only if they both pulled on ropes simultaneously. The chimpanzees did not synchronize their actions until after they received extensive human training, and when they were later given a slightly different transfer task, they all reverted to uncooperative behavior. In more successful experiments with little or no training, most chimpanzee cooperative problem-solving involves individuals learning to refrain from acting until the other is in place and ready to act (Chalmeau 1994; Chalmeau & Gallo 1996) or sometimes even fetching needed partners (Melis et al. 2006a).

Human children begin to collaborate with others with truly joint goals and joint commitments quite early in their lives. For example, Warneken et al. (2006) presented 18- to 24-month-old children and three human-raised juvenile chimpanzees with four collaborative tasks: two instrumental tasks in which there were concrete goals, and two social games in which there were no concrete goals other than collaborative play itself. The adult partners were programmed to stop participating at a certain point in each task as a way of determining the subjects' commitment to the joint activity. Results were clear. In the problem-solving tasks, chimpanzees synchronized their behavior relatively skillfully with that of the humans, often achieving the goals successfully. However, they showed no interest in the social games, basically declining to participate. Most importantly, when the human partner stopped participating, no chimpanzee ever made a communicative attempt to reengage her—even in cases where they were seemingly highly motivated to achieve the goal—thus suggesting that they had not formed a joint goal with the experimenter. In contrast, the human children collaborated enthusiastically in the social games as well as the instrumental tasks. Most importantly, when the adult stopped participating, the children actively encouraged her to reengage by communicating with her in some way, thus suggesting that they had formed with her a joint goal to which they now wanted her to recommit. In a recent study, three-year-old children even showed that they had, and expected their partner to have, a normative commitment to the shared activity such that termination required some kind of acknowledgement or even apology (Gräfenhain et al. 2009).

In addition to a joint goal, a fully collaborative activity requires that there be some division of labor and that each partner understand the other's role. Carpenter et al. (2005) had an adult engage in a collaborative activity with very young children, around 18 months of age, and then repeat that activity, taking over their role—forcing them into a role they had not previously played. Even these very young children readily adapted to their new role, thus suggesting that in their initial joint activity with the adult they had somehow processed her role. Three young, human-raised chimpanzees did not reverse roles with a human partner in the same way (Tomasello & Carpenter 2005). One interpretation is that human infants understand the joint activity from a "bird's-eye view," with the joint goal and complementary roles all in a single representational format, which enables them to reverse roles as needed.

As individuals coordinate their actions with one another in collaborative activities, they also coordinate their attention to things relevant to their joint goal (Bakeman & Adamson 1984). In addition to their joint attention on things, participants in these interactions also have their own perspectives on things. Indeed, Moll and Tomasello (2007) argue that the whole notion of perspective depends on us first having a joint attentional focus, as topic, that we may then view differently (otherwise we just see completely different things). This dual-level attentional structure—shared focus of attention at a higher level, differentiated into perspectives at a lower level—is of course directly parallel to the dual-level intentional structure of the collaborative activity itself—joint goal with individual roles—and ultimately derives from it.

To coordinate their complex collaborative and joint attentional activities, humans have evolved some species-unique forms of communication. Most obvious is, of course, language, but even before this complex form of conventional communication develops, children engage in species-unique forms of gestural communication: specifically, pointing and pantomiming. These nonverbal forms of communication are used to direct other people's attention to something in the external world as a way of informing them about something or requesting help with something. In both cases, the motive for communication involves cooperation, either for requesting help from others or for offering help to others via useful information, in a way that the motives of nonhuman primates do not (Tomasello 2008; see also chapter 29, this volume). Moreover, for pointing and pantomiming to communicate in the complex ways that they do, communicators need to use these gestures in the context of joint attention with the recipient. For example, when a young child points to the end of a tube, as in fig. 32.4, with no shared context, it is not clear what he is attempting to communicate. But in

Fig. 32.4. A human infant points to coordinate a collaborative activity.

the context of the collaborative activity of pulling this tube apart to get a toy contained inside it when the partner is not cooperating, the meaning is immediately clear.

The point is that human pointing and pantomiming are species-unique gestures that communicate in especially powerful ways by taking advantage of a shared joint attentional context, often within a collaborative activity. Given this shared psychological context, a communicator's pointing gesture indicates some aspect of the situation to which he wants his partner to attend. He may be pointing to a piece of paper, the color of the paper, the shape of the paper, or basically any aspect of the paper, depending on the shared context and the reason he is pointing. This means that pointing can actually indicate perspectives on things. Tomasello (1999) argues that the kind of perspective shifting that humans routinely use in communication is internalized during ontogeny to create perspectival cognitive representations, which enable humans to shift flexibly in the way they construe the same situation for different cognitive purposes. Iconic gestures add to this perspectival flexibility the ability to symbolically represent situations that are not currently perceptually present. External icons, such as hands miming the motions of a running antelope, enable humans to create a kind of pretend world that they can then reenact for others, to share, for example, the excitement of this morning's hunting trip (Donald 1991). Symbolic representation, of course, adds a further dimension of flexibility and power to human cognitive representation.

To summarize, human beings collaborate with one another on joint goals and shared plans, and this creates the kind of joint attentional and cooperative context that enables human cooperative communication in the form of the natural referential gestures of pointing and pantomiming. Humans communicate with assumptions of cooperation, most likely because these unique forms of communication originally evolved to coordinate collaborative activities. Internalizing these forms of referential communication leads to humans' species-unique perspectival and symbolic cognitive representations, as individuals, in a sense, communicate perspectives to themselves (Vygotsky 1978).

Cultural Learning, Norms, and Institutions

One of the most important developments in primatology over the last few decades is the discovery that chimpanzees, orangutans, capuchin monkeys, and perhaps other species have a form of culture (chapter 31, this volume). But the cultural behaviors of nonhuman primate groups do not seem to show the key characteristics of human cultural cognition—that is, they do not seem to be collaborative, cumulative, and normative (Tomasello 1990, 1999, 2009; Tennie et al. 2009).

These differences in nonhuman and human culture may be plausibly attributed to differences in underlying processes of social learning. Nonhuman primate cultures depend mainly on noncollaborative processes of social learning in which individual social learners enhance their own fitness by taking advantage of the hard work and experience of others (Whiten & van Schaik 2007; chapter 31, this volume). Human culture also depends on cooperative processes of a unique kind, and these processes play a crucial role in both aspects of cultural and historical change: innovation and transmission. Regarding innovation, it is obvious that in human cultural groups many behavioral novelties are created not by individuals but by groups of individuals working together. Improved ways of hunting large prey, for example, would almost certainly be invented by multiple individuals in a process of collaboration. Even more broadly, when individuals use the cultural artifacts and practices invented by others before them they are, in a sense, collaborating with those others—so that any improvements they make are due to a kind of indirect collaboration, as they build on the innovations of previous inventors. This means that the complex cognitive skills that individual human beings build are thoroughly collaborative enterprises.

Regarding transmission, uniquely human forms of cooperation come to bear in three main ways. First is teaching. Teaching is of course a form of altruism, as an instructor expends some time and energy making sure that a pupil gains a certain piece of knowledge or skill for her benefit (Caro & Hauser 1992; Thornton & Raihani 2008). Teaching is present in one form or another in all human societies

(Kruger & Tomasello 1996), but it is clearly not an everyday activity of any nonhuman primate—though something in this direction may occur on occasion (e.g., see the two single observations of Boesch 1991; chapter 31, this volume). Teaching may be especially important for those kinds of cultural conventions that cannot be invented on one's own but can only be imitated from others. Indeed, Gergely and Csibra (2006) have recently elaborated an account explaining why the existence of relatively "opaque" cultural conventions (whose causal structure is either totally absent or else difficult to see) requires both that human adults be specifically adapted for pedagogy toward children and that human children be specifically adapted for recognizing when adults are being pedagogical. This pedagogical stance is typically indicated by the same behavioral signs as cooperative communication in general, involving such things as eye contact, special tones of voice, and so forth (and indeed teaching may be seen as one manifestation of human cooperative communication in general; Tomasello 2008). Gergely and Csibra emphasize that when children detect pedagogy, they assume that they are supposed to be learning something that is novel to them and that applies to the world in a generalized manner.

The second way in which cooperation affects human cultural transmission involves a special kind of imitation or, more precisely, a special kind of motivation for imitation. Human children imitate not only to acquire more effective behavioral strategies in instrumental situations, but also sometimes for purely social reasons. That is to say, human beings often imitate others simply in order to be like them (Carpenter 2006). The tendency of humans to follow fads and fashions in their cultural group for no apparent instrumental reason is well known and well documented (see any social psychology textbook). The evolutionary basis of this conformist behavior is very likely identification with the group, in the context of cultural group selection which requires, as its fundamental component, very strongly conformist cultural transmission (Richerson & Boyd 2005). This so-called social function of imitation—simply to be like others in the group—is clearly an important part of human culture and cultural transmission, and it enables much more faithful reproduction of behavior in the process of cultural transmission.

The third form of cooperation is the normative dimension of human cultural transmission. Bruner (1990) emphasized that human culture persists, and has the character it does, not just because human children do what others do, but also because adults expect and even demand that they behave in certain ways. Children understand that this is not just the way something usually is done, but the way it *should* be done. In a recent study, Rakoczy et al. (2008) found not only that three-year-old human children copied the way that others did things, but that when those children saw a third party doing them in some other way they objected and told them they were doing it "wrong"—and even taught them the "right" way. Kelemen (1999) has also shown that young children learn very quickly that a particular artifact is "for" a particular function, and that they consider other uses of it to be "wrong." The evolutionary source of this normative dimension to human activities is not immediately clear, although it is presumably bound up with group identity and conformist transmission (this is how we, the members of this group, do things, even if others do them differently), and enforced by sanctions against deviations from the normative standard. Teaching, social imitation, and normativity represent the contribution of humans' special forms of cooperation to the process of cultural transmission across generations. They all assume cooperative motives, group identity, conformity, and even morality, of a type not typically attributed to other primate species. The key point is simply that it is very likely these three cooperative processes, along with humans' ability to invent new things collaboratively with others, that give human cultural traditions their extraordinary stability, cohesion, and cumulativity (Tomasello 2009; Tennie et al. 2009).

An important product of all of this collaborative cultural activity is human language. The evolutionary emergence of this conventional, symbolic means of communication depended first on preexisting forms of cooperative communication such as pointing and pantomiming. Then it built upon human ability to create, by cooperative "agreement," arbitrary conventions for how we refer one another's attention to specific things with specific sounds or gestures. For novices, such as children, participation in this convention depended both on the ability to imitatively learn how to use these communicative acts as others do, and on the knowledge that this is the way all members of the group do it because they all share a common learning history (Tomasello 2008). Conventional symbols thus make human communication and cognition even more deeply cooperative, because even the communicative forms themselves are created by cooperative agreement. In addition, over time, human communities develop not just single symbols to make reference to entities, events, and situations, but also complex constructions for doing so. For example, almost all world languages have set patterns (grammatical constructions) for making reference to common and important events like an agent affecting some object, people transferring objects among one another, objects moving down a path toward a goal, people having feelings about events, and so forth (Goldberg 1995).

Regarding cognition itself, when the human ability to

collaboratively and normatively transmit to others "how we do things" is combined with the ability to symbolically represent the external world in gestures or language, a unique new possibility arises—namely, the creation of social institutions (Searle 1995). Social institutions are, first of all, sets of behavioral practices governed by various kinds of mutually recognized norms and rules. For example, all human cultures engage in mating and marriage in the context of rules for how people in a given culture do it. If one violates those rules, one is sanctioned in some way, and is perhaps even ostracized from the group. But also, as a part of the process in many cases, humans imaginatively create new symbolically defined entities with culturally sanctioned symbolic powers. For example, husbands, wives, and parents are culturally defined roles in the culture which therefore have culturally defined rights and obligations (Searle [1995] refers to this process as the creation of new "status functions"). For another example, all human cultures have rules and norms for exchanging food and other valuable objects. Some objects may be accorded the symbolic status of money (e.g., specially marked paper), which gives them special, culturally backed power in the process of trade. Other sets of rules and norms, at least often in democratic groups, create symbolic leaders, such as chiefs and presidents, who have special rights and obligations for making decisions, or even creating new rules, for a group. Social institutions are created by cooperative agreements and symbolically defined entities and powers, and they then structure much of humans' social and cognitive lives.

Philosophers have long noted that human cognition has a normative dimension, and that this has enormous consequences for how people think. In particular, language is itself a social institution, as it prescribes normatively how to refer other people's attention to things, and to think when we use language that there is a "right and wrong" way to do it—including everything from naming things to reasoning by logic. The role of language in shaping the way humans think is now well established, even in the sense of different languages leading to different ways of construing and reasoning about the world (e.g., Levinson 2003).

The Role of Ontogeny

Clearly, humans are biologically adapted for all of this. This becomes especially clear when we look at autism. Among other things, children with autism are born with a biological deficit in some aspects of shared intentionality, so they cannot take advantage of the cultural world into which they are born, and they do not develop adult human cognitive skills (Hobson 1993). On the other hand, no human could do any of the complex things they do with a biological pre-disposition alone; that is to say, no individual could invent any of the complex cognitive practices and products of culture without a preexisting cultural world in which to grow and learn. A biologically intact human child born outside of any human culture—with no one to imitate, no one to teach her things, no language, no preexisting tools and practices, no symbol systems, no institutions, and so forth—would not develop adult human cognitive skills. Both biology and culture are necessary parts of the process.

Organisms inherit their environment as much as they inherit their genes, so perhaps it is most appropriate to say that human beings biologically inherit the cognitive skills necessary for developing in a cultural environment. Obviously, some kind of social environment is also important in the ontogeny of most other primate species for developing species-typical behaviors of all kinds, and cultural transmission may even play some role as well. But for humans the species-typical sociocultural environment is an absolute necessity for youngsters to develop the culturally specific cognitive skills required for survival in the many very different and sometimes harsh environments that humans inhabit. And so the point is simply that ontogeny plays an especially large and important role in the cognitive development of *Homo sapiens* as compared with other primates. Indeed, perspectival, symbolic, and normative cognitive representations are only possible for individuals growing up in a cultural world and interacting with others who are symbolically communicating different perspectives on things—indeed, the "correct" perspective on things—to them in the first place.

The Evolution of Human Cognition

Great apes' skills of social cognition and culture evolved within the context of the competition and cooperation characteristic of their societies. Competition is perhaps of special importance, and indeed many theorists have argued that the so-called Machiavellian intelligence of nonhuman primates derives primarily from advantages gained by individuals outsmarting group mates in various ways (Humphrey 1976; de Waal 1982; Byrne & Whiten 1988). Great apes also cooperate, of course, in such things as coalitions and alliances, group defense, mutual grooming, and so forth, based in large part on kinship, nepotism, and reciprocity (Silk 2006; chapters 21 and 22, this volume), and these activities may have contributed to apes' remarkable sociocognitive abilities as well.

But in human cognitive evolution, the key role was played by cooperation, and a special form of it at that. Our proposal is (in comparison to other hypotheses that, for example, emphasize intergroup competition or cooperative breeding as the selective force) that what started humans down their unique path to cultural cognition initially was some change in ecology that turned them from mostly individual foragers, like the other great apes (chapter 6, this volume), into obligate collaborative foragers. Whereas other great apes might travel in small groups and sit together while eating, they typically do not procure their food through collaborative activities. Exceptions, such as chimpanzee group hunting of monkeys, are typically additions to the diet and are not obligatory for survival. But, in the current hypothesis, something happened with humans that forced them to either forage collaboratively or perish. One possibility is new competition for their normal means of sustenance. For example, the expansion of grassland and subsequent fragmentation of forest at the end of the Pliocene might have led the radiation of terrestrial monkeys, who would have competed with humans over terrestrial resources (Jablonski 1993). In response to this invasion, perhaps, humans began to exploit food resources requiring complex collaboration and coordination—such things as collaborative hunting of large game and collaborative gathering of embedded plants where the coordinated and self-supporting efforts (including special knowledge and skills) of multiple individuals were required (Hill 2002; Kaplan et al. 2000).

So humans gradually became obligate cooperative foragers. Under these conditions, individuals who were neither appropriately motivated nor skilled in collaborative activities would be at a significant disadvantage. In our view, there were three key steps in the evolutionary process. First, for humans to become truly collaborative foragers, there must have been an initial step that broke them out of the great ape pattern of strong food competition, low tolerance for food sharing, and active food offering mainly under pressure. This great ape pattern can be seen clearly in food acquisition experiments. Melis et al. (2006b) presented pairs of chimpanzees with a board bearing food outside their room that required both to pull simultaneously to get access to food. When all the food was clumped together in the middle of the board, so that there was the problem of sharing at the end, the chimpanzees had considerable difficulty collaborating. When the food was pre-divided for them and placed at the two ends of the board, they had no such difficulty—especially when the individuals in a pair had demonstrated high tolerance for one another in the past. Interestingly, bonobos, who are more tolerant around one another with food, were more successful than chimpanzees at cooperating in the same task when the food was clumped together at the center of the board (Hare et al. 2007)—and human children were not affected by clumped versus pre-divided rewards at all, even sharing equally in

most cases (Warneken et al. 2011). On another dimension of primate temperament, human children, as compared to chimpanzees and orangutans, are especially uncertain when encountering novel people and objects. This often leads to their seeking reassurance from parents and peers in ways that might provide opportunities for social referencing, social learning, and teaching (Herrmann et al. 2007). The proposal is thus that changes in human temperament—towards greater tolerance and social comfort seeking, among other things—were prerequisite for humans beginning down their ultra-cooperative pathway (Hare & Tomasello 2005; Hare 2007; Tomasello 2009; see also chapter 23, this volume).

The temperamental change in humans may have occurred in any one of several possible ways (or in some combination). One possibility is that humans evolved an especially tolerant and prosocial temperament through a process of self-domestication in which aggressive and despotic individuals within a group were systematically punished or shunned—a pattern commonly seen in small-scale societies (Boehm 1999; Wrangham & Pilbeam 2001; Leach 2003). Another possibility is that changes in human temperament in a prosocial direction were brought about by cooperative breeding (Hrdy 2009). In humans, unlike other apes, mothers get the support of alloparents, who contribute to basic child care activities and also engage in a variety of prosocial behaviors such as active food provisioning and teaching of the child. This change in social organization toward pair bonding and cooperative breeding had enormous advantages for child survival and also reduced humans' interbirth interval, thus leading to population growth. In this context, selection presumably favored more tolerant and prosocial individuals for playing the role of helper (who benefits via kin selection and direct/indirect reciprocity). It is of course possible that both self-domestication and cooperative breeding may have played roles. The important point is that there was some initial step away from great apes in human evolution, involving the emotional and motivational side of things, that propelled humans into a new adaptive space in which complex skills and motivations for complex collaborative activities and shared intentionality could be selected.

The second step toward humans' ultracooperativeness was that in this new social context, these tolerant and prosocial individuals would be more likely to be doing the kinds of things together in which cognitive skills for the forming of joint goals, joint attention, cooperative communication, and social learning and teaching would be especially beneficial—such things as cooperative hunting of animals together, cooperative gathering of embedded plants, and so forth. As Alvard (chapter 26, this volume) argues, obligate cooperative foraging poses a basic coordination problem in which individuals have to negotiate a shared goal and somehow communicate about their respective roles. Tomasello (2008) argues that humans' unique skills of cooperative communication, including language, evolved originally to coordinate collaborative foraging activities. A more tolerant motivational/temperamental disposition led to the possibility of individuals acting together with one another in new ways, which in turn set up the conditions for selection for ever more complex cognitive skills of shared intentionality, and for collaborating and communicating in ever more complex ways. Importantly, as argued by Roberts (2005) and Alvard (chapter 26, this volume), the selective context for collaboration of this type is shielded in significant ways from the problems of free-riding. Specifically, when collaborative foraging is obligate, one cannot out-compete others on one's own; each individual has a stake in the well-being of her potential collaborators, since she depends on them; and everyone needs a good reputation as a collaborative partner.

The third step involved group-level processes, including the creation of group-enforced norms and group-constituted social institutions. These required significant sociocognitive coordination skills for forming both mutual expectations and normative rules with others in the group (Tomasello 2009). Then, human groups also began to compete with one another, and this led to processes of cultural group selection. That is, as Richerson and Boyd (2005) have argued, human groups at some point possessed different traditions—and even different norms and institutions—for engaging in various activities, including subsistence. Those groups that passed along culturally "better" traditions, norms, and institutions did better in competition with other groups. And in the new context of obligate cooperative foraging, "better" could easily mean more cooperative. This could then lead to a kind of runaway selection involving a new process of coevolution between culture and cognition: cultural artifacts, norms, and institutions create a new environment to which individuals must adapt. Thus, individuals who could most quickly learn to participate in various collaborative cultural practices and use various cultural artifacts and symbols—through special skills of communication and social learning supported by more sophisticated ways of reading and sharing the intentions of others—were at a selective advantage. Also advantaged were individuals who could most quickly identify with their groups (Richerson and Boyd's "tribal instincts") and negotiate their various social norms (expectations of judging, punishing group mates) for how to interact peaceably and avoid being shunned.

The cognitive skills that human beings develop as a result of living culturally during both phylogeny and ontogeny do not come out of nowhere. Humans have not developed completely new forms of cognition; they have simply taken the great ape version of cognition and "collectivized" it through their skills and motivations of shared intentionality. For example, great apes engage in various group activities, but humans collaborate with others in some special ways (e.g., forming joint goals and commitments with others, or acting in interchangeable roles). Great apes follow the direction of others' gaze, but humans share attention with others recursively (we each know that the other is watching us watch them, and so forth). Great apes also communicate intentionally with others, but humans do so cooperatively and symbolically. Engaging with others—and with their artifacts, norms, and institutions—in these especially collaborative ways leads human children to create some species-unique forms of cognitive representation, namely, perspectival and normative representations. These then enable them to acquire the more specific cognitive skills of their particular culture—from its linguistic symbols to its procedures for procuring and consuming food—and perhaps to contribute to these skills creatively as well.

Human evolution is thus characterized inordinately by niche construction, in the form of cultural practices and products (Odling-Smee et al. 2003), and also in the form of gene-culture coevolution as the species has evolved cognitive skills and motivations enabling it to function effectively in any one of many different self-built cultural worlds (Richerson & Boyd 2005). It is possible that these skills of cultural cognition were still largely absent in *Homo erectus* one to two million years ago, even though the change in foraging strategies may have begun some time earlier. First of all, the relatively rapid brain growth in *Homo erectus* during ontogeny more closely resembled that in modern apes than that in modern humans, and overall modern humans have significantly larger brains than did *Homo erectus* (Coqueugniot et al. 2004). In addition, most evidence suggests that at that time there were very likely no extensive cultural differences between different human groups (Klein 1999). One hypothesis, then, is that humans' special skills of cooperation and shared intentionality—which led to all kinds of specialized cultural practices and products—reached their modern form only with the rise and flourishing of modern humans.

Finally, it is noteworthy that humans also have a species-unique physiological adaptation that may have facilitated their ultracooperativeness. Unlike other primates, humans have eyes with large white sclera, which make it especially easy for an observer to follow the direction of their gaze (Kobayashi & Kohshima 1997). And whereas other great apes mostly follow the head directions of other individuals when they are gazing somewhere, human children mostly follow the eye direction specifically (Tomasello et al. 2007). Such a morphological adaptation could only have evolved in cooperative social groups in which group mates did not too often exploit the gaze direction of others for their own benefit, but instead used it more often in coordinating collaborative and communicative interactions involving joint attention. When this cooperative physical trait evolved in the species is unknown.

Conclusion

In contrast with their nearest great ape relatives, who all live in the general vicinity of the equator, humans occupy an incredibly wide range of environmental niches covering almost the entire planet. To deal with everything from the Arctic to the tropics, humans have evolved a highly flexible suite of cognitive skills. These are not individual cognitive skills that enable individuals to survive alone in the tundra or rain forest, but rather they are sociocognitive skills that enable them to develop, in concert with others in their cultural groups, creative ways of coping with whatever challenges may arise.

In all, then, what most clearly distinguishes the cognition of humans from that of their nearest primate relatives is their sociocognitive adaptations for operating together in cultural groups. Given those adaptations, groups of individuals are able to cooperate to create artifacts and practices that accumulate improvements, or rachet up in complexity, over time as new environmental challenges arise. Since this process creates ever-new cognitive niches for developing youngsters (Tomasello 1999), human children must be equipped to participate in this groupthink with special skills for collaboration, communication, and cultural learning. As part of this process, children construct cognitive representations of the world that incorporate the perspectives and normative judgments of others, and they learn to symbolically represent those perspectives and judgments for others in acts of interpersonal and intrapersonal communication. Humans are adapted for life in a culture, and the particular tools, symbols, and social practices of the cultures into which they are born enable them to construct further cognitive skills for coping with the exigencies of their local environments.

References

Aiello, L. C. & Wheeler, P. 1995. The expensive tissue hypothesis. *Current Anthropolology*, 36, 199–221.

Bakeman, R. & Adamson, L. 1984. Coordinating attention to people and objects in mother-infant and peer-infant interactions. *Child Development*, 55, 1278–1289.

Boehm, C. 1999. *Hierarchy in the Forest: The Evolution of Egalitarian Behavior*. Cambridge, MA: Harvard University Press.

———. 1991. Teaching among wild chimpanzees. *Animal Behaviour*, 41, 530–532.

Boesch, C. & Boesch-Achermann, H. 2000. *The Chimpanzees of the Tai Forest: Behavioural Eecology and Evolution*. Oxford: Oxford University Press.

Boyd, R. & Richerson, P. J. 1996. Why culture is common but cultural evolution is rare. *Proceedings of the British Academy*, 88, 77–93.

Bratman, M. E. 1992. Shared cooperative activity. *Philosophical Review*, 101, 327–341.

Bräuer, J., Call, J. & Tomasello, M. 2007. Chimpanzees really know what others can see in a competitive situation. *Animal Cognition*, 10, 439–448.

Bruner, J. 1990. *Acts of Meaning*. Cambridge, MA: Harvard University Press.

Buttelmann, D., Carpenter, M., Call. J. & Tomasello, M. 2007. Enculturated chimpanzees imitate rationally. *Developmental Science*, 10, F31–38.

Byrne, R. W. & Whiten, A. 1988. *Machiavellian Intelligence: Social Expertise and the Evolution of Intellect in Monkeys, Apes, and Humans*. New York: Oxford University Press.

Call, J. 2004. Inferences about the location of food in the great apes (*Pan paniscus, Pan troglodytes, Gorilla gorilla* and *Pongo pygmaeus*). *Journal of Comparative Psychology*, 118, 232–241.

Call, J., Hare, B. H., Carpenter, M. & Tomasello, M. 2004. "Unwilling" versus "unable": chimpanzees' understanding of human intentional action? *Developmental Science*, 7, 488–498.

Call, J. & Tomasello, M. 2005. Reasoning and thinking in nonhuman primates. In *Cambridge Handbook of Thinking and Reasoning* (ed. by Holyoak, K. J. & Morrison, R. G.), 607–632. Cambridge: Cambridge University Press.

———. 2008. Do chimpanzees have a theory of mind: 30 years later. *Trends in Cognitive Science*, 12, 187–192.

Carpenter, M. 2006. Instrumental, social, and shared goals and intentions in imitation. In *Imitation and the Development of the Social Mind: Lessons from Typical Development and Autism* (ed. by Rogers, S. J. & Williams, J.), 48–70. New York: Guilford.

Carpenter, M., Tomasello, M. & Striano, T. 2005. Role reversal imitation in 12 and 18 month olds and children with autism. *Infancy*, 8, 253–278

Caro, T. M. & Hauser, M. D. 1992. Is there teaching in nonhuman animals? *The Quarterly Review of Biology*, 67, 151–174.

Chalmeau, R. 1994. Do chimpanzees cooperate in a learning task? *Primates*, 35, 385–392.

Chalmeau, R. & Gallo, A. 1996. Cooperation in primates: critical analysis of behavioural criteria. *Behavioural Processes*, 35, 101–111.

Coqueugniot, H., Hublin, J.-J, Veillon, F., Houët, F. & Jacob, T. 2004. Early brain growth in *Homo erectus* and implications for cognitive ability. *Nature*, 431, 299–302.

Crawford, M. P. 1937. The cooperative solving of problems by young chimpanzees. *Comparative Psychology Monographs*, 14, 1–88.

———. 1941. The cooperative solving by chimpanzees of problems requiring serial responses to color cues. *Journal of Social Psychology*, 13, 259–280.

De Waal, F. B. M. 1982. *Chimpanzee Politics*. London: Jonathan Cape.

De Waal, F. B. M., Boesch, C., Horner, V. & Whiten, A. 2008. Comparing social skills of children and apes. *Science*, 319, 569.

Donald, M. 1991 *Origins of the Modern Mind: Three Stages in the Evolution of Culture and Cognition*. Cambridge, MA: Harvard University Press.

Gergely, G., & Csibra, G. 2006. Sylvia's recipe: The role of imitation and pedagogy in the transmission of cultural knowledge. In *Roots of Human Sociality: Culture, Cognition and Interaction* (ed. by Enfield, N. J. & Levinson, S. C.), 229–255. Oxford: Berg Press.

Gilbert, M. 1989. *On Social Facts*. International Library of Philosophy series. Princeton NJ: Princeton University Press.

Goldberg, A. 1995. *Constructions: A Construction Grammar Approach to Argument Structure*. Chicago: University of Chicago Press.

Gräfenhein, M., Behne, T., Carpenter, M. & Tomasello, M. 2009. Young children's understanding of joint commitments to collaborate. *Developmental Psychology*, 45, 1430–43.

Gurven, M. 2004. To give and to give not: The behavioral ecology of human food transfers. *Behavioral and Brain Sciences*, 27, 543–83.

Hanus, D. & Call, J. 2008. Chimpanzees infer the location of a reward on the basis of the effect of its weight. *Current Biology*, 18, R370–372.

Harcourt, A. & de Waal, F. 1992. *Coalitions and Alliances in Humans and Other Animals*. Oxford: Oxford University Press.

Hare, B. 2007. From nonhuman to human mind: What changed and why. *Current Directions of Psychological Science*, 16, 60–64.

Hare, B., Call, J. & Tomasello, M. 2000. Chimpanzees know what conspecifics do and do not see. *Animal Behaviour*, 59, 771–785.

———. 2001. Do chimpanzees know what conspecifics know? *Animal Behaviour*, 61, 139–151.

———. 2006. Chimpanzees deceive a human competitor by hiding. *Cognition*, 101, 495–514.

Hare, B., Melis, A., Woods, V., Hastings, S. & Wrangham, R. 2007. Tolerance allows bonobos to outperform chimpanzees in a cooperative task. *Current Biology*, 17, 619–623.

Hare, B. & Tomasello, M. 2005. The emotional reactivity hypothesis and cognitive evolution. *Trends in Cognitive Sciences*, 10, 464–465.

Herrmann, E., Call, J., Hernández-Lloreda, M. V., Hare, B. & Tomasello, M. 2007. Humans have evolved specialized skills of social cognition: The cultural intelligence hypothesis. *Science*, 317, 1360–1366.

———. 2008. Response to de Waal et al. *Science*, 319, 570.

Herrmann, E., Hernández-Lloreda, M.V., Call, J., Hare, B. & Tomasello, M. 2010. Common and unique components in the cognitive skills of chimpanzees and human children: Evidence from individual differences. *Psychological Science*, 21, 102–110.

Hill, K. 2002. Altruistic cooperation during foraging by the Ache. *Human Nature*, 13, 105–128.

Hill, K. & Hurtado, A. M. 1996. *Ache Life History: The Ecology and Demography of a Foraging People*. New York: Aldine de Gruyter.

Hobson, P. 1993. *Autism and the Development of Mind*. Hillsdale, NJ: Erlbaum.

Hrdy, S. 2009. *Mothers and Others: The Evolutionary Origins of Mutual Understanding*. Cambridge, MA: Harvard University Press.

Humphrey, N. K. 1976. The social function of intellect. In *Growing Points in Ethology*, (ed. by Bateson, P. & Hinde, R. A., 303–321. Cambridge: Cambridge University Press.

Jablonski, N. 1993. *Theropithicus: The Rise and Fall of a Primate Genus*. Cambridge: Cambridge University Press.

Kaminski, J., Call, J. & Tomasello, M. 2008. Chimpanzees know what others know, but not what they believe. *Cognition*, 109, 224–234.

Kaplan, H., Hill, K., Lancaster, J. & Hurtado, A. M. 2000. The theory of human life history evolution: Diet, intelligence, and longevity. *Evolutionary Anthropology*, 156–185.

Kelemen, D. 1999. The scope of teleological thinking in preschool children. *Cognition*, 70, 241–72.

Klein, R. G. 1999. *The Human Career: Human Biological and Cultural Origins*. Chicago: University of Chicago Press.

Kobayashi, H. & Kohshima, S. 1997. Unique morphology of the human eye. *Nature*, 387, 767–768.

Kruger, A. & Tomasello, M. 1996. Cultural learning and learning culture. In D. Olson (Ed.), *Handbook of Education and Human Development: New Models of Teaching, Learning, and Schooling* (ed. by Olson, D.), 369–387. Malden, MA: Blackwell.

Leach, H. 2003. Human domestication reconsidered. *Current Anthropology*, 44, 349–368.

Levinson, S.C. 2003. *Space in Language and Cognition: Explorations in Cognitive Diversity*. Cambridge: Cambridge University Press.

Melis, A., Call, J. & Tomasello, M. 2006. Chimpanzees (*Pan troglodytes*) conceal visual and auditory information from others. *Journal of Comparative Psychology*, 120, 154–162.

Melis, A., Hare, B. & Tomasello, M. 2006a. Chimpanzees recruit the best collaborators. *Science*, 311, 1297–1300.

———. 2006b. Engineering chimpanzee cooperation: Social tolerance constrains cooperation. *Animal Behaviour*, 72, 275–286.

Moll, H. & Tomasello, M. 2007. Co-operation and human cognition: The Vygotskian intelligence hypothesis. *Philosophical Transactions of the Royal Society*, 362, 639–648.

Mulcahy, N. & Call, J. 2006. Apes save tools for future use. *Science*, 312, 1038–1040.

Muller, M. N. & Mitani, J. C. 2005. Conflict and cooperation in wild chimpanzees. *Advances in the Study of Behavior*, 35, 275–331.

Odling-Smee, F. J., Laland, K. N. & Feldman, M. W. 2003. *Niche Construction: The Neglected Process in Evolution*. Princeton, NJ: Princton University Press.

Penn, D. C., Holyoak, K. J. & Povinelli, D. J. 2008. Darwin's mistake: Explaining the discontinuity between human and non-human minds. *Behavioral and Brain Sciences*, 31, 109–178.

Rakoczy, H., Warneken, F. & Tomasello, M. 2008. The sources of normativity: Young children's awareness of the normative structure of games. *Developmental Psychology*, 44, 875–81

Richerson, P. J. & Boyd, R. 2005. *Not By Genes Alone: How Culture Transformed Human Evolution*. Chicago: University of Chicago Press.

Roberts, G. 2005. Cooperation through interdependence. *Animal Behaviour*, 70, 901–908.

Searle, J. 1995. *The Construction of Social Reality*. New York, NY: Free Press.

Silk, J.B. 2006. Practicing Hamilton's Rule: Kin selection in primate groups. In *Cooperation in Primates and Humans: Mechanisms and Evolution* (ed. by Kappeler, P. M. & van Schaik, C. P.), 25–46. Berlin: Springer.

Sterlny, K. 2008. *Nicod Lectures.* Jean Nicod Institute website. Accessed at http://www.institutnicod.org/conf_2008.htm.

Tennie, C., Call, J. & Tomasello, M. 2009. Ratcheting up the ratchet: On the evolution of cumulative culture. *Philosophical Transactions of the Royal Society,* 364, 2405–2415.

Thornton, A. & Raihani, N. J. 2008. The evolution of teaching. *Animal Behaviour,* 75, 1823–1836.

Tomasello, M. 1990. Cultural transmission in the tool use and communicatory signaling of chimpanzees? In *"Language" and Intelligence in Monkeys and Apes: Comparative Developmental Perspectives* (ed. by Parker, S. & Gibson, K.). Cambridge: Cambridge University Press.

———. 1999. *The CulturalOrigins of Human Cognition.* Cambridge, MA: Harvard University Press.

———. 2008. *Origins of Human Communication.* Cambridge, MA: MIT Press.

———. 2009. Postscript: Chimpanzee cultures, 2007. In *The Question of Animal Culture* (ed. by Galef, B. & Laland, K.). Cambridge, MA: Harvard University Press.

Tomasello, M. & Call, J. 1997. *Primate Cognition.* New York: Oxford University Press.

Tomasello, M. & Carpenter, M. 2005. The emergence of social cognition in three young chimpanzees. *Monographs of the Society for Research in Child Development,* 70, (1, serial no. 279).

Tomasello, M., Carpenter, M., Call, J., Behne, T. & Moll, H. 2005. Understanding and sharing intentions: The origins of cultural cognition. *Behavioral and Brain Sciences,* 28, 675–691.

Tomasello, M., Hare, B., Lehmann, H. & Call, J. 2007. Reliance on head versus eyes in the gaze following of great apes and human infants: The cooperative eye hypothesis. *Journal of Human Evolution,* 52, 314–320.

Tomasello, M., Kruger, A. & Ratner, H. 1993. Cultural learning. *Behavioral and Brain Sciences,* 16, 495–552.

Vygotsky, Ł. S. 1978. *Mind and Society: The Development of Higher Mental Processes.* Cambridge, MA: Harvard University Press.

Warneken, F., Chen. F. & Tomasello, M. 2006. Cooperative activities in young children and chimpanzees. *Child Development,* 77, 640–663.

Warneken, F., Lohse, K., Melis, A. P. & Tomasello, M. 2011. Young children share resources equally after collaboration. *Psychological Science,* 22, 267–273.

Whiten, A., & van Schaik, C.P. 2007. The evolution of animal "cultures" and social intelligence. *Philosophical Transactions of the Royal Society B,* 362, 603–620.

Wrangham, R., & Pilbeam, D. 2001. African apes as time machines. In *All Apes Great and Small* (ed. by B. Galdikas, N. Briggs, L. Sheeran, G. Shapiro & J. Goodall), 5–17. New York: Kluwer Academic/Plenum.

Contributors

Susan C. Alberts
Department of Biology
Duke University
Box 90338
Durham, NC 27708
USA

Michael Alvard
Department of Anthropology
Texas A&M University
College Station, TX 77843
USA

Filippo Aureli
Istituto de Neuroetologia
Universidad Veracruzana
Xalapa, Veracruz
Mexico

Josep Call
Department of Developmental and Comparative Psychology
Max Planck Institute for Evolutionary Anthropology
Deutscher Platz 6
D-04103 Leipzig
Germany

Colin A. Chapman
Department of Anthropology and McGill School of Environment
McGill University
Montreal
Quebec H3A 2T7
Canada

Dorothy L. Cheney
Department of Biology
University of Pennsylvania
Philadelphia, PA 19104
USA

Marina Cords
Department of Ecology, Evolution & Environmental Biology
Columbia University
1200 Amsterdam Avenue
New York, NY 10027
USA

Anthony Di Fiore
Department of Anthropology
University of Texas
University Station, C3200
Austin, TX 78712
USA

Melissa Emery Thompson
Department of Anthropology
University of New Mexico
MSC01-1040
Albuquerque, NM 87131
USA

Eduardo Fernandez-Duque
Department of Anthropology
University of Pennsylvania
3260 South Street
Philadelphia, PA 19104
USA

Claudia Fichtel
Behaviorsal Ecology & Sociobiology Unit
German Primate Center
Kellnerweg 4, 37077
Goettingen
Germany

Orlaith N. Fraser
Department of Cognitive Biology
Faculty of Life Sciences
University of Vienna
Althanstrasse 14, A-1090
Vienna
Austria

Ian C. Gilby
Duke University
Department of Evolutionary Anthropology
Box 90383
Durham, NC 27708
USA

Michael Gurven
Department of Anthropology
University of California, Santa Barbara
Santa Barbara, CA 93106
USA

Esther Herrmann
Department of Developmental and Comparative Psychology
Max Planck Institute for Evolutionary Anthropology
Deutscher Platz 6
D-04103 Leipzig
Germany

Maren Huck
Department of Anthropology
University of Pennsylvania
3260 South Street
Philadelphia, PA 19104
USA

Karin Isler
Anthropological Institute & Museum
University of Zurich
Winterthurerstrasse 190
8057-CH Zurich
Switzerland

Keith Jensen
Biological and Experimental Psychology Group
School of Biological and Chemical Sciences
Queen Mary, University of London
Mile End Road
London E1 4NS
United Kingdom

Peter M. Kappeler
Department of Behavioral Ecology and Sociobiology
German Primate Center
Kellnerweg 4D-37077
Göttingen
Germany

Joanna E. Lambert
Department of Anthropology
University of Texas
San Antonio, TX 78249
USA

Kevin E. Langergraber
Department of Anthropology
Boston University
232 Bay Street Road
Boston, MA 02215
USA

Elizabeth V. Lonsdorf
Lester E. Fisher Center for the Study and Conservation of Apes
Lincoln Park Zoo
2001 North Clark Street
Chicago, IL 60614
USA

Frank W. Marlowe
Department of Anthropology
Durham University
Durham DH1 3LE
United Kingdom

Charles Menzel
Language Research Center
Department of Psychology
Georgia State University
Atlanta, GA 30303
USA

John C. Mitani
Department of Anthropology
University of Michigan
1085 South University Avenue
Ann Arbor, MI 48109
USA

Martin N. Muller
Department of Anthropology
University of New Mexico
MSC01-1040
Albuquerque, NM 87131
USA

Julia Ostner
Primate Social Evolution Group
Courant Research Centre Evolution of Social Behaviour
University of Göttingen
Kellnerweg 6 D-37077
Göttingen
Germany

Ryne A. Palombit
Department of Anthropology
Center for Human Evolutionary Studies
131 George Street
Rutgers University
New Brunswick, NJ 08901
USA

Anne Pusey
Duke University
Department of Evolutionary Anthropology
Box 90383
Durham, NC 27708
USA

Jessica M. Rothman
Department of Anthropology
Hunter College
City University of New York
695 Park Avenue
New York, NY 10065
USA

Steven R. Ross
Lester E. Fisher Center for the Study and Conservation of Apes
Lincoln Park Zoo
2001 North Clark Street
Chicago, IL 60614
USA

Laurie R. Santos
Department of Psychology
Box 208205
New Haven, CT 06520
USA

Colleen M. Schaffner
Istituto de Neuroetologia
Universidad Veracruzana
Xalapa, Veracruz
Mexico

Gabriele Schino
Institute of Cognitive Sciences and Technologies
Via Ulisse Aldrovandi 16b
Rome 00197
Italy

Oliver Schülke
Courant Research Centre Evolution of Social Behaviour
University of Göttingen
Kellnerweg 6 D-37077
Göttingen
Germany

Robert M. Seyfarth
Department of Psychology
University of Pennsylvania
3720 Walnut Street
Philadelphia, PA 19104
USA

Joan B. Silk
Department of Anthropology
University of California
Los Angeles, CA 90064
USA

Elisabeth H. M. Sterck
Department of Behavioral Biology
Utrecht University
The Netherlands

Michael Tomasello
Department of Developmental and Comparative Psychology
Max Planck Institute for Evolutionary Anthropology
Deutscher Platz 6
D-04103 Leipzig
Germany

Maria A. van Noordwijk
Anthropological Institute & Museum
University of Zurich,
Winterthurerstrasse 190
8057-CH Zurich
Switzerland

Carel P. van Schaik
Anthropological Institute & Museum
University of Zurich,
Winterthurerstrasse 190
8057-CH Zurich
Switzerland

David P. Watts
Department of Anthropology
Yale University
Box 208277
New Haven, CT 06520
USA

Andrew Whiten
Centre for Social Learning and Cognitive Evolution
School of Psychology
University of St. Andrews
St. Andrews KY16 9JU
Scotland
United Kingdom

Klaus Zuberbühler
School of Psychology
University of St. Andrews
St. Andrews KY16 9JP
Scotland
United Kingdom

Index

Page numbers followed by **f** indicate figures. Page numbers followed by **t** indicate tables.